THE OXFORD HA

MW00846654

PHILOSOPHY OF
SCIENCE

THE OXFORD HANDBOOK OF

PHILOSOPHY OF

SCIENCE

Edited by

PAUL HUMPHREYS

Advisory Editors

ANJAN CHAKRAVARTTY

MARGARET MORRISON

ANDREA WOODY

OXFORD

UNIVERSITY PRESS

OXFORD
UNIVERSITY PRESS

Oxford University Press is a department of the University of Oxford. It furthers the University's objective of excellence in research, scholarship, and education by publishing worldwide. Oxford is a registered trade mark of Oxford University Press in the UK and certain other countries.

Published in the United States of America by Oxford University Press
198 Madison Avenue, New York, NY 10016, United States of America.

Library of Congress Cataloging-in-Publication Data
Names: Humphreys, Paul, editor.
Title: The Oxford handbook of philosophy of science / edited by Paul Humphreys.
Other titles: Handbook of philosophy of science
Description: New York, NY : Oxford University Press, [2016] | Series: Oxford handbooks | Includes bibliographical references and index.
Identifiers: LCCN 2015044838 | ISBN 9780199368815 (hardcover : alk. paper)
ISBN 9780190939397 (paperback : alk. paper)
Subjects: LCSH: Science—Philosophy. | Philosophy and science.
Classification: LCC Q175 .O96 2016 | DDC 501—dc23 LC record available athttp://lccn.loc.gov/2015044838

To all who teach philosophy of science, often under difficult circumstances and without much recognition, this Handbook is dedicated to you.

CONTENTS

PART III NEW DIRECTIONS

ACKNOWLEDGMENTS

This Handbook would not exist without the cooperation of its many authors who dealt with deadlines, referees' comments, revisions, and copy-editing minutiae promptly and with good cheer. It also could not have been carried through to completion without a good deal of sage advice, especially in the formative stages, from the three advisory editors, Anjan Chakravartty, Margaret Morrison, and Andrea Woody, each of whom took time from their already busy schedules to offer suggestions for topics, authors, and overall orientation. I am also grateful for their constructive comments on the Introduction, although the views expressed there are entirely my own. Thanks are due to Akiyoshi Kitaoka, Professor of Psychology at Ritsumeikan University, Kyoto, Japan for permission to use as the cover illustration his beautiful illusion of a Sierpinski gasket constructed from Penrose triangles. The staff at Oxford University Press, especially Molly Davis, Lauren Konopko, and Emily Sacharin were enormously helpful both in the early stages and in production, as were Janish Ashwin and Tharani Ramachandran at Newgen. Peter Ohlin, the philosophy acquisitions editor at the Press, who is a living example of an ideal type, originally persuaded me, against my initial reservations, to take on the task of editing this Handbook. I am now glad that he did.

Contributors

Sibylle Anderl earned a double degree in physics and philosophy from the Technische Universität Berlin, Germany, after which she obtained her PhD in astrophysics and astronomy at the University of Bonn on the topic of interstellar shocks in molecular clouds. Since 2013, she has held a postdoctoral research position at the Institut de Planétologie et d'Astrophysique de Grenoble on star formation and astrochemistry. During her doctoral studies, she helped to establish an interdisciplinary research group that is currently investigating the generation of astrophysical knowledge in modern research.

Claus Beisbart is Professor for Philosophy of Science at the Institute for Philosophy at the University of Bern. He holds a PhD in physics and a second PhD in philosophy from the Ludwig Maximilian University of Munich. His research interests include philosophy of probability, philosophy of physics (particularly of cosmology), and epistemology of modeling and simulation. He has also written about the foundations of ethics, Kant, and public choice theory. He has co-edited the collection *Probabilities in Physics* (with S. Hartmann, Oxford University Press 2011). Since 2015, he is one of the three co-editors of the *Journal for General Philosophy of Science*.

Cristina Bicchieri is the S. J. P. Harvie Professor of Philosophy and Psychology at the University of Pennsylvania and director of the Philosophy, Politics, and Economics program and the Behavioral Ethics Lab (BeLab). She is also a Professor in the Legal Studies department of the Wharton School. Bicchieri received a bachelor's degree in philosophy from the University of Milan and her PhD in Philosophy of Science at Cambridge University in 1984. She was a Harvard Ford fellow in 1980–82. She has been a fellow of Nuffield College (Oxford), the Wissenschaftskolleg zu Berlin, the Swedish Collegium for Advanced Study in the Social Sciences, and the Center for Rationality and Interactive Decision Making, University of Jerusalem. She has worked on problems in social epistemology, rational choice, and game theory. More recently, her work has focused on the nature and evolution of social norms and the design of behavioral experiments to test under which conditions norms will be followed. She published several books and more than 100 articles in top journals. Bicchieri has served as a consultant to UNICEF since 2008, and she advises various nongovernmental organizations and other international organizations and governments on social norms and how to combat harmful social practices. She founded the Penn Social Norms Training and Consulting group in 2014.

Alexander Bird is Professor of Philosophy at the University of Bristol. His work is in the metaphysics and epistemology of science and medicine, with an interest in the history of medical statistics. His published books are *Philosophy of Science* (Routledge and McGill-Queen's 1998), *Thomas Kuhn* (Acumen and Princeton 2000), and *Nature's Metaphysics* (Oxford 2007). His current project, *Knowing Science, Knowing Medicine,* aims to bring insights from general epistemology (the knowledge-first approach in particular) to bear on the philosophy of science and medicine.

Jim Bogen is an emeritus professor of philosophy, Pitzer College, and an adjunct professor of history and philosophy of science, University of Pittsburgh. He is the author and co-author (with Jim Woodward) of several papers on scientific evidence and the distinction between data and phenomena.

Otávio Bueno is Professor of Philosophy and Chair of the Philosophy Department at the University of Miami. His research concentrates in philosophy of science, philosophy of mathematics, philosophy of logic, and epistemology. He has published widely in these areas in journals such as *Noûs, Mind, British Journal for the Philosophy of Science, Philosophy of Science, Philosophical Studies, Studies in History and Philosophy of Science, Analysis, Erkenntnis, Studies in History of Philosophy of Modern Physics,* and *Monist,* among others. He is the author or editor of several books and editor-in-chief of *Synthese.*

Martin Carrier is Professor of Philosophy at Bielefeld University and director of the Institute for Interdisciplinary Studies of Science (I^{2}SoS). His chief area of work is the philosophy of science, in particular historical changes in science and scientific method, theory-ladenness and empirical testability, intertheoretic relations and reductionism, and, presently, the relationship between science and values and science operating at the interface to society. In particular, he addresses changes regarding agenda setting and justificatory procedures imposed on science by the pressure of practice and the commitment to social utility.

Hyundeuk Cheon is an assistant professor in the Institute for the Humanities at Ewha Women's University, Seoul, South Korea. His research interests include philosophy of science, social epistemology, and philosophy of cognitive science. He is currently working on scientific practices and knowledge production seen from a collective-cognitive perspective. Cheon received a PhD in history and philosophy of science from Seoul National University and worked as a research fellow in the Institute for Cognitive Science. He was formerly a visiting scholar at the University of Pittsburgh.

Carl F. Craver is Professor of Philosophy and Philosophy-Neuroscience-Psychology at Washington University in St. Louis. He is the author of *Explaining the Brain* (Oxford University Press 2009) and, with Lindley Darden, of *In Search of Mechanisms: Discoveries across the Life Sciences* (University of Chicago Press 2013) He has published widely in the philosophy of neuroscience, with particular emphasis on causation, discovery, explanation, levels, and reduction. He also works in the neuropsychology of episodic memory.

Heather Douglas is the Waterloo Chair in Science and Society and Associate Professor in the Department of Philosophy at the University of Waterloo, Ontario, Canada. She received her PhD from the History and Philosophy of Science Department at the University of Pittsburgh in 1998. Her work has been supported by the National Science Foundation and the Social Science and Humanities Research Council of Canada. She has published numerous articles as well as a book, *Science, Policy, and the Value-Free Ideal* (University of Pittsburgh Press 2009).

Antony Eagle is Senior Lecturer in Philosophy at the University of Adelaide, having previously taught at the University of Oxford, Princeton University, and the University of Melbourne. His research interests range over topics in metaphysics, epistemology, the philosophy of physics, and the philosophy of language. Current research projects concern persistence, indeterminacy, and the metaphysics and semantics of time. He is the author of a number of articles on the philosophy of probability, and he is the editor of the collection *Philosophy of Probability: Contemporary Readings* (Routledge 2011).

Uljana Feest is professor of philosophy at Leibniz University, Hannover (Germany), where she holds the chair for philosophy of social science and social philosophy. In her research, she specializes in the philosophy and history of the cognitive and behavioral sciences, with special focus on experimental practices of knowledge generation in these fields. In this regard, she is especially interested in the role of operational definitions and operational analysis in psychological experiments.

Ben Fraser is a postdoctoral research fellow in the School of Philosophy at the Australian National University. He works mainly on the evolution of morality, particularly evolutionary debunking arguments, where his research aims to link substantive empirical claims about the nature and origins of moral cognition to metaethical claims about the epistemological status of moral judgments. He is also interested in conceptual issues within biology, especially those to do with costly signaling theory and sexual selection.

Stuart Glennan is Professor of Philosophy and Associate Dean of the College of Liberal Arts and Sciences at Butler University. His research has focused on developing a mechanistic framework to address issues in the philosophy of science and naturalistic metaphysics. He has written on mechanisms, causation, modeling, and explanation, as well as on science and religion and science education. Although his focus is on general philosophy of science, he has particular interests in philosophy of biology and psychology.

Francesco Guala is Associate Professor in the Department of Economics at the University of Milan. A believer in interdisciplinarity, he makes an effort to publish in both philosophical and scientific journals. His first book, *The Methodology of Experimental Economics*, was published by Cambridge University Press in 2005. In 2011, he co-edited with Daniel Steel *The Philosophy of Social Science Reader* (Routledge). His current research is mainly devoted to the study of institutions, making use of empirical and theoretical methods.

Hans Halvorson is Professor of Philosophy, and Associated Faculty of Mathematics, Princeton University. He is also an associate fellow at the Pittsburgh Center for the Philosophy of Science. He has published numerous papers in physics and philosophy journals. His papers have been honored by the Philosopher's Annual, the Philosophy of Science Association, and the Cushing Prize. He was awarded a Mellon New Directions in the Humanities Fellowship to study the application of category-theoretic methods in philosophy of science.

Sven Ove Hansson is professor in philosophy and head of the Department of Philosophy and History, Royal Institute of Technology, Stockholm. He is editor-in-chief of *Theoria*. His research areas include fundamental and applied moral theory, value theory, philosophy of economics, theory of science, epistemology, and logic. He is the author of approximately 300 articles in international refereed journals. His most recent book is *The Ethics of Risk* (Palgrave Macmillan 2013).

Richard Healey has been Professor of Philosophy at the University of Arizona since 1991. He mostly works in the philosophy of science and metaphysics. One aim of his research has been to inquire how far foundational studies of physics shed light on metaphysical topics such as holism, realism, composition, time, and causation. He is now exploring the relations between science and metaphysics from a broadly pragmatist perspective. His *Gauging What's Real* (Oxford University Press 2009) was awarded the 2008 Lakatos prize in philosophy of science. Oxford University Press has also agreed to publish his next book, *The Quantum Revolution in Philosophy*.

Carl Hoefer is ICREA Research Professor, Universitat de Barcelona and Professor of Philosophy, University of Western Ontario. He is a philosopher of physics interested in exploring issues of the metaphysics of nature in light of our best physical theories. He has published numerous papers on topics such as causality, laws of nature, probability and chance, and the natures of space, time, and spacetime. Hoefer has previously held positions at UC Riverside, the London School of Economics, and the Universitat Autònoma de Barcelona.

Paul Humphreys is Commonwealth Professor of Philosophy and Co-director of the Center for the Study of Data and Knowledge at the University of Virginia. He is the author of *The Chances of Explanation* (Princeton 1989), *Extending Ourselves* (Oxford 2004), and *Emergence* (Oxford 2016). His research interests include computational science, data analytics, probability, emergence, and general philosophy of science.

Andreas Hüttemann is Professor for Philosophy at the University of Cologne. Previously, he taught at the Universities of Münster, München (LMU), Bielefeld, and Heidelberg. In 2013, he was elected to the Leopoldina—the German National Academy of Science. He has published mainly on topics relating to metaphysics of science, including *Ursachen* (de Gruyter 2013); *Time, Chance and Reduction* (Cambridge University Press 2010, co-edited with Gerhard Ernst); and *What's Wrong with Microphysicalism?*

(Routledge 2004). He is currently working on a book tentatively entitled *A Minimal Metaphysics for Scientific Practice*.

Muhammad Ali Khalidi teaches philosophy at York University in Toronto. He specializes in general issues in the philosophy of science (especially natural kinds and reductionism) and philosophy of cognitive science (especially innateness, concepts, and domain specificity). His book, *Natural Categories and Human Kinds*, was published by Cambridge University Press in 2013, and he has recently also been working on topics in the philosophy of social science.

Philip Kitcher received his BA from Cambridge University and his PhD from Princeton. He has taught at several American universities and is currently John Dewey Professor of Philosophy at Columbia. He is the author of books on topics ranging from the philosophy of mathematics, the philosophy of biology, the growth of science, the role of science in society, naturalistic ethics, Wagner's *Ring*, to Joyce's *Finnegans Wake*. He has been President of the American Philosophical Association (Pacific Division) and Editor-in-Chief of *Philosophy of Science*. A Fellow of the American Academy of Arts and Sciences, he was also the first recipient of the Prometheus Prize, awarded by the American Philosophical Association for work in expanding the frontiers of science and philosophy. Among his recent books are *Science in a Democratic Society* (Prometheus Books 2011), *The Ethical Project* (Harvard University Press 2014), *Preludes to Pragmatism* (Oxford University Press 2012), and *Deaths in Venice: The Cases of Gustav von Aschenbach* (Columbia University Press 2013). His Terry Lectures were published in the Fall of 2014 as *Life After Faith: The Case for Secular Humanism* (Yale University Press).

Johannes Lenhard has a research specialization in philosophy of science with a particular focus on the history and philosophy of mathematics and statistics. During the past few years, his research has concentrated on various aspects of computer and simulation modeling, culminating in his monograph *Calculated Surprises* (in German). Currently, he is affiliated with the philosophy department at Bielefeld University, Germany. He held a visiting professorship in history at the University of South Carolina, Columbia, long after he had received his doctoral degree in mathematics from the University of Frankfurt, Germany.

Alan C. Love is Associate Professor of Philosophy at the University of Minnesota and Director of the Minnesota Center for Philosophy of Science. His research focuses on conceptual issues in developmental and evolutionary biology, including conceptual change, explanatory pluralism, the structure of scientific questions and theories, reductionism, the nature of historical science, and interdisciplinary epistemology. He is the editor of *Conceptual Change in Biology: Scientific and Philosophical Perspectives on Evolution and Development* (Springer 2015) and of a variety of journal articles and book chapters on topics in philosophy of science and biology.

Aidan Lyon is an assistant professor in philosophy at the University of Maryland. He is also a Humboldt Fellow at the Munich Centre for Mathematical Philosophy and a Research Fellow at the Centre of Excellence for Biosecurity Risk Analysis. His research areas are primarily in Bayesian epistemology, social epistemology, philosophy of explanation, philosophy of mathematics, and biosecurity intelligence and forecasting.

Timothy D. Lyons is Chair of Philosophy and Associate Professor of Philosophy Science at Indiana University-Purdue University Indianapolis. In addition to his co-edited book, *Recent Themes in Philosophy of Science: Scientific Realism and Commonsense* (Springer 2002), his publications include articles in the *British Journal for the Philosophy of Science, Philosophy of Science,* and *Erkenntnis.* He focuses his research on the scientific realism debate. Specifically, he challenges the realist's epistemological tenet while defending the axiological tenet, a position he calls "Socratic scientific realism." He is currently engaged in a multifaceted research project on scientific realism funded by the Arts and Humanities Research Council.

Edouard Machery is Professor in the Department of History and Philosophy of Science at the University of Pittsburgh and an associate director of the Center for Philosophy of Science at the University of Pittsburgh. He is the author of *Doing Without Concepts* (Oxford University Press 2009), as well as the editor of *The Oxford Handbook of Compositionality* (Oxford University Press 2012), *La Philosophie Expérimentale* (Vuibert 2012), *Arguing About Human Nature* (Routledge 2013), and *Current Controversies in Experimental Philosophy* (Routledge 2014). He has been the editor of the Naturalistic Philosophy section of *Philosophy Compass* since 2012.

Margaret Morrison is Professor of Philosophy at the University of Toronto. She has been a fellow of the Wissenchsaftskolleg zu Berlin and is a member of the German National Academy of Sciences (Leopoldina). Recent publications include *Reconstructing Reality: Models, Mathematics, and Simulations* (Oxford University Press 2015) and an edited collection with Brigitte Falkenburg entitled *Why More is Different: Philosophical Issues in Condensed Matter Physics* (Springer 2015). Her interests range over many topics in philosophy of science including emergence in physics, the role of simulation in knowledge production, and issues related to the importance of abstract mathematical models for understanding of concrete systems.

Stathis Psillos is Professor of Philosophy of Science and Metaphysics at the University of Athens, Greece, and a member of the Rotman Institute of Philosophy at the University of Western Ontario (where he held the Rotman Canada Research Chair in Philosophy of Science). He is the author or editor of seven books and of more than ninety papers in learned journals and edited books, mainly on scientific realism, causation, explanation, and the history of philosophy of science.

David B. Resnik is a bioethicist and chair of the Institutional Review Board at the National Institute of Environmental Health Science. He is the author of eight books, the most recent of which is *Environmental Health Ethics* (Cambridge University Press

2012) and more than 190 articles in philosophy and bioethics. He previously taught at the University of Wyoming and the Brody School of Medicine at East Carolina University.

Dean Rickles is Professor of History and Philosophy of Modern of Physics and ARC Future Fellow at the University of Sydney, where he also co-directs the Centre for Time. His main focus of research is space, time, and symmetry, especially within classical and quantum gravity. He has published widely in this area, including several books: *The Structural Foundations of Quantum Gravity* (Oxford University Press 2006, co-edited with S. French and J. Saatsi); *Symmetry, Structure, and Spacetime* (Elsevier 2007); *The Ashgate Companion to Contemporary Philosophy of Physics* (Ashgate 2008); *The Role of Gravitation in Physics* (Max Planck Research Library for the History and Development of Knowledge 2011, co-edited with Cecile DeWitt); and *A Brief History of String Theory: From Dual Models to M-Theory* (Springer 2014).

John T. Roberts is Professor of Philosophy at the University of North Carolina at Chapel Hill. He is the author of *The Law-Governed Universe* (Oxford University Press 2008). Most of his early research focused on laws of nature and related topics such as objective chance. Recently, he has been working on a range of topics including induction, causation, space-time symmetries, the fine-tuning argument, and the alleged paradoxes of time travel. At present he is working on a book about counterfactuals, laws, causation, and dispositions, which will defend a unified normativist account of them all.

Adina L. Roskies is Professor of Philosophy and Chair of the Cognitive Science Program at Dartmouth College. She received a PhD from the University of California, San Diego, in Neuroscience and Cognitive Science in 1995; a PhD from MIT in philosophy in 2004; and an MSL from Yale Law School in 2014. Prior to her work in philosophy, she held a postdoctoral fellowship in cognitive neuroimaging at Washington University and was Senior Editor of the neuroscience journal *Neuron*. Roskies's philosophical research interests lie at the intersection of philosophy and neuroscience and include philosophy of mind, philosophy of science, and neuroethics. Her recent work focuses on free will and responsibility.

Oron Shagrir is a Professor of Philosophy and Cognitive Science and currently the Vice Rector of the Hebrew University of Jerusalem. He works on various topics in philosophy of mind, foundations of computational cognitive and brain sciences, and history and philosophy of computing. He has published in *Mind, Philosophical Studies, Philosophy and Phenomenological Research, Philosophy of Science, The British Journal for the Philosophy of Science*, and *Synthese*. He co-edited (with Jack Copeland and Carl Posy) the volume *Computability: Turing, Gödel, Church, and Beyond* (MIT Press 2013).

Giacomo Sillari is a research assistant at Scuola Normale Superiore in Pisa and also teaches in the Department of Scienze Politiche at the Libera Università Internazionale degli Studi Sociali Guido Carli in Rome. His research lies at the intersection of philosophy and economics, including the epistemic foundations of game theory and the role and nature of social conventions and norms from the viewpoint of game theory. He is a member of the editorial boards of the journals *Topoi* and *Philosophy and Public Issues*.

Bradford Skow is an associate professor of philosophy at the Massachusetts Institute of Technology.

Chris Smeenk is an associate professor in philosophy at Western University and Director of the Rotman Institute of Philosophy. His research has focused primarily on the interplay between theory and evidence in the historical development of physics, in particular gravitational physics and cosmology. This includes projects regarding Newton's methodology in the *Principia Mathematica*, the discovery of general relativity, and recent work in early universe cosmology. He has also worked on topics in general philosophy of science and foundations of physics, such as the status of determinism in general relativity and issues in quantum field theory.

Jan Sprenger completed his PhD in philosophy at the University of Bonn in 2008 after an undergraduate degree in mathematics. Since then, he has been teaching and researching at Tilburg University in the Netherlands and has been the recipient of various Dutch, German, and European grants. Since 2014, he is also Scientific Director of the research center TiLPS. He regularly publishes in journals such as *Philosophy of Science, British Journal for the Philosophy of Science, Mind*, and *Synthese*. He is about to complete a monograph *Bayesian Philosophy of Science* (forthcoming with Oxford University Press) together with Stephan Hartmann.

Mark Sprevak is Senior Lecturer in the School of Philosophy, Psychology, and Language Sciences at the University of Edinburgh. His primary research interests are in philosophy of mind, philosophy of science, metaphysics, and philosophy of language, with particular focus on the cognitive sciences.

P. Kyle Stanford is a professor in the Department of Logic and Philosophy of Science at the University of California at Irvine. He has written extensively on scientific realism and instrumentalism, as well as on a wide range of other philosophical issues, and he is the author of *Exceeding Our Grasp: History, Science, and the Problem of Unconceived Alternatives* (Oxford University Press 2006). His papers concerning scientific realism and instrumentalism have appeared in the *Journal of Philosophy, Philosophy of Science, The British Journal for the Philosophy of Science*, and elsewhere.

Friedrich Steinle is Professor of History of Science at Technische Universität Berlin. His research focuses on the history and philosophy of experiment, on the history of electricity and of color research, and on the dynamics of empirical concepts. His books include *Newton's Manuskript 'De gravitatione'* (Franz Steiner Verlag 1991) and *Explorative Experimente: Ampère, Faraday und die Ursprünge der Elektrodynamik* (Franz Steiner Verlag 2005); he is co-editor of *Experimental Essays: Versuche zum Experiment* (Nomos 1998, with M. Heidelberger), *Revisiting Discovery and Justification* (Springer 2006, with J. Schickore), *Going Amiss in Experimental Research* (Springer 2009, with G. Hon and J.

Schickore), and of *Scientific Concepts and Investigative Practice* (Walter de Gruyter 2012, with U. Feest).

Kim Sterelny is an Australian philosopher based at the Australian National University. He has always worked at the boundaries of philosophy and the natural and social sciences. In recent years, most of his work has been on human evolution, trying to understand the evolution of human social life and of the cognitive capacities that make that life possible, but he retains broad interests in the theoretical and conceptual challenges posed by the life sciences.

Michael Strevens has written books about complex systems, scientific explanation, and the role of physical intuition in science. His other research interests include the psychology of concepts, the philosophical applications of cognitive science, the social structure of science, confirmation theory, causality, the nature of physical probability—and pretty much anything else connected to the workings of science. He is currently Professor of Philosophy at New York University.

Mauricio Suárez is Associate Professor (Profesor Titular) at Madrid's Complutense University; a Research Associate at the Centre for Philosophy of Natural and Social Science, London School of Economics; and presently also a Marie Curie Senior Research Fellow at the School of Advanced Study, London University. Previously he held positions at Oxford, St Andrews, Northwestern, and Bristol Universities, as well as regular visiting appointments at Harvard University. He conducts research on the philosophy of probability and statistics, the metaphysics of causation and chance, and the philosophy of physics. He has also been a leading and pioneering contributor over the years to the interrelated issues of models, fictions, and representations in science. He is the editor of *Fictions in Science: Philosophical Essays on Modelling and Idealisation* (Routledge 2009) and *Probabilities, Causes and Propensities in Physics* (Springer 2011), as well as more than fifty journal articles, book chapters, and edited collections on the topics of his research interest.

Denis Walsh is a professor in the Department of Philosophy, the Institute for the History and Philosophy of Science and Technology, and the Department of Ecology and Evolution at the University of Toronto. Until 2015, he held the Canada Research Chair in Philosophy of Biology. He received a PhD in biology from McGill University and a PhD in Philosophy from King's College London. He is the author of *Organisms, Agency, and Evolution* (Cambridge University Press 2015).

Charlotte Werndl is Professor of Logic and Philosophy of Science at the Department of Philosophy at the University of Salzburg, Austria, and a Visiting Professor at the Department of Philosophy at the London School of Economics (LSE). She is an editor of the *Review of Symbolic Logic* and an associate editor of the *European Journal for Philosophy of Science*. After completing her PhD at Cambridge University, she was a research fellow at Oxford University and an associate professor at LSE. She has published on climate change, statistical mechanics, mathematical knowledge, chaos theory, predictability, confirmation, evidence, determinism, observational equivalence, and underdetermination.

James Woodward is Distinguished Professor in the Department of History and Philosophy of Science at the University of Pittsburgh. Prior to 2010, he was the J. O. and Juliette Koepfli Professor of Humanities at the California Institute of Technology. He works mainly in general philosophy of science, including issues having to do with causation and explanation. He is the author of *Making Things Happen: A Theory of Causal Explanation* (Oxford University Press 2003), which won the 2005 Lakatos Award. He served as president of the Philosophy of Science Association from 2010 to 2012.

CHAPTER 1

INTRODUCTION

New Directions in Philosophy of Science

PAUL HUMPHREYS

1 ORIENTATION

IN its original meaning, a handbook was a ready reference resource small enough to be carried in the hand. As knowledge has grown, that literal meaning has been left behind unless you happen to be reading this on an electronic device. Contemporary philosophy of science is too rich and vibrant a field to be condensed into a single volume, even one containing forty-two substantial essays. This handbook is also more than just a reference work; a primary ambition is to prompt readers to think about the various subject matters in new ways, as well as to convey basic information about the field. To those ends, Section 1 contains broad overviews of the main lines of research and established knowledge in six principal areas of the discipline, Section 2 covers what are considered to be the traditional topics in the philosophy of science, and Section 3 identifies new areas of investigation that show promise of becoming important areas of research. Because the handbook emphasizes topics of broad interest and, where appropriate, new directions in the field, a deliberate decision was made not to cover specialized topics the treatment of which is widely available elsewhere, such as the philosophy of quantum mechanics and of spacetime theories, or topics that are covered in other Oxford Handbooks, such as *The Oxford Handbook of Philosophy of Economics* or *The Oxford Handbook of Philosophy of Time*. Exceptions have been made for fields in which new ideas are being developed at a rapid rate or for which a radically different approach can provide novel insights. As a result, some aspects of quantum and spacetime theories are covered in Chapter 6.

The contents are intended to be useful to philosophers at all levels of expertise and to nonphilosophers who are interested in understanding the current state of the discipline. All the authors in this volume are active researchers in their fields, and each was encouraged to provide an original perspective on the topic at hand. Some chapters are, necessarily, more technical than others and so require more background but, with a little work, anyone with a decent education at the undergraduate level can read any given chapter and benefit. Each article provides reference sources and, in some cases, recommendations for further reading so that those with deeper interests can dig deeper into the topics. For readers who want to follow up on a particular topic, electronic versions of all chapters are available through Oxford Handbooks Online with crosslinks to handbook entries in other fields, with some of those online chapters having content that supplements the print version.

2 UNITY AND DISUNITY

Philosophy of science, as the singular form indicates, has from the earliest stages of its development formulated positions on scientific method, explanation, theory confirmation, and other topics that were intended to apply to all sciences universally. Whether by default, as when natural philosophers took physics and astronomy as models to be emulated in the absence of other systematically successful sciences, or by deliberate policy, as when the Unified Science movement that flourished in the middle third of the twentieth century imposed common standards of formalization and theory assessment across the sciences, the idea was that science had some underlying set of common methods. Today, although what is now called general philosophy of science continues, and its prospects are discussed in Chapter 7, the field has undergone a fragmentation into a number of subdisciplines such as the philosophy of physics, the philosophy of biology, the philosophy of neuroscience, and many others. This is understandable, if only because of easily identifiable differences between fields of investigation. Some of these differences are methodological. Physics appeals to laws of nature or symmetry principles and is heavily mathematical for the most part; biology has few laws, and many of its areas have not yet been subjected to formal treatments. Astronomy is, almost exclusively, a purely observational and theoretical science, unable to exploit experiments as is, for example, chemistry. It is the size and time scales of astronomical systems that make experimental manipulations impossible; as just one example, it is infeasible to experimentally reproduce the processes that led to the formation of the first stars if only because it took hundreds of millions of years after the Big Bang for those stellar objects to form. There is also the uniqueness problem—there is only one observationally accessible universe—a problem that even geophysics does not face as long as one is willing to assimilate geophysics to the more general area of planetary science.

In many cases, differences in subject matter are at least partly responsible for these differences in methods across fields. Some physical and astronomical systems are sufficiently simple that modest idealizations at best are required to bring their theories into contact with real systems. As far as we know, all electrons have the same charge, mass, and spin, thereby allowing genuinely universal claims to be made about them. Although there are research programs that attempt to identify principles of universal biology, they are in their infancy; although they may well succeed in transcending the historical localization of biology to terrestrial systems, perhaps evolutionary contingencies will prevent the biological sciences from ever attaining the universal scope that is characteristic of fundamental physics. A number of features that are characteristic of the philosophy of biology are described in Chapter 4, together with some interesting explorations of how the domain of philosophy of biology has expanded over the course of its history. In the domain of the social sciences, a discussion of how philosophical naturalism has adapted to domain-specific ontologies can be found in Chapter 3. Similarly, most sociological and anthropological principles are tacitly earthbound, with contingent cultural variations restricting generalizations in cases where cultural universals are not available. The prospects for a universal economic theory are perhaps more promising, at least if behavioral economics is excluded.

The resulting proliferation of subdisciplines labeled "the philosophy of X" has produced a fragmentation in our discipline, with the literature in many of the more technical areas becoming inaccessible to those who are not specialists in those areas, a situation that now applies even to subdisciplines in the philosophy of physics and reflects the internal state of the sciences themselves. This specialization produces difficulties for general philosophy of science when providing accounts of such topics as explanation, covered in Chapter 25, that must be reasonably accurate about a variety of fields.

When considering the prospects for general philosophy of science, it is well worth keeping in mind that differences are easy to find, whereas identifying deep unifying features takes sustained research and insight. One should therefore be careful, despite the considerations just noted, not overstate the degree to which science lacks general methods and principles. Complexity theory, discussed in Chapter 33, although lacking a neat definition of its scope, was developed with a goal of applying the same methods across a wide variety of scientific domains, from condensed matter physics through ecology to microeconomics. Overriding the traditional boundaries of scientific fields can provide important insights into situations in which the specific dynamics of the micro-ontology become largely irrelevant, a situation that has long been a feature of statistical mechanics. Indeed, the methods of statistical mechanics have become part of complexity theory itself, and, within the philosophy of physics, statistical mechanics has challenged the long-standing dominance of quantum theory and spacetime theories to become one of the primary areas of research. For different reasons, largely related to its high level of abstraction, probability theory, discussed in Chapter 20, is applicable to almost all fields of scientific endeavor, although since its transition from part of science to a separate field of mathematics with the work of Kolmogorov in the early 1930s, the relation of probability theory to its applications and its ontological status has generated its own philosophical problems. In a related way, methods of confirmation and induction, discussed in Chapter 9, retain a large measure of generality, despite a move toward more local methods of inductive inference.

Philosophical issues arising from computer simulations and from computational science more generally were scarcely discussed until the very end of the past century. In a similar vein to complexity theory, the methods of computational science and especially computer simulations have transcended disciplinary boundaries to a remarkable degree.

This area, which has been part of the philosophy of science for less than three decades, is now a fully fledged subfield and provides additional evidence that there are unifying scientific methods that can be applied to very different subject matters. The novel methodological and epistemological issues that it raises are described in Chapter 34.

Computer science itself, despite its name, has never been fully accepted as a science on a par with physics, biology, and chemistry. Often located in schools of engineering, the discipline covers a wide range of activities, from theoretical work that overlaps with mathematics and mathematical logic to commercial research on memory storage and chip reliability. Perhaps this accounts for the relative lack of interest by philosophers of science in the topic, and they have largely conceded the theoretical aspects of computation to logicians and to philosophers of mind, in the latter case via computational theories of mind. Chapter 2 argues that the concepts of computation and computational complexity are rich sources of philosophical topics, among which are the question of whether computational processes are objective features of the world, the question of whether the standard concept of Turing computability contains an implicit appeal to human computational abilities, and the question of whether every physical process executes at least one computer program. The recommendation in Chapter 2 that philosophers step back from their usual focus on computability *simpliciter* and focus instead on degrees of computational complexity reflects an important division in philosophy of science. The division is between those who prefer to discuss abstract, in principle, possibilities and those who prefer to restrict themselves to situations that have some serious contact with scientific practice. There is no doubt that some results that require abstracting from practicalities are important. When we know that there are in principle undecidable states of certain two-dimensional Ising lattices or that it is impossible even in principle to construct empirically adequate hidden variable versions of quantum mechanics under certain precisely specified restrictions, we thereby know something important about where the limits of scientific knowledge lie.

In contrast, contact with the details of science as it operates under unavoidable constraints is an important corrective to speculative claims, such as how statements representing laws of nature would be formulated at the limit of scientific inquiry. In the absence of evidence that the quantitative truth-likeness of law statements is uniformly converging to a limit, the conception of laws at the epistemological limit is of no help in evaluating contemporary claims about lawhood.

3 REDUCTION RELATIONS

An alternative view to fragmentation is that these differences between fields are only apparent, and all sciences can be reduced in principle to physics, either ontologically or theoretically, although methodological differences may remain. This once widely held view has become increasingly hard to defend, even more so if the "in principle" qualifier is dropped. The options for and alternatives to reduction as a general enterprise are discussed in Chapter 22, with particular attention being paid to prospects for biological reduction. The overall message is that careful analyses show just how difficult theory reduction can be in practice. Some clarification of this claim might be helpful. The program of reducing all of the sciences to physics was from its inception a promissory note, secured by a few apparent successes and an instinctive attachment to physicalism on the part of many philosophers of science back when the philosophy of physics dominated our subject. Local reductions, such as small parts of physical chemistry to physics, have been carried out in great detail. Despite these impressive achievements, projects that attempt a wholesale reduction of one science to another have run into serious difficulties, including the multiple realizability of many "higher level" properties, the fact that some candidates for reduction turn out to be limit cases of the potentially reducing theory, where the limit involves conditions that do not exist in the natural world, and the fact that the predicates of some theories, such as those of psychology, may be social constructions rather than natural kinds.

As with many other topics, a position on reduction carries commitments in other areas in the philosophy of science, such as the nature of theory structure, the form of explanations, the role of idealizations, approximations, and limit results, and whether theoretical or ontological approaches to these questions are preferable. This focus on broader aspects of the reductive process is a characteristic feature of contemporary discussions of reduction in philosophy of science, and its emphasis on the details of actual reduction contrasts sharply not only with the abstract approaches taken in the earlier literature, such as Ernest Nagel's influential account, but also with the traditional attitude toward reduction in the philosophy of mind. The philosophy of neuroscience, a new area that is treated in detail in Chapter 40, allows a naturalistic alternative to those traditional approaches in the philosophy of mind, even if one is not an eliminativist about mental features. It illustrates the need for novel experimental methods suitable for the subject matter at hand, including, in this case, not violating ethical constraints with regard to the subjects, and it poses an especially difficult problem of reduction in matching the often phenomenological and behavioral contents of perceptual experience and psychiatric disorders with the very different classifications of neuroscience.

Thus far, I have emphasized the importance of details, but there is no doubt that abstraction can also reveal important philosophical information. As one example, general dependency relations other than identity and deducibility, such as functional reducibility, or more specific relations than mere part–whole relations, such as mechanistic decomposition, can be valuable in understanding the status of psychological properties. Mechanisms are discussed in Chapter 38, including the prospects for mechanisms as an alternative to law-based accounts of scientific explanation. Laws of nature themselves are covered in Chapter 16.

The failure of reduction is often taken to be a necessary condition for the occurrence of emergence, another area that has seen a resurgence of interest in recent years. Once seen as a home for mysterious entities that could not be brought under the explanatory apparatus of science, the emphasis in emergence research has switched from grand phenomena such as life and consciousness to phenomena within physical, chemical, and biological systems such as universality, phase transitions, and self-organizing systems. Each of these has a well-understood explanatory process underlying the appearance of the emergent phenomenon and is resistant to reduction. This switch from transcendent properties to intradomain properties should make one wary of approaches that take an undifferentiated commitment to physicalism as a starting point for discussions of reduction and associated topics. The realm of the nonmental, which is how the domain of physicalism is usually construed, contains a remarkable variety of ontological and theoretical features, and that variety must be respected rather than lumped into the catchall category of physicalism. To put the point slightly differently, whereas taking physicalism as a starting point for investigations into reduction might have made for a simple initial hypothesis at one time, the weight of evidence now suggests that, with respect to mental properties, starting with the specific ontology of neuroscience but leaving room for adding a finer-grained ontology is preferable. It should be noted that there is no widely agreed definition of emergence, and one of the important tasks for philosophers is to identify different uses of the term. Further discussions of these topics are contained in Chapter 36.

4 Traditional and New Areas

The empirical content of theories, models, and data is a feature that distinguishes science from mathematics and other intellectual endeavors.[1] Yet philosophers of science have widely differing positions on what "empirical content" means. For centuries, empiricism took the form that the only legitimate basis for knowledge, and scientific knowledge in particular, was human perceptual experience. Within philosophy of science, the most vigorous proponents of this view were the logical positivists and logical empiricists, of which the Vienna Circle and the Berlin School, respectively, were the most prominent.[2] The unified science movement was also driven in part by a commitment to empiricism. If evidence for theoretical claims ultimately had to be cashed out in terms of perceptual evidence, then perceptual evidence, at least in the theory-neutral manner in which it was initially regarded, should serve as a subject matter independent epistemological basis for all sciences. Despite impressive efforts to defend it, empiricism is an epistemology that became increasingly out of touch with scientific practice. To reject or to remain neutral about the results from scanning tunneling microscopes that can detect features at sub-nanometer scales, or from X-ray telescopes that provide information about supernova remnants, or from single-molecule real-time DNA sequencing machines that have greater than 99% accuracy rates is to push epistemological risk

---

[1] I set aside here empiricist accounts of mathematics.

[2] As with many such movements, there was a range of opinions on the strictness of the empiricism involved. Rudolf Carnap, for example, was amenable to the use of some instruments in the collection of observations.

aversion beyond reasonable limits. To claim that none of these activities is revelatory of truths about the world is to fail to endorse an enormous range of scientific results for reasons that are oddly anthropocentric. This suggests that specifically scientific epistemologies should be seriously considered as replacements for empiricism. The options for a post-empiricist scientific epistemology are taken up in Chapter 37, and some relations between general epistemology and the philosophy of science are covered in Chapter 11.

One strategy for addressing the deficiencies of empiricism is to switch from the constrained category of observations to the more comprehensive category of data, which can originate from processes far removed from observation and can include experimental, simulated, survey, and many other kinds of data. There is also a profound change occurring within the realm of what has come to be known as big data, a field that ranges from the massive datasets collected at particle accelerators to data harvested from social media sites for commercial purposes. In many of these areas, traditional modeling techniques that use explicitly formulated models are being abandoned in favor of methods that identify patterns and structure in very large datasets, including many that have been collected under uncontrolled conditions and include a certain quantity of contaminated, duplicate, and biased data points. Some of these methods are now being used within the humanities, thus raising the possibility that data science might turn some of the humanities into proto-sciences. These and other aspects of the rise of data are examined in Chapter 35.

Many philosophers of science, even if not empiricists, are suspicious of purely metaphysical investigations. The logical positivist movement was famous for its attempts to eliminate metaphysics from the realm of defensible philosophical activities and, for a while, it succeeded in subduing metaphysical activities. But pure metaphysics gradually revived and produced some distinctive contributions, especially to the analysis of causality, during the 1970s and 1980s. Yet the discomfort with certain styles of metaphysical reasoning has persisted in the philosophy of science, and there has been a growing tension between what is called analytic metaphysics and scientific metaphysics in recent years, with the methods and conclusions of the two approaches becoming increasingly separated from one another. Chapter 17 on metaphysics in science describes in some detail what the appropriate approach to metaphysical claims in science should be.

New areas in the philosophy of science can develop when existing domains of science have been neglected by philosophers, such as with the philosophy of astronomy. In other cases, the science itself develops new approaches that require different philosophical analyses. An example of this is the joint coverage in Chapter 5 of the philosophy of psychology and the philosophy of cognitive science. In the latter, the interplay between neural substrates, representational content, computational and probabilistic processing, and levels of treatment produce topics that are barely recognizable in traditional philosophy of psychology. Finally, completely new scientific disciplines can emerge, such as complexity theory and computational science, that have sui generis features requiring novel treatments.

All sciences have procedures for classification, whether it be the types of particles in the Standard Model of physics; the organization of chemical elements in the familiar periodic table; the arrangement of animals, insects, and plants into species; or the classification of psychiatric disorders. In some disciplines, such as chemistry, the taxonomies are widely accepted; in others, such as psychiatry, they are hotly disputed and not uniform across national boundaries. A central philosophical question is whether a given classification reflects objective divisions in the world—whether they form natural kinds—and, if so, how we pick out those kinds, or whether they are convenient categories that serve the purpose of a compact organizational scheme allowing efficient prediction and description. The topic of scientific realism is obviously in the vicinity of any discussion about natural kinds, as are traditional ontological questions about the nature of properties, and the answers to these questions may well vary across different scientific disciplines. Chapter 19 takes up these and other issues, while Chapter 24 addresses how semantic, cognitive, and experimental approaches to concepts illuminate their use in the sciences.

We have already mentioned complexity theory and computer simulations as new areas in the philosophy of science. A striking example of a very old field that was neglected for decades is the philosophy of astronomy. Philosophers of science have long had an interest in cosmology, and Chapter 39 is devoted to that topic, but little or no attention has been paid to philosophical issues in astronomy, despite the fact that it has made extraordinary advances while remaining a nonexperimental science. That lacuna is remedied here with the first modern reference article on the topic in Chapter 31.

5 On the Value of Philosophy
of Science

There is a further goal that I hope will be achieved by this handbook. I am optimistic that a sympathetic reading of some of these chapters will lessen misunderstandings among nonphilosophers who are curious, and perhaps skeptical, about what twenty-first-century philosophy of science has to contribute to our understanding of various sciences. The work of Karl Popper and Thomas Kuhn, well-known (and rightly so) as it is, long ago ceased to be a topic of central importance as the field moved in other directions. Reading this handbook will bring some understanding of why that is and what are those new directions. There have, of course, been legacies of that era. In the period following the publication of Thomas Kuhn's *The Structure of Scientific Revolutions*, much philosophy of science was influenced by the history and sociology of science. The philosophy of science absorbed lessons from historical studies of science—not the least of which was to avoid anachronistic interpretations of historical episodes—but the majority of research in contemporary philosophy of science is oriented toward the current state of the various sciences. The challenges posed by the Kuhnian era have largely been answered, the result of processes that are insightfully described in Chapter 30, along with suggestions about how to proceed in a post-Kuhnian discipline.

Similarly with the "science wars" of the 1990s, a prominent example of which was Alan Sokal's submission to the postmodern journal *Social Text* of a transparently absurd article on quantum gravity containing numerous errors that would have easily been spotted by a competent referee, had one been used. The article was duly accepted and published. The fallout from this debacle damaged serious philosophy of science because it was often, without justification, lumped together with darker corners of the academic world. There are undoubtedly sociological aspects to the practice of science, and this is reflected in the rise of social epistemology and the fact that large-scale scientific research projects are dependent on social structures. Topics such as these are explored in Chapter 41 on the social organization of science. But such influences should not be exaggerated, and they need not undermine the rationality of the decision-making process in various areas of science.

Regarding the task of demarcating science from non-science (a task that has once again become pressing within biology due to pressure from intelligent design advocates and at the outer limit of mathematical physics where the empirical content of some theories is difficult to discern), Chapter 23 suggests that the well-known demarcation between science and pseudo-science made famous by Popper's work should be replaced with a finer grained set of demarcations between types of knowledge-producing activities—of which science is one—with parts of the humanities, technology, and other areas having their own methods that make them nonscientific rather than unscientific.

The criticism of philosophy of science has parallels in another area. Universities are often criticized by those who have experienced them only from the narrow perspective of an undergraduate education. As with any complex organization or activity, things are seen to be far more complicated when infused with years of further experience, perhaps even wisdom. And so it is with the philosophy of science. From the outside, it can appear disputatious, lacking in progress, concerned with problems remote from experience, and overly simplistic.[3] Some of these criticisms may be appropriate, but it is important that they not revolve around a misunderstanding of how philosophical research is conducted. Disputatious? Yes, definitely, but so are many areas of science. One important example, covered in detail in Chapter 32, concerns the various ways in which the modern synthesis theory of evolution has come under criticism from within biology. Those readers who are concerned by how anti-evolutionist movements have exploited this disagreement can benefit greatly from learning how conflicting views about some body of scientific work can be responsibly assessed from within that science itself. Rational disagreement and a spectrum of different positions are inevitable components of both philosophy and science. This does not mean that all positions have equal merit. One of the great virtues of philosophy of science, properly done, is that when it is wrong, it can shown to be wrong.[4]

---

[3] A large part of what goes on in philosophy is indeed of little interest to those outside the immediate research community, is incremental rather than path-breaking, and is rarely cited. But one can also find plenty of irrelevant, uninspired, and poorly executed scientific research. So the criticism has to be that even the best work in philosophy has the properties attributed to it by the external skeptics.

[4] Persuading the advocates of a false claim to admit this is often more difficult.

What of progress? Chapter 26 assesses the prospects for whether science can be considered to make progress. Does the philosophy of science in turn progress, albeit in a different fashion? In certain respects, the answer is clearly affirmative. Valuable as they were, Mill's methods for classifying and discovering causes have been superseded by far more nuanced accounts of causation, many of which are described in Chapter 8. The analysis of social sciences has been deepened in some areas by the application of game theory, as outlined in Chapter 14, without thereby undermining the usefulness of less formal approaches to the social sciences, many of which are discussed in Chapter 3. A discussion of the role of values within science and of the ways in which ethical considerations affect the human sciences, the topics of Chapters 29 and 12, respectively, would be unlikely to have occurred in the philosophy of science fifty years ago. At a more fine-grained scale, our knowledge of theory structure, described in Chapter 28; of models and theories, discussed in Chapter 18; and of scientific representations more generally, covered in Chapter 21, is far more sophisticated than even ten years ago. The same is true of experiment, the topic of Chapter 13. What is worth keeping in mind is that, within philosophy, progress in understanding can be made without a consensus forming. Well-intentioned thinkers can understand the merits of different positions and yet allocate different weights to the importance of different types of evidence. One can appreciate the force and subtlety of a contemporary Platonist account of mathematical ontology, yet find the inadequacies of Platonistic epistemology to outweigh those virtues.

The charge of being overly simplistic is more difficult to address. The charge is clearly not true of some of the more technical areas of philosophy of science. Chapter 10, on determinism and indeterminism, is a good example of how attention to the details of an abstract formulation of determinism can illuminate a perennial question that is intertwined with topics such as causation, probability, and free will. Sample questions that can be addressed in a formal framework include whether there can be systems for which it is impossible to definitively decide whether the system is deterministic or indeterministic, whatever the degree of resolution of the observational data, and the precise ways in which Newtonian mechanics is nondeterministic. As a second example of the virtues of technical philosophy of science, much attention has been paid to the nature of space and time or, more specifically, the nature of spacetime, but until recently the role of more general spaces in science was largely left unexplored. This deficiency is rectified by Chapter 42, where not only higher dimensional spaces of the kind used in string theory and the abstract spaces of particle physics are considered, but also the more general representational role of spaces in determining domains of possibilities in visual space or color space is explored.

One legacy of the era in which a priori analyses played a dominant role in the philosophy of science is a tendency to overgeneralize the scope of application of some position, such as scientific realism. Even some self-described naturalists were prone to this vice. A famous example is W. V. O. Quine's criterion for ontological commitment. For him, the appearance of an existentially quantified sentence in a logical reconstruction of a scientific theory entailed that anyone advocating that theory was committed to asserting the existence of whatever fell under the scope of that quantifier. But this position fails to recognize the fact that scientists discriminate—and ought to discriminate—between those parts of their theory that they want to interpret realistically and the parts (e.g., those involving severe idealizations) that they do not want to interpret realistically. Planets should be real if anything is, even if how to classify something as a planet is now controversial, but physical systems having a temperature of $0°K$ do not exist, even though the absolute temperature scale includes that point.[5]

So it is possible to be both a scientific realist, in ways that are discussed in Chapter 27, with regard to some types of theory and with regard to some parts of a single theory and to be an anti-realist with regard to others, as discussed in Chapter 15. Even for a single concept, there can be persuasive reasons on both sides of the ontological divide. In some sense, we all suppose money to be real, but the use of virtual currencies and promissory notes in addition to concrete legal tender makes the criteria for reality largely a matter of social convention, for which some commitment to group intentions seems required.

The same point holds with respect to positions in other areas. Some accounts of explanation work well for some domains of science and not for others; some areas of science, such as physics and chemistry, seem to describe natural kinds; others, such as anthropology, deal mostly with social constructs. And so on. This is not evidence for pragmatism but a sharpening of the application criteria for various philosophical positions. Furthermore, just as science uses idealizations to great effect, so one should allow philosophy of science to do the same, within limits. The tension between normative and descriptive philosophy of science is genuine, but philosophy of science does not have to accommodate all historical case studies to be relevant, only those that are representative of identifiable central features of whatever period of science the philosophical claim is about. Science itself is constantly changing; it is unreasonable to expect the methods of eighteenth-century physics to apply to methods currently employed at a major accelerator, for example.

There is one final reason for the existence of different, defensible positions in the philosophy of science. The aims of science—let us assume that a communal activity can have aims—can vary, as can the goals of individual scientists. Those aims can be truth, problem-solving, prediction, control, knowledge, or understanding. The elements of this brief list need not be mutually exclusive, and, if one is a scientific realist, all six could be achievable, with problem-solving resulting from the discovery of truths. But if one were an anti-realist, one might hold in contrast that mere problem-solving, prediction, and control are the only goals of science, with truth, knowledge, and understanding being unattainable. Other positions are possible, such as that one could gain understanding but not knowledge through idealized models that are only approximately true.

---

[5] There is some experimental evidence of negative temperatures, although their interpretation is controversial. This does not affect the point.

At this point, I can turn the reader loose on the rich and rewarding contents of this handbook. Whether you read systematically or selectively, each of the following chapters will provide you with new ways to think about a topic. The material is ample evidence that philosophy of science is a subject with a great past and an even brighter future.

PART I

OVERVIEWS

CHAPTER 2

···

ADVERTISEMENT FOR THE PHILOSOPHY OF THE COMPUTATIONAL SCIENCES

···

ORON SHAGRIR

THE term "the computational sciences" is somewhat ambiguous. At times, it refers to the extensive use of computer models and simulations in scientific investigation. Indeed, computer simulations and models are frequently used nowadays to study planetary movements, meteorological systems, social behavior, and so on. In other cases, the term implies that the subject matter of the scientific investigation or the modeled system itself is a computing system. This use of the term is usually confined to computability theory, computer science, artificial intelligence, computational brain and cognitive sciences, and other fields where the studied systems—Turing machines, desktops, neural systems, and perhaps some other biological ("information processing") systems—are perceived as computing systems. The focus of this chapter is on the latter meaning of "the computational sciences," namely, on those fields that study computing systems.

But, even after clearing up this ambiguity, the task of reviewing the philosophy of the computational sciences is a difficult one. There is nothing close to a consensus about the canonical literature, the philosophical issues, or even the relevant sciences that study computing systems. My aim here is not so much to change the situation but rather to provide a subjective look at some background for those who want to enter the field. Computers have revolutionized our life in the past few decades, and computer science, cognitive science, and computational neuroscience have become respected and central fields of study. Yet, somewhat surprisingly, the philosophy of each computational science and of the computational sciences together has been left behind in these developments. I hope that this chapter will motivate philosophers to pay more attention to an exciting field of research. This chapter is thus an advertisement for the philosophy of the computational sciences.

The chapter comprises two parts. The first part is about the subject matter of the computational sciences: namely, computing systems. My aim is to show that there are varieties of computation and that each computational science is often about a different kind of computation. The second part focuses on three more specific issues: the ontology of computation, the nature of computational theories and explanations, and the potential relevance of computational complexity theory to theories of verification and confirmation.

1 Varieties of Computation

What is the subject matter of the computational sciences? The initial answer is computing systems. But, as it turns out, different computational sciences focus on different kinds of computers, or so I will claim. My strategy of argumentation is to first survey the Gandy-Sieg view that distinguishes between human computation and machine computation. I then provide a critique of this view. Although I commend their distinction between human and machine computation, I point out that we should further distinguish between different kinds of machine computation.

1.1 The Church-Turing Thesis

Until not long ago, people used the term "computers" to designate human computers, where by "human computers" they meant humans who calculate by mechanically following a finite procedure. This notion of procedure, now known as an *effective procedure* or an *algorithm*, has been in use for solving mathematical problems at least since Euclid. The use of effective procedures as a method of proof is found in Descartes, and its association with formal proof is emphasized by Leibnitz. Effective procedures came to the fore of modern mathematics only at the end of the nineteenth century and the first decades of the twentieth century. This development had two related sources. One is the various foundational works in mathematics, starting with Frege and reaching their peak in Hilbert's finitistic program in the 1920s.[1] The other source was the increasing number of foundational decision problems that attracted the attention of mathematicians. The most pertinent problem was the *Entscheidungsproblem* ("decision problem"), which concerns the decidability of logical systems. Hilbert and Ackermann (1928) described it as the most fundamental problem of mathematical logic.

By that time, however, the notion of effective procedure—and hence of logical calculus, formal proof, decidability, and solvability—were not well-defined. This situation was changed when Church, Kleene, Post, and Turing published four pioneering papers on computability in 1936. These papers provide a precise mathematical characterization of the class of effectively computable functions: namely, the functions whose values can

---

[1] See Sieg (2009) for further historical discussion.

be computed by means of an effective procedure. The four characterizations are quite different. Alonzo Church (1936a) characterized the effectively computable functions (over the positives) in terms of lambda-definability, a work he started in the early 1930s and which was pursued by his students Stephen Kleene and Barkley Rosser. Kleene (1936) characterized the general recursive functions based on the expansion of primitive recursiveness by Herbrand (1931) and Gödel (1934).[2] Emil Post (1936) described "finite combinatory processes" carried out by a "problem solver or worker" (p. 103). The young Alan Turing (1936) offered a precise characterization of the computability of real numbers in terms of Turing machines. Church (1936a, 1936b) and Turing (1936) also proved, independently of each other, that the *Entscheidungsproblem* is undecidable for first-order predicate calculus.

It was immediately proved that all four precise characterizations are extensionally equivalent: they identify the same class of functions. The claim that this class of functions encompasses the class of effectively computable functions is now known as the *Church-Turing thesis* (CTT). Church (1936a) and Turing (1936), and to some degree Post (1936), formulated versions of this thesis. It is presently stated as follows:

The Church-Turing Thesis (CTT): Any effectively computable function is Turing-machine computable.[3]

The statement connects an "intuitive" or "pre-theoretic" notion, that of effective computability, and a precise notion (e.g., that of recursive function or Turing machine computability). Arguably, due to the pre-theoretic notion, such a "thesis" is not subject to mathematical proof.[4] This thesis, together with the precise notions of a Turing machine, a recursive function, lambda-definability, and the theorems surrounding them, constitutes the foundations of computability theory in mathematics and computer science.

Although CTT determines the extension of effective computability, there is still a debate about the intension of effective computation, which relates to the process that yields computability. Robin Gandy (1980) and Wilfried Sieg (1994, 2009) argue that effective computation is essentially related to an idealized human agent (called here a "human computer"): "Both Church and Turing had in mind calculation by an abstract human being using some mechanical aids (such as paper and pencil). The word 'abstract' indicates that the argument makes no appeal to the existence of practical limits on time and space" (Gandy 1980, 123–124).[5] This does not mean that only human computers can effectively compute; machines can do this, too. The notion is essentially related to a

---

[2] Church (1936a), too, refers to this characterization. Later, Kleene (1938) expands the definition to partial functions. For more recent historical discussion, see Adams (2011).

[3] In some formulations, the thesis is symmetrical, also stating the "easy part," which says that every Turing-machine computable function is effectively computable.

[4] The claim that CTT is not provable has been challenged, e.g., by Mendelson (1990); but see also the discussion in Folina (1998) and Shapiro (1993).

[5] See also Sieg (1994, 2009) and Copeland (2002a), who writes that an effective procedure "demands no insight or ingenuity on the part of the human being carrying it out."

human computer in that a function is effectively computable *only if* a human computer can, in principle, calculate its values.

This anthropomorphic understanding of effective computation is supported by two kinds of evidence. One pertains to the epistemic role of effective computability in the foundational work that was at the background of the 1936 developments. In this context, it is hard to make sense of a formal proof, decision problem, and so on unless we understand "effectively computable" in terms of human computation. The second source of evidence is the way that some of the founders refer to effective computation. The most explicit statements appear in Post who writes that the purpose of his analysis "is not only to present a system of a certain logical potency but also, in its restricted field, of psychological fidelity" (1936, 105) and that "to mask this identification [CTT] under a definition hides the fact that a fundamental discovery in the limitations of the mathematicizing power of Homo Sapiens has been made" (105, note 8). But the one who explicitly analyzes human computation is Turing (1936), whose work has recently received more careful attention.

1.2   Turing's Analysis of Human Computation

Turing's 1936 paper is a classic for several reasons. First, Turing introduces the notion of an *automatic machine* (now known as a Turing machine). Its main novelty is in the distinction between the "program" part of the machine, which consists of a finite list of states or instructions, and a memory tape that is potentially infinite. This notion is at the heart of computability theory and automata theory. Second, Turing introduces the notion of a universal Turing machine, which is a (finite) Turing machine that can simulate the operations of any particular Turing machine (of which there are infinitely many!); it can thus compute any function that can be computed by any Turing machine. This notion has inspired the development of the general purpose digital electronic computers that now dominate almost every activity in our daily life. Third, Turing provides a highly interesting argument for CTT, one whose core is an analysis of the notion of human computation: the argument is now known as "Turing's analysis."[6]

Turing recognizes that encompassing the effectively computable functions involves the computing processes: "The real question at issue is 'What are the possible processes which can be carried out in computing a number?'" (1936, 249). He proceeds to formulate constraints on all such possible computational processes, constraints that are motivated by the limitation of a human's sensory apparatus and memory. Turing enumerates, somewhat informally, several constraints that can be summarized by the following restrictive conditions:

---

[6] Turing's analysis is explicated by Kleene (1952). It has been fully appreciated more recently by Gandy (1988), Sieg (1994, 2009), and Copeland (2004).

1. "The behavior of the computer at any moment is determined by the symbols which he is observing, and his 'state of mind' at that moment" (250).
2. "There is a bound B to the number of symbols or squares which the computer can observe at one moment" (250).
3. "The number of states of mind which need be taken into account is finite" (250).
4. "We may suppose that in a simple operation not more than one symbol is altered" (250).
5. "Each of the new observed squares is within L squares of an immediately previously observed square" (250).

It is often said that a Turing machine is a model of a computer, arguably, a human computer. This statement is imprecise. A Turing machine is a *letter machine*. At each point it "observes" only one square on the tape. "The [human] computer" might observe more than that (Figure 2.1). The other properties of a Turing machine are more restrictive, too. The next step of Turing's argument is to demonstrate that the Turing machine is no less powerful than a human computer. More precisely, Turing sketches a proof for the claim that every function whose values can be arrived at by a process constrained by conditions 1–5 is Turing-machine computable. The proof is straightforward. Conditions 1–5 ensure that each computation step involves a change in one bounded part of the relevant symbolic configuration, so that the number of types of elementary steps is bounded and simple. The proof is completed after demonstrating that each such step can be mimicked, perhaps by a series of steps, by a Turing machine.

We can summarize Turing's argument as follows:

1. *Premise 1:* A human computer operates under the restrictive conditions 1–5.
2. *Premise 2 (Turing's theorem):* Any function that can be computed by a computer that is restricted by conditions 1–5 is Turing-machine computable.
3. *Conclusion (CTT):* Any function that can be computed by a human computer is Turing-machine computable.

Although Turing's argument is quite formidable, there are still interesting philosophical issues to be addressed: does the analysis essentially refer to humans? And what exactly is a human computer?[7] Other questions concerning computation by Turing machines include: does a Turing machine, as an abstract object, really compute (and, more generally, do abstract machines compute?), or is it a model of concrete objects that compute? We return to some of these questions later. The next step is to discuss machine computation.

---

[7] See Copeland and Shagrir (2013), who distinguish between a cognitive and noncognitive understanding of a human computer.

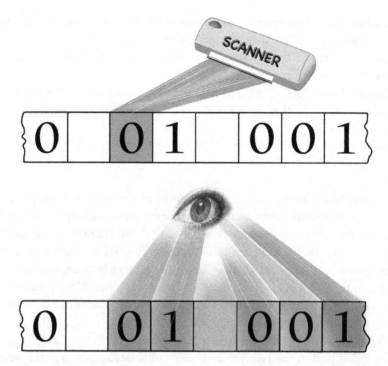

FIGURE 2.1 A human computer vs. a Turing machine. A Turing machine "observes" at each moment only one square on the tape (top). The number of squares observed by a human computer at each moment is also bounded, but it is not limited to a single square (bottom).

1.3 Gandy's Analysis of Machine Computation

It has been claimed that Turing characterized effective computability by analyzing human computation. The question is whether and to what extent this analysis captures the notion of machine computation in general. Gandy observed that "there are crucial steps in Turing's analysis where he appeals to the fact that the calculation is being carried out by a human being" (1980, 124). For example, the second condition in the analysis asserts that there is a fixed bound on the number of cells "observed" at each step. In some parallel machines, however, there is no such limit. In the Game of Life, for instance, such a locality condition is satisfied by each cell: a cell "observes," as it were, only its local bounded environment; namely, itself and its eight neighboring cells. However, the condition is not satisfied by the machine as a whole because all the cells are being observed simultaneously, and there are potentially infinitely many cells.

Gandy moves forward to expand Turing's analysis and to characterize machine computation more generally. He provides an argument for the following claim:

> *Thesis M*: What can be calculated by a machine is [Turing-machine] computable (1980, 124).

Gandy confines the thesis to *deterministic discrete mechanical devices*, which are, "in a loose sense, digital computers" (1980, 126). Thus, Gandy actually argues for the following:

> *Gandy's thesis:* Any function that can be computed by a discrete deterministic mechanical device is Turing-machine computable.

Gandy characterizes discrete deterministic mechanical devices in terms of precise axioms, called Principles I–IV. Principle I ("form of description") describes a deterministic discrete mechanical device as a pair $<S,F>$, where S is a potentially infinite set of states and F is a state-transition operation from S_i to S_{i+1}. Putting aside the technicalities of Gandy's presentation, the first principle can be approximated as:

> I. *Form of description:* Any discrete deterministic mechanical device M can be described by $<S,F>$, where S is a structural class, and F is a transformation from S_i to S_j. Thus, if S_0 is M's initial state, then $F(S_0), F(F(S_0)), \ldots$ are its subsequent states.

Principles II and III place boundedness restrictions on S. They can be informally expressed as:

> II. *Limitation of hierarchy:* Each state S_i of S can be assembled from parts, that can be assemblages of other parts, and so on, but there is a finite bound on the complexity of this structure.
> III. *Unique reassembly:* Each state S_i of S is assembled from basic parts (of bounded size) drawn from a reservoir containing a bounded number of types of basic parts.

Principle IV, called "local causation," puts restrictions on the types of transition operations available. It says that each changed part of a state is affected by a bounded local "neighborhood":

> IV. *Local causation:* Parts from which $F(x)$ can be reassembled depend only on bounded parts of x.

The first three principles are motivated by what is called a discrete deterministic device. Principle IV is an abstraction of two "physical presuppositions": "that there is a lower bound on the linear dimensions of every atomic part of the device and that there is an upper bound (the velocity of light) on the speed of propagation of changes" (1980, 126). If the propagation of information is bounded, an atom can transmit and receive information in its bounded neighborhood in bounded time. If there is a lower bound on the size of atoms, the number of atoms in this neighborhood is bounded. Taking these together, each changed state, $F(x)$, is assembled from bounded, though perhaps overlapping, parts of x. In the Game of Life, for example, each cell impacts the state of several (i.e., its neighboring cells). Those systems satisfying Principles I–IV are known as *Gandy machines*.

The second step in Gandy's argument is to prove a theorem asserting that any function computable by a Gandy machine is Turing-machine computable. The proof is far more complex than the proof of Turing's theorem because it refers to (Gandy) machines that work in parallel on the same regions (e.g., the same memory tape).[8]

Gandy's argument can be summarized as follows:

1. *Premise 1 (Thesis P)*: -"A discrete deterministic mechanical device satisfies principles I–IV" (1980, 126).
2. *Premise 2 (Theorem)*: "What can be calculated by a device satisfying principles I–IV is [Turing-machine] computable" (1980, 126).
3. *Conclusion (Gandy's thesis)*: What can be calculated by a discrete deterministic mechanical device is Turing-machine computable.

Gandy's argument has the same structure as Turing's. The first premise ("Thesis P") posits axioms of computability for discrete deterministic mechanical devices (Turing formulated axioms for human computers). The second premise is a reduction theorem that shows that the computational power of Gandy machines does not exceed that of Turing machines (Turing advanced a reduction theorem with respect to machines limited by conditions 1–5). The conclusion ("Gandy's thesis") is that the computational power of discrete deterministic mechanical devices is bounded by Turing-machine computability (CTT is the claim about the scope of human computability).

The main difference between Gandy and Turing pertains to the restrictive conditions. Gandy's restrictions are weaker. They allow state transitions that result from changes in arbitrarily many bounded parts (in contrast, Turing allows changes in only one bounded part). In this way, Gandy's characterization encompasses *parallel* computation: "if we abstract from practical limitations, we can conceive of a machine which prints an arbitrary number of symbols simultaneously" (Gandy 1980, 124–125). Gandy does set a restriction on the stage transition of each part. Principle IV, of local causation, states that the transition is bounded by the local environment of this part. The Principle is rooted in presuppositions about the *physical world*. One presupposition, about the lower bound on the size of atomic parts, in fact derives from the assumption that the system is discrete. The presupposition about the speed of propagation is a basic principle of relativity. Indeed, Gandy remarked that his thesis P is inconsistent with Newtonian devices: "Principle IV does not apply to machines obeying Newtonian mechanics" (1980, 145).

Gandy's restrictions are weaker than those imposed on human computers. But are they general enough? Does Gandy capture machine computation as stated in his Thesis M? Sieg suggests that he does; he contends that Gandy provides "a characterization of computations by machines that is as general and convincing as that of computations

---

[8] But see Sieg (2009) for a more simplified proof of the theorem.

by human computors given by Turing" (2002, 247). But this claim has been challenged (Copeland and Shagrir 2007): Gandy's principles do not fully capture the notions of algorithmic machine computation and that of physical computation.

1.4 Algorithmic Machine Computation

In computer science, computing is often associated with *algorithms* (effective procedures). Algorithms in computer science, however, are primarily associated with *machines*, not with humans (more precisely, human computers are considered to be one kind of machines). Even CTT is formulated, in most computer science textbooks, in the broader terms of *algorithmic/effective machine computation*.[9] We can thus introduce a modern version of the Church-Turing thesis for an extended, machine version of the thesis:

> *The modern Church-Turing thesis:* Any function that can be *effectively* (algorithmically) computed by a *machine* is Turing-machine computable.

The modern thesis is assumed to be true. The main arguments for the thesis, however, do not appeal to human calculation. The two arguments for the thesis that appear in most textbooks are the argument from confluence and the argument from non-refutation. The argument from non-refutation states that the thesis has never been refuted. The argument from confluence says that many characterizations of computability, although differing in their goals, approaches, and details, encompass the same class of computable functions.[10]

The nature of algorithms is a matter of debate within computer science.[11] But the notion seems broader in scope than that of a Gandy machine. There are asynchronous algorithms that do not satisfy Principle I.[12] We can think of an asynchronous version of the Game of Life in which a fixed number of cells are simultaneously updated, but the identity of the updated cells is chosen randomly. Principle I is also violated by interactive or online (algorithmic) machines that constantly interact with the environment while computing. There might also be parallel synchronous algorithms that violate Principle IV. Gandy himself mentions Markov normal algorithms: "The process of deciding

---

[9] See, e.g., Hopcroft and Ullman (1979, 147), Lewis and Papadimitriou (1981, 223), and Nagin and Impagliazzo, who write: "The claim, called *Church's thesis* or the *Church-Turing thesis*, is a basis for the equivalence of algorithmic procedures and computing machines" (1995, 611).

[10] See Boolos and Jeffrey (1989, 20) and Lewis and Papadimitriou (1981, 223–224). But see also Dershowitz and Gurevich (2008) who use the Turing-Gandy axiomatic approach to argue for the truth of the thesis.

[11] The nature of algorithms is addressed by, e.g., Milner (1971), Knuth (1973, 1–9), Rogers (1987, 1–5), Odifreddi (1989, 3), Moschovakis (2001), Yanofsky (2011), Gurevich (2012), Vardi (2012), and Dean (forthcoming).

[12] See Copeland and Shagrir (2007) and Gurevich (2012) for examples.

whether a particular substitution is applicable to a given word is essentially global" (1980, 145).[13] "Global" here means that there is no bound on the length of a substituted word; hence, "locality" is not satisfied. Another example is a fully recurrent network in which the behavior of a cell is influenced by the information it gets from all other cells, yet the size of the network grows with the size of the problem.

I do not claim that computer science is concerned only with algorithmic machine computation (I return to this point later). The point is that the notion of algorithmic machine computation is broader than that of Gandy machines. This is perhaps not too surprising. As said, Gandy was concerned with deterministic (synchronous) discrete mechanical devices, and he relates *mechanical* to the "two physical suppositions" that motivate Principle IV. Let us turn then to yet another notion, that of physical computation.

1.5 Physical Computation

Physical computation is about physical systems or machines that compute. The class of physical systems includes not only actual machines but also physically possible machines (in the loose sense that their dynamics should conform to physical laws). Gandy machines, however, do not encompass all physical computation in that sense. He himself admits that his characterization excludes "devices which are *essentially* analogue" (Gandy 1980, 125). What is even more interesting is that his *Thesis M* (that limits physical computability to Turing-machine computability) might not hold for physical computing systems. There are arguably physical *hypercomputers*, namely, physical machines that compute functions that are not Turing machine computable.[14]

The claim that physical systems do not exceed the Turing limit is known as the *physical* Church-Turing thesis.[15] Gualtiero Piccinini (2011) distinguishes between bold and modest versions of the thesis. The bold thesis is about physical systems and processes in general, not necessarily the computing ones:

> *The bold physical Church-Turing thesis:* "Any physical process is Turing computable" (746).

The modest thesis concerns physical *computing* processes or systems:

> *The modest physical Church-Turing thesis:* "Any function that is physically computable is Turing computable" (746).

---

[13] An even earlier example is the substitution operation in the equational calculus that is generally recursive yet "essentially global" (Gödel 1934).

[14] For reviews of hypercomputation, see Copeland (2002b) and Syropoulos (2008).

[15] For early formulations, see Deutsch (1985), Wolfram (1985), and Earman (1986).

It turns out that most physical processes are computable in the bold sense that the input-output function that describes their behavior is real-valued Turing comput-able.[16] A well-known exception is presented by Pour-El and Richards (1981). Our focus is the modest thesis. A few counterexamples of physical hypercomputers have been suggested. These are highly idealized physical systems that compute functions that are not Turing machine computable. A celebrated class of counterexamples is that of *supertask machines*. These machines can complete infinitely many operations in a finite span of time. I will discuss here *relativistic machines* that are compatible with the two "physical presuppositions" that underlie Gandy's principle of local causation (the two presuppositions are an upper bound on the speed of signal propagation and a lower bound on the size of atomic parts).[17]

The construction of relativistic machines rests on the observation that there are relativistic space-times with the following property. The space-time includes a future endless curve λ with a past endpoint p, and it also includes a point q, such that the entire stretch of λ is included in the chronological past of q.[18] We can thus have a relativistic machine RM that consists of two communicating standard machines, T_A and T_B. The two standard machines start traveling from p. T_B travels along the endless curve λ. T_A moves along a future-directed curve that connects the beginning point p of λ with q. The time it takes T_A to travel from p to q is finite, while during that period T_B completes the infinite time trip along λ (Figure 2.2).

This physical setup permits the computation of the *halting function*, which is not Turing-machine computable. The computation proceeds as follows (Figure 2.3). T_A receives the input (m,n) and prints "0" in its designated output cell. It then sends a signal with the pertinent input to T_B. T_B is a universal machine that mimics the computation of the mth Turing machine operating on input n. In other words, T_B calculates the Turing-machine computable function $f(m,n)$ that returns the output of the mth Turing machine (operating on input n), if this m^{th} Turing machine halts. T_B returns no value if the simu-lated m^{th} Turing machine does not halt. If T_B halts, it immediately sends a signal back to T_A; if T_B never halts, it never sends a signal. Meanwhile T_A "waits" during the time it takes T_A to travel from p to q (say, one hour). If T_A has received a signal from T_B it prints "1," replacing the "0" in the designated output cell. After an hour (of T_A's time), the out-put cell shows the value of the halting function. It is "1" if the mth machine halts on input n, and it is "0" otherwise.

How is it that RM can compute the halting function? The quick answer is that RM per-forms a supertask: T_B might go through infinitely many computation steps in finite time (from T_A's perspective). But this is not the whole story. A crucial additional feature is

---

[16] There are natural extensions of the notion of Turing computability to real-number functions. Definitions are provided by Grzegorczyk (1957); Pour-El and Richards (1989); Blum, Cucker, Shub, and Smale (1997); and others.

[17] Other examples of supertask machines are the accelerating machines (Copeland 2002c; Earman 1986) and the shrinking machines (Davies 2001).

[18] The first constructions of this setup were proposed by Pitowsky (1990) and Hogarth (1994).

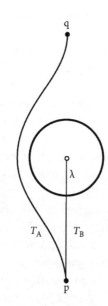

FIGURE 2.2 A setup for relativistic computation (adapted from Hogarth 1994, p. 127; with permission from the University of Chicago Press). In this relativistic Malament-Hogarth space-time the entire stretch of λ is included in the chronological past of q. The machines T_A and T_B start traveling together from a point p. T_A moves along a future-directed curve toward a point q. T_B travels along the endless curve λ. The time it takes for T_A to travel from p to q is finite, while during that period T_B completes an infinite time trip along λ.

that the communication between T_A and T_B is not deterministic in the sense familiar in algorithmic computation. In algorithmic machines, the configuration of each $\alpha + 1$ stage is determined by that of the previous α stage. This condition is explicit in Turing's and in Gandy's analyses. But *RM* violates this condition. Its end stage cannot be described as a stage $\alpha + 1$, whose configuration is completely determined by the preceding stage α. This is simply because there is no such preceding stage α, at least when the simulated machine never halts.[19] *RM* is certainly deterministic in *another* sense: it obeys laws that invoke no random or stochastic elements. In particular, its end state is a deterministic *limit* of previous states of T_B (and T_A).

Does *RM* violate the modest physical Church-Turing thesis? The answer largely depends on whether the machine *computes* and on whether its entire operations are *physically possible*. There is, by and large, a consensus that the relativistic machine is a genuine hypercomputer. The physical possibility of the machine is far more controversial.[20] But even if *RM* is not physically possible, we can use it to make two further points.

---

[19] See Copeland and Shagrir (2007) for an analysis of this point.
[20] See Earman and Norton (1993) for a skeptical note, but see also the papers by István Németi and colleagues (e.g., Andréka et al. (2009)) for a more optimistic outlook. They suggest that physical realizability can be best dealt with in setups that include huge slow rotating black holes.

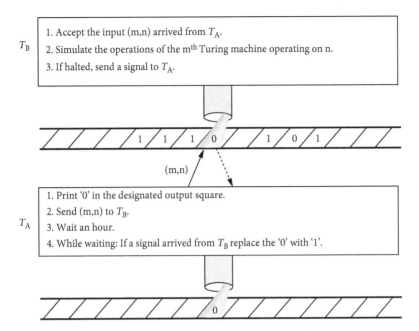

T_B
1. Accept the input (m,n) arrived from T_A.
2. Simulate the operations of the mth Turing machine operating on n.
3. If halted, send a signal to T_A.

(m,n)

T_A
1. Print '0' in the designated output square.
2. Send (m,n) to T_B.
3. Wait an hour.
4. While waiting: If a signal arrived from T_B replace the '0' with '1'.

FIGURE 2.3 Computing the halting function. The relativistic machine RM consisting of two communicating standard (Turing) machines T_A and T_B computes the halting function. T_B is a universal machine that mimics the computation of the mth Turing machine operating on input n. If it halts, T_B sends a signal back to T_A; if it never halts, T_B never sends a signal. Waiting for a signal from T_B, T_A prints "1" ("halts") if a signal is received; otherwise, it leaves the printed "0" ("never halts").

One is that there are "notional," logically possible machines that compute beyond the Turing limit; RM is one such machine, but there are many others. The second point is that the *concept* of physical computation accommodates relativistic computation (and, indeed, hypercomputation): after all, we agree that if *RM* is physically possible, then it does violate the modest Church-Turing thesis.

1.6 On the Relations Between Notions of Computation

Our survey indicates that there are varieties of computation: human, machine, effective, physical, notional, and other kinds of computation. Let us attempt to summarize the relationships among them. Our comparison is between the abstract principles that define each kind of computation ("models of computation"). The inclusion relations are depicted in Figure 2.4. We start with the Turing machine model that applies to (some) letter machines. The human computer model is broader in that it includes all machines that satisfy Turing's conditions 1–5. Gandy machines include even more machines, notably parallel machines that operate on bounded environments. Gandy machines,

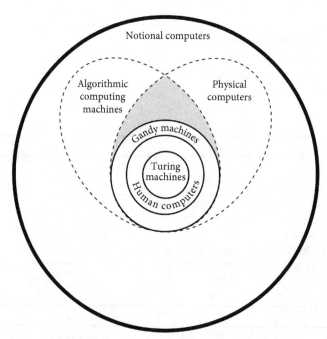

FIGURE 2.4 Relations between classes of computing systems. The relations are measured with respect to models of computation. "Turing machines" are the class of computers that are modeled by some Turing machine. "Human computers" are the class of computers that are limited by Turing's restrictive conditions 1–5. "Gandy machines" are those computers restricted by Principles I–IV. "Algorithmic computing machines" are those machines that compute by means of an algorithm (as characterized in computer science). "Physical computers" refers to computing systems that are physically realizable. "Notional computers" refers to computing systems that are logically possible.

however, do not exhaust two classes of machine computation. Algorithmic computation includes machines that are considered algorithmic in computer science yet are not Gandy machines. Physical computation includes physical machines (e.g., analogue computers) that are not Gandy machines. It seems that the classes of algorithmic computing machines and physical computing machines partially overlap. Algorithmic computation includes logically possible machines that are not physically realizable (possible). Physical computation might include hypercomputers that exceed the computational power of Turing machines (e.g., relativistic machines). Gandy machines are apparently at the intersection of algorithmic and physical computation. Whether they exhaust the intersection is an open question. Last, notional computation includes logically possible machines that are neither algorithmic nor physical; here, we may find some instances of the so-called infinite-time Turing machines.[21]

---

[21] See Hamkins and Lewis (2000).

1.7 What Is the Subject Matter?

What are the computational sciences about? The somewhat complex picture portrayed here indicates that there is no single subject matter covered by the computational sciences. Each "science" might be about a different kind of computation. Early computability theory and proof theory naturally focus on human computation (what can be done by human calculators). Automata theory and many other branches in theoretical computer science have focused on algorithmic machine computation. The Turing-machine model has been so central to these studies because it arguably captures the computational power of all algorithmic machines. But there are certainly some changes in current trends in computer science (and other formal sciences). One trend, which narrows the scope of computation, is to focus on models of physically possible machines.[22] Another trend, which widens the scope, is to go beyond algorithmic computation, and, in this context, I mentioned the interesting recent work on infinite-time Turing machines.

Computational neuroscience and cognitive science are about physical computation in the brain. The working assumption is that neural processes are computing physical processes, and the empirical question is about the functions computed and how they are being computed. The important lesson for philosophers here is that claims about the brain being a Turing machine or about the brain computing Turing-machine computable functions are *not* entailed from the notion of physical computation. Our survey shows that the brain and other physical systems can be computers even if they are not Turing machines and even if the functions they compute are not Turing machine computable at all. Whether or not brains are Turing machines and whether or not they compute Turing computable functions are apparently empirical questions.

2 Ontology, Epistemology, and Complexity

There are many philosophical issues involved in the computational sciences, and I cannot possibly cover all of them. I will briefly touch on three issues: the ontology of computation, the nature of computational theories and explanations, and the relevance of complexity theory to philosophy and specifically to verification and confirmation.

2.1 The Ontology of Computation

There are intriguing questions concerning the ontology of computation: what exactly are algorithms, and how are they related to programs and automata? Are algorithms,

---

[22] An example of this trend is given later in the section on computational complexity.

programs, and automata abstract entities? And what is meant here by abstract? Does abstract mean mathematical? Is abstract opposed to concrete? And how is abstract related to abstraction and idealization?[23] Another set of (related) questions concerns physical computation: what is the difference between a system that computes and one that does not compute? Is the set of physical computing systems a natural kind? Is there only one notion of physical computation, or are there several? And what are the realization or implementation relations between the abstract and the physical?[24] Yet another set of questions concern semantics and information in computing systems.[25]

I focus here on concrete physical computation. The central question is about the nature of physical computation. Looking at the literature, we can identify at least four approaches to physical computation: formal, mechanistic, information-processing, and modeling approaches. The four are not mutually exclusive. An account of physical computation typically includes at least two of the criteria imposed by these approaches. A formal approach proceeds in two steps. It identifies computation with one or another theoretical notion from logic or computer science—algorithm, program, procedure, rule, proof, automaton, and so on. It then associates this theoretical notion with a specific organizational, functional, or structural property of physical systems. An example is Cummins's account of computation. Cummins (1988) associates computation with program execution: "To compute a function g is to execute a program that gives o as its output on input i just in case $g(i) = o$. Computing reduces to program execution" (91). He then associates program execution with the functional property of *steps*: "Program execution reduces to step-satisfaction" (92).[26] I contented that both stages in the formal approach are flawed. The moral of the first part of this chapter is that physical computation is not delimited by the theoretical notions found in logic and theoretical computer science. I also suspect that the difference between computing and noncomputing physical systems is not to be found in some specific functional or structural property of physical systems (Shagrir 2006).

The mechanistic approach identifies computation with *mechanism*, as it is currently understood in the philosophy of science. In this view, computation is a specific kind of mechanistic process; it has to do with the processing of variables to obtain certain relationships among inputs, internal states, and outputs (Piccinini 2007, 2015b; Miłkowski 2013). The information-processing approach, as the name suggests, identifies computation with information processing (Churchland and Sejnowski 1992; Fresco 2014). The difficulty here is to spell out the notion of information in this context. Some claim that there is "no computation without representation" (Fodor 1975, 34; Pylyshyn 1984, 62).[27]

---

[23] See Turner's (2013) entry in the *Stanford Encyclopedia of Philosophy* for a review of some of these issues.

[24] This is one focus of Piccinini's (2015a) entry in the *Stanford Encyclopedia of Philosophy*.

[25] See Floridi (1999) and Fresco (2014, chap. 6) for pertinent discussion.

[26] Fodor (2000) identifies computation with *syntactic properties*, associated in turn with second-order physical properties. Chalmers (2011) identifies computation with the *implementation of automata*, associated in turn with a certain causal-organizational profile. Most accounts of computation fall within this category.

[27] See Sprevak (2010) for an extensive defense of this view.

Others advocate for a nonsemantic notion of information (Miłkowski 2013). The modeling view, which I favor, associates computation with some sort of morphism. According to this view, computation preserves certain relations in the target domain. Take, for example, the oculomotor integrator, which computes the mathematical function of integration. This system preserves the (integration) relation between velocity and positions of the (target) eyes. The distance between two successive eye positions is just the mathematical integration over the eye velocity with respect to time (Shagrir 2010).

A related question is whether computation is a natural kind. This question is important in determining whether the computational sciences are part of the natural sciences or not. This question was raised in the context of the cognitive sciences by Putnam (1988) and Searle (1992). Searle and Putnam argue (roughly) that every physical object can execute any computer program (Searle) or implement any finite state automaton (Putnam). Searle infers from this that computation is not objective, in the sense that it is observer-relative rather than an intrinsic feature of the physical world. He further concludes that "There is no way that computational cognitive science could ever be a natural science, because computation is not an intrinsic feature of the world. It is assigned relative to observers" (212). The argument, if sound, applies to other computational sciences as well.[28]

I would like to make three comments about the argument. First, the argument assumes a formal approach to computation because it identifies computation with programs or automata, although it has some relevance to other approaches too. Second, it is, by and large, agreed that both Putnam and Searle assume a much too liberal notion of program execution and/or implementation of automata. Adding constraints related to counterfactual scenarios, groupings of states and others dramatically narrows the scope of the result.[29] And yet, even when these constraints are in place, almost every physical system executes at least one program and/or implements at least one automaton. What is ruled out is that the system executes/implements all programs/automata. Third, the universal realizability claim (at least under its weaker version, that every physical object executes/implements at least one program/automaton) does not entail that computation is not objective. The fact that every physical object has a mass does not entail that mass is not intrinsic to physics. Similarly, it might well be that executing a program and/or implementing an automaton is objective.

One could still insist that universal realizability does undermine the computational sciences. The assumption underlying these sciences is that we describe some physical systems as computing and others as not. But if every physical system implements an automaton (granted that this relation is objective), then there must be yet another feature of computing systems that distinguishes them from the noncomputing systems. And the worry is that this additional feature is nonobjective. Chalmers (2011) replies to

---

[28] Motivated by other considerations Brian Smith argues that *"Computation is not a subject matter"* (1996, 73).

[29] See Chrisley (1994), Chalmers (1996), Copeland (1996), Melnyk (1996), Scheutz (2001), and Godfrey-Smith (2009).

this worry by noting that universal realizability does not entail that there must be this additional feature and/or that this feature must be nonobjective. He thinks that computation is no more than the objective feature of implementing an automaton. Whether we refer to this computation depends on the phenomena we want to explain. Thus, digestion might be computation, but, presumably, performing this computation is irrelevant to digestion. "With cognition, by contrast, the claim is that it is *in virtue* of implementing some computation that a system is cognitive. That is, there is a certain class of computations such that *any* system implementing that computation is cognitive" (332–333).

My view is that there is some interest-relative element in the notion of computation. This, however, does not entail that the computational sciences do not seek to discover objective facts. These sciences might aim to discover the automata realized by certain physical objects. Moreover, the class of computing systems might have a proper subclass of entirely objective computations. Let us assume that the missing ingredient in the definition of computation is representation. As we know, some people distinguish between derivative and natural representational systems (e.g., Dretske 1988). Partly objective computers are derivative representational systems that (objectively) realize automata. Fully objective computers are natural representational systems that realize automata. Computer science might well be about the first class of systems. Computational cognitive and brain sciences, however, might well be about fully objective computers. One way or another, we cannot infer from the universal realizability result that that cognitive and brain sciences are not natural sciences. The brain might well be a fully objective computer.

2.2 Computational Theories and Explanations

The computational sciences posit computational theories and explanations. What is the nature of these theories and explanations? Again, we do not focus here on the role of computer models and simulations in the scientific investigation. Our query is about the nature of scientific theories and explanations (including models) whose goal is to theorize and explain physical computing systems.

The two main themes that concern philosophers are the nature of computational explanations and the relations between computational explanations of a physical system and other ("more physical") explanations of this system. Robert Cummins (2000) takes computational explanations to be a species of *functional analysis*: "Functional analysis consists in analyzing a disposition into a number of less problematic dispositions such that programmed manifestation of these analyzing dispositions amounts to a manifestation of the analyzed disposition. By 'programmed' here, I simply mean organized in a way that could be specified in a program or flowchart" (125). In some cases, the subcapacities/dispositions may be assigned to the components of the system, but, in other cases, they are assigned to the whole system. An example of the latter is a Turing machine: "Turing machine capacities analyze into other Turing machine capacities" (125). Another example is some of our calculative abilities: "My capacity to multiply

27 times 32 analyzes into the capacity to multiply 2 times 7, to add 5 and 1, and so on, but these capacities are not (so far as is known) capacities of my components" (126).[30]

As for the relations between computational explanations and "lower level" more physical explanations, Cummins and others take the former to be "autonomous" from the latter ones. One reason for the autonomy is the multiple realization of computational properties. Whereas computational properties ("software") are realized in physical properties ("hardware"), the same computational property can be realized in very different physical substrates.[31] Thus, the (computational) explanation referring to computational properties has some generality that cannot be captured by each physical description of the realizers.[32] Another (and somewhat related) reason is that there is a gulf between computational explanations that specify function (or program) and the lower level explanations that specify mechanisms—specifically, components and their structural properties. These are arguably distinct kinds of explanation.[33]

This picture is in tension with the mechanistic philosophy of science. Some mechanists argue that scientific explanations, at least in biology and neuroscience, are successful to the extent that they are faithful to the norms of mechanistic explanations (Kaplan 2011). However, computational explanations are, seemingly, not mechanistic. Moreover, mechanists characterize levels of explanations (at least in biology and neuroscience) in terms of levels of mechanisms, whereas components at different levels are described in terms of a whole–part relationship (Craver 2007). However, it is not clear how to integrate computational explanations into this picture because computational explanations often describe the same components described by lower level mechanistic explanations.

In defending the mechanistic picture, Piccinini and Craver (2011) argue that computational explanations are *sketches* of mechanisms: they "constrain the range of components that can be in play and are constrained in turn by the available components" (303). Computational descriptions are placeholders for structural components or subcapacities in a mechanism. Once the missing aspects are filled in, the description turns into "a full-blown mechanistic explanation"; the sketches themselves can thus be seen as "elliptical or incomplete mechanistic explanations" (284). They are often a guide toward the structural components that constitute the full-blown mechanistic explanations.

Another influential view is provided by David Marr (1982). Marr (1982) argues that a complete cognitive explanation consists of a tri-level framework of computational, algorithmic, and implementational levels. The computational level delineates the

---

[30] As we see, Cummins invokes terminology ("program") and examples of computing systems to explicate the idea of functional analysis. Computations, however, are by no means the only examples of functional analysis. Another example Cummins invokes is the cook's capacity to bake a cake (2000, 125).

[31] Ned Block writes that "it is difficult to see how there could be a non-trivial first-order physical property in common to all and only the possible physical realizations of a given Turing-machine state" (1990, 270–271).

[32] See Putnam (1973) and Fodor (1975). There are many responses to this argument; see, e.g., Polger and Shapiro (2016) for discussion.

[33] See Fodor (1968) and Cummins (2000). But see also Lycan (1987) who argues that the "mechanistic" levels are often also functional.

phenomenon, which is an information-processing task. The algorithmic level characterizes the system of representations that is being used (e.g., decimal vs. binary) and the algorithm employed for transforming representations of inputs into those of outputs. The implementation level specifies how the representations and algorithm are physically realized.

Philosophers disagree, however, about the role Marr assigns to the computational level. Many have argued that the computational level aims at stating the cognitive task to be explained; the explanation itself is then provided at the algorithmic and implementation levels (Ramsey 2007). Piccinini and Craver (2011) argue that Marr's computational and algorithmic levels are sketches of mechanisms. Yet others have associated the computational level with an idealized *competence* and the algorithmic and implementation levels with actual performance (Craver 2007; Rusanen and Lappi 2007). Egan (2010) associates the computational level with an explanatory formal theory, which mainly specifies the computed mathematical function. Bechtel and Shagrir (2015) emphasize the role of the environment in Marr's notion of computational analysis. Although we agree with Egan and others that one role of the computational level is to specify the computed mathematical function, we argue that another role is to demonstrate the basis of the computed function in the physical world (Marr 1977). This role is fulfilled when it is demonstrated that the computed mathematical function mirrors or preserves certain relations in the visual field. This explanatory feature is another aspect of the modeling approach discussed above.

2.3 Complexity: the Next Generation

Theoretical computer science has focused in the past decades not so much on computability theory but on computational complexity theory. They distinguish between problems that are tractable or can be solved efficiently, and those that are not. To make sure, all these functions ("problems") are Turing machine computable. The question is whether there is an efficient algorithm for solving this problem. So far, philosophers have pretty much ignored this development. I believe that, within a decade or so, the exciting results in computational complexity will find their way into the philosophy of science. In "Why Philosophers Should Care About Computational Complexity," Scott Aaronson (2013) points to the relevance of computational complexity to philosophy. In this last section, I echo his message. After saying something about complexity, I will discuss its relevance to scientific verification and confirmation.

An algorithm solves a problem efficiently if the number of computational steps can be upper-bounded by a polynomial relative to the problem size n. The familiar elementary school algorithm for multiplication is efficient because it takes about n^2 steps ($\sim n^2$); n here can refer to the number of symbols in the input. An algorithm is inefficient if the number of computational steps can be lower-bounded by an exponential relative to n. The truth table method for finding whether a formula (in propositional logic) is satisfiable or not is inefficient: it might take $\sim 2^n$ steps (as we recall, the table consists of 2^n rows, where n is the number of different propositional variables in the formula). The difference between efficient and inefficient reflects a gulf in the growth of the computation time it takes to complete the computation "in practice." Let us assume that it takes a machine 10^{-6} seconds to complete one computation step. Using the elementary school algorithm, it would take the machine about 4×10^{-6} seconds to complete the multiplication of two single-digit numbers, and about 1/100 second to complete the multiplication of two fifty-digit numbers. Using the truth table method, it would also take the machine about 4×10^{-6} seconds to decide the satisfiability of a formula consisting of two propositional variables (assuming that n is the number of propositional variables). However, it might take the machine about 10^{14} (more than 100 trillion) years to decide the satisfiability of an arbitrary formula consisting of 100 propositional variables! Not surprisingly, this difference is of immense importance for computer scientists.

A problem is tractable if it can be solved by a polynomial-time algorithm; it is intractable otherwise. It is an open question whether all problems are tractable. The conjecture is that many problems are intractable. Theoreticians often present the question in terms of a "P = NP?" question. P is the class of (tractable) problems for which there is a polynomial-time algorithm; we know multiplication belongs to this class. NP is the class of problems for which a solution can be *verified* in polynomial time.[34] The satisfiability problem (SAT) belongs to NP: we can verify in polynomial time whether a proposed row in the truth table satisfies the formula or not. It is clear that P $\subseteq$ NP. But it is conjectured that there are problems in NP that are not in P. SAT has no known polynomial-time solution, and it is assumed that it is not in P. SAT belongs to a class of *NP-complete* problems, which is a class of very hard problems. If any of these NP-complete problems turns out to belong to P, then P = NP. Our assumption in what follows is that P $\neq$ NP.[35]

Much of the significance of the P = NP question depends on the *extended Church-Turing thesis*. The thesis asserts that the time complexities of any two general and reasonable models of (sequential) computation are polynomially related. In particular, a problem that has time complexity t on some general and reasonable model of computation has time complexity poly(t) in a single-tape Turing machine (Goldreich 2008, 33)[36]; poly(t) means that the differences in time complexities is polynomial. We can think of the thesis in terms of invariance: much as the Turing machine is a general model of algorithmic computability (which is what the Church-Turing thesis is about), the Turing machine is also a general model of time complexity (which is what the extended thesis is about). The computational complexity of a "reasonable" model of computation can be very different from that of a Turing machine, but only up to a point: the time complexities must be polynomially related. What is a *reasonable* model of computation? Bernstein

---

[34] Formally speaking, P stands for polynomial time and NP for nondeterministic polynomial time. NP includes those problems that can be computed in polynomial time by a nondeterministic Turing machine.

[35] The question of whether P = NP is traced back to a letter Gödel sent to von Neumann in 1956; see Aaronson (2013) and Copeland and Shagrir (2013) for discussion.

[36] The most familiar formulation is by Bernstein and Vazirani (1997, 1411).

and Vazirani take "reasonable to mean in principle physically realizable" (1997, 1411). Aharonov and Vazirani (2013, 331) talk about a "physically reasonable computational model." As noted in Section 1, it is interesting to see the tendency of computer scientists today to associate computation with physical computation.

I believe that computational complexity theory has far-reaching implications for the way philosophers analyze proofs, knowledge, induction, verification, and other philosophical problems. I briefly discuss here the relevance of complexity to scientific verification and confirmation (Aharonov, Ben-Or, and Eban 2008; Yaari 2011). So, here goes: quantum computing has attracted much attention in theoretical computer science due to its potentially dramatic implications for the extended Church-Turing thesis. Bernstein and Vazirani (1997) give formal evidence that a *quantum Turing machine* violates the extended Church-Turing thesis; another renowned example from quantum computation that might violate the extended thesis is Shor's factoring algorithm (Shor 1994).[37] More generally, quantum computing theorists hypothesize that quantum machines can compute in polynomial time problems that are assumed to be outside of P (e.g., factorization) and even outside of NP.

This result seems to raise a problem for scientific verification.[38] Take a quantum system (machine) QM. Given our best (quantum) theory and various tests, our hypothesis is that QM computes a function f that is outside NP; yet QM computes f in polynomial time. But it now seems that we have no way to test our hypothesis. If n is even moderately large, we have no efficient way to test whether QM computes $f(n)$ or not. QM proceeds through $\sim n^c$ steps in computing $f(n)$; c is some constant. Yet the scientists use standard methods of verification—"standard" in the sense that they cannot compute problems that are outside P in polynomial time. Scientists calculate the results with paper and pencil[39] or, more realistically, use a standard (Turing) machine TM in testing QM. This method is inefficient because TM will complete its computation in $\sim c^n$ steps, long after the universe ceases to exist. If it turns out that many of the physical (quantum) systems compute functions that are outside NP, the task of their verification and confirmation is seemingly hopeless.[40]

Aharonov, Ben-Or, and Eban (2008) show how to bypass this problem. They have developed a method of scientific experimentation that makes it possible to test non-P hypotheses in polynomial time. The technical details cannot be reviewed here. I will just mention that at the heart of the method is a central notion in computational complexity theory called *interactive proof* (Goldwasser, Micali, and Rackoff 1985). In an interactive

---

[37] Factorization of an integer is the problem of finding the primes whose multiplication together makes the integer. It is assumed that factorization is not in P, although it is surely in NP: given a proposed solution, we can multiply the primes to verify that they make the integer.

[38] I follow Yaari's (2011) presentation.

[39] The (reasonable) assumption is that human computation does not reach problems outside P in polynomial time (Aaronson 2013).

[40] Given that f is not in NP, the verification by TM requires $\sim c^n$ steps. But the same problem arises for problems in NP (Aharonov and Vazirani 2013). Assuming that QM provides only a yes/no answer for SAT, our TM will require $\sim c^n$ steps to verify the hypothesis that QM solves SAT.

proof system, a verifier aims to confirm an assertion made by a prover; the verifier has only polynomial time resources, whereas the prover is computationally powerful. The verifier can confirm the correctness of the assertion by using a protocol of questions by which she challenges the prover (this notion of proof is itself revolutionary because the protocol is interactive and probabilistic.[41]). Aharonov, Ben-Or, and Eban (2008) import the notion to the domain of scientific investigation, in which (roughly) the verifier is the computationally weak scientist and the prover is Mother Nature. Yaari (2011) takes this result a philosophical step further in developing a novel theory of scientific confirmation, called *interactive confirmation*. One obvious advantage of this theory over other theories (e.g., Bayesian) is that interactive confirmation can account for *QM* and similar quantum systems.

ACKNOWLEDGEMENTS

I am grateful to Ilan Finkelstein, Nir Fresco, Paul Humphreys, Mark Sprevak, and Jonathan Yaari for comments, suggestions, and corrections. This research was supported by the Israel Science Foundation, grant 1509/11.

REFERENCES

Aaronson, Scott (2013). "Why Philosophers Should Care about Computational Complexity." In B.J. Copeland, C.J. Posy, and O. Shagrir, *Computability: Turing, Gödel, Church, and Beyond* (Cambridge MA: MIT Press), 261–327.

Adams, Rod (2011). *An Early History of Recursive Functions and Computability from Gödel to Turing* (Boston: Docent Press).

Aharonov, Dorit, Ben-Or, Michael, and Eban, Elad (2008). "Interactive Proofs for Quantum Computations." *Proceedings of Innovations in Computer Science (ICS 2010)*: 453–469.

Aharonov, Dorit, and Vazirani, Umesh V. (2013). "Is Quantum Mechanics Falsifiable? A Computational Perspective on the Foundations of Quantum Mechanics." In B.J. Copeland, C.J. Posy, and O. Shagrir, *Computability: Turing, Gödel, Church, and Beyond* (Cambridge MA: MIT Press), 329–349.

Andréka, Hajnal, Németi, Istvan, and Németi, Péter (2009). "General Relativistic Hypercomputing and Foundation of Mathematics." *Natural Computing* 8: 499–516.

Bechtel, William, and Shagrir, Oron (2015). "The Non-Redundant Contributions of Marr's Three Levels of Analysis for Explaining Information Processing Mechanisms." *Topics in Cognitive Science* 7: 312–322.

Bernstein, Ethan, and Vazirani, Umesh (1997). "Quantum Complexity Theory." *SIAM Journal on Computing* 26: 1411–1473.

Block, Ned (1990). "Can the Mind Change the World?" In G. S. Boolos (ed.), *Method: Essays in Honor of Hilary Putnam* (New York: Cambridge University Press), 137–170.

---

[41] See Aaronson (2013). See also Shieber (2007) who relates interactive proofs to the Turing test.

Blum, Lenore, Cucker, Felipe, Shub, Michael, and Smale, Steve (1997). *Complexity and Real Computation* (New York: Springer-Verlag).

Boolos, George S., and Jeffrey, Richard C. (1989). *Computability and Logic*, 3rd edition (Cambridge: Cambridge University Press).

Chalmers, David J. (1996). "Does a Rock Implement Every Finite-State Automaton?" *Synthese* 108: 309–333.

Chalmers, David J. (2011). "A Computational Foundation for the Study of Cognition." *Journal of Cognitive Science* 12: 323–357.

Chrisley, Ronald L. (1994). "Why Everything Doesn't Realize Every Computation." *Minds and Machines* 4: 403–420.

Church, Alonzo (1936a). "An Unsolvable Problem of Elementary Number Theory." *American Journal of Mathematics* 58: 345–363.

Church, Alonzo (1936b). "A Note on the Entscheidungproblem." *Journal of Symbolic Logic* 1: 40–41.

Churchland, Patricia S., and Sejnowski, Terrence (1992). *The Computational Brain* (Cambridge, MA: MIT Press).

Copeland, B. Jack (1996). "What Is Computation?" *Synthese* 108: 335–359.

Copeland, B. Jack (2002a). "The Church-Turing Thesis." In E. N. Zalta (ed.), *The Stanford Encyclopedia of Philosophy* (http://plato.stanford.edu/archives/sum2015/entries/church-turing/).

Copeland, B. Jack (2002b). "Hypercomputation." *Minds and Machines* 12: 461–502.

Copeland, B. Jack (2002c). "Accelerating Turing Machines." *Minds and Machines* 12: 281–301.

Copeland, B. Jack (2004). "Computable Numbers: A Guide." In B. J. Copeland (ed.), *The Essential Turing* (New York: Oxford University Press), 5–57.

Copeland, B. Jack, Posy, Carl, J., and Shagrir, Oron, eds. (2013). *Computability: Turing, Gödel, Church, and Beyond* (Cambridge MA: MIT Press).

Copeland, B. Jack, and Shagrir, Oron (2007). "Physical Computation: How General are Gandy's Principles for Mechanisms?" *Minds and Machines* 17: 217–231.

Copeland, B. Jack, and Shagrir, Oron (2013). "Turing versus Gödel on Computability and the Mind." In B. J. Copeland, C.J. Posy, and O. Shagrir (eds.), *Computability: Turing, Gödel, Church, and Beyond* (Cambridge MA: MIT Press), 1–33.

Craver, Carl F. (2007). *Explaining the Brain: Mechanisms and the Mosaic Unity of Neuroscience* (New York: Oxford University Press).

Cummins, Robert C. (1988). *Meaning and Mental Representation* (Cambridge MA: MIT Press).

Cummins, Robert C. (2000). "'How Does It Work?' vs. 'What Are the Laws?' Two Conceptions of Psychological Explanation." In F. Keil and R. Wilson (eds.), *Explanation and Cognition* (Cambridge MA: MIT Press), 117–144.

Davies, Brian E. (2001). "Building Infinite Machines." *British Journal for the Philosophy of Science* 52: 671–682.

Dean, Walter (forthcoming). "Algorithms and the Mathematical Foundations of Computer Science." In L. Horsten and P. Welch (eds.), *The Limits and Scope of Mathematical Knowledge* (Oxford University Press).

Dershowitz, Nachum, and Gurevich, Yuri (2008). "A Natural Axiomatization of Computability and Proof of Church's Thesis." *Bulletin of Symbolic Logic* 14: 299–350.

Deutsch, David (1985). "Quantum Theory, the Church-Turing Principle and the Universal Quantum Computer." *Proceedings of the Royal Society of London A* 400: 97–117.

Dretske, Fred (1988). *Explaining Behavior: Reasons in a World of Causes* (Cambridge, MA: MIT Press).

Earman, John (1986). *A Primer on Determinism* (Dordrecht: Reidel).

Earman, John, and Norton, John D. (1993). "Forever Is a Day: Supertasks in Pitowsky and Malament-Hogarth Spacetimes." *Philosophy of Science* 60: 22–42.

Egan, Frances (2010). "Computational Models: A Modest Role for Content." *Studies in History and Philosophy of Science* 41: 253–259.

Floridi, Luciano (1999). *Philosophy and Computing: An Introduction* (London and New York: Routledge).

Fodor, Jerry A. (1968). *Psychological Explanation* (New York: Random House).

Fodor, Jerry A. (1975). *The Language of Thought* (Cambridge, MA: Harvard University Press).

Fodor, Jerry A. (2000). *The Mind Doesn't Work That Way: The Scope and Limits of Computational Psychology* (Cambridge, MA: MIT Press).

Folina, Janet (1998). "Church's Thesis: Prelude to a Proof." *Philosophia Mathematica* 6: 302–323.

Fresco, Nir (2014). *Physical Computation and Cognitive Science* (London: Springer-Verlag).

Gandy, Robin O. (1980). "Church's Thesis and Principles of Mechanisms." In S. C. Kleene, J. Barwise, H. J. Keisler, and K. Kunen (eds.), *The Kleene Symposium* (Amsterdam: North-Holland), 123–148.

Gandy, Robin O. (1988). "The Confluence of Ideas in 1936." In R. Herken (ed.), *The Universal Turing Machine* (New York: Oxford University Press), 51–102.

Gödel, Kurt (1934). "On Undecidable Propositions of Formal Mathematical Systems." In S. Feferman, J. W. Dawson, Jr., S. C. Kleene, G. H. Moore, R. M. Solovay, and J. van Heijenoort (eds.), *Collected Works I: Publications 1929-1936* (Oxford: Oxford University Press), 346–369.

Gödel, Kurt (1956). "Letter to John von Neumann, March 20th, 1956." In S. Feferman, J. W. Dawson, Jr., S. C. Kleene, G. H. Moore, R. M. Solovay, and J. van Heijenoort, (eds.), *Collected Works V: Correspondence, H–Z* (Oxford: Oxford University Press), 372–377.

Godfrey-Smith, Peter (2009). "Triviality Arguments Against Functionalism." *Philosophical Studies* 145: 273–295.

Goldreich, Oded (2008). *Computational Complexity: A Conceptual Perspective* (New York: Cambridge University Press).

Goldwasser, Shafi, Micali, Silvio, and Rackoff, Charles (1985). "The Knowledge Complexity of Interactive Proof Systems." *Proceedings of the Seventeenth Annual ACM Symposium on Theory of Computing (STOC '85)*: 291–304.

Grzegorczyk, Andrzej (1957). "On the Definitions of Computable Real Continuous Functions." *Fundamenta Mathematicae* 44: 61–71.

Gurevich, Yuri (2012). "What Is an Algorithm?" In M. Bieliková, G. Friedrich, G. Gottlob, S. Katzenbeisser, and G. Turán (eds.), *SOFSEM: Theory and Practice of Computer Science, LNCS 7147* (Berlin: Springer-Verlag), 31–42.

Hamkins, Joel D., and Lewis, Andy (2000). "Infinite Time Turing Machines." *Journal of Symbolic Logic* 65: 567–604.

Herbrand, Jacques (1931). "On the Consistency of Arithmetic." In W.D. Goldfarb (ed.), *Jacques Herbrand Logical Writings* (Cambridge, MA: Harvard University Press), 282–298.

Hilbert, David, and Ackermann, Wilhelm F. (1928). *Grundzuge der Theoretischen Logik* (Berlin: Springer-Verlag).

Hogarth, Mark L. (1994). "Non-Turing Computers and Non-Turing Computability." *PSA: Proceedings of the Biennial Meeting of the Philosophy of Science Association* 1: 126–138.

Hopcroft, John E., and Ullman, Jeffery D. (1979). *Introduction to Automata Theory, Languages, and Computation* (Reading, MA: Addison-Wesley).

Kaplan, David M. (2011). "Explanation and Description in Computational Neuroscience." *Synthese* 183: 339–373.

Kleene, Stephen C. (1936). "General Recursive Functions of Natural Numbers." *Mathematische Annalen* 112: 727–742.

Kleene, Stephen C. (1938). "On Notation for Ordinal Numbers." *Journal of Symbolic Logic* 3: 150–155.

Kleene, Stephen C. (1952). *Introduction to Metamathematics* (Amsterdam: North-Holland).

Knuth, Donald E. (1973). *The Art of Computer Programming: Fundamental Algorithms, Vol. 1* (Reading, MA: Addison Wesley).

Lewis, Harry R., and Papadimitriou, Christos H. (1981). *Elements of the Theory of Computation* (Englewood Cliffs, NJ: Prentice-Hall).

Lycan, William (1987). *Consciousness* (Cambridge, MA: MIT Press).

Marr, David C. (1977). "Artificial Intelligence—Personal View." *Artificial Intelligence* 9: 37–48.

Marr, David C. (1982). *Vision: A Computation Investigation into the Human Representational System and Processing of Visual Information* (San Francisco: Freeman).

Melnyk, Andrew (1996). "Searle's Abstract Argument against Strong AI." *Synthese* 108: 391–419.

Mendelson, Elliott. (1990). "Second Thoughts about Church's Thesis and Mathematical Proofs." *Journal of Philosophy* 87: 225–233.

Miłkowski, Marcin (2013). *Explaining the Computational Mind* (Cambridge, MA: MIT Press).

Milner, Robin (1971). "An Algebraic Definition of Simulation between Programs." In D. C. Cooper (ed.), *Proceeding of the Second International Joint Conference on Artificial Intelligence* (London: The British Computer Society), 481–489.

Moschovakis, Yiannis N. (2001). "What is an Algorithm?" In B. Engquist and W. Schmid (eds.), *Mathematics Unlimited—2001 and Beyond* (New York: Springer-Verlag), 919–936.

Nagin, Paul, and Impagliazzo, John (1995). *Computer Science: A Breadth-First Approach with Pascal* (New York: John Wiley & Sons).

Odifreddi, Piergiorgio (1989). *Classical Recursion Theory: The Theory of Functions and Sets of Natural Numbers* (Amsterdam: North-Holland Publishing).

Piccinini, Gualtiero (2007). "Computing Mechanisms." *Philosophy of Science* 74: 501–526.

Piccinini, Gualtiero (2011). "The Physical Church-Turing Thesis: Modest or Bold?" *British Journal for the Philosophy of Science* 62: 733–769.

Piccinini, Gualtiero (2015a). "Computation in Physical Systems." In E. N. Zalta (ed.), *Stanford Encyclopedia of Philosophy* (http://plato.stanford.edu/archives/sum2015/entries/computation-physicalsystems/).

Piccinini, Gualtiero (2015b). *The Nature of Computation: A Mechanistic Account* (New York: Oxford University Press).

Piccinini, Gualtiero, and Craver, Carl F. (2011). "Integrating Psychology and Neuroscience: Functional Analyses as Mechanism Sketches." *Synthese* 183: 283–311.

Pitowsky, Itamar (1990). "The Physical Church Thesis and Physical Computational Complexity." *Iyyun* 39: 81–99.

Polger, Thomas W., and Shapiro, Lawrence A. (2016). *The Multiple Realization Book* (Oxford: Oxford University Press).

Post, Emil L. (1936). "Finite Combinatory Processes—Formulation I." *Journal of Symbolic Logic* 1: 103–105.

Pour-El, Marian B., and Richards, Ian J. (1981). "The Wave Equation with Computable Initial Data Such that Its Unique Solution Is Not Computable." *Advances in Mathematics* 39: 215–239.

Pour-El, Marian B., and Richards, Ian J. (1989). *Computability in Analysis and Physics* (Berlin: Springer-Verlag).

Putnam, Hilary (1973). "Reductionism and the Nature of Psychology." *Cognition* 2: 131–149.

Putnam, Hilary (1988). *Representation and Reality* (Cambridge, MA: MIT Press).

Pylyshyn, Zenon W. (1984). *Computation and Cognition: Toward a Foundation for Cognitive Science* (Cambridge, MA: MIT Press).

Ramsey, William M. (2007). *Representation Reconsidered* (New York: Cambridge University Press).

Rogers, Hartley (1987). *Theory of Recursive Functions and Effective Computability*, 2nd edition (Cambridge, MA: MIT Press).

Rusanen, Anna-Mari, and Lappi, Otto (2007). "The Limits of Mechanistic Explanation in Neurocognitive Sciences." In S. Vosniadou, D. Kayser, and A. Protopapas, (eds.), *Proceedings of European Cognitive Science Conference 2007* (Hove, UK: Lawrence Erlbaum), 284–289.

Scheutz, Matthias (2001). "Causal vs. Computational Complexity?" *Minds and Machines* 11: 534–566.

Searle, John R. (1992). *The Rediscovery of the Mind* (Cambridge MA: MIT Press).

Shagrir, Oron (2006). "Why We View the Brain as A Computer." *Synthese* 153: 393–416.

Shagrir, Oron (2010). "Computation, San Diego Style." *Philosophy of Science* 77: 862–874.

Shapiro, Stewart (1993). "Understanding Church's Thesis, Again." *Acta Analytica* 11: 59–77.

Shieber, Stuart M. (2007). "The Turing Test as Interactive Proof." *Noûs*, 41: 686–713.

Shor, Peter W. (1994). "Algorithms for Quantum Computation: Discrete Logarithms and Factoring." In S. Goldwasser (ed.), *Proceedings of the 35th Annual Symposium on Foundations of Computer Science* (Los Alamitos, CA: IEEE Computer Society Press), 124–134.

Sieg, Wilfried (1994). "Mechanical Procedures and Mathematical Experience." In A. George (ed.), *Mathematics and Mind* (Oxford: Oxford University Press), 71–117.

Sieg, Wilfried (2002). "Calculations by Man and Machine: Mathematical Presentation." In P. Gärdenfors, J. Wolenski, and K. Kijania-Placek (eds.), *The Scope of Logic, Methodology and Philosophy of Science, Volume I* (Dordrecht: Kluwer Academic Publishers), 247–262.

Sieg, Wilfried (2009). "On Computability." In A. Irvine (ed.), *Handbook of the Philosophy of Mathematics* (Amsterdam: Elsevier), 535–630.

Smith, Brian C. (1996). *On the Origin of Objects* (Cambridge, MA: MIT Press).

Sprevak, Mark (2010). "Computation, Individuation, and the Received View on Representation." *Studies in History and Philosophy of Science* 41: 260–270.

Syropoulos, Apostolos (2008). *Hypercomputation: Computing Beyond the Church-Turing Barrier* (New York: Springer-Verlag).

Turing, Alan M. (1936). "On Computable Numbers, with an Application to the Entscheidungsproblem." *Proceedings of the London Mathematical Society, Series 2* 42: 230–265.

Turner, Raymond (2013). "Philosophy of Computer Science." In E. N. Zalta (ed.), *Stanford Encyclopedia of Philosophy* (http://plato.stanford.edu/archives/win2014/entries/computer-science/).

Vardi, Moshe Y. (2012). "What is an Algorithm?" *Communications of the ACM* 55: 5.

Wolfram, Stephen (1985). "Undecidability and Intractability in Theoretical Physics." *Physical Review Letters* 54: 735–738.

Yaari, J. (2011). *Interactive Proofs as a Theory of Confirmation*. PhD thesis (The Hebrew University of Jerusalem).

Yanofsky, Noson S. (2011). "Towards a Definition of Algorithms." *Journal of Logic and Computation* 21: 253–286.

....

CHAPTER 3

PHILOSOPHY OF THE SOCIAL SCIENCES

Naturalism and Anti-naturalism in the Philosophy of Social Science

....

FRANCESCO GUALA

1 INTRODUCTION

....

MY first encounter with the philosophy of social science took place in the mid-1990s, when I was a graduate student at the London School of Economics. The course was taught by different lecturers in each term. On the opening day, the first lecturer explained that there is little consensus regarding the central problems of the philosophy of social science. "What is human action?," in his view, was the most important question, and he told us that the whole term would be devoted to it. He also told us that the problem of human action cannot be solved scientifically. In fact, he claimed, one can do philosophy of social science without paying much attention to what social scientists do.

I was struck by these statements. In the ensuing weeks, the lecturer steered through paradoxes and counterexamples, making his way slowly in the writings of Donald Davidson, Alfred Mele, and other action theorists. The path led us deeply into the philosophy of mind and metaphysics, and on one thing I certainly agreed with him: philosophers' theories of action seemed to be unrelated to the practice of economists, sociologists, or political scientists.

In the second term, however, we were served a different menu. With the new lecturer came a new syllabus and a different approach to the philosophy of social science. The menu included a thorough discussion of modeling in economics, with special emphasis on the problem of idealization. Then, much of the term was devoted to the use of statistics for the identification of causal relations and to the use of causal knowledge for policy-making. The reading material included papers written by scientists, and the students were encouraged to familiarize themselves with the methods used by social scientists in their everyday work.

Thus, I became aware that there are two ways of doing philosophy of social science. The second lecturer believed that research practices in the social sciences raise interesting conceptual and methodological issues—much like physics and biology do. She also recognized that social scientists try to tackle these problems constructively, that they often make progress, and that they have something to teach philosophers on epistemic matters. The implicit message was that philosophers cannot contribute to our understanding of the social world or to the methodology of social science unless they engage seriously with the social sciences.

I learned later that the approach of the second lecturer was a version of philosophical *naturalism*. Naturalism comes in different forms, but, for the purposes of this chapter, I will borrow Harold Kincaid's definition: naturalism "denies that there is something special about the social world that makes it unamenable to scientific investigation, and also denies that there is something special about philosophy that makes it independent or prior to the sciences in general and the social sciences in particular" (Kincaid 2012a, 3). Anti-naturalism, by contrast, affirms both of these theses and in general promotes a style of philosophizing that proceeds autonomously from the (social) sciences.[1]

The divide between naturalists and anti-naturalists used to cut the discipline deeply twenty years ago, and it still does today. In fact, it is difficult to write an impartial overview of the field that does justice to these two ways of doing philosophy of the social sciences. One possible strategy is to take sides and proceed as if those who work on the other side of the divide did not exist. My own instinct then would be to survey the work done by naturalistically minded philosophers on the methodology and ontology of social science. The survey would cover topics such as modeling, causality, mechanistic explanation, theoretical unification, statistical inference, experimentation, simulation, realism, functionalism, individualism, evolution, rational choice, welfare, and justice, among others.[2]

Such a survey, however, would overlook the most distinctive feature of contemporary philosophy of social science: the persistence and influence of anti-naturalism and its recent growth, especially in the area of "social ontology." In this chapter, therefore, I will address the naturalism/anti-naturalism question directly. The next section is devoted to articulating the two positions more precisely. Section 3 introduces a thesis—the dependence of social phenomena on linguistic representation—that is typically used by anti-naturalists to question the possibility of explanation and prediction in social science. Dependence on representation can be interpreted in different ways, and each interpretation raises a different challenge. The causal interpretation is discussed in Sections 3.1 and 3.2, focusing on Ian Hacking's theory of feedback loops and interactive kinds. We will see that feedback loops can be studied scientifically using the tools of game theory

[1] See also Kincaid (1996), Ross (2011).
[2] There are excellent surveys of these topics in Kincaid (2012b) and Jarvie and Zamora Bonilla (2011). For historical reasons related to its separation from the other social sciences, the naturalistic approach has become majoritarian in the philosophy of economics roughly at the same time as in the philosophy of physics and biology. See, e.g., Kincaid and Ross (2009) and Mäki (2012b).

and that causal dependence does not constitute a threat to the explanatory ambitions of social science. Section 4 is devoted to the constitutive interpretation of dependence, defended by influential philosophers like Charles Taylor and John Searle. Section 4.2 introduces the notion of constitutive rule and outlines the epistemic implications of the constitutive view. Section 4.3 questions the constitutive view, illustrating an alternative account of social kinds that does not presuppose their dependence on representation. The chapter ends with some remarks on the "anti-naturalist's predicament," in Section 5 and on the prospects of doing philosophy autonomously from social science.

2 NATURALISM AND ANTI-NATURALISM

Broadly speaking, philosophical naturalists hold that science is the authority on ontological and epistemological matters. This characterization is, however, vague: what is "science" to begin with? Most naturalists agree that the most advanced branches of physics and biology deserve to be taken seriously—but how about psychology, history, or, indeed, social science?

Philosophers have traditionally been deferent toward natural science and diffident toward the social sciences. The mature version of logical positivism that became influential in the middle of the twentieth century, for example, promoted a monistic picture of science shaped on the paradigmatic example of physics. Scientific theories were conceived as axiomatic systems, explanations as deductive or statistical arguments, and evidential support as a logical or probabilistic relation between empirical and theoretical statements. The logical positivist picture was descriptive and normative at the same time: it purportedly described how good science (physics) works and offered a model for less mature scientific disciplines. Most of the gains of philosophical research were to be had in the relatively underdeveloped areas of science, rather than at its most advanced frontier.[3]

Anti-naturalists in the 1950s and 1960s challenged this monistic picture. Philosophers like Alasdair MacIntyre (1957), Peter Winch (1958), and Charles Taylor (1971, 1985) argued forcefully—against the tide—that social scientists should not take natural science as a normative benchmark. They claimed that special features of the social world prevent the successful application of the methods of natural science. Anti-naturalism thus took on a double meaning: it refused to take science as an authority on epistemic matters, and it opposed the "naturalization" of social research on the ideal model of physics.

The decisive blow against the monistic picture, however, was dealt in the 1970s and 1980s by a new generation of philosophers who argued that the logical positivists had not portrayed even the paradigmatic discipline accurately.[4] It also became

---

[3] Hempel's first defense of the deductive-nomological model, for example, occurred in a paper devoted to historical explanation (Hempel 1942).

[4] E.g., Cartwright (1983), Hacking (1983), Salmon (1984), Giere (1988).

clear that many areas of science—like biology, economics, and neuroscience—could not be easily made to fit the positivist straightjacket. This "disunity of science" movement persuaded most philosophers that each scientific discipline uses (and ought to use) methods of inquiry that are suited to its particular topic of investigation. Success depends both on the goals to be achieved by scientists and on contingent matters of fact that may be subject-specific (i.e., characteristic of each domain of inquiry).

The disunity movement opened the way for a new form of naturalism that takes social science seriously. Divergence from the positivistic ideal, one could now argue, may be due to a difference in subject matter rather than to the immaturity of social science. The definition of naturalism given in the previous section in fact is post-positivistic in this respect. Contemporary naturalists like Kincaid deny two theses:

T1: That there is something special about the social world that makes it unamenable to scientific investigation

T2: That there is something special about philosophy that makes it independent or prior to the sciences in general and the social sciences in particular

Naturalists hold that there is no intrinsic obstacle for the social sciences to do the things that the natural sciences do. There is no reason why social scientists should be unable to explain social phenomena, predict them, and intervene to change the course of events in ways that we find desirable.[5] Naturalists, however, are not committed to a specific way in which social phenomena should be explained and predicted. There is no presumption, in particular, that social scientists should use the same theories, models, or inferential techniques that natural scientists use. Even though scientists often apply tools of wide applicability, sometimes they ought to use methods that are especially suited to a specific context of investigation.

According to contemporary naturalists, finally, there is no principled reason to believe that the social sciences are stuck in a blind alley, waiting for philosophers to show the way. Naturalists believe that those who spend their time studying social phenomena are best positioned to assess the efficacy of specific methods and that those philosophers who want to improve the methodology of social science must engage with scientific practice at the same level of analysis.

Anti-naturalists, in contrast, believe that social scientists are oblivious to some overarching philosophical truths. That's why social scientists keep aiming at impossible objectives or keep using inadequate tools in their research. Not only have the social sciences made little progress so far, but they will continue to lag behind the rest of science in the future unless they are radically reformed. The job of the philosopher is to point this out and help social scientists to transform their discipline. According to anti-naturalists,

---

[5] For brevity, in the remainder of the chapter, I will simply refer to these tasks as "prediction and explanation."

therefore, philosophical investigation is prior to scientific research (as stated in T2). John Searle, for example, puts it as follows:

> I believe that where the social sciences are concerned, social ontology is prior to methodology and theory. It is prior in the sense that unless you have a clear conception of the nature of the phenomena you are investigating, you are unlikely to develop the right methodology and the right theoretical apparatus for conducting the investigation. (Searle 2009, 9)[6]

Similarly, according to Raimo Tuomela, the account of the social world that results from philosophical investigation "is meant in part to critically analyze the presuppositions of current scientific research and . . . to provide a new conceptual system for theory-building" (2002, 7).

3 THEORY DEPENDENCE:
THE CAUSAL INTERPRETATION

The prototypical anti-naturalist strategy is to identify a feature of social reality that is overlooked by current social science. There are, of course, different views concerning what the overlooked feature is. But there are also remarkable similarities: an all-time favorite is the capacity of human beings to represent themselves, their own actions, and their own societies in symbolic form. According to Searle, the neglected feature is language:

> Social and political theorists assume that we are already language-speaking animals and then go on to discuss society without understanding the significance of language for the very existence of social reality. (Searle 2009, 2)

Many anti-naturalists agree that symbolic representation plays a crucial role in the creation and reproduction of social phenomena. "The conceptions according to which we normally think of social events . . . enter into social life itself and not merely into the observer's description of it" (Winch 1958, 95). Even our individual identities—who we are and what we do—is shaped by linguistic representation:

> naturalism fails to recognize a crucial feature of our ordinary understanding of human agency, of a person or self. One way of getting at this feature is in terms of

---

[6] Notice that Searle is a self-professed naturalist regarding *natural* science because he believes that social ontology must be consistent with the worldview of contemporary physics and biology (see, e.g., Searle 1995). Since Kincaid's definition of naturalism is discipline-specific, however, Searle falls into the anti-naturalistic camp regarding the philosophy of social science.

the notion of self-interpretation. A fully competent human agent not only has some understanding of himself, but is partly constituted by this understanding. (Taylor 1985, 3)

So, we shall suppose for the sake of the argument that social phenomena are indeed dependent on language or representation. What are the philosophical and scientific implications of this fact?

To begin with, it is useful to distinguish two interpretations of the dependence thesis. According to the first, social phenomena are dependent on representation in a *causal* fashion, whereas according to the second, they are dependent in a more robust *constitutive* or ontological manner. Since different philosophers have endorsed different interpretations, it will be necessary to discuss them both. I will begin with the causal interpretation, which is easier to grasp and analyze. In fact, it turns out that social scientists are aware of the causal dependence of social phenomena on representation and have provided a thorough analysis of the way it works.

The causal dependence of social phenomena on representation has been labeled in many different ways. Ian Hacking (1999) calls it "interactivity"; George Soros (2013) prefers "reflexivity"; the sociologist of science Donald MacKenzie (2006) has revived Austin's "performativity"; and, in a seminal paper, written almost seventy years ago, Robert K. Merton (1948) introduced the notions of "self-fulfilling" and "self-defeating prophecy." All these expressions refer to the same fundamental phenomenon: the concepts that scientists use to theorize can influence the behavior they purport to describe. The causal mechanism involves "feedback loops," a term from cybernetics that recurs frequently in the literature: people are aware of the theorizing that is taking place and adjust their behavior either to conform with the theories or to defeat them. Either way, social behavior is theory-dependent. What people are and what they do depend in part on how they are classified.

Feedback loops have been discussed by philosophers of science at various points in time.[7] But, since the 1990s, the debate has focused on Hacking's theory of "interactive kinds" (1986, 1995, 1999), so I will use it as a paradigmatic example in this section. The moral to be drawn from this debate is that theory-dependence does not constitute an obstacle to the prediction and explanation of social phenomena and does not distinguish the latter from natural phenomena either.

3.1 Interactive Kinds

Hacking describes the phenomenon of theory-dependence as follows:

Responses of people to attempts to be understood or altered are different from the responses of things. This trite fact is at the core of one difference between the natural

---

[7] See, e.g., Grünbaum (1956), Nagel (1960), Buck (1963), Romanos (1973).

and human sciences, and it works at the level of kinds. There is a looping or feed-back effect involving the introduction of classifications of people. New sorting and theorizing induces changes in self-conception and in behaviour of the people classified. These changes demand revisions of the classification and theories, the causal connections, and the expectations. Kinds are modified, revised classifications are formed, and the classified change again, loop upon loop. (Hacking 1995, 370)

"Kind" is an ancient philosophical term. It was originally used by John Locke to translate the Aristotelian notion of genus (*genòs*). Centuries later, it is still used in metaphysics and in philosophy of science, with different connotations. In general, it is used with a realistic slant: the world comes already sorted into kinds before we look at it. The language of kinds is typically used in contrast to the language of sets or classes, collections of things that happen to be grouped for contingent reasons.

An influential conception—first articulated by John Stuart Mill (1843)—relates kinds with scientific inferences and properties. Take cats, for example. Cats have nails, four legs, a liver, and eat mice. They dislike water, consume between twenty and thirty calories per pound of body-weight each day, and have an average life expectancy of fifteen years. Classifying Missy as a cat gives us lots of information because many properties can be inferred from its belonging to the kind *Felis catus*.

An updated version of the Millian conception has been developed by Richard Boyd (1991). Boyd defines kinds as *homeostatic property clusters*. They are "clusters" because the properties of kinds tend to come in a package: they are highly correlated. Having one property makes it highly likely that another property occurs, too. They are "homeostatic" in two senses: (1) the properties are not clustered by chance—there are causal mechanisms that explain why they are correlated—and (2) the correlations are relatively stable. Following Mill, let us call any class that has such properties a *real kind*.[8] Real kinds are important for science because they support inductive inferences and generalizations. They are *projectable*, in Goodman's (1979) sense. The properties and mechanisms of kinds are studied by the sciences, and scientific knowledge makes forecasting possible. If we know some properties of an object, we can infer other properties. Scientific categories like "electron," "uranium," or "competitive price" refer to projectable kinds and real kinds.

Notice that real kinds, according to this conception, may overlap: the same entity may belong to different kinds. Some of these kinds may be natural in the sense of "studied by the natural sciences," and others may be social in the sense of "studied by the social sciences." For instance, I belong to the kind bipedal primate but also to the kinds male, father, husband, Juventus fan. All of these kinds in principle may support inductive inferences and generalizations; so, in principle, it seems that one can be a realist about both natural and social kinds.[9]

---

[8] Philosophers often use the expression "natural kind," but, in the context of social science, that terminology is bound to generate confusion.

[9] For a thorough discussion of pluralism about kinds, see Dupré (1993). Dupré (2001) and Khalidi (2013) discuss the relationship between biological and social kinds.

According to Hacking, however, social and natural kinds are not exactly on a par. There is a difference, and it has to do with the way in which people react to being classified into kinds. Whereas natural entities are *indifferent*, the members of social kinds are not. They interact with classifications—they are *interactive* kinds.

> A cardinal difference between the traditional natural and social sciences is that the classifications employed in the natural sciences are indifferent kinds, while those employed in the social sciences are mostly interactive kinds. The targets of the natural sciences are stationary. Because of looping effects, the targets of the social sciences are on the move. (Hacking 1999, 108)

According to Hacking, theory-dependence is a peculiar social phenomenon, unknown in the natural world.[10] Because interactive kinds lack stability, moreover, they are unlikely to support prediction and explanation in the same way as do natural kinds. Social kinds are not real kinds in the Mill-Boyd sense.

Hacking uses interactivity to defend "an intriguing doctrine" that he calls "dynamic nominalism": *nominalism* because it contends that "our selves are to some extent made up by our naming" (1995, 113), *dynamic* because "the world does not come with a unique pre-packaged structure" (1999, 60). The structure of the world changes because interactive kinds are "on the move." They change over time, and they do so (in part) because of our classifications. This makes scientific inference difficult: projectability requires stability; but if interactive kinds are transient, then their properties are not projectable. Since projectability is the essential feature of real kinds, interactive kinds cannot be real kinds in the Mill-Boyd sense.[11]

3.2 Equilibria

The belief that interactivity makes prediction impossible is as popular as it is false. First, we will see that in many occasions people have incentives to conform with the concepts and predictions that social scientists make. Then, we will see that successful inference is possible even when people want to defeat predictions and classifications.

When people have an incentive to conform with a theoretical category, interactivity may *stabilize* behavior and facilitate inference. Let us illustrate using a simple example adapted from Merton's (1948) essay on self-fulfilling prophecies. Suppose that the

[10] Hacking's demarcation criterion has not had an easy life. The problem is that interactive kinds are undoubtedly common, even in the natural realm: see, e.g., Douglas (1986), Bogen (1988), Haslanger (1995, 2012), Cooper (2004), Khalidi (2010, 2013). The prototypical case—a missile-guiding system—is already in Grünbaum (1956).

[11] Hacking (2002, 2) says that his "dynamic nominalism" might as well be called "dialectical realism." Although he does not explain in which sense a dynamic nominalist could be a realist, I take it that Hacking cannot be a realist in the Mill-Boyd sense.

Table 3.1. Strike–Breaking
as an Equilibrium

	J	B
A	2, 2	0, 0
E	0, 0	1, 1

members of a minority group ("negro workers," in Merton's old-fashioned writing) have a reputation for being unreliable, opportunistic strike-breakers. Let us also suppose that this reputation is reflected in official statistics and social-scientific theorizing. White workers, influenced by these statistics, tend to exclude minority workers from the unions. But because they are excluded, these workers have few opportunities to find a job. If employment is available during a strike, then, the members of the minority group cannot afford to decline the offer. So their reputation as strike-breakers is confirmed: they are strike-breakers because they are believed to be strike-breakers, and they are believed to be strike-breakers because they are strike-breakers.

This reflexive or self-fulfilling prophecy is a *Nash equilibrium*. In the jargon of game theory, an equilibrium is a profile of actions such that the behavior of every individual is optimal given the behavior of the other individuals: each action is a best reply to the actions of others. This means that no one can achieve a better outcome by changing her strategy unilaterally. An important property of this state is that the beliefs of all individuals are true in equilibrium. I choose X because I believe (correctly) that you choose Y, and you choose Y because you believe that I choose X. So, in a Nash equilibrium, the beliefs of the players are consistent and correct. The observation of behavior confirms the expectations of the players, and the expectations induce behavior that is consistent with the expectations. The equilibrium is supported by a feedback loop in Hacking's sense.

The incentives of the players are represented in Table 3.1.[12] The row player stands for the (white) members of the unions: assuming that they will go on strike, the options they face is to either accept (A) or exclude (E) African American workers from the unions. The column player represents African American workers: they can either join the strike (J) or break it (B). There are two possible equilibria: in principle, all the workers would be better off if they went on strike together (AJ). But if Column has a reputation for being a strike-breaker, then Row's rational response is to exclude. Conversely, if Column believes that she is going to be excluded, she is better off breaking the strike. The outcome EB is an equilibrium. The strike-breaker label is a self-fulfilling prophecy, and strike-breaking becomes a real attribute of "negro workers."

---

[12] I follow the usual conventions of game theory: the actions available to each player are represented as rows and columns. The numbers represent payoffs, the first one for the row player and the second one for the column player.

Ron Mallon (2003) has highlighted the analogy between Hacking's interactive kinds and game-theoretic equilibria. The existence of feedback loops, he has argued, is not necessarily an obstacle to prediction. As a matter of fact, the theory-dependence of social phenomena often has the opposite effect than the one suggested by Hacking. A careful analysis of "the looping effect of human kinds shows how the mechanisms involved in creating social roles may act to stabilize those social roles" (Mallon 2003, 340).

But what if people try to *defeat* the theories that have been devised to predict their behavior? This is clearly the most worrying case for the stability of social kinds. Table 3.2 represents a game known as "hide and seek." The analysis of this game caused one of the most spectacular conversions in the history of social science: Oskar Morgenstern gave an early formulation of the problem of theory-dependence in his 1928 dissertation on *Wirtschaftsprognose* (economic forecast). One of his famous examples concerns two individuals—Holmes and Moriarty—playing a game of "hide and seek" (Morgenstern 1928, 98). Holmes is chased by Moriarty, who wants to kill him. If Moriarty predicts that Holmes will take the train to Dover (A), Holmes should stop at an intermediate station (B) to defeat his prediction; but Moriarty should anticipate Holmes's anticipation, prompting Holmes to consider traveling all the way to Dover again, . . . and so forth for an infinite number of iterations.

Morgenstern found this circularity—the fact that any move will provoke a defeating countermove—extremely disturbing, and he used it to draw pessimistic conclusions for the predictive ambitions of social science. But social scientists have gone a long way since then, partly (ironically) thanks to Morgenstern himself (von Neumann and Morgenstern 1944). Any student of game theory now knows that interactions such as these can and often do converge to an equilibrium. Take the game of penalty kicks, a situation that mirrors the Holmes–Moriarty game and that is routinely played by strikers and goal-keepers on European football pitches. Contrary to Morgenstern's intuition, such games have a stable solution (an equilibrium) according to which the strikers shoot stochastically to the right or left of the goal-keeper with a given probability (e.g., Palacios-Huerta 2003). This is a mixed-strategy equilibrium, a pattern that is statistically predictable and that, at the same time, leaves a lot of uncertainty and fun for the fans who watch the game on TV.

These stochastic patterns can be predicted, of course, only if some conditions are satisfied. First, the forecaster must know which options are available to the individuals whose behaviors she is trying to predict. Second, she must know their preferences, or

Table 3.2. Hide and Seek

	A	B
A	1, 0	0, 1
B	0, 1	1, 0

the way in which they will react to the theory. In a pair of classic papers, Simon (1954) and Grunberg and Modigliani (1954) demonstrated that prediction is always possible if conditions of this sort are satisfied. This may seem like a big "if" for, in real life, social scientists do not often have the information that is required for this kind of task. But obviously scientific inferences go wrong when they are not based on enough information, in social and in natural science alike.

So, at best, the theory-dependence argument proves that social science is difficult, which is something most social scientists are happy to concede. It does not prove a priori that scientific inference is impossible or that social kinds are not real kinds in the Mill-Boyd sense. Inferential success is a contingent matter, and science can tell us when and where we are likely to succeed. Theories of interactive kinds do not improve our understanding of social reality and do not pave the way for a new way of doing social research.

To quote Mallon, again:

> Hacking provides little reason to think that such regimes of labeling must always be so causally efficacious as to undermine the associated conception or theory as an instrument of explanation and prediction. Hacking gives us a picture of how instability might work, but no illumination to why it should always work, or in what cases it does work. To recognize that social roles may be unstable is not even to show that they usually are. (Mallon 2003, 346)

4 THEORY-DEPENDENCE: THE CONSTITUTIVE VIEW

The moral so far is that the dependence of social phenomena on representation is not a threat to prediction and explanation if dependence is interpreted causally. For the sake of completeness, we now have to address the other version of the argument, where dependence is interpreted in a more robust ontological fashion. Theory-dependence in this case is a *constitutive,* rather than a causal relation.

Hacking occasionally speaks of "a person being constituted as a certain type of being" (2002, 24), but most of his examples of theory-dependence have an unmistakable causal flavor. To attribute a constitutive interpretation to him would require a considerable stretch of imagination. Other philosophers, however, endorse the constitutive dependence view unambiguously. In "Interpretation and the Sciences of Man," for example, Taylor denounces "the artificiality of the distinction between social reality and the language of description of that social reality. The language is constitutive of the reality, is essential to its being the kind of reality it is" (Taylor 1971, 34).

What does it mean that "the language of description" of social reality "is constitutive of the reality" itself? Taylor refers to the work of John Searle on speech acts, and in particular to his distinction between regulative and constitutive rules.[13] The distinction is well-known: according to Searle, there is an important class of linguistic rules that do not just regulate behavior, but also make the existence of certain social facts possible. The rules of baseball, for example, seem to be conceptually prior to the specific actions that take place on the field during a baseball match. Trying to steal a base or to throw a strike would not make sense in the absence of more fundamental rules that define the nature of the practice that we call "baseball."[14]

All constitutive rules, according to Searle, can be expressed by means of the simple formula "X counts as Y in C," where X is a preinstitutional entity or fact (like sending the ball out of the field), Y is a "status function" (like hitting a home run), and C stands for a domain of application (during a game of baseball). In later writings, the formula has been made somewhat more complicated to acknowledge, for example, that X must satisfy certain conditions or have certain properties in order to count as a genuine instance of Y (Searle 2010, 99). To make these conditions explicit while keeping the beautiful rhythm of the formula, I will use the C-term to refer to these conditions of satisfaction: the grammar of constitutive rules then becomes "X counts as Y if C."

Searle calls all the facts that presuppose the existence of a constitutive rule—like stealing a base or impeaching the president—*institutional* facts. The dependence thesis then takes this form: the existence of an important class of social facts (institutional facts) depends on the existence of a particular set of rules (constitutive rules) and on their acceptance by the members of the relevant community. This dependence is *ontological* or *constitutive* rather than causal.[15]

For example, let us take a set of propositional attitudes like beliefs or expectations that are shared in a community—for simplicity, we shall call them *collective attitudes* (CA).[16] The dependence thesis then states that necessarily, X is Y → [CA(X counts as Y if C) & C].

In other words: if a token entity X has the institutional property Y, then necessarily the members of the relevant social group collectively believe that "X is Y if it satisfies

---

[13] Taylor refers to Searle (1969), but the theory of constitutive rules has been developed in detail especially in later writings (Searle 1995, 2010).

[14] There are many reasons to believe that Searle's distinction between regulative and constitutive rules is spurious (see, e.g., Hindriks 2009; Morin 2013; Guala and Hindriks 2015; Hindriks and Guala 2015); but since this terminology is widely used, I will follow it here.

[15] On the distinction between causal and constitutive dependence, see Boyd (1991), Kukla (2000), and Mäki (2012a). The concept of dependence has generated a sizable literature in the field of metaphysics (see, e.g., Correia 2008, Lowe 2009). Unfortunately metaphysicians have typically discussed types of dependence—like the dependence of a whole on its part—that are not particularly useful for the issues addressed here. Notable exceptions are Thomasson's work on fictional objects (1999, ch. 2) and Epstein's (2015) book on collective entities.

[16] There is a lively debate about the attitudes that are necessary for the constitution of social entities. See, e.g., Gilbert (1989), Bratman (1993, 2014), Searle (1990, 2010), Tuomela (1995, 2002, 2007). On the relationship between game theoretic notions of collective beliefs and philosophical theories of collective intentions, see, e.g., Bacharach (2006), Bardsley (2007), Gold and Sugden (2007), Hakli, Miller, and Tuomela (2011). The reducibility of collective to individual intentions is a thorny issue in the philosophy of action because, according to some authors, preference, beliefs, and common knowledge conditions do

certain conditions C," and these conditions must be satisfied.[17] For instance: money is constituted by our representations in the sense that, for a piece of paper to be money, it is necessary that it satisfies some conditions (such as "being issued by the Central Bank") that are considered sufficient for a thing to count as money within the relevant community.

4.1 Anti-realism and Infallibilism

Making institutional kinds constitutively dependent on representation has some intriguing implications. First, it seems to imply anti-realism in the Mill-Boyd sense. If dependence is understood in a constitutive, noncausal fashion, in fact, it is not clear how institutional kinds may have the properties that real kinds are supposed to have. Recall that, according to the Millian view, real kinds must first and foremost support inductive inferences. But the correlations between properties that make inferences and generalizations possible hold in virtue of contingent facts, in particular the causal mechanisms that, according to Boyd (1991), ensure the "homeostatic" stability of the property clusters. If reflexivity is interpreted causally, along the lines sketched in earlier sections, then we can tell a plausible story about the way in which the creation and maintenance of some behavioral regularities is dependent on human representations. But, under the noncausal or constitutive interpretation, it is not clear how these correlations can be ensured simply by collectively endorsing a certain representation of the relevant entities. The dependence view thus drives a wedge between the criteria for institutional kindhood (dependence on representation) and those for real kindhood in the Mill-Boyd sense (projectability). The conclusion that institutional kinds are not real kinds becomes compelling: they are unified conceptually, by our classifications, rather than by causal mechanisms that guarantee the co-occurrence of their properties. They are unlikely to be projectable and unlikely to be of any use for scientific purposes.[18]

The constitutive view, however, has other surprising consequences. From an epistemic perspective, theory-dependent entities seem to be in an obvious way more transparent and accessible than are theory-independent ones. In a memorable paragraph of

---

not do justice to the normative dimension of institutions (see, e.g., Tuomela 2002, 128–129). For a critical discussion, see Turner (2010).

[17] Another way to put it is that they must have a theory of what it is for something to be an instance of Y. Epstein (2015) argues that collective acceptance and C-conditions play different metaphysical roles (which he calls "anchoring" and "grounding," respectively) and that they should not be conflated within a single relation of dependence.

[18] This conclusion follows from the adoption of a simple "principle of the metaphysical innocence of representation" (Boyd 1991). I agree with Boyd that this principle does not require an explicit defense because "it appears to be underwritten by fundamental conceptions of causal relations common to all the established sciences" (1991, 145). The burden of proof lies with those who intend to deny the metaphysical innocence of representation; minimally, these philosophers owe us a story of how contingent properties and causal relations can be fixed by representation in a noncausal way.

The Construction of Social Reality, for example, Searle claims that we cannot be mistaken about the nature of a social event—a particularly lively cocktail party—because what that social event is (its very nature) depends directly on the way we represent it.

> Something can be a mountain even if no one believes it is a mountain; something can be a molecule even if no one thinks anything at all about it. But for social facts, the attitude that we take toward the phenomenon is partly constitutive of the phenomenon. If, for example, we give a big cocktail party, and invite everyone in Paris, and if things get out of hand, and it turns out that the casualty rate is greater than the Battle of Austerlitz – all the same, it is not a war; it is just one amazing cocktail party. Part of being a cocktail party is being thought to be a cocktail party; part of being a war is being thought to be a war. This is a remarkable feature of social facts; it has no analogue among physical facts. (1995, 33–34)[19]

The point can be extended to all sorts of entities that are dependent on our representations: if a piece of marble is a statue, arguably the art world (critics, historians, collectors) must be aware that it is a statue because being a work of art is partly constituted by the art world's belief that it is a work of art. Or take a banknote: people cannot fail to see that a piece of paper is money, if money is constituted by people's belief that a certain entity is money. Considerations of this type invite the formulation of a general philosophical thesis that we may call *infallibilism about social kinds*: we cannot be wrong about the nature of social kinds that depend constitutively on our representations.

> If we understand institutional entities as dependent on the acceptance of certain constitutive rules laying out (at least) sufficient conditions for their existence, and existing provided something fulfills these conditions, we cannot conceive of investigations into the nature of our own institutional kinds as completely a matter of substantive and fallible discovery. Whereas natural kinds (on a realist view) can exist even if no one knows of their existence or any facts about their nature, institutional kinds do not exist independently of our knowing something about them. Similarly, whereas, in the case of natural kinds, any substantive principles any individual or group accepts regarding the nature of the kind can turn out to be wrong, in the case of institutional kinds those principles we accept regarding sufficient conditions for the existence of these entities must be true. (Thomasson 2003, 589–590)[20]

---

[19] Searle is not always consistent about this matter. In other paragraphs, he makes claims that are at odds with infallibilism; for example: "the process of creation of institutional facts may proceed without the participants being conscious that it is happening according to this form. . . . In the very evolution of the institution [of, say, money] the participants need not be consciously aware of the form of the collective intentionality by which they are imposing functions on objects. In the course of consciously buying, selling, exchanging, etc., they may simply evolve institutional facts" (1995, 47–48). Since this is not an exegetical essay, I will not try to explain how these views may coexist. For a discussion of Searle's "Parisian bash" example, see also Khalidi (2013, chapter 4).

[20] Ruben (1989) defends a similar thesis; I criticize Ruben's and Thomasson's infallibilism in Guala (2010).

Anti-realism and infallibilism are ubiquitous in the *verstehen* and hermeneutic tradition that runs from Wilhelm Dilthey (1927) to Hans-Georg Gadamer (1960), Paul Ricoeur (1965), and Charles Taylor. But they are also common in the Wittgensteinean tradition, for example, in Peter Winch's (1958) influential critique of the naturalistic pretensions of social science and in Michael Dummett's (1978) philosophy of mathematics. Anti-realists and infallibilists usually endorse simultaneously two theses: the first one says that those who share a common culture (a common language and common practices) have a direct and intuitive understanding of social categories. This understanding comes from mastering the relevant social practices—a "way of life"—and is precluded to outsiders. The second thesis says that the intuitive understanding of the insider is the *only* form of understanding that can be attained in social science. Social phenomena cannot be studied empirically as realities that are independent of our thoughts. Indeed, taking the detached point of view of the outsider—as naturalistic social scientists try to do— guarantees that understanding will never be attained. According to Taylor, for example, classic scientific criteria like predictive accuracy are not a good test for social theories:

> We cannot measure such sciences against the requirements of a science of verification: we cannot judge them by their predictive capacity. We have to accept that they are founded on intuitions that all do not share, and what is worse that these intuitions are closely bound up with our fundamental options. (Taylor 1971, 57)

4.2 Money

According to Searle and his followers, such as Taylor and Thomasson, constitutive rules are an essential element of institutional reality. But constitutive dependence is hardly ever mentioned in scientific studies of social institutions. It is possible, of course, that social scientists have failed to appreciate an important metaphysical relationship; but, at the same time, it is also possible that the nature and functioning of institutions is better analyzed in different terms and that the attention that philosophers have paid to constitutive rules is unwarranted. One way to test this hypothesis is to ask whether the existence of some paradigmatic institutions—such as money, private property, or marriage—does indeed require the acceptance of constitutive rules. Since the case of money has become an undisputed classic in social ontology, I will use it as a test case in this section.

Let us begin with some basic social science: economists subscribe to the principle that "money is what money does." The existence of money does not depend on the representation of some things as money, but on the existence of causal mechanisms that ensure that some entities perform money-like functions (checks, debit cards, bank accounts are all money, in economists' sense). What these causal mechanisms are is far from obvious, however, and not a matter of collective acceptance.[21]

---

[21] For a contemporary theoretical analysis of money, see, e.g., Kiyotaki and Wright (1989, 1991). For a comparison of philosophers' and economists' views of money see Tieffenbach (2010), Smit, Buekens, and du Plessis (2011).

Money performs several functions at once. Three of them are usually considered particularly important: money functions as (1) a medium of exchange, (2) a store of value, and (3) a unit of accounting. Economic theory identifies, (1) as the core function of money: a currency, to be money, must first and foremost be used as a medium of exchange. The second function (store of value), however, is strictly linked with it: "if an asset were not a store of value, then it would not be used as a medium of exchange" (Dornbusch and Fischer 1994, 374). The reason is simple: exchanges take place in time. Selling a commodity now with the aim of purchasing another one in the future makes sense only if the revenue from the first trade does not lose its value during the time it takes to effect the second one. Being a store of value, then, is an important precondition for a currency to be money.

Beliefs about the future play a key role in the existence of money. We accept worthless paper bills in exchange for valuable goods or services because we are confident that we will be able to use them later to purchase other commodities. The people who take bills for payment believe that others will take them and so forth. Such beliefs, however, should better not hang up in the air. A primary role of the Central Bank is to sustain these beliefs by enforcing a monopoly on the issuing of money. The Bank prints bills that will be used by everyone because everyone believes that the others believe—and so forth—that they will continue to be used as a medium of exchange. If an entity X (a paper bill) fulfils the condition C (being issued by the Central Bank), then X is usually able to perform the functions of money. But this means only that being issued by the Central Bank makes us very confident that the bill will be accepted in the future—the relation is causal rather than constitutive.

What is so special with a piece of paper that carries the stamp of the Central Bank? The conditions C have a facilitating role: seeing that it carries the stamp of the Central Bank increases our confidence. But people must have good reasons (incentives) to hold that piece of paper.[22] There are various possible incentives, but here I will cite a simple motivation that is crucial in modern societies. The state can guarantee a certain level of demand for the currency via taxation. If the state accepts only paper bills issued by the Central Bank as payment, then, in fact, we can be confident that in the future people will have to hold some official currency for tax purposes. This is true of course only to the extent that the state is strong, stable, and will have the means to collect taxes. So, unsurprisingly, the strength of a currency is strictly linked with the political strength of the state.[23]

The state collects paper bills via taxation and puts them back into the economy by paying salaries to the employees of the public sector. If they meddle with the latter part of the cycle, however, governments and central banks can devalue the currency, creating

---

[22] On the importance of incentives for the existence of money, see Smit et al. (2011). Smit, Buekens, and du Plessis (2014) propose a general incentive-based theory of institutions.

[23] The role played by the state is emphasized by so-called "chartalist" or "claim theories" of money. Early formulations can be found in the writings of Max Weber and Georg Knapp, but the theory still has many disciples today. See, e.g., Wray (1990).

inflation. In extreme cases of hyperinflation an official currency may even become worthless paper and be replaced by informal currencies such as cigarettes or bit-coins. This is because, to function as money, a currency must be a reliable store of value, and this will be true only if its quantity is relatively stable. So it seems that fulfilling the conditions C that people take as sufficient for something to be money is neither necessary nor sufficient. A cigarette can be money even though it has not been issued by the Central Bank, and a bill that has been issued by the Central Bank may fail to work as medium of exchange if the state loses its credibility.

We can now compare the social-scientific view of money with the constitutive view. According to the constitutive view, fulfilling a function Y implies necessarily the acceptance of a constitutive rule. But, as the example of cigarettes shows, for a commodity to function as money it is not necessary that the rule "X counts as money if it has been issued by Philip Morris" is widely accepted. Conversely, the acceptance of a constitutive rule cannot guarantee that any of the functions of money are going to be fulfilled. What matters, ultimately, is the belief that certain entities are going to be accepted as payment by other people in the future.[24] So the real content of terms like "money" is in the actions ("accept it as payment") that are associated with the theoretical term ("money"). Money ultimately is nothing but this set of actions (what money allows us to do) and the related set of expectations. The C conditions are useful in so far as they simplify our decisions: they are coordination devices that help identify quickly and without lengthy inspection an appropriate set of actions in the given circumstances. (Should I accept a piece of paper as payment? Yes, because it has been issued by the Central Bank.) But to focus on the C conditions as what makes something a member of the kind "money" is a perceptual mistake. It mistakes the coordination device for the system of actions and expectations that a social institution is.[25]

5 THE ANTI-NATURALIST PREDICAMENT

The theory dependence of social phenomena is not a threat to explanation and prediction in the social sciences. If dependence is interpreted causally, we have seen that feedback loops may have a stabilizing effect on behavior; moreover, there are always equilibria between actions and representations that make scientific inference possible, at least in principle. If dependence on representation is interpreted constitutively, in contrast, there is no reason to believe that the properties of social kinds support inductive inference. But the theory of constitutive dependence is dubious: there are

[24] In practice, during a crisis, people tend to adopt the currency of another country as "unofficial" money. The fact that a currency—like dollars—is generally used in international trade provides sufficient confidence and sustains its use as a medium of exchange in failing economies.
[25] On the role of coordination devices for the existence of institutions, see Guala and Hindriks (2015) and Hindriks and Guala (2015).

perfectly sound accounts of social institutions that do not presuppose the acceptance of constitutive rules.

The anti-naturalist at this point may reply that we cannot assume that these accounts are correct: to appeal to the authority of social science is to beg the question. But recall that theory dependence was meant to play a foundational role: it was meant to demonstrate that social kinds have peculiar characteristics that prevent the application of standard social science methodology. So it is up to the anti-naturalist to demonstrate a priori that social phenomena are constitutively dependent on representation. If dependence is just a theoretical hypothesis among many, then it cannot play this foundational role.

This is just a symptom of a more general issue: are the grounds of anti-naturalist ontologies robust enough to play a foundational role? Searle (1995, 190–194) says that the theory of constitutive rules is supported by transcendental reasoning. Dependence on representation, he claims, is a condition of intelligibility of ordinary discourse about institutions: "on our normal understanding, statements about money require the existence of representations as part of their conditions of normal intelligibility" (1995, 194).[26] But folk conceptions are fallible: "there is nothing self-guaranteeing about normal understanding. Sometimes we are forced to revise our normal understanding because of new discoveries" (1995, 195). Searle mentions the case of colors: science has shown us that they are extrinsic properties of material objects, contrary to our "normal understanding." But then, clearly, the folk theory of institutions is also revisable in the light of the discoveries of social science.

So we are back at square one: far from demonstrating that current social science is on the wrong track, anti-naturalists start from the assumption that social science is not to be taken seriously. But then one cannot use ontology to show that social science cannot work, unless one has already decided that social science does not work. This is, in a nutshell, the ontologist's predicament.

In the old days of logical positivism, Carnap (1950) had argued convincingly that the metaphysician's ambition to address ontological issues "from the outside"—that is, from a neutral position that does not involve a commitment to any substantial background theory—is misguided. Whether atoms or quarks are the ultimate constituents of matter can only be discussed within the framework of particle physics; whether tables exist can only be asked within the framework of everyday language and folk psychology and so on. Quine (1951/1966) rejected Carnap's sharp contrast between "internal" issues, which can be solved within the framework of a given theory, and external issues (such as the choice between frameworks) that must be resolved on pragmatic grounds. For Quine, pragmatic decisions intervene also in the solution of internal issues, and the relative permeability of frameworks allows that questions of existence be posed across languages and disciplines. This means that ontology is always internal. It always involves a commitment to some (scientific, commonsensical, etc.) theory, and it makes use of whatever

---

[26] Elsewhere, Searle says that "I do not much care if my account of institutional reality and institutional facts matches that ordinary usage" (2005, 18). Again, I will skirt exegetical issues here; Searle (2005) in any case does not tell us what the basis or the criterion is of adequacy for his ontology.

mix of empirical, pragmatic, and logical considerations is considered appropriate to solve a given problem.

Social ontologists with an anti-naturalistic attitude typically work within the theoretical framework of common sense psychology and folk sociology. As Tuomela (2002, 7) has put it, "we have all learned to use this framework as children," and "such information, and examples related to it, form an important data base" for philosophical analysis. The presumption, however, is that a large part of this information is correct, and this presumption needs justification. If the social world was constituted by our representations, then of course we would have a good reason to use the latter as our main source of information. But we have seen that this assumption is questionable and that it is generally rejected by those who study institutions scientifically. Since any ontological investigation must take place within a theoretical framework, the worldview of folk sociology can perhaps *compete* with the worldview of social science, but it cannot aspire to play a foundational role. Ontology is not and can never be prior to science.

REFERENCES

Bacharach, M. (2006). *Beyond Individual Choice* (Princeton, NJ: Princeton University Press).

Bardsley, N. (2007). "On Collective Intentions: Collective Action in Economics and Philosophy." *Synthese* 157: 141–159.

Bogen, J. (1988). "Comments on the Sociology of Science of Child Abuse." *Nous* 22: 65–66.

Boyd, R. (1991). "Realism, Anti-foundationalism, and the Enthusiasm for Natural Kinds." *Philosophical Studies* 61: 127–148.

Bratman, M. (1993). "Shared Intention." *Ethics* 104: 97–113.

Bratman, M. (2014). *Shared Agency* (Oxford: Oxford University Press).

Buck, R. (1963). "Reflexive Predictions." *Philosophy of Science* 30: 359–369.

Carnap, R. (1950). "Empiricism, Semantics and Ontology." *Revue internationale de philosophie* 4: 20–40. Reprinted in Carnap, R., *Meaning and Necessity* (Chicago, Illinois: University of Chicago Press, 1956), 205–221.

Cartwright, N. (1983). *How the Laws of Physics Lie* (Oxford: Clarendon Press).

Cooper, R. (2004). "Why Hacking Is Wrong About Human Kinds." *British Journal for the Philosophy of Science* 55: 73–85.

Correia, F. (2008). "Ontological Dependence." *Philosophy Compass* 3: 1013–1032.

Dilthey, W. (1927). *Der Aufbau der geschichtlichen Welt in den Geisteswissenschaften.* [*Selected Works, Vol. 3. The Formation of the Historical World in the Human Sciences*] (Princeton, NJ: Princeton University Press, 2002).

Dornbusch, R., and Fischer, S. (1994). *Macroeconomics*, 6th edition (New York: McGraw-Hill).

Douglas, M. (1986). *How Institutions Think* (Syracuse, NY: Syracuse University Press).

Dummett, M. (1978). *Truth and Other Enigmas* (Cambridge, MA: Harvard University Press).

Dupré, J. (1993). *The Disorder of Things* (Cambridge, MA: Harvard University Press).

Dupré, J. (2001). *Human Nature and the Limits of Science* (Oxford: Oxford University Press).

Epstein, B. (2015). *The Ant Trap: Rebuilding the Foundations of Social Science* (Oxford University Press).

Gadamer, H. G. (1960). *Truth and Method* (New York: Seabury).

Giere, R. N. (1988). *Explaining Science* (Chicago: University of Chicago Press).

Gilbert, M. (1989). *On Social Facts* (Princeton, NJ: Princeton University Press).

Gold, N., and Sugden, R. (2007). "Collective Intentions and Team Agency." *Journal of Philosophy* 104: 109–137.

Goodman, N. (1979). *Fact, Fiction, and Forecast* (Cambridge, MA: Harvard University Press).

Grünbaum, A. (1956). "Historical Determinism, Social Activism, and Predictions in the Social Sciences." *British Journal for the Philosophy of Science* 7: 236–240.

Grunberg, E., and Modigliani, F. (1954). "The Predictability of Social Events." *Journal of Political Economy* 62: 465–478.

Guala, F. (2010). "Infallibilism and Human Kinds." *Philosophy of the Social Sciences* 40: 244–264.

Guala, F., and Hindriks (2015). "A Unified Social Ontology." *Philosophical Quarterly* 165: 177–201.

Hacking, I. (1983). *Representing and Intervening*. (Cambridge: Cambridge University Press).

Hacking, I. (1986). "Making Up People." In P. Heller, M. Sosna, and D. Wellbery (eds.), *Reconstructing Individualism* (Stanford, CA: Stanford University Press), 222–236). Reprinted in Hacking, 2002.

Hacking, I. (1995). "The Looping Effect of Human Kinds." In A. Premack (ed.), *Causal Cognition: A Multidisciplinary Debate* (Oxford: Clarendon Press), 351–383.

Hacking, I. (1999). *The Social Construction of What?* (Cambridge, MA: Harvard University Press).

Hacking, I. (2002). *Historical Ontology* (Cambridge, MA: Harvard University Press).

Hakli, R., Miller, K., and Tuomela, R. (2011). "Two Kinds of We-Reasoning." *Economics and Philosophy* 26: 291–320.

Haslanger, S. (1995). "Ontology and Social Construction." *Philosophical Topics* 23: 95–125. Reprinted in Haslanger, 2012.

Haslanger, S. (2012). *Resisting Reality: Social Construction and Social Critique* (Oxford: Oxford University Press).

Hempel, C. G. (1942). "The Function of General Laws in History." *Journal of Philosophy* 39: 35–48.

Hindriks, F. (2009). "Constitutive Rules, Language, and Ontology." *Erkenntnis* 71: 253–275.

Hindriks, F., and Guala, F. (2015). "Institutions, Rules, and Equilibria: A Unified Theory." *Journal of Institutional Economics* 11: 459–480.

Jarvie, I., and Zamora Bonilla, J., eds. (2011). *The Sage Handbook of the Philosophy of Social Sciences* (London: Sage).

Khalidi, M. A. (2010). "Interactive Kinds." *British Journal for the Philosophy of Science* 61: 335–360.

Khalidi, M. A. (2013). *Natural Categories and Human Kinds* (Cambridge: Cambridge University Press).

Kincaid, H. (1996). *Philosophical Foundations of the Social Sciences* (Cambridge: Cambridge University Press).

Kincaid, H. (2012a). "Introduction: Doing Philosophy of the Social Sciences." In H. Kincaid (ed.), *The Oxford Handbook of Philosophy of Social Science* (Oxford: Oxford University Press), 3–20.

Kincaid, H., ed. (2012b). *The Oxford Handbook of Philosophy of Social Science* (Oxford: Oxford University Press).

Kincaid, H., and Ross, D., eds. (2009). *The Oxford Handbook of Philosophy of Economics* (Oxford: Oxford University Press).

Kiyotaki, N., and Wright, R. (1989). "On Money as a Medium of Exchange." *Journal of Political Economy* 97: 927–954.

Kiyotaki, N., and Wright, R. (1991). "A Contribution to the Pure Theory of Money." *Journal of Economic Theory* 53: 215–235.

Kukla, A. (2000). *Social Construction and the Philosophy of Science* (London: Routledge).

Lowe, E. J. (2009). "Ontological Dependence." In E. N. Zalta (ed.), *Stanford Encyclopedia of Philosophy*, http://plato.stanford.edu/entries/dependence-ontological/

MacIntyre, A. (1957). "Determinism." *Mind* 66: 28–41.

MacKenzie, D. (2006). *An Engine, Not a Camera* (Cambridge, MA: MIT Press).

Mäki, U. (2012a). "Realism and Anti-realism about Economics." In U. Mäki (ed.), *Handbook of the Philosophy of Science, Vol. 13: Philosophy of Economics* (Amsterdam: Elsevier), 3–24.

Mäki, U., ed. (2012b). *Handbook of the Philosophy of Science, Vol. 13: Philosophy of Economics* (Amsterdam: Elsevier).

Mallon, R. (2003). "Social Construction, Social Roles, and Stability." In F. F. Schmidt (ed.), *Socializing Metaphysics* (Lanham, MD: Rowman and Littlefield).

Merton, R. K. (1948). "The Self-fulfilling Prophecy." *Antioch Review* 8: 193–210.

Mill, J. S. (1843). *A System of Logic* (London: Longmans, Green, Reader and Dyer).

Morgenstern, O. (1928). *Wirtschaftsprognose, eine Untersuchung ihrer Voraussetzungen und Moeglichkeiten* (Vienna: Springer Verlag).

Morin, O. (2013). "Three Ways of Misunderstanding the Power of Rules." In M. Schmitz, B. Kobow, and H. -B. Schmidt (eds.), *The Background of Social Reality* (Berlin: Springer, 185–201).

Nagel, E. (1960). *The Structure of Science* (New York: Harcourt).

Palacios-Huerta, I. (2003). "Professionals Play Minimax." *Review of Economic Studies* 70: 395–415.

Quine, W. O. (1951/1966). "On Carnap's Views on Ontology." Reprinted in *The Ways of Paradox and Other Essay*. (Cambridge, MA: Harvard University Press), 203–211.

Ricoeur, P. (1965). *Interpretation Theory* (Fort Worth: Texas Christian University Press).

Romanos, G. D. (1973). "Reflexive Predictions." *Philosophy of Science* 40: 97–109.

Ross, D. (2011). "Economic Theory, Anti-economics, and Political Ideology." In U. Mäki (ed.), *Handbook of the Philosophy of Science, Vol. 13: Philosophy of Economics* (Amsterdam: Elsevier), 241–286.

Ruben, D. (1989). "Realism in the Social Sciences." In H. Lawson and L. Appignanesi (eds.), *Dismantling Truth* (London: Weidenfeld and Nicolson), 58–75.

Salmon, W. C. (1984). *Scientific Explanation and the Causal Structure of the World* (Princeton, NJ: Princeton University Press).

Searle, J. R. (1969). *Speech Acts: An Essay in the Philosophy of Language* (Cambridge: Cambridge University Press).

Searle, J. (1990). "Collective Intentions and Actions." In P. Cohen, J. Morgan, and M. E. Pollack (eds.), *Intentions in Communication* (Cambridge, MA: MIT Press).

Searle, J. (1995). *The Construction of Social Reality* (London: Penguin).

Searle, J. R. (2005). "What Is an Institution?" *Journal of Institutional Economics* 1: 1–22.

Searle, J. R. (2009). "Language and Social Ontology." In C. Mantzavinos (ed.), *Philosophy of the Social Sciences* (Cambridge: Cambridge University Press), 9–27.

Searle, J. R. (2010). *Making the Social World* (Oxford: Oxford University Press).

Simon, H. (1954). "Bandwagon and Underdog Effects and the Possibility of Election Predictions." *Public Opinion Quarterly* 18: 245–253.

Smit, J. P., Buekens, F., and du Plessis, S. (2011). "What is Money? An Alternative to Searle's Institutional Facts." *Economics and Philosophy* 27: 1–22.

Smit, J. P., Buekens, F., and du Plessis, S. (2014). "Developing the Incentivized Action View of Institutional Reality." *Synthese* 191: 1813–1830.

Soros, G. (2013). "Fallibility, Reflexivity and the Human Uncertainty Principle." *Journal of Economic Methodology* 20: 309–329.

Taylor, C. (1971). "Interpretation and the Sciences of Man." *Review of Metaphysics* 25: 3–51. Reprinted in Taylor, 1985.

Taylor, C. (1985). *Philosophical Papers, Vol. 2: Philosophy and the Human Sciences* (Cambridge: Cambridge University Press).

Tieffenbach, E. (2010). "Searle and Menger on Money." *Philosophy of the Social Sciences* 40: 191–212.

Thomasson, A. (1999). *Fiction and Metaphysics* (Cambridge: Cambridge University Press).

Thomasson, A. (2003). "Realism and Human Kinds." *Philosophy and Phenomenological Research* 68: 580–609.

Tuomela, R. (1995). *The Importance of Us* (Stanford, CA: Stanford University Press).

Tuomela, R. (2002). *The Philosophy of Social Practices* (Cambridge: Cambridge University Press).

Tuomela, R. (2007). *The Philosophy of Sociality* (Oxford: Oxford University Press).

Turner, S. (2010). *Explaining the Normative* (Cambridge: Polity Press).

von Neumann, J., and Morgenstern, O. (1944). *The Theory of Games and Economic Behavior* (Princeton, NJ: Princeton University Press).

Winch, P. (1958). *The Idea of a Social Science and Its Relation to Philosophy* (London: Routledge).

Wray, L. R. (1990). *Money and Credit in Capitalist Economies* (Aldershot: Elgar).

CHAPTER 4

....

PHILOSOPHY OF BIOLOGY

....

BEN FRASER AND KIM STERELNY

1 INTRODUCTION

ONE way of seeing philosophy of biology is as a subfield of philosophy of science more generally. Philosophy of biology in this sense addresses standard topics in the philosophy of science, such as causation or theory reduction, in the context of biological examples, and analyzes and conceptually clarifies issues within biology itself. From this perspective, philosophy of biology takes a particular science, biology, as its subject of philosophical inquiry. Philosophy of biology is distinctive within philosophy of science and plays a special role within it, because biological phenomena are often complex, produced by contingent historical processes, and hence very varied from case to case. Conceptions of confirmation and explanation drawn from the experimental sciences, especially physics and chemistry, do not seem to fit biology well. There is, for example, a long-standing debate as to whether there are any distinctively biological laws of nature. Philosophy of biology thus understood is importantly continuous with biology itself: theoretical biology shades off into philosophy of biology; important contributions to the field are made not only by philosophers but also by biologists themselves.

Another way of seeing philosophy of biology is available, via Peter Godfrey-Smith's distinction between philosophy of science and philosophy of nature (2014: 4). The former, as just noted, takes science—its activities and its products—as targets of philosophical inquiry and analysis. The latter, by contrast, treats science as a resource for further philosophical inquiry, be it ontological, epistemological, or even ethical. Philosophy of nature is a synthesizing discipline, drawing on many fields of science. For example, Richard Dawkins' (1976) *The Selfish Gene* is an essay in philosophy of nature, setting out an overall conception of the evolving, living world and our place in it. Dawkins's conception is drawn from evolutionary biology, but it is interpretation and synthesis, not just report. Likewise, Daniel Dennett's (1995) *Darwin's Dangerous Idea* is a work in philosophy of nature. It is methodologically reflective, but its primary targets are first-order questions about life and cognition.

This chapter covers philosophy of biology in both senses just described. Its structure and contents are as follows. It begins with a brief historical overview of philosophy of biology as an academic discipline. It then surveys a sample of the traditional core issues in the field: units of selection, fitness and the nature of selection, the species problem, systematics, adaptationism, and human nature. Next, we sketch recent expansions of the field to encompass molecular biology, ecology, and conservation biology. Finally, current frontier challenges are identified: the evolution of cooperation and the nature of biological individuality.

2 HISTORY OF PHILOSOPHY OF BIOLOGY

A complete history of philosophy of biology would start with Aristotle, proceed to the early nineteenth century—noting the beginnings of philosophy of science and the foundational contributions of Newton and Bacon—then advance through the late nineteenth century, not neglecting to jump between continents when necessary, paying special attention to Darwin's revelation and its elaboration in the work of Spencer among many others, and end by charting the myriad links between philosophy of biology and biology itself (which of course includes its own numerous subfields) during the twentieth century. Nothing remotely so ambitious will be attempted here.

This section restricts itself to a brief and selective overview of philosophy of biology in the analytic tradition in the mid- to late twentieth century. The fathers of philosophy of biology thus conceived were David Hull and Michael Ruse. Hull's interests were primarily in philosophy of science. As he saw it, the received views of science did not fit the biological sciences. He developed these ideas in papers on species, systematics, human nature, intertheoretic reduction, the nature and importance of genes, and more. Perhaps his most ambitious project—*Science as a Process* (Hull 1988)—was to apply evolutionary models to science itself. Ruse was less focused on detailed issues in philosophy of science, seen through biological examples, and more concerned with showing the importance of evolutionary thinking to more general issues in philosophy and to the humanities more broadly—as in his *Taking Darwin Seriously* (Ruse 1998)—and in connecting the history of biology to philosophy of biology: he argued in his 2009 book *From Monad to Man* that many controversies in evolutionary thinking had deep historical roots and that this history cast a long shadow. Outside academia, he has engaged in often heated and high-profile public controversy over the social, moral, political, and religious significance of evolution.

Hull and Ruse between them introduced many of the central issues concerning philosophers of biology. They also published the first introductory textbooks for the field (Ruse 1973; Hull 1974). Two anthologies of philosophy of biology were published at around the same time. One was edited by evolutionary biologists Francisco Ayala and Theodosius Dobzhansky (1974), the other by philosopher Marjorie Grene and science historian Everett Mendelsohn (1976). These works provided resources for teaching and seed for growing an academic field. Another important early figure was Bill Wimsatt, who was particularly distinctive in taking up the challenges of developmental and molecular biology early and relating these to evolutionary biology. Wimsatt published much of his work in out-of-the-way places, but fortunately the best of it has been collected and supplemented with new papers, in *Re-Engineering Philosophy for Limited Beings* (Wimsatt 2007).

In addition to philosophers interested in biology, philosophy of biology was also nurtured by what might be called the philosophical biologists. These included Richard Dawkins, Michael Ghiselin, Stephen Gould, Richard Lewontin (who hosted many emerging philosophers of biology in his lab), John Maynard Smith, Ernst Mayr, G. C. Williams, and E. O Wilson. These biologists considered the nature of genes and species, the importance of adaptation, and the implications of evolution for morality; sometimes this work was published in philosophy journals, and sometimes it was published with philosophers.

The second generation of philosophers of biology included Robert Brandon, Elliot Sober, Philip Kitcher, and Alex Rosenberg. Brandon made important contributions to debates about fitness; the relations between organism and environment; the levels of selection, reductionism, and species; and the issue of laws in biology. Sober's (1984) *The Nature of Selection* did much to raise the profile of philosophy of biology among both philosophers and biologists. Rosenberg and Kitcher both contributed widely to philosophy of biology, and both weighed in notably on the sociobiology debate (Rosenberg 1980; Kitcher 1984).

Two landmark events further confirmed the arrival of philosophy of biology as an academic discipline. One was the appearance in 1986 of the first issue of *Biology & Philosophy*, an interdisciplinary journal publishing work in philosophy of biology under the editorship of Michael Ruse. Another was the formation in 1989 of the International Society for the History, Philosophy, and Social Studies of Biology, which meets every two years. Philosophy of biology at this point had its own journal, its own professional organization and conference, and a growing group of practitioners. By this time, the leading philosophy of science journals regularly published on core topics in philosophy of biology. It had arrived as a recognized field within philosophy.

3 TRADITIONAL CORE

The initial focus of philosophy of biology was in philosophy of evolutionary biology— probably, in part, for accidental historical reasons. But it was also because evolutionary biologists were themselves puzzled by questions of a mixed empirical-conceptual character. To what extent could human behavior be understood using the same theoretical tools that explain animal behavior? Is the idea that selection acts on genes a new metaphor or a new hypothesis about the evolution of life? Are species real, objective units of the living world, or is that an illusion born of our short lives making it difficult for us to see their cloud-like fading in and out of existence over time? Can groups or species be adapted, or only the organisms from which these collectives are composed? These were initially biologists' questions (Wilson 1975; Gould 1981; Eldredge and Gould 1972; Williams 1966; Dawkins 1976). But because they had this mixed empirical-conceptual character, they and their kin became the staple, the traditional core, of philosophy of biology.

This section reviews these core issues. It opens by covering the debate about the units of selection in evolution, considering genes, groups, species, and cultural variants or "memes." It then discusses a linked set of issues concerning the nature of fitness and of selection itself. Next it details the controversy over adaptationism and the relative importance of selection in explaining features of organisms and large-scale patterns in the history of life. The section then turns to the species problem—that of saying what species are and how to group organisms into them—and the related topic of systematics, or how best to group species into higher level classificatory units. The section closes with a discussion of whether there is an evolved and universal human nature and, if so, in what it might consist.

3.1 Units of Selection

Richard Lewontin (1985: 76) gave an influential statement of the conditions necessary and sufficient for a population to undergo evolution by natural selection. If the members of a population exhibit variation in their traits (be those morphological, physiological, or behavioral), if that variation is at least partially heritable such that offspring resemble their parents more than some random member of the population, and if different variants in the population produce different numbers of offspring, then evolution by natural selection will occur. This three-ingredient "recipe" for evolution by natural selection—variation and heredity plus differential reproduction—is in many respects too simple (Godfrey-Smith 2009a: 20). Godfrey-Smith treats this as a minimal characterization of a Darwinian population and suggests ways to enrich it, but the basic recipe characterizes in a neutral way the kinds of populations that undergo evolution by natural selection. The key question here is what are the targets or units of selection. One obvious answer is that the populations are populations of organisms, and so organisms are the "targets" or "units" of selection. The faster prey organism, the stronger predator—these individuals win out over their slower, weaker rivals and are, therefore, selected. Another appealing answer is that, at least sometimes, groups of organisms are units of selection. Groups that, collectively, better defend a territory or more efficiently use the resources within it are selected over less well coordinated groups. Both these answers, although attractive, have been challenged from the "gene's eye" perspective and in light of an important distinction between vehicles and replicators.

To take the gene's eye view is to consider what characteristics a gene would need in order to increase its frequency over generational time. From this perspective, group selection is problematic, because groups of cooperative individuals (for example) are vulnerable to invasion and undermining from within by individuals with a gene that leads them to take the benefits of others' cooperative behavior while refusing to cooperate themselves. Over generations, such individuals would out-compete and replace cooperators and cooperative groups would collapse (Williams 1966).

It is now widely accepted that this argument is too simple. First, we do not need the gene's eye perspective to see the threat posed by selfish organisms. Second, there are biologically realistic circumstances in which the threat of selfish undermining from within is contained. Indeed, it is now emphasized that the evolution of complex, multicelled individuals depended on selection at the collective level, trumping selection on individual cells, for multicelled individuals evolved from cooperating collectives of individual cells (Maynard-Smith and Szathmary 1995; Okasha 2006; Michod 2007). David Sloan Wilson and Elliot Sober (1994) pioneered the reestablishment within evolutionary biology of "multilevel selection"; selection can and often does act on different levels of biological organization simultaneously. That said, there continues to be debate on whether multilevel selection is really a special form of selection on individual organisms, where individual fitness depends on the individuals intrinsic properties but also on the individual's social context (Kerr and Godfrey-Smith 2002; Okasha 2006), and on the individual's effects on the fitness of his or her close relations (see Okasha and Paternotte's [2014] introduction to the *Biology & Philosophy* special issue on formal Darwinism and the papers therein).

Williams used his attack on group selection to leverage a more radical suggestion. He, and following him Richard Dawkins, argued that the unit of selection was not even the individual organism but instead the gene. Williams's basic thought is that individual organisms are too ephemeral to be targets of selection; they last one generation and are gone. Of course gene tokens are equally ephemeral, but they can form lineages of near-identical copies, and these lineages have the potential to grow deep and bushy over time, through the summed effects of these gene tokens on the organisms that carry them. Organisms do not form lineages of near-identical copies; we are not copies of either of our parents.

Dawkins (1982) distinguished between two kinds of entities: replicators and vehicles (Hull [1981] refined this distinction). Replicators are anything of which copies can be made. Genes are replicators, and of a special sort: they can influence their own chance of being copied. Genes usually exert this influence by building vehicles, entities that mediate the interaction of replicators with the rest of the world. Genes' vehicles are organisms. Dawkins claims that the units of selection must be replicators and that neither individual organisms nor groups of organisms are replicators whereas genes are, so genes are the units of selection (for a recent defence of gene selectionism, see Haig 2012).

It is widely accepted that a small subclass of genes—called ultra-selfish genes—are indeed units of selection. These are genes that improve their prospects of being copied to the next generation by subverting the 50/50 lottery of meiosis in various ways (Trivers and Burt 2008). They do not enhance the fitness of the organism in which they ride. But what of more standard genes: genes for sharp sight or disease resistance? One critical issue is whether genes typically have stable enough effects for us to be able to meaningfully talk of selection in favor of a particular gene lineage. For the effects of a gene on the organism that carries it are very variable: depending on the other genes present, the organism's environment, and the cell in which the gene is expressed. Moreover, counting genes is decidedly tricky, once we see the full complexity of gene regulation (Griffiths and Stotz 2006; Godfrey-Smith 2014). It was once thought that genes were, typically, continuous stretches of DNA that code for a specific protein. It is now known that this is not even approximately true. The DNA stretch is regulated from other places on the chromosome, and, in eukaryotes, the initial transcript made from the DNA includes noncoding sequences that are edited out, with the coding fragments then being reassembled (sometimes in more than one way). In view of the many complications involved in identifying and tracking particular genes and their descendants, Godfrey-Smith argues that genes are by no means a clear case of a Darwinian population.

3.2   Fitness and the Nature of Selection

Whereas the previous section deals with the question of what entities are the target(s) of natural selection, this section considers the nature of selection itself and the closely related issue of how to understand fitness in evolutionary theory. The problem here is that, on one influential and appealing way of understanding selection and fitness, the theory of evolution by natural selection looks to be worryingly empty, an uninformative tautology.

Herbert Spencer (1864) coined the phrase "survival of the fittest" in his attempts to promote and popularize Darwin's theory of evolution by natural selection. Greater fitness is meant to explain why some individuals rather than others survive (and reproduce). Together with the view that the fittest entities are those that produce the most offspring—a common understanding of fitness (for a survey see Mills and Beatty 1979: 265)—this generates a problem: *of course* there is survival of the fittest, since the fittest just are those that in fact survive and reproduce. Spencer's slogan is explanatorily impotent. This is the so-called tautology problem created by defining "fitness" in terms of actual survival and/or reproduction.

The problem has been noted by many, early on by Samuel Butler (1879) and more recently by philosopher Jack Smart, who observed that if "we say that 'even in Andromeda the fittest will survive' we say nothing, for 'fittest' has to be defined in terms of 'survival'" (1959: 366). Smart's statement of the problem is instructive, since the now widely recognized solution is precisely *not* to define fitness in terms of survival and reproduction, or at least of actual survival and reproduction. Instead, fitness should be understood as a *propensity*.

Propensities are probabilistic dispositions. Neither notion here is especially mysterious. Salt has the disposition to dissolve in water. A fair coin has the probabilistic disposition—the propensity—to land heads half the time. On the propensity interpretation of fitness, first defended by Mills and Beatty (1979), an organism's fitness is its propensity to survive and reproduce in a specific set of circumstances. Just as in the coin case, where the object has certain properties that underlie the disposition (the coin has certain structural properties), so too is an organism's fitness a function of its properties or traits, though which traits in particular is a hugely complex and context-sensitive matter. And, just as in the coin case, where a fair coin can nevertheless land heads eight, nine, or even ten times in ten tosses, so too can the fittest organism fail to be best at surviving and reproducing.

Understanding fitness as a propensity blocks the tautology objection. The fitter organism will not necessarily outreproduce all the others; it may, unluckily, die early or not find a mate. And fitness thus interpreted plays a genuinely explanatory role, just as do dispositions elsewhere in science.

Much work has been done on figuring out exactly how to characterize the propensity that is fitness (e.g., Beatty and Finsen 1987; Pence and Ramsey 2013). The difficulties have led some to propose yet other interpretations, such as Rosenberg's (1985) view that fitness is a subjective probability, a measure of human uncertainty about organisms' reproductive success (this view of course comes with problems of its own). There are other suggestions as well: if propensities are intrinsic physical properties of organisms, the propensity interpretation makes it hard to make sense of the idea that an organism's fitness can change, say, as a result of an environmental change. But we often talk about fitness that way (Abrams 2007). However, there is near consensus that fitness is best interpreted as a propensity. The debate from this point on concerns the implications of so understanding fitness for the nature of natural selection itself.

Elliott Sober (1984) drew a parallel between explanations in evolutionary biology (population genetics in particular) and explanations given in Newtonian mechanics: change in populations over time results from the action and interaction of many "forces," including selection, drift, and mutation. On Sober's sort of view, natural selection is a cause of evolutionary change. This causal interpretation of natural selection has been challenged. According to the statistical interpretation, the causes of evolutionary change are individual-level interactions and natural selection is merely a population-level consequence (Walsh, Lewens, and Ariew 2002; Matthen and Ariew 2009). A key issue in the ongoing debate is which interpretation supports the best account of the relationship between natural selection and drift. Proponents of the statistical view say it is theirs (Walsh et al. 2002); proponents of the causal view disagree (Bouchard and Rosenberg 2004; Forber and Reisman 2005; Ramsey 2013). We are inclined to side with the causal view: individual-level interactions and population-level aggregates are both causes, just as individual-level particle motions and pressure (a population-level aggregate) are causes.

3.3 Adaptationism

Darwin's theory of evolution was received in some quarters as a threat to a cherished worldview, since it offered to explain the marvelous functional complexity displayed in nature not as the handiwork of a divine creator but as the product of a gradual process of mindless refinement: features of organisms that appeared intelligently designed were in fact *adaptations* shaped by natural selection.

That adaptation and adaptationist-style explanation are among the most historically and socially significant aspects of Darwin's theory is beyond doubt. What remains controversial, however, is the relative importance of these notions within evolutionary biology itself. How important is natural selection, on the one hand, compared to, on the other, factors like drift, accident, and developmental constraint, when it comes to explaining features of organisms and large-scale patterns in the history of life? The controversy here has come to be known as the "adaptationism" debate.

An evolutionary biologist, Williams (1966: 4) claimed that adaptation is a "special and onerous" concept and warned that it should only be deployed "when really necessary": nonadaptationist options should be considered and exhausted and sufficient positive evidence adduced. In a now classic paper, Stephen Gould and Richard Lewontin (1979) argued that Williams's warning had gone largely unheeded; this paper has continued to be both influential and provocative (it was the focus of a special issue of *Biology & Philosophy* in 2009, the paper's thirtieth anniversary).

Gould and Lewontin charged many practicing biologists with "adaptationism," that is, believing natural selection to be "so powerful and the constraints upon it so few that the direct production of adaptation through its operation becomes the primary cause of nearly all organic form, function, and behaviour" (1979: 150–151). They criticized what they saw as a prominent "adaptationist programme" in biology, consisting of a certain style of argument and investigation: assuming there is an adaptationist explanation for a trait, replacing failed adaptationist explanations with new but still adaptationist candidates, and, importantly, setting the bar for acceptance of adaptationist explanations too low, at mere plausibility (when "plausible stories," as Gould and Lewontin note, "can always be told"). It would be hard to sustain this charge now: evolutionary biology has developed sophisticated methods both from population genetics and, especially, comparative biology to test adaptationist hypotheses.

Much of the debate generated by Gould and Lewontin's paper seemed unproductive, and, in an important paper, Godfrey-Smith (2001) diagnosed much of the problem. Godfrey-Smith identified not one but three kinds of adaptationism and claimed that understanding and resolving the adaptationism debate had been hampered by failure to distinguish these; recently, Tim Lewens (2009) extended Godfrey-Smith's framework. Godfrey-Smith's three adaptationist theses are empirical adaptationism, explanatory adaptationism, and methodological adaptationism. The first claims that natural selection is the most causally important factor in evolution. The second claims that natural selection solves the most interesting problems in biology, namely, accounting for organisms' functional complexity and their fit with their environment. Much of Dawkins's work fits naturally into this category. The third claims that looking for adaptations is a good research strategy. Within biology, this is often known as the "phenotypic gambit": assume that an organism is an ideal problem solver in its environment and then use the mismatch between ideal and actual performance to reveal the constraints on adaptive response.

We take Godfrey-Smith's analysis of the adaptationism question to be an exemplar of philosophy of biology working well. For although there are some natural packages to buy, the three kinds of adaptationism are strictly speaking independent of each other. Pulling them apart, determining what evidence bears on which claims (and whether that evidence is forthcoming), and identifying which form(s) of adaptationism specific figures in the debate hold—all these are valuable services philosophy of biology can perform for biology itself.

3.4 Species

The idea that individual organisms can be grouped together into *species* is both theoretically and practically important, as well as commonsensically appealing. Species both persist and change over evolutionary time. Species also form a fundamental unit of biological classification. So species matter to evolutionary biology and to systematics. Practically, species matter for conservation biology. Determining which species are endangered and which should be prioritized requires first identifying which species exist.

Species thus present an ontological question (what are species?) and a question about classification (on what basis are organisms grouped into species?); depending on the answers here, the commonsensical idea of species may or may not survive scrutiny. There is also a practical question about how to count and prioritize species for purposes of conservation. That last question will be deferred until later (Section 4.2). The focus for now is on the first two.

A longstanding view about the nature of species is typological: species are types of organism in much the same way that chemical elements are types of matter. What defines a species on this view is a distinctive essence, a set of properties necessary and sufficient for being a member of that species. This essentialist kind of typological view of species dates back at least to Aristotle and persists today as a widely held folk view (Griffiths 2002). Species thus conceived were fixed—one species did not change into another—and discontinuous, meaning there were gaps rather than graded variation between species.

Darwin's theory of evolution challenged and eventually led to the displacement of essentialism about species. As Mayr (1959/1997) highlighted, Darwin replaced typological thinking with population thinking. In the latter mode, variation between members of a species is seen as key, and species boundaries are not sharp or permanent. Moreover, species change over time, rather than being eternally fixed. Darwin thus denied both discontinuity and fixity with respect to species.

Philosophers of biology have explicitly argued against species essentialism (Hull 1965; Sober 1980). The question thus arises: If not types defined by essences, what are species? One proposal, first defended by Ghiselin (1974) and Hull (1976), is that species are individuals. Like the most familiar examples of individuals (organisms), a species is spatiotemporally restricted—it has a beginning and (eventually) an end in time and a location in space—and exhibits a kind of integration between its parts. Hull and Ghiselin packaged their view as a rejection of the idea that a specific species—the red kangaroo—is a kind of any description. But it gradually became clear that their essential argument was compatible with thinking of red kangaroos as a natural kind, so long as the distinctive characteristics of the kind are historical and relational. On this view, a particular individual animal is a member of the species, the red kangaroo, in virtue of being genealogically nested within it. Its parents and kin are part of a reproductive network that established as a more or less closed system when the species diverged from its ancestor in Pleistocene Australia. Species are historically defined kinds.

Ghiselin (1974) and Hull's (1976) suggestion that a species as a whole is a lineage with organisms as its parts has shaped much subsequent debate. Considerable effort has been spent trying to say just what kind of integration organisms must exhibit in order to be grouped together into a species. The most natural suggestion to a zoologist was sex. Thus the *biological* species concept groups organisms into species on the basis of reproduction. As first defined by Mayr, species are "groups of actually or potentially interbreeding natural populations which are reproductively isolated from other such groups" (1963: 19). But while this suggestion fits much of the world of multicelled organisms, it does a poor job when asked to handle cases other than multicellular, sexually reproducing animals: plants, bacteria, Achaea, and (obviously) asexual organisms. Prokaryotes create special problems: grouping bacteria into species may well reflect the convenience of microbiologists rather than independent biological reality. But setting aside the problem of bacteria, there have been two main strategies in response to the problems of the biological species concept. One is to recognize that there is no single integration mechanism: there is no single mechanism that unifies a lineage of organisms over time, linking their evolutionary fate and distinguishing this lineage and its members from other lineages. The unifying mechanisms vary over the tree of life: perhaps reproductive isolation is central to mammals, but a shared niche is more central to plants. This strategy thus recognizes a family of species concepts. The second strategy focuses on pattern, not integration mechanism. A species just is an independently evolving lineage, one with an independent evolutionary fate. It is a segment of the tree of life, between its origin as it split from its ancestor and its own real or pseudo-extinction (as it splits into daughters).

The obvious challenge to the pattern view, of course, is to specify what counts as an independently evolving lineage (and, hence, what counts as being a member of this lineage). Different versions of this strategy defend different ways of counting lineages. Geoff Chambers (2012) gives a recent brief review of these issues. Marc Ereshefsky (2001) expresses skepticism about the special status of species (relevant also to the next section). For a selection of recent work on the nature of species and the tree of life, see the special issue of *Biology & Philosophy* on the tree of life, edited by Maureen O'Malley (2010); this too is relevant to the next section.

3.5 Systematics

In traditional taxonomic practice, species are grouped into packets of similar species, collectively known as a genus; similar genera are grouped into a family, families into an order, orders into a class, classes into a phylum, and phyla into a kingdom. This practice posed to its devotees some difficult questions. How similar to the red kangaroo (*Macropus rufus*) does another macropod species have to be to count as being in the same genus, and what similarities count? In the standard works on the mammals of Australia, the swamp wallaby (*Wallabia bicolor*) is not treated as a member of the red kangaroo's genus; indeed, it is in a genus of its own. Why? If one thought that a genus or a family was an objective level of organization in the tree of life, taxonomists would need a principled answer to questions like these. That is not mandatory. One could take this nested system of classification to be just a convenient way of organizing our information about the world's biota. Even though most species have never been formally described, we nonetheless have a vast amount of information about the world's biota, and that information needs to be stored and indexed in a usable way so biologists can retrieve it. Storing information through a series of nested similarities is useful. If I know the genus or the family of a little known species of skink, I can infer, at least as a good working hypothesis, that it will have the characters typical of its genus or family. That will be true, even if there are somewhat different ways of determining the membership of the genus (how similar is similar enough?); it will be true even if there are somewhat different ways of measuring similarity (how important is size? how do we score similarity in skin patterns?). We could choose to treat these higher taxonomic groupings as our constructs, just as principles for shelving books close to one another in the library—subject grouping principles—are our constructs, justified by, and only by, their practical utility to working biologists.

That, however, is not how biological systematics has developed in the past thirty years. So-called evolutionary taxonomy attempted two goals at once: represent in its naming and grouping practices the history of the tree of life and represent biologically significant morphological similarity and divergence. The red kangaroo genus, *Macropus,* named a group of morphologically similar, very closely related species. The genus name was supposed to have genuinely biological significance in two ways: answering both to morphology and recent common descent. Systematics has now essentially abandoned the aim of representing morphological similarity and difference, in part because there seems to be no objective way of specifying and weighting the similarities to count. Systematicists now see their core project as discovering and representing the tree of life. Nothing like a Linnaean hierarchy—species, genus, family, and so on—can be recovered in any natural way from a tree of life (Ereshefsky 2001). In light of that fact, there have been attempts to introduce a new naming practice, PhyloCode, so that taxonomic units are names for well-defined segments of the tree of life (for details, see http://www.ohio.edu/phylocode/).

The project of discovering and representing the tree of life has developed very impressively in the past twenty years on three foundations. One is conceptual: the cladistic insight that similarity is a good guide to shared evolutionary history but only once we discard similarities inherited from ancient ancestors. All the *Macropus* kangaroos share many traits, especially to do with reproduction, by descent from their ancient marsupial ancestor. Those traits can tell us nothing about who is most closely related to whom within the *Macropus* group. Only recent, derived traits shared by some but not all the *Macropus* macropods can tell us that. These cladistic methods were pioneered by German entomologist Willi Hennig (1966) and rely heavily on parsimony. In the simplest versions of the idea, the branching structure of a "cladogram" is drawn in whichever way requires the fewest evolutionary changes to account for the observed distribution of traits. There was considerable philosophical input in developing the core logic behind the reconstruction of evolutionary history (Sober 1991; for an overview, see Haber 2008).

The second foundation was the extraordinary development of computational power and the formal tools to exploit it. Once our biological target group is larger than just a few species and a few traits, finding the simplest tree becomes a terrifyingly difficult search problem, as the space of possible trees becomes vast, and the problem becomes greater still once we go beyond the simplest ways of looking for the most parsimonious tree. Only the development of new formal tools has made this tractable.

The third development is the explosion of genetic data, as it has become possible to sequence large chunks of the genome of many species. DNA sequences are relatively easy to compare, so we can identify historically informative differences between related species. This has proved to be of immense value, especially in recovering ancient evolutionary history, where changes in morphology have masked the deep branching pattern of life. Before genetic data, it had proved impossible to resolve the early history of the radiation of multicelled animals. The history of life is interesting in its own right, but once we have a well-confirmed picture of the history of some branch on the tree of life, that history can be used to test hypotheses about adaptation and constraint. With that data, as many DNA sequences are highly conserved, it has proved possible to resolve that history and use it to test hypotheses about the nature of large-scale, long-term evolutionary change (Bromham 2011).

We should note one more development. Some read recent developments in microbiology as reason to reject the idea that there is a tree of life, denying that the history of life is best represented as a branching, diverging pattern. They argue (i) that the microbial world is ancient, disparate, and numerous: microbes are the dominant players in the world's biota, and (ii) horizontal gene transfer (gene transfer between microbial lineages) is so pervasive that we should not regard microbial lineages as diverging branches, each with an independent evolutionary history (O'Malley 2014; see also O'Malley [2013a,b]). These developments are fascinating, but we are not yet convinced. Consider an individual bacterium—say a specific *E. coli* in our stomach. That bacterium has unbroken lineage of parent–daughter ancestors stretching back to the origins of life, and as we go back in time, sister lineages will diverge from that stem. The further back we go, the more disparate surviving lineages will be from our *E. coli*. At the level of the bacterium itself and its ancestors, there has been no fusion, no merging of formerly independent lineages to form a new organism (there have been such fusions—the eukaryote cell was formed by one—but all agree that these are very rare events). But very likely, some genes in our *E. coli* are not lineal descendants of the genes in the ancient microbe from which the lineage began; they have been picked up from other microbial genes by lateral transfer. To the extent, but only to the extent, that we regard a bacterium as nothing more than a bag of genes, in virtue of this transfer, our bacterial lineage does not have an independent evolutionary history. Since we do not think organisms are just bags of genes, we do not yet see lateral gene transfer as a reason to reject the tree of life as a good representation of the history of life.

3.6 Human Nature

Appeals to human nature have a long history in politics and philosophy. Aristotle based his virtue ethics on a particular conception of human nature—man is uniquely a rational animal—and modern virtue ethicists continue to ground their views in claims about human nature (Nussbaum 1995; Hursthouse 2012). Thomas Hobbes's (1651/1988) belief that unconstrained human nature would lead people to wage a "war of all against all" and lead lives that were "solitary, poor, nasty, brutish, and short" underpinned his belief in the necessity of a social contract. On the other side of the ledger, Jean-Paul Sartre's (1946/1989) denial of the existence of human nature was a foundational element in his formulation of existentialism.

On the conception of human nature in play in these cases human nature is unique, essential, and universal. This should sound suspicious, given what we said earlier about species, and, indeed, one of the participants in that debate applied those general considerations about species to the issue of human nature. Hull (1986) argued that nothing in biology corresponds to traditional notions of human nature; there are no important, unique human universals. Drawing on general principles about species, he argued against in-principle limits on human variation, at and across time. Perhaps all (nonpathological, adult) humans speak a language, are aware of others as intentional agents, and understand social expectations. But there is no guarantee that our future evolutionary trajectory will leave these universals intact. We count as humans not in virtue of our intrinsic characteristics but in virtue of belonging to a certain lineage. But perhaps the in-principle possibility of radical change in the future is of no great consequence. It could be that, in fact, human variation is sharply constrained, and by deep architectural features of the human mind.

Some evolutionary views of human nature make exactly that bet, beginning with Wilson (1975, 1979) and classic sociobiology, and then developed in a more cognitive form by one version of evolutionary psychology. Nativist evolutionary psychologists have argued that a human nature consists of the universally shared genetically coded suite of evolved mental modules. These make up the human mind (Cosmides and Tooby 1992; Barkow et al. 1992; Pinker 1997). Researchers have argued that the extraordinary cognitive efficiency of humans, as they face the very demanding challenges of ordinary life (using language; engaging in complex, coordinated collective action; recognizing and adapting to social expectations; engaging with elaborate material culture), can be explained only on the hypothesis that we have evolved cognitive specialisations that power these skills. Others suggest that this view of our cognitive capacities cannot explain our successful response to evolutionarily novel environments and that instead our minds and social environments are adapted to social learning and collective action. In virtue of these adaptations, unlike other primates, we accumulate cultural information generation by generation (Tomasello 1999, 2014; Richerson and Boyd 2005; Sterelny 2003, 2012; Laland and Brown 2002). We are adapted to cope with novelty, rather than specific environmental challenges. While this too is an evolutionary theory of human nature, it is unclear to us that it would serve as a foundation for moral, social, and political theories.

4 Recent Expansions

The brief survey here is not exhaustive. For example, we do not discuss the relationship between microevolution and macroevolution: whether large-scale patterns in the history of life can be explained wholly in terms of mechanisms acting on local populations over a few generations. It has been suggested that there are mechanisms of species selection (or species sorting) that play an important role in generating these large-scale patterns (perhaps especially through episodes of mass extinction). But rather than further cataloguing the core agenda, we turn now to more recent developments. This section notes several ways in which philosophy of biology has recently been expanded to include issues and fields beyond the traditional core described earlier. It first considers the increasing attention paid to molecular biology, focusing on controversies over the nature of the gene and the role of information in genetics. It then discusses the philosophy of ecology and conservation biology, focusing on issues concerning models and biodiversity.

4.1 Molecular Biology

Classical or Mendelian genetics spoke of "factors" influencing organisms' traits: one factor might affect a plant's flower color, another its seed shape. It did not say what these factors were. Gregor Mendel, after whom the field is named, conducted breeding experiments and observed the pattern of traits appearing in different lineages, concluding that factors are inherited. Mendel's "factors" eventually came to be known as "genes."

The importance of intracellular biochemistry in understanding the nature of genes became apparent in the early twentieth century with the discovery that genes (whatever they were) were located on chromosomes (Morgan et al. 1915). According to the classical gene concept, genes are structures on, or of, chromosomes affecting phenotypic traits (Griffiths and Neumann-Held 1999: 660). Further scientific advances were made in understanding genes with the advent of molecular biology.

Molecular biology is, as the name suggests, the study of biological activity at the molecular level. It overlaps with biology, especially genetics, and chemistry, especially biochemistry. Molecular biologists investigate the structure and function of macromolecules—most famously deoxyribonucleic acid (DNA)—as well as the molecular mechanisms underlying the transcription and translation of DNA and heredity itself.

The field has a short but hectic history. Major events in molecular biology include the identification in 1944 of DNA as the molecule of inheritance, the 1953 discovery by Watson and Crick that DNA has a double-helix structure, Nirenberg and Matthaei's cracking of the "code" by which DNA specifies proteins in 1961, and the 2003 announcement of the sequencing of the entire human genome. With this new molecular-level detail came a new conception of the gene: according to the molecular gene concept, a gene is a segment of DNA that produces a particular molecular product (Griffiths and Neumann-Held 1999: 660).

Molecular biology is relevant to philosophically interesting debates about competing concepts of the gene, as well as the role of information in biology and the issue of reductionism in biology. Taking the latter first, it is now widely thought that if reduction is a matter of restating the theories of one science in terms of another supposedly more general and fundamental one (cf. Nagel 1961), then the prospects for reducing Mendelian to molecular genetics (Schaffner 1967) are poor, given that nothing in the complicated molecular-level machinery subserving inheritance corresponds cleanly to the classical conception of the gene (Hull 1972; Rosenberg 1985).

If reduction is understood differently, however, the reduction of classical to molecular genetics is more promising. In their recent book *Genetics and Philosophy*, Paul Griffiths and Karola Stotz (2013) consider in depth the claim that Mendelian genetics reduces to molecular genetics insofar as the latter elucidates the *mechanisms* underlying the patterns and processes described by the former. Griffiths and Stotz are ultimately unconvinced that this kind of reduction—reduction as mechanistic explanation—can succeed, given that so much explanatory work at the molecular level is done by cellular resources *other than* DNA. However, they do allow that Mendelian and molecular genetics mutually illuminate the same phenomena of interest, biological inheritance, even if the latter is not reducible to or replaceable by the latter. These issues continue to provide an important testing ground for ideas about the nature, role, and success of reduction in science.

More recent philosophical debate about molecular biology has focused on the role played in genetics by *information*, and here our discussion links back to our previous discussion about the gene as unit of selection. In our discussion there, we noted that one of the axes of disagreement concerned the relationship between genes and phenotypes. The gene's eye view of evolution often goes with seeing the genes as playing a special role in development, not just in inheritance, and in turn, that "special role" is often explained in terms of information. Genes program, code for, or carry information about phenotypes. So, for example, Williams suggests that "the gene is a packet of information" (1992: 11). In one unexciting sense, this is uncontroversial. Genes carry causal or covariational information about phenotypes. Carriers of the sickle-cell allele tend to be malaria resistant. But nongenetic factors carry covariational information in the same way. Crocodilians have temperature-dependent sex determination, so the nest temperature to which eggs are exposed carries information about sex. Thus defenders of the idea that genes play a special informational role need, and have developed a richer, semantic sense of information. The most sophisticated version of this view, developed by Nick Shea (2007, 2013), has co-opted ideas from philosophy of cognitive science about information and representation and reengineered them to give an account of the representational content of DNA sequences. The core idea is that specific genes (or gene complexes) are biologically designed to channel development towards a specific phenotype, and this specific evolved function piggybacks on the fact that the DNA transcription-translation system as a whole is an evolved, adapted inheritance system. It does not just happen to reliably support parent–offspring similarities; it is fine-tuned to support similarity-making. Gene sequences carry information for the same reason that word sequences do. They are specific uses of a system designed to send messages (Shea 2007, 2013; Bergstrom and Rosvall 2011). Despite the sophistication of Shea's analysis, this line of thought remains very controversial. Skeptics remain unconvinced about both the role of rich informational notions in biology and of related claims about the special developmental role of the gene. This skeptical view is summed up in their "parity thesis," the idea that there is no qualitative difference between genes and other factors in the developmental matrix (Griffiths and Gray 1994, 2001; Griffiths 2001; Griffiths and Stotz 2013).

4.2 Ecology and Conservation Biology

Ecological systems are complex, unruly, and often unique. The tangled bank of a particular coral reef will resemble other reefs, but there will be differences too, perhaps important ones. This creates an apparently inescapable dilemma exposed in a classic paper by Richard Levins (1966; see also Matthewson and Weisberg 2009; Weisberg 2006). A scientific representation—a model—of such a system cannot simultaneously optimize generality, precision, and realism. A precise, realistic characterization of one reef will mischaracterize one a few kilometers away. A general model of reef dynamics must idealize away many features of real reefs. These may well matter to the future dynamics of these systems. Levins's solution was to recommend the development of a suite of models, each making different trade-offs between precision and generality. We should place most trust in robust results. If all (or most) models identify a link between (say) sea-urchin numbers and coral diversity, we should be inclined to accept that result.

The greater the diversity of models, the greater our confidence in common results. In one of philosophy of science's most memorable phrases, Levins concluded that "truth is the intersection of independent lies" (1966: 423).

In the face of this challenge posed by complex and unique systems, the main role of philosophy of ecology within philosophy of biology has been through its exploration of model-based science (though Wimsatt's [2007] work has been especially important here too, based partly but not wholly in ecology). What are models, and how do they differ from theories? Given that models are inevitably simplified, and often dramatically, they cannot be correct representations of real-world target systems. How then can they explain real-world phenomena? What is it about ecological systems that forces us to trade off between precision, realism, generality? Does the same challenge arise from any complex system? If that were true, given the complexity of, say, the human mind, it would be impossible to build a general and accurate model of facial memory. Alternatively, if we can give a general and accurate model of facial memory, why can we not give a general and accurate model of, for example, invasive species (e.g. a model that would tell us if ocelots would run amok in Australia)?

Rich and nuanced work has been done on all these issues within philosophy of ecology. For example, Weisberg (2013) has developed a taxonomy of models and identified model-building strategies that are driven by characteristics of the target system and distinguished these from strategies that depend on the limits of our knowledge and our computational resources. Peter Godfrey-Smith (2006, 2009b) has built an analysis of models and of the difference between models and theories. On his view, models are indirect representations of target systems; they represent real-world targets via descriptions of fictional systems that resemble these targets. Model representation is often flexible precisely because the character and extent of these resemblances need not be specified in advance of model building.

However, just as evolutionary biology struggles with questions of a mixed conceptual and empirical character, so does ecology. A selection of such questions includes (i) are local ecological communities causally salient, observer-independent systems in the world, or are they our projections onto the natural world? (ii) How does ecosystem ecology, focused on the flow and cycle of materials and energy through a local region, relate to population ecology, which seeks to explain the distribution and abundance of organisms? (iii) What is a niche? In particular, do niches exist independently of the organisms that fill them in a particular system (so a system with a "vacant niche" is invadable)? (iv) Given the differences in physical and temporal scale, how do ecological processes interface with evolutionary ones? Ecological models typically assume populations of a given kind of organism are invariant and unchanging; evolutionary models typically assume, rather than explain, fitness differences (Barker and Odling-Smee 2014). There has been some attention in philosophy of ecology to these issues, and especially the first, on whether ecological systems are real units in nature, and, if so, how their boundaries are to be identified. Greg Cooper (2003) was a pioneer here, and his example has been followed up to some extent (Sterelny 2006; Odenbaugh 2007). Even so, philosophy of ecology is still relatively underdeveloped compared to philosophy of evolutionary biology.

Philosophy of ecology has deep connections with the emerging field of the philosophy of conservation biology. Conservation biology inherits all the difficult empirical problems of ecology—it deals with complex, perhaps historically contingent and hence unique systems—and more, for, as an applied science, conservation biology needs to identify the best practical interventions in contexts not just of limited material resources but also very limited informational resources. We intervene in systems that we at best partially understand. A question also exists about whether the core agenda of conservation biology is genuinely scientific. What should be conserved or restored, and why? Do ecosystems or biological systems have a natural state to which they would return if protected from human destruction? Are ecosystems or communities healthy and functional (or the reverse) in some objective sense, or are these wholly subjective judgements? Readers in the Antipodies will know how historically variable such judgments are. There were sustained attempts to "improve" Australian and New Zealand communities by releasing northern hemisphere fauna: rabbits, foxes, moose, and the like.

Stuart Pimm (1991) showed that biological thinking was profoundly influenced by "balance of nature" concepts, but we now recognize that ecosystems and biological communities are naturally dynamic. There is no privileged original state. Instead, a family of research projects attempts to link the biodiversity of an ecosystem to its stability or productivity. These projects pose very difficult empirical issues: genuine tests require temporal and geographic scales beyond the scope of standard experimental regimes. But they pose difficult conceptual issues too: stability and diversity can both be conceptualized in different ways, and those differences matter (Justus 2008). However, this potential connection between biodiversity and core ecological processes opens up the possibility of treating biodiversity as an objective and causally salient dimension of biological systems, moreover one in which we have a clear prudential stake (Maclaurin and Sterelny 2008).

Sahotra Sarkar (2005, 2012), probably the leading figure in the field, remains unconvinced. For him, the concept of biodiversity is a resource-allocation tool, a means of directing limited resources to one area rather than another. While the tool is constrained by the objective biology of the system, our measure of biodiversity inevitably and rightly reflects both our historically contingent preferences and the fact that we must use proxies, often very crude ones, of the underlying biological processes of interest to us. Yet the concept is useless to conservation biology unless we can measure and compare. Sarkar embraces a conventionalist view of the concept (see Sarkar 2005, 2012). The debate is far from settled; there are, for example, even more sceptical views of the value of thinking about biodiversity as a guide to conservation (Santana 2014).

5 CURRENT FRONTIERS

This section details several current frontier challenges in philosophy of biology. It begins with a topic that is not new to the field but that has recently been taken up in new ways: the evolution of cooperation. It closes by considering the question of what defines an individual, identifying problem cases for traditional accounts and detailing recent attempts to provide a framework that can accommodate the complexity of biological individuality.

5.1 Cooperation

Cooperation has posed a perennial puzzle for biologists and philosophers of biology. The puzzle is to explain how a nature "red in tooth and claw" could permit the startlingly helpful, cooperative, even altruistic behavior many organisms routinely display. Alarm calling, food sharing, collective defense, even suicidal self-sacrifice and the total forego-ing of reproduction—all these phenomena present a prima facie problem for evolution-ary theory, given that selection would seem to strongly disfavor organisms that act so drastically counter to their own fitness interests.

Evolutionary biologists have proposed three main ideas to explain cooperation. The most obvious is that cooperation can be mutually advantageous. A cooperative pack of wolves can kill larger prey, and kill more safely, than a lone wolf. The same mecha-nism works over time: I remove the fleas from your back today; you do the same for me tomorrow, and we both benefit. Cooperation for mutual gain is not intrinsically myste-rious, but it is vulnerable to a famous problem of free-riding. In many contexts: (i) the cooperative behavior has a cost; (ii) the gains of collective action do not require every agent's full participation in the cooperative activity; (iii) there is no automatic mecha-nism that distributes the gains of cooperation in proportion to the effort expended. In these circumstances, there is a temptation to cheat. If the whole pack is not required to kill a moose, each wolf is better off leaving the others to take most of the risk (cf. Scheel and Packer 1991), and cooperation can collapse. There is a large literature on the con-texts in which cooperation through mutual gain is stable despite the risks of cheating—most famously, iterated contexts where cheating is contained by the threat of retaliatory cheating in response (Trivers 1971; Axelrod 1984; Binmore 2006) and on contexts in which cooperation is less vulnerable to cheating but where it still requires coordination and hence signalling (Skyrms 1996, 2004).

The second mechanism depends on the insight that a cooperative trait can increase in frequency if that cooperation is directed to close relatives, for they are likely to carry the same genes, including those that prompt cooperation. This mechanism requires the benefits to the target of cooperation to be greater than the costs to the co-operator, dis-counted by the closeness of the relationship. For example, on average a person shares 50% of his or her genes with a full sibling, so selection favors helping said sister if the benefit to her is more than twice the cost to the individual. This idea was developed by Bill Hamilton (1964). He saw that supposedly fitness-sacrificing behavior can actually pay off in terms not of individual but of *inclusive* fitness, that latter notion being defined in terms of the number of copies of one's genes passed on by oneself *and one's relatives* (hence, biologist J. B. S. Haldane's quip that he would not sacrifice himself for his brother but would do so for two brothers or eight cousins). There is now much controversy about the biological importance of inclusive fitness. It was once thought to be the key to understanding the extraordinarily cooperative behavior of social insects; these have unusual genetic systems, so full sisters are especially closely related. But this is now controversial (see Nowak, Tarnita, and Wilson [2010] and responses in the March 2011 issue of *Nature*; Birch 2014a, 2014b).

The third mechanism takes us back to the core debates about units of selection. One natural explanation is that cooperation evolved through selection in favor of cooperative groups, despite selection favoring selfish individuals in those groups. Debates about this idea have ebbed and flowed since Williams's apparently decisive 1966 critique. This mechanism depends on positive assortment: selection can drive the spread of cooperation by favoring cooperative groups if cooperators interact more with one another than with noncooperators. How strong that positive assortment must be depends on the benefit of cooperator-cooperator interactions, and on the fitness differences between cooperators and defectors in mixed groups. The theory is reasonably well worked out (Okasha 2003, 2006)—how the theory maps onto empirical cases, much less so. The bottom line, though, is that cooperation has been a major driver of evolution. The major transitions in evolution all depend on successful, stable solutions to cooperation problems (Maynard-Smith and Szathmary 1995; Sterelny et al. 2013).

5.2 Individuality

Folk biology recognizes discrete units in the natural world, units that develop, persist by maintaining themselves, and that reproduce: *organisms*, such as horses and humans. Organisms seem the cleanest cases of biological individuals. However, the matter soon gets messy, and the mess matters. Modern evolutionary biology relies heavily on counting—counting offspring, counting members of populations—so knowing how to identify individuals is important. The commonsensically appealing organism-based view of individuality faces many problem cases.

One class of cases is due to the difficulty of distinguishing growth from reproduction. In the case of asexually reproducing plants such as aspens, what looks like the generation of a new individual organism (i.e., reproduction) is in fact the extension of an existing individual (i.e., growth), since the "new" plant shares the root system of the old. Physical links can be broken, and the resultant pieces of the system can then exist independently, creating confusion about how many individuals there are/were. Daniel Janzen (1977) drew a distinction between "ramets" (physiological units) and "genets" (groups of genetically identical ramets) in order to deal with such cases, claiming that the real biological individual was the genet. He thus deemphasized physical cohesion in favor of genetic identity when defining individuals. The problems plants pose for accounts of biological individuality continue to stimulate debate and inform discussion of more general issues in philosophy of biology, such as the units and levels of selection (Clarke 2011).

Another class of problem cases arises when we realize that some apparently unitary organisms are in fact collectives, groups of living things that are themselves individual organisms (or at least organism-like). For instance, lichens are mutualistic associations of fungi and algae, and the Portuguese man o'war is an integrated colony of organisms specialized for movement, defense, feeding, and reproduction. We too fit this bill, given our dependence on a suite of symbiotic gut bacteria. Even less tightly integrated collectives can present as candidates for biological individuality. Eusocial insect colonies, for example, are composed of physiologically distinct if complementarily specialized organisms and have been seen by some as organisms in their own right, so-called super-organisms (Holldobler and Wilson 2008; Ereshefsky and Pedroso 2013; Booth 2014).

More problem cases could be presented, some interesting ones having to do with unusual life cycles. For example, many plants' sexual cycle includes a stage in which the haploid form (an ephemeral, single-cell stage in animals) develops as a complex, multi-celled physiological individual. Do these counts as continuants of the parent or offspring (see Godfrey-Smith 2009a)? The root issue is that genetic shared fate, physiological integration, and functional coadaptation come in degrees and need not coincide. So we are inclined toward developing a *graded* notion of biological individuality.

Peter Godfrey-Smith's (2009a) *Darwinian Populations & Natural Selection* offers a general framework that attempts to capture the complexity of biological individuality, retaining the idea that organisms are important while accommodating the problem cases. A key element of this framework is the notion of a "Darwinian population," a group of entities in which there is variation, heredity, and differential reproductive success (as noted previously, there are complications to be considered here, but these are set aside for current purposes). Another key notion is that of a "Darwinian individual," a member of a Darwinian population.

A crucial feature of Godfrey-Smith's (2009a) framework is that the factors that define a Darwinian population, and hence a Darwinian individual, come in degrees. Take reproduction, for example: this is not a unitary phenomenon but varies along several dimensions, including how narrow a genetic bottleneck divides generations, how much functional integration the reproducing entity exhibits, and how much reproductive specialization it displays. These three dimensions define a space within which different cases can be located. Much the same can be said for other elements necessary for a Darwinian population. Thus Godfrey-Smith offers a multiply graded account of individuality in the Darwinian sense of being a member of a population undergoing natural selection.

This notion of biological individuality is distinct from the organism-based account. Some Darwinian individuals are not organisms, and, perhaps more surprisingly, some organisms are not Darwinian individuals (see Godfrey-Smith 2014: 77–78 for details). But organisms do serve as a useful reference point in the space of Darwinian individuals, against which more exotic cases can be compared. Also, individuality thus understood comes in degrees, allowing us to avoid having to make a choice about what is *the* individual in a given case or *whether or not* a given entity is an individual, while still providing a principled view.

6 SUMMARY

In sum, we suggest that three themes emerge from this survey, despite the fact that it is brief and incomplete. First: the focus on biology makes a difference to how we think of science more generally. It foregrounds the importance of identifying mechanisms, of developing multiple models in the face of complexity, and of idiosyncratic phenomena; generalizations and laws of nature become background. If the take-home message from physics is that the world is deeply strange, the take-home message from biology is that the world is both diverse and untidy. Second, biology is philosophically rich because it continues to pose problems of a mixed empirical-conceptual-theoretical character. Third, in thinking about biological phenomena it is essential to escape the philosopher's preoccupation with discrete categories and necessary and sufficient conditions. Darwinian populations, biological individuals, and bounded ecosystems all come in degrees. That is typical rather than exceptional.

REFERENCES

Abrams, Marshall. (2007). "Fitness and Propensity's Annulment?" *Biology & Philosophy* 22(1): 115–130.

Axelrod, Robert. (1984). *The Evolution of Cooperation* (New York: Basic Books).

Ayala, Francisco, and Dobzhansky, Theodosifus. (1974). *Studies in the Philosophy of Biology* (Berkeley: University of California Press).

Barkow, Jerome, Cosmides, Leda, and Tooby, John, eds. (1992). *The Adapted Mind: Evolutionary Psychology and the Generation of Culture* (Oxford: Oxford University Press).

Beatty, John, and Finsen, Susan. (1987). "Rethinking the Propensity Interpretation of Fitness." In Michael Ruse (ed.), *What Philosophy of Biology Is* (Boston: Kluwer), 18–30.

Bergstrom, C., and M. Rosvall. (2011). "The Transmission Sense of Information." *Biology & Philosophy* 26(2): 159–176.

Binmore, Ken. (2006). *Natural Justice* (Oxford: Oxford University Press).

Birch, Jonathan. (2014a). "Gene Mobility and the Concept of Relatedness." *Biology & Philosophy* 29(4): 445–476.

Birch, Jonathan. (2014b). "Hamilton's Rule and Its Discontents." *British Journal for the Philosophy of Science* 65(2): 381–411.

Booth, Austin. (2014). "Symbiosis, Selection and Individuality." *Biology & Philosophy* 29: 657–673.

Bouchard, Frederic, and Rosenberg, Alex. (2004). "Fitness, Probability, and the Principles of Natural Selection." *British Journal for the Philosophy of Science* 55(4): 693–712.

Bromham, Lindell. (2011). "The Small Picture Approach to the Big Picture: Using DNA Sequences to Investigate the Diversification of Animal Body Plans." In Brett Calcott and Kim Sterelny (eds.), *The Major Transitions in Evolution Revisited* (Cambridge, MA: MIT Press), 271–298.

Butler, Samuel. (1879). *Evolution Old and New; or, the Theories of Buffon, Dr Erasmus Darwin, and Lamarck, as Compared with that of Mr Charles Darwin* (London: Hardwick and Bogue).

Calcott, Brett, and Sterelny, Kim, eds. (2011). *The Major Transitions in Evolution Revisited* (Cambridge, MA: MIT Press).

Chambers, Geoff. (2012). "The Species Problem: Seeking New Solutions for Philosophers and Biologists." *Biology & Philosophy* 27(5): 755–765.

Clarke, Ellen. (2011). "Plant Individuality and Multilevel Selection." In Brett Calcott and Kim Sterelny (eds.), *The Major Transitions in Evolution Revisited* (Cambridge, MA: MIT Press), 227–250.

Cooper, Greg. (2003). *The Science of the Struggle for Existence: On the Foundations of Ecology* (Cambridge, UK: Cambridge University Press).

Cosmides, Leda, and Tooby, John. (1992). *The Adapted Mind* (Oxford: Oxford University Press).

Dawkins, Richard. (1976). *The Selfish Gene* (New York: Oxford University Press).

Dawkins, Richard. (1982). *The Extended Phenotype* (New York: Oxford University Press).

Dennett, Daniel. (1995). *Darwin's Dangerous Idea* (New York: Simon & Schuster).

Eldredge, Niles, and Gould, Stephen. (1972). "Punctuated Equilibrium: An Alternative to Phyletic Gradualism." In T. J. M. Schopf (ed.), *Models in Paleobiology* (San Francisco: Freeman Cooper), 82–115.

Ereshefsky, Marc. (2001). *The Poverty of the Linnaean Hierarchy* (Cambridge, UK: Cambridge University Press).

Ereshefsky, Marc, and Pedroso, Makmiller. (2013). "Biological Individuality: The Case of Biofilms." *Biology & Philosophy* 29: 331–349.

Forber, Patrick, and Reisman, Kenneth. (2005). "Manipulation and the Causes of Evolution." *Philosophy of Science* 72(5): 1113–1123.

Ghiselin, Michael. (1974). "A Radical Solution to the Species Problem." *Systematic Zoology* 23: 536–544.

Godfrey-Smith, Peter. (2001). "Three Kinds of Adaptationism." In S. H. Orzack and E. Sober (eds.), *Adaptationism and Optimality* (Cambridge, UK: Cambridge University Press), 335–357.

Godfrey-Smith, Peter. (2006). "The Strategy of Model-Based Science." *Biology & Philosophy* 21: 725–740.

Godfrey-Smith, Peter. (2009a). *Darwinian Populations and Natural Selection* (Oxford: Oxford University Press).

Godfrey-Smith, Peter. (2009b). "Models and Fictions in Science." *Philosophical Studies* 143: 101–116.

Godfrey-Smith, Peter. (2014). *Philosophy of Biology* (Princeton, NJ: Princeton University Press).

Gould, Stephen. (1981). *The Mismeasure of Man* (New York: Norton).

Gould, Stephen, and Lewontin, Richard. (1979). "The Spandrels of San Marco and the Panglossian Paradigm: A Critique of the Adaptationist Programme." *Proceedings of the Royal Society of London, Series B, Biological Sciences* 205(1161): 581–598.

Grene, Marjorie, and Mendelsohn, Everett. (1976). *Topics in the Philosophy of Biology* (Dordrecht, The Netherlands: D. Riedel).

Griffiths, Paul. (2001). "Genetic Information: A Metaphor in Search of a Theory." *Philosophy of Science* 68(3): 394–412.

Griffiths, Paul. (2002). "What Is Innateness?" *The Monist* 85(1): 70–85.

Griffiths, Paul, and Gray, Russell. (1994). "Developmental Systems and Evolutionary Explanation." *Journal of Philosophy* 91: 277–304.

Griffiths, Paul, and Gray, Russell. (2001). "Darwinism and Developmental Systems." In Susan Oyama, Paul Griffiths, and Russell Gray (eds.), *Cycles of Contingency: Developmental Systems and Evolution* (Cambridge, MA: MIT Press), 195–218.

Griffiths, Paul, and Neumann-Held, Eva. (1999). "The Many Faces of the Gene." *BioScience* 49(8): 656–662.

Griffiths, Paul, and Stotz, Karola. (2006). "Genes in the Post-Genomic Era." *Theoretical Medicine and Bioethics* 27(6): 499–521.

Griffiths, Paul, and Stotz, Karola. (2013). *Genetics and Philosophy: An Introduction* (Oxford: Oxford University Press).

Haber, Matt. (2008). "Phylogenetic Inference." In A. Tucker (ed.), *Blackwell Companion to the Philosophy of History and Historiography* (Malden, MA: Blackwell), 231–242.

Haig, David. (2012). "The Strategic Gene." *Biology & Philosophy* 27(4): 461–479.

Hamilton, William. (1964). "The Genetical Evolution of Social Behavior (I and II)." *Journal of Theoretical Biology* 7: 1–52.

Hennig, Willi. (1966). *Phylogenetic Systematics* (Urbana: University of Illinois Press).

Hobbes, Thomas. (1988). *Leviathan.* Edited by Crawford Macpherson. (London: Penguin). (Original work published 1651)

Holldobler, Bert, and Wilson, E. O. (2008). *The Superorganism* (New York: Norton).

Hull, David. (1965). "The Effects of Essentialism on Taxonomy: Two Thousand Years of Stasis." *The British Journal for the Philosophy of Science* 15(60): 314–326.

Hull, David. (1972). "Reduction in Genetics: Biology or Philosophy?" *Philosophy of Science* 39(4): 491–499.

Hull, David. (1974). *Philosophy of Biological Science* (Englewood Cliffs, NJ: Prentice-Hall).

Hull, David. (1976). "Are Species Really Individuals?" *Systematic Zoology* 25: 174–191.

Hull, David. (1986). "On Human Nature." *PSA: Proceedings of the Biennial Meeting of the Philosophy of Science Association* 2: 3–13.

Hull, David. (1988). *Science as a Process* (Chicago: University of Chicago Press).

Hursthouse, Rosalind. (2012). "Human Nature and Aristotelian Virtue Ethics." *Royal Institute of Philosophy Supplement* 70: 169–188.

Janzen, Daniel. (1977). "What Are Dandelions and Aphids? *American Naturalist* 11: 586–589.

Justus, James. (2008). "Complexity, Diversity, Stability." In Sahotra Sarkar and Anya Plutynski (eds.), *A Companion to the Philosophy of Biology* (Malden, MA: Blackwell), 321–350.

Kerr, Ben, and Godfrey-Smith, Peter. (2002). "Individualist and Multi-Level Perspectives on Selection in Structured Populations." *Biology & Philosophy* 17(4): 477–517.

Laland, Kevin, and Brown, Gillian. (2002). *Sense and Nonsense: Evolutionary Perspectives on Human Behaviour* (Oxford: Oxford University Press).

Lewens, Tim. (2009). "Seven Types of Adaptationism." *Biology & Philosophy* 24: 161–182.

Lewontin, Richard. (1985). *The Dialectical Biologist* (London: Harvard University Press).

Levins, Richard. (1966). "The Strategy of Model Building in Population Biology." *American Scientist* 54: 421–431.

Maclaurin, James, and Sterelny, Kim. (2008). *What Is Biodiversity?* (Chicago: University of Chicago Press).

Matthen, Mohan, and Ariew, Andre. (2009). "Selection and Causation." *Philosophy of Science* 76: 201–224.

Matthewson, John, and Weisberg, Michael. (2009). "The Structure of Tradeoffs in Model Building." *Synthese* 170(1): 169–190.

Maynard-Smith, John, and Szathmary, Eors. (1995). *The Major Transitions in Evolution* (Oxford: Oxford University Press).

Mayr, Ernst. (1963). *Animals and Evolution* (Cambridge, MA: Harvard University Press).

Mayr, Ernst. (1997). *Evolution and the Diversity of Life* (Cambridge, MA: Harvard University Press).

Michod, Richard. (2007). "Evolution of Individuality during the Transition from Unicellular to Multicellular Life." *Proceedings of the National Academy of Sciences USA* 104: 8613–8618.

Mills, Susan, and Beatty, John. (1979). "The Propensity Interpretation of Fitness." *Philosophy of Science* 46: 263–286.

Morgan, Thomas, Sturtevant, Alfred, Muller, Hermann, and Bridges, Calvin. (1915). *The Mechanism of Mendelian Heredity* (New York: Henry Holt).

Nagel, Ernest. (1961). *The Structure of Science* (New York: Harcourt, Brace & World).

Nowak, Martin, Tarnita, Corina, and Wilson, E. O. (2010). "The Evolution of Eusociality." *Nature* 466 (7310): 1057–1062.

Nussbaum, Martha. (1995). "Aristotle on Human Nature and the Foundations of Ethics." In James Edward, John Altham, and Ross Harrison (eds.), *World, Mind, and Ethics: Essays on the Ethical Philosophy of Bernard Williams* (Cambridge, UK: Cambridge University Press), 86–131.

Odenbaugh, Jay. (2007). "Seeing the Forest and the Trees: Realism about Communities and Ecosystems." *Philosophy of Science* 74(5): 628–641.

Okasha, Samir. (2003). "Recent Work on the Levels of Selection Problem." *Human Nature Review* 3: 349–356.

Okasha, Samir. (2006). *Evolution and the Levels of Selection* (Oxford: Oxford University Press).

Okasha, Samir, and Paternotte, Cedric. (2014). "The Formal Darwinism Project: Editor's Introduction." *Biology & Philosophy* 29(2): 153–154.

O'Malley, Maureen. (2013a). "Philosophy and the Microbe: A Balancing Act." *Biology & Philosophy* 28(2): 153–159.

O'Malley, Maureen. (2014). *Philosophy of Microbiology* (Cambridge, UK: Cambridge University Press).

O'Malley, Maureen, ed. (2010). *Special Issue: The Tree of Life. Biology & Philosophy* 25(4).

O'Malley, Maureen, ed. (2013b). *Special Issue: Philosophy and the Microbe. Biology & Philosophy* 28(2).

Pence, Charles, and Ramsey, Grant. (2013). "A New Foundation for the Propensity Interpretation of Fitness." *British Journal for the Philosophy of Science* 64(4): 851–881.

Pimm, Stuart. (1991). *The Balance of Nature?* (Chicago: University of Chicago Press).

Pinker, Steven. (1997). *How The Mind Works* (New York: Norton).

Ramsey, Grant. (2013). "Organisms, Traits, and Population Subdivisions: Two Arguments against the Causal Conception of Fitness?" *British Journal for the Philosophy of Science* 64(3): 589–608.

Richerson, Peter, and Boyd, Robert. (2005). *Not By Genes Alone* (Chicago: University of Chicago Press).

Rosenberg, Alex. (1980). *Sociobiology and the Preemption of Social Science* (Baltimore: Johns Hopkins University Press).

Ruse, Michael. (1973). *The Philosophy of Biology* (London: Hutchinson).

Ruse, Michael. (1998). *Taking Darwin Seriously* (New York: Prometheus).

Ruse, Michael. (2009). *From Monad to Man* (Cambridge, MA: Harvard University Press).

Santana, Carlos. (2014). "Save the Planet: Eliminate Biodiversity." *Biology & Philosophy* 29(6): 761–780.

Sarkar, Sahotra. (2005). *Biodiversity and Environmental Philosophy: An Introduction* (Cambridge, UK: Cambridge University Press).

Sarkar, Sahotra. (2012). *Environmental Philosophy: From Theory to Practice* (Malden, MA: Wiley-Blackwell).

Sartre, Jean-Paul. (1989). "Existentialism Is a Humanism." In Walter Kaufman (ed.), *Existentialism from Dostoyevsky to Sartre* (New York: Meridian). (Original work published 1946), 287–311.

Schaffner, Kenneth. (1967). "Approaches to Reduction." *Philosophy of Science* 34(2): 137–147.

Scheel, David, and Packer, Craig. (1991). "Group Hunting Behaviour of Lions." *Animal Behaviour* 41: 697–709.

Shea, Nick. (2007). "Representation in the Genome and in Other Inheritance Systems." *Biology & Philosophy* 22(3): 313–331.

Shea, Nick. (2013). "Inherited Representations Are Read in Development." *British Journal for the Philosophy of Science* 64(1): 1–31.

Skyrms, Brian. (1996). *The Evolution of the Social Contract* (Cambridge, UK: Cambridge University Press).

Skyrms, Brian. (2004). *The Stag Hunt and the Evolution of Social Structure* (Cambridge, UK: Cambridge University Press).

Sloan-Wilson, David, and Sober, Elliot. (1994). "Reintroducing Group Selection to the Human Behavioral Sciences." *Behavioral and Brain Sciences* 17(4): 585–654.

Sober, Elliott. (1980). "Evolution, Population Thinking, and Essentialism." *Philosophy of Science* 47(3): 350–383.

Sober, Elliot. (1984). *The Nature of Selection* (Chicago: University of Chicago Press).

Sober, Elliot. (1991). *Reconstructing the Past* (Cambridge, MA: MIT Press).

Sterelny, Kim. (2003). *Thought in a Hostile World* (Malden, MA: Blackwell).

Sterelny, Kim. (2006). "Local Ecological Communities." *Philosophy of Science* 72(2): 216–231.

Sterelny, Kim. (2012). *The Evolved Apprentice* (Cambridge, MA: MIT Press).

Sterelny, Kim, Joyce, Richard, Calcott, Brett, and Fraser, Ben, eds. (2013). *Cooperation and Its Evolution* (Cambridge, MA: MIT Press).

Tomasello, Michael. (1999). *The Cultural Origins of Human Cognition.* (Cambridge, MA: Harvard University Press).

Tomasello, Michael. (2014). *A Natural History of Human Thinking* (Cambridge, MA: Harvard University Press).

Trivers, Robert. (1971). "The Evolution of Reciprocal Altruism." *Quarterly Review of Biology* 46: 35–57.

Trivers, Robert, and Burt, Austin. (2008). *Genes in Conflict* (Cambridge, MA: Harvard University Press).

Walsh, Denis, Lewens, Tim, and Ariew, Andre. (2002). "The Trials of Life: Natural Selection and Random Drift." *Philosophy of Science* 69: 452–473.

Weisberg, Michael. (2013). *Simulation and Similarity* (Oxford: Oxford University Press).

Weisberg, Michael, ed. (2006). *Special Issue: Richard Levins' Philosophy of Science. Biology & Philosophy* 21(5).

Williams, George. (1966). *Adaptation and Natural Selection* (Princeton, NJ: Princeton University Press).

Williams, George. (1992). *Natural Selection: Domains, Levels and Challenges* (Oxford: Oxford University Press).

Wilson, E. O. (1975). *Sociobiology: The New Synthesis* (Cambridge, MA: Harvard University Press).

Wilson, E. O. (1979). *On Human Nature* (Cambridge, MA: Harvard University Press).

Wimsatt, Bill. (2007). *Re-Engineering Philosophy for Limited Beings* (Cambridge, MA: Harvard University Press).

CHAPTER 5

....

PHILOSOPHY OF THE PSYCHOLOGICAL AND COGNITIVE SCIENCES

....

MARK SPREVAK

1 INTRODUCTION

....

PHILOSOPHY of the psychological and cognitive sciences is a broad and heterogeneous domain. The psychological and cognitive sciences are rapidly evolving, fragmented, and often lacking in theories that are as precise as one might like. Consequently, philosophers of science have a challenging subject matter: fast moving, variegated, and sometimes slightly fuzzy. Nevertheless, the psychological and cognitive sciences are fertile ground for philosophers. Philosophers can bring their skills to bear with productive effect: in interpreting scientific theories, precisifying concepts, exploring relations of explanatory and logical coherence between theories, and in typing psychological states, processes, and capacities.

In this chapter, I organize work in philosophy of the psychological and cognitive science by the kind of task that philosophers have taken on. The tasks on which I focus are:

1. How should we interpret theories in cognitive science?
2. How should we precisify theoretical concepts in cognitive science?
3. How do theories or methodologies in cognitive science fit together?
4. How should cognitive states, processes, and capacities be individuated?

None of these tasks is distinctively philosophical: not in the sense of being of interest only to philosophers nor in the sense that philosophers would be the best people to solve them. All of the tasks engage a wide range of inquirers; all are carried out to the highest standards by a diverse group of individuals. What marks these tasks out as special is that they are among the problems that tend to interest philosophers and to whose solution philosophers tend to be well placed to contribute. Philosophy of the psychological and cognitive sciences is not defined by questions that set it apart from other forms of inquiry. It is characterized by questions, shared with other researchers, that tend to suit the skills, and attract the interest of, philosophers.

Three qualifications before proceeding. First, this chapter does not attempt to cover non-cognitive research in the psychological sciences (e.g., differential or social psychology). For the purposes of this chapter, "psychological science" and "cognitive science" will be used interchangeably. Second, the term "theory" will be used loosely, and looser than is normal in other areas of philosophy of science. "Theory" may refer to a full-fledged theory or a model, description of a mechanism, sketch, or even a general claim. Finally, the tasks I discuss, and the particular examples I give, only sample work in philosophy of the psychological and cognitive sciences. I focus on a small number of cases that I hope are illustrative of wider, structurally similar projects. The examples are not intended to be a list of the best work in the field, but only examples by which to orientate oneself.

Let us consider the four tasks in turn.

2 INTERPRETING

A crude way to divide up scientific work is between *theory building* and *theory interpretation*. Theory building involves identifying empirical effects and coming up with a theory to predict, test, and explain those effects. Theory interpretation concerns how—granted the empirical prowess or otherwise of a theory—we should understand that theory. Interpretation involves more than assigning a semantics. The job of interpretation is to understand the import of the theory: What is the purpose of the descriptions and physical models used to express the theory? What are the criteria of success for the theory: truth, empirical adequacy, instrumental value, or something else? Which terms in the theory are referring terms? Which kinds of ontological commitments does a theory entail? Which aspects of a theory are essential?

Philosophers have tools to help with theory interpretation. Analogues to the questions above have been pursued for ethical, normative, mathematical, and other scientific discourses. A range of options have been developed concerning theory interpretation, amongst which are versions of realism and instrumentalism. A theory interpretation set a combination of semantic, pragmatic, and ontological parameters regarding a theory. Our job is to see which setting results in the best interpretation of a psychological theory.

These concerns have recently played out with Bayesian models of cognition. Bayesian models predict and explain an impressive array of human behavior. For example, Bayesian models provide a good, predictive model of human behavior in sensory cue integration tasks (Ernst and Banks 2002). In these tasks, subjects are presented with a single stimulus in two different sensory modalities (say, touch and vision) and asked to make judgments about that stimulus by combining information from the two modalities. For example, subjects may be presented with two ridges and asked to decide, using touch and vision, which ridge is taller. Ernst and Banks found that subjects' behavior could be predicted if we assume that the input to each sensory modality is represented by a probability density function and that these representations are combined using the Bayesian calculus to yield a single estimate. Probability density functions and Bayesian computational machinery do a good job of predicting human behavior in sensory cue integration. Bayesian models also appear to explain human behavior. This is because they tie human behavior to the optimal way in which to weigh evidence. Many aspects of human behavior have been modeled with this kind of Bayesian approach, including causal learning and reasoning, category learning and inference, motor control, and decision making (Pouget, Beck, Ma, and Latham 2013).

How should we interpret these Bayesian models?

One option is *realism*. This is known within psychology as the *Bayesian brain hypothesis*. Here, the central terms of Bayesian models—the probability density functions and Bayesian computational methods—are interpreted as picking out real (and as yet unobserved) entities and processes in the human brain. The brain "represents information probabilistically, by coding and computing with probability density functions or approximations to probability density functions," and those probabilistic representations enter into Bayesian, or approximately Bayesian, inference via neural computation (Knill and Pouget 2004, 713). Bayesian models are *both* a theory of human behavior *and* of the neural and computational machinery that underpin the behavior. Realism about Bayesian models is usually qualified with the claim that current Bayesian models of cognition are only approximately true. What current Bayesian models get right is that the brain encodes information probabilistically and that it implements some form of approximate Bayesian inference. The precise format of the probabilistic representations and the precise method of approximate Bayesian inference is left open to future inquiry (Griffiths, Chater, Norris, and Pouget 2012).

Another interpretation option for Bayesian models is *instrumentalism* (Bowers and Davis 2012; Colombo and Seriès 2012; Danks 2008; Jones and Love 2011). According to the instrumentalist, Bayesian models do not aim to describe the underlying neurocomputational mechanisms (other than providing general constraints on their inputs and outputs). The central terms of Bayesian models—probability density functions and Bayesian computational machinery—should be understood, not as referring to hidden neural entities and processes, but as formal devices that allow experimenters to describe human behavioral patterns concisely. The instrumentalist allows that the underlying neural mechanisms could be Bayesian. But this possibility should be distinguished from the content of current Bayesian models. The aim of those models is to predict behavior. The success of Bayesian models in predicting behavior is not evidence that the mechanisms that generate that behavior are Bayesian. One might be inspired by the success of Bayesian models in predicting behavior to entertain the Bayesian brain hypothesis. But inspiration is not evidential support. Bayesian models should be understood as aiming at behavioral adequacy. Their aim is to predict behavior and specify human behavioral competences, not to describe neural or computational mechanisms.

Here we have two incompatible proposals about how to interpret Bayesian models of cognition. The difference between the two proposals matters. Do Bayesian models tell us how our cognitive processes work? According to realism, they do. According to instrumentalism, they do not (or, at least, they only provide constraints on inputs and outputs). How do we decide which interpretation option is correct?

The primary rationale for instrumentalism is epistemic caution. Instrumentalism makes a strictly weaker claim than realism while remaining consistent with the data. There appear to be good reasons for epistemic caution too. First, one might worry about underdetermination. Many non-Bayesian mechanisms generate Bayesian behavior. A lookup table, in the limit, can generate the same behavior as a Bayesian mechanism. Why should we believe that the brain uses Bayesian methods given the vast number of behaviorally indistinguishable, non-Bayesian, alternatives? Second, one might worry about the underspecification of mechanisms in current Bayesian models. The Bayesian brain hypothesis is a claim about neurocomputational mechanisms. There are a huge number of ways in which a behaviorally-adequate Bayesian model could be implemented, both neurally and computationally. Current Bayesian models tend to be silent about their neural or computational implementation in actual brains. Absent specification of the neurocomputational implementation, we should most charitably interpret current Bayesian theories as simply not making a claim about neural or computational mechanisms at all.

What reasons are there for realism? A common inductive inferential pattern in science is to go beyond an instrumentalist interpretation if a theory has a sufficiently impressive track record of prediction and explanation (Putnam 1975). Arguably, Bayesian models do have such a track record. Therefore, our interpretation of Bayesian models of cognition should be realist. The burden is on the instrumentalist to show otherwise: to show that the brain does *not* use probabilistic representations or Bayesian inference. And current scientific evidence gives us no reason to think that the brain does not use Bayesian methods (Rescorla forthcoming).

The disagreement between the realist and the instrumentalist is not epiphenomenal to scientific practice. The choice one makes affects whether, and how, experimental results bear on Bayesian models. For example, if instrumentalism is correct, then no neural evidence could tell in favor (or against) a Bayesian model. The reason is straightforward: the models are not in the business of making any claim about neural implementation, so there is nothing in the model for neural evidence to contest. If realism about Bayesian models is correct, then neural evidence *is* relevant to confirming the Bayesian models. If there is neural evidence that the Bayesian model's probability distributions or methods occur in the brain, then that is evidence in favor of the model. If there is evidence against, that is evidence against the Bayesian model. Bayesian models are evaluated not only by their fit to behavioral data, but also by their neural plausibility.

Bowers and Davis (2012) object that Bayesian theories of cognition are trivial or lacking empirical content. Their objection is that almost any dataset could be modeled as the output of some or other Bayesian model. Specific Bayesian models could be confirmed or disconfirmed by empirical data, but the general Bayesian approach is bound to succeed no matter what. Bayesianism is no more confirmed by behavioral data than number theory is confirmed. If the instrumentalist is right, then Bowers and Davis's objection has bite. Some or another Bayesian model will always fit a behavioral dataset. However, if the realist is right, then it is no longer clear that Bowers and Davis's objection succeeds. Realism raises the stakes of Bayesianism. This opens up the possibility that Bayesianism could be subject to empirical test. Precisely how to test it is not yet obvious. The reason is that currently there are no agreed proposals about the neural implementation of Bayesian models' theoretical posits (probability density functions and Bayesian computational processes). Nevertheless, realism at least opens the door to the possibility of testing Bayesianism. Suppose one were to have a theory of neural implementation in hand. If the brain's measured neural representations and computations—identifiable via the implementation proposal—really do have the properties ascribed by Bayesianism, then the general Bayesian approach would be vindicated. If not—for example, if the brain turns out to employ non-probabilistic representations or to manipulate its representations via a lookup table—then Bayesianism about cognition would be found to be false. A theory of implementation plus neural evidence allows Bayesianism about cognition to be tested.

Interpreting a theory requires making decisions about the theory's goals (truth vs. instrumental accuracy) and how to interpret its theoretical terms (referring vs. formal devices). A final aspect of interpretation is the decision about *which* claims a theory includes. Which parts of a theory are essential, and which are explication or trimming?

Suppose that realism about Bayesian models is correct and therefore that the brain manipulates probabilistic representations. What does a Bayesian model require of these representations? Some things are clearly required: the representations must be probabilistic hypotheses, and they must encode information about the uncertainty of events as well as their truth conditions or accuracy conditions. But to *which* entities and events do these probabilistic hypotheses refer? Do they refer to distal objects in the environment (e.g., tables and chairs) or to mathematical entities (e.g., numerical values of a parameter in an abstract graphical model). In other words, does the correct interpretation of a Bayesian model of cognition make reference to distal objects in the organism's environment, or is the correct interpretation entirely formal and mathematical?

On either view, entertaining a probabilistic hypothesis could enable an organism to succeed. On the first, "distal content", option, this could be because the probabilistic hypotheses are the organism's best guess about the state of its environment, and that appears to be useful information for an organism to consider when deciding how to act in its environment (Rescorla 2015). On this view, neural representations of distal environment stimuli would be an essential part of a Bayesian story, not an optional extra. If this is correct, it would mean that Bayesianism about cognition is incompatible with eliminativism or fictionalism about neural representations of distal objects (Keijzer 1998; McDowell 2010; Sprevak 2013).

On the second, "mathematical content", option, the organism could reliably succeed because the probabilistic hypotheses describe a mathematical structure that is adaptive for the organism to consider in its environment. This need not be because the mathematical structure represents distal properties in the environment. There are other ways than representation in which inferences over a formal structure could reliably lead to successful action. Formal properties may track contingent nomological connections between the organism's environment, body, and sensory system. Consider that it is adaptive for an organism to consider the result of applying a Fourier transform to the values of incoming signals from the auditory nerve, or for an organism to consider the results of applying $\nabla^2 G$ to the values of the incoming signal from its retina. These are useful transformations for an organism to consider even if neither is a representation of a distal environmental property. Egan (2010) argues for a "mathematical content" interpretation of classical computational models in cognitive science. Her reasoning could be extended to Bayesian models. According to such a view, it is adaptive for an organism to consider the mathematical relations described by the Bayesian model even though the terms in that model do not represent distal properties or events in the environment. On this view, representations of distal objects would not be an essential part of a Bayesian story. Distal content could feature in a Bayesian model of cognition, but it is not an essential part of such a model.

My intention here is not to argue for one interpretation rather than another. My intention is only to illustrate that each theory in cognitive science requires interpretation. There are entailments between theory interpretation options. Just as in other areas of science, some aspects of psychological theories should be understood as essential, literal, fact stating, and ontologically committing, whereas other aspects play a different role. It takes care and sensitivity to hit on the correct interpretation, or even to narrow the interpretation options down. This cannot be achieved simply by appealing to the utterances of theory builders because those utterances themselves need interpretation. Theories do not wear their interpretation on their sleeves.

3 PRECISIFYING

Imprecision is bad, not only for its own sake, but also because it permits fallacious inference. If one did not know the difference between the two senses of "bank," one might wrongly infer that since a river has two banks, a river would be a good place to conduct a financial transaction. That no such confusion occurs reflects that the relevant precisification is known to every competent speaker of the language. Unfortunately, the same is not true of every term in psychology. The correct usage of many terms in psychology— "consciousness", "concept", "module"—is murky, even for experts. One task to which philosophers have contributed is to clarify our theoretical terms so that those terms better support reliable, non-trivial, inductive inference. This may involve distinguishing between different things that fall under a term, redefining a term, or sometimes removing a term entirely. This is not mere semantic busy work. Concepts are the building blocks of scientific theories. Fashioning precise and inductively powerful concepts is essential to scientific progress (for more on this, see Machery [ms]).

I discuss here two examples of concepts that philosophers have helped to precisify: the concept of consciousness and the concept of a cognitive module.

The term "consciousness" has its origin in folk use. We might say, "she wasn't conscious of the passing pedestrian," "he was knocked unconscious in the boxing ring," or speak of the "conscious" experience of smelling a rose, making love, or hearing a symphony, making life worth living. Scientific and philosophical work on consciousness only started to make progress when it distinguished different things that fall under the folk term. Precisifying the concept of consciousness has enabled researchers to have a fighting chance to discover the purpose, functional mechanism, and neural basis of consciousness.

A preliminary precisification of consciousness is *arousal*. When we say that someone is conscious, we might mean that she is alert and awake; she is not asleep or incapacitated. When in dreamless sleep, or in a pharmacological coma, a person is unconscious. This sense of "consciousness" is usually accompanied by the assumption that the predicate ". . . is conscious" is a monadic predicate. Someone is simply conscious or unconscious; they need not be conscious *of* something specific. Someone may also be conscious in the sense of being aroused without being capable of consciousness in the sense of being aware of particular stimuli. Patients in a vegetative state show sleep-wake cycles, hence arousal, but they are not aware of particular stimuli. The neural mechanisms that govern consciousness-as-arousal also appear distinct from those that govern consciousness-as-awareness. Arousal is regulated by neural systems in the brainstem, notably, the reticular activating system. In contrast, the neural basis of consciousness-as-awareness appears to be in higher cortical regions and their subcortical reciprocal connections. Consciousness-as-arousal and consciousness-as-awareness have different purposes, relate to different aspects of functional cognitive architecture, and have different neural implementations (Laureys, Boly, Moonen, and Maquet 2009).

Ned Block's concept of access consciousness is one way to further precisify consciousness-as-awareness (Block 1995). A mental representation is defined as access conscious if and only if "it is poised for free use in reasoning and for direct 'rational' control of action and speech" (Block 1995, 382). One indicator of access consciousness is verbal reportability—whether the subject can say he or she is aware of a given mental episode. Reportability is, however, neither necessary nor sufficient for access consciousness. Block's precisification of consciousness-as-awareness highlights a number of other properties of consciousness-as-awareness. First, access consciousness is attributed with a relational predicate: an individual is conscious *of* something. Second, the object of that individual's consciousness is determined by a representation encoded in their brain. Third, access consciousness requires that this representation be "broadcast widely" in their brain: it should be available to central reasoning processes and able to cause a wide variety of behavior, including verbal reports. Fourth, the representation need not have actual behavioral effects; it need only have the disposition to cause appropriate behavioral effects. This catalogue provides a partial functional specification of consciousness-as-awareness. Empirical work has focused on identifying which, if any, neural properties answer to this description (Baars 1997; Dehaene and Changeux 2004).

Despite its virtues, there are idiosyncrasies in Block's precisification of consciousness-as-awareness: Why is rational control necessary for access consciousness? What does it mean for a neural representation to be "poised" to have effects but not actually have them? What does it mean for a representation to "directly" control behavior given that all control of behavior is mediated by other neural systems? The best way to see Block's account is as a stepping stone along the way to an appropriate concept of consciousness-as-awareness.

Forging the right notion of consciousness-as-awareness is not a task for armchair reflection. Precisification does not proceed prior to, or independently of, empirical inquiry. Precisification involves a two-way interaction between empirical hypotheses and consideration of how changes to the concepts that make up the empirical hypotheses would better capture the patterns relevant to scientific psychology. Precisification of a concept must be informed by the utility of the resulting concept to scientific practice. A precisification of consciousness-as-awareness proves its worth by whether the way it groups phenomena pays off for achieving goals in scientific psychology. Groupings that yield reliable inductive inferences or explanatory unification are those that should be favored. For example, a precisification of consciousness-as-awareness should aim to pick out shared facts about purpose, cognitive functional architecture, and neural implementation of consciousness-as-awareness. Recent work suggests that consciousness-as-awareness should be split into smaller concepts as no one concept meets all three of these conditions (Dehaene, Changeux, Naccache, Sackur, and Sergent 2006; Koch and Tsuchiya 2006).

The concepts of consciousness-as-arousal and consciousness-as-awareness are distinct from the concept of phenomenal consciousness. The concept of phenomenal consciousness picks out the qualitative feel—"what it is like"—associated with some mental episodes. It feels a certain way to taste chocolate; it feels a certain way to taste mint; and those two feelings are different. Phenomenal consciousness is characterized purely ostensively and from a subjective, first-person point of view. Consider your mental life, pay attention to the qualitative feelings that accompany certain episodes—those are phenomenally conscious feelings. Given that our concept of phenomenal consciousness is rooted in first-person reflection, it is not surprising that this concept has proven hard to place in relation to scientific concepts related to brain function. Dehaene (2014) suggests that the concept of phenomenal consciousness should for the moment be set aside in pursuing a science of consciousness; phenomenal consciousness is not currently amenable to scientific explanation.

In his introduction to modularity, Fodor (1983) lists nine features that characterize a cognitive module: domain specificity, mandatory operation, limited central accessibility, fast processing, informational encapsulation, shallow outputs, fixed neural architecture, characteristic and specific breakdown patterns, and characteristic ontogenetic pace and sequencing. Whether a mechanism counts as a module depends on whether it meets a weighted sum of these features to an "interesting" extent (Fodor 1983, 37). What counts as interesting, and how different features from the list should be weighed, is left largely unspecified. As one might imagine, there is room for precisifying the concept of modularity in different ways. Fodor claimed that the nine listed properties typically co-occur. Subsequent work has shown that they do not. Moreover, even if they did co-occur, their co-occurrence would not necessarily be of interest to scientific psychology (Elsabbagh and Karmiloff-Smith 2006; Prinz 2006). The concept of a psychological module has since been precisified in different ways and for different purposes. This has been done by giving priority to different properties associated with modularity from Fodor's list.

Fodor himself gives highest priority to two properties from the list: domain specificity and informational encapsulation (Fodor 2000). These properties are distinct. Domain specificity is a restriction on the inputs that a mechanism may receive. A mechanism is domain specific if only certain representations turn the module on, or are processed by the module. For example, in the visual system, a domain-specific module might only process information about retinal disparity and ignore everything else. Informational encapsulation is different. A mechanism is informationally encapsulated if, once the mechanism is processing an input, the information that the mechanism may then draw on is less than the sum total of information in the cognitive system. For example, in the visual system an informationally encapsulated module that processes information about retinal disparity might not be able to draw on the system's centrally held beliefs. Illusions like the Müller–Lyer illusion appear to show that the visual system is, to some extent, informationally encapsulated.

This gets us in the right ball park for precisifying the concept of a cognitive module, but domain specificity and information encapsulation need to be more carefully characterized. Informational encapsulation requires that something like the following three further conditions be met (Samuels 2005). First, the informational encapsulation should not be short-lived; it should be a relatively enduring characteristic of the mechanism. Second, informational encapsulation should not be the product of performance factors, such as fatigue, lack of time, or lapses in attention. Third, informational encapsulation need not shield the mechanism from every external influence. Processing may, for example, be affected by attentional mechanisms (Coltheart 1999). Informational encapsulation requires "cognitive impenetrability." Roughly, this means that although the mechanism may be modulated by certain informational factors (e.g., attention), it cannot be modulated by others (e.g., the high-level beliefs, goals, or similar representational states of the organism). Pinning this down more precisely requires work (see Machery 2015; Firestone and Scholl forthcoming).

Both domain specificity and informational encapsulation admit of degrees. Not just any step down from complete informational promiscuity produces domain specificity or information encapsulation. Moreover, the step down is not merely numerical, but also a matter of kind. The input domain and the informational database should be, in some sense, unified. One should be able to characterize the mechanism as a module for X, where X is some task or process that makes sense as a single unit in the light of concerns about the purpose and cognitive architecture of the organism. Illuminating the precise nature of this constraint on modularity is non-trivial.

The concepts of domain specificity and informational encapsulation offer scope for the development of a palette of distinct precisifications of the concept of a cognition module. If one turns attention to other properties associated with modularity in Fodor's list, more scope for divergent precisifications emerges. One could think of Fodor's criteria as conceptually separable parameters that are important in theorizing about the mind or brain (Elsabbagh and Karmiloff-Smith 2006). Some criteria may be more important for capturing certain kinds of pattern—computational, anatomical, developmental, and so on—than others. Which kind of pattern one wishes to capture depends on the interests of the investigator. Different concepts of modularity may include various combinations of these criteria. How we should separate out and sharpen the parameters of Fodor's characterization into one or more useful working concepts depends on the kinds of pay-off previously described. A precisification of modularity should help inquirers achieve their scientific goals. It should allow us to capture empirical patterns relevant to scientific psychology. Achieving this requires a two-way interaction between empirical inquiry and reflection on how changes to the concepts that make up the empirical hypotheses would allow us to better pick up on significant empirical patterns.

Before moving on, it is worth noting that there is also benefit in having imprecise concepts. Lack of precision in one's concepts may sometimes result in a fortuitous grouping of properties by a hypothesis. And it is not always bad for different research teams to be working with different understandings of theoretical concepts or for their inquiry to proceed with underspecified concepts. The promiscuous inferences that result may sometimes be helpful in generating discovery. Nevertheless, *pace* heuristic benefits, at some stage we need to be aware of exactly which claims are being made, and to understand when a conflict between say, two research teams, is genuine or merely verbal. Eventually, it is in everyone's interest to precisify.

4 Understanding How Things Hang Together

Sellars described the task of philosophy as "to understand how things in the broadest possible sense of the term hang together in the broadest possible sense of the term" (Sellars 1963, 1). One way to do this is to understand the explanatory and logical relations between theories and the ways in which they offer, or fail to offer, each other epistemic support. In this section, we examine two ways in which this is done in philosophy of the psychological and cognitive sciences. First, we examine the relationship between computational and dynamical systems theories in the psychological sciences. Second, we examine the relationship between different levels of inquiry in the psychological sciences.

Computational theories and dynamical systems theories both attempt to explain cognitive capacities. Both aim to explain how, in certain circumstances, we are able to do certain tasks. However, the two theories appear to explain this in different ways.

According to the computational approach, a cognitive capacity should be explained by giving a computational model of that capacity. Pick a cognitive capacity—for example, the ability to infer the three-dimensional shape of an object from information about its two-dimensional shading. An advocate of the computational approach might offer a computation that is able to solve this problem and suggest that this computation, or something like it, is implemented in the brain and causally responsible for the capacity in question (Lehky and Sejnowski 1988). Computational explanations are characterized by appeal to subpersonal representations and formal transformations by mechanisms built from simple components.

Dynamical systems theorists also aim to explain cognitive capacities, but their explanations appear to work differently. A dynamical systems theory involves differential equations relating variables that correspond to abstract parameters and time. The dynamical systems theorist claims that these parameters also correspond to some aspects of neural and bodily activity. Differential equations describe how these parameters interact over time to generate the behavior. Dynamical systems theory explains a cognitive capacity by narrowing in on a dynamical property of the brain or body that is causally responsible for the capacity (Schöner 2008).

Both computational theories and dynamical systems theories have had various successes in explaining cognitive capacities. Aspects of decision-making, such as production of the A-not-B error in infants—the phenomenon of an infant persevering in reaching for box A even though she just saw the experimenter place a desired toy in box B—are well modeled by dynamical systems theory (Thelen, Schöner, Scheier, and Smith 2001). Other cognitive capacities, such as those involved in inferring three-dimensional shape from two-dimensional information recorded by the retina, are well modeled by computation (Qian 1997). Our question is: How do these two theories relate? If psychology employs both, how do the theories fit together?

On the face of it, they appear to be rivals (van Gelder 1995). Each seems to instantiate a rival bet about the nature of the mind: either a cognitive capacity is produced by a computation or it is produced by a dynamical causal relation.

On reflection, this seems too strong. The computational and dynamical systems approach agree on a great deal. They agree that the brain is a dynamical system. It is no part of a computational theory to deny that cognitive capacities could, or should be, explained by a time-based evolution of physical parameters; indeed, the computational approach proposes one class of paths through the space of physical parameters. Computational theories might appear to differ from dynamical systems theories in that only computational theories employ subpersonal representations. However, when explaining a psychological capacity, there are often reasons—independent of the decision to use a computational or dynamical systems theory—to introduce subpersonal representations (Bechtel 1998; van Gelder 1995, 376). Some psychological capacities are "representation hungry"; for example, some capacities require the system to keep track of absent stimuli (Clark and Toribio 1994). Explaining these capacities motivates the introduction of subpersonal representations, no matter whether one places those representations in a computational or dynamical systems context. Furthermore, it is unclear whether subpersonal representations are really a necessary commitment of a computational approach. Not every state in a computation needs to be representational. It is an open question how much, if any, of a mechanism need have representational properties in a computational approach (Piccinini 2008). Appeal to subpersonal representations does not show that computational and dynamical systems theories are incompatible.

Dynamical systems theory appears to differ from a computational approach in that dynamical models give a special role to time. Dynamical models offer descriptions in which time is continuous and the transitions between states are governed by explicitly time-based differential equations. Computational models—for example, Turing machines—use discrete time evolution and are governed by rules that do not mention time. However, not all computational models are like Turing machines. Some computational models involve continuous time evolution and have time-based differential equations as their transition rules. Within this class are connectionist models (Eliasmith 1997) and more recent realistic computational models of neural function (Eliasmith 2013). These computational models assume continuous time evolution, contain parameters that map onto a subset of neural and bodily properties, and use, as their transition rules, differential equations that specify how the parameters evolve over time. These models have all the signature properties of dynamical systems theories.

The distance between computational theories and dynamical systems theories is not as great as it may first appear. In some cases, the two theories converge on the same class of model. This is not to say that every dynamical systems theory is a computational theory or vice versa. It is rather that, in the context of explaining our cognitive capacities, a computational approach and a dynamical systems approach may converge: they may employ the same elements related in the same way—representations, abstract parameters, mapping from those parameters to neural and bodily properties, continuous time evolution, and time-based differential equations. The right way to see the relationship between a computational theory and a dynamical systems theory is not as two rivals, but as two possibly compatible alternatives.

This does not explain how classical, discrete, symbolic-rule-governed computational models relate to the dynamical systems approach. One suggestion is that the latter reduce to the former as a limiting case (for an attempt to show this, see Feldman 2012). And to understand how things hang together generally across theories in the psychological sciences, one needs to understand the relationship between computational theories, dynamical systems theories, and a wide range of other theories: statistical, Bayesian, enactivist, and others. Under which conditions are two theories rivals, compatible alternatives, or do they reduce to one another? Under which conditions do they offer each other epistemic support?

A second way in which to understand how things hang together in the psychological sciences is to understand how investigation at different levels of inquiry coheres. What is the relationship between inquiry at different levels in the psychological sciences? Here, we focus on only two kinds of level: levels of spatial organization inside mechanisms and Marr's levels of computational explanation.

First, let us consider levels of spatial organization inside mechanisms. Different mechanisms exist at different spatial scales. "Higher" level mechanisms involve larger systems and components. "Lower" level mechanisms involve smaller systems and components. We normally explain by describing mechanisms at several levels. For example, we explain the cognitive capacity that a mouse exhibits when navigating a maze to find a reward by appeal to mechanisms inside and around the mouse at multiple spatial scales. The top-level mechanism is the entire mouse engaged in a spatial navigation task. Within this mechanism is a mechanism that involves part of the mouse's brain, the hippocampus, storing a map of locations and orientations inside the maze. Within this mechanism, inside the mouse's hippocampus, is a smaller mechanism storing information by changing weights in the synapses between pyramidal cells. Within this mechanism is a mechanism that produces long-term changes at a synapse by modifying the synapse's N-methyl-D-aspartate (NMDA) receptors. Those NMDA receptors undergo change because of yet smaller mechanisms governing their functioning. A mechanistic explanation of the mouse's cognitive capacity involves appeal to multiple mechanisms at multiple spatial scales and showing how they work in concert to produce the cognitive capacity (Craver 2007). The relationship between higher and lower levels goes beyond the mere mereological whole–part relation. Higher level mechanisms are not just part of, but also are *realized* by lower level mechanisms. Lower level mechanisms form the component parts of higher level mechanisms. Cognitive capacities are explained by showing how mechanisms at different spatial scales, integrated by the mechanistic realization relation, produce the cognitive capacity in question.

Descriptions of mechanisms at different levels of spatial scale have a degree of autonomy from each other. A description of a mechanism at one spatial level may be silent about how that mechanism's component parts work, or about the larger system in which the mechanism is embedded. One might, for example, describe how cells in the mouse hippocampus store a map of locations while remaining silent about the lower level mechanism that produces synaptic change. One might also describe how cells in the hippocampus store a map of locations while remaining silent about how those cells are recruited by the entire mouse to solve the task.

This partial autonomy between descriptions at different spatial levels is not full-blown independence. The autonomy arises because mechanistic realization allows for the (logical) possibility of multiple realization. It is possible for the component parts of a mechanism to be realized in multiple ways, within the constraints that the performance of the mechanism dictates. It is also possible for the mechanism to be embedded in multiple larger contexts, within the constraints that the context should support the operation of the mechanism. Description at a particular level of spatial scale places some constraints on higher or lower level descriptions, but leaves some degree of freedom. This degree of freedom allows different teams of inquirers to focus on discovering mechanisms at different spatial scales in a partially autonomous manner. It allows scientific inquiry in the psychological sciences to split into different fields: biochemistry, cellular biology, neuro-physiology, cognitive neuroscience, and behavioral ecology.

Oppenheim and Putnam (1958) claimed that scientific disciplines are structured into autonomous levels corresponding to the spatial scale of their subject matter. We are now in a position to see what is right about this idea. There is no necessity for scientific inquiry to be structured by levels of spatial scale. Nevertheless, structuring scientific inquiry in the psychological sciences by spatial scale is permissible. It is permissible because the ontologically layered structure generated by the mechanistic realization relation, and the partial autonomy between levels of spatial scale that this provides, allows scientific inquiry to proceed at different spatial scales along relatively separate tracks. The structuring of scientific disciplines by levels of spatial scale that Oppenheim and Putnam describe is a consequence of the ontological layering, and partial autonomy, generated by the mechanistic realization relation.

Marr (1982) introduced a distinct kind of level of inquiry into the psychological sciences. Marr argued that the psychological and cognitive sciences should be divided into three levels of explanation.

Marr's first level is the "computational" level. The aim of inquiry at the computational level is to describe *which* task an organism solves in a particular circumstance and *why* that task is important to the organism. A task should be understood as an extensional function: a pattern of input and output behavior. In order to discover which function an organism computes, we need to understand the ecological purpose of computing that function—why the organism computes this function and what computation of this function would allow the organism to achieve. Without a guess as to the ecological purpose of computing this function, there would be no way of picking out from the vast number of things that the organism does (its patterns of input–output behavior) which are relevant to cognition.

Marr's second level is the "algorithmic" level. The aim of inquiry at the algorithmic level is to answer *how* the organism solves its task. The answer should consist in an algorithm: a finite number of simple steps that take one from input to output. Many different algorithms compute the same extensional function. Therefore, even if we were to know which extensional function the organism computes, the algorithm would still be left open. In order to discover which algorithm an organism uses, researchers look for indirect clues about the information-processing strategies exploited by the organism, such as the organism's reaction times and susceptibility to errors.

Marr's third level is the "implementation" level. The aim of inquiry at the implementation level is to describe how steps in the algorithm map onto physical changes. Even if we were to know both the extensional function and the algorithm, that would still leave open how the algorithm is implemented in physical changes of the organism. The brain is a complex physical system. Without some guide as to which parts of the brain implement which parts of an algorithm, there would be no way to know how the brain enables the organism to solve its task. The implementation level identifies which physical parts are functionally significant: which parts are relevant, and in which ways, to the computation that the organism performs. In the case of an electronic PC, electrical changes inside the silicon chips are functionally significant; the color of the silicon chips or the noise the cooling fan makes are not. Researchers look for implementation level descriptions by using techniques such as magnetic resonance imaging, electroencephalograms, single-cell recording, and testing how performance is affected when physical resources are damaged (e.g., by stroke) or temporarily disabled (e.g., by drugs).

Marr's three levels are not the same as the levels of spatial organization in mechanisms described previously. The levels of spatial organization in mechanisms involve positing an ontological layering relation: psychological mechanisms are related smaller to larger by mechanistic realization; component parts are realized by increasingly smaller mechanisms. One consequence of this ontological layering is that scientific inquiry at different spatial scales can be structured into partially autonomous domains (from biochemistry to behavioral ecology). Structuring scientific inquiry into levels is a consequence, not the principal content, of the claim. In contrast, the principal content of Marr's claim is that scientific inquiry, not ontology, should be structured. Marr divides work in the psychological sciences into three types of inquiry: computational, algorithmic, and implementational. There is no assumption that this division is accompanied by a division in the ontology. Marr's claim is simply that the psychological sciences should pursue three types of question if they are to explain psychological capacities adequately.

Computational, algorithmic, and implementational questions concern a single system at a single level of spatial scale. They also concern a single capacity: which extensional function that capacity instantiates, which algorithm computes that function, and how that algorithm is physically implemented. In contrast, each level of the mechanistic scale concerns a different physical system: the entire organism, the brain, brain regions, neural circuits, individual neurons, and subcellular mechanisms. Each level of scale concerns a different capacity: the capacity of the whole organism to navigate a maze, the capacity of the hippocampus to store spatial information, the capacity of pyramidal synapses to undergo long-term potentiation (LTP), and so on. At each level of spatial scale, and for each corresponding capacity and system, one can ask Marr's questions: Which function does the physical system compute and why? Which algorithm does it use to compute this function? How is that algorithm implemented by physical changes in the system? Marr's questions cut across those concerning mechanisms at different levels of spatial scale.

Marr claimed a degree of autonomy, but not full independence, between his levels. The autonomy that exists between Marr's levels derives from two properties of computation. First, the same extensional function can be computed by different algorithms. Second, the same algorithm can be implemented in different ways. The first property means that proposing a particular function at the computational level does not restrict algorithmic-level inquiry to a particular algorithm. The second property means that proposing a particular algorithm at the algorithmic level does not restrict implementation-level inquiry to a particular physical implementation. Similar degrees of freedom do not hold in reverse. If one were to propose a particular implementation—a mapping from physical activity in the system to steps of some algorithm—then the algorithm that the system employs would be thereby fixed. Similarly, if one were to propose that the system uses a particular algorithm, then the extensional function that the system computes would be thereby fixed. The autonomy between Marr's levels is downwards only: lower levels are partially autonomous from upper levels, but not vice versa.

Downwards autonomy in Marr's scheme, like the autonomy between levels of inquiry about mechanisms at different spatial scales, is only present as a logical possibility. The degree of autonomy is likely to be attenuated in practice. Downwards autonomy for Marr derives from the two logical properties of computation described earlier. However, the psychological sciences are constrained by more than what is logically possible. Their concern is what is reasonable to think given all we know about the brain and agent. The numbers of permissible algorithms and implementations are likely to be significantly less than those that are logically possible when constraints are added about the resources that the brain can employ, the time the system can take to solve its task, assumptions made in other areas of the psychological sciences are taken into account.

The autonomy between Marr's levels of inquiry is different from that between levels of inquiry concerning mechanisms at different spatial scales. The autonomy between Marr's levels of inquiry derives from two properties of computation: that many algorithms compute the same function and that there are many ways to implement the same algorithm. The autonomy between levels of inquiry concerning mechanisms at different spatial scales derives from two properties of the mechanistic realization relation: that it is possible for different mechanisms to produce the same causal power, and that the same mechanism could be embedded in different contexts. The latter two properties give rise to an in principle upward and downward autonomy between levels of spatial scale. Positing a mechanism at a particular level of spatial scale does not fix how its smaller component parts work, nor does it fix how that mechanism is embedded in a larger system. Autonomy between levels of inquiry concerning mechanisms at different spatial scales is bi-directional. This is formally distinct from the downwards-only autonomy associated with Marr's levels of computational explanation.

5 INDIVIDUATING

Disagreements within philosophy of the psychological sciences are sometimes not about the causal flow involved in cognition, but how to individuate that causal flow. Two philosophical camps may agree on the basic causal relations, but disagree about which of the elements in the causal structure are cognitive, representational, perceptual, sensory, doxastic, gustatory, olfactory, and so on. These disagreements may give outsiders the appearance of being merely verbal. But this is rarely the case. The competing sides agree on the meanings of their words. What they disagree about is how cognitive capacities, processes, and states should be individuated. In this section, I look at two examples of this kind of dispute. The first is the disagreement about how to individuate the senses. The second is the disagreement about whether human mental states and processes extend outside our brains and bodies.

The senses are different ways of perceiving, such as seeing, hearing, touching, tasting, and smelling. What makes two senses different? How many senses are there? Under which conditions would an organism have a new sense? The psychological sciences provide data that appear to challenge folk ideas about the senses. Novel sense modalities seem to exist in non-human animals, including magnetic senses, electric senses, infrared senses, and echolocation. Humans seem to have senses for pressure, temperature, pain, balance, and their internal organs in addition to their traditional five senses. Neural processing of sensory information in humans is multimodal; visual areas in the brain are not exclusively visual and integrate information from sound and other stimuli. Blindsight patients appear to have vision without associated visual phenomenology or consciously accessible beliefs. Tactile-visual sensory substitution (TVSS)-equipped patients appear to see via touch. Based on this information, should we revise the folk view that humans have five senses? If so, how? In order to answer this question, we need a way to individuate the senses. Let us look at four contenders.

The first is representation-based. Suppose that each sense has an object or property that is exclusively detected and represented by that sense—its "proper sensible". The proper sensibles of hearing, tasting, smelling, and seeing are sound, flavor, odor, and color respectively. According to the representation-based view, the representations of these proper sensibles—representations that are not generated in any other way—individuate the senses. A sense is individuated by the characteristic representations that the sense produces. A challenge for the view is that it lands us with a new problem: How do we individuate the sensory representations? It is not clear that this is substantially easier than the original problem of how to individuate the senses.

The second approach is experience-based. Hearing, tasting, smelling, seeing, and touch are each associated not only with distinct representations, but also with distinct subjective phenomenal experiences. The phenomenal experiences tend to be similar within sensory modalities and different between sensory modalities. According to the experience-based view, it is because sensory experiences are phenomenally similar to and different from each other that we have distinct senses. A sense is individuated by the types of phenomenal experience to which it gives rise. A challenge for the view is to say what are the relevant similarities and differences in phenomenal experience. Experiences within a sensory modality are not all alike and those between sensory modalities are not all dissimilar. Which phenomenal similarities and differences matter for individuating the senses, and why are they important?

The third approach is stimulus-based. Different senses involve responding to proximal stimuli of different physical kinds. Seeing involves reacting to electromagnetic waves between 380 nm and 750 nm. Hearing involves reacting to air pressure waves in the ear canal. Smelling involves reacting to airborne chemicals in the nose. According to the stimulus-based view, the reason why the senses are distinct is because they involve responses to different physical types of proximal stimulus. A sense is individuated by type of proximal stimulus to which the organism reacts. A challenge for the view is that the same proximal stimulus could be associated with different senses. For example, the same pressure wave in the air may be processed by an organism for hearing and for echolocation.

The final approach is organ-based. Different senses tend to be associated with different sense organs. Seeing involves the eyes, hearing involves the ears, smelling involves the nose. Each sense organ contains physiologically distinct receptor cells. According to the organ-based view, the reason why the senses are distinct is because they employ distinct sense organs. A sense is individuated by its associated sense organ. A challenge for the view is that the same sense organ (e.g., the ear) could be used for two different senses (e.g., hearing and echolocation).

The four proposals prescribe different revisions to folk assumptions about the senses in light of scientific data. The task facing philosophers is to determine which, if any, of these views is correct. Nudds (2004) argues that we should not endorse any of them. His claim is that individuation of the senses is context-dependent. No single account individuates the senses across all contexts. Different criteria apply in different contexts depending on our interests. Macpherson (2011) argues that the senses should be individuated context-independently. She claims that all of the criteria above matter. All contribute jointly to individuating the senses. The four proposals can be used as a multidimensional metric on which any possible sense can be judged. The clustering of an organism's cognitive capacities across multiple dimensions, rather than on a single dimension, determines how its senses should be individuated.

Let us turn to our second example of a dispute about individuation. The hypothesis of extended cognition (HEC) asserts that human mental life sometimes extends outside the brain and takes place partly inside objects in the environment, such as notebooks or iPhones (Clark and Chalmers 1998). Disagreements about whether HEC is true have taken the form of a disagreement about the individuation conditions of human mental states and processes.

The best way to understand HEC is to start with the weaker claim known as *distributed cognition* (Hutchins 1995). An advocate of distributed cognition claims that human cognitive capacities do not always arise solely in, and should not always be explained exclusively in terms of, neural mechanisms. The mechanisms behind human cognitive capacities sometimes include bodily and environmental processes. The brain is not the sole mechanism responsible for our cognitive abilities, but is only part—albeit a large part—of a wider story. The brain recruits environmental and bodily resources to solve problems. Recruiting these resources allows humans to do more than they could otherwise, and to work more quickly and reliably. Brains off-load work onto the body and environment. Sometimes the off-loading is under conscious control: for example, when we consciously decide to use a pen, paper, or our fingers to solve a mathematical problem. Sometimes the off-loading is not under conscious control: for example, when we use our eye gaze to store information (Ballard, Hayhoe, Pook, and Rao 1997; Gray and Fu 2004). Distributed cognition is the claim that *distributed* information-processing strategies figure in the best explanation of, and causal story behind, some human cognitive accomplishments.

HEC is a stronger claim than distributed cognition. According to HEC, not only do brains recruit environmental resources to solve problems, but those non-neural resources, when they have been recruited, also *have mental properties*. Parts of the environment and the body, when employed in distributed cognition strategies, have just as much claim to mental or cognitive status as any neural process. Against this, the hypothesis of embedded cognition (HEMC) accepts the distributed-cognition claim about the exploitation of bodily and environmental resources, but rejects HEC's assertion about the body and environment having mental properties (Rupert 2004, 2013). According to HEMC, non-neural resources, despite figuring in the explanation and causal story behind some cognitive accomplishments, do not have mental properties. Only neural processes have mental or cognitive properties.

How is this about individuation? HEC is, in essence, a claim about the individuation of mental kinds. HEC claims that the causal flow in human cognition should be individuated in such a way that the neural and non-neural parts instantiate a single kind—a mental kind. This is not to say that there are no differences relevant to psychology between the neural and non-neural parts. HEC's claim is merely that the neural and non-neural parts jointly satisfy a condition sufficient for them to instantiate a single mental kind. In contrast, HEMC's claim is that the causal flow in human cognition should be individuated so that the neural and non-neural parts fall under different kinds—mental and non-mental respectively. This is not to say that there are no kinds of interest to psychology that both instantiate. Rather, it is to say that whatever kinds they instantiate, they jointly fail to meet the condition required for them to instantiate a single mental kind. HEC and HEMC disagree, in cases of distributed cognition, about how to individuate mental properties across the causal flow.

Which view is right: HEC or HEMC? To answer this, we need to agree on the minimal condition, mentioned earlier, for a physical process to instantiate a mental kind. There are two main proposals on this score. The first is functionalist. On this view, a physical state or process is mental provided it has the right functional profile. I have argued elsewhere that the functionalist proposal decisively favors HEC (Sprevak 2009). If one does not accept HEC granted functionalism, one concedes the chauvinism about the mind that functionalism was designed to avoid. The second proposal is based on explanatory pay-off to cognitive science. On this view, a physical state or process is mental just in case it fits with best (current or future) cognitive science to treat it as such. I have argued elsewhere that considerations of explanatory value regarding cognitive science are toothless to decide between HEC and HEMC. Cognitive science could continue to be conducted with little or no loss either way (Sprevak 2010). The dispute between HEC and HEMC is a case in point for which a question about individuation of mental states and processes cannot be answered by a straightforward appeal to scientific practice. Work in philosophy of the psychological and cognitive sciences needs to draw on a wide range of considerations to settle such questions about individuation.

6 CONCLUSION

We have surveyed four types of task in philosophy of the psychological and cognitive sciences: How should we interpret our scientific theories? How should we precisify our theoretical concepts? How do our theories or methodologies fit together? How should our cognitive states, processes, and capacities be individuated? We have focused on only some of current work in philosophy of the psychological and cognitive sciences. Work we have not covered includes proposals for psychological mechanisms and architectures (e.g., Apperly and Butterfill 2009; Grush 2004); analysis of scientific methodology in the psychological sciences (Glymour 2001; Machery 2013); examination of key experimental results in the psychological sciences (Block 2007; Shea and Bayne 2010); and analysis of folk psychological concepts (Gray, Gray, and Wegner 2007; Knobe and Prinz 2008).

REFERENCES

Apperly, I. A., and Butterfill, S. A. (2009). "Do Humans Have Two Systems to Track Belief and Belief-Like States?" *Psychological Review* 116: 953–970.

Baars, B. (1997). *In the Theater of Consciousness* (Oxford: Oxford University Press).

Ballard, D. H., Hayhoe, M. M., Pook, P., and Rao, R. (1997). "Deictic Codes for the Embodiment of Cognition." *Behavioral and Brain Sciences* 20: 723–767.

Bechtel, W. (1998). "Representations and Cognitive Explanations: Assessing the Dynamicist's Challenge in Cognitive Science." *Cognitive Science* 22: 295–318.

Block, N. (1995). "On a Confusion About a Function of Consciousness." *Behavioral and Brain Sciences* 18: 227–247.

Block, N. (2007). "Consciousness, Accessibility, and the Mesh Between Psychology and Neuroscience." *Behavioral and Brain Sciences* 30: 481–548.

Bowers, J. S., and Davis, C. J. (2012). "Bayesian Just so Stories in Psychology and Neuroscience." *Psychological Bulletin* 128: 389–414.

Clark, A., and Chalmers, D. J. (1998). "The Extended Mind." *Analysis* 58: 7–19.

Clark, A., and Toribio, J. (1994). "Doing Without Representing?" *Synthese* 101: 401–431.

Colombo, M., and Seriès, P. (2012). "Bayes on the Brain—On Bayesion Modelling in Neuroscience." *British Journal for the Philosophy of Science* 63: 697–723.

Coltheart, M. (1999). "Modularity and Cognition." *Trends in Cognitive Sciences* 3: 115–120.

Craver, C. F. (2007). *Explaining the Brain* (Oxford: Oxford University Press).

Danks, D. (2008). "Rational Analyses, Instrumentalism, and Implementations." In N. Chater and M. Oaksford (eds.), *The Probabilistic Mind: Prospects for Rational Models of Cognition* (Oxford: Oxford University Press), 59–75.

Dehaene, S. (2014). *Consciousness and the Brain: Deciphering How the Brain Codes Our Thoughts* (London: Penguin Books).

Dehaene, S., and Changeux, J. -P. (2004). "Neural Mechanisms for Access to Consciousness." In M. Gazzaniga (ed.), *The Cognitive Neurosciences, III* (Cambridge, MA: MIT Press), 1145–1157.

Dehaene, S., Changeux, J. -P., Naccache, L., Sackur, J., and Sergent, C. (2006). "Conscious, Preconscious, and Subliminal Processing: A Testable Taxonomy." *Trends in Cognitive Sciences* 10: 204–211.

Egan, F. (2010). "Computational Models: A Modest Role for Content." *Studies in History and Philosophy of Science* 41: 253–259.

Eliasmith, C. (1997). "Computation and Dynamical Models of Mind." *Minds and Machines* 7: 531–541.

Eliasmith, C. (2013). *How to Build a Brain: A Neural Architecture for Biological Cognition* (Oxford: Oxford University Press).

Elsabbagh, M., and Karmiloff-Smith, A. (2006). "Modularity of Mind and Language." In K. Brown (ed.), *The Encyclopedia of Language and Linguistics* (Oxford: Elsevier), 8: 218–224.

Ernst, M. O., and Banks, M. S. (2002). "Humans Integrate Visual and Haptic Information in a Statistically Optimal Fashion." *Nature* 415: 429–433.

Feldman, J. (2012). "Symbolic Representation of Probabilistic Worlds." *Cognition* 123: 61–83.

Firestone, C. and Scholl, B. (forthcoming). "Cognition does not affect perception: Evaluating the evidence for 'top-down' effects." *Behavioral and Brain Sciences*.

Fodor, J. A. (1983). *The Modularity of Mind* (Cambridge, MA: MIT Press).

Fodor, J. A. (2000). *The Mind Doesn't Work That Way* (Cambridge, MA: MIT Press).

Glymour, C. (2001). *The Mind's Arrows: Bayes Nets and Graphical Causal Models in Psychology* (Cambridge, MA: MIT Press).

Gray, H. M., Gray, K., and Wegner, D. M. (2007). "Dimensions of Mind Perception." *Science* 315: 619.

Gray, W. D., and Fu, W. T. (2004). "Soft Constraints in Interactive Behavior." *Cognitive Science* 28: 359–382.

Griffiths, T. L., Chater, N., Norris, D., and Pouget, A. (2012). "How the Bayesians Got Their Beliefs (and What Those Beliefs Actually Are): Comment on Bowers and Davis (2012)." *Psychological Bulletin* 138: 415–422.

Grush, R. (2004). "The Emulator Theory of Representation: Motor Control, Imagery, and Perception." *Behavioral and Brain Sciences* 27: 377–442.

Hutchins, E. (1995). *Cognition in the Wild* (Cambridge, MA: MIT Press).

Jones, M., and Love, B. C. (2011). "Bayesian Fundamentalism or Enlightenment? On the Explanatory Status and Theoretical Contributions of Bayesian Models of Cognition." *Behavioral and Brain Sciences* 34: 169–231.

Keijzer, F. A. (1998). "Doing Without Representations Which Specify What to Do." *Philosophical Psychology* 11: 269–302.

Knill, D. C., and Pouget, A. (2004). "The Bayesian Brain: The Role of Uncertainty in Neural Coding and Computation." *Trends in Neurosciences* 27: 712–719.

Knobe, J., and Prinz, J. (2008). "Intuitions About Consciousness: Experimental Studies." *Phenomenology and Cognitive Science* 7: 67–85.

Koch, C., and Tsuchiya, N. (2006). "Attention and Consciousness: Two Distinct Brain Processes." *Trends in Cognitive Sciences* 11: 16–22.

Laureys, S., Boly, M., Moonen, G., and Maquet, P. (2009). "Coma." *Encyclopedia of Neuroscience* 2: 1133–1142.

Lehky, S. R., and Sejnowski, T. J. (1988). "Network Model of Shape-from-Shading: Neural Function Arises from Both Receptive and Projective Fields." *Nature* 333: 452–454.

Machery, E. (2013). "In Defense of Reverse Inference." *British Journal for the Philosophy of Science* 65: 251–267.

Machery, E. (2015). "Cognitive penetrability: A no-progress report." In J. Zeimbekis and A. Raftopoulos (eds.), *The Cognitive Penetrability of Perception* (Oxford: Oxford University Press), 59–74.

Machery, E. ms. "Philosophy Within Its Proper Bounds."

Macpherson, F. (2011). "Individuating the Senses." In F. Macpherson (ed.), *The Senses: Classic and Contemporary Philosophical Perspectives* (Oxford: Oxford University Press), 3–43.

Marr, D. (1982). *Vision* (San Francisco: W. H. Freeman).

McDowell, J. (2010). "Tyler Burge on Disjunctivism." *Philosophical Explorations* 13: 243–255.

Nudds, M. (2004). "The Significance of the Senses." *Proceedings of the Aristotelian Society* 104: 31–51.

Oppenheim, P., and Putnam, H. (1958). "Unity of Science as a Working Hypothesis." In H. Feigl, M. Scriven, and G. Maxwell (eds.), *Concepts, Theories, and the Mind–body Problem.* Minnesota Studies in the Philosophy of Science (Minneapolis: University of Minnesota Press), II: 3–36.

Piccinini, G. (2008). "Computation Without Representation." *Philosophical Studies* 137: 205–241.

Pouget, A., Beck, J. M., Ma, W. J., and Latham, P. E. (2013). "Probabilistic Brains: Knows and Unknowns." *Nature Neuroscience* 16: 1170–1178.

Prinz, J. (2006). "Is the Mind Really Modular?" In R. Stainton (ed.), *Contemporary Debates in Cognitive Science* (Oxford: Blackwell), 22–36.

Putnam, H. (1975). *Mathematics, Matter and Method, Philosophical Papers, Volume 1* (Cambridge: Cambridge University Press).

Qian, N. (1997). "Binocular Disparity and the Perception of Depth." *Neuron* 18: 359–368.

Rescorla, M. (2015). "Bayesian Perceptual Psychology." In M. Matthen (ed.), *The Oxford Handbook of Philosophy of Perception* (Oxford University Press), 694–716.

Rescorla, M. (forthcoming). "Bayesian Sensorimotor Psychology." *Mind and Language.*

Rupert, R. D. (2004). "Challenges to the Hypothesis of Extended Cognition." *Journal of Philosophy* 101: 389–428.

Rupert, R. D. (2013). "Memory, Natural Kinds, and Cognitive Extension; Or, Martians Don't Remember, and Cognitive Science Is Not About Cognition." *Review of Philosophy and Psychology* 4: 25–47.

Samuels, R. (2005). "The Complexity of Cognition: Tractability Arguments for Massive Modularity." In P. Carruthers, S. Laurence, and S. P. Stich (eds.), *The Innate Mind: Vol. I, Structure and Contents* (Oxford: Oxford University Press), 107–121.

Schöner, G. (2008). "Dynamical Systems Approaches to Cognition." In R. Sun (ed.), *Cambridge Handbook of Computational Psychology* (Cambridge: Cambridge University Press), 101–126.

Sellars, W. (1963). *Science, Perception and Reality* (London. Routledge & Kegan Paul).

Shea, N., and Bayne, T. (2010). "The Vegetative State and the Science of Consciousness." *British Journal for the Philosophy of Science* 61: 459–484.

Sprevak, M. (2009). "Extended Cognition and Functionalism." *Journal of Philosophy* 106: 503–527.

Sprevak, M. (2010). "Inference to the Hypothesis of Extended Cognition." *Studies in History and Philosophy of Science* 41: 353–362.

Sprevak, M. (2013). "Fictionalism About Neural Representations." *Monist* 96: 539–560.

Thelen, E., Schöner, G., Scheier, C., and Smith, L. B. (2001). "The Dynamics of Embodiment: A Field Theory of Infant Perseverative Reaching." *Behavioral and Brain Sciences* 24: 1–86.

van Gelder, T. (1995). "What Might Cognition Be, If Not Computation?" *Journal of Philosophy* 91: 345–381.

CHAPTER 6

....

PHILOSOPHY OF THE PHYSICAL SCIENCES

....

CARL HOEFER AND CHRIS SMEENK

1 INTRODUCTION

....

UNDERSTANDING the nature of our knowledge of physics has long been a central topic in philosophy of science. The introduction of relativity theory and quantum mechanics (QM) in the early twentieth century inspired the likes of Reichenbach and Carnap to develop distinctive accounts of epistemology and the structure of theories. Their views about scientific knowledge and later quite different views have often shared one feature, namely taking physics as exemplary. We applaud the recent reversal of this trend, as philosophers considering other areas of science have set aside the physical sciences as a model. Yet there are a variety of philosophical issues that are closely intertwined with the physical sciences. We briefly survey here some aspects of the connection between physics and naturalized metaphysics, starting from a "received view" of the nature and scope of physical theories and exploring how challenges to that view may have ramifications for the philosophical consequences of physics.

2 THE IDEAL VIEW

....

A number of philosophical issues can be thought of as arising in and for the totality of the physical sciences: issues having to do with their aims, the structure of their theories, their presupposed worldview, and the relationship between different theories. To introduce these issues we will begin with a simplified account of the nature of physical theories, illustrated by point-particle mechanics. Drawing on this sketch of a physical theory, we will state an overall position that we will call the *ideal view* (IV).[1] Roughly speaking, this view is a mixture of fundamentalism, scientific realism, and optimism about the simplicity of nature. We expect that certain aspects of the view, or suitably refined versions of it, will appeal to many philosophers of the physical sciences. Yet each of its claims has been challenged in recent philosophical discussions. In ensuing sections, we will review these challenges and assess the prospects for defending the view.

The image of the physical world as having a simple ontology of matter in motion, governed by unchanging laws, comes from classical mechanics. The simplest version of mechanics describes the dynamical evolution of a system of point particles due to forces acting among them. The kinematics of the theory specifies the geometry of motion and the kinds of particles under consideration, whereas the dynamics fixes the forces and their effects. In the Hamiltonian formulation, the state of the system at a given time is represented by a point in phase space Γ, specifying the position and momenta for each of the particles relative to a given reference frame. The interparticle forces are captured in a single function called the *Hamiltonian*, which determines how the particles move via Hamilton's equations. The evolution of the system over time is a trajectory through Γ that specifies the history of the system. The semantics of the theory depends on linking this theoretical description to (idealized) experimental situations. At least in principle, such a link establishes how regions of phase space correspond to answers to experimental questions, such as "The value of property P is given by $X \pm \Delta$ (in appropriate units)." One central project of philosophy of physics is to interpret physical theories in the sense of specifying such semantic links; or, more succinctly, to understand how real physical systems should behave if the theory were true.

This sketch is enough to formulate our foil, a set of four commitments making up the IV. Although the details vary considerably for other physical theories, the structure just described for Hamiltonian mechanics is quite general: many other theories also introduce basic types of entities and their possible states, in concert with modal commitments characterized by laws. In the case of Hamiltonian mechanics, the basic ontology includes particles (possibly of distinct types) and the spacetime through which they move, and Hamilton's equations give the dynamics. The IV adds the following philosophical commitments and aspirations regarding the progress of physics to this account of the structure of theories:

- *Ontology*: theories postulate a set of basic entities, which are the building blocks of all entities falling within the domain of the theory.
- *Laws*: the dynamical laws governing the behavior of the basic entities have non-trivial and non-subjective modal force and explanatory power, and they do not admit exceptions. The laws may either be deterministic, in the sense that a given physical state uniquely fixes a dynamical history, or stochastic, assigning a probability distribution over possible histories. Laws have privilege over other modal

---

[1] The ambiguity in "ideal"—between "perfect, excellent," as opposed to "imaginary, not real or practical"—is intentional and reflects the contrasting assessments of the co-authors.

notions such as cause, disposition, power, and propensity; this means, among other things, that those latter notions need not be invoked in a deep or ineliminable way in presenting or interpreting fundamental physical theories.

- *Fundamentalism*: there is a partial ordering of physical theories with respect to "fundamentality." The ontology and laws of more fundamental theories constrain those of less fundamental theories; more specifically: (1) the entities of a less fundamental theory T_i must be in an appropriate sense "composed out of" the entities of a more fundamental theory T_f, and they behave in accord with the T_f-laws; (2) T_f constrains T_i, in the sense that the novel features of T_i with respect to T_f, either in terms of entities or laws, play no role in its empirical or explanatory success, and the novel features of T_i can be accounted for as approximations or errors from the vantage point of T_f.
- *Status of a final theory*: finding the "final theory," which identifies the fundamental entities and laws and is, in some sense, the ultimate source of physical explanations, is an appropriate aim for physics.

Together, these claims constitute a controversial view about the relationship between physical theories and naturalized metaphysics and about the structure of physical theories. Debates about this view run like a rich vein through much recent work in philosophy of the physical sciences, which we will mine for insights in the sections to follow.

But first we would like to clarify the view in relation to two issues. First, the constraints T_f imposes on T_i according to fundamentalism need not be as restrictive as philosophers often suppose. Fundamentalists, as we have formulated the view, can freely acknowledge that less fundamental theories may have enormous pragmatic value due to their computational tractatibility, simplicity, and other features. They may also play a crucial role in interpreting more fundamental theories. The constraints imposed rather concern ontology and modality. The empirical and explanatory success of T_i must be grounded in the fact that it captures important facts about the deeper structures identified by T_f. Or, in other words, T_i's successes should be recoverable, perhaps as a limiting case within a restricted domain, in terms of T_f's ontology and laws. A second reason why the consistency requirement is weak is because physics is promiscuous when it comes to physical properties. In Hamiltonian mechanics, for example, essentially any mapping from phase space to real numbers qualifies as a "physical property" that can be ascribed to a system. There is no requirement that legitimate properties have a simple definitional link to the basic entities used in setting up the theory. As Wilson (1985) emphasized, the generality with which physics handles properties has been essential to its successful treatment of higher level quantities such as "temperature."[2] Many philosophical

---

[2] Contrary to conventional wisdom among philosophers, temperature is not simply identified with "mean molecular kinetic energy." That relationship only holds in a quite limited case (namely, an ideal gas); for gases with non-negligible interactions, not to mention liquids or solids, the mean kinetic energy depends on temperature and other physical parameters. This complexity blocks a simple "definition" of temperature but poses no obstacle to treating temperature as a physical property.

treatments of the relationship between the "special sciences" and physics overlook the importance of this point, imposing a more tightly constrained relationship between higher level and physical properties than that used in physics itself. Fundamentalists of different persuasions are of course free to introduce and defend different accounts of intertheory relationships, but we do not see stronger versions as following directly from the practice of physics.

Second, the sense in which laws "govern behavior" has to be treated with some care in order to defend the view that the laws do not admit exceptions. A naïve account of laws ties their content directly to behaviors manifested by various systems falling within their domain of applicability—the law of gravity governs the falling of an apple, for example. Yet such systems often manifest behavior that apparently runs counter to these same laws. Does a leaf falling quite differently than the apple somehow render the law of gravity false? The idea that manifest behaviors straightforwardly falsify laws goes wrong in too closely assimilating the content of the laws with that of equations of motion derived from the laws (see, in particular, Smith 2002).[3] The laws in conjunction with a variety of other ingredients, such as fluid resistance in a model of projectile motion, are used to derive specific equations of motion. These other ingredients needed to apply the theory to concrete situations do not have the modal force or status of the laws, and neither do the derived equations of motion. The derived equations for the apple may fail to apply to the leaf, yet that does not show that the laws of mechanics are "false" or admit exceptions. The laws of all existing theories admit exceptions in a very different sense; namely, that there are phenomena falling within overlapping domains of applicability of theories that have not yet been successfully combined. The fundamentalist assumes, however, that it is reasonable to aim for a "final physics" that applies universally, and it is the laws of this theory that truly admit no exceptions.

3 INTERPRETATION

Philosophers have often taken up the challenge of interpreting physical theories, in the sense of explaining what the world would be like if the theory were true (see, e.g., van Fraassen 1989). This project is notoriously challenging in the case of QM, for which there is still no satisfying interpretation despite its empirical success, but a variety of interpretative questions arise for other physical theories as well. An interpretation should characterize the physical possibilities allowed by the theory as well as specifying how the mathematical structures used by the theory acquire empirical content. The standard account of the interpretative project fits well with the IV: the laws of the theory delimit the space of physical possibilities, perhaps regarded as the set

---

[3] Physicists' usage of the term "laws" can confuse the issue; various derived equations that are commonly called "laws" don't deserve the honorific on our view.

of worlds that are physically possible according to the theory. The differences among these possible worlds reflect brute contingencies, such as details about the initial state of the universe. The interpretative project then attempts to explain what these possible worlds would be like, based on the general features of the theory (such as its basic ontology, kinematics, and dynamics). Here, we briefly explore a recent line of criticism of this approach to interpretation, one that argues that interpretations pursued at this abstract level fail to account for the explanatory power and other virtues of our theories.

An interpretation in the traditional sense starts from the general structure of the theory as described earlier: the laws of the theory and its state space. Assigning empirical content to these structures often begins with appeals to preceding theories whose interpretation is taken to be uncontroversial. To discuss one example briefly, the basic laws of general relativity (GR) – Einstein's field equations – determine what sort of gravitational field (= spacetime metric) is compatible with a given large-scale matter distribution (represented by the stress-energy tensor field), and vice versa. But these structures are only connectable to experiment and observation via principles associating the possible trajectories of different kinds of particles or test bodies (massive, massless, force-free) with different kinds of curves in a relativistic spacetime, and the proper time elapsed along a worldline to the time recorded by an ideal clock (see Malament [2007] for a particularly clear presentation). These principles clarify the physical content of the spacetime geometry introduced in GR provided an antecedent understanding of physical trajectories and classification of different types of bodies or particles. The connection between solutions of Einstein's equations and "experiment" or "observation" is mediated by earlier theories that relate data to the determination of a spacetime trajectory—for example, the theories involved in establishing the trajectory of a planet based on a set of astronomical observations.

One recent line of criticism focuses on whether the laws and state space of a theory are appropriate starting points for the interpretative project. Ruetsche (2011), in particular, develops a pragmatist account of interpretation according to which questions of possibility should be posed with respect to particular applications of the theory. Her arguments depend on recondite details of QM applied to ∞ dimensional systems, but, roughly put, she focuses on cases in which the mathematical structures needed to support a schematic representation of a physical system cannot be introduced at the level of generality where philosophers operate. Suppose we require that any bona fide physically possible state of a system must have a well-defined dynamical evolution specifying how it evolves over time. For some models (such as that of ferromagnetism in an infinite spin chain, discussed in her § 12.3), implementing this criterion will depend on the dynamics one chooses—for some (otherwise kosher) states, the dynamics is not well-defined. The particular application one has in mind will determine which dynamics—and hence which states—should be preferred. Based on such cases, Ruetsche argues that interpretive questions need to be indexed to particular contexts of application of a theory rather than at an entirely general level. We expect that Ruetsche's line of argument applies more generally, but making the case would require a more detailed assessment of the parallels between her cases and examples from other domains of physics than we can pursue here.

4 MODALITY

A perennial problem for our understanding of physics is that of adequately describing and accounting for its modal aspects: physical necessity, physical possibility, and physical probability. In this section, we explore the dialectic of debates between two ways of approaching physical modalities: those that make primary the concept of law of nature, and those that demote laws and make primary some other modally loaded notion such as cause, capacity, disposition, or propensity. There is of course a long tradition in the history of natural philosophy of denying that there is any real necessity in physical processes at all and arguing that the necessity we feel that they have is precisely that: merely *felt* necessity, a matter of our psychology and/or sociology rather than a feature of nature itself. Proponents of such views include Hume, Ayer, Goodman, van Fraassen, and Marc Lange, among many others.[4] We will not discuss this tradition, nor the Kantian tradition that seeks to give a transcendental or psychological grounding to modality in nature. Kantian projects arguably fail at giving a modal grounding to the regular behavior in nature, whereas the subjectivist tradition does not even try to offer such a grounding. Both traditions also have had trouble, traditionally, making sense of the modal aspects implicit in actual scientific practice.

4.1 Laws of Nature

The idea that all physical processes occur in conformity with precise, mathematical regularities is one that has been a dominant assumption in Western science since the time of Galileo. But "regularity" and "conformity with" do not adequately capture the strength of the law-idea. Events do not just *happen* to respect the law-regularities; in some sense they have to, it is a matter of (what is usually called) "physical necessity." One way of cashing out that necessity, not widely held now, is to ground it in an omnipotent and omniscient law-giver (or law-enforcer).

Without appealing to the divine, it is hard to explain the necessity of natural laws. One way to go would be to equate physical necessity with metaphysical necessity: the supposed category of necessary truths that follow from the *essences* or *natures* of things. Metaphysical necessity is supposed to be non-logical in character, and how to understand and define it is also a difficult issue, making it a less than ideal resource for explaining or grounding physical necessity. But the basic idea has intuitive appeal: things such

---

[4] This tradition includes modern-day Humeanism of the variety championed by David Lewis; see Loewer (2004*a*) for an excellent overview.

as electrons, neutrinos, or photons behave in the ways that they do because of their fundamental natures; if they behaved differently, we feel inclined to say, then they would not be *the same kinds of things*. A particle that attracted other electrons rather than repelling them would simply not be an *electron*; it would be a positron, perhaps, or some entirely different kind of thing. More generally, this line of thought holds that the universal truths physicists have discovered about fundamental particles and fields are precisely the truths that are, or follow from, their essential natures.

But many philosophers have other intuitions, equally or more strongly held, that go against the metaphysical necessity of laws. These are intuitions concerning the contingency of many physical facts, including many facts that appear in physical laws. A good example is the apparent contingency of the values of most or all natural constants (e.g. c, the speed of light (2.998×10^8 m/s); or G, the gravitational constant (6.67384×10^{-11} $m^3kg^{-1} s^{-2}$)). Physicists can give no good reason why the numerical values of these constants (or their ratios) could not have been slightly—or even greatly—different from what they actually are, and many philosophers would argue that this is for a very good reason: there *is no* such reason; these values are a matter of pure contingency.[5] Other examples often cited in support of contingency involve the mathematical content of the laws themselves: that the gravitational field equations have (or lack) a cosmological constant, that the laws of electromagnetism are Lorentz-covariant rather than Galileo-covariant, and so on. But the consensus in favor of contingency is by no means complete, and many physicists hope that in the light of future theories some of these apparent contingencies may turn out to be necessities given "deeper" level laws. What has never been made clear by even the most hopeful defender of the necessity of physical law is how the necessity could be made complete, i.e., all contingency purged from fundamental physics.[6]

Coming back to the idea of physical necessity being a matter of the essential natures of the basic physical kinds, we note that there is more than a passing resemblance between this "explanation" of physical necessity and the "*virtus dormitiva*" explanation of opium's soporific effects derided by Moliere. The explanation seems to be nothing more than a restatement, in slightly different words, of the claim of necessity. And if

---

[5] In cosmology, anthropic reasoning has at times been invoked to justify constraints on the range of values that certain constants could have (if the universe is to be one with observers). But such arguments, if accepted, merely narrow the range of values that the constants could have; the contingency of the actual values or ratios is still present.

[6] An idea worth mentioning in this connection is this: that the regular behavior we see in the physical world and codify in the laws of physics is in fact something that emerges from pure chaos or randomness at the most fundamental/microscopic level. This idea would give a kind of (probabilistic) relative necessity to the known laws of physics but eliminate laws from the most fundamental level. Examples of how higher level regularities can emerge from chaos or randomness at a lower level have been discussed in depth by Strevens (2003, 2011) and Myrvold (forthcoming) in the context of thermodynamics and statistical mechanics. Filomeno (2014) provides a thorough discussion of speculations by physicists (e.g., Froggatt and Nielsen 1991, Wheeler 1982, Unger and Smolin 2014) regarding such a non–law-grounded emergence of the lawlike behavior codified in quantum theories. But these speculations remain sketchy and incomplete.

we are asked to spell out, for example, what the essential nature of an electron is, there is no way we could do so without invoking the mathematical laws of physics we have discovered so far, which are all we have available to—in part at least—capture the way electrons behave. We find ourselves going around in a very tight circle. Since the laws at least have precise mathematical contents and (given auxiliary assumptions) precise and testable empirical consequences, it is no wonder that many philosophers prefer to stick with mathematical laws themselves and avoid talk of "natures" or related modally loaded notions such as *power, disposition, tendency, cause.* We will return to these related notions in the next subsection; for now, let us proceed assuming that the laws of nature have a *sui generis* form of necessity equivalent neither to brute contingency nor to metaphysical or logical necessity.

In addition to the difficulty of spelling out the nature of natural necessity, laws have been challenged by philosophers who question whether the laws—those we have already found in physics—can even be said to be true, in a respectable sense of "approximately true." We set aside the fact that extant examples of physical laws all come with domain restrictions (e.g., for classical Newtonian mechanics, the restriction to macroscopic bodies and relative speeds $<< c$). Even within their intended domains, laws may seem to fail frequently. A pendulum bob loses momentum and eventually stops even in a vacuum due to friction at the axis. Newton's predictions for Mercury's perihelion fail by a little bit because of GR, we might say, indicating that the earlier applications of Newton's theory extend beyond the domain in which it approximates GR. And many quantum systems treated with Schrödinger's equation fail to come out right unless we introduce apparently ad hoc terms into the Hamiltonian, unmotivated by fundamental theory. Nancy Cartwright is the philosopher who has most forcefully pushed the "laws are false" objection, but many agree with the thrust of her critique.

In the debate over laws' failure to be true initiated by Cartwright (1983), the problem is often presented in terms of physical laws' needing a *ceteris paribus* clause. The laws describe what will happen, *ceteris paribus* (i.e., as long as nothing interferes). Since it is typically impossible to fully cash out what is meant by *ceteris paribus* for any given law, the defenders of laws typically reject the claim that there is such an intended clause appended to law statements. An alternative to accepting the implicit presence of a *ceteris paribus* clause is the perspective mentioned at the end of Section 2: the derivation of empirical equations of motion from general laws is something we do using carefully chosen supplemental conditions (usually mathematically expressable) and having a restricted range of intended application. Outside that range, or when the supplemental conditions do not hold, the equations of motion will not work, but that is not a falsification of the more fundamental general laws.

What tends to be the case when *prima facie* law-failure cases are raised is that we are not faced simply with a failure of the law or laws to accurately describe things. Instead, scientists typically know how to explain the corrections that are needed via de-idealizations of various kinds, introduction of causes from other theories, or from direct observation (which can be mathematized appropriately so as to be plugged into the laws) and the like. The defender of laws and fundamentalism sees here a non-troubling upshot of the complexity of nature, a complexity that the most fundamental laws can, she hopes, fully explain. We introduce an ad hoc friction force to get the pendulum's behavior right, for example, but friction is itself just a consequence of the fundamental quantum laws governing condensed matter.

But a different perspective is possible: one that emphasizes the near-ubiquity of the intrusion of talk of causes and effects in our application of physical laws to real-world systems, both inside and outside the laboratory. In addition to Cartwright, Mathias Frisch has forcefully argued for the importance and ineliminability of causal talk in physics (see Frisch 2009a, 2012). Turning the tables on physical law, some philosophers argue that the notion of *cause* and related notions such as *power, disposition,* or *capacity* are conceptually basic and metaphysically prior, whereas the appearance of precise, mathematical law-governed regularity is nothing more than the upshot of the truism "same cause, same effect."

4.2 Causes, Powers, Dispositions, Etc. *vs* Laws

Russell (1912) famously argued that the notion of cause had no place in modern physics; like the monarchy, it was a relic of medieval times that had long outlived its utility. Much more recently, John Norton has taken up the crusade against causation (2007). Norton characterizes causation as a "folk-science" concept that is certainly useful in daily life and in applications of physics, but one that does not deserve to be thought of as part of the deep structure or content of nature as it is given to us by our physics theories. Mathias Frisch responded to Norton in defense of the ineliminability of causal talk from even fundamental physics (2009a), which led to further exchanges.[7]

Although we think Frisch makes excellent points about the crucial role that causation seems to play at times in physics, there is arguably a fundamental explanatory asymmetry—alluded to already—that favors the priority of laws over the other modal notions (cause, disposition, power, etc.) and helps explain the enduring presence of laws and fundamentalism in the IV. The asymmetry is this: whereas causal talk seems unable to explain the utility and universality of laws in a non-shallow sense, universal and exceptionless laws do seem able to ground the utility of causal talk *and* explain why causes sometimes don't produce their normal effects in a non-shallow sense. Without translation into contentful mathematical statements, talk of causes, powers, and so forth has at most a rough qualitative meaning, and the explanatory power is limited or zero.

---

[7] See Frisch (2009b) and Norton (2009) for this exchange. Smith (2013) defends a neo-Russellian view, considering various cases that Frisch (and others) see as reflecting causal commitments. Smith argues that the conditions thought to reflect causality principles are more aptly regarded as motivated by other concerns, such as imposing conditions needed to ensure the existence and uniqueness of solutions to a given equation.

For example, consider the striking of a match and its subsequent burning. One way to characterize this phenomenon is using disposition-talk. The match-head material has a disposition to ignite when in the presence of sufficient heat (and oxygen); striking the match produces the needed heat via friction; so, if all the necessary triggering conditions are present and no counteracting powers (e.g. falling raindrops or high winds), striking the match produces the effect of ignition. Another way to characterize the phenomenon would be by digging down into the quantum chemistry of the match-head material and treating the combustion phenomenon (chemical reactions binding oxygen and releasing energy in the form of molecular motion, i.e., heat) using mathematical equations and models. Many philosophers would argue that the latter sort of characterization is more explanatory than the former.[8]

The point, again, is that explanatory power may constitute a real asymmetry between laws and (talk of) causes, powers, or dispositions. If we wish, we can talk about what electrons and atoms do using the language of powers and dispositions. But doing so seems to provide only a thin gloss of familiarity and does not come close to providing the detailed predictive power that models constructed from the laws can have. The advocates of causality and powers would surely agree with this, but add that they have every right to help themselves to the mathematical laws and models of physics in formulating their explanations because those laws simply express in precise terms the causal upshot (in certain circumstances) of the causal powers and dispositions of things.

Here, the defender of fundamental laws will object that it is the mathematical laws per se that do the real explanatory work, and many of them cannot plausibly be read as statements about causal powers and so forth. It is fine to say that, for example, *He* nuclei have the *capacity* to bind with two electrons and that those electrons are *disposed* to occupy only certain discrete energy states. But this gives us no handle on when and why the bonds are stable, what the allowed energy levels are, the shapes of the orbitals, and so forth; the Schrödinger equation does give us these things. And the fundamentalist will argue that the Schrödinger equation, Einstein's field equations, Maxwell's equations, and perhaps other fundamental laws cannot be plausibly read as causal power/disposition/capacity statements nor as upshots or consequences of such statements. In order for these laws to be consequences of some (explanatorily) more fundamental causal power/capacity/disposition statements, the latter would have to be expressed in mathematical language yet still clearly be expressions that can be given a causal gloss (as *is* possible for certain laws like Newton's law of gravity and Coulomb's law). But we have no examples of any such more-fundamental statements that could ground our most fundamental laws, such as the Schrödinger equation. So the law fundamentalist sees an explanatory asymmetry in favor of mathematical laws over causes, powers, capacities, and dispositions.

---

[8] See Hoefer (2003) for a separate line of argument in favor of fundamental laws. For a general discussion of reduction, see the chapter by Hütteman and Love in this volume.

Authors such as Cartwright, Frisch, and Woodward (2003) have argued very persuasively that talk of causation and related Aristotelian notions is prevalent throughout the physical sciences, both in practical applications and in theoretical discussions. The priority that the IV gives to laws of nature can be partly defended, but it remains very controversial.

4.3 Probabilistic Law

A question of perennial interest in the physical sciences is that of determinism: do natural systems always evolve in the same way if they start from exactly the same initial conditions? With the rise of classical mechanics and its influence on philosophy in the modern period (from Descartes to Kant), it was common to assume the truth of determinism until the phenomena of QM appeared to inject intrinsic randomness into nature. In the dominant standard interpretation of QM (and quantum field theories), the theory makes only probabilistic predictions in many situations. Via the "Born Rule," the state vector of a quantum system prescribes probabilities for certain events to occur. Taken together, the laws and rules of quantum physics—again, as standardly presented and interpreted—appear to be essentially *stochastic*.

In our laying out of the IV (section 1), we incorporated stochastic laws explicitly as one possible type of physical law, and this is typically what defenders of fundamental laws have done. But it should be noted that some physicists and some philosophers are not happy with the notion of irreducible stochastic laws. There are at least two sorts of root for the discontent. One is simply an aversion to indeterminism: for many, it is hard to give up on the truth of the Principle of Sufficient Reason in some form or other. When things happen, the intuition goes, there is always a reason why they happened thusly, and if we have at the moment only statistical laws to describe these events, that is a defect in our physics that we should seek to correct.[9] Einstein was one physicist (/ philosopher) who took a dim view of stochastic laws for this sort of reason, but there are many others.[10]

---

[9] Although some philosophers and physicists argue against the notion of (irreducibly) chancy fundamental laws, interestingly, a number of physicists have recently argued that experiments establishing violation of Bell inequalities by quantum systems *prove* the existence of intrinsic randomness in nature (e.g., Colbeck and Renner 2012, Pironio, Acín et al. 2010).

Using reasoning similar to that of Bell in his famous 1964 theorem, these authors argue that when pairs of space-like separated events display correlations violating a Bell inequality, then we can exclude the possibility that there is some underlying deterministic account of the correlations, thus leaving pure stochasticity as the only option. Their arguments depend however on assumptions that exclude certain kinds of possible deterministic "hidden variable" alternatives to QM, including theories like Bohmian mechanics, on the basis of the type of nonlocality and contextuality such theories must involve, and many philosophers would find these assumptions too strong to be defensible.

[10] As Arthur Fine (1986) and others have shown, Einstein was more concerned about the "spooky action at a distance" of QM than about its indeterminism. But there are many passages in Einstein's writings and correspondence that show that he was also unhappy about the indeterminism on its own.

A second root of discontent with stochastic laws has to do with puzzlement about their epistemic status and their truth-conditions.[11] Suppose we have a stochastic (fundamental) law that entails that the chance of R_i being the outcome in experimental setup S is $x_i=0.343$. What does this entail about what *will happen* in the world? This is a question that may appear to have no good answer because—strictly speaking and using the understanding of this chance as a *primitive* fact—any actual frequency of R_i outcomes you care to name is physically possible. And, in this sense, the content of the law is no different from that of an alternative law that sets $x_i=0.395$ or $x_i=0.840$. So we have a *prima facie* puzzle about the semantic content of the stochastic law statement. Now, as we all know, if the law says $x_i=0.343$, then that entails that in a set of many S experiments the frequency of R_i outcomes is *unlikely* to be very different from 0.343 and far more unlikely to be close to 0.840 than to 0.395. These claims about what is "likely" or "unlikely" are, of course, themselves probability claims, and if we arrive at them by simply assuming independence of our S experiments and applying the axioms of probability, then these claims are themselves just claims about irreducible objective chances. If we started out wondering what the semantic content of such claims might be, being offered further such claims in response is obviously no help.

The reason why this problem is not more often noted is that when we make our claims about what the outcome frequencies are likely or unlikely to be, we tacitly invoke the *principal principle* (PP), turning the objective chances of certain frequencies' obtaining into a *subjective* probability—that is, a credence, level of expectation, or betting ratio.[12] In so doing we inject some cash value into the claims because (for example) we do things like rejecting hypotheses that have less than 1% or 5% subjective probability. This is the basis of classical statistical testing. So, for example, if our calculations lead us to have subjective probability of less than 1% that the objective chance of R_i is different from $x_i(=0.343)$ by more than 0.05 given the outcomes we have observed, then we reject all possible values for x_i greater than 0.393 or less than 0.293.[13]

So we feel both that we know what the contents of our objective probability claims are and how to test them. But the crucial move was our use of the PP to turn irreducible chance-facts into rational subjective degrees of belief. And it is controversial whether there is any way to argue that PP is justified for primitive or irreducible (alleged) probability statements. For this reason, some philosophers (e.g., Lewis 1994) defend reductive accounts of objective probability and/or probabilistic physical laws. Reductionist approaches will typically deny the possibility of stochastic fundamental laws (e.g. Hoefer 2011) or demote laws themselves to mere regularities without intrinsic modal force (e.g., Humean best-system approaches, such as in Lewis 1994 and Loewer 2004b).

---

[11] See, e.g., Hoefer (2011).

[12] In one common form, the PP simply says that a rational agent who knows that the objective chance of A is x and has no other/better information about whether or not A will obtain should set her subjective credence for A equal to x also. See Lewis (1986).

[13] Or in more Bayesian terms: our subjective credence in the proposition that the objective chance is nearly 0.343 increases and eventually (if we have "reasonable" priors) approaches unity.

When it comes to physical probability, which can be thought of as a species of modality, the same dialectic can arise between those who favor mathematical stochastic laws as primary and those who instead favor notions such as partial cause, probabilistic disposition, or "chance propensity." The concerns raised earlier about the semantic content and epistemology of stochastic laws equally affect the latter notions when they are taken as irreducible primitives.

5 RECOVERY OF THE MANIFEST WORLD

One part of the ambitions encoded in the IV is the idea that physical science can comprehend and account for all features of the external physical world that we all experience. And since at least Eddington's famous discussion of his "two tables"—one the table of everyday experience (solid, continuous, unmoving, impenetrable, colored, etc.), the other the table as apparently described by modern science (a lattice of highly dense points in constant vibratory motion, with much empty space between the atoms or molecules, with no intrinsic color, easily penetrated by many physical things)—an important goal of philosophy of the physical sciences has been to reconcile the world of common sense and daily experience with the world as described by the sciences. This is important in part because we desire to have an overall consistent set of beliefs about the world. But it also has an epistemological side: if the description of the world given by one or more physical theories appears to radically misdescribe the world *in certain special ways*, then that theory or theories may be held to be *epistemically self-undermining*: if we took the theory to be true, we would have to doubt the correctness of the very experiences (of scientists, in laboratories and observatories) that is the only basis for believing the theory in the first place. As we will see, this is a particularly acute concern when it comes to quantum physics.

Space does not permit a detailed exploration of the many issues that can be raised concerning the tension between the manifest image and the scientific image of the physical world; here, we offer some brief introductions and references.

5.1 Flowing Time and Arrow of Time: Apparent Tension with Fundamental Physics

An essential element of our experience of the world is the fact that time seems to *pass* or *flow*. This involves the inexorable movement of events that have already occurred further and further into the past, relative to the moment we call *now*; the equally inexorable approach of future events such as next Monday morning's commute; and the ever-changing nature of the now or the "present." Relatedly, but differently, our experience is full of phenomena that distinguish the past → future direction in time from the future

→ past direction. Memory gives us only information about the past, and knowledge of future events—if we have any—is hard to come by and different in character. Ice cubes melt in warm drinks, leaving a cooler mix, but we never see a cool mix spontaneously turn into a warm drink with an ice cube floating in it. Waves of water or light frequently diverge with circular symmetry from a point-like source, but we never see such waves converging from far away *to* a point. And so forth.

The problem is that both time's flow and time-asymmetry appear to be absent from fundamental physics.[14] Some asymmetric phenomena, such as the melting of ice cubes, can perhaps be handled by imposing a time-asymmetric constraint on initial conditions (e.g., the Boltzmannian statistical mechanics postulate that the universe began in an extremely low-entropy macro state). But such additional postulates are foreign to the fundamental dynamics, and their justification is a matter of controversy (see Albert 2003).

In the case of time's flow and the related notion of "now" or "the present," not only are these notions absent from physical theories, but relativity theory (special and general) appear to be incompatible with any objective "now" or "present." In special relativity (SR), there is no unique way of "foliating" spacetime into slices of space-at-a-time; rather, each inertial reference frame comes with its own way of slicing up spacetime into spaces-at-times. In so far as our manifest notion of the "now" moving into the future requires the now to be universal, SR renders it perspectival and non-unique. And this situation does not change in any important way in GR. Some philosophers argue that we can be content with a "now" that is not spatially extended (or not very far extended). But even if this is accepted, it remains the case that neither SR nor GR contains anything corresponding to the *movement* of the "now": both seem to present spacetime as a four-dimensional (4D) block in which past, present, and future are as indifferently equally real as the left and right sides of your kitchen. Despite this, some philosophers argue that the notion of time's *passing* is both unproblematic and straightforwardly compatible with physics.[15]

5.2 Measurement Problem of QM: Reconciling QM with a Determinate (Single) Macroscopic World

It is well known that quantum theory involves a sort of paradox for which there is as yet no agreed-upon resolution, called the "measurement problem." One way to view the problem is this. QM describes physical systems with mathematical functions (wave functions) that evolve deterministically over time according to the Schrödinger

---

[14] An exception to this claim is the *t*-symmetry violation in parity–non-conserving quantum events such as neutral kaon decay. But such elusive subatomic events are not responsible for the overt macrolevel time asymmetries just listed.

[15] See, e.g., Maudlin (2002) and Norton (2010). For a defense of a spatially restricted objective "now" compatible with relativity theory, see Savitt (2009).

equation. In certain circumstances, we can create systems that have wave functions that are called "superposition states" with respect to some property (position, spin, momentum, energy, etc.), in which the system may not be said to have any definite value of the property in question. When such systems are subjected to measurement, the linearity of the Schrödinger equation entails that the (macroscopic) measurement-indication system should enter into a superposition as well—for example, having neither the property of simply indicating outcome x nor outcome y, but some sort of fusion of both outcomes. But we apparently never see macroscopic systems in superposition states; instead, the measurement yields a single determinate outcome (e.g., x).

Much of the effort devoted by physicists and philosophers to "interpreting" quantum theory over the past 90 years has been directed at resolving the measurement problem. Various solutions are available. The standard "textbook recipe" approach simply denies that macroscopic measurement systems get into superposition states: a single outcome occurs whose probability is given by the Born Rule, and the wave function of the measured system (if it continues to exist) "collapses" into a non-superposition state for the measured property. But this makes the physical theory either inconsistent or incomplete (lacking a precise account of under what circumstances a collapse occurs). On the opposite extreme, the Everett interpretation bites the bullet and asserts that no collapse occurs, macroscopic systems do get into superpositions, and we *do* "see" them—or rather, we ourselves branch into a superposition state with two (or more) copies of our bodies, one seeing result x and another seeing y. This view entails that all possible outcomes do in fact occur in all measurements (and also in many nonmeasurement situations), which creates a serious interpretive puzzle about probability: what meaning can the Born Rule (i.e., the probabilistic predictions of QM) have? Here we see a danger of epistemic self-undermining looming. Quantum theory is trusted by scientists on the basis of experiments having statistical outcomes matching the probabilistic predictions of the theory very precisely. If an interpretation of the theory makes it difficult to understand those probabilistic predictions, then the interpretation undercuts the evidential basis of the theory.[16]

In between the brute collapse account of the standard recipe and the Everett view, there are other approaches to understanding quantum theory that resolve the measurement problem in diverse ways, sometimes by modifying the theory in significant ways. Two examples are the spontaneous collapse theory of Ghirardi, Rimini, and Weber (GRW), and Bohmian mechanics, alternative theories based on modifying standard non-relativistic quantum theory. Although these alternatives arguably resolve the measurement problem and thereby restore the connection between the theory's predictions and the determinate macroscopic world in one respect, there remain further tensions to resolve, as we will now briefly see.

---

[16] See Wallace (2012) for extensive discussion of the Everett interpretation, the probability problem and the related epistemic concern of self-undermining, as well as Greaves and Myrvold (2010).

5.3 Four-Dimensional Space-Time from Quantum Theory

Non-relativistic quantum theory describes systems—for example, a carbon atom—with a wave function, as we noted earlier. The wave function is not a mathematical object defined as a field in ordinary 3D space or 4D spacetime; instead, it is defined in a much higher dimensional space known as "configuration space," which has $3N$ dimensions, where N is the number of quantum particles being described. What is the connection between such a mathematical object and the 4D spacetime (or 3D spaces existing as time passes . . .) we normally think of ourselves as inhabiting? Wave functions—or quantum states more generally, constructed in high- or infinite-dimensional Hilbert spaces— are essential to the description of physical systems in quantum theory. To treat them as mere instruments with no direct representational significance is always possible, but then this leaves the theory itself as a mere instrument, not a description of what reality itself is like or made of. The alternative of reifying in some physical sense configuration space(s) or Hilbert space(s), however, may seem equally unpalatable. If we take such a space or spaces as part of the fundamental ontology and treat the 3D space of experience as illusory or merely an (effectively) emergent structure, then we give ourselves an extremely difficult task: showing how our experience of a 3D world emerges from the high-dimensional underlying quantum reality. And if we postulate that *both* high-dimensional spaces *and* 3D space are real and fundamental, then we give ourselves a problem similar to that which faced Cartesian dualists: giving an account of how the two distinct components of reality can connect, coordinate, and/or influence one another.[17]

This problem of the fundamental dimensionality of physical reality is one currently being much discussed by philosophers of physics (see, e.g., the essays in Ney and Albert 2013). One approach that is being explored is to set aside non-relativistic quantum theory and restrict oneself to quantum field theories that can (perhaps) be interpreted as only postulating fields existing in 4D spacetime.[18]

6 Sufficiency of Fundamental Theories

We now turn to a line of argument challenging the final two commitments of the IV, those that characterize a version of reductionism combined with optimism that a "final theory" will ultimately ground physical explanations. Critics of reductionism have

---

[17] The obvious option, perhaps, is to demote the quantum state itself to the status of a mere predictive mathematical instrument, not something that *directly* represents any structure existing in reality. If one takes this route, then there is no pressure to regard physical reality as having more than the familiar four spatial dimensions. But then one gives up on the scientific realism part of the IV expressed in the *Ontology* commitment.

[18] See Myrvold (2015) and Wallace and Timpson (2010) for examples.

recently emphasized the explanatory autonomy and significance of supposedly less fundamental theories that typically employ theoretical concepts that are not obviously reducible to those of the more fundamental theory. Debates regarding these issues suffer from different formulations of the key concepts—there are a number of different characterizations of "reduction" versus "emergence," as well as "explanation" on the market (see, e.g., the essays in Bedau and Humphreys 2008). Although we cannot go into depth here, the clarifications of the IV are meant to distinguish it from other versions of reductionism that we regard as clearly too strong. For example, the IV is meant to be compatible with theoretical kinds introduced at the level of the less fundamental theory T_i that do not directly map onto those in the more fundamental theory T_f, as illustrated by multiply realized functional kinds such as temperature. Arguments in favor of emergence based on the novelty of theoretical kinds used by T_i are thus not decisive against the IV. Yet the position does still have teeth, in the sense of ruling out stronger senses of emergence that would be established if there are phenomena successfully described by T_i that cannot be accounted for in terms of the ontology and laws of T_f. Recent philosophical debates regarding the viability of even this modest sense of reductionism have been inspired by foundational studies of a wide range of topics and more careful assessment of intertheory relations in the physical sciences.

More fundamental theories are clearly not the source of explanations in the sense of providing a complete, computationally tractable account of all phenomena within their domain. Theories ranging from physical chemistry to condensed matter physics employ semi-empirical methods to determine features of the systems being studied. For example, the frequency of the specific hyperfine transitions in cesium-133 atoms used since 1967 to define the second can, in principle, be calculated within quantum theory. But in fact the value of this frequency is determined experimentally; theory supports the identification of this frequency as a useful invariant quantity, but calculations are nowhere near determining the value of this frequency with sufficient precision. In continuum mechanics, the constitutive equations characterizing how a specific type of material responds to strains are motivated phenomenologically rather than derived. The widespread use of such semi-empirical methods reflects the need to supplement the fundamental theory with empirical inputs in applications. There are a variety of other more intriguing cases in which successful applications apparently require appeal to concepts and mathematical structure of the less fundamental theory. Batterman (2002) discusses, for example, the explanation of universal patterns of fringe spacing and intensities of light observed in a variety of cases, such as dark fringes and supernumerary bows observed in rainbows. A theory called *catastrophe optics* provides successful explanations of these patterns, yet it appeals to mathematical structures of the less fundamental theory (the caustics of ray optics). These aspects of scientific practice challenge any version of reductionism that implies, as Anderson (1972) put it, "the ability to start from those [fundamental] laws and reconstruct the universe." The IV as we formulated it above is not committed to such an implication.

Successful explanations offered by a less fundamental theory T_i that are incompatible with the ontology and laws of T_f would, however, directly challenge the IV, as illustrated by two cases discussed recently by philosophers. First, many systems in classical mechanics are chaotic, in the sense that the trajectories for nearby points in phase space diverge rapidly (exponentially). In the more fundamental theory, QM, the unitarity of the dynamics rules out the divergence of trajectories that is the defining feature of classical chaos. This leads to a quite striking contrast between classical and quantum dynamics. Insofar as the exponential divergence of trajectories is crucial to the successful application of classical mechanics, we have a case where the fundamental theory cannot account for the less fundamental theory's success.[19] Second, thermodynamics (the less fundamental theory in this case) describes phase transitions between distinct phases of matter, such as that between a liquid and gaseous phase of water as the pressure or temperature is slowly changed, in terms of a discontinuity in a thermodynamic quantity. This discontinuity corresponds to a singularity in the partition function in the (more fundamental) statistical mechanical description of the system. Yet the partition function for any system of finite particles is analytic; singularities only arise in the thermodynamic limit, as $N, V \to \infty$ with $\frac{N}{V} = \rho$ (where N is the number of particles and V is the volume). The more fundamental theory hence apparently fails to recover the success of thermodynamics, since the phase transitions occur only in unphysical, idealized systems with an infinite number of particles, rather than in the finite systems of our experience. In both cases, the success of T_i depends on novel features supposedly rendered otiose by T_f.

These are extremely interesting challenges to the IV, but there are (at least) two responses to cases like these open to defenders of the IV. The first challenges the claim that T_i's success actually supports aspects of the theory that are novel with respect to T_f. In general, we do not expect T_f to recover the *exact* structure of T_i—which, after all, has been rejected because it misrepresents some aspects of nature. IV can be preserved if it is possible to construct a simulacrum of T_i's successful results in T_f's language. Callender (2001) pursues this line of thought with regard to the relationship between statistical mechanics and thermodynamics: we should avoid the error of "taking thermodynamics too seriously"; namely, the error of requiring that statistical mechanics reproduces thermodynamics *exactly* rather than being satisfied with an approximation. In the case of phase transitions, a defense of this line of thought requires showing how mathematical treatments of finite systems approximate various results obtained in the thermodynamic limit.[20]

A second line of response considers the nature of the mathematics used by T_i and how it should be interpreted. Belot (2005), for example, argues in reply to Batterman (2002)

---

[19] There is a large technical literature on this topic; Belot and Earman (1997) and Bokulich (2008) survey some of the technical results and provide an entry point to the philosophical issues.

[20] See Batterman (2005a), Butterfield (2011), Kadanoff (2013), and Menon and Callender (2013) for an entry point into these debates.

that the mathematics of wave optics can be understood without appeal to T_i. The caustics and other structures needed to understand phenomena such as universal properties of rainbows (as mentioned earlier) can, in some sense, be discerned within solutions to the wave equations. The debate then turns on the role of the concepts of T_i in interpreting the relevant mathematics; Batterman (2005b) maintains, in response to Belot, that the concepts of ray optics are needed to understand the singular limit of the wave theory and to explain universality. A related line of thought regards the mathematical structures of T_i as having merely instrumental utility. One can evade the apparent conflict between the applied mathematics pressed into service in these cases and a reductionist thesis such as IV by denying that the mathematics has any representational significance. For such an instrumentalist approach, although T_i may provide a valuable inferential pathway to understanding features of T_f, one need not be troubled if the views along the path differ from those at the destination.

We close by raising a different concern: how would a final theory compare to existing physical theories? One obvious contrast concerns the (often implicit) domains of applicability of current theories. Although we expect all current candidates for a fundamental theory to break down at scales where quantum and gravitational effects have to be combined (such as at the Planck scale), presumably the final theory will be truly universal. This may seem like a minimal contrast: we can take existing theories as giving global descriptions of possible worlds distinct from our own but similar within some domains—for example, GR describes a world free from quantum effects, which may capture many larger-scale features of our own world when gravity is the dominant force. This will not work for philosophers who reject the idea that one can delimit the possible worlds allowed by a theory without considering applications (such as Ruetsche 2011, discussed briefly in Section 3). On Wilson's (2006) account, the success of classical mechanics depends on what he calls its "facade"-like structure. Rather than a single axiomatized theory, classical mechanics should be viewed as a collection of different approaches that include different tools for modeling physical systems. Models succeed in giving detailed descriptions of physical systems in part by restricting consideration to a particular length or time scale and by making a variety of assumptions appropriate for that setting. Wilson describes the resulting overall structure as a facade: the locally applicable models may appear complete but in fact have implicit domain restrictions; attempting to provide a complete description leads to jumping over to a different facade based on a distinct approach.[21] The important point for our purposes is that the success of a theory facade does not imply that there is a "globally consistent possible world"

---

[21] He describes, for example, the various ways in which one might describe the physics of billiard ball collisions (see Wilson 2006, chapter 4.vi). One might start with a description of rigid body collisions with results dictated by conservation principles. But this simple description fails to describe energy loss during the collision, distortion of the impacting bodies, and a variety of other effects. Wilson argues that more sophisticated models of collisions based on, for example, continuum mechanics cannot be seen as straightforwardly "completing" or "augmenting" the simple account since these models include physical and mathematical assumptions incompatible with the earlier approach.

described by classical mechanics; instead, we have at best a patchwork of overlapping local models that cover natural phenomena in much the same way as a set of projections in an atlas cover the globe.[22] The final theory envisioned by the IV goes well beyond the natural claim that there is a single "way the world is" to assert that there is a single theory that adequately reflects it, without resorting to a facade-like structure.

REFERENCES

Albert, D. Z. (2003). *Time and Chance* (Boston: Harvard University Press).

Anderson, P. W. (1972). "More Is Different." *Science* 177(4047): 393–396.

Batterman, R. W. (2002). *The Devil in the Details: Asymptotic Reasoning in Explanation, Reduction, and Emergence* (Oxford: Oxford University Press).

Batterman, R. W. (2005a). "Critical Phenomena and Breaking Drops: Infinite Idealizations in Physics." *Studies in History and Philosophy of Science Part B: Studies in History and Philosophy of Modern Physics* 36(2): 225–244.

Batterman, R. W. (2005b). "Response to Belot's 'Whose Devil? Which Details?.'" *Philosophy of Science* 72: 154–163.

Bedau, M. A., and Humphreys, P. E. (2008). *Emergence: Contemporary Readings in Philosophy and Science* (Cambridge, MA: MIT Press).

Belot, G. (2005). "Whose devil? Which details?" *Philosophy of Science* 72(1): 128–153.

Belot, G., and Earman, J. (1997). "Chaos out of order: Quantum mechanics, the correspondence principle and chaos." *Studies in History and Philosophy of Science Part B: Studies in History and Philosophy of Modern Physics* 28(2): 147–182.

Bokulich, A. (2008). *Reexamining the Quantum-Classical Relation* (Cambridge: Cambridge University Press).

Butterfield, J. (2011). "Less Is Different: Emergence and Reduction Reconciled." *Foundations of Physics* 41(6): 1065–1135.

Callender, C. (2001). "Taking Thermodynamics Too Seriously." *Studies in History and Philosophy of Science Part B: Studies in History and Philosophy of Modern Physics* 32(4): 539–553.

Cartwright, N. (1983). *How the Laws of Physics Lie* (Oxford: Clarendon Press).

Colbeck, R., and Renner, R. (2012). "Free Randomness Can Be Amplified." *Nature Physics* 8(6): 450–453.

Filomeno, A. (2014). On the Possibility of Stable Regularities Without Fundamental Laws. Ph.D. thesis (Autonomous University of Barcelona).

Fine, A. (1986). *The Shaky Game: Einstein, Realism, and the Quantum Theory*. Science and Its Conceptual Foundations (Chicago: University of Chicago Press).

Frisch, M. (2009a). "Causality and Dispersion: A Reply to John Norton." *British Journal for the Philosophy of Science* 60(3): 487–495.

Frisch, M. (2009b). "'The Most Sacred Tenet'? Causal Reasoning in Physics." *British Journal for the Philosophy of Science* 60(3): 459–474.

---

[22] Wilson describes the facade-like structure of classical mechanics as a consequence of the way classical mechanics "sits on top of" quantum mechanics, and he does not argue (as far as I am aware) for the more general claim that *all* theories must have a facade-like structure. But an advocate of a more general facade thesis could certainly build on Wilson's discussion of concept use.

Frisch, M. (2012). "No Place for Causes? Causal Skepticism in Physics." *European Journal for Philosophy of Physics* 2(3): 313–336.

Froggatt, C. D., and Nielsen, H. B. (1991). *Origin of Symmetries* (London: World Scientific).

Greaves, H., and Myrvold, W. (2010). "Everett and Evidence." In S. Saunders, J. Barrett, and A. Kent (eds.), *Many Worlds* (Oxford: Oxford University Press), 264–304.

Hoefer, C. (2003). "For Fundamentalism." *Philosophy of Science* 70(5): 1401–1412.

Hoefer, C. (2011). "Time and Chance Propensities." In *The Oxford Handbook of Philosophy of Time* (Oxford: Oxford University Press), 68–90.

Kadanoff, L. P. (2013). "Theories of Matter: Infinities and Renormalization." In R. W Batterman (ed.), *The Oxford Handbook of Philosophy of Physics* (Oxford: Oxford University Press), 141–188.

Lewis, D. (1986). "A Subjectivist's Guide to Objective Chance." In *Philosophical Papers: Vol. II* (Oxford: Oxford University Press), 83–113.

Lewis, D. (1994). "Humean Supervenience Debugged." *Mind* 103(412): 473–490.

Loewer, B. (2004a). "Humean Supervenience" In J. Carroll (ed.), *Readings on Laws of Nature* (Pittsburgh, PA: Pittsburgh University Press), 176–206.

Loewer, B. (2004b). "David Lewis's Humean Theory of Objective Chance." *Philosophy of Science* 71(5): 1115–1125.

Malament, D. B. (2007). "Classical Relativity Theory." In Butterfield, J. and Earman, J. (eds.), *Handbook of the Philosophy of Science. Philosophy of physics. Part A* (Amsterdam: Elsevier), 229–275.

Maudlin, T. (2002). "Remarks on the Passing of Time." *Proceedings of the Aristotelian Society* 102, 259–274.

Menon, T., and Callender, C. (2013). "Ch-ch-changes Philosophical Questions Raised by Phase Transitions." In R. W. Batterman (ed.), *The Oxford Handbook of Philosophy of Physics* (Oxford: Oxford University Press), 189–223.

Myrvold, W. C. (2015). "What Is a Wavefunction?" *Synthese* 1–28.

Myrvold, W. C. (forthcoming). "Probabilities in Statistical Mechanics." In C. Hitchcock and A. Hájek (eds.), *The Oxford Handbook of Probability and Philosophy* (Oxford: Oxford University Press).

Ney, A., and Albert, D. Z. (2013). *The Wave Function: Essays on the Metaphysics of Quantum Mechanics.* (Oxford: Oxford University Press).

Norton, J. D. (2007). "Causation as Folk Science." In H. Price and R. Corry (eds.), *Philosophers' Imprint*, Vol. 3 (Oxford: Oxford University Press).

Norton, J. D. (2009). "Is There an Independent Principle of Causality in Physics?" *British Journal for the Philosophy of Science* 60(3): 475–486.

Norton, J. D. (2010). "Time Really Passes." *Humana.Mente* (online) http://www.humanamente. eu/PDF/Issue13_CompletePDF.pdf.

Pironio, S., Acín, A., et al. (2010). "Random Numbers Certified by Bells Theorem." *Nature* 464(7291): 1021–1024.

Ruetsche, L. (2011). *Interpreting Quantum Theories: The Art of the Possible* (Oxford: Oxford University Press).

Russell, B. (1912). "On the notion of cause." *Proceedings of the Aristotelian Society* 13: 1–26. JSTOR. http://www.jstor.org/stable/4543833.

Savitt, S. F. (2009). "The Transient Nows." In Wayne C. Myrvold and Joy Christian (eds), *Quantum Reality, Relativistic Causality, and Closing the Epistemic Circle*, Vol. 73, The Western Ontario Series in Philosophy of Science (Amsterdam: Springer Netherlands), 349–362.

Smith, S. R. (2002). "Violated Laws, Ceteris Paribus Clauses, and Capacities." *Synthese* 130(2): 235–264.

Smith, S. R. (2013). "Causation in Classical Mechanics." In R. W. Batterman (ed.), *The Oxford Handbook of Philosophy of Physics* (Oxford: Oxford University Press), 107–140.

Strevens, M. (2003). *Bigger Than Chaos: Understanding Complexity Through Probability.* Number 4. (Boston: Harvard University Press).

Strevens, M. (2011). Probability Out of Determinism. In C Beisbart and S. Hartmann (eds.), *Probabilities in Physics* (Oxford: Oxford University Press), 339–364.

Unger, R. M., and Smolin, L. (2014). *The Singular Universe and the Reality of Time* (Cambridge: Cambridge University Press).

van Fraassen, B. (1989). *Laws and Symmetry* (Oxford: Oxford University Press).

Wallace, D. (2012). *The Emergent Multiverse: Quantum Theory According to the Everett Interpretation* (Oxford: Oxford University Press).

Wallace, D., and Timpson, C. G. (2010). Quantum Mechanics on Spacetime I: Spacetime State Realism." *British Journal for the Philosophy of Science* 61(4): 697–727.

Wheeler, J. A. (1982). *Physics and Austerity: Law Without Law: Working Paper* (Austin: Center for Theoretical Physics, University of Texas).

Wilson, M. (1985). "What Is This Thing Called Pain? The Philosophy of Science Behind the Contemporary Debate." *Pacific Philosophical Quarterly* 66(3–4): 227–267.

Wilson, M. (2006). *Wandering Significance: An Essay on Conceptual Behavior* (Oxford: Oxford University Press).

Woodward, J. (2003). *Making Things Happen: A Theory of Causal Explanation.* Oxford Studies in the Philosophy of Science (New York: Oxford University Press).

CHAPTER 7

....

HAVING SCIENCE IN VIEW

*General Philosophy of Science and Its Significance**

....

STATHIS PSILLOS

Quid tandem harmonia?
Similitudo in varietate, seu diversitas identitate compensate.
> "The Confession of a Philosopher," G. W. Leibniz 1672–73

1 PREAMBLE
....

GENERAL philosophy of science (GPoS) is the part of conceptual space where philosophy and science meet and interact. More specifically, it is the space in which the scientific image of the world is synthesized and in which the general and abstract structure of science becomes the object of theoretical investigation.

Yet there is some skepticism in the profession concerning the prospects of GPoS. In a seminal piece, Philip Kitcher (2013) noted that the task of GPoS, as conceived by Carl Hempel and many who followed him, was to offer explications of major meta-scientific concepts such as *confirmation, theory, explanation, simplicity*, and the like. These explications were supposed "to provide general accounts of them by specifying the necessary conditions for their application across the entire range of possible cases" (2013, 187). Yet Kitcher notes, "Sixty years on, it should be clear that the program has failed. We have no general accounts of confirmation, theory, explanation, law, reduction, or causation that will apply across the diversity of scientific fields or across different periods of time" (2013, 188). The chief reasons for this alleged failure are two. The first relates to the diversity of scientific practice: the methods employed by the various fields of natural science

* This essay is dedicated to the *Usual Suspects*: the mighty members of my research group in Athens, who have helped me for years with their ideas and criticism and who motivated me with this essay when I felt I was blocked. Many thanks to Paul Humphreys for useful comments.

are very diverse and field-specific. As Kitcher notes, "Perhaps there is a 'thin' general conception that picks out what is common to the diversity of fields, but that turns out to be too attenuated to be of any great use." The second reason relates to the historical record of the sciences: the "mechanics" of major scientific changes in different fields of inquiry is diverse and involves factors that cannot be readily accommodated by a general explication of the major metascientific concepts (cf. 2013, 189).

Although Kitcher does not make this suggestion explicitly, the trend seems to be to move *from* GPoS *to* the philosophies of the individual sciences and to relocate whatever content GPoS is supposed to have to the philosophies of the sciences.

I think skepticism or pessimism about the prospects of GPoS is unwarranted. And I also think that there can be no philosophies of the various sciences without GPoS. Defending these two claims will be the main target of this chapter. Still, I do not want to contrast GPoS to the philosophies of the individual sciences. As I will show, there is osmosis between them, and this osmosis is grounded on what I will call "Science in general" and the two important functions GPoS plays vis-à-vis Science-in-general: an explicative function and a critical function.

2 WHAT IS SCIENCE?

There have been various public debates about the nature of science, the most prominent being in the early Victorian period with William Whewell and John Stuart Mill as the main protagonists (see Yeo 1993) and in the early decades of the Third Republic in France, with Henri Poincaré among others playing a key part in it (see Paul 1985). But it was in the first half of the twentieth century that the issue of a sharp separation of science from nonscience or pseudoscience became a major philosophical-analytical endeavor.

2.1 Searching for Criteria

The logical positivists aimed to reform philosophy by making it scientific; hence, the issue of what counts as "scientific" was taken to be urgent. Moritz Schlick equated "scientific" with a certain way to be cognitively significant: "A proposition has a statable meaning only if it makes a verifiable difference whether it is true or false" (1932, 88). This came to be known as the *verificationist criterion of meaningfulness*. It rendered science distinct from metaphysics (aka speculative philosophy): metaphysics was meaningless, because unverifiable. But it also *reshaped* philosophy by taking philosophical assertions to be significant insofar as they are analytic: true in virtue of the meanings of their constituent words.

But are scientific assertions, *qua* scientific, verifiable? The answer depends on how exactly we should understand the (modal) concept of verifiability: a statement is strongly verifiable if its truth can be conclusively established in experience, whereas it is weakly verifiable if its truth can be rendered probable by experience. The result is that, depending on what exegesis we may accept, we will get different versions of the criterion of meaningfulness—hence of what counts as "scientific." In the thought of logical positivists, "verifiability" moved from a strict sense of provability on the basis of experience to the much more liberal sense of confirmability. This was a significant shift. Given a claim of provability, many ordinary scientific assertions, for instance, those expressing universal laws of nature, would end up being unverifiable; hence nonscientific. Not so given a claim of confirmability. But, by the same token, and given the typically holistic nature of confirmation, there was no longer a sharp distinction between science and metaphysics (see Hempel 1951; Quine 1951).[1]

Karl Popper (1963) denied that the evidence can have any bearing on the probability of a theory or a hypothesis, but he nonetheless argued that scientific theories can be falsified by the evidence. Popper took (the modal notion of) falsifiability to be the criterion of *demarcation* between science and nonscience or pseudoscience. Scientific theories are supposed to be falsifiable in that they entail observational predictions which can then be tested in order either to corroborate or to falsify the theories that entail them. Nonscientific claims are not supposed to have potential falsifiers: they cannot be refuted. Unlike the logical positivists, Popper did not want to separate science from metaphysics. For him, scientific theories emerge as attempts to concretize, articulate, and render testable metaphysical programs about the structure of the physical world (cf. 1994). Still, there was supposed to be a sharp demarcation between science and pseudoscience.

However, given that in all serious cases of scientific testing, the predictions follow from the conjunction of the theory under test with other auxiliary assumptions and initial and boundary conditions, when the prediction is *not* borne out, it is the whole cluster of premises that gets refuted. Hence, it is not the theory per se that is falsified. It might be that the theory is wrong, or some of the auxiliaries were inappropriate (or both). As a result, any theory can be saved from refutation by making suitable adjustments to auxiliary assumptions. The point here is not that these adjustments are always preferable. They may be ad hoc and without any independent motivation. Hence, the theory might be condemned, as Henri Poincaré (1902, 178) put it, without being, strictly speaking, contradicted by the evidence. The point, rather, is that *qua* a criterion of marking the bounds of the scientific, Popper's falsifiability criterion fails (see my 2012 for details).

Can we find solace in projects such as those associated with Thomas Kuhn (1970) and Imre Lakatos (1970)? Neither of them, to be sure, offered explicit criteria of demarcation. Yet they offered templates as to how science is structured and how it develops over time, which suggested that there are *structural* ways to capture the bounds of science. Kuhn (1977, 277) suggested that "the surest reason" for claiming that some activity (e.g., astrology) is pseudoscientific is precisely that it lacks the right structure; that is, it is not governed by a paradigm-led puzzle-solving normal activity. But it follows that, from the point of view of an existing and established paradigm, a new rival and emerging theory

---

[1] The first to make an effort to distinguish science and metaphysics was Pierre Duhem (1906). He defended the autonomy of science by taking metaphysics to aim at explanation and science at description and classification.

is bound to count as pseudoscientific before it acquires the structure of a paradigm-led puzzle-solving normal activity; which is clearly something we do not want to accept.

Lakatos (1970), too, developed a structural model of science based on the claim that the unit of appraisal is not a single theory but a sequence of theories known as a *scientific research program*. In this way, he aimed to improve on Popper's criterion, which was rightly taken to implausibly imply that theories are falsified and abandoned as soon as they encounter recalcitrant evidence. Lakatos emphasized the role of novel predictions in his own structural model. A research program is progressive as long as it issues in novel predictions, some of which are corroborated. It becomes degenerating when it offers only post hoc accommodations of facts, either discovered by chance or predicted by a rival research program. The price of this way to circumscribe the bounds of science is that there is no way to tell when a research program has reached a terminal stage of degeneration. Even if a research program seems to have entered a degenerating stage, it seems entirely possible (and it did happen with the kinetic theory of gases toward the end of the nineteenth century) that it could stage an impressive comeback in the future.

Expressing the sentiment of despair that was capturing philosophers of science after the so-called *demarcation debacle*, Larry Laudan noted in 1996:

> The failure to be able to explicate a difference between science and nonscience came as both a surprise and a bit of an embarrassment to a generation of philosophers who held, with Quine, that "philosophy of science is philosophy enough." Absent a workable demarcation criterion, it was not even clear what the subject matter of the philosophy of science was. More importantly, the failure of the positivist demarcation project provided an important intellectual rationale for the efforts of relativists in the 1960s and 1970s to argue for the assimilation of science to other forms of belief since the relativists could cite the authority of the positivists themselves as lending plausibility to their denial of any difference that made a difference. (1996, 23)

The problem, indeed, was and *still* is whether science loses any of its intellectual authority if it is not clearly and sharply demarcated from pseudoscience or nonscience. Is it the case that failing to circumscribe the boundaries of science has to lead to an undermining of the objectivity and epistemic reliability of science?

This, note, is a serious and challenging question *within* GPoS. But the answer is negative. As Laudan himself noted (1996, 24), the *epistemic* problem is what makes scientific knowledge reliable and not what makes it scientific. Addressing this problem does not require, for instance, that creationism is proved to be pseudoscientific based on some general and sharp criteria. But it does require engaging with the *epistemic status* of the creationist theories: their relation to evidence, their integrability with other theories we have independent reasons to accept, and so forth. And this presupposes the development of a rather general account of empirical support and confirmation which will make it possible, among other things, to compare evolutionary theory and creationism and to show how and why the latter is epistemically defective. The point I want to stress here is that the very issue of how scientific theories are related to evidence and how theory-appraisal and theory-choice should work is a central concern of GPoS, even in the absence of definite and rigorous ways to characterize creationism or other endeavors as pseudoscience.

2.2 From Essentialism to Family Resemblance

Science does not have an essence waiting to be discovered by conceptual analysis and/ or empirical investigation. An elegant thought, advanced by Massimo Pigliucci (2013), is that "science" is a cluster or family resemblance concept, one whose basic content is captured by a two-dimensional graph in which one dimension is theoretical understanding and the other is empirical knowledge. In this graph, the pseudosciences are supposed to occupy the space near the origin: "they all occupy an area . . . that is extremely low both in terms of empirical content and when it comes to theoretical sophistication" (2013, 24).

However, unless there are objective measures of empirical content and theoretical understanding, it can always be questioned whether paradigmatic cases of pseudoscience are low on both. Pigliucci considers a related objection and argues that, faced with a dilemma of abandoning astrology and creationism or claiming that "established sciences" are not genuinely different from pseudosciences, "the choice is obvious" (2013, 204). Surely it is! But it is so, I claim, because there are plenty of *epistemic* reasons to trust, say, evolutionary biology and astronomy and virtually no reason to trust creationism and astrology. If there is something that genuine scientific theories have in excess over so-called pseudoscientific theories it is that they are supported—objectively speaking—by evidence; they have explanatory content; they cohere better with the rest of the theories in the scientific image of the world. In other words, they score a lot higher in the kind of evidential and nonevidential "metric" that scientists have always employed to appraise theories.[2] Still, Pigliucci's overall approach is commendable because it stresses resemblances and common grounds between the various sciences as being enough for a general characterization of science. James Ladyman (2013, 51) makes this point when he says that "(a) family resemblance certainly exists between the sciences, and the success of fields such as thermodynamics and biophysics shows that science has a great deal of continuity and unity." This "theoretical simplicity and utility of the concept of science" is enough to make it useful even if science is so heterogeneous.

Family resemblance is key to Paul Hoyningen-Huene's (2013) attempt to answer the question "What is science?" by stressing that what makes scientific knowledge distinctive (over other forms of knowledge) is that it is "systematic," where systematicity is analyzed along nine dimensions, including descriptions, explanations, predictions,

---

[2] All this does not imply that the application of scientific methods is algorithmic. In most typical cases, evidence is balanced with considerations concerning theoretical (explanatory) virtues, and this is done by exercising the judgment of scientists. I have elaborated on this claim in my work (2002) and (2012a).

completeness, systematic defense of knowledge claims, epistemic connectedness, and others (2013, 27). Hoyningen-Huene advances his account as a descriptive theory, "describing what exists in science" (2013, 199). But this may well be its chief weakness, since, as Hoyningen-Huene admits, he ends up with a "tenuous sort of unity among all of the sciences" (2013, 209), the reason being that the various ways to understand "systematicity" (by means of the concretizations of *each* of the nine dimensions in *each* of the sciences and/or theories) lead to divergent meanings. It is of little consolation that all these concretizations of "systematicity" in different disciplines, subdisciplines, and areas of research "are connected by multidimensional family resemblance relations" (2013, 209). Ultimately, what makes the relations of family resemblance possible is that all these disciplines are deemed to be members of the same family, viz., *science*, and what makes all these disciplines members of the same family, viz., *science*, is that they are connected by family resemblances.

In Hoyningen-Huene's descriptive approach, the issue of whether scientific theories are supported by evidence is passed over (almost) in silence. One of the dimensions of systematicity—the systematic defense of knowledge claims—is obliquely related to looking for evidence for theories. But he insists that his approach is not evaluative; hence, it is not meant to say whether one theory is supported (or supportable) by evidence more than another nor to compare, in terms of the success of representing reality, science with and other forms of knowledge (including pseudoscience) (2013, 173). This, however, will leave little or no room for a substantial *criticism*—within science and within GPoS—of various theories.

The family resemblance approach has the advantage of avoiding the pitfalls of essentialism, but it has the disadvantage of not adequately explaining where the (family) resemblance lies. In thinking about the question "What is science?" it is important that we rely on our best exemplars of sciences and scientific theories and aim to mold a conception of science that characterizes them. But if we stay at the level of family resemblances, we might end up being too descriptive: science is whatever is being taught in science departments in universities. In my view, it's best to proceed the other way around: *first a conception of science, then the resemblance.*

3 SCIENCE-IN-GENERAL AND GENERAL PHILOSOPHY OF SCIENCE

I want to argue that the object of study of GPoS is *Science-in-general*. And although GPoS is not here to legislate what science ought to be—independently of what is going on (and has been going on) in the various sciences—it is also the case that GPoS is not here to merely describe (or to provide a synopsis of) what the various sciences do. If a purely normative-evaluative perspective is spinning in the void, a purely descriptive perspective does not make room for grounded judgments about the best way to view science and how we characterize the unity that exists among the various sciences despite their differences. In particular, a purely descriptive perspective does not make room for a critique of science—that is, of a critical engagement with science. In engaging with GPoS, we start with some theoretical conception of science and aim for a reflective equilibrium between this conception and actual and historical features appropriate to science and various knowledge-generating practices which are described or characterized as science.

3.1 The Two Functions of GPoS

GPoS fulfills two different but related functions, and it requires both of them in its operation. The first function is *explicative*. It aims to explicate (i.e., to render more precise and more definite) the various concepts that are employed by the various sciences—and hence, to specify their common content as well as their differences and relations. For instance, it is hard to think of science without thinking of it as offering explanations of the phenomena under study, even if there is no overarching account of explanation that covers all paradigmatic sciences or subdisciplines within them. It is equally hard to think of science without thinking of it as offering representations of the phenomena, as relying on models and as advancing theories. And, of course, it is even harder to think of science without some notion of experiment and experimental practice. Even if it turns out that, to give one example among the many, the concept of explanation (or representation, theory, and others) has different explications in the various sciences, all of these explications belong to the same genus, and it is because of this that they are all concepts of explanation and not of something else.[3] The very idea of conceptual pluralism is an idea within GPoS, and its proper development requires GPoS—for admitting that there is no single concept of, say, explanation requires a standpoint from within which all competing concepts of explanation are examined and compared. And this is the standpoint of GPoS.

The second function of GPoS is *critical*. It aims to criticize (in the Kantian sense of passing judgments on) the various conceptions of science as well as the various ways to present science, its methods, and its aims. A key object, for instance, of the critical function of GPoS is to disentangle the part of scientific theories that is up to us and the part that is up to the world; or, in other words, the contribution of the mind and the contribution of the world in our scientific image of the world. Another key object is to discuss the scope and limits of scientific knowledge and the epistemic credentials of the various factors (evidence, theoretical virtues, etc.) that are involved in the acceptance of theories or the relation between philosophy of science and history of science.[4]

---

[3] For a development of this line of thought, especially in connection to the role of explanation in inference to the best explanation (*aka* abduction) see my (2007).

[4] For an account of how these issues were discussed and developed by Henri Poincaré, see my (2014*a*). When it comes to the connection between philosophy of science and history of science, I think that the prototype is still Duhem (1906).

I will discuss these two functions later. In Section 4, I will present four dimensions along which these two functions operate. But before I do this, I want to explicate this idea of Science-in-general as the proper object of GPoS.

3.2 A Mode of Knowledge as Well as a Discipline

In thinking of Science-in-general as the proper object of GPoS, I follow two thinkers. The first is Aristotle. In the opening lines of *Physics* Book 1, Aristotle noted that *episteme* is both a kind of knowledge and a discipline. In fact, this kind of knowledge is what is shared by the various disciplines that are called sciences. Here is how he put it:

> When the objects of an inquiry, in any department, have principles, causes, or elements, it is through acquaintance with these that knowledge and understanding is attained. For we do not think that we know a thing until we are acquainted with its primary causes or first principles, and have carried our analysis as far as its elements. Plainly, therefore, in the science of nature too our first task will be to try to determine what relates to its principles.(Aristotle, 184a10–184a16)[5]

The science of nature ("ἡ περὶ φύσεως ἐπιστήμη" or *scientiae naturalis*, as Thomas Aquinas rendered it in his *Commentary*) is a kind of *episteme*; that is, it is a special way of knowing its subject matter. So, from Aristotle, I take the idea that science is a special mode of knowing the world (or an intended domain of phenomena).

The second thinker is Karl Marx, from whom I form the notion of Science-in-General in analogy with his idea of "production in general," which he put forward in the *Grundrisse*:

> *Production in general* is an abstraction, but a rational abstraction in so far as it really brings out and fixes the common element and thus saves us repetition. Still, this *general* category, this common element sifted out by comparison, is itself segmented many times over and splits into different determinations. Some determinations belong to all epochs, others only to a few. [Some] determinations will be shared by the most modern epoch and the most ancient. No production will be thinkable without them; however even though the most developed languages have laws and characteristics in common with the least developed, nevertheless, just those things which determine their development, i.e. the elements which are not general and common, must be separated out from the determinations valid for production as such, so that in their unity—which arises already from the identity of the subject, humanity, and of the object, nature—their essential difference is not forgotten. (1857–58, 85)

From this rich passage, I take the idea that Science-in-general is a *rational abstraction*. In actual point of fact, there are concrete theories and disciplines with rich and complex

---

[5] In Aristotle (1984, 315).

histories and structures. And yet, they all fall under a general category—*science*—aiming to capture a mode of knowledge of the world, which is subject to different determinations (both conceptually and historically), some of which are general and common to all sciences, while others are different. Neither their unity nor their differences should be neglected. To paraphrase Marx, their unity arises from the identity of the subject (viz., that science is a special mode of knowledge of nature) and from the identity of the object (viz., nature itself). Going above the various sciences to Science-in-general makes it possible to acquire a (revisable and historically conditioned) bird's-eye point of view that is necessary for viewing the various sciences as being parts of a common endeavor to understand the world and to acquire a coherent scientific image of it. As I put it elsewhere (2012b, 101): "Science *as such* is a theoretical abstraction and general philosophy of science is the laboratory of this theoretical abstraction."

The Aristotelian idea that I have used is that the common core of science—what characterizes Science-in-general *qua* an abstraction—is a form of knowledge and a concomitant set of practices and methods that aim to achieve this form of knowledge. This form of knowledge—known as *scientific*—is characterized by a rather rigorous demand for justification as well as for external grounding to (at least some aspects of) reality. The demand for justification renders science an intersubjective enterprise; the demand for external grounding renders it an objective enterprise. The specific determinations of the form of scientific knowledge change over time. And they change: partly because of the critical function of GPoS (until fairly recently performed by practicing scientists themselves as well as by philosophers of science) that unravels problems in the dominant conception of scientific knowledge and its in principle achievability and partly because of changes in the scientific worldview itself.

Here is a quick illustration of the interplay between the abstraction that constitutes Science-in-general and its concrete determinations. The Aristotelian conception of *episteme*—qua certain knowledge of universal and necessary truths founded, ultimately, on induction on the basis of experience and demonstration from first principles—prevailed (although not uncritically and without challenges) for a number of centuries and was criticized and doubly transformed in the seventeenth century by the competing conceptions of Francis Bacon and Rene Descartes.[6] The difference between Descartes and Bacon concerned both the origins of knowledge and the canons of theory appraisal and choice. But neither of them denied that scientific knowledge required strengthened security—although they disagreed on the degree of strength. Briefly put, Bacon was a new inductivist, whereas Descartes was a new "demonstrationist." On Bacon's *new* induction—which was contrasted to Aristotelian induction as this was deployed by the Italian neo-Aristotelians[7]—knowledge starts with experience and the compilation of a detailed natural-experimental history of the phenomena under investigation. It is then acquired by what is in effect an elimination of alternative hypotheses via the careful construction of tables of presences, absences, and concomitant variations. Descartes,

---

[6] For an account of induction in Aristotle and its reception in the Middle Ages, see my (2015).
[7] For an important recent study of this, see Sgarbi (2013).

on the other hand, had no room for induction in his philosophy of nature. But though
he appealed to a demonstrative ideal of scientific knowledge, his conception of demonstration was not Aristotelian syllogistic proof but more akin to what we nowadays
call explanation. "There is a big difference between proving and explaining," as he put
it (1991, 106), meaning that demonstration in science can proceed either from causes to
effects or from effects to causes.

Isaac Newton, as is well known, protested against the use of hypotheses in science,
which he took to be emblematic of the Cartesian approach, and he set forth the famous
methodological rules for doing science. In a letter to Roger Cotes in March 1713, Newton
(2004, 120–121) noted:

> Experimental Philosophy reduces Phenomena to general Rules and looks upon
> the Rules to be general when they hold generally in Phenomena. It is not enough
> to object that a contrary phenomenon may happen but to make a legitimate objec-
> tion, a contrary phenomenon must be actually produced. Hypothetical Philosophy
> consists in imaginary explications of things and imaginary arguments for or against
> such explications, or against the arguments of Experimental Philosophers founded
> upon Induction. The first sort of Philosophy is followed by me, the latter too much by
> Cartes, Leibniz and some others.

Experimental natural philosophy was a new way of acquiring scientific knowledge. It
stressed, among other things, the need to use the phenomena (i.e., empirical laws) as
premises, together with the laws of motion, for the derivation of force laws, and conversely. But, as Newton explained in the *Scholium* of Proposition 69 of Book I of the
Principia, the aim of all this was to argue "more safely" about the physical causes of the
phenomena.[8]

Writing about a century later, Marquis Laplace, one of the greatest Newtonians ever,
took it that he was enlarging Newton's method of experimental natural philosophy by
applying to it the then newly developed mathematical theory of probability. At the very
same time, however, he was exposing it to criticism that was supposed to unravel its
weaknesses: "Yet induction, in leading to the discovery of the general principles of the
sciences, does not suffice to establish them absolutely. It is always necessary to confirm
them by demonstrations or by decisive experiences; for the history of the sciences shows
us that induction has sometimes led to inexact results" (1951, 177). His objective was to
correct this method by strengthening it; that is, by enhancing the degree of certainty it
can attain: "then science acquires the highest degree of certainty and of perfection that it
is able to attain" (1951, 182–183).

There is a clear sense in which all those thinkers (and many who followed them) were
engaged in the same kind of enterprise and doubly so: they were engaged in *science*,
in the sense of the special mode of knowledge noted earlier, *and* they were engaged
in GPoS, too, by developing accounts of the methods appropriate for this kind of

---

[8] For an excellent recent account of Newton's approach to method, see Harper (2011).

knowledge and by criticizing competing accounts. They all had the same starting point, viz., scientific knowledge of the world, and yet the specific determinations of the form of scientific knowledge they favored were different and developed, at least partly, by criticizing competing determinations. The specific determinations of scientific knowledge change over time, and the various relevant conceptions compete with each other, both synchronically and diachronically. Yet, something remains invariant, and this is that science is a special mode of knowledge which is characterized—in its most abstract form—by the demand of increased (but as it turns out never absolute) security and reliability and by the search for methods that make this possible.

3.3 The Conceptual Toolbox of Science-in-General

Science-in-general employs a network of concepts—a toolbox, as it were—in order to characterize the special form of knowledge it is as well as its relation to the world. The knowledge achieved by science offers *explanations* of various phenomena; it unravels *causal connections*; it is supported by the *evidence*. Science employs *theories, hypotheses,* and *principles*; it issues in *predictions*; it relies on *experiments*. Theories *represent* the phenomena and rely on *theoretical concepts* and *models*. The reality science investigates is governed by *laws of nature*. The *entities* there are in the world have *properties* and *powers,* and they stand in various *relations* to each other. This is just a sample of the network of concepts employed by Science-in-general. And though it might not be the case that all of them play a role—or function—in each and every individual science, it is hard to think of any individual science—both in history and as it is practised today—without some (typically most) of these concepts.

   As noted in Section 3.1, one important function of GPoS is to offer explications of these concepts. At one point in the history of GPoS, it was thought that the task of GPoS was to offer formal-syntactic explications. This was the time—associated with some logical positivists like Carnap—of the "logic of science." The idea was—and it was a noble idea—that as it is formally specified when a proposition A entails a proposition B (given a formal language), so it should be formally specified when a proposition A *explains* a proposition B; a proposition A *confirms* a proposition B, and so forth. Part of the reason why this idea was noble was that it was meant to secure some objective content of a concept when it comes to its applications and to leave issues of interpretation open to dispute. This program failed mostly because the content of the concepts under explication was too rich to be formalized. For instance, *explanation* cannot be explicated by being formalized as a species of deduction (see the Deductive-Nomological model); not because there are no deductive-nomological explanations, but because not all explanation is deductive-nomological. But this does not suggest that the task of explication is hopeless. It suggests that, in all probability, the rich conceptual structure of the basic concepts of Science-in-general is not formalizable and that, in all probability, there will be more than one *explicata*. It may even turn out, as Carnap famously argued for the case of probability, that there are two *explicanda* (in his case: two concepts of *probability*, one referring to a measure of confirmation and another to a measure of relative frequency). We should not, I think, equate the failure of formalization with the failure of explication. Most scientific concepts are explicable without being formalizable.

More importantly, explicating a concept is bound to be entangled with the explication of a number of other concepts that fix part of its content or ground its role. For instance, when we try to explicate the concept of *mechanism*, concepts such as *power, explanation, causation, laws of nature,* and others are involved. GPoS in its explicative function works with clusters of concepts and elucidates all of them as networks. The concept of *mechanism*, for instance, has had a diachronic occurrence in science (and in various particular sciences), and there have been (and still are) important controversies about its content and its connections with other concepts such as *causation*. As I have shown in some detail in my own work (2011*b*), the prevalent accounts of mechanism, although motivated by considerations in various particular sciences, are being offered—and justly so—as general *explications* of a central concept that occurs (or can occur) in any theory or science. This case—the explication of mechanism—is a good example of the explicative function of GPoS because explication, though not offering a formalization of a concept, does offer a theoretical account of it; it unravels its inner structure (and hence what kinds of ontic commitments follow from adopting a certain explication as opposed to another); and it makes apparent the connections of this concept with others. In fact, further explicative work suggests that there are two *explicanda* when it comes to *mechanism: mechanical* mechanism and *nonmechanical* mechanism (see my 2011*a* for the details).

4 THE FOUR DIMENSION OF GPoS

Given an understanding of Science-in-general and the two functions of GPoS, we can become more specific about the role and significance of GPoS. There are four major dimensions along which the philosophical study of Science-in-general takes place: epistemic, metaphysical, conceptual, and practical.

The *epistemic dimension* deals with a family of issues that have to do with the epistemic credentials of science and, in particular, with the status of scientific knowledge. Science is not merely a theoretical and experimental practice—it is also (and has always been taken to be) a *mode of knowledge*. It purports to describe and explain the world and employs special methods which offer a systematic understanding of the natural (and the social) phenomena. It relies on theories, hypotheses, and principles that, typically but not invariably, go beyond the observable aspects of the world and describe the world as possessing a hidden-to-the-senses causal-explanatory structure. These theories are not entailed by the available evidence but are supported or licensed by the evidence. The methods employed in science are, typically, ampliative: the output of the application of the method exceeds in content the input of the method; hence, there is supposed to be a problem not just of describing the way these methods work, but also of grounding or justifying their use. And there is the perennial question of how seriously we should take the scientific image of the world as being a true or true-like image and (relatedly) whether we should think of science as aiming to offer a true image of the world in order to have a just view of science.

The *metaphysical* dimension deals with a network of issues that have to do with the implications of the scientific image of the world about the basic ontological categories of the natural (and social) world as well as its connection with the manifest image of the world and its own ontic structure. Scientific theories describe the world as being subject to laws and causal connections between various entities and processes; they postulate various entities, properties, and structures; they deal with natural kinds and species and genera; they purport to unravel the mechanisms that generate or support various functions and behaviors. Two big traditions have fought over the ontic structure of the scientific image of the world: one inflates ontology in order to explain and ground the regularity there is in the world; the other takes regularity as a brute fact; the first posits natural necessities and powers in nature; the second does away with regularity-enforcers and advances a metaphysically thin conception of laws of nature.[9] When it comes to the relation between the scientific image of the world and the manifest image, the lines of controversy concern not only the issue of ontic priority (which is the *real* image of the world?) but also the issue of their connection.

The *conceptual dimension* deals with a cluster of issues that have to do with the ways scientific theories represent the world as well as with the conditions of representational success. Science represents the world via theories, and theories employ a number of representational media, from language, to models, to diagrams, and more. Scientific theories employ, almost invariably, idealizations and abstractions in representing natural phenomena. Scientific concepts acquire their content, to a large extent at least, via the theories in which they occur. But experiments, too, play a significant role in fixing the content of scientific concepts. In fact, theories have to have empirical content, and their abstract (typically but not invariably mathematical) structure has to make contact with the world as this is given to humans in experience. The theory-ladenness of the conceptual structure of scientific theories generates a number of problems that have to do with how best to understand conceptual connections between theories, how best to evaluate the representational content of theories, and how best to understand their relation to experience and experiment.

The *practical dimension* deals with a number of issues that have to do with ethical, social, and other practical—that is, relating to scientific *praxis*—issues. Science is far from being value-free, and the investigation of the place, role, and function of values in science has been an important element of our thinking about science. Values do not function as methods do; yet they are constitutively involved in scientific judgments and in theory-choice and evaluation in science. They are epistemic values as well as social ones, and understanding their interconnections, as well as their role in securing the

---

[9] For a recent attempt to develop this view see my (2014*b*).

objectivity of science, is indispensable for having a view about science and its role in society. Feminist approaches to science have played a key role in uncovering various cognitive and social biases and have promoted the image of a socially responsible science. Issues about the ethics of science, the structure of scientific research, risk-analysis, and the role of science (and of the scientists and the scientific institutions) in policy-making have acquired prominence.

Although it is methodologically useful to separate these four dimensions, in practice, they are intertwined. More importantly, they are all indispensable for having *science* in view. The issues that form each dimension (a sample of which was offered earlier) have a long history and have been subject to important theoretical debates and controversies. Hence a *proper* engagement with GPoS should be both conceptual *and* historical. The rich history of GPoS suggests that the core of the issues that characterize philosophical engagement with science remain the same in *form* throughout history, although their context, content, and the resources available for addressing them have changed over time.

4.1 Unobservables: From Medicine to Atomism

As an example, let me mention the issue of the scientific method(s). Although it is true that there is no account of *the* method that can invariably and accurately characterize all individual sciences and all historical epochs, it is also true that science has been taken to be a *special way of getting to know the world*, which is characterized by two basic features: it is ampliative, *and* it is epistemically warranted. Hence, science is different from mere opinion, and it purports to extend and enlarge our conception of the world as this is given to us by our senses and our memory. Employing methods which succeed in satisfying these two features has always been a key philosophical and scientific endeavor. This endeavor invites two lines of theoretical investigation, as noted earlier. The first is to *devise* and *describe* models of methods that purport to satisfy these desiderata; the second is to show that the thus described models manage to meet these two (prima facie conflicting) desiderata—that is, to *justify* them. This kind of double philosophical endeavor has preoccupied most scientists and philosophers of science ever since Aristotle identified it as an issue in his *Posterior Analytics*. And, no matter how far back we go in time and how far apart in disciplines, we encounter the same *form* of the problem.

Here is a selective, but representative, historical illustration of the idea that the *form* of the problem of the method has a considerable degree of invariance. The starting point of this illustration has, I think, the advantage that it is not widely discussed among philosophers of science. It concerns the debates about method in ancient Greek medicine. For about three centuries (roughly from 300 BCE), there were two competing schools not just for practicing medicine, but for doing science, too. The ancient empirics were a group of doctors in the latter part of the third century BC who took it that, in practicing the art of medicine, doctors should rely on experience (*empeiria*) alone.[10] Empiricism was developed, at least partly, as a reaction to the proliferation of theories in medical practice. Empirics attacked what they took it to be the dominant school in medicine—the so-called rationalists or dogmatists (λογικοί/δογματικοί), who, taking cues from the Stoic theory of signs, argued that there is a special kind of rational inference—called *indicative inference*—which enables the transition from an effect to its *invisible* cause. In particular, the rationalists thought that medicine should be based on understanding and finding the causes of a disease and that this required development of theories about the nature of things and the powers of the causes and of the remedies (cf. Galen, 1985). Indicative inferences were supposed to be grounded on relations of "rational consequence" among distinct existences—relations that were discoverable by means of reason only.

Against all this, empiricists were putting forward the *sola experientia* account of medical knowledge and practice. Medicine, empiricists said, is the accumulation of empirical generalizations—called *theorems*. This accumulation is based on *autopsy* (i.e., on one's own observations) and *history* (i.e., reports on other practitioners' observations). How, then, can there be theoretical novelty in medicine? The thought that prevailed was that this was achieved by *transition to the similar* (την του ομοίου μετάβασιν/de similis transitione). This was taken to be a method (*way*) of invention (οδός ευρεύσεως) (cf. Galen 1985, 5). Yet, there was some active debate about the status of this principle, whose justification is not obvious. The dogmatists were quick to point out that this kind of transition could be justified only if it was accepted that the nature of things was such that they resembled each other (Galen 1985, 70). In response to this, the empiricists insisted that any justification of this principle—that is, of the principle "similar cause, similar effect"—should be based on experience (Galen 1985, 70). They therefore wanted to make it clear that there is no experience-independent justification of the principle that underwrites the transition to the similar.

At stake was the issue of the status and justification of methods that enlarge our image of the world by taking us beyond experience. This issue came into sharp focus in the debate about the status of indicative inference. The key problem was the knowledge of nonapparent things (άδηλα), which the dogmatists thought was necessary for medicine but was going beyond experience. The empiricist doctors joined forces with the skeptic philosophers (in fact, some were both) in order to curtail the rashness of reason, as Sextus (2000, book I §20) put it.

According to Sextus (2000, book II, §§97–98), the dogmatists divided entities into two epistemological categories: (a) pre-evident things (πρόδηλα) and (b) nonevident things (άδηλα). The former are immediately evident in experience without recourse to inference; they come of themselves to our knowledge, as Sextus put it (e.g., that it is day).

---

[10] The founder of the sect (αίρεσις), as it came to be known, was Philinus of Cos (around 260 BC), but the school spread well into the first few centuries AD. The main proponents of empiricism were Serapion (fl. 225 BC) and later Menodotus, Theodas, and Heraclides. Our knowledge of their writings comes mostly from Galen, Sextus Empiricus (who was a skeptic philosopher and an empiricist doctor), and Celsus.

The nonevident things are divided into three subcategories: (b1) those that are nonevident once and for all (καθάπαξ άδηλα; e.g., that the stars are even in number), (b2) those that are nonevident for the moment (πρὸς καιρόν άδηλα), and (b3) those that are nonevident by nature (φύσει άδηλα). The real issue was between b2 and b3, Sextus thought. Temporarily nonevident things have an evident nature but are made nonevident for the moment by certain external circumstances (e.g., for me, now, the city of the Athenians). Naturally nonevident things are those whose nature is such that they cannot be grasped in experience; for example, Sextus says, imperceptible pores—for these are never apparent of themselves but would be deemed to be apprehended, if at all, by way of something else (e.g., by sweating or something similar) (2000, book II §§97–98).

Things that are temporarily nonevident and things that are naturally nonevident are apprehended through signs. There are two kinds of sign. *Recollective* or *commemorative* (υπ ομνηστικά) signs and *indicative* (ενδεικτικά) signs, and, correspondingly, two kinds of inference. The difference between the two concerns the type of things involved in the two inferential procedures. Temporarily nonevident things are known by recollective signs, whereas naturally nonevident things are apprehended via indicative signs (2000, book II §100).

The standard ancient example of a recollective sign is smoke, as in the case "if there is smoke, there is fire." The fire is temporarily nonevident, but knowing that there's no smoke without fire, we can infer that there's fire. So recollective sign-inferences take us from an evident entity to another entity which is temporarily nonevident, but which can be made evident. Indicative-sign inferences take us from an evident thing to another thing which is naturally nonevident. The standard ancient example was the case of sweating as indicative of its nonevident cause (pores in the skin). But there was also another type of dispute: the conclusion of an indicative inference was supposed to be licensed by the fact that the effect (the sign) was *necessarily connected* with the cause (the signified) and flew out of its "proper nature and constitution" (as bodily movements are signs of the soul), as Sextus put it. The commemorative sign-inference, on the other hand, was based on the recollection of the past co-occurrence of the evident effect and the temporarily nonevident cause. Upon the observation of the evident effect, we are led to recall "the thing which has been observed together with it and is not now making an evident impression on us (as in the case of smoke and fire)." So the commemorative inference was supposed to be grounded on past experience and to implicate the memory.

The issue of the distinction between indicative inferences and commemorative ones was centrally disputed among the ancient physicians and philosophers. The empiricists employed special technical terminology to map this distinction. They called indicative sign-inference "analogism" and commemorative sign-inference "epilogism." As in the case of Sextus, the key argument of empiricists in favor of epilogism was that the kind of inference involved in it, being directed toward visible things, is "an inference common and universally used by the whole of mankind" (Galen 1985, 133, 135). Analogism, on the other hand, was not universally accepted because of the invisibility of the things involved in it (Galen 1985, 139; see also Sextus 2000 book II §102).

This kind of debate came to a halt (temporarily) when Galen provided a synthesis of the views of the two sects—a *via media*—by claiming that empiricism should be open to theory, but theory should be open to empirical testing. Reporting his own views, Galen suggested a need for a "reasoned account" to be added to what is known from experience. This reasoned account (a theory) should be tested in experience either by finding confirming instances or by disconfirming it "by what is known in perception" (1985, 89).

And yet, the form (and at least part of the content) of this debate resurfaced in the seventeenth century, in the context of the mechanical conception of nature. In his *Syntagma Philosophicum* (1658/1972), Pierre Gassendi borrowed the distinction between indicative inference and the commemorative one from the ancient doctors and philosophers and argued for principles that can act as a bridge between the macroscopic world and the corpuscularian world of the new mechanical philosophy. For Gassendi, there are circumstances under which the conclusion of indicative inference can be legitimate. This happens, he said, when the sign can exist in one circumstance; that is, when there is only one explanation of the presence of the sign (hence, when there are no competing explanations). Interestingly, this cuts through the visible/invisible distinction. Although Gassendi agreed with the ancient empirics that indicative inferences differ from commemorative ones in the type of entities implicated in them—entities invisible by nature (*occultae as nature*) versus entities temporarily invisible (*res ad tempus occultae*)—he argued that probable knowledge of invisible by nature things (such as the atoms and void) is thereby possible. Note that at the very time that Gassendi was allowing inferential knowledge of unobservables, he was transforming the conception of knowledge appropriate for "*la verite des sciences*" by allowing lesser degrees of certainty than his contemporary Descartes.

The terminology that Gassendi borrowed from Sextus to describe the method that was taken to generate and justify belief in unobservable atoms was likely lost in subsequent discussions, but the form of the issue resurfaced again and again in different contexts—most notably in the debates over atomism and the kinetic theory of gases toward the end of the nineteenth century. There, the issue, as the physical chemist Jean Perrin (1916, xii) put it, was the explanation of "the visible in terms of the invisible" and the grounds for its legitimacy. The fault-line was among those who took it that enlarging the image of the world by positing unobservables was, strictly speaking, beyond the bounds of science and in the vicinity of metaphysics and those who took it that this enlargement was indispensable for understanding the world and licensed by ordinary scientific methods. None of the sides of this debate was monolithic in its approach and arguments. In fact, new arguments turned out to be taken to be relevant and prominent—one of them being what Ludwig Boltzmann (1900) called "the historical principle," viz., that the history of science has shown that theoretical attempts to expand our scientific image of the world by positing unobservable entities have ended up in failure. Another argument, addressed by Pierre Duhem (1906, 88–89) against Gassendi, was that atoms and other unobservables are so unlike ordinary objects that any inference to them should be a matter of the imagination and not of reason. Still, the key issues of the debate were—essentially—invariant.

After his famous work on Brownian motion and its explanation on the basis of the kinetic theory of gases, Jean Perrin—who like Pierre Duhem was a scientist and not a "professional" philosopher—summarized his overall approach as follows (1916, vii): "To divine in this way the existence and properties of objects that still lie outside our ken, *to explain the complications of the visible in terms of invisible simplicity* is the function of the intuitive intelligence which, thanks to men such as Dalton and Boltzmann, has given us the doctrine of Atoms." Perrin's point, as he described in a piece on the method in physical chemistry he published in 1919—in a volume titled "The Methods of the Sciences"—was not to denigrate inductive methods (purportedly staying at the level of sensible entities) nor to promote exclusively deductive methods (aiming at explanation in terms of invisible entities). Rather, as he put it, "the indefinite wealth of nature does not lock itself in a single formula"; hence, cooperation among various methods is required. Still, trying to "eliminate completely invisible elements," like many of his contemporary advocates of energetic wanted to do, would amount to leaving the image of the world unsupported—very much like removing the pillars that support a cathedral (1919, 87).

A number of significant details of this episodic account have been glossed over.[11] But I take it that the intended message is clear enough. This is only an example of how one central problem of GPoS—whether and how science should expand and extend the image of the world by positing unobservable entities—has kept both its form and its significance throughout the centuries and the disciplines. To use current jargon, this might be taken to be the problem of scientific realism, although this is a label we and not most of the protagonists in the various incarnations of this debate are prone to use. It became a standard problem within GPoS in the twentieth century partly because this expansion of the image of the world by positing invisible elements has become the norm in most—if not all—of the sciences. It is noteworthy—and it is occasionally forgotten— that a great deal of the revival of interest in scientific realism in the early 1950s was related to issues in psychology and the status of the so-called intervening variables and hypothetical constructs in moving away from behaviorist approaches to mentality (cf. MacCorquodale and Meehl 1948). But the problem of scientific realism became a standard problem within GPoS for another reason. Precisely because of its generality and resilience, it can provide a conceptual umbrella under which a number of other problems of the scientific image of the world can be subsumed and discussed; for instance, the role of explanation in science, the relation between evidence and theory, the status of scientific truth, the rationality of theory-change, and others.[12]

5 GPoS Versus Philosophy of X?

There is no doubt that the philosophies of the various sciences have flourished over the past forty or so years. There is also no doubt that GPoS was physics-centered until fairly recently. But this trend has been reversed, and GPoS has been more pluralistic in its

---

[11] I have related most of the details in 2011c.

[12] For the issue of scientific realism in the twentieth century, see my (2011a).

relations with the individual sciences. But although, as noted already, the subject matter of GPoS is Science-in-general and not the various individual sciences, GPoS and the philosophies of the various sciences form a seamless web.

What we normally call science X (physics, biology, chemistry, economics, psychology, etc.) is itself a kind of rational abstraction. It's hard to come up with an interesting definition of, say, physics; the subject-matter of physics—as well as the name—has changed over time. The expression "natural philosophy" was used to refer to physics until roughly the middle of the nineteenth century. In 1798, the Académie Royale des Sciences of Paris set the following topic as a prize competition in physics: "the nature, form and used of the liver in the various classes of animals." In 1918, when Max Planck was awarded the Nobel Prize in physics, the citation of the Swedish Academy of Sciences noted that the award was given "in recognition of [Planck's] epoch-making investigations into the quantum theory." In practice, physics comprises a number of fields and disciplines, from mechanics to high-energy physics, to solid-state physics, to physical chemistry, and a good many others. Physics, if anything, is a cluster of sciences and disciplines, as these have evolved over time. Philosophy of physics is the philosophy-of-physics-in-general! And, as we all know, there are various philosophies of sub-X within the philosophy of X: philosophy of spacetime, philosophy of quantum mechanics, philosophy of statistical mechanics, philosophy of string theory, and so on. Philosophy of physics, then, stands to the philosophies of various physical subdisciplines as GPoS stands to philosophies of the various sciences. The same holds for the philosophy of biology, for example. Biology, strictly speaking, is a cluster of sciences or disciplines: ecology, paleontology, synthetic biology, and others. Actually, a new cluster of biological sciences has been formed: "life sciences." The philosophy of biology-in-general stands to the philosophies of various biological subdisciplines as GPoS stands to philosophies of the various sciences. And because there are no impermeable boundaries among the philosophy of X and the philosophies of sub-X, so there are no impermeable barriers between GPoS and the Philosophies of X, where X ranges over individual sciences. This permeability is strengthened by the fact that new hybrid disciplines are being formed (e.g., molecular biology or biophysics). And, with them, there is transfer of methods, theories, and techniques as well as conceptual problems.

The philosophy of X-in-general deals with some issues that are not proper tasks of GPoS. One such issue is the interpretation of the various theories of X; for any theory T in X, there is the question of what the world is like according to T. If there are competing interpretations—or even theories—there is the further issue of how they are related. And there is always the issue of the basic conceptual structure of a theory T in X—that is, of the content of the key explanatory concepts of T. And, of course, there is the issue of how the subdisciplines of X are related to each other. Hence, there are questions and issues that belong—more or less squarely—to the philosophies of X, where X ranges over individual sciences. In the philosophy of biology, for instance, there are issues concerning the concept of reproductive fitness, or the ontological status of the probabilities used in population biology, or the understanding of genes in Mendelian and molecular genetics. In the philosophy of chemistry, there are issues concerning the status of explanations in terms of electron orbitals, whether there are some irreducible chemical laws (e.g., the periodic law of the elements), the issue of "chemical substances" (e.g., whether the world consists of one kind of matter or of a great variety of materials), and the relation between chemistry and physics. In the philosophy of physics, there are the famous issues of the interpretations of quantum mechanics, the nature of space and time, the status of probabilities in statistical mechanics, and many others.

Dealing with philosophical issues that arise within the philosophy of X requires, among other things, an engagement with the sciences or the disciplines themselves that are investigated. The philosophy of X is in many ways the abstract and theoretical end of X itself. But the philosophy of X (or the philosophy of X *in general*, as I would like to put it), *qua* philosophy of science, employs and explores the conceptual resources of GPoS. It performs the two functions of GPoS, the explicative and the critical, at the level of X. In many ways, it deals with problems of GPoS as they are concretized in individual sciences. The idea of explanation in evolutionary theory is a case in point. Elliott Sober (1984), for instance, relied on the composition of forces in dynamics in order to account for the change in gene frequencies over time as the result of different "forces," such as selection, drift, and mutation. And so is the issue of laws of nature: are there laws in biology or chemistry? And the issue of the structure of theories in biology. For instance, evolutionary theory has been taken to support the "semantic view of theories."

In fact, there are areas in which the philosophical investigation of issues in the philosophy of X have exerted serious influence on how the relevant issues are treated within GPoS. One important such issue has been the nature of biological species. These do not conform to standard philosophical approaches to natural kinds, especially when the latter are viewed as possessing essences. A fruitful discussion has opened up about how best to reconceptualize natural kinds or even whether the traditional reliance on natural kinds should be rejected.

But we should not lose sight of the fact that there are philosophical questions that belong—more or less squarely—to GPoS. I have already offered an outline of the issues dealt with in the four dimensions of GPoS. But I want to add two issues that I think are really peculiar to GPoS.

The first, unsurprisingly, is the scientific realism debate. This debate does not concern any science in particular. And this is so because the heart of the problem is about Science-in-general: can and should we trust the scientific image of the world? Think of the challenge of pessimistic induction, for example. This concerns the very idea that science is in a position to offer knowledge of the world. The challenge, as is well known, utilizes the history of science to undercut confidence in the current scientific image of the world. It does that by noting that there is a pattern of theoretical-explanatory failure in the history of science that warrants the claim that, in all probability, current scientific theories will be abandoned and replaced in the years to come despite their impressive empirical success. The details of this debate can be found in my work (1999) and Stanford (2006). The point I want to stress is that this kind of argument is a global argument about science and not about the particular sciences.

In fact, if we were to break it down to subarguments concerning the various individual sciences (e.g., biology, physics, chemistry etc.), we might be able to undermine it. For, even if there is radical theory-change in physics, there is not, it seems, a pattern of radical theory-change in biology or in chemistry. This kind of dividing strategy would possibly warrant the claim that we can be more confident about our current theoretical knowledge in biology than we are in physics. But this kind of strategy would not remove the need to globally address the argument from the pessimistic induction. For unless there are relevant differences in the ways we acquire scientific knowledge in, say, physics and biology, the alleged failures in physics would still undermine the alleged successes in biology or chemistry. The strength of the argument from the pessimistic induction lies precisely in its undermining the relations between explanatory and empirical success and truth. Hence, the attempts to neutralize the pessimistic induction should aim to restore a connection between empirical success and truth, even if this connection is subtler and more sensitive to the history of science than, perhaps, was accepted before.

As I have argued elsewhere (2009, 75–77), when it comes to the realism debate, a key task of GPoS is to address the issue of balancing two types of evidence for or against a scientific theory. The first kind is whatever evidence there is in favor (or against) a specific scientific theory. This evidence has to do with the degree of confirmation of the theory at hand. It is *first-order evidence* and is typically associated with whatever scientists take into account when they form an attitude toward a theory. It can be broadly understood to include some of the theoretical virtues of the theory at hand—of the kind that typically go into plausibility judgments associated with assignment of prior probability to theories. The second kind of evidence (*second-order evidence*) comes from the past record of scientific theories and/or from meta-theoretical (philosophical) considerations that have to do with the reliability of scientific methodology. It concerns not particular scientific theories, but science as a whole. This second-order evidence feeds claims such as those that motivate the pessimistic induction. Actually, this second-order evidence is multifaceted—it is negative (showing limitations and shortcomings) as well as positive (showing how learning from experience can be improved).

This balancing cannot be settled in the abstract—that is, without looking into the details of the various cases. But the need for balancing shows how GPoS and the Philosophies of X can work in harmony. For, in most typical cases, settling issues about the first-order evidence that exists for a theory of X will be a matter of the philosophy of X (subject, as we have noted in Section 2.1, to a general account of the relation between evidence and theory), whereas settling issues about the second-order evidence that exists from the history of science or the conceptual structure of Science-in-general will be a matter of GPoS.

The second issue that predominantly belongs to GPoS concerns the *very idea of a scientific image of the world*. GPoS offers the space in which the various images of the world provided by the individual sciences are fused together into a stereoscopic view of reality. Strictly speaking, the various sciences offer us perspectives on reality. They employ different kinds of taxonomic categories and conceptualize the world by means of different structures of concepts. The category of "chemical bond," for instance, belongs to a different conceptual structure than the category "gene" or "quark." Still, there is a presumption—to say the least—that the various perspectives offered by the various sciences or theories are perspectives *of the same world*. Hence, there is the need to put together the scientific image of the world, to look at the various interconnections among the "partial" images generated by the individual sciences, and to clear up tensions and conflicts. This is precisely the kind of job that GPoS—and only GPoS—can do. It offers a more global (but not absolute) perspective on reality—for seeing the whole picture. Even if there is no way to put together a coherent and unified image of the world, even if, that is, the scientific image is characteristically disunified and disconnected, this can be "seen" only within GPoS.

GPoS offers the toolkit of concepts that are needed for establishing the relations among various partial images. Employing the expression "special sciences" to refer to sciences other than fundamental physics did play a useful role in promoting anti-reductivist approaches to the relations between the various sciences, but it was a misnomer! There are no special sciences—or, all sciences are special in that they offer perspectives on reality. The matter of "putting together" these perspectives is a central task of GPoS, but here again GPoS should be engaged in both of its functions. It should aim to explicate (and examine the conceptual networks) key concepts such as reduction, supervenience, fundamentality, emergence, dependence, and the like. It should aim to criticize the various theoretical accounts concerning the relations among the various sciences—as these were determined both conceptually and historically.

6 Conclusion

I have argued that GPoS can work in harmony with the philosophies of the various sciences, and that, strictly speaking, the philosophies of the various individual sciences require the framework and functions of GPoS. The particular sciences, and the various theories in them, are in many ways dissimilar to each other, both in their historical development and their current conceptual and methodological structures. Still, science need not have an essence for it to be the object of philosophical study. Nor should the philosophical study of science be merely descriptive of whatever is taught in science departments. GPoS has two important functions vis-à-vis Science-in-general: one explicative and another critical. Science-in-general is itself a "rational abstraction," the unity of which arises from the identity of the subject of the various sciences (viz., that science is special *mode of knowledge of nature*) and the identity of the object of the various sciences (viz., *nature itself*). As Leibniz states in the epigram of this chapter: "Similarity in variety, that is, diversity compensated by identity."

REFERENCES

Aristotle. (1984). *Physics*. R. P. Hardie and P. K. Gaye (trans.). *The Complete Works of Aristotle*, J. Barnes (ed.). Vol. 1. (Princeton, NJ: Princeton University Press).

Boltzmann, Ludwig. (1900). "The Recent Development of Method in Theoretical Physics." *The Monist* 11: 226–257.

Descartes, René. (1991). *The Philosophical Writings of Descartes*. Vol. 3: The Correspondence. J. Cottingham, R. Stoothoff, D. Murdoch, and A. Kenny (trans.) (Cambridge: Cambridge University Press).

Duhem, Pierre. (1906). *The Aim and Structure of Physical Theory*. 2nd ed. 1914, P. Wiener (trans.), 1954. (Princeton, NJ: Princeton University Press).

Galen. (1985). *Three Treatises on the Nature of Science*. R. Walzer and M. Frede (trans.). Indianapolis: Hackett.

Gassendi, Pierre. (1658/1972). The Syntagma. In *The Selected Works of Pierre Gassendi*. Craig Brush (trans.). (New York: Johnson Reprints).

Harper, William L. (2011). *Isaac Newton's Scientific Method*. (Oxford: Oxford University Press).

Hempel Carl. (1951). "The Concept of Cognitive Significance: A Reconsideration." *Proceedings and Addresses of the American Academy of Arts and Sciences* 80: 61–77.

Hoyningen-Huene, Paul. (2013). *Systematicity: The Nature of Science*. (New York: Oxford University Press).

Kitcher, Philip. (2013). "Toward a Pragmatist Philosophy of Science." *Theoria* 77: 185–231.

Kuhn, T. S. (1970). *The Structure of Scientific Revolutions*. (2nd enlarged ed., first published 1962). (Chicago: University of Chicago Press).

Kuhn, T. S. (1977). *The Essential Tension. Selected Studies in Scientific Tradition and Change*. (Chicago: University of Chicago Press).

Lakatos, Imre. (1970). "Falsification and the Methodology of Scientific Research Programmes." In I. Lakatos and A. Musgrave (eds.), *Criticism and the Growth of Knowledge*. (Cambridge: Cambridge University Press), 91–196.

Ladyman, James. (2013). "Toward a Demarcation of Science from Pseudoscience." In M. Pigliucci and M. Boudry (eds.), *Philosophy of Pseudoscience: Reconsidering the Demarcation Problem*. (Chicago: University of Chicago Press), 45–59.

Laplace, Pierre Simon. (1951). *A Philosophical Essay on Probabilities*. (New York: Dover).

Laudan, Larry. (1996). *Beyond Positivism and Relativism*. (Boulder, CO: Westview Press).

MacCorquodale, Kenneth, and Meehl, Paul. (1948). "On a Distinction between Hypothetical Constructs and Intervening Variables." *Psychological Review* 55: 95–107.

Marx, Karl. (1857–58). *Grundrisse. Foundations of the Critique of Political Economy (Rough Draft)*. Martin Nicolaus (trans., 1973). (Harmondsworth, UK: Penguin Books).

Newton, Isaac. (2004). *Philosophical Writings*. Andrew Janiak (ed.). (Cambridge: Cambridge University Press).

Paul, Harry W. (1985). *From Knowledge to Power. The Rise of the Science Empire in France, 1860–1939*. (Cambridge: Cambridge University Press).

Perrin, Jean. (1916). *Atoms*. D. L. Hammick (trans.). (London: Constable and Company).

Perrin, Jean. (1919). "Chimie Physique." In B. Baillaud (ed.), *De la Methode dans le Sciences*. (Paris: Librairie Felix Alcan), 65–88.

Pigliucci, Massimo. (2013). "The Demarcation Problem: A (Belated) Response to Laudan." In M. Pigliucci and M. Boudry (eds.), *Philosophy of Pseudoscience: Reconsidering the Demarcation Problem*. (Chicago: University of Chicago Press), 9–28.

Poincaré, Henri. (1902). *La Science et L'Hypothése* (1968 reprint). (Paris: Flammarion).

Popper, Karl. (1963). *Conjectures and Refutations* (3rd ed. revised 1969). (London: RKP).

Popper, Karl. (1994). *The Myth of the Framework: In Defense of Science and Rationality*. London and New York: Routledge.

Psillos, Stathis. (1999). *Scientific Realism: How Science Tracks Truth*. (London and New York: Routledge).

Psillos, Stathis. (2002). "Simply the Best: A Case for Abduction." In A. C. Kakas and F. Sadri (eds.), *Computational Logic: From Logic Programming into the Future* (Berlin-Heidelberg: Springer-Verlag), 605–625.

Psillos, Stathis. (2007). "The Fine Structure of Inference to the Best Explanation." *Philosophy and Phenomenological Research* 74: 441–448.

Psillos, Stathis. (2009). *Knowing the Structure of Nature*. (London: Palgrave-MacMillan).

Psillos, Stathis. (2011a). "Scientific Realism with a Humean Face." In J. Saatsi and S. French (eds.), *The Continuum Companion to Philosophy of Science*. (London: Continuum), 75–95.

Psillos, Stathis. (2011b). "The Idea of Mechanism." In P. McKay, F. Russo, and J. Williamson (eds.), *Causality in the Sciences*. (Oxford: Oxford University Press), 771–788.

Psillos, Stathis. (2011c). "Moving Molecules Above the Scientific Horizon: On Perrin's Case for Realism." *Journal for General Philosophy of Science* 42: 339–363.

Psillos, Stathis. (2012a). "Reason and Science." In C. Amoretti and N. Vassallo (eds.), *Reason and Rationality*. (Frankfurt: Ontos Verlag), 97–115.

Psillos, Stathis. (2012b). "What Is General Philosophy of Science?" *Journal for General Philosophy of Science* 43: 93–103.

Psillos, Stathis. (2014a). "Conventions and Relations in Poincaré's Philosophy of Science." *Methode-Analytic Perspectives* 4: 98–140.

Psillos, Stathis. (2014b). "Regularities, Natural Patterns and Laws of Nature." *Theoria* 79: 9–27.

Psillos, Stathis. (2015). "Induction and Natural Necessity in the Middle Ages." *Philosophical Inquiry* 39: 92–134.

Quine, W. V. (1951). "Two Dogmas of Empiricism." In *From a Logical Point of View* (Cambridge, MA: Harvard University Press).

Schlick, Moritz. (1932). "Positivismus und Realismus." *Erkenntnis* 3: 1–31. Translated as "Positivism and Realism" in A. J. Ayer (ed.), *Logical Positivism*, 1960 (Glencoe, NY: Free Press).

Sextus Empiricus. (2000). *Outlines of Scepticism*. J. Annas and J. Barnes (eds.). (Cambridge: Cambridge University Press).

Sgarbi, Marco. (2013). *The Aristotelian Tradition and the Rise of British Empiricism*. (Dordrecht: Springer).

Sober, Elliott. (1984). *The Nature of Selection: Evolutionary Theory in Philosophical Focus*. (Cambridge, MA: MIT Press).

Stanford, P. Kyle. (2006). *Exceeding Our Grasp: Science, History, and the Problem of Unconceived Alternatives*. (Oxford: Oxford University Press).

Yeo, Richard. (1993). *Defining Science: William Whewell, Natural Knowledge, and Public Debate in Early Victorian Britain*. (Cambridge: Cambridge University Press).

PART II

TRADITIONAL TOPICS

CHAPTER 8

··

CAUSATION IN SCIENCE

··

JAMES WOODWARD

1 INTRODUCTION

THE subject of causation in science is vast, and any article-length treatment must necessarily be very selective. In what follows I have attempted, insofar as possible, to avoid
producing yet another survey of the standard philosophical "theories" of causation and
their vicissitudes. (I have nonetheless found some surveying inescapable—this is mainly
in Section 3.) Instead, I have tried to discuss some aspects of this topic that tend not to
make it into survey articles and to describe some new developments and directions for
future research. My focus throughout is on epistemic and methodological issues as they
arise in science, rather than on the "underlying metaphysics" of the causal relation.

The remainder of this article is organized as follows. After some orienting remarks
(Section 2), Section 3 describes some alternative approaches to understanding causation. I then move on to a discussion of more specific ideas about causation and causal
reasoning found in several areas of science, including causal modeling procedures
(Sections 4 and 5), and causation in physics (Section 6).

2 OVERVIEW

There are controversies in philosophy and philosophy of science not only about which
(if any) account of causation is correct but also about the role of causation (and, relatedly, causal explanation[1]) in various areas of science. For example, an influential strain of
thought maintains that causal notions play little or no legitimate role in physics (Section 6).

---

[1] In what follows I do not sharply distinguish between reasoning involving causation and causal
explanation. Think of a causal explanation as just an assembly of information about the causes of some
explanandum.

There has also been a recent upsurge of interest in (what are taken to be) noncausal forms of explanation, not just in physics but also in sciences like biology. A common theme (or at least undercurrent) in this literature is that causation (and causal reasoning) is less central to much of science than many have supposed. I touch briefly on this issue later but for purposes of this article baldly assert that this general attitude/assessment is wrong-headed, at least for areas of science outside of physics. There are indeed noncausal forms of explanation, but causal reasoning plays a central role in many of areas of science, including the social, behavioral, and biological sciences, as well as in portions of statistics, artificial intelligence, and machine learning. Philosophers of science should engage with this literature rather than ignore it or attempt to downplay its significance.

3 THEORIES OF CAUSATION

3.1 Regularity Theories

The guiding idea is that causal claims assert the existence of (or at least are "made true" by) a regularity linking cause and effect. Mackie's (1974) INUS condition account is an influential example: C^2 causes E if and only if C is a *nonredundant* part (where C is typically but not always by itself *insufficient* for E) of a *sufficient* (but typically not necessary) condition for E. The relevant notions of sufficiency, necessity, and nonredundancy are explicated in terms of regularities: short circuits S cause fires F, because S is a nonredundant part or conjunct in a complex of conditions (which might also include the presence of oxygen O), which are sufficient for F in the sense that $S.O$ is regularly followed by F. S is nonredundant in the sense that if one were to remove S from the conjunct $S.O$, F would not regularly follow, even though S is not strictly necessary for F since F may be caused in some other way—for example, through the occurrence of a lighted match L and O, which may also be jointly sufficient for F. In the version just described, Mackie's account is an example of a *reductive* (sometimes called "Humean") theory of causation in the sense that it purports to reduce causal claims to claims (involving regularities, just understood as patterns of co-occurrence) that apparently do not make use of causal or modal language. Many philosophers hold that reductive accounts of causation are desirable or perhaps even required, a viewpoint that many nonphilosophers do not share.

As described, Mackie's account assumes that the regularities associated with causal claims are deterministic. It is possible to construct theories that are similar in spirit to Mackie's but that assume that causes act probabilistically. Theories of this sort, commonly called *probabilistic* theories of causation (e.g., Eells 1991), are usually formulated

---

[2] I use uppercase letters to describe repeatable *types* of events (or whatever one thinks the relata of causal relationships are) and lowercase letters to describe individual instances or *tokens* of such relata.

in terms of the idea that C causes E if and only if C raises the probability of E in comparison with some alternative situation in which C is absent:

$$Pr\left(E\,/\,C.K\right) > Pr\left(E\,/\,{-}C.K\right) \text{for some appropriate } K. \tag{3.1}$$

(It is far from obvious how to characterize the appropriate K, particularly in non-reductive terms, but I put this consideration aside in what follows.) Provided that the notion of probability is itself understood nonmodally—that is, in terms of relative frequencies—(3.1) is a probabilistic version of a regularity theory. Here what (3.1) attempts to capture is the notion of a *positive* or *promoting* cause; the notion of a *preventing* cause might be captured by reversing the inequality in (3.1).

A general problem with regularity theories, both in their deterministic and probabilistic versions, is that they seem, prima facie, to fail to distinguish between causation and noncausal correlational relationships. For example, in a case in which C acts as a common cause of two joint effects X and Y, with no direct causal connection between X and Y, X may be an INUS condition for Y and conversely, even though, by hypothesis, neither causes the other. Parallel problems arise for probabilistic versions of regularity theories.

These "counterexamples" point to an accompanying methodological issue: causal claims are (at least apparently) often underdetermined by evidence (at least evidence of a sort we can obtain) having to do just with correlations or regularities—there may be a number of different incompatible causal claims that are not only consistent with but even imply the same body of correlational evidence. Scientists in many disciplines recognize that such underdetermination exists and devise ways of addressing it—indeed, this is the primary methodological focus of many of the accounts of causal inference and learning in the nonphilosophical literature. Pure regularity or correlational theories of causation do not seem to address (or perhaps even to recognize) these underdetermination issues and in this respect fail to make contact with much of the methodology of causal inference and reasoning.

One possible response is that causal relationships are just regularities satisfying additional conditions—that is, regularities that are pervasive or "simple" in contrast to those that are not. Pursuing this line of thought, one comes naturally to the view that causal regularities either are or (at least) are "backed" or "instantiated" by *laws* where laws are understood as regularities meeting further conditions, as in the best systems analysis of law (Lewis 1999).

Quite apart from other difficulties, this proposal faces the following problem from a philosophy of science viewpoint: the procedures actually used in the various sciences to infer causal relationships from other sorts of information (including correlations) do not seem to have much connection with the strengthened version of the regularity theory just described. As illustrated later, rather than identifying causal relationships with some subspecies of regularity satisfying very general conditions concerning pervasiveness and so on, these inference techniques instead make use of much more specific assumptions linking causal claims to information about statistical and other kinds of independence relations, to experimentation, and to other sorts of constraints. Moreover, these assumptions connecting correlational information with causal conclusions are not formulated in terms of purely "Humean" constraints. Assuming (as I do in what follows) that one task of the philosopher of science is to elucidate and possibly suggest improvements in the forms that causal reasoning actually takes in the various sciences, regularity theories seem to neglect too many features of how such reasoning is actually conducted to be illuminating.

3.2 Counterfactual Theories

Another natural idea, incorporated into many theories of causation, both within and outside of philosophy, is that causal claims are connected to (and perhaps even reduce to) claims about counterfactual relationships. Within philosophy a very influential version of this approach is Lewis (1973). Lewis begins by formulating a notion of counterfactual dependence between individual events: e counterfactually depends on event c if and only if, (3.2) if c were to occur, e would occur; and (3.3) if c were not to occur, e would not occur. Lewis then claims that c causes e if and only if there is a *causal chain* from c to e: a finite sequence of events $c, d, f. . .e, . . .$ such that d causally depends on c, f on $d, . . .$ and e on f. (Lewis claims that this appeal to causal chains allows him to deal with certain difficulties involving causal preemption that arise for simpler versions of a counterfactual theory.) The counterfactuals (3.2) and (3.3) are in turn understood in terms of Lewis' account of possible worlds: roughly "if c were the case, e would be the case" is true if and only if some possible worlds in which c and e are the case are "closer" or more similar to the actual world than any possible world in which c is the case and e is not. Closeness of worlds is understood in terms of a complex similarity metric in which, for example, two worlds that exhibit a perfect match of matters of fact over most of their history and then diverge because of a "small miracle" (a local violation of the laws of nature) are more similar than worlds that do not involve any such miracle but exhibit a less perfect match. Since, like the INUS condition account, Lewis aspires to provide a reductive theory, this similarity metric must not itself incorporate causal information, on pain of circularity. Using this metric, Lewis argues that, for example, the joint effects of a common cause are not, in the relevant sense, counterfactually dependent on one another and that while effects can be counterfactually dependent on their causes, the converse is not true. Counterfactual dependence understood in this "non-backtracking" way thus tracks our intuitive judgments about causal dependence.

Lewis's similarity metric functions to specify what should be changed and what should be "held fixed" in assessing the truth of counterfactuals. For example, when I claim, that if (contrary to actual fact) I were to drop this wine glass, it would fall to the ground, we naturally consider a situation s (a "possible world") in which I release the glass, but in which much else remains just as it is in the actual world—gravity still operates, if there are no barriers between the glass and ground in the actual world, this is also the case in s and so on.

As is usual in philosophy, many purported counterexamples have been directed at Lewis's theory. However, the core difficulty from a philosophy of science perspective is this: the various criteria that go into the similarity metric and the way in which these trade-off with one another are far too vague and unclear to provide useful guidance for the assessment of counterfactuals in most scientific contexts. As a consequence, although one occasionally sees references in passing to Lewis's theory in nonphilosophical discussions of causal inference problems (usually when the researcher is attempting to legitimate appeals to counterfactuals), the theory is rarely if ever actually used in problems of causal analysis and inference.[3]

Awareness of this has encouraged some philosophers to conclude that counterfactuals play no interesting role in understanding causation or perhaps in science more generally. Caricaturing only slightly, the inference goes like this: counterfactuals can only be understood in terms of claims about similarity relations among Lewisian possible worlds, but these are too unclear, epistemically inaccessible, and metaphysically extravagant for scientific use. This inference should be resisted. Science is full of counterfactual claims, and there is a great deal of useful theorizing in statistics and other disciplines that explicitly understands causation in counterfactual terms but where the counterfactuals themselves are not explicated in terms of a Lewisian semantics. Roughly speaking, such scientific counterfactuals are instead represented by (or explicated in terms of) devices like equations and directed graphs, with explicit rules governing the allowable manipulations of contrary to fact antecedents and what follows from these. Unlike the Lewisian framework, these can be made precise and applicable to real scientific problems.

One approach, not without its problems but that provides a convenient illustration of a way in which counterfactuals are used in a statistical context, is the *potential outcomes* framework for understanding causal claims developed by Rubin (1974) and Holland (1986) and now widely employed in econometrics and elsewhere. In a simple version, causation is conceptualized in terms of the responses to possible treatments imposed on different "units" u_i. The *causal effect* of treatment t with respect to an alternative treatment t' for u_i is defined as $Y_t(u_i) - Y_{t'}(u_i)$ where $Y_t(u_i)$ is the value Y would have assumed for u_i if it had been assigned treatment t and $Y_{t'}(u_i)$ is the value Y would have assumed had u_i instead been assigned treatment t'. (The definition of causal effect is thus explicitly given in terms of counterfactuals.) When dealing with a population of such units, and thinking of $Y_t(u)$ and $Y_{t'}(u)$ as random variables ranging over the units, the average or expected effect of the treatment can then be defined as $E[Y_t(u) - Y_{t'}(u)]$. No semantics of a Lewisian sort is provided for these counterfactuals, but the intuitive idea is that we are to think of $Y_t(u)$ and so on as measuring the response of u in a well-conducted experiment in which u is assigned t—this in turn (in my view) can be naturally explicated by appeal to the notion of an *intervention*, discussed later.

[3] Cf. Heckman (2005).

The absence of a semantics or truth conditions (at least of a sort philosophers expect) for the counterfactuals employed may seem unsatisfying, but in fact the previous characterization is judged by many researchers to be methodologically useful in several ways. For example, the "definitions" given for the various notions of causal effect, even if not reductive, can be thought of as characterizing the *target* to which we are trying to infer when we engage in causal inference. They also draw attention to what Rubin (1974) and Holland (1986) describe as the "fundamental problem" of causal inference, which is that for any given unit one can observe (at most) either $Y_t(u_i)$ or $Y_{t'}(u_i)$ but not both. If, for example, $Y_t(u_i)$ but not $Y_{t'}(u_i)$ is observed, then it is obvious that for reliable causal inference one requires additional assumptions that allow one to make inferences about $Y_{t'}(u_i)$ from other sorts of information—for example, from the responses $Y_{t'}(u_j)$ for other units $u_i \neq u_j$. One can use the potential response framework to more exactly characterize the additional assumptions and information required for reliable inference to causal conclusions involving quantities like $Y_t(u_i)-Y_{t'}(u_i)$, or $E[Y_t(u)-Y_{t'}(u)]$. For example, a sufficient (but not necessary) condition for reliable (unbiased) estimation of $E[Y_t(u)-Y_{t'}(u)]$ is that the treatment variable $T=t, t'$ be independent of the counterfactual responses $Y(u)$. Another feature built into the Rubin-Holland framework is that (as is obvious from the previous definition of individual-level causal effect) causal claims are always understood as having a comparative or contrastive structure (i.e., as claims that the cause variable C taking one value in comparison or contrast to C taking some different value causes the difference or contrast between the effect variable's taking one value rather than another). A similar claim is endorsed by a number of philosophers. As Rubin and Holland argue, we can often clarify the content of causal claims by making this contrastive structure explicit.

Interventionist or *manipulationist* accounts of causation can be thought of as one particular version of a nonreductive counterfactual theory, in some respects similar in spirit to the Rubin-Holland theory. The basic idea is that C (a variable representing the putative cause) causes E (a variable representing the effect) if and only if there is some possible intervention on C such that if that intervention were to occur, the value of E would change. Causal relationships are in this sense relationships that are potentially exploitable for purposes of manipulation. Heuristically, one may think of an intervention as a sort of idealized experimental manipulation that is unconfounded for the purposes of determining whether C causes E. ("Unconfounded" means, roughly, that the intervention is not related to C or E in a way that suggests C causes E when they are merely correlated.) This notion can be given a more precise technical definition that makes no reference to notions like human agency (Woodward 2003). An attraction of this account is that it makes it transparent why experimentation can be a particularly good way of discovering causal relationships.

One obvious question raised by the interventionist framework concerns what it means for an intervention to be "possible." There are many subtle issues here that I lack space to address, but the basic idea is that the intervention operation must be well defined in the sense there are rules specifying which interventions the structure of the system of interest permits and how these are to be modeled. Which interventions are possible in this sense in turn affects the counterfactual and causal judgments we make. Consider a gas is enclosed in a cylinder with a piston that can either be locked in position (so that its volume is fixed) or allowed to move in a vertical direction.[4] A weight rests on the piston. Suppose first (i) the piston is locked and the gas placed in a heat bath of higher temperature and allowed to reach equilibrium. It seems natural to say that the external heat source causes the temperature of the gas, and the temperature and volume together cause the pressure. Correspondingly, if the temperature of the heat bath or the volume had been different, the pressure would have been different. Contrast this with a situation (ii) in which the heat source is still present, the weight is on the piston, the piston is no longer fixed, and the gas is allowed to expand until it reaches equilibrium. Now it seems natural to say that the weight influences the pressure, and the pressure and temperature cause the new volume. Correspondingly, if the weight had been different, the volume would have been different. Thus the way in which various changes come about and what is regarded as fixed or directly controlled and what is free to vary (legitimately) influence causal and counterfactual judgment. The interventionist invocation of "possible" interventions reflects the need to make the considerations just described explicit, rather than relying indiscriminately on the postulation of miracles.

3.3 Causal Process Theories

The theories considered so far are all "difference- making" accounts of causation: they embody the idea that causes "make a difference" to their effects, although they explicate the notion of difference-making in different ways. In this sense, causal claims involve a *comparison* between two different possible states of the cause and effect, with the state of the effect depending on the state of the cause.

By contrast, causal process theories, at least on a natural interpretation, do not take difference-making to be central to causation. In the versions of this theory developed by Salmon (1984) and Dowe (2000), the key elements are *causal processes* and *causal interactions*. In Salmon's version, a causal process is a physical process, such as the movement of a baseball through space, that transmits energy and momentum or some other conserved quantity in a spatiotemporally continuous way. Causal processes are "carriers" of causal influence. A causal interaction is the spatiotemporal intersection of two or more causal processes that involves exchange of a conserved quantity such as energy/momentum, as when two billiard balls collide. Causal processes have a limiting velocity of propagation, which is the velocity of light. Causal processes contrast with pseudo-processes such as the movement of the bright spot cast by a rotating search light inside a dome on the interior surface of the dome. If the rate of rotation and radius of the dome

---

[4] An example of this sort is also described in Hausman, D., Stern, R. and Weinberger, N. (forthcoming). However, I would put the emphasis a bit differently from these authors: it is not that the system lacks a graphical representation but rather that the causal relationships in the system (and the appropriate graphical representation) are different depending on what is fixed, varying and so on.

are large enough, the velocity of the spot can exceed that of light, but the pseudo-process is not a carrier of causal influence: the position of the spot at one point does not cause it to appear at another point. Rather, the causal processes involved in this example involve the propagation of light from the source, which of course respects the limiting velocity c.

Causal processes theories are often described as "empirical" or "physical" theories of causation. According to proponents, they are not intended as conceptual analyses of the notion of causation; instead the idea is that they capture what causation involves in the actual world. It may be consistent with our *concept* of causation that instantaneous action at a distance is conceptually possible, but as matter of empirical fact we do not find this in our world.

According to advocates, the notions of transmission and exchange/transfer of conserved quantities used to characterize causal processes and intersections can be elucidated without reference to counterfactuals or other difference-making notions: whether some process is a causal process or undergoes a causal interaction is a matter that can be read off just from the characteristics of the processes themselves and involves no comparisons with alternative situations. We find this idea in Ney (2009), who contrasts "physical causation," which she takes to be captured by something like the Salmon/ Dowe approach and which she thinks characterizes the causal relationships to be found in "fundamental physics" with difference-making treatments of causation, which she thinks are characteristic of the special sciences and folk thinking about causation.

One obvious limitation of causal process theories is that it is unclear how to apply them to contexts outside of (some parts) of physics in which there appear to be no analogues to conservation laws. (Consider "increases in the money supply cause inflation.") One possible response, reflected in Ney (2009), is to embrace a kind of dualism about causation, holding that causal process theories are the right story about causation in physics, while some other, presumably difference-making approach is more appropriate in other domains. Of course this raises the question of the relationship, if any, between these different treatments of causation. Following several other writers (Earman 2014, Wilson forthcoming), I suggest in Section 6 that intuitions underlying causal processes theories are best captured by means of the contrast between systems governed by hyperbolic and other sorts (elliptical, parabolic) of differential equations. Since all such equations describe difference-making relationships, causal processes should be understood as involving one particular kind of difference-making relationship.

3.4 Causal Principles

So far we have been considering theories that attempt to provide "elucidations" or "interpretations" or even "definitions" of causal concepts. However, a distinguishable feature of many treatments of causation is that they make assumptions about how the notion of causation connects with other notions of interest—assumptions embodying constraints or conditions of some generality about how causes behave (at least typically), which may or may not be accompanied by the claim that these are part of "our concept" of causation. I call these *causal principles*. Examples include constraints on the speed of propagation of causal influence (e.g., no superluminal signaling) and conditions connecting causal and probabilistic relationships—such as the Causal Markov condition described later. Some accounts of causation are organized around a commitment to one or more of these principles—a prohibition on superluminal propagation is assumed in causal process theories, and the Causal Markov condition is assumed in many versions of probabilistic theories. On the other hand, several different interpretive accounts may be consistent with (or even fit naturally with) the same causal principle, so that the same causal principles can "live" within different interpretive accounts. For example, one might adopt an interventionist characterization of causation and also hold that, in contexts in which talk of causal propagation makes sense, causes understood on interventionist lines will obey a prohibition on superluminal signaling. Similarly, one might argue that causes understood along interventionist lines will, under suitably circumscribed conditions, obey the Causal Markov condition (cf. Hausman and Woodward 1999). Philosophers who discuss causation often focus more on interpretive issues than on causal principles, but the latter are centrally important in scientific contexts.

4 CAUSAL MODELING, STRUCTURAL EQUATIONS, AND STATISTICS

Inference (involving so-called causal modeling techniques) to causal relationships from statistical information is common to many areas of science. Techniques are used for this purpose throughout the social sciences and are increasingly common in contemporary biology, especially when dealing with large data sets, as when researchers attempt to infer to causal relationships in the brain from fMRI data or to genetic regulatory relationships from statistical data involving gene expression.

Contrary to what is commonly supposed, such techniques do *not*, in most cases, embody a conception of causation that is itself "probabilistic" and do not provide a straightforward reduction of causation to facts about probabilistic relationships. Instead, we gain insight into these techniques by noting they employ two distinct sorts of representational structures. One has to do with information P about probabilistic or statistical relationships concerning some set of variables V. The other involves devices for representing causal relationships—call these C—among variables in V, most commonly by means of equations or directed graphs. The causal relationships C so represented are not defined in terms of the probabilistic relationships in P—indeed, these causal relationships are typically assumed to be *deterministic*. Instead, the problem of causal inference is conceptualized as a problem of *inferring* from P to C. Usually this requires additional assumptions A of various sorts that go beyond the information in P—examples include the Causal Markov and Faithfulness conditions discussed later. The role of these additional assumptions is one reason why the causal relationships in C should not be thought of as definable just in terms of the relationships in P. Depending on the details of the case, P in conjunction with A may allow for inference to a unique causal structure (the causal structure may be *identifiable* from P, given A) or (the more typical case) the inference may only yield an equivalence class of distinct causal structures.

As a simple illustration, consider a linear regression equation (4.1) $Y = aX + U$. (4.1) may be used merely to describe the presence of a correlation between X and Y but it may also be used to represent the existence of a causal relationship: that a change in X of amount dX causes a change of adY, in which case (4.1) is one of the vehicles for representing causal relationships C referred to earlier. (The convention is that the cause variable X is written on the right-hand side of the equation and the effect variable Y on the left-hand side.) U is a so-called error term, commonly taken to represent other unknown causes of Y besides X (or, more precisely, other causes of Y that do not cause X and that are not caused by X.) U is assumed to be a random variable, governed by some probability distribution. When (4.1) correctly represents a causal relationship, individual values or realizations of U, u_i, combine with realizations of values x_i of X to yield values y_i of Y in accord with (4.1). Note that the stochastic or probabilistic element in (4.1) derives from the fact U is a random variable and not because the relationship between X and Y or between U and Y is itself chancy.

A simple inference problem arising in connection with (4.1) is to infer the value of a from observations of the values of X and Y as these appear in the form of an observed joint probability distribution $Pr(X,Y)$, with $Pr(X,Y)$ playing the role of P in the previous schema. A simple result is that if (4.2) the general functional form (4.1) is correct, and (4.3) the distribution of U satisfies certain conditions, the most important of which is that U is uncorrelated with X, then one may reliably estimate a from $Pr(X,Y)$ via a procedure called ordinary least squares. Here (4.2) and (4.3) play the role of the additional assumptions A, which in conjunction with the information in $Pr(X,Y)$ are used to reach a (causal) conclusion about the value of the coefficient a.

The example just described involves bivariate regression. In multivariate linear regression a number of different variables X_i causally relevant to some effect Y are represented on the right-hand side of an equation $Y = \sum a_i X_i + U$. This allows for the representation of causal relationships between each of the X_i and Y but still does not allow us to represent causal relationships among the X_i themselves. The latter may be accomplished by means of *systems* of equations, where the convention is that one represents that variables X_j are direct causes of some other variable X_k by writing an equation in which the X_j occur on the right-hand side and X_k on the left-hand side. Every variable that is caused by other variables in the system of interest is represented on the left-hand side of a distinct equation. For example, a structure in which exogenous X_1 (directly) causes X_2 and X_1 and X_2 (directly) cause Y might be represented as

$$X_2 = a\,X_1 + U_1 \tag{4.4}$$

$$Y = bX_1 + cX_2 + U_2$$

Here X_1 affects Y via two different routes or paths, one of which is directly from X_1 to Y and the other of which is indirect and goes through X_2. As with (4.1), one problem is that of estimating values of the coefficients in each of the equations from information about the joint probability distribution Pr (Y, X_1, X_2), and other assumptions including assumptions about the distribution of the errors U_1 and U_2. However, there is also the more challenging and interesting problem of *causal structure* learning. Suppose one is given information about Pr (Y, X_1, X_2) but does not know what the causal relationships are among these variables. The problem of structure learning is that of learning these causal relations from the associated probability distribution and other assumptions.

To describe some contemporary approaches to this problem, it will be useful to intro- duce an alternative device for the representation of causal relationships: directed graphs. The basic convention is that an arrow from one variable to another $(X—>Y)$ represents that X is a direct cause (also called a parent $[par]$) of Y. (It is assumed that this implies that Y is some nontrivial function of X and perhaps other variables—nontrivial in that there are at least two different values of X that are mapped into different values of Y.) For example, the system (4.4) can be represented by the directed graph in Figure 8.1.

One reason for employing graphical representations is that it is sometimes reasonable to assume that there are systematic relationships between the graph and dependence/ independence relationships in an associated probability distribution over the variables corresponding to the vertices of the graph. Suppose we have a directed graph G with a probability distribution P over the vertices V in G. Then G and P satisfy the Causal Markov condition (CM) if and only if

> (CM) For every subset **W** of the variables in V, **W** is independent of every other sub- set in V that does not contain the parents of **W** or descendants (effects) of **W**, condi- tional on the parents of **W**.

CM is a generalization of the familiar "screening off" or conditional independence relationships that a number of philosophers (and statisticians) have taken to character- ize the relationship between causation and probability. CM implies, for example, that if two joint effects have a single common cause, then conditionalizing on this common cause renders those effects conditionally independent of each other. It also implies that if X does not cause Y and Y does not cause X and X and Y do not share a common cause, then X and Y are unconditionally independent—sometimes called the principle of the common cause.

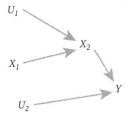

FIGURE 8.1 Graphical Representation of System (4.4).

A second useful assumption, called Faithfulness (F) by Spirtes, Glymour, and Scheines (2000), is

> (F) Graph G and associated probability distribution P satisfy the faithfulness condition if and only if every conditional independence relation true in P is entailed by the application of CM to G.

(F) says that independence relationships in P arise only because of the structure of the associated graph G (as these are entailed by CM) and not for other reasons. This rules out, for example, a causal structure in which X affects Z by two different paths or routes, one of which is direct and the other of which is indirect, going through a third variable Z, but such that the effects along the two routes just happen to exactly cancel each other, so that X is independent of Z.

Although I do not discuss details here, these two assumptions can be combined to create algorithms that allow one to infer facts about causal structure (as represented by G) from the associated probability distribution P. In some cases, the assumptions will allow for the identification of a unique graph consistent with P; in other cases, they will allow the identification of an equivalence class of graphs that may share some important structural features. These conditions—particularly F—can be weakened in various ways so that they fit a wider variety of circumstances, and one can then explore which inferences they justify.

Both CM and F are examples of what I called causal principles. Faithfulness is clearly a contingent empirical assumption that will be true of some systems and not others. For some sorts of systems, such as those involving feedback or regulatory mechanisms in which the design of the system is such that when deviations from some value of a variable that arise via one route or mechanism some other part of the system acts so as to restore the original value, violations of faithfulness may be very common.

The status of CM raises deeper issues. There are a variety of conditions under which systems, deterministic or otherwise, may fail to satisfy CM, but these exceptions can be delimited and arguably are well understood.[5] For systems that are deterministic with additive independent errors (i.e., systems for which the governing equations take the form $X_i = f\left(par\left(X_i\right)\right) + U_i$, with the U_i, independent of each other and each U_i independent of par (X_i)), CM follows as a theorem (cf. Pearl 2000.) On the other hand, our basis for adopting these independence assumptions may seem in many cases to rely on assuming part of CM—we seem willing to assume the errors are independent to the extent that the U_i do not cause one another or $par(X_i)$ and do not have common causes (a version of the common cause principle mentioned earlier). Nonetheless, CM is often a natural assumption, which is adopted in contexts outside of social science including physics, as we note later. In effect, it amounts to the idea that correlations among causally independent ("incoming" or exogenous variables) do not arise "spontaneously" or

---

[5] For discussion see Spirtes, Glymour, and Scheines (2000).

at least that such correlations are rare and that we should try to avoid positing them insofar as this is consistent with what is observed.

5 STABILITY, INDEPENDENCE, AND CAUSAL INTERPRETATION

So far we have not directly addressed the question of what must be the case for a set of equations or a directed graph to describe a causal relationship. My own answer is interventionist: an equation like (4.1) correctly describes a causal relationship if and only if for some range of values of X, interventions that change the value of X result in changes in Y in accord with (4.1). Causal relationships involving systems of equations or more complicated graphical structures can also be understood in interventionist terms. The basic idea is that when a system like (4.4) is fully causally correct, interventions on the each of the variables X_j, setting such variables to some value k may be represented by replacing the equation in which X_j occurs as a dependent variable with the equation $X_j = k$ while the other equations in the system (some of which will have X_j as an independent variable) remain undisturbed, continuing to describe how the system will respond to setting $X_j = k$. Each of the equations thus continues to hold according to its usual interventionist interpretation under changes in others—the system is in this sense *modular*.

Putting aside the question of the merits of this specifically interventionist interpretation, it is useful to step back and ask in a more general way what is being assumed about causal relationships when one employs structures like (4.1) and (4.4). A natural thought is that causal interpretation requires assumptions of some kind (however in detail these may be fleshed out) about the *stability* or *independence* of certain relationships under changes in other kinds of facts. At a minimum, the use of (4.1) to represent a causal relationship rests on the assumption that, at least for some range of circumstances, we can plug different values of X into (4.1) and it will continue to hold for (be stable across or independent of) these different values in the sense that we can use (4.1) to determine what the resulting value of Y will be. Put differently, we are assuming that there is some way of fixing or setting the value of X that is sufficiently independent of the relationship (4.1) that setting X via P does not upset whether (4.1) holds. The modularity condition referred to earlier similarly embodies stability or independence conditions holding *across* equations.

It seems plausible that we rely on such stability assumptions whenever, for example, we interpret (4.1) in terms of a relation of non-backtracking counterfactual dependence of Y on X: what one needs for such an interpretation is some specification of what will remain unchanged (the relationship [4.1], the values of variables that are not effects of X) under changes in X. As suggested earlier, in the contexts under discussion such information is provided by the equational or graphical model employed (and the rules for interpreting and manipulating such models, like those just described), rather than by a Lewis-style similarity metric over possible worlds. Similarly, if one wants to interpret equations like (4.1) in terms of an INUS style regularity account, one also needs to somehow fashion representational resources and rules for manipulating representations that allow one to capture the distinction between those features of the represented system that remain stable under various sorts of changes and those that do not.

These remarks suggest a more general theme, which is that in many areas of science claims about causal relationships are closely connected to (or are naturally expressed in terms of) *independence* claims, including claims about statistical independence, as in the case of CM, and claims about the independence of functional relationships under various changes—the latter being a sort of independence that is distinct from statistical independence. A closely related point is that independence claims (and hence causal notions) are also often represented in science in terms of *factorizability* claims, since factorizability claims can be used to express the idea that certain relationships will be stable under (or independent of) changes in others—the "causally correct" factorization is one that captures these stability facts. As an illustration, CM is equivalent to the following factorization condition:

$$Pr(X_1, \ldots, X_n) = \prod Pr(X_j \,/\, par(Xj)) \tag{5.1}$$

There are many different ways of factoring a joint probability distribution, but one can think of each of these as representing a particular claim about causal structure: in (5.1) each term $Pr(X_j \,/\, par(X_j))$ can be thought of as representing a distinct or independent causal relationship linking X_j and its parents. This interpretation will be appropriate to the extent that, at least within a certain range, each $Pr(X_j \,/\, par(X_j))$ will continue to hold over changes in the frequency of *par* (X_j), and each of the $Pr(X_j \,/\, par(X_j))$ can be changed independently of the others. (The causally correct factorization is the one satisfying this condition.) Physics contains similar examples in which factorization conditions are used to represent causal independence—for example, the condition (clustering decomposition) that the S-matrix giving scattering amplitudes in quantum field theory factorizes as the product of terms representing far separated (and hence causally independent) subprocesses (cf. Duncan 2012, 58).

A further illustration of the general theme of the relationship between causal and independence assumptions is provided by some recent work on the direction of causation in machine learning (Janzing et al. 2012). Suppose (i) Y can be written as a function of X plus an additive error term that is independent of $X : Y = f(X) + U$ with $X \perp\!\!\!\perp U$ (where $\perp\!\!\!\perp$ means probabilistic independence). Then if the distribution is non-Gaussian, there is no such additive error model from Y to X—that is, no model in which (ii) X can be written as $X = g(Y) + V$ with $Y \perp\!\!\!\perp V$. A natural suggestion is that to determine the correct causal direction, one should proceed by seeing which of (i) or (ii) holds, with the correct causal direction being one in which the cause is independent of the error. One can think of this as a natural expression of the ideas about the independence of "incoming" influences embodied in CM. Expressed slightly differently, if $X \rightarrow Y$ is the correct model, but not of additive error form, it would require a very contrived, special (and hence in some sense unlikely) relationship between $P(X)$ and $P(Y/X)$ to yield an additive error model of form $Y \rightarrow X$. Again, note the close connection between assumptions about statistical independence, the avoidance of "contrived" coincidences, and causal assumptions (in this case about causal direction).

In the case just described, we have exploited the fact that at least three causal factors are present in the system of interest—Y, X, and U. What about cases in which there are just two variables—X and Y—which are deterministically related, with each being a function of the other? Janzig et al. (2012) also show that even in this case (and even if the function from X to Y is invertible) it is sometimes possible to determine causal direction by means of a further generalization of the connection between causal direction and facts about independence, with the relevant kind of independence now being a kind of informational independence rather than statistical dependence. Very roughly, if the causal direction runs as $X \rightarrow Y$, then we should expect that the function f describing this relationship will be informationally independent of the description one gives of the (marginal) distribution of X—independent in the sense that knowing this distribution will provide no information about the functional relationship between X and Y and vice versa. Remarkably, applications of this idea in both simulations and real-world examples in which the causal direction is independently known shows that the method yields correct conclusions in many cases. If one supposes that the informational independence of $Pr(X)$ and $X \rightarrow Y$ functions as a sort of proxy for the invariance or stability of $X \rightarrow Y$ under changes in the distribution of X, this method makes normative sense, given the ideas about causal interpretation described previously.

6 CAUSATION IN PHYSICS

Issues surrounding the role of causation/causal reasoning in physics have recently been the subject of considerable philosophical discussion. There is a spectrum (or perhaps, more accurately, a multidimensional space) of different possible positions on this issue. Some take the view that features of fundamental physical laws or the contexts in which these laws are applied imply that causal notions play little or no legitimate role in physics. Others take the even stronger position that because causal notions are fundamentally unclear *in general*, they are simply a source of confusion when we attempt to apply them to physics contexts (and presumably elsewhere as well.) A more moderate position is that while causal notions are sometimes legitimate in physics, they are unnecessary in the sense that whatever scientifically respectable content they have can be expressed without reference to causality. Still others (e. g., Frisch 2014) defend the legitimacy and even centrality of causal notions in the interpretation of physical theories. Those advocating this last position observe that even a casual look at the physics literature turns up plenty of references to "causality" and "causality conditions". Examples include a micro-causality condition in quantum field theory that says that operators at spacelike separation commute and which is commonly motivated by the claim that events at such separation do not interact causally, and the clustering decomposition assumption referred to in Section 5, which is also often motivated as a causality condition. Other examples are provided by the common preference for "retarded" over "advanced" solutions to the equations of classical electromagnetism and the use of retarded rather than advanced Green's functions in modeling dispersion relations. These preferences are often motivated by the claim that the advanced solutions represent "noncausal" behavior in which effects temporally precede their causes (violation of another causality condition) and hence should be discarded. Similarly, there is a hierarchy of causality conditions often imposed in models of general relativity, with, for example, solutions involving closed causal or timelike curves being rejected by some on the grounds they violate "causality." Causal skeptics respond, however, that these conditions are either unmotivated (e.g., vague or unreasonably aprioristic) or superfluous in the sense that what is defensible in them can be restated without reference to any notion of causation.

A related issue concerns whether causal claims occurring in the special sciences (insofar as they are true or legitimate) require "grounding" in fundamental physical laws, with these laws providing the "truth-makers" for such claims. A simple version of such a view might claim that C causes E if and only if there is a fundamental physical law L linking Cs to Es, with C and E "instantiating" L. Although there is arguably no logical inconsistency between this grounding claim, the contention that causation plays no role in physics and the claim that causal notions are sometimes legitimately employed in the special sciences, there is at least a tension—someone holding all these claims needs to explain how they can be true together. Doing this in a plausible way is nontrivial and represents another issue that is worthy of additional philosophical attention.

6.1 Causal Asymmetry and Time-Reversal Invariance

Turning now to some arguments for little or no role for causation in physics, I begin with the implications of time-reversal invariance. With the exception of laws governing the weak force, fundamental physical laws, both in classical mechanics and electromagnetism, special and general relativity, and quantum field theories are time-reversal invariant: if a physical process is consistent with such laws, the "time-reverse" of this process is also permitted by these laws. For example, according to classical electromagnetism, an accelerating charge will be associated with electromagnetic radiation radiating outward from the charge. These laws also permit the time-reversed process according to which a spherically symmetric wave of electromagnetic radiation converges on a single charge which then accelerates—a process that appears to be rare, absent some special contrivances. A widespread philosophical view is that this time-symmetric feature of the fundamental laws is in some way inconsistent with the asymmetry of causal relationships or at least that the latter lacks any ground or basis, given the former. Assuming that an asymmetry between cause and effect is central to the whole notion of causation, this is in turn taken to show that there is no basis in fundamental physics for the application of causal notions.

This inference is more problematic than usually recognized. As emphasized by Earman (2011), the time-reversal invariant character of most of the fundamental equations of physics is consistent with particular solutions to those equations exhibiting various asymmetries—indeed, studies of many such equations show that "most" of their solutions are asymmetric (Earman 2011). An obvious possibility, then, is that to the extent that a causal interpretation of a solution to a differential equation is appropriate at all, it is asymmetries in these solutions that provide a basis for the application of an asymmetric notion of causation. In other words, causal asymmetries are to be found in particular solutions to the fundamental equations that arise when we impose particular initial and boundary conditions rather than in the equations themselves.

To expand on this idea, consider why the convergent wave in the previous example rarely occurs. The obvious explanation is that such a convergent process would require a very precise co-ordination of the various factors that combine to produce a coherent incoming wave. (cf. Earman 2011) On this understanding, the origin of this asymmetry is similar to the origin of the more familiar thermodynamic asymmetries—it would require a combination of circumstances that we think are unlikely to occur "spontaneously," just as the positions and momenta of the molecules making up a gas that has diffused to fill a container are unlikely to be arranged in such a way that the gas spontaneously occupies only the right half of the container at some later time. Of course "unlikely" does not mean "impossible" in the sense of being contrary to laws of nature in either case, and in fact in the electromagnetic case one can readily imagine some contrivance—for example, a precisely arranged system of mirrors and lenses—that produces such an incoming wave.

The important point for our purposes is that the asymmetries in the frequencies of occurrence of outgoing, diverging waves in comparison with incoming, converging waves are not, so to speak, to be found in Maxwell's equations themselves but rather arise when we impose initial and boundary conditions on those equations to arrive at solutions describing the behavior of particular sorts of systems. These initial and boundary conditions are justified empirically by the nature of the systems involved: if we are dealing with a system that involves a single accelerating point charge and no other fields are present, this leads to a solution to the equations in which outgoing radiation is present; a system in which a coherent wave collapses on a point charge is modeled by a different choice of initial and boundary conditions, which leads to different behavior of the electromagnetic field. If we ask why the first situation occurs more often than the second, the answer appears to relate to the broadly statistical considerations described earlier. Thus the time-reversal invariant character of the fundamental equations is consistent with particular solutions to those equations exhibiting various asymmetries, and although there may be other reasons for not interpreting these solutions causally, the time-reversal invariance of the fundamental equations by itself is no barrier to doing so. I suggest that what is true in this particular case is true more generally.

We may contrast this view with the implicit picture many philosophers seem to have about the relationship between causation and laws. According to this picture, the fundamental laws, taken by themselves, have rich causal content and directly describe causal relationships: the "logical form" of a fundamental law is something like

$$\text{All Fs } \textit{cause} \text{ Gs.} \hspace{3cm} (6.1)^6$$

It is indeed a puzzle to see how to reconcile this picture with the time-reversal invariant character of physical laws. Part of the solution to this puzzle is to recognize that the laws by themselves, taken just as differential equations, often do not make causal claims in the manner of (6.1); again, causal description, when appropriate at all, derives from the way in which the equations and the choice of particular initial and boundary conditions interact when we find particular solutions to those equations.

If this general line of thought is correct, several additional issues arise. On the one hand, it might seem that the choice of initial and boundary conditions to model a particular system is a purely empirical question—either one is faced with a coherent incoming wave or not—and distinctively causal considerations need not be thought of as playing any role in the choice of model. One might argue on this basis that once one chooses an empirically warranted description of the system, one may then place a causal gloss on the result and say the incoming wave *causes* the acceleration but that the use of causal language here is unnecessary and does no independent work. On the other hand, it is hard not to be struck by the similarity between the improbability of a precisely coordinated incoming wave arising spontaneously and principles like CM described in Section 4. When distinct segments of the wave front of an outgoing wave are correlated, this strikes us as unmysterious because this can be traced to a common cause—the accelerating charge. On the other hand, the sort of precise coordination of incoming radiation that is associated with a wave converging on a source strikes us as unlikely (again in accord with CM) in the absence of direct causal relations among the factors responsible for the wave or some common cause, like a system of mirrors. On this view, causal considerations of the sort associated with CM play a role in justifying one choice of initial and boundary conditions over another or at least in making sense of why as an empirical mater we find certain sets of these conditions occurring more frequently than others.[7]

6.2 Kinds of Differential Equations

As noted previously, known fundamental physical laws are typically stated as differential equations. Differences among such equations matter when we come to interpret them

---

[6] Davidson (1967) is a classic example in which this commitment is explicit, but it seems to be tacit in many other writers.

[7] For arguments along these lines see Frisch (2014).

causally. From the point of view of causal representation, one of the most important is the distinction between hyperbolic and other sorts (parabolic, elliptical) of partial differential equations (PDEs). Consider a second-order nonhomogeneous partial differential equation in two independent variables

$$Af_{xx} + Bf_{xy} + Cf_{yy} + Df_x + Ef_y + Ff = G$$

where the subscripts denote partial differentiation: $f_{xx} = \partial^2 f / \partial x^2$, and so on. A hyperbolic differential equation of this form is characterized by $B^2 - AC > 0$, while parabolic (respectively, elliptical) equations are characterized by $B^2 - AC = 0$ (respectively, $B^2 - AC < 0$). Hyperbolic PDEs are the natural way of representing the propagation of a causal process in time. For example, the wave equation in one spatial dimension

$$\partial^2 f / \partial t^2 = c^2 \partial^2 f / \partial x^2$$

is a paradigmatic hyperbolic PDE that describes the propagation of a wave through a medium.

The solution domains for hyperbolic PDEs have "characteristic surfaces" or cone-like structures that characterize an upper limit on how fast disturbances or signals can propagate—in the case of the equations of classical electromagnetism these correspond to the familiar light-cone structure. In contrast, elliptical and parabolic PDEs have solution domains in which there is no such limit on the speed of propagation of disturbances. A related difference is that hyperbolic equations are associated with specific domains of dependence and influence: there is a specific region in the solution domain on which the solution at point P depends, in the sense that what happens outside of that region does not make a difference to what happens at P—thus in electromagnetism, what happens at a point depends only on what happens in the backwards light cone of that point. By contrast, elliptical PDEs such as the Laplace equation do not have specific domains of dependence. Instead, the domain of dependence for every point is the entire solution domain. Given this feature and the absence of limiting velocity of disturbance propagation there is, intuitively, no well-defined notion of causal propagation for systems governed by such equations.

Both Earman (2014) and Wilson (forthcoming) suggest that the appropriate way to characterize a notion of causal propagation (and related to this, a Salmon/Dowe–like notion of causal process) is in terms of systems whose behavior is governed by hyperbolic differential equations and that admit of a well-posed initial value formulation. This allows one to avoid having to make use of unclear notions such as "intersection," "possession" of a conserved quantity, and so on. A pseudo-process can then be characterized just by the fact that its behavior is not described by a relevant hyperbolic PDE.

If this is correct, several other conclusions follow. First, recall the supposed distinction between difference-making and the more "scientific" or "physical" notion of causation associated with the Salmon/Dowe account. The hyperbolic PDEs used to characterize the notion of a causal process clearly describe difference-making relationships—one may think of them as characterizing how, for example, variations or differences in initial conditions will lead to different outcomes. So the difference-making/causal process contrast seems ill founded; causal processes involve just one particular kind of difference-making relation—one that allows for a notion of causal propagation.

Second, and relatedly, there are many systems (both those treated in physics and in other sciences) whose behavior (at least in the current state of scientific understanding) is not characterized by hyperbolic PDEs but rather by other sorts of equations, differential and otherwise. Unless one is prepared to argue that only hyperbolic PDEs and none of these other structures can represent causal relationships—a claim that seems difficult to defend—the appropriate conclusion seems to be that some representations of some situations are interpretable in terms of the notion of a causal process but that other representations are not so interpretable.[8] In this sense the notion of a causal process (or at least representations in terms of causal processes) seems less general than the notion of a causal relationship. Even in physics, many situations are naturally described in causal terms, even though the governing equations are not of the hyperbolic sort.

I conclude with a final issue that deserves more philosophical attention and that has to do with the relationships among the mathematical structures discussed in this article. On a sufficiently generous conception of "based," it does not seem controversial that the causal generalizations found in the special sciences (or, for that matter, nonfundamental physics) are in some way or other "based" or "grounded" in fundamental physics. However, this observation (and similar claims about "supervenience," etc.) do not take us very far in understanding the forms that such basing relations might take or how to conceive of the relations between physics and the special sciences. To mention just one problem, many special science generalizations describe the equilibrium behavior of systems—the mathematics used to describe them may involve some variant of the structural equations of Section 4 (which are most naturally interpretable as describing relationships holding at some equilibrium; see Mooij et al. 2013) or else nonhyperbolic differential equations. These generalizations do not describe the dynamical or evolutionary processes that lead to equilibrium. By contrast, most fundamental physical laws, at least as formulated at present, take the form of hyperbolic differential equations that describe the dynamical evolution of systems over time. Understanding how these very different forms of causal representation fit together is highly nontrivial, and for this reason, along with others, the correct stories about the "grounding" of special science causal generalizations in physics is (when it is possible to produce it at all) likely to be subtle and complicated—far from the simple "instantiation" story gestured at earlier. This is just one aspect of the much bigger question of how causal representations and relationships in science at different levels and scales relate to one another.

---

[8] This point is developed at length in Wilson (forthcoming).

Acknowledgements

I would like to thank Bob Batterman, John Earman, Mathias Frisch, Paul Humphreys, Wayne Myrvold, John Norton, and Mark Wilson for helpful comments and discussion.

References

Davidson, D. (1967). "Causal Relations." *Journal of Philosophy* 64: 691–703.

Dowe, P. (2000). *Physical Causation* (Cambridge, UK: Cambridge University Press).

Duncan, A. (2012). *The Conceptual Framework of Quantum Field Theory* (Oxford: Oxford University Press).

Earman, J. (2011). "Sharpening the Electromagnetic Arrow(s) of Time." In C. Callender (ed.), *Oxford Handbook of the Philosophy of Time* (Oxford: Oxford University Press), 485–527.

Earman, J. (2014). "No Superluminal Propagation for Classical Relativistic and Relativistic Quantum Fields." *Studies in History and Philosophy of Modern Physics* 48: 102–108.

Eells, E. (1991). *Probabilistic Causality* (Cambridge, UK: Cambridge University Press).

Frisch, M. (2014). *Causal Reasoning in Physics* (Cambridge, UK: Cambridge University Press).

Hausman, D., Stern, R. and Weinberger, W. (forthcoming) "Systems Without a Graphical Representation." *Synthese*.

Hausman, D., and Woodward, J. (1999). "Independence, Invariance and the Causal Markov Condition." *The British Journal for the Philosophy of Science* 50: 521–583.

Heckman, J. (2005). "The Scientific Model of Causality." *Sociological Methodology* 35: 1–97.

Holland, P. (1986) "Statistics and Causal Inference." *Journal of the American Statistical Association* 81: 945–960.

Janzing, D., Mooij, J., Zhang, K., Lemeire, J., Zscheischler, J., Daniusis, D., Steudel, B., and Scholkopf, B. (2012)."Information-Geometric Approach to Inferring Causal Directions." *Artificial Intelligence* 182–183: 1–31.

Lewis, D. (1973). "'Causation' Reprinted with Postscripts." In D. Lewis, *Philosophical Papers*, vol. 2 (Oxford: Oxford University Press), 32–66.

Lewis, D. (1999). *Papers in Metaphysics and Epistemology* (Cambridge, UK: Cambridge University Press).

Mackie, J. (1974). *The Cement of the Universe* (Oxford: Oxford University Press).

Mooij, J., Janzing, D., and Schölkopf, B. (2013). "From Ordinary Differential Equations to Structural Causal Models: The Deterministic Case." In A. Nicholson and P. Smyth (eds.), *Proceedings of the 29th Annual Conference on Uncertainty in Artificial Intelligence* (Corvallis: AUAI Press), 440–448.

Ney, A. (2009). "Physical Causation and Difference-Making." *The British Journal for the Philosophy of Science* 60(4): 737–764.

Pearl, J. (2000). *Causality: Models, Reasoning, and Inference* (Cambridge, UK: Cambridge University Press).

Rubin, D. (1974). "Estimating Causal Effects of Treatments in Randomized and Nonrandomized Studies." *Journal of Educational Psychology* 66: 688–701.

Salmon, W. (1984). *Scientific Explanation and the Causal Structure of the World* (Princeton, NJ: Princeton University Press).

Spirtes, P., Glymour, C., and Scheines, R. (2000). *Causation, Prediction and Search* (Cambridge, MA: MIT Press).

Wilson, M. (forthcoming). *Physics Avoidance*. Oxford: Oxford University Press.

Woodward, J. (2003). *Making Things Happen: A Theory of Causal Explanation* (New York: Oxford University Press).

CHAPTER 9

....

CONFIRMATION AND INDUCTION

....

JAN SPRENGER

1 THE PROBLEMS OF INDUCTION

....

INDUCTION is a method of inference that aims at gaining empirical knowledge. It has two main characteristics: first, it is *based on experience*. (The term "experience" is used interchangeably with "observations" and "evidence.") Second, induction is *ampliative*, that is, the conclusions of an inductive inference are not necessary but contingent. The first feature makes sure that induction targets empirical knowledge; the second feature distinguishes induction from other modes of inference, such as deduction, where the truth of the premises guarantees the truth of the conclusion.

Induction can have many forms. The simplest one is *enumerative induction*: inferring a general principle or making a prediction based on the observation of particular instances. For example, if we have observed 100 black ravens and no non-black ravens, we may predict that also raven 101 will be black. We may also infer the general principle that all ravens are black. But induction is not tied to the enumerative form and comprises all ampliative inferences from experience. For example, making weather forecasts or predicting economic growth rates are highly complex inductive inferences that amalgamate diverse bodies of evidence.

The first proper canon for inductive reasoning in science was set up by Francis Bacon, in his *Novum Organon* (Bacon 1620). Bacon's emphasis is on learning the cause of a scientifically interesting phenomenon. He proposes a method of *eliminative induction*, that is, eliminating potential causes by coming up with cases where the cause but not the effect is present. For example, if the common flu occurs in a hot summer period, then cold cannot be its (sole) cause. A similar method, although with less meticulous devotion to the details, was outlined by René Descartes (1637). In his *Discours de la Méthode*, he explains how scientific problems should be divided into tractable subproblems and how their solutions should be combined.

Both philosophers realized that without induction, science would be blind to experience and unable to make progress; hence, their interest in spelling out the inductive method in detail. However, they do not provide a foundational *justification* of inductive inference. For this reason, C. D. Broad (1952, 142–143) stated that "inductive reasoning . . . has long been the glory of science" but a "scandal of philosophy." This quote brings us directly to the notorious *problem of induction* (for a survey, see Vickers 2010).

Two problems of induction should be distinguished. The first, fundamental problem is about why we are justified to make inductive inferences; that is, why the method of induction works at all. The second problem is about telling good from bad inductions and developing rules of inductive inference. How do we learn from experience? Which inferences about future predictions or general theories are justified by these observations? And so on.

About 150 years after Bacon, David Hume (1739, 1748) was the first philosopher to clearly point out how hard the first problem of induction is (*Treatise on Human Nature*, 1739, Book I; *Enquiry Concerning Human Understanding*, 1748, Sections IV–V). Like Bacon, Hume was interested in learning the causes of an event as a primary means of acquiring scientific knowledge. Because causal relations cannot be inferred a priori, we have to learn them from experience; that is, we must use induction.

Hume divides all human reasoning into demonstrative and probabilistic reasoning. He notes that learning from experience falls into the latter category: no amount of observations can logically *guarantee* that the sun will rise tomorrow, that lightning is followed by thunder, that England will continue to lose penalty shootouts, and the like. In fact, regularities of the latter sort sometimes cease to be true. Inductive inferences cannot *demonstrate* the truth of the conclusion, but only make it *probable*.

This implies that inductive inferences have to be justified by nondemonstrative principles. Imagine that we examine the effect of heat on liquids. We observe in a number of experiments that water expands when heated. We predict that, upon repetition of the experiment, the same effect will occur. However, this is probable only if nature does not change its laws suddenly: "all inferences from experience suppose, as their foundation, that the future will resemble the past" (Hume 1748, 32). We are caught in a vicious circle: the justification of our inductive inferences invokes the principle of induction itself. This undermines the rationality of our preference for induction over other modes of inference (e.g., counterinduction).

The problem is that assuming the uniformity of nature in time can only be justified by inductive reasoning; namely, our past observations to that effect. Notably, pragmatic justifications of induction, by reference to past successes, do not fly because inferring from past to future reliability of induction also obeys the scheme of an inductive inference (Hume 1748).

Hume therefore draws the skeptical conclusion that we lack a rational basis for believing that causal relations inferred from experience are necessary or even probable. Instead, what makes us associate causes and effects are the irresistible psychological forces of custom and habit. The connection between cause and effect is in the mind rather than in the world, as witnessed by our inability to give a non-circular justification of induction (Hume 1748, 35–38).

Hume's skeptical argument seems to undermine a lot of accepted scientific method. If induction does not have a rational basis, why perform experiments, predict future events, and infer to general theories? Why science at all? Note that Hume's challenge also affects the second problem: if inductive inferences cannot be justified in an objective way, how are we going to tell which rules of induction are good and which are bad?

Influenced by Hume, Karl Popper (1959, 1983) developed a radical response to the problem of induction. For him, scientific reasoning is essentially a deductive and not an inductive exercise. A proper account of scientific method neither affords nor requires inductive inference—it is about *testing hypotheses* on the basis of their predictions:

> The best we can say of a hypothesis is that up to now it has been able to show its worth ... although, in principle, it can never be justified, verified, or even shown to be probable. This appraisal of the hypothesis relies solely upon deductive consequences (predictions) which may be drawn from the hypothesis: There is no need even to mention "induction." (Popper 1959, 346)

For Popper, the merits of a hypothesis are not determined by the degree to which past observations support it but by its performances in severe tests; that is, sincere attempts to overthrow it. Famous examples from science include the Michelson-Morley experiment as a test of the ether theory in physics and the Allais and Ellsberg experiments as tests of subjected expected utility theory in economics. Popper's account also fits well with some aspects of statistical reasoning, such as the common use of null hypothesis significance tests (NHST): a hypothesis of interest is tested against a body of observations and "rejected" if the result is particularly unexpected. Such experiments do not warrant inferring or accepting a hypothesis; they are exclusively designed to disprove the null hypothesis and to collect evidence against it. More on NHST will be said in Section 6.

According to Popper's view of scientific method, induction in the narrow sense of inferring a theory from a body of data is not only unjustified, but even superfluous. Science, our best source of knowledge, assesses theories on the basis of whether their predictions obtain. Those predictions are deduced from the theory. Hypotheses are corroborated when they survive a genuine refutation attempt, when their predictions were correct. Degrees of corroboration may be used to form practical preferences over hypotheses. Of course, this also amounts to learning from experience and to a form of induction, broadly conceived—but Popper clearly speaks out against the view that scientific hypotheses with universal scope are ever guaranteed or made probable by observations.

Popper's stance proved to be influential in statistical methodology. In recent years, philosopher Deborah Mayo and econometrician Aris Spanos have worked intensively on this topic (e.g., Mayo 1996; Mayo and Spanos 2006). Their main idea is that our preferences among hypotheses are based on the *degree of severity* with which they have been tested. Informally stated, they propose that a hypothesis has been severely tested if (1) it fits well with the data, for some appropriate notion of fit; and (2), if the hypothesis were false, it would have been very likely to obtain data that favor the relevant alternative(s) much more than the actual data do.

We shall not, however, go into the details of their approach. Instead, we return to the second problem of induction: how should we tell good from bad inductions?

2 LOGICAL RULES FOR INDUCTIVE INFERENCE

Hume's skeptical arguments show how difficult it is to argue for the reliability and truth-conduciveness of inductive inference. However, this conclusion sounds more devastating than it really is. For example, on a reliabilist view of justification (Goldman 1986), beliefs are justified if generated by reliable processes that usually lead to true conclusions. If induction is factually reliable, our inductive inferences are justified even if we cannot access the reasons for why the method works. In a similar vein, John Norton (2003) has discarded formal theories of induction (e.g., those based on the enumerative scheme) and endorsed a *material* theory of induction: inductive inferences are justified by their conformity to facts.

Let us now return to the second problem of induction; that is, developing (possibly domain-sensitive) *rules of induction*—principles that tell good from bad inductive inferences. In developing these principles, we will make use of the method of *reflective equilibrium* (Goodman 1955): we balance scientific practice with normative considerations (e.g., which methods track truth in the idealized circumstances of formal models). Good rules of induction are those that explain the success of science and that have at the same time favorable theoretical properties. The entire project is motivated by the analogy to deductive logic, in which rules of inference have been useful at guiding our logical reasoning. So why not generalize the project to inductive logic, to rules of reasoning under uncertainty and ampliative inferences from experience?

Inductive logic has been the project of a lot of twentieth-century philosophy of science. Often it figures under the heading of finding criteria for when evidence confirms (or supports) a scientific hypothesis, that is, to explicate the concept of confirmation: to replace our vague pre-theoretical concept, the *explicandum*, with a simple, exact, and fruitful concept that still resembles the explicandum—the *explicatum* (Carnap 1950, 3–7). The explication can proceed *quantitatively*, specifying degrees of confirmation, or *qualitatively*, as an all-or-nothing relation between hypothesis and evidence. We will first look at qualitative analyses in first-order logic since they outline the logical grammar of the concept. Several features and problems of qualitative accounts carry over to and motivate peculiar quantitative explications (Hempel 1945a).

Scientific laws often take the logical form $\forall x: Fx \rightarrow Gx$; that is, all F's are also G's. For instance, take Kepler's First Law that all planets travel in an elliptical orbit around the sun.

Then, it is logical to distinguish two kinds of confirmation of such laws, as proposed by Jean Nicod (1961, 23–25): *L'induction par l'infirmation* proceeds by refuting and eliminating other candidate hypotheses (e.g., the hypothesis that planets revolve around the Earth). This is basically the method of eliminative induction that Bacon applied to causal inference. *L'induction par la confirmation*, by contrast, supports a hypothesis by citing their *instances* (e.g., a particular planet that has an elliptical orbit around the sun). This is perhaps the simplest and most natural account of scientific theory confirmation. It can be expressed as follows:

Nicod condition (NC): For a hypothesis of the form $H = \forall x : Fx \rightarrow Gx$ and an individual constant a, an observation report of the form $Fa \wedge Ga$ confirms H.

However, NC fails to capture some essentials of scientific confirmation—see Sprenger (2010) for details. Consider the following highly plausible adequacy condition, due to Carl G. Hempel (1945a, 1945b):

Equivalence condition (EC): If H and H' are logically equivalent sentences, then E confirms H if and only if E confirms H'.

EC should be satisfied by any logic of confirmation because, otherwise, the establishment of a confirmation relation would depend on the peculiar formulation of the hypothesis, which would contradict our goal of finding a *logic* of inductive inference.

Combining EC with NC leads, however, to paradoxical results. Let $H = \forall x: Rx \rightarrow Bx$ stand for the hypothesis that all ravens are black. H is equivalent to the hypothesis $H' = \forall x: \neg Bx \rightarrow \neg Rx$ that no non-black object is a raven. A white shoe is an instance of this hypothesis H'. By NC, observing a white shoe confirms H', and, by EC, it also confirms H. Hence, observing a white shoe confirms the hypothesis that all ravens are black! But a white shoe appears to be an utterly irrelevant piece of evidence for assessing the hypothesis that all *ravens* are black. This result is often called the *paradox of the ravens* (Hempel 1945a, 13–15) or, after its inventor, *Hempel's paradox*.

How should we deal with this problem? Hempel suggests biting the bullet and accepting that the observation of a white shoe confirms the raven hypothesis. After all, the observation eliminates a potential falsifier. To push this intuition further, imagine that we observe a gray, raven-like bird, and, only after extended scrutiny, we find out that it is a crow. There is certainly a sense in which the crowness of that bird confirms the raven hypothesis, which was already close to refutation.

Hempel (1945a, 1945b) implements this strategy by developing a more sophisticated version of Nicod's instance confirmation criterion in which background knowledge plays a distinct role, the so-called *satisfaction criterion*. We begin with the formulation of *direct* confirmation, which also captures the main idea of Hempel's proposal:

Direct confirmation (Hempel): A piece of evidence E directly Hempel-confirms a hypothesis H relative to background knowledge K if and only if E and K jointly entail the development of H to the domain of E—that is, the restriction of H to the set of individual constants that figure in E. In other words, $E \wedge K \vDash H_{|dom(E)}$ The idea of this criterion is that our observations verify a general hypothesis as restricted to the actually observed objects. Hempel's satisfaction criterion generalizes this intuition by demanding that a hypothesis be confirmed whenever it is entailed by a set of directly confirmed sentences. Notably, Clark Glymour's account of *bootstrap confirmation* is also based on Hempel's satisfaction criterion (Glymour 1980b).

However, Hempel did not notice that the satisfaction criterion does not resolve the raven paradox: $E = \neg Ba$ directly confirms the raven hypothesis H relative to $K = \neg Ra$ (because $E \wedge K \vDash H_{\{a\}}$). Thus, even objects *that are known not to be ravens* can confirm the hypothesis that all ravens are black. This is clearly an unacceptable conclusion and invalidates the satisfaction criterion as an acceptable account of qualitative confirmation, whatever its other merits may be (Fitelson and Hawthorne 2011).

Hempel also developed several *adequacy criteria* for confirmation, intended to narrow down the set of admissible explications. We have already encountered one of them, the EC. Another one, the *special consequence condition*, claims that consequences of a confirmed hypothesis are confirmed as well. Hypotheses confirmed by a particular piece of evidence form a deductively closed set of sentences. It is evident from the definition that the satisfaction criterion conforms to this condition. It also satisfies the *consistency condition* that demands (inter alia) that no contingent evidence supports two hypotheses that are inconsistent with each other. This sounds very plausible, but, as noted by Nelson Goodman (1955) in his book *Fact, Fiction and Forecast*, that condition conflicts with powerful inductive intuitions. Consider the following inference:

Observation: emerald e_1 is green.

Observation: emerald e_2 is green.

. . .

Generalization: All emeralds are green.

This seems to be a perfect example of a valid inductive inference. Now define the predicate "grue," which applies to all green objects if they were observed for the first time prior to time $t =$ "now" and to all blue objects if observed later. (This is just a description of the extension of the predicate—no object is supposed to change color.) The following inductive inference satisfies the same logical scheme as the previous one:

Observation: emerald e_1 is grue.

Observation: emerald e_2 is grue.

. . .

Generalization: All emeralds are grue.

In spite of the gerrymandered nature of the "grue" predicate, the inference is sound: it satisfies the basic scheme of enumerative induction, and the premises are undoubtedly true. But then, it is paradoxical that two valid inductive inferences support flatly opposite conclusions. The first generalization predicts emeralds observed in the future to be green; the second generalization predicts them to be blue. How do we escape from this dilemma?

Goodman considers the option that, in virtue of its gerrymandered nature, the predicate "grue" should not enter inductive inferences. He notes, however, that it is perfectly possible to redefine the standard predicates "green" and "blue" in terms of "grue" and its conjugate predicate "bleen" (= blue if observed prior to t, else green). Hence, any preference for the "natural" predicates and the "natural" inductive inference seems to be arbitrary. Unless we want to give up on the scheme of enumerative induction, we are forced into dropping Hempel's consistency condition and accepting the paradoxical conclusion that both conclusions (all emeralds are green/grue) are, at least to a certain extent, confirmed by past observations. The general moral is that conclusions of an inductive inference need not be consistent with each other, unlike in deductive logic.

Goodman's example, often called the *new riddle of induction*, illustrates that establishing rules of induction and adequacy criteria for confirmation is not a simple business. From a normative point of view, the consistency condition looks appealing, yet it clashes with intuitions about paradigmatic cases of enumerative inductive inference. The rest of this chapter will therefore focus on accounts of confirmation where inconsistent hypotheses can be confirmed simultaneously by the same piece of evidence.

A prominent representative of these accounts is *hypothetico-deductive (H-D) confirmation*. H-D confirmation considers a hypothesis to be confirmed if empirical predictions deduced from that hypothesis turn out to be successful (Gemes 1998; Sprenger 2011). An early description of H-D confirmation was given by William Whewell:

> Our hypotheses ought to foretel [sic] phenomena which have not yet been observed … the truth and accuracy of these predictions were a proof that the hypothesis was valuable and, at least to a great extent, true. (Whewell 1847, 62–63)

Indeed, science often proceeds that way: our best theories about the atmospheric system suggest that emissions of greenhouse gases such as carbon dioxide and methane lead to global warming. That hypothesis has been confirmed by its successful predictions, such as shrinking Arctic ice sheets, increasing global temperatures, its ability to backtrack temperature variations in the past, and the like. The hypothetico-deductive concept of confirmation explicates the common idea of these and similar examples by stating that evidence confirms a hypothesis if we can derive it from the tested hypothesis together with suitable background assumptions. H-D confirmation thus naturally aligns with the Popperian method for scientific inquiry that emphasizes the value of risky predictions, the need to test our scientific hypotheses as severely as possible, to derive precise predictions, and to check them with reality.

An elementary account of H-D confirmation is defined as follows:

Hypothetico-Deductive (H-D) Confirmation: E H-D-confirms H relative to background knowledge K if and only if

1. $H \wedge K$ is consistent,
2. $H \wedge K$ entails E $\left(H \wedge K \vDash E\right)$,
3. K alone does not entail E.

The explicit role of background knowledge can be used to circumvent the raven paradox along the lines that Hempel suggested. Neither $Ra \wedge Ba$ nor $\neg Ba \wedge \neg Ra$ confirms the hypothesis $H = \forall x\colon Rx \rightarrow Bx$, but Ba ("a is black") does so *relative to the background knowledge Ra*, and $\neg Ra$ ("a is no raven") does so *relative to the background knowledge $\neg Ba$*. This makes intuitive sense: only if we know a to be a raven, the observation of its color is evidentially relevant; and, only if a is known to be non-black, the observation that it is no raven supports the hypothesis that all ravens are black in the sense of eliminating a potential falsifier.

Although the H-D account of confirmation fares well with respect to the raven paradox, it has a major problem. *Irrelevant conjunctions* can be tacked to the hypothesis H while preserving the confirmation relation (Glymour 1980a).

Tacking by conjunction problem: If H is confirmed by a piece of evidence E (relative to any K), $H \wedge X$ is confirmed by the same E for an arbitrary X that is consistent with H and K.

It is easy to see that this phenomenon is highly unsatisfactory: assume that the wave nature of light is confirmed by Young's double-slit experiment. According to the H-D account of confirmation, this implies that the following hypothesis is confirmed: "Light is an electromagnetic wave and the star Sirius is a giant bulb." This sounds completely absurd.

To see that H-D confirmation suffers from the tacking problem, let us just check the three conditions for H-D confirmation: assume that some hypothesis X is irrelevant to E, and that $H \wedge X \wedge K$ is consistent. Let us also assume $H \wedge K \vDash E$ and that K alone does not entail E. Then, E confirms not only H, but also $H \wedge X$ (because $H \wedge K \vDash E$ implies $H \wedge K \wedge X \vDash E$).

Thus, tacking an arbitrary irrelevant conjunct to a confirmed hypothesis preserves the confirmation relation. This is very unsatisfactory. More generally, H-D confirmation needs an answer to why a piece of evidence does not confirm every theory that implies it. Solving this problem is perhaps not impossible (Gemes 1993; Schurz 1991; Sprenger 2013), but it comes at the expense of major technical complications that compromise the simplicity and intuitive appeal of the hypothetico-deductive approach of confirmation.

In our discussion, several problems of qualitative confirmation have surfaced. First, qualitative confirmation is grounded on deductive relations between theory and evidence. These are quite an exception in modern, statistics-based science, which standardly deals with messy bodies of evidence. Second, we saw that few adequacy conditions have withstood the test of time, thus making times hard for developing a qualitative *logic* of induction. Third, no qualitative account measures *degree of confirmation* and tells strongly from weakly confirmed hypotheses, although this is essential for a great deal of scientific reasoning. Therefore, we now turn to quantitative explications of confirmation.

3 Probability as Degree of Confirmation

The use of probability as a tool for describing degree of confirmation can be motivated in various ways. Here are some major reasons.

First, probability is, as quipped by Cicero, "the guide to life." Judgments of probability motivate our actions: for example, the train I want to catch will probably be on time, so I have to run to catch it. Probability is used for expressing forecasts about events that affect our lives in manifold ways, from tomorrow's weather to global climate, from economic developments to the probability of a new Middle East crisis. This paradigm was elaborated by philosophers and scientists such as Ramsey (1926), De Finetti (1937), and Jeffrey (1965).

Second, probability is the preferred tool for uncertain reasoning in science. Probability distributions are used for characterizing the value of a particular physical quantity or for describing measurement error. Theories are assessed on the basis of probabilistic hypothesis tests. By phrasing confirmation in terms of probability, we hope to connect philosophical analysis of inductive inference to scientific practice and integrate the goals of normative and descriptive adequacy (Howson and Urbach 2006).

Third, statistics, the science of analyzing and interpreting data, is couched in probability theory. Statisticians have proved powerful mathematical results on the foundations of probability and inductive learning. Analyses of confirmation may benefit from them and have done so in the past (e.g., Good 2009). Consider, for example, the famous De Finetti (1974) representation theorem for subjective probability or the convergence results for prior probability distributions by Gaifman and Snir (1982).

Fourth and last, increasing the probability of a conclusion seems to be the hallmark of a sound inductive inference, as already noted by Hume. Probability theory, and the Bayesian framework in particular, are especially well-suited for capturing this intuition. The basic idea is to explicate degree of confirmation in terms of degrees of belief, which satisfy the axioms of probability. Degrees of belief are changed by *conditionalization* (if E is learned, $p_{new}(H) = p(H|E)$), and the posterior probability $p(H|E)$ stands as the basis of inference and decision-making. This quantity can be calculated via Bayes' theorem:

$$p(H \mid E) = p(H) \frac{p(E \mid H)}{p(E)}.$$

Chapter 20 of this handbook, concerning probability, provides more detail on the foundations of Bayesianism.

We now assume that degree of confirmation only depends on the joint probability distribution of the hypothesis H, the evidence E, and the background assumptions K. More precisely, we assume that E, H, and K are among the closed sentences $\mathfrak{L}$ of a language $\mathcal{L}$ that describes our domain of interest. A Bayesian theory of confirmation can be explicated by a function $\mathfrak{L}^3 \times \mathfrak{B} \to \mathbb{R}$, where $\mathfrak{B}$ is the set of probability measures on the algebra generated by $\mathfrak{L}$. This function assigns a real-valued degree of confirmation to any triple of sentences together with a probability (degree of belief) function. For the sake of simplicity, we will omit explicit reference to background knowledge since most accounts incorporate it by using the probability function $p(\cdot|K)$ instead of $p(\cdot)$.

A classical method for explicating degree of confirmation is to specify *adequacy conditions* on the concept and to derive a *representation theorem* for a confirmation measure. This means that one characterizes the set of measures (and possibly the unique measure) that satisfies these constraints. This approach allows for a sharp demarcation and mathematically rigorous characterization of the explicandum and, at the same time, for critical discussion of the explicatum by means of defending and criticizing the properties that are encapsulated in the adequacy conditions.

The first constraint is mainly of a formal nature and serves as a tool for making further constraints more precise and facilitating proofs (Crupi 2013):

Formality: For any sentences $H, E \in \mathfrak{L}$ with probability measure $p(\cdot)$, $c(H, E)$ is a measurable function from the joint probability distribution of H and E to a real number $c(H,E) \in \mathbb{R}$. In particular, there exists a function $f:[0,1]^3 \to \mathbb{R}$ such that $c(H, E) = f(p(H \wedge E), p(H), p(E))$.

Since the three probabilities $p(H \wedge E), p(H), p(E)$ suffice to determine the joint probability distribution of H and E, we can express $c(H, E)$ as a function of these three arguments.

Another cornerstone for Bayesian explications of confirmation is the following principle:

Final probability incrementality: For any sentences H, E, and $E' \in \mathfrak{L}$ with probability measure $p(\cdot)$,

$$c(H, E) > c(H, E') \quad \text{if and only if} \quad p(H|E) > p(H|E'), \text{ and}$$
$$c(H, E) < c(H, E') \quad \text{if and only if} \quad p(H|E) < p(H|E').$$

According to this principle, E confirms H more than E' does if and only if it raises the probability of H to a higher level. Degree of confirmation co-varies with boost in degree of belief, and satisfactory Bayesian explications of degree of confirmation should satisfy this condition.

There are now two main roads for adding more conditions, which will ultimately lead us to two different explications of confirmation: as *firmness* and as *increase in firmness* (or evidential support). They are also called the *absolute* and the *incremental concept of confirmation*.

Consider the following condition:

Local equivalence: For any sentences H, H', and $E \in \mathfrak{L}$ with probability measure $p(\cdot)$, if H and H' are logically equivalent given E $\left(\text{i.e.}, E \wedge H \vDash H', E \wedge H' \vDash H\right)$, then $c(H, E) = c(H', E)$.

The plausible idea behind local equivalence is that E confirms the hypotheses H and H' to an equal degree if they are logically equivalent conditional on E. If we buy into this intuition, local equivalence allows for a powerful (yet unpublished) representation theorem by Michael Schippers (see Crupi 2013):

Theorem 1: Formality, final probability incrementality, and local equivalence hold if and only if there is a nondecreasing function $g: [0,1] \to \mathbb{R}$ such that for any $H, E \in \mathfrak{L}$ and any $p(\cdot)$, $c(H, E) = g(p(H|E))$.

On this account, scientific hypotheses count as well-confirmed whenever they are sufficiently probable; that is, when $p(H|E)$ exceeds a certain (possibly context-relative) threshold. Hence, all confirmation measures that satisfy the three given constraints are *ordinally equivalent*; that is, they can be mapped on each other by means of a nondecreasing function. In particular, their confirmation rankings agree: if there are two functions g and g' that satisfy Theorem 1, with associated confirmation measures c and c', then $c(H, E) \geq c(H', E')$ if and only if $c'(H, E) \geq c'(H', E')$. Since confirmation as firmness is a monotonically increasing function of $p(H|E)$, it is natural to set up the qualitative criterion that E confirms H (in the absolute sense) if and only if $p(H|E) \geq t$ for some $t \in [0, 1]$.

A nice consequence of the view of confirmation as firmness is that some long-standing problems of confirmation theory, such as the paradox of irrelevant conjunctions, dissolve. Remember that, on the H-D account of confirmation, it was hard to avoid the conclusion that if E confirmed H, then it also confirmed $H \wedge H'$ for an arbitrary H'. On the view of confirmation as firmness, we automatically obtain $c(H \wedge H', E) \leq c(H, E)$. These quantities are nondecreasing functions of $p(H \wedge H'|E)$ and $p(H|E)$, respectively, and they differ the more the less plausible H' is and the less it coheres with H. Confirmation as firmness gives the intuitively correct response to the tacking by conjunction paradox.

It should also be noted that the idea of confirmation as firmness corresponds to Carnap's concept of probability$_1$ or "degree of confirmation" in his inductive logic. Carnap (1950) defines the degree of confirmation of a theory H relative to total evidence E as its probability conditional on E:

$$c(H, E) := p\,(H\,|\,E) = \frac{m(H \wedge E)}{m(E)},$$

where this probability is in turn defined by the measure m that state descriptions of the (logical) universe receive. By the choice of the measure m and a learning parameter λ, Carnap (1952) characterizes an entire continuum of inductive methods from which three prominent special cases can be derived. First, *inductive skepticism*: the degree of confirmation of a hypothesis is not changed by incoming evidence. Second, the rule of *direct inference*: the degree of confirmation of the hypothesis equals the proportions of observations in the sample for which it is true. Third, the *rule of succession* (de Laplace 1814): a prediction principle that corresponds to Bayesian inference with a uniform prior distribution. Carnap thus ends up with various inductive logics that characterize different attitudes toward ampliative inference.

Carnap's characterization of degree of confirmation does not always agree with the use of that concept in scientific reasoning. Above all, a confirmatory piece of evidence often provides a good *argument* for a theory even if the latter is unlikely. For instance, in the first years after Einstein invented the general theory of relativity (GTR), many scientists did not have a particularly high degree of belief in GTR because of its counterintuitive nature. However, it was agreed that GTR was well-confirmed by its predictive and explanatory successes, such as the bending of starlight by the sun and the explanation of the Mercury perihelion shift (Earman 1992). The account of confirmation as firmness fails to capture this intuition. The same holds for experiments in present-day science whose confirmatory strength is not evaluated on the basis of the posterior probability of the tested hypothesis H but by whether the results provide significant evidence in favor of H; that is, whether they are more expected under H than under $\neg H$.

This last point brings us to a particularly unintuitive consequence of confirmation as firmness: E could confirm H even if it *lowers* the probability of H as long as $p(H|E)$ is still large enough. But nobody would call an experiment where the results E are negatively statistically relevant to H a confirmation of H. This brings us to the following natural definition:

Confirmation as increase in firmness: For any sentences H, $E \in \mathcal{L}$ with probability measure $p(\cdot)$,

1. Evidence E *confirms/supports* hypothesis H (in the incremental sense) if and only if $p(H|E) > p(H)$.
2. Evidence E *disconfirms/undermines* hypothesis H if and only if $p(H|E) < p(H)$.
3. Evidence E is *neutral* with respect to H if and only if $p(H|E) = p(H)$.

In other words, E confirms H if and only E raises our degree of belief in H. Such explications of confirmation are also called *statistical relevance accounts* of confirmation because the neutral point is determined by the statistical independence of H and E. The analysis of confirmation as increase in firmness is the core business of Bayesian confirmation theory. This approach receives empirical support from findings by Tentori et al. (Tentori, Crupi, Bonini, and Osherson 2007): ordinary people use the concept of confirmation in a way that can be dissociated from posterior probability and that is strongly correlated with measures of confirmation as increase in firmness.

Confirmation as increase in firmness has interesting relations to qualitative accounts of confirmation and the paradoxes we have encountered. For instance, H-D confirmation now emerges as a special case: if H entails E, then $p(E|H) = 1$, and, by Bayes' theorem, $p(H|E) > p(H)$ (unless $p(E)$ was equal to one in the first place). We can also spot what is wrong with the idea of instance confirmation. Remember Nicod's (and Hempel's) original idea; namely, that universal generalizations such as $H = \forall x: Rx \rightarrow Bx$ are confirmed by their instances. This is certainly true relative to *some* background knowledge. However, it is not true under *all* circumstances. I. J. Good (1967) constructed a simple counterexample in a note for the *British Journal for the Philosophy of Science*: There are only two possible worlds, W_1 and W_2, whose properties are described by Table 9.1.

The raven hypothesis H is true whenever W_1 is the case and false whenever W_2 is the case. Conditional on these peculiar background assumptions, the observation of a black raven is evidence that W_2 is the case and therefore evidence that not all ravens are black:

$$P(Ra.Ba \,|\, W_1) = \frac{100}{1,000,100} < \frac{1,000}{1,001,001} = P(Ra.Ba \,|\, W_2).$$

By an application of Bayes' theorem, we infer $P(W_1|Ra.Ba) < P(W_1)$, and, given $W_1 \equiv H$, this amounts to a counterexample to NC. Universal conditionals are not always confirmed by their positive instances. We see how confirmation as increase in firmness elucidates our pre-theoretic intuitions regarding the theory–evidence relation: the relevant background assumptions make a huge difference as to when a hypothesis is confirmed.

Confirmation as increase in firmness also allows for a solution of the *comparative* paradox of the ravens. That is, we can show that, relative to weak and plausible background assumptions, $p(H|Ra \wedge Ba) < p(H|\neg Ra \wedge \neg Ba)$ (Fitelson and Hawthorne 2011, Theorem 2). By final probability incrementality, this implies that $Ra \wedge Ba$ confirms H more than $\neg Ra \wedge \neg Ba$ does. This shows, ultimately, why we consider a black raven to be more important evidence for the raven hypothesis than a white shoe.

Looking back to qualitative accounts once more, we see that Hempel's original adequacy criteria are mirrored in the logical properties of confirmation as firmness and increase in firmness. According to the view of confirmation as firmness, every consequence H' of a confirmed hypothesis H is confirmed, too (because $p(H') \geq p(H)$). This conforms to

Table 9.1 I. J. Good's (1967) counterexample to the paradox of the ravens

	W_1	W_2
Black ravens	100	1,000
Non-black ravens	0	1
Other birds	1,000,000	1,000,000

Hempel's special consequence condition. The view of confirmation as increase in firmness relinquishes this condition, however, and obtains a number of attractive results in return.

4 Degree of Confirmation: Monism or Pluralism?

So far, we have not yet answered the question of how degree of confirmation (or evidential support) should be quantified. For scientists who want to report the results of their experience and quantify the strength of the observed evidence, this is certainly the most interesting question. It is also crucial for giving a Bayesian answer to the Duhem-Quine problem (Duhem 1914). If an experiment fails and we ask ourselves which hypothesis to reject, the degree of (dis)confirmation of the involved hypotheses can be used to evaluate their standing. Unlike purely qualitative accounts of confirmation, a measure of degree of confirmation can indicate which hypothesis we should discard. For this reason, the search for a proper confirmation measure is more than a technical exercise: it is of vital importance for distributing praise and blame between different hypotheses that are involved in an experiment. The question, however, is which measure should be used. This is the question separating monists and pluralists in confirmation theory: monists believe that there is a single adequate or superior measure—a view that can be supported by theoretical reasons (Crupi, Tentori, and Gonzalez 2007; Milne 1996) and empirical research (e.g., coherence with folk confirmation judgments; see Tentori et al. 2007). Pluralists think that such arguments do not specify a single adequate measure and that there are several valuable and irreducible confirmation measures (e.g., Eells and Fitelson 2000; Fitelson 1999, 2001).

Table 9.2 provides a rough survey of the measures that are frequently discussed in the literature. We have normalized them such that for each measure $c(H, E)$, confirmation amounts to $c(H, E) > 0$, neutrality to $c(H, E) = 0$, and disconfirmation to $c(H, E) < 0$. This allows for a better comparison of the measures and their properties.

Evidently, these measures all have quite distinct properties. We shall now transfer the methodology from our analysis of confirmation as firmness and characterize them in terms of representation results. As before, formality and final probability incrementality will serve as minimal reasonable constraints on any measure of evidential support. Notably, two measures in the list, namely c' and s, are incompatible with final probability incrementality, and objections based on allegedly vicious symmetries have been raised against c' and r (Eells and Fitelson 2002; Fitelson 2001).

Here are further constraints on measures of evidential support that exploit the increase of firmness intuition in different ways:

Disjunction of alternatives: If H and H' are mutually exclusive, then

$$c(H, E) > c(H \vee H', E') \quad \text{if and only if} \quad p(H'|E) > p(H'),$$

Table 9.2 A list of popular measures of evidential support

Difference Measure	$d(H,E) = p(H\mid E) - p(H)$
Log-Ratio Measure	$r(H,E) = \log \dfrac{p(H\mid E)}{p(H)}$
Log-Likelihood Measure	$l(H,E) = \log \dfrac{p(E\mid H)}{p(E\mid \neg H)}$
Kemeny-Oppenheim Measure	$k(H,E) = \dfrac{p(E\mid H) - p(E\mid \neg H)}{p(E\mid H) + p(E\mid \neg H)}$
Rips Measure	$r'(H,E) = \dfrac{p(H\mid E) - p(H)}{1 - p(H)}$
Crupi-Tentori Measure	$z(H,E) = \begin{cases} \dfrac{p(H\mid E) - p(H)}{1 - p(H)} & \text{if } p(H\mid E) \geq p(H) \\[2ex] \dfrac{p(H\mid E) - p(H)}{p(H)} & \text{if } p(H\mid E) < p(H) \end{cases}$
Christensen-Joyce Measure	$s(H,E) = p(H\mid E) - p(H\mid \neg E)$
Carnap's Relevance Measure	$c'(H,E) = p(H \wedge E) - p(H)p(E)$

with corresponding conditions for $c(H, E) = c(H \vee H', E')$ and $c(H, E) < c(H \vee H', E')$.

That is, E confirms $H \vee H'$ more than H if and only if E is statistically relevant to H'. The idea behind this condition is that the sum $(H \vee H')$ is confirmed to a greater degree than each of the parts (H, H') when each part is individually confirmed by E.

Law of Likelihood:

$c(H, E) > c(H', E)$ if and only   if $p(E\mid H) > p(E\mid H')$,

with corresponding conditions for $c(H, E) = c(H', E')$ and $c(H, E) < c(H', E')$.

This condition has a long history of discussion in philosophy and statistics (e.g., Edwards 1972; Hacking 1965). The idea is that E favors H over H' if and only if the likelihood of H on E is greater than the likelihood of H' on E. In other words, E is more expected under H than under H'. The Law of Likelihood also stands at the basis of the *likelihoodist theory of confirmation*, which analyzes confirmation as a comparative relation between two competing hypotheses (Royall 1997; Sober 2008). Likelihoodists eschew judgments on how much E confirms H without reference to specific alternatives.

Modularity: If $p(E|H \wedge E') = p(E|H)$ and $p(E|\neg H \wedge E') = p(E|\neg H)$, then $c(H, E) = c_{|E'}(H, E)$ where $c_{|E'}$ denotes confirmation relative to the probability distribution conditional on E'.

This constraint screens off irrelevant evidence. If E' does not affect the likelihoods of H and $\neg H$ on E, then conditioning on E'—now supposedly irrelevant evidence—does not alter the degree of confirmation.

Contraposition/Commutativity: If E confirms H, then $c(H, E) = c(\neg E, \neg H)$; and if E disconfirms H, then $c(H, E) = c(E, H)$.

These constraints are motivated by the analogy of confirmation to partial deductive entailment. If $H, \vdash E$, then also $\neg E \vdash \neg H$, and if E refutes H, then H also refutes E. If confirmation is thought of as a generalization of deductive entailment to uncertain inference, then these conditions are very natural and reasonable (Tentori et al. 2007).

Combined with formality and final probability incrementality, each of these four principles singles out a specific measure of confirmation up to ordinal equivalence (Crupi 2013; Crupi, Chater, and Tentori 2013; Heckerman 1988):

Theorem 2 (representation results for confirmation measures):

1. If formality, final probability incrementality, and disjunction of alternatives hold, then there is a nondecreasing function g such that $c(H, E) = g(d(H, E))$.
2. If formality, final probability incrementality, and Law of Likelihood hold, then there is a nondecreasing function g such that $c(H, E) = g(r(H, E))$.
3. If formality, final probability incrementality, and modularity hold, then there are nondecreasing functions g and g' such that $c(H, E) = g(l(H, E))$ and $c(H, E) = g'$ $(k(H, E))$. Note that k and l are ordinally equivalent.
4. If formality, final probability incrementality, and commutativity hold, then there is a nondecreasing function g such that $c(H, E) = g(z(H, E))$.

That is, many confirmation measures can be characterized by means of a small set of adequacy conditions. It should also be noted that the *Bayes factor*, a popular measure of evidence in Bayesian statistics (Kass and Raftery 1995), falls under the scope of the theorem since it is ordinally equivalent to the log-likelihood measure l and the Kemeny and Oppenheim (1952) measure k. This is also evident from its mathematical form

$$\mathrm{BF}(H_0, H_1, E) := \frac{p(H_0|E)}{p(H_1|E)} \cdot \frac{p(H_1)}{p(H_0)} = \frac{p(E|H_0)}{p(E|H_1)}$$

for mutually exclusive hypotheses H_0 and H_1 (for which H and $\neg H$ may be substituted).

To show that the difference between these measures has substantial philosophical ramifications, let us go back to the problem of irrelevant conjunctions. If we analyze this problem in terms of the ratio measure r, then we obtain, assuming $H \vdash E$, that for an "irrelevant" conjunct H',

$$r(H \wedge H', E) = p(H \wedge H'|E) / p(H \wedge H') = p(E|H \wedge H') / p(E)$$
$$= 1 / p(E) = p(E|H) / p(E)$$
$$= r(H, E)$$

such that the irrelevant conjunction is supported to the same degree as the original hypothesis. This consequence is certainly unacceptable as a judgment of evidential support since H' could literally be any hypothesis unrelated to the evidence (e.g., "the star Sirius is a giant light bulb"). In addition, the result does not only hold for the special case of deductive entailment: it holds *whenever the likelihoods of H and H ∧ H' on E are the same*; that is, $p(E|H \wedge H') = p(E|H)$.

The other measures fare better in this respect: whenever $p(E|H \wedge H') = p(E|H)$, all other measures in Theorem 2 reach the conclusion that $c(H \wedge H', E) < c(H, E)$ (Hawthorne and Fitelson 2004). In this way, we can see how Bayesian confirmation theory improves on H-D confirmation and other qualitative accounts of confirmation: the paradox is acknowledged, but, at the same time, it is demonstrated how it can be mitigated.

That said, it is difficult to form preferences over the remaining measures. Comparing the adequacy conditions might not lead to conclusive results due to the divergent motivations that support them. Moreover, it has been shown that none of the remaining measures satisfies the following two conditions: (1) degree of confirmation is maximal if E implies H; (2) the a priori informativity (cashed out in terms of predictive content and improbability) of a hypothesis contributes to degree of confirmation (Brössel 2013, 389–390). This means that the idea of confirmation as a generalization of partial entailment and as a reward for risky predictions cannot be reconciled with each other, thus posing a further dilemma for confirmation monism. One may therefore go for pluralism and accept that there are different senses of degree of confirmation that correspond to different explications. For example, d strikes us as a natural explication of increase in subjective confidence, z generalizes deductive entailment, and l and k measure the discriminatory force of the evidence regarding H and $\neg H$.

Although Bayesian confirmation theory yields many interesting results and has sparked interests among experimental psychologists, too, one main criticism has been leveled again and again: that it misrepresents actual scientific reasoning. In the remaining sections, we present two major challenges for Bayesian confirmation theory fed by that feeling: the *problem of old evidence* (Glymour 1980*b*) and the rivaling frequentist approach to learning from experience (Mayo 1996).

5 THE PROBLEM OF OLD EVIDENCE

In this brief section, we expose one of the most troubling and persistent challenges for confirmation as increase in firmness: the problem of old evidence. Consider a phenomenon E that is unexplained by the available scientific theories. At some point, a theory H is discovered that accounts for E. Then, E is "old evidence": at the time when H is developed, the scientist is already certain or close to certain that the phenomenon E is real. Nevertheless, E apparently confirms H—at least if H was invented on independent grounds. After all, it resolves a well-known and persistent observational anomaly.

A famous case of old evidence in science is the Mercury perihelion anomaly (Earman 1992; Glymour 1980b). For a long time, the shift of the Mercury perihelion could not be explained by Newtonian mechanics or any other reputable physical theory. Then, Einstein realized that his GTR explained the perihelion shift. This discovery conferred a substantial degree of confirmation on GTR, much more than some pieces of novel evidence. Similar reasoning patterns apply in other scientific disciplines where new theories explain away well-known anomalies.

The reasoning of these scientists is hard to capture in the Bayesian account of confirmation as increase in firmness. E confirms H if and only if the posterior degree of belief in H, $p(H|E)$, exceeds the prior degree of belief in H, $p(H)$. When E is old evidence and already known to the scientist, the prior degree of belief in E is maximal: $p(E) = 1$. But with that assumption, it follows that the posterior probability of H cannot be greater than the prior probability: $p(H|E) = p(H) \cdot p(E|H) \leq p(H)$. Hence, E does not confirm H. The very idea of confirmation by old evidence, or equivalently, confirmation by accounting for well-known observational anomalies, seems impossible to describe in the Bayesian belief kinematics. Some critics, like Clark Glymour, have gone so far to claim that Bayesian confirmation only describes *epiphenomena* of genuine confirmation because it misses the relevant structural relations between theory and evidence.

There are various solution proposals to the problem of old evidence. One approach, adopted by Howson (1984), interprets the confirmation relation with respect to counterfactual degrees of belief, where E is subtracted from the agent's actual background knowledge. Another approach is to claim that confirmation by old evidence is not about learning the actual evidence, but about *learning a logical or explanatory relation between theory and evidence*. It seems intuitive that Einstein's confidence in GTR increased on learning that it implied the perihelion shift of Mercury and that this discovery was the real confirming event.

Indeed, confirmation theorists have set up Bayesian models where learning $H \vdash E$ increases the probability of H (e.g., Jeffrey, 1983) under certain assumptions. The question is, of course, whether these assumptions are sufficiently plausible and realistic. For critical discussion and further constructive proposals, see Earman (1992) and Sprenger (2015a).

6 Bayesianism and Frequentism

A major alternative to Bayesian confirmation theory is *frequentist inference*. Many of its principles have been developed by the geneticist and statistician R. A. Fisher (see Neyman and Pearson [1933] for a more behavioral account). According to frequentism, inductive inference does not concern our degrees of belief. That concept is part of individual psychology and not suitable for quantifying scientific evidence. Instead of expressing degrees of belief, probability is interpreted as the limiting frequency of an event in a large number of trials. It enters inductive inference via the concept of a sampling distribution; that is, the probability distribution of an observable in a random sample.

The basic method of frequentist inference is hypothesis testing and, more precisely, NHST. For Fisher, the purpose of statistical analysis consists in assessing the relation of a hypothesis to a body of observed data. The tested hypothesis usually stands for the absence of an interesting phenomenon (e.g., no causal relationship between two variables, no observable difference between two treatments, etc.). Therefore it is often called the default or *null hypothesis* (or *null*). In remarkable agreement with Popper, Fisher states that the only purpose of an experiment is to "give the facts a chance of disproving the null hypothesis" (Fisher 1925, 16): the purpose of a test is to find evidence *against* the null. Conversely, failure to reject the null hypothesis does not imply positive evidence for the null (on this problem, see Popper 1954; Sprenger 2015*b*).

Unlike Popper (1959), Fisher aims at experimental and statistical *demonstrations* of a phenomenon. Thus, he needs a criterion for when an effect is real and not an experimental fabrication. He suggests that we should infer to such an effect when the observed data are too improbable under the null hypothesis:

"either an exceptionally rare chance has occurred, or the theory [= the null hypothesis] is not true" (Fisher 1956, 39).

This basic scheme of inference is called *Fisher's disjunction* by (Hacking 1965), and it stands at the heart of significance testing. It infers to the falsity of the null hypothesis as the best explanation of an unexpected result (for criticism, see Royall 1997; Spielman 1974).

Evidence against the null is measured by means of the *p-value*. Here is an illustration. Suppose that we want to test whether the real-valued parameter θ, our quantity of interest, diverges "significantly" from H_0: $\theta = \theta_0$. We collect independent and identically distributed (i.i.d.) data x: $= (x_1, \ldots, x_N)$ whose distribution is Gaussian and centered around θ. Assume now that the population variance σ^2 is known, so $x_i \sim N(\theta, \sigma^2)$ for each x_i. Then, the discrepancy in the data x with respect to the postulated mean value θ_0 is measured by means of the statistic

$$z(x) := \frac{\frac{1}{N}\sum_{i=1}^{N} x_i - \theta_0}{\sqrt{N \cdot \sigma^2}} .$$

We may reinterpret this equation as

$$z = \frac{\text{observed effect} - \text{hypothesized effect}}{\text{standard error}}$$

Determining whether a result is significant or not depends on the p-value or *observed significance level*; that is, the "tail area" of the null under the observed data. This value depends on z and can be computed as

$$p_{obs} := p\left(\left|z(X)\right| \geq \left|z(x)\right|\right),$$

that is, as the probability of observing a more extreme discrepancy under the null than the one which is actually observed. Figure 9.1 displays an observed significance level $p = 0.04$ as the integral under the probability distribution function—a result that would typically count as substantial evidence against the null hypothesis ("$p < .05$").

For the frequentist practitioner, p-values are practical, replicable, and objective measures of evidence against the null: they can be computed automatically once the statistical model is specified, and they only depend on the sampling distribution of the data under H_0. Fisher interpreted them as "a measure of the rational grounds for the *disbelief* [in the null hypothesis] it augments" (Fisher 1956, 43).

The virtues and vices of significance testing and p-values have been discussed at length in the literature (e.g., Cohen 1994; Harlow, Mulaik, and Steiger 1997), and it would go beyond the scope of this chapter to deliver a comprehensive critique. By now, it is

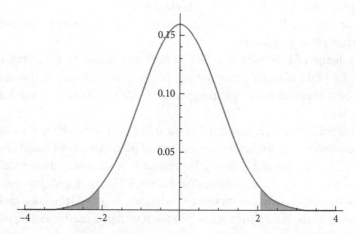

FIGURE 9.1 The probability density function of the null H_0: $X \sim N(0, 1)$, which is tested against the alternative H_1: $X \sim N(\theta, 1)$, $\theta \neq 0$. The shaded area illustrates the calculation of the p-value for an observed z-value of $z = \pm 2.054$ ($p = 0.04$).

consensus that inductive inference based on p-values leads to severe epistemic and practical problems. Several alternatives, such as confidence intervals at a predefined level α, have been promoted in recent years (Cumming 2014; Cumming and Finch 2005). They are interval estimators defined as follows: for each possible value θ' of the unknown parameter θ, we select the interval of data points x that will not lead to a statistical rejection of the null hypothesis $\theta = \theta'$ in a significance test at level α. Conversely, the confidence interval for θ, given an observation x, comprises all values of θ that are *consistent* with x in the sense of surviving a NHST at level α.

We conclude by highlighting the principal philosophical difference between Bayesian and frequentist inference. The following principle is typically accepted by Bayesian statisticians and confirmation theorists alike:

> *Likelihood Principle (LP)*: Consider a statistical model $\mathcal{M}$ with a set of probability measures $p(\cdot|\theta)$ parametrized by a parameter of interest $\theta \in \Theta$. Assume we conduct an experiment $\mathcal{E}$ in $\mathcal{M}$. Then, all evidence about θ generated by $\mathcal{E}$ is contained in the *likelihood function* $p(x|\theta)$, where the observed data x are treated as a constant. (Birnbaum, 1962)

Indeed, in the simple case of only two hypotheses (H and $\neg H$), the posterior probabilities are only a function of $p(E|H)$ and $p(E|\neg H)$, given the prior probabilities. This is evident from writing the well-known Bayes' theorem as

$$p(H\,|\,E) = \left(1 + \frac{p(\neg H)}{p(H)} \frac{p(E\,|\,\neg H)}{p(E\,|\,H)} \right)^{-1}.$$

So Bayesians typically accept the LP, as is also evident from the use of Bayes factors as a measure of statistical evidence.

Frequentists reject the LP: his or her measures of evidence, such as p-values, are based on the probability of results that *could have happened, but did actually not happen*. The evidence depends on whether the actual data fit the null better or worse than most other possible data (see Figure 9.1). By contrast, Bayesian induction is "actualist": the only thing that matters for evaluating the evidence and making decisions is the predictive performance of the competing hypotheses on the actually observed evidence. Factors that determine the probability of possible but unobserved outcomes, such as the experimental protocol, the intentions of the experimenter, the risk of early termination, and the like, may have a role in experimental design, but they do not matter for measuring evidence post hoc (Edwards, Lindman, and Savage 1963; Sprenger 2009).

The likelihood principle is often seen as a strong argument for preferring Bayesian to frequentist inference (e.g., Berger and Wolpert 1984). In practice, statistical data analysis still follows frequentist principles more often than not: mainly because, in many applied problems, it is difficult to elicit subjective degrees of belief and to model prior probability distributions.

7 CONCLUSION

This chapter has given an overview of the problem of induction and the responses that philosophers of science have developed over time. These days, the focus is not so much on providing an answer to Hume's challenge: it is well-acknowledged that no purely epistemic, noncircular justification of induction can be given. Instead, focus has shifted to characterizing valid inductive inferences, carefully balancing attractive theoretical principles with judgments and intuitions in concrete cases. That this is not always easy has been demonstrated by challenges such as the paradox of the ravens, the problem of irrelevant conjunctions, and Goodman's new riddle of induction.

In the context of this project, degree of confirmation becomes especially important: it indicates to what extent an inductive inference is justified. Explications of confirmation can be distinguished into two groups: qualitative and quantitative ones. The first serve well to illustrate the "grammar" of the concept, but they have limited applicability.

We motivated why probability is an adequate tool for explicating degree of confirmation and investigated probabilistic (Bayesian) confirmation measures. We distinguished two senses of confirmation—confirmation as firmness and confirmation as increase in firmness—and investigated various confirmation measures. That said, there are also alternative accounts of inductive reasoning, some of which are nonprobabilistic, such as objective Bayesianism (Williamson 2010), ranking functions (Spohn 1990), evidential probability (Kyburg 1961) and the Dempster-Shafer theory of evidence (Shafer 1976; see also Haenni, Romeijn, Wheeler, and Williamson 2011).

Finally, we provided a short glimpse of the methodological debate between Bayesians and frequentists in statistical inference. Confirmation theory will have to engage increasingly more often with debates in statistical methodology if it does not want to lose contact with inductive inference in science—which was Bacon's target in the first place.

ACKNOWLEDGMENTS

I would like to thank Matteo Colombo, Vincenzo Crupi, Raoul Gervais, Paul Humphreys, and Michael Schippers for their valuable feedback on this chapter. Research on this chapter was supported through the Vidi project "Making Scientific Inferences More Objective" (grant no. 276-20-023) by the Netherlands Organisation for Scientific Research (NWO) and ERC Starting Investigator Grant No. 640638.

SUGGESTED READINGS

For qualitative confirmation theory, the classical texts are Hempel (1945a, 1945b). For an overview of various logics of inductive inference with scientific applications, see Haenni et al. (2011). A classical introduction to Bayesian reasoning, with comparison to frequentism, is given by Howson and Urbach (2006). Earman (1992) and Crupi (2013) offer comprehensive reviews of Bayesian confirmation theory, and Good (2009) is an exciting collection of essays on induction, probability, and statistical inference.

REFERENCES

Bacon, F. (1620). *Novum Organum; Or, True Suggestions for the Interpretation of Nature* (London: William Pickering).

Berger, J., and Wolpert, R. (1984). *The Likelihood Principle* (Hayward, CA: Institute of Mathematical Statistics).

Birnbaum, A. (1962). "On the Foundations of Statistical Inference." *Journal of the American Statistical Association* 57(298): 269–306.

Broad, C. D. (1952). *Ethics and the History of Philosophy* (London: Routledge).

Brössel, P. (2013). "The Problem of Measure Sensitivity Redux." *Philosophy of Science* 80(3): 378–397.

Carnap, R. (1950). *Logical Foundations of Probability* (Chicago: University of Chicago Press).

Carnap, R. (1952). *Continuum of Inductive Methods* (Chicago: University of Chicago Press).

Cohen, J. (1994). "The Earth Is Round (p<.05)." *Psychological Review* 49: 997–1001.

Crupi, V. (2013). "Confirmation." In E. Zalta (ed.), *The Stanford Encyclopedia of Philosophy*. Retrieved on December 28, 2015, at http://plato.stanford.edu/entries/confirmation/.

Crupi, V., Chater, N., and Tentori, K. (2013). "New Axioms for Probability and Likelihood Ratio Measures." *British Journal for the Philosophy of Science* 64(1): 189–204.

Crupi, V., Tentori, K., and Gonzalez, M. (2007). "On Bayesian Measures of Evidential Support: Theoretical and Empirical Issues." *Philosophy of Science* 74: 229–252.

Cumming, G. (2014): "The New Statistics: Why and How." *Psychological Science* 25: 7–29.

Cumming, G. and Finch, S. (2005). "Inference by Eye: Confidence Intervals and How to Read Pictures of Data." *American Psychologist* 60(2): 170–180.

De Finetti, B. (1937). "La Prévision: ses Lois Logiques, ses Sources Subjectives." *Annales de l'institut Henri Poincaré* 7: 1–68.

De Finetti, B. (1974). *Theory of Probability*. Vol. 1 (New York: John Wiley & Sons).

de Laplace, P. S. (1814). *A Philosophical Essay on Probabilities* (Mineola, NY: Dover).

Descartes, R. (1637). *Discours de la méthode* (Leiden: Jan Maire).

Duhem, P. (1914). *La Théorie Physique: Son Objet, Sa Structure* (Paris: Vrin).

Earman, J. (1992). *Bayes or Bust? A Critical Examination of Bayesian Confirmation Theory* (Cambridge, MA: MIT Press).

Edwards, A. (1972). *Likelihood* (Cambridge: Cambridge University Press).

Edwards, W., Lindman, H., and Savage, L. J. (1963). "Bayesian Statistical Inference for Psychological Research." *Psychological Review* 70: 193–242.

Eells, E., and Fitelson, B. (2000). "Measuring Confirmation and Evidence." *Journal of Philosophy* 97(12): 663–672.

Eells, E., and Fitelson, B. (2002). "Symmetries and Asymmetries in Evidential Support." *Philosophical Studies* 107(2): 129–142.

Fisher, R. (1956). *Statistical Methods and Scientific Inference* (New York: Hafner).

Fisher, R. A. (1925). *Statistical Methods for Research Workers* (Edinburgh: Oliver & Boyd).

Fitelson, B. (1999). "The Plurality of Bayesian Measures of Confirmation and the Problem of Measure Sensitivity." In *Philosophy of Science*, Vol. 66 (Chicago: University of Chicago Press), S362–S378.

Fitelson, B. (2001). *Studies in Bayesian Confirmation Theory.* PhD thesis, University of Wisconsin–Madison.

Fitelson, B., and Hawthorne, J. (2011). "How Bayesian Confirmation Theory Handles the Paradox of the Ravens." In J. H. Fetzer and E. Eells (eds.), *The Place of Probability in Science* (New York: Springer), 247–275.

Gaifman, H., and Snir, M. (1982). "Probabilities Over Rich Languages, Testing and Randomness." *Journal of Symbolic Logic* 47(3): 495–548.

Gemes, K. (1993). "Hypothetico-Deductivism, Content and the Natural Axiomatisation of Theories." *Philosophy of Science* 60: 477–487.

Gemes, K. (1998). "Hypothetico-Deductivism: The Current State of Play; the Criterion of Empirical Significance: Endgame." *Erkenntnis* 49(1): 1–20.

Glymour, C. (1980a). "Hypothetico-Deductivism Is Hopeless." *Philosophy of Science* 47(2): 322–325.

Glymour, C. (1980b). *Theory and Evidence* (Princeton, NJ: Princeton University Press).

Goldman, A. I. (1986). *Epistemology and Cognition* (Cambridge, MA: Harvard University Press).

Good, I. (2009). *Good Thinking* (Mineola, NY: Dover).

Good, I. J. (1967). "The White Shoe Is a Red Herring." *British Journal for the Philosophy of Science* 17(4): 322.

Goodman, N. (1955). *Fact, Fiction and Forecast* (Cambridge, MA: Harvard University Press).

Hacking, I. (1965). *Logic of Statistical Inference* (Cambridge: Cambridge University Press).

Haenni, R., Romeijn, J. -W., Wheeler, G., and Williamson, J. (2011). *Probabilistic Logic and Probabilistic Networks* (Berlin: Springer).

Harlow, L. L., Mulaik, S. A., and Steiger, J. H. (1997). *What If There Were No Significance Tests?* (Mahway, NJ: Erlbaum).

Hawthorne, J., and Fitelson, B. (2004). "Re-Solving Irrelevant Conjunction with Probabilistic Independence." *Philosophy of Science* 71: 505–514.

Heckerman, D. (1988). "An Axiomatic Framework for Belief Updates." In J. F. Lemmer and L. N. Kanal (eds.), *Uncertainty in Artificial Intelligence 2* (Amsterdam: North-Holland), 11–22.

Hempel, C. G. (1945a). "Studies in the Logic of Confirmation [I]." *Mind* 54(213): 1–26.

Hempel, C. G. (1945b). "Studies in the Logic of Confirmation [II]." *Mind* 54(214): 97–121.

Howson, C. (1984). "Bayesianism and Support by Novel Facts." *British Journal for the Philosophy of Science* 34: 245–251.

Howson, C., and Urbach, P. (2006). *Scientific Reasoning: The Bayesian Approach*, 3rd ed. (La Salle, IL: Open Court).

Hume, D. (1739). *A Treatise of Human Nature* (Oxford: Clarendon Press).

Hume, D. (1748). *Enquiry Concerning Human Understanding* (Oxford: Clarendon Press).

Jeffrey, R. C. (1965). *The Logic of Decision*, 2nd ed. (Chicago: University of Chicago Press).

Jeffrey, R. C. (1983). "Bayesianism with a Human Face." In J. Earman (ed.), *Testing Scientific Theories* (Minneapolis: University of Minnesota Press), 133–156.

Kass, R. E., and Raftery, A. E. (1995). "Bayes Factors." *Journal of the American Statistical Association* 90: 773–795.

Kemeny, J. G., and Oppenheim, P. (1952). "Degree of Factual Support." *Philosophy of Science* 19: 307–324.

Kyburg, H. E. (1961). *Probability and the Logic of Rational Belief* (Middletown, CT: Wesleyan University Press).

Mayo, D. G. (1996). *Error and the Growth of Experimental Knowledge* (Chicago: University of Chicago Press).

Mayo, D. G., and Spanos, A. (2006). "Severe Testing as a Basic Concept in a Neyman-Pearson Philosophy of Induction." *British Journal for the Philosophy of Science* 57: 323–357.

Milne, P. (1996). "log[P(h/eb)/P(h/b)] Is the One True Measure of Confirmation." *Philosophy of Science* 63: 21–26.

Neyman, J., and Pearson, E. S. (1933). "On the Problem of the Most Efficient Tests of Statistical Hypotheses." *Philosophical Transactions of the Royal Society A* 231: 289–337.

Nicod, J. (1961). *Le problème logique de l'induction* (Paris: Presses Universitaires de France).

Norton, J. D. (2003). "A Material Theory of Induction." *Philosophy of Science* 70(4): 647–670.

Popper, K. (1954). "Degree of Confirmation." *British Journal for the Philosophy of Science* 5: 143–149.

Popper, K. R. (1959). *The Logic of Scientific Discovery* (London: Hutchinson).

Popper, K. R. (1983). *Realism and the Aim of Science* (Towota, NJ: Rowman & Littlefield).

Ramsey, F. P. (1926). "Truth and Probability." In D. H. Mellor (ed.), *Philosophical Papers* (Cambridge: Cambridge University Press), 52–94.

Royall, R. (1997). *Statistical Evidence: A Likelihood Paradigm* (London: Chapman & Hall).

Schurz, G. (1991). "Relevant Deduction." *Erkenntnis* 35: 391–437.

Shafer, G. (1976). *A Mathematical Theory of Evidence* (Princeton, NJ: Princeton University Press).

Sober, E. (2008). *Evidence and Evolution: The Logic Behind the Science* (Cambridge: Cambridge University Press).

Spielman, S. (1974). "The Logic of Tests of Significance." *Philosophy of Science* 41(3): 211–226.

Spohn, W. (1990). "A General Non-Probabilistic Theory of Inductive Reasoning." In R. D. Shachter, T. S. Levitt, J. Lemmer, and L. N. Kanal (eds.), *Uncertainty in Artificial Intelligence* 4, 149–158 (Amsterdam: Elsevier).

Sprenger, J. (2009). "Evidence and Experimental Design in Sequential Trials." *Philosophy of Science* 76: 637–649.

Sprenger, J. (2010). "Hempel and the Paradoxes of Confirmation." In D. M. Gabbay, S. Hartmann, and J. Woods (eds.), *Handbook of the History of Logic*, Vol. 10 (Amsterdam: North-Holland), 235–263.

Sprenger, J. (2011). "Hypothetico-Deductive Confirmation." *Philosophy Compass* 6(7): 497–508.

Sprenger, J. (2013). "A Synthesis of Hempelian and Hypothetico-Deductive Confirmation." *Erkenntnis* 78: 727–738.

Sprenger, J. (2015a). "A Novel Solution of the Problem of Old Evidence." *Philosophy of Science* 82: 383–401.

Sprenger, J. (2015b). "Two Impossibility Results for Measures of Corroboration." Forthcoming in *The British Journal for Philosophy of Science*.

Tentori, K., Crupi, V., Bonini, N., and Osherson, D. (2007). "Comparison of Confirmation Measures." *Cognition* 103: 107–119.

Vickers, J. (2010). "The Problem of Induction." In E. N. Zalta (ed.), *The Stanford Encyclopedia of Philosophy*. Fall 2010 edition. Retrieved on December 28, 2015, at http://plato.stanford.edu/entries/induction-problem/.

Whewell, W. (1847). *Philosophy of the Inductive Sciences, Founded Upon Their History* (London: Parker).

Williamson, J. (2010). *In Defence of Objective Bayesianism* (Oxford: Oxford University Press).

CHAPTER 10

....

DETERMINISM AND INDETERMINISM

....

CHARLOTTE WERNDL

1 INTRODUCTION

....

DETERMINISM reigns when the state of the system at one time fixes the past and future evolution of the system. The question of determinism can be asked about real systems (i.e., whether the state of a real system at one time fixes the state of the system at all times) or about models of real systems (i.e., whether the state of a model at one time fixes the state of the model at all times). *Indeterminism* amounts simply to the negation of determinism. Of course one usually uses models to arrive at claims about the deterministic characters of real systems, but, as we will see, the relationship between deterministic systems and models is not at all straightforward.

This article focuses on three major themes in the recent debate on determinism in the philosophy of science. Throughout the article, emphasis is placed not just on summarizing the debates but also on presenting some novel criticism and arguments. The first major theme is determinism, indeterminism, and observational equivalence. Here I critically discuss various notions of observational equivalence between deterministic and indeterministic systems and whether there is underdetermination between deterministic and indeterministic models (Section 2). The second major theme is whether Newtonian mechanics is indeterministic and how scientists' debates in the nineteenth century differ from the contemporary debate (Section 3). The third major theme is how probabilities can arise in deterministic systems. Here I stress the usefulness of the method of arbitrary functions for understanding deterministic probabilities (Section 4). The article ends with a conclusion (Section 5).

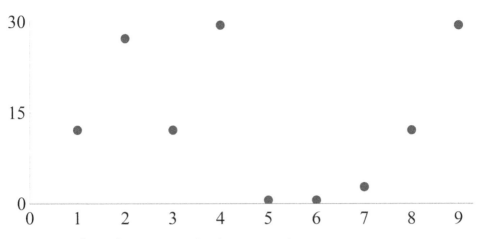

FIGURE 10.1 Observed temperature in London over nine days.

2 DETERMINISM, INDETERMINISM, AND OBSERVATIONAL EQUIVALENCE

2.1 Deterministic and Indeterministic Models

Consider the evolution of the temperature in London (which is assumed to take values between 0 and 30 degrees). A meteorologist measures the temperature over nine days and obtains a sequence of observations as shown in Figure 10.1. Meteorologists aim to find a model that reproduces these observations and correctly predicts the future temperature values. In this context, the question arises whether the temperature evolution is best described by a deterministic or an indeterministic model, and one might think that the observations allow for only one or the other. However, in several cases, including the example of the evolution of the temperature, both a deterministic and an indeterministic model is possible (i.e., there is observational equivalence between deterministic and indeterministic models). This raises the question regarding which model is preferable and whether there is underdetermination. To tackle these questions, I first introduce deterministic and indeterministic models.

The deterministic models we focus on are measure-theoretic deterministic models (M, T_t, p)[1]. Here M is the set of possible states (the *phase space*), where $m \in M$ represents the state of the system. The functions $T_t: M \to M$ are the *evolution equations*, telling one that a state $m \in M$ evolves to $T_t(m)$ after t time steps $(t \in \mathbb{Z})$. p is a probability measure,

---

[1] For technical details see Werndl (2009a, 2011). For simplicity, I focus on models with discrete time, but all that is said carries over to models with continuous time (cf. Werndl 2011).

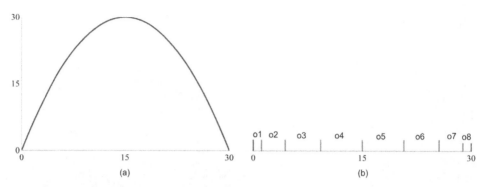

FIGURE 10.2 (a) the logistic map $T(m)$; (b) the observation function Φ_8.

assigning a probability to regions of M.[2] The *solution* through m represents a possible path of the deterministic system over time. Formally, it is the bi-infinite sequence $\left(\ldots T_{-2}(m),\, T_{-1}(m),\, m,\, T_1(m),\, T_2(m)\ldots\right)$. Clearly, because T_t are functions, the models (M, T_t, p) are deterministic: the initial state m fixes the past and future evolution of the system. These deterministic models are among the most important models in science (e.g., including all Newtonian models of energy-conserving systems).

In the observation of a deterministic system, a value is observed that is dependent on, but usually different from, the actual value (since observations cannot be done with infinite precision). Formally, an observation corresponds to an *observation function*, that is, a function $\Phi : M \to M_O$, where $\Phi(m)$ represents the observed value (M_O is the set of all possible observation values). Since we can make observations with only finite precision, in what follows it is assumed that observation functions take only finitely many values.

Lorenz (1964) used the logistic map to model the evolution of the temperature. Since this is a very simple model, I use it for illustration purposes. More specifically, the phase space M of the logistic map is $[0,30]$ (representing the temperature values between 0 and 30). The temperature at day $t+1$ is obtained from the temperature m at day t by the equation (cf. Figure 10.2):

$$T(m) = 4m\left(1 - \frac{m}{30}\right). \tag{1}$$

Hence the evolution equations are given by $T_t(m) := T^t(m)$ (where T_t is the t-th iterate of m).[3] The probability measure assigns to a region A of $[0; 30]$ the value:

---

[2] There are various interpretations of this probability measure from being a physical quantity that describes the probability of finding a system in a certain region of phase space to the long-run average of the proportion of time a solution spends in a certain region (cf. Lavis 2011).

[3] The logistic map is only forward deterministic. That is, the state of the model at one time determines all the future states but not the past states. Nothing hinges on this, and all the results presented in this section carry over to systems that are only forward deterministic.

$$p(A) = \int_A \frac{30}{\pi \sqrt{\dfrac{m}{30}\left(1 - \dfrac{m}{30}\right)}} \, dm. \tag{2}$$

Consider the observation function Φ_8 of the logistic map with eight values $\Phi_8\,(m)$ = $o1$ = 0.5709505 for $0 \le m < 1.14181$, $\Phi_8\,(m)$ = $o2$ = 2.767605 for $1.14181 \le m < 4.3934$, $\Phi_8\,(m)$ = $o3$ = 6.82605 for $4.3934 \le m\ 9.2587$, $\Phi_8\,(m)$ = $o4$ = 12.12935 for $9.2587 \le m\ 15$, $\Phi_8\,(m)$ = $o5\ 17.87015$ for $15 \le m < 20.7403$, $\Phi_8\,(m)$ = $o6$ =23.17345 for $20.7403 \le m <$ 25.6066, $\Phi_8\,(m)$= $o7$ =27.2324 for $25.6066 \le m < 28.8582$, and $\Phi_8\,(m)$ = $o8$ = 29.4291 for $28.8582 \le m < 30$ Figure 10.2) shows this observation function. Suppose that the initial temperature is 0.3355. Then the first nine iterates coarse-grained by the observation function Φ_8 are: $o4$, $o7$, $o4$, $o8$, $o1$, $o1$, $o2$, $o4$, $o8$. This is the sequence shown in Figure 10.1. Consequently, the time series of Figure 10.1 can be derived from observations of the logistic system.

The indeterministic models we focus on are *stochastic models* $\{Z_t\}$ with a finite number of outcomes E (representing systems that evolve according to probabilistic laws). Here Z_t represents the outcome of the system at time t. Probability distributions characterize the probabilistic behavior of the stochastic system: for example, the probability distribution $P\,(Z_t = e)$ gives one the probability that the outcome of the system is e at time t; joint probability distributions $P\,(Z_s = e$ and $Z_t = f)$ tell one the probabilities of outcomes at different times; conditional probability distributions P $(Z_s = e$ given that $Z_t = f)$ tell one the probability that the outcome is e at s given that it was f at t (for any s and t). A *realisation* represents a possible evolution of the stochastic system over time. Formally, it is a bi-infinite sequence $\left(...Z_{-2},\ Z_{-1},\ Z_0,\ Z_1,\ Z_2 ...\right)$. A stochastic model is indeterministic in the sense that, given the initial outcome, there are several outcomes that might follow (and these possibilities are measured by probabilities). Most indeterministic models in science are stochastic models. For stochastic systems observations are also modeled by *observation functions*, that is, functions $\Gamma : E \rightarrow E_O$ where $\Gamma\,(e)$ represents the observed value (E_0 is the set of all possible observed values).

Probably the best-known stochastic models are *Bernoulli models* (representing a sequence of identically distributed random experiments, where the outcomes are independent of each other, as for a sequence of coin tosses or a sequence of throwing dice). Also widely used in science are Markov models (representing a sequence of identically distributed random experiments, where the next outcome depends only on the previous outcome). Consider the following specific Markov model $\{V_t\}$: there are eight possible states $o1 = 0.570905$ $o2 = 2.767605$, $o3 = 6.82605$, $o4 = 12.12935$, $o5 = 17.87015$, $o6 = 23.17345$, $o7 = 27.2324$, $o8 = 29.4291$, which each have probability 1/8. Each state can be followed by two other states, and the probability that a state is followed by any the two other states is 1/2. More specifically, $o1$ can be followed by $o1$ or $o2$, $o2$ by $o3$ or $o4$, $o3$ by $o5$ or $o6$, $o4$ by $o7$ or $o8$, $o5$ by $o7$ or $o8$, $o6$ by $o5$ or $o6$, $o7$ by $o3$ or $o4$, $o8$ by $o1$ or $o2$. For one of the realizations of $\{V_t\}$ the entries from time 0 to 9 are: $o4$, $o7$, $o4$, $o8$, $o1$, $o1$, $o2$, $o4$, $o8$.

This is the sequence shown in Figure 10.1. Therefore, the time series shown in Figure 10.1 can also derive from a Markov model. Recall that the very same time series can also arise from the logistic system. This raises the question of observational equivalence, to which we now turn.

2.2 Observational Equivalence

Observational equivalence occurs when *the deterministic model, relative to an observation function* Φ, *and the stochastic model, relative to an observation function* Γ *give the same predictions* (formally, this kind of observational equivalence is called manifest isomorphism—cf. Werndl 2009a). What it means to "give the same predictions" needs further elaboration. The predictions obtained from a stochastic model are the probability distributions over its realizations coarse-grained by Γ. Recall that a probability measure p is defined on the phase space of a deterministic model. Therefore, the predictions derived relative to an observation function Φ are the probability distributions over the solutions coarse-grained by Φ. Therefore "give the same predictions" means that (i) the possible values of the observation function Γ of the stochastic model and of the observation function Φ of the deterministic model are the same, and (ii) the probability distributions over the realizations of the stochastic model coarse-grained by Γ and the probability distributions over the solutions of the deterministic model coarse-grained by Φ are the same.

Given a deterministic model (M, T_t, p) and an observation function $\Phi: M \to M_O$, can an observationally equivalent stochastic model be found? Yes: $\{Z_t\} := \{\Phi(T_t)\}$ is a stochastic model, constructed by applying the observation function to the deterministic model. The possible values of $\{\Phi(T_t)\}$ are the same as the possible observed values of (M, T_t, p). Further, the realizations of $\{\Phi(T_t)\}$ and the solutions of (M, T_t, p) coarse-grained by Φ have the same probability distributions. Hence *there is observational equivalence* between the stochastic model $\{\Phi(T_t)\}$ (assuming that all values can be observed, i.e. that Γ is the identity function) and the deterministic model (M, T_t, p) relative to Φ. One might wonder whether $\{\Phi(T_t)\}$ has only trivial probabilities (0 and 1) because it derives from applying an observation function to a deterministic model. However, importantly, this is *not* so. It can be shown that in several cases the stochastic model $\{\Phi(T_t)\}$ is *nontrivial* (i.e., there are probabilities assigned to outcomes that are strictly between 0 and 1; Werndl 2009a, 2011).

This result can be illustrated with the example of the evolution of the temperature. We know that we can describe the evolution of the temperature by the logistic map (Equation 1). The set of possible values of the observation function Φ_8 is the same as the set of all possible outcomes of the stochastic model $\{V_t\} := \{\Phi_8(T_t)\}$. The probability distributions of this stochastic model are determined by applying Φ_8 to the logistic map and hence are identical to those of $\{\Phi_8(T_t)\}$. Instances of these identical probability distributions are $p\big(\Phi_8\big(T_t\big)=o1\big)=P\big(Z_t=o1\big)$ or $p\big(\Phi_8(T_{t+1})=o2$ given that $\Phi_8\big(T_t\big)=o1\big)=P\big(Z_{t+1}=o2$ given that $Z_t=o2\big)$ for all $t \in \mathbb{N}$ Hence the conclusion is that the logistic map, relative to Φ_8, and the stochastic model $\big\{\Phi_8\big(T_t\big)\big\}$ (when all values are observed), are observationally equivalent. Indeed, $\big\{\Phi_8\big(T_t\big)\big\}$ is the Markov model $\{V_t\}$. Thus the logistic map, relative to Φ_8, and the Markov model $\{V_t\}$, are observationally equivalent. So we have found an explanation of why the time series of Figure 10.1 can arise from both the logistic map and $\{V_t\}$.

In an insightful paper, Berlanger (2013) investigates whether manifest isomorphism can serve as a *purely mathematical* notion of observational equivalence. He concludes that the answer is negative, because the specific choice of the observation function will depend on the context and the physical situation at hand. While I agree with his conclusion, I do not think that this shows that there is anything wrong with manifest isomorphism. Observational equivalence is about observations. Hence it is only desirable that manifest isomorphism not be a purely mathematical notion and that the physical situation at hand will influence the choice of the observation function. Importantly, there are investigations of physical phenomena that can be regarded as instances of manifest isomorphism. This is certainly the case. For instance, chaos theory is about deterministic systems that are nevertheless unpredictable and show irregular and random behaviour (cf. Werndl 2009b). Also, in the context of chaos theory many scientists have reported that they first described a physical phenomenon with a stochastic model only to find later that the data can also be regarded as deriving from a deterministic system (e.g., Shaw 1984).

The results I have presented so far show only that there can be observational equivalence between deterministic and stochastic models. Yet one might still doubt that *stochastic models arising in scientific theorizing* (in short: stochastic models in science) *can be observationally equivalent to deterministic models arising in scientific theorizing* (in short: deterministic models in science). If such doubts were justified, then one could divide the probability distributions found in science into two groups: the ones deriving from observations of deterministic systems in science and the ones deriving from observations of stochastic systems in science. Then one might argue that if the observed probability distributions are of the type of stochastic models in science, this amounts to evidence for a stochastic model, and if they are of the type of deterministic models, this provides evidence for an underlying deterministic system. Clearly, such an argument works only if stochastic models in science cannot be observationally equivalent to deterministic models in science.

Indeed, this is what Kolmogorov believed. More specifically, Kolmogorov introduced the Kolmogorov-Sinai entropy to measure the amount of information produced by a stochastic model and a deterministic model, and he expected that deterministic models in science would have positive entropy and that stochastic models in science would have zero entropy. But when Kolomogorov tried to prove this conjecture, he failed. A few years later it was found that many deterministic systems in science, including

Newtonian systems, have positive Kolmogorov-Sinai entropy (cf. Sinai 1989: 835–837; Werndl 2011). In conclusion, Kolmogorov's attempt of separating deterministic models in science from stochastic models in science failed. Indeed, *many deterministic models in science, including Newtonian models, are observationally equivalent (i.e., manifestly isomorphic) to stochastic models in science.* Returning to our example of the evolution of the temperature, the logistic map (a deterministic model in science) relative to the observation function Φ_8 is observationally equivalent to the Markov model $\{V_t\}$ (Markov models are widely used in science). Relative to the coarser observation function $\Phi_4(m)$, where $\Phi_4(m) = 2.1967$ for $0 \leq m < 4.3934$, $\Phi_4(m) = 9.6967$ for $4.3934 \leq m < 15$, $\Phi_4(m) = 20.3033$ for $15 \leq m < 25.6066$, $\Phi_4(m) = 27.8033$ for $25.6066 \leq m \leq 30$, the logistic map is even observationally equivalent to a Bernoulli model with two outcomes (i.e., to a series of coin tosses; Werndl 2009a, 2011).

One might think that only if certain coarse observation functions are applied to deterministic models in science can one obtain stochastic models in science and that fine-enough observations of deterministic systems in science should yield probability distributions that do not derive from stochastic models in science. In other words, one might doubt that deterministic models in science are observationally equivalent to stochastic models in science *at every observation level.*

The idea of observational equivalence at every observation level can be spelled out in various ways (cf. Werndl 2011). Here I discuss only results about the most commonly used notion, referring to $\left(\varepsilon_1, \varepsilon_2\right)$-*congruence,* which was introduced by the mathematician Ornstein (for technical details, see Ornstein and Weiss 1991; Werndl 2009a, 2011). For a sufficiently small $\varepsilon_1 \geq 0$ one will not be able to distinguish states of the deterministic system that are less than the distance ε_1 apart. Further, suppose that for sufficiently small $\varepsilon_2 \geq 0$ one will not be able to distinguish differences in probabilities of less than ε_2. Then a deterministic model and a stochastic model are $\left(\varepsilon_1, \varepsilon_2\right)$-congruent (i.e., give the same predictions at level $\left(\varepsilon_1, \varepsilon_2\right)$) iff there is a one-to-one correspondence between the solutions of the deterministic model and the realizations of the stochastic model such that the state of the deterministic model and the outcome of the stochastic model are at all time points within distance ε_1 except for a set of probability smaller than ε_2.

The following can be shown for many deterministic models in science, including our example of the logistic map (the evolution of the temperature) and several Newtonian models: for any arbitrary $\varepsilon_1 > 0$ and $\varepsilon_2 > 0$ there is a Markov model, which is $\left(\varepsilon_1, \varepsilon_2\right)$-congruent to the deterministic model (Ornstein and Weiss 1991; Werndl 2009a). From this we can draw the conclusion that the doubts raised earlier cannot be substantiated: deterministic models in science can indeed be observationally equivalent at every observation level to stochastic models in science.

Berlanger (2013) has argued that $\left(\varepsilon_1, \varepsilon_2\right)$-congruence is not sufficient for observational equivalence because the set of points where the models differ by ε_2 (the ε_2-set) is restricted only in its probability measure and *not in its distribution.* He goes on to construct examples of $\left(\varepsilon_1, \varepsilon_2\right)$-congruent models that differ at regular time intervals and argues that they are not observationally equivalent because they differ systematically and detectably.

Berlanger's argument is correct. It shows that in order to arrive at a valuable notion of observational equivalence, $(\varepsilon_1, \varepsilon_2)$-congruence needs to be *strengthened* by adding the condition that the ε_2-set is *distributed randomly* (to match our expectations of random experimental noise). Indeed, for the examples discussed by Ornstein and Weiss (1991) and Werndl (2009, 2011), it is easy to see that they cannot differ at regular time intervals as in Berlanger's counterexample.[4] Still, there remains the question whether, for the examples of $(\varepsilon_1, \varepsilon_2)$-congruence discussed in the literature, the ε_2-set is really distributed in a way that matches our expectations of random noise.

In our context it is important that we can show that several deterministic models in science (including the logistic map) are $(\varepsilon_1, 0)$-congruent to Markov models for every $\varepsilon_1 \geq 0$ Here Berlanger's concerns do not arise because there is no exceptional set of positive measure ε_2 where the models differ. To conclude, there are indeed deterministic models in science that are observationally equivalent at every observation level to stochastic models in science.

2.3 Choice and Underdetermination

As we have shown, there are cases where deterministic and stochastic models are observationally equivalent. Let one of the cases be given where the deterministic model (M, T_t, p) relative to Φ and the stochastic model $\{\Psi(T_t)\}$ relative to Γ are observationally equivalent. Here there is a choice between different models, and the question arises: Is the deterministic model or the stochastic model preferable relative to evidence? If the data equally supports the deterministic and the stochastic model, there is *underdetermination*. To illustrate this with our example: here the question arises whether to choose the logistic map or the Markov model $\{\Phi_8(T_t)\}$ to describe the evolution of the temperature in London.

Suppes (1993) and Suppes and de Barros (1996) argue that there is underdetermination in these cases. Yet more care is needed. In particular, in order to answer the question regarding which model is preferable, one needs to specify the class of observations under consideration. The two main cases are (i) the *currently possible* observations given the available technology (which is the kind of choice arising in practice) and the (ii) the observations that are *possible in principle* (where it is assumed that there are no limits, in principle, on observational accuracy; cf. Werndl 2013b).

Let us first consider case (ii): the observations that are possible in principle. Here one quickly sees that *there is no underdetermination*. If always finer observations can be made, then the deterministic model is preferable (since only the deterministic

---

[4] These examples are chaotic (strongly mixing), which implies that regular differences as in Berlanger's counterexample are impossible.

model allows that always finer observations can be made). On the other hand, suppose the possible observations show that there are no other states apart from those corresponding to the values of a certain observation function Ψ. Then the stochastic model $\{Z_t\}=\{\Psi(T_t)\}$ is preferable because only this model has no more states. Hence there is no underdetermination. Winnie (1998) and Wuthrich (2011) also present an argument along these lines to argue against Suppes's (1993) underdetermination thesis.

Let us now turn to case (i): the currently possible observations. To avoid a trivial answer, assume that Φ is at least as fine as the currently possible observations and that hence it is not possible to find out whether there are more states than the ones corresponding to the values given by Ψ. In other words, the predictions of the deterministic model and stochastic model $\{Z_t\}=\{\Psi(T_t)\}$ agree at all currently possible observation levels. To provide an example, if $\Psi := \Phi_8$ corresponds to an observation at least as fine as the currently possible observations, then the logistic map and $\{\Psi_8(T_t)\}$ will give the same predictions at all currently possible observation levels.

Werndl (2013b) argues that underdetermination can still be avoided in the case most commonly discussed in the literature, that is, the choice between deterministic models derived from Newtonian theory and stochastic models obtained by applying observation functions to these deterministic models. Her argument involves the idea of *indirect evidence*, which is best explained with an example. Galileo's theory is only about the motion of bodies on Earth, and Kepler's theory is only about the motion of planets. So data about planets cannot be derived from Galileo's theory (with the help of standard auxiliary assumptions). Still, data about planets support Kepler's theory, and, with Newtonian mechanics as a bridge, they provide indirect evidence for Galileo's theory. As emphasized by Laudan and Leplin (1991), indirect evidence can block underdetermination. For instance, suppose there is a hypothesis H from which the same predictions are derivable as from Galileo's theory but that does not follow from Newtonian mechanics (or another general theory). In this case there is no underdetermination because only Galileo's theory (but not H) is supported by indirect evidence. Along these lines, Werndl (2013b) argues that for deterministic models derived from Newtonian mechanics there is indirect evidence from other similar Newtonian models. Yet for the stochastic models there is no indirect evidence. Hence the deterministic models are preferable, and there is no underdetermination. To illustrate this argument with our simple example of the evolution of the temperature: suppose that the logistic map were derivable from the generally well-confirmed theory of fluid dynamics but the stochastic model is not derivable from such a theory. Then the deterministic model would receive indirect evidence from other similar models of fluid mechanics but the stochastic model would not. Hence in this case the deterministic model would be preferable.

We now turn to the second main theme of this article, where the concern is an altogether different version of indeterminism that does not involve any probability distributions.

3 INDETERMINISM IN NEWTONIAN PHYSICS

3.1 Examples of Indeterminism

In past decades the question of whether or not the equations of Newtonian physics are deterministic has received much attention in the philosophy of science community. Contra to popular belief that Newtonian physics is deterministic, the answer to this question is not clear-cut.

Two kinds of examples have been discussed in the literature that are taken to show that Newtonian mechanics is indeterministic. The first class of examples are systems *where the initial conditions of the bodies do not uniquely fix their solutions.* For example, Norton (2003, 2008) discusses a system where a point particle of unit mass is moving on a dome of the shape shown in Figure 10.3. The dome is rotationally symmetric about the origin $r = 0$ which corresponds to the highest point of the dome. The shape of the dome is specified by

$$h\left(r\right) = \frac{2}{3g}\, r^{\frac{3}{2}},$$

(3)

describing how far the surface of the dome lies below the highest point, where r is the radial distance coordinate at the surface of the dome. The mass is accelerated by the gravitational force along the surface. At any point of the surface the gravitational force tangential to the surface is directed radially outward and is assumed to be given by

$$F = \frac{d\left(gh\right)}{dr} = r^{\frac{1}{2}}$$

(4)

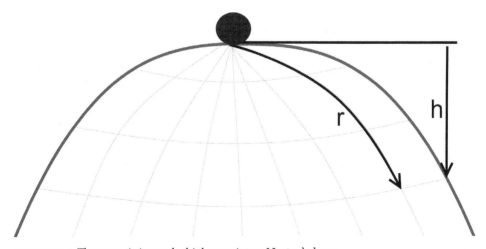

FIGURE 10.3 The mass sitting at the highest point on Norton's dome.

(there is no tangential force at $r = 0$). Recall Newton's second law of motion: $F = m \cdot a$ (i.e., the force equals the mass times the acceleration). When this law is applied to the radial acceleration d^2r/dt, one obtains

$$\frac{d^2r}{dt} = r^{\frac{1}{2}}. \tag{5}$$

If the mass sits initially (at $t = 0$) at the highest point of the dome, there is an obvious solution given by $r(t) = 0$ for all times t. Yet, unexpectedly, there is another class of solutions given by

$$r(t) = \frac{1}{144}(t-T)^4 \text{ for } t \geq T \tag{6}$$

$$r(t) = 0 \text{ for } t \leq T, \tag{7}$$

where $T \geq 0$ is an arbitrary constant.[5] Hence the evolution of the mass sitting initially at the highest point of the dome is not determined: it can stay on the highest point forever or start moving down the surface at an arbitrary point in time.

Mathematicians have of course investigated under what conditions unique solutions exist for differential equations. In this context, a crucial condition is *local Lipschitz continuity*. Intuitively speaking, a local Lipschitz continuous function is limited in how fast it can change. More formally: a function $F(r)$ is *locally Lipschitz* continuous when, for every initial state x in the domain, there is a neighbourhood U of x such that for the restriction of F to U it holds that for all r, s in U:

$$|F(r) - F(s)| \leq K|r - s|. \tag{8}$$

The Picard-Lindelöf theorem from the theory of ordinary differential equations is a key theorem about the existence and uniqueness of differential equations. It states that if the force F is locally Lipschitz continuous on its domain, then there is a unique maximally extended solution to the equation of motion (cf. Arnold 1992). Most differential equations used in science fulfill the conditions of the Picard-Lindelöf theorem. Since for Norton's dome there is a failure of uniqueness, Norton's dome is not locally Lipschitz continuous. More specifically, for Norton's dome $F(r) = r^{\frac{1}{2}}$ (cf. Equations 4 and 5), and

---

[5] Equations (6) and (7) solve Newton's second law (5) because

$$\frac{d^2r}{dt} = \frac{d^2\left(\frac{1}{144}(t-T)^4\right)}{dt} = \frac{4*3}{144}(t-T)^2 = \frac{1}{12}(t-T)^2 = \left(\frac{1}{144}(t-T)^4\right)^{\frac{1}{2}} \text{ for } t \geq T, \quad \text{and}$$

$$\frac{d^2r}{dt} = \frac{d^2 0}{dt} = 0 = 0^{\frac{1}{2}} = r^{\frac{1}{2}} \text{ for } t \leq T.$$

for $F(r) = r^{\frac{1}{2}}$ condition (8) fails for $x = 0$ when the particle is at rest at the highest point of the dome.

The second kind of example taken to show that Newtonian mechanics is indeterministic concern *space invaders*. There is no upper bound to the speed of particles in Newtonian mechanics; hence there is the possibility of space invaders (i.e., particles zooming in from spatial infinity in a finite amount of time). For instance, Xia (1992) proved that there can be space invaders for $n \geq 5$ particles where each of the n particles is subject to the gravitational influence of the other $n - 5$ particles and there are no other forces present. Hence for a universe that is empty at t_0 there are the following two possibilities. First, obviously it is possible that the universe remains empty forever. However, second, it is also possible that after t_0 there are, say, five particles in the universe with interactions as discussed in Xia (1992). Hence Newtonian particle mechanics is an indeterministic theory: the fate of the universe after t_0 is not determined; it can stay empty forever or be occupied by five particles.

3.2 Is Newtonian Physics Indeterministic?

Based on these examples, various authors such as Earman (1986) and Norton (2003, 2008) have argued that Newtonian physics is indeterministic. Others have objected to this conclusion and defended the claim that Newtonian physics is deterministic. For instance, Korolev (2007) argues for local Lipschitz continuity as an implicit assumption of Newtonian physics. In Korolev one also finds the idea that improper idealizations lead to indeterminism and that once these improper idealizations are abandoned, the resulting systems will be deterministic. For instance, if the dome were not completely rigid (and in reality it certainly is not completely rigid), then it would deform in a way that guarantees local Lipschitz continuity and hence would prevent indeterminism. Zinkernagel (2010) claims that Norton's dome arises from an incorrect application of the first law. In essence, what Zinkernagel requires is that the first law be understood in a way that every force has a first "cause" and that forces are not turned on smoothly from zero to non-zero magnitude. Applied to Norton's dome, this yields the conclusion that the particle at rest at the top of the dome must stay at rest, because for the other solutions the forces are turned on smoothly from zero to non-zero magnitude.

Overall, I would agree with Fletcher (2012) that the question "Is Newtonian mechanics indeterministic?" is too simplistic. The answer to this question depends on what one takes Newtonian mechanics to be, and there is no unequivocal answer to this question (cf. Malament 2008; Wilson 2009). There are various different conceptions of Newtonian physics, which are all useful in their own way, and none is a priori privileged over the others. For instance, while an applied mathematician may consider only forces with certain continuity properties, a physicist may focus on a class of properties of models that can be investigated through experiments. There are still interesting scientific and philosophical questions about determinism to be answered: for instance, whether a certain configuration of matter allows for space invaders or what role Lipschitz indeterminism plays in fluid dynamics. Yet these questions are about precisely specified versions of Newtonian physics rather than about "the Newtonian physics".

Nevertheless, much has been learned from this debate. The analysis of the various indeterministic systems and the detailed investigations into the reasons why determinism obtains or fails has been highly insightful. To provide a few examples, we have learned that the violation of Lipschitz continuity is benign in the sense that the force on a ball rolling off the table is not Lipschitz at the point where it loses contact (Fletcher 2012). Zinkernagel's (2010) analysis has shown us that are two different understandings of Newton's first law (one in which it is required that every force has a first "cause" and one in which this is not the case) and that some cases of indeterminism can be avoided by requiring that every force has a first "cause." Furthermore, we have learned that trying to exclude the indeterministic examples by forbidding certain kinds of idealizations is unlikely to succeed, as virtually all of the idealizations are used elsewhere without any complaint (Fletcher 2012; Wilson 2009).

3.3 Determinism and Indeterminism: Past and Present

Although examples of indeterminism in Newtonian physics have attracted much attention in the philosophy community in the past decades, these examples are nothing new. They were discussed in the nineteenth century by scientists such as Poisson, Duhamel, and Bertrand. In an interesting paper, van Strien compares the current debates to those in the nineteenth century and argues

> nineteenth century conceptions of determinism were essentially different from the contemporary conception of determinism in classical physics. Contemporary philosophers of physics largely regard determinism as a property of the equations of physics, specially as the statement that for each system there are equations of motion with unique solutions for given initial conditions. However, I show that in the nineteenth century, this claim was not strongly established, and that the authors that I discuss from this period treated determinism in an essentially different way. Specifically, from their arguments it appears that they thought that determinism could hold even in cases in which the equations of physics did not have a unique solution for given initial conditions. [. . .] Apparently, for these nineteenth century authors, whether or not there was determinism in physical reality did not necessarily depend on whether the equations of physics had unique solutions. This indicates that for them, determinism was not an idea based on the properties of the equations of physics, but rather an a priori principle that was possibly based on metaphysical considerations about causality or the principle of sufficient reason; rather than a result derived from science, determinism was a presupposition of science, that had to be upheld even if it was not reflected in the equations. (2014: 168)

Van Strien makes two main claims. The first is that nineteenth-century conceptions of determinism were *essentially different* from those in use today. The second is that determinism was not a result derived from science but an a priori *principle* possibly based on metaphysical considerations about causality or the principle of sufficient reasons. We discuss these claims in turn.

I agree with van Strien that the nineteenth-century discussion has a very different focus, but I would not say that conceptions of determinism were different. As discussed in the introduction, determinism is the idea that the state of the system at one time determines the future and past evolution of the system, and this idea can be applied either to models and equations or to physical systems. So I would rather say that while all had the same idea of determinism in mind, the focus is now different: the debate in the nineteenth century centered on determinism in real physical systems, but the current debate focuses on determinism in equations or models (as opposed to real systems).

Regarding the second claim: van Strien is right to emphasize that metaphysical considerations about causality or the principle of sufficient reason possibly (even likely) influenced the thinking of scientists in the nineteenth century. Still, I would not go so far as to claim that determinism was an a priori principle. It seems likely that part of scientists' reason in the nineteenth century to upheld determinism was *the empirical success of deterministic equations.* In other words, there was (and still is) no evidence that the indeterminism showing up in examples such as Norton's dome or the space invaders is a real feature of physical systems (cf. Fletcher 2012; Wilson 2009). Instead, there was a vast amount of evidence confirming deterministic equations, and it is likely that this contributed to the general belief that physical systems are governed by deterministic laws.[6]

We now turn to the third main topic of this article: the question of deterministic probabilities.

4 DETERMINISTIC PROBABILITIES

4.1 The Method of Arbitrary Functions

Philosophers often puzzle how stable *ontic probabilities* (i.e., probabilities that are real features of the world) can arise out of *deterministic equations.* The *method of arbitrary functions* is philosophically important because it shows how this is possible. It has an illustrious history and has been advocated by, among others, Hopf, Poincare, Reichenbach, and von Kries (cf. von Plato 1983).

The following is a simple example introducing the method of arbitrary functions (cf. Strevens 2011). Consider a very simple wheel of fortune (i.e., a wheel painted in equal

---

[6] As Wilson (2009) describes in detail, physicists coming across indeterministic equations often find that there are gaps in the mathematical description of the physical system and that, once these gaps are closed, determinism is regained.

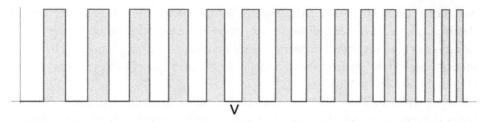

FIGURE 10.4 Conversion of initial velocities v into grey and white outcomes.

numbers of small, equal-sized white and grey sections). The wheel is spun with a certain
initial velocity, and when it comes to rest a fixed pointer indicates the outcome (white
or grey). Our immediate judgement is that the probability of the outcome "grey" is 1/2,
even though the wheel is governed by deterministic equations.[7]

This judgement can be substantiated by analyzing the wheel in more detail. The first
component we must consider is how the *dynamics* of the wheel converts initial velocities
into outcomes. Figure 10.4 shows the conversion of initial velocities into grey and white
outcomes. The crucial feature that emerges here is what Strevens (2003) calls *microcon-
stancy* (i.e., for small ranges of initial velocities the proportion of initial velocities leading to the outcome "grey" is 1/2 and, likewise, the proportion of initial velocities leading
to the outcome "white" is 1/2.) The second component we have to look at is the *preparation of the wheel of fortune in a certain initial velocity*. This preparation is modeled
by a probability distribution p over initial velocities. Of course we usually know very
little about this initial distribution, and different ways of spinning the wheel (by different persons or the same person in different contexts) will correspond to different initial
probability distributions. Yet this does not matter. All we need is the plausible assumption that the class of probability densities p we might possibly employ are what Strevens
(2003) calls *macroperiodic* (i.e., do not fluctuate drastically on a very small region).
Then, as illustrated by Figure 10.5 for two different initial probability densities, the probabilities for the outcomes "grey" and "white" will approximately be 1/2.[8] In sum, there
are stable probabilities for the wheel of fortune even though the wheel is governed by
deterministic equations. What explains these stable probabilities is that (i) the dynamics
is microconstant and (ii) the class of possible probability distributions describing the
preparation of the wheel are macro-periodic.

The method of arbitrary functions is particularly relevant when there is a *class of possible initial densities* (e.g., because the method of preparation differs from scientist to
scientist).[9] It is not meant to apply to all situations where there are ontic probabilities

---

[7] If quantum effects crop up, the wheel is governed by equations that are approximately deterministic
(everything that is said in this article carries over to this case).

[8] The term "arbitrary functions" is fitting in the sense that a large class of initial probability densities
assigns probability 1/2 to the outcomes "grey" and "white." Yet it is also misleading in the sense that only
certain initial densities lead to these probabilities. Hence at least some plausibility arguments need to be
given for the assumption that the initial densities are macroperiodic.

[9] Of course the class of possible densities can also arise from purely natural processes.

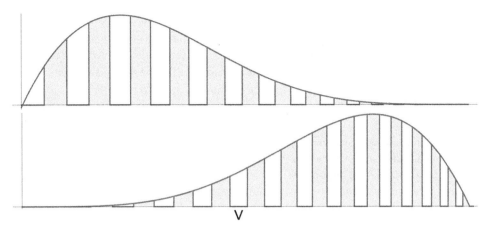

FIGURE 10.5 Probabilities of "grey" and "white" for two different initial probability densities.

but only to certain cases. The prime examples to which the method of arbitrary functions has been successfully applied are games of chance (cf. Strevens 2003; von Plato 1983). Applying this method to probabilities in statistical mechanics, ecology, and the social sciences has also been suggested (the method has not been proven to apply to realistic systems of these disciplines since the relevant mathematics is extremely difficult; Abrams 2012; Strevens 2003; Werndl 2010). In particular, although occasionally discussed (e.g., Myrvold 2012, in press), the method of arbitrary functions deserves more attention in *statistical mechanics*. A common complaint in statistical mechanics has been that different ways of preparing, say, a gas, result in different initial probability distributions and that thus identifying the microcanonical measure (restricted to a macroregion) with the initial probability density is beside the point (Leeds 1989; Maudlin 2007; Werndl 2013a). Clearly, the method of arbitrary functions is ideally suited to make sense of the idea that there is a class of possible initial probability distributions that all lead to (approximately) the same macroscopic behavior.

We now turn to the question of how to interpret the probabilities that arise by applying the method of arbitrary functions and, related to this, how to interpret the initial probability distributions.

4.2 Interpretational Issues

Myrvold (2012, in press) suggests applying the method of arbitrary functions to statistical mechanics. He interprets the initial probability distributions as representing agents' possible rational credences about finding the system in a certain initial state. He argues that the resulting probabilities of outcomes that arise out of the dynamics are approximately equal to the measure assigned to these outcomes by the microcanonical measure. Hence the probabilities obtained by the method of arbitrary functions combine epistemic (credences) and physical considerations (the deterministic dynamics). To reflect this, Myrvold

(2012, in press) calls these probabilities *epistemic chances*. Myrvold's discussion is insightful. In particular, his point that all that is needed is that the standard probability measures in statistical mechanics assign probabilities to outcomes that are effectively the same as the correct probabilities (but need not be exactly the same) cannot be stressed enough.

The only concern I have is that credences cannot be invoked to explain how systems behave as they do. Myrvold anticipates this concern when he writes

> There is a connection, however, between the epistemic considerations we have invoked, and what would be required of an explanation of relaxation to equilibrium. The processes that are responsible for relaxation to equilibrium are also the processes that are responsible for knowledge about the system's past condition of nonequilibrium becoming useless to the agent. (in press, 33)

There certainly is this link. Despite this, in my view, the initial probability distributions of the method of arbitrary functions should be primarily understood not as credences but as ontic probabilities describing the preparation of the system (since there is a fact of the matter in which states a system is prepared; cf. Rosenthal 2012). The probabilities of the method of arbitrary functions arise from these ontic initial probability distributions, while the epistemic chances arising from credences as propounded by Myrvold are *derivative*.

Rosenthal (2010, 2012) explores the intriguing idea that for the method of arbitrary functions probability is defined in terms of the *phase space structure* (various small contiguous regions, on each of which the proportion of the outcomes is the same). The idea is that for such a phase space structure the probability of an outcome simply amounts to the proportion of initial conditions of phase space that lead to this outcome. To illustrate this idea with the wheel of fortune: the probability of "grey" is 1/2 because half of the initial conditions lead to the outcome "grey" (cf. Figure 10.2). This idea is doomed to failure because the phase space structure *plus* certain initial probability distributions are needed to arrive at stable probabilities (Rosenthal also seems to acknowledge this). For instance, it is possible to construct a machine that prepares the wheel of fortune with an initial speed that always leads to the outcome "grey." Hence in this case the probability of "grey" will not be 1/2.

Instead, what seems promising is the idea that the initial probability distributions are negligible in the following sense: whenever there are stable probabilities at the macroscopic level arising *for various different ways of preparing the system*, then one can expect that the probabilities derive from the *phase space structure*. The underlying reasoning is that if the same probabilities arise for various ways of preparing a system, then one can expect that the initial probability distributions are macroperiodic. If the initial probability distributions were not macroperiodic, slight changes in the way the system is prepared would lead to different macroscopic probabilities, contradicting the existence of stable probabilities.[10]

---

[10] As Rosenthal (2010, 2012) remarks, this argument assumes that our usual phase spaces and their metric properties are privileged (in the sense that probabilities over distorted representations of standard physical quantities cannot be expected to be macroperiodic).

Note that whenever the structure of phase space is discussed, the assumption is that the phase space is composed of various small contiguous regions, on each of which the proportion of the outcomes is the same (e.g., Abrams 2012; Rosenthal 2010, 2012; Strevens 2003). However, for stable probabilities to arise, various small contiguous regions are not necessary. Indeed, there are cases where there is just one such region. For instance, consider a funnel located precisely above a central nail (i.e., a Galton board with just one level). When the ball is poured into the funnel, the ball either bounces to the left or to the right of the nail. Assuming that there is only a vertical velocity component when pouring the ball into the funnel, the phase space consists of just one contiguous region with initial conditions corresponding to two outcomes (landing "right" and landing "left" of the nail). Still, there are stable probabilities for the outcomes "right" and "left" (both 1/2) because the class of initial probability distributions describing the various ways of preparing the system contains only (approximately) symmetric distributions.

Strevens (2011) takes a different route. He claims that for nearly all long series of trials of the system the initial conditions form macroperiodically distributed sets. As Strevens stresses, this means that the initial distributions represent only actual occurrences of initial states and have nothing to do with *probabilities*. He regards this as desirable because then, under his interpretation of probability based on the method of arbitrary functions, probabilities arise from *nonprobabilistic facts*. At the same time, by appealing to a robustness condition (that nearly all long series of trials produce macroperiodically distributed sets), he avoids the major pitfalls of finite frequentism.

Streven's proposal is worthwhile. Still, there are some problems. First, Strevens makes the "nearly all"-condition more precise by stating that the actual distributions are macro-periodic in nearly all relevantly close possible worlds, and he appeals to the Lebesgue measure to quantify this claim. There is the worry that it is nontrivial to justify the Lebesgue measure as the correct measure over close possible worlds (cf. Rosenthal 2010, 2012). Further, it is not clear to me whether it is formally possible to assign a measure to "ways of altering the actual world" (the exact formal details are not spelled out by Strevens). Also, consider again the initial velocities of the wheel of fortune. When the wheel is spun by the same person in a specific context again and again, the frequency distribution of the initial velocities approximates the shape of a certain density. For these reasons, scientists usually postulate a probability density describing the probability of preparing the wheel with a certain initial velocity. This probability density is predictively useful: it is (usually) found to give accurate predictions about the future frequencies of initial velocities produced by the person. However, because under Strevens's account the initial velocities are nothing more than actual occurrences of initial states, there cannot be any such predictive power.

In my view, there is a way to avoid these problems; namely, I suggest interpreting the initial distributions simply as *probability distributions that are physical, biological, and other quantities characterizing the particular situation at hand* (as suggested by Szabò 2007; see also Sober 2010). That is, the concept of probability reduces to ordinary physical, biological, and other quantities, and the word "probability" is a collective term whose precise meaning depends on the context of its application. Accordingly, the probabilities arising from the method of arbitrary functions also correspond to physical, biological, and other quantities that characterize particular situations. These probabilities are ontic, and, as desired, they support counterfactuals about future predictions.[11]

A final remark: Werndl (2013a) put forward an account of how to understand typicality measures in statistical mechanics. The main idea is that there is a class of initial probability distributions of interest that corresponds to the possible ways of preparing a system (elements of this class are assumed to be translation-continuous or translation-close). The typicality measure can then be used to make claims about all initial probability distributions of interest; for example, if the claim is that typical initial conditions show thermodynamic-like behavior, this implies that for all initial probability distributions of interest the probability of thermodynamic-like behavior is close to one. Werndl does not mention the method of arbitrary functions. Still, formally, the justification of the typicality measures involves (next to other ingredients) the application of the method of arbitrary functions to outcomes that have probability close to one. The slogan "typicality is not probability" is correct here in the sense that the typicality measure is a measure at the microlevel and does not amount to a probability over the initial states (such as the probability of finding the system in a certain state).

4.3 Puzzles about Deterministic Probabilities Resolved

Philosophers have often doubted that there can be ontic probabilities in a deterministic world (see, e.g., Schaffer [2007] for a recent paper expressing such doubts). As our discussion has shown, the method of arbitrary functions provides an explanation of how probabilities can arise out of determinism.

A prominent worry about deterministic probabilities is that they lead to a violation of the Principal Principle, which establishes a connection between credences and chances (e.g., Schaffer 2007). According to this principle, a rational agent's credence in the occurrence of an event E should equal the chance of E as long as the agent has no inadmissible knowledge about the truth of E. Formally: for all events E, all P and all K

$$cr_t\left(E\mid P \& K\right)= p \qquad (9)$$

where cr_t stands for the credence of the agent at time t, P is the proposition that the chance of E is p, and K is any admissible proposition. The crucial question here is what an "admissible proposition" amounts to. Lewis (1986) suggests that historical information about the exact state of a system up to time t, as well as information about the laws

[11] If the initial distributions are understood as summarizing actual frequencies, the method of arbitrary functions is also compatible with an account of probabilities as Humean chances (Frigg and Hoefer 2010, 2014). As for all Humeans, it is not trivial to say what it means to strike the best balance between simplicity, strength, and fit. The account I propose does not have to deal with such difficulties.

of nature, are always admissible. But given deterministic laws, this implies that the exact future state of the system can be predicted. Hence the credences in Equation (9) can only be 0 and 1, and for nontrivial deterministic probabilities p this leads to a violation of the Principal Principle. Commonly, the conclusion drawn from this is that there exist no nontrivial deterministic probabilities (since a violation of the Principal Principle is deemed to be unacceptable).

This argument is too quick, and there are better ways to characterize an admissible proposition. Following Frigg and Hoefer (2013), let a proposition K be *admissible* with respect to event E and chance setup S iff "K contains only the sort of information whose impact on reasonable credence about E, if any, comes entirely by way of impact on credence about the chances of those outcomes." With this modified notion of admissibility, as desired, the Principal Principle (Equation 9) comes out as true. As Glynn (2010) stresses, when making rational decisions, it would be a real loss if we could not rely on macroscopic probabilities such as those of the outcome "tail" for a coin toss, the outcome "black" for a wheel of fortune, and so on. Glynn argues that chances are level-relative and that only the initial history and laws at the specific *level of reality* should count as admissible (this amounts to a special case of Frigg and Hoefer's general definition of "admissibility").

Probabilities arising from deterministic equations for games of chance, in statistical mechanics, and so on are often said to be *epistemic* in the sense that if we had precise knowledge about the initial conditions and the deterministic laws, then we would not need them. Our discussion puts these claims into the right perspective. The probabilities of the method of arbitrary functions are *not* epistemic in the sense that they are credences. The only sense in which they are epistemic is that they are relative to a certain *level of reality*. If the system were described at the level where determinism reigns, we would indeed not need any probabilities (cf. Frigg and Hoefer 2010; Glynn 2010; Rosenthal 2010).

5 CONCLUSION

This article centered on three major themes in the recent discussion on determinism in philosophy of science. The first major theme was determinism, indeterminism, and observational equivalence. Here I first critically discussed various notions of observational equivalence and then presented results about the observational equivalence between deterministic and indeterministic models. I also put forward an argument on how to choose between deterministic and indeterministic models involving the idea of indirect evidence. The second major theme concerned the question of whether Newtonian physics is indeterministic. I argued that the answer to this question depends on what one takes Newtonian mechanics to be. Further, I discussed how contemporary debates on this issue differ from those of scientists in the nineteenth century. In particular, I pointed out that the focus of scientists in the nineteenth century was on investigating whether determinism holds for real systems rather than just for certain equations or models. The third major theme was how the method of arbitrary functions can be used to make sense of deterministic probabilities. Here I discussed various ways of interpreting the initial probability distributions, and I argued that they are best understood as physical, biological, and other quantities characterizing the particular situation at hand. I also emphasized that the method of arbitrary functions deserves more attention than it has received so far and that it is among the most promising contenders for understanding probabilities in certain fields (e.g., classical statistical mechanics).

The topic of determinism is a very old one, but it is still very much alive: Much has been learned in recent decades, but a lot still remains to be discovered.

REFERENCES

Abrams, M. (2012). "Mechanistic Probability." *Synthese* 187(2): 343–375.

Arnold, V. I. (1992). *Ordinary Differential Equations* (Berlin: Springer).

Berlanger, C. (2013). "On Two Mathematical Definitions of Observational Equivalence: Manifest Isomorphism and Epsilon-Congruence Reconsidered." *Studies in History and Philosophy of Science, Part B* 44(2): 69–76.

Earman, J. (1986). *A Primer on Determinism* (Berlin and New York: Springer).

Frigg, R., and Hoefer, C. (2010). "Determinism and Chance from a Humean Perspective." In F. Stadler (ed.), *The Present Situation in the Philosophy of Science* (Dordrecht, The Netherlands: Springer), 351–372.

Frigg, R. and Hoefer, C. (2013). "The Best Humean System for Statistical Mechanics. *Erkenntnis*. Advance online publication. doi:10.1007/s10670-013-9541-5

Fletcher, S. (2012). "What Counts as Newtonian System—The View from Norton's Dome." *European Journal for the Philosophy of Science* 2(3): 275–297.

Glynn, L. (2010). "Deterministic Chance." *The British Journal for the Philosophy of Science* 61(1): 51–80.

Korolev, A. V. (2007). "Indeterminism, Asymptotic Reasoning, and Time Irreversibility in Classical Physics." *Philosophy of Science* 74(5): 943–956.

Laudan, L., and Leplin, J. (1991). "Empirical Equivalence and Underdetermination." *The Journal of Philosophy* 88: 449–472.

Leeds, S. (1989). "Malament and Zabell on Gibbs Phase Space Averaging." *Philosophy of Science* 56: 325–340.

Lewis, D. (1986). "A Subjectivist's Guide to Objective Chance." In R. C. Jeffrey (ed.), *Studies in Inductive Logic and Probability*, Vol. 2 (Berkeley: University of California Press), 83–132.

Lorenz, E. (1964). "The Problem of Deducing the Climate from the Governing Equations." *Tellus* 16(1): 1–11.

Malament, D. (2008). "Norton's Slippery Slope." *Philosophy of Science* 75(5): 799–816.

Maudlin, T. (2007). "What Could Be Objective about Probabilities?" *Studies in History and Philosophy of Modern Physics* 38: 275–291.

Myrvold, W. (2012). "Deterministic Laws and Epistemic Chances." In Y. Ben-Menahem and M. Hemmo (eds.), *Probability in Physics* (New York: Springer), 73–85.

Myrvold, W. (in press). "Probabilities in Statistical Mechanics." In C. Hitchcock and A. Hajek (eds.), *Oxford Handbook of Probability and Philosophy* (Oxford: Oxford University Press).

Norton, J. (2003). "Causation as Folk Science." *Philosopher's Imprint* 3(4): 1–22.

Norton, J. (2008). "The Dome: An Unexpectedly Simple Failure of Determinism." *Philosophy of Science* 75(5): 786–798.

Ornstein, D., and Weiss, B. (1991). "Statistical Properties of Chaotic Systems." *Bulletin of the American Mathematical Society* 24: 11–116.

Rosenthal, J. (2010). "The Natural-Range Conception of Probability." In G. Ernst and A. Hüttemann (eds.), *Time, Chance and Reduction: Philosophical Aspects of Statistical Mechanics* (Cambridge, UK: Cambridge University Press), 71–91.

Rosenthal, J. (2012). "Probabilities as Ratios of Ranges in Initial State Spaces." *Journal of Logic, Language and Information* 21: 217–236.

Schaffer, J. (2007): "Deterministic Chance?" *The British Journal for the Philosophy of Science* 58: 113–140.

Shaw, R. (1984). *The Dripping Faucet* (Santa Cruz: Aerial Press).

Sinai, Y. G. (1989). "Kolmogorov's Work on Ergodic Theory." *The Annals of Probability* 17: 833–839.

Sober, E. (2010). "Evolutionary Theory and the Reality of Macro Probabilities." In E. Eells and J. Fetzner (eds.), *Probability in Science* (Heidelberg: Springer), 133–162.

Strevens, M. (2003). *Bigger than Chaos* (Cambridge, MA: Harvard University Press).

Strevens, M. (2011). "Probability Out of Determinism." In C. Beisbart and S. Hartmann (eds.), *Probabilities in Physics* (Oxford: Oxford University Press), 339–364.

Suppes, P. (1993). "The Transcendental Character of Determinism." *Midwest Studies in Philosophy* 18: 242–257.

Suppes, P., and de Barros, A. (1996). "Photons, Billiards and Chaos." In P. Weingartner and G. Schurz (eds.), *Law and Prediction in the Light of Chaos Research* (Berlin: Springer), 189–201.

Szabò, L. E. (2007). "Objective Probability-Like Things With and Without Indeterminism." *Studies in History and Philosophy of Modern Physics* 38: 626–634.

von Plato, J. (1983). "The Method of Arbitrary Functions." *The British Journal for the Philosophy of Science* 34: 37–47.

Van Strien, M. (2014). "The Norton Dome and the Nineteenth Century Foundations of Determinism." *Journal for General Philosophy of Science* 45(1): 167–185.

Werndl, C. (2009a). "Are Deterministic Descriptions and Indeterministic Descriptions Observationally Equivalent?" *Studies in History and Philosophy of Modern Physics* 40: 232–242.

Werndl, C. (2009b). "What Are the New Implications of Chaos for Unpredictability?" *The British Journal for the Philosophy of Science* 60: 195–220.

Werndl, C. (2010). "The Simple Behaviour of Complex Systems Explained? Review of 'Bigger Than Chaos, Understanding Complexity Through Probability' (Michael Strevens)." *The British Journal for the Philosophy of Science* 61(4): 875–882.

Werndl, C. (2011). "On the Observational Equivalence of Continuous-Time Deterministic and Indeterministic Descriptions." *European Journal for the Philosophy of Science* 1(2): 193–225.

Werndl, C. (2013b). "On Choosing Between Deterministic and Indeterministic Models: Underdetermination and Indirect Evidence." *Synthese* 190(12): 2243–2265.

Werndl, C. (2013a). "Justifying Typicality Measures of Boltzmannian Statistical Mechanics and Dynamical Systems." *Studies in History and Philosophy of Modern Physics* 44(4): 470–479.

Wilson, M. (2009). "Determinism and the Mystery of the Missing Physics." *The British Journal for the Philosophy of Science* 60(1): 173–193.

Winnie, J. (1998). "Deterministic Chaos and the Nature of Chance." In J. Erman and J. Norton (eds.), *The Cosmos of Science—Essays of Exploration* (Pittsburgh: Pittsburgh University Press), 299–324.

Wuthrich, C. (2011). "Can the World Be Shown to be Indeterministic After All?" In C. Beisbart and S. Hartmann (eds.), *Probabilities in Physics* (Oxford: Oxford University Press), 365–389.

Xia, Z. (1992). "The Existence of Noncollision Singularities in Newtonian Systems." *Annals of Mathematics* 135: 411–468.

Zinkernagel, H. (2010). "Causal Fundamentalism in Physics." In M. Suarez, M. Dorato, and M. Redei (eds.), *EPSA Philosophical Issues in the Sciences: Launch of the European Philosophy of Science Association* (Dordrecht, The Netherlands: Springer), 311–322.

CHAPTER 11

....

EPISTEMOLOGY AND PHILOSOPHY OF SCIENCE

....

OTÁVIO BUENO

1 INTRODUCTION

....

IT is a sad fact of contemporary epistemology and philosophy of science that there is very little substantial interaction between the two fields. Most epistemological theories are developed largely independently of any significant reflection about science, and several philosophical interpretations of science are articulated largely independently of work done in epistemology. There are occasional exceptions, of course. But the general point stands.

This is a missed opportunity. Closer interactions between the two fields would be beneficial to both. Epistemology would gain from a closer contact with the variety of mechanisms of knowledge generation that are produced in scientific research, with attention to sources of bias, the challenges associated with securing truly representative samples, and elaborate collective mechanisms to secure the objectivity of scientific knowledge. It would also benefit from close consideration of the variety of methods, procedures, and devices of knowledge acquisition that shape scientific research. Epistemological theories are less idealized and more sensitive to the pluralism and complexities involved in securing knowledge of various features of the world. Thus, philosophy of science would benefit, in turn, from a closer interaction with epistemology, given sophisticated conceptual frameworks elaborated to refine and characterize our understanding of knowledge, justification, evidence, and the structure of reasons, among other key epistemological notions.

In this chapter, I argue for a closer involvement between epistemology and philosophy of science by examining two areas in which a close interaction would be beneficial to both: approaches to knowledge in traditional epistemology and in philosophy of science and the roles played by instruments in the production of scientific knowledge (considering, in particular, how these roles can be illuminated by certain forms of epistemological theorizing, such as internalism). I consider each of them in turn.

2 APPROACHES TO KNOWLEDGE: FROM EPISTEMOLOGY TO PHILOSOPHY OF SCIENCE AND BACK

Epistemologists and philosophers of science have been concerned with the proper understanding of knowledge. In the hands of epistemologists, knowledge is approached primarily from the perspective of a conceptual analysis—as emerging from the attempt to introduce individually necessary and collectively sufficient conditions to characterize knowledge. The traditional tripartite characterization of knowledge in terms of justified true belief, which can be traced back all the way to Plato, received a significant blow with Ed Gettier's (1963) counterexamples, which questioned the sufficiency of the tripartite account by providing clear cases in which an agent satisfied the three conditions for knowledge—thus having a justified true belief—but failed to have knowledge. In response to the counterexamples, a variety of different accounts have been proposed, from causal accounts (Goldman 1967) through accounts that do not allow for unjustified false beliefs (Littlejohn 2012) all the way to accounts that resist the need to provide an analysis of knowledge (Williamson 2000). It seems safe to think, after decades of work by epistemologists on this issue, that the Gettier problem is unlikely to go away (Zagzebski 1994).

Although searching for a proper conceptual analysis of knowledge may help to clarify the concept, philosophers of science approach the issue differently. They tend not to provide an account of knowledge, but to examine ways of producing and assessing knowledge, however such knowledge is ultimately established. Different scientific fields invoke distinct resources to achieve the goal of advancing knowledge in the relevant field. There is very little in common between particle physics and clinical research, and, not surprisingly, techniques and procedures used in one are largely irrelevant to the other: particle accelerators are crucial to implement the research in the case of the former, whereas randomized controlled trials are central to the latter. Philosophers of science tend to examine critically the ways in which these techniques and procedures are invoked to secure knowledge, how they work, their limitations, and the possibilities they open. Epistemic resources are explored in the particularities of the fields under study. In order to do that, it is not required to have a conceptual analysis of knowledge; it is enough to have an understanding of what is involved in securing, obtaining, and assessing the relevant information in a given domain.

Is a conceptual analysis of knowledge necessary to be able to implement this evaluation? It is not. One can assess how information is obtained and assessed in a particular field independently of taking any stand on the success or failure of the project of conceptually analyzing knowledge. We often use concepts whose conceptual analyses are not available and, in many cases, cannot be available given their indefinability. Consider, for instance, the concept of *identity*. It cannot be defined since any attempt to define this concept ultimately presupposes identity, at least in the meta-language. After all, the sentence expressing identity requires the identity of the variables used to properly characterize the notion. This can be easily noted in the expression of Leibniz identity laws (in second-order logic), $x = y$ if, and only if, $\forall P\,(Px \leftrightarrow Py)$, which requires the *same* variables on both sides of the equivalence connectives (for a discussion and references to the relevant literature, see Bueno 2014 and 2015). Despite this fact, it would be a mistake to suppose that the concept of *identity* is somehow poorly understood and cannot be used in a variety of contexts. There is typically no difficulty in determining whether two objects are the same—unless their identity conditions are not properly specified. But this is not a problem for the concept of identity; instead, it is a difficulty that emerges from the lack of identity conditions for the objects in question.

Similarly, despite not having a conceptual analysis of knowledge, it is possible to determine the conditions under which, in a given field, knowledge is obtained and assessed. There are, of course, proposals in epistemology that argue that one should take the concept of knowledge as primitive rather than try to analyze it in more basic terms. These are knowledge-first approaches (Williamson 2000). In Timothy Williamson's approach, knowledge is then formulated, although not defined, as the most general factive mental state. This account provides an interesting alternative to attempts to define knowledge and it redirects some of the epistemological work away from the goal of identifying more basic concepts in terms of which knowledge should be characterized to examining the epistemological landscape by starting with the concept of knowledge in the first place. Epistemologically, it is a suggestive approach.

It is unclear, however, what implications the approach would have to the philosophical understanding of the sciences. First, it is not obvious that the concept of knowledge is, in fact, presupposed in scientific practice. Whether or not knowledge is presupposed to make assertions, it is safe to maintain that, to the extent that scientific practice is involved at all with knowledge, this involvement is one in which knowledge needs to be established rather than being assumed from the start.

There are, of course, those within epistemology who balk at the claim that science is in the business of producing knowledge. As part of the development of his contextualist account of knowledge, David Lewis notes:

> The serious business of science has to do not with knowledge per se; but rather, with the elimination of possibilities through the evidence of perception, memory, etc., and with the changes that one's belief system would (or might or should) undergo under the impact of such eliminations. (Lewis 1996, p. 234)

This remark seems right (although somewhat one-sided). The elimination of possibilities is indeed an integral part of scientific practice, particularly in the context of obtaining and assessing evidence. In contrast, knowledge per se plays no comparable role. Despite being crucial to epistemological views, knowledge is not what, ultimately, scientific practice is involved with. (Although Lewis doesn't note this, the sciences also open venues for epistemic possibilities—highlighting, in some cases, possibilities that we can only be aware of given the theoretical and experimental practices they enable.)

Despite that, certain philosophical conceptions about the sciences do assign a key role to knowledge. In certain philosophical views (particularly realist ones), knowledge plays a key role. Knowledge (in particular, objective knowledge) may be taken as something science aims to reach in the long run by searching for the truth (Popper 1972). Alternatively, knowledge can be inherently invoked in the characterization of scientific progress (Bird 2007).

Both proposals, however, face difficulties. The notion of truth approximation in the way conceptualized by Popper cannot be implemented because false theories are equally far away from the truth (for a critical discussion, see Miller 1994). As a result, because, on the standard conception, truth is part of objective knowledge, one has no epistemic indication as to how close or far away one is from having such knowledge.

The characterization of progress in terms of knowledge is similarly problematic for it presupposes that knowledge has already been obtained. But how can it be determined that such knowledge has indeed been established in the first place? One could argue that we know more now than we knew centuries ago, and thus, given the intervening scientific progress, not only do we have knowledge, we have increasingly more of it. But this, once again, presupposes that we have the relevant knowledge to begin with, and from an epistemological viewpoint it is questionable whether one is entitled to start with this assumption. There are many possibilities of mistakes and mismatches that can undermine presumed knowledge claims, and unless one is in a position to rule out the relevant ones, one cannot properly have knowledge. As a result, to start by assuming knowledge is more troublesome than it may initially seem.

Note that the concern here is not that skepticism is prima facie correct (although, of course, it may well be). The issue is not that we don't have any knowledge of the world, but rather that particular knowledge claims may be contested because they can be undermined. Some local, restricted form of skepticism may play an important role in scientific practice. But an open-ended, global, Cartesian skepticism, according to which we don't know anything, is not part of that practice. Global skeptical arguments, such as brain-in-a-vat arguments (Nozick 1981), have no purchase in scientific activity and are properly ignored. According to these arguments:

(P_1) If I know P (i.e., some perceptually salient fact about the world), I know that I'm not a brain in a vat.
(P_2) I don't know that I'm not a brain in a vat.
Therefore, I don't know P.

(P_1) is motivated from the fact that my presumed knowledge of some perceptually salient fact about the world, P, undermines my being a brain in a vat because if I were a brain in a vat I wouldn't have the knowledge in question. Thus, if I know P, I know that I'm not a brain in a vat. (P_2) is motivated from the fact that my perceptual experiences would be the same even if I were a brain in a vat. As a result, it is unclear how I could have the knowledge that I'm not a brain in a vat. These considerations challenge that I may know anything at all about the world based on my perceptual experiences. However, there is a serious concern about the coherence of the situation entertained in this scenario. After all, the perceptual experiences one has are presumably the result of one's interactions with the world, which are undermined by the brain-in-a-vat scenario.

Rather than engaging with this sort of global skeptical challenge, scientific practice takes for granted that there is a world to be studied, although the details and properties of that world are precisely the kind of issue scientific research aims to elucidate. We will examine, in the next section, some conditions developed in scientific practice to this effect.

In response, Lewis provides an account of knowledge that emphasizes the importance of eliminating possibilities while still granting the difficulty of expressing the proposal. After all, the moment we note that we are properly ignoring certain possibilities, we are *ipso facto* attending—and hence, no longer ignoring—such possibilities. In his unique style, Lewis makes the point in this way:

> S knows that P iff P holds in every possibility left uneliminated by S's evidence— Psst!—except for those possibilities that we are properly ignoring. (Lewis 1996, p. 232)

For Lewis, in providing an account of knowledge, two constraints need to be met. First, knowledge should be *infallible* on pain of not really being knowledge. Second, we should be in a position to assert that we have lots of knowledge; that is, skepticism should turn out to be *false*. In providing his form of contextualism, Lewis is trying to walk the thin line between infallibilism and skepticism. His starting point is Peter Unger's work, which takes infallibilism to be a requirement for knowledge. However, Unger ends up in skepticism, given that infallible knowledge is not possible (see Unger 1975). Lewis shares Unger's infallibility requirement, but tries to avoid skepticism. As Lewis insists:

> If you are a contented fallibilist, I implore you to be honest, be naive, hear it afresh. "He knows, yet he has not eliminated all possibilities of error." Even if you've numbed your ears, doesn't this overt, explicit fallibilism still sound wrong? (Lewis 1996, p. 221)

Given that, on Lewis's account, if you know P, then P holds in every possibility left uneliminated by your evidence (except for the possibilities you're properly ignoring), then your knowledge is indeed infallible. After all, you have eliminated all possibilities of error, and those possibilities that you haven't eliminated, you can properly ignore.
For Lewis, however, infallibilism doesn't lead to skepticism. Because skepticism is ultimately *false*—well, at least when we are not doing epistemology! In epistemological contexts, when we cannot properly ignore "uneliminated possibilities of error everywhere"

(Lewis 1996, pp. 230–231), ascriptions of knowledge are hardly ever true. That's how epistemology destroys knowledge (Lewis 1996, p. 231). But this is only temporarily:

> The pastime of epistemology does not plunge us forevermore into its special context. We can still do a lot of proper ignoring, a lot of knowing, a lot of true ascribing of knowledge to ourselves and others, the rest of the time. (Lewis 1996, p. 231)

In other words, in nonepistemological contexts, where we can properly ignore various possibilities of error, we do end up knowing a lot of things. In these contexts, it would be a mistake to claim that we lack knowledge. After all, as long as what we claim to know holds in every possibility that is not eliminated by the available evidence (except for the possibilities that we are properly ignoring), the proposed characterization of knowledge is satisfied. As a result, Lewis is entitled to claim that we know; in fact, we know lots of things. In these contexts, skepticism—conceived as the claim that we don't have knowledge—is therefore false, and Lewis achieves his goal.

Does this strategy successfully deal with the skeptic? I don't think it does. In this context, sensitivity to details of scientific practice would have been helpful. In epistemological contexts, as Lewis acknowledges, the game is handed over to the skeptic because we cannot rule out all possibilities of error, and, as a result, we cannot claim to have knowledge on Lewis' account. Similarly, in nonepistemological contexts, the skeptic can still raise the issue as to whether the ignored possibilities are indeed *properly* ignored. And in all the cases in which such possibilities are *not* properly ignored, we still don't have knowledge. But when are possibilities properly ignored?

Lewis introduces three main rules for how possibilities can be properly ignored:

(a) The *Rule of Reliability*: "Consider processes whereby information is transmitted to us: perception, memory, and testimony. These processes are fairly reliable. Within limits, we are entitled to take them for granted. We may properly presuppose that they work without a glitch in the case under consideration. Defeasibly—*very* defeasibly—a possibility in which they fail may properly be ignored" (Lewis 1996, p. 229).

(b) The *Rules of Method*: "We are entitled to presuppose—again, very defeasibly—that a sample is representative; and that the best explanation of our evidence is the true explanation" (Lewis 1996, pp. 292–230).

(c) The *Rule of Conservatism*: "Suppose that those around us normally do ignore certain possibilities, and it is common knowledge that they do. . . . Then—again, very defeasibly!—these generally ignored possibilities may properly be ignored" (Lewis 1996, p. 230).

But these rules are problematic, particularly if they are used in response to skepticism. Employed in this way, they end up begging the question against the skeptic. (a′) With regard to the rule of reliability, perception, memory, and testimony are reliable processes for the most part. However, this doesn't entitle us to suppose that we can ignore the possibilities in which they fail, even defeasibly. The skeptic would insist that, despite the fair reliability of these processes, the rule of reliability doesn't entitle us to ignore the cases in which these processes can be mistaken (i.e., mislead us in significant ways). After all, in all of the cases in which the processes are mistaken, we will fail to have knowledge. Clearly, cases such as these cannot be properly ignored, not even defeasibly.

Part of the difficulty here is to determine under what conditions we can properly ignore the cases in which these information transmission processes fail. The rule of reliability recommends that we uniformly ignore all such cases (although defeasibly so). But this uniform rejection ends up including those cases in which the skeptic is right in insisting that we would be mistaken in trusting these processes of transmission of information. After all, perception, memory, and testimony are not always trustworthy. To ignore the problematic, unsuccessful cases given that, for the most part, these processes work well is clearly an invalid inference. Moreover, the move ends up begging the question against the skeptic since it assumes, without argument, that perception, memory, and testimony are in fact reliable. Whether and under what conditions these processes can be so taken is what the skeptic is challenging. Thus, invoked as a response to skepticism, the rule of reliability is clearly inadequate.

The contextualist strategy is to insist that we can ignore the cases in which perception, memory, and testimony fail. However, to do that, the contextualist simply ignores along the way the challenge raised by the skeptic. Because it just ignores the skeptic, the strategy can't provide an adequate response to skepticism.

(b′) With regard to the rules of method, two points should be made: (i) The skeptic would challenge that one is entitled to presuppose that a sample is representative. In fact, if anything, samples tend to be *un*representative in all sorts of ways. Convenient ways of getting a sample typically are incompatible with a random selection, and the sample size is often not large enough to be representative. Moreover, usual randomization devices, such as a roulette wheel, are not truly random because, strictly speaking, we would need an infinite list of outputs to generate a truly random process. In practice, we end up working with samples that are not randomized and so are not strictly speaking representative. And even if a sample is obtained via some almost randomized method, the way in which we get to particular units in the sample is often not done randomly at all, but rather by invoking some convenient type of method. Suppose that Joe Smith is randomly selected to be part of the sample of a population we are interviewing for a polling study. But when we get to Joe Smith's house, he has just moved out, and the current owner of the place has no idea of Joe's whereabouts. Since we are already there, we just interview the owner instead of Joe. That's convenient, no doubt; but at that point, a truly random sample is left behind. Cases such as this are, of course, ubiquitous, and they bring in additional sources of bias.In other words, as opposed to what the reliability rule states, samples tend *not* to be representative. If you simply get a sample, more likely than not, it will turn out to be unrepresentative. As a result, one needs to work extraordinarily hard to obtain a representative sample. To assume that samples are, for the most part, reliable is to ignore, without argument, the huge difficulties of actually obtaining a sample that is in fact representative. As a result, to use the reliability rule as part of an argument against skepticism is entirely ineffective. The skeptic is entitled to complain that the outcome of the reliability rule (that samples are representative and thus can be trusted) is something that needs to be earned rather than simply assumed.

(ii) Clearly, a skeptic would also challenge that the best explanation of our evidence is the true explanation. There are several arguments as to why inference to the best explanation is not generally reliable. Consider, for instance, the bad lot argument (van Fraassen 1989). Suppose we have various explanations of the same phenomena to choose from. Inference to the best explanation would lead us to conclude that the best explanation is the true one. However, unbeknownst to us, the lot of explanations we were considering was a very bad one, and none of the available explanations, including the best one, was even remotely close to the truth. Clearly, inference to the best explanation is not reliable in this context.

The trouble is that we typically don't know whether, among the lot of available explanations, the true explanation is included. And so the situation we are typically in is one to which the bad lot argument clearly applies. Unless we can rule out the bad lot possibility whenever we infer the truth of the best explanation, it is not clear how such an inference could be reliable.

In response, one could suppose that the true explanation is included in the lot of available explanations. This move is an attempt to ensure the reliability of the inference to the best explanation rule. After all, if the true explanation is in the lot of admissible explanations, by inferring the best explanation one minimizes the chance of inferring something false from true premises.

However, this move still doesn't work. To guarantee reliability, it is not enough to include the true explanation among the lot of available explanations. One needs also to assume that the best explanation is the true one. Suppose, for example, that the true explanation is too complicated, computationally intractable, and extraordinarily difficult to understand. It is unlikely that anything having these features would count as the best explanation, even if it were true. Moreover, if we assume that the true explanation is in the relevant lot of explanations, we end up trivializing the rule. If we already know that a certain lot includes the true explanation, we might just as well infer the true explanation directly!

(c′) With regard to the rule of conservatism, clearly, a skeptic would challenge the idea that one can properly ignore the possibilities that those around us normally do ignore. It is precisely by systematically *violating* this rule that scientific change takes place. Consider the case of splitting the atom in the 1930s. Consider also the case of the moving earth in the context of the formulation of the Copernican theory.

In response, the Lewisian can always note that Lewis was very careful in formulating each rule with a built-in defeasibility condition: "We are entitled to presuppose—again, *very defeasibly*—that a sample is representative . . .." But note that this is, first, a significant concession to the fallibilist! Second, this eventually grants the point to the skeptic. If these possibilities cannot be properly ignored after all, one cannot still claim to know.

As these considerations illustrate, attention to details of scientific practice would have made the difficulties involved in this defense of contextualism in epistemology more transparent.

3 INSTRUMENTS AND SCIENTIFIC KNOWLEDGE

Associated with changes in particular contexts of knowledge assessment we find a phenomenon that has challenged various realist views about the development of science: the incommensurability problem. [1] There are many ways of understanding this problem (see, e.g., Feyerabend 1981; Hoyningen-Huene 1993; Kuhn 1970; Sankey 1994; Siegel 1980 and the references quoted in these works). For the purposes of the present work, I take it to be the problem to the effect that there is no common standard of evaluation to determine the reference of certain theoretical terms. Depending on the context one considers (such as particular domains of application of a theory), there are different ways of determining the reference of such terms. And if that reference is opaque, it is unclear how one could provide realist interpretations of the theories that use these theoretical terms. If there is no fact of the matter as to whether, say, "mass" refers to Newtonian mass or to relativistic mass, it is unclear how one could claim that we should read such theories realistically. It seems that a commitment either way would be just arbitrary. As a result, the incommensurability phenomenon raises problems for different forms of realism about science.

It may be thought that a contextualist approach to knowledge would simply reinforce a relativist view regarding scientific practice, according to which anything goes. But this is not the case. In fact, the contextualist conception will grant that incommensurability will inevitably emerge, but it will resist any relativist consequence. Central to this move, I argue, is a particular understanding of the functions that instruments play in scientific practice and the epistemic features they share with observation.

But can instruments be used to overcome the problems that emerge from incommensurability? In principle, it might be thought that the answer should be negative. After all, the argument goes, the use of instruments relies on various theoretical assumptions, and the latter will just carry over the incommensurability found at the level of theories.

---

[1] In this section, I rework some parts of Bueno (2012). But here I examine, in particular, the significance of epistemological considerations for the proper understanding of scientific instruments.

In other words, since theories are routinely invoked in the construction and use of scientific instruments, it would not be possible to use the latter to resist the effects of incommensurability. After all, these effects will emerge just as vividly in the theories that underlie the instruments as in the theories that inform the relevant parts of scientific practice. Bringing in the instruments will be of no help.

This line of argument, however, can be resisted in multiple ways. First, it is quite controversial to claim that theories need to be adopted in the construction and use of scientific instruments. In several instances, some theories are indeed used in the *construction* of instruments, but these theories turn out to be false—and the instruments work just as adequately. This is the case, for instance, in the use of Kepler's optics in the construction of the first microscopes in the seventeenth century. Despite the falsity of Kepler's theory, the telescope ended up becoming a major tool for the study of astronomical phenomena (for a provocative discussion, see Feyerabend 1974). With the falsity of the theories used in the construction of the instruments, one cannot claim that the problem of incommensurability seriously threatens these theories. After all, the fact that the theories in question weren't even (taken to be) true indicates a less than full commitment to them. In the presence of better alternatives, one may just simply rule out these theories.

Second, it is quite controversial to claim that theories need to be invoked in the *use* of instruments (see Hacking 1983). Although some theories may need to be employed in the case of sophisticated use of certain instruments such as electron microscopes, this need not be the case for every instrument in every context. There's a relative autonomy of the instruments with respect to theories (see Hacking 1983, again, as well as Azzouni 2000; Baird 2004; Humphreys 2004). And some optical microscopes clearly fit into this category.

It is more difficult, however, to make this case for complex tools such as electron microscopes. After all, to obtain particularly sophisticated results in electron microscopy, not only does one need to know in great detail how such instruments work, but one also needs to master a great deal about the quantum mechanics that goes into the instrument's construction. In these cases, can we say that the incommensurability issues that emerge in the context of quantum mechanics will also carry over to the use of an electron microscope? Yes, we can. But the incommensurability won't pose a problem here—at least not for the empiricist. After all, the use of these instruments is ultimately based on a robust group of observational practices, and these practices are invariant under theory transformations. What are these practices? (Observational practices are linked to Azzouni's gross regularities [Azzouni 2000]. However, they are importantly different from the latter because observational practices are restricted to observations in a way that gross regularities need not be.)

Observational practices are certain procedures that are ultimately grounded on observation. Observation, of course, is not a static, passive process of simply opening one's eyes (Hacking 1983). It is rather a much more active process of interaction with different, observable aspects of the world, one that relies on all our sense modalities and explores the triangulation provided by the interconnection between them. Processes of this kind involve the following epistemic features (which are freely based on and adapted from Azzouni 2004 and Lewis 1980):

(i) *Counterfactual dependence*: These are the two key conditions from which the remaining three listed below follow. (a) Had the scene before our eyes been different, the image it yields would have been correspondingly different (within the sensitivity range of our senses). (b) Had the scene before our eyes been the same, the resulting image would have been correspondingly the same (again, within our senses' sensitivity range). In other words, the images that we form are sensitive to the scenes before us. The particular form this sensitivity takes is expressed in terms of the robustness, refinement, and tracking conditions that are entailed by counterfactual dependence.

(ii) *Robustness*: The scene before our eyes produces an image independently of us. There are two related senses of "independence" here: (a) Given the sensory faculties we have, the image that will be produced is not of our own making. Unless we are hallucinating, we won't see a pink elephant in front of us if none is there (counterfactual dependence condition (i)(a)). (b) What we see doesn't depend on our beliefs regarding the scene before us (the two counterfactual dependence conditions apply independently of our beliefs about the objects we experience). Our beliefs may change our *interpretation* (our understanding) of what we see, though. Someone who doesn't have the concept of a *soccer ball* won't see a ball *as* a soccer ball, even though the person will see the round object. It is the latter sense that is relevant here.

(iii) *Refinement*: We can refine the image of what we see, say, by getting closer for a better look. As we approach the target object, we compare the resulting images of the object with the previous ones (obtained when the object was seen further away). The counterfactual dependence conditions remain invariant as we move closer to the object, thus enabling the comparison. This improvement in the image allows us to discriminate better what is seen—even though there are well-known physical limits to how far the discrimination can go.

(iv) *Tracking*: Two forms of tracking are enabled by the counterfactual dependence condition: given the sensitivity of our perceptual apparatus to the environment around us, we can track objects in *space* and in *time*. We can move around an object to detect differences in it (spatial tracking). We can also stay put and observe changes in a given object over time (temporal tracking). In both cases, the differences in the resulting images, allowed for by the invariance in the counterfactual dependence conditions, are contrasted and compared. As a result, we are able to track the scene before our eyes in space and time.

Based on these four conditions (or on conditions of this sort), observational practices are elaborated. These practices explore the significant epistemic features of observation.

With *counterfactual dependence*, we have the sort of sensitivity to the scene before our eyes that we would expect in the case of good detection systems. Changes in the scene correspond to changes in our observations. In this way, what we observe is sensitive and responsive to the environment.

With *robustness*, we obtain the sort of independence that is significant for the objectivity of observation. It is not up to us what we will observe when we open our eyes even though our observations depend on concepts that we invoke to interpret, categorize, and understand what we see.

With *refinement*, we can improve our observations. From an epistemic point of view, this increases our trust in what we observe. This doesn't mean, of course, that observations are always reliable. They aren't, and we learn when to trust the results of observation and when not to. Since this is accomplished by using our own senses (what else could we use?), refinement clearly plays an important role in this process. After all, refinement allows us to determine in more fine-grained ways how the images we form depend on our position with respect to the scene before our eyes.

Tracking finesses the counterfactual dependence condition, and it spells out the particular type of correlation that is established between the scene before our eyes and what we observe. The sensitivity that tracking establishes is to both the presence of certain details in the scene before our eyes and to the absence of other details there.

Observational practices are grounded on these four epistemic conditions, and they explore the latter to guarantee reliable results. These practices ground theoretical and experimental activities in scientific practice. The central idea, briefly, is that the observational features of our experience are grounded on observational practices, and these, in turn, are invariant under theoretical reinterpretations. As a result, the presence of incommensurable theories won't affect them. This is because, although observation is undeniably theory laden, it has certain aspects that are preserved even when theories change. These robust features of observation emerge from the four epistemic features discussed earlier. And reliable instruments, it is not difficult to see, exemplify these features. After all, the same support that the four epistemic conditions offer to observation extends very naturally to certain instruments and the stability of the results the latter offer.

Now, the presence of incommensurability challenges the realist understanding of scientific results because it calls into question the existence of common standards to assess the adequacy of rival theories. That is, with incommensurability in place, it is not clear that there are standards that do not favor one theory in contrast with another. This questions the realist's ability to choose between rival theories without assuming the standards that favor one of them (e.g., by implicitly assuming the framework of one of the theories). Since the realist needs to decide if one of theses theories (if any) is true (or approximately true), he or she needs to be in a position to provide noncircular grounds to prefer one theory over the other.

But are we inevitably led to embrace relativism if we accept the incommensurability thesis? As will become clear shortly, I don't think so. The proposal advanced here supports the incommensurability thesis but resists the move to relativism. There may not be common standards in theory evaluation, but we can't conclude that anything goes— that any theory is just as good as any other. Observational regularities, even those based on some instruments, would be preserved—they are quasi-true and thus preserved through theory change. And we can choose among rival theories based on their consistency with these observational regularities. Although this criterion will not uniquely single out one theory (which is not a problem, of course, for the anti-realist), it rules out some theories. As a result, we have theoretical pluralism without relativism.

How exactly do instruments get into the picture? First, they are devised in order to extract observational regularities from things that are unobservable. And by preserving the observational regularities, one can choose between different theories (those that accommodate such regularities), even though the results may have different interpretations in different theories.

Second, by invoking the four epistemic conditions, we can neutralize the familiar troubles posed by incommensurability. After all, the robustness of the access to the sample that is provided by the relevant instruments prevents conceptual changes from affecting that access. Even though theories may be needed to interpret certain results from microscopes, the practice with these instruments doesn't change with the change of theories. Priority is still given to the rich texture of data obtained from the instruments.

Note, however, that the results that the instruments generate don't change with changes in the relevant theories. As will become clear, this is the outcome of the four epistemic conditions mentioned earlier. The conditions establish a strong connection between the sample and the instruments. And this connection, as we'll see, is not theory-dependent—even though theories are involved at various points in the construction and use of the instruments in question. Hence, the type of the detection of the objects in the sample is invariant under transformations of the theories that are used.

But how can instruments provide a common ground—and a common standard—in theory evaluation? How exactly can instruments be used to avoid the relativistic consequences of the incommensurability thesis? The proposal here, I insist, is not to *undermine* the latter thesis. It is not even clear how exactly that could be done. After all, in theory evaluation, one does end up invoking some standard or other in order to judge the adequacy of the theories involved. And, typically, the standards that directly support the theory one already favors would be invoked in defense of that theory. But it doesn't follow from this that there are no objective standards of theory assessment. Different scientific theories are usually expected to conform to the evidence. And, even if such theories may interpret the evidence differently, the results conveyed by that evidence are stable and systematic enough that the theories couldn't simply ignore them. The results—typically produced by various kinds of scientific instruments—provide stable, theory-neutral grounds for theory assessment.

The results generated by scientific instruments, I noted, are relatively theory-neutral in the sense that relatively few theories are needed to interpret them. Usually, only those theories that were used to construct the instruments are invoked in interpreting the results obtained with these instruments. Additional theories are not required—practice with the images in question and with the instruments are central here, though. This means that instruments have a relative independence from theories, which means that the results that they offer can be used to adjudicate between rival theories without favoring any one of the theories in particular.

Let me elaborate on this point. Instruments can provide a common ground between rival theories in their evaluation process for the following reasons:

(1) As just noted, instruments are relatively theory-independent, and thus no theory is favorably supported by an instrument alone. This condition is sufficient to guarantee that instruments yield a common ground in theory assessment. As we saw, the reason why instruments are only relatively rather than absolutely theory-independent is because one needs theories to construct sophisticated instruments, and it is often crucial to invoke suitable theories to use the instruments to their maximum limit.

For instance, without knowledge of the tunneling effect in quantum mechanics, one cannot construct—let alone imagine the construction of—a scanning tunneling microscope (STM). This doesn't mean, of course, that knowledge of this effect is sufficient for the construction of that instrument. Clearly, it isn't: the tunneling effect was known since the 1920s, but the STM wasn't actually built until the 1980s. In addition to theoretical knowledge, a considerable dose of engineering and practical skills is required. Moreover, to use an STM effectively and obtain detailed results about the surface structure of the sample, it is crucial that one knows not only the way the instrument operates, but also the theories that are relevant in bringing about the phenomena under study—particularly, quantum mechanics. Once again, knowledge of the *relevant* theories is not sufficient to guarantee successful results given that it is central that one has the required skills to obtain the results. And that is certainly *not* simply a matter of having theoretical knowledge (see Baird 2004; Hacking 1983; Humphreys 2004).

But if theories are involved in the construction process and in the use of certain instruments, isn't the epistemic status of these instruments necessarily compromised? After all, it might be argued, because the instruments rely on these theories, the former can't yield results that conflict with the latter. I don't think this is correct, though. First, I mentioned earlier that knowledge of the *relevant* theories is involved in the successful use of certain instruments. Now, what are the relevant theories? This is obviously a context-dependent matter. Usually, the relevant theories are those that are explicitly invoked by those who have constructed and by those who regularly use the instruments in question. But reference to these theories won't settle the matter. After all, one can successfully construct an instrument using a false theory. This was actually the case with the construction of the first telescopes in the seventeenth century based on Kepler's optics. The use of that theory was clearly relevant. However, one can't claim that, from an epistemological point of view, the theory is responsible for the results of the instrument. In the end, what matters is how reliable, reproducible, and stable are the results generated by the instruments. The epistemological status of the theories used won't affect that.

Second, this feature also highlights another sense in which instruments are only relatively theory-independent: although in sophisticated cases theories are used in the construction of instruments, these theories do not support the instruments in question.

In fact, the theories sometimes are not even consistent—as the case of Kepler's optics illustrates.

(2) Instruments can provide a common ground in theory assessment for a second reason. Instruments navigate across theoretical boundaries without disturbing the theories involved. Consider, for instance, an electron microscope (whether a transmission or a scanning electron microscope). This type of tool is regularly used in a variety of fields, ranging from chemistry through molecular biology to physics, and so it needs to bridge the theoretical gap between quite different disciplines. In practice, the instrument is simply used in the way it needs to be used in each field. Not surprisingly, different practitioners use the instrument in different ways depending on their specific research needs. And that is accomplished independently of the quite dramatic differences between the various theories invoked in each case. Again, this doesn't mean that theories play no role in the practice with this instrument. It just means that different users will engage with different theories depending on the application at hand. As a result, also in this sense, instruments yield a common ground in theory evaluation.

(3) Moreover, the results obtained from the instruments are robust (in the sense discussed earlier with the strong epistemic conditions of access they provide to the sample). When working correctly, instruments systematically track the stuff in the sample. This will generate suitable, robust data to adjudicate between rival theories. (Of course, different theories may be used to interpret the data differently. But the body of data itself can be used to assess the theories in question.)

Now, since instruments provide such a common ground, can they also be used as a common standard to adjudicate between rival theories? I think they can. Being relatively theory-independent and producing robust results, instruments can be used to assess the adequacy of scientific theories. Moreover, instruments are often used to yield novel results that, in turn, will require explanation; in terms of the adequacy of the latter, one can choose among competing theories.

For these reasons, we can rule out relativism even acknowledging the presence of incommensurability. Instruments—at least good ones—provide strong epistemic access and are relatively theory-neutral. The fact that the results from instruments track the evidence (in particular, relevant features of the sample) provides grounds to choose a theory that entails that piece of evidence.

Someone may resist this move. After all, the argument goes, to use instruments and their results in the adjudication between rival theories, one has to adopt a methodological standard to the effect that if there is an inconsistency between the results generated by certain instruments and the consequences derived from relevant theories, we need to revise something in the resulting package. The problem here is that, for instruments to play such a role, their results need to be stable over radical theory change. It turns out, however, that the meaning of the notions that are used to characterize and describe the results of instruments changes with shifts in the theories one considers. As a result, incommensurability applies to the notions that are invoked in the description of the results of instruments. The result is therefore clear: relativism.

In response, we can simply block the last step in the argument. Even if we grant the presence of incommensurability in this context, as we should, this is not enough to entail relativism. The nature of the access provided by the instruments is central for that. Consider the *robustness* condition: the results we obtain from an instrument are not dependent on our beliefs about the sample under study. Whatever we believe about what might be going on in the sample is irrelevant to what is actually happening. The interaction between the sample and the instrument—once the latter is suitably calibrated and the former is properly prepared—is a process that depends on the instrument and the sample. It is in this sense that the robustness condition is an independence condition.

Now, when we *interpret* the results of an instrument, our beliefs about what might be happening in the sample are obviously relevant and so are the theories we may invoke in making sense of the data. But the process of interpretation of experimental results is a very different process from the one in which these results are obtained. This doesn't mean that no interpretation or no theorizing goes on when experimental results are being obtained and constructed. For example, the statistical manipulation of data is clearly in place, and it constitutes an extremely significant form of theorizing in the construction of experimental results. However, before any such statistical techniques can be invoked, it is crucial that raw data are generated. The robustness of each particular piece of datum is central.

And this robustness is a crucial feature of the strategy of resisting relativism. The robustness is another way of supporting the relative theory-independence of the instrument's results. Moreover, the theories involved in the actual construction of the data, being statistical in character, are not contentious in the context of assessing rival *physical* theories. After all, such statistical theories can be applied only if proper application conditions are satisfied. But as long as these theories are correctly applied, rival physical theories can be discriminated even if the relevant statistical theories may be different for the different physical theories. In this way, statistical theories also provide a common ground between rival physical hypotheses. Clearly, this sort of common ground is not strong enough to overcome the incommensurability between the theories in question because the standards are not general enough to effectively undo the effects of incommensurability. However, the presence of theories in the construction of experimental results doesn't undermine the robustness of such results, which can then be used to assess between rival theories. As a result, relativism doesn't emerge here.

Consider, also, the *counterfactual dependence* conditions: had the sample been different (within the sensitivity range of the instrument), the image produced would have been correspondingly different; had the sample been the same (within the instrument's sensitivity range), the image produced would have been correspondingly the same. This establishes a clear dependence between the sample and the images that are generated. Clearly, there might be a variety of different theories involved in the description of the particular form of interaction between the sample and the instrument. And, as an outcome of the presence of these theories, incommensurability may emerge. However, the counterfactual dependence between the sample and the instrument prevents us from getting relativistic consequences from the incommensurability. After all, the relation between the sample and the instrument holds independently of the theories involved. It is a matter of the relation between the two (sample and instrument) alone.

A similar point emerges when we consider the third epistemic feature of scientific instruments: the *refinement* condition. In a series of instruments, it is possible to improve the resolution of the images that are generated. This indicates that the access to the sample provided by each instrument, although robust, can yield further information about the sample in the end. Once again, in the process of improving the instruments, in addition to careful practice, theories can play a significant role. These theories can suggest new avenues to explore the interactions between the instrument and the sample. They can also suggest new mechanisms that might be active in the sample, which in turn may indicate new possible forms of intervening on the latter. Obviously, theories are not sufficient to achieve these results because systematic engineering techniques and, sometimes, better instruments are also needed.

Given the presence of various sorts of theories throughout the process of refinement, it might be thought that the incommensurability involved will prevent instruments from being used effectively in theory selection. But this is not the case. After all, the theories employed in the development of the instruments need not be involved in the actual implementation of the latter. Moreover, once the instruments are constructed, the results that are obtained are largely independent from the theories in question. And with the refinement of the instruments, more fine-grained results can be generated. In this way, even more detailed ways of testing theories and yielding novel effects can be produced. As a result, once again, incommensurability won't generate relativistic conclusions. Because the results are largely independent of the theories, the incommensurability won't raise a problem for the objectivity of the results.

Finally, consider the *tracking* condition, which, given the counterfactual dependence conditions, specifies the particular form of the dependence between the instrument and the sample: the instrument is sensitive to the sample and tracks its main features—spatially and, in some cases, temporally. Now the tracking relation between the instrument and the sample depends ultimately on what goes on between the two of them. Thus, the fact that some theories are invoked in the construction and use of certain instruments doesn't affect the outcome of the instrument. The images that the latter yield are sensitive to what is going on—to the best of our knowledge—in the sample. Of course, since we may not *know* (or may not be justified in believing) that the counterfactual (and, hence, the tracking) conditions are actually satisfied in the case of certain events, we cannot know for sure what is really going on in the sample. But *if* the conditions are met, then the instrument is indeed sensitive to the presence of the sample and the details it has. In other words, the fact that theories are involved doesn't affect the counterfactual dependence (or the tracking). As a result, due to the presence of various theories, we may well have incommensurability. But, once again, for the reasons just discussed, this won't yield relativism.

Note that, at this point, being able to know whether the counterfactual conditions actually apply enters the considerations. An externalist about instruments—in analogy with an externalist in the theory of knowledge (BonJour and Sosa 2003)—insists that as long as the counterfactual and the tracking conditions are met, instruments will be reliable. There's no need to know (or be justified in believing) that these conditions are indeed satisfied. The reliability of scientific instruments depends only on their properties and the relations they bear to the environment. Our epistemic access to the relevant epistemic conditions is irrelevant. In contrast, an internalist about instruments—again, in analogy with the epistemological internalist—argues that the satisfaction of the counterfactual conditions is not enough: one also needs to know (or, at least, be justified in believing) that these conditions are indeed met. Otherwise, we will be unable to know whether the instrument in question is indeed reliable or not. After all, suppose we don't know (or have reason to believe) that the counterfactual conditions are indeed satisfied, and suppose that these conditions fail; we may then take an image produced by an instrument as depicting a feature that the sample actually lacks or may fail to identify a feature of the sample because no such a corresponding feature is depicted in the image. Admittedly, if the counterfactual conditions are met, we shouldn't be prone to these mistakes. But unless we are able to know (or be justified in believing) that the conditions are in place, we can't rule these errors out.

These considerations indicate the significance that internalist and externalist considerations play in the proper formulation of suitable epistemic conditions for scientific instruments. By engaging with this aspect of epistemological theorizing, philosophers of science can be provided a better account of the epistemology of instruments, which produce some of the most significant pieces of knowledge in scientific practice.

ACKNOWLEDGMENTS

My thanks go to Paul Humphreys for extremely helpful and perceptive comments on an earlier draft of this chapter. They led to significant improvements.

REFERENCES

Azzouni, J. (2000). *Knowledge and Reference in Empirical Science* (London: Routledge).
Azzouni, J. (2004). *Deflating Existential Consequence* (New York: Oxford University Press).
Baird, D. (2004). *Thing Knowledge: A Philosophy of Scientific Instruments* (Berkeley: University of California Press).
Bird, A. (2007). "What Is Scientific Progress?" *Noûs* 41: 92–117.
BonJour, L., and Sosa, E. (2003). *Epistemic Justification: Internalism vs. Externalism, Foundations vs. Virtues* (Oxford: Blackwell).

Bueno, O. (2012). "Inconmensurabilidad y Dominios de Aplicacíon." In Lorenzano and Nudler (eds.) (2012), 27–65.

Bueno, O. (2014). "Why Identity Is Fundamental." *American Philosophical Quarterly* 51: 325–332.

Bueno, O. (2015). "Can Identity Be Relativized?" In Koslow and Buchsbaum (eds.), 2015, 253–262.

DeRose, K., and Warfield, T. (eds.) (1999): *Skepticism: A Contemporary Reader* (New York: Oxford University Press).

Feyerabend, P. (1974). *Against Method* (London: New Left Books).

Feyerabend, P. (1981). *Realism, Rationalism and Scientific Method.* Philosophical Papers, Vol. 1. (Cambridge: Cambridge University Press).

Gettier, E. (1963). "Is Justified True Belief Knowledge?" *Analysis* 26: 144–146.

Goldman, A. (1967). "A Causal Theory of Knowing." *Journal of Philosophy* 64: 355–372.

Hacking, I. (1983). *Representing and Intervening* (Cambridge: Cambridge University Press).

Hoyningen-Huene, P. (1993). *Reconstructing Scientific Revolutions: Thomas S. Kuhn's Philosophy of Science.* A. Levin (trans.) (Chicago: University of Chicago Press).

Humphreys, P. (2004). *Extending Ourselves: Computational Science, Empiricism, and Scientific Method* (New York: Oxford University Press).

Koslow, A., and Buchsbaum, A. (eds.). (2015). *The Road to Universal Logic*, Vol. II (Dordrecht: Birkhäuser).

Kuhn, T. (1970). *The Structure of Scientific Revolutions*, 2nd ed. (Chicago: University of Chicago Press).

Lewis, D. (1980). "Veridical Hallucination and Prosthetic Vision." *Australasian Journal of Philosophy* 58: 239–249. (Reprinted, with a postscript, in Lewis, 1986, 273–290.)

Lewis, D. (1986). *Philosophical Papers*, Vol. II (Oxford: Oxford University Press).

Lewis, D. (1996). "Elusive Knowledge." *Australasian Journal of Philosophy* 74: 549–567. (Reprinted in DeRose and Warfield [eds.] 1999, 220–239.)

Littlejohn, C. (2012). *Justification and the Truth-Connection* (Cambridge: Cambridge University Press).

Lorenzano, P., and Nudler, O. (eds.). (2012). *El Camino desde Kuhn: La Inconmensurabilidad Hoy* (Madrid: Editorial Biblioteca Nueva).

Miller, D. (1994). *Critical Rationalism: A Restatement and Defence* (La Salle: Open Court).

Nozick, R. (1981). *Philosophical Explanations* (Cambridge: Harvard University Press).

Popper, K. (1972). *Objective Knowledge* (Oxford: Clarendon Press).

Sankey, H. (1994). *The Incommensurability Thesis* (Aldershot: Avebury).

Siegel, H. (1980). "Objectivity, Rationality, Incommensurability, and More." *British Journal for the Philosophy of Science* 31: 359–384.

Unger, P. (1975). *Ignorance: A Case for Scepticism* (Oxford: Oxford University Press).

van Fraassen, B. (1989). *Laws and Symmetry* (Oxford: Clarendon Press).

Williamson, T. (2000). *Knowledge and Its Limits* (Oxford: Oxford University Press).

Zagzebski, L. (1994). "The Inescapability of Gettier Problems." *Philosophical Quarterly* 44: 65–73.

CHAPTER 12

ETHICS IN SCIENCE

DAVID B. RESNIK

1 INTRODUCTION

ETHICS questions, problems, and concerns arise in many different aspects of scientific research (Shrader-Frechette, 1994; Resnik, 1998a, Macrina, 2005; Shamoo and Resnik, 2009). Ethical issues, such as reporting misconduct, sharing data, assigning authorship, using animals or humans in experiments, and deciding whether to publish results that could be used by others in harmful ways impact the day-to-day activities of scientists and frequently draw the attention of the media, policymakers, and the public (Steneck, 2007; National Academy of Sciences, 2009; Briggle and Mitcham, 2012). These issues occur in many different types of research disciplines, including biology, medicine, physics, chemistry, engineering, psychology, sociology, anthropology, economics, mathematics, and the humanities (including philosophy). Research ethics is a type of professional ethics, similar to medical or legal ethics (Shamoo and Resnik, 2009). Philosophers have tended to focus on fundamental questions concerning the relationship between science and ethics, such as whether value judgments influence concept formation and theory-choice (Rudner, 1953; Kuhn, 1977; Longino, 1990; Kitcher, 1993; Harding, 1998; Elliott, 2011). While these issues are important, they are not the main concern of this chapter, which will focus, for the most part, on practical and policy issues related to conduct of science.

2 HISTORICAL BACKGROUND

Questions about the ethics of science are not a uniquely modern concern. In *Reflections on the Decline of Science in England*, Charles Babbage (1830) scolded his colleagues for engaging in deceptive research practices he described as trimming, cooking, and forging. Trimming, according to Babbage, involves reporting only the data that support one's hypothesis or theory. Cooking involves conducting an experiment that is designed only to obtain a specific result consistent with one's hypothesis. Since the experiment is rigged in advance, it is not a genuine test of a hypothesis. Forging involves making up or fabricating data. One of the first major ethical scandals involving science began in 1912, when Charles Dawson claimed to have discovered parts of a skull in the Piltdown gravel bed near Surrey, England, which he said was a "missing link" between humans and apes. Though paleontologists doubted that the skull was genuine because it was inconsistent with the hominid fossil record, the skull was not definitely proven to be a fake until 1953, when laboratory tests indicated that the bones had been artificially aged with chemicals. The artifact was actually composed of a human skull, an orangutan jaw, and chimpanzee teeth (Shamoo and Resnik, 2009).

Ethical issues came to the forefront in World War II. American and British physicists were concerned that Germany would develop an atomic bomb that it could use to win the war. Albert Einstein wrote a letter to President Roosevelt warning him about this threat and urging him to support scientific research to develop an atomic bomb. Roosevelt took this advice and initiated the top secret Manhattan Project in 1941. Robert Oppenheimer led a group of scientists, including Enrico Fermi and Richard Feynman, who worked on developing an atomic bomb in Los Alamos, NM. Germany surrendered to the Allied forces in May 1945, but Japan continued to fight. In August 1945, the US dropped atomic bombs on Hiroshima and Nagasaki, and the Japanese surrendered shortly thereafter. During the war, physicists wrestled with their ethical responsibilities related to the atomic bomb. After the war ended, Robert Oppenheimer and other scientists led an "atoms for peace" movement that sought to stop the proliferation of nuclear weapons and encourage the use of nuclear energy for non-violent purposes, such as electric power generation (Resnik, 1998a; Briggle and Mitcham, 2012).

During the war crimes trials at Nuremberg, Germany, the world learned about horrific experiments that German physicians and scientists had conducted on concentration camp prisoners. Most of these experiments caused extreme suffering, injury, and death. The prisoners did not consent to these experiments. Some experiments involved exposing subjects to extreme environmental conditions, such as low air pressure or freezing cold temperatures. Other experiments involved injuring subjects with shrapnel or bullets to study wound healing. One of the most notorious researchers, Josef Mengele, injected chemicals into children's eyes in an attempt to change their eye color; amputated limbs and organs without anesthesia, and subjected prisoners to electroconvulsive therapy. In 1947, at the conclusion of the war trials, the judges promulgated the Nuremberg Code, a set of ten principles for the conduct of ethical experimentation involving human beings. The Code requires that subjects provide their informed consent to participate in research, that experiments should be expected to yield socially valuable results that cannot be obtained by other means, and that experiments minimize suffering and injury. In 1964, the World Medical Association adopted the Helsinki Declaration to provide ethical guidance for medical research, and in the 1970s and 1980s the United States adopted laws and regulations governing research with human subjects (Shamoo and Resnik, 2009).

By the 1980s, most scientists acknowledged that ethical issues arise in the conduct of science, but they viewed these as having mostly to do with science's interactions with society. Science itself was viewed as objective and ethically pristine. All of this changed, however, when several scandals emerged involving fraud or allegations of fraud in the conduct of federally funded research. The most famous of these became known as the Baltimore Affair, after Nobel Prize–winning molecular biologist David Baltimore, who was associated with the scandal. In 1986, Baltimore, Thereza Imanishi-Kari, and four co-authors associated with the Whitehead Institute published a paper in *Cell* on using gene transfer methods to induce immune reactions (Weaver et al., 1986). The study was funded by the National Institutes of Health (NIH). Margot O'Toole, a postdoctoral fellow in Imanishi-Kari's lab, had trouble repeating some of the experiments reported in the paper and she asked for Imanishi-Kari's lab notebooks. When O'Toole could not reconcile data recorded in the lab notebooks with the results reported in the paper, she accused Imanishi-Kari of misconduct. After an internal investigation found no evidence of misconduct, the Office of Scientific Integrity (OSI), which oversees NIH-funded research, began investigating the case. A congressional committee looking into fraud and abuse in federal agencies also investigated the case, and the *New York Times* reported the story on its front pages. In 1991, the OSI concluded that Imanishi-Kari falsified data and recommended she be barred from receiving federal funding for ten years. However, this finding was overturned by a federal appeals panel in 1996. Though Baltimore was not implicated in misconduct, his reputation was damaged and he resigned as President of Rockefeller University due to the scandal. Imanishi-Kari acknowledged that she was guilty of poor recordkeeping but not intentional wrongdoing (Shamoo and Resnik, 2009).

Other scandals that made headlines during the 1980s included a dispute between Robert Gallo, from the NIH, and Luc Montagnier, from the Pasteur Institute, over credit for the discovery of the human immunodeficiency virus (HIV) and patent claims concerning a blood test for the HIV; a finding by the OSI that Harvard Medical School postdoctoral fellow John Darsee had fabricated or falsified data in dozens of published papers and abstracts; and a finding by the NIH that University of Pittsburgh psychologist Stephen Breuning had fabricated or falsified data on dozens of grant applications and published papers. Breuning was convicted of criminal fraud and sentenced to sixty days in prison and five years of probation (Shamoo and Resnik, 2009).

In response to these scandals, as well as growing concerns about the impact of financial interests on the objectivity and integrity of research, federal agencies, chiefly the NIH and National Science Foundation (NSF), took additional actions to address ethical problems in the conduct of science (Steneck, 1999). In 1989, the NIH required that all extramurally funded graduate students receive instruction in the responsible conduct of research (RCR). It later extended this requirement to include post-doctoral fellows and all intramural researchers. In 2009, the NSF expanded its RCR instructional requirements to include all undergraduate or graduate students receiving NSF research support. During the 1990s, federal agencies adopted and later revised policies pertaining to research misconduct. In 2000, federal agencies agreed upon a common definition of research misconduct as fabrication of data, falsification of data, or plagiarism (FFP) (Office of Science and Technology Policy 2000). During the 1990s, federal agencies adopted policies concerning the disclosure and management of financial interests related to research. The federal government also reorganized the OSI and renamed it the Office of Research Integrity (ORI). The ORI expanded the scope of its mission beyond oversight and investigation and started sponsoring research on research integrity and conferences on scientific ethics (Steneck, 1999).

Although the United States is the world's leader in research ethics oversight, regulation, and policy, many other countries have now adopted rules and guidelines pertaining to scientific integrity and RCR instruction. Scientific journals, professional associations, and universities also have developed rules and policies that cover a number of different topics, including the ethical conduct of research involving humans or animals; authorship, publication, and peer review; reporting and investigating misconduct; and data ownership and management (Resnik and Master, 2013).

3 Philosophical Foundations of Science's Ethical Norms

Science's ethical norms are standards of behavior that govern the conduct of research. Many of science's norms are embodied laws and regulations, institutional policies, and professional codes of conduct. Science's ethical norms have two distinct foundations. First, ethical norms in science can be viewed as rules or guidelines that promote the effective pursuit of scientific aims. When scientists fabricate or falsify data, they propagate errors and undermine the search for truth and knowledge. Conflicts of interest in research are an ethical concern because they can lead to bias, fraud, or error (Resnik, 2007). Ethical norms also indirectly foster the advancement of scientific aims by helping to promote trust among scientists, which is essential for collaboration, publication, peer review, mentoring, and other activities. Unethical behavior destroys the trust that is vital to social cooperation among scientists (Merton, 1973; Whitbeck, 1995).

Ethical norms also help to foster the public's support for science. Science takes place within a larger society, and scientists depend on the public for funding, human research participants, and other resources (Ziman, 2000). Unethical behavior in science can undermine funding, deter people from enrolling in studies, and lead to the enactment of laws that restrict the conduct of research. To receive public support, scientists must be accountable to the public and produce results that are regarded as socially valuable (Resnik, 1998a). Since science's ethical norms stem from the social aspects of research, they can be viewed as part of the social epistemology of science (Longino, 1990; Resnik, 1996; Goldman, 1999).

Second, science's ethical norms are based on broader, moral norms. For example, rules against data fabrication and falsification can be viewed as applications of the principle of honesty, a general rule that applies to all moral agents (Resnik, 1998a). Rules for assigning credit in science can be viewed as applications of a moral principle of fairness, and rules pertaining to the treatment of human subjects in research can be viewed as applications of respect for autonomy, social utility, and justice (Shamoo and Resnik, 2009). Although science's ethical norms are based on moral norms, science's norms are not the same as the norms that apply to all people. For example, honesty in science differs from honesty in ordinary life because it is more demanding. It is morally acceptable, one might argue, to stretch the truth a little bit when telling a friend about a fish one caught at a lake. However, even a little bit of stretching of the truth is unethical in science because minor changes in data can impact results and mislead the scientific community. Further, honesty in science is different from honesty in ordinary life because scientific honesty is defined by technical rules and procedures that do not apply to ordinary life. For example, falsification of data includes unjustified exclusion of data in research. Some data may be excluded only when they are due to experimental error or are statistical outliers. To determine whether a data point from a scientific study can be excluded, one must therefore have an expert understanding of the field of research.

4 Science's Ethical Norms

Ethical norms that are widely recognized by working scientists and endorsed by philosophers and other scholars include (Resnik, 1998a; Shamoo and Resnik, 2009; Elliott, 2012):

Honesty. Do not fabricate or falsify data. Honestly report the results of research in papers, presentations, and other forms of scientific communication.

Openness. Share data, results, methods, and materials with other researchers.

Carefulness. Keep good records of data, experimental protocols, and other research documents. Take appropriate steps to minimize bias and error. Subject your own work to critical scrutiny and do not overstate the significance of your results. Disclose the information necessary to review your work.

Freedom. Support freedom of inquiry in the laboratory or research environment. Do not prevent researchers from engaging in scientific investigation and debate.

Due credit. Allocate credit for scientific work fairly.

Respect for colleagues. Treat collaborators, students, and other colleagues with respect. Do not discriminate against colleagues or exploit them.

Respect for human research subjects. Respect the rights and dignity of human research subjects and protect them from harm or exploitation.

Animal welfare. Protect and promote the welfare of animals used in research.

Respect for intellectual property. Do not plagiarize or steal intellectual property. Respect copyrights and patents.

Confidentiality. Maintain the confidentiality of materials that are supposed to be kept confidential, such as articles or grants proposals submitted for peer review, personnel records, and so on.

Legality. Comply with laws, regulations, and institutional policies that pertain to research.

Stewardship. Take proper care of research resources, such as biological samples, laboratory equipment, and anthropological sites.

Competence. Maintain and enhance your competence in your field of study. Take appropriate steps to deal with incompetence in your profession.

Social responsibility. Conduct research that is likely to benefit society; avoid causing harm to society. Engage in other activities that benefit society.

A few comments about this above list of scientific norms are in order. First, the list is not intended to be complete. There may be other norms not included on this list. However, the list probably includes most of science's most important norms.

Second, most of these norms imply specific rules and guidelines that are used to apply the norms to specific situations. For example, respect for human research subjects implies rules pertaining to informed consent, risk minimization, confidentiality, and so on. Due credit implies rules concerning authorship on scientific papers, and animal welfare implies rules for minimizing pain and suffering to animals, and so on.

Third, the norms pertain not only to individuals but also to institutions and organizations, which play a key role in promoting and enforcing ethical conduct in research (Shamoo and Resnik, 2009). For example, university contracts with pharmaceutical companies can help to ensure that researchers have access to data, and that companies cannot suppress the publication of data or results. University policies pertaining to the reporting and investigation of illegal or unethical activity can help ensure the researchers comply with the law and abide by ethical standards. Journal policies concerning authorship can play an important role in sharing credit fairly in scientific publications. Conflict of interest rules adopted by funding agencies can help ensure that grants are reviewed fairly, without bias.

Fourth, sometimes scientific norms conflict with each other or with other moral or social norms. For example, openness may conflict with the respect for human research subjects when scientists are planning to share private data about individuals. One way of handling this conflict is to remove information that can identify individuals (such as name or address) from the data. Alternatively, one can share data by requiring recipients to sign an agreement in which they promise to keep the data confidential. When a conflict among norms arises in a particular situation, scientists should use their reasoning and good judgment to decide how to best resolve the conflict in light of the relevant information and available options (Shamoo and Resnik, 2009). The possibility of conflicts among ethical norms in science implies that RCR instruction must involve more than teaching students how to follow rules: it must also help students learn to use their reasoning and judgment to deal with ethical dilemmas (Shamoo and Resnik, 2009). Ethical dilemmas in research will be discussed at greater length below.

Fifth, science's ethical norms can be studied from an empirical or conceptual perspective. Empirical approaches to scientific norms attempt to describe the norms that are accepted by scientists and explain how they function in the conduct of research. Sociologists, psychologists, and historians have gathered data on scientific norms and developed explanatory hypotheses and theories. For example, Brian Martinson and colleagues surveyed 3,247 NIH-funded scientists at different stages of their careers concerning a variety of behaviors widely regarded as unethical or ethically questionable. They found that 0.3% admitted to falsifying or cooking data in the last three years; 0.3% admitted to ignoring major aspects of human subjects requirements; 1.4% said they had used someone else's ideas without permission or giving proper credit; 10% said they had inappropriately assigned authorship credit; and 27.5% admitted to keeping poor research records (Martinson et al., 2005). In the 1940s, Robert Merton (1973) described four different scientific norms that he regarded shaping scientific behavior: universalism, communism, disinterestedness, and organized skepticism. He later added originality to his list of norms (Ziman, 2000).

Conceptual approaches examine the justification, definition, and meaning of scientific norms. Conceptual approaches attempt to evaluate and criticize norms that are accepted by the scientific community rather than simply describe them. Philosophers, ethicists, and working scientists have taken a conceptual approach to scientific norms. Philosophers have tended to focus on epistemic norms, such as simplicity, explanatory power, testability, empirical support, and objectivity, rather than ethical ones (Quine and Ullian, 1978; Kitcher, 1993; Thagard, 1993). Only a handful of philosophers have conducted extensive studies of science's ethical norms (Shrader-Frechette, 1994; Resnik, 1998a; Whitbeck, 1998). Working scientists have tended to address specific normative issues in research, such as authorship (Rennie et al., 1997), misconduct (Kornfeld, 2012), conflict of interest (Morin et al., 2002), data integrity (Glick and Shamoo, 1993), and human subjects research (Hudson et al., 2013).

Sixth, variations in scientific practices raise the issue of whether there is a core set of norms that applies to all sciences at all times. For example, different scientific disciplines have different practices concerning authorship (Shamoo and Resnik, 2009). In some biomedical fields, authorship is granted for providing biological samples for analysis; in other fields, providing biological samples, without making any other contribution, would not merit authorship (Shamoo and Resnik, 2009). There are also variations in informed consent practices for research. In the United States and other industrialized nations, the research subject must provide consent. If the subject is incapable of providing consent, the subject's legal representative may provide it. However, in some African communities tribal elders make all the important decisions, including consent for medical treatment or research. In other communities, a woman does not provide consent for herself, but consent may be provided by her husband (if she is married) or an older relative (such as her father) if she is not (Shamoo and Resnik, 2009).

There are also variations in scientific practices across time. Robert Millikan won the Nobel Prize in Physics in 1923 for measuring the smallest electrical charge (i.e., the charge on an electron) and for his work on the photoelectric effect. To measure the smallest electrical charge, Millikan sprayed oil drops through electrically charge plates.

When a drop was suspended in the air, the electrical force pulling the drop up was equivalent to the force of gravity pulling it down. By calculating these forces, Millikan was able to determine the smallest possible charge. In 1978, historian of science Gerald Holton (1978) examined Millikan's laboratory notebooks for these experiments and compared them to the data presented in the published paper. In the paper, Millikan said that he had reported all the data. However, he did not report 49 out of 189 observations (26%). Holton also discovered that Millikan had graded his observations as "good," "fair," and "poor." Other physicists who have attempted to measure the smallest electrical charge have obtained results very close to Millikan's. Even though Millikan got the right answer, some commentators have argued that Millikan acted unethically by excluding observations that should have been reported. At the very least, he should have mentioned in the paper that not all observations were reported and explained why (Broad and Wade, 1983). Others have argued that Millikan should be judged by the ethical standards of his own time. While it would be standard practice today to explain why some data were excluded, this was not the practice in Millikan's time. Science has become more honest, open, and rigorous since then (Franklin, 1981).

How can one account for these social and historical variations in scientific norms? Two different questions arise. The first is empirical: are there variations in the norms accepted by scientists? The answer to this question appears to be in the affirmative, given the examples previously discussed. The second is normative: should there be variations in scientific norms? One might argue that variations in scientific norms are ethically acceptable, provided that they are based on different interpretations of core ethical standards (Macklin, 1999). This would be like claiming that variations in laws pertaining to murder in different societies are acceptable even though a general prohibition against murder should apply universally. For example, in all fields of research, authorship should be awarded for making a significant intellectual contribution to a research project, even though different disciplines may interpret "significant contribution" differently. Informed consent is an ethical norm that applies to all human subjects research even though it may be interpreted differently in different cultures. Honesty applies to all fields of research, but different disciplines may have different understandings of what it means to fabricate or falsify data. Honesty was still an important concern during Millikan's time even though the standards for reporting data may not have been the same as they are today (Shamoo and Resnik, 2009).

5 ETHICAL DILEMMAS AND ISSUES IN RESEARCH

As noted previously, ethical dilemmas can arise when scientific norms conflict with each other or with other social or moral norms. The following are some of the ethical dilemmas and issues that frequently arise in research.

Research Misconduct. Research misconduct raises a number of issues. The first one is definitional: what is research misconduct? In general, research misconduct can be viewed as behavior that is widely regarded as highly unethical. In many cases, a finding of misconduct has legal and career consequences. For example, an individual who is found to have committed misconduct while conducting federally funded research may be barred from receiving federal funding for a period of time. Individuals who are found to have committed misconduct may also have their employment terminated by their institution. In extreme cases, misconduct could lead to prosecution for criminal fraud. As noted earlier, US federal agencies have defined misconduct as fabrication, falsification, or plagiarism. However, before the agencies agreed upon this common definition, many agencies also include "other serious deviations" from accepted scientific practices in the definition of misconduct. Many US universities still include this category of behavior in the definition of misconduct. Additionally, other countries define misconduct differently. While most include FFP in the definition of misconduct, some include significant violations of rules pertaining to research with humans or animals in the definition (Resnik, 2003). One might object that the US government's definition is too narrow to deal with all the highly unethical behaviors that undermine the integrity of scientific research. However, if the definition of misconduct is too broad, it may be difficult to enforce, due to lack of clarity and the sheer number of violations that need to be handled (Resnik, 2003).

Another difficult issue confronts those who suspect that someone has committed misconduct. Although scientists have an ethical obligation to report suspected misconduct in order to help protect the integrity of research, honoring this obligation often comes at some personal cost. Whistleblowers will usually need to expend considerable time and effort in providing testimony, and they may face the threat of retaliation or harassment. They may develop a reputation as a troublemaker and have difficulty finding employment (Malek, 2010). Of course, not reporting misconduct may also have adverse consequences. If an individual is involved in a research project in which they suspect a collaborator of misconduct, they could be implicated if the misconduct is discovered by someone else. Even if the misconduct is not discovered, the individual would have to live with the fact that they allowed fraudulent or erroneous research to be published.

A third issue concerns the distinction between misconduct and scientific disagreement (Resnik and Stewart, 2012). Sometimes scientists accuse their peers of misconduct because they disagree with the assumptions their peers have made or the methods they have used. These sorts of disagreements frequently occur in research. Science sometimes makes significant advances when researchers propose radical ideas or use untested methods (Kuhn, 1962). Of course, radical ideas and untested methods may also lead researchers astray. In either case, the best way to handle these sorts of disagreements is through honest and open scientific debate, rather than by accusing one's peers of misconduct. The US federal definition of misconduct distinguishes between misconduct and scientific disagreement and error. Misconduct involves intentional or reckless deviation from commonly accepted scientific standards (Office of Science and Technology Policy, 2000).

Sharing Data and Materials. Scientists frequently receive requests to share data, methods, and materials (such as chemical reagents or biological samples). As noted earlier, the norm of openness obliges scientists to share their work. Sharing is important in science to make prudent use of resources and achieve common goals. Sharing is also essential to scientific criticism and debate. Knowledge can be obtained more readily when scientists work together (Munthe and Welin, 1996). However, sharing may conflict with the desire to achieve scientific credit or priority (Shamoo and Resnik, 2009). If a chemistry laboratory shares preliminary data with another laboratory working on the same topic, the other laboratory may publish first and win the race for priority. If a molecular biology laboratory has devoted considerable time and effort toward developing a transgenic mouse model of a human disease, it may be hesitant to share the mouse with other laboratories unless it can receive credit for its work. A social scientist who has amassed a considerable amount of demographic and criminological data concerning a population may not want to share this data with other researchers if she is concerned that they could use it to publish papers on topics on which she is planning to publish. Scientists have developed different strategies for dealing with these dilemmas. For example, most scientists do not share preliminary data. Data is shared widely only after publication. Many scientists may also reach agreements with researchers who receive data or materials. The molecular biology lab that developed the transgenic mouse could collaborate with other laboratories that use the mouse and receive some credit. The sociologists with the large database could share the data with the understanding that the recipients would not publish papers on certain topics.

Authorship. As discussed previously, authorship raises ethical issues. Indeed, authorship disputes are some of the most common ethical controversies in everyday science (Shamoo and Resnik, 2009). Although deciding the authorship of a scientific paper is not as important to society as protecting human subjects from harm, it is very important for scientists, since careers may hang in the balance. The adage "publish or perish" accurately describes the pressures that academic researchers face. The number of authors per scientific paper has increased steadily since the 1950s (Shamoo and Resnik, 2009). For example, the number of authors per paper in four of the top medical journals increased from 4.5 in 1980 to 6.9 in 2000 (Weeks et al., 2004). Some papers in experimental physics list more than 1,000 authors (King, 2012). One of the reasons why the numbers of authors per paper is increasing is that science has become more complex, collaborative, and interdisciplinary in the last few decades. While this is clearly the case, the desire for career advancement undoubtedly helps drive this trend (Shamoo and Resnik, 2009).

Two different authorship issues frequently arise. First, sometimes researchers are not listed as authors on a paper even though they have made a significant intellectual contribution. Students and technicians are especially vulnerable to being exploited in this way, due to their lack of power. Indeed, some senior scientists believe that technicians should not be included as authors because they are paid to do technical work and do not need publications (Shamoo and Resnik, 2009). This attitude in unjustified, since authorship should be based on one's contributions, not on one's professional status. Sometimes individuals who have made significant contributions to the paper are not listed as authors in order to make the research appear unbiased. Pharmaceutical companies frequently hire professional writers and statisticians to help prepare articles submitted for publication involving research they have funded. One study found that 7.9% of articles published in six leading medical journals had ghost authors (Wislar et al., 2011).

A second type of ethical problem related to authorship occurs when individuals are listed as authors who have not made a significant intellectual contribution to the research. This phenomenon, known as honorary authorship, is fairly common: one study found that 17.6% of papers published in six leading medical journals have at least one honorary author (Wislar et al. 2011). There are several reasons why honorary authorship is fairly common. First, some laboratory directors insist that they be named as an author on every paper that comes out of their lab, even if they have not made a significant intellectual contribution to the research. Second, some researchers demand that they receive authorship credit for supplying data or materials, even if they make no other contribution to the research. Third, some researchers include well-known scientists on the authorship list in order to enhance the status of the paper and increase its chances of being read or taken seriously. Fourth, some researchers have informal agreements to name each other as authors on their papers, so they can increase their publication productivity (Shamoo and Resnik, 2009).

Most scientific journals have adopted authorship policies to deal with the previously mentioned problems. Policies attempt to promote two important values: fair assignment of credit and accountability (Resnik, 1997). Assigning credit fairly is important to make sure that researchers receive the rewards they deserve. Accountability is important for ensuring that individuals who are responsible for performing different aspects of the research can explain what they did, in case questions arise concerning the validity or integrity of the results. If reviewers or readers detect some problems with the data, it is important to know who conducted the experiments that produced the data. Many journals follow authorship guidelines similar to those adopted by the International Committee of Medical Journal Editors (ICMJE). According to the ICJME (2013) guidelines authorship should be based on:

> Substantial contributions to the conception or design of the work; or the acquisition, analysis, or interpretation of data for the work; AND
> Drafting the work or revising it critically for important intellectual content; AND
> Final approval of the version to be published; AND
>   Agreement to be accountable for all aspects of the work in ensuring that questions related to the accuracy or integrity of any part of the work are appropriately investigated and resolved.

The ICMJE also recommends that each author be accountable for the work he or she has done and should be able to identify parts of the research for which other authors are accountable. Individuals who meet some but not all of the authorship criteria should be listed in the acknowledgments section of the paper (ICMJE, 2013). Some journals go a step farther than the ICMJE recommendations and require authors to describe their specific contributions to the research (Shamoo and Resnik, 2009).

Conflict of Interest. A conflict of interest (COI) in research is a situation in which an individual has financial, professional, or other interests that may compromise his or her judgment or decision-making related to research. COIs are ethically problematic because they can lead to bias, misconduct, or other violations of ethical or scientific standards, and can undermine the public's trust in research (Resnik, 2007). The most common types of interests that create COIs include financial relationships with companies that sponsor research, such as funding, stock ownership, or consulting arrangements; and intellectual property related to one's research. An institution can also have a COI if the institution or its leaders have interests that could compromise institutional decision-making. Non-financial COIs can occur when a scientist has a professional or personal relationships with a researcher whose work he or she is reviewing. For example, if a scientist reviews the grant application from another researcher at the same institution, the scientist would have a COI. (COIs in peer review will be discussed more below.)

Scientific journals, federal agencies, and research institutions have rules that require COI disclosure. Such disclosure can help to minimize ethical problems by providing interested parties with information that can be useful in evaluating research (Resnik and Elliott, 2013). For example, if the author of a review article on hypertension claims that a drug made by company X is superior to the alternatives, and the author has a significant amount of stock in company X, then a reader of the article may be skeptical of the author's conclusions. Disclosure can also help promote public trust in research. If a researcher has an undisclosed COI that later comes to light, the revelation may cause members of the public to distrust the researcher because they feel that they have been deceived. By disclosing COIs, problematic relationships are out in the open, for all to see and assess.

Sometimes disclosures may not be a strong enough response to a COI. If a COI is particularly egregious or difficult to manage, the best option may be to prohibit the COI. For example, most grant agencies do not allow reviewers to evaluate research proposals from colleagues at the same institution or from former students or supervisors. However, sometimes prohibiting a COI has adverse consequences for science or society that are worse than the consequences of allowing the COI. For example, suppose that a speech pathology researcher, who is a stutterer, has invented a device to help control stuttering. He has several patents on the invention, which he has transferred to the university in exchange for royalties. He has also formed a start-up company, with university support, to develop, manufacture, and market the device. The researcher and the university both own a significant amount of stock in the company. The researcher is planning to conduct a clinical trial of the device on campus.

One could argue that this COI should be prohibited, given this complex web of individual and institutional interests in the research. However, prohibiting the COI might adversely impact the development of this device, because the researcher is best qualified to conduct the clinical trial. A clinical trial of the device conducted by another researcher on another campus is not a feasible option. Since development of the device can benefit science and the public, the best option might be to permit the trial to take place on campus but require that it has additional oversight from a group of independent investigators, who could make sure that scientific and ethical standards are upheld (Resnik, 2007).

Peer Review. Peer review is a key part of science's self-correcting method (Ziman, 2000). Journals and research funding organizations use peer review to ensure that papers (or research proposals) meeting appropriate scientific and ethical standards. Peer review can improve the quality of papers and proposals and promote scientific rigor and objectivity. The *Philosophical Transactions of the Royal Society of London* instituted the first peer review system in 1665, but the practice did not become common in science until the mid-twentieth century (Resnik, 2011). When a paper is submitted to a journal, it will usually be reviewed by an editor. If the editor determines that the paper falls within the journal's scope and is has the potential to make an important contribution to the literature, he or she will send it to two or more reviewers who have expertise in the paper's topic area. After receiving reports from the reviewers, the editor will decide whether it should be published, revised, or rejected. The entire peer review process is supposed to be confidential in order to protect the author's unpublished work.

Although most scientists believe that the peer review system is an indispensable part of the evaluation of research, peer review is far from perfect. Studies have shown that reviewers often fail to read manuscripts carefully, provide useful comments, or catch obvious errors. Reviewers may be influenced by various biases, such as the author's gender, ethnicity, geographic location, or institutional affiliation (Resnik, 2011). Studies have also shown that reviewers often have very different assessments of the same paper: one may recommend that it be rejected, while another may recommend acceptance without any changes (Resnik, 2011). Finally, research indicates that a number of ethical problems can impact peer review: reviewers may breach confidentiality; unnecessarily delay the review process in order to publish an article on the same topic or stifle competitors; use data or methods disclosed in a manuscript without permission; make personal attacks on the authors; require unnecessary references to their own publications; and plagiarize manuscripts (Resnik et al., 2008).

Scientists and scholars have proposed some reforms to improve the reliability and integrity of peer review. First, reviewers should inform editors if they have any COIs related to the paper. For example, a reviewer would have a conflict of interest if he or she is from the same institution as the author, has a significant professional or personal relationship to the author, or has a financial interest related to the manuscript, such as ownership of stock in a company sponsoring the research or a competitor. The editors can review disclosures and decide whether the reviewer should review the paper. In some cases, it may be desirable to have a reviewer review a paper despite a COI, due to the limited availability of other experts to review the paper (Shamoo and Resnik, 2009).

A second reform is to experiment with alternative methods of peer review. Most scientific disciplines use a single-blind approach: the reviewers are told the identities of the authors but not vice versa. Some journals use a double-blind approach: neither the reviewers nor the authors are told the others' identity. One of the advantages of this approach is that it could help to minimize gender, ethnic, geographic, or institutional biases, as well as COIs. However, studies have shown that more than half of the time reviewers are able to accurately identify authors when they are blinded (Resnik, 2011). Hence, a double-blind system may give editors and authors a false sense of security. Some journals have begun using an un-blinded (or open) approach: both authors and reviewers are told the others' identity. The chief advantage of this approach is that it will hold reviewers accountable. Reviewers will be less likely to plagiarize manuscripts, unnecessarily delay publication, or commit other ethical transgressions, if they know that the authors can hold them accountable. They may also be more likely to read the manuscript carefully and provide useful comments. One of the problems with the open approach is that scientists may not want to reveal their identities to the authors out of fear of possible retribution. Reviewers also may not give candid comments if their identities will be revealed. Since the current system has numerous flaws, journals should continue to experiment with different methods of peer review and other reforms (Resnik, 2011).

Research with Human Subjects. Research with human subjects raises many different ethical issues. Since entire volumes have been written on this topic, only a couple of issues will be discussed in this chapter (see Shamoo and Resnik, 2009, Emanuel et al., 2011). Various ethical guidelines, such as the Nuremberg Code and Helsinki Declaration (mentioned above), and legal regulations, such as the US Common Rule (Department of Health and Human Services, 2009), require researchers to take steps to protect human subjects, such as obtaining informed consent, minimizing risks, and protecting confidentiality. However, these regulations and guidelines do not cover every situation and are subject to interpretation. Hence, ethical dilemmas can still arise even when researchers comply with applicable rules.

Most of the ethical dilemmas in research with human subjects involve a conflict between protecting the rights and welfare of research participants and advancing scientific knowledge and benefiting society. A recurring example of this dilemma occurs when researchers use placebo control groups in clinical trials. The placebo effect is a well-documented phenomenon in which individuals tend to respond better when they think they are getting an effective treatment. To control for this effect, researchers can use a placebo group when testing a new drug or other treatment. When this happens, neither the subjects nor the investigators will be told who is receiving a placebo or the experimental treatment. To reduce bias, subjects are randomly assigned to different groups (Emanuel and Miller, 2001).

Using a placebo control group does not raise any significant ethical issues when there is no effective treatment for a disease. However, it does create an ethical dilemma if there is an effective treatment, because some subjects would be foregoing treatment in order to help advance medical research. Foregoing treatment could adversely impact their health and violate the physician's ethical obligation to provide his or her patients with medical treatment. Some commentators have argued that placebos should never be used when there is an effective treatment, because promoting the welfare of the subjects should take priority over advancing research. If there is an effective treatment for a disease, then the experimental treatment should be compared to a currently accepted effective treatment. Others have argued that it is more difficult to conduct scientifically rigorous clinical trials that compare two treatments because such studies will usually find small differences between groups. In statistics, the sample size needed to obtain significant results is inversely proportional to the effect size: the smaller the effect, the larger the sample needs to be. By using a placebo control group in a clinical trial, one can conduct a smaller study with scientific rigor. Smaller studies are easier to execute and take less time and money than larger ones. They also expose fewer subjects to risks (Emanuel and Miller, 2001). For these reasons, some commentators argue that placebos can be used even when effective treatments exist in two situations: (a) foregoing an effective treatment is not likely to cause subjects any significant pain or permanent harm and (b) the new treatment has the potential to significantly benefit society. For example, a placebo group could be used to study the effectiveness of a pain-control medication for moderate arthritis, because subjects could forego arthritis treatment during a clinical trial without significant pain or permanent harm. Placebo control groups could also be used in testing a treatment that has potential to significantly benefit society, such as an inexpensive method for preventing mother-child transmission of HIV (Resnik, 1998b).

Another situation that creates a conflict between the rights/welfare of human subjects and the advancement of knowledge/social benefit is the use of deception in behavioral research. Subjects in an experiment may alter their behavior if they know what is being studied. Psychologists sometimes use some form of deception to minimize this type of bias. The most famous experiments involving deception were conducted in the 1960s by Stanley Milgram, a Harvard psychologist. Milgram wanted to understand why people follow orders given by those they view as authorities. This was an important topic to address, especially since many German soldiers, prison guards, and government officials who committed atrocities during World War II claimed they were only following orders. Milgram's experiments involved three people: a scientist, a helper, and a test-taker. The test-taker was hooked up to some electrodes connected to a machine that would produce a shock. The helper would ask the test-taker a series of questions. If the test-take got the wrong answer, the scientist instructed the helper to administer a shock. The test-taker would react by expressing discomfort or pain. The shock would be increased each time, eventually reaching levels classified as "dangerous" or "potentially life-threatening." Milgram found that a majority of the helpers continued to give shocks when instructed to do so, even when the shocks reached "life-threatening" levels. They also gave shocks even when the test-takers pleaded with the scientist to end the experiment. Whenever the helpers expressed doubts about giving a shock, the scientists would ask them to continue and stressed the importance of completing the experiment. Fortunately, no one got a real shock; the test-takers were faking pain and discomfort. The point of the experiment was to see whether the helpers would obey an authority (the scientist), even when they were asked to harm another person. Deception was a necessary part of the experiment, because it would not have been possible to test the helpers' tendency to obey authority if they had known that the test-takers were not getting shocked (Milgram, 1974).

Milgram debriefed the test-takers after the experiments were over by telling them what had really taken place. Some of them were distraught after the experiments ended. Although they were glad that they hadn't hurt anyone, they were upset that they had been willing to harm someone in order to complete the experiment. They found out something about their character they did not want to know. Many of them were upset that they had been manipulated to answer a scientific question. Milgram (1974) has defended the use of deception in these experiments on the grounds that (a) it was necessary to answer an important scientific question; (b) it did not significantly harm the subjects; (c) the subjects did consent to be in an experiment; and (d) the subjects were debriefed. However, critics have argued that the experiments significantly harmed some of the subjects and that the knowledge Milgram sought could have been gotten without resorting to deception. For example, researchers could have developed a role-playing game with different subjects placed in positions of authority. By observing people play the game, they would learn about obedience to authority (Shamoo and Resnik, 2009).

Research with Animals. The most important ethical question with research involving animals is whether it should be conducted at all. Scientists who work with animals regard this research as necessary to produce knowledge that benefits people. Animal research can provide us with a basic knowledge of physiology, anatomy, development, genetics, and behavior that is useful for understanding human health and treating human diseases. Animal experiments also play an important role in the testing of new drugs, medical devices, and other medical products. Indeed, research laws and guidelines require that investigators provide regulatory authorities data from animal studies before they can test a new drug in human beings (Shamoo and Resnik, 2009). Although animal research has produced many important benefits for human societies, it has been embroiled in controversy for many years. Animal rights activists have expressed their opposition to animal research by protesting on university campuses, disrupting experiments, infiltrating laboratories, turning laboratory animals loose in the wild, and threatening researchers. Most universities now tightly control access to animal research facilities to protect experiments, laboratory animals, and researchers (Shamoo and Resnik, 2009).

Peter Singer (1975), Tom Regan (1983), and other philosophers have developed arguments against animal experimentation. Singer's argument is based on three assumptions. First, many species of animals used in laboratory research can suffer and feel pain. Second, animal pain and suffering should be given equal moral consideration to human pain and suffering. Third, the right thing to do is to produce the greatest balance of benefits/harms for all beings deserving of moral consideration (utilitarianism). From these three assumptions, Singer concludes that animal experimentation is unethical because the pain and suffering it causes to animals do not outweigh the benefits it produces for people (Singer, 1975). Observational research on animals (i.e., studying them in the wild) may be acceptable, but research that involves inflicting pain or suffering or causing disability or death should not be allowed.

Regan (1983) approaches the topic from a moral rights perspective, not a utilitarian one. He argues that the function of moral rights is to protect interests: the right to life protects one's interest in life; the right to property protects one's property interest; etc. According to Regan, animals have rights because they have interests, such as the interest in living, obtaining food, and avoiding suffering. Although research with human subjects may adversely impact their interests, this research does not violate their rights because people can make an informed decision to sacrifice some of their interests in order to benefit science and society. Animals cannot give informed consent, however. Hence, research on animals is unethical because it violates their rights. Like Singer, Regan would have no problems with studying animals in the wild, but he would object to controlled experiments that cause pain, suffering, and death.

While evaluating Singer's and Regan's arguments is beyond the scope of this chapter (see LaFollette and Shanks, 1996; Garrett, 2012), it is worth noting that other issues arise in animal research besides the fundamental question of whether it should be done at all. Scientists who work with animals and policymakers have developed laws and guidelines for protecting animal welfare in research. These rules deal with issues such as feeding, living conditions, the use of analgesia and anesthesia, and euthanasia. Some widely recognized ethical principles that promote animal welfare in research are known as the three Rs: reduction (reduce the number of animals used in research wherever possible); refinement (refine animal research techniques to minimize pain and suffering); and replacement (replace animals experiments with other methods of obtaining knowledge, such as cell studies or computer models, wherever possible). One of the more challenging ethical issues for scientists with no moral objections to experimenting with animals involves the conflict between minimizing pain/suffering and advancing scientific knowledge (LaFollette and Shanks, 1996). For example, consider an experiment to study a new treatment for spinal injuries. In the experiment, researchers partially sever the spinal cords of laboratory mice. The experimental group receives the treatment; the control group receives nothing. The scientists measure variables (such as the ability to walk, run a maze, etc.) that reflect the animals' recovery from the spinal injury in both groups. Because analgesia medications could interfere with the healing process and confound the data, they will not be administered to these animals. Thus, it is likely that the animals will experience considerable pain or discomfort in this experiment. Does the social value of this experiment—to develop a treatment for spinal injuries—justify inflicting harm on these animals?

Social Responsibility. The final group of issues considered in this chapter pertains to scientists' social responsibilities. Scientists interact with the public in many different ways: they give interviews to media, provide expert testimony in court, educate the public about their work, and advocate for policies informed by their research. Some of the most challenging issues involve conducting and publishing potentially harmful research. As noted earlier, the nuclear weapons research conducted at Los Alamos was kept secret to protect national security interests. Military research raises many issues that will not be considered here, such as whether scientists should work for the military and whether classified research should be conducted on university campuses (see Resnik, 1998b). This chapter will focus on issues related to conducting and publishing non-military research that may be used for harmful purposes.

Biomedical researchers have published numerous studies that have potential public health benefits but also may be used by terrorists, criminals, or rogue nations to develop weapons that could pose a grave threat for national security, society, and the economy (also known as "dual use research"). Some examples include studies that could be used to develop smallpox viruses that can overcome the human immune system's defenses (Rosengard et al., 2002); a demonstration of how to make a polio virus from available genetic sequence data and mail-order supplies (Cello et al., 2002); a paper describing how to infect the US milk supply with botulinum toxin (Wein and Liu, 2005); and research on genetically engineering the H5N1 avian flu virus so that it can be transmissible by air between mammals, including humans (Imai et al., 2012; Russell et al., 2012).

Dual-use research raises several concerns for scientists, institutions, journals, and funding agencies: (1) should the research be conducted?, (2) should the research be publicly funded?, and (3) should the research be published? As noted earlier, freedom, openness, and social responsibility are three of science's ethical norms. Restricting research to protect society from harm creates a conflict between freedom/openness and social responsibility (Resnik, 2013). To manage this conflict, one needs to consider the potential benefits and harms of the research and the impact on the scientific community (such as a chilling effect) of proposed restrictions. Because freedom and openness are vital to the advancement of scientific knowledge, arguments to restrict publication must overcome a high burden of proof. The default position should be that unclassified research will normally be published in the open literature unless there is a significant chance that it can be used by others for harmful purposes. If restrictions on publication are warranted, the research could be published in redacted form, with information necessary to repeat experiments removed. The full version of the publication could be made available to responsible scientists (Resnik,

2013). Another option would be to classify the research. The decision not to fund or defund research does not need to overcome as high a burden of proof as the decision to restrict publication, because freedom of inquiry does not imply a right to receive funding. Researchers who cannot obtain funding for their work from a government agency are still free to pursue other sources of funding, such as private companies or foundations.

6 CONCLUSION

Research ethics is of considerable concern to scientists, the media, policymakers, and the public. The ethical issues that scientists face are important, challenging, complex, and constantly evolving. To date, philosophers have focused mostly on fundamental questions concerning the relationship between science and human values and have had little to say about the day-to-day ethical questions and problems that confront scientists. Hopefully, this trend will change and more philosophers will take an interest in investigating and analyzing ethical issues in science. The benefits for science, society, and the philosophical profession could be significant.

ACKNOWLEDGMENTS

This article is the work product of an employee or group of employees of the National Institute of Environmental Health Sciences (NIEHS), National Institutes of Health (NIH). However, the statements, opinions, or conclusions contained therein do not necessarily represent the statements, opinions, or conclusions of NIEHS, NIH, or the United States government.

REFERENCES

Babbage, C. (1830) [1970]. *Reflections on the Decline of Science in England* (New York: Augustus Kelley).

Broad, W., and Wade, N. (1983). *Betrayers of Truth: Fraud and Deceit in the Halls of Science* (New York: Simon and Schuster).

Briggle, A., and Mitcham C. (2012). *Ethics in Science: An Introduction* (New York: Cambridge University Press).

Cello, J., Paul A., Wimmer, E. (2002). "Chemical Synthesis of Poliovirus cDNA: Generation of Infectious Virus in the Absence of Natural Template." *Science* 297(5583): 1016–1018.

Department of Health and Human Services. (2009). "Protection of Human Subjects." Code of Federal Regulations 45, Part 46.

Elliott, K. C. (2011). *Is a Little Pollution Good for You?: Incorporating Societal Values in Environmental Research* (New York: Oxford University Press).

Emanuel, E. J., Grady, C. C., Crouch, R. A., Lie, R. K., Miller, F. G., and Wendler, D. D., eds. (2011). *The Oxford Textbook of Clinical Research Ethics* (New York: Oxford University Press).

Emanuel, E. J., and Miller, F. G. (2001). "The Ethics of Placebo-Controlled Trials—A Middle Ground." *New England Journal of Medicine* 345(12): 915–919.

Franklin, A. (1981). "Millikan's Published and Unpublished Data on Oil Drops." *Historical Studies in the Physical Sciences* 11: 185–201.

Garrett, J. R., ed. (2012). *The Ethics of Animal Research: Exploring the Controversy* (Cambridge, MA: Massachusetts Institute of Technology Press).

Glick J. L., Shamoo, A. E. (1993). "A Call for the Development of "Good Research Practices" (GRP) Guidelines." *Accountability in Research* 2(4): 231–235.

Goldman, A. (1999). *Knowledge in a Social World* (New York: Oxford University Press).

Harding, S. (1998). *Is Science Multicultural?: Postcolonialisms, Feminisms, and Epistemologies* (Bloomington, IN: Indiana University Press).

Holton, G. (1978). "Subelectrons, Presuppositions, and the Millikan-Ehrenhaft Dispute." *Historical Studies in the Physical Sciences* 9: 166–224.

Hudson, K. L., Guttmacher, A. E., and Collins, F. S. (2013). "In Support of SUPPORT—A view from the NIH." *New England Journal of Medicine* 368(25): 2349–2351.

Imai, M., Watanabe, T., Hatta M., Das, S. C., Ozawa, M., Shinya, K., Zhong, G., Hanson, A., Katsura, H., Watanabe, S., Li, C., Kawakami, E., Yamada, S., Kiso, M, Suzuki, Y., Maher, E. A., Neumann, G., Kawaoka, Y. (2012). "Experimental Adaptation of an Influenza H5 HA Confers Respiratory Droplet Transmission to a Reassortant H5 HA/H1N1 Virus in Ferrets." *Nature* 486(7403): 420–428.

International Committee of Medical Journal Editors. (2013). Recommendations for the Conduct, Reporting, Editing and Publication of Scholarly Work in Medical Journals. Available at: http://www.icmje.org/. Accessed: August 31, 2013.

King, C. (2012). Multiauthor Papers: Onward and Upward. ScienceWatch Newsletter, July 2012. http://archive.sciencewatch.com/newsletter/2012/201207/multiauthor_papers/. Accessed: August 31, 2013.

Kitcher, P. (1993). *The Advancement of Science* (New York: Oxford University Press).

Kornfeld, D. S. (2012). "Perspective: Research Misconduct: The Search for a Remedy." *Academic Medicine* 87(7): 877–882.

Kuhn, T. S. (1962). *The Structure of Scientific Revolutions*. (Chicago: University of Chicago Press).

Kuhn, T. S. (1977). *The Essential Tension* (Chicago: University of Chicago Press).

LaFollette, H., and Shanks, S. (1996). *Brute Science: Dilemmas of Animal Experimentation* (New York: Routledge).

Longino, H. (1990). *Science as Social Knowledge* (Princeton, NJ: Princeton University Press).

Macklin, R. (1999). *Against Relativism: Cultural Diversity and the Search for Ethical Universals in Medicine* (New York: Oxford University Press).

Macrina, F. (2005). *Scientific Integrity*. 3rd ed. (Washington, DC: American Society of Microbiology Press).

Malek, J. (2010). "To Tell or Not to Tell? The Ethical Dilemma of the Would-Be Whistleblower." *Accountability in Research* 17(3): 115–129.

Martinson, B. C, Anderson, M. S., and de Vries, R. (2005). "Scientists Behaving Badly." *Nature* 435(7043): 737–738.

Merton, R. (1973). *The Sociology of Science: Theoretical and Empirical Investigations* (Chicago: University of Chicago Press).

Milgram, S. (1974). *Obedience to Authority* (New York: Harper and Row).

Morin, K., Rakatansky, H., Riddick, F. A. Jr., Morse, L. J., O'Bannon, J. M., Goldrich, M. S., Ray, P., Weiss, M., Sade, R. M., and Spillman, M. A. (2002). "Managing Conflicts of Interest in the Conduct of Clinical Trials." *Journal of the American Medical Association* 287(1): 78–84.

Munthe, C., Welin, S. (1996). "The Morality of Scientific Openness." *Science Engineering Ethics* 2(4): 411–428.

National Academy of Sciences. (2009). *On Being a Scientist: A Guide to Responsible Conduct in Research*. 3rd. ed. (Washington, DC: National Academy Press).

Office of Science and Technology Policy. 2000. "Federal Research Misconduct Policy." *Federal Register* 65(2350): 76260–76264.

Quine, W. V., and Ullian, J. S. (1978). *The Web of Belief*. 2nd ed. (New York: McGraw-Hill).

Regan, T. (1983). *The Case for Animal Rights* (Berkeley: University of California Press).

Rennie, D., Yank, V., and Emanuel, L. (1997). "When Authorship Fails. A Proposal to Make Contributors Accountable." Journal of the American Medical Association 278(7): 579–585.

Resnik, D. B. (1996). "Social Epistemology and the Ethics of Research." *Studies in the History and Philosophy of Science* 27(4): 566–586.

Resnik, D. B. (1997). "A Proposal for a New System of Credit Allocation in Science." *Science and Engineering Ethics* 3(3): 237–243.

Resnik, D. 1998a. *The Ethics of Science* (New York: Routledge).

Resnik, D. B. (1998b). "The Ethics of HIV Research in Developing Nations." *Bioethics* 12(4): 285–306.

Resnik, D. B. (2003). "From Baltimore to Bell Labs: Reflections on Two Decades of Debate about Scientific Misconduct." *Accountability in Research* 10(2): 123–135.

Resnik, D. (2007). *The Price of Truth* (New York: Oxford University Press).

Resnik, D. (2009). *Playing Politics with Science* (New York: Oxford University Press).

Resnik, D. B. (2011). "A Troubled Tradition." *American Scientist* 99(1): 24–28.

Resnik, D. B. (2013). H5N1 "Avian Flu Research and the Ethics of Knowledge." Hastings Center Report 43(2): 22–33.

Resnik, D. B.. and Elliott, K. C. (2013). "Taking Financial Relationships into Account When Assessing Research." *Accountability in Research* 20(3): 184–205.

Resnik, D. B., Gutierrez-Ford, C., and Peddada, S. (2008). "Perceptions of Ethical Problems with Scientific Journal Peer Review: An Exploratory Study." *Science and Engineering Ethics* 14(3): 305–310.

Resnik, D. B., and Master, Z. (2013.) "Policies and Initiatives Aimed at Addressing Research Misconduct in High-Income Countries." *PLoS Med* 10(3): e1001406.

Resnik, D. B., and Stewart, C. N. Jr. (2012). "Misconduct Versus Honest Error and Scientific Disagreement." *Accountability in Research* 19(1): 56–63.

Rosengard, A. M, Liu, Y., Nie, Z., and Jimenez, R. (2002). "Variola Virus Immune Evasion Design: Expression of a Highly Efficient Inhibitor of Human Complement." *Proceedings of the National Academy of Sciences* 99(13): 8808–8813.

Rudner, R. (1953). "The Scientist Qua Scientist Makes Value Judgments." *Philosophy of Science* 21(1): 1–6.

Russell, C. A., Fonville, J. M., Brown, A. E., Burke, D. F, Smith, D. L., James, S. L., Herfst, S., van Boheemen, S., Linster, M., Schrauwen, E. J., Katzelnick, L., Mosterín, A., Kuiken, T., Maher, E., Neumann, G., Osterhaus, A. D., Kawaoka, Y., Fouchier, R. A., and Smith, D. J. (2012). "The Potential for Respiratory Droplet-Transmissible A/H5N1 Influenza Virus to Evolve in a Mammalian Host." *Science* 336(6088): 1541–1547.

Shamoo, A. E., and Resnik, D. B. (2009). *Responsible Conduct of Research*. 2nd ed. (New York: Oxford University Press).

Shrader-Frechette, K. (1994). *Ethics of Scientific Research* (Lanham, MD: Rowman and Littlefield).

Singer, P. (1975). *Animal Liberation* (New York: Random House).

Steneck, N. H. (1999). "Confronting Misconduct in Science in the 1980s and 1990s: What Has and Has Not Been Accomplished?" *Science and Engineering Ethics* 5(2): 161–176.

Steneck, N. H. (2007). *ORI Introduction to the Responsible Conduct of Research*. Rev. ed. (Washington, DC: Department of Health and Human Services).

Thagard, P. (1993). *Computational Philosophy of Science* (Cambridge, MA: Massachusetts Institute of Technology Press).

Weaver, D., Reis M. H., Albanese, C., Costantini F., Baltimore, D., and Imanishi-Kari, T. (1986). "Altered Repertoire of Endogenous Immunoglobulin Gene Expression in Transgenic Mice Containing a Rearranged Mu Heavy Chain Gene." *Cell* 45(2): 247–259.

Weeks, W., Wallace, A., and Kimberly, B. (2004). "Changes in Authorship Patterns in Prestigious US Medical Journals." *Social Science and Medicine* 59(9): 1949–1954.

Wein, L., and Liu, Y. (2005). "Analyzing a Bioterror Attack on the Food Supply: The Case of Botulinum Toxin in Milk." *Proceedings of the National Academy of Sciences* 102(28): 9984–9989.

Whitbeck, C. (1995). "Truth and Trustworthiness in Research." *Science and Engineer Ethics* 1(4): 403–416.

Whitbeck, C. (1998). *Ethics in Engineering Practice and Research* (New York: Cambridge University Press).

Wislar, J. S., Flanagin, A., Fontanarosa, P. B., and Deangelis, C. D. (2011). "Honorary and Ghost Authorship in High Impact Biomedical Journals: A Cross Sectional Survey." *British Medical Journal* 343: d6128.

Ziman, J., 2000. *Real Science: What It Is, And What It Means* (Cambridge: Cambridge University Press).

CHAPTER 13

··

EXPERIMENT

··

ULJANA FEEST AND FRIEDRICH STEINLE

1 INTRODUCTION

EXPERIMENTS are a central and important aspect of science. Hence, one might expect that they would be accorded a central status in the philosophy of science, on a par with topics like explanation, confirmation, causation, and laws of nature. The fact that this was—through the better part of the twentieth century—not the case, has been much remarked upon ever since the early 1980s began to see a renewed interest in the philosophy of experimentation. Ian Hacking's 1983 *Representing and Intervening* (Hacking 1983) is commonly regarded as having been a milestone in this regard. Other important early publications include Bruno Latour and Steve Woolgar's *Laboratory Life* (Latour and Woolgar 1979), Harry Collins's *Changing Order: Replication and Induction in Scientific Practice* (Collins 1985), Steve Shapin and Simon Schaffer's *Leviathan and the Air Pump* (Shapin and Schaffer 1985), Peter Galison's *How Experiments End* (Galison 1987), and Allan Franklin's *The Neglect of Experiment* (Franklin 1986). By the late 1980s, the expression "new experimentalism" had been coined to describe this new movement (Ackermann 1989).[1]

The new interest in experiments was by no means restricted to philosophy, but was rather part of the more general "practical turn" in the history of science, sociology of science, and science studies. Even though other important works and collections followed (e.g., Gooding, Pinch and Schaffer 1989; Gooding, 1990; Mayo, 1996; Rheinberger, 1997; Heidelberger and Steinle 1998), scientific experiments remained fairly marginal as a

---

[1] One less well-known earlier strand originated already in the 1920s, with perhaps the first book devoted to the philosophy of experiment ever by the German philosopher and physicist Hugo Dingler who emphasized the constructive role of agency in experiments (Dingler 1928). Dingler later discredited himself by his fierce opposition to the theory of relativity that he dismissed as "Jewish." Members of the "Erlangen school" in postwar Germany developed on Dingler's ideas but did not gain international visibility.

topic of analysis in mainstream philosophy of science through the 1990s. This was perhaps in part due to the interdisciplinary nature of this field of study, which focused on material aspects of scientific research and highlighted the process of discovery as worthy of analysis. It may also have had something to do with the fact that beyond the catchy phrase that "experiments have many lives of their own" (Hacking 1983), new experimentalists were not united by a single question, but were rather making contributions to a variety of other discussions, such as discussions about scientific realism, theory-ladenness, social constructionism, and skepticism.

Within the past fifteen years, things have changed, however (e.g., Radder 2003), and there can be little doubt that there is now a greater philosophical interest in scientific experiments. A good deal of this more recent interest in experiments is due to the growing diversification of philosophy of science itself, including an increasing focus on scientific practice (with topics like scientific models, simulations, and concept formation). With regard to *experimental* practice, this focus is especially noticeable within philosophical analyses of various special sciences, such as biology, neuroscience, and economics (e.g., Weber, 2005; Guala, 2005; Steel, 2008), which frequently look at methodological issues that arise in the sciences themselves. However, there are also important recent contributions coming from the history and philosophy of physics (e.g., Steinle 2016a).

This article is organized into three groups of topics. In Section 1, we recount some debates that continue to be important but were especially dominant in the 1980s and 1990s: scientific realism, theory-ladenness, skepticism, and robustness. Section 2 provides an overview of the literature concerning exploratory experiments, concept formation, and the iterative process of knowledge generation, which came to prominence since the 1990s. In Section 3, we look at the ways in which these topics figure in some of the more recent works within various philosophies of the special sciences, in particular focusing on questions of causal inferences and the extrapolation of experimental results to the world. We conclude (in Section 4) by laying out some open questions and future directions of research.

2 THE NEW EXPERIMENTALISM: BETWEEN SCIENTIFIC REALISM AND SKEPTICISM

One theme that unites advocates of (new) experimentalism since the 1980s is a rejection of what they regard as the theory-centeredness of traditional philosophy of science. The traditional approach (as construed by experimentalists) asks how various theories are related to one another, both across time and across different "levels" or disciplines, and how theories can be confirmed by experimental data. Experimentalists point out that this approach to science not only excludes processes of discovery and theory construction from philosophical analyses, but also neglects to reflect on the important roles of instruments in scientific experiments and on the skills required to use them. The theory-dominated view suggests that experiments only play the subservient role of supplying the data that put theories to the test. It was this latter assumption, in particular, that advocates of new experimentalism revolted against in the early 1980s, arguing instead that the function of experiments cannot be reduced to testing theories and that experiments ought to be analyzed on their own terms. Ironically, different camps within the experimentalist movement drew rather different conclusions from this assertion: while some have argued that experiments are capable of producing results that can hold up independently of theories, others have suggested that experiments pose unique skeptical challenges. These two points need not be mutually exclusive, but will be reviewed separately here.

2.1 Realism and the Relationship Among Entities, Theories, and Phenomena

In his seminal *Representing and Intervening*, Ian Hacking (1983) points to a number of ways in which experiments can be independent of theories. The book also has a larger agenda, however: to argue for a position he dubbed "entity realism," which holds that there can be good reasons to say about particular entities (e.g., electrons) that they exist without committing to the claim that any particular theory of the entity is true. With this position, Hacking responds to well-known problems with realism about theories (in particular, the problems raised by the pessimistic meta-induction) while holding on to (and trying to justify) the basic realistic intuitions that are at the heart of much experimental work.

Now, given that entities like electrons have long been treated as typical examples of *theoretical* entities, it may seem counterintuitive to be a realist about such an entity without thereby endorsing as true a theory of this entity. If true, this objection threatens to have two far-reaching consequences for experimental practice; namely, (1) that claims about the existence of the entity cannot be separated from other assumptions (and hence are underdetermined by the evidence), and (2) observational reports of the entity are informed by theoretical assumptions that are built into the method of measurement (see Duhem 1906/1962). In response to these problems, Hacking provides two arguments in support of his claim. The first is that it is possible to detect a given entity by means of different instruments, not all of which presuppose the same theory (cf. Hacking 1981). Second, he argues that instruments sometimes make use of the known causal properties of entities, even if there are competing theoretical accounts of those entities. Hacking illustrates this point by means of the use of electrons and positrons in solid-state physics, where electrons are sprayed onto metal spheres. This point is summarized in the well-known phrase that "if you can spray them they are real" (Hacking 1983, 24).

Some commentators have questioned Hacking's assumption that it is possible to obtain knowledge about entities independently of the theoretical assumptions required to individuate them in the first place. In response to this point, Theodore Arabatzis

(2012) has attempted to develop Hacking's experimentalism by bringing it together with causal theories of reference as proposed by authors like Putnam and Kripke. Drawing on his historical research about the history of the electron (e.g., Arabatzis 2006), Arabatzis coined the expression "hidden entity" and argues that it is possible to distinguish between "levels" of theorizing: in an experimental context, different scientists can agree on low-level causal knowledge about the entities they are concerned with. This knowledge, Arabatzis argues, is provided by shared experiments across different theoretical accounts of an entity, which fix the reference of the entity-term in question.

A second class of arguments against Hacking's entity realism stresses that experimental manipulability, while perhaps providing reasons for believing in the existence of particular entities, is not sufficient to establish entity realism as such. Although some passages in Hacking's writings seem to suggest that he wants to argue that the existence of entities can be inferred from success in manipulation, Morrison (1990) points out that it is contradicted by Hacking's own assertion that his is not an epistemological but a metaphysical account. However, some commentators point out that Hacking's metaphysical entity realism does seem to presuppose an epistemological argument. In this vein, Reiner and Pierson (1995) argue that Hacking's position (despite his claims to the contrary) is an instance of an IBE account and runs into similar problems as other such accounts. Morrison (1990) argues that Hacking's realism should be construed as a transcendental one, in that it explicates a commitment that must be presupposed for the very activity of experimentation.

Leaving aside the question of the scope and shape of Hacking's entity realism, we wish to emphasize one key feature of his approach, which is similar to that of others within the experimentalist camp: namely, that it attempts to draw a wedge between theories and experimentally produced data, thereby hoping to deflate worries typically associated with incommensurability, underdetermination, and theory-ladenness of observations. Bogen and Woodward (1988) make a similar point by singling out phenomena as independent from theories and data. With this move, they call into question the simplistic dichotomy between theories and data that has characterized much of traditional philosophy of science, drawing attention to the local and idiosyncratic character of experimental data. Bogen and Woodward's account importantly brings to the fore questions about the status and quality of experimental evidence and raises questions about the causal and evidential relationship between "local" experimental effects and the more context-independent phenomena under investigation.

2.2 Skepticism and Epistemological Strategies

As we saw in the previous section, some authors have argued that the local and material practices of experimentation and instrument use provide experimentally produced knowledge with a certain degree of autonomy from our (ever-changing) scientific theories, thereby making it more secure. Ironically, other authors have drawn the exact opposite conclusion, arguing instead that experimental practices (and in particular practices of instrument use) add an element of epistemic uncertainty to science that ought to make us skeptical of experimental results (and of science in general). The most prominent advocate of this line of thought is Harry Collins (e.g., Collins 1985). Collins has developed and modified his position since the 1980s, but since his early formulations have come to be canonical of a particular form of skepticism, we recount them here.

Collins points to cases where the existence of particular phenomena or entities is predicted by theories, but the entities in question are very hard to detect by specific instruments or experimental setups (his famous example is that of gravitational waves as predicted by Einstein). In such cases, Collins argues, any claim to have an instrument that allows the detection will inevitably face a problem of circularity: the proper functioning of the instrument has to be justified by its ability to detect the entity in question, and the claim to have detected the entity will be justified by appeal to the proper functioning of the instrument. One step toward breaking this circle might be to show that the finding is replicable. In response, Collins points out that experimental practices involve a tacit element that cannot be fully explicated by means of rules, and he concludes from this that it is in principle impossible to decide whether an experiment is an exact replication of another one and whether a given instrument or experiment has been used/conducted correctly, because any attempt to explicate how to apply rules of replication will formulate more rules. This leads to what Collins has dubbed the "experimenter's regress," which (according to him) can only be resolved by social power dynamics.

Unsurprisingly, this argument did not sit well with philosophers of science, who saw it as an attack on the rationality of science. Collins's most prominent critic was Allan Franklin (e.g., 1999), who presented an alternative reconstruction of Collins's case study involving the purported detection of weak gravitational radiation by Joseph Weber. While Collins concludes that Weber was not strictly disproved by his peers, Franklin argues that there were nevertheless many rational reasons for distrusting, and ultimately rejecting, Weber's findings. Franklin builds here on the concept of "epistemological strategies" (first introduced in Franklin 1986, 1989) and uses it to take on the challenge of "[h]ow [to] distinguish between a result obtained when an apparatus measures or observes a quantity and a result that is an artifact created by the apparatus" (Franklin 1986, 165). Whereas Hacking had suggested that scientists use the strategy of using multiple instruments that rely on different background theories, Franklin recognizes that this strategy is not always an option and adds more strategies, such as (1) indirect validation (if the instrument or experimental apparatus in question can detect several similar entities), (2) intervention and prediction, (3) argument from the consistency of data, (4) calibration and experimental checks (calibrate instrument by showing that it can reliably detect already known phenomena), (5) the elimination of alternative explanations, and more (see Franklin 1989).

2.3 Experimental Errors and Robustness

Franklin's notion of epistemological strategy is characteristic of a pragmatic (and practice-inspired) attitude that permeates much of the philosophy of experimentation.

Epistemological strategies are discovered by analyzing the methods scientists actually use to establish the validity of their findings. Other authors have studied how the validity of a given finding can be established as separate from the background conditions presupposed in experimental inquiry. What is at stake here is essentially the Duhemian problem of underdetermination. Two key concepts that deserve special mention here are (1) Deborah Mayo's concept of error-probing and (2) William Wimsatt's concept of robustness. In her 1996 *Error and the Growth of Experimental Knowledge*, Mayo argues that experimental scientists have at their disposal an array of strategies ("methodological rules"). These are methods of probing for (and removing) a variety of "canonical" experimental errors: "The overarching picture . . . is of a substantive inquiry being broken down into inquiries into more or less canonical errors in such a way that methodological strategies and tools can be applied in investigating those errors" (Mayo 1996, 18). Mayo specifically focuses her attention on statistical significance tests and emphasizes that these error-statistical methods are actually used by practicing scientists. Moreover, she targets Bayesian confirmation theory as relying on a highly abstract vision of the relationship between theory and evidence and as providing an unrealistic picture of the use of statistical methods in scientific practice. According to her account, then, a hypothesis can be adopted or retained if it passes a "severe test"; that is, a test where the probability of a false hypothesis passing is very low.

Mayo's analysis has made important contributions but has also been critically discussed. For example, Giora Hon (1998) argues that it is too narrow in that it does not succeed in offering a full-blown philosophy of experimental error or even of the category of error as illuminating the experimental process. In turn, Hon has proposed what he takes to be a more comprehensive framework to understand kinds of experimental inferences and how they can go wrong (Hon 1989). Aiming at an analysis of the role scientific errors actually play in scientific practice, Jutta Schickore (2005) has pointed out that even though Mayo claims to show how scientists can learn from error, errors are—in her framework—merely something to be eliminated, not something that can contribute to knowledge generation. By contrast, Schickore argues that "[e]rrors are not merely uncovered to be cast aside. Rather, they become objects of systematic exploration and discourse" (2005, 554–555), thereby playing a positive epistemic role.

One salient class of experimental errors are experimental artifacts. The central question here is whether there are principled ways of determining whether an experimental finding is "genuine" or an artifact of some flaw in the experiment or instrument. This question is frequently discussed in relation to Wimsatt's notion of robustness (Wimsatt 1981). Wimsatt's notion is closely related to the idea (already touched on above) that an experimental finding or measurement can be considered as warranted if it can be established by means of "multiple determination"; that is, by means of different instruments that do not rely on the same theory.[2] The concept of robustness has since come to acquire a broader meaning in that some authors use it to describe a finding being "solidified" in a broader sense through a process of "mutual adjustment of different factors, such

---

[2] In the contemporary literature, the term "robustness" is also (and primarily) used to describe models. We restrict ourselves to experiments here.

as data and data processing, experiment, equipment, and theory" (Soler 2012), akin to Hacking's thesis of the self-vindication of laboratory science (Hacking 1992). However, this broader meaning defeats the philosophical purpose of the original notion, namely to free specific experimental findings of their entanglement with theories and instruments (Coko and Schickore 2013). Nicolas Rasmussen (1993), using the example of the bacterial mesosome (which turned out to be an artifact), argues that, in scientific practice, experimental results cannot be disentangled in a way that would enable robustness in Wimsatt's sense. Rasmussen views the mesosome case as illustrating an experimenters' regress, but other authors have offered competing accounts of this episode in the history of biology. For example, Sylvia Culp argues that the mesosome was shown to be an artifact precisely because the body of data pointing in this direction was more robust than the data suggesting otherwise (e.g., Culp 1995).[3]

In a recent article, Schickore and Coko (2014) point to a surprising divergence in the literature about the role of multiple determinations in scientific practice. While some authors claim that it is ubiquitous, others say that it is not. Schickore and Coko argue that although all of the authors in question draw on real cases to make their argument, some do so in a way that abstracts away from the details, whereas others proceed in a more detail-oriented fashion. This difference corresponds to different meta-philosophical aims, namely, either to elucidate reasoning patterns under ideal epistemic situations or to provide analyses of the "messiness" of actual research.

3 EXPLORATORY EXPERIMENTS AND THE ITERATIVE PROCESS OF SCIENTIFIC RESEARCH

A somewhat different approach to experiment focuses less on the validity of experimental results and more on the generation of particular (types of) experimental findings. Some (but not all) of the writings in this tradition tend to be more historical in that they provide analyses of the research process itself. They analyze how objects of research are constituted in the research process and emphasize that this research is ongoing and iterative.

3.1 Experimental Systems and Epistemic Things

Within the history and philosophy of biology, Hans-Jörg Rheinberger's (1997) framework for the analysis of experimental knowledge generation has been especially

---

[3] But see Hudson (1999) and Weber (2005) for competing accounts of how the mesosome controversy was resolved. For additional texts about the data-instrument circle, see Bechtel (1994), Culp (1994). For a comprehensive overview of the literature about multiple determination, see Coko and Schickore (2013).

influential. Although his terminology, situated in the French tradition of history and philosophy of science, is not easily accessible for Anglophone philosophers, we would like to highlight two aspects of his work.

First, according to Rheinberger, when scientists investigate a given object of research, this object is, by definition, ill-understood. Rheinberger refers to such objects as "epistemic things" to emphasize that they are "things" toward which scientists direct their epistemic attention without having a very clear idea of what their important features are. He contrasts them with what he calls "technical objects" that are more clearly understood and no longer objects of researchers' curiosity. The analytical distinction between epistemic things and technical objects is "functional rather than structural" and serves the purpose of elucidating the research process, not its results. Epistemic things are what they are precisely "by virtue of their opacity" or "preliminarity" (Rheinberger 2005, 406–407). Often enough, in the process of research, they become clarified and transparent at some point, which is equivalent to being turned from epistemic things into technical objects. For Rheinberger, then, the crucial questions concern the conditions under which such things are constituted as objects of scientific study and what the dynamics are by which they get consolidated. Rheinberger's concept has been sharply criticized by David Bloor (2005) since it deliberately leaves open the question of reference—indeed, in Bloor's view, this makes Rheinberger a linguistic idealist or even brings him, by disregard of reference, close to being a social constructivist. In his response, Rheinberger emphasizes that the category of epistemic things is intended as a tool to "investigate the process of research itself," as Bloor rightly observes, and that, in the course of that process, the precise meaning of reference "remains elusive" (Rheinberger 2005, 407).

Other authors, too, have inquired into Rheinberger's use of the expression "epistemic thing" since he at times seems to assert that things can exist in a vague and preliminary fashion. Feest (2011) has tried to disambiguate this notion by suggesting that it is concepts, not things, that are preliminary, and that it is scientists' knowledge of things, not the things themselves, that can be vague. Chang suggests that Rheinberger's notion might be understood "[a]long vaguely Kantian lines," according to which "metaphysical objects-in-themselves are forever out of our reach, while epistemic objects constitute nature-as-phenomena" (Chang 2011, 412), although he also points out that Rheinberger's epistemic things are deeply historical.

Second, Rheinberger's central unit of analysis is that of an experimental system. This concept is meant to express the fact that a researcher "does not, as a rule, deal with isolated experiments in relation to theory" (Rheinberger 1997). They are, according to him, "systems of manipulation," creating the conceptual and practical space within which epistemic things are investigated. Rheinberger thereby distances himself from approaches that attempt to locate the constitution of scientific objects solely in the realm of theories or Kuhnian paradigms. Experimental systems encompass, in a way that can perhaps not be further disentangled, all elements of experimental research practice such as material resources, instrumentation, personal preferences, theoretical and conceptual structures, and sheer historical chance. In a similar vein, but with a view toward large-scale, global developments, Andrew Pickering (1995) has framed the notion of the "mangle of practice" that again emphasizes the close entanglement of material, conceptual, theoretical, social, and cultural aspects in the development of experimental science (see also Rheinberger's incisive comment [1999]). Going rather in the opposite direction, and with the brick model of Peter Galison (1988) in mind, Klaus Hentschel (1998) has attempted to further differentiate such experimental systems, drawing on Hacking's former taxonomy of experiments and significantly expanding it.

3.2 Exploratory Experiments and Concept Formation

One version of the idea that experiments can make epistemic contributions that are not focused on (or informed by) theories was introduced in 1997 by Richard Burian and Friedrich Steinle independently of each other. They identified a type of experiment that they dubbed "exploratory experiment." Although the expression had been used previously (Gooding 1990; Sargent 1995, with reference to Boyle), Steinle and Burian used it in a novel way to describe a type of experiment that consists in systematic variation of experimental variables with the aim of phenomenologically describing and connecting phenomena in a way that allows a first conceptual structure of a previously poorly structured research field. According to Steinle, exploratory experiments typically take place during historical episodes in which no theories or appropriate conceptual frameworks are available and can result in the formation of such frameworks.

Both Burian and Steinle base their claims on historical studies of experimental research practices: Burian analyzes an episode of histochemical research of protein synthesis in the 1940s that provided, as a result, a first map outlining the basic elements of protein biosynthesis processes in the cell (Burian 1997). Steinle focuses on early electromagnetism in 1820–21, with Ampère developing new spatial concepts, and Faraday, for electromagnetic induction, forming the new concept of magnetic curves, in both cases while aiming at laws. Faraday characterized his approach as "putting facts most closely together" (Steinle 1997, 2016a). Both authors emphasize that exploratory experimentation is not random trial and error, but constitutes a systematic procedure and is of utmost epistemic importance in that it is essential in providing conceptual foundations for all further research.

Like Rheinberger's, the analytical frameworks proposed by Burian and Steinle have a negative and a positive aspect to them. On the negative side, this work is clearly aligned with the rejection of a theory-centered vision of experimentation. But, in doing so, they also shift their interest toward the *processes* (as opposed to the *products*) of research and knowledge generation. Former generations would have spoken of the context of discovery, but it is exactly within the turn to the practices of science that the usefulness of the dichotomy of discovery and justification has been questioned (Schickore and Steinle, 2006).

It bears stressing that the idea of non–theory-centered experiments is not original to twentieth-century new experimentalism. The earlier tradition would at least include Francis Bacon with his idea of experiment providing ample material for filling what he called "tables" of presence, absence, and of different degrees; the Accademia del Cimento with its idea of experimentally establishing laws; Robert Boyle with his "Experimental Histories" and his success in founding the law of expansion of gases directly on experiment; and Edme Mariotte with his concept of "principes de l'éxpérience" or laws that are directly based on experimental research. In the eighteenth century, d'Alembert characterized the procedure of "physique éxperimentale" as "piling up as many facts as possible, putting them into a most natural order, and relating them to a certain number of general facts of which the other are the consequences" and thereby took recourse to experimental practice that he saw around him in researchers like Reaumur or Dufay (Steinle 2016b). In the nineteenth century, J. S. Mill spelled out, in his 1843 *System of Logic,* his four (inductive) experimental methods. Different as those approaches were, and different as are the ways in which they sustain modern philosophical analysis, they show that for most of the time during which experimental research has been discussed, there was an awareness (albeit not a detailed explication) of the existence of experiments that are not driven by theory. It was the anti-inductivistic turn in the early twentieth century, by authors like Duhem or Popper, that discredited those approaches all in all as naïve and mistaken and made the whole tradition disappear in the philosophy of science until the rise of New Experimentalism.

The idea of exploratory experimentation has been developed further in the past decade, based on historical case studies in the physical and biological sciences. At least three significant issues were addressed in those studies. First, in countering the supposition that exploratory experiments resembled radical Baconianism, which constructs all of science—so to speak—from the bottom up, Laura Franklin (2005) distinguishes between a theoretical background—which can direct or constrain an experiment— and a local hypothesis under test. Drawing on case studies from molecular biology, she argues that scientists will refer to experiments as exploratory when there is an absence of theory in the latter sense but that this does not rule out the existence of theories in the former sense. Much in the same direction, Karaca (2013), analyzing particle physics, differentiates between "theory-driven experiments" and "theory-laden experiments." Whereas exploratory experimentation is always, and inevitably, theory laden, this does not mean it is already theory-driven.

Second, further historical studies have led to the articulation of a variety of forms of exploratory experimentation (Burian, 2007; Waters, 2007; Peschard, 2012). In order to deal with that variety, Maureen O'Malley (2007) points to the alternatives of introducing further types of experiment or widening the notion of exploratory experimentation. Kevin Elliot (2007) goes for the second option and answers the ensuing question of what keeps the family of exploratory experimentation together by emphasizing the epistemological criterion of not being theory-driven and the methodological characteristic of having variation of parameters as a central procedure. With those in mind, he provides a full scheme of varying characteristics of exploratory experimentation. To his two criteria, we might add two further points: the flexibility and openness of the conceptual framework dealt with and the characteristic that it is not individual experiments but rather chains, series, or networks of experiments that allow conclusions in exploratory experimentation.

Third, the finding that exploratory experiments can result in the conceptual restructuring of a research field raises questions about the nature and circumstances of concept formation. Assuming that all experiments require some conceptual presuppositions, the question is how exactly, within the course of the investigative process, the very concepts that frame the experimental work at the outset can be revised. This question has not been addressed in depth so far, although Gooding (1990) has proposed an analysis that draws on the role of agency in experiment and concept formation. There is, moreover, the question of the dynamic processes that prompt scientists to form and eventually adopt new concepts. Revising existing concepts and creating new ones is a risky process that calls into question much previous knowledge, thus raising the question of what the typical situations are that drive researchers to go down this risky path. In Kuhnian terms, one might describe the situation as one of deliberately putting the existing paradigm in question. As Steinle argues, this uncomfortable situation is usually not sought out: rather, concept revision is done as a reaction to specific epistemic challenges. The historical observation that conceptual revisions often occur in the course of exploratory experimentation should not mislead us into the conclusion that this was the initial goal (Steinle 2012).

A major recent challenge concerning exploratory experimentation is the question of Big Data: how can exploratory research work in fields that come with large amounts of data, what forms can such research take, and what new possibilities and restrictions come with that? Those topics have been partly addressed by Franklin (2005) for genomics and by Karaca (2013) for high energy physics but have still to be further analyzed, all the more so because the whole topic of the characteristics of data-driven science has only recently been taken up (Leonelli 2014).

3.3 Concepts, Instruments, Theories, and the Iterative Nature of Science

In addition to analyzing the formation of concepts in experiments, there are also intriguing philosophical questions with regard to more developed concepts and theories. Scientific concepts and theories do not come with instructions on how to apply them in particular experiments. Two questions follow from this: first, how do abstract theories or concepts get "translated" into specific experimental contexts? Second, assuming that we have a specific method to measure a particular property, how do we know that this method can be extended to a different context, and how do we know that a different method measures the same property? The physicist Percy Bridgman (1927) famously answered the latter question by suggesting that every method of measurement defines a new concept. Although this statement was long taken to stand for a rather simple-minded theory of meaning, several authors have argued that Bridgman—in the wake of the Einsteinian revolution—merely meant to caution scientists not to assume that a particular theoretical concept could be generalized from one specific context of application (measurement situation) to another (Chang 2009).

The recent revival of interest in the position commonly attributed to Bridgman—operationalism or operationism (see, e.g., Chang, 2004; Chang, 2009; Feest, 2005)—arises from the recognition that whenever an experiment is designed (be it to test a hypothesis or to explore a given object of research), this requires an "operationalization," that is, a specification of operations that, when performed, will provide insights into the question at hand. This draws attention to the fact that experimental data production is often guided by substantial presuppositions about the subject matter, raising questions about the ways in which operationalizations are constructed. Chang (2004, 206) suggests that this happens in a two-step process, which involves turning an abstract concept into an "image" and then matching it with a physical system. There are in principle two kinds of responses to this analysis. One is to raise skeptical concerns about the ability of experiments to deliver abstract and general knowledge given the potentially arbitrary nature of operationalizations and operational definitions and the specific circumstances of their implementations. The other response is to aim for a more detailed analysis of the ways in which operational definitions can produce knowledge.

Hasok Chang has coined the notion of the iterative nature of experimental research to point to the dynamic historical processes whereby presuppositions, theories, standards, and instruments go through iterative cycles of mutual adjustment in the process of scientific research. Kevin Elliott (2012) distinguishes between epistemic and methodological iterations, where the former concern the revisiting of knowledge claims, and the latter concern the revisiting of modes of research. In turn, Feest (2010, 2011) also emphasizes the dynamic and iterative nature of scientific knowledge generation. Like Chang, she emphasizes the importance of operationalism for an understanding of scientific research, arguing that operational definitions should be regarded as instruments that can also be revised or discarded in the process. Related to those questions, the problem of concept formation and revision in experimental practice has been taken up recently with particular focus on the role(s) of concepts in scientific research practice (Feest and Steinle, 2012).

4 EXPERIMENTAL INFERENCES IN THE LAB AND THE EXTRAPOLATION OF EXPERIMENTAL FINDINGS

The literature covered in the previous section uses case studies to analyze the historical conditions that enable successful experimental practice. In recent philosophy of science, however, there have also been advances along more normative lines, scrutinizing the types of (often causal) inferences that are drawn from experimental data. Although these kinds of inquiries, of course, go back as far as J. S. Mill's four methods, James Woodward's (2005) notion of an "ideal intervention" has been particularly influential in the recent literature, although it must be noted that this notion figures as part of his account of causality, not as a method of discovering causes (see also Woodward 2009). In this section, we take a look at some recent work on experimental inferences, especially with regard to the social and behavioral sciences.

4.1 Inferences Within Experiments and Background Knowledge

In the literature about causal inference, a standard question is how to distinguish between spurious correlations and genuine causal relationships. Building on work by Judea Pearl (1988), Spirtes, Scheines, and Glymour (1993) have, for the past twenty-five years, developed a formal theory of the discovery of causal relations that relies on Bayes nets to enable causal inferences and predictions. Our concern here is not with the adequacy of formal approaches, but with the challenges faced by experimental approaches. First, whereas (for example) in the case of smoking, both variables (smoking and the possession of ashtrays) can be directly manipulated, there are many scientific cases where this is not the case. This is especially obvious in the behavioral sciences, where the variables in question are often latent (i.e., not accessible to direct intervention) and hence are manipulated indirectly by way of an experimental instruction. Second, even if we can identify the variable to be manipulated (e.g., the visual cortex), it may not be possible to manipulate it in isolation (cf. Bogen 2002). Woodward (2008) talks of "fat-handed interventions" here. Third, even in cases where it is possible to isolate and manipulate the to-be-manipulated variable, the question arises of what kinds of causal inferences can be drawn from the effects of doing so, given (a) the limited knowledge about other intervening variables that may play a role in the effect and (b) the possibility that the system might reorganize itself in a way that compensates for the disruption caused by the intervention (cf. Mitchell 2009).[4] Fourth, even if inferences from effects to causes within a given experiment are warranted, it is not clear whether the inference in question can be generalized to other contexts (other laboratories or real-life applications).

Dealing with experimental economics, Francesco Guala (2012) spells out some aspects of these problems, referring to them as inferences *within* an experiment and inferences *from* the experiment, respectively. As to the former, he points out that the key idea of experimentation is *control*, which he distinguishes into (1) control over a variable that is changed or manipulated by the experimenter and (2) control over other (background) conditions or variables that are set by the experimenter. Ideally, the latter is accomplished by holding background conditions stable by varying the variable of interest to observe changes in another variable of interest (as worked out by Graßhoff 2011). In the social and behavioral sciences, however, it is extremely complicated to tease the to-be-manipulated variable apart from the background conditions of the experiment,

---

[4] This problem is closely related to the issue of modularity (see Steel 2008; Woodward 2008).

both theoretically and practically. As Guala (2005) argues, very small differences in the experimental setup can potentially trigger entirely different processes. Similarly, Jacqueline Sullivan (2009) points out that even small differences in experimental protocols may tap into entirely different phenomena. She concludes that we should—at present—be very skeptical of the comparability of experimental results of different neurobiological experiments. Guala's proposed solution to this problem is slightly more optimistic in that he suggests that inductive support in experiments has to be viewed as "a three-place relation" among *H, e,* and background knowledge rather that a two-place relation between *H* and *e* (Guala 2005, 109) and that such background knowledge is in principle attainable by a gradual and piecemeal process that includes error elimination. He claims that inferences from experimental data to causes cannot be justified in isolation, but that scientists' knowledge of the context and circumstances of research is required.

The problem of how to ensure the soundness of causal inferences hence turns on the question of how to ensure that a particular experimental design succeeds in controlling all the relevant variables. Differently put, just as scientists need to make sure that a given result is not an artifact of an instrument, they also need to make sure that it is not an artifact of the experimental design. Within the methodological and philosophical literature about experiments in psychology and the social sciences, this issue is typically discussed under the general rubric of "validity." While this concept has in recent years also been adopted by philosophers of science, there is some confusion about the question of whether "validity" (cf. Jiménez-Buedo 2011) refers to the outcomes of inferences or to the apparatus that forms some of the background for the inference. We also need to distinguish between cases in which the background in question is (a) a scientific instrument/test that has been imported into the experiment or (b) an experimental design that has been specifically devised for the purposes of a particular study. The distinction is relevant because instruments, once calibrated, become tools that remain the same, whereas types of experiments evolve with the research in question.

In the social sciences, the term "validity" was first introduced in the 1950s (cf. Cronbach and Meehl 1955) and pertained to psycho diagnostic tests. It was not until a decade later that methodologists used the term to consider the quality of *experiments.* In this context, the psychologists Campbell and Stanley (1966) popularized the distinction between internal and external validity (i.e., between whether the inferences drawn from a particular experiment may be considered as sound within the highly confined and artificial conditions of the experiment or whether they can be generalized to the outside world, both in a research context and an applied context; e.g., Cronbach and Shapiro 1982).

4.2 Inferences from Experiments to the World

The term "external validity" has become customary in some of the philosophical literature (see Jiménez-Buedo and Miller 2010), although strictly speaking it runs together several distinct problems: (1) the representativeness of the sample, (2) the relevance of the experimental preparation of experimental subjects in the lab to the real-world phenomenon of interest, and (3) the representativeness of the experimental environment. We prefer to use the more general term "extrapolation" to cover the various questions posed by inferences (a) from the lab to the world and (b) from an experimental intervention to an application of the results in a different context.

The simplicity and manageability of small-scale experiments makes them scientifically attractive for psychologists and social scientists. It is precisely this simplicity, however, that also raises concerns about extrapolation because, after all, the real world is neither small nor simple. Daniel Steel (2008) has coined the expression "extrapolators circle" to indicate that a reasoned judgment of whether a given experiment adequately captures the relevant features of the target system seems to require precisely the kind of knowledge that the experiment aims to produce in the first place. Guala argues that the soundness of inferences within the lab improves as our knowledge of the background improves, but such knowledge becomes possible precisely because we have simplified the background so as to make it epistemically manageable. As a result, there appears to be a tradeoff between the soundness of our inferences within the confines of the experiment and the soundness of our extrapolations from the experiment; that is, between internal and external validity (Guala 2005, 144; see also Sullivan 2009, 534, for a related argument).

On Guala's account (2005, 198), the problem of extrapolation can ultimately be overcome in a similar fashion as the problems of inferences within experiments; namely, by way of gaining a better understanding of the (real-world) background. By contrast, Steel (2008) argues that this type of answer (which treats our assumptions about the relevant background as empirical hypotheses) is incapable of providing a principled account of the conditions under which such inferences from experiments (or models) to real-life settings are warranted. On Steel's account, causal structures or mechanisms can be individuated by means of what he calls the "disruption principle," which states that "interventions on a cause make a difference to the probability of the effect only if there is an undisrupted mechanism running from the cause to the effect" (2008, 7).

Viewed in this way, experiments have a lot in common with models: they simulate real-world scenarios with the aim of making inferences about the real world. Hence, they raise similar questions as those raised by models, such as determining the relevant similarities between model and target. This suggests that the (huge) literature about scientific models that has emerged in the past couple of decades will likely be relevant to the issue of extrapolation from experiments.[5]

4.3 Experiment, Application, Policy

In the past decade, the question of extrapolation from scientific models and experiments has become especially virulent in the course of discussions about applied research

---

[5] Guala (2005) suggests that experiments, like models, have the status of "mediators."

and evidence-based policy decisions. Importantly, Nancy Cartwright (2007; see also Cartwright and Hardie 2012) asks how it can be determined that a policy decision that worked in one context can be expected to have an equally positive effect in a different context. As Cartwright and Hardie (2012, 6) put it, "It is a long road from 'it worked somewhere' to the conclusion you need—'it will work here.'" In particular, they argue against the view that randomized controlled studies (RTCs) can answer this question, arguing instead that RCTs are at best going to provide more evidence to the effect that a given causal intervention "works here," not that it will work somewhere else (2012, 8). The authors argue that it is a big challenge to figure out the right level of description for both causes and effects in the source experiment. One example discussed by the authors concerns the policy to educate pregnant mothers in developing countries in an attempt to reduce child mortality, a policy that had been successful in one context, but not in another.

In addition to the general inferential problems just mentioned, the recent literature about causal inferences in contexts of applied experimental research has also focused on different types of causal inferences and the distinct problems of extrapolation they pose. For example, a recent publication (Osimani and Mignini 2015) argues that, in pharmacological drug trials, researchers want to know not only whether a given drug is effective (i.e., cures the affliction in question), but also whether it is safe (i.e., does not have any harmful side effects). Both these questions can be studied in the laboratory, but, they argue, in the case of the former, issues of extrapolation arise (will the drug be effective in the target population?), whereas in the latter case, there is no problem of extrapolation because knowing that the drug can be harmful in any context is sufficient. As this example shows, applied experimental research leads to questions about values and inductive risk and the novel ethical-cum-epistemological issues that arise when commercial interests enter the realm of the scientific experiment (see, e.g., Douglas, 2000; Wilholt, 2009; Carrier, 2011). This opens up important new avenues of research for the philosophy of experiment.

5 CONCLUSION

In conclusion, we would like to highlight some issues at the heart of current and future philosophical work about experiments:

- As we have pointed out, the renewed interest in experiments in the 1970s and 1980s initially came out of an interdisciplinary field of the history, philosophy and social studies of science. In the meantime, experiments have become a more accepted part of mainstream philosophy of science. However, we think that scholarly investigations that integrate different disciplinary perspectives will continue to add original impulses to the philosophy of experimentation.
- There is a general agreement at this point that the role of experiments cannot be reduced to that of theory-testing. Corresponding to this de-emphasis of

theory-testing in the philosophy of experimentation, the notion of scientific theory has come under scrutiny in other areas of philosophy of science as well. This can be seen, for example, in the philosophy of neuroscience, with its emphasis on mechanistic models, and in debates about the status of climate models and computer simulations. This raises the question of whether some of the initial critiques of the theory-centered vision of science have become obsolete. At the very least, it appears that some of the epistemological questions in relation to experiments have shifted.

- The shift of emphasis from theories to (mechanistic) models in the life sciences also changes some questions we can ask about the role of experiments in relation to models. As shown in Section 3.1, some of the epistemological problems that beset experiments (e.g., what kinds of inferences one can draw from them) are very similar to problems with models. Exploring this relationship further strikes us a worthwhile topic of future study.

- There has recently been a lot of interest in experimental methods in the behavioral sciences, for example, pertaining to inferences from experimentally induced functional imaging data. Whereas some authors (e.g., Klein 2010) have remarked on the exploratory function of such data, there is little contact between this literature and more general philosophical debates about scientific (including exploratory) experiments. Similarly, it strikes us as worthwhile to investigate possible points of contact between recent debates about psychology's "replication crisis" and the literature about the problem of experimental replication.

- A further question concerns a more encompassing analysis of the notion of scientific and, in particular, experimental *practice*. Although some philosophers (like A. Franklin) have highlighted the importance of epistemological strategies or methodological rules, others (like Collins) point to the tacit skills required for scientific practice (including the skills required to apply rules correctly). With few exceptions (e.g., Rouse 2002), different aspects of experimental practice have not yet received much attention from the side of philosophers. Related to this point, the role of concepts in (experimental) scientific practice deserves further attention (Feest and Steinle 2012).

- Further, while the work by Bogen and Woodward about the distinction between data and phenomena has generated a lot of interest over the years, there are still many competing interpretations of how phenomena and data should properly be conceived, and how they figure in experimental science. With regard to this, we specifically want to highlight the issue of inferences from local and idiosyncratic experimental data to less context-specific phenomena (Section 3.2).

- Finally, the issue of applied experimental research touched on in section 3.3 strikes us as extremely important, especially in the light of recent interest in questions concerning the role of science in a liberal democracy.

ACKNOWLEDGEMENTS

We would like to thank Jutta Schickore and Paul Humphreys for helpful comments on previous versions of this article.

REFERENCES

Ackermann, R. (1989). "The New Experimentalism." *British Journal of Philosophical Sciences* 40: 185–190.

Arabatzis, T. (2006). *Representing Electrons: A Biographical Approach to Theoretical Entities* (Chicago: University of Chicago Press).

Arabatzis, T. (2012). "Experimentation and the Meaning of Scientific Concepts." In U. Feest and F. Steinle (eds.), *Scientific Concepts and Investigative Practice* (Berlin: De Gruyter).

Bechtel, W. (1994). "Deciding on the Data. Epistemological Problems Surrounding Instruments and Research Techniques in Cell Biology." In R. M. Burian, M. Forbes, and D. Hull (eds.), *PSA: Proceedings of the 1994 Biennial Meeting of the Philosophy of Science Association. Vol. 2: Symposia and invited papers* (East Lansing, MI: Philosophy of Science Association), 167–178.

Bloor, D. (2005). "Toward a Sociology of Epistemic Things." *Perspectives on Science* 13(3): 285–312.

Bogen, J. (2002). "Epistemological Custard Pies from Functional Brain Imaging." *Philosophy of Science* 69(3): 59–71.

Bogen, J., and Woodward, J. (1988). "Saving the Phenomena." *Philosophical Review* 97: 303–352.

Bridgman, P. W. (1927). *The Logic of Modern Physics* (New York: Macmillan).

Burian, R. M. (1997). "Exploratory Experimentation and the Role of Histochemical Techniques in the Work of Jean Brachet, 1938–1952." *History and Philosophy of the Life Sciences* 19: 27–45.

Burian, R. M. (2007). "On MicroRNA and the Need for Exploratory Experimentation in Post-Genomic Molecular Biology." *History and Philosophy of the Life Sciences* 29(3): 285–312.

Campbell, D. T., and Stanley, J. C. (1966). *Experimental and Quasi-Experimental Designs for Research* (Chicago: Rand McNally).

Carrier, M. (2011). "Knowledge, Politics, and Commerce. Science Under the Pressure of Practice." In M. Carrier and A. Nordmann (eds.), *Science in the Context of Application* (Dordrecht/London: Springer), 11–30.

Cartwright, N. (2007). *Hunting Causes and Using Them: Approaches in Philosophy and Economics* (Cambridge: Cambridge University Press).

Cartwright, N., and Hardie, J. (2012). *Evidence-Based Policy: A Practical Guide to Doing It Better* (Oxford/New York: Oxford University Press).

Chang, H. (2004). *Inventing Temperature: Measurement and Scientific Progress* (New York: Oxford University Press).

Chang, H. (2009). "Operationalism. The Stanford Encyclopedia of Philosophy": http://plato.stanford.edu/archives/fall2009/entries/operationalism/

Chang, H. (2011). "The Persistence of Epistemic Objects Through Scientific Change." *Erkenntnis* 75: 413–429.

Coko, K., and Schickore, J. (2013). "Robustness, Solidity and Multiple Determinations." *Metascience* 22(3): 681–683.

Collins, H. M. (1985). *Changing Order: Replication and Induction in Scientific Practice* (Beverly Hills/London: Sage).

Cronbach, L. J., and Meehl, P. E. (1955). "Construct Validity in Psychological Tests." *Psychological Bulletin* 52(4): 281–302.

Cronbach, L. J., and Shapiro, K. (1982). *Designing Evaluations of Educational and Social Programs* (San Francisco: Jossey-Bass).

Culp, S. (1994). "Defending Robustness. The Bacterial Mesosome as a Test Case." In D. Hull, M. Forbes, and R. M. Burian (eds.), *PSA 1994: Proceedings of the 1994 Biennial Meeting of the Philosophy of Science Association* (East Lansing, MI: Philosophy of Science Association), 46–57.

Culp, S. (1995). "Objectivity and Experimental Inquiry. Breaking Data-Technique Circles." *Philosophy of Science* 62: 430–450.

Dingler, H. (1928). *Das Experiment: Sein Wesen und seine Geschichte* (München: Reinhardt).

Douglas, H. (2000). "Inductive Risk and Values in Science." *Philosophy of Science* 67(4): 559–579.

Duhem, P. (1906/1962). *The Aim and Structure of Physical Theory* (Atheneum, NY: Atheneum).

Elliott, K. C. (2007). "Varieties of Exploratory Experimentation in Nanotoxicology." *History and Philosophy of the Life Sciences* 29(3): 313–336.

Elliott, K. C. (2012). "Epistemic and Methodological Iteration in Scientific Research." *Studies in History and Philosophy of Science* 43: 376–382.

Feest, U. (2005). "Operationism in Psychology—What the Debate Is About, What the Debate Should Be About." *Journal for the History of the Behavioral Sciences* 41(2): 131–149.

Feest, U. (2010). "Concepts as Tools in the Experimental Generation of Knowledge in Cognitive Neuropsychology." *Spontaneous Generations: A Journal for the History and Philosophy of Science* 4(1): 173–190.

Feest, U. (2011). "Remembering (Short-Term) Memory: Oscillations of an Epistemic Thing." *Erkenntnis* 75(3): 391–411.

Feest, U., and Steinle, F., eds. (2012). *Scientific Concepts and Investigative Practice* (Berlin: De Gruyter).

Franklin, A. (1986). *The Neglect of Experiment* (Cambridge: Cambridge University Press).

Franklin, A. (1989). "The Epistemology of Experiment." In D. C. Gooding, T. Pinch, and S. Schaffer (eds.), *The Uses of Experiment. Studies in the Natural Sciences* (Cambridge: Cambridge University Press), 437–460.

Franklin, A. (1999). *Can That Be Right?: Essays on Experiment, Evidence, and Science.* Boston Studies in the Philosophy of Science, vol. 199. (Dordrecht/Boston: Kluwer).

Franklin, L. R. (2005). "Exploratory Experiments." *Philosophy of Science* 72: 888–899.

Galison, P. (1987). *How Experiments End* (Chicago: University of Chicago Press).

Galison, P. (1988). "History, Philosophy, and the Central Metaphor." *Science in Context* 2(1): 197–212.

Gooding, D. C. (1990). *Experiment and the Making of Meaning: Human Agency in Scientific Observation and Experiment* (Dordrecht: Kluwer).

Gooding, D. C., Pinch, T., and Schaffer, S., eds. (1989). *The Uses of Experiment: Studies in the Natural Sciences.* (Cambridge: Cambridge University Press).

Graßhoff, G. (2011). "Inferences to Causal Relevance from Experiments." In D. Dieks, W. J. Gonzalez, S. Hartmann, et al. (eds.), *Explanation, Prediction, and Confirmation* (Dordrecht: Springer), 167–182.

Guala, F. (2005). *The Methodology of Experimental Economics* (Cambridge: Cambridge University Press).

Guala, F. (2012). "Handbook of the Philosophy of Science." In U. Mäki (ed.), *Handbook of the philosophy of science. Vol. 13: Philosophy of Economics* (Boston: Elsevier/Academic Press), 597–640.

Hacking, I. (1981). "Do We See Through a Microscope?." *Pacific Philosophical Quarterly* 62: 305–322.

Hacking, I. (1983). *Representing and Intervening: Introductory Topics in the Philosophy of Natural Science* (Cambridge: Cambridge University Press).

Hacking, I. (1992). "The Self-Vindication of the Laboratory Science." In A. Pickering (ed.)., *Science as Practice and Culture* (Chicago: University of Chicago Press), 29–64.

Heidelberger, M., and Steinle, F. (eds). (1998). *Experimental Essays—Versuche zum Experiment* (Baden-Baden: Nomos Verlag).

Hentschel, K. (1998). "Feinstruktur und Dynamik von Experimentalsystemen." In M. Heidelberger and F. Steinle (eds)., *Experimental Essays—Versuche zum Experiment* (Baden-Baden: Nomos Verlag), 325–354.

Hon, G. (1989). "Towards a Typology of Experimental Errors. An Epistemological View." *Studies in History and Philosophy of Science* 20: 469–504.

Hon, G. (1998). "'If This Be Error.' Probing Experiment with Error." In M. Heidelberger and F. Steinle (eds.), *Experimental Essays—Versuche zum Experiment* (Baden-Baden: Nomos Verlag), 227–248.

Hudson, R. (1999). "Mesosomes: A Study in the Nature of Experimental Reasoning." *Philosophy of Science* 66: 289–309.

Jiménez-Buedo, M. (2011). "Conceptual Tools for Assessing Experiments: Some Well-Entrenched Confusions Regarding the Internal/External Validity Distinction." *Journal of Economic Methodology* 18(3): 271–282.

Jiménez-Buedo, M., and Miller, L. M. (2010). "Why a Trade-Off? The Relationship between the External and Internal Validity of Experiments." *Theoria* 69: 301–321.

Karaca, K. (2013). "The Strong and Weak Senses of Theory-Ladenness of Experimentation: Theory-Driven versus Exploratory Experiments in the History of High-Energy Particle Physics." *Science in Context* 26(1): 93–136.

Klein, C. (2010). "Philosophical Issues in Neuroimaging." *Philosophy Compass* 5(2): 186–198.

Latour, B., and Woolgar, S. (1979). *Laboratory Life: The Social Construction of Scientific Facts* (Beverly Hills/London: Sage).

Leonelli, S. (2014). "What Difference Does Quantity Make? On the Epistemology of Big Data Biology." *Big Data and Society* 1(1): 1–11.

Mayo, D. G. (1996). *Error and the Growth of Experimental Knowledge* (Chicago: University of Chicago Press).

Mitchell, S. D. (2009). *Unsimple Truths: Science, Complexity, and Policy* (Chicago: University of Chicago Press).

Morrison, M. (1990). "Theory, Intervention and Realism." *Synthese* 82: 1–22.

O'Malley, M. A. (2007). "Exploratory Experimentation and Scientific Practice: Metagenomics and the Proteorhodopsin Case." *History and Philosophy of the Life Sciences* 29(3): 337–358.

Osimani, B., and Mignini, F. (2015). "Causal Assessment of Pharmaceutical Treatments. Why Standards of Evidence Should Not be the Same for Benefits and Harms?" *Drug Safety* 38: 1–11.

Pearl, J. (1988). *Probabilistic Reasoning in Intelligent Systems.* (San Mateo, CA: Morgan and Kaufman).

Peschard, I. (2012). "Modeling and Experimenting." In P. Humphreys and C. Imbert (eds.), *Models, Simulations, and Representations* (New York: Routledge), 42–61.

Pickering, A. (1995). *The Mangle of Practice: Time, Agency and Science* (Chicago: University of Chicago Press).

Radder, H., ed. (2003): *The Philosophy of Scientific Experimentation*. (Pittsburgh, PA: University of Pittsburgh Press).

Rasmussen, N. (1993). "Facts, Artifacts, and Mesosomes. Practicing Epistemology with the Electron Microscope." *Studies in History and Philosophy of Science* 24: 227–265.

Reiner, R., and Pierson, R. (1995). "Hacking's Experimental Realism: An Untenable Middle Ground." *Philosophy of Science* 62: 60–69.

Rheinberger, H. -J. (1997). *Towards a History of Epistemic Things: Synthesizing Proteins in the Test Tube* (Stanford, CA: Stanford University Press).

Rheinberger, H. -J. (1999). "Reenacting History. Review of: Pickering, Andrew: The Mangle of Practice, 1995." *Studies in the History and Philosophy of Science* 30: 163–166.

Rheinberger, H. -J. (2005). "A Reply to David Bloor." *Perspectives on Science* 13: 406–410.

Rouse, J. (2002). *How Scientific Practices Matter: Reclaiming Philosophical Naturalism* (Chicago: University of Chicago Press).

Sargent, R. -M. (1995). *The Diffident Naturalist: Robert Boyle and the Philosophy of Experiment* (Chicago: University of Chicago Press).

Schickore, J. (2005). "'Through Thousands of Errors We Reach the Truth'—But How? On the Epistemic Roles of Error in Scientific Practice." *Studies in History and Philosophy of Science Part A* 36(3): 539–556.

Schickore, J., and Coko, K. (2014). "Using Multiple Means of Determination." *International Studies in the Philosophy of Science* 27(3): 295–313.

Schickore, J., and Steinle, F., eds. (2006). *Revisiting Discovery and Justification. Historical and Philosophical Perspectives on the Context Distinction* (Dordrecht: Springer).

Shapin, S., and Schaffer, S. (1985). *Leviathan and the Air-pump: Hobbes, Boyle and the Experimental Life* (Princeton, NJ: Princeton University Press).

Soler, L. (2012). *Characterizing the Robustness of Science: After the Practice Turn in Philosophy of Science* (Dordrecht/New York: Springer).

Spirtes, P., Glymour, C. N., and Scheines, R. (1993). *Causation, Prediction, and Search* (New York: Springer).

Steel, D. (2008). *Across the Boundaries: Extrapolation in Biology and Social Science* (Oxford/New York: Oxford University Press).

Steinle, F. (1997). "Entering New Fields. Exploratory Uses of Experimentation." *Philosophy of Science* 64 (Supplement): 65–74.

Steinle, F. (2012). "Goals and Fates of Concepts: The Case of Magnetic Poles." In U. Feest and F. Steinle (eds.), *Scientific Concepts and Investigative Practice* (Berlin: De Gruyter), 105–125.

Steinle, F. (2016a). *Exploratory Experiments: Ampère, Faraday, and the Origins of Electrodynamics* (Pittsburgh: University of Pittsburgh Press). English version of Steinle, F. (2005). *Explorative Experimente. Ampère, Faraday und die Ursprünge derElektrodynamik* (Stuttgart: Steiner).

Steinle, F. (2016b). "Newton, Newtonianism, and the Roles of Experiment." In H. Pulte and S. Mandelbrote (eds.), *The Reception of Isaac Newton in Europe* (New York: Continuum Publishing).

Sullivan, J. (2009). "The Multiplicity of Experimental Protocols. A Challenge to Reductionist and Non-Reductionist Models of the Unity of Neuroscience." *Synthese* 167: 511–539.

Waters, C. K. (2007). "The Nature and Context of Exploratory Experimentation: An Introduction to Three Case Studies of Exploratory Research." *History and Philosophy of the Life Sciences* 29(3): 275–284.

Weber, M. (2005). *Philosophy of Experimental Biology* (New York: Cambridge University Press).

Wilholt, T. (2009). "Bias and Values in Scientific Research." *Studies in History and Philosophy of Science* 40(1): 92–101.

Wimsatt, W. C. (1981). "Robustness, Reliability, and Overdetermination." In M. B. Brewer and B. E. Collins (eds.), *Scientific Inquiry and the Social Sciences* (San Francisco: Jossey-Bass), 124–163.

Woodward, J. (2005). *Making Things Happen: A Theory of Causal Explanation* (New York/Oxford: Oxford University Press).

Woodward, J. (2008). "Invariance, Modularity, and All That. Cartwright on Causation." In N. Cartwright, S. Hartmann, C. Hoefer et al. (eds.), *Nancy Cartwright's Philosophy of Science* (New York: Routledge), 198–237.

Woodward, J. (2009). "Agency and Interventionist Theories." In H. Beebee, C. Hitchcock, and P. C. Menzies (eds.), *The Oxford Handbook of Causation* (Oxford/New York: Oxford University Press), 234–262.

CHAPTER 14

GAME THEORY

CRISTINA BICCHIERI AND GIACOMO SILLARI

1 GAMES OF COMPLETE INFORMATION

A game is an abstract, formal description of a strategic interaction. Any strategic inter-action involves two or more decision-makers (players), each with two or more ways of acting (strategies), such that the outcome depends on the strategy choices of all the play-ers. Each player has well-defined preferences among all the possible outcomes, enabling corresponding utilities (payoffs) to be assigned. A game makes explicit the rules gov-erning players' interaction, the players' feasible strategies, and their preferences over outcomes.

1.1 Normal Form

A possible representation of a game is in *normal form*. A normal form game is com-pletely defined by three elements that constitute the *structure of the game*: a list of players $i = 1, \dots, n$; for each player i, a finite set of pure strategies S_i; and, for each player i, a payoff function u_i that gives player i's payoff $u_i(s)$ for each n-tuple of strategies $s = (s_1, \dots, s_n)$, where u_i is a typical Von Neumann-Morgenstern utility function, that is $u_i : \times_{j=1}^{n} S_j \to \mathbb{R}$. A player may choose to play a *pure* strategy, or, instead, she may choose to randomize over her pure strategies; a probability distribution over pure strategies is called a *mixed strategy* and is denoted by σ_i. Each player's randomization is assumed to be indepen-dent of that of his opponents, and the payoff to a mixed strategy is the expected value of the corresponding pure strategy payoffs. A different interpretation of mixed strategies, based on the idea that players do not (always) randomize over their feasible action, is that the probability distribution σ_i represents i's *uncertainty* about what other players will do. Thus, a mixed strategy is thought of as a player's conjecture about other players' plans of action. The conjectures depend on players' private information, which remains

	C	D
C	3,3	0,4
D	4,0	1,1

FIGURE 14.1 The Prisoner's Dilemma.

unspecified in the model. A problem with this interpretation is that if there are reasons behind the choices a player makes, they should be included in the model because they are likely to be payoff relevant.

The three elements defining a normal form game can be arranged and easily read in a matrix, as in the example of a two-person game in Figure 14.1:

The 2×2 matrix in Figure 14.1 depicts the two-player normal form representation of the famous Prisoner's Dilemma game, where C stands for "cooperate" and D for "defect." The numbers in the cell of the matrix denote players' payoffs: the first number is the payoff for the Row player, the second for the Column player. Each player picks a strategy independently, and the outcome, represented in terms of players' payoffs, is the joint product of these two strategies. Notice that in the game in Figure 14.1, each player is better off defecting *no matter* what the other player does. For example, if Column cooperates, Row gets a payoff of 4 by defecting and a payoff of 3 by cooperating, whereas if Column defects, Row gains a payoff of 1 by defecting and of 0 by cooperating. When, regardless of what other players do, a strategy yields a player a payoff strictly inferior to some other strategy, it is called a strictly *dominated* strategy. When a strategy yields the same payoff as another undominated strategy, but it has an inferior payoff against at least one opponent's strategy, it is called a *weakly dominated strategy*.

The game in Figure 14.1 is one of *complete information* in that the players are assumed to know the rules of the game (which include players' strategies) and other players' payoffs. If players are allowed to enter into binding agreements before the game is played, we say that the game is *cooperative*. On the other hand, *non-cooperative games* make no allowance for the existence of an enforcement mechanism that would make an agreement binding on the players. What strategies should rational players choose? In the prisoner's dilemma, surely players will choose D because choosing D guarantees them a higher payoff no matter what the other player chooses. In other words, players have an *unconditional* preference to choose D. What if players do not have strictly dominated strategies, as, for instance, in the game shown in Figure 14.2?

In this game, players want to *coordinate* their actions in that their interests are aligned and therefore their preferences for action are *conditional* on the action of the other player. Thus Row will like to play L if Column also plays L, and Column will prefer R if Row chooses R. That is to say, L is, for both players, the *best reply* to L and R the best reply to R. When best replies "meet" like that, then players are content with their action, meaning that they would not want to unilaterally change their course of action if they

	L	R
L	3,3	0,0
R	0,0	1,1

FIGURE 14.2 A coordination game.

had a chance. They won't unilaterally deviate because, since each one is playing a best reply to the other's action, neither player could improve by changing action: both players are maximizing their payoffs given what the other is doing. When players jointly maximize their utility, they are playing the *Nash equilibrium* of the game (cf. Nash 1996). Informally, a Nash equilibrium specifies players' actions and beliefs such that (1) each player's action is optimal given his beliefs about other players' choices, and (2) players' beliefs are correct. Thus, observing an outcome that is not a Nash equilibrium entails either that a player chose a suboptimal strategy or that some players "misperceived" the situation.

More formally, a Nash equilibrium is a vector of strategies $\left(\sigma_1^*,\ldots,\sigma_n^*\right)$, one for each of the n players in the game, such that each σ_i^* is optimal given (or is a *best reply* to) σ_{-i}^*, where the subscript $-i$ denotes all players except i. That is

$$u_i\left(\sigma_i^*,\sigma_{-i}^*\right)\geq u_i\left(\sigma_i,\sigma_{-i}^*\right) \text{ for all mixed strategies of player } i, \ \sigma_i.$$

Note that optimality is only conditional on a fixed σ_{-i}^*, not on *all* possible σ_{-i}. A strategy that is a best reply to a given combination of the opponents' strategies may fare poorly vis á vis a different strategy combination.

In a game like that depicted in Figure 14.3, Row gains a payoff of 1 if the toss of two coins results in two heads or two tails and loses 1 otherwise, and vice-versa for Column.

This game has no Nash equilibrium in pure strategies. Nash proved that—provided certain restrictions are imposed on strategy sets and payoff functions—every game has at least one equilibrium in mixed strategies. In a mixed strategy equilibrium, the equilibrium strategy of each player makes the other indifferent between the strategies on which she is randomizing. In particular, the game in Figure 14.3 has a unique equilibrium in which both players randomize between their strategies with probability ½. Then, if the first player plays $\sigma_1 = \left(1/2\ H, 1/2\ T\right)$, her expected payoff is $1/2\ 1 + 1/2 - 1 = 0$ regardless of the strategy choice of the second player. A Nash equilibrium in mixed strategy is a *weak* equilibrium in that unilateral deviation does not guarantee a gain, although it does not entail a loss either.

Matching Pennies	H	T
H	1, −1	−1, 1
T	−1, 1	1, −1

FIGURE 14.3 Matching Pennies.

The players (and the game theorist) predict that a specific equilibrium will be played just in case they have enough information to infer players' choices. The standard assumptions in game theory are

CK1: The structure of the game is common knowledge.
CK2: The players are rational (i.e., they are expected utility maximizers), and this is common knowledge.

The concept of *common knowledge* was introduced by Lewis (1969) in his study on convention, which is arguably the first major philosophical work in which game theory plays a central role as a modeling tool. Simply put, the idea of common knowledge is that a certain proposition p is common knowledge among two players if both of them know p, both of them know that they know p, and so on ad infinitum (for some of the technicalities, see Section 3 herein or Vanderschraaf and Sillari 2014). CK1 and CK2 may allow players to predict their opponents' strategy. For example, in the Prisoner's Dilemma game of Figure 14.1, CK1 would ensure that players know that C is strictly dominated, whereas CK2 would then make players choose the utility-maximizing strategy D (and predict that other players will also choose D). However (cf. Bicchieri 1993), these CK assumptions do not always guarantee that a prediction be feasible. For instance, even if the game has a unique equilibrium, the set of strategies that a player may choose from under CK1 and CK2 need not contain only equilibrium strategies. Moreover, predictability is problematic in case (cf. the game in Figure 14.2) of multiple Nash equilibria. Let us consider another example of a game with multiple equilibria. Suppose two players have to divide $100 between them. They must restrict their proposals to integers, and each has to independently propose a way to split. If the total proposed by both is equal or less than $100, each gets what she proposed, otherwise they get nothing. This game has 101 Nash equilibria. Is there a way to predict which one will be chosen? In real life, many people would go for the 50/50 split. It is simple and seems equitable. In Schelling's words, it is a *focal point* (Schelling 1960). Unfortunately, mere salience is not enough to provide a player with a reason for choice. In our example, only if it is common knowledge that the 50/50 split is the salient outcome does it becomes rational to propose $50. Game theory, however, filters out any social or cultural information regarding strategies, leaving players with the task of coordinating their actions on the sole basis of common knowledge of rationality (and of the structure of the game). CK1 and CK2 are of no use to prune a player's strategy set: rational play remains indeterminate. There are two main avenues to cope with the problem of indeterminacy. One, that we will explore in Section 5, is to think of a game as played by a population of players subject to evolutionary pressures: studying such evolutionary dynamics will be key to solving indeterminacy.

A different approach to the problem of indeterminacy is to start by considering the set of Nash equilibria and ask whether some of them should be eliminated because they are in some sense "unreasonable." In a sense, this approach *strengthens* in various ways the requirement CK2, with the aim of excluding unreasonable strategies and being left

	C	D
A	2, 4	0, 1
B	0, 1	0, 1

FIGURE 14.4 A game with a weak equilibrium.

with a truly rational selection. This is the approach taken by the *refinement* program (Kohlberg 1990, van Damme 1987). Consider the game in Figure 14.4:

The game has two Nash equilibria in pure strategies: *(A,C)* and *(B,D)*. The equilibrium *(A,C)* is *Pareto-dominant*, since it gives both players a higher payoff than any other equilibrium in the game. However, common knowledge of rationality and of the structure of the game does not force Column player to expect Row player to eliminate the weakly dominated strategy *B*, nor is Row forced to conclude that Column will discard *D*. Prudence, however, may suggest that one should never be too sure of the opponents' choices. Even if players have agreed to play a given equilibrium, some uncertainty remains. If so, we should try to model this uncertainty in the game. Selten's insight was to treat perfect rationality as a limit case (Selten 1965). His "trembling hand" metaphor presupposes that deciding and acting are two separate processes in that, even if one decides to take a particular action, one may end up doing something else by mistake. An equilibrium strategy should be optimal not only against the opponents' strategies, but also against some very small probability $\sigma > 0$ that the opponents make "mistakes." Such an equilibrium is *trembling-hand perfect*. Is the equilibrium *(B,D)* perfect? If so, *B* must be optimal against *C* being played with probability σ and *D* being played with probability $1-\sigma$ for some small $\sigma > 0$. But, in this case, the expected payoff to *A* is 2σ, whereas the payoff to *B* is 0. Hence, for all $\sigma > 0$, *A* is a better strategy choice. The equilibrium *(B,D)* is not perfect, but *(A,C)* is. A prudent player therefore would discard *(B,D)*. In this simple game, checking perfection is easy because only one mistake is possible. With many strategies, there are many more possible mistakes to take into account. Similarly, with many players, we may need to worry about who is likely to make a mistake.

1.2 Extensive Form

A different, possibly richer representation of a game is the *extensive form*. In the extensive form, we specify the following information: a finite set of players $i = 1,\dots,n$; the order of moves; the players' choices at each move and what each player knows when she has to choose. The order of play is represented by a game tree T, which is a finite set of partially ordered nodes $t \in T$ satisfying a precedence relation <. A *subgame* is a collection of branches of a game such that they start from the same node, and the branches and the node together form a game tree by itself. A tree representation is sequential because it shows the order in which actions are taken by the players. It is quite natural to think of sequential-move games as being ones in which players choose their strategies one after the other and of simultaneous-move games as ones in which players choose their strategies at the same time. What is important, however, is not the temporal order of events, but whether players know about other players' actions when they have to choose their own. In the normal form representation, players' information about other players' choices is not represented. This is the reason why a normal form game could represent any one of several extensive form games. When the order of play is irrelevant to a game's outcome, then restricting oneself to the normal form is justifiable. When the order of play is relevant, however, the extensive form must be specified.

In an extensive form game, the information a player has when she is choosing an action is explicitly represented using *information sets*, which partition the nodes of the tree. If an information set contains more than one node, the player who has to make a choice at that information set will be uncertain as to which node she occupies. Not knowing which node one occupies means that the player does not know which action was chosen by the preceding player. If a game contains information sets that are not singletons, the game is one of *imperfect information*. Consider, for instance, following game in Figure 14.5:

This is an extensive form representation of the Prisoner's Dilemma. In the Prisoner's Dilemma, each player acts in ignorance of the other player's choice. Such ignorance is reproduced in Figure 14.5, introducing the *information set* that contains player 2's nodes. The meaning of the set is that player 2 is unable to distinguish between the left node (reached after player 1 has chosen C) and the right node (after player 1 has chosen D).

In an extensive form game, a *strategy* for player *i* is a complete plan of action that specifies an action at every node at which it is *i*'s turn to move. Note that a strategy specifies actions even at nodes that will never be reached if that strategy is played. Consider the game in Figure 14.6:

It is a finite game of perfect information in which player 1 moves first. If he chooses D at his first node, the game ends and player 1 nets a payoff of 1, whereas player 2 gets 0. But choosing D at the first node is only part of a strategy for player 1. For example, it can be part of a strategy that recommends "play D at your first node, and x at your last node." Another strategy may instead recommend playing D at his first node and y at his last decision node. Although it may seem surprising that a strategy specifies actions even at

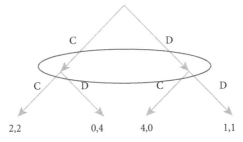

FIGURE 14.5 The extensive form representation of the Prisoner's Dilemma.

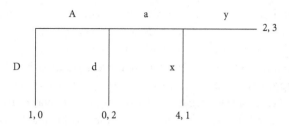

FIGURE 14.6 An extensive game of perfect information.

nodes that will not be reached if that strategy is played, we must remember that a strategy is a full *contingent* plan of action. For example, the strategy *Dx* recommends playing *D* at the first node, thus effectively ending the game. It is important, however, to be able to have a plan of action in case *D* is not played. Player 1 may, after all, make a mistake and, because of 2's response, find himself called to play at his very last node. In that case, having a plan helps. Note that a strategy cannot be changed during the course of the game. Although a player may conjecture about several scenarios of moves and counter-moves before playing the game, at the end of deliberation a strategy must be chosen and followed through the game.

The game of Figure 14.6 has two Nash equilibria in pure strategies: *(Dx,d)* and *(Dy,d)*. Is there a way to solve the indeterminacy?

Suppose player 1 were to reach his last node. Since he is by assumption rational, he will choose *x*, which guarantees him a payoff of 4. Knowing (by assumption) that 1 is rational, player 2—if she were to reach her decision node—would play *d*, since by playing *a* she would net a lower payoff. Finally, since (by assumption) player 1 knows that 2 is rational and that she knows that 1 is rational, he will choose *D* at his first decision node. The equilibrium *(Dy,d)* should therefore be ruled out because it recommends an irrational move at the last node. In the normal form, both equilibria survive. The reason is simple: Nash equilibrium does not constrain behavior out of equilibrium. In our example, if 1 plans to choose *D* and 2 plans to choose *d*, it does not matter what player 1 would do at his last node because that node will never be reached.

The sequential procedure we have used to conclude that only *(Dx,d)* is a reasonable solution is known as *backward induction*. In finite games of perfect information with no ties in payoffs, backward induction always identifies a unique equilibrium. The premise of the backward induction argument is that mutual rationality and the structure of the game are common knowledge among the players. It has been argued by Binmore (1987), Bicchieri (1989, 1993), and Reny (1992) that, under certain conditions, common knowledge of rationality leads to inconsistencies. For example, if player 2 were to reach her decision node, would she keep thinking that player 1 is rational? How would she explain 1's move? If 1's move is inconsistent with common knowledge of rationality, player 2 will be unable to predict future play; as a corollary, what constitutes an optimal choice at her node remains undefined. As a consequence of these criticisms, the usual premises of backward induction arguments have come to be questioned (cf. Basu 1990; Bonanno 1991; Pettit and Sugden 1989). There are a number of further equilibrium refinements for games in extensive form. Their abundance makes it impossible to delve into details here. The interested reader can consult Bicchieri (1993, chapter 3).

1.3 Extensive Form Refinements

We return now to the discussion of the refinement program. We have seen how, in the normal form, Selten's trembling-hand perfection requires players to check how a strategy will perform were another player to take an action that has zero probability in equilibrium. In the extensive form representation, players ask what would happen off-equilibrium, at points in the game tree that will never be reached if the equilibrium is played. In both cases, the starting point is an equilibrium, which is checked for stability against possible deviations. By its nature, the Nash equilibrium concept does not restrict action choices off the equilibrium path because those choices do not affect the payoff of the player who moves there. For example, the equilibrium (Dy, d) in the game in Figure 14.6 lets player 1 make an irrational choice at the last node because that choice is not going to affect his payoff (which is determined by his choosing D at the beginning of the game). However, the strategy of a player at an off-equilibrium information set can affect what other players choose in equilibrium. Suppose the players consider agreeing to play (Dy, d). In order to choose D, player 1 must decide what would happen were he to play A instead. To decide whether D is a rational move, 1 has to think about player 2's choice at an off-equilibrium node. His conclusion about 2's choice will affect his own choice. But player 2's choice will depend on how she interprets 1's off-equilibrium move. Player 1, in turn, must be able to anticipate 2's interpretation of his deviating from the equilibrium path. For example, if 2 were to interpret the deviation as a mistake, would she still play her part in the equilibrium (Dy, d) and choose d? If she expects y to be played at the last node, a is a best reply. But, at his last node, why would rational player 1 choose y? Is the agreement to play (Dy, d) reasonable?

The earliest refinement proposed to rule out implausible equilibria in extensive games of perfect information is *subgame perfection* (Selten 1965). A Nash equilibrium is subgame-perfect if its component strategies (when restricted to any subgame) remain a Nash equilibrium of the subgame. The equilibrium (Dy, d) is not subgame-perfect: in the subgame starting at the last node, y is a dominated strategy. Note that the backward induction equilibrium is always subgame-perfect.

Consider as another example the *ultimatum game*, introduced by Güth et al. (Güth, Schmittberger, and Schwarze 1982). In the ultimatum game, two players, called Proposer and Respondent, bargain over a sum of money, say $10. proposer offers an allocation of the money, and responder may *accept* that allocation (and, in this case, the payoffs are as specified by the proposer's allocation) or *reject* it (and, in this case, the payoff for both players is zero). A simplified version of the game, sometimes called the "mini-ultimatum game," is illustrated in Figure 14.7:

Here, the only possible allocations are (5,5) and (8,2). The former is always accepted, whereas the latter can be rejected by player 2. This game has two Nash equilibria (*offer 5, reject*) and (*offer 2, accept*). The former, however, is not subgame-perfect because it

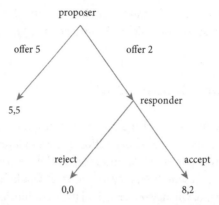

FIGURE 14.7 The "mini-ultimatum game".

requires the irrational move "reject" in player 2's unreached node. Subgame perfection, however, applies only (nontrivially) to games that have proper subgames. Any Nash equilibrium of a game without proper subgames is trivially subgame-perfect (because the whole game can be considered a subgame), but, in this case, the criterion does not help in resolving the indeterminacy.

As I mentioned at the outset, the refinement program attempts to establish stability criteria for Nash equilibrium. It presupposes that players will choose to play a Nash equilibrium after having eliminated several alternative equilibria on the ground that they are unreasonable. What counts as a reasonable equilibrium, however, depends on how off-equilibrium behavior is interpreted. This, in turn, hinges on players' out-of-equilibrium beliefs. A formalization of the off-equilibrium deliberation process requires the use of counterfactuals (from the viewpoint of playing a given equilibrium, deviations are contrary-to-fact events). Some work in this direction has been done by Selten and Leopold (1982), Samet (1996), Stalnaker (1994), and Skyrms (1990). So far, we have developed no comprehensive theory of out-of-equilibrium behavior that indicates, for example, when a deviation should be interpreted as a signal and when as a mistake. Such theory would supply substantive (as opposed to merely formal) rationality criteria for players' beliefs and would thus expand the traditional notion of practical rationality to include an epistemic component. This theoretical inadequacy undermines the deductive goal of inferring (and predicting) equilibrium play from rationality principles alone.

1.4 An Application: Bicchieri's Theory of Social Norms

The norm-based utility function introduced by Bicchieri (2006) tries to capture, through a game-theoretic model, the idea that, when a norm exists, individuals will show different "sensitivities" to it, and this should be reflected in their utility functions (cf. also Bicchieri and Chavez 2010; Bicchieri and Sontuoso 2014; Bicchieri and Zhang 2012). Consider a typical *n*-person (normal-form) game. For the sake of formal treatment, we represent a norm as a (partial) function that maps what the player expects other players to do into what the player "ought" to do. In other words, a norm regulates behavior conditional on other people's (expected) behaviors. Denote the strategy set of player i by S_i, and let $S_{-i} = \prod_{j}^{j=i} S_j$ be the set of strategy profiles of players other than i. Then a norm for player i is formally represented by a function $N_i : L_{-i} \to S_i$, where $L_{-i} \subseteq S_{-i}$. Two points are worth noting. First, given the other players' strategies, there may or may not be a norm that prescribes how player i ought to behave. So L_{-i} need not be—and usually is not—equal to S_{-i}. In particular, L_{-i} could be empty in the situation where there is no norm whatsoever to regulate player i's behavior. Second, there could be norms that regulate joint behaviors. A norm, for example, that regulates the joint behaviors of players i and j may be represented by $N_{ij} : L_{-i,-j} \to S_i \times S_j$. Since we are concerned with a two-person game here, we will not further complicate the model in that direction.

A strategy profile $s = (s_1, \ldots, s_n)$ instantiates a norm for j if $s_{-j} \in L_{-j}$; that is, if N_j is defined at s_{-j}. It violates a norm if, for some j, it instantiates a norm for j but $s_j \neq N_j(s_{-j})$. Let π_i be the payoff function for player i. The norm-based utility function of player i depends on the strategy profile s and is given by

$$U_i(s) = \pi_i(s) - k_i max \; s_{-j} \in L_{-j} max_{m \neq j} \left\{ \pi_m \left(s_{-j}, N_j(s_{-j}) \right) - \pi_m(s), 0 \right\},$$

where $k_i \geq 0$ is a constant representing i's sensitivity to the relevant norm. Such sensitivity may vary with different norms; for example, a person may be very sensitive to equality and much less so to equity considerations. The first maximum operator takes care of the possibility that the norm instantiation (and violation) might be ambiguous in the sense that a strategy profile instantiates a norm for several players simultaneously. We conjecture, however, that this situation is rare, and, under most situations, the first maximum operator degenerates. The second maximum operator ranges over all the players other than the norm violator. In plain words, the discounting term (multiplied by k_i) is the maximum payoff deduction resulting from all norm violations.

In the ultimatum game, the norm we shall consider is the norm that prescribes a fair amount that the proposer ought to offer. To represent it, we take the norm functions to be the following: the norm function for the proposer, N_1, is a constant N function, and the norm function for the responder, N_2, is nowhere defined. If the responder (player 2) rejects, the utilities of both players are zero:

$$U_{1,reject}(x) = U_{2,reject}(x) = 0$$

Given that the proposer (player 1) offers x and the responder accepts, the utilities are the following:

$$U_{1,accept}(x) = M - x - k_1 \; max \; (N - x, 0)$$

$$U_{2,accept}(x) = x - k_2 \; max \; (N - x, 0)$$

where N denotes the fair offer prescribed by the norm, and k_i is non-negative. Note, again, that k_i measures how much Proposer dislikes to deviate from what he takes to be the norm. To obey a norm, "sensitivity" to the norm need not be high. Fear of retaliation may make a proposer with a low k behave according to what fairness dictates but, absent such risk, her disregard for the norm will lead her to be unfair.

Again, Responder should accept the offer if $U_{2,accept}(x) > U_{2,reject} = 0$, which implies the following threshold for acceptance: $x > k_2 N / (1+k_2)$. Obviously, the threshold is less than N: an offer of more than what the norm prescribes is not necessary for the sake of acceptance.

For Proposer, the utility function is decreasing in x when $x \geq N$; hence, a rational proposer will not offer more than N. Suppose $x \leq N$. If $k_1 > 1$, the utility function is increasing in x, which means that the best choice for the proposer is to offer N. If $k_1 < 1$, the utility function is decreasing in x, which implies that the best strategy for the proposer is to offer the least amount that would result in acceptance; that is (a little bit more than) the threshold $x > k_2 N / (1+k_2)$. If $k_1 = 1$, it does not matter how much Proposer offers provided the offer is between $k_2 N / (1+k_2)$ and N.

It should be clear at this point that the ks measure people's sensitivity to various norms. Such sensitivity will usually be a stable disposition, and behavioral changes may thus be caused by changes in focus or in expectations. A theory of norms can explain such changes, whereas a theory of inequity aversion à la Fehr and Schmidt (1999) does not.

2 EPISTEMIC FOUNDATIONS
OF GAME THEORY

An important development of game theory is the so-called *epistemic* approach. In the epistemic approach to game theory, strategic reasoning is analyzed on the basis of hypotheses about what players know about the game, about other players' knowledge, and about other players' rationality. Since Aumann's formalization, the idea of common knowledge and the analysis of what players choose depending on what their beliefs about each other are began to play an increasingly important role in game theory. In particular, one can evaluate solution concepts by examining the epistemic assumptions and hypotheses in terms of which they can be derived (cf. Battigalli and Bonanno 1999). Such epistemic hypotheses are treated formally using the tools provided by *interactive epistemology* (cf. Aumann 1999).

To formalize players' knowledge states, one considers a space set Ω whose elements are *possible worlds*. An *event* is then represented by a subset of Ω. For example, the event "it is sunny in Philadelphia" is represented by the set of all possible worlds in which it is sunny in Philadelphia. For each player, there exists an *information function* that partitions the space set. Intuitively, a player cannot distinguish among worlds belonging to the same cell of her information partition. Thus, in a possible world ω, player i knows an event E if and only if the set E (of possible worlds in which E obtains) includes the cell of her information partition containing ω. The intuition behind this is that if a player cannot distinguish among all the worlds in which E is true, then he knows that E is the case. It is possible to define a knowledge function K_i for each player i so that, when given E as an argument, it returns as a value the set of those worlds such that, for each one of them, the cell of i's information partition that contains it is a subset of E. That is to say, $K_i E$ is the event that i *knows* E. By imposing certain conditions on the K_i's one can force the epistemic functions to have certain properties. For example, by requiring that $K_i E$ be a subset of E, one requires that what players know is true, since in every possible world in which $K_i E$ obtains, E obtains as well; similarly, by requiring that $K_i K_i E$ be a subset of $K_i E$, one establishes that players know what they know, and by requiring that $K_i \neg K_i E$ be a subset of $\neg K_i E$ that they know what they do not know (where "$\neg$" is the usual set-theoretical operation of complementation). The first condition is often referred to as *truth axiom*, the second as *positive introspection axiom*, the third as *negative introspection axiom*. Notice that this setup has an equivalent formulation in terms of modal logics (cf. Fagin, Halpern, Moses, and Vardi 1995; Meyer and van der Hoek 2004). To see the equivalence of the two approaches, consider that modal formulas express propositions whose semantic interpretation is given in terms of Kripke structures of possible worlds. It is then possible to establish a correspondence between formulas of the modal logic and events in the approach described earlier. In a Kripke model, then, an event corresponds to the set of those possible worlds that satisfy the formula expressing the proposition associated with that event.

Knowledge functions can be iterated; thus, they can represent mutual and higher order knowledge, and Aumann (1976) provides a mathematical definition of the idea of common knowledge in the setup just sketched. A proposition p is *common knowledge* between, say, two players i and j if and only if the set of worlds representing p includes the cells of i's and j's partitions *meet*, which contain p, where the *meet* of two partitions is the finest common coarsening of them. An application of the definition is the theorem proved in the same article, in which it is shown that if players have common priors, and their posteriors are common knowledge, then the posteriors are equal, even if the players derived them by conditioning on different information. Or, in other words, one cannot "agree to disagree." Although (Aumann 1976) formalizes Lewis's definition of common knowledge, it is debated whether Aumann's seminal definition is a faithful rendition of Lewis's informal characterization of common knowledge (cf. Cubitt and Sugden 2003; Paternotte 2011; Sillari 2005, 2008b; Vanderschraaf 1998).

It is often argued that these models pertain to idealized agents and that they fail to adequately capture the epistemic attitudes of realistic agents. On this issue, the interested reader can consult Sillari (2006, 2008a), where a solution based on so-called *awareness structures* is examined from both a philosophical and a logical point of view. In a nutshell, game theoretic models enriched with (un)awareness represent how individual players *perceive* the game: it could be, for instance, that player j is not simply uncertain as to whether i's strategy set includes strategy s_i, but in fact is entirely in the dark about the very availability of such a s_i for i. This would be naturally a very important aspect of the interaction from a strategic point of view. Awareness thus allows us to represent strategic interactions more realistically. The large literature on the application of *awareness structures* to game theory is well summarized in Schipper (2014).

In such a framework, it is possible to investigate which strategy profiles are compatible with certain epistemic assumptions about the players. For example, CK1 and CK2 from Section 1 imply that players would never choose strictly dominated strategies. The first contributions in this sense are Pearce (1984) and Bernheim (1984), in which a procedure is devised to eliminate players' strategies that are not *rationalizable* (i.e., not supported by internally consistent beliefs about other players' choices and beliefs). In general, certain epistemic conditions may be proved to justify all the strategy profiles yielded by a certain solution concept—and only those—hence providing an *epistemic foundation* for that solution concept. For example, in Aumann and Brandenburger (1995) it is proved that, for two-person games, mutual knowledge (i.e., first-order knowledge among all the players) of the structure of the game, of rationality, and of the players' strategies implies that that the strategies constitute a Nash equilibrium of the game.

2.1 Correlated Equilibrium

We have assumed so far that players' strategies are independent, as though each player receives a private, independent signal and chooses a (mixed) strategy after having observed her own signal. However, signals need not be independent. For example, players can *agree* to play a certain strategy according to the outcome of some external observed event (e.g., a coin toss). If the agreement is self-fulfilling, in that players have no incentive to deviate from it, the resulting strategy profile is a strategic equilibrium in correlated strategies or, in short, a correlated equilibrium (cf. Aumann 1974, 1987). For any Nash equilibrium in mixed strategies, a correlation device can be set such that it indicates a probability distribution over the possible outcomes of the game yielding such an equilibrium profile. If the correlation signal is common knowledge among the players, we speak of *perfect* correlation. However, players may correlate their strategies according to *different* signals (less than perfect correlation). The idea is that players have information partitions whose cells comprehend more than one possible outcome because they ignore which signals are received by other players. To represent the fact that players receive different signals (i.e., they ignore which strategies will be chosen by other players), it is required that in every cell of the information partition of player i, her strategy does not change. It is then possible to calculate the expected payoff of playing the strategy indicated by the correlation device versus the expected payoff yielded by playing a different strategy. If the players have no incentive not to play the indicated strategy, the profile yielded by the correlation device is an equilibrium. Correlation by means of private signals may yield outcomes that are more efficient than those of the optimal Nash equilibrium. An important philosophical application of correlated equilibrium is found in Vanderschraaf (1998, 2001), in which convention as defined in Lewis (1969) is shown to be the correlated equilibrium of a coordination game. Gintis (2009) puts correlated equilibrium at the very core of his unified theory of social sciences.

3 RPEATED GMES

We saw in Section 1 that, in the Prisoner's Dilemma, rational players choose to defect. Since mutual defection is the socially inefficient outcome of the interaction, this observation has prompted many philosophers to speculate that it is in fact *irrational* to defect in Prisoner's Dilemma and, furthermore, that agents are *morally* required to cooperate. Game-theoretic rationality is understood as maximization of payoffs, and payoffs are expressed in Von Neumann-Morgenstern utility; thus, game-theoretic rationality amounts to consistency of preferences. Given such a strict definition of rationality, cooperation is ruled out as a viable outcome between rational players in Prisoner's Dilemma. This, however, does not entail that game theory is unfit to examine, both from a descriptive and a justificatory point of view, the issues of cooperation and reciprocity. In fact, game theory can provide us with a thorough understanding of the characteristics of cooperative social arrangements (or, at least, of toy models of social institutions), as well as with a description of how such arrangements may come about. The crucial idea is that it is wrong to assume that cooperation and morality can be examined through the analysis of *one-shot* games and, in particular, prisoner's dilemma interactions. In order to understand how cooperation is possible, we need to look at the same agents repeatedly interacting in Prisoner's Dilemma-like situations. Were we to find that self-sustaining cooperative equilibria exist, we would be observing an important first step toward a game-theoretic justification and description of the mechanisms of reciprocity in society. First, we need to distinguish repeated interaction with a *finite* horizon and repeated interaction with an *infinite* horizon. In the case of the repeated Prisoner's Dilemma, in the former case, we only have one, inefficient equilibrium of the repeated game, whereas in the latter case we have a huge set of equilibrium outcomes—in fact, a huge set of subgame perfect equilibrium outcomes.

3.1 Finitely Repeated Games

In repeated interactions of the Prisoner's Dilemma, cooperative outcomes are viable outcomes. My cooperating today may bring about your cooperation tomorrow. Or the threat of your responding to my defection today by retaliating against me tomorrow could prevent me from choosing to defect. However, we have to be careful in considering the nature of the repeated interaction. In fact, if the interaction has a well-defined horizon, and this is known by the parties, then cooperation is impossible for rational agents who know that they are rational, just as it is impossible in the one-shot case. Informally, the argument goes as follows: in a non-anonymous repeated Prisoner's Dilemma interaction, a player has an incentive to cooperate *insofar as* her cooperation may lead the other player to cooperate in turn. Of course, if the game is repeated, say, ten times, at round ten of play, no player has an incentive to cooperate because round ten is the final round of the overall interaction and players know this. Hence, at round

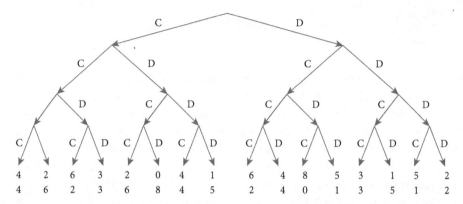

FIGURE 14.8 Extensive form representation of a twice-repeated Prisoner's Dilemma.

ten, both players will defect. This entails that, whatever a player's action at round nine is, the other player is going to defect at round ten. Hence, cooperation at round nine cannot possibly bring about cooperation at round ten, and hence both players defect at round nine. Cooperation at round eight is ruled out by similar reasoning. Repeating the argument backward for all rounds, it follows that players will defect at round one and at all subsequent rounds. This should not come as a surprise. A repeated interaction with a finite horizon of n rounds is but a game of perfect information. In fact, we can give an extensive form representation of such an interaction: see Figure 14.8 for the game in Figure 14.5 and the case in which $n = 2$, where the payoffs are calculated as the simple *sum* of the payoffs received in each of the two Prisoner's Dilemma iterations.

As we know from Section 1, all perfect information games are solvable by backward induction, and the finitely repeated Prisoner's Dilemma represents no exception. It is easy to see that, in the twice-repeated game, D still strictly dominates C, and backward induction solves the game showing that defection at every round is the Nash equilibrium of the interaction.

Formally, to define a repeated game Γ, we first consider a game, called the *stage game*, where I is, as usual, the set of players, A_i is the set of actions available in the stage game to player i, and v_i is the payoff function in the stage game for player i. The stage game remains fixed for the entire duration of the repeated game (sometimes called a *super-game*). At each repetition, each player has the same actions available to her, and each outcome yields the same payoffs. We call the players' pure strategies in the stage game *actions* to differentiate them from the players' pure strategies in the supergame. Given that the elements of the stage game are set, what are the elements of the supergame? We will assume that the players of the supergame are the same at each stage game (i.e., players do not change during the repeated interaction). What are the overall pure strategies available to them? What are their overall payoffs?

A pure strategy determines a player's action in all contingencies (even those that do not materialize); therefore, a pure strategy in the supergame is a function of the history of play. Let $a^t = \left(a_1^t, \ldots, a_n^t\right)$ be the action profile played at round t, and define a^0 to be Ø. A *history* of play at t is $h^t = \left(a^0, \ldots, a^{t-1}\right)$. Let H^t be the set of all possible histories of play at t. We then define $H = \bigcup_{t=0}^{\infty} H^t$ to be the set of all possible histories. As in extensive form games, a pure strategy for player i is a function from H to A_i specifying an action following each history of play. Each pure (or, in fact, mixed) strategy profile s for the repeated game induces an infinite sequence of action profiles for the stage game as follows: let $a = (a^0, a^1, \ldots)$ denote an infinite sequence of action profiles and let a^t as defined earlier denote the history at t based on a; then $a(s)$ is induced by s recursively:

$$a^0\left(s\right) = \left(s_1\left(\varnothing\right), \ldots, s_n\left(\varnothing\right)\right)$$

$$a^1\left(s\right) = \left(s_1\left(a^0\left(s\right)\right), \ldots, s_n\left(a^0\left(s\right)\right)\right)$$

$$a^2\left(s\right) = \left(s_1\left(a^0\left(s\right), a^1\left(s\right)\right), \ldots, s_n\left(a^0\left(s\right), a^1\left(s\right)\right)\right)$$

$$\ldots$$

Intuitively, at stage 0, the actions induced by s are played. At stage 1, the action profile a^0 constituting history h^0 is observed, and the action profile specified by s is played. At stage 2, the sequence of action profiles $(a^0, a^1 = h^1)$ is observed, and the action profile specified by s is played, and so on. Therefore $a(s)$ gives rise to an infinite stream of payoffs. In particular, for player i, we have that $a(s)$ gives rise to the stream

$$v_i\left(a^0\left(s\right)\right), v_i\left(a^1\left(s\right)\right), v_i\left(a^2\left(s\right)\right), \ldots$$

The overall payoff for i is based on the average of the payoffs for each stage game: $\sum_{t=0}^{T} \dfrac{v_i\left(a^t\left(s\right)\right)}{T}$, where a^t is the action profile induced by s at t and T is the total number of repetitions. However, the sum has to be conveniently discounted. There are two reasons why we need to think in terms of discounted sums. The first is that we realistically assume that players are not endowed with infinite patience and that a payoff of x utils today is worth to them more than a payoff of x utils a number of rounds in the future. The second is that we may realistically assume that there is no guarantee that the stage game just played will be repeated at the next round, but rather there is some (determinate) probability that it will be repeated. The players' patience and the probability of repetition are understood in terms of a discount factor applied to the stream of payoffs. Suppose that players discount the future at a constant rate of $0 \leq \delta \leq 1$ each round. Suppose the game is repeated T times. Player i's utility will then be

$$u_i\left(s\right) = \sum_{t=0}^{T} \frac{v_i\left(a^t\left(s\right)\right)\delta^{t-1}}{T}.$$

What if the game is infinitely repeated? In that case we have that $V_i = \sum_{t=0}^{\infty} \delta^{t-1} v_i\left(a^t(s)\right)$.
A stream of payoffs $(c,c,c,...) = \dfrac{c}{1-\delta}$ is as good as V_i for i if $c = (1-\delta)V_i$. Thus, we define

$$u_i(s) = (1-\delta)\sum_{\infty}^{t=0} v_i\left(a(s)\right)$$

3.2 Indefinitely Repeated Prisoner's Dilemma

Depending on how patient the players are or, equivalently, on how probable continuation of play is, we see that there are in fact a myriad possible cooperative outcomes of the Prisoner's Dilemma when the horizon of the repetitions is infinite or indefinite. Assume that at each round there is a probability p that the next round will take place. Intuitively, cooperation is now justifiable since there is a chance that cooperation at the current stage will affect the other player's behavior in the next stage.

As an example, consider the following strategy for the indefinitely repeated Prisoner's Dilemma, traditionally called *grim trigger*. The player adopting the trigger strategy begins by cooperating at stage one, and she cooperates at each round if she has observed her opponent cooperate in the previous round. Were the opponent ever to defect, the trigger strategy switches to defection and never contemplates cooperation again. What happens if a player using the trigger strategy meets another player adopting the same strategy? They both start off by cooperating, and, since cooperation is observed at each round, they continue cooperating at each subsequent iteration of the stage game. Is this sustainable for rational players? It depends on the probability that the game will end at each stage. To see why cooperation can be achieved, suppose that, after each repetition, there is a probability of 1/3 that this is the last stage. Then, the probability that the game will persist until at least the Nth stage is $(2/3)^N$. Note that the probability that the game is eternal is zero, since $(2/3)^N \to 0$ as $N \to \infty$. Now consider the following "trigger" strategy S: "Play C as long as the other plays C, but if he plays D, play D forever after." If both players adopt S, they will cooperate forever. Player i's expected payoff of playing C at each stage is $3 + 3(2/3) + ... + 3(2/3)^{N-1} + 3(2/3)^N + 3(2/3)^{N+1} + ...$ Can a player gain by deviating? Suppose player i plays D at the Nth stage. If the other player plays S, the deviant player will at most get the following payoff $D = 3 + 3(2/3) + ... + 3(2/3)^{N-1} + 4(2/3)^N + 1(2/3)^{N+1} + ...$ For a deviation to be unprofitable, it must be the case that the expected payoff of continuous cooperation (enforced by S) is greater or equal to the expected payoff of, say, deviating at the Nth stage. In our example, $C - D = (3-4)(2/3)^N + (3-1)(2/3)^{N+1} + (3-1)(2/3)^{N+2} + ...$, which amounts to $(2/3)(-1+4)$. Since this value is greater than zero, this means that the expected payoff of continuous cooperation is better than the expected payoff of deviating at the Nth stage. Therefore (S, S) is a Nash equilibrium. In fact, we know that in our game any feasible expected payoff combination that gives each player more than 1 can be sustained in equilibrium. This result is known as the *folk theorem*, so called because it is uncertain who proved it first but it has become part of the folk literature of game theory. What is rational in a one-shot interaction may not be rational when the game is repeated, provided certain conditions obtain. The conclusion we may draw is that phenomena such as cooperation, reciprocation, honoring commitments, and keeping one's promises can be explained as the outcome of rational, self-interested choices. In repeated interactions, even a person who cares only about material, selfish incentives can see that it is in his interest to act in a cooperative way.

4 EVOLUTIONARY GAME THEORY

A Nash equilibrium need not be interpreted as a unique event. If we think of it as an observed regularity, we want to know by what process such an equilibrium is reached and what accounts for its stability. When multiple equilibria are possible, we want to know why players converged to one in particular and then stayed there. An alternative way of dealing with multiple equilibria is to suppose that the selection process is made by nature.

Evolutionary theories are inspired by population biology (e.g., Maynard Smith 1982). These theories dispense with the notion of the decision-maker, as well as with best responses/optimization, and use in their place a natural selection, "survival-of-the-fittest" process together with mutations to model the frequencies with which various strategies are represented in the population over time. In a typical evolutionary model, players are preprogrammed for certain strategies and are randomly matched with other players in pairwise repeated encounters. The relative frequency of a strategy in a population is simply the proportion of players in that population who adopt it. The theory focuses on how the strategy profiles of populations of such agents evolve over time, given that the outcomes of current games determine the frequency of different strategies in the future.

As an example, consider the game in Figure 14.9 and suppose that there are only two possible behavioral types: "hawk" and "dove."

A hawk always fights and escalates contests until it wins or is badly hurt. A dove sticks to displays and retreats if the opponent escalates the conflict; if it fights with another dove, they will settle the contest after a long time. Payoffs are expected changes in fitness due to the outcome of the game. Fitness here means just reproductive success (e.g., the expected number of offspring per time unit). Suppose injury has a payoff in terms of loss of fitness equal to C, and victory corresponds to a gain in fitness B. If hawk meets hawk, or dove meets dove, each has a 50% chance of victory. If a dove meets another dove, the winner gets B and the loser gets nothing, so the average increase in fitness for a dove meeting another dove is $B/2$. A dove meeting a hawk retreats, so her fitness is unchanged, whereas the hawk gets a gain in fitness B. If a hawk meets another hawk, they escalate until one wins. The winner has a fitness gain B, the loser a fitness loss C. So the average increase in fitness is $(B-C)/2$. The latter payoff is negative, since we assume the cost of injury is greater than the gain in fitness obtained by winning the contest. We assume that players will be randomly paired in repeated encounters, and in each encounter they will play the stage game of Figure 14.9.

If the population were to consist predominantly of hawks, selection would favor the few doves because hawks would meet mostly hawks and end up fighting with an average loss in fitness of $(B-C)/2$, and $0 > (B-C/2)$. In a population dominated by doves, hawks would spread, since every time they meet a dove (which would be most of the time), they would have a fitness gain of B, whereas doves on average would only get $B/2$. Evolutionary game theory wants to know how strategies do on average when games are played repeatedly between individuals who are randomly drawn from a large population. The average payoff to a strategy depends on the composition of the population, so a strategy may do very well (in term of fitness) in one environment and poorly in another. If the frequency of hawks in the population is q and that of doves correspondingly $(1-q)$, the average increase in fitness for the hawks will be $q(B-C)/2+(1-q)B$, and $(1-q)B/2$ for the doves. The average payoff of a strategy in a given environment determines its future frequency in the population. In our example, the average increase in fitness for the hawks will be equal to that for the doves when the frequency of hawks in the population is $q = B/C$. At that frequency, the proportion of hawks and doves is stable. If the frequency of hawks is less that B/C, then they do better than doves and will consequently spread; if their frequency is larger than B/C, they will do worse than doves and will shrink.

Note that if $C > B$, then $(B-C)/2 < 0$, so the game in Figure 14.9 has two pure-strategy Nash equilibria: (H, D) and (D, H). There is also a mixed strategy equilibrium in which Hawk is played with probability $q=B/C$ and Dove is played with probability $(1-q)=C-B/C$. If the game of Figure 14.9 were played by rational agents who *choose* which behavior to display, we would be at a loss in predicting their choices. From common knowledge of rationality and of the structure of the game, the players cannot infer that a particular equilibrium will be played. In the hawk/dove example, however, players are not rational and do not choose their strategies. So, if an equilibrium is attained, it must be the outcome of some process very different from rational deliberation. The process at work is natural selection: high-performing strategies increase in frequency whereas low-performing strategies' frequency diminishes and eventually goes to zero.

dove/hawk	H	D
H	(B-C)/2, (B-C)/2	B, 0
D	0, B	B/2, B/2

FIGURE 14.9 The Hawk–Dove game.

We have seen that in a population composed mostly of doves, hawks will thrive, and the opposite would occur in a population composed mainly of hawks. So, for example, if "hawks" dominate the population, a mutant displaying "dove" behavior can invade the population because individuals bearing the "dove" trait will do better than hawks. The main solution concept used in evolutionary game theory is the *evolutionarily stable strategy* (ESS) introduced by Maynard Smith and Price (1973). A strategy or behavioral trait is evolutionarily stable if, once it dominates in the population, it does strictly better than any mutant strategy, and hence it cannot be invaded. In the hawk/dove game, neither of the two pure behavioral types is evolutionarily stable because each can be invaded by the other. We know, however, that a population in which there is a proportion $q = B/C$ of hawks and $(1-q) = C-B/C$ of doves is stable. This means that the type of behavior that consists in escalating fights with probability $q = B/C$ cannot be invaded by any other type; hence, it is an ESS. An ESS is a strategy that, when it dominates the population, is a best reply against itself. Therefore, an evolutionarily stable strategy such as *(B/C, C–B/C)* is a Nash equilibrium. Although every ESS is a Nash equilibrium, the reverse does not hold; in our stage game, there are three Nash equilibria, but only the mixed strategy equilibrium *(B/C, C–B/C)* is an ESS.

Evolutionary games provide us with a way of explaining how agents that may or may not be rational and—if so—subject to severe information and calculation restrictions achieve and sustain a Nash equilibrium. Philosophical implications and applications can be found in Skyrms (1990, 1996, 2004). When there exist evolutionarily stable strategies (or states), we know which equilibrium will obtain without the need to postulate refinements in the way players interpret off-equilibrium moves. Yet we need to know much more about the processes of cultural transmission and to develop adequate ways to represent payoffs so that the promise of evolutionary games is actually fulfilled.

References

Aumann, R. (1974). "Subjectivity and Correlation in Randomized Strategies." *Journal of Mathematical Economics* 1: 67–96.

Aumann, R. (1976). "Agreeing to Disagree." *The Annals of Statistics* 4: 1236–1239.

Aumann, R. (1987). "Correlated Equilibrium as an Expression of Bayesian Rationality." *Econometrica* 55: 1–18.

Aumann, R. (1999). "Interactive Epistemology I: Knowledge." *International Journal of Game Theory* 28: 263–300.

Aumann, R., and Brandenburger, A. (1995). "Epistemic Conditions for Nash Equilibrium." *Econometrica* 63: 1161–1180.

Basu, K. (1990). "On the Non-existence of a Rationality Definition for Extensive Games." *International Journal of Game Theory* 19: 33–44.

Battigalli, P., and Bonanno, G. (1999). "Recent Results on Belief, Knowledge and the Epistemic Foundations of Game Theory." *Research in Economics* 503: 149–226.

Bernheim, B. (1984). "Rationalizable Strategic Behavior." *Econometrica* 52: 1007–1028.

Bicchieri, M. C. (1989). "Self-Refuting Theories of Strategic Interaction: A Paradox of Common Knowledge." *Erkenntnis* 20: 69–85.

Bicchieri, M. C. (1993). *Rationality and Coordination*. (Cambridge: Cambridge University Press).

Bicchieri, M. C. (2006). *The Grammar of Society*. (Cambridge: Cambridge University Press).

Bicchieri, C. and Chavez, A. (2010). "Behaving as Expected: Public Information and Fairness Norms." *Journal of Behavioral Decision Making* 23(2): 161–178. Reprinted in M. Baurmann, G. Brennan, R. E. Goodin, and N. Southwood (eds.), *Norms and Values. The Role of Social Norms as Instruments of Value Realisation* (Baden-Baden: Nomos Verlag).

Bicchieri, C., and Sontuoso, A. (2014). "I Cannot Cheat on You After We Talk." In M. Peterson (ed.), *The Prisoner's Dilemma* (Cambridge: Cambridge University Press), 101–115.

Bicchieri, C., and Zhang, J. (2012). "An Embarrassment of Riches: Modeling Social Preferences in Ultimatum Games." In U. Maki (ed.), *Handbook of the Philosophy of Science*, Volume 13: Philosophy of Economics (Amsterdam: Elsevier).

Binmore, K. (1987). "Modeling rational players: Part I." *Economics and philosophy* 3(2): 179–214.

Bonanno, G. (1991). "The Logic of Rational Play in Games of Perfect Information." *Economics and Philosophy* 7: 37–61.

Cubitt, R., and Sugden, R. (2003). "Common Knowledge, Salience, and Convention: A Reconstruction of David Lewis's Game Theory." *Economics and Philosophy* 19: 175–210.

Fagin, R., Halpern, J., Moses, Y., and Vardi, M. Y. (1995). *Reasoning About Knowledge*. (Cambridge, MA: MIT Press).

Fehr, E., and Schmidt, K. M. (1999). "A Theory of Fairness, Competition, and Cooperation." *Quarterly Journal of Economics* 114(3): 817–868.

Gintis, H. (2009). *The Bounds of Reason: Game Theory and the Unification of the Behavioral Sciences*. (Princeton, NJ: Princeton University Press).

Güth, W., Schmittberger, R., and Schwarze, B. (1982). "An Experimental Analysis of Ultimatum Bargaining." *Journal of Economic Behavior and Organization* 3(4): 367–388.

Kohlberg, E. (1990). "Refinement of Nash Equilibrium: The Main Ideas." In T. Ichiischi, A. Neyman, and Y. Tauman (eds.), *Game Theory and Applications*. (San Diego: Academic Press), 3–45.

Lewis, D. K. (1969). *Convention*. (Cambridge, MA: Harvard University Press).

Maynard Smith, J. (1982). *Evolution and the Theory of Games* (Cambridge: Cambridge University Press).

Maynard Smith, J., and Price, G. R. (1973). "The Logic of Animal Conflict." *Nature* 146: 15–18.

Meyer, J. -J., and van der Hoek, W. (2004). *Epistemic Logic for AI and Computer Science, new edition*. (Cambridge University Press: Cambridge).

Nash, J. (1996). *Essays on Game Theory*. (Cheltenham, UK: Edward Elgar).

Paternotte, C. (2011) "Being Realistic About Common Knowledge: A Lewisian Approach." *Synthese* 183(2): 249–276.

Pearce, D. (1984). "Rationalizable Strategic Behavior and the Problem of Perfection." *Econometrica* 52: 1029–1050.

Pettit, P., and Sugden, R. (1989). "The Backward Induction Paradox." *Journal of Philosophy* 4: 1–14.

Reny, P. (1992). "Rationality in Extensive Form Games." *Journal of Economic Perspectives* 6: 429–447.

Samet, D. (1996). "Hypothetical Knowledge and Games with Perfect Information." *Games and Economic Behavior* 17(2): 230–251.

Schelling, T. (1960). *The Strategy of Conflict*. (Cambridge: Harvard University Press).

Schipper B. C. (2014). "Awareness." In H. van Ditmarsch, J. Y. Halpern, W. van der Hoek, and B. Kooi (eds.), *Handbook of Epistemic Logic* (London: College Publications), 77–146.

Selten, R. (1965). "Spieltheoretische Behandlung eines Oligopolmodels mit Nachfragetragheit." *Zeitschrift fur die gesamte Staatwissenschaft* 121: 301–324.

Selten, R., and Leopold, U. (1982). "Subjunctive Conditionals in Decision and Game Theory." *Philosophy of Economics*. (Berlin/Heidelberg: Springer), 191–200.

Sillari, G. (2005). "A Logical Framework for Convention." *Synthese* 147: 379–400.

Sillari, G. (2006). "Models of Awareness." *Proceedings of The 7th Conference on Logic and the Foundations of Game and Decision Theory (LOFT)*, 209–218.

Sillari, G. (2008*a*). "Quantified Logic of Awareness and Impossible Possible Worlds." *The Review of Symbolic Logic* 1(4): 514–529.

Sillari, G. (2008*b*). "Common Knowledge and Convention." *Topoi* 27(1–2): 29–39.

Skyrms, B. (1990). *The Dynamics of Rational Deliberation*. (Cambridge: Harvard University Press).

Skyrms, B. (1996). *Evolution of the Social Contract*. (Cambridge: Cambridge University Press).

Skyrms, B. (2004). *The Stag Hunt and the Evolution of Social Structure*. (Cambridge: Cambridge University Press).

Stalnaker, R. (1994). "On the evaluation of solution concepts." *Theory and decision* 37(1): 49–73.

Van Damme, E. (1987). *Stability and Perfection of Nash Equilibria*. (Berlin: Springer).

Vanderschraaf, P. (1998). "Knowledge, Equilibrium, and Convention." *Erkenntnis* 49: 337–369.

Vanderschraaf, P. (2001). *Learning and Coordination: Inductive Deliberation, Equilibrium and Convention* (New York: Routledge).

Vanderschraaf, P., and Sillari, G. (2014). "Common Knowledge." In E. N. Zalta (ed.), *The Stanford Encyclopedia of Philosophy* (Spring 2014 Edition): http://plato.stanford.edu/archives/spr2014/entries/common-knowledge/.

CHAPTER 15

....

INSTRUMENTALISM

Global, Local, and Scientific

....

P. KYLE STANFORD

[A]ll thought processes and thought-constructs appear a priori to be not essentially rationalistic, but biological phenomena. . . .. Thought is origi- nally only a means in the struggle for existence and to this extent only a biological function.

—Hans Vaihinger, *The Philosophy of 'As If'* (xlvi)

1 PRELUDE: INSTRUMENTALISM, THE VERY IDEA

....

THE leading idea of instrumentalism is that ideas themselves, as well as concepts, theo- ries, and other members of the cognitive menagerie (including the idea of instrumen- talism itself, of course) are *most* fundamentally tools or instruments that we use to satisfy our needs and accomplish our goals. This does not imply that such ideas, theo- ries, and the like cannot also be truth-apt or even true, but simply that we misunder- stand or overlook their most important characteristics—including the most important questions to ask about them—if we instead think of them most fundamentally as candi- date descriptions of the world that are simply true or false. Indeed, the American prag- matist John Dewey originally coined the term "instrumentalism" to describe his own broad vision of human beings as creatures whose cognitive activities are much more deeply entangled with our practical needs and our attempts to successfully navigate the world and its challenges than we usually recognize, creatures whose efforts to engage the world intellectually must proceed using cognitive tools that are no less a product of and no less conditioned by our long history of seeking to meet such practical needs and objectives than are the arms, legs, and eyes we use to find food or shelter. A natural contrast here is with the tradition of Cartesian rationalism in the Early Modern period, which seemed (at least in caricature) to presuppose (and then discover to its evident surprise) that we are most essentially creatures of pure disembodied intellect and that the most fundamental question to ask about our ideas or beliefs is therefore whether or not they accurately describe or represent how things stand not only in our immediate physical environment, but also in far more remote realms of concern like pure mathematics and theology. Like empiricists, pragmatists view this rationalist tradition as having gone wrong at the very first step, having tried to derive substantive conclusions about how things stand in the world around us from the ideas we encounter in introspection without first asking just what those ideas are and how we came to have them in the first place. But where the empiricists simply proposed a competing (and deservedly influential) conception of where our ideas or beliefs come from and how they provide us with knowledge of the world when they do, pragmatists went on to defend a fundamentally and systematically distinct conception of the very point, purpose, role, and/or function of cognitive entities like ideas or theories and cognitive states like belief in the first place.

Indeed, the most enduring legacy of American pragmatism has been an influential philosophical account of truth that embodies this broad view of ideas and beliefs as instruments for satisfying our needs and goals. As pragmatist thinkers went on to emphasize, however, the centrality and significance they ascribed to understanding the role that such ideas or beliefs play in guiding our practical interactions with the world does not *compete* with the possibility that those same ideas and beliefs might "correspond to" or "agree with" reality. What they argued instead was that such verbal formulae serve simply to mask or even obscure the need for investigating, as William James was fond of putting the point, what truth is "known-as." Such pragmatists held that "truth" is simply the name we give to what *works* for us in the cognitive arena, to the beliefs, ideas, theories, or other cognitions that do or would enable us to most effectively and efficiently satisfy our needs and realize our practical goals, whether or not we have yet managed to identify which particular cognitions those are. The verbal formula of "correspondence to" or "agreement with" reality certainly represents another way to pick out such ideas and beliefs, but it is extraordinarily misleading and unhelpful as a philosophical *theory* of truth because it makes a mystery out of both the nature of and our access to this supposed "correspondence" and, in the process, serves to obscure the central roles that thinking and talking about truth and falsity actually play in our cognitive engagement with the world. Such pragmatists argued that what we point to as evidence of the falsity of a belief invariably turns out to be one or more ways in which it fails to fully satisfy one or more of an extremely broad spectrum of our practical needs, concerns, and interests, including the need to effectively integrate that belief with others to guide our actions. Thus, when James famously argues that "the true" is only the expedient in our way of thinking, he hastens to add

> expedient in almost any fashion; and expedient in the long run and on the whole, of course; for what meets expediently all the experience in sight won't necessarily

meet all further experiences equally satisfactorily. Experience, as we know, has ways of *boiling over*, and making us correct our present formulas. ([1907] 1978, 106)

Accordingly, true beliefs are not those that correspond to states of the world that are somehow independent of how we conceive of or conceptually engage with it (a correspondence whose very intelligibility seems open to question) but are instead those that correspond to the members of a privileged collection of beliefs that are specified or picked out in a distinctive way. Just how this privileged collection should be picked out was a matter of considerable and enduring controversy: C. S. Peirce suggested, for instance, that it was those that would be embraced by an ideal set of inquirers at the end of an idealized inquiry, whereas James himself held that it was the set of beliefs that no further experience would ever incline us to abandon. But, most fundamentally, pragmatists regarded the truth or falsity of any given belief as a matter of the correspondence between that belief and the members of a set of such beliefs that would maximally satisfy our embedded, situated, and unavoidably human needs and desires rather than the match between that belief and some raw, unconditioned, or unconceptualized reality. Of course, the pragmatists' philosophical opponents immediately accused them of simply conflating what is useful or pleasing to us in the way of belief with what is true, and the rest is history.

2 INSTRUMENTALISM GOES LOCAL: DEBATES CONCERNING SCIENTIFIC REALISM

Note that this pragmatist version of instrumentalism is a global doctrine: it asserts a distinctive view of ideas, beliefs, concepts, and the like *in general*. But some philosophers have been strongly attracted by the idea that we might embrace more *local* versions of the fundamental instrumentalist conception of cognitive entities or states, seeing it or something very like it as articulating the right view to take of just some specific class or category of those entities and states. In particular, the idea of embracing a localized form of instrumentalism has been persistently attractive to critics of the "scientific realist" view that the incredible practical and epistemic achievements of our best scientific theories should lead us to think that those theories must be at least probably and/or approximately true. Debates concerning scientific realism are as old as science itself, but in our own day those who resist such realism are typically (although not especially helpfully) characterized as "antirealists." This heterogeneous category includes a motley collection of suspicious characters, undesirables, and degenerates with a wide variety of grounds for doubting whether we should or must join the realist in regarding even our best scientific theories as even approximately true. But prominent among them are what I will call "scientific instrumentalists" who argue that we should instead regard *scientific theories in particular* merely as powerful cognitive instruments or tools.

The influential attempts of logical positivist and logical empiricist thinkers to articulate such scientific instrumentalism in the early and middle decades of the twentieth century often did so by proposing a distinctive analysis of the semantic content or role of theoretical discourse in science. Ernst Mach suggested, for example, that the point of such theoretical discourse was simply to "replace, or *save*, experiences, by the reproduction and anticipation of facts in thought" ([1893] 1960, 577), and a law of nature such as Snell's "law of refraction is a concise, compendious rule, devised by us for the mental reconstruction of" large numbers of such observable facts or experiences ([1893] 1960, 582). The early Rudolph Carnap argued explicitly that the very meaning of theoretical scientific claims was simply exhausted by what we usually think of as the observable implications of those claims, and he devoted considerable effort and ingenuity to the attempt to actually carry out a convincing general reduction of the language of theoretical science to such a privileged phenomenological or observational basis. But these efforts rapidly encountered a daunting collection of both technical and philosophical obstacles, and this reductive project was ultimately abandoned even by its original architects including, most influentially, Carnap himself.

Later logical empiricist thinkers would respond to the failure of this attempted reduction by proposing alternative forms of scientific instrumentalism that nonetheless persisted in attributing a distinctive semantic role or linguistic function specifically to the claims of theoretical science. One particularly influential such alternative proposed, for example, that theoretical scientific claims were not even *assertoric*, insisting that such claims instead functioned simply as "inference tickets" allowing us to infer some observable states from others (or the truth of some observational claims from others), rather than themselves asserting anything at all or (therefore) even possessing truth values. Ernst Nagel famously argued, however, that this somewhat desperate semantic maneuver simply eviscerated any distinction between scientific realism and instrumentalism, suggesting that there was a "merely verbal difference" between the claim that a theory functions as a reliable "inference ticket" between some observable states and others and the supposedly competing realist contention that the theory in question is simply true (1961, 139).

Another alternative sought to avoid such counterintuitive construals of the semantic content of theoretical claims by proposing instead that although such claims are genuinely assertoric and their meaning is not reducible to that of claims about observations or observation statements, they can nonetheless be eliminated without loss from our scientific discourse. This proposal was supported by an influential theorem of William Craig (1953) showing that if we start with any recursively axiomatized first-order theory (T) and an effectively specified subvocabulary of that theory (O) that is exclusive of and exhaustive with the rest of the theory's vocabulary, we can then effectively construct a further theory (T') whose theorems will be all and only those of the original theory containing no nonlogical expressions in addition to those in the specified subvocabulary.

As Carl Hempel went on to point out in connection with his influential "theoretician's dilemma," if we restrict the relevant subvocabulary of T to its "observational" terms, Craig's Theorem thus establishes that there is a "functionally equivalent" alternative to T that eliminates all nonobservational vocabulary but nonetheless preserves any and all deductive relationships between observation sentences expressed by T itself. In that case, Hempel noted, "any chain of laws and interpretive statements establishing [definite connections among observable phenomena] should then be replaceable by a law which directly links observational antecedents to observational consequents" (Hempel [1958] 1965, 186).

The significance of this result was immediately challenged, however, once again most famously by Ernst Nagel, who pointed out that the axioms of any such "Craig-transform" T' would be infinite in number (no matter how simple the axioms of T), would correspond one-to-one with all of the true statements expressible in the language of T, and could actually be constructed only after we already knew all of those true statements expressible using the restricted observational subvocabulary of T. In more recent decades, the challenges facing such semantic and/or eliminative forms of instrumentalism have only increased in severity and number: philosophers of science have come to recognize an increasingly wide range of profound differences between actual scientific theories and the sorts of axiomatic formal systems to which tools like Craig's Theorem can be naturally applied, and such phenomena as the "theory-ladenness of observation" have generated considerable skepticism regarding any attempt to divide the language or vocabulary of science into "theoretical" and "observational" categories in the first place.

Although this history makes the prospects for attempting to develop scientific instrumentalism by means of a distinctive semantic or eliminative analysis of the theoretical claims of science appear exceedingly dim, this strategy always represented just one possible way of articulating the fundamental instrumentalist idea that our best scientific theories are cognitive tools or instruments rather than accurate descriptions of otherwise inaccessible domains of nature. More recently, philosophers of science attracted by this fundamental idea have largely abandoned dubious proposals concerning the meaning of our theoretical discourse or the eliminability of that discourse from science altogether and instead tried to develop scientific instrumentalism by suggesting that although the claims of our best scientific theories mean just what they seem to and cannot be eliminated from science, we nonetheless do not have sufficient grounds for *believing* many of those claims when they are so regarded. That is, whether motivated by pessimistic inductions over the history of science, worries about the underdetermination of theories by the evidence, or something else altogether, such distinctively *epistemic* versions of scientific instrumentalism argue that we need not *believe* everything that our best scientific theories (really do) say about the world in order to use them effectively as tools for navigating that world and guiding our practical interactions with it. (For a broad discussion of the most influential motivations for such epistemic instrumentalism, see Stanford [2006, chap. 1].)

Such epistemic scientific instrumentalists cannot, however, see themselves as simply applying the pragmatist's global instrumentalist attitude in a more local or restricted way. The global instrumentalist holds that cognitive entities like ideas and theories are best conceived *quite generally* as tools or instruments we use to make our way in the world, and she insists that this conception does not compete with the possibility that those same cognitive entities might be true. By contrast, the epistemic scientific instrumentalist *denies* that the admittedly instrumentally useful theories of contemporary science are also true, or at least that we have rationally compelling reasons for believing that they are (a subtlety I will henceforth leave aside for ease of exposition)—indeed, it is the scientific *realist* who holds that many or all of the theories of contemporary science are *both* instrumentally powerful *and* (at least approximately) true! Thus, where the global instrumentalist could happily concede that many of the beliefs concerning which she advocated her instrumentalism could also be correctly (although less helpfully) characterized as "corresponding to the world" or "agreeing with reality," the epistemic scientific instrumentalist insists instead that we should think of a particular set of our scientific beliefs simply as useful tools or instruments rather than thinking that they are true, and therefore the scientific instrumentalist cannot accept the global instrumentalist's view that the correspondence formula is simply an especially unhelpful or obscure way to pick out the most instrumentally powerful of these ideas or claims.

It would seem, then, that the epistemic scientific instrumentalist must face a question that simply never arose for the global instrumentalist: she will have to identify precisely *which* ideas, claims, or theories are those she regards as *merely* instrumentally useful rather than also corresponding to or agreeing with reality. But it might also seem that she has a natural and obvious response to this demand: after all, she is a *scientific* instrumentalist, so she might suggest that it is all and only the claims of *science* that she regards as merely instrumentally useful in this way. Unfortunately, this proposal cannot pick out the class of claims toward which she advocates her distinctive epistemic form of instrumentalism because that very instrumentalism recommends that we make effective use of our best scientific theories in practical contexts, and it would seem that to do so *just is* to believe at least some of what they tell us about the world. That is, it would seem that when we put our best scientific theories to good instrumental use we do so *by* believing the claims they make concerning such matters as how much fuel the rocket will need to reach orbit, which drug will prevent transmission of the disease, and how existing weather patterns will change in response to global warming. The epistemic scientific instrumentalist therefore cannot regard the claims of science generally as merely instrumentally useful because she cannot make effective instrumental use of her best scientific theories without simply believing at least some of what they say about the world to be true.

Recognizing this problem suggests a natural refinement of this proposal, however. We might suggest instead that epistemic scientific instrumentalists accept the *predictions* and *recipes for intervention* offered by our best scientific theories, but not the *descriptions* of otherwise inaccessible parts of nature that they offer. Indeed, this proposal seems to capture the broad flavor of a number of prominent and influential forms of epistemic scientific instrumentalism. Thomas Kuhn famously denies, for example, that successive theoretical representations of some natural domain provide "a better representation of what nature is really like," but nonetheless holds that a later theory will typically be "a better instrument for discovering and solving puzzles," offering more impressive "puzzle-solutions and . . . concrete predictions" ([1962] 1996, 206) than its historical predecessors. Similarly, Larry Laudan argues that the scientific enterprise is progressive because our theories improve over time in their ability to solve empirical and conceptual problems, but he nonetheless forcefully denies that this is because such theories are more closely approximating the truth about nature itself (1977, 1996). And Bas van Fraassen's influential constructive empiricism (1980) holds that we should take our best scientific theories to be "empirically adequate," meaning simply that the claims they make about observable matters of fact are true. To whatever extent solving Kuhn's puzzles, addressing Laudan's problems, or exhibiting van Fraassen's empirical adequacy involve predicting and intervening in the world around us, these suggestions would seem to embody the broad idea that what we should believe are the predictions and recipes for intervention provided by our best scientific theories but not the descriptions of otherwise inaccessible parts of nature that they offer.

Notwithstanding the widespread intuitive appeal of this proposal, however, it likewise fails to distinguish those claims that the epistemic scientific instrumentalist regards as merely instrumentally useful from those that she instead believes to be true. One important reason for this failure is that many of what we regard as a scientific theory's empirical predictions simply are descriptive claims about parts or aspects of nature that are difficult to investigate directly, a problem articulated in a characteristically elegant and enigmatic way by Howard Stein in paraphrasing Eugene Wigner's observation that one also "uses quantum theory, for example, to calculate the density of aluminum" (1989, 49). To illustrate Stein's point using a different example, we might note that some contemporary cosmological theories seek to explain the present rate of expansion of the universe by positing a field of "dark energy," and among the most important predictions they make are those that specify the characteristics of that hypothesized field. Perhaps even more importantly, however, the predictions and recipes for intervention generated by our best scientific theories concerning perfectly familiar entities and events like eclipses, earthquakes, and extinctions are made using precisely the same descriptive apparatus with which those theories characterize the world more generally. That is, what our best scientific theories actually predict are such phenomena as the occlusion of one celestial body by another, the shifting of the Earth's tectonic plates, or the elimination of all organisms belonging to a particular phylogenetic group, and such predictions cannot be treated as having a more secure claim to truth than the relevant theory's own description of nature. If we do not believe what a theory says earthquakes or eclipses are, how are we to even understand its predictions concerning when and where the next earthquake or eclipse will occur? Nor is it open to us to try to evade the problem by seeking to couch our predictions and recipes for intervention in a mythical "observation language" of instrument-needle readings and colored patches in the visual field supposedly devoid of any theoretical commitment whatsoever. Not only did the attempt to articulate or develop such a pure language of observation come to ruin (see earlier discussion), but even if we had such a language it would not suffice to characterize the earthquakes, eclipses, extinctions, and wide range of further empirical phenomena with respect to which the instrumentalist herself takes our best scientific theories to serve as effective tools for prediction and intervention.

It thus turns out to be considerably more difficult than we might have initially suspected for the epistemic scientific instrumentalist to specify just those claims she regards as merely instrumentally useful rather than true. But even if this problem can somehow be solved, another looms that would seem at least as difficult to surmount because critics of epistemic scientific instrumentalism have repeatedly suggested that there is simply no room to distinguish a sufficiently sophisticated commitment to the instrumental utility of our best scientific theories across the full range of instrumental uses to which we put them from the realist's own commitment to the truth of those same theories. Thus, to convince us that she is offering a coherent and genuinely distinct alternative to scientific realism, it seems that the epistemic scientific instrumentalist will have to be able to precisely specify not only which scientific claims are those toward which she adopts an instrumentalist attitude, but also what difference it makes for her to be an instrumentalist rather than a realist concerning those claims. The next section will examine this latter demand in greater detail before I go on to suggest that both of these foundational challenges can indeed be overcome if the epistemic scientific instrumentalist avails herself of what might seem a surprising source of assistance in characterizing the distinctive attitude she recommends toward some of even the most successful contemporary scientific theories.

3 FACING THE MUSIC: WHAT DIFFERENCE DOES IT MAKE?

The need for the scientific instrumentalist to clearly articulate the difference between regarding a given scientific claim or theory as a useful tool or instrument and simply believing that same claim or theory to be true arises largely in response to the persistent suggestion that any apparent substantive difference between these two possibilities simply dissolves under further scrutiny. Earlier, we saw Nagel raise this charge against the "inference ticket" version of semantic scientific instrumentalism popular with many of his contemporaries, but much the same criticism has been raised against epistemic versions of scientific instrumentalism as well. Paul Horwich (1991), for example, points out that some philosophical accounts of the nature of belief simply characterize it as the mental state responsible for use, and he suggests that epistemic instrumentalists are not entitled to conclude that their own position is really any different from that of their realist opponents until they show why such accounts of belief itself are mistaken. A much more detailed argument is offered by Stein (1989), who argues that once we refine both realism and instrumentalism in ways that are independently required to render them at all plausible in the first place, no room remains for any real difference between the resulting positions. He proposes that realists must give up both the idea that scientific theorizing can achieve reference or truth of any metaphysically transcendent or noumenal variety and the idea that any property of a scientific theory can explain its empirical success without simply describing the uses to which the theory itself has been put. For their part, he argues, instrumentalists must recognize that the instrumental functions of a scientific theory include not only calculating experimental outcomes but also representing phenomena adequately and in detail throughout the entire domain of nature to which that theory can be usefully applied and (especially) serving as our primary resource for further extending our inquiry into that domain successfully. But, he suggests, this process of sophisticating realism and instrumentalism in ways that are independently required to make each view plausible or appealing simultaneously eradicates any substantive difference between them.

The most detailed and systematic version of this challenge, however, is offered by Simon Blackburn (1984, 2002), who uses Bas van Fraassen's (1980) influential constructive empiricism as his representative form of scientific instrumentalism. Instead of believing our best scientific theories to be true, van Fraassen's constructive empiricist instead simply "accepts" them as "empirically adequate," which is to say that she believes their claims concerning observable matters of fact while remaining agnostic concerning their further claims regarding the unobservable. Blackburn quite rightly points out, however, that the acceptance van Fraassen recommends involves much more than simply using theories to predict observable outcomes:

> The constructive empiricist is of course entirely in favor of scientific theorising. It is the essential method of reducing phenomena to order, producing fertile models, and doing all the things that science does. So we are counselled to *immerse* ourselves in successful theory. . ... Immersion will include acceptance as empirically adequate, but it includes other things as well. In particular it includes having one's dispositions and strategies of exploration, one's space of what it is *easy* to foresee and what difficult, all shaped by the concepts of the theory. It is learning to speak the theory as a native language, and using it to structure one's perceptions and expectations. It is the possession of habits of entry into the theoretical vocabulary, of manipulation of its sentences in making inferences, and of exiting to empirical prediction and control. Van Fraassen is quite explicit that all of this is absolutely legitimate, and indeed that the enormous empirical adequacy of science is an excellent argument for learning its language like a native. . ... Immersion, then, is belief in empirical adequacy plus what we can call being "functionally organized" in terms of a theory. (Blackburn 2002, 117–119)

Blackburn thus sees van Fraassen's enthusiasm for our "immersion" in our best scientific theories as seeking to capture the wide and heterogeneous range of ways in which we make effective instrumental use of those theories, just as Stein suggested we must in order to render any form of scientific instrumentalism attractive. Like Stein, however, Blackburn further suggests that once the full range of such instrumentally useful functions is recognized, no room remains for any substantive difference between the constructive empiricist's "immersion" in or "animation" by our best scientific theories and the scientific realist's own commitment to the truth of those same theories:

> The problem is that there is simply no difference between, for example, on the one hand being animated by the kinetic theory of gases, confidently expecting events to fall out in the light of its predictions, using it as a point of reference in predicting and controlling the future, and on the other hand believing that gases are composed of moving molecules. There is no difference between being animated by a theory according to which there once existed living trilobites and believing that there once existed living trilobites. . . . What can we do but disdain the fake modesty: "I don't really believe in trilobites; it is just that I structure all my thoughts about the fossil record by accepting that they existed"? (Blackburn 2002, 127–128)

Here, Blackburn articulates the central challenge in an especially perspicuous way: once instrumentalists like van Fraassen have formulated the acceptance of, immersion in, or animation by a scientific theory in a way that recognizes the full range of useful instrumental functions that such theories perform for us, how will the acceptance, immersion, or animation they recommend be any different from simply believing those same theories to be true?[1]

Blackburn goes on to argue that although there are indeed genuine forms of variation in the character of our embrace of particular scientific theories that might seem to be appealing candidates for capturing the contrast between realist and instrumentalist commitments, none of these is available to van Fraassen to use in distinguishing the constructive empiricist's attitude from that of her realist counterpart. We might naturally distinguish, for example, the past and present empirical adequacy of a theory from its complete or total empirical adequacy, but van Fraassen's fully immersed constructive empiricist is no less committed to the ongoing or future empirical adequacy of a theory she accepts than she is to its past and present empirical adequacy. Although we might well have reasons to doubt that some particular theory that has been empirically adequate to date will remain so in the future, any room we recognize for drawing such a distinction will have to be reconstructed from *within* the constructive empiricist's own more general commitment to the empirical adequacy of the theories she accepts and therefore cannot constitute the *difference* between the constructive empiricist's commitments and those of her realist opponent. And the same would seem to apply to any potential variation in our commitment to the *ongoing* ability of a given scientific theory to solve Kuhn's puzzles or Laudan's empirical and theoretical problems.

---

[1] At times, van Fraassen seems to suggest that such "immersion" is required only for working scientists themselves, rather than for philosophical interpreters of scientific activity. But he nonetheless insists that such immersion remains perfectly consistent with adopting the constructive empiricist's instrumentalism, arguing that even the working scientist's "immersion in the theoretical world-picture does not preclude 'bracketing' its ontological implications" (1980, 81). Moreover, many aspects of the immersion van Fraassen recommends to working scientists are matters on which the philosophical interpreter cannot afford to remain agnostic in any case, such as the propriety of using a theory as the foundation for our further investigation.

Similarly, although it is perfectly natural to contrast full and unreserved acceptance of a theory with acceptance in a more cautious or tentative spirit, Van Fraassen's fully immersed constructive empiricist embraces a given scientific theory's empirical adequacy no less fully or confidently than the realist embraces its truth, so this sort of difference between more and less cautious acceptance cannot be what distinguishes realism from constructive empiricism. That is, the constructive empiricist does not believe any less confidently than the realist, but instead believes with equal confidence only a theory's claims about observable phenomena. Once again, although we might well have good reasons to embrace *either* the truth *or* the empirical adequacy of some particular theory with varying degrees of confidence, any such room for variation in the strength of our conviction would have to be recognized from *within* both the realist's and constructive empiricist's respective forms of commitment to a theory (i.e., to all of its claims or only to its claims about observable phenomena) and therefore cannot constitute the difference between them. And, once again, it seems that we will need to recognize the same room for variation in the degree or extent of our confidence in a theory's ability to solve Kuhnian puzzles or Laudanian problems.

It would seem, then, that epistemic forms of scientific instrumentalism face formidable obstacles not only, as we saw earlier, in precisely specifying those claims toward which such an instrumentalist attitude is appropriate, but also in recognizing the full range of ways in which we rely on our best scientific theories instrumentally without simply collapsing any distinction between such an instrumentalist attitude and realism itself. I now want to propose, however, that by taking advantage of what might seem a surprising source of assistance, the epistemic scientific instrumentalist can articulate the difference between the realist's epistemic commitments and her own in a way that addresses both of these fundamental challenges in a convincing fashion.

4 Singing a Different Tune: Scientific Realism and Instrumentalism Revisited

We might begin by noting that the fundamental idea that some scientific theories are useful conceptual instruments or tools despite not being even approximately true is one that the scientific realist needs no less than the instrumentalist; after all, this represents the realist's *own* attitude toward a theory like Newtonian mechanics. That is, the realist flatly rejects the claims of Newtonian mechanics concerning the fundamental constitution and operation of nature: she denies that space and time are absolute, she denies that gravitation is a force exerted by massive bodies on one another, and so on. But she knows perfectly well that we routinely make use of Newtonian mechanics to send rockets to the moon and, more generally, to make predictions and guide our interventions concerning the behavior of billiard balls, cannonballs, planets, and the like under an extremely wide (although not unrestricted) range of conditions.

We might begin by asking, then, what the realist means when she herself claims that Newtonian mechanics constitutes a useful conceptual tool or instrument that is not even approximately true. The answer, presumably, is that although she does not accept the theory's account of the fundamental constitution of nature, she nonetheless knows how to apply the theory just as a true believer would to a wide range of entities and phenomena whose existence she thinks can be established and that she thinks can be accurately characterized in ways that simply do not depend on Newtonian mechanics itself. That is, she can make use of other theories that she *does* regard as accurately describing the physical domain, as well as her own perceptual experience and perhaps other epistemic resources besides, to generate an *independent* conception of the billiard balls, cannonballs, and rockets to which she can then apply Newtonian mechanics, deploying the theoretical machinery of masses, forces, inelastic collisions, and the like to guide her prediction and intervention with respect to those independently characterized entities, processes, and phenomena. Such phenomena need not be observable, of course, as she knows how a Newtonian would characterize subatomic particles and their gravitational attractions in terms of masses and forces just as well as billiard balls and planets. And over *whatever* domain she believes the theory to be an instrumentally reliable conceptual tool, she can apply it just as a Newtonian would to guide her prediction and intervention concerning such *independently characterized* entities, events, and phenomena while nonetheless insisting that the theoretical description Newtonian mechanics gives of those entities, events, and phenomena is not even approximately true.

Indeed, the realist presumably takes this very same attitude toward other empirically successful theories of past science that are fundamentally distinct from contemporary theoretical orthodoxy. Of course, in the case of Newtonian mechanics, she can specify quite precisely just where she expects Newtonian mechanics to fail in application (and by how much), but this feature of the example is incidental, as is the fact that Newtonian mechanics is still actually used in a wide variety of engineering and practical contexts. What matters is that the realist herself regards Newtonian mechanics as a practically useful cognitive tool or instrument despite not being even approximately true, and it seems that she must regard this as an apt characterization of other empirically successful past theories that have been subsequently abandoned, whether or not they are still actually used and whether or not she can specify with mathematical precision what she expects the limits of the range or extent of their instrumental utility to be.

But, of course, this very same strategy is available to the scientific instrumentalist for characterizing her own attitude toward those theories she regards as "mere instruments." She, too, can characterize billiard balls, cannonballs, and planets and form straightforwardly factual beliefs about them by relying on whatever sources of information she has concerning them that are simply independent of Newtonian mechanics or any other theory toward which she adopts an instrumentalist stance. At a minimum, of course, she can rely on the evidence of her senses concerning such entities and phenomena. But, crucially, the same strategy remains open to her even if she accepts W. V. O. Quine's influential suggestion that the familiar middle-sized objects of our everyday experience are no less "theoretical" entities hypothesized to make sense of the ongoing stream of experience around us than are atoms and genes and that it is only by "the positing of the bodies of common sense" ([1960] 1976, 250) that we come to have any coherent picture of the world around us in the first place. If so, the instrumentalist will then simply need to decide just *which* theories are those toward which she will adopt an instrumentalist attitude, and the characteristics relevant to making this determination will surely depend on the more general reasons she has for adopting an instrumentalist attitude toward some or all of even our best scientific theories in the first place. If Quine is right, it may well be that a localized instrumentalism concerning any and all "theories" whatsoever is not a coherent possibility, but the epistemic scientific instrumentalist remains free to commit herself to realism concerning some theories (e.g., the hypothesis of the bodies of common sense) and instrumentalism concerning others in just the same way we found necessary in order to make sense of the realist's own commitments.

These reflections suggest that it was a mistake all along not only to hold the epistemic scientific instrumentalist responsible for defending the coherence of some exotic and unfamiliar cognitive attitude that she alone adopts toward a subset of scientific claims, but also to think of her as adopting this attitude toward any and all theories or theoretical knowledge as such. *Both* realists and instrumentalists regard some theories (e.g., the hypothesis of the bodies of common sense) as providing broadly accurate descriptions of entities and events in the natural world, and both regard some theories (e.g., Newtonian mechanics) merely as useful instruments for predicting and intervening with respect to entities, events, and phenomena as they can be characterized independently of those very theories. The thinkers we have traditionally called "instrumentalists" have simply been those prepared to take the latter attitude toward a much wider range of theories than their "realist" counterparts, including most saliently those contemporary scientific theories for which we are not currently in a position to articulate even more instrumentally powerful successors. That is, we have tended to reserve the term "instrumentalist" for someone who is willing to regard even an extremely powerful and pragmatically successful theory as no more than a useful instrument even when she knows of no competing theory that she thinks does indeed represent the truth about the relevant natural domain. But we all take instrumentalist attitudes toward some theories and not others, and it is the very same attitude that the realist herself adopts toward Newtonian mechanics (and other instrumentally powerful past scientific theories) that the instrumentalist is putting into wider service: scratch a scientific realist and watch an instrumentalist bleed!

In some cases, of course, scientific theories posit the existence of entities, processes, or phenomena to which we simply have no routes of epistemic access that are independent of the theory itself. For example, contemporary particle physics does not allow quarks to be isolated and therefore posits "gluons" to bind quarks within a proton, but our only point of epistemic contact with gluons or reason for thinking that they exist is the theory's insistence that *something* must play this role. Accordingly, an instrumentalist concerning particle physics will not believe any of its substantive claims concerning the existence and/or properties of gluons, although she will nonetheless typically be willing to make use of many of those claims in the course of arriving at new beliefs concerning entities, processes, and phenomena that she can characterize in ways that do not depend on contemporary particle physics or any other theories toward which she adopts an instrumentalist attitude.

Accordingly, although this account neither appeals to a mythical observation language devoid of any theoretical commitment whatsoever nor ascribes any foundational epistemic role to observability as such, it nonetheless recognizes that the empiricist's cherished epistemic resources of observation and perception more generally will often figure prominently among the ways we characterize those entities, processes, and phenomena (like earthquakes, eclipses, or extinctions) concerning which we think a given scientific theory is able to provide effective instrumental guidance. On this view of the matter, a scientist who finds a new way of detecting entities or phenomena posited by a theory, of creating them in the laboratory, or of demonstrating their causal influence on other entities or phenomena has achieved something extremely important even by the lights of those who are instrumentalists concerning the theory in question, for she has expanded the range of *independent* empirical phenomena concerning which we may regard that theory as an effective guide to prediction and intervention, sometimes in ways that are largely of theoretical interest and sometimes in ways that serve as the foundation for extraordinary technological and practical achievements. Thus, the tracks in a cloud chamber, the patterns on an electrophoresis gel, and the distinctive sour taste ascribed to acids by early chemists are all phenomena whose existence and central features can be characterized in ways that are, although not free of any theoretical commitments altogether, nonetheless independent of the commitments of the particular theories in whose terms scientific realists interpret them. If we are instead instrumentalists concerning any or all of those theories, these points of epistemic contact will help *constitute* our independent epistemic grasp of the entities, events, and phenomena concerning which we think the theory in question offers effective prediction, intervention, and instrumental guidance quite generally.

It is not hard to imagine, however, an objector who insists that a subtle incoherence lurks at the heart of the proposed parallel between the epistemic scientific instrumentalist's attitude toward some of even the most successful contemporary scientific theories and the realist's own attitude toward a theory like Newtonian mechanics. In the latter case, she might suggest, the merely instrumental character of the realist's commitment to the theory simply *consists in* her unwillingness to make use of Newtonian mechanics with unrestricted scope. She will instead use theories like general and special relativity to make predictions and guide her interventions when even very small errors might be consequential, in cases where the approximate predictive equivalence of the two theories is either unknown or is known to fail, and to ground her further theoretical investigation and exploration of the relevant natural domain. But such restrictions of scope cannot capture the difference between realism and instrumentalism regarding the best of our own contemporary scientific theories because, in such cases, we do not have any competing theory to whose truth (or even just general applicability) we are more fully committed that we might fall back to in these ways and/or under these circumstances. Thus, the objection goes, an instrumentalist attitude characterized by means of such a parallel will once again simply collapse back into realism itself in just those cases that actually divide scientific realists and instrumentalists.

This suggestion, however, ignores further crucial differences between the scientific realist's attitude toward the most powerful and successful theory we have concerning a given domain of nature and the form that even an extremely robust commitment to the mere instrumental utility of that same theory might take. Consider, for example, those scientific instrumentalists whose conviction is inspired in one way or another by reflection on the historical record of scientific inquiry itself. Such instrumentalists typically do *not* share the realist's expectation that the most powerful and successful theory we now have concerning a given domain of nature will retain that position indefinitely as our inquiry proceeds. Instead, such an instrumentalist expects that, in the fullness of time, even that theory will ultimately be replaced by a fundamentally distinct and still more instrumentally powerful successor that she is in no position to specify or describe in advance.

This expectation is, of course, connected to the two grounds we saw Blackburn recognize as intuitively plausible candidates for the difference between realism and instrumentalism that he argued were simply not available to van Fraassen's constructive empiricist: the distinction between a theory's empirical adequacy to date and its final or complete or future empirical adequacy and the distinction between embracing a theory fully and without reservation and embracing it in a more tentative or cautious spirit. Blackburn argued (quite rightly) that van Fraassen cannot characterize his constructive empiricist's instrumentalism in these terms because the constructive empiricist is no less committed to a theory's future empirical adequacy than to its past and present empirical adequacy, and she embraces this complete empirical adequacy of the theory (i.e., the truth of its claims about observable states of affairs) with no less confidence or conviction than the realist embraces its truth simpliciter; but the historically motivated scientific instrumentalist we are now considering simply does not share these commitments. She fully expects even the best conceptual tool we currently possess for thinking about a given natural domain to be ultimately discovered not to be fully empirically adequate and/or for future inquirers to eventually replace that tool with another that is even more instrumentally powerful and yet distinct from it in ways sufficiently fundamental as to prevent that successor from being counted as simply a more sophisticated, more advanced, or more completely developed version of existing theoretical orthodoxy. This instrumentalist's commitment to the ongoing instrumental utility of our best current theory is therefore *not* a commitment to its complete and total instrumental utility, and it is indeed systematically more cautious and tentative than that of the realist who believes that the theory itself is at least approximately true and therefore will not ultimately be replaced in this manner.

Blackburn may be right, then, to suggest that there is no room for a difference between van Fraassen's constructive empiricism and realism itself, but the grounds on which he rests this judgment help us to see why there are indeed profound differences between the provisional embrace of even the most powerful and impressive scientific theory we have concerning a given natural domain by a more historically motivated form of scientific instrumentalism and the scientific realist's own attitude toward that same theory. Moreover, these differences in turn produce a further and equally profound divergence concerning the actual pursuit of scientific inquiry itself: a scientific instrumentalist of this historically motivated variety will be systematically more sanguine than her realist counterpart concerning the investment of time, attention, energy, taxpayer dollars, and other limited resources in attempts to discover and develop theoretical alternatives that diverge in fundamental ways from or even directly contradict the most powerful and impressive scientific theory we have in a given natural domain. Although the realist might encourage such exploration as a way of further developing our current theory in an existing domain, the prospect that such exploration and development will actually discover a fundamentally distinct alternative theory that ultimately overturns or replaces current theoretical orthodoxy is one that it seems she must regard as remote. The instrumentalist thus has all the same motivations that the realist has for investing in the search for fundamentally distinct and even more instrumentally powerful successors to our best scientific theories and at least one more that is far more compelling: in stark contrast to the realist, she fully expects this search to ultimately attain its intended object. Such an instrumentalist does not say, with Blackburn, "I don't really believe in genes, or atoms, or gluons; it is just that I structure all my thoughts by accepting that they exist." Her thoughts, her expectations, and even her pursuit of scientific inquiry itself are all structured quite differently than they would be if she believed that our best current theories of inheritance or of the minute constitution of matter were even approximately true.

5 Conclusion: Reprise and Coda

We may now return at long last to the two fundamental challenges that it seemed any epistemic version of scientific instrumentalism must face: the need to specify precisely which claims are those toward which it recommends an instrumentalist attitude and the need to articulate how adopting such an attitude toward any given scientific theory would substantially differ from the realist's own belief in the truth of that same theory. We have just seen how the second of these challenges can be answered by recognizing that epistemic scientific instrumentalists are simply adopting the same attitude that the realist herself takes toward a theory like Newtonian mechanics toward a much wider range of theories than realists themselves do, including some or all of the most instrumentally powerful and successful theories of contemporary science. But seeing how this challenge can be met makes clear that it was unnecessary (and perhaps always hopeless) to try to *divide* the claims of any given scientific theory into those we must believe in order to make effective instrumental use of that theory and those we need not. This is certainly not what the realist does in the case of Newtonian mechanics. Instead, she treats that theory as a mere tool or instrument for predicting and intervening with respect to entities, events, and phenomena as they can be conceived in ways that do not depend on Newtonian mechanics itself, thus making use of not only the evidence of her senses but also of other theories concerning which she is not an instrumentalist (such as the hypothesis of the bodies of common sense, among others). Thus, it is always with respect to some independent conception of the world and its inhabitants that a scientific theory that is a "mere" conceptual tool or instrument exhibits its (sometimes remarkable) instrumental utility, and this simply does not require that the constituent claims of that theory should or even can be neatly separated into distinct categories consisting of those we must believe in order to make effective instrumental use of the theory and those we need not.

I conclude by pointing out the coherent (perhaps even attractive) possibility of combining such localized epistemic scientific instrumentalism with the global or pragmatic variety with which we began. Someone holding this distinctive combination of views would share the global instrumentalist's insistence that ideas, beliefs, theories, and cognitive entities or states *quite generally* are tools or instruments for accomplishing the full range of our practical and pragmatic goals as effectively and efficiently as possible, and she will agree that this characterization simply does not compete with thinking of some of those same beliefs or cognitive states as also corresponding to or accurately describing the world itself. However, she will also deny that some of even the best scientific theories of our own day in fact correspond to or agree with reality in this way, meaning simply that she doubts that these particular theories will or would persist throughout the entire course of further inquiry. Although she grants that our current theory concerning some particular scientific domain represents the best cognitive or conceptual tool we presently have for guiding our prediction, intervention, and other practical engagement with entities, events, and phenomena in that domain, she nonetheless fully expects that cognitive tool to be replaced in the course of further inquiry by other, even more empirically impressive and instrumentally powerful successors that are fundamentally distinct from it. That is, she thinks that any such theories will not or would not ultimately be retained in the description of nature adopted by idealized inquirers at the end of a suitably idealized inquiry (Peirce), the set of beliefs that no further experience would lead us to abandon if we adopted it now (James), or in whatever way her favored version of global instrumentalism picks out the special class of beliefs constituting the truth about the world, and, by her lights, this is just what it is for such theories to turn out not to be true. Thus, although global and scientific instrumentalism are distinct and separable views, their combination holds evident attractions for those who find themselves with deep reservations about both the coherence of the classical scientific realist's correspondence conception of truth *and* her conviction that the most empirically successful and instrumentally powerful scientific theories of the present day are or must be at least probably, approximately true.

Acknowledgments

My thanks to Jeff Barrett; Pen Maddy; Arthur Fine; Jim Weatherall; David Malament; Aldo Antonelli; Paul Humphreys; Ludwig Fahrbach; Yoichi Ishida; Mark Newman;

Michael Poulin; students in graduate seminars at the University of Pittsburgh Department of History and Philosophy of Science and the University of California, Irvine, Department of Logic and Philosophy of Science; audiences at the University of Michigan, Cambridge University, and the Australian National University during my stay there as a Visiting Fellow; and many others I have inexcusably forgotten for useful discussion and suggestions regarding the material in this chapter.

Suggested Reading

Beyond those mentioned in the text, classic discussions of instrumentalist themes can be found in:

Duhem, P. ([1914] 1954). *The Aim and Structure of Physical Theory*. Translated by P. Weiner. (Princeton, NJ: Princeton University Press).
Poincare, H. ([1905] 1952). *Science and Hypothesis*. (New York: Dover).

And more recently influential treatments can be found in:

Fine, A. (1986). "Unnatural Attitudes: Realist and Instrumentalist Attachments to Science." *Mind* 95: 149–179.
Kitcher, P. (1993). *The Advancement of Science: Science Without Legend, Objectivity Without Illusions*. (New York: Oxford University Press).
Psillos, S. (1999). *Scientific Realism: How Science Tracks Truth*. (New York: Routledge).
Stanford, P. K. (2006). *Exceeding Our Grasp: Science, History, and the Problem of Unconceived Alternatives*. (New York: Oxford University Press).

References

Blackburn, S. (1984). *Spreading the Word*. (Oxford: Clarendon).
Blackburn, S. (2002). "Realism: Deconstructing the Debate." *Ratio* 15: 111–133.
Craig, W. (1953). "On Axiomatizability Within a System." *Journal of Symbolic Logic* 18: 30–32.
Hempel, C. ([1958] 1965). "The Theoretician's Dilemma: A Study in the Logic of Theory Construction." Reprinted in C. Hempel (1965), *Aspects of Scientific Explanation* (New York: Free Press), 173–226.
Horwich, P. (1991). "On the Nature and Norms of Theoretical Commitment." *Philosophy of Science* 58: 1–14.
James, W. ([1907] 1978). *Pragmatism: A New Name for Some Old Ways of Thinking* and *The Meaning of Truth: A Sequel to Pragmatism*. (Cambridge, MA: Harvard University Press).
Kuhn, T. S. ([1962] 1996). *The Structure of Scientific Revolutions*, 3rd ed. (Chicago: University of Chicago Press).
Laudan, L. (1977). *Progress and its Problems: Towards a Theory of Scientific Growth*. (Berkeley: University of California Press).
Laudan, L. (1996). *Beyond Positivism and Relativism*. (Boulder, CO: Westview).
Mach, E. ([1893] 1960). *The Science of Mechanics*, 6th ed. Translated by T. J. McCormack. (La Salle, IL: Open Court).
Nagel, E. (1961). *The Structure of Science*. (New York: Harcourt, Brace, and World).

Quine, W. V. O. ([1960] 1976). "Posits and Reality." Reprinted in W. V. O. Quine (1976), *The Ways of Paradox and Other Essays*. 2nd ed. (Cambridge, MA: Harvard University Press), 246–264.

Stanford, P. K. (2006). *Exceeding Our Grasp: Science, History, and the Problem of Unconceived Alternatives*. (New York: Oxford University Press).

Stein, H. (1989). "Yes, but . . . Some Skeptical Remarks on Realism and Anti-realism." *Dialectica* 43: 47–65.

Vaihinger, H. ([1924] 1965). *The Philosophy of 'As If': A System of the Theoretical, Practical, and Religious Fictions of Mankind*. (London: Routledge and Kegan Paul).

van Fraassen, B. (1980). *The Scientific Image*. (Oxford: Clarendon).

CHAPTER 16

....

LAWS OF NATURE

....

JOHN T. ROBERTS

1 INTRODUCTION

....

THE thought that the universe is somehow governed by laws of nature, and that one of the central jobs of science is to discover these laws, has played a central, structuring role in much scientific thought since the seventeenth century (see Milton 1998; Feynman 2010: 1-1, 1-2). But what does this thought amount to?

In most cases, what are called "laws of nature" are generalizations or regularities in the course of events. But not every generalization or regularity counts as a law. For example:

(C) Throughout space and time, no massive particle ever travels faster than c (the speed of light).

This fact, presumably, is a law of nature (barring possible new revolutionary developments that this article ignores). On the other hand, there must be some speed d such that this is a fact too:

(D) Throughout space and time, no dodo ever travels faster than d.

Because the dodos are no more, and there were only ever a finite number of them, one of them must have achieved the dodo speed record; let d be a speed just a bit faster than that record. (C) and (D) are parallel in their syntactic and semantic features, and yet it seems obvious that one of them is a law of nature (or, at least, it might well be one) whereas the other is plainly not: (D) is just a matter of happenstance. Had the dodos not gone extinct when they did, or had history's fastest dodo been just a little more highly motivated on her best day, (D) might well have been false. Moreover, (D) certainly would have been false had the dodos managed to hang on until the days of jet airplanes and one of them taken a ride on the Concorde.

The parallelism between (C) and (D) makes it evidently hopeless that lawhood can be defined in syntactic and semantic terms—for it does not look as if (C) and (D) differ in any relevant semantic or syntactic respect, but they do differ with respect to lawhood. (Petty objection: This example ignores the frame-relativity of speed. Obvious reply: Revise (C) and (D) so that they say "no two massive particles ever achieve a mutual relative speed greater than . . ." and "no two dodos ever achieve a mutual relative speed greater than . . . ." The worry then dissolves at the cost of making the example more complicated.) This example (which is a descendant of a more famous one involving metallic spheres, due to Reichenbach 1947: 368) usefully brings out four important features of laws of nature and the philosophical puzzles they raise.

First, lawhood is a feature that is properly attributed to some, but not all, true regularities. The fundamental philosophical problem about laws can be put in this form: What makes the difference between these two kinds of regularities? That is, what criterion separates the laws from the other regularities, and (equally importantly) why is that criterion so important that we mark it with the exalted name "law of nature" and treat the regularities that satisfy the criterion in all the special ways that laws get treated (see following discussion)? Incidentally, in the philosophical literature about laws of nature, all true regularities that are not laws of nature are called "accidental regularities." This is unfortunate, because many of them are not accidental at all, in the ordinary sense of the English word. But this is a piece of terminology that has become stubbornly standard.

Second, laws have a counterfactual-supporting power that accidental regularities often lack. Had we tried our best to accelerate a massive particle to a speed greater than c, we would have failed, no matter what. Why? Because (C) would still have been true, and it follows logically from (C) and the antecedent of this counterfactual that we would have failed. By contrast, had we tried our best to accelerate a dodo to a speed greater than d, then—who knows? We might well have succeeded, in which case (D) would have been false. The point is that (D) does not enjoy a presumption of still-getting-to-be-true in our counterfactual speculation, whereas (C) does. This point generalizes: Whenever we reason hypothetically about non-actual scenarios, it seems that we hold the actual laws of nature constant (so far as it is logically possible to do so), whereas accidental regularities do not always seem to enjoy this privilege.

Third, laws seem to enjoy the power to underwrite scientific explanations in a way that accidental regularities often do not. If we wanted an explanation of why a certain massive particle failed to accelerate to a speed greater than c—despite the massive amounts of energy poured into an effort to accelerate it, say—then the fact that (C) is a law seems relevant to the explanation we seek. If, however, we wanted an explanation of why some particular dodo failed to travel at a speed greater than d—despite its great natural athleticism and its years of training—then (D) would plainly be irrelevant to this explanation. If someone pointed to (D) in order to try to explain this dodo's failure, we would likely reply that (D) is really another way of putting what we want explained—why did no dodo every break the d barrier, and in particular why did this one never break it, despite its great efforts?

Fourth, the fundamental philosophical problem about laws of nature is not one of those topics in the philosophy of science where the philosophical work tends to be "close to the science." No scientist would ever for a moment be tempted to wonder whether (D) was a law of nature. Indeed, most likely, no competent scientist would ever waste a moment on (D) at all, except to be polite to some philosopher. Scientists know laws, or good guesses at laws, when they see them and proceed to use them accordingly; they do not (except in rare cases—witness the odd history of "Bode's law") recognize the truth of some regularity in the course of nature and then spend time wondering and debating whether or not it should be considered a law of nature. So what philosophers working on the topic of lawhood are searching for—namely, the criterion for distinguishing laws from accidental regularities—is not something that typically rises to the level of consciousness in the course of the work of scientists. It is not something that scientists themselves appeal to, or debate about, or problematize, or depend upon in arguments that they give. By contrast with many topics in general philosophy of science—such as causation, probability, confirmation, explanation, and mechanism—there do not seem to be many relevant clues in the details of contemporary scientific work. (There may be some exceptions to this though; see Roberts 2008: 12–24.) By comparison with recent work by philosophers of science on those other topics, the recent literature on laws of nature seems to have more overlap with pure metaphysics and less with philosophy of the particular sciences, and I believe that this is the explanation.

2 THE FUNDAMENTAL PROBLEM

Our philosophical problem can be motivated and defined as follows: Much (though, to be sure, not all) modern scientific work is organized around the idea that the course of events is somehow "governed" by a set of "laws of nature" that it is the business of science to uncover, or to uncover closer and closer approximations to. What are called "laws of nature" are (what are believed to be) regularities that hold universally in the course of events—but not all such regularities are called "laws of nature." Scientists have no trouble recognizing the ones that should be so-called; these are generally the ones discovered in a certain way—usually ones discovered via inductive reasoning and believed to be integrated in some way into a more general theory. Thus, for example, John Arbuthnot's discovery that in each year from 1629 to 1710 more boys than girls were born in London (Kitcher 2001: 71) might easily have provided an inductive basis for a generalization that more boys are always born in London, but unless and until some basis for integrating this into a general theory was found, it is difficult to imagine anyone calling it a "law of nature." And, in fact, when a Darwinian theoretical explanation of it was discovered much later by R. A. Fisher, the explanation was found to depend not only on some general laws such as the principle of natural selection but also on some crummy contingent facts—such as the fact that among humans in urban English conditions, males were the more vulnerable sex. This undermined the thought that Arbuthnot's regularity was a law, precisely because it did not follow from the central organizing principles of evolutionary theory. Once a principle is considered a law of nature, it is granted certain privileges: it is held constant in counterfactual reasoning, and it is granted a certain power to underwrite scientific explanations. Our problem, fundamentally, is to explain why—assuming as we do that modern science embodies a justified method (or family of methods) for acquiring knowledge about the natural world—scientists are justified in treating regularities discovered in this way as if they deserved these privileges.

These days, almost all philosophers in the analytical tradition are realists and will be quick to rewrite the problem I have just stated like this: "To explain what it is about certain regularities in the course of natural events that really does set them apart, making them deserving of being called 'laws of nature,' and of being accorded the counterfactual and explanatory privileges of laws, and why the marks scientists look for count as evidence that a regularity belongs to this special class." Pragmatists and radical empiricists will demur (see, e.g., Blackburn 1986; van Fraassen 1989), and it might turn out that this realist hegemony is a fashion that will pass; nevertheless, the remainder of this article speaks the language of realism. So our question is: What is it that distinguishes the laws from the accidental regularities, makes the laws deserve their counterfactual and explanatory privileges, and makes the things that scientists take to be evidence that something is a law evidence that it is indeed a law?

Numerous theories have been proposed to answer this question. Today the most influential ones are the best-system account, according to which the laws are those regularities that belong to the system of truths that achieves the best combination of strength and simplicity (Lewis 1973); the contingent-necessitation account, according to which laws are metaphysically contingent relations among universals that ensure that certain patterns hold among the particulars that instantiate those universals (Armstrong 1983; Dretske 1977; Tooley 1977); the dispositional essentialist account, according to which natural properties have essential features that include their conferring certain dispositions on their instances and laws of nature are simply expressions of the regularities in behavior that result from these dispositions (Ellis 2001; Mumford 2004; Bird 2007); and the no-theory theory of laws, according to which laws cannot be reduced to anything more basic (Carroll 1994; Maudlin 2007). Two more recent theories include one that holds that counterfactuals are more basic than laws, and laws are simply those truths that are most counterfactually robust (Lange 2009), and one that holds that laws are relativized to theories and that the laws of a theory are the principles of that theory that guarantee the reliability of measurement methods (Roberts 2008).

Rather than surveying these theories one by one, this article proceeds by examining a number of subsidiary questions about the nature of laws that have attracted a great deal of attention in the recent literature, the examination of which will shed light on the advantages and liabilities of these theories.

3 Do We Need a Philosophical Theory of Laws?

Do we need a theory of lawhood at all? Why not simply adopt laws as a metaphysical primitive in our philosophy of science? The latter course is endorsed by Carroll (1994) and Maudlin (2007).

It might be objected that if we do take laws as a primitive, then we will not be able to explain why they play all the roles that that they do, since they simply will not have enough built into them. But this would be a mistake: To take the concept of a law of nature as a primitive does not mean making no assumptions about them; in Euclidean geometry, points and lines are primitives, but there are several fruitful axioms about them. A primitivist about laws can postulate a number of basic connections between lawhood and other basic concepts that account for why laws support counterfactuals, explain their instances, get inductively supported by observed instances, and so on.

Primitive posits certainly can explain things. But do laws, taken as a primitive posit, explain what they are supposed to explain? Consider two cases. In the first case, our livestock has been disappearing at night, and strange tracks have been found on the ground in the morning. We come up with a theory: a heretofore unknown predatory animal—which he hereby dub "the Gruffalo," knowing nothing else about it—has been stealing our livestock and leaving the tracks. The Gruffalo is a primitive of this theory; we had no concept of it before, and we cannot define it in terms of more basic vocabulary. Appealing to the Gruffalo, it seems, we can explain what we want to explain. Given the truth of our theory, we understand what is going on perfectly—for we have seen animals attack and carry off prey before, so we have a model of the kind of causal process in question that is supposed to be doing the explaining here.

In the second case, our livestock has been disappearing but here we posit a primitive entity—"the Bruffalo"—about which we say only that its nature is such as to explain the disappearance of our animals at night. Appealing to the Bruffalo, can we explain what we want explained? It seems clear that the answer is "no": a primitive posit cannot explain a phenomenon simply by being postulated to have the power to explain that very phenomenon. It must explain by being postulated to have some feature in the light of which (and together with our background knowledge) sense can be made of the target explanandum—and simply saying "this thing's nature, *ex hypothesi,* is to explain that target explanandum" does not count.

Now consider again laws, taken as a primitive, not explicated in term of anything more basic. Certain things are postulated about them—for example, that they are capable of explaining their instances. Unlike in the case of the Gruffalo, we do not already have a model of the way in which the explaining gets done. The law is a "we-know-not-what" that somehow explains its instances. Newton's second law, for example, is a we-know-not-what that somehow has the power to make it the case that $f = ma$; its existence thus has the power to explain why $f = ma$. Is this not a posit of the same sort as the Bruffalo? It is introduced to explain something, but it does not do the explaining by having particular features that help us make sense of the target explanandum but rather by its simply being stipulated that it has the power to explain this explanandum. This, it seems, is not how primitive theoretical posits play their role in explanation. For this reason, it seems unlikely that laws of nature are best viewed as primitive posits in our total theory of the world.

4 THE QUESTION OF HEDGED LAWS

I mentioned previously that laws are regularities that are strictly true through space and time, but not all agree with this. Numerous philosophers have argued that laws of nature are typically not strict regularities but rather hedged regularities, or ceteris paribus laws: they come with an (implicit or explicit) qualifying clause, which means that they can tolerate some exceptions without being false. What kinds of exceptions, and how many? These are not made explicit; if they were, the law could be restated in a strict form (e.g., Cartwright 1983; Pietroski and Rey 1995; Lange 2002).

This view faces two formidable problems (Earman and Roberts 1999). The first is the problem of disconfirmation: Falsificationism is dead (and good riddance), but there is nevertheless something suspicious about a supposed item of scientific knowledge that is such that we cannot specify an observation that, alone or in conjunction with auxiliary hypotheses, would even disconfirm it. But hedged laws seem to be in just this boat. Suppose I hypothesize that, ceteris paribus, Fs are G; then you discover an F that is non-G. Is your discovery evidence against my hypothesis? How can we tell? It might be one of those exceptions that my hypothesized CP-law can tolerate with no trouble, and it might not be; if we knew anything relevant to the probability that it was in one class or the other, then it would not be a hedged law after all (or at least it would not be *this* hedged law), for in that case I would have to have hypothesized something more contentful about the circumstances under which an F must be G. So it seems that with a genuine ceteris paribus law, there is no such thing as evidence that would unequivocally disconfirm it, even to a small degree. The idea that natural science legitimately has truck with such items is suspect.

The second is the problem of nontrivial content. Impatient skeptics of hedged laws sometimes ask, "What could it mean to say that, ceteris paribus, Fs are G, except that every F is a G unless it isn't?" This rhetorical question is not the knock-down argument it is often presented as. But what is the answer to it? A standard answer is that the hedged law says that a *typical* F is G *unless it gets interfered with*. But what makes an F "typical," and what counts as "interfering" with an F? If we can determine that, then we can plug our answers in and thus rewrite the hedged law out without any hedges, so it turns out that the ceteris paribus clause was after all only an abbreviation for qualifications that we were too lazy to state the first time round. On the other hand, if we cannot determine it, then why should we believe that the hedged law statement really makes any genuine claim about the natural world at all? (This is sometimes called "Lange's dilemma," after Lange 1993.)

However, there are influential arguments in favor of hedged laws. One, originating with Cartwright (1983), is that the function of laws of nature is to state the causal powers of different kinds of entities and that the regularities we use to express those laws state only what those causal powers can make happen when nothing else interferes with them. But, of course, in the real world, something else is almost always interfering.

A second argument, originating perhaps with Fodor (1991), is that the laws cited by psychology and other sciences concerning multiply realizable kinds are almost certainly bound to fail to be exceptionless. But it is a widespread view that scientific explanations require laws. Hence it seems that one must admit the existence of hedged laws, because the only two alternatives are to (a) reject the widespread view just mentioned, which seems radically revisionary and (b) deny that the sciences other than physics can explain anything, which is ludicrous. This second argument seems compelling. Of course, one might reasonably doubt that all explanations require laws, but it is hard to deny that sciences like biology and psychology uncover broad generalizations that have explanatory power yet are not exceptionless, so (even if one were to insist that these are not laws "strictly speaking") the difficulty here cannot be avoided forever.

One strategy for dealing with this problem, without running afoul of either the problem of disconfirmation or Lange's dilemma, is to understand *hedged law statements* as devices we use for selecting particular laws of nature that do not fully specify the contents of those laws. The contents of the laws in question are fully precise and exceptionless, but we do not need to be in a position to specify their contents fully in order to pick them out. (There is ample precedent for this: for example, the full content of Newton's law of gravitation presumably includes the precise value of the gravitational constant, but every statement of this law to date has failed to indicate this value precisely, providing instead only a finite number of significant figures. Thus we can refer to this august law and specify enough of its content to be useful to us without specifying its entire content.) To be more specific, there is some strict law that covers the Fs that we, as a matter of fact, have in mind, or are likely to apply the law in question to, that guarantees that they will (mostly) be Gs, though we are not now in a position to say exactly what that law says. One proposal along these lines (Strevens 2012) is that when scientists propose a hedged law (Strevens actually says "causal generalization"), they typically have in mind a particular causal mechanism M that they figure stands behind this generalization; they might not know enough about this mechanism to describe its workings in much detail, but (if it in fact exists) they are able to refer to it. "Ceteris paribus, Fs are G" is true just in case, when the normal operating conditions of this mechanism M are in place, F brings about G by way of M (see Strevens 2012: 658–660). A similar account I have recently defended (Roberts 2014) has it that "Ceteris paribus, Fs are G" is a true hedged law statement just in case there exists some property K such that (i) it is a law that most F and Ks are G, and (ii) (it is true but not necessarily a law that) almost all actual Fs to which our scientific community will apply *this very hedged law statement* for purposes of prediction and explanation are K.

Proposals such as these get around the two problems for hedged laws considered previously, for it is both clear what it would it take for a hedged law statement to be false on such a proposal and possible to specify observations that very clearly would count as evidence against any given hedged law statement on such a proposal (e.g., we observe lots and lots of Fs, under conditions that do not surprise us and strike us as typical, and none of them are G). Furthermore, these proposals enable us to explain how law statements in the life sciences can be mostly hedged and yet still provide explanatory power: so long as we know that there is some law that guarantees that most of the robins' eggs around here are bluish-green (for example), we seem to have satisfied whatever nomological requirement on explanation it seems fair to impose on an explanation, whether or not we can state that law with complete precision. However, proposals like these do not tell us much about what the laws themselves are like. The proposal of Strevens (2012) does refer to causal mechanisms, but it is not committed to any particular metaphysical account of what they are like; the proposal of Roberts (2014) is thoroughly neutral on the metaphysics of laws. So, despite Cartwright's famous argument that the hedged character of laws is a clue to their being ultimately about powers, it seems that this hedged character tells us nothing about the nature of lawhood at all.

5 THE QUESTION OF METAPHYSICAL NECESSITY

Are the laws of nature metaphysically necessary? Or is the kind of necessity they enjoy (variously called *physical, natural,* and *nomic* necessity) a status somehow intermediate between metaphysical necessity and brute fact?

The traditional view is that the laws must be metaphysically contingent, because they are discovered empirically. (And besides, surely God could have made the universe with different laws had He wanted to!) But we have learned from Kripke (1980) that the metaphysically necessary need not be knowable a priori.

Moreover, an attractive new theory of properties has become very influential that would seem to imply that the laws must be necessary: this is the dispositional essentialist account, according to which natural properties have essential features, and among these features are their causal and nomic relations to other properties (Shoemaker 1980; Ellis 2001; Mumford 2004; Bird 2007). The chief arguments in favor of this theory of properties deal mainly with the shortcomings of its main competitor, quidditism, according to which a property is simply a "suchness" with no essential features at all. If quidditism is correct, then the identity or distinctness of two properties occurring in two different possible worlds is a matter of brute fact. This means that there can be two possible worlds that differ only in that two properties (e.g., mass and charge) have their causal and nomic roles reversed, or in that the causal and nomic roles of properties have been permuted, and this seems to lead to grave problems—ontological, semantic, and epistemological (Shoemaker 1980; Bird 2007). On the other hand, dispositional essentialism is free of such worries, and it implies that the actual nomic relations among properties must hold in all possible worlds. For example, the laws relating mass and force must be obeyed by all possible instances of mass and force (otherwise they simply would not be instances of mass and force), and in any possible worlds where there are no instances of mass or force these laws will be vacuously true. Hence the metaphysical necessity of laws falls out of an attractive theory of properties. Moreover, this theory of laws seems to succeed in explaining why laws support counterfactuals and explanations as well as they do.

However, it should be noted that there is middle ground between quidditism and dispositional essentialism. We could take the view that each natural property has some of its actual causal-nomic profile essentially but not all of it. So, for example, it might not be essential to mass that things with mass attract one another in accordance with Newton's exact law (an inverse square law with a specific fundamental constant G) but nevertheless essential to mass that any two bodies that have mass attract one another and that the attraction be greater the greater their masses. This necessity might be conceptual in nature: perhaps no property in any other possible world that failed to satisfy that weaker description could count as the referent of *mass,* though a property that obeyed a law like Newton's but with a slightly different constant could—and similarly with other natural properties. This would rule out possible worlds that differ only via permutations of the properties but would leave many of the laws of nature metaphysically contingent.

It might be objected that to make this move would be to give an account only of our concepts of properties like mass—the nature of *mass itself* would be left quite out of account; what is it that makes some property in some other world the same property as the one we call "mass" in this word? But we might simply decline to reify properties in a way that makes this meaningful. Questions about their transworld identity might not have serious metaphysical answers, over and above what we stipulate when we specify the possible situations we want to talk about, as Kripke (1980) suggests concerning persons.

Moreover, there is an important argument for the view that not all laws are metaphysically necessary: there are some counterfactuals that seem to be supported by the laws but that are not supported by the metaphysical necessities. If there were a table in here, it would be held to the floor by the inverse-square gravitational force. Why? Well, it would be made of matter, which has mass, and so on.[1] But there are presumably metaphysically possible worlds where there is no matter but instead schmatter, which is characterized not by mass but by schmass, which obeys not the familiar inverse-square law but instead an inverse-2.00000000001 law. Why should we think that if there were a table in here, we would not be in one of those worlds? For it is not metaphysically impossible for the world to contain things with schmass rather than mass, after all. The right answer to this strange question seems to be: tables are, by definition, made of the mattery stuff, and, in this world, it is physically (though not metaphysically) necessary that the mattery stuff is matter, which has mass, and so on. So the laws support the counterfactual that if there were a table in here, it would have mass and therefore would be held down by the inverse-square law. But this answer requires it to follow from the law that the mattery stuff has mass, and this is metaphysically contingent. Therefore, there is at least one metaphysically contingent law (or, at any rate, there is at least one metaphysically

---

[1] In the argument to follow, I assume either that inertial mass and gravitational mass are the same property or that "mass" refers to the latter.

contingent regularity that has the counterfactual-supporting power of the laws—and once this has been established, why quibble over the word "law"?)

6 THE QUESTION OF HUMEANISM

Humeanism about laws is the thesis that the laws of nature are reducible to (and therefore supervene upon) the great spatiotemporal distribution of particular matters of actual fact—the so-called "Humean mosaic." This is composed of some kind of space-time structure, together with what actually goes on throughout it—which, if anything like our modern physical theories is correct, can be characterized by the values of some set of mathematical variables defined over the space-time. These might be scalar and vector fields defined over space-time points, but they might be something a lot more complicated—more complex objects defined over sets of points, or regions, or sets of regions; the space-time structure might have more than four dimensions, and so on. But the crucial point is that there is nothing essentially modal, or counterfactual, or dispositional in this world: it is wholly a world of things actually happening, actually going on, here or there. There are fully actual occurrences, and there are regularities and patterns in those, and that is it: the laws of nature must be somehow reducible to this structure. (The same goes for causation, counterfactuals, chances, powers, and so forth, but here we are concerned only with laws, fortunately!) "No necessary connexions between distinct existences," as the great Hume said. (The thesis of "Humean supervenience" was introduced and defined prosaically by Lewis [1973] and revised by Loewer [1997] and Earman and Roberts [2005]. But to my mind, the worldview of contemporary Humeanism is most thoroughly captured by Beebee [2001, 2011] and by a tragically still-unpublished but very well-known paper by Hall [n.d.]).

The literature is full of arguments against Humeanism. First, there are numerous putative counterexamples—descriptions of pairs of putative possible worlds that share their Humean bases but have different laws of nature. The simplest of these is perhaps the pair of lonesome-particle worlds, consisting of nothing but a single particle, moving uniformly through the void forever. It conforms to the laws of Newtonian mechanics, with gravity as the only force. But it also conforms to the laws of nineteenth-century classical physics, with Newtonian mechanics and gravitation theory wedded with Maxwell's electrodynamics. Surely (the argument goes) there is a possible world where the laws are those of the first theory, and there is nothing but a single particle moving uniformly, as well as a possible world where the laws are those of the second theory, and there is nothing but a single such (electrically neutral) particle? Yet these worlds would pose a counterexample to Humean supervenience and thus to Humean reductionism. A determined Humean can grit his teeth and reject one of these worlds as merely epistemically possible, but Carroll (1994) has posed a much more sophisticated counterexample in the same spirit, which will be much more painful for any would-be Humean to reject.

Beebee (2001) argues persuasively, however, that all of these arguments depend on taking for granted what she calls the Governing Conception of Laws—the idea that the laws of nature govern the course of nature in some robust sense. She argues that an alternative Non-Governing Conception—according to which what is special about the laws is that they are a family of principles that collectively provide a uniquely compact and coherent summary of natural history—is available and does justice to the role played by laws in the practice of science. She also argues that anyone who has cast their lot with Humeanism has, in effect, already rejected the Governing Conception in favor of the Non-Governing Conception, and so all such counterexample-based arguments fail to have any rational force against them.

Another common argument against Humeanism is that Humean laws could not possibly explain their instances, as laws of nature are supposed to do. For on a Humean view, the laws themselves are constituted by the great Humean mosaic, which is itself constituted by all of the particular events that occur in the universe; if a law were called upon to explain one of these events, then, we would have a vicious explanatory circle. This argument is currently the subject of considerable controversy. Loewer (2012) argues that the alleged explanatory circle is merely apparent, since the laws scientifically explain the particular events, whereas the particular events metaphysically explain the laws, and these are totally different kinds of explanation. Lange (2013) counters that whatever helps to constitute the explanans in a scientific explanation also helps to explain its explanandum, and this blocks Loewer's reply. My own view, however, is that the argument against the Humean is not conclusive, because we have a clear and relevant precedent for cases in which an element in a pattern can be explained by that pattern, even though it helps to constitute the pattern: namely, aesthetic explanations, in which the placement of one element within a work of art is explained by the fact that the work belongs to a certain style or genre and that, given the arrangement of other elements, the element in question must be placed as it is in order for the work to conform to the conventions of that style or genre. (For example, we can explain why a certain movement ends on a certain chord by pointing out that it is a work of sonata form, and given the other notes occurring in the piece, it must end with that chord—even though its being a work of sonata form is partly constituted by the fact that it ends with that chord.) We could also explain the placement of the element causally, by appeal to the psychological states of the artist, but that is not what I have in mind; there is a different kind of explanation, namely an account that provides the audience with a different kind of understanding of the work, namely aesthetic understanding. Similarly, a Humean might argue, there is a kind of understanding—quite different from what we get from a causal explanation—that we get by learning to place a particular event within the larger pattern constituted by a set of, say, Newtonian laws—even if it turns out that the Newtonian laws are party constituted by that event itself. The Humean who adopts this reply to the argument ends up with a view on which covering-law explanations are more like aesthetic explanations than causal explanations—a surprising thesis but one that seems in the spirit of Humeanism.

Yet another common argument against Humeanism alleges that only a non-Humean conception of laws of nature is consistent with the justification of induction. If there are no necessary connections between distinct existences, as the Humean alleges, then the universe is a mosaic of loose and separate elements, and what happens in one part of the mosaic affords no clue as to what happens in another part. To draw any inference at all—even a merely probable inference—from the part of the mosaic we have seen to the part we have not seen requires some premise to the effect that there is some connection between the two, such as a non-Humean law might provide but that nothing in a Humean universe could supply. The problem with this argument is that there is no good reason to think that the non-Humean is any better off than the Humean, with respect to solving the problem of induction. Hume himself already anticipated the argument and thoroughly answered it near the end of Section IV of the *Enquiry Concerning Human Understanding* (Hume 1993: 23–24); the crucial point has been recast in a contemporary idiom by Beebee (2011): the problem is that even if there are necessary connections of a non-Humean sort, this still does not provide us with any guarantee that the patterns we have seen in our data so far are representative of what we will see in the future. One reason why this is so is that the non-Humean laws may well change in the future—or even if the laws themselves cannot change, they might involve temporal parameters in such a way that the superficial behavior they result in is quite different tomorrow than it was today.

So much for arguments against Humeanism. But are there any good arguments in favor of Humeanism? I think that one pretty good argument for Humeanism goes like this: imagine that we are magically afforded a god's-eye glimpse of the entire Humean base of the universe, with supernatural secretarial help for organizing it. We would in principle be able to pick out the regularities that, given this massive body of data, look most to be the laws of nature. Let's call those *the ideally apparent laws*. On the other hand, there are the regularities that are, in fact, the laws of nature. From a Humean point of view, the ideally apparent laws of nature are, presumably, the very same things as the laws of nature. But from the non-Humean point of view, they may very well not be. Both sides will agree that, if through some stroke of luck we found ourselves in the ideal epistemic position, our best guess at what the laws of nature were would be these ideally apparent laws. But while the Humeans say we would be bound to be right, the non-Humeans say there is a chance that we would be wrong. But now suppose that the ideally apparent laws and the "real laws" are in fact different—in that case, what are the "real laws" to us? Why should we give a hoot about them? Given a choice, right now, between knowing what these non-Humean "real laws" are and what the mere ideally apparent laws are, which would we choose to know? For all purposes of prediction, manipulation of nature, building better engines, building better telescopes, curing disease, finding explanations that give satisfaction to our feelings of curiosity, and so on, which should we expect to serve us better? Even if the non-Humean "real laws" are the Great Truth, they are, in our hypothesized situation, a Hidden Truth that will ex hypothesi never reveal itself in any manifest form. In short, for all scientific and practical purposes, would we not prefer to know the ideally apparent laws? And, in that case, why not reappropriate the term "laws of nature" for the regularities we are really interested in finding—the ideally apparent laws—the things that the Humeans have been calling "laws of nature" all along?

One might object that a parallel argument could be constructed that would lead us to the conclusion that we ought to be phenomenalists about the material world, or crude behaviorists about other minds, or instrumentalists about the theoretical posits of science, and in general that we ought to take any class of things we believe in and reduce it to some construct out of the stuff that we use as evidence for it. But we have long since learned to resist arguments like that, and similarly we should resist this one. But this one is special. If we gave a similar argument for phenomenalism about tables and chairs, we would have to start out assuming that our evidence for tables and chairs comes from sense impressions. But it does not; we see tables and chairs. If we gave a similar argument for crude behaviorism about other minds, we would have to start out assuming that we draw inferences about the mental states of others from their behavior. But that is false: we can see and hear others' feelings in their faces and voices and postures and gaits; we can know what others are thinking because they tell us (and not because we infer it from their "verbal behavior"). With laws it is different: scientists can build instruments that let them see (or at least measure and detect) electrons; they cannot build a device for seeing laws—the very idea that they might is a category mistake.

However, the best argument for Humeanism, in my opinion, would come from a particular Humean reductive account of laws that captured all the roles that laws are supposed to play. With some such account in hand, the Humean could issue the challenge: "Why shouldn't these items—which everybody agrees exist anyhow—qualify just as well as candidate referents for our term 'laws of nature' as any of the more ontologically robust items posited by non-Humeans?" Moreover, unless some detailed account of exactly how the laws can somehow emerge out of the primordial soup that is the Humean base, Humeanism is at best a promissory note.

The literature is full of claims that the best Humean approach to laws is the "best-system" approach, inspired by J. S. Mill and F. Ramsey and articulated by Lewis (1973, 1994); alternative versions have been offered by Loewer (1997), Halpin (1999), Roberts (1999), Braddon-Mitchell (2001), and Cohen and Callender (2009), inter alia. Details vary, but the basic template is the thought that the laws are the regularities that belong to the best system of true propositions describing the actual Humean mosaic, where the "best" such system is the one that achieves the best combination of some short list of virtues, such as strength and simplicity. The virtues in question are supposed to be ones that scientists themselves appeal to in theory-choice, and the standards for measuring them are to be borrowed from scientific practice itself. A neat thing about this view is that it purports to look to the epistemic standards scientists use to discover laws and reverse-engineer them to find out what they are good standards for discovering; whatever *that* turns out to be, it identifies with the laws. So the laws turn out to be the ultimate goal of scientific inquiry, and it makes perfect sense that scientists should treat the regularities that have met their epistemic standards so far as deserving the privileges due to the laws (including counterfactual and explanatory robustness). Different versions of the best-system account differ on how to deal with worries about how the systems should be formulated, how to measure simplicity, whether laws can have exceptions, whether there is only one set of laws of nature or whether laws are relativized to different fields of science, and whether alien scientists might work with radically different standards than ours, among others.

The frequent claims that the best-system approach is the best Humean approach to laws are rarely if ever supported by explicit comparisons of the best-system approach with other Humean accounts of laws (or even accompanied by acknowledgements that other Humean approaches exist). In fact, the best-system approach suffers from some serious difficulties (this article discusses only one but for more, see Hall [n.d.] and Roberts [2008: 16–24]). Where are the standards of strength, simplicity, and the manner of trading them off against each other supposed to come from? Defenders of the best-system approach usually point to the practice of scientists in their theory choices. But the choices scientists confront require them to trade off between simplicity and fit with the existing data; they are never in a situation where they are given all of the facts and must choose among systems of truths by trading off between simplicity and strength. In fact, these two choice situations have completely different logical structures: in the case scientists face, one is typically choosing among mutually inconsistent hypotheses, each of which fails to fit all of the existing data points; in the choice situation envisioned by the best-system approach, one is choosing among systems that are mutually consistent (because they are all true), each of which fits the facts (again, because they are all true). Thus a little reflection shows that nothing in the actual practice of science resembles the choice among systems that the best-system approach requires (ironically, since one of the virtues most frequently claimed for this approach is its close fit with scientific practice; e.g., Cohen and Callender 2009: 3). So where the standards for choosing a "best" system are to come from is a mystery.

So what might a better Humean account look like? One possibility is that, drawing on the projectivist moment in Hume, we might develop a quasi-realist account of laws on which law-talk primary serves the function of expressing attitudes that we take toward certain regularities but is usefully regimented into a discourse that mimics that of realist-friendly subject matters so that we can coordinate our attitudes (e.g., Blackburn 1986; Ward 2002—but note that Ward holds that his version of projectivism about laws is inconsistent with the doctrine of Humean supervenience).

Another Humean alternative I have defended at length (Roberts 2008) is a contextualist account. First, lawhood is relativized to a (scientific) theory—essentially, lawhood is not a property that regularities can have or lack but rather a role that a proposition can play within a theory. Next, "L is a law of nature" is true at world w as asserted in context k if L is a law relative to some theory T that is both true at w and salient in k. Boiling things down considerably: in typical contexts, the default salient theory is the total body of scientific theory accepted (and not later rejected) by the relevant scientific experts of the community to which the conversationalists belong (where this community is considered as extended in time and including stages later than the context itself). So, "L is a law of nature," as spoken by me in a typical context, is true just in case L plays the law role within some true theory that is or will be accepted by our scientific community. But what is the "law role"? Roughly, L plays the law role within theory T just in case L is the regularity that expresses the reliability of some method that is a legitimate measurement procedure, according to that theory. Why measurement procedures? It is very plausible that for anything to count as a legitimate measurement procedure, it must be guaranteed reliable by the laws of nature, and, conversely, any empirical procedure for finding out the value of a variable that was guaranteed reliable by the laws of nature would seem to be a legitimate measurement procedure. So we have a neat connection between two problems: that of distinguishing the laws from the accidental regularities and that of distinguishing the legitimate measurement procedures from the empirical finding-out procedures that just happen always to work as a matter of contingent fact—solve one and we have solved the other. My proposal is the order of explanation runs from measurements to laws: it is because a method is a legitimate measurement method that the regularity that expresses its reliability counts as a law, not the other way round. This proposal grounds the modal status of the laws of nature in the epistemic-normative status of the legitimate measurement methods. I argue at length elsewhere (Roberts 2008: 355–361) that this view is consistent with Humean supervenience and that its contextualism enables it to get around the counterexamples considered previously: in each case, we have a single possible world (e.g., a world containing nothing but a single particle, traveling uniformly through the void), within which two different theories are both true (e.g., Newton's theory and Newton's theory plus Maxwell's), within which different propositions play the law role. In the course of the argument, we are first maneuvered into considering the world in a context within which the salient theory is one of these two theories ("this description of events is consistent with the Newton-only theory, so surely there is a possible lonesome-particle world where the Newton-only theory is true"), and in this context it is correct to say that the laws of the world under consideration are the laws of the first theory; later, we are maneuvered into a context within which the second theory is the salient one ("but there is surely also a possible lonesome-particle world at which the laws are those of Newton together with those of Maxwell"), and in that context it is correct to say that the laws of the world under consideration are those of the second theory. We naturally conclude that we have two different worlds here, since they have incompatible laws, failing to notice that we have shifted contexts in the middle of the discussion in a way that renders our two statements about the laws perfectly compatible with one another.

7 THE QUESTION OF GOVERNING

The laws of nature are said to "govern" the universe, but is this to be taken seriously, or is just a manner of speaking?

On a widespread view, the only accounts of lawhood according to which laws genuinely govern are non-Humean, non-essentialist accounts. For on Humean accounts, laws are just patterns in the course of events, not anything that enforces those patterns; on the other hand, essentialist accounts hold that laws merely describe the powers inherent in active beings that make their ways through the world by themselves at their own direction, and, though natural history is predictable, it is ungoverned. It is only the remaining accounts—such as the contingent necessitation account of Armstrong (1983), Dretske (1977), and Tooley (1977), or the primitivist accounts of Carroll (1994) and Maudlin (2007)—that leave the laws any real work to do in "pushing things around" in the world. And only if the laws push things around are they aptly described as "governing." Otherwise they sit somehow serenely outside the hustle-bustle of the universe, like Leibniz's God, and radically unlike Newton's Lord God Pantokrator.

I think this widespread line of thought contains multiple confusions. First, if saying that the laws "govern" the universe is supposed to have serious content, then presumably it should help us understand what the relation between the laws and the course of nature is. But if "governing" means something like "pushing around," or playing an active role in directing, then it cannot help us understand this. Suppose that the law that Fs are G were something that intervened actively in the world, seeing to it that whenever something is an F, it is also G. The very presence of this law would then have to be a state of affairs that somehow necessitated the G-ness of any given F. How are we to understand this necessitation? If the necessitation is merely logical, then we do not have active pushing-around; but if the necessitation is something more robust, then we have the same problem we had originally all over again: for now we are faced with a "meta-law" that whenever it is a law that every F is a G, this law makes each F be G. Presumably, if we think that the laws truly govern the universe, then we want *this* law to govern as well, and now we seem to be off on an infinite regress. So governing-as-pushing-around does not help us understand the relation between laws the course of nature at all.

Second, in politics we speak of law-governed nations and of the rule of law. But it is not the laws themselves that actively intervene and see to it that the propositions that we call the laws (e.g., "No person shall commit murder") remain true; that is done instead by the government and its agencies. We say it is a *law that* you cannot commit murder, and that the laws govern—but things like *that you cannot commit murder* are not the agents of the governing but rather the contents of the governing; similarly, to say that the course of nature is governed by certain laws—namely P, Q, and R—should not be taken to imply that P, Q, and R are items with the power to make anything do anything. Instead, they are the contents of the governing: Nature is somehow governed, and it is governed to conform to the propositions that P, that Q, and that R—whether there is any agent that does this governing is another question altogether.

So whether the laws of nature govern the universe does not depend on whether the laws are things of some sort that have the power to boss the universe or its inhabitants around (they are not); it depends on whether it makes sense to think of the universe as *governed,* and, if so, whether the laws of nature are the *contents of the governing.* One way to approach this question is to take a "governed" universe to be one in which certain nontrivial omnitemporal propositions are inevitably true and to take those propositions to be the contents of the governing.

Plausibly, a proposition is inevitably true just in case it is necessarily true in a robust sense that passes a test proposed in Lange (2009). Lange suggests that a pair of modal operators <"necessarily," "possibly"> that are interdefinable in the usual way ("possibly" = "not necessarily not") and satisfy the axioms of some respectable modal logic nevertheless should count as corresponding to a genuine kind of necessity and possibility only if they satisfy the following formula:

> For all A such that possibly A, and all C such that necessarily C: Had it been the case that A, then it would (still) have been the case that C.

In other words, what is necessary is that which would still have been true, no matter what possible circumstances had obtained. (Any stronger demand would surely be unfair—for even the necessary might have failed, had the impossible happened!)

Putting this together: the laws govern if the universe is governed and they are the contents of the governing, which they are if they are inevitably true, which they are if they are necessary in a genuine sense, which they are if they collectively satisfy this principle:

> For all A such that A is consistent with the laws, and all C such that C is a law: Had it been the case that A, then it would (still) have been the case that C.

Not all accounts of laws of nature satisfy this principle (e.g., Lewis 1973). However, there are many that do (e.g., Carroll 1994; Bird 2007; Lange 2009; Roberts 2008). One of these (Roberts 2008) is a Humean account, and another (Bird 2007) is an essentialist account. So I conclude that on a reasonably charitable way of understanding it, the doctrine that laws of nature govern the universe is one that is available to Humeans, non-Humeans, essentialists, and non-essentialists alike.

REFERENCES

Armstrong, D. M. (1983). *What is a Law of Nature?* (Cambridge, UK: Cambridge University Press).

Beebee, H. (2001). "The Non-Governing Conception of Laws of Nature." *Philosophy and Phenomenological Research* 61(3): 571–594.

Beebee, H. (2011). "Necessary Connections and the Problem of Induction." *Nous* 45(3): 504–527.

Bird, A. (2007). *Nature's Metaphysics* (Oxford: Oxford University Press).

Blackburn, S. (1986). "Morals and Modals." In G. MacDonald and C. Wright (eds.), *Fact, Science and Morality* (Oxford: Blackwell), 119–142.

Braddon-Mitchell, D. (2001). "Lossy Laws." *Nous* 35(2): 260–277.

Carroll, J. (1994). *Laws of Nature* (Cambridge, UK: Cambridge University Press).

Cartwright, N. (1983). *How the Laws of Physics Lie* (Oxford: Oxford University Press).

Cohen, J., and Callender, C. (2009). "A Better Best System Account of Lawhood." *Philosophical Studies* 145(1): 1–34.

Dretske, F. I. (1977). "Laws of Nature." *Philosophy of Science* 44(2): 248–268.

Earman, J., and Roberts, J. T. (1999). "*Ceteris Paribus*, There Is No Problem of Provisos." *Synthese* 118(3): 439–478.

Earman, J., and Roberts, J. T. (2005). "Contact With the Nomic: A Challenge for Deniers of Humean Supervenient, Part One." *Philosophy and Phenomenological Research* 71(1): 1–22.

Ellis, B. (2001). *Scientific Essentialism* (Cambridge, UK: Cambridge University Press).

Feynman, R. (2010). *The Feynman Lectures on Physics*, Vol. I (New York: Basic Books).

Fodor, J. A. (1991). "You Can Fool All of the People Some of the Time, Everything Else Being Equal: Hedged Laws and Psychological Explanation." *Mind* 100(397): 19–34.

Hall, N. (n.d.). "Humean Reductionism about Laws of Nature." http://philpapers.org/rec/HALHRA

Halpin, J. (1999). "Nomic Necessity and Empiricism." *Nous* 33(4): 630–643.

Hume, D. (1993). *An Enquiry Concerning Human Understanding*. Edited by E. Steinberg (Indianapolis: Hackett).

Kitcher, P. (2001). *Science, Truth and Democracy* (Oxford: Oxford University Press).

Kripke, S. (1980). *Naming and Necessity* (Princeton, NJ: Princeton University Press).

Lange, M. (1993). "Natural Laws and the Problem of Provisos." *Erkenntnis* 38(2): 233–248.

Lange, M. (2002). "Who's Afraid of Ceteris Paribus Laws? or, How I Learned to Stop Worrying and Love Them." *Recentness* 57(3): 407–423.

Lange, M. (2009). *Laws and Lawmakers* (Oxford: Oxford University Press).

Lange, M. (2013). "Grounding, Scientific Explanation, and Humean Laws." *Philosophical Studies* 164(1): 255–261.

Lewis, D. (1973). *Counterfactuals* (Cambridge, MA: Harvard University Press).

Lewis, D. (1994). "Humean Supervenience Debugged." *Mind* 103(412): 473–490.

Loewer, B. (1997). "Humean Supervenience." *Philosophical Topics* 24: 101–126.

Loewer, B. (2012). "Two Accounts of Laws and Time." *Philosophical Studies* 160(1): 115–137.

Maudlin, T. (2007). *The Metaphysics within Physics* (Oxford: Oxford University Press).

Milton, J. R. (1998). "Laws of Nature." In D. Garber and M. Ayers (eds.), *The Cambridge History of Seventeenth-Century Philosophy*, Vol. 1 (Cambridge, UK: Cambridge University Press), 680–701.

Mumford, S. (2004). *Laws in Nature* (London: Routledge).

Pietroski, P., and Rey, G. (1995). "When Other Things Aren't Equal: Saving Ceteris Paribus Laws from Vacuity." *British Journal for the Philosophy of Science* 46(1): 81–110.

Reichenbach, H. (1947). *The Elements of Symbolic Logic* (New York: Macmillan).

Roberts, J. T. (1999). "'Laws of Nature' as an Indexical Term: A Reinterpretation of the Best-System Analysis." *Philosophy of Science* 66 (Suppl. 3): S502–S511.

Roberts, J. T. (2008). *The Law-Governed Universe* (Oxford: Oxford University Press).

Roberts, J. T. (2014). "CP-Law Statements as Vague, Self-Referential, Self-Locating, Statistical, and Perfectly in Order." *Erkenntnis* 79(10): 1775–1786.

Shoemaker, S. (1980). "Causality and Properties." In P. van Inwagen (ed.), *Time and Cause* (Dordrecht: Reidel), 109–135.

Strevens, M. (2012). "'Ceteris Paribus' Hedges: Causal Voodoo that Works." *Journal of Philosophy* 109(11): 652–675.

Tooley, M. (1977). "The Nature of Laws." *Canadian Journal of Philosophy* 7: 667–698.

van Fraassen, B. (1989). *Laws and Symmetries* (Oxford: Oxford University Press).

Ward, B. (2002). "Humeanism without Humean Supervenience." *Australasian Journal of Philosophy* 60: 203–223.

FURTHER READING

Carroll, J. W. (2004). *Readings on Laws of Nature* (Pittsburgh: University of Pittsburgh Press).

CHAPTER 17

....

METAPHYSICS IN SCIENCE

....

RICHARD HEALEY

1 INTRODUCTION

SCIENCE may have been conceived in pre-Socratic metaphysics, but it has proved to be an unruly offspring. Science has transformed, if not solved, some metaphysical problems while posing new ones. Metaphysical ideas such as those of the ancient atomists have sometimes proved helpful in developing new scientific theories. But the widespread agreement on the empirically grounded progress achieved in science has often been contrasted with what seem to be abstruse and interminable disputes over metaphysical theses. In the twentieth century Karl Popper sought to demarcate scientific from metaphysical and other claims by appealing to their empirical falsifiability, while Rudolf Carnap and other logical positivists dismissed metaphysical claims as cognitively meaningless since they are neither empirically verifiable nor true by virtue of meaning.

While sympathetic to Carnap's view of philosophy as the logical analysis of science, W. V. O. Quine raised influential objections to the positivists' views of meaning and evidence. Instead he advocated a holistic view of meaning and hypothesis confirmation that many have taken to imply that metaphysics is continuous with and indistinguishable in its methods from highly theoretical science. For this and other reasons, metaphysics is once again actively pursued, even by those unwilling to accept Quine's holism.

I offer a brief assessment of the contemporary relation between philosophy of science and resurgent metaphysics only after examining the impact of science on a number of more specific metaphysical issues.

2 WHAT IS EVERYTHING MADE OF?

It has been suggested that metaphysics consists in an attempt to give a general description of ultimate reality. But that leaves unclear why science or, at any rate, the more theoretical parts of science should not count as "metaphysics" by this definition (see Van Inwagen 2007). Indeed, the atomic theory began as one answer offered by pre-Socratic philosophers to the question "What is everything made of?"

The idea that everything is ultimately composed of some fundamental kind of thing continues to figure prominently in contemporary analytic metaphysics. Debates continue, about which things are fundamental and by virtue of what they compose anything else. These debates are occasionally conducted at a level of formal generality to which empirical considerations seem irrelevant, as when properties or facts are contenders for a fundamental ontology and the composition relation is wholly characterized by axioms of mereology (see Varzi 2003). But if ultimate reality has any connection to the empirical world, scientific findings cannot be ignored in metaphysics.

According to our best current science, everything in the physical world is not composed of atoms or their component particles. Light is not composed of atoms, nor are neutrinos or other elementary particles, including the recently discovered Higgs boson. Together this "ordinary matter" is believed to amount to a mere 5% of the universe's non-gravitational mass/energy content. Our best current physical theories of it are quantum field theories. But while interacting quantum field theories are key components of what is often called the Standard Model of elementary particles, they do not literally describe particles, and neither physicists nor philosophers have reached a consensus on their fundamental ontology.[1]

Contemporary physics employs a variety of different composition/decomposition relations, including aggregation, superposition, and the subsystem/supersystem relation. They differ in their formal properties and cannot be combined into any single relation of physical composition that might give significance to the question "What is everything made of?" More positively, advances in physics have suggested new candidates for a fundamental ontology and also for how a fundamental ontology might give rise to everything else in the world.

Scientific theorizing has influenced metaphysical debate about the nature and existence of space and time at least since Newton. The locus of the debate altered after relativity theory was taken to unify these into a new category of space-time. While some argued that space-time exists independently of matter, Einstein came to believe that his general theory of relativity represented space-time as having no existence independent of the gravitational field metric (see Einstein 1916/1961: fifth appendix). This is just one example of how changed conceptions of space, time, and matter have transformed the traditional metaphysical debate between substantivalism and relationism about space and time.

These changes continue, as string theories suggest physical space may have ten or more dimensions, loop quantum gravity (and other proposals) explore the possibility that space-time is not fundamental but emerges in some limit from a quantum theory set in some quite different space, and certain proposed revisions of quantum theory

---

[1] Healey (2013, Section 6) contains references to work on the ontology of quantum field theories and discusses different composition relations employed in physics.

posit as fundamental a (configuration) space of an enormous number of dimensions (see Albert 1996).

Relativity theory may be seen to favor a fundamental ontology of events or processes in space-time rather than spatially located material objects persisting through time. A physicist will sometimes use the term "event" in a context where a philosopher would use "space-time point" to refer to a smallest part of space-time. But material objects are nothing over and above a set of discrete spatiotemporally localized events—"flashes"— in one objective collapse theory proposed as a conceptually unproblematic improvement on current formulations of quantum theory.[2]

Atomic theory provided a paradigm for physical composition in which spatial juxtaposition and possibly some interatomic force combined atoms as constituent parts of a material object capable of persisting through time. Contemporary science offers an alternative model in which a less fundamental object may emerge as a more or less persistent dynamic pattern constituted by the behavior of more fundamental objects: examples include water waves, Jupiter's red spot, and phonons and other varieties of quasi-particles in condensed matter physics. Some contemporary Everettians maintain that material objects, human and other organisms, and indeed the world in which we find ourselves all emerge as dynamic patterns in the quantum state of the universe (along with many other similarly emergent worlds, which together form the multiverse; see Wallace 2012).

3 IDENTITY AND INDIVIDUALITY

The nature and identity conditions of individuals have emerged as important foundational topics in contemporary science, both biological and physical. In evolutionary biology the thesis that species are individuals has challenged the traditional idea that species are natural kinds, while the nature of individuality has become a central concern of biologists and philosophers investigating major transitions such as that involved in the coalescence of cells into multicellular organisms.

Physicists and philosophers have queried the status of quantum particles as individuals, as well as the applicability of Leibniz's Principle of the Identity of Indiscernibles (PII): similar issues arise for space-time points in the general theory of relativity. Resolving these issues is one goal of the currently influential program of structural realism in metaphysically oriented philosophy of science.

While living organisms are almost universally recognized as individuals in biology, the extraordinary diversity of life presents biologists with problem cases in which it is not clear whether something is an organism or how organisms should be individuated. Viruses, ant colonies, and the 1,500-year-old "humongous fungus" found to have

---

[2] Maudlin (2011: ch. 11) is a good introduction to this proposal.

taken over a large area of northern Michigan are three examples. Van Inwagen (1995) argued that living organisms (unlike ordinary physical objects) are the only composite things that exist, leaving it to biology to say what these are. Not only organisms but also organs, genes, cells, mitochondria, groups, and species all lay claim to the status of biological individuals (see Wilson and Barker 2013). Common to all of these is the idea that an individual is something spatial that persists through time in interaction with its environment.

Appealing to the role of the species concept in evolutionary biology, Ghiselin (1997) and Hull (1989) argued that species are individuals rather than classes. This conclusion has been resisted even by many who accept the importance of population thinking and reject biological essentialism (see Kitcher 1984; Crane 2004 is one metaphysician's response).

Leibniz appealed to biological individuals to illustrate his famous PII when he told a story of an "ingenious gentleman" who thought he could find two identical leaves in a princess's garden; after an extensive search, the gentleman failed because "he was convinced by his eyes that he could always note the difference." The gentleman would have done better to look to fundamental physics in his search for indistinguishable but distinct individuals. Physicists use the term "identical" to refer to particles (or other physical items) that share all the same intrinsic physical properties, such as mass and charge. According to current theory, any two electrons (for example) are identical in this sense though not (of course) in the strict sense of being a single electron.

In classical physics a pair of "identical" electrons could be distinguished by their extrinsic properties since their spatiotemporal trajectories must differ. But electrons do not have spatiotemporal trajectories according to (orthodox) quantum theory, and the two electrons in a helium atom (for example) cannot be distinguished from each other either intrinsically *or extrinsically*, nor can two photons in a laser beam. But electrons and photons have different ways of being "identical." When taken to represent a collection of "identical" bosons (such as photons), a quantum wave-function remains the same if they are interchanged, while that representing a collection of "identical" fermions (such as electrons) is multiplied by –1. The distinction between fermions and bosons has important physical manifestations.[3]

We often think of elementary particles, including electrons, as ultimate constituents of matter (see Section 2). If they are not distinguishable individuals, then it is not obvious how they could compose spatially distinct material objects or living organisms. Quantum field theory suggests a different perspective on electrons: photons and other so-called elementary particles as excited states ("quanta") of some more fundamental field—the electron, electromagnetic field, or other fields of the Standard Model of high energy physics. Teller (1995) argues that "identical" quanta of such a field cannot be labeled, counted, or individuated in any other way (he says they lack "primitive

[3] The fermionic character of electrons accounts for the exclusion principle that determines basic features of the periodic table of chemical elements, while bosonic helium becomes a superfluid at very low temperatures.

thisness"). Huggett (1997) agrees with the last claim but rejects Teller's arguments for the first two: quanta could exist as metaphysically distinct and persist (for a while) even if we could not distinguish or trace them over time and they could not be modally switched. Maybe that is enough to confer individuality.

If quantum particles are individuals, then they appear to refute Leibniz's PII. Some have attempted to rescue PII, while others have concluded that quantum particles are not individuals.[4]

One can try to rescue PII by appealing to distinguishing features of "identical" quantum particles not captured by orthodox quantum theory: such features are usually referred to as hidden variables. For example, according to Bohmian mechanics (which is arguably empirically equivalent to orthodox quantum mechanics), each electron (or other massive particle) has a different, continuous trajectory. Or one can adopt a weakened form of PII, according to which two individuals may be distinguished not by any intrinsic or extrinsic properties but by bearing one another an irreflexive relation.[5]

An alternative is that quantum particles exist but are not individuals because they lack self-identity. It is not immediately clear what this means: certainly it violates the Quinean dictum "no entity without identity." Nevertheless, French and Krause (2006) set out to develop a formal framework that would enable one to speak intelligibly about (quantum) objects that are not individuals.

But the quantum fields of the Standard Model interact, and it seems that neither particles nor quanta can be recovered as an ontology of an interacting quantum field theory (see Fraser 2008; Ruetsche 2011). If so, currently fundamental physics does not itself describe elementary particles—either as individuals or as nonindividuals. Their status as emergent entities may incline the metaphysician seeking ultimate constituents to suspend inquiries or to pursue them elsewhere.

One place she might look is to the points or regions of space-time (see Stachel 2002; Wallace and Timpson 2010). General relativity may be interpreted as a theory of the structure of space-time, and algebraic quantum field theory is most naturally understood as attributing quantum states to regions of space-time. Permuting points in a manifold representing space-time has been argued to leave everything the same (i.e., to represent no distinct possible world/state of the world). This raises the issue of whether space-time points (or regions) are individuals.[6]

While it is not obvious that general relativity attributes any properties to space-time *points* (see Section 6), it certainly attributes properties (corresponding to values of the magnitudes whose limits define curvature, for example) to compact *regions*, and algebraic quantum field theory attributes states to them. Our universe is so complex that in

---

[4] French (2011) is a good introduction with many references.
[5] See Saunders (2003), who takes the electrons in a helium atom to bear one another a characteristically quantum irreflexive relation we might call "having opposite spins" even though neither has a determinate spin! Others have extended this line of argument to arbitrary finite sets of "identical" bosons as well as fermions.
[6] Section 8 revisits the issue of the individuality of space-time points and quantum particles.

any accurate model these properties may serve to individuate those regions in accordance with PII. This would not be true in a simpler, more symmetric universe, but philosophers of science will gladly leave such universes to pure metaphysicians.

4 TIME AND CHANGE

Metaphysicians have offered rival accounts of change ever since Parmenides denied its possibility. If change involves something's having incompatible properties at different times, then such an account will involve assumptions about things, properties, and time. Relativity theory revised basic assumptions about all three, and current attempts to unify general relativity with quantum theory intimate yet more radical revisions.

In relativity theory, the basic distinction between temporal and spatial structure is drawn locally—by the light-cone structure at each point in space-time. Temporal order and duration are defined with respect to a time-like curve—a continuous, future-directed curve in space-time threading the interior of the light cones at each point. An ideal clock would measure duration along the time-like curve that traces out its history. To define a global notion of time, we need to move off this curve.

Consider first the Minkowski space-time of special relativity. Suppose we start with a single time-like curve—the potential world-line of a freely moving, everlasting, ideal clock. Then there is one particularly natural way of arriving at a global temporal structure. At each instant on this curve there is a uniquely privileged three-dimensional (3D) Euclidean space, every point of which can be assigned the same time. These instantaneous spaces stack up parallel to one another in space-time and so define time order and duration for events and processes everywhere in space-time, as well as giving sense to the idea of points of space enduring through this global time.

What we have arrived at is simply the time of an inertial frame in which our ideal clock is at rest. If we wished, we could choose to understand change as simply something's having incompatible properties at different values of this time. But of course that would be to single out this frame from all other inertial frames as the arbiter of genuine change. Minkowski space-time satisfies the special principle of relativity precisely because it provides no basis for privileging a particular inertial frame in this way.[7]

There are no global inertial frames in a generic general-relativistic space-time; however, in a neighborhood of any space-time point p there is a continuous time-like curve in whose neighborhood one can define what is called a normal frame. Such a frame still specifies a privileged family of instantaneous 3D spaces and says what events in different instantaneous spaces occur at the same enduring place. But the family may not be defined globally—a curve in an instantaneous space may simply end at a singularity, or the 3D spaces may intersect one another far enough away from the initial curve. Also,

---

[7] There is then no way to state the metaphysical doctrine of presentism, according to which the only things in the world that exist are those that exist now.

the instantaneous spaces will not, in general, be Euclidean or even share the same geometry. Still, we do have a well-defined frame in the neighborhood of p in which we can now define change.

To summarize, to introduce notions of duration and identity of spatial position at different times into a relativistic space-time one must add at least a time-like curve, and extending notions of simultaneity and sameness of place off a single time-like curve requires additional structure. This makes change in relativity frame-dependent. So change is real but relational in relativity just like distance, duration, and velocity.

This conclusion has been disputed, and interesting physical arguments have been given against the reality of change and even against the reality of time itself. Gödel (1949) constructed a novel model of general relativity and used it to argue that time is not real.[8] He proposed as a necessary condition for time's reality that there be an "objective lapse of time," requiring a succession of universe-wide moments. Taking his model universe to be physically possible, he noted that it contained not even one such moment and argued that time is not real since the reality of time could not hinge on contingent features of our actual universe.

In attempting to quantize general relativity, physicists have selected classical magnitudes to promote to quantum observables by requiring that they be invariant under the theory's basic symmetries. That has been understood either as requiring genuine physical magnitudes to be diffeomorphism invariant or as requiring them to be gauge invariant in a Hamiltonian formulation of general relativity. Ordinary magnitudes such as the curvature and space-time metric are neither, and in a standard interpretation of the Hamiltonian formulation of general relativity no gauge invariant magnitude can change. Earman (2002) used these considerations to construct an argument for the unreality of change (also see Maudlin 2003; Healey 2004).

Barbour (2000) argued that time is unreal based on his assumption that a quantum theory of gravity will posit a collection of "momentary" configurations of the universe (for reactions see Butterfield 2002; Healey 2002; Ismael 2002). Any temporal order or duration would have to emerge from the internal features of these configurations, but these are insufficient to determine a unique temporal ordering. The time order, and temporal duration, of general relativity would emerge only as an approximation in a suitable classical limit. More recent collaborative work (Barbour et al. 2013) proposes a near empirical equivalent to general relativity that reintroduces a universal temporal order as a step toward a quantum theory of gravity.

5 DETERMINISM

Determinism and materialism (or its contemporary descendant, physicalism) are metaphysical theses brought into prominence with the rise of modern science. Each can be

---

[8] Malament (1985) is a good introduction to Gödel's model. Yourgrau (1991) and Earman (1995: 194–200) offer contrasting evaluations of Gödel's argument.

formulated as a claim that certain physical features of the world determine some distinct class of worldly features. Determinism holds that the complete physical state of the world at one time determines its complete physical state at any other time, that is, its complete (future and past) physical history.[9] Physicalism holds that the complete physical history of the world determines its entire history.

In each case, the determination relation may be understood in modal terms. Determinism says that given the complete physical state of the world at one time and the laws of physics, only one complete physical history of the world is possible. Physicalism says that given the complete physical history of the world only one complete history of the world is possible. The following famous passage from Laplace (1820/1951) may be read as implying both theses:[10]

> An intellect which at a certain moment would know all forces that set nature in motion, and all positions of all items of which nature is composed, if this intellect were also vast enough to submit these data to analysis, it would embrace in a single formula the movements of the greatest bodies of the universe and those of the tiniest atom; for such an intellect nothing would be uncertain and the future just like the past would be present before its eyes. (p. 4)

The basis for his confidence was the Newtonian classical mechanics Laplace himself had so successfully applied to the solar system. This confidence appears engagingly naive in the light of subsequent scientific developments.

Assuming classical mechanics, there are models of somewhat artificial but quite simple worlds whose complete physical state at one time is compatible with more than one complete physical state at other times.[11] While any nonidealized classical model of our world would be vastly more complex, the absence of an upper limit on velocities in classical mechanics means no evidence could decide between a model in which determinism holds and one in which it does not.[12]

Classical mechanics was superseded as fundamental physics by relativity and quantum theory. Special relativity improves the prospects for determinism because its fixed light-cone structure defines a limiting velocity. But general relativity clouds them once more. Determinism cannot even be defined in a space-time in which not every local moment of time can consistently be extended to distant regions.[13] We have strong

---

[9] Here are two related formulations: the world's complete physical state at/up to any time t determines its complete physical state at every later time.

[10] By embellishing his formulation with the metaphor of a vast intellect, Laplace introduces a distracting epistemic element into what is basically a metaphysical claim.

[11] Hoefer (2003) gives a simple introduction to this issue, as well as everything else in this section. Earman (1986) offers a more comprehensive treatment of physical determinism.

[12] Earman (1986: 33–39) gives further details of so-called space-invaders—particles that do not exist before time t then appear from arbitrarily far away, traveling here at an arbitrarily fast speed and slowing as they approach.

[13] Only in a space-time foliated by space-like Cauchy surfaces can one define the state of the world at each moment: there are no space-like Cauchy surfaces in Gödel space-time.

evidence that our universe contains black holes where general relativity predicts a space-time singularity is concealed behind an event horizon. Determinism fails in a rotating black hole that may form from the collapse of a massive, spinning star but not for our region of the universe outside its event horizons. It would fail more dramatically if any singularities were not concealed behind an event horizon, a possibility yet to be ruled out theoretically. Anything might emerge from such a naked singularity even given the entire history of the world prior to its formation.

The alleged determinism of classical physics is popularly contrasted with the supposed indeterminism of quantum physics, although some prominent approaches seek to reconcile quantum probabilities with an underlying determinism.

Probability enters quantum theory via the Born rule. As traditionally formulated, this specifies the probability that a measurement of a magnitude on a system will yield one outcome rather than another, given the quantum state of that system. If a quantum system is neither measured nor interacts with another quantum system then its quantum state evolves deterministically, in accordance with the Schrödinger equation (or some linear relativistic analog). But measurement typically requires reassignment of the quantum state, where the postmeasurement state is not determined by the premeasurement state but depends on the probabilistic outcome of the measurement. If this quantum state describes a fundamental aspect of the physical world, then quantum measurement induces a stochastic physical change in that aspect, so the world is fundamentally indeterministic.

Understood this way, the Born rule and associated indeterminism apply only on measurement. This is widely regarded as unacceptably vague, and since it introduces 'measurement' as primitive, quantum theory cannot provide a noncircular explication of the term.[14] Slightly modified dynamical equations have been proposed for a system's quantum state that would encompass its (mostly) linear evolution for an isolated system like an atom as well as the nonlinear, stochastic behavior manifested by enormously larger systems, including apparatus used to record the outcome of the interaction involved when such a microscopic system is measured. These proposals result in alternative indeterministic "collapse" theories. Although each differs in some predictions from quantum theory, those differences are currently too small to be experimentally distinguished (for further details, see Ghirardi 2011). No current "collapse" theory can be considered a serious alternative to the interacting quantum field theories of the Standard Model.

Bohm independently rediscovered ideas proposed earlier by de Broglie and further developed them into an alternative interpretation of quantum mechanics—perhaps better thought of as an alternative theory. According to Bohmian mechanics, all particles have, at all times, a definite position and velocity. In addition to the Schrödinger equation, this theory posits a guidance equation that determines, on the basis of the joint quantum state and initial positions of n particles, what their future positions and

---

[14] For further development of this objection and possible reactions (including the Ghirardi–Rimini–Weber collapse theory and Bohmian mechanics) see Bell (2004: 213–231).

velocities should be. Bohmian mechanics is deterministic: the particles' quantum state evolves deterministically with no collapse, and the guidance equation specifies deterministically evolving trajectories for all the particles. The Born rule applies directly to the actual positions of the particles: to justify its application to a measurement on a subset of them whose outcome is recorded in the positions of others it is necessary to analyze the kind of interaction that would require. As long as the statistical distribution of initial positions of particles is chosen so as to meet a "quantum equilibrium" condition, Bohmian mechanics is empirically equivalent to standard quantum mechanics. Critics object that this equivalence cannot be extended to a fully relativistic Bohmian version of interacting quantum field theories.

Everett (1957) proposed a (re)formulation of quantum theory according to which the quantum state of the universe always evolves deterministically. It appears to evolve indeterministically only because every physical observation is carried out in a world within the universe, with a different outcome in different emergent worlds. Critics have objected that Born probabilities make no sense when every possible outcome of a quantum measurement actually occurs (in some world). But by adapting decision theory to a branching multiverse, contemporary Everettians have offered a derivation of the Born rule, as well as an account of how observations of measurement outcomes provide evidence for Everettian quantum theory. Wallace (2012) has also offered an account of what constitutes a world in Everettian quantum theory and of how different worlds emerge within a single universe. Besides its potential for combining empirical indeterminism with underlying determinism, Everettian quantum theory also prompts a reevaluation of the metaphysics of constitution and continued identity for worlds, material objects, and persons.[15]

6 Physicalism, Supervenience, and Holism

The previous section expressed physicalism as the thesis that the complete physical history of the world determines its entire history. For various reasons this should be taken only as a rough, preliminary statement.[16] But it does serve to introduce three respects in which developments in physics bear on a metaphysical thesis of physicalism.

What is a world, and what counts as its history? Laplace could assume a unique Newtonian world of particles and forces existing in 3D space and persisting through infinite absolute time. The analog to the complete physical history of a relativistic world is the entire physical contents of four-dimensional space-time. The dimensionality and

---

[15] Together, Saunders et al. (2010) and Wallace (2012) provide a comprehensive survey of the contemporary debate on Everettian quantum theory.

[16] Stoljar (2009) gives some reasons and offers improvements on this statement.

even the existence of space-time (except as emergent structure) remain open questions in projected quantum theories. According to contemporary Everettians, our familiar physical world emerges as just one branch of the universal quantum state, along with a vast multiverse of similarly real worlds, all within a single space-time that may itself come to be seen as emergent from some more basic physical structure.

How might the world's physical history determine its complete history? The idea is that there can be no difference in the world's history without a difference in its physical history. But this raises several questions. What makes a difference in world history physical? What kind of modality is conveyed by the word "can," and what grounds the necessary connection between the world's physical features and its nonphysical features?

Defining the physical by reference to current physics may well render physicalism trivially false, since a future physics will quite likely specify new physical features of the world not determined by features specified by current physics. However, defining the physical by reference to some hypothetical "final" completed physics renders its content inscrutable.

Since an important application of physicalism is to the mental, one option is to define the physical as whatever is not mental (where this is independently characterized). To proceed this way is to place a bet on the development of physics by assuming that future physics will not appeal directly to such mental features.

The necessity by which the physical determines the nonphysical is presumably not grounded in laws of physics, which relate purely physical features of the world. This appears to make physicalism a purely metaphysical thesis. But Quine (1978) argued that the truth of physicalism derives from the nature of physics itself, as the discipline a major purpose of which is to compile a catalog of elementary physical states sufficient to render physicalism true.

Lewis (1986) sought to defend the tenability of what he called Humean supervenience:

> the doctrine that all there is to the world is a vast mosaic of local matters of particular fact, just one little thing and then another. (. . .) We have geometry: a system of external relations of spatiotemporal distance between points. Maybe points of space-time itself, maybe point-sized bits of matter or aether or fields, maybe both. And at those points we have local qualities: perfectly natural intrinsic properties which need nothing bigger than a point at which to be instantiated. For short: we have an arrangement of qualities. And that is all. There is no difference without difference in the arrangement of qualities. All else supervenes on that. (pp. ix–x)

He noted that this is not exactly physicalism ("materialism"), because it does not require these local qualities to be physical, but doubted that anything but qualities like those recognized by current physics could make it true.

The following principle is a natural restriction of Humean supervenience to physics:

Spatiotemporal Separability (SS): Any physical process occupying spacetime region R supervenes upon an assignment of qualitative intrinsic physical properties at space-time points in R.

Contemporary physics is not easily reconciled with SS. Even classical physics posits magnitudes like velocity and mass density that apparently depend on qualities whose instantiation requires at least a *neighborhood* of the point at which the magnitude is defined (see Butterfield 2006). Quantum theory arguably involves attribution of physical properties or relations to systems with "entangled" states that do not supervene on the physical properties of their subsystems together with their spatiotemporal relations (see Healey 1991; Maudlin 2007). Such failure would be an instance of a kind of metaphysical holism within physics (see Healey 2008). The two photons constituting an entangled pair have been detected over 100 kilometers apart: whenever its subsystems are not spatiotemporally coincident, the persistence of such a system would constitute a nonseparable physical process, in violation of SS. Wallace and Timpson (2010) locate a radical failure of SS even in a quantum field theory, which postulates no persisting systems.

7 CAUSATION, LAWS, AND CHANCE

Section 5 introduced the thesis of determinism by quoting a passage from Laplace (1820/1951). The following sentence immediately precedes that passage: "We ought to regard the present state of the universe as the effect of its antecedent state and as the cause of the state that is to follow."

This squares with the idea that causes are antecedent conditions that bring about their effects in accordance with some law of nature—a popular idea among logical empiricist philosophers of science (see, e.g., Hempel 1965). It also squares with a metaphor they would have resisted—of the world as a giant machine that generates world history from some initial state through the operation of fundamental dynamical laws it is the business of physics to discover.[17] With the advent of quantum theory, the idea and metaphor came to be relaxed to admit the possibility of laws that are stochastic rather than deterministic so causation may be "chancy." But there is still a widespread view that science should explore the causal structure of the world and that it is by appeal to that structure that science is able to explain phenomena (see Salmon 1984; Dowe 2000).

Following Lewis's influential critique of regularity analyses of causation and advocacy of counterfactual analyses, analytic metaphysics has downplayed the importance of laws in contributing to our understanding of the causal relation (see Menzies 2014; Schaffer 2014). Maudlin (2007) argues against this trend. Taking his cue from physics, he offers accounts of counterfactuals and causation in terms of laws taken as primitive. Steven Weinberg is one physicist who seems ready to accept the primitive status of fundamental physical laws.[18]

---

[17] See Maudlin (2007) who calls these FLOTEs (Fundamental Laws of Temporal Evolution).

[18] Weinberg (1996) says, "What I mean when I say that the laws of physics are real is that they are real in pretty much the same sense (whatever that is) as the rocks in the fields."

Russell's (1913) famous argument that a law of causality plays no role in fundamental physics continues to provoke debate (see Price and Corry 2007; Frisch 2012). But some physicists still seek explicitly causal explanations and look for a way to express the fundamental causal structure of theoretical physics (see Bell 2004, especially 245–246). This is in tension with some naturalist philosophers' attempts to understand causation not as a relation to be analyzed physically or metaphysically but as a notion that is best understood as arising only from the perspective of a physically situated agent.

Beside their scientific importance in fueling what some call a second quantum revolution, Bell's theorem and his subsequent arguments (see Bell 2004) form a locus of philosophical inquiry: the experimental violation of associated Bell inequalities even led Shimony to speak of experimental metaphysics (Cushing and McMullin 1989: 64, and elsewhere). Bell himself argued that quantum mechanics is not a locally causal theory and cannot be embedded in such a theory (2004, 230–244), and that "certain particular correlations, realizable according to quantum mechanics, are locally inexplicable. They cannot be explained, that is to say, without action at a distance", pp. 151–152. Some have gone further, claiming that experimental results confirming violation of Bell inequalities show that the world is nonlocal (see Albert and Galchen 2009: 36; Maudlin 2011: 13; Goldstein et al. 2011: 1).

Bell's condition of local causality was designed to permit the application to probabilistic theories of the intuitive principle that "The direct causes (and effects) of events are near by, and even the indirect causes (and effects) are no further away than permitted by the velocity of light" (2004: 239).

Here is the condition:

> *Local Causality:* A theory is said to be locally causal if the probabilities attached to values of local beables in a space–time region 1 are unaltered by specification of values of local beables in a space–like separated region 2, when what happens in the backward lightcone of 1 is already sufficiently specified, for example by a full specification of all local beables in a space–time region 3. (pp. 239–240)

Region 3 is a thick slice right across the backward light cone of 1. A "local beable" is just some magnitude to which the theory ascribes a value in the relevant region.

Based on this condition Bell writes down an equation like this expressing the probabilistic independence of the value of magnitude A in 1 on the values of magnitudes B,b in 2, given the value of a in 1 and sufficient specification of magnitude values λ in region 3:

$$PI: prob(A|B,a,b,\lambda) = prob(A|a,\lambda).$$

For certain systems in certain states there are magnitudes A,B,a,b whose observed values violate a Bell inequality implied by this equation (together with the assumption that the probability of λ does not depend on a,b—magnitudes whose values we can freely choose, unlike those of A,B).

Maudlin and others argue that violation of probabilistic independence here entails some causal dependence of the value of A in 1 on the values of magnitudes B,b in 2, even though this would require superluminal causal influence. On these grounds they argue that the world is nonlocal. Note that quantum theory plays no role in the argument: observations merely bear out the theory's predictions in this situation.

Healey (2014) rejects this conclusion. He argues that the failure of *PI* refutes neither *Local Causality* nor the intuitive principle on which Bell based *PI*. The condition of *Local Causality* is not applicable to quantum theory since it postulates no unique probability whose value is capable of being altered: quantum theory correctly attaches not one but *two* probabilities to values of local beables in 1, neither of which is altered by specification of values of local beables in a space-like separated region 2 (one probability already requires this specification; the other is uniquely determined by what happens in the backward light cone of 1).

Since there is no universal time in a relativistic theory, any probabilities it specifies for a localized event cannot be specified at such a time but only at a space-time point—the idealized location of a hypothetical situated agent who should set his or her credence equal to that probability. $prob(A|a,\lambda)$ specifies the probability for such an agent situated in the backward light cone of 1, while $prob(A|B,a,b,\lambda)$ specifies the probability for such an agent situated in the forward light cone of 2 but not 1. Their different space-time situations give each access to different evidence concerning the value of A, so they should set their credences to match different probabilities (both supplied by quantum theory). Here as elsewhere quantum theory provides good advice about a local world.

8 STRUCTURE AND FUNDAMENTALITY

If metaphysics consists in an attempt to give a general description of ultimate reality, do the more theoretical parts of science count as metaphysics? Physicists and metaphysicians have each claimed to be investigating the fundamental structure of reality, and fundamental physics is often allotted a prominent role in contemporary philosophy (e.g., Ladyman and Ross 2007; Maudlin 2007; Sider 2011; Kutach 2013). However, a closer examination reveals that "fundamental" and "structure" are terms often used differently in pursuit of different intellectual projects about whose respective value strong opinions have been expressed.

Physicists typically think of fundamental physics as a work in progress satisfying two criteria:

1. It is unsettled because it lies at or beyond the cutting edge of current research.
2. It concerns everything in the universe at all scales (of energy, time, length, etc.).

When philosophers appeal to fundamental physics, they often have in mind some completed theory or theories capable of providing a detailed and very good (if not yet perfectly accurate) description of the natural world.[19] Contemporary analytic metaphysicians often assume such theories will be of a certain form (e.g., they will contain laws describing properties of objects). But any ascription of content tends to be either purely schematic or based on physics that physicists would regard as no longer fundamental (and often simply mistaken). As we see in Section 9, this has provoked a reaction from some philosophers of science, especially philosophers of physics interested in clarifying the conceptual foundations of particular physical theories.

Structural realism emerged as a mediating position in the dispute between scientific antirealists and traditional scientific realists (Worrall 1989). Following Ladyman's (1998) essay, it has acquired the character of a metaphysical view of how fundamental science is able to reveal the structure of physical reality (see Ladyman 2014; French 2014). In this guise the structure that is revealed is more fundamental than any objects or properties one may attempt to abstract from it. This view appears to mesh better with a semantic than a syntactic conception of scientific theories.

By contrast, the structure Sider (2011) takes as the domain of metaphysics, though worldly, is characterized in terms modeled on the syntactic categories of modern logic. Sider holds not only that some (but not other) predicates pick out natural properties but also that there are fundamental objects, since the "joint-carving" metaphor can be given literal content when applied to quantifiers and names. Maudlin reminds us of Bertrand Russell's (1923: 84) warning that almost all thinking that purports to be philosophical or logical consists in attributing to the world properties of language and appeals to gauge theory to argue that "if one believes that fundamental physics is the place to look for the truth about universals (or tropes or natural sets), then one may find that physics is telling us there are no such things" (2007: 96).

Healey (2007) and Arntzenius (2012) disagree: each draws different metaphysical conclusions from gauge theory. Healey argues that (at least classical) gauge theories are best understood as attributing nonseparable features (holonomy properties) to the world, while Arntzenius maintains that the "internal spaces" on which gauge potentials are defined as connections are one of many examples of physical spaces, each just as real as ordinary 3D space.

Ladyman distinguished epistemic from ontological structural scientific realism and advocated the latter: "structural realism amounts to the claim that theories tell us not about the objects and properties of which the world is made, but directly about structure and relations" (1998: 422). In extreme slogan form, ontological structural realism is the view that structure is all there is.[20]

The view has spawned a cottage industry in recent philosophy devoted to attempts to clarify, develop, attack, defend, and apply it to particular scientific theories.[21] Two

---

[19] Ladyman and Ross (2007) provide a notable exception by endorsing a version of (2).

[20] Contrast this slogan with the opening sentences(?) of Sider (2011): "Metaphysics, at bottom, is about the fundamental structure of reality . . . Not about what there is. Structure."

[21] Ladyman (2014) is a good place to begin.

prominent applications are to general relativity and to the status of "identical" quantum particles (see Stachel 2002; Pooley 2006; Ladyman and Ross 2007).

Earman and Norton (1987; see also Norton 2011) developed the so-called hole argument (which influenced Einstein's route to general relativity) into an argument against a form of space-time substantivalism—the view that points of space-time exist independently of anything else that may be located in space-time. Noting the analogy between the permutations involved in a "hole diffeomorphism" applied to a general relativistic space-time manifold and permutations of the quantum particles discussed in Section 3, Stachel formulated a principle of

> *General Permutability (GP)*: Every permutation *P* of the *a* entities in *S*, *R(a)*, *R(Pa)* represents the same possible state of the world. (2002: 242)

Here *S* is a set of *a* entities and *R* a collection of relations among them. Accepting the analogy, Ladyman and Ross (2007) maintain that the ontological problems of quantum mechanics and general relativity in respect of identity and individuality both demand resolution by ontological structural realism. Pooley (2006) rejects the analogy, as well as the demand.

If the *a* entities are mathematical elements of a mathematical model (points of a differential manifold or numerical labels of quantum particles) we use to represent the relevant physical system, then GP may be taken to express a redundancy in the models of a physical theory with no metaphysical import. But if a permutation is understood to apply to physical entities so represented (space-time points or quantum particles), then GP faces a dilemma. Either there are no such individuals or their permutation does *not* represent the same possible state of the world. Teller (2001) considers ways of responding to the hole argument that deny the possibility of the permuted state but concludes that none could secure the individuality of quantum particles.

Although common and relatively harmless in science, the possibility of such model/world ambiguity is a present danger for those who would draw metaphysical conclusions from science. A critic might issue an updated version of Russell's warning (to which Maudlin drew our attention—see previous page): almost all thinking that purports to be naturalistic metaphysics consists in attributing to the world properties of mathematical scientific models.

9 THE STATUS OF METAPHYSICS

Metaphysics has long been a controversial branch of philosophy. Declared dead by the logical positivists, it was reborn in the last quarter of the twentieth century and now asserts the full privileges of adulthood. However, with privilege comes responsibility, and contemporary analytic metaphysics has come under attack from two directions for shirking its responsibilities.

Externally, prominent scientists have criticized philosophers for ignoring scientific developments bearing on metaphysical questions.[22] They may be criticized in turn for ignoring what philosophers of particular sciences (including physics, neuroscience, and biology) have had to say about many of those developments.[23] Criticisms of contemporary analytic metaphysics from within philosophy must be taken more seriously. Some are accompanied by suggestions as to how metaphysicians might more profitably redirect their philosophical energies.

Ladyman and Ross (2007) is a prominent example of an internal attack. The authors maintain that

> standard analytic metaphysics (or "neo-scholastic" metaphysics as we call it) contributes nothing to human knowledge and, where it has any impact at all, systematically misrepresents the relative significance of what we do know on the basis of science. (vii)

They propose to confine metaphysics to a particular kind of radically naturalistic metaphysics (naturalistic metaphysics$_{ETMG}$). Naturalistic metaphysics$_{ETMG}$ attempts to unify hypotheses and theories taken seriously by contemporary science: moreover, these are to include at least one specific hypothesis drawn from fundamental physics, but no hypothesis that current science declares beyond our capacity to investigate should be taken seriously: "no alternative kind of metaphysics can be regarded as a legitimate part of our collective attempt to model the structure of objective reality" (2007: 1). So neoscholastic metaphysics is dead, but scientifically informed naturalistic metaphysics$_{ETMG}$ rises as a phoenix from its ashes![24]

For Huw Price, contemporary analytic metaphysics is a mere apparition:

> the ghost of a long-discredited discipline. Metaphysics is actually as dead as Carnap left it, but—blinded, in part, by these misinterpretations of Quine—contemporary philosophy has lost the ability to see it for what it is, to distinguish it from live and substantial intellectual pursuits. (2011: 282)

Rather than resurrect some naturalistically purified metaphysics, he proposes subject naturalism as metaphysics' legitimate heir. The subject naturalist investigates such

---

[22] At a 2011 conference Stephen Hawking asked, "Why are we here? Where do we come from? Traditionally, these are questions for philosophy, but philosophy is dead. Philosophers have not kept up with modern developments in science. Particularly physics."

[23] Price's (1989) letter to *Nature* querying one of Hawking's own arguments is just one instance: issues of the journals *Studies in History and Philosophy of Modern Physics* and *Biology & Philosophy* are full of others.

[24] Readers will find in Ross, Ladyman, and Kincaid (2013) a variety of reactions and alternatives to naturalistic metaphysics$_{ETMG}$ offered by authors' sympathetic to naturalistic metaphysics of one kind or another.

topics as time, causation, probability, modality, and moral objectivity not to see how they figure in the structure of objective reality but to understand the function and origin of our concepts of them. The investigation is naturalistic because it is constrained by our best scientific accounts of the world and our situation in it. It is even a scientific investigation, although since its successful pursuit will dissolve philosophical problems Price calls it "philosophical anthropology."

So far this section has focused on the status of metaphysics in contemporary professional philosophy. But scientific progress has itself often depended on acceptance or rejection of very general assumptions that might reasonably be counted as metaphysical.[25] The atomic hypothesis is a classic example. Despite its important influence on Newton, he was able to conceive a mechanics that transcended the limits of the "mechanical philosophy" in part through his alchemical investigation into what Hume called the "secret springs" of nature. Some current projects in quantum gravity are motivated by appeals to metaphysical views of space and time that hark back to Leibniz, Mach, and Poincaré (see Barbour 2000; Smolin 2013, especially ch. 14).

Friedman (2001) makes a case that certain general principles have served an important function in rationally mediating progress through Kuhnian scientific revolutions. He thinks of these as constitutive principles analogous to those synthetic a priori principles Kant located in Newtonian mechanics. Since they define the space of real (mathematical and physical) possibilities within which properly empirical laws can be formulated, conformity to these principles is more than physically necessary. But for Friedman they have a merely *relativized* a priori status, which means they may come to be questioned or abandoned as science progresses.

An important manifestation of the resurgence of analytic metaphysics has been a renewed respect for a category of metaphysical necessity, intermediate between physical and logical necessity. But the notion of metaphysical necessity remains obscure if the correlative modality of metaphysical possibility cannot simply be identified with some kind of conceivability. By proposing that metaphysical possibility amounts to conceivable physical possibility, Leeds (2001) arrives at a very similar position as Friedman. For each of them, metaphysics is not insulated from but in a sense parasitic upon science because science importantly influences what we conceive as possible.

ACKNOWLEDGMENTS

Thanks to Paul Humphreys, Jenann Ismael, David Glick, and Paul Teller for helpful comments on an earlier draft of this article.

---

[25] Though dated, Burtt (2003) is still worth reading.

References

Albert, D. (1996). "Elementary Quantum Metaphysics." In J. T. Cushing, A. Fine, and S. Goldstein (eds.), *Bohmian Mechanics and Quantum Theory: An Appraisal* (Berlin: Springer), 277–284.

Albert, D., and Galchen, R. (2009). "Was Einstein Wrong? A Quantum Threat to Special Relativity." *Scientific American* 300: 32–39.

Arntzenius, F. (2012). *Space, Time and Stuff* (Oxford: Oxford University Press).

Barbour, J. (2000). *The End of Time* (Oxford: Oxford University Press).

Barbour, J., Koslowski, T., and Mercati, F., et al. (2013). "The Solution to the Problem of Time in Shape Dynamics." http://arxiv.org/abs/1302.6264

Bell, J. S. (2004). *Speakable and Unspeakable in Quantum Mechanics* (rev. ed.) (Cambridge, UK: Cambridge University Press).

Burtt, E. A. (2003). *The Metaphysical Foundations of Modern Science* (Minneola, NY: Dover).

Butterfield, J. (2002). "Critical Notice." *British Journal for the Philosophy of Science* 53: 289–330.

Butterfield, J. (2006). "Against *Pointillisme* about Mechanics." *British Journal for the Philosophy of Science* 57: 709–753.

Crane, J. (2004). "On the Metaphysics of Species." *Philosophy of Science* 71: 156–173.

Cushing, J., and McMullin, E. (1989). *Philosophical Consequences of Quantum Theory: Reflections on Bell's Theorem* (Notre Dame, IN: University of Notre Dame Press).

Dowe, P. (2000). *Physical Causation* (Cambridge, UK: Cambridge University Press).

Earman, J. (1986). *A Primer on Determinism* (Dordrecht, The Netherlands: Reidel).

Earman, J. (1995). *Bangs, Crunches, Whimpers and Shrieks* (Oxford: Oxford University Press).

Earman, J. (2002). "Thoroughly Modern McTaggart: or What McTaggart Would Have Said If He Had Learned the General Theory of Relativity." *Philosophers' Imprint* 2, http://hdl.handle.net/2027/spo.3521354.0002.003

Earman, J., and Norton, J. (1987). "What Price Spacetime Substantivalism?" *British Journal for the Philosophy of Science* 38: 515–525.

Einstein, A. (1961). *Relativity: The Special and the General Theory* (New York: Bonanza). (Original work published 1916)

Everett, H. (1957). "'Relative State' Formulation of Quantum Mechanics." *Reviews of Modern Physics* 29: 454–462.

Fraser, D. (2008). "The Fate of 'Particles' in Quantum Field Theories with Interactions." *Studies in History and Philosophy of Modern Physics* 39: 841–859.

French, S. (2011). "Identity and Individuality in Quantum Theory." In *Stanford Encyclopedia of Philosophy*, http://plato.stanford.edu/entries/qt-idind/

French, S. (2014). *The Structure of the World: Metaphysics and Representation* (Oxford: Oxford University Press).

French, S., and Krause, D. (2006). *Identity in Physics* (Oxford: Clarendon).

Friedman, M. (2001). *Dynamics of Reason* (Stanford: CSLI).

Frisch, M. (2012). "No Place for Causes? Causal Skepticism in Physics." *European Journal for Philosophy of Science* 2: 313–336.

Ghirardi, G. C. (2011). "Collapse Theories." In *Stanford Encyclopedia of Philosophy*, http://plato.stanford.edu/entries/qm-collapse/

Ghiselin, M. (1997). *Metaphysics and the Origin of Species* (Albany: State University of New York Press).

Gödel, K. (1949). "A Remark about the Relationship between Relativity Theory and Idealistic Philosophy." In P. A. Schilpp (ed.), *Albert Einstein: Philosopher-Scientist*, Vol. 2 (Lasalle, IL: Open Court), 555–562.

Goldstein, S. et al. (2011). "Bell's Theorem." *Scholarpedia* 6: 8378. http://www.scholarpedia.org/article/Bell%27s_theorem

Healey, R. A. (1991). "Holism and Nonseparability." *Journal of Philosophy* 88: 393–421.

Healey, R. A. (2002). "Can Physics Coherently Deny the Reality of Time?" In C. Callender (ed.), *Time, Reality and Experience* (Cambridge, UK: Cambridge University Press), 293–316.

Healey, R. A. (2004). "Change Without Change and How to Observe it in General Relativity." *Synthese* 141: 1–35.

Healey, R. A. (2007). *Gauging What's Real* (Oxford: Oxford University Press).

Healey, R. A. (2016). "Holism and Nonseparability in Physics." In *Stanford Encyclopedia of Philosophy*, http://plato.stanford.edu/entries/physics-holism/

Healey, R. A. (2013). "Physical Composition." *Studies in History and Philosophy of Modern Physics* 44: 48–62.

Healey, R. A. (2014). "Causality and Chance in Relativistic Quantum Field Theories." *Studies in History and Philosophy of Modern Physics*, 48: 156–167.

Hempel, C. (1965). *Aspects of Scientific Explanation* (New York: Free Press).

Hoefer, C. (2003). "Causal Determinism." In *Stanford Encyclopedia of Philosophy*, http://plato.stanford.edu/entries/determinism-causal/

Huggett, N. (1997). "Identity, Quantum Mechanics and Common Sense." *The Monist* 80: 118–130.

Hull, D. (1989). *The Metaphysics of Evolution* (Stony Brook: State University of New York Press).

Ismael, J. (2002). "Rememberances, Mementos and Time-Capsules." In C. Callender (ed.), *Time, Reality and Experience* (Cambridge, UK: Cambridge University Press), 317–328.

Kitcher, P. (1984). "Species." *Philosophy of Science* 51: 308–333.

Kutach, D. (2013). *Causation and its Basis in Fundamental Physics* (Oxford: Oxford University Press).

Ladyman, J. (1998). "What Is Structural Realism?" *Studies in History and Philosophy of Science* 29: 409–424.

Ladyman, J. (2014). "Structural Realism." In *Stanford Encyclopedia of Philosophy*, http://plato.stanford.edu/entries/structural-realism/

Ladyman, J., and Ross, D. (2007). *Every Thing Must Go* (Oxford: Oxford University Press).

Laplace, P. (1951). *Philosophical Essay on Probabilities* (New York: Dover). (Original work published 1820).

Leeds, S. (2001). "Possibility: Physical and Metaphysical." In C. Gillett and B. Loewer (eds.), *Physicalism and its Discontents* (Cambridge, UK: Cambridge University Press), 172–193.

Lewis, D. (1986). *Philosophical Papers*, Vol. 2 (Oxford: Oxford University Press).

Malament, D. (1985): "'Time Travel' in the Gödel Universe." in *PSA: Proceedings of the Biennial Meeting of the Philosophy of Science Association 1984*, Vol. 2 (Chicago: University of Chicago Press), 91–100.

Maudlin, T. (2003). "Thoroughly Muddled McTaggart" [with a Response by John Earman]. *Philosophers' Imprint* 2: 1–23. http://quod.lib.umich.edu/cgi/p/pod/dod-idx/thoroughly-muddled-mctaggart-or-how-to-abuse-gauge-freedom.pdf?c=phimp;idno=3521354.0002.004

Maudlin, T. (2007). *The Metaphysics within Physics* (Oxford: Oxford University Press).

Maudlin, T. (2011). *Quantum Non-Locality and Relativity* (3rd ed.) (Chichester, UK: Wiley-Blackwell).

Menzies, P. (2014). "Counterfactual Theories of Causation." In *Stanford Encyclopedia of Philosophy*, http://plato.stanford.edu/entries/causation-counterfactual/

Norton, J. (2011). "The Hole Argument." In *Stanford Encyclopedia of Philosophy*, http://plato.stanford.edu/entries/spacetime-holearg/

Pooley, O. (2006). "Points, Particles and Structural Realism." In D. Rickles, S. French, and J. Saatsi (eds.), *The Structural Foundations of Quantum Gravity* (Oxford: Oxford University Press), 83–120.

Price, H. (1989). "A Point on the Arrow of Time." *Nature* 340: 181–182.

Price, H. (2011). *Naturalism without Mirrors* (Oxford: Oxford University Press).

Price, H., and Corry, R. eds. (2007). *Causation, Physics and the Constitution of Reality* (Oxford: Clarendon).

Quine, W. V. O. (1978). "Otherworldly" [Review of N. Goodman, *Ways of World-making*]. *New York Review of Books*, November 23.

Ross, D., Ladyman, J., and Kincaid, H., eds. (2013). *Scientific Metaphysics* (Oxford: Oxford University Press).

Ruetsche, L. (2011). *Interpreting Quantum Theories* (Oxford: Oxford University Press).

Russell, B. (1913). "On the Notion of Cause." *Proceedings of the Aristotelian Society* 13: 1–26.

Russell, B. (1923). "Vagueness." *Australasian Journal of Philosophy and Psychology* 1: 84–92.

Salmon, W. (1984). *Scientific Explanation and the Causal Structure of the World* (Princeton, NJ: Princeton University Press).

Saunders, S. (2003). "Physics and Leibniz's Principles." In K. Brading and E. Castellani (eds.), *Symmetries in Physics* (Cambridge, UK: Cambridge University Press), 289–307.

Saunders, S., Barrett, J., Kent, A. and Wallace, D., eds. (2010). *Many Worlds?* (Oxford: Oxford University Press).

Schaffer, J. (2014). "The Metaphysics of Causation." In *Stanford Encyclopedia of Philosophy*, http://plato.stanford.edu/entries/causation-metaphysics/

Sider, T. (2011). *Writing the Book of the World* (Oxford: Oxford University Press).

Smolin, L. (2013). *Time Reborn* (Boston: Houghton Mifflin).

Stachel, J. (2002). "The Relations between Things versus the Things between Relations." In D. B. Malament (ed.), *Reading Natural Philosophy* (Chicago: Open Court), 231–266.

Stoljar, D. (2009). "Physicalism." In *Stanford Encyclopedia of Philosophy*, http://plato.stanford.edu/entries/physicalism/

Teller, P. (1995). *An Interpretive Introduction to Quantum Field Theory* (Princeton, NJ: Princeton University Press).

Teller, P. (2001). "The Ins and Outs of Counterfactual Switching." *Nous* 35: 365–393.

Van Inwagen, P. (1995). *Material Beings* (Cornell, NY: Cornell University Press).

Van Inwagen, P. (2007). "Metaphysics." In *Stanford Encyclopedia of Philosophy*, http://plato.stanford.edu/entries/metaphysics/

Varzi, A. (2003). "Mereology." In *Stanford Encyclopedia of Philosophy*, http://plato.stanford.edu/entries/mereology/

Wallace, D. (2012). *The Emergent Multiverse* (Oxford: Oxford University Press).

Wallace, D., and Timpson, C. (2010). "Quantum Mechanics on Spacetime I." *British Journal for the Philosophy of Science* 61: 697–727.

Weinberg, S. (1996). "Sokal's Hoax." *New York Review of Books*, August 8.

Wilson, R., and Barker, M. (2013). "The Biological Notion of Individual." In *Stanford Encyclopedia of Philosophy*, http://plato.stanford.edu/entries/biology-individual/

Worrall, J. (1989). "Structural Realism: The Best of Both Worlds?" *Dialectica* 43: 99–124.

Yourgrau, P. (1991). *The Disappearance of Time* (Cambridge, UK: Cambridge University Press).

CHAPTER 18

MODELS AND THEORIES

MARGARET MORRISON

1 INTRODUCTION: CONTEXT AND BACKGROUND

Two particularly vexing questions for philosophy of science are "What is a model?" and "What is a theory?" I am not going to pretend to answer either of these questions here, partly because there is not just one answer but many, depending on which philosophical position one adopts. In everyday parlance one often hears the phrase "It's only a model" to describe a hypothesis or claim that is tentative or yields some less than accurate information about a system/phenomena of interest. The contrast here is with the notion of theory as a body of knowledge that is better confirmed, justified, and more widely accepted. That use of the terms "model" and "theory" is typically found in practical contexts where the participants are less concerned with defining what a model is and more concerned with constructing good ones. Philosophers, on the other hand, are often worried about how terms are used, an exercise that typically involves some degree of conceptual analysis with epistemological and/or ontological implications. For example, if we characterize models as abstract entities we are immediately faced with the question of how they convey concrete information or how they relate to or represent the physical world. If models are understood as describing or representing fictional situations then similar problems arise in attempting to explain how they aid us in understanding. These are some of the philosophical problems surrounding the nature of models—problems that also have analogues or counterparts in the domain of theories.

The existence of many different kinds of models—physical models, mathematical models, fictional models, toy models, and so on—gives rise to different ways of characterizing them in both ontological and epistemological terms, something about which there is a certain amount of philosophical disagreement. My aim here is not to provide an extensive description of these various types of models but rather to address some of the problematic issues regarding their interpretation, their relation to theory, and how they convey information.[1] Here again, however, I want to stress that I do not necessarily see any way of providing generalizations that might cut across these different types of models, generalizations that would enable us to say how a model provides information about the world. Instead, we need to examine the model and its intricacies in order to determine how it functions in an informative capacity.

One of the consequences of the emphasis on models in philosophy of science has been a shift away from focusing on the nature of theories. Perhaps the most prominent account of theories, the semantic view, is, in most of its guises, not about theories at all but about models because the former are defined solely in terms of the latter (van Fraassen 1980, 1989, 2008; Giere 1988, 2004; Suppes 1961, 1962, 1967, 2002; Suppe 1989). Add to this the fact that a good deal of the literature on the practice of modeling, an approach sometimes referred to as the "scientific" account (Achinstein 1964; Hesse 1953, 1970; Kuhn 1972; Cartwright 1999; Cartwright, Suarez, and Shomar 1995; Morrison 1998, 1999; Morgan and Morrison 1999) and many more since (Frigg 2002; Suarez 2003, to name a few) has emphasized varying degrees of independence from theory, with the consequence that very little attention has been paid to the role theory plays in articulating scientific knowledge. The other focus of this chapter is an attempt to redress that imbalance by emphasizing the importance of both theories and models and how each functions in the construction and acquisition of scientific knowledge. Accomplishing the latter involves examining some of the difficulties encountered by various accounts of theory structure, in particular the problems faced by the semantic view. That we need a more robust notion of theory becomes clear once we look, even superficially, at much of the structure of modern science, especially physics.[2] Moreover, a cursory examination reveals that models and theories function at a different level of representation and explanation. Indeed, no general account of "scientific knowledge" can be articulated without some notion of theory that differentiates it from models.

Much of the criticism of the semantic view by the proponents of the "scientific" account has centered on its inability to account for the ways models are constructed and function in practical contexts. Although the semanticists stress that their view is primarily a logico-philosophical account of theory structure, they also emphasize its ability to capture scientific cases. Because of this dual role, it is important to evaluate the model-theoretic features of the account to determine its success in clarifying the nature of theory structure as well as its merits in dealing with the "scientific" or practical dimensions of modeling.

Some may think that the fate of the semantic view and its different instantiations (van Fraassen 1980; da Costa and French 2003; Suppes 2002; Giere 1988) has been debated so much in the literature that there is little left to say.[3] But with philosophy there is always

---

[1] For an extensive treatment of different kinds of models, see Frigg and Hartmann (2012).

[2] I focus primarily on physics since that is the science where the notion of a "theory" is most clearly articulated and where the distinction between theories and models is perhaps most obvious.

[3] The partial structures approach of da Costa and French (2003) is an extension of Suppes' account with the addition of a structure where only partial relations are defined on the individuals. The advantage

more to say, and, indeed, the debate is very much alive.[4] Van Fraassen describes the tension between the semantic view and its "scientific" rivals in the following way:

> Structural relations among models form a subject far removed from the intellectual processes that lead to those models in the actual course of scientific practice. So, what is important depends on what is of one's interest. Accordingly, if one's interest is not in those structural relations but in the intellectual processes that lead to those models, then the semantic view of theories is nowhere near enough to pursue one's interest.
>
> Both interests are important . . . leaving out either one we would leave much of science un-understood. But the basic concepts of theory and model—as opposed to the historical and intellectual dealing with formulations of theories, or on the contrary, construction of models—do not seem to me very different in the two approaches to understanding science. (2008: 311)

The quotation seems to suggest that (a) each view addresses different issues and hence there is no real tension between them (both are needed and each fulfills a different role) and (b) because the concepts "theory" and "model" have similar meanings in the different contexts, each view simply attends to different aspects of the concept or invokes a different type of analysis.

While it is surely right that one approach is more interested in model construction as opposed to structural features of models, where "structural" here should be understood in the model-theoretic sense, it is not clear that the differences can be so easily reconciled or that there is a similarity of concepts that facilitates the kind of different "perspectives" van Fraassen suggests. Moreover, it is certainly not the case that the scientific or practice-based approach aligns models with theory in the way the semanticists do. Hence some clarification is in order.

To some extent we can go along happily using, constructing, and discussing theories and models without having an exact definition of either. But if we are going to engage in the philosophical job of clarification, then we need some way to specify the content of each (something the semantic view claims to address). If we think it is the job of theories to tell us what the world is possibly like, then we need some way of differentiating what a theory is *about* (i.e., its content) from the various assumptions required for its *application* in particular contexts. One way to remedy this problem is to differentiate models and theories based on different notions of *representation* and *explanation* appropriate to each. My own view is that part of that differentiation will involve the notion of a theoretical core—a set of fundamental assumptions that constitute the basic content of the theory, as in the case of Newton's three laws and universal gravitation. This core constrains not only the behavior but also the representation of objects governed by the theory, as well as the construction of models of the theory. In other words, the laws of

---

of this is that it can purportedly reflect the fact that we may lack knowledge about the domain we wish to model.

[4]   See, for example, Halverson's recent critiques (2012, 2013) of the semantic approach.

motion and UG allow us to have a kind of physical representation of how certain types of objects behave in certain circumstances. The basic motions described by the laws are then reconceptualized in more concrete terms when detailed models are constructed, models that allow us to represent certain behaviors as instances of particular kinds of motion. In that sense, theory does play a unifying role that is not present in a simple family of models. This is not to say that theory determines the way models are constructed but only that the way we model phenomena is often constrained by theoretical laws or principles that are part of the larger scientific context.[5] I say more about this later.

I begin with some history of the semantic and syntactic view of theories and the role of models in each of these accounts. I then go on to discuss different ways of thinking about models that have emerged out of the practice-based accounts, some of the epistemological and ontological problems associated with modeling, and why difficulties in formulating general claims about the model-world relation do not preclude philosophically interesting analyses of those issues.

2 THEORIES, MODEL THEORY, AND MODELS

The semantic view was formulated as an alternative to the difficulties encountered by its predecessor, the syntactic, or received view. The former defines theories in terms of models while the later defines models in terms of theories, thereby making models otiose. The history and variations associated with these views is a long and multifaceted story involving much more technical detail than I can (or a reader would want me to) rehearse here.[6] What these views do have in common, however, is the goal of defining what a *theory* is, a definition that speaks to, albeit in different ways, the notion of a model and its function.

On the syntactic view the theory is an uninterpreted axiomatized calculus or system—a set of axioms expressed in a formal language (usually first-order logic)—and a model is simply a set of statements that interprets the terms in the language. While this is one way of interpreting the language, the more common approach has been to use correspondence rules that connected the axioms (typically the laws of the theory) with testable, observable consequences. The latter were formulated in what was referred to as the "observation language" and hence had a direct semantic interpretation. Many

---

[5] Weisberg (2013) differentiates between models and theories on the basis of practices associated with each. He claims that modeling and theorizing involve different procedures for representing and studying real-world systems, with the former employing a type of *indirect* representation and analysis. While this distinction may be helpful in identifying certain features or stages of theorizing and/or certain types of modeling, as different categories for knowledge acquisition/production, theories and models cannot, in the end, be differentiated in terms of direct versus indirect representation. In many cases there is a great deal of overlap in the activities involved in both theorizing and modeling, and while the products may be different, many of the processes are not.

[6] For an extended discussion see da Costa and French (2003).

of the problems associated with the syntactic view involved issues of interpretation—specifying an exact meaning for correspondence rules, difficulties with axiomatization, and the use of first-order logic as the way to formalize a theory. But there were also problems specifically related to models. If the sets of statements used to interpret the axioms could be considered models, then how should one distinguish the intended from the unintended models? Other criticisms of models were put forward by Carnap (1939: 68) who claimed that models contributed nothing more than aesthetic or didactic value, and Braithwaite (1953) who saw models as aids for those who found it difficult to "digest" an uninterpreted calculus. He also claimed that in some cases models were a *hindrance* to our understanding of physical theory since they had the potential to lead us down the road to falsity.

The difficulties associated with axiomatization and the identification of a theory with a linguistic formulation gave rise to the semantic view whose advocates (Suppes 1961, 1967, 2002; Suppe 1989; Giere 1988) appeal, in a more or less direct way, to the notion of model defined by Tarski. Although van Fraassen (1980) opts for the state-space approach developed by Weyl and Beth, the underlying similarity in these accounts is that the model supposedly provides an interpretation of the theory's formal structure but is not itself a linguistic entity. Instead of formalizing the theory in first-order logic, one simply defines the intended class of models directly.

Suppes' version of the semantic view includes a set-theoretic axiomatization that involves defining a set-theoretical predicate, (i.e., a predicate like "is a classical particle system" that is definable in terms of set theory), with a model for the theory being simply an entity that satisfies the predicate. He claims that the set-theoretical model can be related to what we normally take to be a physical or scientific model by simply interpreting the primitives as referring to the objects associated with a physical model. Although he acknowledges that the notion of a physical model is important in physics and engineering, the set-theoretical usage is the fundamental one. It is required for an exact statement of any branch of empirical science since it illuminates not only "the exact statement of the theory" but "the exact analysis of data" as well (2002: 24). Although he admits that the highly physical or empirically minded scientists may disagree with this, Suppes also claims that there seems to be no point in "arguing about which use of the word model is primary or more appropriate in the physical sense" (22).[7]

Van Fraassen specifically distances himself from Suppes' account, claiming that he is more concerned with the relation between physical theories and the world than with the structure of physical theory (1980: 67). To "present a theory is to specify a family of structures, its models" (64); and, "any structure which satisfies the axioms of a theory [. . .] is called a model of that theory" (43). The models here are state-spaces with trajectories and constraints defined in the spaces. Each state-space can be given by specifying a set of variables with the constraints (laws of succession and coexistence) specifying

---

[7] What this suggests, then, is that as philosophers our first concern should be with the exact specifications of *theoretical structure* rather than how the models used by scientists are meant to deliver information about physical systems.

the values of the variables and the trajectories their successive values. The state-spaces themselves are mathematical objects, but they become associated with empirical phenomena by associating a point in the state-space with a state of an empirical system.[8] Giere's approach to models, while certainly identified as "semantic," does not specifically emphasize this nonlinguistic aspect, but he also claims that his notion of a model, which closely resembles the scientist's notion, "overlaps nicely with the usage of the logicians" (1988: 79).[9]

The relevant logician here is Tarski, who defines a model as "a possible realization in which all valid sentences of a theory T are satisfied is called a model of T" (1953: 11). In 1961 and later in 2002, we find Suppes claiming that "the concept of model in the sense of Tarski may be used without distortion and as a fundamental concept" in the disciplines of mathematical and empirical sciences (2002: 21). He claims that the meaning of the concept "model" is the same in these disciplines with the only difference to be found in their *use* of the concept. In contrast to the semantic view, Tarski defines a theory as a set of sentences and says the role of the model is to provide the conditions under which the theory can be said to be true. Hence the ultimate goal is defined in terms of truth and satisfaction, which are properties of *sentences* comprising the theory. The importance of the model is defined solely in terms of its relation to the sentences of the theory, and in that sense it takes on a linguistic dimension.

The point of Tarski's definition is that it rests on a *distinction* between models and theories, something the semanticists essentially reject. For them models are not *about* the theory as they are for Tarski; the theory is simply defined or identified by its models. For example, Suppes' account lacks a clearly articulated distinction between the primitives used to define particle mechanics and the realization of those axioms in terms of the ordered quintuple.[10] Moreover, if theories are defined as families of models there is, strictly speaking, nothing for the model to be true of, except all the other models. In other words, the models do not provide an interpretation of some distinct theory but stand on their own as a way of treating the phenomena in question. While there may

---

[8] But once this occurs the state-space models take on a linguistic dimension; they become models of the theory in its linguistic formulation. Similarly, in Suppes' account, when it comes to specifying the set theoretical predicate that defines the class of models for a theory, we do need to appeal to the specific language in which the theory is formulated. And, in that context, which is arguably the one in which models become paramount, they cease to become nonlinguistic entities. But as long as no specific language is given priority at the outset, we can talk about models as nonlinguistic structures.

[9] At the risk of being unfair to Giere, I emphasize here that I am referring to his earlier views on models and the semantic account, some of which he may no longer explicitly hold.

[10] We can axiomatize classical particle mechanics in terms of the five primitive notions of a set P of particles, an interval T of real numbers corresponding to elapsed times, a position function s defined in the Cartesian product of the set of particles and the time interval, a mass function m and a force function F defined on the Cartesian product of the set of particles, the time interval and the positive integers (the latter enter as a way of naming the forces). A realization or model for these axioms would be an ordered quintuple consisting of the primitives $P = <P,T,s,m,f>$. We can interpret this to be a physical model for the solar system by simply interpreting the set of particles as the set of planetary bodies, or the set of the centers of mass of the planetary bodies.

be nothing wrong with this in principle, it does create a rather peculiar scenario: there is no way of identifying what is fundamental or specific about a particular theoretical framework since, by definition, all the paraphernalia of the models are automatically included as part of the theory. But surely something like perturbation theory, as a mathematical technique, should not be identified as part of quantum mechanics any more than the differential calculus ought to be included as part of Newton's theory.

Moreover, if a theory is just a family of models, what does it mean to say that the model/structure is a realization of the theory? The model is not a realization of the theory because there is no theory, strictly speaking, for it to be a realization of. In other words, the semantic view has effectively dispensed with theories altogether by redefining them in terms of models. There is no longer anything to specify as "Newtonian mechanics" except the models used to treat classical systems. While there may be nothing wrong with this if one's goal is some kind of logical/model-theoretic reconstruction—in fact it has undoubtedly addressed troublesome issues associated with the syntactic account— but, if the project is to understand various aspects of how models and theories are related, and how they function, then reducing the latter to the former seems unhelpful. And, as a logical reconstruction, it is not at all clear how it has enhanced our understanding of theory structure—one of its stated goals. Replacing theories with models simply obviates the need for an account of theories without showing how or why this improves things.

Both van Fraassen's focus on state-spaces and Giere's emphasis on hypotheses in the application of models to the world do speak to aspects of the scientific practice of modeling. As I noted earlier, the state-space approach typically involves representing a system or a model of a system in terms of its possible states and its evolution. In quantum mechanics a state-space is a complex Hilbert space in which the possible instantaneous states of the system may be described by unit vectors. In that sense, then, we can see that the state-space approach is in fact a fundamental feature of the way that theories are represented. The difficulty, however, is that construed this way the theory is nothing more than the different models in the state-space. While we might want to refer to the Hilbert space formulation of QM for a mathematically rigorous description, such a formalism, which has its roots partly in functional analysis, can be clearly distinguished from a theory that gives us a more "physical" picture of a quantum system by describing its dynamical features.

Similarly, on Giere's (2004) account we have nothing that answers to the notion of "theory" or "law." Instead we have principles that define abstract objects that, he claims, do not directly refer to anything in the world. When we refer to something as an "empirical" theory, this is translated as having models (abstract objects) that are structured in accordance with general principles that have been applied to empirical systems via hypotheses. The latter make claims about the similarity between the model and the world. The principles act as general templates that, together with particular assumptions about the system of interest, can be used for the construction of models. Although the notion of a general principle seems to accord well with our intuitive picture of what constitutes a theory, the fact that the principles refer only to models and do not describe anything in the world means that what we often refer to as the "laws of physics" are principles that are true only of models and not of physical objects.

> If we insist on regarding principles as genuine statements, we have to find something that they describe, something to which they refer. The best candidate I know for this role would be a highly abstract object, an object that, by definition, exhibits all and only the characteristics specified in the principles. So the principles are true of this abstract object, though in a fairly trivial way. (Giere 2004: 745)

This leaves us in the rather difficult position of trying to figure out what, if any, role there could be for what we typically refer to as quantum theory, Newtonian mechanics, evolutionary theory, and so on. Or, alternatively, what is gained by the kind of systematic elimination of theory that characterizes the various formulations of the semantic view.

Giere claims that one of the problems that besets the notion of "theory" is that it is used in ambiguous and contradictory ways. While this is certainly true, there are other, equally serious problems that accompany the semantic reconstruction. One of these, mentioned at the outset, involves specifying the content of the theory if it is identified strictly with its models. Another, related issue concerns the *interpretation* of that content. In particular, the models of many of our theories typically contain a good deal of excess structure or assumptions that we would not normally want to identify as part of a theory. Although van Fraassen claims that it is the task of theories to provide literal descriptions of the world (1989: 193), he also recognizes that models contain structure for which there is no real world correlate (225–228). However, the issue is not simply one of determining the referential features of the models, even if we limit ourselves to the empirical data. Instead I am referring to cases where models contain a great deal of structure that is used in a number of different theoretical contexts, as in the case of approximation techniques or principles such as least time. Because models are typically used in the *application* of higher level laws (that we associate with theory), the methods employed in that application ought to be distinguished from the content of the theory (i.e., what it purports to say about physical systems).[11]

Consider the following example (discussed in greater detail in Morrison 1999). Suppose we want to model the physical pendulum, an object that is certainly characterized as empirical. How should we proceed when describing its features? If we want to focus on the period, we need to account for the different ways in which it can be affected by air, one of which is the damping correction. This results from air resistance acting on the pendulum ball and the wire, causing the amplitude to decrease with time while increasing the period of oscillation. The damping force is a combination of linear and quadratic damping. The equation of motion has an exact solution in the former case

---

[11] One might want to argue that the way around this problem is to distinguish between models as abstract entities and model descriptions where the latter contain mathematical details that do not relate specifically to a conceptualization of the system being modeled. What this does, however, is to replace one problem with a similar one of specifying the nature of the model. See Weisberg (2013) for more on the model/model-description distinction.

but not in the latter case, since the sign of the force must be adjusted each half-period to correspond to a retarding force. The problem is solved using a perturbation expansion applied to an associated analytic problem where the sign of the force is not changed. In this case, the first half-period is positively damped and the second is negatively damped with the resulting motion being periodic. Although only the first half-period corresponds to the damped pendulum problem, the solution can be reapplied for subsequent half-periods. But only the first few terms in the expansion converge and give good approximations—the series diverges asymptotically, yielding no solution.

All of this information is contained in the model, yet we certainly do not want to identify the totality as part of the theory of Newtonian mechanics. Moreover, because our treatment of the damping forces requires a highly idealized description, it is difficult to differentiate the empirical aspects of the representation from the more mathematically abstract ones that are employed as calculational devices. My claim here is not just that the so-called empirical aspects of the model are idealized since all models and indeed theories involve idealization. Rather, the *way* in which the empirical features are interconnected with the nonempirical makes it difficult to isolate what Newtonian mechanics characterizes as basic forces.[12]

Why do we want to identify these forces? The essence of Newtonian mechanics is that the motion of an object is analyzed in terms of the forces exerted on it and described in terms of the laws of motion. These core features are represented in the models of the theory, as in the case of the linear harmonic oscillator which is derived from the second law. Not only are these laws common to all the models of Newtonian mechanics, but they constrain the kind of behavior described by those models and provide (along with other information) the basis for the model's construction. Moreover, these Newtonian models embody different kinds of assumptions about how a physical system is constituted than, say, the same problem treated by Lagrange's equations. In that sense we identify these different core features as belonging to different "theories" of mechanics.

The notion that these are different theories is typically characterized in terms of the difference between forces and energies. The Newtonian approach involves the application of forces to bodies in order to see how they move. In Lagrange's mechanics, one does not deal with forces and instead looks at the kinetic and potential energies of a system where the trajectory of a body is derived by finding the path that minimizes the action. This is defined as the sum of the Lagrangian over time, which is equal to the kinetic energy minus the potential energy. For example, consider a small bead rolling on a hoop. If one were to calculate the motion of the bead using Newtonian mechanics, one would have a complicated set of equations that would take into account the forces

---

[12] Giere (1988, ch. 3) has also discussed the pendulum model as a way of illustrating the role that idealizations play in models and to show the importance of models in applying laws to particular problem situations. I have no disagreement with Giere's description of models per se—only with the idea that theories should be construed as families of models and that Newton's laws, for instance, should be described as "principles" (2004). It is not clear to me what the advantages of this type of characterization are, except to avoid using the word "theory."

that the hoop exerts on the bead at each moment. Using Lagrangian mechanics, one looks at all the possible motions that the bead could take on the hoop and finds the one that minimizes the action instead of directly calculating the influence of the hoop on the bead at a given moment.

As discussed earlier, when we model the physical pendulum using Newtonian mechanics, the calculation of the forces involved becomes very complex indeed. The *application* of the theory to physical phenomena requires special assumptions about how these systems/phenomena are constituted, as well as calculational methods for dealing with those assumptions. In other kinds of situations we frequently need to incorporate rigid body mechanics into our models in order to deal with rigid bodies defined as systems of particles. In these cases we assume that the summation $\Sigma F = \Sigma ma$ over a system of particles follows directly from $F=ma$ for the motion of a single particle. That is, we simply assume this is valid despite the fact that there are limiting processes involved in the mathematics when modeling rigid bodies as continuous distributions of matter. Another problem is that the internal electromagnetic forces cannot be adequately modeled as equal and opposite pairs acting along the same line, so we simply assume that the equations are valid with the summation confined to external forces. Although the models require rigid body mechanics, no one would suggest that this is an essential feature of Newtonian theory, nor in the pendulum example, the specific methods for perturbation expansions. Even if we take account of Nancy Cartwright's (1999) point that theories or theoretical laws do not literally describe anything, we would still want to distinguish between what I call the fundamental core of Newton's theory (laws of motion and UG) and the models and techniques used in the application of those laws.

If we identify a theory with a core set of laws/equations, no such difficulties ensue. For example, when asked for the basic structure of classical electrodynamics, one would immediately cite Maxwell's equations. These form a theoretical core from which a number of models can be specified that assist in the application of these laws to specific problem situations. Similarly, an undisputed part of the theoretical core of relativistic quantum mechanics is the Dirac equation. Admittedly there may be cases where it is not obvious that such a theoretical core exists. Population genetics is a good example. But even here one can point to the theory of gene frequencies as the defining feature on which many of the models are constructed. My point is simply that by defining a theory solely in terms of its many models, one loses sight of the theoretical coherence provided by core laws, laws that may not *determine* features of the models but certainly constrain the kind of behaviors that the models describe. Indeed, it is the identification of a theoretical core rather than all of the features contained in the models that enables us to claim that a set of models belongs to Newtonian mechanics. Moreover, nothing about this way of identifying theories requires that they be formalized or axiomatized.

Although the semanticists claim that their account extends to scientific models, questions arise regarding its ability to incorporate the many different roles models play and the various ways they are constructed within the broader scientific context. More recently, van Fraassen claimed that the criticisms of the semantic view by those interested in the more practical issues related to modeling were due not so much to a different concept of models but to "different interests in the same subject" (2008: 309). He cites Suarez and Cartwright (2007), who distinguish the theory-driven (structural) approach to models characteristic of the semantic view with the "practical" account by appealing to a process-product distinction. The structural approach focuses on the product, the models themselves, without any attention to the processes involved in the model's construction. While van Fraassen agrees that this distinction forms the natural divide between the two approaches, he does not directly address Suarez's and Cartwright's main point, which is: any assessment of the degree to which the structuralist's models are models of theory *requires* that we pay attention to the processes by which they were constructed. In that sense, we cannot simply say that the two approaches to models are simply the result of different interests. For semanticists to claim that their models are models of a theory requires that they address practical issues about processes.

However, once we acknowledge that our discussion of models needs to move beyond structures, there seems no reason to retain the semantic view. Not only does it suffer from difficulties in specifying the theory-model relationship, but identifying a model with a structure in the logician's sense, as a structure in which the sentences of the theory are true, restricts its function in ways that make it overly theory dependent. Of course the irony here is that no definition of a theory is ever given beyond the notion of a model.[13] As models of a "theory," the structures are constrained by the laws of logic and model theory, which typically fail to capture the rather loose connections between theories and models that are more indicative of scientific practice (see Morrison 1990).[14]

3 MODELS, REPRESENTATION, AND EXPLANATION: SOME PRACTICAL ISSUES

Moving the focus away from the semantic to the scientific approach to modeling, the question I next address is how to understand the relation between models and the systems they represent. In other words, rather than provide a characterization or classification of the "ontology" of models—toy, mathematical equations, scale models, fictional models, and so on, I focus on something that is common to all—their idealized or abstract nature—and how, despite this feature, they are able to deliver information

[13] van Fraassen (2008: 309) acknowledges that what a theory is has never been entirely settled in the literature on the semantic view. While several of its proponents have offered accounts in terms of models and their application (Giere 1988), van Fraassen himself emphasized that theories are the kinds of things that are believed, disbelieved, and so on but identifies them ultimately with families of models.

[14] The partial structures view of da Costa and French (2003) was formulated as an attempt to overcome some of these difficulties. Unfortunately space prevents me from discussing the various aspects of this view, but, as I have argued elsewhere (Morrison 2007), not only does their formulation fail to solve the problem(s); it introduces ambiguities that further complicate the issue.

about concrete systems.[15] Two important categories for assessing information regarding the model-world relation are representation and explanation, categories that are also important in the characterization of theories. Cartwright (1999), for example, claims that in virtue of their abstractness theories simply do not represent anything, a view that is similar to Giere's account discussed earlier. Contrary to this, I claim that both models and theories have a representational function, but one that is different in each case. That is not to say, however, that we can formulate a general notion of representation or explanation that is common to all models (or theories), but rather that in answering questions about the relation of models to the world, representation and explanation are important categories on which to base our analyses.[16]

As I mentioned, one of the defining features of a model is that it contains a certain degree of representational inaccuracy.[17] In other words, it *is* a model *because* it fails to accurately represent its target system. Sometimes we know specifically the type of inaccuracy the model contains because we have constructed it precisely in this way for a particular reason. Alternatively, we may be simply unsure of the kind and degree of inaccuracy because we do not have access to the system that is being modeled and thus no basis for comparison.

One way of thinking about the representational features of a model is to think of it as incorporating a type of picture or likeness of the phenomena in question, a notion that has had a long and distinguished history in the development of the physical sciences. The idea of "likeness" here, though, is somewhat misleading and by no means straightforward. For example, many scale models of the solar system were constructed both before and after Copernicus, some in order to demonstrate that a planetary conjunction would not result in a planetary collision. One can think of these as representing, in the physical sense, both the orbits and the relation of the planets to each other in the way that a scale model of a building represents certain relevant features of the building itself. Nineteenth-century physics is replete with both the construction of and demand for models as a way of developing and legitimating physical theory. In the initial stages of the development of electrodynamics, Maxwell relied heavily on different mechanical models of the aether (some of which were illustrated in a pictorial fashion) as an aid to formulating the field equations—models that bore no relation to what he thought the aether could possibly be like. What these models did was represent a mechanical system of rotating vortices whose movements would set up electric currents that satisfied certain equations. While no one thought that the aether consisted of vortices, the model represented the way in which electric currents could arise in a mechanical system. Once the field equations were in place, Maxwell abandoned these pictorial mechanical models

---

[15] We often see this relation described in terms of the model being isomorphic to the world, something that is, strictly speaking, meaningless. The isomorphism relation is a mapping between structures; it is a category mistake to think a model can be isomorphic to the world construed as a physical thing.

[16] For more on this see Morrison (2015).

[17] I should mention here that the notion of representation used by the semanticists—isomorphism—does not really incorporate this notion of inaccuracy. I say more about isomorphism later.

and chose to formulate the theory in the abstract formalism of Lagrangian mechanics, which itself functions as a kind of mathematical model for different types of physical systems, both classical and quantum.

Maxwell's work was severely criticized by Lord Kelvin, who maintained that the proper understanding of nature required physical, mechanical models that could be manipulated as a way of simulating experiments. In other words, Kelvin's notion of a mechanical model was something that could be built, not simply drawn on paper, and both he and FitzGerald constructed such aether models as a way of representing the propagation of electromagnetic waves.[18] But here again the important point is that no one really believed that the aether itself bore a similarity to these models; rather they were useful because they represented a mechanical system that behaved according to the electromagnetic equations and because they led to modifications of some of Maxwell's mathematics. Add to this the Rutherford and Bohr models of the atom, the shell and liquid drop models of the nucleus, and we begin to see how models, functioning as both mathematical and physical representations, have emerged as important sources of knowledge.[19] What is important in each of these examples is that each of the models represents the target system in some particular way, either by approximating it as a likeness/similarity or as a type of analogue system that obeys the same equations.

The semanticists have another way of thinking about representation, one grounded in the notion of a structural mapping (e.g., an isomorphism). Since the semanticists' models are defined in terms of nonlinguistic structures, how should we understand their representational capacity? The response given by van Fraassen is that the empirical substructures of the model are candidates for "the *direct representation* of the observable phenomena" (1980: 64, italics added). But how does this "direct representation" take place when one cannot have an isomorphism between phenomena and structures?

The answer involves the notion of "appearances," which include the "structures described in experimental and measurement reports" (van Fraassen 1980: 64). A theory is said to be empirically adequate if it has some model such that all appearances are isomorphic to empirical substructures of that model. For Suppes, the situation is similar in that isomorphism enters in a central way. He makes use of representation theorems as a way of characterizing the models of a theory. For example, a representation theorem for a theory means that we can pick out a certain class of models of the theory that exemplifies, up to isomorphism, every model of the theory. So if M is the set of all models of some theory T and S is a subset of M, then a representation theorem for M with respect to S would be the claim that for every model m in M there is a model in S isomorphic to

[18] For more on Maxwell's use of models see Morrison (2000) and for criticisms of Maxwell's models by Kelvin see Smith and Wise (1989).

[19] My intention here is not to draw a sharp distinction between mathematical and physical models per se because in many cases physical models, like FitzGerald's aether models, have a mathematical interpretation as well. I do not deny that there may be important ways in which mathematical and physical models differ, but those differences are not relevant for the discussion here.

m.[20] I do not discuss the relation between this type of representation and the more pictorial type mentioned earlier except to say that it is analogous to the relation discussed by Suppes between his use of the term "model" (i.e., the set theoretic one) and the more general notion of a physical model. While it may be possible to capture the pictorial notion in terms of the more formal one, doing so will undoubtedly add an extra layer of "structure" that seems unnecessary if our goal is to represent a physical system for the purposes of understanding, say, its possible causal connections.

It is important to note that scientific representation characterized in terms of isomorphism is not without its critics. Suarez (2003) in particular has argued against both isomorphism and similarity as the constituents of scientific representation.[21] He sees the emphasis on similarity and isomorphism as indicative of a reductive naturalistic approach that ignores scientists' purposes and intentions, thereby relegating the latter as nonessential features of representation. While I agree with many of Suarez's points, my own view is that the poverty of similarity and isomorphism as characterizations of representation stems, ultimately, from adopting the semantic or model theoretic approach to theories. If one chooses to interpret theories as families of models/structures, then one is all but forced to rely on isomorphism as *the* way to flesh out the notion of representation.

On Giere's account of the semantic view, which also characterizes models as structures, the model-world relation is explicated via theoretical hypotheses; however, those hypotheses are useful only insofar as they tell us the extent to which the model is similar to the physical system that is modeled. But if we dissociate models from the notion of a formal structure, then we are free to talk about similarity as a basis for representation in some contexts but not in others. The advantage here is that in some cases models will be similar to the systems they model, making the concept useful but without committing us to similarity as *the* way of understanding how models represent physical systems. This becomes important when we realize that often a motivating aspect of model construction is a lack of knowledge of the system/phenomena being modeled (as in the Maxwell case), and in those situations similarity as a representational tool is of little use. My point then is that the difficulties with isomorphism and similarity as accounts of representation arise partly from problems inherent in the notions themselves (especially in the case of similarity) but primarily from the account of models that necessitate their use. Consequently, an abandonment of the semantic view will also liberate us from reliance on isomorphism as the way of representing scientific phenomena.

As I mentioned, the goal in promoting a notion of theory that extends beyond a "collection of models", especially where representation is at issue, requires that we be able to

[20] Suppes (1967) also uses the notion of definition as a way to think about the representation of concepts of the theory. The first definition in a theory is a sentence of a certain form that establishes the meaning "of a new symbol of the theory in terms of the primitive concepts of the theory" (53). The point of introducing new symbols is to facilitate deductive investigation of the theory's structure, not to add to the structure. In that sense then the definitions are not a crucial aspect of the models.

[21] See also Frigg (2002).

capture certain "general" features used to classify physical systems; features that are not always obvious or easily extractable from collections of models. To clarify consider the following: If we take Newtonian theory to encompass *all* the models used in its application, then how do we determine which features of this collection have the appropriate representational status; that is, what features of the models are singled out as *essential* for representing physical systems as basically Newtonian? If we can pick out the common elements that make the models Newtonain (e.g., laws of mechanics), then are we not just isolating certain features that, taken together, specify a core notion of theory from within the family of models—something the semanticists are at pains to avoid and something they claim is not needed.

On this interpretation, one of the roles of theory—one that is not successfully captured by a family of models—is to provide a *general* representation of an entire class of phenomena, where "representation" should be understood in the nontechnical sense as a way of simply describing the phenomena as exhibiting certain kinds of behavior (e.g., quantum, classical, relativistic, etc.). Part of the description involves the constraints placed on that behavior by the laws of the theory. The Schrodinger equation, for example, describes the time evolution of quantum systems, and it together with other principles like the uncertainty and exclusion principles and Planck's constant form the core of our theory of quantum mechanics. The models of nonrelativistic quantum mechanics all obey these constraints in their role of filling in the details of specific situations in order to apply these laws in cases like the infinite square well potential. This is not to say that this is the *only* role for models; indeed I have spent a good deal of time arguing for their autonomy in various circumstances. Rather, my point is that theory too plays an important role in the way we understand and represent physical systems; viewed in this way, we can see how the generality expressed by theories can be understood in terms of their representational power.

Some of the literature on models suggests that they can function as the only vehicle for scientific representation.[22] Cartwright, for instance, has claimed that "the fundamental principles of theories in physics do not represent what happens"; only models represent in this way "and the models that do so are not already part of any theory" (1999: 180).[23] The reason for this is that theories use abstract concepts, and although these concepts are made more concrete by the use of interpretive models (e.g., the harmonic oscillator, the Coulomb potential, etc.), they are still incapable of describing what happens in actual situations. For that we need representative models that go beyond theory. These latter models, which account for regular and repeatable situations, function like "blueprints for nomological machines" (Cartwright 1999: 180).[24]

---

[22] Although Suppes' use of representation theorems refers to theories, that reference is strictly to the set of models of the theory.

[23] Not being an advocate of the semantic view Cartwright typically speaks of models in the sense used by physicists: physical models that are either theoretical or phenomenological or a mixture of both. Although she does not develop a full account of representation, it seems clear that what she has in mind is more closely aligned with the kind of representation afforded by, for example, the Bohr model of the atom.

[24] For a discussion of this notion see Cartwright (1999).

Cartwright's argument trades on the assumption that the abstractness inherent in the description of the harmonic oscillator, for example, renders it incapable of describing any real physical phenomena.[25] While theories undoubtedly contain what Cartwright calls abstract concepts, it is important to recognize that these concepts and their interpretive models are part of the way theories classify particular kinds of behavior—by representing it as an instance of, say, of harmonic motion. No actual system looks like the model of the harmonic oscillator, but the model may still represent basic features of harmonic motion or tell us important things about it. For example, when we use the model to analyze the motion of an object attached to a spring that obeys Hooke's law, we find that the period of the oscillator is completely independent of the energy or amplitude of the motion—a result that is not true of periodic oscillations under any other force law. Of course if the frequency varied significantly with the amplitude, the situation would become much more complicated, but most vibrating systems behave, at least to some approximation, as harmonic oscillators with the properties described by the model.

The important point here is how to think about representation. We use models to represent physical phenomena in more or less abstract ways. Some models build in features that describe the system in more detail than others; but regardless of the degree of detail we typically intend the model to refer to a physical system/phenomena *understood* in a certain way.[26] When we say that a diatomic molecule can be *considered* as two point masses connected by a spring and can undergo quantized oscillations, we are making a point about how quantum mechanics (and the model) predicts that the energy levels of a harmonic oscillator are equally spaced with an interval of h times the classical frequency and have a minimum value (the zero-point energy). It also follows from this that photons absorbed and emitted by the molecule have frequencies that are multiples of the classical frequency of the oscillator. While the model may be highly abstract, it nevertheless refers to basic features of a physical system and in that sense functions in a representational capacity.[27] Theory too can function in this way. It can represent physical systems/phenomena by isolating and highlighting certain basic features, behaviors, or causes from which more specialized applications can be then be derived or constructed via models. For example, Cooper pairing (the pairing of electrons) in the Bardeen-Cooper-Schrieffer (1957) theory of superconductivity acts as a concrete representation insofar as it specifies exactly what processes give rise to superconductivity in metals.

---

[25] Cartwright (1989) makes a distinction between abstract and idealized models where the latter involves an idealized description of a physical object, like a frictionless plane, that can be made more concrete by adding back certain features while the former describes an object that cannot be made more realistic simply by adding correction factors.

[26] *How* exactly it does this is a question that most likely involves matters related to cognitive psychology. My claim here is that theories and abstract models do have a representational function, contrary to Cartwright's point that they do not.

[27] Another important concern here is the role of mathematics. Since most models in physics, biology, economics, and many other disciplines are mathematical, the question arises as to the representational relation between mathematics and physical phenomena. This is a separate topic that requires an investigation into the applicability of mathematics.

How that occurs is described in more detailed terms using a model that takes account of a variety of physical processes, a model that nevertheless incorporates a number of highly abstract and unrealistic assumptions.

In some other works (Morrison 1998, 1999, 2015), I claim that one of the important features of models is their explanatory power—their ability to provide more or less detailed explanations of specific phenomena/behaviors. But one of the reasons models have explanatory power is because they provide representations of the phenomena that enable us to understand why or how certain processes take place. Although my claim here is that theories can explain/represent the general features of a physical system, the important difference between their representational function and that of models is that the latter, unlike the former, tend not to be generalizable across different kinds of phenomena. For example, the Moran model in population genetics deals strictly with overlapping generations, while the notion of gene frequencies, which is arguably the theoretical foundation of population genetics, applies to all evolving populations.

To return to our original question about how models deliver concrete information, the answer is: in a variety of different ways, depending on the type of model we have and what we want our model to do. But in each and every instance, the model can only deliver information insofar as it furnishes a type of representation that we can use for explanatory or predictive purposes. Mathematical models will fulfill this role in ways that are totally different from scale models. The information we extract from fictional models will require us to specify the ways in which the fiction represents the system of interest; and each model will do that in a different way. While all models embody unrealistic or idealized assumptions, characterizing them generally as "fictions" fails to differentiate the ways in which different types of models enable us to understand the world and, moreover, has the potential to undermine their status as sources of concrete knowledge (Morrison 2015).

Although there is a great deal to be said about the construction and function of models, together with the way they are utilized, my own view is that none of these considerations can be codified into a "theory of models." Attempting to do so is, in some ways, to misconstrue the role models play in contributing to human knowledge, not only in the sciences but across many diverse disciplines.

REFERENCES

Achinstein, Peter. (1964). "Models, Analogies and Theories." *Philosophy of Science* 31(4): 328–350.

Bardeen, J., Cooper, L. and Schrieffer, J. R. (1957). "Theory of Superconductivity." *Physical Review* 108, 1175–1204.

Braithwaite, R. (1953). *Scientific Explanation* (Cambridge, UK: Cambridge University Press).

Carnap, R. (1939). *International Encyclopedia of Unified Science*, Vol. 1, No. 3: *Foundations of Logic and Mathematics* (Chicago: University of Chicago Press).

Cartwright, N. (1999). "Models and the Limits of Theory: Quantum Hamiltonians and the BCS Models of Superconductivity." In M. Morgan and M. Morrison, (eds.), *Models as Mediators: Perspectives on Natural and Social Science* (Cambridge, UK: Cambridge University Press), 241–281.

Cartwright, N. (1989). *Nature's Capacities and their Measurement* (Oxford: Oxford University Press).

Cartwright, N., Suarez, M., and Shomar, T. (1995). "The Tool Box of Science." In W. E. Herfel et al. (eds.), *Theories and Models in Scientific Processes* (Amsterdam: Rodopi), 137–149.

da Costa, N. C. A., and French, S. (2003). *Science and Partial Truth: A Unitary Approach to Models and Scientific Reasoning* (Oxford: Oxford University Press).

Frigg, R. (2002). "Models and Representation: Why Structures Are Not Enough." CPNSS Discussion Paper Series, DP MEAS 25/02 (London: LSE Centre for Philosophy of Natural & Social Science).

Frigg, R., and Hartmann, S. (2012). "Models in Science." In *Stanford Encyclopedia of Philosophy.* http://plato.stanford.edu/entries/models-science/

Giere, R. (1988). *Explaining Science: A Cognitive Approach* (Chicago: University of Chicago Press).

Giere, R. (2004). "How Models Are Used to Represent Reality." *Philosophy of Science* 71: 742–752.

Halverson, H. (2012). "What Scientific Theories Could Not Be." *Philosophy of Science* 79: 183–206.

Halverson, H. (2013). "The Semantic View, If Plausible, Is Syntactic." *Philosophy of Science* 80: 475–478.

Hesse, M. (1953). "Models in Physics." *British Journal for the Philosophy of Science* 4: 198–214.

Hesse, M. (1970). *Models and Analogies in Science* (Oxford: Oxford University Press).

Kuhn, T. (1972). *The Essential Tension* (Chicago: University of Chicago Press).

Morgan, M., and Morrison, M. (1999). *Models as Mediators: Perspectives on Natural and Social Science* (Cambridge, UK: Cambridge University Press).

Morrison, M. (1990). "Unification, Realism and Inference." *The British Journal for the Philosophy of Science* 401: 305–332.

Morrison, M. (1998). "Modelling Nature: Between Physics and the Physical World." *Philosophia Naturalis* 38: 65–85.

Morrison, M. (1999). "Models as Autonomous Agents." In M. Morgan and M. Morrsion (eds.), *Models as Mediations: Essays in the Philosophy of the Natural and Social Sciences* (Cambridge, UK: Cambridge University Press), 38–65.

Morrison, M. (2000). *Unifying Scientific Theories: Physical Concepts and Mathematical Structures* (Cambridge, UK: Cambridge University Press).

Morrison, M. (2007). "Where Have All the Theories Gone?" *Philosophy of Science* 74: 195–228.

Morrison, M. (2015). *Reconstructing Reality: Models, Mathematics and Simulations* (Oxford: Oxford University Press).

Smith, C., and Wise, N. (1989). *Energy and Empire* (Cambridge, UK: Cambridge University Press).

Suarez, M. (2003). "Scientific Representation: Against Similarity and Isomorphism." *International Studies in the Philosophy of Science* 17: 225–244.

Suarez, M., and Cartwright, N. (2007). "Theories: Tools versus Models." *Studies in History and Philosophy of Science, Part B* 39: 62–81.

Suppe, Frederick. (1989). *The Semantic Conception of Theories and Scientific Realism* (Urbana: University of Illinois Press).

Suppes, P. (1961). "A Comparison of the Meaning and Use of Models in the Mathematical and Empirical Sciences." In H. Freudenthal (ed.), *The Concept and Role of the Model in Mathematics and Natural and Social Sciences* (Dordrecht: The Netherlands: Reidel), 163–177.

Suppes, P. (1962). "Models of Data." In Ernest Nagel, Patrick Suppes, and Alfred Tarski (eds.), *Logic, Methodology and Philosophy of Science: Proceedings of the 1960 International Congress* (Stanford, CA: Stanford University Press), 252–261.

Suppes, P. (1967). "What Is a Scientific Theory?" In S. Morgenbesser (ed.), *Philosophy of Science Today* (New York: Basic Books), 55–67.

Suppes, P. (2002). *Representation and Invariance of Scientific Structures* (Stanford, CA: CSLI).

Tarski, A. (1953). *Undecidable Theories* (Amsterdam: North Holland).

van Fraassen, B. (1980). *The Scientific Image* (Oxford: Oxford University Press).

van Fraassen, B. (1989). *Laws and Symmetries* (Oxford: Oxford University Press).

van Fraassen, B. (2008). *Scientific Representation: Paradoxes of Perspective* (Oxford: Oxford University Press).

Weisberg, M. (2013). *Simulation and Similarity* (Oxford: Oxford University Press).

CHAPTER 19

....

NATURAL KINDS

....

MUHAMMAD ALI KHALIDI

1 PRELIMINARY CHARACTERIZATION

CONSIDER the following scientific claims:

1. The Higgs boson can decay into two tau leptons.
2. There are 100 million black holes in the Milky Way galaxy.
3. Protons can transform into neutrons through electron capture.
4. Gold has a melting point of 1064° C.
5. Viruses multiply by attaching themselves to host cells.
6. The Eurasian wolf (*Canis lupus lupus*) is a predator and a carnivore.
7. Schizophrenics experience auditory hallucinations.

Some of the terms in these sentences designate individuals (e.g., *Milky Way*) or properties (e.g., *melting point of 1064° C*), but others refer to what philosophers have dubbed "kinds" or "natural kinds" (e.g., *proton, gold, Eurasian wolf*). Though scientists may not explicitly discuss natural kinds, they employ terms that denote neither individuals nor properties and would thereby seem committed to the existence of such things. To be sure, some of these terms can also be construed as referring to properties (e.g., the property of being a proton, atom of gold, or virus), but, in most of these cases, the term in question stands for a collection of properties that tend to be co-instantiated rather than a single property. For protons, these properties would be positive charge of 1.6×10^{-19} C, mass of 1.7×10^{-27} kilograms, and spin of ½. These three properties jointly specify what it is to be a member of the kind *proton*. Moreover, although all three properties are found together in all and only protons, they do not always co-occur in nature but are dissociable, since the charge of a proton is also carried by a pion (a π^+ meson), which has a different mass, and the spin of a proton is shared with an electron, which has a different charge. But when they are co-instantiated in the same individual, a number of other properties also tend to be instantiated. That would seem to be a central characteristic of natural kinds: they are associated with collections of properties that, when co-instantiated, result in the instantiation of yet other properties. Though a natural kind may occasionally be identifiable with a single property, that property must lead to the manifestation of a number of other properties for there to be a genuine distinction between kinds and properties. Many properties, even some of those properties associated with natural kinds, do not lead to the instantiation of a whole range of other properties, for instance the property of having a mass of 1.7×10^{-27} kilograms.

Natural kinds are said to have members, namely those individuals that possess the properties associated with the kind. All the kinds mentioned so far have as their members particular objects or entities (e.g., proton particle, atom of gold, virion), but there are also kinds of process or event, such as *radioactive decay, oxidization, volcanic eruption, supernova, species extinction, phosphorylation,* and *mitosis.* These are apparently also kinds, although their members are particular events or processes rather than objects or entities.

Now that we have a preliminary characterization of natural kinds, two philosophical questions can be raised about them (cf. Hawley and Bird 2011: 205; Khalidi 2013: 12):

1. Kindhood: *What* are kinds? Are kinds anything over and above their members? In other words, are they mere collections of individuals or are they universals or abstract entities in their own right?
2. Naturalness: *Which* kinds are natural? What makes a given kind natural? Is any assortment of properties a natural kind? Is any set of individuals, however arbitrary, equivalent to a natural kind?

These two questions are relatively independent, in the sense that the answer to one does not seem to preempt the answer to the other—unless one answers Question 2 by denying that there is any difference between natural and arbitrary categories, in which case it would be odd if not inconsistent to maintain that kinds are universals or abstract entities distinct from their members.

The first question takes us deep into the realm of metaphysics. The kindhood question is continuous with the traditional debate between realists and mominalists, between those philosophers who posit universals that correspond to properties and kinds and those who think that reality consists only of particulars and collections of such particulars. Universals can be transcendent (in the traditional Platonic conception), abstract entities that are not spatiotemporal, or immanent (in the view of some contemporary philosophers such as Armstrong [1989]), wholly present in each of their particular instances. The problems with both realism and nominalism are familiar and need not be rehearsed in detail here. Realists are committed to the existence of problematic entities with nonstandard attributes. If they are transcendent entities (like Plato's forms), then they exist outside of space-time and cannot be directly experienced. But if they are immanent (like Armstrong's universals), then they are capable of being wholly present in many different particulars at the same time. Both views posit entities that are very different from the familiar particulars that populate the universe. Moreover, realists about kinds face special problems not faced by realists about properties, since they must grapple with the question of whether kinds are universals over and above property universals or whether they are mere conjunctions of property universals, and each view comes with its difficulties. Meanwhile, nominalists encounter puzzles that arise from the fact that claims made about kinds cannot always be reduced to claims about their particular members (e.g., there are six kinds of quark, or there are more extinct than extant biological species). If such claims are not to be expunged from our discourse, they owe us an account of them.

The second question pertains more centrally to the philosophy of science. Here the issue is: Which categories correspond to natural kinds and which are just arbitrary; which categories capture something genuine about the universe and which are merely gerrymandered and unreflective of reality? Presumably, science has a say in the answer to this question, since it aims to isolate nonarbitrary categories that pertain to the very nature of reality and to classify individuals into categories that reveal something about the universe itself. The paradigmatic natural kinds are often thought to be chemical elements, chemical compounds, and biological species. With the proliferation of scientific disciplines and subdisciplines, the candidates have multiplied, and this is perhaps what makes the problem more acute for contemporary philosophers. Do all these scientific categories correspond to natural kinds, and, if not, what would rule some of them out? And are there other natural kinds, apart from those countenanced by science?

Before delving into the question of naturalness in more detail, it will be useful to examine some historical and contemporary approaches to the topic of natural kinds. This is not just a matter of antiquarian interest, since "natural kind" is a philosophical term of art, and, hence, what its originators meant by it is a significant constraint on what it can be taken to mean. More to the point, these early discussions contain important insights and raise a number of issues that continue to figure prominently in debates about natural kinds.

2 HISTORICAL ANTECEDENTS

Discussion of natural kinds is implicit in the work of many philosophers of earlier eras, but the philosophical doctrine of "kinds" or "natural groups" appears to have been overtly articulated in the mid-nineteenth century first by William Whewell in *The Philosophy of the Inductive Sciences* (1840; 2nd ed. 1847) and then by John Stuart Mill in *A System of Logic* (1843; 8th ed. 1882). Neither Whewell nor Mill spoke of "natural kinds" but mostly just of "kinds." In addition to occasionally using the terminology of "kinds," Whewell also discusses "natural groups" and "natural classes," distinguishing them from artificial groups and classes. Meanwhile, Mill sometimes uses the expressions "real kind" and "true kind"; he also refers to "natural categories" and "natural groups."

Whewell holds that we classify things into kinds or classes based on similarity or resemblance. But resemblance among particulars can be analyzed further in terms of the possession of common properties, namely those properties that "enable us to assert true and general propositions" (1847: 486). Thus resemblance, and hence the classification of particulars into kinds, may only be ascertained after extensive scientific study of a domain, when we have been able to frame general propositions. Moreover, the general propositions that we assert about a kind or natural group need to agree with one another, leading to a "consilience" of properties. The true mark of a natural classification scheme is that it sorts individuals into groups that share properties that, although apparently disparate, are discovered to be consilient. Using mineralogy as his example, Whewell states that classifying minerals according to hardness, specific gravity, color, lustre, and other observable properties leads to the same classes as those obtained on the basis of chemical constitution, and this is a sign that we have a natural rather than an artificial system of classification. As he puts it: "each of these arrangements is true and natural, then, and then only, when it coincides with the other . . . such classifications have the evidence of truth in their agreement with one another" (1847: 541). The notion of "consilience of inductions" is an important feature of Whewell's philosophy of science, and it is central to his conception of natural groups and scientific classification schemes. He also sometimes refers to the idea of the coincidence among different sets of properties by the term "affinity". Significantly, Whewell denies that all natural groups or classes have definitions; instead, they are often organized around a type or specimen. But even though natural groups may not always be definable, he does not think that makes them arbitrary; natural classes are given "not by a boundary line without, but by a central point within" (1847: 494). Another innovation of Whewell's is that he thinks that natural groups may sometimes not be sharply divided and that there may be individuals intermediate between two natural groups, or groups intermediate between superordinate groups (1847: 495).

Mill agrees with the identification of kinds with a number of co-instantiated properties or attributes. This means that a multitude of general assertions can be made about them and they can serve as the basis for inductive inference, since we infer from the presence of some of those properties the presence of others. So far, Mill's view concurs broadly with Whewell's, but he qualifies this characterization in at least two crucial ways. First, he thinks that the properties associated with kinds must be "inexhaustible" or "indefinite" in number, and second, he holds that they must not follow or be deducible from one another by an "ascertainable law" (1882: I vii 4). On the first count, some of Mill's critics have wondered why the number of properties involved must be indefinite in number. After all, if a number of different properties are jointly instantiated in natural kinds and they enable us to make inductive inferences and project to new instances, it does not seem necessary to require that they be inexhaustible. The second claim is also controversial, since, as critics such as C. S. Peirce (1901) observed, the very point of scientific inquiry is presumably to discover the laws that connect the properties associated with kinds. Sometimes Mill suggests that the properties of kinds are linked as a mere matter of coexistence or by "uniformities of coexistence"—in other words, a kind of ultimate brute fact (1882: III xxii 2). However, at other times he seems to allow the properties associated with kinds to be causally linked: "The properties, therefore, according to which objects are classified, should, if possible, be those which are causes of many other properties: or at any rate, which are sure marks of them" (1882: IV vii 2; cf. 1882: III xxii 2). Moreover, he suggests that uniformities of coexistence, unlike causal uniformities ("uniformities of succession"), only account for the co-instantiation of the ultimate properties in nature, "those properties which are the causes of all phenomena, but are not themselves caused by any phenomenon" (1882: III xxii 2). This would suggest that coexistence obtains for the properties of the most fundamental kinds, whereas causality might obtain among the properties of other kinds.

A further complication in Mill's view is that at times he identifies kinds with natural groups or groups that feature in scientific classification schemes, but at other times he states that real kinds are a subset of such groups. After equating kinds with natural groups, he goes on to say that although all plant species are real kinds, very few of the higher taxa, such as genera and families, can be said to be kinds (1882: IV vii 4). What would disqualify such classes from corresponding to kinds? One additional condition that he places on kinds, as opposed to groups in natural classification schemes, is that there be an "impassable barrier" between them and that they be entirely distinct from one another (1882: IV vii 4). In contrast with Whewell, Mill thinks that there cannot be intermediate members between real kinds; as he puts it, they must be separated by a "chasm" rather than a "ditch" (1882: I vii 4). In pre-Darwinian biology, species were commonly thought to satisfy this condition but not higher taxa. Mill also disagreed with Whewell about types, denying that natural groups could be delimited by typical members and considering definitions to be important to the delimitation of natural groups (1882: IV vii 4). Finally, Mill affirms that different sciences may classify the same individuals in different ways, since these classifications may serve different purposes; for example a geologist categorizes fossils differently from a zoologist (1882: IV vii 2).

The logician John Venn built on the insights of Whewell and Mill, emphasizing the importance of natural groups for probabilistic inference and statistical generalization.[1] For Venn, who was primarily interested in probability and induction, natural kinds or natural groups are the main grounds of statistical inference. He writes in the *Logic of Chance* (1866; 3rd ed. 1888): "Such regularity as we trace in nature is owing, much more than is often suspected, to the arrangement of things in natural kinds, each of them containing a large number of individuals" (1888: III, sect. 3). It is these regularities that enable us to project from one instance of a kind to another, or to make general statements about them. Venn was particularly interested in cases in which causal links between properties were not strict and where accurate predictions could not be made about single cases but only about large numbers of them. When the properties associated with natural kinds are not linked to them invariably but only with a certain frequency, "there is a "limit" to which the averages of increasing numbers of individuals tend to approach" (1888: XVIII sect 16).

---

[1] Hacking (1991: 110) credits the first usage of the expression "natural kind" to Venn. But in using that term, Venn seems to have thought he was following Mill's own usage. There is a possibility that Venn misremembered Mill's exact expression and inserted the modifier "natural" inadvertently.

In a later work, *The Principles of Empirical or Inductive Logic* (1889; 2nd ed. 1907), Venn distinguished "natural substances" from "natural kinds," the former being chemical elements and compounds and the latter biological species. He also held that many conventional actions or institutional phenomena share the features of natural kinds. In all these cases, the particulars falling under the kinds possess a number of common attributes, although "attributes are not things which can be counted like apples on a tree" (1907: 333). Still, we choose the attributes upon which to base our classifications in such a way that one attribute will point to a number of others as well. For example, if we classify plants not on the basis of color but on the structure of the seed, we will find that a number of other attributes will be associated with that attribute (Venn 1907: 335). Venn also seems to endorse Mill's requirement that natural kinds ought to be separated from one another by a sharp divide (335). However, he notes that these two requirements may not always coincide, because there could be classifications that are based on maximizing the number of coincident properties, which, however, do not lead to clean breaks between classes (336). Like Mill, Venn subscribes to the interest-relativity of classification schemes, averring that each will serve a specific purpose (562).

Peirce's views on natural kinds emerged initially in opposition to Mill's. As already mentioned, he objects strenuously to Mill's apparent contention that the properties of natural kinds should not be linked by law. Instead, he claims that scientists aim precisely at accounting for the connections between the various properties associated with natural kinds. He also proposes an alternative definition of a "real kind": "Any class which, in addition to its defining character, has another that is of permanent interest and is common and peculiar to its members, is destined to be conserved in that ultimate conception of the universe at which we aim, and is accordingly to be called 'real.'"(1901). There are two features of this definition that are worth underlining. First, Peirce is rather minimalist in allowing a kind to have just two properties pertaining to it, presumably properties that are logically independent though linked by scientific law. Second, he builds into his definition a conception of the end of inquiry by suggesting that we cannot definitively rule on kinds until we have an ultimate scientific theory. Elsewhere, Peirce (1902/1932) linked natural kinds to final causation, characterizing a "real" or "natural class" as "a class of which all the members owe their existence as members of the class to a common final cause." Like Whewell but unlike Mill, Peirce (1902/1932) also denies that there are always sharp divisions between natural classes and holds that not all natural classes are definable, maintaining that there may be no essential properties that are both necessary and sufficient for membership in a natural class.

After these seminal discussions of natural kinds from the mid-nineteenth century into the first decade of the twentieth, the topic was seldom discussed until the 1970s, with three notable exceptions: the work of C. D. Broad, Bertrand Russell, and W. V. Quine. One can discern a definite trajectory in the views of these three philosophers, as they represent increasing disillusionment with the notion of natural kinds and growing skepticism regarding their utility for science.

Broad invokes natural kinds in discussing induction by enumeration, saying that such inductions are plausible when it comes to natural kinds (e.g., crows, swans, pieces of silver) but not when it comes to non-natural kinds (e.g., billiard balls, counters in a bag). But Broad's main innovation is the introduction of a striking and influential image for conceiving of natural kinds (although it is not clear whether it was proposed independently by others). Imagine an n-dimensional state space of n properties (e.g., color, temperature, hardness), each of which can take on various values, and plot each individual in the universe as a point in this space depending on the values that it has for each property. If we carry out this exercise, what we find is "a 'bunching together' of instances in the neighborhood of certain sorts of states," rather than a uniform distribution (Broad 1920: 25). In other words, some combinations of properties are very common in the universe whereas others are rare or nonexistent. To make the illustration more vivid, Broad asks us to imagine a fluid spread out in this space such that its density at any point represents the number of individuals with that combination of properties:

> We should find a number of blobs in the space surrounding certain points. These blobs would be very dark near their centres and would shade off very quickly in all directions as we moved away from these centres. In the regions between the blobs there would be practically no dots at all, and such as there were would be extremely faint. (1920: 26)

As well as serving as a compelling representation of natural kinds, this analogy contains a couple of important claims, first that members of natural kinds need not have exactly identical properties, just similar properties whose values cluster around a "type," and second that even though intermediate instances between two or more natural kinds may be very rare, they are not nonexistent. Both points were mentioned in connection with Whewell's views on kinds and were denied by Mill. Moreover, Broad stresses that the existence of "blobs" is a contingent feature of the universe, and the fact that not all combinations of properties are allowable is due to the operation of causal laws. In the end, he concludes that the notion of natural kinds is intertwined with that of causation (1920: 44). In making inductive inferences, we simply assume that natural kinds exist, but this assumption is not capable of proof, although it forms the basis of our scientific inquiries (1920: 42).

Russell begins by characterizing natural kinds as "a class of objects all of which possess a number of properties that are not known to be logically interconnected" (1948: 278). He also states that there are certain laws that make some combinations of properties more stable than others and affirms that these "functional laws of correlation" are "probably more fundamental than natural kinds" (390). Moreover, in physics, the properties associated with atomic elements are no longer thought to coexist "for no known reason" but are rather now held to be due to "differences of structure." This means that the only natural kinds whose properties are correlated for no known reason (according to the physics of his day) are electrons, positrons, neutrons, and protons, although Russell states that eventually these "may be reduced to differences of structure" (1948: 390). In short, Russell's treatment seems to contain the same tension found in Mill, in considering natural kinds to be associated with properties that are correlated due to laws of coexistence while also allowing them to be correlated due to causal laws. But in contrast to most of the philosophical tradition, Russell seems to think that natural kinds will be superseded as science comes to discover the laws that account for the coincidence of their properties and that natural kinds will be a "temporary phase" on the road to the discovery of scientific laws that relate properties directly to one another.

On this last point, Quine would seem to agree with Russell, although he does not cite him in this context. In fact, Quine has two main objections to the utility of the notion of natural kinds in modern science. First, he thinks that the notion is mired in the obscure idea of similarity and cannot therefore serve as a sound basis for scientific theorizing. Indeed, matters are even worse since one cannot even define "kind" precisely in terms of similarity due to various logical complications. Second, Quine thinks that in certain domains, similarity can be bypassed in favor of the matching of components or the coincidence of properties; in chemistry, for instance, "Molecules will be said to match if they contain atoms of the same elements in the same topological combinations" (1969: 135). But if one understands similarity in this way, then one has effectively traded a generic notion of similarity for one that applies specifically to this case and relies ultimately on identity of molecules. It is not clear from Quine's analysis why such exact measures of property coincidence would not lead to retaining natural kinds rather than rendering them extinct. Perhaps his point is that there is no need to posit kinds once one has laws that directly link properties, or perhaps he thinks that the similarity standards operative in different branches of science do not have anything interesting in common, so there is no generic notion of natural kind, just the kinds posited in each specific science.

Having examined the tradition from Whewell to Quine, several themes emerge in discussions of natural kinds, some of which represent points of consensus while others are matters of contention. There seems to be broad agreement that natural kinds collect together particulars that are similar in certain respects, where similarity may be understood in terms of sharing properties as posited by an established scientific theory (rather than superficial similarity). This means that categories corresponding to natural kinds support inductive inference and enable us to make true generalizations. Moreover, that is because natural kinds are associated with collections of properties that are not logically linked but are correlated either as a matter of brute coexistence or as a matter of causality or both. There is some disagreement as to whether these correlated properties should be indefinite in number (Mill), or just a large number (Whewell, Venn, Broad, Russell), or even as few as two (Peirce). There is no consensus on the definability of natural kinds, with some arguing that they are definable (Mill) and others that they are not always so (Whewell, Venn, Peirce, Broad). There is also controversy as to whether natural kinds must be separated from one another by a chasm with no intermediate members (Mill, Venn) or whether they can shade off into one another without discrete breaks (Whewell, Broad). Finally, most of these authors are pluralists and allow the possibility that classification by natural kinds may be discipline-relative, with different scientific disciplines sometimes classifying the same individuals differently.[2]

---

[2] This retelling of the history of the natural kind concept can be contrasted with the ones given in Hacking (1991, 2007) and Magnus (2013), where more discontinuity is found in the tradition of philosophical discussion of natural kinds.

3 CONTEMPORARY APPROACHES

The two most influential approaches to natural kinds in the recent philosophy of science are essentialism and the homeostatic property cluster (HPC) theory.

3.1 Essentialism

Essentialism is an ancient doctrine in philosophy, but it has been revived in contemporary discussions of natural kinds due to considerations deriving from the usage of certain terms in natural language. In discussions of meaning and reference, Kripke (1971/1980) and Putnam (1975) both noted parallels between proper names (e.g., "Aristotle") and natural kind terms (e.g., "gold," "water," "tiger"). Their basic observation was that the two types of terms could be used to refer successfully by a speaker even though the speaker was partly ignorant of or misinformed about the properties associated with the referent.[3] Reference in these cases is not to whatever possesses the properties associated by the speaker with the referent but to the referent's real essence. In the case of natural kinds, successful reference by a speaker depends not on possessing a description that uniquely identifies the natural kind but (at least in part) on causal contact with an exemplar of that kind (contact that may be mediated by other speakers). For example, the term "gold" can be used to refer successfully even though a speaker might be mistaken about the properties of gold, or even if something of a different kind has those same properties (e.g., pyrites). Furthermore, a natural kind term like "gold" is a "rigid designator" in the sense that it would refer to the same kind of thing in every possible world in which that substance exists. Hence, what the term "gold" refers to, in the actual world or in other possible worlds, is any exemplar that has the real essence of gold, the substance that was initially dubbed by speakers when the term "gold" was introduced.

This account of reference presupposes that there is an essence associated with each natural kind to which our terms ultimately point, even though we may be unaware of that essence and it may take an undetermined amount of time for us to discover it. If by "essence" is meant something loosely like the true nature of the thing, then that presupposition might not be a very controversial one. But numerous philosophers who built on the work of Kripke and Putnam have come to associate essences with various more substantive features (and some of these features are at least implicit in the work of Kripke and Putnam themselves and may be required by their theory of reference). Thus

---

[3] However, Putnam's extension of these claims from proper names to natural kind terms was somewhat more cautious. He writes, for example: "it is instructive to observe that nouns like 'tiger' or 'water' are very different from proper names. One can use the proper name 'Sanders' correctly without knowing anything about the referent except that he is called 'Sanders'—and even that may not be correct" (1975: 166).

the claim that natural kinds have essences has come to encompass a number of theses, including some or all of the following:

1. Necessary and sufficient properties: An essence consists of a set of properties, possession of which is singly necessary and jointly sufficient for membership in a natural kind.
2. Modally necessary properties: An essence consists of a set of properties that are (a) necessarily associated with the natural kind (i.e., associated with it in all possible worlds) and/or such that (b) members of the kind possess them necessarily (i.e., possess them in every possible world).[4]
3. Intrinsic properties: An essence consists of a set of properties that are intrinsic to members of the kind (rather than relational, extrinsic, functional, etiological, and so on).
4. Microstructural properties: An essence consists of a set of properties, all of which are microstructural in nature (i.e., pertain to the microconstituents of the kind and their arrangement in a certain configuration, such as the structural formula of a chemical compound).

Sometimes additional claims are made regarding essences:

5. Essential properties issue in sharp boundaries between natural kinds.
6. Essential properties result in a hierarchy of natural kinds that cannot crosscut one another.
7. There is a bedrock of most fundamental essences that result in a basic level of natural kinds (infima species).

Although essentialism continues to have many adherents and may indeed be the default position among contemporary philosophers, there is increasing disenchantment with it, especially among philosophers of science. One main source of dissatisfaction relates to a perceived misalignment between the claims of essentialism and the results of science. Essentialism is controversial across a wide swathe of sciences, particularly the "special sciences" that pertain to various macro-domains (e.g., fluid mechanics, stellar astronomy, biochemistry, geology), in which many kinds are functionally characterized rather than intrinsically and in terms of macro-properties rather than micro-structure. Moreover, it has become increasingly clear that the key essentialist features are not found even in many of the paradigmatic instances of natural kinds. Of course, it is always open in principle to save the essentialist thesis in question by ruling that the alleged instances are not in fact natural kinds. But if this occurs for many if not most of the paradigmatic natural kinds, it may be more reasonable to reject the thesis and preserve the natural kind.

It is not difficult to see that the essentialist theses are violated by a range of categories that are strong candidates for natural kinds, such as the atoms of chemical elements.

---

[4] Some essentialists would say *nomologically* possible worlds, but others think that the laws of nature are inherent in the essences, so they would say *metaphysically* possible worlds (Ellis 2001).

Consider thesis 2b, the claim of modal necessity. Since elemental atoms can decay into atoms of other elements, they can become atoms of other kinds (e.g., an atom of uranium-234 decays into an atom of thorium-230). This implies that they do not always belong to the same kind in the actual world. Hence one cannot maintain that atoms of chemical elements belong to those elemental kinds necessarily or in all possible worlds. Although they could still be said to satisfy 2a, since the properties associated with each kind of chemical element can be said to be associated with that kind in every possible world, this thesis seems to be satisfied trivially by categories that do not correspond to natural kinds. For example, the category of things that have a *mass of less than one kilogram* is associated with that very property in every possible world, yet no one would claim that it corresponds to a natural kind. Similar problems arise when it comes to biological species. Consider theses 1 and 3, the claims that essential properties are necessary and sufficient for membership in a kind and intrinsic to the kind, respectively. There is no intrinsic set of properties common to all and only members of a particular biological species (e.g., certain strands of DNA or a collection of genes, much less phenotypic features), so these claims are not jointly true of species. Moreover, if the essence of a biological species is taken to be historical rather than intrinsic (as many philosophers of biology would claim), then it would satisfy 1 but not 3 (or 4). In response to such problems, some essentialist philosophers have ruled that species are not natural kinds (Wilkerson 1993; Ellis 2001), while others have concluded that species have extrinsic historical essences, thereby giving up on theses 3 and 4 (Griffiths 1999; Okasha 2002), and yet others have challenged the consensus of biologists and philosophers of biology, maintaining that species do have intrinsic essences (Devitt 2008). Meanwhile, Laporte (2004) argues that 2b is also violated for biological species. In sum, it is clear that even some of the paradigmatic natural kinds fail to satisfy at least one of 1 through 4, let alone 5 through 7.

In addition to these doubts about specific essentialist theses, questions have also been raised about the cogency of deriving essentialist metaphysical conclusions from observations concerning the use of terms in natural language. If the use of certain general terms presupposes that kinds have essences, this presupposition may just reflect a human cognitive bias that is displayed in features of natural language (Leslie 2013). In response, it might be said that the presuppositions that underwrite our linguistic usage are themselves firm metaphysical intuitions (e.g., about what would be the same substance in other possible worlds). But not only are these intuitions largely unjustified postulates, they are not universally shared, as many early critics of Kripke and Putnam pointed out (e.g., Mellor 1977, Dupré 1981).

3.2 Homeostatic Property Clusters

A different account of natural kinds has been proposed in a number of papers by Richard Boyd (1989, 1991, 1999a, 1999b), whose theory adheres more closely to the details of at least some scientific categories. Boyd begins from the insight that many kinds in the special sciences correspond to property clusters. In describing them as "clusters," he is arguing for a frequent co-occurrence among the properties while allowing for a certain looseness of association among them. This constitutes a departure from essentialism as ordinarily understood, and the break with essentialism is confirmed by Boyd when he writes: "The natural kinds that have unchanging definitions in terms of intrinsic necessary and sufficient conditions . . . are an unrepresentative minority of natural kinds (perhaps even a minority of zero)" (1999a: 169). Instead of a set of properties that are singly necessary and jointly sufficient for membership in the kind, Boyd holds that members of a kind may share a largely overlapping set of properties. That is because many natural kinds consist of properties loosely maintained in a state of equilibrium or homeostasis. This is achieved by means of an underlying mechanism that ensures that many of these properties are instantiated together, although it may also be that "the presence of some of the [properties] favors the others" (Boyd 1989: 16). Either way, we denote this loose collection or cluster of properties by some kind term, which picks out either all or most of the properties in the cluster and possibly some of the underlying mechanisms as well. Although Boyd (1999a: 141) sometimes notes that the HPC theory can be considered the essence of a natural kind and allows that natural kinds have "definitions" provided by the cluster, this is clearly not a version of essentialism in the ordinary sense. Some philosophers who have endorsed the HPC account of natural kinds think that it is compatible with essentialism (Griffiths 1999), but most have regarded it as an alternative to essentialist accounts.

Boyd emphasizes that the fact that properties cluster loosely, rather than by way of necessary and sufficient conditions, is a contingent fact about the world rather than a reflection of our ignorance. It arises because in the special sciences, there are few if any exceptionless laws to be found. Instead, there are "causally sustained regularities" that apply to natural kinds and specify their properties in such a way as to allow for exceptions and instances in which properties are imperfectly correlated (Boyd 1999a: 151–152, 1999b: 72). The fact that such generalizations are not exceptionless or eternal by no means suggests that they are merely accidental, since they are sustained by underlying causal structures (Boyd 1999a: 152). Although Boyd does not appear to say this in so many words, clustering occurs because many causal connections in the universe are not strict, and this is due to the fact that some causal processes can interfere with other causal processes, leading to the imperfect co-instantiation of groups of properties. That is presumably why we find loose clusters of causal properties rather than exactly the same set of properties in every individual member of a kind.

An important aspect of Boyd's account is what he calls the "accommodation thesis," namely the claim that we devise our categories in such a way as to accommodate the causal structure of the world. The fact that some inductive generalizations are true and the terms that feature in those generalizations are projectible (Goodman 1954/1979) is a reflection of this causal structure. Boyd sums up his conception of accommodation as follows:

> We are able to identify true generalizations in science and in everyday life because we are able to accommodate our inductive practices to the causal factors that sustain them. In order to do this—to frame such projectable generalizations at all—we require a vocabulary . . . which is itself accommodated to relevant causal structures. (1999a: 148)

Boyd never meant his account to apply across the board to all natural kinds but at best to a range of natural kinds in the special sciences, which as he says "study complex structurally or functionally characterized phenomena" (1989: 16). Although he often mentions biological examples like species and higher taxa, he also refers in passing to natural kinds in geology, meteorology, psychology, and the social sciences. Like many philosophers in the tradition of theorizing about natural kinds, Boyd affirms "disciplinary or sub-disciplinary pluralism about natural kinds" (1999b: 92), such that natural kinds in one discipline may not be such in another. But he also makes the point that some natural kinds facilitate inductions and feature in explanations "beyond the domain of a single scientific discipline" (Boyd 1999b: 81).

Despite the fact that Boyd's account seems to accord with a range of natural kinds in the special sciences, particularly the biological sciences, some philosophers have found it wanting precisely when it comes to capturing the nature of these kinds. It has been argued that Boyd's focus on interbreeding as the homeostatic mechanism that preserves the stability of biological species does not do justice to the historical nature of species. Moreover, given polymorphisms within biological species, there is no cluster of properties common to all members of a species that are kept in equilibrium (Ereshefsky and Matthen 2005). Other philosophers have defended the HPC account by allowing a number of different mechanisms to account for the stability of species (Wilson, Barker, and Brigandt 2009). However, even if Boyd's account aptly characterizes a range of natural kinds in the special sciences, where homeostatic mechanisms are plentiful and many systems have elaborate feedback mechanisms that maintain them in a state of equilibrium, it may not account for all such kinds (let alone nonspecial science kinds such as those of elementary particle physics). The HPC account certainly represents an advance over essentialist views in its fidelity to scientific practice, but it may not go far enough in the direction of naturalism.

4 A NATURALIST ALTERNATIVE

What would a naturalist approach to natural kinds look like? A naturalist approach is one that takes its cue from science rather than specifying certain conditions that natural kinds should meet on a priori grounds or on the basis of conceptual analysis. To determine which categories correspond to natural kinds, we should ascertain which categories matter to science. Does this mean that our default assumption should be that all categories that occur in our current best scientific theories correspond to natural kinds? There are two caveats here. The first is that scientific theories are corrigible, so that our current best theories might be revised and at least some of their categories revoked or abandoned. A definitive judgment concerning the identity of natural kinds can only be made when and if the categories of science are finalized (Khalidi 2013). The second caveat is that not every category invoked in scientific inquiry, much less every one mentioned in a scientific journal article, plays an indispensable epistemic role in the relevant science (Magnus 2012). Some categories may be redundant, or adopted for nonscientific purposes, or incidental to the actual practice of scientific inquiry, or holdovers from discredited scientific theories.

On this way of characterizing naturalism, it is not a position universally shared by contemporary philosophers. Some philosophers would insist, rather, that philosophy ought to specify the conditions that natural kinds should meet, and then science will tell us which entities in the world satisfy those conditions.[5] The first step presupposes that philosophy has access to conceptual or a priori truths independently of the deliverances of science. That is an assumption that many contemporary naturalists would dispute. But does naturalism then imply that philosophy simply registers what science has to say and that it has no original contribution to make on this and other matters? No, because, as already suggested, a critical philosophical stance can help to determine whether some scientific categories are redundant, dispensable, inconsistent, or incoherent.

A number of contemporary philosophers have adopted such an approach, identifying natural kinds with categories that play an epistemic or investigative role in science, sometimes referring to them as "epistemic kinds" or "investigative kinds" (Kornblith 1993; Brigandt 2003; Griffiths 2004; Craver 2009; Weiskopf 2011; Magnus 2012; Slater 2014; Franklin-Hall 2015). One common denominator among most (but not all; see Ereshefsky and Reydon 2014) recent approaches to natural kinds is the claim that categories corresponding to natural kinds are projectible, play a privileged role in induction, and feature in explanatory arguments. In fact, this approach is already found in Boyd's account of natural kinds and is implicit in his "accommodation thesis." As Boyd puts it: "It is a truism that the philosophical theory of *natural* kinds is about how classificatory schemes come to contribute to the epistemic reliability of inductive and explanatory practices" (1999a: 146; emphasis in original). Thus certain categories play a role in scientific discovery, explanation, and prediction and figure in successful inductive inferences. A number of nontrivial generalizations can be made using such categories, and any new member that is classified under such a category is likely to have all or many of the other associated features. Moreover, there are many such features associated with the kind, not just a few, and we tend to discover more as time goes by (although not necessarily an indefinite number, as Mill held).

Rather than rest with a purely epistemic understanding of natural kinds, one can ask, in line with Boyd's accommodation thesis: What are the ontological underpinnings of the epistemic role of natural kind categories? According to the tradition beginning with Whewell and Mill, natural kinds are associated with a cluster of properties that when co-instantiated lead to the instantiation of a multitude of other properties. If kinds are just clusters of properties, the question arises: Why do properties cluster? If we think of some of the most fundamental constituents of the universe, such as quarks or leptons, there does not seem to be an answer to the question as to why they have the basic properties that they do. At least in the current state of scientific theorizing, there is no widely

---

[5] See Papineau (2009) for a characterization of the contrast between the naturalist approach and the alternative.

accepted explanation of their having just those combinations of mass, charge, spin, and other quantum numbers. An explanation may yet be found, but it may also be that for the most fundamental kinds (whatever they may turn out to be) the co-occurrence of their "core" properties is just a matter of brute coexistence. However, when it comes to other natural kinds, even the chemical elements, there is a causal account of the clustering of properties. An atom of carbon-12, for example, has six protons, six neutrons (each of which consist of three quarks), and (in a nonionized state) six electrons, and we have scientific theories that tell us why those particular combinations of particles are stable, namely the theories of strong forces and electroweak forces. Hence there is a causal account of the co-instantiation of the properties of carbon atoms. Moreover, many of the macro-level properties of large assemblages of carbon atoms can also be accounted for with reference to the micro-level properties of carbon atoms. Here too there are causal theories that explain why these micro-properties, when aggregated in certain ways and in certain contexts, will yield just those macro-properties. For a different isotope of carbon, carbon-14, the different macro-properties (e.g., half-life) will be accounted for with reference to slightly different micro-properties. Once the "core" properties of natural kinds cluster together, the "derivative" properties follow as a matter of causality. Even with regard to the properties of elementary particles, their quantum numbers may be co-instantiated by way of mere coexistence, but their derivative properties are causally linked to those core properties. For instance, the fact that electrons curve in the opposite direction as positrons in an electromagnetic field, or that they leave a visible track in a cloud chamber, follows causally from the more basic properties of electrons.

So far we have canvassed two reasons for the co-instantiation of the core properties of natural kinds—brute coexistence and causal relations—just as Mill seems to have held. Another reason that is sometimes given for the clustering of properties in nature as well as in the social world is the presence of a copying mechanism. Ruth Millikan (1999, 2005) distinguishes "eternal kinds" from "copied kinds" (or "historical kinds") on the grounds that the latter share certain properties not as a result of natural law but rather as a consequence of a copying process that ensures that many of the same properties are instantiated in individual members of those kinds. As she puts it, each member of the kind "exhibits the properties of the kind because other members of that same historical kind exhibit them" (1999: 54). This is supposed to obtain for biological kinds such as species (e.g., *tiger*), as well as for many social kinds (e.g., *doctor*) and artifactual kinds (e.g., *screwdriver*). Moreover, members of such kinds may not all share exactly the same set of properties and may not be separated from members of other kinds by sharp boundaries. They belong to the same kind because a copying process has produced the kind members from the same models, they have been produced in response to the same environment, and some function is served by members of the kind (Millikan 2005: 307–308). However, although Millikan makes much of this division between two kinds of kinds, causal relations are also central to copied kinds, since the copying process that gives rise to all members of the kind would not copy them had they not performed a certain function in a particular environment. Moreover, there would appear to be some kinds that combine features of both natural and copied kinds, for example DNA molecules.

Millikan also refers to copied kinds as "historical kinds," but there is a distinction to be made between copied kinds and etiological kinds, which are individuated solely or largely in terms of a shared causal history. Members of some kinds from a variety of sciences are grouped together because of the fact that they are the result of the same or exactly similar causal process. This does not necessarily entail being copied in response to the same environment or serving the same function. Consider the cosmological kind *cosmic microwave background radiation*, which includes all radiation left over from the Big Bang. Collections of photons belonging to this kind are intrinsically indistinguishable from others that have the same frequency yet are distinguished from them because of their common origin. Similarly, members of a geological kind like *sedimentary rock* are grouped together and distinguished from other rocks on the basis of the process that gave rise to them. Some of them may share causal properties as well, but they are primarily individuated on the basis of the causal process that produced them. Etiological kinds can therefore be distinguished from other kinds on the grounds that they possess a shared causal history, although they do not necessarily share synchronic causal properties (at any rate, not ones that distinguish them from some nonmembers). They are also causally individuated, but with reference to diachronic rather than synchronic properties. Some philosophers may consider etiological kinds not to be natural kinds, but many categories drawn from a range of sciences are at least partly etiological, including biological species. If they are considered natural kinds, then they are also causally individuated, although on the basis of diachronic rather than synchronic causality. Hence, apart from those fundamental kinds whose "core" properties cluster as a matter of coexistence, it appears as though the common ontological ground for natural kinds is causal.

5 REMAINING QUESTIONS AND FUTURE DIRECTIONS

A contrast is often drawn between interest-relative classification and objective or human-independent classification. However, if the previously discussed considerations are correct, this contrast is overblown. It is true that some interests are involved in constructing our categories and delimiting their boundaries, but if the interests are epistemic—that is, they aim to acquire knowledge of the world by locating categories that are projectible, support inductive generalizations, have explanatory value, and so on—then they ought to enable us to identify genuine features of the world. As long as the interests guiding our taxonomic schemes are epistemic, they should point us in the direction of real distinctions in nature. Nevertheless, naturalism is compatible with the claim that there may be a variety of genuinely epistemic interests, which may pertain to different scientific disciplines or subdisciplines. This may result in different taxonomic schemes that classify the same individuals in a crosscutting manner. Two individuals, i_1

and i_2, may be classified in some taxonomic category K_1 by one branch of science, while i_2 may be classified with some other individual i_3 in category K_2 by another branch of science, where neither category K_1 nor K_2 includes the other (Khalidi 1998). Although crosscutting taxonomies reflect different interests, they do not necessarily undermine realism; provided they are epistemic interests they will presumably uncover different aspects of the causal structure of the universe. But one may still harbor doubts as to whether even our most considered epistemic categories generally cleave to the joints of nature. These doubts may loom larger if we allow that some of our categories are delimited for reasons of human convenience, ease of measurement, relevance to human well-being, and other idiosyncratic considerations. Hence it may be argued that even at the idealized end of inquiry, some details concerning the boundaries of our categories or their specific features may be determined in part by our human predilections. It is an open question as to whether and to what extent such considerations play a role in demarcating our categories and hence to what extent these categories correspond to nature's own divisions.

The naturalist conception of natural kinds is an avowedly pluralist one, countenancing a more diverse set of natural kinds than some traditional philosophers might have allowed. But there is a more radical variety of pluralism about natural kinds that needs to be mentioned, namely pluralism about the notion of natural kind itself. Despite the unitary causally based account of natural kinds offered in the previous section, it may be possible to maintain that there is no single notion of natural kind that is applicable to all the diverse branches of science and across all scientific domains. Just as some philosophers have argued for a pluralist conception of causality, with several types of causal relation, each playing a central role in different areas of science, others have held that diverse factors ground natural kinds in different domains (Dupré 2002). This position was already prefigured in the discussion of Quine's views on natural kinds, and it remains to be seen whether all the natural kinds identified by science can be made to fit a single overarching (causal) template.

The naturalist characterization of natural kinds is in principle amenable to the existence of natural kinds in the social sciences, and this accords with the thinking of at least some nineteenth-century philosophers (Mill, Venn, and Peirce, all countenance social kinds). But there are some obstacles to assimilating social kinds to natural kinds. Some philosophers claim that social kinds are ontologically subjective, depending for their very existence on human mental attitudes (Searle 1995). Others argue that they are different from natural kinds because they are interactive and can change in response to our attitudes toward them, in what Hacking (1999) has dubbed the "looping effect" of social or human kinds. Perhaps most important, some philosophers hold that the difference between natural kinds and social kinds is that the latter are fundamentally evaluative or normative in nature (Griffiths 2004). These are all significant challenges to the incorporation of social kinds into natural kinds, and there is more work to be done to ascertain whether these features obtain for all social kinds and whether they preclude their being natural kinds. Meanwhile, the term "natural kind" should not be taken to preempt this issue, with its implied contrast between the natural and the social. Since it should be an open question as to whether social kinds can be natural kinds, it might have been better to adopt the term "real kind" instead of "natural kind." This would also have helped dispense with the problematic assumption that natural kinds cannot be artificially produced, since it is clear that many human-made kinds are good candidates for natural kinds, for example synthetic chemical compounds, artificially selected organisms, and genetically engineered organisms.

If one thinks of natural kinds as corresponding to clusters of causal properties, the possibility arises that there may be more such clusters than human beings are able to observe or apprehend. This means that although our most established scientific categories will indeed correspond to natural kinds, there may be a multitude of other natural kinds that we have not succeeded in capturing. The naturalist account of natural kinds is realist, but it allows that there may be more kinds of things in the universe than are dreamt of in a completed science. It may be that, given human frailties and limitations, the kinds that we identify are a small fraction of those that exist in the universe. This consequence may be thought to undermine realism, since it leads to too many natural kinds, only some of which we will actually identify. But it is not clear why allowing an abundance of natural kinds and admitting our inability to identify them all would impugn their reality. An alternative that is sometimes suggested consists in identifying the natural kinds with the most fundamental kinds in nature: quarks, leptons, and bosons (at least according to our current theories). There are a couple of problems with this view. First, it is possible that there is no fundamental level in nature but that the universe consists of strata of evermore fundamental entities (Schaffer 2003). Second, it leaves many of the interesting questions that exercise philosophers as well as working scientists unaddressed, namely questions about which categories and systems of classification we should adopt beyond the most fundamental level. The question of natural kinds is not just of philosophical interest; it occurs at least implicitly in many scientific controversies. Questions often arise over the validity of certain categories or their legitimacy, such as *Asperger syndrome, race,* or *social class.* These questions appear not just in psychiatry and the social sciences but also in the natural sciences. Consider: Is *dark matter* a natural kind? That is, is there a uniform kind of thing that is a (hitherto undiscovered) kind of matter and is characterized by a number of properties that give rise to a multitude of others? Or are there perhaps two or a few such kinds? Or is "dark matter" just a catchall term for whatever it is that accounts for a discrepancy in our calculations or has certain gravitational effects, without having a common set of properties? Presumably, these questions are at the heart of contemporary physical and cosmological inquiries. If we answer the last question in the affirmative, then one way of putting our conclusion would be: *dark matter* is not a natural kind.

References

Armstrong, D. (1989). *Universals: An Opinionated Introduction.* (Boulder, CO: Westview Press).

Boyd, R. (1989). "What Realism Implies and What It Does Not." *Dialectica* 43: 5–29.

Boyd, R. (1991). "Realism, Anti-Foundationalism, and the Enthusiasm for Natural Kinds." *Philosophical Studies* 61: 127–148.

Boyd, R. (1999a). "Homeostasis, Species, and Higher Taxa." In R. A. Wilson (ed.), *Species: New Interdisciplinary Essays* (Cambridge, MA: MIT Press), 141–186.

Boyd, R. (1999b). "Kinds, Complexity, and Multiple Realization." *Philosophical Studies* 95: 67–98.

Brigandt, I. (2003). "Species Pluralism Does Not Imply Species Eliminativism." *Philosophy of Science* 70: 1305–1316.

Broad, C. D. (1920). "The Relation between Induction and Probability (Part II)." *Mind* 29: 11–45.

Craver, C. (2009). "Mechanisms and Natural Kinds." *Philosophical Psychology* 22: 575–594.

Devitt, M. (2008). "Resurrecting Biological Essentialism." *Philosophy of Science* 75: 344–382.

Dupré, J. (1981). "Natural Kinds and Biological Taxa." *Philosophical Review* 90: 66–90.

Dupré, J. (2002). "Is 'Natural Kind' a Natural Kind Term?" *Monist* 85: 29–49.

Ellis, B. (2001). *Scientific Essentialism* (Cambridge, UK: Cambridge University Press).

Ereshefsky, M., and Matthen, M. (2005). "Taxonomy, Polymporphism, and History: An Introduction to Population Structure Theory." *Philosophy of Science* 72: 1–21.

Ereshefsky, M., and Reydon, T. A. C. (2014). "Scientific Kinds." *Philosophical Studies* 167: 1–18.

Franklin-Hall, L. (2015). "Natural Kinds as Categorical Bottlenecks." *Philosophical Studies* 172: 925–948.

Goodman, N. (1979). *Fact, Fiction, and Forecast* (4th ed.) (Cambridge, MA: Harvard University Press). (Original work published 1954)

Griffiths, P. E. (1999). "Squaring the Circle: Natural Kinds with Historical Essences." In R. A. Wilson (ed.), *Species: New Interdisciplinary Essays* (Cambridge, MA: MIT Press), 209–228.

Griffiths, P. E. (2004). "Emotions as Natural and Normative Kinds." *Philosophy of Science* 71: 901–911.

Hacking, I. (1991). "A Tradition of Natural Kinds." *Philosophical Studies* 61: 109–126.

Hacking, I. (1999). *The Social Construction of What?* (Cambridge, MA: Harvard University Press).

Hacking, I. (2007). "Natural Kinds: Rosy Dawn, Scholastic Twilight." *Royal Institute of Philosophy Supplement* 82: 203–239.

Hawley, K., and Bird, A. (2011). "What Are Natural Kinds?" *Philosophical Perspectives* 25: 205–221.

Khalidi, M. A. (1998). "Natural Kinds and Crosscutting Categories." *Journal of Philosophy* 95: 33–50.

Khalidi, M. A. (2013). *Natural Categories and Human Kinds: Classification Schemes in the Natural and Social Sciences* (Cambridge, UK: Cambridge University Press).

Kornblith, H. (1993). *Inductive Inference and Its Natural Ground* (Cambridge, MA: MIT Press).

Kripke, S. (1980). *Naming and Necessity* (Cambridge, MA: Harvard University Press). (Original work published 1971)

LaPorte, J. (2004). *Natural Kinds and Conceptual Change* (Cambridge, UK: Cambridge University Press).

Leslie, S. J. (2013). "Essence and Natural Kinds: When Science Meets Preschooler Intuition." *Oxford Studies in Epistemology* 4: 108–166.

Magnus, P. D. (2012). *Scientific Inquiry and Natural Kinds: From Planets to Mallards* (New York: Palgrave Macmillan).

Magnus, P. D. (2013). "No Grist for Mill on Natural Kinds." *Journal for the History of Analytical Philosophy* 2: 1–15.

Mellor, D. H. (1977). "Natural Kinds." *British Journal for the Philosophy of Science* 28: 299–312.

Mill, J. S. (1882). *A System of Logic* (8th ed.) (New York: Harper & Brothers).

Millikan, R. (1999). "Historical Kinds and the 'Special Sciences.'" *Philosophical Studies* 95: 45–65.

Millikan, R. (2005). "Why Most Concepts Aren't Categories." In H. Cohen and C. Lefebvre (eds.), *Handbook of Categorization in Cognitive Science* (Amsterdam: Elsevier), 305–315.

Okasha, S. (2002). "Darwinian Metaphysics: Species and the Question of Essentialism." *Synthese* 131: 191–213.

Papineau, D. (2009). "Naturalism." In *Stanford Encyclopedia of Philosophy* (Spring 2009 ed.), http://plato.stanford.edu/archives/spr2009/entries/naturalism/

Peirce, C. S. (1901). "Kind." In J. M. Baldwin (ed.), *Dictionary of Philosophy and Psychology* (New York: Macmillan). http://www.jfsowa.com/peirce/baldwin.htm

Peirce, C. S. (1932). "The Classification of the Sciences." In C. Hartshorne and P. Weiss (eds.), *Collected Papers of Charles Sanders Peirce*, Vol. I (Cambridge, MA: Harvard University Press). (Originally published 1902). http://www.textlog.de/4255.html

Putnam, H. (1975). "The Meaning of 'Meaning.'" In *Philosophical Papers*, Vol. 2: *Mind, Language, and Reality* (Cambridge, UK: Cambridge University Press), 215–271.

Quine, W. V. (1969). "Natural Kinds." In W. V. Quine, *Ontological Relativity and Other Essays* (New York: Columbia University Press), 114–138.

Russell, B. (1948). *Human Knowledge: Its Scope and Limits* (London: Routledge).

Schaffer, J. (2003). "Is There a Fundamental Level?" *Noûs* 37: 498–517.

Searle, J. (1995). *The Construction of Social Reality* (New York: Free Press).

Slater, M. (2014). "Natural Kindness." *British Journal for the Philosophy of Science*. Advance online publication. doi: 10.1093/bjps/axt033

Venn, J. (1888). *The Logic of Chance* (3rd ed.) (London: Macmillan).

Venn, J. (1907). *The Principles of Empirical or Inductive Logic* (2nd ed.) (London: Macmillan).

Weiskopf, D. (2011). "The Functional Unity of Special Science Kinds." *British Journal for the Philosophy of Science* 62: 233–258.

Whewell, W. (1847). *The Philosophy of the Inductive Sciences*, vol. 1 (2nd ed.) (London: John W. Parker).

Wilkerson, T. E. (1993). "Species, Essences, and the Names of Natural Kinds." *Philosophical Quarterly* 43: 1–19.

Wilson, R. A., Barker, M. J., and Brigandt, I. (2009). "When Traditional Essentialism Fails: Biological Natural Kinds." *Philosophical Topics* 35: 189–215.

CHAPTER 20

....

PROBABILITY

....

ANTONY EAGLE

1 INTRODUCTION

....

WHEN thinking about probability in the philosophy of science, we ought to distinguish between probabilities *in* (or *according to*) theories, and probabilities *of* theories.[1]

The former category includes those probabilities that are assigned to events by particular theories. One example is the quantum mechanical assignment of a probability to a measurement outcome in accordance with the Born rule. Another is the assignment of a probability to a specific genotype appearing in an individual, randomly selected from a suitable population, in accordance with the Hardy-Weinberg principle. Whether or not orthodox quantum mechanics is correct, and whether or not any natural populations meet the conditions for Hardy-Weinberg equilibrium, these theories still assign probabilities to events in their purview. I call the probabilities assigned to events by a particular theory *t*, *t-chances*. The *chances* are those *t*-chances assigned by any true theory *t*. The assignment of chances to outcomes is just one more factual issue on which different scientific theories can disagree; the question of whether the *t*-chances are the chances is just the question of whether *t* is a true theory.

Probabilities of theories are relevant to just this question. In many cases, it is quite clear what *t*-chances a particular theory assigns, while it is not clear how probable it is that those are the chances. Indeed, for any contentful theory, it can be wondered how probable it is that its content is correct, even if the theory itself assigns no chances to outcomes.

Perhaps these two seemingly distinct uses of probability in philosophy of science can ultimately be reduced to one underlying species of probability. But, at first glance, the prospects of such a reduction do not look promising, and I will approach the two topics

---

[1] I use *theory* as a catchall term for scientific hypotheses, both particular and all-encompassing; I hereby cancel any implication that theories are *mere* theories, not yet adequately supported by evidence. Such well-confirmed theories as general relativity are still theories in my anodyne sense.

separately in the following two sections of this chapter. In the last section, I will turn to the prospect of links between probabilities in theories and probabilities of theories.

2 PROBABILITIES IN THEORIES

2.1 Formal Features

Scientific theories have two aims: prediction and explanation. But different theories may predict and explain different things, and generally each theory will be in a position to address the truth of only a certain range of propositions. So, at least implicitly, each theory must be associated with its own space of (possible) *outcomes*, a collection of propositions describing the events potentially covered by the theory. We can assume that, for any theory, its space of outcomes forms an *algebra*. That is, (1) the trivially necessary outcome $\top$ is a member of any space of outcomes; (2) if p is an outcome, then so is its negation $\neg p$; and (3) if there exists some countable set of outcomes $\{p_1, p_2, \ldots\}$, then their countable disjunction $\bigvee_i p_i$ is also an outcome. Each of these properties is natural to impose on a space of outcomes considered as a set of propositions the truth value of which a given theory could predict or explain.

A theory includes chances only if its principles or laws make use of a *probability function P* that assigns numbers to every proposition p in its space of outcomes O, in accordance with these three axioms, first formulated explicitly by Kolmogorov (1933):

1. $P(\top) = 1$;
2. $P(p) \geq 0$, for any $p \in O$;
3. $P(\bigvee_i p_i) = \Sigma_i P(p_i)$ for any countable set of mutually exclusive outcomes $\{p_1, \ldots\} \subseteq O$.

To these axioms is standardly added a stipulative definition of *conditional probability*. The notation $P(p|q)$ is read *the probability of p given q*, and defined as a ratio of unconditional probabilities: when $P(q) > 0$, $P(p|q) =_{df} \dfrac{P(p \wedge q)}{P(q)}$

A theory that involves such a function P meets a formal mathematical condition on being a chance theory. (Of course, a theory may involve a whole family of probability functions, and involve laws specifying which probability function is to apply in particular circumstances.) As Kolmogorov noted, the formal axioms treat probability as a species of *measure*, and there can be formally similar measures that are not plausibly chances. Suppose some theory involved a function that assigned to each event a number corresponding to what proportion of the volume of the universe it occupied—such a function would meet the Kolmogorov axioms, but "normalized event volume" is not chance. The formal condition is necessary, but insufficient for a theory to involve chances (Schaffer 2007, 116; but see Eagle 2011*a*, Appendix A).

2.2 The Modal Aspect of Chance

What is needed, in addition to a space of outcomes and a probability measure over that space, is that the probability measure play the right role in the theory. Because the outcomes are those propositions whose truth value is potentially predicted or explained by the theory, each outcome—or its negation, or both—is a *possibility* according to the theory. Because scientific theories concern what the world is like, these outcomes are objectively (physically) possible, according to the associated theory. It is an old idea that probability measures the "degree of possibility" of outcomes, and chances in theories correspondingly provide an objective measure of how possible each outcome is, according to the theory. This is what makes a probability function in theory T a chance: that T uses it to quantify the possibility of its outcomes (Mellor 2005, 45–48; Schaffer 2007, 124; Eagle 2011*a*, §2).

Principles about chance discussed in the literature can be viewed as more precise versions of this basic idea about the modal character of chances. Consider the Basic Chance Principle (BCP) of Bigelow, Collins, and Pargetter (1993), which states that when the T-chance of some outcome p is positive at w, there must be a possible world (assuming T is metaphysically possible) in which in which p occurs and in which the T-chance of p is the same as in w, and which is relevantly like w in causal history prior to p:[2]

> In general, if the chance of A is positive there must be a possible future in which A is true. Let us say that any such possible future grounds the positive chance of A. But what kinds of worlds can have futures that ground the fact that there is a positive present chance of A in the actual world? Not just any old worlds ... [T]he positive present chance of A in this world must be grounded by the future course of events in some A-world sharing the history of our world and in which the present chance of A has the same value as it has in our world. That is precisely the content of the BCP. (Bigelow, Collins, and Pargetter 1993, 459)

So the existence of a positive chance of an outcome entails the possibility of that outcome, and indeed, the possibility of that outcome under the circumstances which supported the original chance assignment. Indeed, Bigelow et al. argue that "anything that failed to satisfy the BCP would not deserve to be called *chance*."

Does the BCP merely precisify the informal platitude that p's having some chance, in some given circumstances, entails that p is possible in those circumstances? The BCP

---

[2] In the same vein is the stronger (assuming that the laws entail the chances) "Realization Principle" offered by Schaffer (2007, 124): that when the chance of p is positive according to T, there must be another world alike in history, sharing the laws of T, in which p.

entails that when an outcome has some chance, a very specific metaphysical possibility must exist: one with the same chances supported by the same history, in which the outcome occurs. This means that the BCP is incompatible with some live possibilities about the nature of chance. For example, consider *reductionism* about chance, the view that the chances in a world supervene on the Humean mosaic of occurrent events in that world (Lewis 1994; Loewer 2004). Reductionism is motivated by the observation that chances and frequencies don't drastically diverge, and indeed, as noted earlier, that chances predict frequencies. Reductionists explain this observation by proposing that there is a nonaccidental connection (supervenience) between the chances and the pattern of outcomes, including outcome frequencies: chances and long-run frequencies don't diverge because they can't diverge. Plausible versions of this view require supervenience of chances at some time t on the total mosaic, not merely history prior to t (otherwise chances at very early times won't reflect the laws of nature but only the vagaries of early history). Consider, then, a world in which the chance of heads in a fair coin toss is 1/2; the chance of 1 million heads in a row is $1/2^{\left(10^6\right)}$. By the BCP, since this chance is positive, there must be some world w in which the chance of heads remains 1/2, but where this extraordinary run of heads occurs. But if the prior history contains few enough heads, the long run of heads will mean that the overall frequency of heads in w is very high—high enough that any plausible reductionist view will deny that the chance of heads in w can remain 1/2 while the pattern of outcomes is so skewed toward heads. So reductionists will deny the BCP for at least some outcomes: those that, were they to occur, would *undermine* the actual chances (Ismael 1996; Lewis 1994; Thau 1994). Reductionists will need to offer some other formal rendering of the informal platitude. One candidate is the claim—considerably weaker than BCP—that when the actual chance of p is positive at t, there must be a world perfectly alike in history up to t in which p.[3]

2.3 Chance and Frequency

Chancy theories do not make all-or-nothing predictions about contingent outcomes. But they will make predictions about outcome *frequencies*. It makes no sense to talk about "the frequency" of an outcome because a frequency is a measure of how often a certain type of outcome occurs in a given trial population; different populations will give rise to different frequencies. The frequencies that matter—those that are predicted by chance theories—are those relative to populations that consist of genuine repetitions of the same experimental setup. The population should consist of what von Mises called "mass phenomena or repetitive events . . . in which either the same event repeats itself again and again, or a great number of uniform elements are involved at the same

---

[3] Since reductionists don't accept that chance can float free from the pattern of occurrences, independent restriction to a world with the same chances is redundant where it is not—as in the case above—impossible.

time" (von Mises 1957, 10–11). Von Mises himself held the radical view that chance simply reduced to frequency, but one needn't adopt that position in order to endorse the very plausible claim that chance theories make firm predictions only about "practically unlimited sequences of uniform observations."

How can a chance theory make predictions about outcome frequencies? Not in virtue of principles we've discussed already: although a probability function over possible outcomes might tell us something about the distribution of outcomes in different possible alternate worlds, it doesn't tell us anything about the actual distribution of outcomes (van Fraassen 1989, 81–86). Because chances do predict frequencies, some additional principle is satisfied by well-behaved chances. One suggestion is the Stable Trial Principle (STP):[4]

> If (i) A concerns the outcome of an experimental setup E at t, and (ii) B concerns the same outcome of a perfect repetition of E at a later time t', then $P_{tw}(A) = x = P_{t'w}(B)$. The STP predicts, for instance, that if one repeats a coin flip, the chance of heads should be the same on both trials. (Schaffer 2003, 37)

If chances obey the STP for all possible outcomes of a given experimental setup, they will be *identically distributed*. Moreover, satisfaction of the STP normally involves the trials being *independent* (dependent trials may not exhibit stable chances because the chances vary with the outcomes of previous trials). So if, as Schaffer and others argue, STP (or something near enough) is a basic truth about the kinds of chances that appear in our best theories, then the probabilities that feature in the laws of our best theories meet the conditions for the *strong law of large numbers*: that almost all (i.e., with probability 1) infinite series of trials exhibit a limit frequency for each outcome type identical to the chance of that outcome type.[5]

If a theory entails that repeated trials are stable in their chances—that the causal structure of the trials precludes "memory" of previous trials and that the chance setup can be more or less insulated from environmental variations—then the chance, according to that theory, of an infinite sequence of trials not exhibiting outcome frequencies that reflect the chances of those outcomes is zero. The converse to the BCP—namely, that there should be a possible world in which p only if p has some positive chance—is false: it is possible that a fair coin, tossed infinitely many times, lands heads every time. But something weaker is very plausible (see the principle HCP discussed later): that if the

---

[4] Here "P_{tw}" denotes the probability function at time t derived from the laws of w.

[5] Stepping back from the earlier remarks, the STP may be satisfied by processes that produce merely *exchangeable* sequences of outcomes (de Finetti 1937). In these, the probability of an outcome sequence of a given length depends only on the frequency of outcome types in that sequence and not their order. For example, sampling *without replacement* from an urn gives rise to an exchangeable sequence, although the chances change as the constitution of the urn changes. If a "perfect repetition" involves drawing from an urn of the same prior constitution, then this may be a stable trial even though successive draws from the urn have different probabilities. It is possible to prove a law of large numbers for exchangeable sequences to the effect that almost all of them involve outcome frequencies equal to the chances of those outcomes.

chance of p being the outcome of a chance process is 1, then we ought to expect that p would result from that process. Given this, an infinite sequence of stable trials would be expected to result in an outcome sequence reflecting the chances.

So some chance theories make definite predictions about what kinds of frequencies would appear in an infinite sequence of outcomes. They make no definite prediction about what would happen in a finite sequence of outcomes. They will entail that the chance of a reasonably long sequence of trials exhibiting frequencies that are close to the chances is high, and that the chance increases with the length of the sequence. But because that, too, is a claim about the chances, it doesn't help us understand how chance theories make predictions. If the goals of science are prediction and explanation, we remain unclear on how chancy theories achieve these goals.

Our actual practice with regard to this issue suggests that we implicitly accept some further supplementary principles. Possibility interacts with the goals of prediction and explanation. In particular: the more possible an outcome is, the more frequently outcomes of that type occur. If a theory assigns to p a high chance, it predicts that events relevantly like p will frequently occur when the opportunity for p arises. Chance, formalizing degree of possibility, then obeys these principles:

- HCE: If the chance of p is high according to T, and p, then T explains why p.[6]
- HCP: If the chance of p is high according to T, then T predicts that p, or at least counsels us to expect that p.

These principles reflect our tacit opinion on how chance theories predict and explain the possible outcomes in their purview. So, if a probability measure in some theory is used in frequency predictions, that suggests the measure is a chance function. Similarly, if an outcome has occurred, and theoretical explanations of the outcome cite the high value assigned to the outcome by some probability measure, then that indicates the measure is functioning as a chance because citing a high chance of some outcome is generally explanatory of that outcome. So, chance theories will predict and explain observed frequencies, so long as the series of repeated stable trials is sufficiently long for the theory to entail a high chance that the frequencies match the theoretical chances.

Emery (2015, 20) argues that a basic characteristic of chance is that it explains frequency: a probability function is a chance function if high probability explains high frequency. Assuming that high frequency of an outcome type can explain why an instance of that type occurred, HCE is a consequence of her explanatory criterion for chance.

2.4 Classical and Propensity Chances

Two issues should be addressed before we see some chance theories that illustrate the themes of this section. The first concerns the modal aspect of chances. I've suggested

---

[6] In the sense of *explains* on which *T explains why p* doesn't entail or presuppose *T*.

that chance theories involve a space of possible outcomes, and that chances measure possibilities. These are both core commitments of the *classical theory of chance*. But that theory adds a further element; namely, that "equally possible" outcomes should be assigned the same chance:

> The theory of chance consists in reducing all the events of the same kind to a certain number of cases equally possible, that is to say, to such as we may be equally undecided about in regard to their existence, and in determining the number of cases favorable to the event whose probability is sought. (Laplace 1951, 6–7)

This further element is not now widely accepted as part of the theory of chance. It is presupposed in many textbook probability problems that do not explicitly specify a probability function: for example, we might be told the numbers of red and black balls in an urn, and asked a question about the probability of drawing two red balls in a row with replacement. The only way to answer such a question is to assume that drawing each ball is a case "equally possible," so that the facts about the number of different kinds of ball determine the probability. The assumption seems right in this case, though that appears to have more to do with the pragmatics of mathematics textbooks than with the correctness of the classical theory (the pragmatic maxim being that one ought to assume a uniform distribution over elementary outcomes by default, and accordingly mention the distribution explicitly only when it is non-uniform over such outcomes). But the classical theory fails to accommodate any chance distribution in which elementary outcomes, atoms of the outcome space, are not given equal probability, such as those needed to model weighted dice. Since the space of outcomes for a weighted die is the same as for a fair die, chance cannot supervene on the structure of the space of outcomes in the way that the classical theory envisages. Although chance has a modal dimension, it is unlike alethic modals, where the modal status of a proposition as necessary, or possible, can be read straight off the structure of the space of possible outcomes. And, for that reason, the classical theory of chance is untenable. That is not to say that symmetries play no role in the theory of chance; it is clearly plausible in many cases that empirically detected symmetries in the chance setup and the space of outcomes play a role in justifying a uniform distribution (North 2010; Strevens 1998). But the fact that symmetrical outcomes, where they exist, can justify equal probabilities for those outcomes falls well short of entailing that they always exist, or that they always justify equal probabilities.

The other issue I wish to take up here concerns the formal aspect of chances. I've assumed that chances must meet the probability axioms, and the pressing question on that assumption was how to discriminate chances from other mathematically similar quantities. But having picked on the connections between chance, possibility and frequency to characterize the functional role of chance, a further question now arises: *must something that plays this functional role be a probability?*

Given that chances predict and explain outcome frequencies in repeated trials, an attractive thought is that they manage to do so by being grounded in some feature of the trials constituting a tendency for the setup involved to cause a certain outcome. It is standard to call such a tendency a *propensity* (Giere 1973; Popper 1959). The idea that chances reflect the strength of a causal tendency both promises to explain the modal character of chances—because if a chance setup has some tendency to produce an outcome, it is not surprising that the outcome should be a possible one for that setup—and also to provide illumination concerning other aspects of the chance role. The STP, for example, turns out to be a direct consequence of requiring that repeated trials be alike in their causal structure.

This invocation of tendencies or propensities is implicitly conditional: a propensity for a certain outcome, given a prior cause. So it is unsurprising that many propensity accounts of chance take the fundamental notion to be conditional, recovering unconditional chances (if needed) in some other way (Popper 1959; see also Hájek 2003). The chance of complications from a disease could be understood as the chance of the complication, given that someone has the disease; or, to put it another way, the propensity for the complication to arise from the disease. The idea of conditional probability antedates the explicit stipulative definition given earlier, and causal propensities seem promising as a way of explicating the pre-theoretical notion.

But as Humphreys pointed out (1985; see also Milne 1986), grounding conditional chances in conditional propensities makes it difficult to understand the existence of non-trivial chances without such underlying propensities. Yet such chances are ubiquitous, given that chances are probabilities. The following theorem—*Bayes' theorem*—is an elementary consequence of the definition of conditional probability:

$$P(q \mid p) = \frac{P(p \mid q) P(q)}{P(p)}.$$

We say that p is *probabilistically independent* of q if $P(p \mid q) = P(p)$. It follows from this definition and Bayes' theorem that p is probabilistically independent of q if and only if q is independent of p. If p is causally prior to q, then although there may be a propensity for p to produce q, there will generally be no propensity for q to produce p. If there is no propensity for q to produce p, then the chance of p given q should be just the unconditional chance of p because the presence of q makes no difference: p should be probabilistically independent of q. But it will then follow that q is independent of p too, even though there is a propensity for p to produce q. Humphreys offers an illustrative example: "heavy cigarette smoking increases the propensity for lung cancer, whereas the presence of (undiscovered) lung cancer has no effect on the propensity to smoke" (Humphreys 1985, 559). In this case, causal dependence is not symmetric. There is a causal dependence of lung cancer on smoking—a propensity for smoking to produce lung cancer. This propensity grounds a chancy dependence of lung cancer on smoking. But there is no causal dependence of smoking on lung cancer; so, on the propensity view, there should not be a dependence in the chances. These two results are not compatible if chances are probabilities because probabilistic dependence is symmetrical:

a necessary condition for probability theory to provide the correct answer for conditional propensities is that any influence on the propensity which is present in one direction must also be present in the other. Yet it is just this symmetry which is lacking in most propensities. (Humphreys 1985, 559)

Humphreys uses this observation in arguing that chances aren't probabilistic, because conditional chances are grounded in conditional propensities. But even conceding that some chances are grounded in propensities, many are not. Consider the conditional chance of rolling a 3 with a fair die, given that an odd number was rolled. There is no causal influence between rolling an odd number and rolling a three—whatever dependence exists is constitutive, not causal. These conditional chances exist without any causal propensity to ground them. Given this, the lack of causal dependence of smoking on lung cancer, and the corresponding absence of a propensity, cannot entail that there is no chance dependence between those outcomes. We could retain the role of causal tendencies in explaining certain conditional chances without violating the probability calculus. To do so, we must abandon the idea that every conditional chance should be grounded in a causal tendency. Retaining the formal condition that chances be probabilities certainly puts us in a better position to understand why it is probability functions that appear in our best scientific theories, rather than whatever mathematical theory of "quasi-probabilities" might emerge from executing Humphreys's program of re-founding the theory of chances on causal propensities. (For an example of what such a theory might look like, see Fetzer 1981, 59–67.) Many successful theories presuppose that chances are probabilities, and dropping that presupposition involves us in a heroic revisionary project that we might, for practical reasons, wish to avoid if at all possible.

2.5 Examples of Chance in Scientific Theory

I've suggested that theories involve chances only when they involve a probability assignment to a theoretically possible outcome that is used by that theory in prediction and explanation of outcome frequencies. So understood, a number of current scientific theories involve chances. Examination of the grounds for this claim involves unavoidable engagement with the details of the theories; and interestingly enough, the way that different theories implement a chance function does not always follow a common pattern. I give more details in the longer online version of this chapter.

What is common to implementations of chance in physics, from quantum mechanics to classical statistical mechanics, is the use of a *possibility space* (see Rickles, this volume: §3.1). Regions of this space are possible outcomes, and the (one appropriately normalized) volumes of such regions get a natural interpretation as chances if the theory in question gives the appropriate theoretical role to such volumes. What we see in quantum and classical theories is that an assignment of numbers that is formally a probability follows more or less directly from the basic postulates of the theories and that those numbers are essentially used by those theories to predict and explain outcomes—including frequency outcomes—and thus play the chance role.[7]

So, in quantum mechanics, the wavefunction—which represents the quantum state of a system at some time—determines a probability, via the Born rule, for every possible outcome (region of the quantum possibility space), and it is those probabilities that give the theory its explanatory and predictive power because it is those probabilities that link the unfamiliar quantum state with familiar macroscopic measurement outcomes. These probabilities satisfy the role of chance and so are chances. Many questions remain about how precisely the measurement outcomes relate to the possibility space formalism, but no answer to such questions suggests that a quantum explanation of a measurement outcome might succeed even without citing a probability derived using the Born rule (Albert 1992; Ney 2013; Wallace 2011).

The story is similar in classical statistical mechanics (Albert 2000, 35–70; Meacham 2005; Sklar 1993). The interesting wrinkle in that case is that, although it postulates probabilities that play an explanatory role—especially with respect to thermodynamic phenomena—and that cannot be excised from the theory without crippling its explanatory power (Eagle 2014, 149–154; Emery 2015, §5), classical statistical mechanics is not a *fundamental* theory. Whereas thermodynamic "generalisations may not be fundamental . . . nonetheless they satisfy the usual requirements for being laws: they support counterfactuals, are used for reliable predictions, and are confirmed by and explain their instances" (Loewer 2001, 611–612). The statistical mechanical explanation of thermodynamics doesn't succeed without invoking probability, and those probabilities must therefore be chances.

2.6   The Question of Determinism

Chances arise in physics in two different ways: either from fundamentally indeterministic dynamics or derived from an irreducibly probabilistic relationship between the underlying physical state and the observable outcomes, as in statistical mechanics—and, indeed, in many attractive accounts of quantum mechanics, such as the Everett interpretation (Greaves 2007; Wallace 2011). But many have argued that this second route does not produce genuine chances and that indeterministic dynamical laws are the only way for chance to get in to physics:

> To the question of how chance can be reconciled with determinism . . . my answer is: it can't be done. . . . There is no chance without chance. If our world is deterministic there are no chances in it, save chances of zero and one. Likewise if our world somehow contains deterministic enclaves, there are no chances in those enclaves. (Lewis 1986, 118–120)

---

[7]   Indeed Ismael (2009) argues that every testable theory comes equipped with a probability function which enables the theory to make predictions from partial information, and that such probability functions are perfectly objective features of the content of the theory.

There is a puzzle here (Loewer 2001, §1). For it does seem plausible that if a system is, fundamentally, in some particular state, and it is determined by the laws to end up in some future state, then it really has no chance of ending up in any other state. And this supports the idea that nontrivial chances require indeterministic laws. On the other hand, the predictive and explanatory success of classical statistical mechanics, and of deterministic versions of quantum mechanics, suggests that the probabilities featuring in those theories are chances.

This puzzle has recently been taken up by a number of philosophers. Some have tried to bolster Lewis's incompatibilist position (Schaffer 2007), but more have defended the possibility of deterministic chance. A large number have picked up on the sort of considerations offered in the last section and argue that because probabilities in theories like statistical mechanics behave in the right sort of way, they are chances (Emery, 2015; Loewer 2001). Often these sorts of views depend on a sort of "level autonomy" thesis, that the probabilities of nonfundamental theories are nevertheless chances because the nonfundamental theories are themselves to a certain degree independent of the underlying fundamental physics and so cannot be trumped by underlying determinism (Glynn 2010; List and Pivato 2015; Sober 2010).

It is difficult, however, to justify the autonomy of higher level theories because every occurrence supposedly confirming them supervenes on occurrences completely explicable in fundamental theories. If so, the underlying determinism will apparently ensure that only one course of events can happen consistent with the truth of the fundamental laws and that therefore nontrivial probabilities in higher level theories won't correspond to genuine possibilities. If chance is connected with real possibility, these probabilities are not chances: they are merely linked to epistemic possibility, consistent with what we know of a system. In responding to this argument, the crucial question is: is it possible for determinism to be true and yet more than one outcome be genuinely possible? The context-sensitivity of English modals like *can* and *possibly* allows for a sentence like *the coin can land heads* to express a true proposition even while the coin is determined to land tails as long as the latter fact isn't contextually salient (Kratzer 1977). We may use this observation, and exploit the connection between chance ascriptions and possibility ascriptions, to resist this sort of argument for incompatibilism (Eagle 2011a) and continue to regard objective and explanatory probabilities in deterministic physics as chances.

3 Probabilities of Theories

Many claims in science concern the evaluation, in light of the evidence, of different theories or hypotheses: that general relativity is better supported by the evidence than its nonrelativistic rivals, that relative rates of lung cancer in smokers and nonsmokers confirm the hypothesis that smoking causes cancer, or that the available evidence isn't decisive between theories that propose natural selection to be the main driver of population genetics at the molecular level and those that propose that random drift is more important. Claims about whether evidence *confirms, supports*, or is *decisive between* hypotheses are apparently synonymous with claims about how probable these hypotheses are in light of the evidence.[8] How should we understand this use of probability?

One proposal, due to Carnap, is to invoke a special notion of "inductive probability" in addition to chance to explain the probabilities of theories (Carnap 1955, 318). His inductive probability has a "purely logical nature"—the inductive probability ascribed to a hypothesis with respect to some evidence is entirely fixed by the formal syntax of the hypothesis and the evidence. Two difficulties with Carnap's suggestion present themselves. First, there are many ways of assigning probabilities to hypotheses relative to evidence in virtue of their form that meet Carnap's desiderata for inductive probabilities, and no one way distinguishes itself as uniquely appropriate for understanding evidential support. Second, and more importantly, Carnap's proposal leaves the cognitive role of confirmation obscure. Why should we care that a theory has high inductive probability in light of the evidence, *unless* having high inductive probability is linked in some way to the credibility of the hypothesis?

Particularly in light of this second difficulty, it is natural to propose that the kind of probability involved in confirmation must be a sort that is directly implicated in how credible or believable the hypothesis is relative to the evidence. Indeed, we may go further: the hypothesis we ought to accept outright—the one we ought to make use of in prediction and explanation—is the one that, given the evidence we now possess, we are most confident in. Confirmation should therefore reflect confidence, and we ought to understand confirmation as implicitly invoking probabilistic levels of confidence: what we might call *credences*.

3.1 Credences

A credence function is a probability function, defined over a space of propositions: the belief space. Unlike the case of chance, propositions in the belief space are not possible outcomes of some experimental trial. They are rather the possible objects of belief; propositions that, prior to inquiry, might turn out to be true. An agent's credence at a time reflects their level of confidence then in the truth of each proposition in the belief space.

There is no reason to think that every proposition is in anyone's belief space. But since a belief space has a probability function defined over it, it must meet the algebraic conditions on outcome spaces, and must therefore be closed under negation and disjunction. This ensures that many unusual propositions—arbitrary logical compounds of propositions the agent has some confidence in—are the objects of belief. It also follows from the logical structure of the underlying algebra that logically equivalent propositions are

---

[8] Given the focus of this chapter, I am setting aside those philosophical views that deny that claims about support or confirmation can be helpfully understood using probability (Popper 1963). See Sprenger (this volume) for more on confirmation.

the same proposition. The laws of probability ensure that an agent must be maximally confident in that trivial proposition which is entailed by every other. Since every propositional tautology expresses a proposition entailed by every other, every agent with a credence function is certain of all logical truths. (Note: they need not be certain of a given sentence that it expresses a tautology, but they must be certain of the tautology expressed.) These features make having a credence function quite demanding. Is there any reason to think that ordinary scientists have them, that they may be involved in confirmation?

The standard approach to this question certainly accepts that it is psychologically implausible to think that agents explicitly represent an entire credence function over an algebra of propositions. But a credence function need not be explicitly represented in the brain to be the right way to characterize an agent's belief state. If the agent behaves in a way that can be best rationalized on the basis of particular credences, that is reason to attribute those credences to her. Two sorts of argument in the literature have this sort of structure. One, given mostly by psychologists, involves constructing empirical psychological models of cognitive phenomena, which involves assigning probabilistic degrees of belief to agents. The success of those models then provides evidence that thinkers really do have credences as part of their psychological state (Perfors 2012; Perfors et al. 2011).

3.2 Credences from Practical Rationality

The second sort of argument, given mostly by philosophers and decision theorists, proposes that certain assumptions about rationality entail that, if an agent has degrees of confidence at all, these must be credences. The rationality assumptions usually invoked are those about *rational preference* (although recent arguments exist that aim to avoid appealing to practical rationality: see Joyce 1998). If an agent has preferences between options that satisfy certain conditions, then these preferences can be represented as maximizing subjective expected utility, a quantity derived from a credence function and an assignment of values to outcomes. Any such *representation theorem* needs to be supplemented by an additional argument to the effect that alternative representations are unavailable or inferior, thus making it plausible that an agent really has a credence function when they can be represented as having a credence function. These additional arguments are controversial (Hájek 2008; Zynda 2000).

There are many representation theorems in the literature, each differing in the conditions it imposes on rational preference (Buchak 2013; Jeffrey 1983; Maher 1993; Savage 1954). Perhaps the simplest is the *Dutch book argument* (de Finetti 1937; Ramsey 1926). The argument has two stages: first, show that one can use an agent's preferences between options of a very special kind—bets—to assign numbers to propositions that reflect degrees of confidence; and, second, show that those numbers must have the structure of probabilities on pain of practical irrationality—in particular, that non-probabilistic credences commit an agent to evaluating a set of bets that collectively guarantee a sure loss as fair.

Ramsey suggests the fundamental irrationality of such a valuation lies in its inconsistent treatment of equivalent options. Such a valuation

> would be inconsistent in the sense that it violated the laws of [rational] preference between options... if anyone's mental condition violated these laws, his choice would depend on the precise form in which the options were offered him, which would be absurd. (Ramsey 1926, 78)

The Dutch book argument, like other representation theorems, is suggestive but far from conclusive. But even philosophers can avail themselves of the psychological style of argument for credences, arguing that, because approaching epistemic rationality via credences allows the best formal systematization of epistemology and its relation to rational action, that is reason to suppose that rational agents have credences:

> A remarkably simple theory—in essence, three axioms that you can teach a child—achieves tremendous strength in unifying our epistemological intuitions. Rather than cobbling together a series of local theories tailored for a series of local problems—say, one for the grue paradox, one for the raven paradox, and so on—a single theory in one fell swoop addresses them all. While we're at it, the same theory also undergirds our best account of rational decision-making. These very successes, in turn, provide us with an argument for probabilism: our best theory of rational credences says that they obey the probability calculus, and that is a reason to think that they do. (Eriksson and Hájek 2007, 211).

3.3 Bayesian Confirmation Theory

The basic postulate of *Bayesian confirmation theory* is that evidence e confirms hypothesis h for A if A's credence in h given e is higher than their unconditional credence in h: $C_A(h|e) > C_A(h)$. That is, some evidence confirms a hypothesis just in case the scientist's confidence in the hypothesis, given the evidence, is higher than in the absence of the evidence. Note that, by Bayes' theorem, e confirms h if and only if $C_A(e|h) > C_A(e)$: if the evidence is more likely given the truth of the hypothesis than otherwise. In answering the question of what scientists should believe about hypotheses and how those beliefs would look given various pieces of evidence, we do not need a separate "logic of confirmation": we should look to conditional credences in hypotheses on evidence.

This is overtly agent-sensitive. Insofar as confirmation is about the regulation of individual belief, that seems right. But although it seems plausible as a sufficient condition on confirmation—one should be more confident in the hypothesis, given evidence confirming it—many deny its plausibility as a necessary condition. Why should it be that evidence can confirm only those hypotheses that have their credence level boosted by the evidence? For example, couldn't it be that evidence confirms a hypothesis of which I am already certain (Glymour 1981)?

Some Bayesians respond by adding additional constraints of rationality to the framework. Credences of a rational agent are not merely probabilities: they are probabilities that also meet some further condition. It is fair to say that explicit constructions of such further conditions have not persuaded many. Most have made use of the classical principle of indifference (see Section 2.4), but such principles are overly sensitive to how the hypothesis is presented, delivering different verdicts for equivalent problems (Milne 1983; van Fraassen 1989). It does not follow from the failure of these explicit constructions that anything goes: it just might be that, even though only some credences really reflect rational evaluations of the bearing of evidence on hypotheses, there no recipe for constructing such credences without reference to the content of the hypotheses under consideration. For example, it might be that the rational credence function to use in confirmation theory is one that assigns to each hypothesis a number that reflects its "intrinsic plausibility . . . prior to investigation" (Williamson 2000). And it might also be that, when the principle of indifference is conceived as involving epistemic judgments in the evaluation of which cases are "equally possible," which give rise to further epistemic judgments about equal credences, it is far weaker and less objectionable (White 2009).

Moreover, there are problems faced by "anything goes" Bayesians in explaining why agents who disagree in credences despite sharing all their evidence shouldn't just suspend judgment about the credences they ought to have (White 2005). The reasonable thing to do in a case like that, it's suggested, is for the disagreeing agents to converge on the same credences, adopt indeterminate credences, or do something like that— something that involves all rational agents responding in the same way to a given piece of evidence. But how can Bayesians who permit many rational responses to evidence explain this? This takes a more pressing form too, given the flexibility of Bayesian methods in encompassing arbitrary credences: why is it uniquely rational to follow the scientific method (Glymour 1981; Norton 2011)? The scientific method is a body of recommended practices designed to ensure reliable hypothesis acceptance (e.g., prefer evidence from diverse sources as more confirmatory or avoid ad hoc hypotheses as less confirmatory). The maxims of the scientific method summarize techniques for ensuring good confirmation, but if confirmation is dependent on individual credence, how can there be one single scientific method? The most plausible response for the subjective Bayesian is to accept the theoretical possibility of a plurality of rational methods but to argue that current scientific training and enculturation in effect ensure that the credences of individual scientists do by and large respond to evidence in the same way. If the scientific method can be captured by some constraints on credences, and those constraints are widely endorsed, and there is considerable benefit to being in line with community opinion on confirmation (as there is in actual scientific communities), that is a prudential reason for budding scientists to adopt credences meeting those constraints.

Whether we think that scientists have common views about how to respond to evidence as a matter of a priori rationality or peer pressure doesn't matter. What ultimately matters is whether the Bayesian story about confirmation delivers plausible cases where, either from the structure of credences or plausible assumptions about "natural" or widely shared priors, we can derive that the conditional credence in h given e will exceed the unconditional credence if and only if e intuitively confirms h. Howson and Urbach (1993, 117–164) go through a number of examples showing that, in each case, there is a natural Bayesian motivation for standard principles of scientific methodology (see also Horwich 1982). I treat some of their examples in the online version of this chapter (see also Sprenger, this volume). The upshot is that Bayesian confirmation theory, based on the theory of credences, provides a systematic framework in which proposed norms of scientific reason can be formulated and evaluated and that vindicates just those norms that do seem to govern actual scientific practice when supplied with the kinds of priors it is plausible to suppose working scientists have.

3.4 The Problem of Old Evidence

The Bayesian approach to confirmation faces further problems. In the online version of this chapter, I also discuss apparent difficulties with the Bayesian justification of abductive reasoning. But here I will only discuss the so-called *problem of old evidence*.

The problem stems from the observation that "scientists commonly argue for their theories from evidence known long before the theories were introduced" (Glymour 1981). In these, cases there can be striking confirmation of a new theory, precisely because it explains some well-known anomalous piece of evidence. The Bayesian framework doesn't seem to permit this: if e is old evidence, known to be the case already, then any scientist who knows it assigns it credence 1. Why? Because credence is a measure of epistemic possibility. If, for all A knows, it might be that $\neg e$, then A does not know that e. Contraposing, A knows e only if there is no epistemic possibility of $\neg e$. And if there is no epistemic possibility of $\neg e$, then $C_A(\neg e) = 0$, and $C_A(e) = 1$.[9] If $C_A(e) = 1$, however, it follows from Bayes' theorem that $C_A(h \mid e) = C_A(h)$, and therefore e does not confirm h for anyone who already knows e.

The most promising line of response might be to revise orthodox Bayesianism and deny that known evidence receives credence 1.[10] But can we respond to this objection without revising the orthodox picture? Perhaps by arguing that old evidence should not confirm because confirmation of a theory should involve an increase in confidence. But it would be irrational to increase one's confidence in a theory based on evidence one already had: either you have already factored in the old evidence, in which case it would be unreasonable to count its significance twice, or you have not factored in the old evidence, in which case you were unreasonable before noticing the confirmation. Either

---

[9] This argument is controversial. Because someone can reasonably have different fair betting rates on p and q, even when they know both, the betting dispositions interpretation of credence does not require that each known proposition has equal credence, so they needn't each have credence 1.

[10] Other revisionary responses include claiming that something is learned and confirmatory: namely, that h entails e (Garber 1983). Note that an agent who genuinely learns this is subject to Dutch book before learning it (since they are uncertain of a trivial proposition) and so irrational; it cannot be a general account of confirmation by old evidence for rational agents.

way, the only reasonable position is that confirmatory increases in confidence only arise with new evidence. If there seems to be confirmation by old evidence, that can only be an illusion, perhaps prompted by imperfect access to our own credences.

3.5 Believing Theories

The orthodox response to the problem of old evidence, as well as the problem itself, presupposes a certain view about how the acquisition of confirmatory evidence interacts with credences. Namely: if h is confirmed by e, and one acquires the evidence that e, one should come to be more confident in h. The simplest story that implements this idea is that if A's credence is C, and e is the strongest piece of evidence A receives, their new credential state should be represented by that function C^+ such that for every proposition in the belief space p, $C^+(p) = C(p|e)$. This procedure is known as *conditionalization* because it involves taking one's old conditional credences given e to be one's new credences after finding out that e. If conditionalization describes Bayesian learning, then the Bayesian story about confirmation can be described as follows: e confirms h for A just in case A will become more confident in h if they learn e.

There are a number of controversial issues surrounding conditionalization, when it is treated as the one true principle for updating belief in light of evidence. For example, since it is a truth of probability theory that if $C(p) = 1$, then for every q, $C(p|q) = 1$, conditionalization can never lower the credence of any proposition of which an agent was ever certain. If conditionalization is the only rational update rule, then forgetting or revision of prior certainty in light of new evidence is never rational, and this is controversial (Arntzenius 2003; Talbott 1991).

But, regardless of whether conditionalization is always rational, it is sometimes rational. It will often be rational in the scientific case, where evidence is collected carefully and the true propositional significance of observation is painstakingly evaluated, so that certainty is harder to obtain and occasions demanding the revision of prior certainty correspondingly rarer. In scientific contexts, then, the confirmation of theory by evidence in hand will often lead to increased confidence in the theory, to the degree governed by prior conditional credence. This is how Bayesian confirmation theory (and the theory of credence) proposes to explain probabilities of scientific hypotheses.

4 THE PRINCIPAL PRINCIPLE

4.1 Coordinating Chance and Credence

We've discussed chance in theories, and credibility of theories. Is there any way to connect the two? Lewis (1986) offered an answer: what he called the *Principal Principle* (PP) because "it seem[ed] to [him] to capture all we know about chance." The discussion in Sections 2.2–2.3 already shows that we know more about chance than this; nevertheless the PP is a central truth about the distinctive role of credence about chance in our cognitive economy.

We need some notation. Suppose that C denotes some reasonable initial credence function, prior to updating in the light of evidence. Suppose that $[\![P(p)=x]\!]$ denotes the proposition that the real chance of p is x. That is, it is the proposition that is true if the true theory t is such that the t-chance of p is x. Suppose that e is some admissible evidence, evidence

> whose impact on credence about outcomes comes entirely by way of credence about the chances of those outcomes.(Lewis 1986, 92; see also Hoefer 2007, 553)

Using that notation, the PP can be expressed:

$$C\big(p\,|\,[\![P(p)=x]\!]\wedge e\big)=x.$$

Given a proposition about chance and further evidence that doesn't trump the chances, any reasonable conditional credence in p should equal the chance.[11]

The PP doesn't say that one should set one's credences equal to the chances; if the chances are unknown, one cannot follow that recommendation; and yet, one can still have conditional credences recommended by PP. It does follow from the PP that (1) if you were to come to know the chances and nothing stronger and (2) you update by conditionalization from a reasonable prior credence, then your credences will match the chances.[12]

But even before knowing the chances, the PP allows us to assign credences to arbitrary outcomes, informed by the credences we assign to the candidate scientific hypotheses. For (assuming our potential evidence is admissible and that we may suppress mention of e), the theorem of total probability states that, where Q is a set of mutually exclusive and jointly exhaustive propositions, each with non-zero credence, then for arbitrary p, $C(p)=\sum_{q_i\in Q}C(p|q_i)C(q_i)$. If Q is the set of rival scientific hypotheses about the chances (i.e., each $P_x=[\![P(p)=x]\!]$), then the PP entails that $C(p)=\sum_{P_x\in Q}C(p|P_x)C(P_x)=\sum_{P_x\in Q}x\cdot C(P_x)$.

That is, the PP entails that one's current credence in p is equal to one's subjective expectation of the chance of p, weighted by one's confidence in various hypotheses about

---

[11] Lewis's own formulation involves reference to time-dependency of chance, a reference that, in my view, is a derivative feature of the dependence of chance on the physical trial and that I suppress here for simplicity (Eagle 2014).

[12] The PP is a schema that holds for any proposition of the form $[\![P(p)=x]\!]$, not just the true one. So, whatever the chances might be, there is some instance of PP that allows updating on evidence about the actual chances.

the chance. You may not know which chance theory is right, but if you have a credence distribution over those theories, then the chances in those theories are already influencing your judgments about possible outcomes. Suppose, for example, you have an unexamined coin that you are 0.8 sure is fair and 0.2 sure is a double-headed trick coin. On the former hypothesis, $P(heads) = 0.5$; on the latter, $P(heads) = 1$. Applying the result just derived, $C(heads) = 1 \times 0.2 + 0.5 \times 0.8 = 0.6.$, even while you remain in ignorance of which chance hypothesis is correct.

The PP is a *deference* principle, one claiming that rational credence defers to chances (Gaifman 1988; Hall 2004). It has the same form as other deference principles, such as deference to expert judgment (e.g., adopt conditional credences in rain tomorrow equal to the meterologist's credences), or to more knowledgeable agents (e.g., adopt conditional credences given e equal to your future credences if you were to learn e, the Reflection Principle of van Fraassen [1984]). It is worth noting that the proposition about chance in the PP is not a conditional credence; this seems to reflect something about chance, namely, that the true unconditional chances are worth deferring to, regardless of the (admissible) information that the agent is given (Joyce 2007).

Lewis showed that, from the PP, much of what we know about chance follows. For example, if it is accepted, we needn't add as a separate formal constraint that chances are probabilities (Section 2.1). Suppose one came to know the chances, had no inadmissible evidence, and began with rational credences. Then, in accordance with the PP, one's new unconditional credences in possible outcomes are everywhere equal to the chances of those outcomes. Since rational credences are probabilities, so too must chances be (Lewis 1986, 98). This and other successes of the PP in capturing truths about chance lead Lewis to claim that

> A feature of Reality deserves the name of chance to the extent that it occupies the definitive role of chance; and occupying the role means obeying the [PP]...(Lewis 1994, 489)

One major issue remains outstanding. How can we link up probabilities *in* and probabilities *of* theories? In particular: can we offer an explanation of how, when a theory makes predictions about frequency, the observation of frequencies in line with prediction is confirmatory? Of course it *is* confirmatory, as encapsulated in the direct inference principle HCP from Section 1.3. But it would be nice to offer an explanation.

A chance theory t, like the theory of radioactive decay, makes a prediction f that certain frequencies have a high chance according to the theory. Let us make a simplifying assumption that we are dealing with a collection of rival hypotheses that share an outcome space. Then, we may say that t predicts f if and only if $P_t(f)$ is high. Notice that such a prediction doesn't yet permit the machinery of Bayesian confirmation theory to show that the observation of f would confirm t. While the t-chance of f is high, there is no guarantee yet that anyone's credence in f given t is correspondingly high. So, even if $P_t(f)$ is considerably greater than $C_A(f)$, that doesn't yield confirmation of t by f for A unless $C_A(f|t) \approx P_t(f)$.

Suppose t is some theory that entails a claim about the chance of a certain frequency outcome f. Presumably, for any plausible t, the rest of t is compatible with that part of t that is about the chance of f; so factorize t into a proposition $[\![P(f)=x]\!]$ and a remainder t'. Suppose that one doesn't have other information e that trumps the chances. Then, the PP applies to your initial credence function (assuming your current credence C was obtained by conditionalizing on e), so that $C(f|t) = C_{initial}(f|t \wedge e) = C_{initial}(f | [\![P(f)=x]\!] \wedge t' \wedge e) = x$. But since $[\![P(f)=x]\!]$ is a consequence of t, $P_t(f) = x$, and we thus have the desired equation, $C(f|t) = P_t(f)$, which allows the frequency predictions of a theory to confirm it—or disconfirm it (Howson and Urbach 1993, 342–347; Lewis 1986, 106–108).

Not all is smooth sailing for the PP. Consider the case of undermining discussed in Section 2.2. There, a reductionist chance theory assigned some positive chance to a subsequent pattern of outcomes that, if it occurred, would undermine the chance theory. Since the actual chance theory t is logically incompatible with the chance theory that holds in the possibility where the undermining pattern exists, it is not possible for t to be true and for the undermining future u to occur. Given logical omniscience, $C(u|t) = 0$. But by the PP, $C(u|t) = C(u | [\![P_t(u)=x]\!] \wedge e) = x > 0$. Contradiction: the existence of undermining futures, which seems to follow from reductionism, and the PP are jointly incompatible. There is a large literature examining the prospects for the PP and for reductionism and canvassing various related principles that are not susceptible to this problem (Hall 1994; Ismael 2008; Joyce 2007; Lewis 1994). I discuss some of this further in the online version of this chapter.

SUGGESTED READING

Briggs (2010) covers material similar to that found in this chapter, and some readers may find a second opinion useful. Implicit in much of the chapter were references to various substantive theories of the truth-conditions for probability claims, historically known as "interpretations" of probability: Hájek (2012) contains a much fuller and more explicit account of the various positions on this issue. Lewis (1986) is the classic account of the Principal Principle, the relationship between credence and chance. The foundations of the theory of credences can be found in a wonderful paper (Ramsey 1926); the best account of the application of the theory of credences to confirmation is Howson and Urbach (1993). There are many textbooks on the philosophy of probability; none is bad, but one that is particularly good on the metaphysical issues around chance is Handfield (2012). But perhaps the best place to start further reading is with Eagle (2011b), an anthology of classic articles with editorial context: it includes the items by Ramsey, Lewis, and Howson and Urbach just recommended.

REFERENCES

Albert, David Z. (1992). *Quantum Mechanics and Experience*. (Cambridge, MA: Harvard University Press).

Albert, David Z. (2000). *Time and Chance*. (Cambridge, MA: Harvard University Press).

Arntzenius, Frank. (2003). "Some Problems for Conditionalization and Reflection," *Journal of Philosophy* 100: 356–370.

Bigelow, John, Collins, John, and Pargetter, Robert. (1993). "The Big Bad Bug: What are the Humean's Chances?" *British Journal for the Philosophy of Science* 44: 443–462.

Briggs, Rachael. (2010). "The Metaphysics of Chance." *Philosophy Compass* 5: 938–952.

Buchak, Lara. (2013). *Risk and Rationality*. (Oxford: Oxford University Press).

Carnap, Rudolf. (1955). *Statistical and Inductive Probability*. (Brooklyn: Galois Institute of Mathematics and Art). Reprinted in Eagle (2011b), 317–326; references are to this reprinting.

de Finetti, Bruno. (1937). "Foresight: Its Logical Laws, Its Subjective Sources." In Henry E. Kyburg, Jr., and Howard E. Smokler (eds.). *Studies in Subjective Probability, [1964]*. (New York: Wiley), 93–158.

Eagle, Antony. (2011a). "Deterministic Chance." *Noûs* 45: 269–299.

Eagle, Antony (ed.). (2011b). *Philosophy of Probability: Contemporary Readings*. (London: Routledge).

Eagle, Antony. (2014). "Is the Past a Matter of Chance?" In Alastair Wilson (ed.), *Chance and Temporal Asymmetry* (Oxford: Oxford University Press), 126–158.

Emery, Nina. (2015). "Chance, Possibility, and Explanation." *British Journal for the Philosophy of Science* 66: 95–120.

Eriksson, Lina, and Hájek, Alan. (2007). "What Are Degrees of Belief?" *Studia Logica* 86: 183–213.

Fetzer, James H. (1981). *Scientific Knowledge: Causation, Explanation, and Corroboration*. (Dordrecht: D. Reidel).

Gaifman, Haim. (1988). "A Theory of Higher Order Probabilities." In Brian Skyrms and William Harper (eds.), *Causation, Chance and Credence, vol. 1* (Dordrecht: Kluwer), 191–219.

Garber, Daniel. (1983). "Old Evidence and Logical Omniscience in Bayesian Confirmation Theory." In John Earman (ed.), *Minnesota Studies in Philosophy of Science, volume 10: Testing Scientific Theories*. (Minneapolis: University of Minnesota Press), 99–132.

Giere, Ronald N. (1973). "Objective Single-Case Probabilities and the Foundations of Statistics." In P. Suppes, L. Henkin, G. C. Moisil, et al. (eds.), *Logic, Methodology and Philosophy of Science IV*. (Amsterdam: North-Holland), 467–483.

Glymour, Clark. (1981). "Why I am not a Bayesian." In C. Glymour (ed.), *Theory and Evidence*. (Chicago: University of Chicago Press), 63–93.

Glynn, Luke. (2010). "Deterministic Chance." *British Journal for the Philosophy of Science* 61: 51–80.

Greaves, H. (2007). "Probability in the Everett Interpretation." *Philosophy Compass* 2: 109–128.

Hájek, Alan. (2003). "What Conditional Probability Could Not Be." *Synthese* 137: 273–323.

Hájek, Alan. (2008). "Arguments for—or Against—Probabilism?" *British Journal for the Philosophy of Science* 59(4): 793–819.

Hájek, Alan. (2012). "Interpretations of Probability." In Edward N. Zalta (ed.), *The Stanford Encyclopedia of Philosophy* (Winter 2012 Edition) http://plato.stanford.edu/archives/win2012/entries/probability-interpret/.

Hall, Ned. (1994). "Correcting the Guide to Objective Chance." *Mind* 103: 505–518.

Hall, Ned. (2004). "Two Mistakes About Credence and Chance." In Frank Jackson and Graham Priest (eds.), *Lewisian Themes*. (Oxford: Oxford University Press), 94–112.

Handfield, Toby. (2012). *A Philosophical Guide to Chance*. (Cambridge: Cambridge University Press).

Hoefer, Carl. (2007). "The Third Way on Objective Probability: A Sceptic's Guide to Objective Chance." *Mind* 116: 549–596.

Horwich, Paul. (1982). *Probability and Evidence*. (Cambridge: Cambridge University Press).

Howson, Colin, and Urbach, Peter. (1993). *Scientific Reasoning: The Bayesian Approach* 2nd ed. (Chicago: Open Court).

Humphreys, Paul. (1985). "Why Propensities Cannot be Probabilities." *Philosophical Review* 94: 557–570.

Ismael, Jenann. (1996). "What Chances Could Not Be." *British Journal for the Philosophy of Science* 47: 79–91.

Ismael, Jenann. (2008). "Raid! Dissolving the Big, Bad Bug." *Noûs* 42: 292–307.

Ismael, Jenann. (2009). "Probability in Deterministic Physics." *Journal of Philosophy* 106: 89–108.

Jeffrey, Richard C. (1983). *The Logic of Decision* 2nd ed. (Chicago: University of Chicago Press).

Joyce, James M. (1998). "A Nonpragmatic Vindication of Probabilism." *Philosophy of Science* 65: 575–603.

Joyce, James M. (2007). "Epistemic Deference: the Case of Chance." *Proceedings of the Aristotelian Society* 107: 187–206.

Kolmogorov, A. N. (1933). *Grundbegriffe der Wahrscheinlichkeitrechnung, Ergebnisse Der Mathematik und ihrer Grenzgebiete*, no. 3 (Springer, Berlin); translated as (1956). *Foundations of the Theory of Probability*, 2nd ed. (New York: Chelsea).

Kratzer, Angelika. (1977). "What 'Must' and 'Can' Must and Can Mean." *Linguistics and Philosophy* 1: 337–355.

Laplace, Pierre-Simon. (1951). *Philosophical Essay on Probabilities*. (New York: Dover).

Lewis, David. (1986). "A Subjectivist's Guide to Objective Chance." In D. Lewis (ed.), *Philosophical Papers*, vol. 2. (Oxford: Oxford University Press), 83–132.

Lewis, David. (1994). "Humean Supervenience Debugged." *Mind* 103: 473–490.

List, Christian, and Pivato, Marcus. (2015). "Emergent Chance." *Philosophical Review* 124: 59–117.

Loewer, Barry. (2001). "Determinism and Chance." *Studies in History and Philosophy of Modern Physics* 32: 609–620.

Loewer, Barry. (2004). "David Lewis's Humean Theory of Objective Chance." *Philosophy of Science* 71: 1115–1125.

Maher, Patrick. (1993). *Betting on Theories*. (Cambridge: Cambridge University Press).

Meacham, Christopher J. G. (2005). "Three Proposals Regarding a Theory of Chance." *Philosophical Perspectives* 19(1): 281–307.

Mellor, D. H. (2005). *Probability: A Philosophical Introduction*. (London: Routledge).

Milne, P. (1983). "A Note on Scale Invariance." *British Journal for the Philosophy of Science* 34: 49–55.

Milne, P. (1986). "Can There Be a Realist Single-Case Interpretation of Probability?" *Erkenntnis* 25: 129–132.

Ney, Alyssa. (2013). "Introduction." In Alyssa Ney and David Z. Albert (eds.), *The Wave Function*. (New York: Oxford University Press), 1–51.

North, Jill. (2010). "An Empirical Approach to Symmetry and Probability." *Studies in History and Philosophy of Modern Physics* 41: 27–40.

Norton, John. (2011). "Challenges to Bayesian Confirmation Theory." In P. S. Bandypadhyay and M. R. Forster (eds.), *Handbook of the Philosophy of Science, vol. 7: Philosophy of Statistics.* (Amsterdam: North-Holland), 391–439.

Perfors, Amy. (2012). "Bayesian Models of Cognition: What's Built in After All?." *Philosophy Compass* 7(2): 127–138.

Perfors, Amy, Tenenbaum, Joshua B., Griffiths, Thomas L., et al. (2011). "A Tutorial Introduction to Bayesian Models of Cognitive Development." *Cognition* 120: 302–321.

Popper, Karl. (1959). "A Propensity Interpretation of Probability." *British Journal for the Philosophy of Science* 10: 25–42.

Popper, Karl. (1963). *Conjectures and Refutations.* (New York: Routledge).

Ramsey, F. P. (1926). "Truth and Probability." In D. H. Mellor (ed.), *Philosophical Papers, 1990.* (Cambridge: Cambridge University Press), 52–94.

Savage, Leonard J. (1954). *The Foundations of Statistics.* (New York: Wiley).

Schaffer, Jonathan. (2003). "Principled Chances." *British Journal for the Philosophy of Science* 54: 27–41.

Schaffer, Jonathan. (2007). "Deterministic Chance?" *British Journal for the Philosophy of Science* 58: 113–140.

Sklar, Lawrence. (1993). *Physics and Chance.* (Cambridge: Cambridge University Press).

Sober, Elliott. (2010). "Evolutionary Theory and the Reality of Macro Probabilities." In E. Eells and J. H. Fetzer (eds.), *The Place of Probability in Science.* (Dordrecht: Springer), 133–161.

Strevens, Michael. (1998). "Inferring Probabilities from Symmetries." *Noûs* 32: 231–246.

Talbott, W. J. (1991). "Two Principles of Bayesian Epistemology." *Philosophical Studies* 62: 135–150.

Thau, Michael. (1994). "Undermining and Admissibility." *Mind* 103: 491–503.

van Fraassen, Bas C. (1984). "Belief and the Will." *Journal of Philosophy* 81: 235–256.

van Fraassen, Bas C. (1989). *Laws and Symmetry.* (Oxford: Oxford University Press).

von Mises, Richard. (1957). *Probability, Statistics and Truth.* (New York: Dover).

Wallace, David. (2011). *The Emergent Multiverse.* (Oxford University Press).

White, Roger. (2005). "Epistemic Permissiveness." *Philosophical Perspectives* 19: 445–459.

White, Roger. (2009). "Evidential Symmetry and Mushy Credence." In T. S. Gendler and J. Hawthorne (eds.), *Oxford Studies in Epistemology.* (Oxford: Oxford University Press), 161–186.

Williamson, Timothy. (2000). *Knowledge and Its Limits.* (Oxford: Oxford University Press).

Zynda, Lyle. (2000). "Representation Theorems and Realism About Degrees of Belief." *Philosophy of Science* 67: 45–69.

CHAPTER 21

....

REPRESENTATION
IN SCIENCE

....

MAURICIO SUÁREZ

1 HISTORICAL INTRODUCTION

....

SCIENTIFIC representation is a latecomer in philosophy of science, receiving considerable attention only in the past fifteen years or so. In many ways this is surprising since
the notion of representation has long been central to philosophical endeavors in the
philosophies of language, mind, and art. There are a number of historical reasons for
the earlier neglect. At least initially this may be due to the logical empiricists' stronghold
upon the field and their mistrust of notions of correspondence and direct reference—
it was perhaps assumed that representation was one of those notions. In addition, in
what has come to be known as the "received view" (the name stuck even though it was
only "received" in the 1970s), scientific knowledge is articulated and embodied entirely
within scientific theories, conceived as linguistic or propositional entities. Models play
only a heuristic role, and the relation between scientific claims and theories on the one
hand and the real-world systems that they are putatively about on the other, is therefore
naturally one of description rather than representation (Bailer-Jones 2009). Although
the distinction is not entirely sharp, a critical difference between description and representation concerns the applicability of semantic notions such as truth, which are
built into descriptions but seem prima facie ill-suited for representations as has been
emphasized by, for example, Ronald Giere (Giere 1988, ch. 4). In focusing on the role
that language plays in science, the logical empiricists and their successors may thus have
implicitly privileged theoretical description.

The analysis of the logical structure of scientific theories remained a central concern
for philosophers of science during the postwar years, to the detriment of any thorough
attention to the role that models, model-building, and other genuinely representational
entities and activities play in science. The rejection of the "received" or syntactic view in
favor of a "semantic" conception in the 1980s did not in the first instance improve things

much, since the focus continued to be the analysis of scientific *theory*. But the semantic view arguably contained the seeds of a deeper change of outlook since the central claim that theories could be understood as collections of models surreptitiously shifted attention from description onto representation. In particular Bas Van Fraassen's version of the semantic view in terms of state spaces and Ronald Giere's in terms of cognitive models both emphasized how theories are tools for representation rather than description. These authors came to the conclusion that the linguistic analysis of scientific theory was of limited interest and emphasized instead the representational roles of models (Giere 1988, ch. 3; Van Fraassen 1980, ch. 3) There is no doubt that the development of this non-linguistic version of the semantic view was an essential step in the upsurge of representation in philosophy of science.

Nevertheless, the most important historical route to the notion of representation—and also the main source of current interest in it within the philosophy of science—is contributed by what we may call the modelling tradition or the "modelling attitude" (see Suárez 2015). This is the historical series of attempts by both philosophers and practicing scientists to understand and come to terms with model-building, analogical reasoning, and the role that images, metaphors, and diagrams play in modelling. The tradition has an interesting history too, beginning much further back in the works of philosophically informed scientists in the second half of the nineteenth century, such as William Thomson, James Clerk Maxwell, Heinrich Hertz, and Ludwig Boltzmann. Boltzmann's widely read article in *Encyclopedia Britannica*, in particular, signals the belle époque of this "modelling attitude," which has done nothing but continue to flourish and inform much scientific practice during the twentieth century (in spite of philosophical detractors such as Pierre Duhem, who famously disparaged it in Duhem 1906/1954). Within the philosophy of science, it was mainly opponents to logical empiricist reconstructions of knowledge (both Carnapian and post-Carnapian) who pursued this modelling tradition and continued to emphasize the essential role of models, model-building, and analogical reasoning in the sciences. Thus Norman Campbell's masterly *Physics: The Elements* (Campbell 1920) had considerable influence in advancing the case for modelling amongst mainly British scholars such as Max Black (1962) and Mary Hesse (1966). Rom Harré (1960) was also influenced by Stephen Toulmin's (1960) vindication of theories as maps.

Within the sciences the modelling attitude has arguably been the prevalent methodology for acquiring predictive knowledge and control of natural and social systems ever since. Within philosophy of science, however, it has enjoyed more varied fortunes: after a peak of interest in the 1960s, the focus on models and representation waned considerably again, only to fully reemerge in the late 1990s around what is known as the "mediating models" movement. This was a movement of scholars based at the London School of Economics, the Tinbergen Institute, and the Wissenschaftkolleg in Berlin who developed and advanced a case for models and their role in scientific inquiry during the 1990s. The mediating models movement did not just look back in order to vindicate Mary Hesse's work in the 1960s but also proposed a view entirely of its own, according to which models are autonomous entities that mediate between theory and the world.

Models are neither simply inductive generalizations of data, nor are they merely elementary structures of theory. Rather they are independent entities, very much endowed with a life of their own and playing out a variety of roles in inquiry, which prominently include "filling in" theoretical descriptions for their concrete application (Morgan and Morrison 1999). The view lends itself to a certain sort of instrumentalism about theories in the building of models (Cartwright et al. 1994; Suárez 1999), which paves the way for an understanding of models as representations.

The contemporary discussions of representation thus emerge from two somewhat distinct currents of thought: the semantic approach and the mediating models movement. Some of the defenders of mediating models reject the semantic view but mainly on account of its construal of what models are (under the constraint that models are whatever provides identity conditions for theories); but it is nonetheless the case that scholars working in either movement emphasize the view that models are genuinely representational.[1] One of the most significant pioneering papers (Hughes 1997) is in fact the result of exposure to both movements. Hughes was already a leading defender of the semantic view (which he had successfully applied to problems in the philosophical foundations of quantum theory in Hughes [1989]) when he went on to become a prominent contributor to the mediating models movement (Hughes 1999). It is important to bear this dual heritage in mind since it goes some way towards explaining some of the inner tensions and open disagreements that one finds nowadays in this area.

2   ELEMENTS OF REPRESENTATION

There are many different types of representations in the sciences, in areas as diverse as engineering, mathematical physics, evolutionary biology, physical chemistry, and economics. Modelling techniques in these areas also vary greatly, as do the typical means for a successful application of a model. This is prima facie a thorny issue for a theory of representation, which must provide some account of what all these representations have in common. Nevertheless we may consider just one particular model, for our purposes, particularly since it has been widely discussed in the literature as a paradigmatic example, namely the "*billiard ball* model of gases."

2.1 Sources and Targets

The billiard ball model is a central analogy in the kinetic theory of gases developed in the second half of the nineteenth century (Brush 2003). Perhaps its first

---

[1] The notable exception may be "structuralism," which holds to a set-theoretical version of the semantic conception, arguably in conjunction with a kind of nonrepresentationalism regarding scientific theory (Balzer et al. 1989).

appearance—certainly the most celebrated one—in the philosophical literature occurs in Mary Hesse's work (Hesse 1962, pp. 8ff) where the model is employed to distinguish what Hesse calls the "negative," "positive," and "neutral" analogies:[2]

> When we take a collection of billiard balls in random motion as a model for a gas, we are not asserting that billiard balls are in all respects like gas particles, for billiard balls are red or white, and hard and shiny, and we are not intending to suggest that gas molecules have these properties. We are in fact saying that gas molecules are *analogous* to billiard balls, and the relation of analogy means that there are some properties of billiard balls which are not found in molecules. Let us call those properties we know belong to billiard balls and not to molecules the *negative analogy* of the model. Motion and impact, on the other hand, are just the properties of billiard balls that we do want to ascribe to molecules in our model, and these we can call the *positive analogy* [. . .] There will generally be some properties of the model about which we do not yet know whether they are positive or negative analogies [. . .] Let us call this third set of properties the *neutral analogy*.

In order to unravel the implications of this quote for representation, we need to first draw a number of distinctions. Let us refer to the system of billiard balls as the *source* and the system of gas molecules as the *target*. We say that billiard balls represent gas molecules if and only if the system of billiard balls is a representational source for the target system of gas molecules.[3] The extensions of "source" and "target" are then picked out implicitly by this claim; that is, any pair of objects about which this claim is true is a <source, target> pair. We can then list the properties of the source object as $\left\{P_1^s, P_2^s, \ldots, P_i^s, \ldots, P_j^s, \ldots, P_n^s\right\}$ and those of the target object as $\left\{P_1^t, P_2^t, \ldots, P_i^t, \ldots, P_j^t, \ldots, P_n^t\right\}$. The claim then is that some of these properties are identical: $P_1^s = P_1^t, P_2^s = P_2^t, \ldots, P_i^s = P_i^t$. Hence $\left\{P_1^t, P_2^t, \ldots, P_i^t\right\}$ constitute the positive analogy. What are we saying about the remaining properties? On Hesse's account we minimally know that some properties of the source system of billiard balls are not at all present in the system of gas molecules; thus, for example, $\left\{P_{i=1}^s, \ldots P_j^s\right\}$ are not identical to any of the properties of gas molecules $\left\{P_1^t, P_2^t, \ldots, P_i^t, \ldots P_j^t, \ldots, P_n^t\right\}$. These properties of billiard balls constitute then the negative analogy, while the remaining properties $\left\{P_{j+1}^s, \ldots, P_n^s\right\}$ constitute the neutral analogy (since we do not know whether they are in the positive or negative analogies). However, this is a purely epistemic criterion, and we may suppose that all properties of billiard balls are objectively

---

[2] Hesse (1966, pp. 8ff.) traces the model back to Campbell (1920), although it is significant that the term "billiard ball model" does not appear there once—and in fact neither does "billiard balls" but only the more generic "elastic balls of finite diameter." The first explicit appearance of the full billiard balls analogy that I am aware of occurs in Sir James Jean's influential textbook (Jeans 1940, p. 12ff). This is a significant point in historical scholarship, since the billiard ball analogy itself cannot have played the heuristic role that Hesse ascribes to it in the development of the theory; however, such historical considerations need not detain us here.

[3] The claim may be generic for any pair of types of such systems or particular for one particular system of billiard balls with respect to a particular set of gas molecules enclosed in a container.

in either the positive or negative analogy: they are either really shared by both balls and molecules or they are not, regardless of how accurate our knowledge of these facts is.

However, note that Hesse frames her notion of analogy in terms of the properties of the source: the criterion is that some of these properties obtain in the target and some do not. This criterion is symmetrical as long as "analogy" is understood as the sharing of identical properties; but even under this assumption, it does not provide some important information regarding the relation between the properties involved in the negative analogy. For it could be that properties are absent altogether in the target, or that they are explicitly denied in the target. And this would amount to a major difference. (In other words, it could be that there is some $P^t_{j+x} \in \{P^t_j, \ldots, P^t_n\}$, such that $P^t_{j+x} = \neg P^s_{j+w}$, for some $P^s_{j+w} \in \{P^s_j, \ldots, P^s_n\}$). For there is a difference between saying that gas molecules are unlike billiard balls in that they are soft (the property applies properly, but it is plainly contradicted) and saying that they are unlike billiard balls in that they are neither hard nor soft (the property just does not apply and there is no genuine contradiction). The latter statement seems more appropriate in the case of hardness/softness. But it need not always be more appropriate to deny the application of a property for the negative analogy. For example, when it comes to elasticity everything is quite different: billiard balls are after all only *imperfectly* elastic (Jeans 1940, pp. 15–16) Thus there are different ways in which properties can be objectively in the negative analogy. Billiard balls are hard and shiny, and these properties are simply inapplicable to gas molecules. They are both in the negative analogy "by absence." But the ball's imperfect elasticity is certainly applicable— even though denied in the molecules, which are presumed to be completely elastic. The inelastic character of the billiard balls is in the negative analogy "by denial." As we shall see, the difference between absence and denial turns out to be of some significance for discussions of scientific representation.

2.2 Means and Constituents

Another important distinction that the example brings into relief is one between means and constituents of a representation (Suárez 2003). The billiard balls provide a representation of the molecules in the gas and as such share a number of properties with them. So one can reason from the properties of billiard balls in the model to the properties of molecules. The positive analogy therefore provides the material *means* that allow representations to do their work. But it would go beyond this minimal statement regarding the function of the positive analogy to assert that the positive analogy *constitutes* the representational relation. This is not merely a statement regarding the functional grounds of the representation, but rather the nature, essence, or definition of the representation. There is a world of a difference between stating the positive analogy as what allows the representation to do its work (i.e., the reason the representation is useful, accurate, predictive, explanatory, relevant, etc.) and stating it as what the representation is essentially (i.e., the very constituent representational relation, its defining condition). We may summarize them as follows:

Means: R is the means (at a particular time and context) of the representation of some target *t* by some source *s* if (i) R (*s, t*) and (ii) some user of the representation employs R (at that particular time and in that context) in order to draw inferences about *t* from *s*.

Constituent: R constitutes (at all times and in every context) the representation of some target *t* by some source *s* if and only if (i) R (*s, t*) and (ii) for any source–target pair (*S, T*): *S* represents *T* if and only if R (*S, T*).

It is important to note both the inverted order of the quantifiers and the temporal index in the definition of the means. In other words, the constituent of a representation is an essential and perduring relation between sources and targets, while the means are those relations that at any given time allow representation-users to draw inferences from the source about the target. It turns out to be an interesting question whether these coincide in general or whether there is any need for a constituent at all. Note that this issue is independent of whether or not there are objective positive and negative analogies between sources and targets. Hence a fundamental question for theories of representation is whether Hesse's positive analogy belongs to the constituents of representations (in which case it is essential to the representation itself) or to their means (in which case it is not so essential). The answer ultimately depends upon whether representation in general is substantive or deflationary.

3 THEORIES OF REPRESENTATION

A theory of scientific representation will aim to provide some philosophical insight into what all representations have in common, as well as what makes representations scientific. There are at least two different ways to go about providing such insight, corresponding roughly to an analytical or a practical inquiry into representation. An analytical inquiry attempts to provide a definition of representation that does justice to our intuitions and its central applications. A practical inquiry, by contrast, looks into the uses of the representations directly and attempts to generalize, or at least figure out what a large number of its key applications may have in common. And, correspondingly, a theory or account of scientific representation may be substantive (if it aims for an analytical inquiry into its constituents) or deflationary (if its aim is instead to provide the most general possible account of its means in practice). In addition, a substantive theory may be primitivist or reductive, depending on whether it postulates representation as an explanatory primitive or attempts to analyze it away in terms of some more fundamental properties or relations.[4]

---

[4] It is logically possible for a deflationary theory to also be reductive—that is, to deny that representation is any substantial explanatory relation but reduce every singular means of representation to some further nonsubstative properties. See Section 6.2. for further discussion.

3.1 Substantive and Deflationary Accounts

On a substantive account every representation is understood to be some explanatory property of sources and targets or their relation. Since on this analysis this property or relation is constitutive of representation, it is perdurably always there, as long as representation obtains. Its explanatory power suggests that it is the relation between the objects that make up possible sources and targets—or the type of properties shared, or the type of relation—and not the singular properties of the concrete source–target pair or their singular relations. The latter simply instantiate the type and may vary greatly from application to application. By contrast, on a deflationary account there may be nothing but the singular properties of the concrete source–target pair in the particular representation at hand. On such an account nothing can explain why those singular properties (or relations), which are in fact used, are the appropriate ones for representation—since there is, on a deflationary account, no constitutive explanatory relation of representation that they may instantiate. On the contrary, on a typical deflationary account, more generally, the appropriate properties are exactly the ones that get used in the particular application at hand, and there is no reason why they are appropriate other than the fact that they do so get used.

Let me try to unravel this distinction a little further in analogy with theories of truth in metaphysics (as suggested in Suárez 2004, p. 770; for an accessible review to theories of truth see Blackburn and Simmons 1999). A substantive approach to truth assumes that there is a particular type of relation between propositions and facts (or between propositions and other propositions, or between propositions and utility functions) such that any proposition that stands in that relation to a fact (or to a set of other propositions, or some value of some agent's utility function) is true. This relation may be correspondence between propositions and facts, or coherence with other propositions, or the maximization of some utility function of an agent. On any of these substantive accounts, there is an explanation for why a proposition is true rather than false, namely that it so happens to hold such a relation (correspondence, coherence, utility) to something else. By contrast, on a deflationary account, nothing explains the truth of a proposition, since there is nothing substantive that it is true in "virtue of" that could possibly explain it. To put it very bluntly, a proposition's truth is rather determined by its functional role in our linguistic practice. In other words, "truth" picks out not a natural kind out there in the world but a function in speech and discourse.

In a similar vein, substantive theories of representation understand it to be a substantive and explanatory relation akin to correspondence (or coherence or utility). Deflationary theories, by contrast, claim it to pick out nothing other than a functional role in scientific practice. Sections 5 and 6 review a few of each of those types of theories. A few distinctions must first be introduced.

3.2 Reductive and Nonreductive Accounts

The other important distinction already mentioned concerns reductive versus nonreductive theories of representation. A reductive theory is one that aims to reduce representation to something else—and most substantive accounts of representation will be reductive. Again, one may think of the analogous case for theories of truth, where substantive accounts on the whole attempt to reduce the "truth" property to a cluster of further properties that includes correspondence (or coherence or utility). What explains the truth or falsehood of some proposition is then whether or not such properties obtain, since truth is just the obtaining of those properties. Similarly, a substantive account of representation will reduce it to some substantive and explanatory relation such as, for example, similarity. What explains representation is then the obtaining of this substantive property, since representation just is that property. On a reductive substantive theory, explanation comes for free.

A nonreductive account of representation, by contrast, will not attempt to reduce representation to anything else—it will not suppose that there exists any underlying property that explains away representation. Rather, on this account representation is irreducible. Many deflationary accounts are nonreductive: they assume that representation cannot be analyzed away in terms of other properties. However, not all nonreductive accounts are deflationary; some are *primitivist*: they accept that the concept of representation cannot be reduced further, but they assume that this is so because it is an explanatory primitive—it can and usually is invoked in order to explain other concepts (e.g., it can be used to explain empirical adequacy, since this may be defined as accurate representation of the observable phenomena). In other words, a primitivist account accepts that representation cannot be reduced any further but not because it lacks substance.[5]

Hence, from a logical point of view, the substantive/deflationary and reductive/nonreductive distinctions are orthogonal. Any of the four combinations is logically possible. However, from a more practical point of view, substantive accounts of representation naturally line up with reduction, while deflationary accounts often, but not always, go along with the claim that representation cannot be further reduced or analyzed.

4 THE ANALOGY WITH ART AND AESTHETICS

I have so far been elaborating on a novel analogy between theories of representation and theories of truth. There is a yet more conspicuous analogy running through the contemporary literature on representation linking it to discussions in aesthetics and the philosophy of art.[6] This analogy is often employed because similar views regarding

---

[5] It is unclear if there exists any primitivist account of truth, but Williamson's (2000) account of knowledge is certainly primitivist.

[6] The origins of the analogy are not always attributed correctly, or at all. It was originally introduced in Suárez (1999), which also contains some of the examples from art that have gone on to be discussed regularly. Van Fraassen (1994) is an acknowledged ancestor; although it does not present the analogy explicitly, it already discusses artistic representation in this context.

artistic representation have been discussed for a long time (for an excellent treatment, see Lopes 1996). Some consensus has been reached there regarding the strength of certain arguments against substantive theories of scientific representation and in particular resemblance theories. In this section I briefly review and expand on those arguments with an eye to an application later on in the context of scientific representation. The critical assumption must be that if representation is a substantive relation, then it must be the same in both art and science (although the means of representation and the constraints on its application may vary greatly in both domains). Hence if representation is substantive, it is so in both domains, while if it is deflationary, then this must also be the case in both domains. This has plausibility because in both domains the representation in question is objectual; for example, objects (models) stand for other objects (systems).

I review here only one important type of argument against resemblance theories of artistic representation, originally due to Nelson Goodman (1968), which may be referred to as the *logical argument*. According to this argument, resemblance cannot constitute the relation of representation because it does not have the right logical properties for it. Resemblance is reflexive, symmetrical, and transitive, but representation in general is none of this. This is best illustrated by a couple of paintings. Thus Velázquez's *Portrait of Innocent X* depicts the Pope as he was sitting for Velázquez, but it certainly does not depict itself, and the Pope certainly does not represent the canvas. (Goodman uses the example of the portrait of the Duke of Wellington to illustrate the same "Innocent" point.) Francis Bacon's *Study after Velázquez's Portrait of Pope Innocent X* (1953) is a formidable depiction of the Velázquez canvas, but it would be wrong to say that it depicts the Pope himself— thus exhibiting a failure of transitivity. Yet to the extent that the Bacon resembles the Velázquez it also resembles the Pope; the Pope resembles the Velázquez just as much as is resembled by it; and, certainly, the canvas maximally resembles itself. In other words, resemblance is an equivalence relation: it is impossible to explain these failures of reflexivity, symmetry, and transitivity if representation really is resemblance. Therefore resemblance [Res] cannot *constitute* representation in the sense of the definition: it is not the relation R such that "(i) [for the canvas (s) and the Pope (t)]: R (s, t) and (ii) for any source–target pair (S, T): S represents T if and only if R (S, T)." Here, only the first condition is fulfilled but not the second. Now, certainly resemblance can *overlap* with representation—and indeed the Pope and the Velázquez canvas do (presumably!) resemble each other. Moreover, resemblance can be the effective means by which a viewer infers for example the color of the Pope's clothes from the color of the Velázquez canvas. So indeed the definition of means is satisfied at that very time for that particular viewer, since "(i) [for the relation Res of resemblance, the canvas (s) and the Pope (t):] Res (s, t); and (ii) some user of the representation employs Res (at that particular time and in that context) in order to draw inferences about t from s."

In other words, the analogy with artistic representation provides a logical argument against a substantive theory of the constituents of representation in general as resemblance (or indeed as any other relation that has the logical properties of reflexivity, symmetry, and/or transitivity). It shows that representation in general is not a logical equivalence relation.

5 SUBSTANTIALISM

Two main substantive accounts have been discussed in the literature, and they are both reductive in character. One of these accounts attempts to reduce representation to similarity relations between sources and targets while the other attempts a reduction to isomorphism between their structures. I argue in the following that both reductive attempts fail and that it is an informative exercise to determine exactly why. Nevertheless, it bears reminding ourselves at this stage that one live option for the substantivist—one that moreover has not been sufficiently discussed in the literature so far—is to go "primitivist" at this point and deny that representation can in fact be reduced to any other property. I do not discuss this option here for the reasons of plausibility already mentioned, but it would certainly be a logically admissible way out of some of the arguments in this section.

5.1 Similarity

The connection between similarity and representation has been emphasized before (Aronson et al. 1995; Giere 1988; Godfrey-Smith 2006; Weisberg 2012), and in most of these cases it is at least plausible to suppose that the background assumption has been one of reduction. In other words, all these accounts may be understood to be approximations to the following reductive theory of representation:

[*sim*]: *A* represents *B* if and only if *A* and *B* are similar

Any theory that has this form is a substantive account of the constituents of scientific representation that reduces it to the relation of similarity between sources and targets. In other words, according to theories like this, the underlying similarity between sources and targets explains their representational uses. Nevertheless, the theories will differ in the different degrees of sophistication and complexity of their accounts of similarity. The simplest account understands similarity as the mere sharing of properties, and representation then boils down to the sharing of properties between representational sources and targets. Suppose then that the properties of the model source are given by $P^s = \{P^s_1, P^s_2, \ldots, P^s_n\}$, and the properties of the target object as $P^t = \{P^t_1, P^t_2, \ldots, P^t_m\}$. The simplest account then assumes that the source represents the target if and only if they share a subset of their properties. In other words, there are some

$\left\{P_1^s, P_2^s, \ldots, P_i^s\right\} \in P^s$, with $i \le n$, and some $\left\{P_1^t, P_2^t, \ldots, P_i^t\right\} \in P^t$, with $i \le m$, such that $\left\{P_1^s = P_1^t, P_2^s = P_2^t, \ldots, P_i^s = P_i^t\right\}$.[7] On this account, the complete objective positive analogy constitutes the representational relation between the billiard ball model and the gas molecules, while the negative analogy is a list of those properties of the model that play no genuine representational role. Hence only those properties of billiard balls that are shared with gas molecules, such as putatively elasticity and motion, are genuinely representational and can be said to be in the representational part of the model.

The simplicity of this account has a number of advantages, including its intuitiveness and fit with our ordinary or unreflective ways of talking about similarity. It moreover also fits in very well with Mary Hesse's discussion of analogy and her example of the kinetic theory of gases. But it has a number of problems too, which follow fairly straightforwardly from our earlier discussions. The most obvious problem is brought home by the analogy with art. The logical argument applies here in full force since similarity so simply construed is reflexive and symmetrical (and transitive over the properties shared by the intermediate targets). In other words, the simple account of similarity provides the wrong reduction property for representation.

The problem may be arguably confronted by more sophisticated definitions of similarity. I focus here on Michael Weisberg's (2012) recent account, which is heavily indebted to seminal work by Tversky and collaborators on the psychology of similarity judgments (Tversky 1977). On the Tversky–Weisberg account, a source s represents a target t if and only if their comparative similarity is large, where the degree of comparative similarity is measured in accordance to the following function: $Sim(s, t) = \theta \cdot f(S \cup T) - \alpha \cdot f(S - T) - \beta \cdot f(T - S)$, where S and T are the full set of salient and relevant features of source and target and θ, α, and β are relative weights that typically depend on context. This similarity measure is a function of shared features of source and target, which takes into account both features of the source missing in the target (i.e., Hesse's negative analogy) but also features of the target missing in the source. Moreover, and this is what is really nice about the Tversky–Weisberg account, it renders similarity nonsymmetric: $Sim(s, t)$ need not equal $Sim(t, s)$ because s and t may be endowed with a different range and number of salient features or attributes. So the Tversky–Weisberg account provides an answer to the "*Innocent*" point and gets around the awkward conclusion that a model represents a target only to the extent that it is represented by it.

This more complex proposal is promising, but it may be challenged on a number of grounds. First, the Tversky–Weisberg ingenious measure of similarity—while going further than any similarity measure relying only on Hesse's positive and negative analogies—nonetheless cannot capture the impact of "negative analogies by denial,"

---

[7] There is an issue, which I gloss over in the discussion in the main text, about how to define properties in such a way that these identity statements obtain. For instance, a purely extensional definition in principle does not help, since the entities that come under the extension of these properties are typically very different in the source and the target. I am therefore assuming that there is a different account of properties that makes sense of the aforementioned identity statements—otherwise this particular theory of representation as similarity would not even get off the ground.

that is, the properties of the target that are explicitly denied in the source. We saw that Hesse does not distinguish those properties of the model source that fail to apply to the target (arguably color and shine) from those other properties of the source that are explicitly denied in the target (limited elasticity, escape velocity). The former properties may perhaps be ignored altogether—since they do not play a role in the dynamical processes that ensue in either billiard balls or, naturally, gas molecules. The latter properties cannot, however, be so dismissed—so it seems altogether wrong to claim that they are not part of the representation. In other words the similarity proposal does not account for a typical way in which models go wrong or misrepresent their targets. The form of misrepresentation that involves "lying," "simulating," or "positively ascribing the wrong properties" is not describable under this account, which prima facie—given how pervasive "simulation" is in model-building—constitutes a problem for the account.[8]

Second, the logical argument has not been completely answered since Tversky–Weisberg similarity continues to be reflexive—and so is therefore representation so construed, which seems just wrong. Also, notice the emphasis on contextual relevance. The sets of features of s, t to be taken into account are relative to judgments of relevance in some context of inquiry. In other words, there are no context-independent or context-transcendent descriptions of the features of sources and targets. Third, and related to this, notice that the idea of context-relative description presupposes that some antecedent notion of representation is already in place, since it assumes that sources and targets are *represented as* having particular sets of features in context.

5.2 Isomorphism

The other substantive theory appeals to the notion of isomorphism and its cognates. Once again there have been a number of different approaches (Bartels 2006; Mundy 1986; Pincock 2012; Suppes 2000; Swoyer 1991), but they all seem to have at their foundation a commitment to reducing representation to some kind of morphism relation between the structures that are instantiated by both the source and the target. Therefore these views are all approximations to the following theory:

[iso]: A represents B if and only if the structures S_A and S_B exemplified by A and B are isomorphic: $S_A \equiv S_B$.

The definition of isomorphism is then given as follows. Two structures $S_A = \langle D_A, P_j^n \rangle$ and $S_B = \langle D_B, T_j^n \rangle$, where P_j^n and T_j^n are n-place relations; are isomorphic ($S_A = S_B$) if and only if there is a one-to-one and onto mapping $f: D_A \rightarrow D_B$, such that for any

---

[8] One may suppose that this problem may be overcome by incorporating a fourth factor in the measure of similarity successfully representing the weight of features of the source that are explicitly denied in the target. It remains to be seen if this is possible, but at any rate the other three objections would remain.

n-tuple $(x_1, \ldots, x_n)$, where each $x_i \in D_A : P_j^n[x_1, \ldots, x_n]$ only if $T_j^n[f^\circ(x_1), \ldots, f^\circ(x_n)]$; and for any n-tuple $(y_1, \ldots, y_n)$, where each $y_i \in D_B : T_j^n[y_1, \ldots, y_n]$ only if $P_j^n[f^{-1}(y_1), \ldots, f^{-1}(y_n)]$. In other words, an isomorphism is a *relation preserving* mapping between the domains of two extensional structures, and its existence proves that the *relational framework* of the structures is the same.[9]

This theory is another substantive account of the constituent of representation, now as the relation of isomorphism between instantiated structures of the source and target pair. On this theory what explains representation is the conjunction of the obtaining of the relation of isomorphism and the appropriate relation of instantiation.[10]

Once again there are different versions of the theory appealing to different types of morphism relation, ranging from the strongest form (isomorphism) to weaker versions in this order: partial isomorphism, epimorphism, and homomorphism. I will not enter a detailed discussion of the differences, except to state that the logical argument applies to some extent to all of them. For instance, isomorphism and partial isomorphism are reflexive, symmetrical, and transitive; epimorphism and homomorphism are reflexive and transitive for the same set of relations, and so on. More important still, for our purposes, is the fact that none of these approaches can accommodate the two senses of negative analogy. For there are no resources in any of these approaches to accommodate in particular the explicit denial of a relation in the target system that has been asserted in the source system. That is, whatever minimal morphism there is between two structures S_A and S_B, the relations that stand functionally related in the morphism cannot deny each other; that is, $P_j^n[x_1, \ldots, x_n]$ and $T_j^n[f^\circ(x_1), \ldots, f^\circ(x_n)]$ cannot deny each other. At best in an epimorphism or homomorphism there will be some relations that are not asserted in either source or model: some $P_j^n[x_1, \ldots, x_n]$ will thus have no "correlate." But none of the known morphisms is able to account for a structural mapping where $P_j^n[x_1, \ldots, x_n]$ is explicitly denied in its correlate, that is, where

$T_j^n[f^\circ(x_1), \ldots, f^\circ(x_n)] = \neg P_j^n[x_1, \ldots, x_n]$. And this, as we saw, is precisely what happens in negative analogies "by denial," for instance when we model the perfectly elastic collisions of gas molecules by means of imperfectly elastic billiard balls. The only structural rendition of such analogies is one that does not insist on any transfer of relevant

---

[9] Or, as sometimes said in the literature, it is an expression of "structural identity." But this is equivocal and should really be avoided, since the domains D_A, D_B are typically not identical but made up of very different elements. In other words, except in the reflexive case, isomorphism is certainly not an expression of numerical identity, as a cursory inspection of the definition makes obvious: "Two structures are isomorphic if and only if there is a mapping between their domains" can only be true if there exist two numerically distinct structures S_A and S_B endowed with their own distinct domains D_A and D_B.

[10] Thus what explains a particular item of representation is a product relation X.Y where X is isomorphism and Y is instantiation. There is an issue here with multiple instantiation, since X.Y = Z can be reproduced by an in principle infinite number of distinct such products, for example, X'.Y' = X.Y = Z. Given that the extant leeway for instantiation turns out to be very large—most objects dramatically underdetermine their structure—it follows that what precise isomorphism is constitutive of any particular representation is also almost always undefined.

structure. Since most scientific modelling has this "simulative" character, so representation is not a structural relation even when model sources and targets are structures or may be described as possessing or instantiating them.

6 Deflationism

Let us now look at deflationary theories. These are theories that do not presuppose that representation is substantive or can otherwise be reduced to substantive properties or relations of sources and targets. The analogy with theories of truth was already noticed, and this also holds for deflationary views. Thus deflationary views of truth come in a couple of different forms, including what I call redundancy and use-based theories, and I argue in the following that deflationary theories of representation also take similar forms. In particular I develop a "redundancy" version of R. I. G. Hughes's Denotation-Demonstration-Interpretation (DDI) model of representation and a "use-based" version of my own inferential conception.

6.1 The DDI Account

Redundancy theories of truth originate in Frank Ramsey's work, in particular in "Facts and Propositions" (Ramsey 1927). The basic thought is that to assert of some proposition P that it is true is to assert nothing over and above P itself. The predicate "true" is instead redundant, in the sense that to predicate of any proposition that it is "true" adds nothing to the content of that proposition. There is no substantive property that all true propositions share. The ascription of the predicate "true" to a proposition is rather taken to possess only a kind of honorific value: it merely expresses the strength of someone's endorsement of a particular proposition. The use of the predicate may have other functions—for instance it helps in generalization and quotation—as in "everything Ed says is true"—but even then it does not express any substantive property: it does not establish that everything Ed says is true in virtue of any substantive property that all the propositions that he utters share. Truth is, if it is a property at all, a *redundant* property. For my purposes here I focus on the part of the redundancy theory that most closely approaches the view that the terms "truth" and "falsity" do not admit a theoretical elucidation or analysis but that, since they may be eliminated in principle—if not in practice—by disquotation, they do not in fact require such an analysis. I take this implicitly to mean that there are no nontrivial necessary and sufficient conditions for these concepts.

The transposition of all this discussion to theories of scientific representation is, I argue, the claim that representation is itself a redundant concept in the same way: it expresses some honorific value that accrues to a particular use that agents make of some model, but it does not in itself capture any relational or otherwise property of sources or targets. The use of the term in addition signals further commitments, which I study in greater depth in the next section and which are mainly related to the source's capacity to generate surrogate inferences regarding the target. But here again—as in the case of truth—such commitments do not signal that the term "representation" picks out any substantive property. What they rather signal is that the term has no analysis in terms of nontrivial necessary and sufficient conditions. Concomitantly, we cannot work out explanations for the diverse applications of this term on the basis of any substantive definition or precise application conditions, since it has none.

Perhaps the most outstanding example of a redundancy account of scientific representation is Hughes's (1997) DDI model.[11] On Hughes's account, representation typically although not necessarily involves three separate speech acts: first, the denoting of some target by a model source; second, the demonstration internal to the model of some particular result; and third, the interpretation of some aspects of the target as aspects of the source and, concomitantly, the transposition of the results of the demonstration back in terms of the target system with ensuing novel predictions, and so on. The most important aspect of the DDI model for our purposes is Hughes's claim not to be "arguing that denotation, demonstration and interpretation constitute a set of speech-acts individually necessary and jointly sufficient for an act of theoretical representation to take place" (Hughes 1997, p. 329). In other words, the DDI account is proposed explicitly as a deflationary account of representation along the lines of the redundancy theory: It refrains from defining the notion of representation, and it is not meant as an explanatory account of use. It is instead a description of three activities, or speech acts, that are typically enacted in scientific modelling.[12] It connects to practice in the sense that these items provide a description of three typical norms in the practice of representation.

The main example of the application of a DDI model is Galileo's model of motion along an inclined plane (Hughes 1997, pp. 326–329), where a purely geometrical demonstration in geometry is taken to represent inertial motion on an inclined plane. Hughes shows convincingly how to apply the DDI three-part speech act theory to this model and how much hinges in particular on the intermediate demonstration in the model. The DDI is of course supposed rather generally, so let us apply it to our main example, the billiard ball model in the kinetic theory of gases.

The denotation part of this model is prima facie straightforward: billiard balls are taken within the model to denote gas molecules. (There are some complications that arise when one considers the denotation of particular properties, particularly those in the negative analogy "by absence," but we may leave property denotation aside for the purposes of the discussion.) As for the demonstration and interpretation stages,

---

[11] Van Fraassen (2008) too endorses a redundancy sort of deflationism regarding representation, but the details of his specific account are less developed.

[12] Hughes's original DDI model appealed essentially to the relation of denotation. I have argued that this appeal is problematic from a deflationary point of view but that the original account may be amended into a fully deflationary one by replacing denotation with something I call denotative function—a notion suggested by some of Catherine Elgin's writings (e.g., Elgin 2009)—which is not a relation but a feature of representational activity (Suárez 2015).

Campbell (1920, pp. 126–128) helpfully separates what he calls the "hypothesis" of the theory from the "dictionary" provided by the model, which provides ideal grounds to apply both the demonstration and interpretation requirements in the DDI account. Thus among the hypothesis of the theory we find all the relevant mathematical axioms for the mechanical system, including those relating to the constants (l, m, v), the $3n$-dependent variables (x_s, y_s, z_s) for the n-system of elastic balls, and the equations that govern them. The demonstrations will be carried out at this level via these mathematical equations. The dictionary brings in a link to the physics of gas molecules, by establishing for example that (i) l is the length of the cubical vessel in which the "perfect gas" is contained, (ii) m is the mass of each molecule and nm the total mass of the gas, (iii) $\frac{1}{\alpha} mv^2$ is the absolute temperature T of the gas, and (iv) p_i is the pressure on the wall i of the container for $i = x, y, z$, which for the given interval of time comprised between t and $t + \gamma$ is given by $p_i = \lim_{\gamma \to \infty} \sum_{s=1}^{s=n} \frac{1}{\gamma} \Delta m \frac{d(x_s, y_s, z_s)}{dt}$. Then, using the equations in the hypothesis of the theory we may calculate the pressure to be $p_i = \frac{1}{3l^3} nmv^2$, for any value of i. This constitutes the demonstration step referred to previously. Now interpreting this result back into the physical description of the gas by means of the dictionary, we obtain $p_i = \frac{\alpha \cdot n}{3} \frac{T}{V}$, which is an expression of Boyle's and Gay-Lussac's law since $\frac{\alpha \cdot n}{3}$ is a constant.

Hence the mechanical model generates, via the three steps of denotation (of gas molecules by infinitely elastic mechanical billiard balls), demonstration (by means of the hypothetical laws of the model), and interpretation (back in terms of thermodynamic properties of the gas) just the kind of prediction that is central to the kinetic theory of gases. The DDI account thus shows it to provide a *representation* of the gas and its properties.

6.2 The Inferential Conception

A different deflationary account is provided by the inferential conception (Suárez 2004). It differs minimally from both Hughes's DDI account and other deflationary views in leaning more toward use-based deflationism.[13] The difference between these two sorts of deflationism, however minimal, can again be illustrated by means of the analogy with theories of truth. All deflationary theories of some concept X deny that there is a definition of the concept that explains its use. Redundancy theories, as we saw, deny that X may be defined altogether; use-based theories admit that X may be defined and may have possession conditions, but they deny that the use of X is thereby explained.

---

[13] Ron Giere's (2004) recent four-place pragmatic account is also arguably of this use-based deflationary kind.

There is a sense then in which use-based views are reductive but not substantive. They certainly aim to "anchor" the concept in features of practice, and, depending on how they go about this, they may in fact be reducing the concept. For instance, Giere (2004, p. 743) comes very close to providing necessary and sufficient conditions for sources to represent targets in terms of similarities, agents, and purposes—and may be considered reductive. The inferential conception that I have defended is more explicit in providing merely very general or necessary conditions. It does not aim at a reduction, but it does aim to link representation to salient features of representational practice—in a way that dissolves a number of philosophical conundrums regarding the notion.

On the inferential conception [inf] a source s represents a target t only if (i) the representational force of s points to t and (ii) s allows an informed and competent agent to draw specific inferences regarding t (Suárez 2004, p. 773). There are several important caveats and consequences to this definition, which may be summarized as follows:

(1) [inf] may be objected to on grounds of circularity, given the reference to representational force in part (i). However, [inf] does not define representation, and, at any rate, representational force is a feature of practice, not itself a "concept."

(2) The term "representational force" is generic and covers all intended uses of a source s to represent a target t, including but not restricted to the notion of denotation whenever that may obtain. In particular, while sources of fictional objects (such as the models of the ether at the end of the nineteenth century) do not denote them, they may well have representational force pointed at them.

(3) Inferences are "specific" if they lead to features of t that do not follow from the mere fact that s represents t. So, for instance, that someone's name is Sam follows from the fact that "Sam" is used to denote him, so it is not specific—and the inference is therefore unable to ground representation.

(4) A object or model is a representational source for some target within some representational practice as long as the practice in question involves norms of correct surrogate inference—it is not required in addition that any of the consequences of these inferences be to true conclusions. This allows the inferential conception to account for all instances of idealization and misrepresentation, including negative analogies "by denial."

(5) The dynamics for developing scientific representations, on the inferential account, responds to the productive interplay of (i) representational force and (ii) inferential capacities.

The last point is particularly important to the prospects of the inferential conception, since any adequate account of scientific representation must do justice to this dynamical aspect. It arguably distinguishes the inferential conception from its more static substantive competitors (Suárez 2004, p. 773–774).[14] It seems apposite to

---

[14] It moreover distinguishes it from other accounts that rhetorically claim to be in the same inferential spirit yet ultimate presuppose that all inference rides on some substantial structural relation. Contrary to appearances, those accounts are, unlike the inferential conception, neither deflationary nor dynamical.

end this article with an outline of the application of the inferential conception to the same example of the billiard ball model in the kinetic theory of gases that has occupied us before. The main extension of the theory (discussed by both Campbell [1920], pp. 134–135, and Jeans [1940], pp. 170–174) concerns viscosity in a gas. It does so happen that layers of the gas slide past each other as would be expected in a viscous liquid. This affects the relationship between molecular velocities and temperature, to the point that the coefficient α inserted into the hypothesis and related to temperature in the dictionary (see equation iii earlier relating absolute temperature to molecular velocity) must be made to depend on the value of the temperature itself. In fact in a thorough treatment of the phenomenon, the assumption that the gas molecules are perfect elastic spheres must be relaxed (see Jeans 1940, p. 171) to deal with the fact that viscosity does not depend in actual fact—as may be established experimentally—on the size and shape of the container but only on the temperature and density of the gas.

In other words, the inferential capacities of the original version of the model lead, via its representational force, to predictions that in turn motivate an alteration in the model. This new model's reinforced inferential capacities avoid the potential refutation by adjusting the representational force of some of the elements in the model, notably viscosity, and so on. The conceptual development of the kinetic theory of gases is therefore accounted for in terms of the playing out of an inbuilt tension amongst the two essential surface features of scientific representation. The point certainly calls for further development and generalization, but the thought is that a deflationary account of scientific representation possesses the resources to naturally account for the heuristics that drive the dynamics of modelling practice.

7 CONCLUSIONS

Scientific representation continues to be a very lively area of research within contemporary philosophy of science. This article is intended to provide a review of some of the most significant work carried out in the area recently while offering some critical commentary, guide, and direction. The first section introduced the topic of scientific representation from a historical perspective. The second section reviewed common terminology and drew some significant conceptual distinctions. In the third section accounts of representation were divided along two orthogonal dimensions into reductive–nonreductive and substantive–deflationary. I argued that while there is no logical compulsion, it stands to reason that reductive accounts will be typically substantive while nonreductive ones will tend toward deflationism. Section 4 developed an analogy with artistic representation. In Section 5 the major substantive accounts of representation were reviewed and it was argued that they all confront important challenges. Section 6 discussed and endorsed two deflationary approaches, the DDI model and the inferential conception. Some promising avenues for future work in both deflationary approaches were suggested in Section 7.

ACKNOWLEDGMENTS

Support from the Spanish government (DGICT Grants FFI2011-29834-C03-01 and FFI2014-57064-P) and the European Commission (FP7-PEOPLE-2012-IEF: Project number 329430) is gratefully acknowledged. This piece is best read in conjunction with the entry on Scientific Representation that I wrote for Oxford Bibliographies Online (Suárez, 2014), which provides additional bibliographical information.

REFERENCES

Aronson, J. E. Way, and Harré, R. (1995). *Realism Rescued* (London: Duckworth).

Bailer-Jones, D. (2009). *Scientific Models in Philosophy of Science* (Pittsburgh: Pittsburgh University Press).

Balzer, W., Moulines, C. U., and Sneed, J. (1989). *An Architectonic for Science: The Structuralist Program* (Dordrecht: Springer).

Bartels, A. (2006). "Defending the Structural Concept of Representation." *Theoria* 55: 7–19.

Black, M. (1962). *Models and Metaphors* (Ithaca, NY: Cornell University Press).

Blackburn, S., and Simmons, K., eds. (1999). *Truth* (Oxford: Oxford University Press).

Boltzmann, L. (1902). "Models." In *Encyclopaedia Britannica* (Edinburgh, UK: Black).

Brush, S. G. (2003). *The Kinetic Theory of Gases* (London: Imperial College Press).

Campbell, N. R. (1920). *Physics: The Elements* (Cambridge: Cambridge University Press).

Cartwright, N., Shomar, T., and Suárez, M. (1994). "The Toolbox of Science." *Poznan Studies in the Philosophy of the Sciences and the Humanities* 44: 137–149.

Duhem, P. (1954). *The Aim and Structure of Physical Theory* (Princeton, NJ: Princeton University Press). (Original work published 1906 in French).

Elgin, C. (2009). "Exemplification, Idealization and Understanding," in M. Suárez (Ed.), *Fictions in Science: Philosophical Essays on Modelling and Idealization* (New York: Routledge, 2009), pp. 77–90.

Giere, R. (1988). *Explaining Science* (Chicago: University of Chicago Press).

Giere, R. (2004). "How Models Are Used to Represent Reality." *Philosophy of Science* 71: 742–752.

Godfrey-Smith, P. (2006). "The Strategy of Model-Based Science." *Biology and Philosophy* 21: 725–740.

Goodman, N. (1968). *Languages of Art: An Approach to a Theory of Symbols* (Indianapolis: Bobbs-Merrill).

Harré, R. (1960). *An Introduction to the Logic of the Sciences* (London: Macmillan Press).

Hesse, M. (1966). *Models and Analogies in Science* (Notre Dame: University of Notre Dame Press).

Hughes, R. I. G. (1989). *The Structure and Interpretation of Quantum Mechanics* (Cambridge, MA: Harvard University Press).

Hughes, R. I. G. (1997). "Models and Representation." *Philosophy of Science* 64: 325–336.

Hughes, R. I. G. (1999). "The Ising Model, Computer Simulation and Universal Physics," in M. Morgan and M. Morrison (Eds.), *Models as Mediators* (Cambridge: Cambridge University Press), pp. 97–145.

Jeans, J. (1940). *Kinetic Theory of Gases* (Cambridge: Cambridge University Press).

Lopes, D. (1996). *Understanding Pictures*. Oxford: Oxford University Press.

Morgan, M., and Morrison, M., eds. (1999). *Models as Mediators* (Cambridge: Cambridge University Press).

Mundy, B. (1986). "On the General Theory of Meaningful Representation." *Synthese* 67: 391–437.

Pincock, C. (2012). *Mathematics and Scientific Representation* (Oxford: Oxford University Press).

Ramsey, F. P. (1927). "Symposium: Facts and Propositions," in *Proceedings of the Royal Society of London Supplementary Volumes*, vol. 7 (London: The Royal Society), pp. 153–206.

Suárez, M. (1999), "Theories, Models and Representations," in L. Magnani et al. (Eds.), *Model-Based Reasoning in Scientific Discovery* (New York: Kluwer Academic), pp. 83–95.

Suárez, M. (2003). "Scientific Representation: Against Similarity and Isomorphism." *International Studies in the Philosophy of Science* 17: 225–244.

Suárez, M. (2004). "An Inferential Conception of Scientific Representation." *Philosophy of Science* 71: 767–779.

Suárez, M. (2014). "Scientific Representation." Oxford Bibliographies Online. doi: 10.1093/OBO/9780195396577-0219

Suárez, M. (2015). "Deflationary Representation, Inference, and Practice." *Studies in History and Philosophy of Science* 49: 36–47.

Suppes, P. (2000). *Representation and Invariance of Scientific Structures*. Stanford CSLI series (Chicago: University of Chicago Press).

Swoyer, C. (1991). "Structural Representation and Surrogative Reasoning." *Synthese* 87: 449–508.

Toulmin, S. (1960). *The Philosophy of Science* (New York: Harper Torchbooks).

Tversky, A. (1977). "Features of Similarity." *Psychological Review* 84: 327–352.

Van Fraassen, B. (1980). *The Scientific Image* (Oxford: Oxford University Press).

Van Fraassen, B. (1994). "Interpretation of Science, Science as Interpretation," in J. Hilgevoord (Ed.), *Physics and Our View of the World* (Cambridge: Cambridge University Press), pp. 188–225.

Van Fraassen, B. (2008). *Scientific Representation* (Oxford: Oxford University Press).

Weisberg, M. (2012). *Simulation and Similarity* (Oxford: Oxford University Press).

Williamson, T. (2000). *Knowledge and Its Limits* (Oxford: Oxford University Press).

CHAPTER 22

....

REDUCTION

....

ANDREAS HÜTTEMANN AND ALAN C. LOVE

REDUCTION and reductionism have been central philosophical topics in analytic philosophy of science for more than six decades. Together they encompass a diversity of issues from metaphysics (e.g., physicalism and emergence) and epistemology (e.g., theory structure, causal explanation, and methodology). "Reduction" usually refers to an asymmetrical relationship between two items (e.g., theories, explanations, properties, etc.) where one item is *reduced* to another. The nature of this relationship has been characterized in distinct ways for different sets of items: reductive relations between theories have often been understood in terms of logical derivation, whereas reductive relations between properties have sometimes been understood in terms of identity or supervenience. Depending on how the relationship is characterized, one can speak of successful reductions when the asymmetrical relationship is established or manifest, and unsuccessful reductions when for some reason the relationship does not hold. "Reductionism" usually refers to a more general claim or assumption about the existence, availability, or desirability of reductions in an area of research. For example, if particular kinds of reductions are established routinely in molecular biology and constitute a major part of its explanatory capacity, then reduction*ism* can be used as a descriptive label (molecular biology pursues reductionist explanations or has a reductionist methodology).

Arguments for and against reductionism often trade in several different meanings of reduction simultaneously or assume that a particular construal is primary, but the relationship between different conceptions of reduction is complex, and there are few if any entailment relations among them. The specialization trend in philosophy of science, whereby philosophers have increasingly concentrated on actual scientific practices and controversies, has shifted debates away from providing a uniquely correct account of *what* reduction is and toward *why* or *how* scientists pursue reductions in more localized contexts. As a consequence, a rift has grown between metaphysical and epistemological questions because the former often invoke "in principle" considerations about what a completed science *would* be able to say, whereas the latter emphasizes "in practice" considerations that temper the metaphysical inferences one might draw, in part because we find both successful and unsuccessful reductions of different kinds in scientific practice.

Thus an argument establishing a particular relationship of reduction between two items in an area of science does not necessarily generalize to an argument in favor of reductionism for all items in that area of science or for other areas of science.

It is impossible to adequately summarize the many detailed analyses of reduction and reductionism in different sciences, even when only focused on the natural sciences. Our goal is not to be comprehensive but rather to provide an introduction to the topic that illuminates how contemporary discussions took their shape historically, especially through the work of Ernest Nagel, and limn the contours of concrete cases of reduction in specific natural sciences. To this end, we begin with a discussion of the unity of science and the impulse to accomplish compositional reduction in accord with a layer-cake vision of the sciences (Section 1). Next, we review Nagel's seminal contributions on theory reduction and how they strongly conditioned subsequent philosophical discussions (Section 2). Then we turn to detailed issues that arise from analyzing reduction in different sciences. First, we explore physical science by explicating different accounts of reduction (e.g., limit reduction and part-whole reduction) and probing their applicability to cases from condensed matter physics and quantum mechanics (Section 3). Second, we explore biological science by rehearsing how an antireductionist consensus grew out of the juxtaposition of genetics with a refined Nagelian view of theory reduction and then subsequently dissipated into a myriad of perspectives on explanatory reduction that apply across the life sciences, including the growth of mechanism approaches (Section 4). In conclusion, we argue that the epistemological heterogeneity and patchwork organization of the natural sciences encourages a pluralist stance about reduction (Section 5).

1 Unity of Science and Microreduction

A common thread running through different accounts of reduction is a concern with coherence between two or more domains. This has appeared frequently under the guise of unification, such as how different theories fit together and provide a more unified explanation. Whether or not this unification obtained seemed to speak to perennial questions: Are living systems anything over and above physical constituents with suitable organization? Are tables and chairs simply swarms of subatomic particles? Many logical empiricists of the mid-twentieth century officially bracketed such metaphysical questions and approached the issues indirectly. They focused on how these questions would be formulated within the framework of scientific theories: Is biology reducible to physics? Is the macroscopic behavior of objects reducible to the behavior of their microscopic constituents? The working assumption was that it was desirable to counterbalance the increasing specialization of the sciences with a meta-scientific study "promoting the integration of scientific knowledge" (Oppenheim and Putnam 1958, 3). This integration was advanced under the auspices of the "Unity of Science" as an organizing principle pertaining to "an ideal state of science" and "a pervasive trend within science, seeking the attainment of that ideal" (4). The "unity" invoked relied on a complex conception of reduction and a vision of reductionism, and the diversity of issues now seen in philosophical discussions can be understood as a slow disentangling of these interwoven claims.

Logical empiricists conceived of reduction as a form of progress: "The label 'reduction' has been applied to a certain type of progress in science . . . replacement of an accepted theory . . . by a new theory . . . which is in some sense superior to it" (Kemeny and Oppenheim 1956, 6–7). Thus the unity of science was underwritten by the progressive reduction of one theory to another (i.e., a form of reductionism). Successful reductions contribute to the goal of unification. The conception of reduction appealed to—"microreduction"—involves establishing explicit compositional relations between the objects in one science (e.g., molecules in chemistry) and its components in another science (e.g., subatomic particles in physics). This picture relies on a hierarchy of levels so that every whole at one level can be decomposed into constituents at a lower level: social groups, multicellular organisms, cells, molecules, atoms, and elementary particles. Theories of objects and their constituents were related in microreductions, which encourages a layer-cake view of the sciences as offering theories at corresponding levels of the compositional hierarchy: sociology, organismal biology, molecular biology, chemistry, and physics. Therefore a microreduction reduces one branch of science with a theory of objects in its domain (e.g., chemistry) to another branch of science with a theory of the constituents of those objects in its domain (e.g., physics).

This perspective sets out a clear philosophical agenda: characterize the nature of theory reduction so that assessments can be made of the success (or failure) of microreductions at different hierarchical levels with respect to the trend of attaining the ideal of unification (i.e., reductionism). Associated tasks include explicating the structure of scientific theories, translating the distinct vocabulary of one science into another, and spelling out the derivation of the higher-level science from the lower-level one. Given the predominant view of explanation as a logical derivation of an *explanandum* phenomenon from universal laws and initial conditions (Hempel and Oppenheim 1965[1948]), a successful microreduction was explanatory. Although it might not occur in practice ("whether or not unitary [completed, unified] science is ever attained"; Oppenheim and Putnam 1958, 4), the "in principle" vision was bracing: a fundamental theory of the ultimate constituents of the universe from which we can derive all the theories and laws that pertain to its more complex wholes. An epistemological consequence would be a reduced total number of laws, "making it possible, in principle, to dispense with the laws of [the higher-level science] and explain the relevant observations by using [the lower-level science]" (7).

There are a variety of different claims about reduction contained within this global vision of the unity of science based on microreduction: (a) reduction is a type of progress, (b) reduction is a relation between theories, (c) reduction involves logical derivation, (d) reduction is explanatory, and (e) reduction is compositional (part-whole). Much of the debate about reduction from the 1960s through the 1980s can be described as a disentangling and evaluation of these claims: (a) How should we understand scientific progress? Is it always reductionist in character? (b) What is the structure of theories? Are theories the only thing that can be reductively related? (c) Does reduction always involve logical derivation or should we understand it differently? (d) Is reduction always explanatory? How should we understand what counts as an explanation? (e) Is reduction always compositional or can it sometimes be, for example, causal?

Within the debates surrounding these questions there are at least three identifiable trends: (a) conceptualizing scientific progress required far more than the concepts of reduction developed by logical empiricists; (b) questions about the nature of theory reduction came to predominate through discussions of theory structure (Suppe 1977), whether theories relate logically or otherwise, the empirical inadequacy of a layer-cake view of the sciences, and an unraveling of the consensus about a deductive-nomological view of explanation (Woodward 2011); and, (c) increasing attention to examples from nonphysical sciences called into question whether reduction should be seen as a relation between theories and pertain to composition (Hüttemann and Love 2011; Kaiser 2012; Love and Hüttemann 2011). We ignore (a) in what follows and concentrate on (b) and (c) by reviewing Nagel's influential conception of theory reduction (Section 2), exploring the development of ideas about reduction and explanation in physical science (Section 3), and analyzing biological reasoning that challenged the applicability of theory reduction and encouraged different formulations of explanatory reduction (Section 4).

Oppenheim and other logical empiricists were already aware of a proliferation of notions of reduction: "the epistemological uses of the terms 'reduction', 'physicalism', 'Unity of Science', etc. should be carefully distinguished from the use of these terms in the present paper" (Oppenheim and Putnam 1958, 5). In order to increase clarity regarding the diversity of meanings for reduction and reductionism that have been treated, we offer the following coarse-grained classification scheme (Brigandt and Love 2012, Sarkar 1992).

Metaphysical reduction: This refers to theses like metaphysical reduction*ism* or physicalism, which claim that all higher-level systems (e.g., organisms) are constituted by and obtain in virtue of nothing but molecules and their interactions. Associated concepts include supervenience (no difference in a higher-level property without a difference in some underlying lower-level property), identity (each higher-level token entity is the same as an array of lower-level token entities), emergence (a higher-level entity is not reducible to its constituent molecules and their interactions), and downward causation (a higher-level entity has the capacity to bring about changes in lower-level entities, which is not reducible to its lower-level constituents).

Epistemological reduction: This refers to claims about representation and explanation or methodology. For representation and explanation, issues revolve around the idea that knowledge about higher-level entities can somehow be reduced to knowledge about lower-level entities. Dominant themes include what form the knowledge takes, such as whether it is theories structured in a particular way or other units (e.g., models or concepts), what counts as an explanation, and whether this is manifested similarly in different areas of science. For methodology, issues revolve around the most fruit-ful tactics for scientific investigation. Should experimental studies always be aimed at uncovering lower-level features of higher-level entities, such as by decomposing a complex system into parts (Bechtel and Richardson 1993)? Although a track record of success might suggest an affirmative answer, the exclusive use of reductionist research strategies may lead to systematic biases in data collection and explanatory models (Wimsatt 2007).

The remainder of our discussion focuses on epistemological reduction with special refer-ence to representation and explanation, beginning with Nagel's influential treatment.

2 NAGEL'S ACCOUNT OF REDUCTION

Debates about theory reduction concern the relation of two theories that have an overlapping domain of application. This is typically the case when one theory suc-ceeds another and thereby links reduction to progress, such as by requiring that the new theory predict or explain the phenomena predicted or explained by the old theory (Kemeny and Oppenheim 1956). Ernest Nagel's (1961, ch. 11) account of reduction was the most influential attempt to spell out this idea. It remains the shared background for any contemporary discussion of theory reduction. Nagel conceived of reduction as a special case of explanation: "the explanation of a theory or a set of experimental laws established in one area of inquiry, by a theory usually though not invariably formu-lated for some other domain" (1961, 338). Explanation was understood along the lines of the deductive-nomological model: "a reduction is effected, when the experimental laws of the secondary science . . . are shown to be logical consequences of the theoreti-cal assumptions . . . of the primary science" (352). Nagel gave two formal conditions that were necessary for the reduction of one theory to another. The first was the condition of *connectability*. If the two theories in question invoke different terminology, connections of some kind need to be established that link their terms. For example, "temperature" in thermodynamics does not appear in statistical mechanics. These connections became known as *bridge laws* and were frequently assumed to be biconditionals that express synthetic identities (e.g., the temperature of an ideal gas and the mean kinetic energy of the gas).

The second condition was *derivability*. Laws of the reduced theory must be logically deducible from laws of the reducing theory. A putative example is provided by the wave theory of light and Maxwell's theory of electromagnetism. Once it was estab-lished that light waves are electromagnetic radiation of a certain kind (a bridge law), *connectability* was fulfilled and the wave theory of light could be reduced to Maxwell's theory. Everything the wave theory had to say about the propagation of light could (seemingly) be deduced from Maxwell's theory plus the bridge law (*derivability*). The wave theory becomes a part of Maxwell's theory that describes the behavior of a certain kind of electromagnetic radiation (light waves), though expressed in a different terminology that can be mapped onto Maxwell's terminology (Sklar 1967). Provided this account is correct, the wave theory of light can be reduced to Maxwell's theory of electromagnetism. Thus successful "Nagelian reductions" absorb, embed, or integrate the old theory into the successor theory and reduce the number of independent laws or assumptions that are necessary to account for the phenomena (i.e., just those of the new theory). Although the successful predictions of the old theory concerning the phenomena are retained, one can dispense with its assumptions or laws, at least in principle.

Nagel and other logical empiricists were aware that their conception of reduction was a kind of idealization, but a number of key problems soon surfaced. The first was meaning incommensurability. If the meaning of a theoretical term is partially determined by its context, then terms in the old and new theories will have different meanings. Thus a prerequisite for the condition of derivability is violated (Feyerabend 1962). However, this criticism rested on the controversial assumption that the meaning of theoretical terms derived from the context of the entire theory (meaning holism). The second problem was the absence of bridge laws in the form of biconditionals, especially in biology and psychology. A third problem was the absence of well-structured theories with universal laws; Nagel's account cannot explicate reduction in sciences where theories and laws play a less significant role (see Section 4). Finally, the new theory typically does not recover the predictions of the old theory—and for good reasons, because the new theory often makes *better* predictions. This last objection was particularly influential in debates about theory reduction in physics (see Section 3).

Despite these problems, Nagel's account of reduction served as the primary reference point for almost all debates on reduction and reductionism. This is true for debates about whether chemistry can be reduced to physics (Weisberg, Needham, and Hendry 2011, Hendry and Needham 2007) and whether classical genetics can be reduced to molecular genetics (see Section 4). These debates followed a standard pattern. First, they assumed that if some sort of reductionist claim is true, then the pertinent concept of reduction was Nagelian. Second, it was discovered that Nagelian theory reduction failed to capture the relations between different theories, models, or representations in the field of science under scrutiny. As a consequence, various sorts of antireductionist or nonreductionist claims became popular, but it became apparent that such purely negative claims fail to do justice to the relations under investigation in the natural sciences, which encouraged the development of new conceptions of reduction.

Debates about reduction in philosophy of mind, which were tracked widely by philosophers who did not necessarily work on specific natural sciences, serve as an illustration of how Nagelian theory reduction shaped discussions. When it became apparent that claims about mental states, properties, or events could not be reduced *sensu* Nagel to neurobiology or allied sciences, various forms of nonreductive physicalism became popular. This in turn led to the development of alternative conceptions of reduction in order to characterize the relation between different forms of representations. For example, it has been argued that the essential desideratum in philosophy of mind is not Nagelian reduction but rather the explanation of bridge laws. What is needed is an explanation of *why* pain is correlated with or identical to the stimulation of certain nerve fibres, which is left unexplained in Nagelian reduction. So-called *functional reduction*, if achieved, would give us a reductive explanation of pain in terms of underlying physical or neurobiological features and thereby provide evidence for a metaphysical reduction of the mental to the physical (Chalmers 1996, Levine 1993, Kim 1998, 2005). Functional reduction consists of three steps: (1) the property M to be reduced is given a functional definition of the following form: having M = having some property P (in the underlying reduction domain) such that P performs causal task C; (2) the properties or mechanisms in the reduction domain that perform causal task C are discovered; and, (3) a theory that explains how the realizers of M perform task C is constructed (Kim 2005, 101–102).

This account of functional reduction is abstract and not well connected to the actual scientific practices of neurobiology or well anchored in empirical details of concrete examples. In this respect, the discussion in philosophy of mind differs from developments surrounding reduction in philosophy of science. This divergence in part explains why the revival of talk about mechanisms in neuroscience has been associated with a rejection of reduction (see Section 4).

3   REDUCTION IN PHYSICS

3.1   Amending Nagel's Model

When one theory succeeds another in an area of science, the theories typically make contradictory claims. This can be illustrated by what Newtonian mechanics (NM) and the special theory of relativity (STR) have to say about the dependence of momentum on velocity. Figure 22.1 depicts how, relative to some fixed observer, there is an increasing divergence in the predictions made by NM and STR as the ratio of the velocity of a particle to the velocity of light approaches 1. This divergence indicates that STR makes *better* predictions in certain domains, which is part of the reason why it was accepted as a successor of NM. But if NM and STR make contradictory claims, then they are logically incompatible; NM cannot be deduced from STR. Because STR makes better and contradictory predictions compared to NM, Nagelian theory reduction is incapable of adequately describing the reductive relations between STR and NM.

In the light of this objection, Kenneth Schaffner (1967, 1969, 1976, 1993) revised and developed the theory reduction framework into the general reduction model (GRM). Schaffner acknowledged that the old or higher-level theory typically could not be deduced from the succeeding or lower-level theory. However, he argued that a suitably corrected version of the higher-level theory should be the target of a deduction from the lower-level theory, assuming there are bridge laws that facilitate connectability.

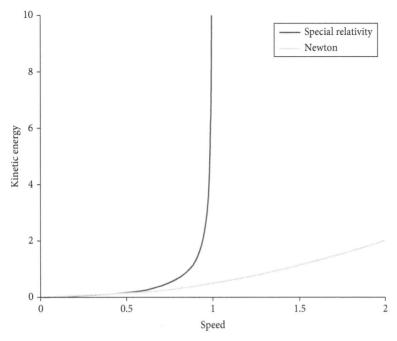

FIGURE 22.1 Kinetic energy as a function of the velocity of a particle relative to some observer (measured in v/c), as predicted by Newtonian mechanics and the special theory of relativity.

Attribution: By D. H (Own work using Gnuplot) [CC BY-SA 3.0 (http://creativecommons.org/licenses/by-sa/3.0)], via Wikimedia Commons (http://upload.wikimedia.org/wikipedia/commons/3/37/Rel-Newton-Kinetic.svg)

This corrected version of the higher-level theory needs to be *strongly analogous* to the original higher-level theory (Schaffner 1993, 429). Strong analogy or "good approxima-tion" (Dizadji-Bahmani et al. 2010) allows for some divergence in predictions, as well as some amount of meaning incommensurability, but the details of the analogical relation are left unspecified (Winther 2009). More recently, Dizadji-Bahmani and colleagues (2010) have defended a generalized Nagel-Schaffner account (GNS) where the higher-level theory is corrected and the lower-level theory is restricted by the introduction of boundary conditions and auxiliary assumptions. Only then are bridge laws utilized to logically deduce the former from the latter. The GNS account retains the Nagelian idea that reduction consists in subsumption via logical deduction, though it is now a *corrected* version of the higher-level theory that is reduced to a *restricted* version of the lower-level theory.

One might be skeptical about how well GRM or GNS capture actual cases of theory relations because, as Figure 22.1 illustrates, NM is not a good approximation of STR. GRM and GNS are limited to cases in which pairs of theories have largely overlapping domains of application and are simultaneously held to be valid (Dizadji-Bahmani et al. 2010). Most of the cases that Nagel and Schaffner had in mind do not fall in this range. Additionally, the authors' paradigm case—the reduction of thermodynamics to statistical mechanics—generates problems for this account (see Section 3.2). At best, Nagel's model of reduction, where a higher-level level theory is absorbed into a lower-level theory and cashed out in terms of logical deduction, only applies to a restricted number of theory pairs.

3.2 Limit-Case Reduction

Even though it is false that predictions from NM and STR are *approximately* the same, it is nevertheless true that *in the limit* of small velocities, the predictions of NM approximate those of STR. While Nagelian reduction does *not* apply in the case of NM and STR, physicists use a notion of reduction according to which STR reduces to NM under certain conditions (Nickles 1973).[1] Nickles refers to this latter concept as *limit-case reduction* in order to distinguish it from Nagel's concept of theory reduction. Limit-case reduction often permits one to skip the complexities of STR and work with the simpler theory of NM, given certain limiting conditions.

Limit-case reduction is very different from Nagelian reduction. Not only does it obtain in the converse direction (for Nagelian reduction, NM reduces to STR; for limit-case reduction, STR reduces to NM), but limit-case reduction is also a much weaker concept. Successful Nagelian reduction shows that the old theory can be *embedded* entirely in the new theory, whereas limit-case reduction focuses on two theories that make different predictions about phenomena that converge under special circumstances. Thus limit-case reduction is typically *piecemeal*; it might be possible for one pair of equations from STR and NM to be related by a limit-case reduction, while another pair of equations fails. Further, even though both accounts involve derivation, they differ on what is derived. On Nagel's account, the *laws* of the old or higher-level theory have to be logically deducible from the new theory. For limit-case reduction, the classical equation (as opposed to a particular value) is derived from the STR equation, but this sense of "derivation" refers to the process of obtaining a certain result by *taking a limit process*. So, strictly speaking, it is not the classical *equation* that is logically derived from STR but rather *solutions* of the new equations that are shown to coincide with solutions of the old equations in the limit, and solutions of the new equations are shown to differ from those of the old theory only minimally in the neighborhood of the limit (e.g., Ehlers 1997).

Limit-case reduction aims to explain the past success as well as the continued application of a superseded theory from the perspective of a successor theory. It is a *coherence* requirement that the successes of the old theory should be recoverable from the perspective of the new theory (Rohrlich 1988). Although there are many detailed studies of the relations between thermodynamics and statistical mechanics (e.g., Sklar 1993, Uffink 2007), NM and the general theory of relativity (e.g., Earman 1989, Ehlers 1986, Friedman 1986, Scheibe 1999), and classical mechanics and quantum mechanics (e.g.,

---

[1] "[I]n the case of (N)M and STR it is more natural to say that the *more* general STR reduces to the *less* general (N)M in the limit of low velocities. Epitomizing this intertheoretic reduction is the reduction of the Einsteinian formula for momentum, $p = m_0 v / \sqrt{(1 - (v/c)^2)}$, where m_0 is the rest mass, to the classical formula $p = m_0 v$ in the limit as $v \rightarrow 0$" (Nickles 1973, 182).

Scheibe 1999, Landsman 2007), we confine ourselves to some observations about the limit-process that are relevant for all of these cases.

1. If limit-case reductions aim to explain the past successes of an old theory, then not all limits are admissible. The equations of STR may reduce to those of NM in the limit $c \to \infty$ (i.e., the solutions of STR and NM coincide in the limit $c \to \infty$), but this kind of limit does not explain why the old theory was successful. As a matter of fact, c is constant and the *actual* success of the old theory cannot be accounted for in terms of solutions that converge only under *counterfactual* circumstances (Rohrlich 1988). For limit-case reduction to explain the success of the old theory and yield the coherence in question, the limits have to be specified in terms of parameters that can take different values in the actual world, such as v/c.

2. Limit processes presuppose a "topological stage" (Scheibe 1997). A choice of topology is required to define the limit because it assumes a concept of convergence. This choice could be nontrivial, and whether or not a pair of solutions counts as similar may depend on it. Some recent work focuses on developing criteria for this choice (Fletcher 2015).

3. Limit processes may involve idealizations. The "thermodynamic limit" assumes that the number of particles in a system goes to infinity. This issue has been discussed extensively with respect to phase-transitions and critical phenomena (Batterman 2000, 2002, 2011, Butterfield 2011, Morrison 2012, Norton 2012, Menon and Callender 2013). Thermodynamics typically describes systems in terms of macroscopic quantities, which often depend only on other macroscopic quantities, not the microphysical details. From the perspective of statistical mechanics, this can be explained only in the thermodynamic limit (i.e., for systems with an infinite size) because it is only in this limit that the sensitivity to microphysical details disappears. Explaining the nonfluctuating quantities in terms of infinite system size is an idealization because real systems have only finitely many constituents. However, this is unproblematic if one can approach the limit smoothly; that is, neighboring solutions for large yet finite systems differ minimally from solutions in the thermodynamic limit. Thus, in this case, statistical mechanics can explain why thermodynamic descriptions apply to large systems—the appeal to infinity is eliminable (Hüttemann et al. 2015).

Other cases may be more problematic for limit-case reduction. Thermodynamically, phase transitions and critical phenomena are associated with non-analyticities in a system's thermodynamic functions (i.e., discontinuous changes in a derivative of the thermodynamic function). Such non-analyticities cannot occur in *finite* systems as described by statistical mechanics because it allows for phase transitions only in *infinite* particle systems (see, e.g., Menon and Callender 2013). It has been argued that if the limit is singular, then solutions in the limit differ significantly from the neighboring (i.e., finite-system) solutions, and these fail to display phase transitions. Thus the appeal to infinity appears to be ineliminable for the explanation of the observed phase transitions (Batterman

2011), though this claim has been disputed (Butterfield 2011, Norton 2012, Menon and Callender 2013). The question then is: Under what conditions does the appeal to (infinite) idealizations undermine limit-case reduction? Disagreement partially depends on the conditions that successful reductions are supposed to fulfill. Does it suffice for the new theory to explain *the success* of the old theory, or should it explain the old theory itself, which may require the logical deduction of the equations of the old theory (Menon and Callender 2013)?

3.3 Reductive Explanations in Physics

It is sometimes argued that quantum mechanics and quantum entanglement in particular tell us that, "reductionism is dead . . . the total physical state of the joint system cannot be regarded as a consequence of the states of its (spatially separated) parts, where the states of the parts can be specified without reference to the whole" (Maudlin 1998, 54). This claim concerns neither Nagelian reduction nor limit-case reduction because it is not about pairs of theories. The claim is rather that, within one and the same theory (i.e., quantum mechanics), the state of the compound system cannot be explained in terms of the states of the parts. Maudlin's anti-reductionist claim does not concern the failure of a particular theory reduction but rather the failure of a kind of part-whole explanation.

In contrast to philosophy of biology (see Section 4), part-whole explanations have not been discussed in detail within philosophy of physics. One attempt to characterize part-whole explanations in physical science follows suggestions made by C. D. Broad (1925). According to this conception, a compound (whole) system's behavior can be explained by its parts if it can be explained in terms of (a) general laws concerning the behavior of the components considered in isolation, (b) general laws of composition, and, (c) general laws of interaction. Many macroscopic features like specific heat and the thermal or electrical conductivity of metals or crystals can be explained according to this model (Hüttemann 2004, 2005). Quantum entanglement is perhaps the most interesting case because the reductive explanation of the whole in terms of its parts clearly fails (Humphreys 1997, Hüttemann 2005, Maudlin 1998).

4 BIOLOGY

4.1 From Nagelian Theory Reduction to an Antireductionist Consensus

Historically, discussions of reductionism in the life sciences included extended arguments about vitalism, the claim that nonphysical or nonchemical forces govern biological systems. The story is more complex than a bald denial of physicalism, in part because what counts as "physicalism" and "vitalism" differ among authors and over time. Some late-eighteenth-century authors took a vitalist position that appealed to distinctly biological (i.e., natural) forces on analogy with NM, whereas organicists of the early twentieth century focused on organization as a nonreducible, system-level property of organisms. Many questions in the orbit of vitalism are reflected in contemporary discussions, but here we concentrate on how different aspects of reduction gained traction when the molecularization of genetics was juxtaposed with a revised form of Nagelian theory reduction. The manifold difficulties encountered in applying theory reduction to the relationship between classical and molecular genet-ics encouraged new approaches to explanatory reduction that were forged on a wide variety of biological examples where theories were much less prominent, especially within the ambit of different approaches to scientific explanation (e.g., mechanism descriptions).

A central motivation for Schaffner's refinement of the Nagelian model (see Section 2) was the putative success of molecular biology in reducing aspects of traditional fields of experimental biology to biochemistry. Although this was a work in progress, it was assumed that a logical derivation of classical genetics from a finished theory of biochemistry was in principle possible and would eventually be achieved. Schaffner's account and the case of genetics was the touchstone of discussions about reduc-tion in philosophy of biology for several decades. Most of the reaction was critical, spawning an "antireductionist consensus." Three core objections were leveled against Schaffner's GRM:

1. Molecular genetics appeared to be replacing classical genetics, which implied that the reductive relations among their representations were moot (Ruse 1971, Hull 1974). A suitably corrected version of the higher-level theory seemingly yields a different theory that has replaced classical genetics in an organic, theory-revision process (Wimsatt 2007, ch. 11).
2. Nagel and Schaffner assumed that a theory is a set of statements in a formal lan-guage with a small set of universal laws (Kitcher 1984). The knowledge of molec-ular genetics does not correspond to this type of theory structure, which called into question the presumed logical derivation required to accomplish a reduction (Culp and Kitcher 1989, Sarkar 1998).
3. GRM focused on formal considerations about reduction rather than substantive issues (Hull 1976, Sarkar 1998, Wimsatt 2007, ch. 11). This was made poignant in the acknowledgment that GRM was peripheral to the actual practice of molecular genetics (Schaffner 1974). If reduction is a logical relation between theories that is only "in principle" possible, why should we think GRM captures the progressive success of molecular genetics in relation to classical genetics?[2]

---

[2] Other formal approaches to theory structure also failed to capture the actual practices of scientific reasoning in genetics, thereby running afoul of the latter two objections (Balzer and Dawe 1986a, 1986b).

4.2 Models of Explanatory Reduction

These three objections—the difference between reduction and replacement in the context of theories changing over time, the mismatch between GRM theory structure assumptions and the knowledge practices of geneticists, and the gap between in principle formal problems and in practice substantive issues—collectively spurred new approaches to reduction in biology that were more sensitive to case studies of knowledge development, more empirically adequate with respect to the actual reasoning practices observed, and more responsive to substantive issues about reduction in practice (Kaiser 2011). These models of *explanatory* reduction differ from theory reduction in at least two salient ways: (a) they permit a variety of features as relata in reductions, such as subsets of a theory, generalizations of varying scope, mechanisms, and individual facts; and (b) they foreground a feature absent from the discussion of GRM, the idea that a reduction explains the whole in terms of its parts (Winther 2011).

One model of explanatory reduction that was animated by all three objections and exemplifies different reductive relata is the difference-making principle—gene differences cause differences in phenotypes (Waters 1990, 1994, 2000). Waters identifies this as a central principle of inference in both classical genetics and molecular genetics. An explanatory reduction is achieved between them because the causal roles of genes in instantiations of the inference correspond in both areas of genetics. Another model—explanatory heteronomy—requires that the *explanans* include biochemical generalizations, but the *explanandum* can be generalizations of varying scope, mechanisms, and individual facts that make reference to higher-level structures, such as cells or anatomy (Weber 2005). One of the most prominent models of explanatory reduction to emerge in the wake of theory reduction is mechanistic explanation (Darden 2006, Craver 2007, Bechtel 2011, Bechtel and Abrahamsen 2005, Glennan 1996), but whether it should be categorized as explanatory reduction is unclear (see Section 4.3).

Many models of explanatory reduction focus on how a higher-level feature or whole is explained by the interaction of its lower-level constituent parts. These approaches stress the importance of decomposing complex wholes into interacting parts of a particular kind (Kauffman 1971, Bechtel and Richardson 1993, Wimsatt 2007, ch. 9, Sarkar 1998). There is no commitment to the wholes and parts corresponding to different sciences or theories in order to achieve a compositional redescription or causal explanation of a higher-level state of affairs in terms of its component features (Wimsatt 2007, ch. 11, Hüttemann and Love 2011). These models avoid the basic objections facing GRM and fit well with the absence of clearly delineated theories in genetics, the emphasis on a whole being explained in terms of its parts in molecular explanations, and the piecemeal nature of actual scientific research. Molecular biology can offer reductive explanations despite the fact that many details are left our or remain unexplained.

4.3 Mechanistic Explanation and Reduction

Although several philosophers drew attention to the fact that biologists use the language of "mechanism" regularly and emphasize explanation in terms of decomposing a system into parts and then describing how these parts interact to produce a phenomenon (Kauffman 1971, Wimsatt 2007, ch. 9, 11), this was largely ignored because the conception of explanation diverged from the predominant deductive-nomological framework (Hempel and Oppenheim 1965[1948]). Explanatory power derived from laws in this framework, and their absence from a mechanism description meant they were interpreted as either temporary epistemic formulations or, to the degree that they were explanatory, reliant on "laws of working" (Glennan 1996, Schaffner 1993). These laws of working were presumed to be a part of a lower-level theory that would (in principle) reductively explain the features of higher-level entities.

One of the preeminent reasons offered for a mechanisms approach was its ubiquity in practice, both past and present (Darden 2006, Machamer, Darden, and Craver 2000). Reduction and replacement fail to capture the relations between classical and molecular genetics. These sciences deal with different mechanisms that occur at different points of time in the cell cycle—classical genetics focuses on meiosis, whereas molecular genetics focuses on gene expression—and involve different entities and activities, such as chromosomal behavior for classical genetics and nucleotide sequences for molecular genetics. Mechanisms approaches share the values of sensitivity to actual knowledge development, empirical adequacy with respect to scientific practices, and awareness of substantive rather than formal issues. Discussions about mechanistic explanation derive from attention to large areas of successful science, especially molecular biology and neurobiology, where standard conceptions of theory structure, explanation, and reduction seem ill suited to capture actual scientific practices. In this sense, they are motivated by the mismatch between GRM theory structure assumptions and the knowledge practices of geneticists: "these models do not fit neuroscience and molecular biology" (Machamer, Darden, and Craver 2000, 23).

Should we understand mechanisms as a variant on explanatory reduction? In one sense, the answer is "no" because of a stress on the multilevel character of mechanism descriptions. Instead of logical relations between two theories or levels, the entire description of the mechanism, which involves entities and activities operating at different levels, is required.[3]

And yet almost all approaches to mechanistic explanation share the idea of explaining by decomposing systems into their constituent parts, localizing their characteristic activities, and articulating how they are organized to produce a particular effect. Mechanistic explanations illustrate and display the generation of specific phenomena

---

[3] "Higher-level entities and activities are . . . essential to the intelligibility of those at lower levels, just as much as those at lower levels are essential for understanding those at higher levels. It is the integration of different levels into productive relations that renders the phenomenon intelligible and thereby explains it" (Machamer et al. 2000, 23).

by describing the organization of a system's constituent components and activities. Additionally, entities and activities at different levels bear the explanatory weight unequally, and it becomes important to look at abstraction and idealization practices involved in representing mechanisms (Brigandt 2013b, Levy and Bechtel 2013, Love and Nathan 2015). These show patterns of reasoning where some kinds of entities and activities are taken to be more explanatory than others. To the degree that these are lower-level features, a form of explanatory reduction may be occurring. Discussions of "bottoming out" are germane to sorting out this possibility. The existence of "components that are accepted as relatively fundamental" (Machamer et al. 2000, 13) provides a clear rationale for why biologists often label mechanistic descriptions as reductive. That one science takes restricted types of entities or activities as fundamental, and another science takes different types of entities or activities as fundamental, does not mean reduction is inapplicable.

Whether mechanistic explanation circumvents discussions of explanatory reduction is an open question (Craver 2005). A core reason for the difficulty in answering the question is that mechanisms approaches are sometimes advanced as a package deal; they not only avoid reduction*ism* but also provide a novel conception of how knowledge is structured and account for how scientific discovery operates. But once a shift has been made from theory reduction to explanatory reduction, many of the issues comprising Schaffner's GRM package become disaggregated. The crucial issue becomes characterizing reduction so as to better identify what assumptions are and are not being made about associated issues, such as theory structure or explanation (Sarkar 1998).

4.4 Standard Objections to Models of Theory and Explanatory Reduction

Although Schaffner's GRM faced a variety of specific objections, there are two standard objections to both theory and explanatory reduction that routinely arise: (a) context: the effects of lower-level entities and their interactions depend on the context in which they occur, which leads to one-many relations between lower-level features and higher-level features; and (b) multiple realization: higher-level features can be implemented by different kinds of lower-level features, so that many-one relations between lower-level features and higher-level features obtain. We cannot do justice to the complexity of these standard objections, but it is important to describe their main contours (see Brigandt and Love 2012).

Classical geneticists were aware of the fact that a phenotype is brought about by the interaction of several classical genes—the same allele may lead to two different phenotypes if occurring in two individuals with a different genotype (Waters 2004). There are many situations where the relationship between lower-level features and higher-level features is context dependent (Gilbert and Sarkar 2000, Hull 1974, Wimsatt 1979, Burian 2004). These different contexts include the developmental history, spatial region, and physiological state of a cell or organism. Reduction seems to fail because there is a one-many relation between lower-level and higher-level features, both compositionally and causally. The context of the organized whole is somehow primary in understanding the nature and behavior of its constituent parts.

Proponents of theory reduction have replied that a molecular reduction can take the relations of parts and the context of lower-level features into account. For example, one could specify the relevant context as initial conditions so that the higher-level features can be deduced in conjunction with premises about lower-level features. This strategy is subject to an objection plaguing Schaffner's GRM—scientists simply do not do this. The logical derivation of theory reduction would require that a representation of higher-level features be deduced from premises containing any and all of the lower-level context (internal or external) that is causally relevant. This may be possible in principle but not in practice. Models of explanatory reduction avoid this concern entirely. Explanations can highlight one among many causes, relegating everything else to the background, which is often held fixed in experimental studies. Explanations can appeal to a gene as a salient causal factor relative to a context even if the other genes involved in the phenotype are unknown and the cellular context of the gene has not yet been understood (Waters 2007). If biologists discover that the same mechanism produces different effects in distinct contexts, and only one of these effects is the target of inquiry, then the relevant aspects of the context can be included (Delehanty 2005). In this respect, models of explanatory reduction have a clear advantage over models of theory reduction.

Turning to multiple realization, the fact that higher-level features can be implemented by different kinds of lower-level features (many-one relations) also seems to challenge reductions. Knowledge of the lower-level features alone is somehow inadequate to account for the higher-level feature. For example, higher-level wholes can be composed of different lower-level component configurations or produced through causal processes that involve different interactions among lower-level components (Brigandt 2013a). Schaffner defends GRM against this objection by emphasizing that it is sufficient to specify one such configuration of lower-level features from which the higher-level feature can be derived (Schaffner 1976). This reply is inadequate because scientists usually attempt to explain types of higher-level phenomena rather than tokens; for example, not *this* instantiation of classical genetic dominance in pea plants but classical genetic dominance in sexually reproducing multicellular organisms. Token-token reduction may be possible but is relatively trivial epistemologically, whereas type-type theory reduction is empirically false due to multiple realization (Fodor 1974, 1997, Kimbrough 1978).

One inference that has been drawn from this result is that higher-level theories are justified in abstracting away from irrelevant variation in lower-level features to arrive at explanatory generalizations that involve natural kinds at higher levels (Kitcher 1984, Strevens 2009), though lower-level features may be informative for exceptions to a higher-level generalization. Models of explanatory reduction must deal with many-one relations between lower-level features and higher-level features, but it is a potentially manageable problem. One mitigation strategy is to challenge the commitment to unification that implies there is an explanatory loss in appealing to the multiply realized "gory details" (Waters 1990). Scientists find explanations of higher-level phenomena in terms of disjunctive lower-level types preferable in many cases. More generally, scrutinizing this lower-level heterogeneity facilitates a fine-grained dissection of compositional and causal differences that are otherwise inexplicable at the higher-level (Sober 1999). Thus models of explanatory reduction can manage multiple realizability objections through an emphasis on different explanatory virtues. Sometimes a lower-level explanation is better relative to one virtue (e.g., specificity), while a higher level explanation is preferable relative to another (e.g., generality).

4.5 Case Studies: Molecular, Developmental, and Behavioral Biology

Despite the emphasis on the relationship between classical and molecular genetics, philosophers of biology have analyzed reduction in other domains, such as evolutionary biology (Beatty 1990, Dupré 1993, Rosenberg 2006, Okasha 2006). Three brief examples from molecular, developmental, and behavioral biology are valuable in surfacing further conceptual issues relevant to models of explanatory reduction.

All models of explanatory reduction require representing the phenomena under investigation, just not always in terms of a theory (Sarkar 1998). Mathematical equations, scale miniatures, and abstract pictorial diagrams are examples. Every epistemological reduction involves a representation of the systems or domains to be related. Discussions surrounding the standard objections of one-many and many-one relationships in biological systems largely ignore questions related to representation, such as idealization or approximation, even though these have an impact on arguments about reduction. For example, hierarchical levels ("higher" and "lower") can be represented in different ways. Questions of one-many or many-one relations between different levels can be answered differently depending on how a hierarchy is represented (Love 2012). The decomposition of a system into parts depends on the principles utilized, such as function versus structure, and these can yield competing and complementary sets of part representations for the same system (Kauffman 1971, Wimsatt 2007, ch. 9, Bechtel and Richardson 1993, Winther 2011). Therefore, questions of representational choice and adequacy need to be addressed prior to determinations of whether reductive explanations succeed or fail. One can have successful reductions of features of complex wholes to constituent parts under some representations and simultaneously have failures of reduction under different representations, a situation that scientists methodologically exploit for the purpose of causal discovery (Wimsatt 2007, ch. 12).

In addition to hierarchy, another representational issue salient in biological explanations is temporality (Hüttemann and Love 2011). In most discussions of explanatory reduction, no explicit distinction has been drawn between compositional or spatial relations (arrangements) and causal or temporal relations (dynamics). Spatial composition questions have dominated, but biological models are frequently temporal if not explicitly causal (Schaffner 1993).[4] This is because a key aim is to explain how the organizational relations between parts change over time. Scientific explanations commonly invoke dynamic (causal) processes involving entities on several levels of organization (Craver and Bechtel 2007). Temporality takes on special significance in developing organisms where interactions over time among parts bring about new parts and new interactions (Parkkinen 2014). An orientation toward compositional reduction has encouraged an assumption of no change in the constituency of a whole being related to its parts.

Protein folding within molecular biology illustrates the importance of temporality (Love and Hüttemann 2011, Hüttemann and Love 2011). Functional proteins are folded structures composed of amino acid components linked together in a linear chain. If we ask whether the folded protein is mereologically composed of its amino acid parts given current representations in molecular biology, then we get an affirmative answer for an explanatory reduction with respect to composition. But if we ask whether the linear amino acid chain folds into a functional protein (a causal process with a temporal dimension) purely as consequence of its linked amino acid parts, then the answer is less clear. Empirical studies have demonstrated that other folded proteins (i.e., wholes) are required to assist in the proper folding of newly generated linear amino acid chains (Frydman 2001). The significance of temporality and dynamics is foregrounded precisely because the linked amino acid components alone are sufficient constitutionally but insufficient causally (Mitchell 2009), and the relations concern only molecular biological phenomena (as opposed to higher levels of organization, such as cells or anatomy).

Once the distinction between composition and causation is drawn, another issue related to representation becomes visible: reductive causal explanations that involve appeals to more than one type of lower-level feature (Love 2015, forthcoming). Explanations in developmental biology can be interpreted as reductive with respect to the difference-making principle (Waters 2007). Genetic explanations identify changes in the expression of genes and interactions among their RNA and protein products that lead to changes in the properties of morphological features during ontogeny (e.g., shape or size), while holding a variety of contextual variables fixed. Another type of reductive explanation invokes mechanical forces due to the geometrical arrangements of mesoscale materials, such as fluid flow (Forgacs and Newman 2005), which also can be interpreted as ontogenetic difference makers. Instead of preferring one reductive explanation to another or viewing them as competitors, many biologists seek to represent the combined dynamic of both types of lower-level features to reductively explain the manifestation of higher-level features of morphology: "an increasing number of examples point to the existence of a reciprocal interplay between expression of some developmental genes and the mechanical forces that are associated with morphogenetic movements" (Brouzés and Farge 2004, 372, Miller and Davidson 2013).

---

[4] A very different distinction utilizing time separates the historical succession of theories via reduction—diachronic reduction—from attempts to relate parts to wholes, such as in explanatory reduction or interlevel theory reduction —synchronic reduction (Rosenberg 2006, Dupré 1993).

Finding philosophical models for the explanatory integration of genetics and physics is an ongoing task, but the ability to represent these causal relations in temporal periodizations is a key element of explanatory practice (Love forthcoming). This type of situation was not recognized in earlier discussions because reduction was conceptualized in terms of composition rather than causation and as a two-place relation with a single, fundamental lower level. A reductive explanation of a higher-level feature in terms of two different kinds of lower-level features was unimagined (and ruled out) within theory reduction because of the layer-cake view of theories corresponding to distinct levels of organization. It was ignored in most models of explanatory reduction, which focused on dyadic relations between classical genetics and molecular genetics or morphology and molecules.

The possibility of reductive explanations involving both molecular genetics and physics is a reminder that we routinely observe the coordination of a multiplicity of approaches, some reductive and others not, in biological science. An illuminating example is the study of human behavioral attributes such as aggression and sexuality (Longino 2013). Several different approaches to these behaviors can be distinguished, such as quantitative behavioral genetics, social-environmental analysis, molecular behavioral genetics, and neurobiology. Some of these are reductive in the sense of identifying lower-level features (e.g., genetic differences) to account for higher-level features (i.e., behavioral differences); others are not (e.g., social-environmental differences). Reductive relationships exist among the approaches themselves, such as the relationship between quantitative behavioral genetics and molecular behavioral genetics. When examined closely, Longino shows that different approaches conceptualize the higher-level phenomenon of behavior differently (e.g., patterns of individual behavior or tendencies in a population) and parse the space of causal possibilities differently (e.g., allele pairs, neurotransmitter metabolism, brain structure, and parental income). Each approach, reductive or otherwise, is limited; there is no fundamental level in the science of behavior and no single hierarchy of parts and wholes in which to organize the approaches.[5] There can be multiple successes and failures of different kinds of reductive explanation within the study of aggression and sexuality, but these arise from different representational assumptions and explanatory standards—multiple concepts of reduction are required simultaneously.

5 CONCLUSION

A major theme emerging from the previous discussion is that the epistemological heterogeneity and patchwork organization of the natural sciences requires an array of

---

[5] "Each approach offers partial knowledge of behavioral processes gleaned by application of its investigative tools. In applying these tools, the overall domain is parsed so that effects and their potential causes are represented in incommensurable ways. We can (and do) know a great deal, but what we know is not expressible in one single theoretical framework" (Longino 2013, 144).

concepts to capture the diversity of asymmetrical, reductive relations found in the sciences, in addition to symmetrical, coordinative relations. This theme has been tracked in the growing specialization within philosophy of science, but it also has nurtured a growing rift between metaphysical questions about reductionism and epistemological questions about reduction. The "in practice" successes or failures of particular explanatory reductions do not yield straightforward building blocks for various projects in metaphysics that frequently demand more universal claims about the existence, availability, or desirability of reductions (i.e., forms of reductionism). A shift toward in practice considerations does not mesh tightly with metaphysical projects, such as deciding whether a higher-level feature is emergent and not reducible, and therefore the significance of debates about mechanistic explanation and models of explanatory reduction may appear irrelevant to topics of discussion in philosophy of mind or metaphysics.

We offer a procedural recommendation by way of a conclusion: given the existence of many senses of reduction that do not have straightforward interrelations, terminology such as "reductionist versus anti-reductionist" should be avoided. It is more perspicuous to articulate particular metaphysical and epistemological notions of reduction and then define acceptance or rejection of those notions or their failure or success in specific areas of science. This means it will be possible to argue for the success and failure of different types of reduction simultaneously within a domain of inquiry (Hüttemann and Love 2011). There is a strong rationale for talking about different kinds of reduction rather than in terms of a unified account of reduction or overarching dichotomies of reductionism versus anti-reductionism. Once we incorporate distinctions regarding different types of epistemological reduction (e.g., Nagel reduction, limit-case reduction, and part-whole reduction), the different interpretations of these types (e.g., the difference-making principle versus mechanisms as types of explanatory reduction), the different kinds of explanatory virtues that operate as standards (e.g., logical derivation, specificity, and generality), and the different kinds of representational features involved (spatial composition, causal relationships, or temporal organization), it is problematic to seek a single conception of reduction that will do justice to the diversity of phenomena and reasoning practices in the sciences. No global notion of reduction accurately characterizes what has occurred in the past or is currently happening in all areas of scientific inquiry. A pluralist stance toward reduction seems warranted (Kellert et al. 2006).

ACKNOWLEDGMENTS

Order of authorship is alphabetical. Andreas Hüttemann is supported in part by funding from the Deutsche Forschungsgemeinschaft (Research Group: Causation and Explanation [FOR 1063]). Alan Love is supported in part by a grant from the John Templeton Foundation (Integrating Generic and Genetic Explanations of Biological Phenomena, ID 46919). We would like to thank Paul Humphreys and Samuel Fletcher for helpful comments.

REFERENCES

Balzer, W., and Dawe, C. M. (1986a). "Structure and Comparison of Genetic Theories: (1) Classical Genetics." *The British Journal for Philosophy of Science* 37: 55–69.

Balzer, W., and Dawe, C. M. (1986b). "Structure and Comparison of Genetic Theories: (2) The Reduction of Character-Factor Genetics to Molecular Genetics." *The British Journal for Philosophy of Science* 37: 177–191.

Batterman, R. W. (2000). "Multiple Realizability and Universality." *The British Journal for Philosophy of Science* 51: 115–145.

Batterman, R. W. (2002). *The Devil in the Details: Asymptotic Reasoning in Explanation, Reduction, and Emergence* (Oxford: Oxford University Press).

Batterman, R. W. (2011). "Emergence, Singularities, and Symmetry Breaking." *Foundations of Physics* 41: 1031–1050.

Beatty, J. (1990). "Evolutionary Anti-Reductionism: Historical Reflections." *Biology and Philosophy* 5: 197–210.

Bechtel, W. (2011). "Mechanism and Biological Explanation." *Philosophy of Science* 78: 533–557.

Bechtel, W., and Abrahamsen, A. (2005). "Explanation: A Mechanist Alternative." *Studies in the History and Philosophy of Biology and Biomedical Sciences* 36: 421–441.

Bechtel, W., and Richardson, R. (1993). *Discovering Complexity: Decomposition and Localization as Strategies in Scientific Research* (Princeton, NJ: Princeton University Press).

Brigandt, I. (2013a). "Explanation in Biology: Reduction, Pluralism, and Explanatory Aims." *Science & Education* 22: 69–91.

Brigandt, I. (2013b). "Systems Biology and the Integration of Mechanistic Explanation and Mathematical Explanation." *Studies in History and Philosophy of Biological and Biomedical Sciences* 44: 477–492.

Brigandt, I., and Love, A. C. (2012). "Reductionism in Biology." In E. N. Zalta (ed.), *Stanford Encyclopedia of Philosophy*, http://plato.stanford.edu/entries/reduction-biology/.

Broad, C. D. (1925). *Mind and its Place in Nature* (London: Routledge & Kegan Paul).

Brouzés, E., and Farge, E. (2004). "Interplay of Mechanical Deformation and Patterned Gene Expression in Developing Embryos." *Current Opinion in Genetics & Development* 14: 367–374.

Burian, R. M. (2004). "Molecular Epigenesis, Molecular Pleiotropy, and Molecular Gene Definitions." *History and Philosophy of the Life Sciences* 26: 59–80.

Butterfield, J. (2011). "Less Is Different: Emergence and Reduction Reconciled." *Foundations of Physics* 41: 1065–1135.

Chalmers, D. (1996). *The Conscious Mind: In Search of a Fundamental Theory* (New York: Oxford University Press).

Craver, C. F. (2005). "Beyond Reduction: Mechanisms, Multifield Integration and the Unity of Neuroscience." *Studies in the History and Philosophy of Biological and Biomedical Sciences* 36: 373–395.

Craver, C. F. (2007). *Explaining the Brain: Mechanisms and the Mosaic Unity of Neuroscience* (New York: Oxford University Press).

Craver, C. F., and Bechtel, W. (2007). "Top-Down Causation Without Top-Down Causes." *Biology & Philosophy* 22: 547–563.

Culp, S., and Kitcher, P. (1989). "Theory Structure and Theory Change in Contemporary Molecular Biology." *The British Journal for the Philosophy of Science* 40: 459–483.

Darden, L. (2006). *Reasoning in Biological Discoveries: Essays on Mechanisms, Interfield Relations, and Anomaly Resolution* (New York: Cambridge University Press).

Delehanty, M. (2005). "Emergent Properties and the Context Objection to Reduction." *Biology & Philosophy 20*: 715–734.

Dizadji-Bahmani, F., Frigg, R., and Hartmann, S. (2010). "Who's Afraid of Nagelian Reduction?" *Erkenntnis 73*: 393–412.

Dupré, J. (1993). *The Disorder of Things: Metaphysical Foundations of the Disunity of Science* (Cambridge, MA: Harvard University Press).

Earman, J. (1989). *World Enough and Spacetime: Absolute versus Relational Theories of Space and Time* (Cambridge, MA: MIT Press).

Ehlers, J. (1986). "On Limit Relations Between and Approximate Explanations of Physical Theories." In B. Marcus, G. J. W. Dorn, and P. Weingartner (eds.), *Logic, Methodology and Philosophy of Science, VII* (Amsterdam: Elsevier), 387–403.

Ehlers, J. (1997). "Examples of Newtonian Limits of Relativistic Spacetimes." *Classical and Quantum Gravity 14*: A119–A126.

Feyerabend, P. K. (1962). "Explanation, Reduction and Empiricism." In H. Feigl and G. Maxwell (eds.), *Scientific Explanation, Space, and Time* (Minneapolis: University of Minnesota Press), 28–97.

Fletcher, S. (2015). "Similarity, Topology, and Physical Significance in Relativity Theory." *The British Journal for the Philosophy of Science.* doi: 10.1093/bjps/axu044

Fodor, J. A. (1974). "Special Sciences (Or: the Disunity of Sciences as a Working Hypothesis)." *Synthese 28*: 77–115.

Fodor, J. A. (1997). "Special Sciences: Still Autonomous After All These Years." In J.A. Tomberlin (ed.), *Mind, Causation and World–Philosophical Perspectives,* Vol. 11 (Oxford: Blackwell), 149–163.

Forgacs, G., and Newman, S. A. (2005). *Biological Physics of the Developing Embryo* (New York: Cambridge University Press).

Friedman, M. (1986). *Foundations of Space-Time Theories: Relativistic Physics and Philosophy of Science* (Princeton, NJ: Princeton University Press).

Frydman, J. (2001). "Folding of Newly Translated Proteins in Vivo: The Role of Molecular Chaperones." *Annual Review of Biochemistry 70*: 603–647.

Gilbert, S. F., and Sarkar, S. (2000). "Embracing Complexity: Organicism for the 21st Century." *Developmental Dynamics 219*: 1–9.

Glennan, S. (1996). "Mechanisms and the Nature of Causation." *Erkenntnis 44*: 49–71.

Hempel, C. G., and Oppenheim, P. (1965 [1948]). "Studies in the Logic of Explanation." In *Aspects of Scientific Explanation and other Essays in the Philosophy of Science* (New York: Free Press), 245–290.

Hendry, R., and Needham, P. (2007). "Le Poidevin on the Reduction of Chemistry." *The British Journal for the Philosophy of Science 58*: 339–353.

Hull, D. L. (1974). *Philosophy of Biological Science* (Englewood Cliffs, NJ: Prentice-Hall).

Hull, D. L. (1976). "Informal Aspects of Theory Reduction." *PSA: Proceedings of the Biennial Meeting of the Philosophy of Science Association 1974*: 653–670.

Humphreys, P. (1997). "How Properties Emerge." *Philosophy of Science 64*: 1–17.

Hüttemann, A. (2004). *What's Wrong with Microphysicalism?* (London: Routledge).

Hüttemann, A. (2005). "Explanation, Emergence, and Quantum Entanglement." *Philosophy of Science 72*: 114–127.

Hüttemann, A., Kühn, R., and Terzidis, O. (2015). "Stability, Emergence and Part-Whole-Reduction." In *Why More Is Different: Philosophical Issues in Condensed Matter Physics and Complex Systems*, edited by, 169–200. Dordrecht: Springer.

Hüttemann, A., and Love, A. C. (2011). "Aspects of Reductive Explanation in Biological Science: Intrinsicality, Fundamentality, and Temporality." *The British Journal for Philosophy of Science* 62: 519–549.

Kaiser, M. I. (2011). "The Limits of Reductionism in the Life Sciences." *History and Philosophy of the Life Sciences* 33(4): 453–476.

Kaiser, M. I. (2012). "Why It Is Time to Move beyond Nagelian Reduction." In D. Dieks, W.J. Gonzalez, S. Hartmann, M. Stöltzner, and M. Weber (eds.), *Probabilities, Laws, and Structures* (Berlin: Springer), 245–262.

Kauffman, S. A. (1971). "Articulation of Parts Explanations in Biology and the Rational Search for Them." *PSA: Proceedings of the Biennial Meeting of the Philosophy of Science Association* 1970 8: 257–272.

Kellert, S. H., Longino, H. E., and Waters, C. K. (2006). "Introduction: The Pluralist Stance." In S. H. Kellert, H. E. Longino and C. K. Waters (eds.), *Scientific Pluralism* (Minneapolis: University of Minnesota Press), vii–xxix.

Kemeny, J. G., and Oppenheim, P. (1956). "On Reduction." *Philosophical Studies* 7: 6–19.

Kim, J. (1998). *Mind in a Physical World* (Cambridge, MA: MIT Press).

Kim, J. (2005). *Physicalism, or Something Near Enough* (Princeton, NJ: Princeton University Press).

Kimbrough, S. O. (1978). "On the Reduction of Genetics to Molecular Biology." *Philosophy of Science* 46: 389–406.

Kitcher, P. (1984). "1953 and All That: A Tale of Two Sciences." *Philosophical Review* 93: 335–373.

Landsman, N. P. (2007). "Between Classical and Quantum." In J. Earman and J. Butterfield (eds.), *Handbook of the Philosophy of Physics* (Amsterdam: Elsevier), 417–553.

Levine, J. (1993). "On Leaving Out What It's Like." In M. Davies and G. W. Humphreys (eds.), *Consciousness: Psychological and Philosophical Essays* (Oxford: Blackwell), 121–136.

Levy, A., and Bechtel, W. (2013). "Abstraction and the Organization of Mechanisms." *Philosophy of Science* 80: 241–261.

Longino, H. E. (2013). *Studying Human Behavior: How Scientists Investigate Aggression and Sexuality* (Chicago: University of Chicago Press).

Love, A. C. (2012). "Hierarchy, Causation and Explanation: Ubiquity, Locality and Pluralism. " *Interface Focus* 2: 115–125.

Love, A. C. (2015). "Developmental Biology." In E. N. Zalta (ed.), *Stanford Encyclopedia of Philosophy*, http://plato.stanford.edu/entries/biology-developmental/.

Love, A. C. (forthcoming). "Combining Genetic and Physical Causation in Developmental Explanations." In C. K. Waters and J. Woodward (eds.), *Causal Reasoning in Biology* (Minneapolis: University of Minnesota Press).

Love, A. C., and Hüttemann, A. (2011). "Comparing Part-Whole Reductive Explanations in Biology and Physics." In D. Dieks, J. G. Wenceslao, S. Hartmann, T. Uebel, and M. Weber (eds.), *Explanation, Prediction, and Confirmation* (Berlin: Springer), 183–202.

Love, A. C., and Nathan, M. D. (2015). "The Idealization of Causation in Mechanistic Explanation." *Philosophy of Science* 82: 761–774.

Machamer, P., Darden, L., and Craver, C. F. (2000). "Thinking about Mechanisms." *Philosophy of Science* 67: 1–25.

Maudlin, T. (1998). "Part and Whole in Quantum Mechanics." In E. Castellani (ed.), *Interpreting Bodies* (Princeton, NJ: Princeton University Press), 46–60.

Menon, T., and Callender, C. (2013). "Turn and Face the Strange ... Ch-ch-changes: Philosophical Questions Raised by Phase Transitions." In R. Batterman (ed.), *The Oxford Handbook of Philosophy of Physics* (Oxford: Oxford University Press), 189–223.

Miller, C. J., and Davidson, L. A. (2013). "The Interplay Between Cell Signalling and Mechanics in Developmental Processes." *Nature Reviews Genetics* 14: 733–744.

Mitchell, S. D. (2009). *Unsimple Truths: Science, Complexity, and Policy* (Chicago and London: University of Chicago Press).

Morrison, M. (2012). "Emergent Physics and Micro-Ontology." *Philosophy of Science* 79: 141–166.

Nagel, E. (1961). *The Structure of Science: Problems in the Logic of Scientific Explanation* (New York: Harcourt, Brace & World).

Nickles, T. (1973). "Two Concepts of Intertheoretic Reduction." *Journal of Philosophy* 70: 181–201.

Norton, J. (2012). "Approximation and Idealization: Why the Difference Matters." *Philosophy of Science* 79: 207–232.

Okasha, S. (2006). *Evolution and the Levels of Selection* (New York: Oxford University Press).

Oppenheim, P., and Putnam, H. (1958). "Unity of Science as a Working Hypothesis." In H. Feigl, M. Scriven, and G. Maxwell (eds.) *Concepts, Theories, and the Mind-Body Problem* (Minneapolis: University of Minnesota Press), 3–36.

Parkkinen, V-P. (2014). "Developmental Explanations." In M. C. Galavotti, D. Dieks, W. J. Gonzalez, S. Hartmann, T. Uebel, and M. Weber (eds.), *New Directions in the Philosophy of Science: The Philosophy of Science in a European Perspective*, Vol. 5 (Berlin: Springer), 157–172.

Rohrlich, F. (1988). "Pluralistic Ontology and Theory Reduction in the Physical Sciences." *The British Journal for the Philosophy of Science* 39: 295–312.

Rosenberg, A. (2006). *Darwinian Reductionism: Or, How to Stop Worrying and Love Molecular Biology* (Chicago: University of Chicago Press).

Ruse, M. (1971). "Reduction, Replacement, and Molecular Biology." *Dialectica* 25: 39–72.

Sarkar, S. (1992). "Models of Reduction and Categories of Reductionism." *Synthese* 91: 167–194.

Sarkar, S. (1998). *Genetics and Reductionism* (Cambridge, UK: Cambridge University Press).

Schaffner, K. F. (1967). "Approaches to Reduction." *Philosophy of Science* 34: 137–147.

Schaffner, K. F. (1969). "The Watson-Crick Model and Reductionism." *The British Journal for the Philosophy of Science* 20: 325–348.

Schaffner, K. F. (1974). "The Peripherality of Reductionism in the Development of Molecular Biology." *Journal of the History of Biology* 7: 111–139.

Schaffner, K. F. (1976). "Reductionism in Biology: Prospects and Problems." *PSA: Proceedings of the Biennial Meeting of the Philosophy of Science Association* 1974 32: 613–632.

Schaffner, K. F. (1993). *Discovery and Explanation in Biology and Medicine* (Chicago: University of Chicago Press).

Scheibe, E. (1997). *Die Reduktion Physikalischer Theorien: Teil I, Grundlagen und Elementare Theorie* (Berlin: Springer).

Scheibe. E. (1999). *Die Reduktion Physikalischer Theorien: Teil II, Inkommensurabilitat und Grenzfallreduktion* (Berlin: Springer).

Sklar, L. (1967). "Types of Inter-Theoretic Reduction." *The British Journal for Philosophy of Science* 18: 109–124.

Sklar, L. (1993). *Physics and Chance: Philosophical Issues in the Foundations of Statistical Mechanics* (New York: Cambridge University Press).

Sober, E. (1999). "The Multiple Realizability Argument against Reductionism." *Philosophy of Science* 66: 542–564.

Strevens, M. (2009). *Depth: An Account of Scientific Explanation* (Cambridge, MA: Harvard University Press).

Suppe, F., ed. (1977). *The Structure of Scientific Theories,* 2nd ed. (Urbana: University of Illinois Press).

Uffink, J. (2007). "Compendium of the Foundations of Classical Statistical Physics." In J. Earman and J. Butterfield (eds.), *Handbook of the Philosophy of Physics* (Amsterdam: Elsevier), 923–1074.

Waters, C. K. (1990). "Why the Antireductionist Consensus Won't Survive the Case of Classical Mendelian Genetics." *PSA: Proceedings of the Biennial Meeting of the Philosophy of Science Association,* Vol. 1: Contributed Papers: 125–139.

Waters, C. K. (1994). "Genes Made Molecular." *Philosophy of Science* 61: 163–185.

Waters, C. K. (2000). "Molecules Made Biological." *Revue Internationale de Philosophie* 4: 539–564.

Waters, C. K. (2004). "What Was Classical Genetics?" *Studies in the History and Philosophy of Science* 35: 783–809.

Waters, C. K. (2007). "Causes that Make a Difference." *Journal of Philosophy* 104: 551–579.

Weber, M. (2005). *Philosophy of Experimental Biology* (New York: Cambridge University Press).

Weisberg, M., Needham, P., and Hendry, R. (2011). "Philosophy of Chemistry." In E. N. Zalta (ed.), *Stanford Encyclopedia of Philosophy,* http://plato.stanford.edu/entries/chemistry/.

Wimsatt, W. C. (1979). "Reductionism and Reduction." In P. D. Asquith and H. E. Kyburg (eds.), *Current Research in Philosophy of Science* (East Lansing, MI: Philosophy of Science Association), 352–377.

Wimsatt, W. C. (2007). *Re-Engineering Philosophy for Limited Beings: Piecewise Approximations to Reality* (Cambridge, MA: Harvard University Press).

Winther, R. G. (2009). "Schaffner's Model of Theory Reduction: Critique and Reconstruction." *Philosophy of Science* 76: 119–142.

Winther, R. G. (2011). "Part-Whole Science." *Synthese* 178: 397–427.

Woodward, J. (2011). "Scientific Explanation." In E. N. Zalta (ed.), *Stanford Encyclopedia of Philosophy,* http://plato.stanford.edu/archives/win2011/entries/scientific-explanation/.

CHAPTER 23

....

SCIENCE AND NON-SCIENCE

....

SVEN OVE HANSSON

1 INTRODUCTION

....

DISCUSSIONS about the definition or delimitation of science are often couched in terms of its "demarcation." Moreover the phrases "demarcation of science" and "demarcation of science from pseudoscience" are in practice taken to be synonymous. But this is an oversimplified picture. Everything that can be confused with science is not pseudoscience, and science has nontrivial borders for instance to metaphysics, ethics, religion, the arts, technology, and many types of nonscientific systematized knowledge. They give rise to a wide variety of area-specific boundary issues, such as where to draw the borders between religious studies and confessional theology, between political economics and economic policy, between musicology and practical musicianship, and between gender studies and gender politics.

In this chapter the term *nonscientific* is used to cover everything that is not science. The term *unscientific* tends to imply some form of contradiction or conflict with science. *Antiscientific* and *pseudoscientific* both convey a strong disagreement with science, but they refer to different types of such disagreements. An antiscientific statement is inimical to science, whereas a pseudoscientific statement professes adherence to science but misrepresents it. *Quasi-scientific* is very close in meaning to pseudoscientific.

2 THE CONCEPT OF SCIENCE

....

Classifications of human knowledge have a long history. In particular, the typology of knowledge was a popular theme among learned writers in the Middle Ages. A large number of classification schemes have survived. They usually organized disciplines in groups and subgroups, thus giving rise to a tree-like structure. Such *divisiones scientiarum* (*divisiones philosophiae*) served to identify areas deemed worthy of scholarly efforts and/or suitable for inclusion in educational curricula (Ovitt 1983; Dyer 2007). But no agreement was reached on the overall term to be used to cover all types of knowledge, and there is no sign that much importance was attached to the choice of such a term. *Scientia* (science), *philosophia* (philosophy), and *ars* (art) were all common for that purpose.

Etymologically one might expect a clear distinction between the three terms. *Scientia* is derived from *scire* (know), which was used primarily about knowledge of facts. *Philosophia* is of Greek origin and literally means "love of wisdom," but it was often interpreted as systematic knowledge and understanding in general, covering both empirical facts and more speculative topics such as existence and morality. Cicero influentially defined it as "the knowledge of things human and divine and of the causes by which those things are controlled" (*c.* 44 BC/1913: 2.5.). *Ars* refers to skills, abilities, and craftsmanship. It did not have the special association with aesthetic activities that our term "art" has. The arts emphasized in most knowledge classifications were the liberal arts, a collection of skills that had been identified already in antiquity as suitable for free men and for being taught in schools. (Chenu 1940; Tatarkiewicz 1963: 233). Since the early Middle Ages, the liberal arts were divided into two groups: the trivium, which consisted of logic, rhetoric, and grammar, and the quadrivium, which consisted of arithmetic, geometry, astronomy, and music. However, when the term *ars* was used in knowledge classifications, it usually included other arts that were not parts of the trivium or quadrivium.

All three terms, *scientia*, *philosophia*, and *ars*, were used interchangeably as umbrella terms for all knowledge. Some authors used *scientia* as the most general term and *philosophia* as a second-level term to denote some broad category of knowledge disciplines. Others did it exactly the other way around, and still others used *scientia* and *philosophia* as synonyms. Similarly, *ars* was sometimes used to cover all the disciplines and sometimes to cover some broad subcategory of them. This terminological confusion persisted well into the sixteenth and seventeenth centuries (Ovitt 1983; Freedman 1994; Covington 2005). For a modern reader, it may be particularly surprising to find that in the Middle Ages, "philosophy" included all kinds of knowledge, as well as practical craftsmanship. Starting at the end of the fifteenth century it became common to exclude the crafts (the mechanical arts) from philosophy, but as late as in the eighteenth century the word "philosophy" was commonly used to denote all kinds of knowledge (Tonelli 1975; Freedman 1994).

The English word "science" derives from *scientia* and originally had an equally wide meaning. It could refer to almost anything that one has to learn in order to master it—including everything from scholarly learning to sewing to horseback riding. In the 1600s and 1700s the meaning of the term was restricted to systematic knowledge, and during the 1800s it was further restricted to denote the new, more empirical type of knowledge that was then emerging in the academic area previously called "natural philosophy" (Layton 1976). The word still has a strong association with the study of nature. For instance, a science class is self-evidently taken to be devoted to natural science. But today the term refers not only to the disciplines investigating natural phenomena but also to those studying individual human behavior and some of those studying human societies and human creations. Several of the latter disciplines are not counted as sciences and instead are referred to as *humanities*. Hence, according to the conventions of the English language, political economy is a science (one of the social sciences) but classical philology and art history are not.

The closest synonym to "science" in German, *Wissenschaft*, also originally meant "knowledge." However, *Wissenschaft* has a much broader meaning than "science." It includes all the academic specialties, including the humanities. The same applies to the corresponding words in Dutch and the Nordic languages, such as *wetenschap* (Dutch), *vitenskap* (Norwegian), and *vetenskap* (Swedish). This linguistic divergence gives rise to a fundamental issue that we must resolve before considering the details of how to distinguish between science and non-science: Is it the outer limits of that which is called "science" in English or that which is called *Wissenschaft* in German that we should characterize? The fact that English is the dominant academic language does not contribute much to solving this issue. We should look for a delimitation that is based on important epistemic criteria. In my view, this will lead us to use a concept whose delimitation is closer to the German term, which includes the humanities, than to the English term, which excludes them. This is because the sciences and humanities—the *Wissenschaften*—form a community of knowledge disciplines that is unified by epistemic criteria (Hansson 2007b). In particular they have four important characteristics in common.

First, they are united by all taking as their primary task to provide *the most reliable information* currently obtainable in their respective subject areas. The reliability referred to here is intersubjective, as is the knowledge required to be that is produced in the knowledge disciplines. In this it differs from statements arising in many other types of human discourse. For instance, when discussing religious worldviews or aesthetic standpoints one may arguably be justified in saying, "That may be true for you, but this is true for me." In the sciences and humanities, such relativity to individuals is not accepted. Neutrinos do not have mass for one but not the other. The Bali tiger is not extinct for some but not for others. Immanuel Kant was not born in 1724 for one person and in some other year for another. (Of course in all three cases this obtains only provided that we use the key terms concordantly; for instance in the last example we must use the same calendar). This means that science offers us a common worldview or, perhaps more precisely, common parts of our worldviews. In order to achieve this common worldview, scientific knowledge must be intersubjectively reliable (i.e., it must be reliable for all of us). Ultimately it has to strive for knowledge that corresponds as well as possible (given the limits of our linguistic means) to how things really are. This means that its (elusive) aim is to obtain objective knowledge (Douglas 2004).

Second, the knowledge disciplines are characterized by a strong drive for *improvement through critical appraisals and new investigations*. They do not claim to have certain knowledge in their respective domains. Instead, their claim is to possess both the best (but imperfect) knowledge that is currently available and the best means to improve it. What makes the knowledge disciplines superior to their rivals is not infallibility—which they do not possess—but strong internal mechanisms for uncovering and correcting their own mistakes. The resulting improvements can reach their very foundations, not least in methodological issues.

Third, the knowledge disciplines are connected by an informal but nevertheless well worked-out division of intellectual labor and by *mutual respect for each other's competences*. For instance, biologists studying animal movement rely on the mechanical concepts and theories developed by physicists. Astronomers investigating the composition of interstellar matter rely on information from chemistry on the properties of molecules, and so on. This mutual reliance is equally self-evident across the supposed barrier between the sciences (in the traditional, limited sense) and the humanities. An astronomer wishing to understand ancient descriptions of celestial phenomena has to rely on the interpretations of these texts made by classical scholars; a historian must ask medical scientists for help in identifying the diseases described in historical documents; art historians turn to chemists for the identification of pigments used in a painting and to physicists and chemists for radiocarbon dating of its canvas; and so on.

Fourth, there is a strong and rapidly growing *interdependence* among the knowledge disciplines. Two hundred years ago, physics and chemistry were two independent sciences with only a few connections. Today they are closely knit together not least by integrative subdisciplines such as physical chemistry, quantum chemistry, and surface science. The interconnections between biology and chemistry are even stronger. The interdependencies between natural sciences and the humanities are also rapidly growing. Game theory and the neurosciences are bringing biology and the social sciences closer to each other. Recently, methods and concepts from studies of biological evolution (such as the serial founder effect) have been applied successfully to the development of human societies and to the development of languages tens of thousands of years before the written evidence (Henrich 2004; Pagel, Atkinson, and Meade 2007; Lycett and von Cramon-Taubadel 2008; Atkinson 2011). These and many other bonds between the natural sciences and the humanities have increased dramatically in the half century that has passed since C. P. Snow's (1959/2008) pessimistic prediction of a widening gap between them.

The communality among the knowledge disciplines can also be expressed in terms of their value commitments. Robert K. Merton (1942/1973) summarized the ethos of science in the form of four institutional imperatives. *Universalism* demands that truth claims are evaluated according to preestablished, impersonal criteria. Their acceptance or rejection should not depend on the personal or social qualities of their protagonists. According to *communality* (infelicitously called "communism" by Merton), the findings of science belong to the community and not to individuals or groups. *Disinterestedness* means that institutional controls curb the effects of personal or ideological motives that individual scientists may have. Finally, *organized skepticism* requires that science allows detached scrutiny of beliefs that are supported by other institutions such as religious and political organizations. These values are crucial for creating an intellectual environment of collective rationality and mutual criticism (Settle 1971). Importantly, they apply equally to the humanities and the sciences.

For lack of a better term, I use the phrase "science(s) in a broad sense" to refer to the community of knowledge disciplines. The name is not important, but it is important to rid ourselves of the increasingly irrelevant distinction between the sciences and the humanities and base our analysis on more epistemically relevant categories.

3 PRACTICAL ARTS

One of the most widespread types of nonscientific knowledge is the knowledge that we all have about how to achieve things in practical life. This includes trivial knowledge, such as how to open a door and make coffee. It also includes advanced knowledge that takes years to master and is typically part of the special competences of a profession. Aristotle referred to such activities as productive arts and defined them as the making of things "whose origin is in the maker and not in the thing made; for art is concerned neither with things that are, or come into being, by necessity, nor with things that do so in accordance with nature (since these have their origin in themselves)" (350 BC/1980: VI: 4).

The productive arts included most of the activities undertaken for a living by the lower and middle classes. In antiquity such arts were contrasted with the liberal arts and called illiberal, vulgar, sordid, or banausic (*artes illiberales, artes vulgares, artes sordidae, artes banausicae*). These were all derogative terms, indicating the inferior social status of these activities and reflecting a contemptuous view of manual work that was predominant in classical Greece (Van Den Hoven 1996: 90–91; Ovitt 1983; Tatarkiewicz 1963; Whitney 1990). In the Middle Ages, the most common term was "mechanical arts" (*artes mechanicae*), which was introduced by Johannes Scotus Eriugena (*c.* 815–*c.* 877) in his commentary on Martianus Capella's allegorical text on the liberal arts, *De nuptiis Philologiae et Mercurii* (On the Marriage of Philology and Mercury; Noble 1997).

Today, medieval treatises on mechanical arts are often seen as precursors of later discussions of technology. The modern meaning of the word "mechanical" contributes to this interpretation. However, although the word has its origins in a Greek root that relates to machines, in the Middle Ages it acquired the meaning "concerned with manual work; of the nature of or relating to handicraft, craftsmanship, or artisanship." The old sense of "relating to machines" had disappeared by this period, and it is probably a learned reconstruction from the sixteenth century (*Oxford English Dictionary*).

A famous list of seven mechanical arts (or, more accurately, groups of arts) was provided in the late 1120s by Hugh of Saint Victor:

1. *lanificium*: weaving, tailoring
2. *armatura*: masonry, architecture, warfare
3. *navigatio*: trade on water and land
4. *agricultura*: agriculture, horticulture, cooking
5. *venatio*: hunting, food production

6. *medicina*: medicine and pharmacy
7. *theatrica*: knights' tournaments and games, theatre

(Hoppe 2011: 40–41)

The reason Hugh summarized all the practical arts under only seven headings was obviously to achieve a parallel with the seven liberal arts. Hugh emphasized that, just like the liberal arts, the mechanical ones could contribute to wisdom and blessedness. He also elevated their status by making them one of four major parts of philosophy (the others were theoretical, practical, and logical knowledge; Weisheipl 1965: 65). After Hugh it became common (but far from universal) to include the mechanical arts in classifications of knowledge (Dyer 2007).

Hugh's list confirms that the concept of practical arts was much wider than our modern concept of technology. Only about half of the items on his list would be classified as technology today. Warfare, trade, hunting, medicine, games, and theatre playing are the clearest examples of items not so classified.

The distinction between liberal and mechanical arts continued to be in vogue in the early modern era. It had an important role in the great French *Encyclopédie* (published 1751–1772) that pioneered the incorporation of the mechanical arts into the edifice of learning. In the preface Jean Le Rond d'Alembert (1717–1783) emphasized that the mechanical arts were no less worthy pursuits than the liberal ones (d'Alembert 1751: xiij).

Medieval texts on the mechanical arts are often treated as beginnings of a discussion of the science–technology relationship, but they were in fact concerned with the much more general issue of the relationship between theoretical and practical knowledge. One interesting example is a discussion by the English philosopher Robert Kilwardby (1215–1279). He emphasized that a distinction must be made between science in a broad sense (which he called "speculative philosophy") and the practical skills, but he also pointed out that they are dependent on each other in a fundamental way. He said, "the speculative sciences are practical and the practical ones are speculative" (quoted in Whitney 1990: 120).

In his *Opera Logica* (1578) the Italian philosopher Jacopo Zabarella (1533–1589) discussed the same issue but reached a different conclusion. In his view, the productive arts can learn from science but not the other way around (Mikkeli 1997: 222). The topic discussed by these scholars seems to have been lost in modern philosophy. However, the German historian of technology Otto Mayr has proposed that research should be conducted on "historical interactions and interchanges between what can roughly be labelled 'theoretical' and 'practical' activities, that is, between man's investigations of the laws of nature and his actions and constructions aimed at solving life's material problems" (1976: 669). This topic is interesting not only historically; it has philosophical aspects that are well worth investigating. Such investigations would amount to a resumption of a topic that has largely been lost, perhaps in part for terminological reasons. It can be seen as a double extension of the current discussion of the relationship between science and technology, involving both an extension from the standard to the broad concept of science and an extension from technology to the much broader notion of practical arts. We can also describe it as the philosophical investigation of relationships between human knowledge and human activity. It includes, but is not restricted to, the relation between theoretical and practical rationality.

Today the relationship between practical and theoretical knowledge is more complex than in medieval times. One important reason for this is that practical activities are now based to a large degree on science. To mention just one example, clinical medicine is increasingly science based (Hansson 2014b). We therefore must distinguish between those parts or aspects of the practical arts that are science based and those that are not. Similarly, we can distinguish between factual knowledge with and without a scientific base. This gives rise to the division shown in Table 23.1.

Category 1 in the table, scientific factual knowledge, is the type of knowledge that science, conducted for "its own sake," aims at producing. Note that the term "science" is used here in a broad sense that includes the humanities. Category 2, nonscientific factual knowledge, includes most of our knowledge about the world as we encounter it in our daily lives. Such knowledge need not be less well founded or less certain than scientific knowledge, but it is not based on the systematic collective process that produces scientific knowledge. The remaining two categories both refer to practical knowledge. Category 3, science-based action knowledge, can be delineated more precisely as scientific knowledge about how to achieve certain practical results. Modern science-based medical knowledge is a prime example, as is much of modern technological knowledge. This category of knowledge is often described as "applied science," but the term is misleading. Science-based action knowledge is largely founded on scientific investigations specifically tailored to obtain action guidance, such as clinical trials and other directly action-guiding experiments, rather than on knowledge obtained for other purposes (Hansson 2007a; 2015). Much research remains to be done to help us understand the intricate relationship between science and practical knowledge.

Category 4, action knowledge not based on science, includes many cases of trivial knowledge but also highly systematic and sophisticated forms of knowledge that may take many years to acquire. The art of violin-playing is one example. Although marginally informed by scientific results from acoustics, musicology, and the behavioral sciences, this is still a predominantly experience-based art. Many other "practical arts" have the same character.

Table 23.1 Four Types of Knowledge

	Factual knowledge	Action knowledge
Science	(1) Scientific factual knowledge	(3) Science-based action knowledge
Non-science	(2) Nonscientific factual knowledge	(4) Action knowledge not based on science

Practical knowledge, not least in the crafts, is often expressed in the form of action rules or "rules of thumbs" that are constructed to be easy to memorize and apply (Norström 2011). The practical knowledge of electricians is a good example. There are many rules for how to connect wires. These rules are based on a mixture of theoretical justification, practical experience, and conventions. For instance, there are good science-based reasons to have a uniform color code for all earth wires, but the choice of green and yellow for that purpose is a convention that cannot be derived from physics. In his or her everyday work, the electrician applies the rules, not the underlying theory. Such bodies of knowledge with mixed origins should be the subject of in-depth epistemological investigations, not least due to the challenges they pose to the delimitation of science versus nonscientific forms of knowledge. The next two sections deal with two more specific forms of practical knowledge, namely technology and the (fine) arts.

4 TECHNOLOGY

The word "technology" is of Greek origin, based on *techne,* which means art or skill and *–logy,* which means "knowledge of" or "discipline of." The word was introduced into Latin by Cicero (Steele 1900: 389; Cicero 44–46 BC/1999: 4:16.) However, it does not seem to have been much used until Peter Ramus (1515–1572) started to use it in the sense of knowledge about the relations among all *technai* (arts). In 1829 the American physician and scientist Jacob Bigelow published *Elements of Technology,* in which he defined technology as "the principles, processes, and nomenclatures of the more conspicuous arts, particularly those which involve applications of science" (Tulley 2008). Already in the late seventeenth century "technology" often referred specifically to the skills and devices of craftspeople. (Sebestik 1983). This sense became increasingly dominant, and in 1909 *Webster's Second New International Dictionary* defined technology as "the science or systematic knowledge of industrial arts, especially of the more important manufactures, as spinning, weaving, metallurgy, etc." (Tulley 2008). Technology had acquired a more limited sense, referring to what is done with tools and machines.

In English the word "technology" also acquired another meaning: increasingly it referred to the actual tools, machines, and procedures used to produce material things, rather than to knowledge about these things. This usage seems to have become common only in the twentieth century. The earliest example given in the *Oxford English Dictionary* is a text from 1898 about the coal-oil industry, according to which "a number of patents were granted for improvements in this technology, mainly for improved methods of distillation" (Peckham 1898: 119). Today this is the dominant usage (Mertens 2002). However, this is not true of all languages. For instance, French, German, Dutch and Swedish all have a shorter word (*technique, Technik, techniek, teknik*) that refers to the actual tools, machines, and practices. In these languages, the word corresponding to "technology" (*technologie, Technologie, technologie, teknologi*) is more often than in English used to denote knowledge about these practical arts rather than the arts and their material devices themselves; however, due to influence from English, the use of "technology" in the sense of tools, machines, and practices is common in these languages as well. According to the *Svenska Akademiens Ordbok*, the Swedish counterpart of the *Oxford English Dictionary*, this usage seems to have become common in Swedish in the 1960s.

The meaning of "technology" has evolved with the development of the tools, machines, and other artefacts that we use. Two major extensions of its meaning took place in the second half of the twentieth century. First, with the development of computer and information technology, a wide range of programming and other software-related activities became recognized as technological. Second, through the equally rapid development of biotechnology, many activities based on biological knowledge are now considered technological, and new types of artefacts, namely various forms of modified biological organisms, are regarded as technological products.

Higher education in what is today called "technology" has its origin in the polytechnical schools of the nineteenth century, the first being the École Polytechnique in Paris (founded in 1794). In the nineteenth and early twentieth centuries, schools of engineering fought to obtain the same status as universities. To prove their case, they had to base their teaching of technology on a scientific basis. Two different strategies were adopted to achieve this. One was to use results from the natural sciences to investigate the workings of machines and other technological constructions. Formulas from mechanical science were used to characterize the movements of machine parts, and the theory of electromagnetism was applied in the construction of electric machines and appliances. New disciplines, such as structural mechanics, were developed that broadened the basis of this type of calculations. The other strategy was to apply scientific method directly to technological constructions. Machines and machine parts were built and measurements were made on alternative constructions in order to optimize their performance. (Faulkner 1994; Kaiser 1995). In many cases, this was the only way to solve practical technological problems (Hendricks, Jakobsen, and Andur Pedersen 2000). For instance, the processes studied in wind tunnels were usually too complex to allow for a mathematical solution. Direct testing of technological constructions has continued to be an essential part of scientific engineering. For example, without crash tests, automobile safety would have been much worse. Even when a construction is based on relatively simple, well-known principles, it must be tested in practice, as exemplified by endurance tests of furniture and household appliances.

Whereas the first of these two strategies can be described as "applied natural science," the second does not fit well into that description. Due to the essential role of this type of investigation in all branches of technology, it would be grossly misleading to describe technology as "applied" natural science. Engineers who develop or evaluate new technology have much more to do than apply results from the natural sciences (Layton, 1978). We can use the term "technological science" to denote the scientific study of technological devices and constructions. As I have proposed elsewhere (Hansson 2007c), technological science should be treated as a major species of science along with natural, social, medical, and other science, rather than (as is common) as a subspecies of the natural sciences. This treatment is justified by at least six major differences between technological and natural science (Hansson 2007c):

1. The primary study objects of technological science have been constructed by humans, rather than being objects from nature.
2. Design is an important part of technological science. Technological scientists not only study human-made objects; they also construct them.
3. The study objects are largely defined in functional, rather than physical, terms. For instance, a device with the function to crack nuts is a nutcracker, irrespective of its physical structure. A device whose function is to turn over the upper layer of the soil is a plow, irrespective of its construction, etc. In the natural sciences, functional definitions are much less common. (Kroes and Meijers 2002; Hansson 2006; Vermaas and Houkes 2006; Kroes 2012).
4. The conceptual apparatus of the technological sciences contains a large number of value-laden notions. Examples are "user friendly," "environmentally friendly," and "risk" (Layton 1988; Hansson 2014a).
5. There is less room than in the natural sciences for idealizations. For instance, physical experiments are often performed in a vacuum in order to correspond to theoretical models in which the impact of atmospheric pressure has been excluded, and for similar reasons chemical experiments are often performed in the gas phase. In the technological sciences, such idealizations cannot be used since the technical constructions have to work under common atmospheric conditions.
6. In their mathematical work, technological scientists are satisfied by sufficiently good approximations. In the natural sciences, an analytical solution is always preferred if at all obtainable.

Previously, technology was sometimes described by philosophers as applied natural science (Bunge 1966), but today there is consensus that such a description of technology is grossly misleading (Mitcham and Schatzberg 2009; Bunge 1988; Hansson 2007c). In recent years, several authors have claimed that it is more accurate to describe science as applied technology than the other way around (Lelas 1993). However, the science–technology relationship is much too complex to be captured with reductive terms such as "application." The prevalence of specifically technological science is a further complication that has usually not been taken into account in discussions of the science–technology relationship. It is probably useful to treat that relationship as a species of the more general relationship between science (in a broad sense) and the practical arts that was introduced in the previous section. However, technology has features of its own that need special studies, not least its elements of design and the role of functional terms.

5 THE FINE ARTS

Sculpture, painting, music, drama, and literature (in the form of tales) all go back to prehistoric ages. However, they do not seem to have been separated out as a special kind of human endeavor distinct from practical, intellectual, or religious activities. There was, for instance, no division between the part of a potter's work that consists in making the pottery durable and fit for use and that which consists in making it appealing to the eye. The occupations that we today call artistic, such as sculpture, painting, and music, were not grouped together or treated differently from other qualified manual trades.

We have already noted that the Latin *ars*, from which our "art" originates, originally denoted skills and abilities in general, and there was no other term in antiquity for what we today call "art." The concept had simply not been invented. The same was true in the Middle Ages. There was no concept of fine arts in the knowledge classifications. Poetry was grouped with grammar and rhetoric, the visual arts with the other manual crafts, and music with mathematics and astronomy (Kristeller 1980: 174). This classification of music referred primarily to music theory, in which analogies were drawn between musical harmony and the supposed harmony of the astronomical universe (James 1995). There was as yet no idea of a common trait combining poetry, painting, and music into a special category of human activities. Painters and sculptors were counted as manual craftsmen, and they were organized in the same types of guilds as other manual workers (Hauser 1968: 222–230). Musicians were largely organized in the guilds of watchmen, and many musicians performed double duties as musicians and nightwatchmen (Raynor 1972: 55–69).

In the Renaissance, the status of what we now call artists was gradually improved. An important step was taken in 1563 by painters, sculptors, and architects in Florence. They dissociated themselves from the guild system and formed instead an *Accademia delle Arti del Disegno* (Academy of the Arts of Drawing). However, treatises of poetics, painting, and music continued to deal with only one of these subjects, without comparing it to the others. It was not until the eighteenth century that a literature emerged in which the fine arts were compared with each other and discussed on the basis of common principles. The term "fine arts" (in French, *beaux arts*) was introduced to denote painting, sculpture, architecture, music, and poetry and sometimes other art forms such as gardening, opera, theatre, and prose literature (Kristeller 1980: 163–165). A decisive step in forming the concept was taken by Charles Batteux (1713–1780), professor of philosophy in Paris. In his book from 1746, *Les beaux arts réduits à un même principe* (The Fine Arts Reduced to a Single Principle), he for the first time clearly separated the fine arts such as music, poetry, painting, and dance from the mechanical arts. The single principle referred to in the title was imitation of nature, which he claimed to be the common principle of the fine arts (Batteux 1746: 5–7).

In this way, two classes of the practical or productive arts were singled out according to their purposes: technology that strives to satisfy our material needs and the fine arts that pursue beauty as an end in itself (cf. Fourastié 1977). This distinction was well received by Batteux's contemporaries. In his introduction to the *Encyclopédie*, d'Alembert made extensive use of Batteux's terminology. According to d'Alembert, the fine arts are much less rule-bound than theoretical learning and therefore much more dependent on individual inspiration or genius (1751: xiij).

The distinction between art and technology is by no means simple. Many technological activities, not least in the design and construction of new technological objects, have considerable aesthetic components. At the same time, many objects of art can be described as technological. The concept of art (in the modern sense of "fine arts") is notoriously difficult to define. Most of these definitional problems need not concern us here since the problematic outer limits of art are usually not limits with science on the other side. However, recently the phenomenon of artistic research may have opened a new area of contact between science and art (Lesage 2009). The concept is by no means well defined, but it covers artistic practices that aim at the production of new knowledge—but knowledge of an artistic rather than a scientific kind. In most cases artistic research is clearly distinguishable from scientific research (by which I mean research in the sciences and humanities), but there may be some doubtful cases and some overlaps. It is still too early to say how this new field will develop. It may possibly have the potential to give rise to extensive contacts and overlaps with science (in the broad sense), similar to those of technology, but that remains to be seen.

6 PSEUDOSCIENCE

"Pseudoscience" and "pseudoscientific" are unavoidably defamatory words (Laudan 1983: 118; Dolby 1987: 204). Etymology provides us with an obvious starting point for a definition. "Pseudo-" means "false." The *Oxford English Dictionary* defines pseudoscience as "a pretended or spurious science," and many writers on pseudoscience have emphasized that pseudoscience is non-science posing as science. The foremost modern classic on the subject (Gardner 1957) bears the title *Fads and Fallacies in the Name of Science*. According to Brian Baigrie, "what is objectionable about these beliefs is that they masquerade as genuinely scientific ones" (1988: 438). These and many other authors assume that to be pseudoscientific, a claim or a teaching has to satisfy the following two criteria according to which pseudoscience is similar to science in one respect and dissimilar in another:

1. It is not scientific (the dissimilarity condition).
2. Its major proponents try to create the impression that it is scientific (the similarity condition).

These two criteria can be taken as a first approximation of a delimitation of pseudoscience, but they both need some adjustment. Beginning with the dissimilarity condition, it should be noted that in this context, the term "scientific" refers in practice to science in the broad sense proposed above in Section 2, although this is seldom explicitly stated. The reason for this is that many teachings taken to be paradigm examples of pseudoscience are misrepresentations, not of science in the conventional sense but of the humanities. Holocaust deniers, ancient astronaut theorists, fabricators of Atlantis myths, proponents of fringe theories on Shakespearian authorship, promoters of the Bible code, and many others are primarily distorters of historical and literary scholarship, although in many of these cases, neglect or misrepresentation of natural science adds to the confusion.

Second, the term "not scientific" in the dissimilarity condition should not be interpreted as "nonscientific" but as "unscientific" (i.e., in conflict with science). My views on the quality of the works of different contemporary painters are nonscientific for the simple reason that they refer to issues of taste that cannot be resolved by science. However, since they are compatible with science, it would be misleading to call them unscientific. Our usage of the term "pseudoscientific" should therefore be restricted to issues that belong to the area of science (in the broad sense). It is also useful to specify "unscientific" in terms of the fundamental criterion of reliability referred to in Section 2. An investigation does not qualify as pseudoscience merely by lacking in scientific fruitfulness or importance. (Such failings are better referred to as "bad science.") It has to fail in terms of reliability (epistemic warrant). We can summarize this by saying that pseudoscience is characterized by suffering from such a severe lack of reliability in scientific issues that it cannot at all be trusted.

The similarity condition is also in need of modification. To begin with, there are ways to imitate science that we do not label as pseudoscientific. For instance fraud in otherwise legitimate branches of science is an unscientific practice with a high degree of scientific pretense, but it is seldom if ever called "pseudoscience." The reason for this seems to be the lack of a *deviant doctrine*. Isolated breaches of the requirements of science are not commonly regarded as pseudoscientific. Pseudoscience, as it is commonly conceived, involves a sustained effort to promote teachings that do not have scientific legitimacy at the time. Most fraudulent scientists avoid doing this. Instead they present (faked) results that are in conformity with the predictions of established scientific theories, thereby reducing the risk of disclosure.[1] For a phenomenon to be pseudoscientific, it must be part of a striving to promote standpoints not supported by (good) science. We can therefore modify the similarity condition and specify that a pseudoscientific standpoint must be *part of a nonscientific doctrine that is claimed to be scientific* (Hansson 1996).

There is an obvious tension between this modified version of the similarity condition and the conventional view of science. Most philosophers of science, and most scientists, regard science as constituted by methods of inquiry rather than by particular doctrines. However, this modification of the similarity condition is adequate (although it makes the condition less oriented toward similarities) since pseudoscience typically involves a representation of science as a closed and finished doctrine rather than as a methodology for open-ended inquiry. One very clear example is homeopathy, which was conceived before it was known that all material substance consists of molecules. Since homeopathy

[1] Obviously an activity can be pseudoscientific and fraudulent at the same time. There is a long history of fraudulent science being performed to obtain support of paranormal claims (Alcock 1981; Randi 1982). Another prominent example is Andrew Wakefield's fraudulent vaccine study that was part of a pseudoscientific campaign against vaccination (Deer 2011). Probably, fakers promoting pseudoscientific doctrines run a larger risk of disclosure than fakers who refrain from doing so.

is a closed doctrine, it is still built on ideas that are incompatible with this information, in sharp contrast with scientific medicine, which had no difficulty in assimilating this information and making effective use of it (Singh and Ernst 2008).

There is a further problem with the similarity condition. Must a doctrine be presented as scientific in order to be pseudoscientific? Many astrologists describe their teachings as science. But there are also astrologists who take an antiscientific stance, repudiate science, and claim that astrology is much better than science. Does this exempt their astrological teachings from being labeled as pseudoscientific? Most critics of pseudoscience would probably be unwilling to exempt them. In practice, the term "pseudoscience" is widely used about doctrines conflicting with science that are advanced as better alternatives to science rather than as scientific. Hence, Grove included among the pseudoscientific doctrines those that "purport to offer alternative accounts to those of science or claim to explain what science cannot explain" (1985: 219). Similarly, Lugg maintained that "the clairvoyant's predictions are pseudoscientific whether or not they are correct" (1987: 227–228), despite the fact that most clairvoyants do not profess to be practitioners of science. In order to comply with common usage, we therefore must further modify the similarity condition and include statements that are parts of unscientific doctrines, irrespectively of whether these doctrines are presented as scientific. This leads us to the following, much-modified version of the definition of pseudoscience that we started with. A claim or a teaching is pseudoscientific to the extent that it

1. suffers from such a severe lack of reliability in scientific issues that it cannot at all be trusted.
2. is part of a doctrine that conflicts with (good) science (Hansson 1996, 2009, 2013a).

As emphasized in Section 2, the successfulness of science depends largely on its self-corrections, and these often concern fundamental methodological issues. Therefore, the methodological criteria and requirements of science change considerably over time. In consequence, what was good science at one time can be bad science, perhaps even pseudoscience, at another. For a concrete example, consider the clinical experiment in the treatment of pneumonia that was reported by Joseph Dietl to the Viennese physicians' association in 1849. He treated three groups of pneumonia patients in different ways. One group received blood-letting and a second an emetic. These were the two standard treatments at that time. The third group received no specific treatment. Mortality was 20.4% in the first group, 20.7% in the second, and 7.4% in the third (Dietl 1849). This was exceptionally good research at the time, and it provided much of the impetus for the abolition of bloodletting from medical practice. However, his study employed a method that would have been unacceptable today. The three groups were pneumonia patients admitted to the hospital in three consecutive years. Today, randomization would have been required, and a study following Dietl's protocol would run a considerable risk of being labeled as pseudoscientific.

Unfortunately, most of the discussion about the demarcation between science and pseudoscience has centered on specific methodological features, such as the falsifiability of hypotheses (Popper 1962, 1994), the progress of research programs (Lakatos 1970, 1974a, 1974b, 1981), integrability into other sciences (Reisch 1998), and so on. (See Hansson [2008] for details on these and other demarcation criteria.) Such methodological criteria are often highly useful for judging the quality of research, but due to the continuous transformation of scientific methodology, they are too time-bound to be suitable as demarcation criteria for pseudoscience.

7 RELIGION

A religion is a system of beliefs that includes claims about supernatural entities, such as a God or gods, spirits, and human souls surviving physical life. It is not always obvious whether a particular system of beliefs is religious or not, and self-attributions may not be accurate. In secularized societies such as Western Europe and antireligious societies such as China, religious beliefs have sometimes been presented as science in order to further their acceptance. One example of this is anthroposophy, a German variant of theosophy that has extensive teachings on reincarnation and recognizes many supernatural beings but denies being a religion and instead describes itself as "spiritual science." In China, Qigong has been cleansed of spiritual elements and presented as a science-based practice. Conversely, in some social circumstances there are advantages to being identified as a religion. Scientology, whose extensive pseudoscientific teachings contain comparatively few religious elements, puts much emphasis on being a religion and a church. Such a classification has substantial advantages, legally and taxation-wise, for the movement in question. As these examples show, in scholarly studies of a movement, its own answer to the question whether it is a religion may need a critical appraisal.

Is there a conflict between science and religion? The answer to that question in practice depends on the extent and nature of the connections between the spiritual beings included in the religious belief system and the material world studied by science. If the spiritual entities are assumed to live a separate existence, unconnected to the world we experience with our senses, then science has nothing to say about them. A religion with such limited claims has no conflicts with science. However, most religious views involve some sort of interaction between the spiritual and the physical spheres. In some religions, spiritual beings are believed to be almost constantly interfering in the lives of human beings. In other religions, such as the three Abrahamic ones, the two spheres are believed to be mostly independent, but on rare occasions the spiritual sphere can interfere with the physical one. Such occurrences are called "miracles." Miracles are an essential component of the faith of some believers; for others they are unimportant or not even part of their belief system. Since miracles are physical occurrences, they can run into conflict with science in at least two ways: scientists can show that a professed miracle did not take place or (if its occurrence is accepted) explain what happened in nonsupernatural terms (Nickell 2007).

In societies where science has a high status, some religious proponents have tried to solve or preempt such conflicts by promoting incorrect claims in scientific issues. Religiously motivated pseudoscience can be divided into four main categories according to the types of claims that are defended (Hansson 2013b):

1. Alleged scientific proof of the literal veracity of Scriptures or traditional dogmas. The foremost example is creationism and its variant intelligent design that are both constructed to promote a view of the origin of species that is incompatible with science but compatible with a literalist interpretation of Scripture (Ruse 1996; Young and Edis 2006). Misrepresentations of archaeological evidence are commonly made in order to obtain confirmations of scriptural narrations. Major forms are so-called biblical archaeology, the search for Noah's Ark, and Mormon archaeology (Davis 2004). Pseudoscientific claims that stigmatize homosexuality usually have a background in attempts to defend religious dogma (Grace 2008).
2. Alleged scientific proof of the authenticity of venerated physical objects. This is particularly common in Christianity due to the importance of such physical objects (relics) in several major variants of that religion. The most famous example is the so-called shroud of Turin that is claimed by "sindonologist" pseudoscientists to be the true burial shroud of Jesus Christ, although radiocarbon dating by three laboratories leaves no doubt that this piece of cloth dates from the period between 1260 and 1390. Pseudoscientific claims have also been made about a large number of other relics (Nickell 2007).
3. Alleged scientific proof of afterlife or the existence of nonhuman spiritual beings. The two major forms are spiritualism (Brandon 1983) and the presentation of purported memories from previous lives as proof of reincarnation (Edwards 2002).
4. Alleged scientific proof of divine interference or the spiritual powers of humans. A large variety of physical phenomena explainable as legerdemain have been used to prove the special powers of (self-)chosen individuals, usually not in connection with traditional religions. Attempts to prove their genuineness have often resulted in pseudoscience (Alcock 1981).

The distinction between science and religion is easy to draw in principle. But in practice it is often difficult due to the prevalence of religiously motivated pseudoscience.

8 CONCLUSION

The message conveyed in this chapter can be summarized as follows:

- The traditional focus on the demarcation between science and pseudoscience is much too limited, since science has philosophically interesting borders to many human endeavors that are not pseudoscientific.

- In discussions on the relations between science and various nonscientific activities, it is preferable to use a widened concept of science that also includes the humanities. The distinction between the sciences and the humanities is an idiosyncrasy of the English language that has become increasingly inadequate.
- One of the most interesting border issues concerns the relationships between science (in this widened sense) and the productive arts (i.e., knowledge about how to achieve various practical results). This is an extensive topic that includes problems such as the science–technology relationship, the relationship between practical and theoretical rationality, the epistemology of practical rule knowledge, and the nature of scientific procedures such as action-guiding experiments that aim at practical rather than theoretical knowledge.

Science is a socially embedded activity. We should not expect to understand its inner workings without also understanding how it interacts with other human endeavors.

References

Alcock, James E. (1981). *Parapsychology, Science or Magic? A Psychological Perspective* (Oxford: Pergamon).

Aristotle. (1980). *Nichomachean Ethics*. Translated by W. D. Ross (Oxford: Clarendon). (Original work *c.* 350 BC)

Atkinson, Quentin D. (2011). "Phonemic Diversity Supports a Serial Founder Effect Model of Language Expansion from Africa." *Science* 332: 346–349.

Baigrie, B. S. (1988). "Siegel on the Rationality of Science." *Philosophy of Science* 55: 435–441.

Batteux, Charles. (1746). *Les beaux arts réduits à un même principe* (Paris: Durand).

Brandon, Ruth. (1983). *The Spiritualists: The Passion for the Occult in the Nineteenth and Twentieth Centuries* (London: Weidenfeld and Nicolson).

Bunge, Mario. (1966). "Technology as Applied Science." *Technology and Culture* 7: 329–347.

Bunge, Mario. (1988). "The Nature of Applied Science and Technology." In V. Cauchy (ed.), *Philosophy and Culture: Proceedings of the XVIIth Congress of Philosophy*, vol. II (Montréal: Éd. Montmorency), 599–604.

Chenu, Marie-Dominigue. (1940). "Arts 'mecaniques' et oeuvres serviles." *Revue des sciences philosophiques et theologiques* 29: 313–315.

Cicero, Marcus Tullius. (1913). *De officiis*. Translated by Walter Miller. Loeb Classical Library (Cambridge, MA: Harvard University Press). (Original work *c.* 44 BC)

Cicero, Marcus Tullius. (1999). *Epistulae ad Atticum*, Vols 1–4. Edited and translated by D. R. Shackleton Bailey. Loeb Classical Library (London: Harvard University Press). (Original work *c.* 44–46 BC)

Covington, Michael A. (2005). "Scientia Sermocinalis: Grammar in Medieval Classifications of the Sciences." In Nicola McLelland and Andrew Linn (eds.), *Flores Grammaticae: Essays in Memory of Vivien Law* (Münster: Nodus Publikationen), 49–54.

d'Alembert, Jean le Rond. (1751). "Discours préliminaire." In Denis Diderot and Jean le Rond d'Alembert (eds.), *Encyclopédie, ou dictionnaire raisonné des sciences, des arts et des métiers, par une sociétéde gens de lettres*, Vol. 1 (Paris: Briasson).

Davis, Thomas. (2004). *Shifting Sands: The Rise and Fall of Biblical Archaeology* (New York: Oxford University Press).

Deer, Brian. (2011). "How the Vaccine Crisis Was Meant to Make Money." *British Medical Journal* 342: c5258.

Dietl, Józef. (1849). *Der Aderlass in der Lungenentzündung* (Wien: Kaulfuss Witwe, Prandel).

Dolby, R. G. A. (1987). "Science and Pseudoscience: The Case of Creationism." *Zygon* 22: 195–212.

Douglas, Heather. (2004). "The Irreducible Complexity of Objectivity." *Synthese* 138: 453–473.

Dyer, Joseph. (2007). "The Place of Musica in Medieval Classifications of Knowledge." *Journal of Musicology* 24: 3–71.

Edwards, Paul. (2002). *Reincarnation: A Critical Examination* (Amherst, MA: Prometheus).

Faulkner, Wendy. (1994). "Conceptualizing Knowledge Used in Innovation: A Second Look at the Science-Technology Distinction and Industrial Innovation." *Science, Technology and Human Values* 19: 425–458.

Fourastié, Jean. (1977). "Art, Science and Technique." *Diogenes* 25: 146–178.

Freedman, Joseph S. (1994). "Classifications of Philosophy, the Sciences, and the Arts in Sixteenth- and Seventeenth-Century Europe." *Modern Schoolman* 72: 37–65.

Gardner, Martin. (1957). *Fads and Fallacies in the Name of Science* (New York: Dover).

Grace, André P. (2008). "The Charisma and Deception of Reparative Therapies: When Medical Science Beds Religion." *Journal of Homosexuality* 55: 545–580.

Grove, J. W. (1985). "Rationality at Risk: Science against Pseudoscience." *Minerva* 23: 216–240.

Hansson, Sven Ove. (1996). "Defining Pseudoscience." *Philosophia Naturalis* 33: 169–176.

Hansson, Sven Ove. (2006). "Defining Technical Function." *Studies in History and Philosophy of Science* 37: 19–22.

Hansson, Sven Ove. (2007a). "Praxis Relevance in Science." *Foundations of Science* 12: 139–154.

Hansson, Sven Ove. (2007b). "Values in Pure and Applied Science." *Foundations of Science* 12: 257–268.

Hansson, Sven Ove. (2007c). "What is Technological Science?" *Studies in History and Philosophy of Science* 38: 523–527.

Hansson, Sven Ove. (2008). "Science and Pseudo-Science." In *Stanford Encyclopedia of Philosophy*, http://plato.stanford.edu/entries/pseudo-science

Hansson, Sven Ove. (2009). "Cutting the Gordian Knot of Demarcation." *International Studies in the Philosophy of Science* 23: 237–243.

Hansson, Sven Ove. (2013a). "Defining Pseudoscience—and Science." In Massimo Pigliucci and Maarten Boudry (eds.), *The Philosophy of Pseudoscience* (Chicago: Chicago University Press), 61–77.

Hansson, Sven Ove. (2013b). "Religion and Pseudoscience." In A. Runehov and L. Oviedo (eds.), *Encyclopedia of Sciences and Religions* (Dordrecht, The Netherlands: Springer), 1993–2000.

Hansson, Sven Ove. (2014a). "Values in Chemistry and Engineering." In Peter Kroes and Peter-Paul Verbeek (eds.), *The Moral Status of Technical Artefacts* (New York: Springer), 235–248.

Hansson, Sven Ove. (2014b). "Why and For What Are Clinical trials the Gold Standard?" *Scandinavian Journal of Public Health* 42(Suppl. 13): 41–48.

Hansson, Sven Ove. (2015). "Experiments before Science: What Science Learned from Technological Experiments." In Sven Ove Hansson (ed.), *The Role of Technology in Science: Philosophical Perspectives* (New York: Springer), 81–110.

Hauser, Arnold. (1968). *The Social History of Art*, Vol. 1: *From Prehistoric Times to the Middle Ages* (London: Routledge and Kegan Paul).

Hendricks, Vincent Fella, Jakobsen, Arne, and Andur Pedersen, Stig. (2000). "Identification of Matrices in Science and Engineering." *Journal for General Philosophy of Science* 31: 277–305.

Henrich, Joseph. (2004). "Demography and Cultural Evolution: How Adaptive Cultural Processes Can Produce Maladaptive Losses: The Tasmanian Case." *American Antiquity* 69: 197–214.

Hoppe, Brigitte. (2011). "The Latin Artes and the Origin of Modern Arts." In Maria Burguete and Lui Lam (eds.), *Arts: A Science Matter*, Vol. 2 (Singapore: World Scientific), 35–68.

Hoven, Birgit Van Den. (1996). *Work in Ancient and Medieval Thought: Ancient Philosophers, Medieval Monks and Theologians and their Concept of Work, Occupations and Technology*. Dutch Monographs on Ancient History and Archaeology 14 (Amsterdam: Gieben).

James, Jamie. (1995). *The Music of the Spheres* (London: Abacus).

Kaiser, W. (1995). "Die Entwicklung der Elektrotechnik in ihrer Wechselwirkung mit der Physik." In L. Schäfer and E. Ströker (eds.), *Naturauffassungen in Philosophie, Wissenschaft, Technik, Band III; Aufklärung und späte Neuzeit* (München: Verlag Karl Alber Freiburg), 71–120.

Kristeller, Paul Oskar. (1980). *Renaissance Thought and the Arts. Collected Essays* (Princeton, NJ: Princeton University Press).

Kroes, Peter. (2012). *Technical Artefacts: Creations of Mind and Matter* (New York: Springer).

Kroes, Peter, and Meijers, Anthonie. (2002). "The Dual Nature of Technical Artifacts— Presentation of a New Research Programme." *Techné: Research in Philosophy and Technolog* 6(2): 4–8.

Lakatos, Imre. (1970). "Falsification and the Methodology of Research Program." In Imre Lakatos and Alan Musgrave (eds.), *Criticism and the Growth of Knowledge* (Cambridge, UK: Cambridge University Press), 91–197.

Lakatos, Imre. (1974a). "Popper on Demarcation and Induction." In P. A. Schilpp (ed.), *The Philosophy of Karl Popper*. Library of Living Philosophers 14, Book 1 (La Salle, IL: Open Court), 241–273.

Lakatos, Imre. (1974b). "Science and Pseudoscience." *Conceptus* 8: 5–9.

Lakatos, Imre. (1981). "Science and Pseudoscience." In S. Brown et al (eds.), *Conceptions of Inquiry: A Reader* (London: Methuen), 114–121.

Laudan, Larry. (1983). "The Demise of the Demarcation Problem." In R. S. Cohan and L. Laudan (eds.), *Physics, Philosophy, and Psychoanalysis* (Dordrecht, The Netherlands: Reidel), 111–127.

Layton, Edwin T. (1976). "American Ideologies of Science and Engineering." *Technology and Culture* 17: 688–701.

Layton, Edwin T. (1978). "Millwrights and Engineers, Science, Social Roles, and the Evolution of the Turbine in America." In W. Krohn, E. T. Layton, and P. Weingart (eds.), *The Dynamics of Science and Technology: Sociology of the Sciences*, Vol II (Dordrecht, The Netherlands: Reidel), 61–87.

Layton, Edwin T. (1988). "Science as a Form of Action: The Role of the Engineering Sciences." *Technology and Culture* 29: 82–97.

Lelas, Srdjan. (1993). "Science as Technology." *British Journal for the Philosophy of Science* 44: 423–442.

Lesage, Dieter. (2009). "Who's Afraid of Artistic Research? On Measuring Artistic Research Output." *Art and Research: A Journal of Ideas, Contexts and Methods* 2(2).

Lugg, Andrew. (1987). "Bunkum, Flim-Flam and Quackery: Pseudoscience as a Philosophical Problem." *Dialectica* 41: 221–230.

Lycett, Stephen J., and von Cramon-Taubadel, Noreen. (2008). "Acheulean Variability and Hominin Dispersals: A Model-Bound Approach." *Journal of Archaeological Science* 35: 553–562.

Mayr, Otto. (1976). "The Science-Technology Relationship as a Historiographic Problem." *Technology and Culture* 17: 663–673.

Mertens, Joost. (2002). "Technology as the Science of the Industrial Arts: Louis-Sébastien Lenormand (1757–1837) and the Popularization of Technology." *History and Technology* 18: 203–231.

Merton, Robert K. (1942). "Science and Technology in a Democratic Order." *Journal of Legal and Political Sociology* 1: 115–126. Reprinted as "The Normative Structure of Science," In Robert K. Merton (ed.), *The Sociology of Science. Theoretical and Empirical Investigations* (Chicago: University of Chicago Press, 1973), 267–278.

Mikkeli, Heikki. (1997). "The Foundation of an Autonomous Natural Philosophy: Zabarella on the Classification of Arts and Sciences." In Daniel A. Di Liscia, Eckhard Kessler, and Charlotte Methuen (eds.), *Method and Order in Renaissance Philosophy of Nature: The Aristotle Commentary Tradition* (Aldershot, UK: Ashgate), 211–228.

Mitcham, Carl, and Schatzberg, Eric. (2009). "Defining Technology and the Engineering Sciences." In A. Meijers (ed.), *Handbook of the Philosophy of Science*, Vol. 9: *Philosophy of Technology and Engineering Sciences* (Amsterdam: Elsevier), 27–63.

Nickell, Joe. (2007). *Relics of the Christ* (Lexington: University Press of Kentucky).

Noble, David F. (1997). *The Religion of Technology: The Divinity of Man and the Spirit of Invention* (New York: Alfred A. Knopf).

Norström, Per. (2011). "Technological Know-How from Rules of Thumb." *Techné: Research in Philosophy and Technology* 15: 96–109.

Ovitt, George, Jr. (1983). "The Status of the Mechanical Arts in Medieval Classifications of Learning." *Viator* 14: 89–105.

Pagel, Mark, Atkinson, Quentin D., and Meade, Andrew. (2007). "Frequency of Word-Use Predicts Rates of Lexical Evolution Throughout Indo-European History." *Nature* 449: 717–720.

Peckham, S. F. (1898). "The Genesis of Bitumens, as Related to Chemical Geology." *Proceedings of the American Philosophical Society* 37: 108–139.

Popper, Karl. (1962). *Conjectures and Refutations. The Growth of Scientific Knowledge* (New York: Basic Books).

Popper, Karl (1994). "Falsifizierbarkeit, zwei Bedeutungen von." In Helmut Seiffert and Gerard Radnitzky (eds.), *Handlexikon zur Wissenschaftstheorie* (2nd ed.) (München: Ehrenwirth GmbH Verlag), 82–86.

Randi, James. (1982). *The Truth about Uri Geller* (Buffalo, NY: Prometheus).

Raynor, Henry. (1972). *A Social History of Music* (London: Barrie & Jenkins).

Reisch, George A. (1998). "Pluralism, Logical Empiricism, and the Problem of Pseudoscience." *Philosophy of Science* 65: 333–348.

Ruse, Michael, ed. (1996). *But Is It Science? The Philosophical Question in the Creation/Evolution Controversy* (Buffalo, NY: Prometheus).

Sebestik, Jan. (1983). "The Rise of the Technological Science." *History and Technology* 1: 25–43.

Settle, Tom. (1971). "The Rationality of Science versus the Rationality of Magic." *Philosophy of the Social Sciences*, 1: 173–194.

Singh, Simon, and Ernst, Edzard. (2008). *Trick or Treatment: The Undeniable Facts about Alternative Medicine* (London: Bantam).

Snow, C. P. (2008). *The Two Cultures* (Cambridge, UK: Cambridge University Press). (Original work published 1959)

Steele, R. B. (1900). "The Greek in Cicero's Epistles." *American Journal of Philology* 21: 387–410.

Tatarkiewicz, Wladyslaw. (1963). "Classification of Arts in Antiquity." *Journal of the History of Ideas* 24: 231–240.

Tonelli, Giorgio. (1975). "The Problem of the Classification of the Sciences in Kant's Time." *Rivista Critica di Storia Della Filosofia* 30: 243–294.

Tulley, Ronald Jerome. (2008). "Is There Techne in My Logos? On the Origins and Evolution of the Ideographic Term—Technology." *International Journal of Technology, Knowledge and Society* 4: 93–104.

Vermaas, Pieter E., and Houkes, Wybo. (2006). "Technical Functions: A Drawbridge between the Intentional and Structural Natures of Technical Artefacts." *Studies in History and Philosophy of Science Part A*, 37: 5–18.

Weisheipl, James A. (1965). "Classification of the Sciences in Medieval Thought." *Mediaeval Studies* 27: 54–90.

Whitney, Elspeth. (1990). "Paradise Restored: The Mechanical Arts from Antiquity Through the Thirteenth Century." *Transactions of the American Philosophical Society* 80: 1–169.

Young, M., and Edis, T., eds. (2006). *Why Intelligent Design Fails. A Scientific Critique of the New Creationism* (New Brunswick, NJ: Rutgers University Press).

Recommended Reading

Hansson, Sven Ove, ed. (2015). *The Role of Technology in Science: Philosophical Perspectives* (New York: Springer).

Kroes, Peter. (2012). *Technical Artefacts: Creations of Mind and Matter* (New York: Springer).

Meijers, Anthonie, ed. (2009). *Handbook of the Philosophy of Science*, Vol. 9: *Philosophy of Technology and Engineering Sciences* (Amsterdam: Elsevier).

Pigliucci, Massimo, and Boudry, Maarten, eds. (2013). *The Philosophy of Pseudoscience* (Chicago: Chicago University Press).

Skeptical Inquirer. (1976–). Buffalo, N.Y.: Committee for Skeptical Inquiry.

CHAPTER 24

....

SCIENTIFIC CONCEPTS

....

HYUNDEUK CHEON AND EDOUARD MACHERY

PHILOSOPHERS of science have often been interested in characterizing the concepts scientists use, such as the concepts of gene, natural selection, space, and time. Their interest is sometimes normative: they want to assess these concepts (e.g., Machery 2009, on the concept of concept in cognitive psychology). Sometimes, their interest is ameliorative: they want to reform the concepts scientists currently rely on (Reichenbach 1938; Carnap 1950). Sometimes, it is instrumental: characterizing a scientific concept may be important for understanding the explanatory success and limitation of a scientific tradition, or it can be a case study for understanding conceptual change or progress in science. Finally, sometimes, it is simply descriptive: Sometimes, philosophers of science just want to know how scientists think of a given scientific topic and why they do so.

In this chapter, we review the philosophical literature on scientific concepts, with an eye toward highlighting some exciting recent developments. We focus on three distinct approaches to studying scientific concepts. The *semantic approach* is concerned with the semantic content of scientific concepts (e.g., NATURAL SELECTION) and with the meaning of the predicates expressing them (e.g., "natural selection"); the *cognitive approach* views concepts as psychological entities and brings the psychological research on lay concepts to bear on theorizing about scientific concepts; the *experimental approach* brings experimental tools drawn from psychology or sociology to the study of scientific concepts.[1] Section 1 focuses on the semantic approach, Section 2 on the cognitive approach, Section 3 on the experimental approach.

1 THE SEMANTIC APPROACH

....

1.1 The Incommensurability Thesis

Philosophers of science have been particularly interested in the semantic content of scientific concepts or, equivalently for our purposes, in the meaning of the terms expressing

---

[1] We use small caps to name concepts.

them. The main impetus behind this interest is the connection between the semantic content of concepts and the incommensurability thesis that emerged from Hanson's (1958), Kuhn's (1962), and Feyerabend's (1962) work (for discussion, see, e.g., Devitt 1979; Fine 1975; Kitcher 1978; Scheffler 1967; Shapere 1966). The incommensurability thesis says, roughly, that when scientists are committed to different scientific theories (e.g., different physical theories), they mean different things or have different concepts despite using the same terminology. For instance, according to this thesis, "mass" expresses two different concepts when used by scientists before and after Einstein's formulation of special relativity in 1905 (Kuhn 1962, 101–102). Feyerabend expressed the incommensurability thesis as follows (1962, 74):

> The "inertial law" of the impetus theory is incommensurable with Newtonian physics in the sense that the main concept of the former, viz., the concept of impetus, can neither be defined on the basis of the primitive descriptive terms of the latter, nor related to them via a correct empirical statement. The reason for this incommensurability was also exhibited: although [the inertial law], taken by itself, is in quantitative agreement both with experience and with Newton's theory, the "rules of usage" to which we must refer in order to explain the meanings of its main descriptive terms contain the law (7) [viz.: "motion is a process arising from the continuous action of a source of motion or a 'motor,' and a 'thing moving'"] and, more especially, the law that constant forces bring about constant velocities. Both of these laws are inconsistent with Newton's theory.

The incommensurability thesis seems to have unsettling consequences. If two scientists mean different things despite using the same words, they do not seem to disagree when one asserts a given sentence (e.g., a sentence expressing a scientific law) while the other asserts its negation. Appearances notwithstanding, physicists before and after Einstein's formulation of special relativity do not really disagree when they respectively assert and deny the sentence, "Mass is constant." If scientists committed to different theories do not really disagree, the replacement of one scientific theory by another cannot constitute progress since progress requires the replacement of falsehoods by truths or at least of less verisimilar beliefs by more verisimilar beliefs. So, if the incommensurability thesis is true, scientists committed to different theories seem not to really disagree, and science does not really progress!

1.2 The Semantic Content of Scientific Concepts

Whether the incommensurability thesis holds depends in part on what determines the semantic content of scientific concepts.[2] If the content of a scientific concept stands in a bijective relation with the class of inferences whose premises or conclusions involve that concept (what we will call "concept use"), then a term like "mass" expresses different

---

[2] Alternatively, one can grant that the semantic content of scientific concepts changes when scientific theories change while insisting that commensurability and progress simply require stable reference despite conceptual change.

concepts when used by scientists committed to Newtonians mechanics and special relativity theory. The incommensurability thesis follows.

More plausibly, instead of standing in a bijective relation with use, the semantic content of scientific concepts could supervene on their use, in which case a term like "mass" need not express different concepts when used by scientists committed to Newtonian mechanics and special relativity theory, although it may (depending on how different Newtonian mechanics and special relativity theory really are and on the exact relation between the content of concepts and their use). On some views, the semantic content of a scientific concept stands in a bijective relation with its uses that are determined by the central principles of the theory this concept belongs to. Since two distinct theories can share the same central principles, scientists committed to one of them can mean the same as those committed to the other. How to identify the central principles of a theory may, of course, be unclear (Feyerabend 1962).

Finally, the semantic content of scientific concepts could be entirely independent of their use, and scientists committed to different theories could, as a result, share the same concepts. According to externalist theories of meaning, the semantic content of scientific concepts is identical to the properties the concepts happen to be somehow associated with (Boyd 1984; Putnam 1975; Scheffler 1967, chapter 3). As Putnam put it (1975, 237),

> It is beyond question that scientists use terms as if the associated criteria were . . . approximately correct characterizations of some world of theory-independent entities, and that they talk as if later theories in a mature science were, in general, better descriptions of the same entities that earlier theories referred to. In my opinion, the hypothesis that this is right is the only hypothesis that can account for the communicability of scientific results.

If the semantic content is determined by factors external to scientists' minds, then the content of concepts need not change when scientific theories change, and the incommensurability thesis does not follow from scientific change.[3]

1.3 The Semantic Approach: A Dead End?

To address the challenges raised by the incommensurability thesis, we thus need to identify the correct view about the semantic content of scientific concepts (but see footnote 2 as an alternative strategy). Unfortunately, we argue in the remainder of Section 1 that the prospects for this task are dim.

It would not do to simply appeal to the dominant theory of content determination in the philosophy of mind or philosophy of language because no consensus has emerged

---

[3] Of course, were reference to change with at least some scientific change, the incommensurability thesis would threaten again.

among philosophers of mind and language on this question (e.g., Block 1986; Fodor 1998; Peacocke 1992). In fact, the very same philosophical options about the determination of the content of concepts are found in these different areas of philosophy and in the philosophy of science.

More plausibly, one could appeal to case studies in the history of science and in contemporary science to determine which of the views about the determination of the content of scientific concepts is right. Case studies could perhaps tell us how variation (including change) in the semantic content of scientific concepts and in scientific theories are related to one another. For instance, Brigandt (2010) defends his account of the semantic content of concepts by reviewing the history of the concept of gene in the twentieth century and the apparent disagreement about genes in contemporary biology. On his view, we can explain—and even see the rationality of—the variation and stability in the semantic content of the concepts of gene if we distinguish the reference of these concepts, their use (what he calls their "inferential role"), and their explanatory goal. The explanatory goal of a concept is the set of explananda for which scientists developed a particular scientific concept.

We are skeptical of the prospects of identifying the correct theory about the content of scientific concepts on the basis of case studies in the history of science and in contemporary science (for a similar skepticism about the role of the history of science on characterizing the reference of scientific concepts, see Fine 1975). First, the history of science can typically be reconstructed in quite different ways, thus lending support to different views about the semantic content of scientific concepts. Variation in the content of MASS in physics was one of the stock examples of Kuhn (1962) and Feyerabend (1962) when they defended the incommensurability thesis (see also Field 1973). In contrast to the claim that the content of MASS changed when the central principles of Newtonian mechanics were replaced by those of special relativity theory, Earman (1977) has argued that both theories in fact characterize mass by means of identical principles. As he put it (1977, 535–536):

> [T]hree principles of NM [Newtonian mechanics] appear as
> (N1) m_N is a scalar invariant
> (N2) $P_N = m_N V_N$
> (N3) $F_N = m_N A_N$
> where m_N, P_N, V_N, F_N, A_N are, respectively, the Newtonian mass, the Newtonian four-momentum, four-velocity, four-force, and four-acceleration. In SRT [special relativity theory] there are exact analogues (RI), (R2), (R3) of (N1), (N2), (N3) with proper mass m_o in place of m_N, the relativistic four-momentum P_R in place of P_N, etc.

If semantic content supervenes on concept use—in particular if it is determined by central principles of scientific theories—and if Earman is right about the central principles of Newtonian mechanics and special relativity theory (respectively, (RI), (R2), (R3), and (N1), (N2), (N3)), then MASS has plausibly the same semantic content when used by a Newtonian physicist and by a physicist committed to special relativity theory. If Kuhn and Feyerabend are right, then MASS has plausibly a different semantic content, and the incommensurability thesis is correct. We doubt that there is any way to decide between competing reconstructions of this and other key episodes in the history of science.

Furthermore, even if the previous problem were somehow solved, case studies in the history of science and in contemporary science could only tell us how scientific concepts happen to be used (e.g., what inferences scientists draw with what concepts, what principles involving particular concepts they endorse, and what objects concepts happen to be applied to) and perhaps what explanatory goals happen to be associated with them. (But even that may be optimistic: for instance, Brigandt says little about how explanatory goals are to be identified and individuated—i.e., what makes two explanatory goals distinct.) Another step is required to relate concept use and scientists' explanatory goals to semantic content, and nothing in the historical record determines how this step is to be taken. A historian of science or a philosopher examining contemporary science can almost freely stipulate which concept uses or which explanatory interests are constitutive of the semantic content of concepts, and different stipulations are compatible with the historical record. Brigandt (2010), for instance, does not really argue that explanatory goals and concept use determine two dimensions of the semantic content of concepts; he merely stipulates it. The same is true of Kuhn and Feyerabend.

That there is little to gain for philosophers of science from focusing on the semantic content of concepts is not very surprising since this conclusion is in line with Quine's compelling skeptical arguments about meaning (e.g., Quine 1951, 1960). Furthermore, to examine scientific change descriptively and normatively, philosophers of science need not frame the question in semantic terms. Whether or not the semantic content of scientific concepts supervenes on their use, and, if it does, whatever aspect of their use is relevant, historians and philosophers of science can still describe how concepts' use—including the inferences scientists draw, the propositions they accept, and the entities scientists apply their concepts to—varies across time or scientific traditions and research programs. Furthermore, variation, including change, and stability in concept use can be assessed for its rationality, whether or not semantic content supervenes on it. We conclude that the focus on the semantic content of scientific concepts was a dead end for theorizing about concepts. In the remainder of this chapter, we look at alternative approaches.

2 THE COGNITIVE APPROACH

2.1 Scientific Concepts in Light of Cognitive Science

For a long time, philosophical discussion about scientific concepts has been divorced from the psychology of concepts, even though some historicist philosophers of science have relied on Gestalt psychology to support their views about science (e.g., Hanson 1958; Kuhn 1962), and it was not until the 1980s that philosophers exploited the burgeoning psychological literature on concepts (Nersessian 1984, 1989; Thagard 1990). By paying attention to empirical studies of concepts, they attempted to shed a new light on conceptual change in science. Thagard summarized the underlying idea as follows (1990, 256): "the nature of concepts and conceptual change is in fact an important epistemological topic and . . . drawing on ideas from the cognitive sciences can provide an account of conceptual change for epistemology and the philosophy of science." We review this approach in this section.

Inspired by the cognitive science of concepts, Thagard proposed to think of concepts as complex structures like frames or schemata. A *frame* or a schema is a data structure representing a property, a situation, or the members of a given class. A frame organizes the information about its extension by specifying *slots* and *values* for each slot (Minsky 1975). A slot represents a particular aspect or dimension of the world (a determinable) and accepts a range of values (determinates). In a frame, each slot has a default value that is used if there is no available information about an instance of the frame. For example, the frame WHALE might have a slot for size with the value LARGE, and a slot for locomotion with the value SWIM.[4]

On Thagard's view, frames encode different types of information. The frame WHALE may encode information about kind relations (e.g., that whales are a kind of mammal, blue whale is a subkind of whale, etc.), part–whole relations (e.g., that whales have fins, bones, and a tail as parts), and "rules" that enable scientists to infer, explain, or make analogies (e.g., the rule that if something is a whale, then it swims). The constituents of a concept are not all regarded as equally important for the identity of this concept. Kind relations and part–whole relations are more important than the rules, and rule change is thus a weaker form of conceptual change than change in part–whole relations or kind relations.

In a similar spirit, Andersen, Barker, and Chen (2006) have appealed to Barsalou's frame model to understand the cognitive processes underlying Kuhnian revolutions. Barsalou (1992) distinguishes the *attributes* and *values* represented by frames. This distinction is closely akin to the slot–value distinction in Minsky's frame theory. An attribute corresponds to a type of property possessed by the members of a class, and a value corresponds to the properties possessed by some members of this class. For instance, COLOR is an attribute of the frame APPLE, whereas RED is a value for this attribute. A frame defines a conceptual field, that is, a set of concepts that share the same attributes but have different values for these attributes. The frame MAMMAL defines the conceptual field constituted by mammal concepts such as DOG, CAT, COW, etc. These mammal concepts share the same attributes but have different values for those attributes. Similarly, the frame FRUIT allows for a variety of possibilities (its attributes can have different values), and a particular set of values (e.g., COLOR: RED; SIZE: SMALL, TASTE: SWEET, SHAPE: SPHERICAL) can represent a concept APPLE. Thus, a frame can represent several

---

[4] Since the slots and values themselves can be thought of as concepts, we write their names in small caps.

concepts that fall under a superordinate concept, and it embodies the hierarchical relations between concepts.

In Barsalou's frame theory, constraints govern the relations between attributes and the relations between values. The constraints among attributes are called "structural invariants." For BIRD, if something (a particular object or a subkind) has a value for the attribute BEAK, it also must have the attribute NECK. Other constraints imply probabilistic or deterministic co-variation between values. For BIRD, if the value for the attribute FOOT is CLAWED, then the value for the attribute BEAK is more likely to be POINTED than ROUND; if the value for FOOT is WEBBED, then the value for the attribute GAIT is more likely to be WADDLING than RUNNING. These patterns of co-variation might reflect physical and biological facts.

Barsalou's frame model provides a useful framework for characterizing scientific concepts and conceptual change in science. As we have seen, frames can represent hierarchies of concepts. The frame DUCK and the frame BIRD have the same attributes (e.g., SOUNDS), but the attributes of DUCK have typically a single value (supposing ducks are represented as making a single kind of sounds), whereas the attributes of BIRD have various values (since different kinds of birds make different sounds). Thus, frames are well-suited to express the inclusion of DUCK into BIRD. This feature of Barsalou's frame theory is useful to understand an important idea found in Kuhn's later writings, namely, revolutionary changes in scientific taxonomy: A scientific revolution involves a change in the hierarchy embodied by scientists' frames.

In addition, constraints between values represent scientists' beliefs about empirical regularities and their ontological commitments (what kind of things they believe reality encompasses).[5] Some combinations of values are excluded because scientists believe that they are not found in nature on theoretical or empirical grounds. For example, the fact that a value of an attribute (e.g., BEAK: ROUND) is associated with the value of another attribute (e.g., FOOT: WEBBED) represents scientists' belief that no bird species has both a round beak and clawed feet.

Taxonomic change is understood to be the result of incremental changes in frames. Some empirical anomalies might be accommodated by altering the combinations of values while retaining the list of attributes; others may require making changes at the attribute level so that the attributes of pre-revolution frames differ from those of post-revolution frames (for an analysis of the Copernican revolution in this framework, see Andersen et al. 2006, chapter 5). On this view, extreme differences between scientific theories and the concepts embedded in them (before and after a scientific revolution) result from incremental, gradual processes, a proposal that is consistent with the detailed historical case studies showing that scientific revolutions are not sudden, discontinuous, holistic changes but instead historically extended processes (for the historical analysis of the Copernican revolution, see Barker 1999; Barker and Goldstein 1988).

---

[5] Hoyningen-Huene (1993) coined the expressions "knowledge of regularity" and "knowledge of quasi-ontology."

A cognitive-scientific approach is not only useful to understand the nature of scientific concepts, but also how new conceptual structures emerge from old ones. In light of the cognitive science of mental models, Nersessian (2008) has suggested that scientific reasoning often consists in constructing, evaluating, and modifying mental and physical models. Models, including mental models, are understood to be analog representations, representing salient spatiotemporal relations and causal structures. In model-based reasoning, inference is characterized as the manipulation of models rather than of propositions, and models are manipulated by means of simulation (Johnson-Laird 1983).

In science, conceptual structures are created when new models develop out of older models. Nersessian proposes that analogical reasoning plays a crucial role in this creation. Cognitive scientists have long been aware that analogy plays a role in creative problem-solving in ordinary and scientific practices, and it has thus been extensively studied, but Nersessian's take on analogy-based creation of new conceptual structures is original. Her proposal is based on her detailed, illuminating description of how Maxwell developed a new mathematical representation of electromagnetism by constructing a series of models (Nersessian 2002, 2008; for discussion, see Cheon and Machery 2010). In cognitive science, analogy is often thought to consist in applying some known facts from a source domain to a target domain about which there is only partial knowledge, but Maxwell's "physical analogy" is more complicated. A physical analogy uses a set of mathematical relationships in a source domain to understand a target domain. In the case of Maxwell, the source domain was fluid mechanics (a special instance of continuum mechanics) and the target domain electromagnetism. Nersessian's main insight is that a physical analogy does not consist in the direct mapping from the source domain to the target domain; rather, the mapping needs to be built. The resources from a source domain provide constraints for building intermediate models (e.g., the vortex fluid model, the vortex-idle wheel model, and the elastic vortex-idle wheel model), which are themselves progressively elaborated. In this intermediate processes, both models and analogical sources are explored and tested carefully to determine their usefulness in understanding the target domain. This process is iterative: additional intermediate models are produced by taking into account features of the source and target domains, and are tested, as a result of which new intermediate models are developed. This iterative process generates new conceptual structures, such as Maxwell's model of electromagnetism.

2.2 Concept Heterogeneity

2.2.1 *Concept Heterogeneity in Lay Cognition*

Despite its significant contribution to understanding scientific concepts, much of the philosophical literature inspired by cognitive science has selectively relied on a subset of the vast literature on concepts. Philosophers of science first adopt a particular theory of concepts (e.g., frame model), then apply it to scientific concepts. Unfortunately, this strategy fails to consider the cognitive science of concepts in its entirety and overlooks an embarrassing situation in the psychology of concepts: different theories of concepts are still in competition, and there is no consensus on what concepts are.

In the psychological literature, three proposals are dominant (for review, see Machery 2009). According to the *prototype approach*, concepts (called "prototypes") represent the typical or diagnostic properties of their extension (Rosch 1978; Rosch and Mervis 1975); according to the *exemplar approach* (Medin and Schaffer 1978; Nosofsky 1986), a concept is really a set of representations of particular instances (called "exemplars"); finally, according to the *theory approach* (Carey 1985, 2009; Keil 1989; Murphy and Medin 1985), a concept (called a "theory") represents its extension as falling under explanatory and causal generalizations. Remarkably, no theory of concepts has managed to provide a satisfactory explanation of all the experimental findings that are relevant to human conceptual cognition. Some theories can account for some data, which are hard to be explained by another theory, and vice versa.

There are several ways of making sense of this situation, among which the most notable is Machery's (2009, 2010) *heterogeneity hypothesis*. It can be decomposed into the following tenets[6]:

- *Multiplicity thesis*: For each category, an individual typically has multiple kinds of information, including prototypes, exemplars, and theories.
- *Few-commonalities thesis*: These co-referential bodies of information do not constitute a homogeneous class about which scientifically relevant generalizations can be formulated.
- *Anti-hybrid thesis*: Different bodies of information are constitutive parts of a concept when those are connected and coordinated (individuation conditions); however, evidence suggests that prototypes, exemplars, and theories are not coordinated; therefore, they are not parts of a concept.[7]
- *Anti-natural-kind thesis*: Therefore, the notion of concept fails to pick out a natural kind.
- *Eliminativist thesis*: The notion of concept ought to be eliminated from the theoretical vocabulary of cognitive science.

Machery's heterogeneity hypothesis stands in contrast with the "received view" in cognitive science that concepts form a natural kind whose members share many interesting properties. In the end, he suggests that cognitive scientists would be better off without the very notion of concept.

The eliminativist thesis could be resisted. As Machery acknowledges, a term's failure to pick out natural kinds is not sufficient for its elimination from the scientific vocabulary.

---

[6] This decomposition differs from Machery's original formulation.

[7] Roughly, two bodies of information are connected if and only if retrieving one primes the retrieval of the other; two bodies of information are coordinated if and only if, if they happen to underwrite contradictory judgments, one of them defeats the other.

Elimination of a scientific term is recommended just in case it would be more costly than beneficial to keep it. Thus, one might opt to reject the eliminativist thesis while admitting the anti-natural-kind thesis on the pragmatic ground that, everything considered, the notion of concept is still useful. Call this position "concept pragmatism." Instead, one might argue that concepts are indeed a natural kind in a robust sense, thus undercutting the motivation for the eliminativist thesis. To do so, one can deny either the few-commonalities thesis or the anti-hybrid thesis. According to the first option, concepts share in fact several, if not many, characteristics. They might form a superordinate natural kind, which comprises prototypes, exemplars, and theories as subordinate kinds (Samuels and Ferreira 2010; Weiskopf 2009). This view can be called "concept pluralism." According to the second option, hybrid theories are correct (Rice forthcoming; Vicente and Manrique 2016): in this "concept hybridism," prototypes, exemplars, and theories should be viewed as parts of concepts rather than as concepts in their own right.

We cannot settle this issue in this limited space. At minimum, however, what we ought to learn from the landscape in the psychology of concepts is a common ground shared by eliminativism, concept pragmatism, concept pluralism, and concept hybridism: the multiplicity thesis itself. We argue in the remainder of Section 2 that this thesis is important for scientific concepts, too.

2.2.2 *Concept Heterogeneity in Science*

If scientific concepts are not different in kind from lay concepts, as proponents of the cognitive approach have productively assumed, we can extend the multiplicity thesis to scientific concepts: Scientific concepts may be structurally heterogeneous perhaps because each scientific concept consists of summary information (the kind of information that is supposed to be constitutive of prototypes), information about individual cases (the kind of information that is supposed to be constitutive of exemplars), and causal-explanatory information (the kind of information that is supposed to be constitutive of theories), each of which is involved in most, if not all, cognitive processes underlying scientific practices (for a fuller exposition, see Cheon ms.). We call this "the *structural heterogeneity thesis.*"

One may be puzzled by the claim that scientific concepts are theories because concepts are constituents of scientific theories. Theoretical information (e.g., that having wings enables birds to fly) encoded in concepts (e.g., BIRD) is often referred to as a *mini-theory* (Murphy and Medin 1985), but a mini-theory is not identical to a full-blown theory like fluid mechanics, which explains why having wings allows birds to fly in the sky. Thus, mini-theories are parts of concepts, and concepts including mini-theories are constituents of scientific theories.

One might wonder whether scientific concepts really encode prototypical or exemplar information. However, this is the case of many important scientific concepts that have both a theoretical and an observational nature (e.g., SPECIES or PLANET). Furthermore, even theoretical concepts can encode empirical information about models or instances to which the concepts are applied. For example, HARMONIC OSCILLATOR may include nomological information embedded in Hooke's law as well as information about pendulums or springs on which weights are hanging.

Finally, one may object that scientific concepts differ from ordinary ones on the grounds that, unlike ordinary concepts, scientific concepts are definitions; but, in fact, few scientific concepts seem to be definable. In particular, the logical-positivist project to define theoretical concepts in terms of observational ones has failed.

In fact, the structural heterogeneity thesis provides a new explanation of an intriguing phenomenon: productive scientific research without agreed-upon definitions. This phenomenon can be illustrated by the "species problem" in the philosophy of biology. Despite the importance of the concept of species in biology, biologists have failed to agree on the definition of species, and biologists endorse different concepts of species (e.g., the biological species concept, the ecological species concept, and the phylogenetic species concept). Whereas philosophers have been troubled by this situation and have attempted to identify the correct definition of species, biologists seem unfazed by this lack of definition. The structural heterogeneity thesis can account for biologists' attitude. The crucial idea is that biologists have both prototypical and theoretical information about what species are. Biologists often agree about what a paradigmatic species looks like (Amitani 2010). This is evidence that biologists share a common prototype of species, which is satisfied by "good species" and suffices for productive research and scientific progress. On the other hand, disagreement among biologists about what species are can be explained in terms of differences in their theoretical information about species (e.g., interbreeding ability, niche, or monophyletic taxon). Of course, this explanation raises a range of fascinating questions that call for further investigation. In particular, one would want to better understand the dynamics between the diverse informational structures constitutive of concepts (prototypical, exemplar, and theoretical information).

3 THE EXPERIMENTAL APPROACH

If scientific concepts are psychological entities, then they can plausibly be studied by means of the experimental methods that psychologists use to study lay concepts. Experimentally minded philosophers of science have recently embraced this approach (for review, Griffiths and Stotz 2008; Machery 2016). This section describes and defends the goals of this new approach and briefly illustrates it with two examples.

3.1 Goals and Strengths of the Experimental Approach

Studying scientific concepts experimentally does not consist in asking scientists to make explicit their own concepts (e.g., GENE or INNATENESS) since scientists may not be good at describing their own concepts. Similarly, because lay people are not good at describing their own concepts, they are not asked to make these explicit when lay concepts are studied experimentally. Rather, the suggestion is to run elicitation studies: scientists are to be put in controlled situations where they have to use their concepts (just like lay people when lay concepts are examined), and their token concept uses (i.e., their judgments) provide evidence about the nature of the underlying concepts.

The study of scientific concepts by means of experimental methods can serve to test competing accounts of these concepts, including those that philosophers of science have proposed. It can also serve to study what Stotz and Griffiths have called the "conceptual ecology" of scientific concepts, that is, the epistemic needs that influence scientific concepts. Scientists modify concepts flexibly and adaptively to respond to theoretical and experimental needs, which can vary across scientific fields, communities, and traditions. One approach for studying the conceptual ecology of scientific concepts is to examine how a given term (e.g., "gene" or "innate") expresses different concepts across fields, communities, or traditions and how differences between these concepts relate to differences among epistemic needs across fields, communities, or traditions. Finally, the experimental study of scientific concepts can serve to formulate and test hypotheses about the relationship between particular scientific concepts and lay concepts when those are related. For instance, just like lay people, psychologists and linguists deploy the concept of innateness in their effort to characterize human cognitive capacities, and this common use raises the question of how the scientific concept or concepts and the lay concept of innateness are related. Studies of particular concepts may serve to test more general hypotheses about the relations between scientific and lay concepts.

The experimental study of scientific concepts possesses a number of strengths. Large representative samples can be collected, scientists can be put in situations where they use the very concepts at work in their scientific research, and variation in conceptualization can be easily examined. The controlled nature of experimental set-ups also allows philosophers of science to investigate hypotheses about scientific concepts that can't be assessed by examining the natural occurrences of the terms expressing them.

3.2 Two Examples: GENE and INNATENESS

Stotz and Griffiths's groundbreaking research on the concepts of gene kickstarted the experimental study of scientific concepts in philosophy (Stotz and Griffiths 2004; Stotz, Griffiths, and Knight 2004).[8] They describe their project as follows (Stotz and Griffiths 2004, 5):

> philosophical analyses of the concept of the gene were operationalized and tested
> using questionnaire data obtained from working biologists to determine whether

---

[8] Psychologists interested in scientific education have extensively studied how scientific concepts can be taught and how scientific learning interacts with learners' intuitive theories. Experimentally minded philosophers of science would greatly benefit from getting acquainted with this research tradition (e.g., Koponen and Kokkonen 2014; Shtulman 2009).

and when biologists conceive genes in the ways suggested. These studies throw light on how different gene concepts contribute to biological research. Their aim is not to arrive at one or more correct "definitions" of the gene, but rather to map out the variation in the gene concept and to explore its causes and its effects.

Stotz and Griffiths put forward three hypotheses about the variation of the concept of gene across molecular biology, developmental biology, and evolutionary biology. One of their hypotheses proposed that "molecular biologists [should] emphasize the investigation of the intrinsic, structural nature of the gene and . . . be reluctant to identify a gene only by its contributions to relatively distant levels of gene expression. Conversely, evolutionary biologists should be more interested in genes as markers of phenotypic effects and reluctant to treat two similar DNA sequences as the same gene when they lead to different outcomes for the larger system in which they are embedded" (Stotz and Griffiths 2004, 12).

To test their hypotheses, they presented eighty biologists from the University of Sydney with a three-part questionnaire (for detail, see Stotz et al. 2004). The first part was used to identify participants' area of research; the second asked explicit questions about what genes are, what their function is, and the utility of the concept of gene; the third part was the elicitation study itself. In this third part, participants were given individuation questions: they had to decide whether two stretches of DNA were the same or a different gene. There, they made use of their concept of gene instead of attempting to define what they took genes to be.

Stotz and Griffiths found that, for some of their questions, scientists answered differently when asked direct questions about what genes are and when asked to use their concept of gene. Evolutionary biologists did not emphasize the contribution of genes to phenotypic traits in the former kind of situation (contrary to what had been expected), whereas they did in the latter kind of situation (in line with Stotz and Griffiths's third prediction). Differences between research communities were also found, although the results turned out to be often difficult to interpret. Although their findings remain tentative, they provide at least a proof of possibility that experimental studies of concepts are possible and enlightening. They gave evidence that scientists sometimes conceptualize entities (genes in this case) in two different ways—their explicit concept differs from an implicit concept—and they provided evidence that the implicit concept can be influenced by field-specific epistemic needs.

Knobe and Samuels (2013) compared how lay people and scientists conceive of innateness (following up on Griffiths, Machery, and Linquist 2009). They put forward three competing hypotheses about the relation between scientific concepts (e.g., the scientific concepts of species) and lay concepts (e.g., the lay concept of species):

- *The overwriting hypothesis*: The acquisition of a scientific concept leads to the elimination of the lay concept.
- *The overriding hypothesis*: The lay concept coexists with the acquired scientific concept, but scientists rely on the latter in typical scientific contexts.

- *The filtering hypothesis*: Scientists do not acquire a distinct concept, but they reject the judgments the lay concept underwrites when those violate general scientific principles (such as "Do not allow your judgments to be affected by your moral values").

Lay people and scientists were presented with vignettes describing the development of a biological trait, and they were asked whether this trait is innate. Surprisingly, just like lay people, scientists judged that a trait can be innate despite being learned (as predicted by Griffiths and colleagues' model of the lay concept of innateness). Furthermore, moral considerations influenced lay people's and scientists' innateness judgments: valuable traits (e.g., the capacity "to solve very complicated math problems") were judged more innate than disvalued traits; neutral traits were judged more innate when good environmental factors rather than bad environmental factors contributed to their development. Scientists and lay people alike were also able to overcome these tendencies when they were experimentally primed to make reflective judgments about innateness. In one experimental study, instead of being presented with only one vignette (e.g., a vignette describing the development of a valuable trait), participants were presented with a pair of vignettes (e.g., one describing the development of a valuable trait and one describing the development of a disvalued trait). Being confronted with the pair of vignettes led scientists and lay people to be more reflective, which limited the influence of moral considerations on their innateness judgments and prompted them to sharply distinguish learned and innate traits.

Knobe and Samuels's innovative research provides stimulating data that bear on the relations between scientific and lay concepts. Unfortunately, since lay people and scientists behaved similarly in all the studies they report, these data undermine the three hypotheses they put forward about these relations, all of which predict a difference between lay people's and scientists' judgments. At least for the case of the concept of innateness—and plausibly more generally—novel hypotheses about scientific concepts and their relation to lay concepts, to be tested by means of experiments, are called for.

3.3 A Defense of the Experimental Approach

In response to the proposal to study scientific concepts experimentally, Waters (2004) has argued that experimental studies of concepts cannot serve to test philosophical analyses of scientific concepts. In a nutshell, for Waters, the philosophical analysis of scientific concepts is at least partly a normative enterprise, whereas experimental studies of scientific concepts can only be descriptive and thus cannot serve to test philosophical analyses. Following the first section of Reichenbach's *Experience and Prediction* (1938), Waters proposes that philosophers of science engage in "rational reconstruction" of scientific concepts, where rationally reconstructing a scientific concept of x consists in presenting some propositions about x (or inference schemas about x) in a way that accords with logical and epistemological principles. The rational reconstruction of a concept need not be identical to what scientists unreflectively mean by a predicate expressing this concept. To adjudicate between competing rational reconstructions of a concept, Waters begins by describing Reichenbach's postulate of correspondence: the correct rational reconstruction of a concept is the one that scientists recognize as capturing what they themselves mean "properly speaking" when they use the predicate expressing that concept. As Waters put it (2004, 38):

> Epistemological descriptions may depart from the way scientists actually talk (in order to heed epistemological standards such as consistency and logical complete-ness), but the descriptions must satisfy the following condition: scientists would agree that the accounts accurately describe what they mean, "properly speaking."

In contrast, experimental studies are typically limited to studying how scientists actually talk or use a concept, and thus they cannot test proposed philosophical analyses of concepts. In the last section of his article, Waters ends up rejecting the idea that the postulate of correspondence is the only constraint on rational reconstruction on the grounds that scientists' own epistemic values should not provide the final say about what counts as a successful rational reconstruction. In the remainder of this section, we mostly discuss the postulate of correspondence.

Waters's attack against experimental conceptual analysis suffers from two main flaws that we discuss in turn. First, even if conceptual analysis were a form of rational reconstruction, it would not follow that candidate analyses could not be tested experimentally. Rational reconstructions reconstruct the way scientists use particular concepts. If one can show experimentally that a candidate rational reconstruction of a given concept x has nothing or little to do with scientists' unreconstructed use of x, then this gives us a strong reason to assume that the reconstruction is erroneous (for related ideas, see Schupbach forthcoming; Shepherd and Justus 2014). Second, Waters's alternative to experimental conceptual analysis—that is, his account of rational reconstruction—is unsatisfactory because, we argue, neither a judgment that a candidate rational reconstruction captures the content of a scientific concept properly speaking nor a judgment that it fails to do so can adjudicate between competing reconstructions. Rationally reconstructed concepts are bound to be used in ways that differ substantially from unreconstructed concepts since their use is supposed to obey logical and epistemological principles; as a result, scientists may deny that the proposed use captures what they mean, properly speaking. But this denial need not amount to a failure of the candidate reconstruction; rather, scientists may simply not be very good at seeing what they themselves mean, properly speaking. Scientists' agreement that a candidate reconstruction captures what they mean properly speaking does not fare better because a concept can typically be reconstructed in several incompatible ways and because scientists would probably agree with several of these. As noted, Waters ends up denying that the postulate of correspondence is the only constraint bearing on rational reconstruction, but his amendment to Reichenbach's approach is of little help since no constraint is described in detail.

To summarize, even if Waters were right that conceptual analysis in the philosophy of science is rational reconstruction, this would not undermine the utility of the type of experimental surveys conducted by Stotz and Griffiths. Furthermore, he has not presented a compelling characterization of the rational reconstruction of scientific concepts.

4 CONCLUSION

This chapter has reviewed three distinct approaches to study scientific concepts. Philosophers of science have often focused on the semantic content of scientific concepts or, equivalently for our purposes, on the meaning of scientific terms, motivated to a large extent by the challenges raised by the incommensurability thesis. In line with Quinean concerns about meaning, we have expressed doubts about the prospects of identifying the correct theory about what determines the semantic content of scientific concepts, and we have suggested that the description and assessment of scientific change and variation need not be framed in semantic terms. We then turned to the cognitive approach that treats scientific concepts as cognitive entities (data structures) and appeals to the cognitive psychology of concepts to understand scientific change. We have highlighted the importance of taking into account the recent developments in the psychology of concepts, highlighting the significance of the heterogeneity hypothesis for scientific concepts. Finally, if concepts are cognitive entities, then they plausibly can be studied experimentally. The last section reviewed this new experimental approach, highlighting its potential and defending it against criticisms.

REFERENCES

Amitani, Y. (2010). *The Persistence Question of the Species Problem.* Unpublished Ph.D. Dissertation, University of British Columbia.

Andersen, H., Barker, P., and Chen, X. (2006). *The Cognitive Structure of Scientific Revolutions* (New York: Cambridge University Press).

Barker, P. (1999). "Copernicus and the Critics of Ptolemy." *Journal for the History of Astronomy* 30, 343–358.

Barker, P., and Goldstein, B. R. (1988). "The Role of Comets in the Copernican Revolution." *Studies in the History and Philosophy of Science* 19, 299–319.

Barsalou, L. W. (1992). "Frames, Concepts, and Conceptual Fields." In A. Lehrer and E. F. Kittay (eds.), *Frames, fields, and contrasts* (New York: Lawrence Erlbaum), 21–74.

Block, N. (1986). "Advertisement for a Semantics for Psychology." *Midwest Studies in Philosophy* 10, 615–678.

Boyd, R. (1984). "The Current Status of Scientific Realism." In J. Leplin (ed.), *Scientific Realism* (Berkeley: University of California Press, 41–82.

Brigandt, I. (2010). "The Epistemic Goal of a Concept: Accounting for the Rationality of Semantic Change and Variation." *Synthese* 177, 19–40.

Carey, S. (1985). *Conceptual Change in Childhood* (Cambridge, MA: MIT Press).

Carey, S. (2009). *The Origin of Concepts* (Oxford: Oxford University Press).

Carnap, R. (1950). *Logical Foundations of Probability* (Chicago: University of Chicago Press).

Cheon, H. (ms). Structural Heterogeneity of Scientific Concepts.

Cheon, H., and Machery, E. (2010). "Review of *Creating Scientific Concepts*, by Nancy J. Nersessian." *Mind* 119, 838–844.

Devitt, M. (1979). "Against Incommensurability." *Australasian Journal of Philosophy* 57, 29–50.

Earman, J. (1977). "Against Indeterminacy." *The Journal of Philosophy* 74, 535–538.

Feyerabend, P. K. (1962). "Explanation, Reduction and Empiricism." In H. Feigl and G. Maxwell (eds.), *Minnesota Studies in the Philosophy of Science (Vol. 3): Scientific Explanation, Space and Time* (Minneapolis: University of Minnesota Press), 28–97.

Field, H. (1973). "Theory Change and the Indeterminacy of Reference." *The Journal of Philosophy* 70, 462–481.

Fine, A. (1975). "How to Compare Theories: Reference and Change." *Nous* 9, 17–32.

Fodor, J. A. (1998). *Concepts: Where Cognitive Science Went Wrong* (Oxford: Clarendon Press).

Griffiths, P. E., Machery, E., and Linquist, S. (2009). "The Vernacular Concept of Innateness." *Mind and Language* 24, 605–630.

Griffiths, P. E., and Stotz, K. (2008). "Experimental Philosophy of Science." *Philosophy Compass* 3, 507–521.

Hanson, R. N. (1958). *Patterns of Discovery* (Cambridge: Cambridge University Press).

Hoyningen-Huene, P. (1993). *Reconstructing Scientific Revolutions: Thomas S. Kuhn's Philosophy of Science* (Chicago: University of Chicago Press).

Johnson-Laird, P. N. (1983). *Mental Models: Towards a Cognitive Science of Language, Inference, and Consciousness* (Cambridge: Harvard University Press).

Keil, F. C. (1989). *Concepts, Kinds, and Cognitive Development* (Cambridge: MIT Press).

Kitcher, P. (1978). "Theories, Theorists and Theoretical Change." *The Philosophical Review* 87, 519–547.

Knobe, J., and Samuels, R. (2013). "Thinking Like a Scientist: Innateness as a Case Study." *Cognition* 128, 72–86.

Koponen, I. T., and Kokkonen, T. (2014). A Systemic View of the Learning and Differentiation of Scientific Concepts: The Case of Electric Current and Voltage Revisited." *Frontline Learning Research* 2, 140–166.

Kuhn, T. S. (1962). *The Structure of Scientific Revolutions* (Chicago: University of Chicago Press).

Machery, E. (2009). *Doing Without Concepts* (New York: Oxford University Press).

Machery, E. (2010). "Précis of *Doing Without Concepts*." *Behavioral and Brain Sciences* 33, 195–206.

Machery, E. (2016). "Experimental Philosophy of Science." In W. Buckwalter and J. Sytsma (eds.), *Routledge Companion to Experimental Philosophy* (New York: Routledge), 475–490.

Medin, D. L., and Schaffer, M. M. (1978). "Context Theory of Classification Learning." *Psychological Review* 85, 207–238.

Minsky, M. (1975). "A Framework for Representing Knowledge." In P. Winston (ed.), *The Psychology of Computer Vision* (New York: McGraw-Hill).

Murphy, G. L., and Medin, D. L. (1985). "The Role of Theories in Conceptual Coherence." *Psychological Review* 92, 289–316.

Nersessian, N. J. (1984). *Faraday to Einstein: Constructing Meaning in Scientific Theories* (Dordrecht: Kluwer Academic).

Nersessian, N. J. (1989). "Conceptual Change in Science and in Science Education." *Synthese* 80, 163–183.

Nersessian, N. J. (2002). "Maxwell and 'the Method of Physical Analogy': Model-Based Reasoning, Generic Abstraction, and Conceptual Change." In D. Malament (ed.), *Reading Natural Philosophy* (Chicago: Open Court), 129–166.

Nersessian, N. J. (2008). *Creating Scientific Concepts*. (Cambridge, MA: MIT Press).

Nosofsky, R. M. (1986). "Attention, Similarity, and the Identification–Categorization Relationship." *Journal of Experimental Psychology: General* 115, 39.

Peacocke, C. (1992). *A Study of Concepts*. (Cambridge, MA: MIT Press).

Putnam, H. (1975). "The Meaning of 'meaning.'" In H. Putnam (ed.), *Mind, Language, and Reality* (Cambridge: Cambridge University Press), 215–271.

Quine, W. V. O. (1951). "Two Dogmas of Empiricism." *The Philosophical Review* 60, 20–43.

Quine, W. V. O. (1960). *Word and Object* (Cambridge, MA: MIT Press).

Rice, C. (forthcoming). "Concepts as Pluralistic Hybrids." *Philosophy and Phenomenological Research*.

Reichenbach, H. (1938). *Experience and Prediction: An Analysis of the Foundations and the Structure of Knowledge* (Chicago: University of Chicago Press).

Rosch, E. (1978). "Principles of Categorization." In E. Rosch and B. Lloyd (eds.), *Cognition and Categorization* (New York: Lawrence Erlbaum), 27–48.

Rosch, E., and Mervis, C. B. (1975). "Family Resemblances: Studies in the Internal Structure of Categories." *Cognitive Psychology* 7, 573–605.

Samuels, R., and Ferreira, M. (2010). "Why Don't Concepts Constitute a Natural Kind?" *Behavioral and Brain Sciences* 33, 222–223.

Scheffler, I. (1967). *Science and Subjectivity* (Indianapolis, IN: Bobbs-Merrill).

Schupbach, J. N. (forthcoming). "Experimental Explication." *Philosophy and Phenomenological Research*.

Shapere, D. (1966). "Meaning and Scientific Change." In R. Colodny (ed.), *Mind and Cosmos: Essays in Contemporary Science and Philosophy* (Pittsburgh, PA: University of Pittsburgh Press), 41–85.

Shepherd, J., and Justus, J. (2014). "X-phi and Carnapian Explication." *Erkenntnis* 80, 381–402.

Shtulman, A. (2009). "Rethinking the Role of Resubsumption in Conceptual Change." *Educational Psychologist* 44, 41–47.

Stotz, K., and Griffiths, P. E. (2004). "Genes: Philosophical Analyses Put to the Test." *History and Philosophy of the Life Sciences* 26, 5–28.

Stotz, K., Griffiths, P. E., and Knight, R. D. (2004). "How Scientists Conceptualize Genes: An Empirical Study." *Studies in History and Philosophy of Biological and Biomedical Sciences* 35, 647–673.

Thagard, P. (1990). "Concepts and Conceptual Change." *Synthese* 82, 255–274.

Vicente, A., and Martinez Manrique, F. (2016). "The Big Concepts Paper: A Defense of Hybridism." *The British Journal for the Philosophy of Science* 67, 59–88.

Waters, C. K. (2004). "What Concept Analysis in Philosophy of Science Should Be (and Why Competing Philosophical Analyses of Gene Concepts Cannot Be Tested by Polling Scientists)." *History and Philosophy of the Life Sciences* 26, 29–58.

Weiskopf, D. (2009). "The Plurality of Concepts." *Synthese* 169, 145–173.

CHAPTER 25

....

SCIENTIFIC EXPLANATION

....

BRADFORD SKOW

1 IDENTIFYING THE SUBJECT

....

THE title of this chapter is misleading. "Scientific explanation" is the traditional name for a topic that philosophers of science are supposed to have something to say about. But it is a bad name for that topic. For one thing, theories of scientific explanation don't have an activity that is exclusive to scientists as their subject matter. Nonscientists do it, too, all the time; they just have less specialized knowledge to use and direct their attention to less complicated phenomena. This suggests that we drop the adjective "scientific" and call the topic "explanation." But the word "explanation" all by itself is also a bad name for the topic. What philosophers have called theories of explanation are meant to say something about what is happening in scenarios like these:

(1) A physicist explains why the planetary orbits are stable.
(2) A father explains to his older child why his younger child is crying.

There is the word "explains," so what is the problem? The problem is that what philosophers have called theories of explanation are not intended to have anything to say about what goes on when

(3) A biologist explains what the theory of evolution says

or when

(4) A policeman explains where the train station is.

The difference is clear: (1) and (2) are cases where someone explains *why* something is the case, whereas in (3) and (4) someone explains *what* or *where*.

So, maybe a better name for what philosophers of science are after is a "theory of explaining why." But I have my doubts even here.[1] Explaining why P is a speech act, and when someone successfully performs this speech act they, among other things, convey to their audience the answer to the question why P. But it seems that one can answer the question why P without explaining why P. It seems, for example, that I could tell you why the planetary orbits are stable without explaining to you why the planetary orbits are stable. So there are two components to explaining why: (i) answering the relevant why-question and (ii) doing this by performing the speech-act of explaining. How important is the second part to the philosophy of science? Does it matter to the philosophy of science what makes explaining different from other speech-acts?

Some theories of explanation do give the speech-act of explaining an important role to play, but in this survey I will focus just on the first component of explaining why. A theory that addresses that component is a theory of answers to why-questions.[2]

Why-questions are important in everyday life and in the sciences. The natural curiosity that we are all born with and that most of us retain throughout our lives is in part a curiosity about why things happen. And scientists do not just happen to go around answering why-questions; answering why-questions is one of the aims of science. It was a great scientific advance when physicists finally figured out why hydrogen has the spectrum it does. And the fact that they needed to use quantum mechanics to answer this question told in favor of that theory (which at the time was still new).[3]

So what does it take for a body of information to be an answer to the question why P? One very common form answers take is "P because Q." This is not the only form answers to why-questions take: "The chicken crossed the road to get to the other side" is an answer to "Why did the chicken cross the road?" But I lack the space here to go into the relationship between because-answers and to-answers, and, anyway, a great deal of science aims at because-answers not to-answers.[4]

---

[1] I develop these doubts in more detail in chapter 2 of Skow (2016).

[2] Van Fraassen (1980) identifies theories of explanation with theories of answers to why-questions. Most other philosophers deny that their theory is meant to cover all and only answers to why-questions. Hempel (1965) tried to distinguish between "explanation-seeking" why-questions and "evidence-seeking" why-questions and claimed that his theory was only about the first kind of why-question. I doubt that there is any distinction of this sort to be made; see chapter 2 of Skow (2016). Achinstein (1983) gave the speech-act of explaining an important role in his theory. Other philosophers think a theory of explanation should say something about explaining how and so should say something about how-questions, not just why-questions: see Hempel (1965: section 5.2), which cites Dray (1957); see also Cross (1991). But I doubt that there is a principle by which we can broaden a theory of explanation to cover how-questions and still exclude who-, what-, when-, and where-questions.

[3] This is controversial; one form of scientific anti-realism says that science aims to systematically describe what happens, not to say anything about why things happen.

[4] Biology and psychology are major exceptions. We often explain people's behavior in terms of their purposes and at least seem to explain the behavior of "lower" forms of life in terms of the functions that behavior serves. I regret that I cannot address the controversies around these kinds of explanation.

Having focused again on a narrower target, the question becomes: what have philosophers of science had to say about what it takes for it to be the case that *P* because *Q*? We want to see how the theories fill in the schema

(5) Necessarily, *P* because *Q* if and only if. . . [5]

Let us start at the beginning.

2 The Deductive-Nomological Model

The "fountainhead" of almost all contemporary thinking about explanation is Carl Hempel's deductive-nomological (DN) "model" of explanation.[6] Put into the form (5), the theory says

(6) *P* because *Q* if and only if (i) it is true that *Q*, (ii) the fact that *Q* entails the fact that *P*, (iii) the fact that *Q* has at least one law of nature as a conjunct, and (iv) the fact that *Q* would not entail the fact that *P* if the laws were removed.

A more traditional way to state the DN model is to say that "explanations are arguments" of a certain kind. The conjunction of the argument's premises expresses the fact that *Q*; the argument's conclusion expresses the fact that *P*; clauses (i) and (ii) correspond to the claim that the argument is sound; and clause (iii) corresponds to the claim that at least one of the argument's premises must be (or express) a law of nature.

There are plenty of examples that Hempel's theory seems to get exactly right. Suppose you drop a rock from 1 meter above the ground, and it hits the ground at a speed of .45 meters per second. Why does it hit with that speed? The following seems like a good way to answer. Use Newton's second law and the law of universal gravitation to deduce the speed at impact from the height of the fall. Here, the facts cited in the answer include laws of nature (Newton's), and the fact being explained is deduced from them, just as Hempel's theory requires. And this is not an isolated example. Hang out with the right sort of physicists or applied mathematicians and you will spend all day solving differential equations. It looks like the point of doing this is often to answer why-questions. Why is such and such system in such and such state at time T? Solve the differential equation describing its behavior over time, plug in its state at the initial time, and you have

---

[5] Depending on one's metaphilosophical views, one might want something stronger than a necessarily true biconditional. I won't worry about those distinctions here. I will also also suppress the "necessarily" in what follows.

[6] First presented in 1948, in a paper jointly authored with Paul Oppenheim (Hempel and Oppenheim 1948). The most elaborate development and defense is in Hempel (1965). Wesley Salmon called it the "fountainhead" (1989, 12).

thereby deduced its state at T from, among other things, a law (the equation). You have also thereby, it certainly seems, explained why it is in that state.

Hempel's DN model was not meant to apply to "statistical explanations." Still, even in its intended domain it was shown to be false in the 1960s. Here are two of the many well-known counterexamples:

- *The ink bottle*: Jones knocks an ink bottle with his knee. It topples over and spills ink all over the rug. Then the rug is stained because Jones knocked the ink bottle. But "Jones knocked the ink bottle" does not contain a law of nature.[7]
- *The flagpole*: A flagpole is 10 meters high and it casts a shadow 10 meters long. The angle that the line passing through the sun, the top of the flagpole, and the top of the shadow makes with the earth is 45 degrees. The fact that the shadow is 10 meters long and that the sun is in that position, together with the laws of geometric optics, entails that the flagpole is 10 meters high. But it is false that the flagpole is 10 meters high because this fact obtains. It is not 10 meters high because it casts a 10-meter long shadow.[8]

A diagnosis of where Hempel went wrong is not hard to find. In the ink bottle, we find out why the rug is stained by learning what caused the stain, even though no laws are mentioned.[9] In the flagpole, we see the height of the flagpole deduced from a law without any mention of any the causes of its having that height, but we do not learn why the flagpole is that high. The obvious moral is that an answer to a why-question needs to say something about causes but does not need to mention any laws.

3 CAUSAL THEORIES OF EXPLANATION

The moral needs to be qualified and refined if it is to have any chance of being true. First, we are clearly here focusing only on why-questions that ask why some given event occurred. Only events can have causes, so it only makes sense to impose a causal requirement on answers to why-questions about events. This limits the scope of the theory; it makes sense to ask why Galileo's law of free fall is true or why every polynomial over the complex numbers has a root even though these are not questions about events.

Second, "an answer to the question why E occurred must say something about the causes of E" is too vague. A philosophical theory needs to say something much more precise and detailed.

---

[7] This example is originally due to Scriven (1959, 456). See also Scriven (1962).

[8] This example is originally due to Bromberger. He never published it, although he did discuss similar examples in published work (see Salmon 1989, 72).

[9] In response to Scriven's example, Hempel held that laws were still in some sense implicit in the explanation (Hempel 1965). Woodward examines and opposes this reply in chapter 4 of his book (2003).

There have been several attempts to do so. A naïve causal theory of explanation might just say that to answer the question why E occurred it is necessary and sufficient to describe at least one cause of E. Philosophers have articulated more detailed and nuanced causal theories of explanation. Wesley Salmon's theory (1984) and David Lewis's theory (1986) are two prominent examples. Lewis's can be stated with fewer technical notions, so let us take a brief look at it.[10]

It is actually not easy to put Lewis's theory in the form (5). Lewis's theory is instead best understood as a theory of complete answers, and of partial answers, to why-questions. The distinction between complete answers and partial answers should be familiar. Suppose I ask you who is taking your class. You have twenty students, but you just say that Jones is enrolled. You have given me a partial answer. If you had said the names of all your students and told me that those were all the students, you would have given me a complete answer.

Lewis's theory is best understood as having two parts; one says what it takes for something to be a complete answer to a why-question, the other what it takes to be a partial answer:

(7) A proposition P is the complete answer to the question why event E occurred if and only if P is a maximally specific proposition about E's causal history. (It says what E's causes are, what the causes of its causes are, and so on.)

(8) A proposition P is a partial answer to the question why event E occurred if and only if P is about E's causal history (it need not be maximally specific).

A window breaks, and we want to know why. On Lewis's view, "Because someone threw a rock at it" is only a partial answer. For although the proposition that someone threw a rock is about the causal history of the breaking, it is not maximally specific. It does not tell us who threw the rock or about any of the other causes of the breaking (e.g., the causes of the throw). "Because Adam threw a rock at it" is also a partial answer but is more complete. It pins down more facts about the causal history than the first answer does.

Obviously, if (7) is right then complete answers to why-questions are impossible to write down. None of us has ever heard one. We trade only in partial answers.

Is some causal theory of explanation—Lewis's, Salmon's, or some other—correct? A lot of alleged counterexamples turn on controversial claims about causation and apply only to some causal theories of explanation. A survey of all of them is too big a project for this chapter; here is just one example. I hit a vase, and it cracks. It cracked, in part, because it was fragile. But fragility is a disposition, and some philosophers hold that dispositions cannot cause anything. If that is right, then the naïve causal theory of

---

[10] There are also a lot more objections to Salmon's than to Lewis's, in part because Salmon's is tied to one particular (and particularly implausible) theory of causation. See Hitchcock (1995) for some persuasive objections to Salmon's theory. (David Lewis, of course, also had a theory of causation—a few in fact—but one need not accept his views about causation to accept his theory of explanation.)

explanation is in trouble. But this kind of example may not refute the naïve theory; there are theories of causation on which dispositions can be causes.[11] And, anyway, even if the naïve theory is in trouble, Lewis's theory, for example, may not be. One could make the case that the fact that the vase is fragile tells us something about the causal history of the breaking. (Certainly the set of "possible causal histories" of a nonfragile vase is different from the set of possible causal histories of a fragile vase. It "takes more" to break a nonfragile vase.)

Lewis's theory is relatively undemanding; the class of facts that are about the causal history of some event E will typically be quite large, much larger than the class of facts that describe some particular cause of E. Still, some explanations appear to fail to meet even Lewis's undemanding criteria. Elliott Sober (1983) drew attention to "equilibrium explanations." R. A. Fisher (1931) explained why the current sex ratio among humans is one-to-one. In broad outline, his answer was that if the sex ratio had ever departed from one-to-one, individuals who overproduced the minority sex would have higher fitness, and so the sex ratio would move back toward one-to-one. Elliott Sober argued that equilibrium explanations are noncausal. Here is Sober's summary:

> a causal explanation . . . would presumably describe some earlier state of the population and the evolutionary forces that moved the population to its present [sex ratio] configuration. . . .. Where causal explanation shows how the event to be explained was in fact produced, equilibrium explanation shows how the event would have occurred regardless of which of a variety of causal scenarios actually transpired. (202)

If Sober had had Lewis's theory of explanation in mind, he might have continued: someone X could know every detail of every cause of the current sex ratio without knowing anything about what the sex ratio would have been if those causes had not occurred. And, the argument goes, if X were to learn more about what the sex ratio would have been, if those causes had not occurred, then after learning that information X would know more about why that ratio is 1:1. If all this is right, then it looks like Lewis's theory is false.[12]

Another important class of examples worth mentioning are examples of "mathematical explanations of physical phenomena." Why are the periods of the life cycles of cicadas prime numbers? Members of a cicada species emerges once every N years to mate and then die. If a predator species also has a periodic life cycle, and its life cycle in years is some number that divides N, then predators of those species will be around whenever the cicadas emerge. Predation will then reduce the size of the next cicada generation. It is therefore advantageous for the cicada to have a life cycle period that, when measured in years, has as few divisors as possible. Periods that, when measured in years, are prime

---

[11] On the question of whether the fragility explanation is causal see Lange (2013, 494), who gets the example from Jackson and Pettit (1992). For a discussion of whether dispositions can be causes see Choi and Fara (2014).

[12] Few now accept that Fisher's explanation is a "noncausal explanation." Strevens, for example, argues in detail that equilibrium explanations count as causal on his theory (2008, 267–272).

numbers, are therefore the best candidates. It looks here like the fact that prime numbers have the smallest number of divisors helps answer a why-question. Some doubt that this fact could count as causal-explanatory information.[13]

I think the jury is still out on whether a causal theory of explanation is defensible.[14] But if no such theory is true, what is the alternative? Some opponents of causal theories see some other common factor in the alleged counterexamples we have surveyed. I'll say something about one attempt to isolate that factor—unificationist approaches to explanation—later. Another conclusion would be to think that the class of answers to why-questions exhibits a great deal of diversity and little unity. If that's right, then what philosophers should be doing is cataloguing that diversity, identifying the different varieties of answer and putting them on display.

4 Probabilities in Explanation

Alongside his DN model of explanation, Hempel developed what he called the *inductive-statistical* (IS) model of explanation. What kinds of answers to why-questions is this model meant to apply to?

Suppose Jones has strep throat, and his doctor gives him penicillin. The penicillin works, and Jones recovers within a week. Why did he recover? It certainly looks like the fact that he took (an appropriate dose of) penicillin should be part of the answer. But taking penicillin does not guarantee recovery from strep in a week; only, say, 90% of those who take it recover that fast.[15] So this cannot be a DN explanation citing the "law" that everyone who suffers from strep and takes penicillin recovers in a week. For there is no such law. Hempel's IS model is meant to apply here instead:

- *P* because *Q* if and only if (i) it is true that *Q*, (ii) the probability of *P* given *Q* is high, and (iii) the fact that *Q* has at least one law as a conjunct.

The IS model, like the DN model, requires there to be a law, but relaxes the requirement that the fact that *Q* entail the fact that *P*, substituting a high probability requirement. If we assume Hempel's "explanations are arguments" point of view, then IS explanations are good inductive arguments with law-statements among their premises, where a good inductive argument is one in which the conclusion has high probability given the

---

[13] This example was introduced into the literature by Baker (2005). Lange (2013) discusses a host of examples that he calls "distinctively mathematical" and defends the claim that they are noncausal explanations.

[14] I defended a causal theory similar to Lewis's in Skow (2014), and I and propose and defend a new causal(ish) theory in Skow (2016).

[15] I made this number up.

premises. In the penicillin example, Hempel was willing to count as a law the fact that the probability of recovery from strep in 1 week when given penicillin is 90%.

The IS model fails. Even if taking penicillin raised the probability of recovery within a week from, say, 10% to 20%, it would still be true that Jones recovered in a week because he took penicillin. In the wake of the IS model's failure, several other theories have tried to state just what facts about probabilities must, or may, appear in an explanation. Wesley Salmon's *statistical relevance* theory of explanation was an important successor (Salmon, Jeffrey, and Greeno 1971). It was meant to apply to why-questions of the form "Why is this x, which is A, also B?" ("Why is Jones, who had strep, a person who recovered from strep in one week?") Roughly speaking, Salmon required an answer to this question to (i) cite a factor C that is "statistically relevant" to being B—the probability that x is B, given that x is A and C, must be different from the probability that x is B, given that it is A; and also (ii) state just what these probabilities are.[16]

Is there any need to have a separate theory of statistical explanations? The naïve causal theory of explanation does not need to be amended or augmented to deal with the penicillin example. True, penicillin doesn't guarantee recovery—it just makes recovery more likely. Nevertheless, Jones's taking penicillin still caused his recovery, and so the naïve theory will say that the proposition that Jones took penicillin answers the question why he recovered. No statistics need to appear in the answer.

One might reply that if (i) one answers the question why some event E occurred by citing a cause of E, and (ii) there is an interesting statistical relationship between E-type events and C-type events,[17] then one's answer could be improved by citing that statistical relationship. The suggestion is that "Because he took penicillin, and penicillin raises the chance of recovery in a week to 90%" is a better answer to the question why Jones recovered than is "Because he took penicillin." If this is right, then causal theories of explanation do need to say something special about statistical explanations. But one might doubt that the longer explanation is a better answer. Does the fact that penicillin raised the chance of recovery really add anything to the fact that the penicillin caused the recovery? The statistics might be helpful if we do not yet know whether penicillin caused the recovery, but, once we do know this, the facts about the statistics can seem irrelevant. One might in fact hold that, if some causal theory of explanation is true, then no true answer to a why-question about an event mentions probabilities or chances.[18]

One complication here is the possibility that causation can be analyzed in terms of probabilities. Perhaps for C to be a cause of E is for the probability of E given C to be related in the right way to other probabilities. For example, one might hold that for C to be a cause of E is for the probability of E given C to be higher than the unconditional

---

[16] I have omitted many details of the theory.

[17] That Suzy threw a rock at the window did not guarantee that the window would break. But there is no interesting statistical relationship between rock-throwings and window-breakings.

[18] Compare Humphreys (1989, 109) and Kitcher (1989, 422). The restriction to why-questions about events is important. Even if answers to why-questions about events should not cite statistics, it might still be that answers to other kinds of why-questions should.

probability of E. If this view is right, then maybe the fact that C is a cause of E is the very same fact as some fact about probabilities. Then one cannot say that an answer to a why-question about an event should cite facts about causes but not facts about probabilities.

In fact, if to be a cause of E is to raise the probability of E, then the rough version of Salmon's statistical relevance theory that I have been working with is (almost) equivalent to the naïve causal theory of explanation.[19] It seems to me that insofar as the statistical relevance theory (in its more complete version) gets things right, it is because it is using statistical relationships as a stand-in, or surrogate, for causal relationships. The idea that causation can be analyzed in terms of probabilities, however, faces a lot of challenges.[20]

There is a need to have a separate theory of statistical explanations if there are "noncausal" statistical explanations. On some interpretations, quantum mechanics says that radioactive decay is a genuinely indeterministic process. When an atom decays by, say, emitting a beta particle, the laws of quantum mechanics, plus the exact condition of the atom and its environment just before the decay, did not determine that it would decay. Now suppose we ask why the atom decayed. Peter Railton held that we can answer this question by deducing the probability of decay from the laws of quantum mechanics and the prior condition of the atom and its environment (Railton 1978). This probability doesn't seem to have anything to do with the causes of the decay (it is not clear that the decay has any causes). So, if Railton is right that facts about the probability help explain why the atom decayed, then we have a "noncausal" role for probabilities in explanations.[21] Probabilities also play an important role in the answers to why-questions that statistical mechanics offers. Some have held that the information about probabilities in these explanations is also not causal-explanatory information.[22]

5 THEORIES OF CAUSAL EXPLANATION

Coming up with a theory that applies generally to all why-questions, a filling-in of (5)

(5) P because Q if and only if . . .

[19] Salmon allows conditions that lower the probability of E to explain E, even though they do not count as causes on the current proposal.
[20] See Hitchcock (2012) for an overview. Cartwright (1979) and chapter 4 of Spirtes, Glymour, and Scheines (2001) contain arguments that statistical relationships between events underdetermine causal relationships.
[21] This is a big "if"; Kitcher, for example, denies that quantum mechanics explains why the atom decayed (Kitcher 1989, 451). I should say that Railton himself thought explanations like this were causal.
[22] Chapter 10 of Strevens (2008) attempts to locate a noncausal explanatory role for probabilities. Statistical mechanics is a popular hunting ground for those looking for noncausal explanations. For some examples, see Batterman (1992) and Reutlinger (2014).

that covers all cases, is a daunting project. We have already seen some philosophers narrow their sights and just focus on questions that ask why some event occurred. Some philosophers narrow their sights even more. They say that they aim to state and defend a theory of causal explanation of events. Instead of saying what it takes for something to be an explanation "full stop," or even an explanation of an event, the project is that of saying what it takes for something to be a *causal* explanation of an event. James Woodward's book *Making Things Happen* (2003) and Michael Strevens's book *Depth* (2008) are two important recent works of this kind.

Each book builds its theory of causal explanation on a theory of causation. Woodward starts with a counterfactual theory of causation. His basic insight is that causation is a kind of conditional counterfactual dependence: C is a cause of E if and only if, holding fixed the occurrence of certain facts, had C not occurred, E would not have occurred. Woodward has a lot to say about just how this insight is to be developed and understood: a lot to say about what facts may be held fixed, and a lot to say about how to understand the way he is using counterfactual "had C not occurred, E would not have occurred." Since this is a chapter on explanation, not on causation, I must pass over these details.[23]

What does Woodward say it takes to causally explain why some event occurred? In one place, he summarizes his theory like this: "[causal] explanation is a matter of exhibiting systematic patterns of counterfactual dependence" (191). So, on Woodward's view, citing a cause of E is a way to causally explain why E occurred because, he claims, citing a cause of E conveys information about what E counterfactually depends on. But if Woodward is right, then even after I have told you that C caused E, I can give you still more causal-explanatory information, *even if I do not tell you anything about E's other (token) causes.* For I might tell you more about the way in which E depends on C. Suppose that there is a pot of water on my stove, over a lit burner, and that the water is at a steady 110° Fahrenheit. Suppose that it is the knob's being turned halfway that caused the water to be at that temperature. Then I can give some causal-explanatory information about the temperature of the water by citing this cause and saying that the knob is turned halfway. On Woodward's way of thinking about causation I have, in effect, told you that there is *some* other position of the knob P such that, if the knob had been at P (instead of at the half-way mark), then the water would have been at a different temperature. Obviously, I could provide you with more of this kind of information. I could be more specific about the way the temperature depends on the knob's position. For example, I could say that if the knob were turned all the way, the water would be boiling. Woodward holds that this information is also causal-explanatory information about the

---

[23] Woodward's theory of causation makes use of "structural equations" for modeling causal facts. Two important technical works in this tradition are Pearl (2009) and Spirtes, Glymour, and Scheines (2001). I have described Woodward's theory of token causation between events, but more central to his theory is his theory of type-level causation between variables (a variable is a "generalized event"; events occur, or fail to occur, in each possible world, whereas variables take on one of many possible values in each possible world). Yablo (2002) is an important defender of a conditional dependence theory of token causation that does not rely on structural equations.

water's temperature. In general, information about how the event being explained is situated in a range of counterfactual alternatives is causal-explanatory information. Bare claims about the token causes of the event are one way to provide this kind of information, but not the only way.

Let me now turn to Strevens's theory. Strevens thinks that there are two kinds of causation. On the one hand, there is the low-level, microphysical production of precisely specified events, like that event consisting in *these* atoms being in exactly *these* positions at this time. On the other hand, there is high-level causation between imprecisely specified events, as when Suzy's throw causes the window to break. These events are imprecise because there are a huge number of microphysical ways Suzy could have thrown and a huge number of ways in which the window could break.

Strevens analyzes high-level causation in terms of low-level causation; he says relatively little about what low-level causation is. Here, in brief outline, is how the analysis goes. He agrees with the common thought that the causes of an event "make a difference" to whether that event occurred. But he does not understand difference-making in counterfactual terms. Instead, he starts with the notion of a set of facts "causally entailing" another fact. Set S causally entails F if and only if, first, S entails F, and second, this logical relation "mirrors" the low-level, microphysical production of F (or of the "maximal precisification of F") by the members of S. Suppose we have such an S and F. Strevens then subjects the set S to a process of "abstraction." Roughly speaking, this process weakens the members of S as much as possible, consistent with S still causally entailing F. So we might start with the fact that I threw a rock with a mass of 50 grams at that window, the fact that it had an initial velocity of 10 m/s, the fact that there was nothing around to interfere with the rock's flight, the fact that the window was made of normal glass and was 5 mm thick, and the fact that certain laws of mechanics are true. This set causally entails that the window breaks. But none of the facts mentioned is a difference-maker; they are "too specific." After Strevens's process of abstraction does its work on this set, the fact that the rock had a mass of 50 grams and was thrown at 10 m/s will have been replaced by the (weaker) fact that it was thrown with an initial momentum in a certain range. (It will be the range containing exactly those initial momenta that causally entail that a window of that sort will break under these conditions; the momentum .5 kg m/s will fall in this range.) *This* fact, about the initial momentum, is *not* "too specific" to be a difference-maker; it is just specific enough and so counts as a cause of the breaking.[24]

So we start with a set of facts S that causally entails the occurrence of some event E; Strevens' abstraction procedure returns a set of weaker facts S^* that (i) also causally

[24] Stephen Yablo was a pioneer of the idea that the causes of E are the events that are (i) "causally sufficient" for the occurrence of E that are also (ii) "not too specific" (Yablo 1992). The idea that the causes of E are facts that play an essential role in entailing the occurrence of E was a key component of J. L. Mackie's theory of causation (Mackie 1974). Officially Strevens's process is a process of "optimization," which has a process of abstraction as a part. I do not have space to go into the other components of the process.

entails that E occurs, but now also (ii) contains facts that correspond to causes of E.[25] What then is Strevens's theory of causal explanation? He says that the deduction of the occurrence of E from S^* is a "standalone" causal explanation of the occurrence of E. Other kinds of causal explanations are to be understood in terms of standalone explanations; they are the basic building blocks of his theory.

6 Unificationist Theories of Explanation

The original goal was to say what it takes for it to be the case that P because Q. Causal theories of explanation lower their sights and focus only on explanations of events. Theories of causal explanation lower their sights further and focus only on one particular kind of explanation of events. Unificationist theories of explanation aim back up at the original target. Their slogan is: to explain why is to unify.

So what is unification, and how might the notion of unification figure in a theory of answers to why-questions? Michael Friedman and Philip Kitcher are the founding fathers of the unificationist approach to explanation (see Friedman 1974; Kitcher 1981, 1989). For reasons of space I will just discuss Kitcher's theory. Central to Kitcher's theory is the notion of a set of argument-forms unifying a set of propositions. He illustrates this notion with an example:

> The unifying power of Newton's work consisted in its demonstration that one pattern of argument could be used again and again in the derivation of a wide range of accepted sentences. (1981, 514)

You want to prove that apples take so-and-so many seconds to fall from a tree of such-and-such a height? Newton showed that you could prove this by setting up a differential equation governing the apple's motion, use his three laws and the law of universal gravitation, and then solve that equation. You want to prove some facts about the moon's orbit? You can prove them using the same method. You want to prove that the tide is higher when the moon is overhead? Same form of argument again. In Kitcher's view, that's what it is for Newton's theory to unify all of these phenomena.

We need a more precise statement of Kitcher's notion of unification. The things to which Kitcher's notion of unification apply are sets of argument-forms. Different sets of argument-forms might unify a body of fact to different degrees. Explanation is going to go with being a "best" unifier, so we need to know: given a set of propositions K, what is it for one set of argument-forms to unify K better than another? Only sets of

---

[25] For example, the fact that this window broke at T corresponds to the event that consists in the breaking of the window.

argument-forms that "generate K from one of its subsets" are even in the running. (These are sets of argument-forms for which there is some subset P of K—the "premise-set"— such that, for each proposition X in K, there is a valid argument for X from premises in P that instantiates one of the forms.) Among these, Kitcher says, sets of argument-forms that use smaller premise-sets unify better, other things being equal; smaller sets of argument-forms unify better, other things being equal; and sets of argument-forms each of whose members have a relatively small set of instances unify better, other things being equal. These criteria pull against each other, and it is not obvious how to arrive at an all-things-considered verdict. But let us pass over these details. Kitcher's theory of explanation makes use of this notion of unification in the following way:

- An argument is an answer to the question why P[26] if and only if it is a sound argument with P as its conclusion and it instantiates one of the argument-forms that belongs to the set of argument-forms that best unifies the set of all "Humean" truths.[27] (To save writing, let us say that an argument-form that appears in the set of argument-forms that best unifies is a "good" argument-form.)

Kitcher takes from Hempel the idea that explanation is deduction. But for Hempel explanation is local: if you can deduce P from some laws, then you have answered the question why P is true no matter what the answers to other why-questions look like. For Kitcher, explanation is global or holistic: roughly speaking, a deduction of P from some laws gets to be an answer to the question why P only if arguments of the same form can be used to deduce many other truths.[28] Kitcher thinks the global nature of his theory is a virtue. The standard counterexamples to the DN model, he claims, involve a deduction of some proposition from some laws where the deduction does not instantiate one of the good argument-forms. We can deduce the height of the flagpole from the length of its shadow and the laws of optics. But, he argues, the set of argument-forms that best unifies the set of all truths will not include an argument-form for deducing phenomena from their effects; the set of argument-forms that best unifies the phenomena will certainly need an argument-form for deducing phenomena from their causes, and there do not seem to be any phenomena that can be deduced from the effects but not their causes.

Some have worried that Kitcher's theory, even if it says the right thing about the flagpole example, says it for the wrong reasons. If flagpoles were in the habit of springing up uncaused, then maybe we could unify the phenomena better if we had an argument-form for deducing events from their effects. Even in a possible world like that, though,

---

[26] Kitcher's theory is slightly easier to digest if put in this form, rather than form (5).

[27] Among the non-Humean truths are those of the form "P because Q." See section 8.3 of Kitcher (1989) for a more complete characterization of the Humean truths, as well as Kitcher's reasons for excluding the non-Humean ones. It is worth noting that Kitcher's restriction means that, on his view, questions of the form "Why is it the case that P because Q?" do not have answers. This is a flaw since "higher order" explanations do exist. My statement of Kitcher's view is simplified in a few ways that will not matter for my discussion.

[28] For the reason why this is rough, see the "natural worry" later.

it does not seem that the flagpole is 10 meters high because it casts a shadow 10 meters long.[29]

Kitcher had an unusual view about causation. He held that *what it is* for *C* to be a cause of *E* is for *C* to appear in the correct answer to the question why *E* occurred (for a more precise statement of his view see Kitcher [1989, 497]; see also Kitcher [1986, 229]). So Kitcher would say that the world we are imagining, in which the length of a flagpole's shadow explains why the flagpole is as tall as it is, is thereby also a world in which the length of the shadow is a cause of the flagpole's height. (For what it's worth, I do not find this defense plausible.[30])

Here's another natural worry about Kitcher's theory: doesn't it entail that the world is necessary unified? That would be bad, because surely it is possible for the world not to be unified. But to become an objection, this worry needs more development. Kitcher's theory appears to allow for possible worlds in which the set of argument-forms that best unifies the Humean truths contains a huge number of argument-forms, each of which is instantiated only a few times. That is, it allows for worlds in which the set of argument-forms that best unifies the Humean truths unifies them only to a tiny degree.[31]

The specific details of Kitcher's theory aside, there is a general feature of unification-ist theories that bothers me. Unification, it seems to me, should be analyzed in terms of explanation, not the other way around. It may be that there are many different kinds of unification,[32] but still, one central way for a theory to unify, I think, is for it to provide answers to many different why-questions. This conception of unification cannot appear in a theory of answers to why-questions because its appearance there would make the theory circular.

If there is an objection to Kitcher's theory here, though, it is a frustratingly subtle one. Kitcher said that Newton's theory unified a lot of phenomena by allowing for deductions of all those phenomena, deductions that all instantiated the same argument-form. I want to say that I disagree; what it was for Newton's theory to unify all of those phenomena was, instead, for his theory to figure in the answer to the question why *X* occurred, for each of those phenomena *X*. But just what do Kitcher and I disagree about here? Kitcher will say that, since his theory of explanation is correct and since the argument-form Newton's theory uses is a good one, the deductions instantiating it *do* answer the rel-evant why-questions about those phenomena. So my objection here is not that Kitcher cannot say that Newton's theory answers a lot of why-questions. Nor is my objection

[29] For objections to Kitcher like this see Woodward (2003, 361) and Paul and Hall (2013, 91–92).

[30] Generically, empiricists hold that both causation and explanation must be analyzed in terms of more epistemologically unproblematic notions. Kitcher's analyses of explanation in terms of unification and of causation in terms of explanation are empiricist in spirit. Perhaps a committed empiricist will find his view more plausible than I do. Kitcher's view is not, however, the only empiricist option; there are empiricist theories of causation that do not analyze it in terms of explanation.

[31] Section 8.2 of Kitcher (1989) is headed by the question "What If the World Isn't Unified?," but it ends up addressing instead the objection that, on his view, a cause of some event *E* might not explain *E*. Kitcher presents the views about causation that I just mentioned in response to this objection.

[32] That there are many kinds of unification is a theme of Morrison (2000).

that Kitcher fails to provide correct necessary and sufficient conditions for unification.[33] Instead, my objection is to his "order of analysis." Kitcher says that Newton's theory unified *in virtue of the fact that* it made it possible to deduce many phenomena using a good argument-form. I say that even if Newton's theory did make this possible, the theory did not unify those phenomena in virtue of making this possible. It unified in virtue of the fact that it answered many why-questions. Kitcher may be right about which theories unify, but I think he is wrong about what unification is. And again, if I am right about what unification is, then the notion of unification cannot appear in the correct theory of answers to why-questions.

7 RADICAL CONTEXT SENSITIVITY?

In *The Scientific Image* (1980) Bas Van Fraassen defended a view about explanation that entails that there can be no interesting filling-in of schema (5). It is worth discussing here because, if true, it shows that every theory I have mentioned goes wrong right from the beginning.

Van Fraassen's theory makes use of the notion of context-sensitivity. The pronoun "I" is a paradigm case of a context-sensitive word, so it can serve as an illustration. There is no one fixed thing that the word "I" refers to. When I use it, it refers to Bradford Skow; when my oldest brother uses it, it refers to Erik Skow; you get the idea. Clearly, the rule is: on any occasion when "I" is spoken, it refers to the person speaking.[34] Now the fact that a certain person is speaking is fact about the context in which "I" is being used. The context also includes facts about who is listening, where and when the conversation takes place, and also the beliefs and presuppositions of the parties to the conversation.[35] So, more generally, what "I" refers to depends on the context. That is what it is to say that "I" is context-sensitive. "Zebra," by contrast, is not context-sensitive. No matter who is speaking, or where, or what the parties to the conversation believe, "zebra" applies to all and only the zebras.

Another example of context-sensitivity, one that may be more helpful for what is to follow, is the word "tall." Suppose Jones is 6 feet tall. Then, in a context in which we have been talking about the heights of various philosophers, "Jones is tall" is true. Jones has the property that "tall" expresses in that context. But in a context in which we have been

---

[33] That is, my objection *here* is not that Kitcher fails to provide correct necessary and sufficient conditions. I do happen to think that his theory fails to do this. (I think that his theory of explanation is wrong, and I think that his conditions on unification are correct only if his theory of explanation is right.) But I do not have space here to develop this objection in a way that does not make it look like it begs the question.

[34] Ignoring exceptional cases, as when I am reading out loud from your diary.

[35] There is a lot of debate in the philosophy of language about just what contexts are. See Stalnaker (2014).

discussing basketball players, "tall" expresses a different property, and "Jones is tall" is false.[36]

Van Fraassen claimed that why-questions are context-sensitive. To explain his view it will help to temporarily make a distinction. There is a familiar distinction between a declarative sentence and a proposition. Propositions are the "semantic values" of declarative sentences: a declarative sentence, like "Jones is human," expresses a proposition. But there are other kinds of sentences in addition to declarative ones. Our interest is in interrogative sentences. Sometimes "question" is used as a synonym for "interrogative sentence," but it helps to distinguish between them. Let us say that questions are the semantic values of interrogative sentences. Then, questions stand to interrogatives as propositions stand to declaratives.

Van Fraassen's thesis, then, is that why-interrogatives are context-sensitive. They express different why-questions in different contexts.

He claims that they are context-sensitive in two ways. First, he holds that why-questions are contrastive. That is, an interrogative of the form

- Why is it the case that P?

in a given context, expresses the question why it is the case that P rather than A_1, or A_2, ..., or A_n. What goes in for "A_1," "A_2," and so on depends on the context. The proposition that A_1, the proposition that A_2, and so on, are alternatives to the proposition that P: each is incompatible with P and with all the others.

If this is right, if why-interrogatives are context-sensitive in this way, then which sentence expresses a true answer to a why-interrogative must be context-sensitive in a similar way. And there is good evidence that both of these claims are right. Van Fraassen asks us to think about Adam. I might ask a question using the sentence, "Why did Adam eat the apple?" And I might ask this in a context in which the salient alternative course of action was for Adam to eat nothing. Then, it seems, in this context, I asked the question why Adam ate the apple rather than nothing at all. In another context, the most salient alternative might have been for Adam to eat the pear instead. Then a use of "Why did Adam eat the apple?" serves to ask the question why Adam ate the apple rather than the pear. In the first context, the answer to "Why did Adam eat the apple?" is "Because he was hungry." This answer is not correct, however, in the second context. In the second context, "Because he was hungry" expresses the proposition that Adam ate the apple rather than the pear because he was hungry. And that is false. (The apple wouldn't do any more to alleviate Adam's hunger than the pear, and Adam knew this.)

This kind of context-sensitivity in "because" statements is no real threat to the project of finding an interesting way to complete (5). We just need to make a small change to the project. We just need to make the "hidden contrasts" explicit in our schema. The goal becomes to fill in the right-hand side of

---

[36] This is an oversimplified model of the way in which "tall" is context-sensitive, but the simplifications do not matter here.

(9) *P* rather than *A1*, or *A2*, or . . . or *An*, because *Q* if and only if . . .

Many philosophers have thought that causation is also a contrastive relation: a fully explicit claim about causation looks something like: "*C* rather than *C** caused *E* rather than *E**." Adam's being hungry, rather than satiated, caused him to eat the apple, rather than nothing at all. If this is the right way to think about causation, then those who want to develop a causal theory of explanation will not be bothered by the move from (5) to (9).[37]

But Van Fraassen thought that the context-sensitivity of why-interrogatives, and therefore of because-sentences, went beyond the role of context in supplying alternatives to the "*P*" in "why *P*?" He held that context also supplied a "relation of relevance." In brief, and ignoring from now on the fact that why-interrogatives and because-statements are contrastive, he thought the following about the semantics of because-sentences:

(10) "*P* because *Q*" is true in context *C* if and only if (i) it is true that *P*, and it is true that *Q*, and (ii) the fact that *Q* is "relevant-in-*C*" to the fact that *P*.

We can think of claim (10) in this way: just as we all think that "tall" expresses different properties in different contexts, Van Fraassen thought that "because" expresses different relations (between facts) in different contexts.

How radical a claim (10) is depends on how many "candidate relevance relations" you think there are. One could accept (10) and still make the modest claim that there are only, say, two candidate relevance relations. In any context, one or the other of them is the semantic value of "because." But Van Fraassen was a radical. He appeared to believe

(11) For any two propositions there is a candidate relevance relation that the first bears to the second.

If (11) is true then there is not much to say about the nature of explanation beyond (10). Those who think that only causes explain were right—about one relevance relation; those who think that *F* can explain G even if *F* does not cause G are also right—about a different relevance relation.

But (11) is hard to believe. Consider "The flagpole is 10 meters high because it casts a shadow 10 meters long," or "The economy is weak because Saturn is in retrograde motion." These sound false. But if (11) is right, then there is a conversational context we can get into in which these sentences express truths. (Importantly, this is so even if we restrict our attention to contexts in which we still believe that the flagpole's shadow does

---

[37] For arguments in favor of taking causation to be contrastive see Hitchcock (1996*a*) and Schaffer (2005). On contrasts in explanation see, for example, Hitchcock (1996*b*) and Lipton (2004). The causal claim I wrote down is "contrastive on both sides" whereas in schema (9) there are only contrasts on the left-hand side. Die-hard contrastivists will say that (9) should be contrastive on both sides. I don't have space to go into these details here.

not cause its height and in which Saturn's motion does not have any causal influence on the economy.) In his book, Van Fraassen tried to produce a context in which "The flag-pole is 10 meters high because it casts a shadow 10 meters long" expresses a truth, but he did not convince many people.[38] I suspect that the most that Van Fraassen establishes is a modest form of context-sensitivity, one that is no threat to traditional theories of explanation.

It is worth mentioning that one can believe that there are many explanatory relevance relations and that context determines which one is operative for a given use of "because" while still having a unified theory of explanation. One could say that every relevance relation is a restriction of some one single ur-relevance relation. A causal theory of explanation, for example, could say that the ur-relevance relation is causation. Some contexts select a relevance relation that only some of E's causes bear to E. Different causes will be relevant in different contexts. This kind of context-sensitivity is not radical and is no threat to the project of completing (5).

8 OTHER TOPICS

There are many topics I have not been able to cover, so I will end by just throwing out some references about three of them. First: we explain not just physical phenomena but also mathematical facts. Since mathematical facts cannot be caused, it is unclear whether the work that has been done on causal explanation sheds any light on explanation in mathematics. Steiner (1978) can be read as an attempt to find a surrogate for causal explanation in mathematics. A recent and much deeper investigation of explanation in mathematics is found in Lange (2014).

Second: a distinctively metaphysical kind of explanation—"grounding," or "in virtue of" explanation—is currently the focus of a lot of attention in metaphysics. Some of that work is relevant to the philosophy of science also, since some scientific explanations are of this kind. One example is an explanation of the fact that it is 72° Fahrenheit in this room that cites the motions of the room's air molecules. Rosen (2010) and Fine (2012) are standard references on grounding and explanation.

Third and finally: some explanations are better than others. There are two obvious ways in which one may be better: an explanation may be better by being more complete, and an explanation may be better by being more relevant in context. (If what I want are the psychological factors that explain your behavior, then an explanation citing those factors is better in this sense.) Are there any other ways for one explanation to be better than another? We seem to have a word for one: depth. One explanation may be deeper than another, and being deeper seems like a way of being better. If this is right, then a complete theory of explanation should include a theory of depth. What does it take for

---

[38] The most influential response to Van Fraassen's views on explanation is found in Kitcher and Salmon (1987).

one explanation to be deeper than another? Weslake (2010) surveys some answers and proposes his own.

References

Achinstein, P. (1983). *The Nature of Explanation* (Oxford: Oxford University Press).

Baker, A. (2005). "Are There Genuine Mathematical Explanations of Physical Phenomena?" *Mind* 114: 223–238.

Batterman, R. (1992). "Explanatory Instability." *Nous* 26: 325–348.

Cartwright, N. (1979). "Causal Laws and Effective Strategies." *Nous* 13: 419–437.

Cross, C. (1991). "Explanation and the Theory of Questions." *Erkenntnis* 34: 237–260.

Dray, W. (1957). *Laws and Explanation in History* (Oxford: Oxford University Press).

Choi, Sunglow and Fara, Michael. (2014). "Dispositions." *The Stanford Encyclopedia of Philosophy* (Spring 2014 Edition), Edward N. Alta (ed.), URL = <http://plato.stanford.edu/archives/spr2014/entries/dispositions/>.

Fine, K. (2012). "Guide to Ground." In F. Correia and B. Schnieder (eds.), *Metaphysical Grounding: Understanding the Structure of Reality* (Cambridge: Cambridge University Press), 37–80.

Fisher, R. (1931). *The Genetical Theory of Natural Selection* (New York: Dover).

Friedman, M. (1974). "Explanation and Scientific Understanding." *Journal of Philosophy* 71, 5–19.

Hempel, C. (1965). "Aspects of Scientific Explanation." In *Aspects of Scientific Explanation and Other Essays in the Philosophy of Science* (New York: The Free Press), 331–496.

Hempel, C., and Oppenheim, P. (1948). "Studies in the Logic of Explanation." *Philosophy of Science* 15, 135–175.

Hitchcock, C. (1995). "Salmon on Explanatory Relevance." *Philosophy of Science* 62, 304–320.

Hitchcock, C. (1996a). "Farewell to Binary Causation." *Canadian Journal of Philosophy* 26, 267–282.

Hitchcock, C. (1996b). "The Role of Contrast in Causal and Explanatory Claims." *Synthese* 107, 395–419.

Hitchcock, Christopher. (2012). "Probabilistic Causation." *The Stanford Encyclopedia of Philosophy* (Winter 2012 Edition), Edward N. Zalta (ed.), URL = <http://plato.stanford.edu/archives/win2012/entries/causation-probabilistic/>.

Humphreys, P. (1989). *The Chances of Explanation* (Princeton, NJ: Princeton University Press).

Jackson, F., and Pettit, P. (1992). "In Defense of Explanatory Ecumenism." *Economics and Philosophy* 8, 1–21.

Kitcher, P. (1981). "Explanatory Unification." *Philosophy of Science* 48, 507–531.

Kitcher, P. (1986). "Projecting the Order of Nature." In R. E. Butts (ed.), *Kant's Philosophy of Science* (Dordrecht: D. Reidel Publishing), 201–235.

Kitcher, P. (1989). "Explanatory Unification and the Causal Structure of the World." In P. Kitcher and W. Salmon (eds.), *Scientific Explanation. Volume 13 of Minnesota Studies in the Philosophy of Science.* (Minneapolis: University of Minnesota Press), 410–505.

Kitcher, P., and Salmon, W. (1987). "Van Fraassen on Explanation." *Journal of Philosophy* 84, 315–330.

Lange, M. (2013). "What Makes a Scientific Explanation Distinctively Mathematical?" *British Journal for the Philosophy of Science* 64, 485–511.

Lange, M. (2014). "Aspects of Mathematical Explanation: Symmetry, Unity, and Salience." *Philosophical Review* 123, 485–531.

Lewis, D. (1986). "Causal Explanation." In *Philosophical Papers, Volume II* (Oxford: Oxford University Press).

Lipton, P. (2004). *Inference to the Best Explanation*, 2nd ed. (New York: Routledge).

Mackie, J. L. (1974). *The Cement of the Universe* (Oxford: Oxford University Press).

Morrison, M. (2000). *Unifying Scientific Theories* (Cambridge: Cambridge University Press).

Paul, L. A., and Hall, N. (2013). *Causation* (Oxford: Oxford University Press).

Pearl, J. (2009). *Causality*, 2nd ed. (Cambridge: Cambridge University Press).

Railton, P. (1978). "A Deductive-Nomological Model of Probabilistic Explanation." *Philosophy of Science* 45, 206–226.

Reutlinger, A. (2014). "Why Is There Universal Macro-Behavior? Renormalization Group Explanation As Non-causal Explanation." *Philosophy of Science* 81, 1157–1170.

Rosen, G. (2010). "Metaphysical Dependence: Grounding and Reduction." In B. Hale and A. Hoffmann (eds.), *Modality: Metaphysics, Logic, and Epistemology*. (Oxford: Oxford University Press), 109–136.

Salmon, W. (1984). *Scientific Explanation and the Causal Structure of the World* (Princeton, NJ: Princeton University Press).

Salmon, W. (1989). *Four Decades of Scientific Explanation* (Pittsburgh, PA: University of Pittsburgh Press).

Salmon, W., Jeffrey, R., and Greeno, J. (1971). *Statistical Explanation and Statistical Relevance* (Pittsburgh, PA: University of Pittsburgh Press).

Schaffer, J. (2005). "Contrastive Causation." *Philosophical Review* 114, 297–328.

Scriven, M. (1959). "Truisms as the Grounds for Historical Explanation." In P. Gardiner (ed.), *Theories of History* (New York: The Free Press), 443–475.

Scriven, M. (1962). "Explanations, Predictions, and Laws." In H. Feigl and G. Maxwell (eds.), *Minnesota Studies in the Philosophy of Science III* (Minneapolis: University of Minnesota Press), 170–230.

Skow, B. (2014). "Are There Non-Causal Explanations (of Particular Events)?" *British Journal for the Philosophy of Science* 65, 445–467.

Skow, B. (2016). *Reasons Why* (Oxford: Oxford University Press).

Spirtes P., Glymour, C., and Scheines, R. (2001). *Causation, Prediction, and Search*, 2nd ed. (Cambridge, MA: MIT Press).

Sober, E. (1983). "Equilibrium Explanation." *Philosophical Studies* 43, 201–210.

Stalnaker, R. (2014). *Context* (Oxford: Oxford University Press).

Steiner, M. (1978). "Mathematical Explanation." *Philosophical Studies* 34, 135–151.

Strevens, M. (2008). *Depth* (Cambridge, MA: Harvard University Press).

Van Fraassen, B. C. (1980). *The Scientific Image* (Oxford: Oxford University Press).

Weslake, B. (2010). "Explanatory Depth." *Philosophy of Science* 77, 273–294.

Woodward, J. (2003). *Making Things Happen* (Oxford: Oxford University Press).

Yablo, S. (1992). "Mental Causation." *Philosophical Review* 101, 245–280.

Yablo, S. (2002). "De Facto Dependence." *Journal of Philosophy* 99, 130–148.

CHAPTER 26

SCIENTIFIC PROGRESS

ALEXANDER BIRD

1 INTRODUCTION

"What Des-Cartes did was a good step. You have added much several ways, & especially in taking the colours of thin plates into philosophical consideration. If I have seen further it is by standing on the shoulders of Giants." Newton's famous lines come in a letter to Hooke concerning research into the nature of light, carried out by Newton, Hooke, and Descartes. Those words have come to represent the idea of scientific progress—the idea that science is a collective enterprise in which scientists add to the edifice upon which their colleagues and predecessors have been laboring. While the metaphor is an old one, it is only in the early modern period that thinkers came to view history and culture—science in particular—in terms of progress. Descartes (1637/1960: 85) himself, in his *Discourse on Method*, having remarked on the state of knowledge in medicine—that almost nothing is known compared to what remains to be known—invites "men of good will and wisdom to try to go further by communicating to the public all they learn. Thus, with the last ones beginning where their predecessors had stopped, and a chain being formed of many individual lives and efforts, we should go forward all together much further than each one would be able to do by himself." Descartes's desire found its expression in the scientific societies of the time, such as the Académie des Sciences (1666) and the Royal Society (1660), and in learned journals, such as the *Journal des Sçavans* and the *Philosophical Transactions of the Royal Society*. Zilsel (1945) finds late renaissance artisans publishing knowledge of their crafts in the spirit of contributing incrementally to the public good and understanding. But it is in the work of Francis Bacon that progress as an ideal for science is first promoted as such. In *The Advancement of Learning* (1605) and *The New Organon* (1620), Bacon lays down the growth of knowledge as a collective goal for scientists, knowledge that would lead to social improvement also. And in his *New Atlantis* (1627) Bacon articulates a vision of a society centered upon Salomon's House, a state-supported college of scientists working on cooperative projects and the model for the learned academies founded in the seventeenth century.

With the notion of scientific progress in hand, several questions naturally arise: What *is* scientific progress? How does one promote scientific progress? How can we detect or measure scientific progress? Is there in fact progress in science?[1] This article concentrates principally on the first question, but the various questions are related. For example, Descartes and Bacon accepted that science is the sort of activity where progress can be made by collective activity, in particular by adding to the achievements of others. This contrasts with the idea of progress in the arts. Although controversial, a case can be made that progress in the arts is made when the horizons of an art form are expanded through the exploration of new expressive possibilities. If so, that progress is typically individual and not straightforwardly cumulative—new artworks do not pick up where their predecessors left off. Furthermore, it seems plausible that the very point of scientific activity is to make progress, whereas it is rather less plausible that the raison d'etre of the arts is related to progress of any kind. If this is correct, the scientific progress is the sort of thing that is best promoted collectively, can be added to incrementally, and is linked intimately to the aim of science.

2 PROGRESS AND THE AIM OF SCIENCE

How does a conception of progress relate to the goals of an activity? The simple view of progress here is that if an activity A aims at goal X, then A makes progress insofar as it gets closer to achieving X or does more of X or does X better (depending on what X is and how it is specified). However, it is worth noting that an activity can be said to show progress along dimension D, even if that activity does not have any goal, let alone a consciously entertained or publicly agreed one, to which D is related. Progress in the arts, as mentioned, is like this. We might judge that political institutions have made progress if they are more responsive to the desires and needs of citizens, even if we do not suppose that this is the aim of those institutions. In such cases we judge progress by some appropriate standard, typically concerning a good that relates to the nature of the activity in question. For example, an artist might aim at producing a certain kind of aesthetic experience in an audience. An appropriate standard for judging progress in the arts is therefore the expansion of the possibilities for producing aesthetic experiences. Note also that a collective enterprise may make progress and may have a goal even if it is not the intention of any individual to contribute to that progress or to promote that goal. The artist may be focused solely on the experience of his or her audience and have no specific intention of widening the expressive possibilities of the art form. Nonetheless, if the work in fact does broaden the range of expressive possibilities, then he or she may thereby be contributing to artistic progress. Likewise, a business may have the aim of

---

[1] Niiniluoto (1980: 428) raises three questions corresponding to first, third, and fourth of these.

making a profit, even though most of its employees (perhaps even all of them) do not have this as an individual goal.

Thus we can talk of progress in science even if we do not attribute an aim to science and do not attribute the same aim to scientists. That said, most commentators do attribute a goal to science and tend to assume that scientists have that goal also. So the three principal approaches to scientific progress relate to three views of the aim of science, in accordance with the simple view of progress, as follows:

(a)   Science aims at solving scientific problems. Science makes progress when it solves such problems.
(b)   Science aims at truth. Science makes progress when it gets closer to the truth.
(c)   Science aims at knowledge. Science makes progress when it adds to the stock of knowledge.

Some well-known views of the aim of science might appear not to fit any of these. Bas van Fraassen (1980: 12) holds empirical adequacy to be the aim of science. But this can be accommodated within (b) insofar as the aim of empirical adequacy is the aim of achieving a certain kind of truth, the truth of a theory's observational consequences.

All three approaches accept that the goals of science can be achieved collectively and in particular in an incremental and (normally) cumulative way, just as described by Descartes. Problem-solving takes place within a tradition; problems solved by a scientist will add to the problems solved by the tradition. A theory developed by one scientist may be improved by a successor, thus bringing it closer to the truth. Scientists add to the stock of knowledge generated by predecessors (including by improving their theories).

3   PROGRESS AS SOLVING SCIENTIFIC PROBLEMS

Scientists engage with scientific problems. Such problems arise from a tradition of solving problems. Consider this example: Given that we can account for the motions of the known planets within the framework of Newtonian mechanics and theory of gravitation, how should we account for the motion of the moon? This problem troubled astronomers and mathematicians for several decades in the eighteenth century and spurred research. Alexis Clairaut eventually showed that the difficulty in reconciling observations with Newtonian theory was due to the use of unsuitable approximations. In solving this problem, Clairaut thereby made progress in astronomy.

The principal proponents of the problem-solving approach to progress are Thomas Kuhn (1970) and Larry Laudan (1977). Kuhn uses the terminology of "puzzle-solving," but this has the same intent as Laudan's "problem-solving." For both, scientific activity takes place within a tradition of research. Research traditions are characterized by shared commitments that form a background that gives significance to research problems and guides the testing and evaluation of theories (i.e., proposed solutions to problems). For example, those in a common research tradition will share basic ontological commitments (beliefs about what sorts of entities exist—e.g., in medicine whether there are humors or germs), background theories (e.g., Newtonian mechanics), mathematical techniques (e.g., the calculus), and methods of theory assessment (e.g., significance testing in classical statistics). Kuhn used the terms "paradigm" and "disciplinary matrix" to refer to research traditions just described.

Kuhn emphasized a particularly important component of a research tradition: shared exemplars of scientific problems and their solutions that act as a model for future problem-solving within the tradition. (Kuhn also used the term "paradigm" to refer to such exemplars.) These paradigms-as-exemplars are Kuhn's answer to the question about how science is able to make progress. For individuals, they explain the cognitive psychology of problem-solving. Training with exemplars allows scientists to see new problems as similar to old ones and as such requiring a similar approach to solving them. For research communities, shared exemplars and disciplinary matrices permit the individuals and groups within the community to agree on fundamentals and thus agree on the problems that need to be solved, the means of solving them, and, for the most part, whether a proposed solution is satisfactory. Research carried out in this way, governed by a shared disciplinary matrix, Kuhn calls "normal science." Kuhn contrasts normal science with "pre-paradigm" (or "pre-normal") science, a period of science characterized by a multiplicity of schools that differ over fundamentals. Pre-paradigm science fails to make progress because intellectual energy is put into arguing over those fundamental disagreements rather than into solving agreed puzzles according to agreed standards. For mature sciences, a more important contrast is "extraordinary" (or "revolutionary") science, which occurs when the research tradition finds itself unable to solve particularly significant problems. Under such circumstances a change of paradigm is needed. New kinds of problem-solution are required, ones that differ in significant ways from the exemplars that had previously dominated the field.

According to Laudan (1981: 145), "science progresses just in case successive theories solve more problems than their predecessors." Kuhn and Laudan regard progress during normal science as unproblematic—during normal science solutions to problems are generally accepted and add to the sum of problems solved. Extraordinary science is less straightforward, because the rejection of a paradigm (i.e., a rupture within the research tradition) will often mean that some of the problems previously regarded as solved (by that paradigm, within that tradition) are no longer regarded as such. For example, Descartes's vortex theory of planetary motion explained why the planets moved in the same plane and in the same sense (rotational direction). The successor theory, Newton's inverse square law of gravitational force, solved various problems (e.g., why the planets obeyed Kepler's laws) but lacked an explanation for the aforementioned problems to which Descartes's theory offered a solution. Hence the adoption of Newton's theory required relinquishing certain apparent problem-solutions. Such reductions in problem-solving ability are known as "Kuhn-losses." The existence of Kuhn-losses makes assessment of progress more complicated, because, according to Laudan's account, the change in question will be progressive only if the Kuhn-losses are compensated for by a greater number of successful problem-solutions provided by the new paradigm. That requires being able to individuate and count problems. Furthermore, presumably some problems are more significant than others, and we would therefore want to weight their solutions accordingly.

Laudan (1981: 149) admits that the best way of carrying out this individuation of problems is not entirely clear, but he argues that all theories of progress will come up against that or an analogous difficulty—individuating confirming and disconfirming instances. We shall see that this is a rather less pressing matter for the other views of progress. Kuhn, on the other hand, accepts that there is no definitive way of making such assessments. The significance of a problem is determined by the tradition or paradigm that gives rise to it. There is no uncontentious way of making such assessments across paradigms: this is the problem of incommensurability. Thus there can be rational disagreement between adherents of two different paradigms. But that does not mean that Kuhn denies that there is progress through revolutions. On the contrary, Kuhn (1970: 160–173) is clear that there is progress. Scientists operating within the new paradigm must be able to say, from their perspective at least, that the new paradigm retains much of the problem-solving power of its predecessor and that the reduction in the number of problems solved (the Kuhn-losses) are outweighed by the ability of the new paradigm to solve the most pressing anomalies of the predecessor while offering the promise of new problems and solutions to come. Incommensurability means that during a revolution this assessment—the assessment that moving to the new paradigm is progressive relative to retaining the old paradigm—is not rationally mandated and can be rationally disputed.

What counts a solving a problem? For that matter, what counts as a problem or a puzzle in the first place? A crucial feature of the problem-solving approach is that the scientific tradition or paradigm determines what a problem is and what counts as a solution. For Kuhn puzzles may be generated by a paradigm in a number of ways. For example, if a theory at the heart of a paradigm involves an equation with an unknown constant, then one puzzle will be to determine the value of that constant; subsequently, determining its value with greater precision or by a different means will also be puzzles. Most puzzles acquire their status by similarity to exemplar puzzles, and solutions are accepted as such on the basis of similarity to exemplary solutions. For Laudan also, problems are determined by the tradition and its current leading theories—for example a (conceptual) problem exists if the theory makes assumptions about the world that run counter to prevailing metaphysical assumptions. In the simplest cases, an empirical problem is solved by a theory if the theory entails (along with boundary conditions) a statement of the problem—there is no requirement that the theory should be true. I have called this feature of the problem-solving approach "internalist" in the epistemological sense (Bird 2007: 69). Internalist epistemologies are those that maintain that the epistemological status (e.g., as justified) of a belief should be accessible to the subject. The problem-solving approach is internalist in that it provides, as Laudan (1981: 145) says, an account of the goal of science (and so of progress) such that scientists can determine whether that goal is being achieved or approached—the community is always in a position to tell whether a scientific development has the status of being progressive.

A consequence of this way of thinking about problems and solutions is that entirely false theories may generate problems and solutions to those problems. It implies that a field might be making progress (not merely appearing to make progress) by the accumulation of falsehoods, so long as those falsehoods are deducible from the principal theories in the field. For example, according to the dominant theory among alchemists, all matter is made up of some combination of earth, air, fire, and water; therefore it should be possible to transmute one substance into another by appropriate changes to the proportions of these elements. Consequently, a leading problem for alchemists was to discover a mechanism by which one metal could be transmuted into another. Alchemists provided solutions in terms of the dominant theory, i.e., solutions referred to the four elements and the four basic qualities (moist, cold, dry, and warm). Neither the problem nor its solutions are genuine, but that does count against progress being made in alchemy according to the problem-solving approach. Early writers in the Hippocratic tradition disagreed about the correct humors. One—the author of *On Diseases, IV*—proposes blood, phlegm, bile, and water. This sets up an asymmetrical relationship with the elements—the humors and elements overlap in one (water) but not the others. By replacing water with black bile, a symmetrical relationship between the humors, elements, and qualities could be restored. Furthermore, according to the author of *On the Nature of Man*, a relationship between the humors and the seasons could be established, explaining the prevalence of phlegm in winter for example (Nutton 1993: 285). The problem-solving approach takes such developments as progressive, despite their being devoid of any truth.

Another implication of the problem-solving approach is that the rejection of any theory without a replacement counts as regressive, so long as the theory did raise and solve some problems. Laudan, Kuhn, and others point out that because theories are not rejected outright with nothing to replace them, this scenario does not normally arise. However, even if rare, it does occur. In 1903 René Blondlot believed that he had discovered a new form of radiation, which he named "N-rays" ("N" for Nancy, where he was born and worked). N-rays attracted much interest until the American physicist R. W. Wood visited Blondlot's laboratory and surreptitiously removed a crucial prism from Blondlot's experiment. Blondlot nonetheless continued to see the rays. After Wood's intervention it became accepted that N-rays simply did not exist; scientists gave up research in N-rays and the community repudiated the earlier "discoveries". During the intervening period, 1903–1905, more than 300 articles on the topic were published (Lagemann 1977), discussing the nature of the rays and their properties (e.g., which materials could generate or transmit the rays). N-ray research was carried out as normal science—research problems were posed and solved. So, according to the problem-solving approach, N-ray research was progressive until 1905, and, because its repudiation after Wood involved a loss of problems and solutions without any compensating gain, that repudiation must count as regressive. This conclusion will strike many as just wrong—N-ray research did not add anything to progress, but its rejection did perhaps add something.

Laudan regards the disconnection between progress and truth as an advantage of his account. Truth, says Laudan (1981: 145), is transcendent and "closed to epistemic access." Kuhn (1970: 206) also argues that truth is transcendent. Note that to assert this is to assume antirealism. Scientists gather evidence and provide arguments as to why that evidence gives us reason to believe, to some degree or other, the relevant hypotheses. Scientific realists claim that, in many mature sciences at least, this process does in fact lead to beliefs that are true or close to the truth. Antirealists deny this. Laudan's claim that truth is transcendent, if correct, implies that the scientists are on a hiding to nothing. Their evidence and arguments cannot be leading us toward the truth, for if they did, then truth would not be transcendent—it would be epistemically accessible through the process of scientific research.

So scientific realists have two reasons to reject the problem-solving approach. First, the motivation ("truth is transcendent") that Laudan gives for the internalism of that approach assumes antirealism. Second, the approach gives perverse verdicts on particular scientific episodes: if scientists in the grip of a bad theory add further falsehoods (false problem-solutions), that counts as progressive; if, on the other hand, they realize their error and repudiate those falsehoods, that counts as regressive.

4 Progress as Increasing Nearness to the Truth

A natural response to the realist's objections to the problem-solving approach is to seek an account of progress in terms of truth—what I have called the "semantic" approach to progress (Bird 2007: 72). If the problem-solving approach takes progress to consist of later theories solving more problems than their predecessors, then a truth approach might take progress to consist in later theories containing more truth than their predecessors. This would occur, for example, when later theories add content to earlier theories and that new content is true.

However, the view of progress as accumulating true content has not found favor even among realists. The principal reason is the recognition that theories are often not in fact true. A good theory may in fact be only a good approximation to the truth, and a better successor theory may cover the same subject matter but be a better approximation to the truth without in fact being strictly true itself. So realists have tended to favor accounts of progress in terms of increasing *truthlikeness* (nearness to the truth, verisimilitude), not (directly) in terms of truth itself. Indeed, what David Miller (1974: 166) regards as the "problem of verisimilitude" is the problem of saying how it is that there is progress, for example in the sequence of theories of motion from Aristotle to Oresme to Galileo to Newton, even though all the theories are false. It would seem that we do have an intuitive grasp of a nearness to the truth relation. In the following cases, T_2 is intuitively nearer to the truth than T_1 and T_4 is nearer to the truth than T_3: Although it is not clear whether T_4 is nearer to the truth than T_2, there is no reason why we should expect any verdict on such a question; the predicate "nearer to the truth than" delivers only a partial order.

The value of Avogadro's constant, N_A	
truth	$N_A = 6.022 \times 10^{-23}$ particles per mole
T_1 (Perrin 1908)[2]	$N_A = 6.7 \times 10^{-23}$ particles per mole
T_2 (Rutherford 1909)	$N_A 6.16 \times 10^{23}$ particles per mole
Nature (e.g., chemical formula) of water	
truth	water is H_2O
T_3 (Aristotle)	water is an element
T_4 (Dalton)	water is a compound with the formula HO

It may seem obvious that someone who prefers the increasing nearness to the truth approach should provide an account of what nearness to the truth is and when one theory is nearer to the truth than another. However, it is unclear to what extent this is an essential task. If the various available accounts of truthlikeness are all unsatisfactory, the truthlikeness account of progress is severely undermined. On the other hand, consider the simple accumulation of truth view with which I started this section. Would it be a fair criticism of that view that philosophers do not have an agreed, unproblematic account of what *truth* is? Is the latter a problem for scientific realism, insofar as the latter also appeals to a notion of truth? Presumably not— we can and do use the concept of truth unproblematically. So the issue may come down to this: Is the concept of truthlikeness a theoretical or technical one introduced for the philosophical purposes discussed earlier? Or is it a concept with a pre-philosophical use, one that has clear application in a sufficient range of important cases? If the former, then the lack of a clear philosophical account of truthlikeness does cast doubt on the coherence of the concept and its use to account for progress. If the latter, then (like the concepts of truth, cause, knowledge, etc.) the truthlikeness concept has a legitimate philosophical application even in the absence of a satisfactory analysis.[3]

Either way, it is instructive to look at one of the key debates in the early development of theories of truthlikeness. The basic idea behind Popper's (1972: 52) account of verisimilitude is that theory A is closer to the truth than B when A gets everything

---

[2] See Becker (2001) for details.
[3] I am now inclined to think that our ability to give a clear and unambiguous judgment in the cases just mentioned is evidence in favor of the second view, whereas I previously (Bird 2007) took the first.

right that B gets right and some other things right besides, without getting anything wrong that B does not also get wrong.[4] Now this turns out to fail even in standard cases where A is false but is clearly closer to the truth than B (e.g., provides a closer numerical approximation of some constant; Tichý 1974). It can be shown that A will have some false consequences that are not consequences of B. So one might be tempted to drop the second condition: so long as the truths are accumulating, we can ignore the fact that there might be some accumulation of false consequences also. But a (prima facie) problem with ignoring the falsehoods is that a change to a theory might add to the true consequences in a small way but add to the false consequences in a major way. One might believe that such a change should be regarded as regressive. Focusing just on true consequences would also imply that a maximally progressive strategy would be to adopt an inconsistent theory, because it has *every* truth among its consequences.

A different approach initiated by Tichý and others (Hilpinen 1976; Niiniluoto 1977) considers how much the possible worlds described by theories are like the actual world. Tichý imagines a language with three atomic propositions to describe the weather: "it is hot," "it is rainy," and "it is windy". Let us assume that all three propositions are true. Say Smith believes "it is not hot and it is rainy and it is windy" whereas Jones believes "it is not hot and it is not rainy and it is not windy". We can say that Jones is further from the truth than Smith since there are three respects in which a world may be like or unlike the actual world, and the possible world described by Jones's theory differs from the actual world in all three respects (so the distance from the actual world is 3), while the possible world described by Smith's theory differs from the actual world in one respect (so distance = 1). The world of Jones's theory is thus more distant from the actual world than the world of Smith's theory, and so the latter theory is closer to the truth than the former. Miller (1974), however, considers an alternative language with the atomic propositions "it is hot", "it is Minnesotan", and "it is Arizonan". "It is Minnesotan" is true precisely when "it is hot ↔ it is rainy" is true, and "it is Arizonan" is true precisely when "it is hot ↔ it is windy" is true. In this language the truth is expressed by "it is hot and it is Minnesotan and it is Arizonan," while Smith's beliefs are expressed by "it is not hot and it is not Minnesotan and it is not Arizonan" and Jones's beliefs are expressed by "it is not hot and it is Minnesotan and it is Arizonan". Given this language for the description of possible worlds, the world described by Smith's theory differs from the actual world in three respects (distance = 3), and the world of Jones's theory differs in just one respect (distance = 1). So Jones's theory is closer to the truth than Smith's. It would thus appear that this approach makes verisimilitude relative to a language. That is a problematic conclusion for the scientific realist, who wants progress and so (on this approach) verisimilitude to be objective. One might tackle this problem in various ways. For example, one might say that some languages are objectively better at capturing the natural structure of the world. (While I think that

---

[4] More formally: A is closer to the truth than B iff B's true consequences are a subset of the true consequences of A and A's false consequences are a subset of the false consequences of B and one or both of these two subset relations is a proper subset relation.

is correct, it seems insufficient to solve the problem entirely.) Or one might try quite different approaches to truthlikeness. (See Oddie [2014] for a clear and detailed survey of the options.) Eric Barnes (1991) proposes a more radical response that notes an epistemic asymmetry between the beliefs expressed in the two sets of vocabularies; this, I propose in the next section, is instructive for the purposes of understanding progress.

The weather according to Smith and Jones (original)			
truth	hot	rainy	windy
Jones	not hot	not rainy	not windy
Smith	not hot	rainy	windy
The weather according to Smith and Jones (redescribed)			
truth	hot	Minnesotan	Arizonan
Jones	not hot	Minnesotan	Arizonan
Smith	not hot	not Minnesotan	not Arizonan

What is more obviously problematic for the truthlikeness approach are cases in which our judgments of progress and of truthlikeness are clear but divergent. I have suggested that we can imagine an example of this by considering again the case of Blondlot and N-rays (Bird 2007, 2008).[5] Observation of new effects, with new apparatus, particularly at the limits of perceptual capacities, is often difficult, so it is not entirely surprising that scientists could honestly think they were observing something when they were not. Almost all the work on the subject was carried out by French scientists—although some British scientists claimed to detect N-rays, most scientists outside France (and quite a few in France) could not detect them. It is plausible that the French N-ray scientists were motivated by national sentiment (Lagemann 1977), consciously or unconsciously proud that that there were now French rays to stand alongside the X-rays (German), cathode rays (British), and canal rays (German again) discovered and investigated in the preceding decades.[6] Blondlot had an excellent reputation—he was a member of the

---

[5] See Rowbottom (2008, 2010) for an opposing view and further discussion of this case.

[6] X-rays were first observed by Röntgen in 1895, while cathode rays were first observed by another German, Johann Hittorf, in 1869 and named as such by a third, Eugen Goldstein. Nonetheless, cathode rays were investigated primarily by British physicists—they were produced in the eponymous tube devised by William Crookes, and their nature, streams of electrons, was identified by J. J. Thompson in 1897, confirming the British (Thompson and Schuster) hypothesis against the German (Goldstein, Hertz, and Wiedemann) one that they were a form of electromagnetic radiation. Canal rays, also known as anode rays, were also produced in a Crookes tube and were discovered by Goldstein in 1886 and investigated by Wien and Thompson. Until N-rays, the French has no horse in the physical ray race.

Académie des Sciences; others who worked on N-rays were well respected, including Augustin Charpentier and Jean Becquerel (son of Henri Becquerel) (Klotz 1980). No doubt these reputations also had a persuasive role—in that context the observation or otherwise of N-rays became a test of a scientist's skills, not of the theory. Whatever the explanation, it is clear that the scientific justification for belief in N-rays was at best limited. Now let us imagine, counter-to-fact, that at some later date it is discovered that there really are rays answering to many of Blondlot's core beliefs about N-rays; many of Blondlot's beliefs (and those of the scientific community in France) turn out to be correct or highly truthlike, although those scientists held those beliefs for entirely the wrong reasons (their techniques could not in fact detect the genuine rays). In this case, according to the truthlikeness view of progress, Blondlot's work contributed to progress (by adding true and highly truthlike propositions to the community's beliefs) whereas Wood's revelation was regressive (because it removed those beliefs). That seems entirely wrong—even if Blondlot got things right (by accident), the episode is clearly one of pathological science, not of progressive science—excepting Wood's intervention, which got things back on track.

The preceding example shows that the truthlikeness account of progress conflicts with our judgments concerning particular instances. The case is partly hypothetical—in science it is difficult to get things right (factually correct or highly truthlike) for the wrong reasons. There are nonetheless some real cases. For example, it was believed in the Renaissance period that infants and young children had more body water as a proportion of total body weight than older children and adults. The source of this belief was a physiological theory based on the doctrine of the humors. As mentioned previously, the body was supposed to be governed by four humors (bile, black bile, blood, and phlegm); ill health was held to be a matter of an imbalance or corruption of the humors. The natural balance was supposed to change with time, and since the humors were characterized by their proportions of the four basic qualities (moist, dry, warm, and cold), the changing balance of the humors implied change in the natural proportion of these qualities with age. The prevailing view was that the predominant humor in childhood is blood, which is moist and warm (Newton 2012). Children are moist because they are formed from the mother's blood and the father's seed, both of which are moist and warm. Childhood is also influenced by the moon, which was held to be warm and moist. In youth, however, the dominant humor is choler, which is warm and dry. And so in growing to maturity a child loses moisture, and indeed this drying continues with age, explaining the wrinkled skin of the elderly in contrast to the soft skin of babies and infants. When humoral theory was dismissed in the late eighteenth and nineteenth centuries, these opinions fell into abeyance. It turns out that modern physiology has shown that body water is proportionally highest in neonates and then declines (Friis-Hansen et al. 1951; Schoeller 1989). So the Renaissance doctors, basing their opinion in the false humoral theory, fortuitously had correct opinions about the variation of moisture in the body with age. According to the "increasing truthlikeness" view, when true implications of humoral theory were drawn, science thereby made progress, but it regressed when these beliefs were dropped along with the humoral theory. On the contrary, even the fortuitously true consequences of humoral theory were no contribution to scientific progress insofar as they were based principally on that theory and not on appropriate evidence.

5 PROGRESS AS THE ACCUMULATION OF KNOWLEDGE

The third approach to scientific progress is an *epistemic* one—scientific progress is the accumulation of scientific knowledge; as Sir William Bragg put it: "If we give to the term Progress in Science the meaning which is most simple and direct, we shall suppose it to refer to the growth of our knowledge of the world in which we live" (1936: 41). As Mizrahi (2013) shows, Bragg was far from alone in his opinion—scientists in general take knowledge to be at the core of scientific progress. It is also my own preferred view.

Let us return to Miller's example of Smith and Jones who are making assertions about the weather. In Miller's version of the story, they inhabit a windowless, air-conditioned room, thus they are making guesses. Jones guesses that it is not hot and it is not rainy and it is not windy, while Smith guesses that it is not hot and it is rainy and it is windy. The truth is that it is hot and it is rainy and it is windy, so none of Jones's guesses are correct whereas two of Smith's are correct. If we further assume that Smith's guesses are made after Jones's, then that would look like progress according to the truthlikeness view. But, as argued in the preceding section, lucky guesses are no contribution to progress. Now let us imagine, as does Barnes (1991: 315), a variation whereby Smith and Jones form their beliefs not by guessing but by investigating indicators of temperature, precipitation, and airspeed in their vicinity, using methods that are usually very reliable by normal scientific standards. However, things go awry in the case of all Jones's investigations, leading to his false beliefs. They go awry for Smith also in the case of temperature, but his methods for measuring precipitation and airspeed are reliable and working fine. Accordingly, Smith knows that it is rainy and that it is windy. Now consider the beliefs of Smith and Jones when expressed using the language of "hot," "Minnesotan," and "Arizonan." The oddity is that now Jones has two correct beliefs: that the weather is Minnesotan and that it is Arizonan. Let us have a closer look at how it is that Jones has these two true beliefs. The belief that it is Minnesotan is logically equivalent to the belief that it is hot ↔ it is raining, which in turn is equivalent to the belief **DIS** that either it is both hot and rainy or it is neither hot nor rainy. Jones believes this proposition because he believes that it is not hot and not rainy. So he believes the second disjunct of **DIS**. Note that this disjunct of **DIS** is false. Since **DIS** is a simple consequence of that disjunct, Jones believes the whole of **DIS**. However, the whole of **DIS** is true, since the first disjunct is true—a disjunct that Jones believes to be false. So the position is that Jones believes a true proposition, **DIS**, because it is a disjunction and he believes one of the disjuncts—the one that happens to be false. As Barnes (1991: 317) points out, Jones's belief has the structure of a standard

---

Why Jones believes that the weather is Minnesotan

Jones believes

it is Minnesotan

which is equivalent to

DIS:

it is hot and rainy	OR	it is not hot and it is not rainy
true		*false*
not believed by Jones		*believed by Jones*

Gettier case: a case of a true and justified belief but where the truth and justification come apart. The justification (in this case Jones's normally reliable investigation that has gone awry) is attached to the false disjunct of **DIS** whereas the truth comes from the other disjunct of **DIS** for which Jones has no justification (in fact he thinks it is false). As a standard Gettier case, we must deny that Jones has knowledge in this case, although he has fortuitously obtained a true belief.

The problem we faced was that it looked as if truthlikeness depends on choice of language. Combined with the "progress is increasing truthlikeness" view, that implies that Smith makes progress relative to Jones when we express their beliefs using normal language and that Jones makes more progress when their beliefs are expressed using the terms "Minnesotan" and "Arizonan". Barnes's important observation shows that the change of language makes no corresponding difference regarding *knowledge*: a shift of language reveals that Jones has some true beliefs, but it does not render those beliefs epistemically successful—they are not knowledge. So that in turn means that while the truthlikeness view faces problems, the epistemic view can simply bypass them.[7] Fortuitously true beliefs notwithstanding, Jones knows nothing substantive about the weather whereas Smith does, and therefore Smith contributes to progress and Jones does not.

A suitable response therefore proposes that progress should be seen in terms of increasing knowledge, not nearness to the truth. This, the epistemic view, gives the right results in the other cases considered also. If by fluke Blondlot had some true beliefs about N-rays, they would not have been knowledge and so would have made no contribution to progress. Renaissance doctors did not know that infants have high body water, although that is true, because they had no way of determining that—their

---

[7] Barnes himself thinks that we can use the epistemic features of such cases to identify an appropriately privileged language—appropriate to that epistemic circumstance. The language issue strikes me as a red herring and that Barnes's contribution is more effective without the detour via language.

beliefs were based on the radically false humoral theory. So there was no progress when they came to have those beliefs either. Likewise, normal science based on a radically false theory, such as the theory of four elements and four basic qualities, is not progressing, even if it appears, from the practitioners' point of view, to be providing successful solutions to their problems. That is because in such cases there is no addition of knowledge.

The view that progress consists in the accumulation of knowledge faces an objection relating to that which stimulated the truthlikeness approach. A straightforward accumulation of truth view suffered from the fact that there could be progress in a sequence of theories, all of which are false. The idea of truthlikeness, which could increase through that sequence, appeared to solve the problem. If we now switch to the epistemic approach we face a difficulty that does not appear to have such a straightforward solution. Knowledge entails truth, so the community cannot know any proposition in a series of false propositions: however close to the truth p maybe, if p is strictly false, then it cannot be known that p. Nor, however, is there an obvious analogue to truthlikeness for knowledge: a state that is much like knowledge except that the content of that state is false. It might be that we could attempt a formal definition of such a state—call it "approximate knowing". Approximate knowing would be a certain kind of belief. For example, it might be proposed that the state of approximate knowing that p is a belief that p that is justified, where the justification is appropriately linked to the fact that p has high truthlikeness. Note that if an approach of this sort says anything detailed about what approximate knowing is (such as in this proposal), then it is likely to draw on a particular definition of (non-approximate) knowledge, substituting truthlikeness for truth. Yet accounts and definitions of knowledge are notoriously subject to counterexample. So it may be that the most we can confidently say about this state is that it is like knowing except for a false content. Those, like Williamson (1995), who take knowledge to be a factive mental state, different in kind from belief, will note that approximate knowing is not a state of the same kind as knowing.

While I do not rule out the idea of approximate knowing, a better approach starts by noting that when a theory T is accepted, what scientists believe is true is not limited to a precise statement of T—indeed they might not believe that precise statement. Scientists will believe many other propositions that are related to T, most importantly its most obvious and salient logical consequences. For example, Dalton not only believed (a) that water is HO; he also believed (b) that water is a compound of hydrogen and oxygen and (c) that water is made up of molecules, each containing a small number of atoms of hydrogen and oxygen, and so forth. Rutherford almost certainly did not believe (d) that Avogadro's constant has the precise value of $6.16 \times 10^{23} \text{mol}^{-1}$, but he did likely believe (e) that the value of Avogadro's constant is approximately $6.16 \times 10^{23} \text{mol}^{-1}$, or (f) that the value of Avogadro's constant lies between $6.0 \times 10^{23} \text{mol}^{-1}$ and $6.3 \times 10^{23} \text{mol}^{-1}$. While (a) and (d) are false propositions, (b), (c), (e), and (f) are true propositions. Given the evidence and methods that Dalton and Rutherford employed, it is plausible that Dalton knew (b) and (c) and Rutherford knew (e) and (f). Thus insofar as their predecessors did not have this knowledge, then Dalton and Rutherford contributed to progress, according to the epistemic approach. The relevant consequences are logically weaker and so less informative that the propositions (a) and (c) from which they are derived: in the case of (e) and (f), the consequences are inexact propositions, while in the case of (b) and (c), the consequences are exact propositions but omit details contained in the original proposition (a).

This approach to defending the epistemic view of progress, in the case of false but progressive theories, requires that we are able to find (true) propositions that are consequences of such theories and that are now known by the community but that were not known previously. It is highly plausible that this condition can be met.[8] As the previous examples suggest, if a proposition is strictly false but highly truthlike, then there will exist related nontrivial true propositions. At the very least "T is highly truthlike" will be such a proposition (for false but highly truthlike theory T). But in many cases we will be able to find propositions more precise than this.[9]

The next question concerns belief—do the scientists in question believe the true proposition in question? In some cases this may be clear enough—Dalton clearly believed (a) and also (b) and (c). But in other cases it might be less clear, often because the scientist does not believe the principal truthlike but false proposition. The most difficult cases are those in which the theory in question is a model that scientists know from the outset to be strictly false. Nonetheless, in such a case, if the model does contribute to progress, then it will have some significant element of truth or truthlikeness (at least relative to its predecessors), and that will have been achieved by suitable gathering of evidence and reasoning. If so, the scientists promoting the model should be able to say something about the respects in which the model matches reality, even if, at a minimum, that is just to say that the model delivers approximately accurate predictions within certain ranges of the relevant parameters. They will typically have some idea regarding some of the implications of theory that these are supported by the evidence and reasoning whereas others are not. For example, the simple kinetic theory of gases is clearly false, since it assumes that gas molecules have no volume. So no-one ever believed the theory in toto, but scientists do believe nontrivial implications of the theory: that gases are constituted by particles; that the temperature of a gas is in large part a function of the kinetic energy

---

[8] See Niiniluoto (2014) for criticisms of this claim—what follows, I hope, provides some answers to those criticisms.

[9] Is "T is highly truthlike" a proposition of science? The answer might depend on one's view of the truth predicate and its relatives, such as "truthlike." For example, one might think that one's use involves semantic ascent: "T" is about the world whereas "T is true" concerns words (and their relation to the world). On the other hand, a redundancy theorist about truth will think that "T is true" says exactly what "T" says and is equally about the world. Presumably analogous remarks could be make about "truthlike." In my view, even if the truth and truthlike predicates do involve semantic ascent, they can still be used in scientific propositions: they still make claims about the world, even if they say something else as well. For example, if an economist says, "The simple supply and demand model is a good approximation to the truth as regards many commodity markets, but it is not a good approximation for markets with asymmetric information, such as the markets for complex financial securities and health insurance" then he or she is saying something *both* about the world and about a theory; the assertion does not seem any the less a scientific statement because of that.

of the particles; and that the ideal gas equation holds with a high degree of approximation for gases at moderate pressure and temperature.[10]

So false theories, even those known to be false, can contribute to progress on the epistemic view because they often have significant true content or true implications that are believed by scientists on the basis of good evidence and reasoning (i.e., these implications are known propositions). Thus the epistemic approach is able to accommodate progress made through false theories. At the same time, it delivers more accurate verdicts regarding cases of accidentally true unjustified (or partially) theories, which gives it a distinct advantage, in my opinion. On the other hand, it does need to confront the widely held view that truth is the aim of inquiry and belief and that the justification element of knowledge is not constitutive of the aim of science (nor of scientific progress) but is rather merely instrumental in achieving that aim (Rowbottom 2010; Niiniluoto 2011).

6 PROGRESS AND THE AIM OF SCIENCE

As Niiniluoto (2011) puts it, "Debates on the normative concept of progress are at the same time concerned with axiological questions about the aims and goals of science." At the outset I linked approaches to progress with views regarding the aim of science. Laudan (1977, 1981) and Kuhn (1970) think that science aims at solving problems, and so success in so doing is their standard of progress. Oddie (2014) accepts that truth is the aim of inquiry, motivating accounts of truthlikeness.[11] Barnes (1991) and I take knowledge to be the goal of science and accordingly hold scientific progress to be the addition of scientific knowledge.[12] Note that because a goal can have subsidiary goals as means, those who hold that science aims at truth or truthlikeness can agree that science aims at problem-solving, since solving a problem will be a route (albeit a fallible one) to the truth about the matter in question. Likewise the view that knowledge is the goal of science will imply that problem-solving and truth are legitimate goals for science, since pursuing these will further the pursuit of knowledge. So the difference between the views must turn on differences between these goals as ultimate or constitutive goals.

---

[10] Note that belief here does not need to be occurrent; it may be dispositional. This is not to say that scientists will believe all the significant true implications of a truthlike model—many scientists did not believe the ontological implications of the kinetic theory, developed in the 1850s and 1860s, until the twentieth century. It may be a matter of further discovery that a certain element of a model is highly truthlike rather than merely instrumental; such discoveries will themselves be contributions to progress, since now those elements will themselves, by being believed, become knowledge.

[11] I note that in the earlier (2007) version of his article, Oddie begins "*Truth* is the aim of inquiry," whereas he now (2014) commences more circumspectly "*Truth* is widely held to be the constitutive aim of inquiry."

[12] The debate about the aim of science naturally relates to debates concerning the nature and aim of belief; cf. Velleman (2000), Wedgwood (2002), and Owens (2003).

Although the problem-solving view of the constitutive aim of science is consistent with scientific realism, it is not a view that will appeal to the realist. If that view were correct, then a piece of science would have achieved its goal if it solves some problem in the (internalist) sense of Kuhn and Laudan but is nonetheless false. The only way of making that view acceptable is to reject the notion of truth (or truthlikeness) or at least adopt a strong kind of skepticism that would make the goal of truth utopian. If one is a realist and thinks that truth is achievable, then one will be inclined to reject the idea that a problem-solution that *appears* to solve a problem but is false does satisfy the aim of science.

It might appear difficult to draw a distinction between the goals of truth and those of knowledge. After all, if one uses rational means to pursue the first goal and one does achieve it, then one will typically have achieved the second also. There are differences, however. If one could believe a contradiction and all of its consequences, then one would thereby believe all truths (and all falsehoods). That would achieve the aim of science to a maximal degree, if that aim is truth, whereas it would not achieve the aim of science, if that aim is knowledge. If to avoid this it is added that one's beliefs must be logically consistent, one could adopt a policy of believing propositions at random, so long as they are consistent with one's existing beliefs. This would lead to a large quantity of true belief but to no knowledge. Neither of these policies, although productive of truth if successfully adopted, would count as promoting the goal of science. If that verdict is correct, then the view that science aims at knowledge has an advantage. Alternatively, the truth aim would have to be supplemented by an avoidance of error aim, in which case there is a question about how to balance these aims: To what extent does a possible gain in truth outweigh a risk of falsity? The epistemic account does not have to face such a question, since risky belief, even if true, fails to be knowledge.

7 CONCLUSION: REALISM, PROGRESS, AND CHANGE

A simplistic picture of the history of science presents science as a highly reliable source of truth—science progresses because it is able to add new truths to the stock of old truths. While it not clear that any scholar really believed this to be true, it may be the impression given by some whiggish, heroic histories of science. Furthermore, optimistic accounts of the scientific method or of inductive logic seem to suggest science should progress by accumulating truth. The history of science portrayed by Kuhn and others, with its periodic reversals of belief, rejects such a view. And if realist philosophies of science say that the history of science ought to be like this, then those philosophies may be rejected too.

If one draws antirealist conclusions from this history, the view of progress as increasing problem-solving power looks attractive. However, such a view does have to take a stand on what is happening during those moments of rupture when previous problem-solutions are rejected—episodes of Kuhn-loss. One may argue that there is an overall increase in problem-solving power, but to do so implies that there is a weighting of problems and their solutions, so that the new problem-and-solution combinations are worth more than those that are lost. Furthermore, Kuhn is clear that old problems-plus-solutions may be rejected on the promise of more and better to come. Problem-solving power is clearly a dispositional concept on this view. If so, it may not be as transparent as Laudan would like whether an episode of scientific change is progressive.

Realists will hold that this approach to progress suffers from not distinguishing apparent from real progress. That there is little or no difference is a consequence of the internalism espoused by Laudan—it should be possible to tell directly whether a problem has been solved by a proposed solution and progress thereby made. But to those who avail themselves of the concept of truth, a sequence of false solutions to a pseudo-problem (a problem founded on a false assumption) cannot be progress, however convincing such solutions are to those working within that paradigm.

The realist can maintain that the existence of falsehoods in the history of science is consistent with realism, so long as those falsehoods are in due course replaced by truths, without truths being replaced by falsehoods. Nonetheless, the favored variant on this for realists has been to conceive of progress in terms of truthlikeness. A sequence of theories may be progressive even if false when each is closer to the truth than its predecessor. This view can avoid much of the difficulty raised by Kuhn-loss, so long as the new problem-solutions are closer to the truth than those they replace. That seems to be the case in the standard examples. Although there was Kuhn-loss in the rejection of Descartes's vortex theory, his problem-solutions were badly false (it is not because gravity operates via a vortex that the planets rotate in a plane and in the same direction). So the loss of those "solutions" should be nothing for a truth/truthlikeness account of progress to worry about. That said, there are some cases of truth-loss (e.g., that concerning the moisture content of infants' bodies, when the humoral theory was abandoned) that per se would be regarded as regressive by the truth/truthlikeness standard. One might think that this is wrong, because it is progressive to cease believing in a proposition, true or false, if one discovers that one's previous reasons for believing it are themselves mistaken. The truth/truthlikeness view might attempt to retrieve the situation by arguing that in such cases the truth-loss is outweighed by truth-gain, which would then require weighting some truths as more significant than others. More problematic are hypothetical cases in which there is no corresponding significant gain (such as the example in which N-rays turn out to exist and Blondlot was right but for the wrong reasons).

Neither Kuhn-loss nor truth-loss are, per se, problematic for the epistemic account. The episodes discussed, actual and hypothetical, are not regressive, since the beliefs rejected do not amount to knowledge. A regressive episode for the epistemic view is one in which there is knowledge loss—a scientific change that involves scientists knowing less at the end than at the beginning. Realists will expect such episodes to be few and far between, for if scientists have acquired enough evidence and good reasons so that they know that p, it will be fairly unlikely that evidence and reasons will come along that will persuade them to give up belief in *p*. That said, we can conceive of hypothetical cases; if a mass of misleading evidence is generated, then the community might lose its knowledge. But even such cases need not trouble the epistemic account of progress, for such episodes really would be regressive. Thus the epistemic account can claim both to match our pretheoretic judgments about which (real or hypothetical) episodes are progressive relative to their rivals while also finding it more straightforward to deliver the verdict that science has indeed generally been progressive.

References

Barnes, E. (1991). "Beyond Verisimilitude: A Linguistically Invariant Basis for Scientific Progress." *Synthese* 88: 309–339.

Becker, P. (2001). "History and Progress in the Accurate Determination of the Avogadro Constant." *Reports on Progress in Physics* 64: 1945–2008.

Bird, A. (2007). "What Is Scientific Progress?" *Noûs* 41: 64–89.

Bird, A. (2008). "Scientific Progress as Accumulation of Knowledge: A Reply to Rowbottom." *Studies in History and Philosophy of Science, Part A* 39: 279–281.

Bragg, W. (1936). "The Progress of Physical Science." In J. Jeans, W. Bragg, E. V. Appleton, E. Mellanby, J. B. S. Haldane, and J. S. Huxley (eds.), *Scientific Progress* (London: Allen & Unwin), 39–77.

Descartes, R. (1960). *Discourse on Method*. Translated by A. Wollaston (Harmondsworth, UK: Penguin). (Original work published 1637.)

Friis-Hansen, B. J., M. Holiday, T. Stapleton, and W. M. Wallace (1951). "Total Body Water in Children." *Pediatrics* 7: 321–327.

Hilpinen, R. (1976). "Approximate Truth and Truthlikeness." In M. Przelecki, A. Szaniawski, and R. Wójcicki (eds.), *Formal Methods in the Methodology of the Empirical Sciences* (Dordrecht, The Netherlands: Reidel), 19–42.

Klotz, I. M. (1980). "The N-Ray Affair." *Scientific American* 242: 168–175.

Kuhn, T. S. (1970). *The Structure of Scientific Revolutions* (2nd ed.) (Chicago: University of Chicago Press).

Lagemann, R. T. (1977). "New Light on Old Rays." *American Journal of Physics* 45: 281–284.

Laudan, L. (1977). *Progress and its Problems: Toward a Theory of Scientific Growth* (Berkeley: University of California Press).

Laudan, L. (1981). "A Problem-Solving Approach to Scientific Progress." In I. Hacking (ed.), *Scientific Revolutions* (Oxford: Oxford University Press), 144–155.

Miller, D. (1974). "Popper's Qualitative Theory of Verisimilitude." *British Journal for the Philosophy of Science* 25: 166–177.

Mizrahi, M. (2013). "What Is Scientific Progress? Lessons from Scientific Practice." *Journal for General Philosophy of Science* 44: 375–390.

Newton, H. (2012). *The Sick Child in Early Modern England, 1580–1720* (Oxford: Oxford University Press).

Niiniluoto, I. (1977). "On the Truthlikeness of Generalizations." In R. E. Butts and J. Hintikka (eds.), *Basic Problems in Methodology and Linguistics* (Dordrecht, The Netherlands: Reidel), 121–147.

Niiniluoto, I. (1980). "Scientific Progress." *Synthese* 45: 427–462.

Niiniluoto, I. (2011). "Scientific Progress." In E. N. Zalta (Ed.) *The Stanford Encyclopedia of Philosophy* (Summer 2011 ed.). http://plato.stanford.edu/entries/scientific-progress/

Niiniluoto, I. (2014). "Scientific Progress as Increasing Verisimilitude." *Studies in History and Philosophy of Science, Part A* 46: 73–77.

Nutton, V. (1993). "Humoralism." In W. F. Bynum and R. Porter (eds.), *Companion Encyclopedia of the History of Medicine,* Vol. I (London: Routledge), 281–291.

Oddie, G. (2014). "Truthlikeness." In E. N. Zalta (Ed.) *The Stanford Encyclopedia of Philosophy* (Summer 2014 ed.). http://plato.stanford.edu/entries/truthlikeness/

Owens, D. J. (2003). "Does Belief Have an Aim?" *Philosophical Studies* 115: 283–305.

Popper, K. (1972). *Objective Knowledge* (Oxford: Clarendon).

Rowbottom, D. (2008). "N-Rays and the Semantic View of Scientific Progress." *Studies in History and Philosophy of Science, Part A* 39: 277–278.

Rowbottom, D. P. (2010). "What Scientific Progress Is Not: Against Bird's Epistemic View." *International Studies in the Philosophy of Science* 24: 241–255.

Schoeller, D. A. (1989). " Changes in Total Body Water with Age." *The American Journal of Clinical Nutrition* 50: 1176–1181.

Tichý, P. (1974). "On Popper's Definitions of Verisimilitude." *British Journal for the Philosophy of Science* 25: 155–160.

van Fraassen, B. (1980). *The Scientific Image* (Oxford: Oxford University Press).

Velleman, J. D. (2000). "On the Aim of Belief." In J. D. Velleman, *The Possibility of Practical Reason* (Oxford: Oxford University Press), 244–281.

Wedgwood, R. (2002). "The Aim of Belief." *Philosophical Perspectives* 16: 267–297.

Williamson, T. (1995). "Is Knowing a State of Mind?" *Mind* 104: 533–565.

Zilsel, E. (1945). "The Genesis of the Concept of Scientific Progress." *Journal of the History of Ideas* 6: 325–349.

CHAPTER 27

SCIENTIFIC REALISM

TIMOTHY D. LYONS

1 INTRODUCTION

CONTEMPORARY scientific realism embraces two core tenets, one axiological and the other epistemological. According to the axiological tenet, the primary aim of science is truth—truth, for the realists, being no less attributable to assertions about unobservables than assertions about observables. According to the epistemological tenet, we are justified in believing an "overarching empirical hypothesis," a testable meta-hypothesis, about scientific theories. For standard scientific realism, that meta-hypothesis is that our successful scientific theories are (approximately) true. The justification for believing this meta-hypothesis resides in the "no-miracles argument": were our successful scientific theories not at least approximately true, their success would be a miracle. In other words, rejecting miracles as explanatory, the (approximate) truth of our theories is the only, and so the best, explanation of their success. Because both sides of the debate have taken the axiological tenet to stand or fall with the epistemological tenet, it is the debate over the epistemological tenet that is at the forefront of the literature. Accordingly, this article is concerned primarily with epistemic scientific realism, in particular, clarifying and reinforcing the two primary nonrealist objections against it. Only briefly, at the end, do I return to the axiological tenet.

2 THE HISTORICAL ARGUMENT AGAINST SCIENTIFIC REALISM

There are two primary arguments against epistemic scientific realism. In the next section, I address the argument from the underdetermination of theories by data. Here

I discuss the historical argument against scientific realism,[1] which takes seriously the assertion that the scientific realist's meta-hypothesis is empirically testable. At its core lies a list of successful theories that cannot, by present lights, be construed as (even approximately) true. Contemporary versions have generally drawn on Larry Laudan's (1981) well-known list of such theories (e.g., theories positing phlogiston, caloric, the luminiferous ether, etc.). The most prevalent version of this argument is known as "the pessimistic meta-induction": because many past successful theories have turned out to be false, we can inductively infer that our present-day theories are likely false. Accordingly, realists respond that there are insufficiently many positive instances to warrant the conclusion: referencing Laudan's historical argument, Stathis Psillos writes, "This kind of argument can be challenged by observing that the inductive basis is not big and representative enough to warrant the pessimistic conclusion" (1999: 105). Here I discuss two alternative *noninductive* variants of the historical argument, both of which take the form of a *modus tollens*: the first I articulated earlier (Lyons 2002), and the second I introduce here.

2.1 The First Meta–*Modus Tollens*

As with the standard pessimistic induction, the core premise of the noninductive historical *modus tollens* is a list of successful theories that are not, by present lights, approximately true. In this historical argument, the list of false successful theories demonstrates not that our contemporary theories are (likely) false but that the realist meta-hypothesis is false, in which case we cannot justifiably believe it. In fact, I contend (Lyons 2001, 2002) that this is how we should understand the structure of the historical argument in Laudan (1981). (However, notably, in texts published both before [1977: 126] and after [1983: 123] "A Confutation of Convergent Realism" [1981], Laudan explicitly embraces the standard pessimistic meta-induction.) As we have seen, in its standard form the empirical meta-hypothesis the realist says we can justifiably believe is "our successful theories are approximately true." And the standard argument for that thesis is that it would be a miracle were our theories to be as successful as they are, were they not at least approximately true. The choice this realist gives us is between approximate truth and miracles, the latter being no genuine option at all. The historical threat, then, briefly stated yet properly construed, is as follows

1. If (a) that realist meta-hypothesis were true, then (b) we would have no successful theories that cannot be approximately true. (If we did, each would be a "miracle," which no one of us accepts.)
2. However, (not-b) we do have successful theories that cannot be approximately true: the list (of "miracles").

[1] See Mill (1859/1998), Laudan (1981), Sklar (1981), Rescher (1987), Worrall (1989), Leplin (1997), Chakravartty (1998), Psillos (1999), Sankey (2001), Lange (2002), Lyons (2002), Kitcher (2001), Chang (2003), Stanford (2003), Magnus and Callender (2004), Parsons (2005), Lyons (2006), Stanford, (2006), Doppelt (2007), Schurz (2009), Lyons (2009a, 2009b), Harker, (2010, 2013), Vickers (2013), Peters (2014).

3.   Therefore, (not-a) the realist meta-hypothesis is false. (And the no-miracles argument put forward to justify that meta-hypothesis is unacceptable.)

As mentioned, the historical premise has been traditionally understood to provide positive instances toward a nonrealist conclusion that our successful *scientific* theories are (likely) false. However, we now see that premise as providing counterinstances to the realist meta-hypothesis that our successful theories are approximately true. We are making no meta-induction toward an affirmation that our scientific theories are (likely) false but rather a meta–*modus tollens* that reveals the falsity of the realist meta-hypothesis (and the unacceptability of the no-miracles argument). Not only, then, are we not justified in believing the realist's meta-hypothesis; that hypothesis cannot even be accepted as a "fallible" or "defeasible"—let alone a "likely"—conjecture. It is simply false. One virtue attributable to this variant of the historical argument is that it no longer hinges on the quantitative strength of an inductive basis, and efforts to weaken that basis will not eliminate the threat. Notice also, for instance, that the *modus tollens* requires no commitment as to which specific scientific theories among the mutually contradictory successful theories are the false ones—the falsity may lie in the past theories, in the contemporary theories that directly contradict them, or both. On this view, the quest to empirically increase the quantity of instances is not to provide strength for an inductive inference. It is rather to secure the soundness of the *modus tollens*, to secure the truth of the pivotal second premise, the claim that there are counterinstances to the realist meta-hypothesis. Put another way, increasing counterinstances serves to strengthen and secure the nonampliative, nonuniversal falsifying hypothesis "some successful theories are false." An additional role served by increasing the counterinstances (implicit in the previous parenthetical notes) pertains to the very argument realists invoke to justify believing their meta-hypothesis: the greater the quantity of "miracles," the more obvious it becomes that the core claim of the no-miracles argument is false. Increasing the collection of what become for the realist inexplicable successes, we are forced to reject the realist claim to offer the only, or even merely the best, explanation of those successes.[2] Moreover, for each counterinstance, it is now the realist who fails to live up to the realist's own much-touted demand for explanation. (Later we will see that increasing the historical counterinstances serves another, still noninductive purpose, in a second *modus tollens*.) Most crucially, this construal of the historical argument renders impotent numerous defenses against it. (Notably, among plenty of others, Lange [2002], Kitcher [2001], Magnus and Callender [2004], Parsons [2005], and Stanford [2006: 10–11][3] offer other recent arguments against the inductive variant of the historical argument, each of which misses the mark against this meta–*modus tollens*.)

---

[2] I discuss nonrealist explanations of success in "Explaining the Success of a Scientific Theory" (Lyons 2003).

[3] Stanford offers instead his "new induction." In short, because past scientists failed to think of alternatives, contemporary scientists fail as well.

One, albeit awkward, response to this *modus tollens* is to deny my claim that the realist invokes a meta-hypothesis. In fact, surprisingly, on both sides of the contemporary realism debate realism is construed as claiming we can justifiably believe the best explanation of natural phenomena. However prevalent this construal may be, it is the result of confusion regarding both the phenomena that realism explains and what it is that explains those phenomena. With the *modus tollens* now in hand, we can make clear just how untenable realism is rendered by this misconstrual. Of course the phenomena calling for explanation are not those of the natural world, so to speak; the latter are what scientific theories purport to explain. What realism purports to explain is, rather, a property of scientific theories. Nor, of course, is that which does the explaining itself a scientific theory; what the latter explains are natural phenomena. Just as that which is explained by realism is a property of scientific theories, so is that which does the explaining: one property of scientific theories explains another.

With regard to the property to be explained: in order to justify belief that bears on the content of science, realists must appeal to a property of scientific theories, which, in some epistemically relevant way, distinguishes those theories from any other arbitrarily selected collection of statements—say, collections of possible statements about the observable and unobservable that are contrary to what is accepted by contemporary science. And here one might think, yes, that property can simply be that of being the best explanation for a given set of phenomena. However, first the explanation needs to be "good enough" (Lipton, 2004); that is, it needs to meet the criteria scientists employ for acceptance. Second, even if scientists employ strict criteria, realists cannot simply map their beliefs onto scientific acceptance. Since to believe P is to believe that P is true, to believe the accepted explanation for a set of phenomena is to believe that it is true. Thus we have already landed on a meta-hypothesis "*accepted* theories are true." Moreover, with the proper understanding of the historical argument, we realize that this meta-hypothesis is unequivocally refuted by the meta–*modus tollens,* at nearly every turn in the history of science. Success, then, cannot be equated with superiority over available competitors, even when that superiority is determined by strict criteria for scientific acceptance. The property that realism purports to explain must go even beyond those criteria: it must be a surprising property—for instance, predictive success—such that, without some other property (e.g., truth), it would be a miracle were the theory to have it.

Regarding the property that does the explaining: the *modus tollens* forces us to see that it cannot simply be the truth of the theory. Not only is "*accepted* theories are true" refuted at nearly every historical turn; it turns out there are numerous counterinstances to even "theories achieving novel predictive success are true." Although Laudan was not concerned with novelty when formulating his list, numerous novel successes are listed in Lyons (2002, 70–72): there I show that, along with false theories outside of Laudan's list, those theories positing phlogiston, caloric, ether, and so on *did* achieve novel predictive success. In fact, even if the realist could somehow justify a commitment to only our most successful contemporary theories, the latter theories are not such that they can be believed outright: despite their individual successes, general relativity and quantum field theory, for instance, contradict one another; they cannot both be, strictly speaking, true—at best, at least one is only approximately so. So realists are forced to invoke as the explanatory property not truth per se but something along the lines of *approximate* truth. A meta-hypothesis is required and—despite a desire for a "face-value" realism, wherein we simply believe our favorite theories—that meta-hypothesis must be something similar to "predictively successful theories are (at least approximately) true."

Because it is only contemporary theories about which most realists want to be realists, and because our realist inclinations may blind us to the relevance of the conflict between general relativity and quantum field theory, we are prompted here to bolster the point that there is a conflict. I submit that the clash between those theories is taken as a given and in fact is taken to be *the* driving problem in contemporary theoretical science, by any number of physicists—including those in as much disagreement as, on one hand, Brian Greene and Leonard Susskind, who advocate string theory, and, on the other, Lee Smolin and Peter Woit, who argue against it.[4] Additionally, profoundly relevant but generally overlooked in philosophical circles is the dramatic conflict arising from the conjunction of those otherwise successful theories and the data regarding the acceleration of the universal expansion, the value of λ. In short, that conjunction results in what may well be the greatest clash between prediction and data in the history of science: the predicted value for λ is inconceivably greater than what the data permit, off by "some 120 orders-of-magnitude!" (Frieman et al. 2008: 20), letting physicists speak for themselves. This "blatant contradiction with observations . . . indicates a profound discrepancy"—"between General Relativity (GR) and Quantum Field Theory (QFT)—a major problem for theoretical physics" (Ellis et al. 2011: 1). Predicting, for instance, that "there would be no solar system" (2), this "profound contradiction" amounts to a "major crisis for theoretical physics" (10). (See also Murugan et al. 2012: 2–3.) Now contrast this "major crisis"—along with, perhaps, the galactic discrepancies that have prompted the positing of dark matter—against the comparatively negligible discrepancies that ultimately led to the overthrow of Newtonian physics by general relativity: the mildly aberrant behavior of Mercury, the less than determinate results (as it now turns out) of the Eddington expedition. . . . Again, the primary point is that, given the relation between general relativity and quantum field theory, at least one of these otherwise predictively successful theories cannot be strictly speaking true; at best at least one is only approximately true. Since believing that T is approximately true is to believe that T is strictly speaking false, believing that T is approximately true is altogether distinct from simply believing T. We must wholly discard the thesis that realists simply believe the best explanation of phenomena, along with any lip service to face-value realism. Recognizing the untenability of these notions, it is clear that our realist cannot avoid a meta-hypothesis. And, with the *modus tollens* in hand, we see that this meta-hypothesis must be one that invokes as its explanatory correlate something along the lines of *approximate* truth and, as the correlate to be explained, a property more restrictive than scientific acceptability,

---

[4] Ioan Muntean has brought to my attention a formal proof of the clash between quantum field theory and general relativity in Weinberg and Witten (1980).

for instance, predictive success—hereafter meaning "novel success," where the data predicted were not used in developing the theory.[5]

2.2 The Second Meta–*Modus Tollens*

We can now introduce a second, new historical meta–*modus tollens*. This variant embraces lessons from the underdetermination argument, discussed later, and from the insight provided by twentieth-century philosophy of science that evaluations of empirical theories are comparative. Here, however, what are to be compared are meta-hypotheses. Claiming that we are justified in believing the meta-hypothesis "predictively successful theories are approximately true," the scientific realist is committed to claiming justification for believing that meta-hypothesis and the no-miracles premise *over* those ampliative meta-hypotheses that oppose them. By the latter, I mean to include the following "ContraSR" meta-hypotheses:

- "predictively successful theories are *statistically unlikely* to be approximately true"
- "predictively successful theories are *not-even-approximately-true.*"

Asking whether we are justified in believing the realist's meta-hypothesis and no miracles premise over ContraSRs, we must ask: Which are in *better* accord with the data? In line with what we have seen so far, if any data could stand as correlatively precise positive instances for a ContraSR, they would have to be successful theories that are *not* approximately true. That given, and given the list of predictively successful theories that are not approximately true, the ContraSRs must be credited with such positive instances (which, as noted later, need not be literally "confirming" instances). And, in the historical argument, these positive instances are identified empirically, wholly independent of any presupposition that such a ContraSR is true. Moreover, without granting victory in

---

[5] For the sake of simplicity and clarity, here I use this standard realist meta-hypothesis as my foil for articulating the form and implications of the primary arguments against realism. However, the most sophisticated variant of realism focuses not on theories but only those theoretical constituents "responsible for" success (e.g., Psillos 1999). The meta-hypothesis becomes "those theoretical constituents that are deployed in the derivation of successful novel predictions are at least approximately true." Testing this revision constitutes a research program in itself, taking us well beyond Laudan's list. Although my goal here is not to offer historical case studies, we can note that such a program has at least been launched. Invoking the first *modus tollens* and exploring the theoretical constituents genuinely deployed by, for instance, Kepler, Newton, Leverrier, and Adams, I detail numerous counterinstances to this sophisticated meta-hypothesis (Lyons 2002, 2006), also included in a more recent list offered in Vickers (2013). (See also Lyons 2013, footnote 13.) It turns out that theories—and in particular "theoretical constituents *responsible for*"—achieving novel successes include the usual suspects, those positing phlogiston, caloric, ether, and so on but also less familiar theories such as Rankine's thermodynamic vortex theory, as well as familiar yet overlooked theories, such as Kepler's theory of the *anima motrix*. Although, here I do not use the deployment realist's far more cumbersome meta-hypothesis as my foil, the points in this article hold no less if we pivot our discussion around it.

advance to the realist's hypothesis, we have no data that stand as correlatively precise negative instances of a ContraSR. By contrast, the realist's meta-hypothesis has no correlatively precise positive instances without already establishing it. And it has a set of correlatively precise negative instances, the list, again without requiring any presupposition of a ContraSR. Given the list of successful theories that are not approximately true, the ContraSRs are rendered in their relation to the data superior to the realist hypothesis, and we are not justified in believing the latter over these ContraSRs. We are not, therefore, justified in believing the scientific realist's meta-hypothesis.

One reason I have included two nonidentical meta-hypotheses as ContraSRs is to emphasize that none of this has anything to do with believing one or the other of the two. And here we must not mislead ourselves to think that there is anything intrinsically ampliative in the relevant notions of "superiority" and "relation to the data." Even if most swans are white and only a few are black, it is unproblematic to say that black swans stand as correlatively precise positive (but not literally "confirming") instances of the false hypothesis "all swans are black"; and this unproblematic assertion requires no induction back to that false hypothesis. Likewise with the ContraSRs (irrespective of whether they are true or false): it is unproblematic to say that predictively successful theories that are not approximately true stand as correlatively precise positive (but not literally "confirming") instances of the meta-hypothesis "predictively successful theories are statistically unlikely to be approximately true" or "predictively successful theories are not-even-approximately-true." This unproblematic assertion also requires no ampliative inference back to either of these ContraSRs. No induction is involved in the task of tallying correlatively precise positive/negative instances (which are not, in themselves, literally confirming or refuting) or the task of comparing hypotheses in light of those results. Whether the same can be said for the act of *choosing between* the meta-hypotheses is a distinct question and not at issue here: for the recognition of the superiority of ContraSRs in respect to the data over the realist's meta-hypothesis does nothing in itself to necessitate, in turn, an inference to the truth of, a belief in, or even the tentative choosing or acceptance of any such empirically ampliative meta-hypothesis about successful theories. Nonetheless, employing the data, we find that we are not justified in believing the realist's meta-hypothesis. In fact, the argument just presented can be expressed as follows:

1. (a) We are justified in believing the realist meta-hypothesis (given the evidence) only if (b) we *do not* have greater evidence for those that oppose it (ContraSRs).
2. However, (not-b) we *do* have greater evidence for those that oppose it (for ContraSRs): the list; (and *as above, this claim involves no induction to any ContraSR).*
3. Therefore, (not-a) it is not the case that we are justified in believing the realist's meta-hypothesis (given the evidence).

This is another *modus tollens*, related to but distinct from the original introduced previously. There we saw that increasing the evidence against the realist meta-hypothesis secures soundness, the second premise of the first *modus tollens*. We now see a further reason for increasing the evidence and strengthening that premise. Via ContraSR meta-hypotheses, increasing the quantity of the items on the list increases the counterevidential weight against the realist meta-hypothesis—again necessitating no inductive (or other kind of) inference to a conclusion that our contemporary successful scientific theories are statistically unlikely to be approximately true or that they are not-even-approximately true, and so on.

To close this section on the historical argument, let us say, for the moment, that in response to the second premise of the first *modus tollens*, the realist chooses to embrace the following as the meta-hypothesis we can justifiably believe: "predictively success-ful theories are statistically likely to be approximately true." Not only would a retreat to this statistical meta-hypothesis diminish (but not eliminate) the testability of the real-ist meta-hypothesis; it would also wholly concede to miracles and to a failure on the part of the realist to explain successful theories. More to the point, the first ContraSR, "predictively successful theories are statistically unlikely to be approximately true," is bolder than but entails the negation of this new realist statistical hypothesis; that given, realists would find themselves in exactly the same situation as with their original meta-hypothesis: we have greater evidence for a bolder hypothesis that entails its negation. Hence, such a weakening of the realist hypothesis makes no difference in light of this new, second *modus tollens*.[6]

3 THE ARGUMENT FROM THE UNDERDETERMINATION OF THEORIES BY DATA

3.1 The Competitor Thesis

Beyond the historical argument, the other central challenge to realism is the argu-ment from the underdetermination of theories by data.[7] Its core premise is a *competitor*

---

[6] Encouraged now by the possibility of invoking statistical likelihood, the realist may be tempted to move it, so to speak, outside of the meta-hypothesis and claim that we can justifiably believe that the meta-hypothesis "predictively successful theories are approximately true" is statistically likely. However, this meta-hypothesis has already been rendered false by my first *modus tollens*; that given, along with the fact that "statistically likely" means "statistically likely to be true," we clearly cannot justifiably believe that this meta-hypothesis is statistically likely. It is simply false.

[7] See Mill (1843), Duhem (1914/1991), Quine (1951), Boyd (1973), van Fraassen (1980), Sklar (1981), Churchland (1985), Ben-Menahem (1990), Horwich (1991), Laudan and Leplin (1991), Leplin (1997), Kukla (1998), Psillos (1999), Okasha (2002), Lipton (2004), Chakravartty (2008), Godfrey-Smith (2008), Wray (2008), Lyons (2009b), Manchak (2009), Hoyningen-Huene (2011), Worrall (2011), Tulodziecki (2012), Lyons (2013).

thesis. In basic terms: our successful theories have empirically equivalent, yet incompatible, competitors. From that, the conclusion of the underdetermination argument is drawn. In basic terms: despite their empirical success, we cannot justifiably believe that our favored theories, rather than their competitors, are (approximately) true. I argued earlier that, contrary to much of the literature, the historical argument is not properly construed as inductive; I suggest that, likewise, the argument from underdetermination calls for clarification.

The first step is to isolate that question regarding competitors that is genuinely at issue in the scientific realism debate. That question is not "*Which* theories can we justifiably believe?" or, even more carefully, "*Which* theories are those among the class of theories such that we can justifiably believe that they are approximately true?" Since answers to these questions concede that there are such theories, the questions themselves grant a victory to realism in advance of being answered. Toward the identification of the proper question, we are prompted to clarify our terminology. Although our theories may have many alternatives (including alternative formulations), we can take "distinct alternatives" to denote alternatives such that, if those alternatives are approximately true, our favored theories cannot be approximately true: were the alternatives approximately true, they would render our preferred theories patently false. Realism, claiming we can *justifiably believe* the meta-hypothesis "our successful theories are approximately true" requires that we can *justifiably deny* the approximate truth of any such distinct alternatives to our favored theories. Taking "competitors" to mean distinct alternatives that are *empirically on par* with our favored theories, we see that the relevant and legitimate question in the realism debate is *whether* our successful scientific theories have competitors *such that we cannot justifiably deny that they are approximately true.*

With the proper competitor question in hand, the second step is to recognize that the realist cannot answer the argument from underdetermination merely by denying that our favored theories have empirically *equivalent* competitors. In fact, nonrealists other than, say, Bas van Fraassen, such as Laudan (1981), Sklar (1981), and Stanford (2006), recognize that, at least historically, the most genuine threats to scientific realism are empirically distinct. Just as theories that are empirically distinct from our contemporary theories occupy the core premise of the historical argument, such theories can occupy the competitor premise of the underdetermination argument. Clarifying "empirically on par" to allow for this, we can finalize our previous definition of "competitors": competitors are distinct alternatives whose empirical predictions accord with the observed data predicted by our favored theories. Since it is so commonly assumed that realism is threatened only by underdetermination arguments that invoke empirically equivalent competitors, we are prompted to go beyond Laudan, Sklar, and Stanford and make clear just why such an assumption arises and why it is false.

It arises because, in contrast with an empirically distinct competitor, a competitor that is empirically equivalent to our favored theory is *empirically ineliminable*: no matter how many empirical predictions are tested, there is no point in the future at which such a competitor can be empirically eliminated; and, given the fact of empirical ineliminability, our favored scientific theories face *temporally unrestricted* underdetermination.

Empirical equivalence suffices for empirical ineliminability, which in turn suffices for temporally unrestricted underdetermination. It is for this reason, I suggest, that the literature is replete with concern about empirical equivalence. Nonetheless, I contend that, as long as we find the right set of empirically distinct theories, that demand is irrelevant. That is, although empirical equivalence is sufficient for empirical ineliminability and so temporally unrestricted underdetermination, it is not a necessary condition for either.

First, an obvious point, whose importance is unencumbered by its obviousness: until distinguishing tests have been performed, the empirical distinguishability of competitors does nothing whatsoever to license belief in the (approximate) truth of our favored theories. Because realism is, in every instance, about belief (in approximate truth) at time t, potential future tests of competitors (at t-plus-a-century or even t-plus-a-decade) offer no positive evidence in favor of our preferred theories over those competitors. This fact alone strikes at the claim that we can be justified in believing (in the approximate truth of) our favored theories—empirically distinct competitors, which are *not yet* eliminated, posing no less a threat than empirically equivalent competitors. Second, if the situation is such that our theories will always have indefinitely many competitors, the distinguishing tests can never be performed for the entire set: in that case, our favored scientific theories are and will always be faced with empirically ineliminable competitors. Even if each competitor is individually empirically distinct from a given favored theory, and thus in principle individually empirically eliminable, the underdetermination that results from indefinitely many such competitors remains temporally unrestricted. The epistemic threat is no less severe than that posed by empirically equivalent competitors. With these points, I suggest that, just as the pessimistic induction is a straw objection to, so a distraction in favor of, scientific realism, so too are irrelevant competitor questions and the superfluous demand for empirical equivalence.

We can now endeavor to identify classes of genuine competitors and so secure the relevant competitor thesis. Toward that end, it is notable that, although the historical and underdetermination arguments have generally been treated as distinct, important relations hold between them.[8] Here I develop an empirically and historically informed competitor thesis and thus a foundational premise for an argument from underdetermination. However, I endeavor to avoid any inductive inference. The general strategy I offer for doing so is as follows: we empirically identify historically exemplified competitor relations between past and present theories, and, in particular, we isolate those competitor relations that extend to any theory related to phenomena in the way that scientific theories are noncontentiously required to be related to phenomena—according

---

[8] Stanford (2006), drawing on Sklar, brings this important point to the fore, offering a historical *induction* to ground what I've referred to as a competitor thesis (see footnote 2). However—even setting aside the fact that Stanford does not argue for indefinitely many competitors in the situations he specifies—I show (Lyons 2013) that his argument poses no threat to contemporary realism: it fails to concern itself with the type of theories invoked in the realist's meta-hypothesis (e.g., those making successful novel predictions, a problem initially pointed to in Lyons 2006: 544, footnote 10); it rests on a problematic thesis regarding the *failure of scientists,* and it relies on not one but two dubious inductions. The arguments I articulate here face none of these problems.

to both realists and antirealists. From among the latter subset of competitor relations, we identify those that can be instantiated in indefinitely many ways. In what follows, I specify two such competitor relations, demonstrating along the way that we have a historically informed yet wholly noninductive way to realize that any theories we may favor have indefinitely many competitors.

3.2 Competitors, Set 1

Toward the empirical identification of one such competitor relation between past and present theories, we can look first to those theories on "the list" employed in the historical argument. We can select as T such a historically successful but now rejected theory. According to contemporary science, T (approximately) predicts that a certain set of observed phenomena obtains. However, again according to contemporary science, in certain situations, the phenomena behave in a manner that significantly diverges from T's predictions. Here it is crucial to recognize that, since we are employing contemporary science to articulate the details of the divergence and since "scientific seriousness" is relevant to the realism debate (and its proper competitor question) only insofar as it pertains to whether or not a theory could be approximately true, the competitors we are considering cannot be excluded on the grounds that they are, for instance, "Cartesian fantasies." Contemporary science itself reveals a competitor (CT), which, though contradicting T, shares those predictions successfully made by T. The following is expressed by CT:

> The phenomena are (approximately) as T predicts, except in situations S, in which case the phenomena behave in manner M.

Instantiating this expression as a relation that obtains between past and present successful theories, one can insert for T, say, Kepler's deep theory of the *anima motrix*, which includes the following posits, each of which is patently false by contemporary lights: the sun is a divine being at the center of the universe; the natural state of the planets is rest; there is a nonattractive emanation, the *anima motrix*, coming from the sun that pushes the planets forward in their paths; the planets have an inclination to be at rest and to thereby resist the solar push, and this contributes to their slowing speed when more distant from the sun; the force that pushes the planets is a "directive" magnetic force, and so on. One can include for S occasions in which, say, Jupiter approaches Saturn, and occasions in which a planet's orbit is particularly close to the sun, and so on, adding as M, say, "non-Keplerian perturbations" and "the advancement of Mercury's perihelion." Using contemporary science to articulate S and M in their full particulars, these will be awkward and utterly nonsimple assertions, and the realist will be tempted to claim that such a competitor (in this case, a competitor to Kepler's theory) lacks certain explanatory virtues. However, because this competitor is an expression of successful contemporary science and because the realist cannot sacrifice the possibility of taking successful theories of contemporary science to be approximately true, the realist must concede that the absence of such virtues does nothing to provide grounds for denying the approximate truth of such a competitor. We see, then, that this expression reveals that there are rivals for the truth of such past theories. However, that expression also reveals that there are such rivals to any contemporary theory that accounts for some set of phenomena. Instantiating T with any accepted contemporary theory, the remaining clauses can include any S that has not (yet) been acknowledged as obtaining and any M that significantly differs from the behavior that T describes. We can add the fact that, according to science itself, the data set we have is infinitesimally small compared to the totality of events in the 13.8-billion-year-old universe. These points noted, we recognize that there are indefinitely many options and combinations, all of which will share T's predictions about observed phenomena. Although a very small sample of these competitors may be subject to future empirical elimination, indefinitely many such competitors will remain at any time.

3.3 Competitors, Set 2

Inspired by such historical considerations, we can move forward to identify another set of competitor relations and so another set of competitors. To similarly ensure that the next competitor set is empirically informed by science itself, we recognize first that, from science itself, including both past and present science, one can extract the assertion of situations in nature such that they

(a) appear in some experiential context, or were even taken at some stage in the history of science, to exemplify invariable and foundational continuities of the natural world, but

(b) are deemed by (e.g., later or contemporary) science to be no more than the residual effects of, and contingent on, what are ultimately uncommon conditions.

After noting a few cases in which this situation is exemplified, I unpack the way in which the prevalence in contemporary science of such posited special conditions reveals an additional set of competitors to the theories of contemporary science itself. In what follows, I use "Φ" to mean "an invariable and foundational continuity of the natural world." From the Newtonian picture, the posit that the sidereal period of a celestial object is proportional to the radius cubed constitutes no Φ but instead captures a consequence of the special condition of the low mass of the planets in our solar system in relation to that of the sun. Likewise from the Newtonian picture, no Φ is captured by the posit that the rate of acceleration for all objects falling toward earth is 9.8 meters per second per second: on one hand, this posit neglects the possibility of a terminal velocity; on the other, for instance, conjoining it to Newton's posit that the moon is in freefall, entails a sudden and extraordinary collision between the earth and moon. Consider as well, say, the assertion that the fluids argon, H_2O, and freon R-113, come to a "rolling" boil at 185.85°,

100°, and 47.6°, respectively. By contemporary lights, rather than capturing intrinsic and invariable properties of argon, H_2O, and freon R-113, this claim describes mere by-products of special conditions—for example, terrestrial convection currents acting on the molecules in the fluid, the currents themselves the result of the earth's mass, the proximity of the fluid to the earth's center of mass, and so on. (In fact, in microgravity experiments aboard the space shuttle, freon R-113 has been observed to boil at a lower temperature and to result not in a "rolling" boil but rather a single bubble coalescing in the middle of the fluid; Merte et al. 1996). That the earth is at rest, that the sun passes over us, that iron filings and compass needles are drawn in the direction of Polaris, that the sun emits primarily yellow light—in themselves, and by present lights, these claims approximate no Φs but stand only as artifacts of certain conditions: respectively, for instance, the nondiscernible nature of minimally accelerating motion, the location from which we observe the earth–sun relation, a temporary phase in the cycle of geomagnetic reversals, the particular dispersal of photons that results given the oxygen, nitrogen, carbon molecules in the atmosphere, and so on.

With each of these posits, it is not that a few exceptional phenomena deny their candidacy as Φs; rather, according to contemporary science, these posits capture only what ultimately amount to uncommon situations. Extending well beyond these examples, contemporary science insists that there are innumerably many other situations such that, though they might appear (or might have once appeared) to exemplify Φs, they are nothing more than the residual consequences of particular conditions. Acknowledging the ubiquitous positing of such special situations in contemporary science affords the recognition of a second collection of competitors to our favored theories. These competitors posit that the observed phenomena predicted by our favored theory, T, and its auxiliaries are, themselves, merely the result of unique conditions. Our favored theory's empirical claims constitute the description found in manner M (a description of no Φ but only a special situation), brought about by condition C (the particularly unique condition). Competitor:

- Φ obtains.
- Φ allows for the presence and absence of condition C and, in itself, patently contradicts T.
- *Condition C obtains* (according to the theory complex within which a description of Φ is embedded) *in spatiotemporal location* l, *and causes observable entities E at* l *to behave (approximately) in manner M, as T claims.*

Any number of cosmically rare conditions can be posited for C, whose particular effect is to bring about a set of phenomena that are in approximate accord with the confirmed predictions of our favored theory. Among them are dimensions intersecting, relations between our universe and others in the multiverse, stages in our universe's expansion, perhaps even relations between our galaxy and the Great Attractor, or our solar system and a galactic center, and so on. In fact there is one such condition that is available at nearly any level of nature: C can simply be a *threshold for emergence*—met by any variable regarding populations/quantities of objects, masses, charges, relations between entities, and so on—where the descriptions we favor describe no more than rare emergent properties. The properties we attribute to observed phenomena have come about only because a particular threshold was reached. And upon either surpassing the narrow limits of, or dropping below, that threshold, those properties will no longer persist. Specifying such a condition allows for indefinitely many competitors that diverge dramatically from and hence patently contradict T, all of which will share T's predictions about observed phenomena: nearly any randomly chosen self-consistent set of descriptions can qualify as descriptions of Φs that are posited in the competitors to govern phenomena in the absence of C. Given the possibilities expressed in quantum mechanics and cosmology—with particle/wave duality of "entities," wormholes, Kaluza-Klein theories, branes, holographic universes, and the like—we have great leeway for competitors that strike us psychologically as absurd. From the standpoint of these competitors, our favored T's success is no more than a by-product of condition C; and T need not, from the standpoint of these competitors, describe, *to any stretch of "approximation,"* any actual Φs.

Notice that these competitors explain their predicted observable events that are unexpected by our favored theory: in the absence of condition C, a condition perhaps deeply embedded in our *extraordinarily limited experience* of the 13.8-billion-year-old universe, the phenomena predicted by our favored T will no longer obtain (and those events predicted by Φ will). Notice also that, just as there is no generally accepted causal mechanism for certain unique conditions posited in contemporary science (e.g., vacuum fluctuations or, to use an example mentioned earlier, geomagnetic reversals), the realist demand for approximate truth cannot require that the competitors specify a causal mechanism for condition C. More generally, we cannot justifiably deny the approximate truth of these competitors merely because they do not posit causal mechanisms. Nor can we justifiably deny their approximate truth on the grounds that they violate a principle of the uniformity of nature or that they require that the world be nonsimple. These competitors are subject to no such charges.

We have now seen two sets of genuine competitors, alternatives that can render our favored theories such that they are not even approximately true and that nonetheless enjoy the same success as our favored theories. Taking both sets into account, we need concede little more than that some such competitors may be incomplete. However, because that particular concession is taken to be applicable to even our best theories, general relativity and quantum field theory among them, a charge of incompleteness cannot justify a denial of their approximate truth. Of course there are no data available that favor these competitors over our current theories, but nor are there data to eliminate them in favor of our preferred theories. And that is what is key. Moreover, in line with points made earlier, the nonrealist is no epistemic realist about competitors: acknowledging that there are *competitors that assert* that, say, S, M, and/or C obtain is wholly distinct from *asserting* that S, M, and/or C in fact obtain. And liberating ourselves from the pessimistic induction to the falsity of present-day theories, as we have, the nonrealist is wholly liberated from any denial of the truth of our favored present theories. With this we have two empirically informed, yet noninductively grounded, methods for revealing the following competitor thesis, which provides a positive answer to the competitor question that is relevant to the scientific realism debate: our favored theories have genuine competitors such that we cannot justifiably deny that they are approximately true.

4 Taking Stock: The *Modi Tollentis* Against Scientific Realism, Where History and Underdetermination Meet

I now consider how our competitor thesis can be put to work in a set of arguments against epistemic scientific realism. Note first that, although we arrived at the competitor thesis by way of analyzing, for instance, historically exemplified relations between past and present theories, because that thesis explicitly pertains to competitors that are not past theories, when we consider it only in explicit terms, it is not historical. Closing this section, I revisit the two *modus tollens* arguments we considered earlier. Going beyond those arguments for now, our historically grounded but ultimately ahistorical competitor thesis allows me to introduce a third—this time ahistorical—*modus tollens* argument. We have seen that epistemic realists claim we can justifiably believe the empirical meta-hypothesis, "predictively successful theories are approximately true." As noted earlier, taking "distinct alternative" to denote a theory such that, if it is (approximately) true, our favored theory cannot be (approximately) true (e.g., as specified earlier, alternatives whose posited Φs are not described by T to any stretch of "approximation") and taking "competitor" to mean a distinct alternative whose empirical predictions accord with the observed data predicted by T, it is again clear that epistemic realism *requires* justification for denying that successful theories have any genuine competitors that are (approximately) true. We can call this requirement the epistemic realist's noncompetitor condition (see Lyons 2009b). Recognizing that epistemic realism entails this epistemic noncompetitor condition, we have the material for a third *modus tollens* argument against realism. Its basic structure is as follows:

1. If (a) epistemic realism holds, then (b) the epistemic realist's noncompetitor condition holds.
2. However, (not-b) the epistemic realist's noncompetitor condition does not hold.
3. Therefore, (not-a) epistemic realism does not hold.

More carefully,

1. If (a) we can justifiably believe the realist's meta-hypothesis "our predictively successful theories are approximately true," then (b) we can justifiably *deny* that our successful theories have approximately true competitors.
2. However, (not-b) we cannot justifiably deny that our predictively successful theories have approximately true competitors; in fact, on the contrary, given the

previous argument, our predictively successful theories have indefinitely many competitors whose approximate truth we cannot justifiably deny.

3.   Therefore, (not-a) it is not the case that we can justifiably believe the realist's meta-hypothesis "our predictively successful theories are approximately true."

I have suggested that there are important relations between the historical argument and the argument from underdetermination, one example being the empirically/historically informed approach we employed toward revealing the two sets of competitors, the latter of which substantiate our competitor thesis: our favored theories have indefinitely many competitors whose approximate truth we cannot justifiably deny. With the content and structure of the arguments against realism now clarified, I add here the bold posit that this empirically informed, yet ultimately ahistorical, competitor premise can be conjoined to the empirical historical premise ("the list") in the first and second *modus tollens* arguments. That is, recognizing that indefinitely many competitors share the predictive success of our favored theories, yet nonetheless patently contradict our favored theories (and one another), we must grant that indefinitely many successful theories are patently false. The first two *modus tollens* arguments, no longer limited to being historical, receive a drastic increase in the quantity of items on "the list" of counterinstances, giving them a vastly stronger, more encompassing core premise. Additionally, given the dramatic rise in the quantity of "miracles," we significantly buttress our earlier conclusion that the no-miracles justification for believing the realist meta-hypothesis— "it would be a miracle were our theories as successful as they are were they not at least approximately true"—is utterly false.[9]

Adding the competitors to our counterinstances, it is important to note that they cannot be excluded from that list for failing to make novel predictions. A full competitor's predictions are what they are irrespective of when and whether they are actually derived. (Consider for instance the frequency and importance of *discoveries* that an already accepted theory entails a given consequence.) Just as the nonarticulated nature of a full competitor does nothing to prohibit its approximate truth, it does not affect its predictions. Now the sole purpose of the previous two competitor expressions is to reveal that there are competitors, and of course those competitors that are never generated cannot be excluded for using the data in their generation. Nonetheless, for the sake of illustration, consider a computer employing the previous expressions to generate sets of competitors to a proposed theory, T. In this case, we have a set of theories, all of which contradict one another but make some of the same predictions. And any novel success attributable to T is equally attributable to every competitor in the class of those that do not diverge in respect to T's successful novel predictions. These points noted, the competitors of concern cannot be excluded from the list of counterinstances for failing to make successful novel predictions. I submit that, irrespective of the realist appeal

---

[9] Since, for instance, few of the competitors constitute instances of our theorizing, pessimistic inductions that require uniformity in the falsity of our theorizing do not inherit this support.

to novelty, innumerably many competitors genuinely qualify as counterinstances to the realist meta-hypothesis.[10]

Finally, this conjunction of an ahistorical competitor thesis to the list of theories in the historical argument makes newly salient an important explanatory relation between the two primary arguments against epistemic realism: the fact that, historically, we find theories that are predictively successful but not even approximately true is wholly unsurprising, and even expected, given the fact of underdetermination and, in particular, given the fact that every theory has indefinitely many competitors. Now seen in their deductive light, I suggest that both the evidence and these arguments against epistemic realism are far stronger than both the epistemic realist's meta-hypothesis and the no-miracles argument that is meant to provide justification for believing that meta-hypothesis.

5 Socratic Scientific Realism

Despite such threats to epistemic realism, taking a cue from Nicholas Rescher (1987), I argue that the nonrealist is mistaken to discard the realist's most fundamental tenet: science seeks truth. I advocate a wholly *non*epistemic, purely axiological scientific realism (see Lyons 2001, 2005, 2011, 2012). This nonepistemic realism endeavors to overcome our 2,500-year-old obsession with belief[11] and to treat meta-hypotheses about science, including "science seeks truth," the same way it treats scientific theories—not as objects of belief but as tools for inquiry, tools to be deployed in the quest for truth. I take the key idea here—discarding belief in the quest for truth—to have a long pedigree. Beyond occasionally receiving lip-service as the proper scientific attitude, Rescher, Russell, Popper, and Peirce number among those offering comments that accord with it. However, I am unsure just who, if anyone, has embraced it as fully as I aspire to.[12]

[10] Notably the *deployed* constituents of concern in footnote 5 also have competitors. Cutting to the chase, consider our computer invoking, for instance, the second competitor expression as a method for generating competitors. Among the posits genuinely deployed toward that end are the following, each of which is patently false by present lights: our favored theory, T, is false; only a subclass of T's observable consequences obtain, and (nearly any sized set of specific) observable events outside of that subclass fail to accord, and can significantly clash, with observable events described by T; Φ obtains, as an invariable and foundational continuity of the natural world; Φ allows for the presence and absence of condition C; condition C obtains; it obtains in spatiotemporal location l; condition C is what allows for the subclass of T's observable consequences that do obtain, and so on. Given indefinitely many possibilities for each of the variables, we have indefinitely many competitors even to those constituents that are deployed in the derivation of novel successes.

[11] That is, nonsyntactic belief.

[12] Though Rescher, Russell, and Peirce, for instance, may well advocate epistemic humility, they do not follow through on bracketing belief. And as soon as Popper is comfortable discussing truth, he introduces the epistemic realist's thesis that corroboration indicates verisimilitude; moreover he at least appears to advocate believing the truth of singular empirical predictions. (Additionally, I do not embrace what I take to be the other key components of Popperianism, e.g., the denial that science employs induction, an obsession with a demarcation criterion and falsifiability, the claim that scientists seek to falsify their theories, the demand for content-retention across theory change, etc.)

Historically tracing back the injunction "seek truth without claiming to possess it," I suppose the figure whose voice we would hear—even if he turned away from the physical world—is Socrates. This is one factor leading me toward the name I give my position in this article: *Socratic scientific realism*.

The compulsion to believe may well be what has diverted standard scientific realists from refining their empirically uninformative axiological meta-hypothesis, thereby prohibiting them from a comprehensive account of science. Specifically, realists, including Rescher, have failed to specify just which subclass of true statements science seeks. To at least indicate the core idea of my axiological postulate, it is that *science seeks to increase the XT statements in its theory complexes*. XT statements are those whose truth is *experientially concretized*—that is, true statements whose truth is made to deductively impact, is deductively pushed to and enters into, documented reports of specific experiences. This postulate shifts from "truth" to a *subclass* of true statements, from theories to complexes/systems, and from endeavoring to simply *attain* the truth to endeavoring to *increase the quantity* of true statements in the specified subclass. Moreover, explicitly including "increase" in the postulated goal, evaluation is unambiguously comparative, theory complex against theory complex. And although realist truth is not contingent on the system of statements in which a statement is embedded, the experiential concretization of a statement's truth is. Most significantly, I have shown elsewhere (Lyons 2005) that the *actual* achievement of this state, an increase in the XT statements of a theory complex, *requires* the achievement of a set of syntactic theoretical desiderata, the importance of which both sides of the debate agree: namely, an increase in empirical accuracy and consistency and an increase in, or at least the retention of, breadth of scope, testability, and number of forms of simplicity. I make no claim that we can justifiably believe that the posited primary goal has been achieved. However, since it cannot be achieved without these *otherwise potentially disparate theoretical virtue*s, my axiological meta-hypothesis offers both an explanation and, crucially, a justification for their mutual pursuit: if we do not have these necessary conditions for our primary goal, an increase in experientially concretized truth, we know we do not have what we seek. Given this relation, I contend, this meta-hypothesis lives up to what it demands: for instance, it is informative toward our understanding of "inference to the best explanation," now liberated from epistemic baggage; it provides a better account of science than nonrealist axiologies (e.g., Laudan's meta-hypothesis that science seeks problem-solving effectiveness and van Fraassen's meta-hypothesis that the aim of science is empirical adequacy; Lyons, 2005); and finally, and more generally, it dramatically improves the realist's ability to account for what is going on in science. The battle cry of Socratic scientific realism is the following: science seeks truth and does so rationally, irrespective of whether we can justifiably believe we have achieved it.

ACKNOWLEDGEMENTS

Research for this article was supported by the Arts and Humanities Research Council, United Kingdom.

REFERENCES

Ben-Menahem, Y. (1990). "Equivalent Descriptions." *The British Journal for the Philosophy of Science 41*: 261–279.

Boyd, R. (1973). "Realism, Underdetermination and the Causal Theory of Reference." *Nous 7*: 1–12.

Chakravartty, A. (1998). "Semirealism." *Studies in History and Philosophy of Science 29*: 391–408.

Chakravartty, A. (2008). "What You Don't Know Can't Hurt You: Realism and the Unconceived." *Philosophical Studies 137*: 149–158.

Chang, H. (2003). "Preservative Realism and Its Discontents: Revisiting Caloric." *Philosophy of Science 70*: 902–912.

Churchland, P. M. (1985). "The Ontological Status of Observables: In Praise of the Superempirical Virtues." In P. M. Churchland and C. A. Hooker (eds.), *Images of Science* (Chicago: University of Chicago Press), 35–47.

Doppelt, G. (2007). "Reconstructing Scientific Realism to Rebut the Pessimistic Meta-Induction." *Philosophy of Science 74*: 96–118.

Duhem, P. (1991). *The Aim and Structure of Physical Theory*. 2d ed. Translated by P. Weiner (Princeton, NJ: Princeton University Press). (Original work published 1914)

Ellis, G., Elst, H., Murugan, J., and Uzan, J. (2011). "On the Trace-Free Einstein Equations as a Viable Alternative to General Relativity." *Classical and Quantum Gravity 28*: 225007–225017.

Frieman, J., Turner M., and Huterer, D. (2008). "Dark Energy and the Accelerating Universe." *Annual Review of Astronomy and Astrophysics 46*: 385–432.

Godfrey-Smith, P. (2008). "Recurrent Transient Underdetermination and the Glass Half Full." *Philosophical Studies 137*: 141–148.

Harker, D. (2010). "Two Arguments for Scientific Realism Unified." *Studies in History and Philosophy of Science 41*: 192–202.

Harker, D. (2013). "How to Split a Theory: Defending Selective Realism and Convergence without Proximity." *The British Journal for Philosophy of Science 64*: 79–106.

Horwich. P. (1991). "On the Nature and Norms of Theoretical Commitment." *Philosophy of Science 58*: 1–14.

Hoyningen-Huene, P. (2011). "Reconsidering the Miracle Argument on the Supposition of Transient Underdetermination." *Synthese 180*: 173–187.

Kitcher, P. (2001). "Real Realism: The Galilean Strategy." *The Philosophical Review 110*: 151–197.

Kukla, A. (1998). *Studies in Scientific Realism* (New York: Oxford University Press).

Lange, M. (2002). "Baseball, Pessimistic Inductions, and the Turnover Fallacy." *Analysis 62*: 281–285.

Laudan, L. (1977). *Progress and its Problems* (New York: Oxford University Press).

Laudan, L. (1981). "A Confutation of Convergent Realism." *Philosophy of Science 48*: 19–49.

Laudan, L. (1983). "The Demise of the Demarcation Problem." In R. S. Cohen and L. Laudan (eds.), *Physics, Philosophy and Psychoanalysis* (Dordrecht: Kluwer), 111–127.

Laudan, L., and Leplin, J. (1991). "Empirical Equivalence and Underdetermination." *Journal of Philosophy 88*: 449–472.

Leplin, J. (1997). *A Novel Defense of Scientific Realism* (New York: Oxford University Press).

Lipton, P. (2004). *Inference to the Best Explanation*. 2d ed. (New York: Routledge).

Lyons, T. (2001). *The Epistemological and Axiological Tenets of Scientific Realism*. Ph.D. diss., University of Melbourne.

Lyons, T. (2002). "Scientific Realism and the Pessimistic Meta-*Modus Tollens*." In S. Clarke and T. D. Lyons (eds.), *Recent Themes in the Philosophy of Science: Scientific Realism and Commonsense* (Dordrecht: Springer), 63–90.

Lyons, T. (2003). "Explaining the Success of a Scientific Theory." *Philosophy of Science* 70(5): 891–901.

Lyons, T. (2005). "Toward a Purely Axiological Scientific Realism." *Erkenntnis* 63: 167–204.

Lyons, T. (2006). "Scientific Realism and the *Stratagema de Divide et Impera*." *The British Journal for the Philosophy of Science* 57(3): 537–560.

Lyons, T. (2009a). "Criteria for Attributing Predictive Responsibility in the Scientific Realism Debate: Deployment, Essentiality, Belief, Retention. . . ." *Human Affairs* 19: 138–152.

Lyons, T. (2009b). "Non-Competitor Conditions in the Scientific Realism Debate." *International Studies in the Philosophy of Science* 23(1): 65–84.

Lyons, T. (2011). "The Problem of Deep Competitors and the Pursuit of Unknowable Truths." *Journal for General Philosophy of Science* 42(2): 317–338.

Lyons, T. (2012). "Axiological Realism and Methodological Prescription." In H. W. de Regt, S. Hartmann, and S. Okasha (eds.), *EPSA Philosophy of Science: Amsterdam 2009, European Philosophy of Science Association* (Dordrecht: Springer), 187–197.

Lyons, T. (2013). "The Historically Informed *Modus Ponens* against Scientific Realism: Articulation, Critique, and Restoration." *International Studies in Philosophy of Science* 27(4): 369–392.

Magnus, P. D., and Callender, C. (2004). "Realist Ennui and the Base Rate Fallacy." *Philosophy of Science* 71: 320–338.

Manchak, J. (2009). "Can We Know the Global Structure of Spacetime?" *Studies in History and Philosophy of Modern Physics* 40: 53–56.

Merte, H., Lee, H., and Keller, R. (1996). "Report on Pool Boiling Experiment Flown on STS-47 (PBE-IA), STS- 57 (PBE-IB), and STS-60 (PBE-IC)." NASA Contractor Report CR-198465, Contract NAS 3-25812.

Mill, J. S. (1843). *A System of Logic* (London: John W. Parker).

Mill, J. S. (1998). *On Liberty* (New York: Oxford University Press). (Original work published 1859)

Murugan, J., Weltman, A., and Ellis, G. (2012). "The Problem with Quantum Gravity." In J. Murugan, A. Weltman, and G. F. R. Ellis (eds.), *Foundations of Space and Time: Reflections on Quantum Gravity* (New York: Cambridge University Press), 1–7.

Okasha, S. (2002). "Underdetermination, Holism and the Theory/Data Distinction." *Philosophical Quarterly* 52: 302–319.

Parsons, K. (2005). *Copernican Questions: A Concise Invitation to the Philosophy of Science* (New York: McGraw-Hill).

Peters, D. (2014). "What Elements of Successful Scientific Theories Are the Correct Targets for 'Selective' Scientific Realism?" *Philosophy of Science* 81(3): 377–397.

Psillos, S. (1999). *Scientific Realism: How Science Tracks Truth* (New York: Routledge).

Quine, W. (1951). "Two Dogmas of Empiricism." *The Philosophical Review* 60: 20–43.

Rescher, N. (1987). *Scientific Realism: A Critical Reappraisal* (Dordrecht: Kluwer).

Sankey, H. (2001). "Scientific Realism: An Elaboration and a Defence." *Theoria* 98: 35–54.

Schurz, G. (2009). "When Empirical Success Implies Theoretical Reference: A Structural Correspondence Theorem." *The British Journal for Philosophy of Science* 60: 101–133.

Sklar, L. (1981). "Do Unborn Hypotheses Have Rights?" *Pacific Philosophical Quarterly* 62: 17–29.

Stanford, P. K. (2003). "No Refuge for Realism: Selective Confirmation and the History of Science." *Philosophy of Science 70*: 913–925.

Stanford, K. (2006). *Exceeding Our Grasp: The Problem of Unconceived Alternatives* (New York: Oxford University Press).

Tulodziecki, D. (2012). "Epistemic Equivalence and Epistemic Incapacitation." *The British Journal for the Philosophy of Science 63*: 313–328.

van Fraassen, B. (1980). *The Scientific Image* (Oxford: Oxford University Press).

Vickers, P. (2013). "Confrontation of Convergent Realism." *Philosophy of Science 80*: 189–211.

Weinberg, S., and Witten, E. (1980). "Limits on Massless Particles." *Physics Letters B*: 59–62.

Worrall, J. (1989). "Structural Realism: The Best of Both Worlds?" *Dialectica 43*: 99–124.

Worrall, J. (2011). "Underdetermination, Realism and Empirical Equivalence." *Synthese 180*: 157–172.

Wray, K. B. (2008). "The Argument from Underconsideration as Grounds for Anti-Realism: A Defence." *International Studies in the Philosophy of Science 22*: 317–326.

....

CHAPTER 28

SCIENTIFIC THEORIES

....

HANS HALVORSON

WHAT is a scientific theory? Several philosophers have claimed that this question is *the* central philosophical question about science. Others claim still that the answer one gives to this question will fundamentally shape how one views science. In more recent years, however, some philosophers have become tired of this focus on theories and they have suggested that we stop trying to answer this question. In this chapter, I will canvass and critically scrutinize the various answers that have been given to the question, "what is a scientific theory?" Then I will consider a recent argument against trying to answer this question. Finally, I will address the question of the utility of formal models of scientific theories.

1 THE ONCE-RECEIVED VIEW OF THEORIES

....

In the 1960s and 1970s, the vogue in philosophy of science was to identify problematic assumptions made by logical positivists and to suggest that a better analysis of science would be possible once these assumptions were jettisoned. Of particular relevance for our discussion was the claim that the logical positivists viewed theories as "linguistic" or "syntactic" entities. Where did the 1960s philosophers get this idea? And is there any justice to their claim? (The references here are too numerous to list. Some of the most important include Achinstein 1968; Putnam 1962; Suppe 1972, 1974; Suppes 1964; and van Fraassen 1972.)

The story here is complicated by the history of formal logic. Recall that axiomatic systems of formal logic were common currency among philosophers of science in the 1920s, culminating in Rudolf Carnap's *The Logical Syntax of Language* (Carnap 1934). At that time there was no such thing as formal semantics; instead, semantic investigations were considered to be a part of psychology or even of the dreaded metaphysics. Thus, when philosophers in the 1920s placed emphasis on "syntax," they really meant to place emphasis on mathematical rigor. Indeed, what we now call "model theory" would almost certainly have been considered by Carnap et al. as a part of logical syntax. But more about this claim anon.

In any case, in his 1962 critique of the "received view (RV)" of scientific theories, Hilary Putnam describes the view as follows: "(RV) A scientific theory is a partially interpreted calculus."

What is meant by this notion? First of all, a "calculus" is a set of rules for manipulating symbols. What Putnam has in mind here is something like the "predicate calculus," which involves a set L of symbols (sometimes called a *signature*), a list of formation rules, and a list of transformation rules. The notion of "partial interpretation" is a bit more difficult to specify, a likely result of the fact that, in the 1940s and 1950s, philosophers were still coming to terms with understanding model theory. In fact, in his critique of the RV, Putnam lists three ways of trying to understand partial interpretation and rejects all three as inadequate.

The idea behind partial interpretation, however, is clear: some scientific theories rely heavily on various mathematical calculi, such as the theory of groups, tensor calculus, or differential equations. But the statements of mathematics don't, by themselves, say anything about the physical world. For example, Einstein's field equations

$$R_{ab} - \frac{1}{2} g_{ab} \, R = T_{ab},$$

will mean nothing to you unless you know that R_{ab} is supposed to represent space-time curvature. Thus, Einstein's theory is more than just mathematical equations; it also includes certain claims about how those equations are linked to the world of our experience.

So, for the time being, it will suffice to think of "partial interpretation" as including at least an intended application of the formalism to the empirical world. We can then spell out the RV further as follows:

(RV) A scientific theory consists of two things:

1. A formal system, including:
    (a) symbols,
    (b) formation rules, and
    (c) deduction rules.
2. Some use of this formal system to make claims about the physical world and, in particular, empirically ascertainable claims.

Perhaps the closest thing to an explicit assertion of RV is found in (Carnap 1939a, 171–209; see also Feigl 1970; Hempel 1970; Nagel 1961). Before proceeding, note that Carnap himself was quite liberal about which kinds of formal systems would be permitted under the first heading. Unlike Quine, Carnap didn't have any problem with second-order quantification, intensional operators, nonclassical logics, or infinitary logics. (When I need to be more precise, I will use $L\omega\omega$ to indicate the first-order predicate calculus, where only finite conjunctions and disjunctions are permitted.)

The RV has been the focus of intense criticism from many different angles. In fact, it seems that between 1975 and 2010, beating up on RV was the favorite pastime of many philosophers of science. So what did they think was wrong with it?

The most obvious criticism of the RV, and one which Carnap himself anticipated, is that "scientific theories in the wild" rarely come as axiomatic systems. It is true that Carnap's proposal was based on some very special cases (e.g., Einstein's special theory of relativity, which admits of at least a partial axiomatization in $L\omega\omega$). And it is not at all clear that other interesting scientific theories could be reconstructed in this way—not even Einstein's general theory of relativity, nor quantum mechanics (QM), not to speak of less-formal theories such as evolutionary biology. (The problem with the former two theories is that they seem to require at least second-order quantification, for example, in their use of topological structures.) So why did Carnap make this proposal when it so obviously doesn't fit the data of scientific practice?

Here, we must remember that Carnap had a peculiar idea about the objectives of philosophy. Starting with his book *The Logical Structure of the World* (Carnap 1928), Carnap aimed to provide a "rational reconstruction" of the knowledge produced by science. (For an illuminating discussion of this topic, see Demopoulos 2007.) Carnap's paradigm here, following in the footsteps of Russell, was the nineteenth-century rigorization of mathematics afforded by symbolic logic and set theory. For example, just as nineteenth-century mathematicians replaced the intuitive idea of a "continuous function" with a precise logically constructed counterpart, so Carnap wanted to replace the concepts of science with logically precise counterparts. In other words, Carnap saw the objective of philosophical investigation as providing a "nearest neighbor" of a scientific concept within the domain of rigorously defined concepts. For Carnap, if it was possible to replace individual scientific concepts with precise counterparts, then it was a worthy aim to formalize an entire domain of scientific knowledge.

Carnap's ideas about "rational reconstruction" and of "explication" are worthy of a study in their own right. Suffice it to say for now that any serious discussion of Carnap's views of scientific theories needs to consider the goals of rational reconstruction. (A nice discussion of these issues can be found in the introduction to Suppe 1974.)

I've explained then why Carnap, at least, wanted to replace "theories in the wild" with formal counterparts. Many people now prefer to take a different approach altogether. However, in this chapter, I will mostly consider descendants of Carnap's view—in particular, accounts of scientific theories that attempt to provide at least some formal precision to the notion.

But even among philosophers who agree with the idea of explicating "scientific theory," there are still many objections to RV. (For a rather comprehensive listing of purported difficulties with RV, see Suppe 1974 and Craver 2008). Rather than review all of these purported difficulties, I will focus on what I take to be misunderstandings.

1. *Fiction: RV treats scientific theories as linguistic entities.*

    Fact: The RV gives the "theoretical definition" in terms of something that is often called a "formal language." But a formal language is really not a language at all because nobody reads or writes in a formal language. Indeed, one of the primary features of these so-called formal languages is that the symbols don't have any meaning. Thus, we might as well stop talking about "formal language" and re-emphasize that we are talking about *structured sets*, namely, sets of symbols, sets of terms, sets of formulas, and the like. There is nothing intrinsically linguistic about this apparatus.

2. *Fiction: RV confuses theories with theory formulations.*

   Fact: To my knowledge, no advocate of RV ever claimed that the language of formulation was an essential characteristic of a theory. Rather, one and the same theory can be formulated in different languages. The failure to distinguish between theories and theory formulations is simply a failure to understand the resources of symbolic logic. All that is needed to make this distinction is an appropriate notion of "equivalent formulations," where two formulations are equivalent just in case they express the same theory. (For one reasonable account of equivalent theory formulations, see Glymour 1971 and Barrett and Halvorson 2015).

   The confusion here lies instead with the supposition that two distinct theory formulations, in different languages, can correspond to the same class of models—a supposition that has been taken to support the semantic view of theories. This confusion will be unmasked in the subsequent section.

3. *Fiction: RV is inadequate because many interesting theories cannot be formulated in the first-order predicate calculus.*

   Fact: I've already noted that Carnap, at least, was not committed to formulating theories in first-order logic. But even so, it is not clear what is meant by saying that "the theory T cannot be formulated in first-order logic" when T hasn't already been described as some sort of structured collection of mathematical objects. Consider, for example, Einstein's general theory of relativity (GTR). Is it possible to formulate GTR in a syntactic approach? First of all, this question has no definitive answer—at least not until GTR is described in enough mathematical detail that its properties can be compared with properties of first-order axiomatizable theories. Moreover, it won't do to point out that, as it is standardly formulated, GTR involves second-order quantification (e.g., in its use of topological structure). Some theories formulated in second-order logic *also* admit a first-order axiomatization; or, in some cases, although the original theory might not admit a first-order axiomatization, there might be a similar replacement theory that does. One example here is the (mathematical) theory of topological spaces. Whereas the definition of the class of topological spaces requires second-order quantification, the theory of "locales" can be axiomatized in (infinitary) first-order logic. And, indeed, several mathematicians find the theory of locales to be a good replacement for the theory of topological spaces. (In fact, there is something like a first-order axiomatization of GTR; see Reyes 2011.) It is another question, of course, why a philosopher of science would want to try to axiomatize a theory when that would involve translating the theory into a completely alien framework. For example, I suspect that little insight would be gained by an axiomatization of evolutionary biology. But that's not surprising at all: evolutionary biology doesn't use abstract theoretical mathematics to the extent that theories of fundamental physics do. Perhaps there is a simple solution to the supposed dilemma of whether philosophers ought to try to axiomatize theories: let actual science be our guide. Some sciences find axiomatization useful, and some do not. Accordingly, some philosophers of science should be concerned with axiomatizations, and some should not.

1.1 Correspondence Rules

The most important criticism of RV regards the related notions of correspondence rules, coordinative definitions, bridge laws, and partial interpretation. Each of these notions is meant to provide that additional element needed to differentiate empirical science (applied mathematics) from pure mathematics.

The notion of a coordinative definition emerged from late nineteenth-century discussions of the application of geometry to the physical world. As was claimed by Henri Poincaré, the statements of pure mathematical geometry have no intrinsic physical meaning—they are neither true nor false. For example, the claim that

(P) "The internal angles of a triangle sum to 180 degrees,"

says nothing about the physical world, at least until the speaker has an idea in mind about the physical referents of the words "line," "triangle," and the like. A *coordinative definition*, then, is simply a way of picking out a class of physical things that correspond to the words or symbols of our mathematical formalism.

One example of a coordinative definition is a so-called operational definition. For example, Einstein defined the simultaneity of events in terms of observations of light signals sent from those events. Early logical positivists such as Hans Reichenbach took Einstein's definition of simultaneity as a paradigm of good practice: taking a theoretical concept—such as simultaneity—and defining it in terms of simpler concepts. (Granted, concepts such as "seeing two images at the same time" are not as simple as they might at first seem!)

As the logical positivists came to rely more on symbolic logic for their explications, they attempted to explicate the notion of coordinative definitions within a logical framework. The key move here was to take the language L of a theory and to divide it into two parts: the observation language L_O and the theoretical language L_P. By the late 1930s, the RV included this dichotomization of vocabulary, and efforts were focused on the question of how the terms in L_P "received meaning" or "received empirical content."

It is the current author's belief that Carnap and others erred in this simplistic method for specifying the empirical content of a scientific theory. However, I do *not* grant that the notion of empirical content cannot be specified syntactically, as has been suggested by van Fraassen (1980), among others. (For example, the distinction might be drawn among equivalence classes of formulas relative to interderivability in the theory T; or the distinction might be drawn using many-sorted logic. The formal possibilities here seem hardly to have been explored.) Be that as it may, it was the simplistic way of specifying empirical content that was criticized by Putnam and others. With a devastating series of examples, Putnam showed that L_O terms can sometimes apply to unobservable objects, and L_P terms are sometimes used in observation reports. But let's set those criticisms aside for the moment and consider a second problem. Even if the vocabulary L could legitimately be divided into observational and theoretical components, there remains the question of how the theoretical vocabulary ought to be related to the observation vocabulary.

In the years between 1925 and 1950, Carnap gradually loosened the restrictions he placed on the connection between theoretical (L_P) and observational (L_O) vocabulary. In the earliest years, Carnap wanted every theoretical term to be *explicitly defined* in terms of observation terms. That is, if $r(x)$ is a theoretical predicate, then there should be a sentence $\phi(x)$ in the observation language such that

$$T' \vdash \forall x \big(r(x) \leftrightarrow \phi(x) \big).$$

That is, the theory T implies that a thing is r if and only if that thing is ϕ; that is, it provides a complete reduction of r to observational content. (Here, Carnap was following Russell's (1914) proposal to "construct" the physical world from sense data.)

However, by the mid-1930s, Carnap had become acutely aware that science freely uses theoretical terms that do not permit complete reduction to observation terms. The most notable case here is disposition terms, such as "x is soluble." The obvious definition,

x is soluble $\equiv$ if x is immersed, then x dissolves,

fails, because it entails that any object that is never immersed is soluble (see Carnap 1936; 1939b). In response to this issue, Carnap suggested that disposition terms must be connected to empirical terms by means of a certain sort of partial, or conditional, definition. From that point forward, efforts focused on two sorts of questions: were reduction sentences too conservative or too liberal? That is, are there legitimate scientific concepts that aren't connected to empirical concepts by reduction sentences? Or, conversely, is the requirement of connectability via reduction sentence too permissive?

The final, most liberal proposal about coordinative definitions seems to come from Hempel (1958). Here, a theory T is simply required to include a set C of "correspondence rules" that tie the theoretical vocabulary to the observational vocabulary. Around the same time, Carnap put forward the idea that theoretical terms are "partially interpreted" by means of their connection with observation statements. However, as pointed out by Putnam (1962), Carnap doesn't provide any sort of precise account of this notion of partial interpretation. Indeed, Putnam argues that the notion doesn't make any sense. Ironically, Putnam's argument has been challenged by one of the strongest critics of the RV (Suppe 1971).

Thus, the so-called syntactic approach to theories was subjected to severe criticism and was eventually abandoned. But I've given reason to think that a more sophisticated syntactic approach might be possible and that such an approach would have all the advantages of the semantic approach to theories. In the next section, I'll also explain why I'm not convinced that semantic approaches have any intrinsic advantage over syntactic approaches. In fact, as I will explain in Section 5, the best versions of the syntactic and semantic approaches are formally dual to each other and provide essentially the same picture of the structure of scientific theories.

2 THE SEMANTIC VIEW OF THEORIES

What I have been calling the "once RV" of theories is often called the "syntactic view" of theories—emphasizing that theories are formulated by means of logical syntax. According to Fred Suppe, the syntactic view of theories died in the late 1960s, after having met with an overwhelming number of objections in the previous two decades. At the time when Suppe wrote of the death of the syntactic view, it was unclear where philosophy of science would go. Several notable philosophers—such as Feyerabend and Hanson—wanted to push philosophy of science away from formal analyses of theories. However, others such as Patrick Suppes, Bas van Fraassen, and Fred Suppe saw formal resources for philosophy of science in other branches of mathematics, most particularly set theory and model theory. Roughly speaking, the "semantic view of theories" designates proposals to explicate theory-hood by means of the branch of mathematical logic called *model theory*. I will talk about the semantic view in this section, and I will discuss Suppes's set-theoretic account of theories in the following section.

Recall that the study of mathematical models (i.e., model theory) came alive in the mid–twentieth century with the work of Tarski and others. For philosophy of science, this advance was particularly significant because the early positivists had banished semantical words such as "meaning," "reference," or "truth." With the invention of formal semantics (i.e., model theory) these words were given precise explications.

However, philosophers of science were not all that quick to make use of model theory. One of the primary pioneers along these lines was the Dutch logician and philosopher Evert Beth. (A partial account of Beth's contribution to the development of the semantic view can be found in van Fraassen 1970, 1972.) We will, however, proceed ahistorically and present the mature version of the semantic view.

Let L be a signature. An L-*structure* M (alternatively: an *interpretation* of L) is defined to be a set S and an assignment of elements of L to relations or functions on Cartesian products of S. For example, if c is a constant symbol of L, then c^M is an element of S. As is described in any textbook of model theory, the interpretation M extends naturally to assign values to all L-terms and then to all L-formulas, in particular, L-sentences. In fact, if an L-formula $\phi(x_1, \ldots, x_n)$ has n free variables, then it will be assigned to a subset of the n-fold Cartesian product of S. As a special case, a sentence ϕ of L (which has zero free variables) is assigned to a subset of the singleton set; that is, either the singleton set itself (in which case we say that "ϕ is true in M") or the empty set (in which case we say that "ϕ is false in M").

An L-structure is sometimes misleadingly called a "model." This terminology is misleading because the technically correct phrase is "model of Σ" where Σ is some set of sentences. In any case, we have the technical resources in place to state a preliminary version of the semantic view of theories:

(SV) A scientific theory is a class of L-structures for some language L.

Now, proponents of the semantic view will balk at SV for a couple of different reasons. First, semanticists stress that a scientific theory has two components:

1. A theoretical definition and
2. A theoretical hypothesis.

The theoretical definition, roughly speaking, is intended to replace the first component of Carnap's view of theories. That is, the theoretical definition is intended to specify some abstract mathematical object—the thing that will be used to do the representing. Then the theoretical hypothesis is some claim to the effect that some part of the world can be represented by the mathematical object given by the theoretical definition. So, to be clear, SV here is only intended to give one half of a theory, viz. the theoretical definition. I am not speaking yet about the theoretical hypothesis.

But proponents of the semantic view will balk for a second reason: SV makes reference to a language L. And one of the supposed benefits of the semantic view was to free us from the language-dependence implied by the syntactic view. So, how are we to modify SV in order to maintain the insight that a scientific theory is independent of the language in which it is formulated?

I will give two suggestions, the first of which I think cannot possibly succeed. The second suggestion works, but it shows that the semantic view actually has no advantage over the syntactic view in being "free from language dependence."

How then to modify SV? The first suggestion is to formulate a notion of "mathematical structure" that makes no reference to a "language." At first glance, it seems simple enough to do so. The paradigm case of a mathematical structure is supposed to be an ordered n-tuple $\langle S, R_1, \ldots R_n \rangle$, where S is a set, and $R_1, \ldots, R_n$ are relations on S. (This notion of mathematical structure follows Bourbaki 1970.) Consider, for example, the proposal made by Lisa Lloyd:

In our discussion, a model is not such an interpretation [i.e., not an L-structure], matching statements to a set of objects which bear certain relations among themselves, but the set of objects itself. That is, models should be understood as structures; in the cases we shall be discussing, they are mathematical structures, i.e., a set of mathematical objects standing in certain mathematically representable relations. (Lloyd 1984, 30)

But this proposal is incoherent. Let a be an arbitrary set, and consider the following purported example of a mathematical structure:

$$M = \langle \{a, b, \langle a, a \rangle\}, \{\langle a, a \rangle\} \rangle.$$

That is, the base set S consists of three elements a, b, $\langle a, a \rangle$, and the indicated structure is the singleton set containing $\langle a, a \rangle$. But what *is* that structure? Is that singleton set a monadic property? Or is that singleton a binary relation? (The former is a structure for a language L with a single unary predicate symbol; the latter is a structure for a language L' with a single binary relation symbol.) The simple fact is that in writing down M as an ordered n-tuple, we haven't really described a mathematical structure. Thus, a mathematical structure cannot simply be "a set of mathematical objects standing in certain mathematically representable relations."

To press the point further, consider another purported mathematical structure:

$$N = \langle \{a, b, \langle a, b \rangle\}, \{\langle a, b \rangle\} \rangle.$$

Are M and N isomorphic structures? Once again, the answer is underdetermined. If M and N are supposed to be structures for a language L with a single unary predicate symbol, then the answer is Yes. If M and N are supposed to be structures for a language L' with a single binary relation symbol, then the answer is No.

Thus, it is not at all clear how SV is supposed to provide a "language-free" account of theories. The key, I suggest, is to define a reasonable notion of equivalence of theory formulations—a notion that allows one and the same theory to be formulated in different languages. But that same stratagem is available for a syntactic view of theories. Thus, "language independence" is not a genuine advantage of the semantic view of theories as against the syntactic view of theories. If the semantic view has some advantages, then they must lie elsewhere.

What, then, *are* the purported advantages of the semantic view of theories? What is supposed to recommend the semantic view? Here, I will enumerate some of the advantages that have been claimed for this view and provide some critical commentary.

1. *Claim: Scientists often work with heterogeneous collections of models that aren't all models of a single syntactically formulated theory.*

   This claim may well be true, but scientists engage in a lot of activities that don't involve constructing or evaluating theories. It seems that what might be suggested here is to consider the *ansatz* that the primary objective of science is the construction and use of models. I myself am loath to jump on the bandwagon with this assumption; for example, mathematical physicists don't actually spend much time building models—they are busy proving theorems.

2. *Claim: Scientists often deal with collections of models that are not elementary classes (i.e., aren't the collection of models of some set of first-order sentences).*

This claim is strange because it seems to indicate that scientists work with classes of L-structures (for some language L) that are not elementary classes (i.e., not the classes of models of a set of first-order sentences). I happen to know of no such example. Certainly, scientists work with classes of models that are not in any obvious sense elementary classes but largely because they haven't been given a precise mathematical definition.

What about mathematical structures like Hilbert space, which are used in QM? Isn't it obvious that the theory of Hilbert spaces is not elementary (i.e., there is no set Σ of first-order axioms such that the models of Σ are precisely the Hilbert spaces)?

The problem mentioned earlier is still present. Although Hilbert spaces are fully legitimate citizens of the set-theoretic universe, the class of Hilbert spaces is not in any obvious way a class of L-structures for some language L.

What, then, are we to do in this case? QM is formulated in terms of Hilbert spaces, and so physics needs Hilbert spaces. If the RV can't countenance Hilbert spaces, then so much the worse for the RV, right?

I grant the legitimacy of this worry. However, there is a problem here not just for the RV, but also for any philosopher looking at QM: although physicists use Hilbert spaces for QM, it is not at all clear what "theory" they are proposing when they do so. That is, it is not clear what assertions are made by QM. Or to restate the problem in a Quinean fashion, it is not clear what the domain of quantification in QM is.

One way to understand the task of "interpreting" QM is finding the correct way to formulate the theory syntactically (i.e., finding the correct predicates, relations, and domain of quantification).

3. *Claim: Scientists often want to work with an intended model, but the syntactic approach always leaves open the possibility of unintended models.*

This point invokes the following well-known fact (the Löwenheim-Skølem theorem): for any theory T in $L\omega\omega$ (assuming a countably infinite language), if T has a model of cardinality κ, then T has models of all smaller and larger infinite cardinalities. For example: if T is the first-order theory of the real numbers (say, in the language of ordered fields), then T has a countable model Q. But Q is not the model we intend to use if we believe that space has the structure of the continuum! Once again, this "problem" can be simply dealt with by means of various technical stratagems (e.g., infinite conjunctions). But even if we remain in $L\omega\omega$, it is not clear what this criticism was really meant to show. On the one hand, the point could be that scientists need to discriminate between models that cannot be discriminated via first-order logic; for example, a scientist might want to say that M is a good or accurate model of some phenomenon, and N is *not* good or accurate, even though M and N are elementarily equivalent (i.e., they agree in the truth values they assign to first-order sentences). The key word here is "need." I will grant that sometimes scientists' preferences don't respect elementary equivalence—that is, they might prefer one model over an elementarily equivalent model. But I'm not sure that this preference would have anything to do with them believing that their preferred model provides a more accurate representation of reality. They might well think that the differences between these models are irrelevant! Suppose, however, that we decide that it *does* make a difference to us—that we want to be able to say: "the world might be like M, but it is not like N, even though M and N are elementarily equivalent." If we want to do that, must we adopt the semantic view of theories? Not necessarily: we could simply adopt a stronger language (say, an infinitary logic or second order logic) that allows us to discriminate between these models. If we claim that M has some feature that N lacks, then that's because we are implicitly relying on a more expressive language than $L\omega\omega$. In many paradigm cases, the distinctions can be drawn by means of second-order quantification or even just with infinitary connectives and first-order quantification. The honest thing to do here would be to display our ontological commitments clearly by means of our choice of language.

4. *Claim: The semantic view is more faithful to scientific practice.*

"[A semantic approach] provides a characterization of physical theory which is more faithful to current practice in foundational research in the sciences than the familiar picture of a partially interpreted axiomatic theory" (van Fraassen 1970). This criticism is partially dealt with by the fact that the syntactic view wasn't supposed to provide a completely accurate description of what's going on in science—it was supposed to provide an *idealized* picture. In the earliest years of logical positivism, Carnap explicitly described the relation of philosophy of science to science on analogy to the relation of mathematical physics to the physical world. The point of this comparison is that mathematical physics makes idealizing assumptions so as to provide a tractable description that can be used to do some theoretical work (e.g., to prove theorems). In the same way, we can think of a syntactically formulated theory as an idealized version of a theory that provides *some*, but not complete, insight into the nature of that theory.

Perhaps the most helpful thing to say here is to echo a point made by Craver (2008) and also by Lutz (2015): the syntactic and semantic views are attempts to capture aspects of scientific theorizing, not the essence of scientific theorizing. Consequently, these two approaches need not be seen as competitors.

2.1 The Problem of Representation

Here's an initial puzzle about the semantic view of theories: if a theory is a collection of models, then what does it mean to *believe* a theory? After all, we know what it means to believe a collection of sentences; but what would it mean to believe a collection of models? At first glance, it seems that the semantic view of theories commits a basic category mistake.

Semanticists, however, are well aware of this issue, and they have a simple answer (albeit an answer that can be sophisticated in various ways).

> (*B*) To believe a theory (which is represented by a class Σ of models), is to believe that the world is isomorphic to one of the models in Σ.

(See van Fraassen [1980, 68–69; 2008, 309]; note that van Fraassen himself claims that "belief" is not typically the appropriate attitude to take toward a theory. The more appropriate attitude, he claims, is "acceptance," which includes only belief in that theory's empirical adequacy.) This idea seems simple enough . . . until you start to ask difficult questions about it. Indeed, there are a number of questions that might be raised about B.

First, "isomorphism" is a technical notion of a certain sort of mapping between mathematical structures. The physical world, however, is presumably not itself a mathematical structure. So what could be meant here by "isomorphism"? In response to this sort of question, some semanticists have suggested replacing the word "isomorphic" with the word "similar" (see Giere 1988, 83). Presumably, the thought here is that using a less precise word will invite less scrutiny. In any case, a nontrivial amount of recent literature in philosophy of science has been devoted to trying to understand what this notion of similarity is supposed to be.

Second, the technical notion of isomorphism presupposes a specific category of mathematical structures. For example, consider the real numbers ℝ, and suppose that the intended domain of study X is supposed to be isomorphic to ℝ. What would that tell us about X? Would it tell us something about the cardinality of X? Would it also tell us something about the topological structure of X? How about the order structure of X? And what about the smooth (manifold) structure of X? The point here is that ℝ belongs to several different categories—sets, groups, topological spaces, and the like—each of which has a different notion of isomorphism.

Third, and related to the second point, how are we to tell the difference between a genuinely significant representational structure in a model and a surplus, nonrepresentational structure? For example, suppose that we represent the energy levels of a harmonic oscillator with the natural numbers ℕ. Well, which set did we mean by ℕ? Did we mean the Zermelo ordinals, the von Neumann ordinals, or yet some other set? These sets have different properties (e.g., the Zermelo ordinals are all singleton sets, the von Neumann ordinals are not). Does the "world" care whether we represent it with the Zermelo or von Neumann ordinals (cf. Benacerraf 1965)?

The ironic thing, here, is that one the primary motivations for the semantic view of theories was to move away from problems that philosophers were creating for themselves and back toward genuine problems generated within scientific practice. But the "problem of representation"—the question of how a mathematical model might be similar to the world—is precisely one of those problems generated internal to philosophical practice. That is, this problem was not generated internal to the practice of *empirical science* but internal to the practice of *philosophical reflection on science*. To my eye, it is a sign of deterioration of a philosophical program when it raises more questions than it answers, and it appears that the semantic view has begun doing just that.

2.2 Criticisms of the Semantic View

Critics of the syntactic, or received, view of theories have typically called for its rejection. In contrast, the semantic view of theories has mostly been subjected to internal criticisms, with calls for certain modifications or reforms. Many of these discussions focus on the notion of "models." While the earliest pioneers of semantic views were using the technical notion of "model" from mathematical logic, several philosophers of science have argued for a more inclusive notion of scientific models.

There have been a few external criticisms of the semantic view of theories, but most of these have been shown to rest on misunderstandings. For example, a classic criticism of the semantic view is that whereas one can believe in a theory, one cannot believe in a collection of models. But semanticists have been very clear that they see belief as involving the postulation of some notion of similarity or resemblance between one of the models and the intended domain of study.

One reason, however, for the paucity of criticism of the semantic view is that philosophers' standards have changed—they no longer demand the same things from an account of theories that they demanded in the 1960s or 1970s. Recall, for example, that Putnam demanded that the RV give a precise account of partial interpretation. When he couldn't find such an account, he concluded that scientific theories could not be what the RV said they were. The semantic view seems not to have been subjected to such high demands.

Consider, for example, the claim that the RV of theories entails that theories are language-dependent entities, whereas the semantic view of theories treats theories as language-independent. This comparison is based on a misunderstanding: classes of models *do* depend on a choice of language. Consider, for example, the following question: what is the class of groups? One might say that groups are ordered quadruples $\langle G, o, e, i \rangle$ where o is a binary function on G, i is a unary function on G, e is an element of G, and the like. Alternatively, one might say that groups are ordered triples $\langle G, o, i \rangle$, where o is a binary function on G, and so forth. Or, for yet another distinct definition, one might say that a group is an ordered triple $\langle G, e, o \rangle$, where e is an element of G, and so forth. Note that none of these classes is the same. For example, $\langle G, o, i \rangle$ cannot be a member of the last named class for the simple reason that i is a unary function on G. (For further discussion of this sort of criticism, see Halvorson 2012, 2013; Glymour 2013; van Fraassen 2014.)

3 THE SET-THEORETIC VIEW OF THEORIES

Beginning in the 1950s and 1960s, Patrick Suppes developed a distinctive view of scientific theories as set-theoretic structures. In one sense, Suppes's view is a semantic view insofar as mathematical semantics involves looking for structures inside the universe of sets. However, Suppes's approach differs in emphasis from the semantic view. Suppes doesn't talk about *models* but about *set-theoretic predicates* (STPs). He says, "To axiomatize a theory is to define a predicate in terms of the notions of set theory" (Suppes 1999, 249). (For the most comprehensive summary of Suppes' views on this subject, see Suppes 2002.)

Recall that Zermelo-Frankel (ZF) set theory is a first-order theory in a language with a single binary relation symbol ∈, where $x \in y$ intuitively means that the set x is an element of the set y. Following the typical custom, I'll make free use of the definable symbols ∅ (empty set), ⊆ (subset inclusion), and ordered n-tuples such as $\langle x_1, \ldots, x_n \rangle$.

In short, an STP is just an open formula $\Phi(x)$ in the language of set theory. Thus, for example, the predicate "x is an ordered pair" can be defined by means of the formula

$$\Phi(x) \equiv \exists y \exists z \left(x = \langle y, z \rangle \right).$$

Similarly, the predicate "y is a function from u to v" can be defined by means of a formula saying that y is a subset of ordered pairs of elements from u, v such that each element of u is paired with at most one element of v.

These STPs also allow us to define more complicated mathematical structures. For example, we can say that a set x is a "group" just in case $x = \langle y, z \rangle$, where y is a set and z is a function from $y \times y$ to y satisfying certain properties. The result would be a rather complicated set-theoretic formula $\Gamma(x)$, which is satisfied by all and only those ordered pairs that are groups in the intuitive sense.

What are the advantages of the STP approach? First of all, it is more powerful than the syntactic approach—at least if the latter is restricted to first-order logic. On the one hand, there are STPs for second-order structures, such as topological spaces. On yet another hand, STPs can select intended models. For example, there is an STP $\Phi(x)$ that holds of x just in case x is countably infinite.

Thus, it seems that the models of almost any scientific theory could be picked out by means of an STP. But just because you *can* do something doesn't mean that you *should* do it. What advantage is there to formulating a scientific theory as an STP? Do such formulations provide some insight that we were looking for? Do they answer questions about theories? Moreover, what is the success record of Suppes's proposal?

Let's consider how Suppes's proposal fares in answering the sorts of questions philosophers of science might have about theories:

1. *Is the set-theoretic approach of any use for understanding the relations between models of a single theory?*

    One advantage of a model-theoretic approach to theories is that model theory provides uniform definitions of the notions of *embeddings* and *isomorphisms* between models (e.g., if M, N are L-structures, then a map $j: M \to N$ is called an *embedding* just in case it "preserves" the interpretation of all nonlogical vocabulary). Similarly, $j: M \to N$ is an *isomorphism* if it is a bijection that preserves the interpretation of the nonlogical vocabulary. This single definition of isomorphism of L-structures then specializes to give the standard notion of isomorphism for most familiar mathematical structures such as groups, rings, vector spaces, and the like. It is this feature of model theory that makes it a fruitful mathematical discipline: it generalizes notions used in several different branches of mathematics. Now if A and B both satisfy an STP $\Gamma(x)$, then when is A embeddable in B, and when are A and B isomorphic? A cursory scan of the literature shows (surprisingly!) that no such definitions have been proposed.

2. *Is the set-theoretic approach of any use for answering questions about relations between theories?*

Under what conditions should we say that two STPs $\Phi(x)$ and $\Psi(x)$ describe *equivalent* theories? For example, according to the folklore in mathematical physics, Hamiltonian and Lagrangian mechanics are equivalent theories—although this claim has recently been contested by philosophers (see North 2009). Suppose then that we formulated an STP $H(x)$ for Hamiltonian mechanics and another $L(x)$ for Lagrangian mechanics. Could these set-theoretic formulations help us clarify the question of whether these two theories are equivalent?

Similarly, under what conditions should we say that the theory described by $\Phi(x)$ is *reducible to* the theory described by $\Psi(x)$? For example, is thermodynamics reducible to statistical mechanics?

For a rather sophisticated attempt to answer these questions, see Pearce (1985), which develops ideas from Sneed and Stegmüller. Length considerations will not permit me to engage directly with these proposals. What's more, I have no intention of denigrating this difficult work. However, a cursory glance at this account indicates that it requires translating theories into a language that will be foreign to most working scientists, even those in the exact sciences. The problem here is essentially information overload: giving an STP for a theory often means giving more information than a working scientist needs.

It is sometimes touted as a virtue of the set-theoretic approach that predicates in the language of set theory can be used to pick out intended models (and to rule out those unintended models, such as end extensions of Peano arithmetic). But this advantage comes at a price: there are also predicates of the language of set-theory to which scientists would show complete indifference. Consider an example: there are many sets that satisfy the STP "x has the structure of the natural numbers." Moreover, these different sets have different set-theoretic properties (e.g., one might contain the empty set and another might not). That is, there is an STP $\Phi(x)$ such that $\Phi(M)$ but $\neg\Phi(N)$, where M and N are sets that both instantiate the structure of the natural numbers. Thus, while STPs can be used to rule out unintended models, they also seem too fine-grained for the purposes of empirical science.

The criticisms I've made here of Suppes's approach can be considered as complementary to those made by Truesdell (1984). In both cases, the worry is that the set-theoretic approach requires translating scientific theories out of their natural idiom and into the philosopher's preferred foundational language (viz. set theory). Obviously, set theory plays a privileged role in the foundations of mathematics. For example, we call a mathematical theory "consistent" just in case it has a set-theoretic model; in other words, just in case it can be translated into set theory. But the goal of philosophy of science today is not typically to discover whether scientific theories are consistent in this sense;

the goal, more often, is to see how scientific theories work—to see the inferential relations that they encode, to see the strategies for model building, and the like—and how they are related to each other. The set-theoretic approach seems to provide little insight into those features of science that are most interesting to philosophers.

4 FLAT VERSUS STRUCTURED VIEWS
OF THEORIES

Typically, the syntactic versus semantic debate is taken to be the central question in the discussion of the structure of scientific theories. I maintain, however, that it is a distraction from a more pressing question. The more pressing question is whether scientific theories are "flat" or whether they have "structure." Let me explain what I mean by this.

The syntactic view of theories is usually formulated as follows:

A theory is a set of sentences.

This formulation provides a *flat* view: a theory consists of a collection of things and not in any relations between those things or structure on those things. In contrast, a *structured* view of theories might say that a theory consists of both sentences and, for example, inferential relations between those sentences.

The flat versus structured distinction applies not just to the syntactic view of theories, but also to the semantic view of theories. A flat version of the semantic view might be formulated as:

A theory is a set (or class) of models.

In contrast, a structured version of the semantic view might say that a theory is a set of models *and* certain mappings between these models, such as elementary embeddings.

Both the syntactic and the semantic views of theories are typically presented as flat views. In the latter case, I suspect that the flat point of view is accidental. That is, most proponents of the semantic view are not ideologically committed to the claim that a theory is a bare set (or class) of models. I think they just haven't realized that there is an alternative account.

But, in the case of the syntactic view, some philosophers have ideological commitments to a flat view—a commitment that derives from their rejection of "intensional" concepts. The most notable case here is Quine. Quine's criticism of the analytic-synthetic distinction can also be seen as a criticism of a structured view of theories. On a structured syntactic view of theories, what matters in a theory is not just the sentences it contains, but also the relations between the sentences (e.g., which sentences are logical consequences of which others). But, in this case, commitment to a theory would involve claims about inferential relations, in particular, claims about which sentences are logical consequences of the empty set. In other words, a structured syntactic view of theories needs an analytic-synthetic distinction!

Quine's powerful criticisms of the analytic-synthetic distinction raise worries for a structured syntactic picture of theories. But is all well with the unstructured, or flat, syntactic view? I maintain that the unstructured view has *severe* problems that have never been addressed. First of all, if theories are sets of sentences, then what is the criterion of equivalence between theories? A mathematically minded person will be tempted to say that between two *sets*, there is only one relevant condition of equivalence: namely, equinumerosity. But certainly we don't want to say that two theories are equivalent if they have the same number of sentences! Rather, if two theories are equivalent, then they should have some further structure in common. What structure should they have in common? I would suggest that, at the very least, equivalent theories ought to share the same inferential relations. But, if that's the case, then the content of a theory includes its inferential relations.

Similarly, Halvorson (2012) criticizes the flat semantic view of theories on the grounds that it would trivialize the notion of theoretical equivalence.

5 A Category-Theoretic Approach to Theories

As mentioned in Section 3, Suppes's set-theoretic approach to theories was criticized by the historian of physics Clifford Truesdell. Interestingly, a student of Truesdell's became similarly agitated about the use of set-theoretic models for physical systems, seeing these models as obscuring the salient features of physical systems. This student was so firmly convinced that a new approach was needed that he devoted his career to developing an alternative foundation of mathematics, a "category-theoretic" foundation of mathematics.

Truesdell's student was William Lawvere, who has gone on to be one of the most important developers of category theory (see Lawvere 2007). In retrospect, it is shocking that almost no philosophers of science followed the developments of Lawvere and his collaborators. To the best of my knowledge, there are at most a half-dozen papers in philosophy of science that make use of category-theoretic tools.

So what is a category? One way of seeing it is that a category is a set where the elements can stand in a variety of relations to each other. In a bare set, two elements are either equal or unequal—there is no third way. But for two elements a, b in a category, there might be several different "paths" from a to b, some of which might be interpreted as showing that a and b are the same and others showing that some other relation holds between them (see Figure 28.1).

Slightly more rigorously, a category C consists of a collection C_0 of "objects," and a collection C_1 of "arrows" between these objects. The collection of arrows is equipped with a partial composition operation: if the head of one arrow meets the tail of another, then

FIGURE 28.1 An impressionistic picture of one particular category. The nodes are the elements (or objects) of the category, and the lines are the paths (or arrows) between elements. We have labeled a couple of the nodes as familiar groups, since the collection of groups (objects) and group homomorphisms (arrows) is a paradigmatic category.

the two can be composed to give another arrow. Furthermore, for each object, there is an identity arrow going from that object to itself.

If you're familiar with any abstract mathematical structures, then you are familiar with some categories. For example, the category **Grp** of groups has groups as objects and group homomorphisms as arrows. Similarly, the category **Man** of manifolds has differential manifolds as objects and smooth mappings as arrows.

Some mathematicians and philosophers have made bold claims about category theory; for example, that it should replace set theory as the foundation of mathematics. For the purposes of this chapter, I won't need to take any position on that debate. Nonetheless, there are good reasons to think that category theory provides a better language than set theory for philosophers of science. Whereas set theory is best at plumbing the depths of individual mathematical structures (e.g., the real number line), category theory excels at considering large collections of mathematical structures and how they relate to each other. But isn't this point of view preferable for the philosopher of science? Isn't the philosopher of science's goal to see how the various components of a theory hang together and to understand relations between different theories?

It was completely natural, and excusable, that when Suppes and collaborators thought of mathematical structures they thought of sets. After all, in the early 1950s, the best account of mathematical structure was that given by Bourbaki. But mathematics has evolved significantly over the past sixty years. According to current consensus in the mathematical community, the best account of mathematical structure is provided by category theory:

> The concept of mathematical structures was introduced by N. Bourbaki ... The description which Bourbaki used was, unfortunately, rather clumsy. Simultaneously, a more abstract (and more convenient) theory of categories was introduced by S. Eilenberg and S. Mac Lane. (Adámek 1983, ix)

Thus, if philosophers of science want to stay in touch with scientific practice, they need to think about the category-theoretic account of mathematical structure.

Categories have the capacity to represent theories as *structured* things—either structured syntactic things or structured semantic things. From the syntactic point of view, a theory can be thought of as a category whose objects are sentences and whose arrows are inferential relations between those sentences. From the semantic point of view, a theory can be thought of as a category whose objects are models and whose arrows are mappings between those models. In both cases, category theory provides a particularly natural way to develop a structured view of theories.

Taking a category-theoretic view of theories has many advantages and suggests many future projects for technically oriented philosophers of science:

1. Recall that flat views of theories have trouble providing a reasonable notion of theoretical equivalence. A category theoretic approach to theories fixes this problem. First of all, with regard to the syntactic point of view, the best account of theoretical equivalence is the notion of "having a common definitional extension," as explained by Glymour (1971). It is not difficult to see, however, that two theories have a common definitional extension just in case there are appropriate mappings between them that preserve inferential relations (see Barrett and Halvorson 2015).

2. To my knowledge, the semantic view of theories offers no resources for answering interesting question about relations between different scientific theories. For example, are Hamiltonian and Lagrangian mechanics equivalent theories?

    The reason that the semantic view cannot answer this questions is that if theories are treated as flat, structureless classes of models, then there is nothing interesting to say about relations between theories. It is only when theories are treated as structured things that there are interesting mathematical questions about equivalence, reduction, and other relations between theories.

    Once again, category theory can help here. If theories are categories, then a mapping between theories is a mapping between categories, something known as a "functor." And there are interesting properties of functors that look very much like the relations between theories that interest philosophers of science. For example, a functor that is full, faithful, and essentially surjective looks like a good candidate for an equivalence of theories (Weatherall 2015).

3. Some philosophers have claimed that there is no significant difference between the syntactic and semantic views of theories (see Friedman 1982; Halvorson 2012); or, at least, that in typical cases we can freely switch back and forth between syntactic and semantic representations of theories. In fact, category theory provides the means to make this claim rigorous.

    First of all, given a syntactically formulated theory T, we can construct a class $M(T)$ of models. In fact, $M(T)$ is more than just a class—it is a category, since there is a natural notion of arrows between models, namely, elementary embeddings. Thus, M can in fact be thought of as a mapping from syntactically formulated theories to semantically formulated theories.

    Is there a map N going in the reverse direction? Clearly, it cannot be the case that for any class Σ of models, there is a corresponding syntactically formulated theory $N(\Sigma)$. What's more, it is clear that the input for N shouldn't be bare classes of models because one and the same class of models can correspond to different syntactically formulated theories (see Halvorson 2012). Thus, the input for N should be *categories* of models.

In short, there are interesting technical questions about moving from categories of models to syntactically formulated theories. The idea here is that when a category is sufficiently "nice," then it is the category of models of a syntactically formulated theory. (These issues are discussed at length in Awodey and Forssell [2013], Makkai [1987], and Makkai and Reyes [1977] among other places in the literature on categorical logic.)

6 THE NO-THEORY VIEW

Some philosophers of science will think that we've been wasting a lot of time splitting hairs about whether theories are syntactic things, or semantic things, or something else. And, obviously, I somewhat agree with them. As indicated in the previous section, the best versions of the syntactic and semantic views of theories are dual to each other, and both analyses are helpful in certain philosophical discussions. However, some philosophers would go further and say that we needn't ever use the word "theory" in our philosophical discussions of science. In this section, I intend to address this intriguing suggestion.

Rather than try to deal with every critique of "theory," I'll zoom in on a recent discussion by Vickers (2013) in the context of trying to understand how scientists deal with inconsistencies. According to Vickers, whenever we might ask something about a certain theory, we can just ask the same question about a set of propositions: "Why not simply talk about sets of propositions?" (28). At this stage, it should be clear to what problems this suggestion might lead. First of all, we have no direct access to propositions—we only have access to sentences that express those propositions. Thus, we only know how to talk about sets of propositions by using sets of sentences. Second, why should we think that when a scientist puts forward a theory, she's only putting forward a *set* of propositions? Why not think that she means to articulate a *structured set* of propositions (i.e., a set of propositions with inferential relations between them)? Thus, we see that Vickers's stance is not neutral on the major debates about scientific theories. It is not neutral about the semantic versus syntactic point of view, and it is not neutral on flat versus structured points of view.

This is not to say that we can't go a long way in philosophical reflection on science without answering the question, "what is a theory?" Indeed, Vickers's book is a paradigm of philosophical engagement with issues that are, or ought to be, of real concern to working scientists and to scientifically engaged laypersons. But it seems to me that Vickers doesn't practice what he preaches. He disclaims commitment to any theory of theories, while his actual arguments assume that scientists are rational creatures who enunciate propositions and believe that there are inferential relations between these propositions. Thus, I believe that Vickers's work displays the fruitfulness of a structured syntactic view of theories.

7 Why a Formal Account of Theories?

Clearly, this chapter has leaned heavily toward formal accounts of theories. Other than the author's personal interests, are there good reasons for this sort of approach? Once again, we are faced here with the question of what philosophers of science are hoping to accomplish, and, just as there are many different sciences with many different approaches to the world, I suggest that there are many different legitimate ways of doing philosophy of science. Some philosophers of science will be more interested in how theories develop over time, and (*pace* the German structuralist school) formal analyses have so far offered little insight on this topic. Other philosophers of science are interested in how theory is related to experiment, and the more accurate and detailed an account of real-world phenomena we give, the more difficult it will be to illuminate issues via an abstract mathematical representation.

Thus, if a formal approach to scientific theories has utility, it has a limited sort of utility—in precisely the same way that mathematical physics has a limited sort of utility. Mathematical physics represents a sort of limiting case of scientific inquiry, where it is hoped that pure mathematical reasoning can provide insight into the workings of nature. In the same way, formal approaches to scientific theories might be considered as limiting cases of philosophy of science, where it is hoped that pure mathematical reasoning can provide insight into our implicit ideals of scientific reasoning, the relations between theories, and the like. For some philosophers, this sort of enterprise will be seen as not only fun, but also illuminating. But, just as there are scientists who don't like mathematics, it can be expected that there will be philosophers of science who don't like formal accounts of theories.

8 Further Reading

For another perspective on scientific theories, see Craver (2008), especially its account of mechanistic models.

For a historically nuanced look at the logical empiricists' account of theories, see Mormann (2007). The most elaborate reconstruction of this approach is given in the introduction to Suppe (1974). For a recent defense of some parts of the RV, see Lutz (2012, 2014).

For a somewhat biased account of the development of the semantic view of theories, see Suppe (2000). For another account of this approach, from one of its most important advocates, see van Fraassen (1987). For discussion of the "problem of representation" in the semantic view of theories, see Frigg (2006) and Suárez (2010). For a recent discussion of the syntactic versus semantic debate, see Lutz (2015).

For a detailed look at Suppes's set-theoretic approach, see Suppes (2002). For a comparison of semantic and set-theoretic views of theories, see Przełecki (1974) and Lorenzano (2013).

ACKNOWLEDGMENTS

Thanks to Thomas Barrett, Robbie Hirsch, Dimitris Tsementzis, Jim Weatherall, and Bas van Fraassen for sharpening my views on this subject and for feedback on an earlier draft. Thanks to Jeff Koperski for bringing Truesdell's critique of Suppes to my attention.

REFERENCES

Achinstein, Peter. (1968). *Concepts of Science* (Baltimore, MD: Johns Hopkins Press).

Adámek, Jiří. (1983). *Theory of Mathematical Structures* (New York: Springer).

Awodey, Steve, and Forssell, Henrik. (2013). "First-Order Logical Duality." *Annals of Pure and Applied Logic* 164 (3): 319–348.

Barrett, Thomas, and Halvorson, Hans. (2015). "Quine and Glymour on theoretical equivalence." Forthcoming in *Journal of Philosophical Logic*.

Benacerraf, Paul. (1965). "What Numbers Could Not Be." *Philosophical Review* 74: 47–73.

Bourbaki, Nicholas. (1970). *Théorie des Ensembles* (Paris: Hermann).

Carnap, Rudolf. (1928). *Der logische Aufbau der Welt* (Berlin: Springer Verlag).

Carnap, Rudolf. (1934). *Logische Syntax der Sprache* (Berlin: Springer Verlag).

Carnap, Rudolf. (1936). "Testability and Meaning." *Philosophy of Science* 3 (4): 419–471.

Carnap, Rudolf. (1939a). "Foundations of Logic and Mathematics." In Otto Neurath, Rudolf Carnap, and Charles Morris (eds.), *International Encyclopedia of Unified Science* (Chicago: University of Chicago Press), 2: 139–212.

Carnap, Rudolf. (1939b). "Logical Foundations of the Unity of Science." In Otto Neurath, Rudolf Carnap, and Charles Morris (eds.), *International Encyclopedia of Unified Science* (Chicago: University of Chicago Press), 1: 42–62.

Craver, Carl F. (2008). "Structures of Scientific Theories." In Peter Machamer and Michael Silbertstein *(eds.), The Blackwell Guide to the Philosophy of Science* (New York: Wiley), 55–79.

Demopoulos, William. (2007). "Carnap on the rational Reconstruction of Scientific Theories." In Michael Friedman and Richard Creath (eds.), *The Cambridge Companion to Carnap* (New York: Cambridge University Press), 248–272.

Feigl, Herbert. (1970). "The 'Orthodox' View of Theories: Remarks in Defense as Well as Critique." In Michael Radner and Stephen Winokur (eds.), *Analyses of Theories and Methods of Physics and Psychology*. Minnesota Studies in the Philosophy of Science (Minneapolis: University of Minnesota Press), 4: 3–16.

Friedman, Michael. (1982). "Review of *The Scientific Image*." *Journal of Philosophy* 79 (5): 274–283.

Frigg, Roman. (2006). "Scientific Representation and the Semantic view of Theories." *Theoria* 21 (1): 49–65.

Giere, R. N. (1988). *Explaining Science: A Cognitive Approach* (Chicago: University of Chicago Press).

Glymour, Clark. (1971). "Theoretical Realism and Theoretical Equivalence." In Roger C. Buck and Robert S. Cohen *(eds.), PSA: Proceedings of the Biennial Meeting of the Philosophy of Science Association* (New York: Springer), 275–288.

Glymour, Clark. (2013). "Theoretical Equivalence and the Semantic View of Theories." *Philosophy of Science* 80 (2): 286–297.

Halvorson, Hans. (2012). "What Scientific Theories Could Not Be." *Philosophy of Science* 79 (2): 183–206.

Halvorson, Hans. (2013). "The Semantic View, If Plausible, Is Syntactic." *Philosophy of Science* 80 (3): 475–478.

Hempel, Carl Gustav. (1958). "The Theoretician's Dilemma: A Study in the Logic of Theory Construction." In Herbert Feigl, Michael Scriven, and Grover Maxwell (eds.), *Concepts, Theories, and the Mind-Body Problem*. Minnesota Studies in Philosophy of Science (Minneapolis: University of Minnesota Press), 2: 37–98.

Hempel, Carl Gustav. (1970). "On the "Standard Conception" of Scientific Theories." In Michael Radner and Stephen Winokur (eds.), *Analyses of Theories and Methods of Physics and Psychology* (Minneapolis: University of Minnesota Press), 4: 142–163.

Lawvere, F. William. (2007). "Interview with F. William Lawvere" (December). http://www.cim.pt/files/publications/b23www.pdf.

Lloyd, Elisabeth. (1984). "A semantic approach to the structure of evolutionary theory." PhD diss., Princeton University.

Lorenzano, Pablo. (2013). "The Semantic Conception and the Structuralist View of Theories: A Critique of Suppe's Criticisms." *Studies in History and Philosophy of Science Part A* 44 (4): 600–607.

Lutz, Sebastian. (2012). "On a Straw Man in the Philosophy of Science: A Defense of the Received View." *HOPOS* 2 (1): 77–120.

Lutz, Sebastian. (2014). "What's right with a syntactic approach to theories and models?" *Erkenntnis* 79 (8): 1475–1492.

Lutz, Sebastian. (2015). "What Was the Syntax-Semantics Debate in the Philosophy of Science About?" Forthcoming in *Philosophy and Phenomenological Research*.

Makkai, Michael. (1987). "Stone Duality for First Order Logic." *Advances in Mathematics* 65 (2): 97–170.

Makkai, Michael, and Reyes, Gonzalo E. (1977). *First Order Categorical Logic* (New York: Springer).

Mormann, Thomas. (2007). "The Structure of Scientific Theories in Logical Empiricism." In Alan Richardson and Thomas Uebel (eds.), *The Cambridge Companion to Logical Empiricism* (New York: Cambridge University Press), 136–162.

Nagel, Ernest. (1961). *The Structure of Science* (New York: Harcourt, Brace & World).

North, Jill. (2009). "The "Structure" of Physics: A Case Study." *Journal of Philosophy* 106: 57–88.

Pearce, David. (1985). *Translation, Reduction and Equivalence* (Frankfurt: Peter Lang).

Przełecki, Marian. (1974). "A Set Theoretic versus a Model Theoretic Approach to the Logical Structure of Physical Theories." *Studia Logica* 33 (1): 91–105.

Putnam, Hilary. (1962). "What Theories Are Not." In Ernest Nagel, Patrick Suppes, and Alfred Tarski (eds.), *Logic, Methodology and Philosophy of Science: Proceedings of the 1960 International Congress* (Stanford, CA: Stanford University Press), 240–251.

Reyes, Gonzalo. (2011). "A Derivation of Einstein's Vacuum Field Equations." In Bradd Hart (ed.), *Models, Logics, and Higher-Dimensional Categories* (Providence, R.I.: American Mathematical Society), 245–261.

Russell, Bertrand. (1914). "The Relation of Sense-Data to Physics." *Scientia* 16: 1–27.

Suárez, Mauricio. (2010). "Scientific Representation." *Philosophy Compass* 5 (1): 91–101.

Suppe, Frederick. (1971). "On Partial Interpretation." *Journal of Philosophy* 68: 57–76.

Suppe, Frederick. (1972). "What's Wrong with the Received View on the Structure of Scientific Theories?" *Philosophy of Science* 39: 1–19.

Suppe, Frederick. (1974). *The Structure of Scientific Theories* (Urbana: University of Illinois Press).

Suppe, Frederick. (2000). "Understanding Scientific Theories: An Assessment of Developments, 1969–1998." *Philosophy of Science 67 supplement*: S102–S115.

Suppes, Patrick. (1964). "What Is a Scientific Theory?" In Sydney Morgenbesser (ed.), *Philosophy of Science Today* (New York: Basic Books), 55–67.

Suppes, Patrick. (1999). *Introduction to Logic* (Mineola NY: Dover Publications).

Suppes, Patrick. (2002). *Representation and Invariance of Scientific Structures* (Palo Alto, CA: Center for the Study of Language & Information).

Truesdell, Clifford. (1984). "Suppesian Stews." In *An Idiot's Fugitive Essays on Science* (New York: Springer), 503–579.

van Fraassen, Bas. (1970). "On the Extension of Beth's Semantics of Physical Theories." *Philosophy of science* 37: 325–339.

van Fraassen, Bas. (1972). "A Formal Approach to the Philosophy of Science." In Robert Colodny (ed.), *Paradigms and Paradoxes* (Pittsburgh, PA: University of Pittsburgh Press), 303–366.

van Fraassen, Bas. (1980). *The Scientific Image* (New York: Oxford University Press).

van Fraassen, Bas. (1987). "The Semantic Approach to Scientific Theories." In N. J. Nersessian (ed.), *The Process of Science* (New York: Springer), 105–124.

van Fraassen, Bas. (2008). *Scientific Representation: Paradoxes of Perspective* (New York: Oxford University Press).

van Fraassen, Bas. (2014). "One or Two Gentle Remarks About Hans Halvorson's Critique of the Semantic View." *Philosophy of Science* 81 (2): 276–283.

Vickers, Peter. (2013). *Understanding Inconsistent Science* (New York: Oxford University Press).

Weatherall, James. (2015). "Are Newtonian Gravitation and Geometrized Newtonian Gravitation Theoretically Equivalent." Forthcoming in *Erkenntnis*.

CHAPTER 29

VALUES IN SCIENCE

HEATHER DOUGLAS

1 INTRODUCTION

SCIENCE takes place in society. Although this might seem obvious, incorporating an understanding of this into philosophy of science has driven dramatic changes in the understanding of values in science in recent years. Seeing science as fully embedded in and responsible to the broader society that supports it, the scientific endeavor becomes fully value-laden. This change is not just for the description of science performed by all-too-human actors who inevitably bring their values with them into their science. The change also applies to science in the ideal. However, the distinctive nature of science must also be described and protected from abuse or the misuse of values in science. The issues for philosophers thus become (1) What are the implications of the embedding of science in society? (2) When and how are values legitimate? (3) Which values are legitimate? And (4), what is the ideal for science embedded in society?

The changes that this conceptual shift have wrought take place against the backdrop of the value-free ideal. This chapter will first describe the value-free ideal before moving on to the challenges philosophers have leveled against it. It will then describe the terrain of how to conceive of the role of values in science once the value-free ideal is rejected *qua* ideal. As we will see, challenging questions of the place and role of science in society come to the fore, with important implications for the philosophy of science.

2 THE VALUE-FREE IDEAL

The value-free ideal has historical roots stretching back centuries (Proctor 1991), but the contemporary formulation dates from the Cold War era (Douglas 2009: ch. 3). Precursors to the contemporary value-free ideal, including Max Weber, did not articulate the full-fledged value-free ideal, suggesting instead that values served to help focus scientists on what was significant and that science could not proceed without values (Douglas 2011). That values shape what scientists chose to explore was, and remains, an uncontroversial way in which values influence science. That scientists' interests and concerns help direct scientists' attention in some directions and not others seemed unproblematic and was never contested by philosophers of science. But, by 1950, this was thought to be part of the "logic of discovery" in science rather than the "logic of justification." If values played a role in the logic of justification, the worry was (and remains) that science could be corrupted.

It was not until a debate over the role of values in science erupted in the 1950s that the value-free ideal emerged in its fully articulated form. The debate began with the work of C. West Churchman (1948) and Richard Rudner (1953) and their acknowledgment that, in addition to values shaping the direction of research, values were required to set what was thought to be sufficient evidence. Rudner pointedly argued that because science was not a deductive process and there was always an inductive gap between the evidence and the scientific claims being made ("no scientific hypothesis is ever completely verified"), one had to decide "that the evidence is sufficiently strong or that the probability is sufficiently high to warrant the acceptance of the hypothesis." (Rudner 1953, 2) Making this decision involved considering "the importance, in the typically ethical sense, of making a mistake in accepting or rejecting the hypothesis" (Rudner 1953, 2). This meant that social and ethical values had a legitimate role in deciding whether a scientific claim is sufficiently justified—the determination of sufficient justification can include social and ethical values weighing the importance of a mistake and whether we have enough evidence to consider such a risk worth worrying about. For example, if we think a false-positive type of error is more worrisome, we might demand more evidence. If we think a false-negative type of error is more worrisome, we might demand less before finding a claim "sufficiently warranted." Whether one or the other is more worrisome is a value judgment, one that often depends on social and ethical valuations of the consequences of error. What counts as sufficient warrant thus depends on the claim and the context in which it is made and whether there are serious implications of error in that context.

Such a view brought values, however, into the heart of doing science. And, as such, there was substantial resistance to it. Philosophers like Richard Jeffrey and Isaac Levi resisted this line of argument, albeit in different ways. Jeffrey argued that scientists did not need to accept and reject hypotheses; instead, he thought they should just assign probabilities to them (Jeffrey 1956). Rudner had anticipated this approach, however, and had argued that even in assigning probabilities, one needed to decide that the probability was sufficiently warranted. The problem of assessing sufficiency might be mitigated, but it was not eliminated.

Levi took a different tack, arguing that, as a normative matter, scientists should not consider the broader societal implications or the context of their work when assessing scientific claims. Levi suggested that certain "canons of inference" should be used when examining evidence and its relationship to hypotheses and that such canons were entirely sufficient for addressing the inferential gap that confronted scientists (Levi

1960). For Levi, such canons demanded a uniform response from scientists when confronting judgments about the strength of evidence.

It was this line of argument that helped to formulate the value-free ideal for science. The canons of inference were to include such considerations as simplicity, scope, and explanatory power, aspects that were soon to be called "epistemic" or "cognitive" values. The value-free ideal was the idea that social and ethical values should influence only the external aspects of science—such as which projects were undertaken and which methodologies were thought to be ethically acceptable—but that in the heart of science, at the moment of inference, no social and ethical values were to have any role whatsoever. Rather, only the canons of inference (the epistemic values) should shape scientists' judgments.

It was in Thomas Kuhn's 1977 paper, "Objectivity, Value Judgment, and Theory Choice," that such canons of inference came to be called "values." Kuhn argued that one cannot generate an algorithm or set of rules for scientific theory choice and that the process of judgment regarding theories is better characterized as a value judgment, one informed by the characteristics desired of good scientific theories, characteristics that in Kuhn's view included "accuracy, consistency, scope, simplicity, and fruitfulness" (Kuhn 1977, 322). By establishing a shared set of values used to inform theory choice, science could remain objective even if *some* values were essential to science. In addition, scientists could disagree about which theories to adopt (as historical work showed they did) and still be working within the canons of scientific inference because some scientists might emphasize particular values over others or have different interpretations of what instantiated the values properly. What made science objective, on this picture, was an adherence to excluding social and ethical values from the inferential aspects of science. It was this sense of which values should be allowed in (the epistemic values only) at the crucial point in science (when making inferences about which theories to accept) that constituted the value-free ideal (see also Laudan 1984; McMullin 1983).

3 CHALLENGES TO THE VALUE-FREE IDEAL

The value-free ideal, a somewhat inaccurate label for the epistemic-values-only-in-scientific-inference ideal, has been challenged in a number of ways. I describe three main approaches to critiquing the ideal: the descriptive challenge, the boundary challenge, and the normative challenge. These are not meant to be fully distinct challenges, but the thrust of each of the three leads in different directions.

3.1 The Descriptive Challenge

The first challenge began in the 1980s, as feminist scholars of science noted that science did not appear to be value-free. Although some scientific claims under scrutiny could be shown to be due to poor methodology, others could not. Instead, they appeared to be the result of either a lack of sufficient alternative explanations or theories, or the result of insufficiently examined background assumptions. Feminist philosophers argued that even good science, science thought to be exemplary, appeared value-laden in important ways. (Fausto-Sterling 1985; Harding 1986, 1991; Keller and Longino 1996; Nelson 1990) Their work showed how fields that carefully gathered evidence, that followed the traditional practices of scientific inference, still managed to produce blatantly sexist results. Examples such as early primatology (Haraway 1989), anthropological accounts of human development (Longino 1990), and biological accounts of sperm–egg interactions (Martin 1996) were shot through with sexist presuppositions that blinded scientists to alternative explanations of phenomena and thus led to acceptance of views later shown to be inadequate. The scientists who did such work were (generally) not committing fraud or allowing social values to shape their inferences in obvious ways, but, once the science was scrutinized, it seemed anything but value-free.

The problems unearthed by this work showed the pervasive difficulty of eliminating values from the scientific process and the results produced—you could only test whichever theories were available, and you could only test them using your background assumptions. What theories you had available and which background assumptions you held often reflected your values. Values could not be removed from science on this account.

This line of argument is also tied to the understanding of science as underdetermined by the available evidence (in the tradition of Duhem, Neurath, and Quine; see Brown 2013a), particularly transient underdetermination (Anderson 2004; Biddle 2013) Although all of the evidence may one day be in and make clear what we should think, as actual epistemic actors, we are not in that position. The evidence does not clearly determine which claims are the right ones, nor does it indicate that we have all the plausible options on the table for consideration, nor even whether our background assumptions are adequate. If a surprising result does arise, we don't know precisely where the problem lies. Against this backdrop of uncertainty, we still make judgments about what is scientifically supported. This has been called the "gap argument" for the role of values in science (Brown 2013a; Elliott 2011a; Intemann 2005).

The recognition of this gap between theory and evidence, and the possibility that it is filled by values, has led many feminists to call for increased diversity in science and better scrutiny of theories within science. If background assumptions used by scientists often "encode social values" (Longino 1990, 216), we need better scientific practices to uncover such encoding and help us generate background assumptions with better values. Such work led philosophers to consider more carefully the social structure of the scientific community and scientific practices, to see if improving those could assist in both ferreting out these problematic values and ensuring better science. Feminist philosophers invigorated the field of social epistemology, which examines more closely the social conditions of knowledge production and what kinds of conditions we should consider good for reliable knowledge production (Longino 1990, 2002). I will call this the descriptive challenge to the value-free ideal: that science, even science done well, is not value-free, and so the ideal is irrelevant. This approach turns elsewhere, to social structures, for ideals for science.

Helen Longino, for example, has argued for a set of principles that should structure epistemic communities for robust knowledge production. An epistemic community should instantiate public forums for criticism, prima facie equality of intellectual authority, responsiveness to criticism, and shared values in order to be able to produce claims worth calling "knowledge" (Longino 1990, 2002). Miriam Solomon, in contrast, focuses on the range of "decision-vectors" (influences on decisions scientists make) and argues for two principles to guide the distribution of effort within a scientific community: efforts should be distributed equitably (i.e., proportionally) with respect to empirical successes of different theories, and effort should be distributed equally with respect to nonempirical factors (i.e., there should be no clustering of effort in science for purely social or political reasons) (Solomon 2001). Both of these approaches value diversity in science for creating critical pressure on existing scientific approaches and for bringing new ideas into the fore, thus fostering debate and discussion (see also Intemann 2009).

This emphasis on the social structures of science produced by the descriptive challenge has illuminated much about science. By examining how scientific communities should be structured, and in particular how such structures would reveal the hidden values embedded in science, social epistemological approaches have much to teach us about values in science. But such approaches on their own are not enough to take down the value-free ideal *qua* ideal, nor enough to replace it. One could still argue that the goal of science should be the removal of values from science, and social approaches help us ferret out those values, thus leaving the ideal intact. If, on the other hand, one embraced rejecting the ideal, the social-level norms (about the structure of epistemic communities) do not tell individual actors what counts as a good argument within those communities. Some guidance on the legitimate (and illegitimate) grounds for inference in science is still needed so that actors can know what is (and what is not) appropriate argumentation. Section 4 on alternative ideals takes up this challenge.

3.2 The Boundary Challenge

In addition to the descriptive challenge, philosophers critiqued the value-free ideal on another front: on the plausibility that a clear distinction between epistemic and nonepistemic values could be made. The value-free ideal, in its post-Levi/Kuhn form, requires a clear distinction between the values that are allowed to influence theory choice (often called epistemic values) and those that are not (the nonepistemic social and ethical values). If this distinction cannot withstand scrutiny, then the value-free ideal collapses.

Several philosophers have challenged the boundary between epistemic and nonepistemic values. Phyllis Rooney, for example, argued that when we examine the epistemic values espoused by scientists at different historical moments, we can often recognize the nonepistemic values influencing what is thought to be purely epistemic (Rooney 1992). In examining episodes in the history of physics, Rooney finds cultural or religious values shaping supposedly epistemic values as reasons for accepting or rejecting theories. Longino argued that the selection of epistemic values by Kuhn and others was arbitrary and, indeed, that other values could also be argued to be important criteria for theory selection, including ones that seemed opposite to the canonical ones (Longino 1995, 1996). She proposed that, in contrast to the traditional epistemic values of consistency, simplicity, scope, and fruitfulness, equally plausible are such values as novelty, applicability, and ontological heterogeneity, values that arise out of feminist criticisms of science, criticisms that are motivated by a concern for social justice. For example, feminist scientists value novelty in scientific theories because the traditional theories were so often laden with sexism. Feminist scientists also value ontological heterogeneity so that scientists are more likely to recognize the diversity among the individuals they are studying (Longino 1996, 45–47). How the value-free ideal is to be maintained as a coherent ideal when the boundary between epistemic and nonepistemic values appears to be porous is not clear.

Arguments about how to understand the epistemic values in science have become more complex in recent years. As will be discussed later, examinations of the traditional set of epistemic values within a context of rejecting the value-free ideal have opened up new possibilities and insights about epistemic values. It is possible to have differences in purpose and emphasis among values, including among epistemic and nonepistemic values, without having a clear distinction that can bear the weight of the value-free ideal.

3.3 The Normative Challenge

The third challenge to the value-free ideal confronts the ideal more directly. Rather than undermine its plausibility or feasibility, this approach argues that the ideal, *qua* ideal, is the wrong ideal for science. This argument draws from the insights of Churchman, and Rudner and from even earlier work by William James (Magnus 2013), but provides a stronger articulation of and basis for those insights. I will call this the normative challenge. (This has been called the "error argument"; see Brown 2013*a*; Elliott 2011*a*.) Because this approach challenges the ideal *qua* ideal, it must replace the value-free ideal with an alternative ideal by which to manage values in science.

There are several conceptual bases that combine in the normative challenge to the value-free ideal. First, as with the descriptive challenge and arguments from underdetermination, the endemic uncertainty in science is central. That we never have conclusive proof for our scientific theories or complete evidence for our hypotheses means that there is always some uncertainty regarding scientific knowledge. (This was also a key premise in Rudner's argument.) It is because of this uncertainty that there is an inductive gap involved in every scientific claim and an "inductive risk" that accompanies every decision to make that claim—a risk that one makes an inaccurate claim or fails to make an accurate one (Hempel 1965).

Second, the epistemic authority of science and of scientists in society is acknowledged. Scientists do not just do science for themselves, but also for the broader society that supports scientific research. Scientists are generally granted—and indeed we think they should be granted—a prima facie epistemic authority regarding their research areas. As the bearers of our most reliable source of knowledge, scientists should speak and (mostly) expect to be believed. They should also expect their claims about the world to serve as a further basis for decision-making. If one examines the history of science advising, scientists were brought into the heart of government decision-making increasingly throughout the twentieth century, across developed democratic countries, albeit through varied institutional mechanisms (Douglas 2009, ch. 2; Lentsch and Weingart 2009, 2011). The rise of scientific epistemic authority is both descriptively manifested and to be normatively desired.

This baseline epistemic authority brings with it general responsibilities to be neither reckless nor negligent in one's actions (Douglas 2003, 2009, ch. 4). Although this is a responsibility every person has, scientists must consider in particular the impact of their authoritative statements and whether they think they have sufficient evidence for their claims. The inductive risk scientists bear when deciding on a claim must be confronted. To do so requires considering the context(s) in which scientific claims will likely be used and what the consequences of making an error would be (whether it is a false-positive or false-negative error). Evidential sufficiency is not set at a fixed level across all possible contexts (Holter 2014; Miller 2014a; Wilholt 2013). Although good, solid evidence is generally needed to support a scientific claim (a coin toss will not do), how much evidence is enough, what constitutes sufficient evidence, will vary depending on what is at stake (see also Elliott and Resnik 2014).

For example, if one is testing a new treatment for a disease, the level of evidential sufficiency will vary depending on whether the disease has an alternative, mostly successful treatment or whether there is no existing treatment, as well as on how deadly or harmful the disease is. If the disease is highly fatal and there is no treatment, a lower threshold of evidential sufficiency is often warranted before we claim the treatment has enough evidence to support its implementation. If the disease is mild and/or we have successful treatments, we can legitimately demand a stricter standard of safety and efficacy before releasing it on the market. Even disciplines have different levels of sufficient evidence attached to them. High-energy physics, with theories that are far removed from practical application and thus have few impacts of withholding acceptance (low false-negative impacts), can demand very high evidential thresholds to avoid the embarrassment (and potential for wasted effort) of prematurely accepted (and later proved false) claims (Staley 2014). Other areas of science, both because of the technical difficulties of achieving very high levels of statistical significance and because the costs of error are more evenly distributed, legitimately accept a lower threshold (often—but not universally—two standard deviations) for evidential sufficiency.

In addition to the flexibility in standards of evidential sufficiency found in scientific practice, the argument from inductive risk also demonstrates the importance of inductive risk at more stages in the research process than the decision of whether the evidence is sufficient to support the conclusions of a research project. Inductive risk also plays an important role earlier in the scientific process, when data must be characterized. Even here, scientists can be confronted with ambiguous events that they must decide what to do with (Miller 2014a). Should they discard them (potentially lowering the power of their study)? Should they characterize them one way or another? Should they give up on the study until a more precise methodology can be found? Each of these choices poses inductive risks for the scientist, a chance that a decision could be the wrong one and thus will incur the consequences of error.

For example, Douglas (2000) describes a case of scientists attempting to characterize the changes in rat liver slides. The rats had been dosed with an important environmental contaminant (dioxin) for 2 years at various levels and then killed and autopsied. Changes to the livers, and whether the rats' livers exhibited carcinogenic changes, were assessed visually. But some of the slides were interpreted differently by different toxicologists. (Douglas [2000] discusses three different sets of evaluations taking place over the course of 12 years.) There were clear inductive risks to describing a liver slide as having a malignant versus a benign change or even no substantial change at all. Although one might balk at the need for interpretation of the slides and the expert disagreement regarding it, doing the study over again and culturing the liver tissues would be prohibitively expensive (costing millions of dollars). Weighing the consequences of error could help scientists decide how much evidence was enough—or how sure they needed to be—to assign any particular characterization.

The value-free ideal is thus too restrictive in excluding values from moments of inference in science. It is also too permissive in that the epistemic values the value-free ideal allows are given too much free rein. Consider, for example, the epistemic value of simplicity. Should scientists select a theory because it is simple or elegant? Only if the world actually were simple or elegant, which it only sometimes is. Other times it is complex and messy. To allow such a value to play the same role as evidence gives too much weight to such a value and moves science away from its core empiricism (Douglas 2009, ch. 5). Because scientists have responsibilities to be neither reckless nor negligent, and thus to consider and weigh the consequences of mistakes, and because scientists always confront inductive risk in doing studies and making claims, the value-free ideal is the wrong ideal. It is also the wrong ideal because it allows epistemic values too much ability to influence science. Taken together, these arguments seem destructive of the value-free ideal *qua* ideal.

Nevertheless, we cannot do without some ideal for values in science. The descriptive challenge and resultant emphasis on social structure in science are insufficient sources of alternative ideals. In addition to norms for social structures, we need norms to address the concern that values could be used in place of evidence. It would be deeply problematic for the epistemic authority of science if we were to decide that what best characterizes the rat liver slide is what we would like it to be or we were to decide that a drug is safe and effective (or not) because we would like it to be (or not). In order to understand what is or is not a good use of values in scientific reasoning, we need an alternative ideal for values in science.

4 ALTERNATIVE IDEALS FOR VALUES IN SCIENCE

There are several contenders for a replacement ideal once the value-free ideal is set aside. As noted earlier, allowing any values to play any role whatsoever seems to open the door to wishful thinking and a complete undermining of the value of science to society. What to say in the face of these concerns has brought out a number of distinct arguments. Some of these ideals depend on rethinking the terrain of values that can influence science, particularly the nature of epistemic values.

4.1 Rethinking Epistemic Values in Science

As noted in section 3.2, some philosophers have challenged the boundary between epistemic and nonepistemic values as being too porous to support the value-free ideal. But with the value-free ideal set aside, there has been renewed interest in a possible topography of values to articulate a new ideal.

One approach has been to redefine what is meant by epistemic and to acknowledge the context dependence of epistemic values. In his 2010 paper, Daniel Steel argues that epistemic values should be defined as those values that "promote the attainment of truth" (Steel 2010, 15). Some of these values Steel suggests are "intrinsic," that is, "manifesting that value constitutes an attainment of or is necessary for truth" (Steel 2010, 15). Examples of such values include empirical adequacy and internal consistency. This is a rather small set of values, as others such as Laudan (2004) and Douglas (2009) (where they are called "epistemic criteria") have noted. Extrinsic epistemic values, on the other hand, "promote the attainment of truth without themselves being indicators or requirements of truth" (Steel 2010, 18). Examples include simplicity, external consistency, testability, and open discourse in science. Whether such values actually promote the attainment of truth will depend on the context. For example, external consistency promotes truth only when the additional theories one is being consistent with are true. With this account of epistemic values in hand, Steel argues that nonepistemic values are allowable in science only so long as they do not "hinder or obstruct the attainment of truth" (Steel 2010, 25).

Douglas (2013), on the other hand, has articulated a different terrain for the traditional epistemic values. Starting with a core set of values (similar to Steel's intrinsic epistemic values), one can then make further distinctions among the remaining values from the Kuhnian set. For example, one can distinguish between values that are instantiated by the claims or theories on their own and values that are instantiated by a scientific claim and the evidence that supports it. Douglas (2013) notes that conflating these two different ways in which a value is instantiated in science has led to some confusion. For example, simplicity applied to and exemplified in a theory on its own is properly not epistemic, but instead at best a reflection of the fact that simplicity makes theories easier to work with (Douglas 2009, 107). On the other hand, simplicity instantiated by a theory in relationship to evidence (e.g., a best fit curve) is in fact a good reason to adopt a theory and is properly epistemic (i.e., indicative of truth) (Douglas 2013; Forster and Sober 1994). With this distinction in hand, one can distinguish between epistemic values (reliable guides for inference) and cognitive values (attributes of theories that make them productive or easier to use). These complexities in the terrain for epistemic values are then reflected in replacement ideals.

4.2 Distinguishing Roles for Values in Science

Douglas (2009) both argues against the value-free ideal for science and proposes an alternative ideal to properly constrain values in science. The alternative ideal depends on a distinction between two kinds of roles for values in science: a direct role and an indirect role. In the direct role, the values serve as reasons in themselves for the choices being made and directly assess those choices. For example, when a scientist decides to pursue a particular project because she is interested in the subject, her interest in the subject is a value judgment that serves in a direct role for the decision to pursue it. She pursues a project *because* of her values. Similarly, when a scientist decides not to use a particular methodology because it would cause extreme suffering in the primates on which he is experimenting, his moral concern for suffering in primates is a value that serves in a direct role in his decision not to use that methodological approach. He decides not to use that method *because* of his values. Douglas (2009) argues that the direct role is acceptable (indeed, often laudable) in deciding which projects to pursue, in making methodological choices, and in deciding what to do with the science produced (i.e., in the "external" phases of science).

This direct role is to be distinguished from an indirect role for values in science. In the indirect role, the values help determine whether the evidence or reasons one has for one's choice are sufficient (sufficiently strong or well supported) but do not contribute to the evidence or reasons themselves. In the indirect role, values assess whether evidence is sufficient by examining whether the uncertainty remaining is acceptable. A scientist would assess whether a level of uncertainty is acceptable by considering what the possible consequences of error would be (false positives or false negatives) by using values to weigh those consequences. In this way, the indirect role captures the concerns over inductive risk by using values to weigh the consequences of error in order to assess evidential sufficiency. The values do not contribute to the weight of evidence because that would be the values serving in a direct role. Douglas (2009) argues that in the "internal" phases of science, in evidence characterization and interpretation (i.e., when scientists decide what to make of the available evidence), (nonepistemic) values should play only an indirect role.

In short, the alternative to the value-free ideal is that (nonepistemic) values should be constrained to the indirect role when scientists are characterizing phenomena and assessing hypotheses with respect to available evidence. If values were allowed to play a direct role, the values would be reasons to accept a hypothesis or to characterize phenomena, and this would shift science away from empiricism. Most problematically, values in a direct role during evidential assessment would be equivalent to allowing wishful thinking into the heart of science. If values could play a direct role in the assessment of evidence, a preference for a particular outcome could act as a reason for that outcome or for the rejection of a disliked outcome. In the indirect role, acting to assess evidential sufficiency, values do not have such potency.

Kevin Elliott (2011b, 2013) has raised a conceptual challenge for the direct versus indirect role distinction. He has argued that it is not clear what the distinction refers to, whether it is a distinction about the logical role for values or a distinction about which kinds of consequences are important: intended or unintended. Conceptually, the distinction is first and foremost about the logical role for values in science, in particular about assessing whether evidence is sufficient for a claim (or specific phenomena are sufficiently captured by a description). One relevant consideration for assessing evidential sufficiency is whether the uncertainty regarding the claim is acceptable. In assessing the acceptability of uncertainty, one considers the two directions of error, false positive or false negative, and the potential consequences of these errors. Values are used to weigh the seriousness of these consequences.

One may wonder, however, what the role is for the epistemic values (as described earlier—the genuinely epistemic values such as internal consistency, empirical adequacy, or explanatory power of a theory with respect to evidence). These values should be conceived of as part of the epistemic evaluation of science—they help us assess how strong the evidence is in relationship to the claim being made or whether the claim is even an adequate contender for our consideration (Douglas 2009, 2013). As such, they are central to the assessment of *how much* uncertainty we have (and less about the *acceptability* of such uncertainty, where inductive risk considerations arise). They have what Matthew Brown has called "lexical priority," along with the evidence (Brown 2013a). Once these values have been utilized to assess how much uncertainty we think there is, the other values (social, ethical, *and cognitive*) must compete to help weigh whether the evidence (and its relationship to the theory) is enough. It is here that inductive risk is crucial, that the indirect role is central, and that a direct role is prohibited.

One can ask further, however, whether this distinction between the roles for values in science is useful in assessing what scientists do. Because the distinction between a direct role (the value serves as a reason for a choice) and an indirect role (values are used to assess the sufficiency of evidence) depends on the reasoning of the scientist, and the actual reasoning is not always transparent—even to the scientists themselves—the distinction may not be helpful in policing acceptable and unacceptable roles for values in science in practice (Elliott 2011b; Hicks 2014; Steel and Whyte 2012). Daniel Steel and Kyle Powys Whyte have argued that their alternative ideal, that "non-epistemic values should not conflict with epistemic values [both intrinsic and extrinsic, i.e., any value that is truth-promoting] in the design, interpretation, or dissemination of research that is practically and ethically permissible," does better than an ideal that is based on the roles that values play in the reasoning of scientists (Steel and Whyte 2012, 164). Their essay examines the methodological choices and interpretations of evidence in environmental justice research. (Unfortunately, the test is not as incisive as one could hope for because both direct and indirect roles for values are allowable for methodological choices in science—see later discussion.) But, in order to apply this ideal, we need to know which extrinsic epistemic values are in play in a given context, which means knowing which ones will promote the attainment of the truth. Depending on where the truth lies, this can change from context to context and be only apparent in hindsight, well after the dust has settled (Elliott 2013).

One need not intuit, however, a scientist's reasoning in each particular instance for the role-based ideal to be useful. (No ideal for reasoning can meet such a standard.) Instead, one can look at patterns of argumentation to see whether, for example, scientists respond in appropriate ways to new pieces of evidence (Douglas 2006). If values serve in an appropriate indirect role, new evidence (particularly evidence that reduces uncertainty) should shift a scientist's claims. If a scientist remains intransigent, either he must espouse particularly strong values (a strong concern for the impacts of a false positive, for example, with no concern for a false negative), or his values are playing an improper direct role. In either case, there are grounds for rejecting that particular scientist's views. If we disagree with their weighing of inductive risk, we should not rely on them (Douglas 2015). If they are using values in an improper role, we should not rely on them either. Finally, the purpose of an ideal is as much about guiding our own reasoning as for policing the behavior of others. We need some sense of when and how using values is appropriate in science, without the benefit of hindsight, before we know where the truth lies.

4.3 Arguing for the Right Values in Science

Another approach taken to the problem of a replacement ideal has been to argue for scientists adopting the right values. For example, in her book *Philosophy of Science After Feminism*, Janet Kourany argues for an ideal of socially responsible science (Kourany 2010). In this ideal, science should meet jointly both epistemic and social standards. It is science instantiating the right values that is important to Kourany, whether those values influence the external or internal parts of science. As Kourany argues, the ideal of socially responsible science "maintains that sound social values as well as sound epistemic values must control every aspect of the research process" (Kourany 2013, 93–94). One can view Kourany's position as requiring "the joint necessity of evidence and values" (Brown 2013b, 68). Getting the values right is thus central to having a responsible science. Kourany argues for more intensive discussion both within scientific specialties and with the groups affected by scientific communities for developing the codes that will guide responsible science.

Kevin Elliott (2013) has argued that rather than focus on the different roles that values can play in science, we should focus on the various goals scientists have in addition to epistemic goals and ensure those goals are both clearly articulated and that the values employed further them (see also Elliott and McKaughan 2014). As he says, "a particular value can appropriately influence a scientist's reasoning in a particular context only to the extent that the value advances the goals that are prioritized in that context" (Elliott 2013, 381). If a scientist prefers a speedy test to an accurate test, then she should make that clear and use the faster test. Such an approach "allows non-epistemic values to count as reasons for accepting a model or hypothesis as long as they promote the goals of the assessment" (Elliott 2013, 381).

Daniel Hicks further argues that "philosophers of science should undertake a deeper engagement with ethics" in order to grapple with a fuller picture of the aims of science and the aims of other endeavors of which science is a part (Hicks 2014, 3; see also Anderson 2004). Hicks thinks we should analyze whether certain values, particularly nonepistemic values, should have priority in a particular context based on what the constitutive goals are for that context. Thus, in analyzing the machinations of the pharmaceutical industry, Hicks suggests that placing profit motive over good science only appears to make sense for the industry but is, in fact, self-undermining because the ultimate aim of the industry is not profit but good health (Hicks 2014). Distorting science is wrong not because of the harm it does to science or our epistemic enterprise, but because doing so subverts the correct order of values, placing profits over health. It is this sort of analysis that Hicks thinks we can use to assess whether the influence of values is legitimate or not in science—analyzing whether we have the right values prioritized.

Matthew Brown has argued for a return to a more pragmatic image of inquiry (drawn from John Dewey), one that considers the incorporation of values in science a necessary and ongoing feature (Brown 2012). Brown further urges us to be wary of accepting the "lexical priority of evidence," both because evidence may in fact be suspect and because value judgments are often the result of careful inquiry and reflection, even grounded in empirical information, rather than mere subjective preferences (Anderson 2004; Brown 2013a). Brown urges that we coordinate evidence, values, and theory in our inquiries and that all must fit together well in a way that "resolves the problem that spurred the inquiry" at the start of the process (Brown 2013a, 837). Success is defined as resolving the problem that spurred the inquiry, and it assures us that the parts of inquiry have ultimately worked properly.

One might wonder whether these approaches undermine the particular empirical character of science by suggesting that values can play an even broader array of roles in scientific inference as long as they are the right values. For example, even with the requirement of joint satisfaction of epistemic and ethical considerations for theory acceptance, we can be faced with situations where the evidence strongly supports a theory we find unwelcome because of our value commitments. Should we reject such a theory solely because it does not live up to our value judgments (even though the epistemic reasons for it are strong)? Whether such approaches can avoid the problems of wishful thinking (or excessively motivated reasoning) remains to be seen. Regardless, if one examines the broader set of contexts in which values influence science, getting the values right proves to be paramount.

5 More Locations for Values in Science

Thus far, we have discussed the value-free ideal, its challengers, and contenders for replacements, including further examinations of the nature of epistemic and cognitive values. This debate has focused primarily on the assessment of evidence and the crucial moments of inference in science. But values play a role in science at far more than these "internal" aspects of science. Values also help shape the direction of research, the methodologies pursued, and the ways in which research is disseminated and applied. Whether or not one judges the challenges to the value-free ideal as successful, the debate about values in inference only scratches the surface of the full range of ways in which values influence scientific practice. And this broader array has proved to crucially influence what evidence we have to examine and which theories or hypotheses get pursued. Regardless of the proper place of values in inference, the broader array of locations of values in science deeply shapes what we claim to know (Okruhlik 1994). Depending on what research we do and how we collect evidence, our inferences will be substantially altered. Considering this broader array raises an interesting set of challenges for philosophers of science. Restricting roles for values across this array is inappropriate—values play both direct and indirect roles legitimately in the locations discussed herein. Whether the emphases on diversity arising from the social epistemic approaches to science are sufficient remains an open question.

5.1 Values in Research Direction

Values deeply shape which projects scientists pursue. From societal decisions of which projects to fund, to corporate decisions for which projects to pursue (and how to pursue them), to individual decisions for where to place effort, the evidence we eventually have to consider depends on scientists', institutional, and societal values. The decisions here can be fraught. For example, decisions on whether (and under what conditions) to pursue potentially dangerous research must weigh the potential epistemic and societal benefits of pursuit against the potential societal risks of pursuit. (Such research is often called "dual-use" but sometimes the danger has little to do with intentional uses.) In examples such as "gain-of-function" research on pathogenic viruses, such weightings involve fraught and highly charged value-based debates (Casadevall, Howard, and Imperiale 2014; Douglas 2014a, 975–977; Evans 2014; Lipsitch 2014).

Additional examples of the importance of values (and institutional structures instantiating values) in shaping scientific research agendas are found in the work of Hugh Lacey and Philip Kitcher, among others. Although a staunch proponent of the value-free ideal for scientific inference (for whether to accept or reject scientific theories), Lacey has argued for greater attention to which research projects are taken up, noting that what looks like a good project can be deeply shaped by (potentially problematic and one-sided) social values (Lacey 1999, 2005). He has focused in particular on agricultural research, comparing the intensive efforts to create mass-marketed genetically modified organisms (GMOs) with more locally based agro-ecology research and finds the imbalance of efforts problematic. Kitcher has elucidated tensions regarding whether some research should be pursued at all, questioning whether democratic societies should support research that seeks to empirically undermine presumptions of shared equality central to democratic systems (Kitcher 2001). These questions of which knowledge is worth pursuing (as evaluated from within liberal democracies) remain pressing.

5.2 Values and Methodological Choices

In addition to which projects to pursue, among the most important decisions scientists make concern the methodologies they use. These can be general questions, such as whether to utilize human subjects in a controlled setting or whether to observe humans and attempt to find a naturally occurring control cohort. These can be very specific, detailed questions, such as whether to utilize a particular source of information (e.g., death certificates versus medical records, or census tracts versus zip codes; see Steel and Whyte 2012) as being sufficiently appropriate or reliable for a study. At both the specific and general levels of questions, values play a vital and potent role.

The importance of values in methodological choices can be readily seen in the ethical prohibitions we have for some options (Douglas 2014*b*). Despite the knowledge we might gain, we rightly do not think it acceptable to round people up involuntarily and submit them to various experimental conditions. Regardless of whether we think the conditions might harm or help the people subjected to them, to do so without their informed consent is deeply unethical. This reflects a strongly held (and well-justified) concern for human autonomy that the value of knowledge does not trump. Beyond human subject research, ethical considerations have the potential to restrict other areas of research as well, arising, for example, from ethical concern over inflicting unnecessary suffering on animals. What counts as an acceptable or necessary level of suffering is contentious.

The ability of social and ethical values (such as the concern over human autonomy or animal suffering) to restrict the avenues of research open to scientists speaks to the potency of values in methodological choices. Values can legitimately play a powerful and direct role in deciding which research projects can and cannot be pursued. But, at the same time, we do not want social and ethical values to determine the epistemic outcomes of research. We know there are techniques scientists can employ to "game" the research process, skewing results in a particular favor. How should the pervasive, potent, and potentially problematic influence of values here be managed?

This is not merely an academic question. As detailed by Michaels (2008) and McGarity and Wagner (2008), scientists have perfected ways of structuring research projects so that they appear to be an open-ended process of inquiry to the casual observer but are, in fact, set up to generate a particular result. Torsten Wilholt, for example, describes a case where researchers were ostensibly testing for the estrogenic potential of particular synthetic substances (due to concern over hormone disruption) but ensured a negative outcome to the study by testing the substances on a mouse strain known for estrogen insensitivity (Wilholt 2009). Other ways of ensuring particular outcomes include using an insufficient exposure or dosage level so that no difference between the dosed and control group appears, doing multiple trials and publishing only the successful ones (an approach trial registries are attempting to thwart), and using inappropriate testing conditions so that results are irrelevant to actual practice (Wickson 2014).

So, although values can legitimately restrict the methodological choices of scientists, we don't want such values to shape research so that genuine tests are avoided. For this decision context, discussing roles for values is not helpful. What is needed is a clear examination of the values employed and whether they are the right ones.

If researchers choose to use a noncoercive methodology when working with human subjects for ethical reasons, we should applaud them not because of the role the values are (or are not) playing, but because the researchers are displaying or using the right values. And if researchers deliberately select a research method so that particular desired results are produced (i.e., the results are practically guaranteed by the methods chosen so no genuine test is possible), we should excoriate them not for the role values are playing but for the values on display; in this case, an abuse of the authority of science while giving short shrift to the value of genuine discovery, for whatever other interest or gain being sought.

Yet there is still difficult conceptual terrain to be explored. As Steel and Whyte (2012) note, it is acceptable for ethical and moral considerations to trump epistemic ones in shaping methodological choices. How extensive are these considerations? How are we to weigh, in cases where the values are contested—for example, in potentially dangerous research (because the methods are risky to humans or to the environment) or in cases where animal suffering must be inflicted in order to pursue a particular research program—which is more important, the value of the knowledge to be gained or the ethical issues at stake? Values in methodological issues force us to confront these difficult problems.

5.3 Values Everywhere Else

Values in inference, in decisions about which projects to pursue, and in decisions about which methodologies to employ do not exhaust the places where values influence science. Consider these additional locations:

- *The language we choose to use can embed values in science.* Sometimes these choices hinder or hamper the ability of scientists to do good work (as in the case of using the term "rape" to describe duck behavior, Dupré 2007), and sometimes the value-inflected language choice can make research epistemically and socially stronger (as in the decision to call a collection of behaviors "spousal abuse," Root 2007). This issue does not belong to social science in particular; natural science domains can also have value-inflected language debates (Elliott 2011a). An analysis of how values embedded in language, or in the "thick concepts" that are often central topics of investigation, is needed, including the ideals that should guide their use. (Alexandrova 2014).

- *Values can play an important role in the construction and testing of models.* As Biddle and Winsberg (2010) note, models, especially models of great complexity, often have a particular historical development that reflects certain priorities (e.g., scientists want to reflect temperature patterns more than precipitation patterns), which then become embedded in how the models work. These priorities are value judgments, and they are value judgments that subsequently affect how different modules in the models combine and how we estimate uncertainties regarding the models. Winsberg (2012) further argues that we cannot unpack all the value judgments embedded in models, nor can we capture them all in uncertainty estimates, nor can we expect scientists to be fully aware of what they are doing. What this means for the use of models—particularly complex computer models—in scientific assessments of public import remains unclear, except that more work needs to be done (Parker 2014).

- *Values are crucial in deciding how to use and disseminate science.* Even if potentially dangerous research is pursued, debates about whether to disseminate the knowledge gleaned are as fraught with value judgments as decisions about whether to pursue that research. Despite a baseline Mertonian norm of openness, some research should potentially be kept confidential (Douglas 2014a). This can be true because of a desire to avoid weapons proliferation or because of a desire to protect the privacy of individuals involved in the research.

6 IMPLICATIONS FOR SCIENCE IN SOCIETY

What is perhaps most interesting about the wide-ranging discussion concerning values in science is how important the conceptualization of science and society is for what one thinks the norms should be. The normative challenge to the value-free ideal—that what counts as sufficient evidence could vary by context and the implications of error—is perfectly rational. Indeed, it has been noted by epistemologists such as John Heil (1983) and the burgeoning interest in pragmatic encroachment among epistemologists (Miller 2014b). If one is to object to a role for values in science or to the presence of particular values in science, one has to do so on the grounds that the role of science in society makes the presence of values problematic (e.g., Betz 2013). And this is where the debate is properly joined. Given the rationality of and need for values in science, how should we conceive of the relationship between science and society? Does the importance of science in society place additional constraints or requirements on the presence of values in science?

The inductive risk argument gains its traction because of the view of the responsibilities of scientists articulated in Douglas (2003, 2009). Is this the right view? More specifically, do particular institutional contexts in which scientists operate (e.g., within advising structures, academia, private industry, government science) alter the nature of the role responsibilities scientists have, even in ways that alter their general responsibilities? If institutional demands clash with general moral responsibilities, what should scientists do?

The relationship between science and democratic societies also needs further exploration (following on the work of Brown [2009] and Kitcher [2001, 2011]). Should scientists' judgments be free of social and ethical values to preserve democratic ideals (as Betz [2013] has argued), or should they be value explicit (Douglas 2009; Elliott and Resnik 2014) or embedded in value-laden collaborations to preserve democratic ideals (Douglas 2005; Elliott 2011a)? Stephen John has argued that having varying standards for scientists' assertions depending on the expected audience is an unworkable approach (John 2015) and has argued instead for uniform high evidential standards for the public assertions of scientists. Is this the right standard, or should scientists' assertions be more nuanced, carrying with them clear expressions of why scientists think the evidence is enough, so that audience members can interpret them properly? Would such an approach be more workable than the complex institutional structures John proposes (e.g., distinguishing between public and private assertions, generating different norms for each, and structuring institutions accordingly)? Furthermore, what is the correct basis for trust in science by the broader society, and what role should values play in science given that basis (Elliott 2006; Wilholt 2013)?

In short, what the values in science debate forces us to confront is the need to theorize—in a grounded way by assessing actual science policy interfaces, actual practices of science communication, and actual research decision contexts—what the proper role is of science in society. It is exactly where philosophy of science should be.

ACKNOWLEDGMENTS

My thanks to Ty Branch, Matt Brown, Kevin Elliott, Dan Hicks, Paul Humphreys, and Ted Richards for their invaluable advice and assistance with this paper. Deficiencies remaining are mine alone.

REFERENCES

Alexandrova, A. (2014). "Well-being and Philosophy of Science." *Philosophy Compass* 10(3), 1–13.

Anderson, E. (2004). "Uses of Value Judgments in science: A General Argument, with Lessons from a Case Study of Feminist Research on Divorce." *Hypatia* 19(1), 1–24.

Betz, G. (2013). "In Defence of the Value Free Ideal." *European Journal for Philosophy of Science* 3(2), 207–220.

Biddle, J. (2013). "State of the Field: Transient Underdetermination and Values in Science." *Studies in History and Philosophy of Science Part A* 44(1), 124–133.

Biddle, J., and Winsberg, E. (2010). "Value Judgements and the Estimation of Uncertainty in Climate Modeling." In P. D. Magnus and J. Busch (eds.), *New waves in philosophy of science* (New York: Palgrave Macmillan), 172–197.

Brown, M. B. (2009). *Science in Democracy: Expertise, Institutions, and Representation* (Cambridge, MA: MIT Press).

Brown, M. J. (2012). "John Dewey's Logic of Science." *HOPOS: The Journal of the International Society for the History of Philosophy of Science* 2(2), 258–306.

Brown, M. J. (2013a). "Values in Science Beyond Underdetermination and Inductive Risk." *Philosophy of Science* 80(5), 829–839.

Brown, M. J. (2013b). "The Source and Status of Values for Socially Responsible Science." *Philosophical Studies* 163(1), 67–76.

Casadevall, A., Howard, D., and Imperiale, M. J. (2014). "An Epistemological Perspective on the Value of Gain-of-Function Experiments Involving Pathogens With Pandemic Potential." *mBio* 5(5), e01875–e01914.

Churchman, C. W. (1948). "Statistics, Pragmatics, Induction." *Philosophy of Science* 15(3), 249–268.

Douglas, H. (2000). "Inductive Risk and Values in Science." *Philosophy of Science* 67(4), 559–579.

Douglas, H. (2003). "The Moral Responsibilities of Scientists (Tensions Between Autonomy and Responsibility)." *American Philosophical Quarterly* 40(1), 59–68.

Douglas, H. (2005). "Inserting the Public into Science." In S. Massen and P. Weingart (eds.), *Democratization of Expertise? Exploring Novel Forms of Scientific Advice in Political Decision-Making* (Dordrecht: Springer), 153–169.

Douglas, H. (2006). "Bullshit at the Interface of Science and Policy: Global Warming, Toxic Substances and Other Pesky Problems." In G. L. Hardcastle and G. A. Reisch (eds.), *Bullshit and Philosophy* (Chicago: Open Court Press), 215–228.

Douglas, H. (2009). *Science, Policy, and the Value-Free Ideal* (Pittsburgh, PA: University of Pittsburgh Press).

Douglas, H. (2011). "Facts, Values, and Objectivity." In I. C. Jarvie and J. Zamora-Bonilla (eds.), *The Sage Handbook of the Philosophy of Social Science* (London: Sage Publications), 513–529.

Douglas, H. (2013). "The Value of Cognitive Values." *Philosophy of Science* 80(5), 796–806.

Douglas, H. (2014a). "The Moral Terrain of Science." *Erkenntnis* 79(5), 961–979.

Douglas, H. (2014b). "Values in Social Science." In N. Cartwright and E. Montuschi (eds.), *Philosophy of Social Science: A New Introduction* (Oxford: Oxford University Press), 162–182.

Douglas, H. (2015). "Politics and Science Untangling Values, Ideologies, and Reasons." *The Annals of the American Academy of Political and Social Science* 658(1), 296–306.

Dupré, J. (2007). "Fact and Value." In H. Kincaid, J. Dupré, and A. Wylie (eds.), *Value-Free Science?: Ideals and Illusions* (Oxford: Oxford University Press), 27–41.

Elliott, K. C. (2006). "An Ethics of Expertise Based on Informed Consent." *Science and Engineering Ethics* 12(4), 637–661.

Elliott, K.C. (2011a). *Is a Little Pollution Good for You? Incorporating Societal Values in Environmental Research* (New York: Oxford University Press).

Elliott, K. C. (2011b). "Direct and Indirect Roles for Values in Science." *Philosophy of Science* 78(2), 303–324.

Elliott, K. C. (2013). "Douglas on Values: From Indirect Roles to Multiple Goals." *Studies in History and Philosophy of Science Part A* 44(3), 375–383.

Elliott, K. C., and McKaughan, D. J. (2014). "Nonepistemic Values and the Multiple Goals of Science." *Philosophy of Science* 81(1), 1–21.

Elliott, K. C., and Resnik, D. B. (2014). "Science, Policy, and the Transparency of Values." *Environmental Health Perspectives* 122(7), 647–650.

Evans, N. G. (2014). "Valuing Knowledge: A Reply to the Epistemological Perspective on the Value of Gain-of-Function Experiments." *mBio* 5(5), e01993-14.

Fausto-Sterling, A. (1985). *Myths of Gender: Biological Theories About Women and Men* (New York: Basic Books).

Forster, M., and Sober, E. (1994). "How to Tell When Simpler, More Unified, or Less Ad Hoc Theories Will Provide More Accurate Predictions." *British Journal for the Philosophy of Science* 45(1), 1–35.

Haraway, D. J. (1989). *Primate Visions: Gender, Race, and Nature in the World of Modern Science* (New York/London: Routledge).

Harding, S. G. (1986). *The Science Question in Feminism* (Ithaca, NY: Cornell University Press).

Harding, S. G. (1991). *Whose Science? Whose Knowledge?* (Ithaca, NY: Cornell University Press).

Heil, J. (1983). "Believing What One Ought." *Journal of Philosophy* 80(11), 752–765.

Hempel, C. G. (1965). "Science and Human Values." In C.G. Hempel (ed.), *Aspects of Scientific Explanation* (New York: Free Press), 81–96.

Hicks, D. J. (2014). "A New Direction for Science and Values." *Synthese* 191(14), 3271–3295.

Holter, B. (2014). "The Epistemic Significance of Values in Science." University of Calgary: http://hdl.handle.net/11023/1317

Intemann, K. (2005). "Feminism, Underdetermination, and Values in Science." *Philosophy of Science* 72(5), 1001–1012.

Intemann, K. (2009). "Why Diversity Matters: Understanding and Applying the Diversity Component of the National Science Foundation's Broader Impacts Criterion." *Social Epistemology* 23(3–4), 249–266.

Jeffrey, R. C. (1956). "Valuation and Acceptance of Scientific Hypotheses." *Philosophy of Science* 23(3), 237–246.

John, S. (2015). "Inductive Risk and the Contexts of Communication." *Synthese* 192(1), 79–96.

Keller, E. F., and Longino, H. E. (eds.). (1996). *Feminism and Science* (Oxford: Oxford University Press).

Kitcher, P. (2001). *Science, Truth, and Democracy* (Oxford: Oxford University Press).

Kitcher, P. (2011). *Science in a Democratic Society* (Amherst, NY: Prometheus Books).

Kourany, J. (2010). *Philosophy of Science After Feminism* (Oxford: Oxford University Press).

Kourany, J. A. (2013). "Meeting the Challenges to Socially Responsible Science: Reply to Brown, Lacey, and Potter." *Philosophical Studies* 163(1), 93–103.

Kuhn, T. S. (1977). "Objectivity, Value Judgment, and Theory Choice." In T. Kuhn (ed.), *The Essential Tension* (Chicago: University of Chicago Press), 320–339.

Lacey, H. (1999). *Is Science Value Free?: Values and Scientific Understanding* (New York: Routledge).

Lacey, H. (2005). *Values and Objectivity in Science: The Current Controversy About Transgenic Crops* (Lanham, MD: Rowman and Littlefield).

Laudan, L. (1984). *Science and Values: The Aims of Science and their Role in Scientific Debate* (Oakland: University of California Press).

Laudan, L. (2004). "The Epistemic, the Cognitive, and the Social." In P. Machamer and G. Wolters (eds.) *Science, Values, and Objectivity* (Pittsburgh, PA: University of Pittsburgh Press), 14–23.

Lentsch, J., and Weingart, P. (eds.). (2009). *Scientific Advice to Policy Making: International Comparison* (Farmington Hills, MI: Barbara Budrich Publishers).

Lentsch, J., and Weingart, P. (eds.). (2011). *The Politics of Scientific Advice: Institutional Design for Quality Assurance* (Cambridge: Cambridge University Press).

Levi, I. (1960). "Must the Scientist Make Value Judgements?" *Journal of Philosophy* 57(11), 345–357.

Lipsitch, M. (2014). "Can Limited Scientific Value of Potential Pandemic Pathogen Experiments Justify the Risks?" *mBio* 5(5), e02008-14.

Longino, H. E. (1990). *Science as Social Knowledge: Values and Objectivity in Scientific Inquiry* (Princeton, NJ: Princeton University Press).

Longino, H. E. (1995). "Gender, Politics, and the Theoretical Virtues." *Synthese* 104(3), 383–397.

Longino, H. E. (1996). "Cognitive and Non-Cognitive Values in Science: Rethinking the Dichotomy." In L. H. Nelson and H. Nelson (eds.), *Feminism, Science, and the Philosophy of Science* (Dordrecht: Kluwer), 39–58.

Longino, H. E. (2002). *The Fate of Knowledge* (Princeton, NJ: Princeton University Press).

Magnus, P. D. (2013). "What Scientists Know Is Not a Function of What Scientists Know." *Philosophy of Science* 80(5), 840–849.

Martin, E. (1996). "The Egg and the Sperm: How Science Has Constructed a Romance Based on Stereotypical Male-Female Roles." in E. F. Keller and H. Longino (eds.), *Feminism and Science* (Oxford: Oxford University Press), 103–117.

McGarity, T. O., and Wagner, W. E. (2008). *Bending Science: How Special Interests Corrupt Public Health Research* (Cambridge, MA: Harvard University Press).

McMullin, E. (1983). "Values in Science." In P. D. Asquith and T. Nickles (eds.), *PSA: Proceedings of the Biennial Meeting of the 1982 Philosophy of Science Association vol. 1* (East Lansing, MI: Philosophy of Science Association), 3–28.

Michaels, D. (2008). *Doubt Is Their Product: How Industry's Assault on Science Threatens Your Health* (Oxford: Oxford University Press).

Miller, B. (2014a). "Catching the WAVE: The Weight-Adjusting Account of Values and Evidence." *Studies in History and Philosophy of Science Part A* 47, 69–80.

Miller, B. (2014b). "Science, Values, and Pragmatic Encroachment on Knowledge." *European Journal for Philosophy of Science* 4(2), 253–270.

Nelson, L. H. (1990). *Who Knows: From Quine to a Feminist Empiricism* (Philadelphia, PA: Temple University Press).

Okruhlik, K. (1994). "Gender and the Biological Sciences." *Canadian Journal of Philosophy* 24(supp1), 21–42.

Parker, W. (2014). "Values and Uncertainties in Climate Prediction, Revisited." *Studies in History and Philosophy of Science Part A* 46, 24–30.

Proctor, R. (1991). *Value-free Science?: Purity and Power in Modern Knowledge* (Cambridge, MA: Harvard University Press).

Rooney, P. (1992). "On Values in Science: Is the Epistemic/Non-Epistemic Distinction Useful?" In K. Okruhlik, D. Hull and M. Forbes (eds.), *PSA: Proceedings of the Biennial Meeting of the Philosophy of Science Association, Volume One* (Chicago: Philosophy of Science Association), 13–22.

Root, M. (2007). "Social Problems." In H. Kincaid, J. Dupré, and A. Wylie (eds.), *Value-Free Science?: Ideals and Illusions* (New York: Oxford University Press), 42–57.

Rudner, R. (1953). "The Scientist *Qua* Scientist Makes Value Judgments." *Philosophy of Science* 20(1), 1–6.

Staley, K. (2014). "Inductive Risk and the Higgs Boson," presented at Philosophy of Science Association Meetings, November 2014, Chicago Illinois.

Steel, D. (2010). "Epistemic Values and the Argument from Inductive Risk." *Philosophy of Science* 77(1), 14–34.

Steel, D., and Whyte, K. P. (2012). "Environmental Justice, Values, and Scientific Expertise." *Kennedy Institute of Ethics Journal* 22(2), 163–182.

Solomon, M. (2001). *Social Empiricism* (Cambridge, MA: MIT Press).

Wickson, F. (2014). "Good Science, Bad Researchers and Ugly Politics: Experimenting with a Norwegian Model," presented at the American Association for the Advancement of Science Annual Meeting, Feburary 2015, Chicago Illinois.

Wilholt, T. (2009). "Bias and Values in Scientific Research." *Studies in History and Philosophy of Science Part A* 40(1), 92–101.

Wilholt, T. (2013). "Epistemic Trust in Science." *British Journal for the Philosophy of Science* 64(2), 233–253.

Winsberg, E. (2012). "Values and Uncertainties in the Predictions of Global Climate Models." *Kennedy Institute of Ethics Journal* 22(2), 111–137.

PART III

NEW DIRECTIONS

CHAPTER 30

AFTER KUHN

PHILIP KITCHER

1 INTRODUCTION

Books in the philosophy of science do not typically sell more than a million copies. Kuhn's *The Structure of Scientific Revolutions* (Kuhn 1962) is the great exception, a work that has attracted readers from a wide variety of fields, as well as many outside the academy. Public figures have testified to its influence on them, and it has contributed a vocabulary that pervades discussion of changes in view. References to paradigms trip off many tongues and flow from many pens, often from people who have never read the original and almost always from those who are ignorant of Kuhn's own later ambivalence about his central term. His book has inspired a variety of intellectual trends, including some about which he voiced skepticism or distaste. Ian Hacking pithily sums up its importance at the beginning of his preface to the fiftieth anniversary edition (Kuhn 2012): "Great books are rare. This is one. Read it and you will see."

The monograph was originally intended as a contribution to the *International Encyclopedia of Unified Science*, a series of short works on topics in the philosophy of the sciences aimed at presenting the ideas of logical empiricism (as it evolved out of logical positivism) and allied themes in American pragmatism (initially an equal partner in the enterprise but later playing a reduced role). Kuhn's assigned task was to provide a concise picture of the historical development of the sciences, and, although his manuscript was considerably longer than the standard limit, the editors were happy to publish it as the final volume in the series. Commentators have sometimes been puzzled by the editorial enthusiasm, viewing Kuhn's discussions as central to the demolition of the logical empiricist project. Recent studies of the filiations between Kuhn and such logical empiricists as Carnap and Frank have dissolved the perplexity (Earman 1993; Reisch 1991, 2005).

Given Kuhn's topic, it would be natural to suppose that his work has decisively reshaped both the history of science and the philosophy of science. Although there is a sense in which that assumption is correct, the impact of the monograph on the two fields it straddles is rather different from the changes it has wrought elsewhere—one will not find many (perhaps not any) historians or philosophers of science using Kuhn's central concepts to elaborate their views. Strictly speaking, Kuhn was the first and last Kuhnian. Yet, even though historians have been reluctant to parcel up the development of scientific fields into large discontinuous blocks, each dominated by a paradigm, today's history of science has been shaped by his emphases on the tacit knowledge possessed by scientists and on the reasonableness of those who resisted what are now conceived as major scientific advances. Within philosophy, there has been no such similar reshaping. Throughout most of the philosophy of the special sciences and much of general philosophy of science, Kuhn is rarely cited. The late 1960s witnessed attempts to build interdisciplinary study of "history and philosophy of science," but the venture has fallen out of fashion, with historians and philosophers agreeing to go their separate ways. Nevertheless, later sections of this article will attempt to identify areas of contemporary philosophy of science in which Kuhn's work has prompted new questions and launched significant inquiries that had been unjustly neglected. Professional philosophers of science would benefit from taking Hacking's advice.

2 THE BOGEYMAN

In the first years after its publication, *Structure* did not sell particularly well, but by the late 1960s it had attracted attention both from scientists and philosophers. Apparently these two audiences read different books that happened to share a title. Practicing scientists expressed their enthusiasm for a discussion that, unlike so much philosophy of science, seemed to understand the character of scientific work. The treatment of normal science as puzzle-solving resonated with their professional experience, as it continues to do for young scientists who read *Structure* for the first time. By contrast, philosophers of science reacted negatively, often harshly, to the "subjectivism" and "relativism" they discerned in the treatment of scientific revolutions. Israel Scheffler's (1967) was only one of a series of books by authors who saw themselves as fighting to restore a threatened order.

Scientists and philosophers focused on different parts of the same book. The former viewed the earlier sections (I–VIII) as supplying a rich account of scientific practice, far more subtle and nuanced than the pervasive tales of encounters between hypothesis *h* and evidence *e*. Their reading of the later sections was informed by a serious engagement with Kuhn's framework. Philosophical readers tended to be more impatient, seeing those early sections as a labored restatement of familiar, relatively trivial features of scientific research in contrived terminology. The real philosophical action lay in the sections on revolutions (IX–XIII), where provocations leapt from the page. After a revolution, "scientists are responding to a different world" (Kuhn 1962: 111); champions of different paradigms "necessarily talked through each other, and their debate was entirely inconclusive" (132); to change allegiance from one paradigm to another involves "a conversion experience that cannot be forced" (151); revolutionary change requires decisions that "can only be made on faith" (158). Despite the fact that Kuhn often wrote about the "reasonableness" of scientists and their decisions, even in the thick of scientific revolutions, there were enough remarks akin to those just cited to inspire charges that he was advocating a form of relativism threatening the justified authority of the sciences.

Kuhn strenuously denied the charges but to no avail. His defense was compromised by a common philosophical practice of associating him with one of his erstwhile Berkeley colleagues, a philosopher of science who also saw history as profoundly relevant, who also deployed the term "incommensurability" in talking about major scientific change and who gladly (ironically? playfully?) accepted labels Kuhn resolutely rejected ("relativist," "anarchist," "Dadaist"). Paul Feyerabend's seminal essays of the 1960s (Feyerabend 1962, 1965) already indicated the positions he would later espouse (Feyerabend 1970, 1975), and even before the publication of *Structure* he had opposed what he saw as Kuhn's conservatism. For most philosophers, however, the differences between the two were invisible. Late 1960s philosophy of science constructed a bogeyman, sometimes Kuhn-Feyerabend, sometimes Feyerabend-Kuhn. The bogeyman was so alarming that it had to be demolished. Articulating the ideas of the thinkers behind it and coming to a serious assessment of their positions would have to wait a decade, or three.

Kuhn-Feyerabend was supposed to have launched an attack on the rationality of science by revealing obstacles to rational decision at crucial moments in the history of the sciences. Using Kuhn's preferred vocabulary, different paradigms were claimed to be "incommensurable": this term, drawn from mathematics, was viewed as signaling the impossibility of any common measure for rival paradigms and thus as dooming any attempt at objective comparison of them. Hence Kuhn-Feyerabend claimed that scientific revolutions could not rationally be resolved.

Structure offers three different notions of incommensurability between paradigms. Before we distinguish these, it is helpful to consider Kuhn's conception of a paradigm. As has often been noted, Kuhn appears to use "paradigm" in a number of different ways, and a probing critique of his usage distinguished twenty-one senses (Masterman 1970; Kuhn clearly saw Masterman's analysis as an unusually constructive early response to his work—see his 2000: 167ff [originally in Lakatos and Musgrave 1970]). It is useful to simplify Masterman's distinctions by recognizing two major foci. Paradigms enter *Structure* as concrete scientific achievements (like Newton's account of gravitation and the orbits of the planets), serving members of the pertinent scientific community as exemplars to be imitated in their own work. Learning "how to go on" from a paradigm, the apprentice scientist acquires a body of tacit knowledge that is typically not fully articulated. Much later in *Structure*, however, Kuhn hopes to analyze the debates between competing paradigms in terms of the concepts, values, techniques, and methods that practitioners deploy, and this requires him to make that tacit knowledge explicit. Consequently, paradigms become identified with a rich set of elements that constitute the normal scientific practice they guide—the constellation Kuhn came to call the "disciplinary matrix." Making the tacit knowledge explicit is essential to his arguments about the kinds of incommensurability figuring in scientific revolutions. We should always recall, however, that charting those kinds of incommensurability in revolutionary debates is not something fully available to the participants in those debates—much of the knowledge remains tacit. In presenting the disciplinary matrix, an analyst makes explicit what the members of a scientific community find latent in the concrete scientific achievements that guide them.

Central to *Structure*, and to Kuhn's later work, is the notion of *conceptual* incommensurability. Kuhn was much impressed by the linguistic shifts that occur in scientific revolutions: remarking on the transition from Newton to Einstein, he claims that "the whole conceptual web whose strands are space, time, matter, force, and so on, had to be shifted and laid down again on nature whole" (Kuhn 1962: 149). Such shifts produce languages that are at cross purposes, that divide up natural phenomena in ways that do not map smoothly onto one another, and, as Kuhn later came to see it, that are not intertranslatable (Kuhn 2000: ch. 2). Incommensurability of this sort underlies the thesis that "communication across the revolutionary divide is inevitably partial" (Kuhn 1962: 149) and that proponents of different paradigms talk past one another. Philosophers obsessed with Kuhn-Feyerabend responded to the phenomenon of conceptual incommensurability by trying to show that it would not constitute an insuperable barrier to discussion and debate. Once that issue had been settled, they lost interest. As we shall see, Kuhn did not.

A second, related notion is that of *observational* incommensurability. Here Kuhn began from the insight (already developed in Sellars 1953 and Hanson 1958) that observation presupposes categories: seeing is always "seeing as" (Kuhn borrowed from Wittgenstein, as well as from Hanson). When champions of different paradigms stand in the same place, they may observe different things: as they watch the sunrise, Tycho sees the sun beginning its diurnal round, whereas Kepler perceives the horizon dipping to bring a static sun into view. To switch to another paradigm thus requires a "gestalt switch" after which a scientist "works in a different world" (Kuhn 1962: 135). Once again, those concerned with Kuhn-Feyerabend's attack on scientific rationality sought to show how researchers adopting different paradigms could share a body of observational evidence on the basis of which they might hope to resolve their disagreements. Typically they also aimed to rebut the apparently profligate multiplication of worlds.

Kuhn's third notion of incommensurability, *methodological* incommensurability, appeared to threaten common standards for assessing whatever evidence might be available. The notion of a paradigm as a concrete scientific achievement leads very naturally to the thought that paradigms structure a domain of scientific research, picking out certain types of problems as central: the inspiring work urges scientists to go and do likewise, effectively giving priority to a class of questions typified by the one originally resolved. Clashing paradigms are likely to disagree on which problems are important to solve, and Kuhn emphasizes that no paradigm ever resolves all the issues taken to lie in its domain. Hence he envisages a protracted dispute in which proponents of rival paradigms insist that their favorite settles the truly central questions (or at least some of them) and that the alternative fails to do so. We might say that the rivals have different methodological standards for evaluating the track records of the alternative normal scientific traditions. Methodological differences can arise when scientists share common aims but disagree on the best means for achieving them. But the kind of methodological incommensurability that particularly interested Kuhn, reflected in language he later brought to the discussion of scientific objectivity, supposes that the ends of the rival traditions differ—they espouse different *values* (Kuhn 1977: 320–339). Philosophers replying to the threat from Kuhn-Feyerabend aimed to show either how the methodological differences had been exaggerated or that enough common standards (values) remained to permit an orderly resolution of major scientific debates.

The bogeyman kept a significant number of philosophers busy for about two decades, until philosophers of science (and some scientists) became aware that scholars in other disciplines, particularly in the anthropology and sociology of science, were saying more radical things. By the mid-1980s, Kuhn's "relativism" (which he continued to disavow) had come to seem rather tame. We return to the ensuing "science wars" later. For the moment, the goal is to understand how the Kuhn-Feyerabend "challenge to rationality" was overcome.

3 The Bogey Laid (or Back to Kuhn?)

The easiest issue to resolve is the apparent obstacle posed by observational incommensurability. Assume not only that there are occasions on which the defenders of rival paradigms will naturally respond to the same phenomena by using different categories, but that this *always* happens. The Sellars-Hanson-Kuhn point correctly denies the existence of a neutral observation language, one that would be free of all theoretical presuppositions and thus available for formulating the evidence with respect to *any and every* paradigm. The fact that there is no tool available for performing all jobs does not imply, however, that there are jobs for which no tool is available—the nonexistence of a neutral observation language does not commit us to concluding that, in an instance of interparadigm debate, we cannot find a way of presenting the observational evidence that is neutral between the conflicting paradigms. The imagined case of Tycho and Kepler watching the dawn shows clearly how such local relative neutrality is possible. Both observers can retreat from their favored theoretical idiom, talking about the relative separation between sun and horizon, without presupposing which is genuinely moving. They can achieve descriptions of what they each see, descriptions both can accept, and thus find common starting points for reasoning about what is occurring. Those descriptions are not completely theoretically neutral, for they have presuppositions (e.g., that the observed yellow disk is a distant object). Agreement is reached because the presuppositions are shared. Moreover, that basic point indicates a general solution. In interparadigm debates, observational evidence is assembled by retreating from the usual rich theoretical descriptions to a level at which presuppositions are held in common.

Philosophers troubled by the specter of Kuhn-Feyerabend can leave observational incommensurability at this point, having overcome one obstacle to rational scientific decision. Others, intrigued by Kuhn's remarks about changing worlds, might press further (as the final section of this article attempts to do). The second potential barrier to orderly scientific change stems from conceptual incommensurability. Here the task for the would-be champion of rationality is to explain how scientists divided by their allegiance to rival paradigms can communicate with one another. A familiar riposte to Kuhn's claims about incommensurability (one he understandably found irritating) asks how the historian, speaking the language of contemporary science, is able to understand, and reconstruct, the content of sciences that lie on the far side of some revolutionary divide. Kuhn's explicit response to this, developed very clearly in some of the papers collected in Kuhn (2000), is that the languages of such sciences are not intertranslatable but that users of the languages can grasp the content of one another's claims by "going bilingual."

That response concedes enough to dissolve the threat philosophers discerned in *Structure*. Consider the following examples. Scientists adopting different stances in the Copernican revolution use "planet" in different ways: for heliocentrists, there is one more planet (Earth) than geocentrists would allow. Does that make for communicative troubles? Hardly. Perhaps there will be some initial stage of mutual confusion, but the rivals can quickly come to recognize the simple difference and make allowances.

The chemical revolution poses a more complex challenge. Traditional approaches to combustion supposed that bodies capable of burning share a common "principle," phlogiston, which is emitted to the atmosphere when they burn. Lavoisier, by contrast, saw combustion as a process in which something (oxygen) is absorbed from the atmosphere. In an important experiment, Lavoisier's scientific opponent (but personal friend) Joseph Priestley isolated a new gas (or "air") by gently heating the result of initially burning mercury. On Priestley's account, the mercury first emitted its phlogiston to form the "calx of mercury"; when this was later heated gently the phlogiston was reabsorbed, leaving ambient air that was "dephlogisticated." Priestley investigated the properties of this gas, eventually breathing it (it felt light to his lungs). Lavoisier, who learned how to do the experiment from reading Priestley's account of it, identified the gas as oxygen. Kuhn is entirely correct to suppose that the languages used by Priestley and Lavoisier are at cross purposes with one another. When Priestley talks of phlogiston, he is, from Lavoisier's viewpoint, failing to refer to anything in nature. What then could "dephlogisticated air" be? If we think of it as what we get when phlogiston is removed from the air then, because there is no phlogiston, there is no such thing (just as "the mother of Santa Claus" does not denote anyone). But reading Priestley's explanation of how he did his experiments, Lavoisier can recognize that his colleague is attaching the term "dephlogisticated air" to a gas he collected in a vessel, and on Lavoisier's own theory that gas is oxygen. Despite the fact that the languages are at odds, the two men can communicate.

Kuhn is quite right to declare that some scientific languages are not mutually intertranslatable, at least given one conception of translation. If translation requires a uniform process of substituting terms in one language for terms in another, we cannot translate Priestley's language into Lavoisier's or vice versa. That standard for translation sets the bar very high. As professional translators often lament, there are terms even in closely related languages for which a similar problem arises: the languages divide up the world differently. Once this point has emerged, a philosophical response to Kuhn-Feyerabend is complete. Linguistic cross-cutting poses no bar to rational scientific discussion. Yet, once again, Kuhn saw more significance in the phenomenon. We investigate in an upcoming section the ways in which conceptual incommensurability became central to the research of his later life.

A third hurdle remains. Does methodological incommensurability doom attempts to restore the rationality of scientific change? Some philosophers have supposed it does (Doppelt 1978) and have embraced what they take to be "Kuhnian relativism"; others have agreed that Kuhn is committed to relativism but have tried to show how methodological incommensurability can be tamed (Laudan 1977, 1984). Certainly, if the task is to rebut Kuhn-Feyerabend, the problem of methodological incommensurability poses the largest challenge.

The course of the chemical revolution shows quite clearly how proponents of rival paradigms weighed different problems differently. The experimental work initiated by Priestley and his fellow traditionalists (Henry Cavendish prominent among them), as well as by Lavoisier, generated a corpus of results about chemical reactions, challenging the chemical community to find ways of identifying the constituents of the substances involved so that they could provide a systematic account of them (Holmes [1985] provides a comprehensive historical narrative, on the basis of which Kitcher [1993: 272–290] offers a philosophical analysis). Throughout a period of nearly two decades, champions of the rival approaches disagreed not only about the compositions of various reactants but also concerning which problems should be addressed and which reactions analyzed. Between 1775 and the early 1790s, Lavoisier's new chemistry was able to amass an increasing number of impressive analyses, including some that handled reactions the phlogistonian traditionalists had previously hailed as central. While this extensive problem-solving was occurring, Priestley and his allies were facing recurrent difficulties in offering analyses of reactions they had once claimed to be able to understand. Small wonder, then, that the 1780s witnessed a parade of chemists who, one by one or in small clusters, abandoned the traditional approach for Lavoisier's.

This is no isolated episode. The same pattern is readily discernible in a smaller historical change, namely the nineteenth-century shift in paleontology caused by "the Great Devonian Controversy" (Rudwick 1985; analyzed in Kitcher 1993: 211–218; Kuhn [2000: 288] singles out Rudwick's account as a "splendid" achievement). Here too we find proponents of different perspectives starting from clashing methodological judgments (assigning different value to particular problems, unsolved or apparently resolved), as well as the same gradual shift of allegiances—this time to a compromise position, initially beyond the horizon of either paradigm—during the course of an eight-year debate.

When methodological standards (or value judgments) conflict, the differences can sometimes be resolved as the rival perspectives are applied to a growing body of factual information. The history of science recapitulates phenomena from everyday life. People engaged in a common project—for example, to buy some needed device or to select someone for a particular position—may start with different priorities but come to see that qualities they had hoped to find must be abandoned to achieve the common ends at which their venture is aimed. It is a mistake—born, perhaps, of a lingering positivism, whose credo is *de pretiis non est disputandum?*—to regard conflicts in values or methodological standards as inevitably blocking possibilities of reasonable and orderly change in view.

Kuhn made no such mistake. If we take his resistance to charges of relativism seriously, there is an obvious way to interpret *Structure*. The three forms of incommensurability interfere with the speedy resolution of interparadigm debates—and the fact that the commitments of normal scientists are largely pieces of tacit knowledge adds to the difficulties; we cannot expect them to sort through their partial communications, their observational disagreements, or their clashing values with the efficiency later analysts might bring to the debate. Yet that is not to suppose that revolutions close through the triumph of unreason.

The possibility of rational change without "instant rationality" can be illustrated by juxtaposing two passages from *Structure* that have often inclined Kuhn's critics to accuse him of inconsistency. Toward the end of chapter 12, he remarks:

> The man who embraces a new paradigm at an early stage must often do so in defiance of the evidence provided by problem-solving. He must, that is, have faith that the new paradigm will succeed with the many large problems that confront it, knowing only that the older paradigm has failed with a few. A decision of that kind can only be made on faith. (Kuhn 1962: 158)

Two paragraphs later, he seems to withdraw the provocative claim about faith, opting for a far more traditional view of scientific change.

> This is not to suggest that new paradigms triumph ultimately through some mystical aesthetic. . . .. if a paradigm is ever to triumph it must gain some first supporters, men who will develop it to the point where hardheaded arguments can be produced and multiplied. And even those arguments, when they come, are not individually decisive. Because scientists are reasonable men, one or another argument will ultimately persuade many of them. But there is no single argument that can or should persuade them all. Rather than a single group conversion, what occurs is an increasing shift in the distribution of professional allegiances. (158)

How can he have it both ways?

The previous discussion of how the obstacles to rationality erected by Kuhn-Feyerabend can be overcome—and how they are overcome in the chemical revolution and the "Great Devonian Controversy"—shows directly how Kuhn's allegedly incompatible claims fit together. At the beginning of the debate, the older paradigm has succeeded in solving many puzzles, but it has encountered one that resists standard methods of treatment. The early advocates of a new paradigm take overcoming that difficulty to be crucial and take their undeveloped proposal to have resources eventually to replicate the successes of the older approach. They do so on the basis of confidence, or faith, and it is good for the community that some are prepared to make a leap of this sort. If their confidence proves justified, they will turn back challenges from traditionalists that the new paradigm has failed to solve specific problems with which its predecessor succeeded, and different traditionalists are likely to view different problems as the crucial ones. As the problems fall one by one, some traditionalists switch to the new paradigm—we see that parade of "conversions" found in the chemical revolution and in the Devonian controversy.

The rebuttal of Kuhn-Feyerabend outlined here is articulated in more detail in Kitcher (1993); other philosophers might offer much the same story, using a different background framework (e.g., Laudan 1984; Lakatos 1970). Extending the reading of the two passages from *Structure* discussed in the preceding paragraphs to a full interpretation of the book shows that the critique of Kuhn-Feyerabend was worked out by Kuhn himself. Indeed, that critique was given in the book that helped to build the bogeyman.

4 THE ROAD TO THE "SCIENCE WARS"

As Barry Barnes (1982) makes very clear, *Structure* provided resources for an important development in the sociology of science. During the 1970s, a group of scholars at the University of Edinburgh, of whom Barnes, David Bloor, Donald MacKenzie, and Steven Shapin were the most prominent members, initiated the Strong Programme in the Sociology of Knowledge (SSK). At the core of SSK was a commitment to symmetry in explanation. That commitment, made clear in Bloor (1976), starts from the important insight that all kinds of beliefs, those rejected as false or irrational as well as the ones accepted as true or well justified, require causal explanation. Add the thought that human capacities for observing and thinking do not vary extensively across space, time, and cultures. The conclusion will be that scholars should not suppose that the processes leading scientists to judgments rejected by their fellows are likely to be inferior to those underlying the majority view. History's losers are as epistemically virtuous as the winners.

So far the line of reasoning does have a Kuhnian pedigree. The style of historical explanation exemplified by *Structure* does reject the idea of dismissing the scientists whose views were replaced in scientific revolutions as dogmatic or unintelligent (those who condemned Copernicus as "mad" for taking the earth to be a planet were not "either just wrong or quite wrong"; Kuhn 1962: 149). Yet Kuhn, unlike Kuhn-Feyerabend, was not committed to denying any epistemically important differences at all stages of revolutionary debate. Central to his picture of a procession of reasonable scientists switching their allegiance is the thought of an *evolving* controversy in which evidential considerations come to make a difference.

SSK and its descendant positions in the sociology of science went on to challenge that thought. Inspired sometimes by the Duhem-Quine thesis ("Any hypothesis can be held true, come what may") and sometimes by a reading of Wittgenstein's discussion of following a rule ("Anything can be the next term of any finite sequence"), many sociologists of science maintained that epistemic symmetry *always* prevails. No matter how many observations are made, it will still be logically possible—and "therefore" equally reasonable—to hang on to one's prior views.

Confining the power of the evidence paves the way for other factors to play a role in the resolution of revolutions. Sociology of science can tell a positive story of scientific change by emphasizing the intertwining of factual claims with social and political values. Instead of viewing the history of the sciences as driven by objective evidence, thus generating an account of nature that constrains proposals for ordering society, nature and society may be seen as constructed simultaneously. In what is probably the most influential study in the historical sociology of science, Shapin and Simon Schaffer (1985) review the seventeenth-century debate between Hobbes and Boyle over the status of the air pump. They show very clearly how both Hobbes and Boyle were committed to distinct webs of mutually reinforcing assumptions about the natural phenomena, the appropriate methods for investigation, and the character of a healthy political order. Because (allegedly) they *could not* resolve their dispute on the basis of any amount of experimental findings (Shapin and Schaffer appeal here to the Duhem-Quine thesis), Hobbes was no tedious crank but rather a thinker entitled to his position, a man who lost because a different vision of political order became popular. Officially advocating complete epistemic symmetry between Hobbes and Boyle, the authors slightly overstate their position in the last sentence of the book: "Hobbes was right."

Other sociologists saw the same features not merely as affecting large historical changes but as omnipresent in scientific research (Collins 1985). All scientific decision-making came to be seen as thoroughly contingent, a matter of which positions could "recruit the most allies," whether in society or in nature (Latour 1988). Bruno Latour characterized the vocabulary used in appraising scientific judgments ("purely rational," "totally absurd," and the like) as "*compliments* or *curses*," functioning in the same way as "swear words" for workmen who "push a heavy load" (Latour 1987: 192). We are free to make whatever evaluations we like, to see Galileo as "courageously rejecting the shackles of authority, arriving at his mathematical law of falling bodies on purely scientific grounds" or as "a fanatic fellow traveler of protestants [who] deduces from abstract mathematics an utterly unscientific law of falling bodies" (Latour 1987: 191–192). Nor was this the end of the line. Latour and some of his allies went on to plead for further forms of symmetry, offering heady metaphysical accounts of the processes underlying the joint construction of society and nature (Latour 1992, 1993, 2005).

Kuhn wanted none of this. Reacting to the (relatively restrained) arguments of Shapin and Schaffer, whose book he found "fascinating," Kuhn protested the strategy of dismissing the experimental evidence:

> what [Shapin and Schaffer] totally miss is the vastly greater explanatory power that comes, including that straightforwardly with the Puy-de-Dôme experiment and many others. So that there is every rational reason to switch from one of these ways of doing it to the other, whether you think you've shown that there is a void in nature or not.(Kuhn 2000: 316–317)

His critique recapitulates the view from *Structure*, and it is aligned with the judgments of philosophers who have discussed the significance of the Duhem-Quine thesis—simply because it is logically possible to maintain one's old belief does not mean that it is reasonable or rational to do so (for a thorough elaboration of the point, see Laudan and Leplin 1991). It is entirely consistent to accept the Kuhnian themes that originally inspired SSK, to recognize that the beliefs of both winners and losers need causal explanations, that those explanations find very similar capacities and processes at work at many stages of scientific controversies, and yet to suppose that there can come a later moment where the mass of evidence acquired makes further allegiance to a long-held position unreasonable. While human beings tend to share the same cognitive capacities, life inside and outside the classroom teaches us that they have quite diverse opportunities for evaluating theses about nature and sometimes quite different abilities to think things through.

Although scientists have generally not taken much interest in the history, philosophy, or sociology of science, the provocative character of the claims made by proponents of SSK and their more radical followers aroused considerable ire in some scientific quarters. The "science wars" broke out in 1994, with the publication of Gross and Levitt (1994) and Alan Sokal's revelation that the journal *Social Text* had published an article (Sokal 1996) deliberately written as a pastiche of work in science studies. The ensuing controversy pits some scientists and their philosophical allies against scholars in the sociology and anthropology of science, joined by some literary and cultural theorists; historians of science sometimes expressed their sympathy for the latter group.

By the turn of the millennium, the dispute had petered out, with a tacit agreement that individual disciplines should go their separate ways. General philosophy of science largely returned to the problems Hempel had set as the core of the logical empiricist agenda, concentrating on providing formal accounts of confirmation, theory, law, and explanation, almost as if the Kuhnian invitation to rethink the image of science in light of its history had never been issued. Even when philosophers reject core assumptions of logical empiricism—typically by liberalizing, or even discarding, the constraints the logical empiricists had imposed—they usually continue to seek general analyses of a small family of metascientific notions. History of science, meanwhile, became focused on questions that appeared of little direct relevance to philosophy, and the sociology of science became ever more preoccupied with theoretical issues about the exact forms of symmetry and relativism to be commended. Those who sought middle ground found themselves marginalized (Kitcher 1998).

5 ENDURING LEGACIES

Attempts to engage in interdisciplinary science studies were the most obvious expressions of the revolutionary impact of *Structure*. Although they broke down, some themes and developments in contemporary philosophy of science are plainly indebted to Kuhn. This section offers a brief survey.

One obvious response to *Structure*, sometimes but not always motivated by the urge to demolish Kuhn-Feyerabend, consisted in efforts to elaborate rival accounts of scientific change. In effect, some philosophers saw Kuhn as having added a new, large problem to the logical empiricist agenda: How do the sciences evolve? This was a problem to be solved by using examples from the history of science to generate and test analyses of the considerations governing reasoned resolution of revolutions and of scientific progress (the latter issue being one whose difficulty Kuhn highlighted [1962: final chapter]. The enterprise was begun at the important conference held in 1965 at the London School of Economics focused on *Structure* (Lakatos and Musgrave 1970), particularly in Lakatos's seminal account of a theory of scientific change (Lakatos 1970). Lakatos began from some central Popperian concepts and theses but extended these quite radically to cope with the kinds of historical episodes to which Kuhn had drawn attention.

During the 1970s and 1980s, other philosophers were inspired to emulate Kuhn's search for larger patterns in the history of science. Although they sought alternatives to the concept of a paradigm, both Larry Laudan and Dudley Shapere were convinced that an adequate treatment of scientific practice and its evolution required identifying a far richer framework than the theories (conceived as axiomatic systems) popular among logical empiricists (Laudan 1977, 1984 ; Shapere 1984). They attempted to offer alternatives to Kuhn's analyses of "disciplinary matrices" (Kuhn 1970; see Kuhn 2000: 168), in which he included not only shared generalizations (the constituents of logical empiricists' theories) but also shared models and—significantly—shared values. From a quite different, formal perspective, Wolfgang Stegmüller built on the work of Joseph Sneed to offer a reconstruction of a Kuhnian approach to science and scientific change. Kuhn's remarks about proposals that diverge substantially from his own are uncharacteristically (and unwarrantedly) harsh—for example, the dismissal of Laudan's work (Kuhn 2000: 311). He shows more sympathy for the Sneed-Stegmüller program, although claiming that it must overcome the very large problem of accounting for revolutionary change (Kuhn 2000: ch. 7, especially 192–195).

Further exploration of the components of scientific practice and the dynamics of scientific change is undertaken in Kitcher (1993) and Friedman (2001), both works attempting to elaborate Kuhnian themes (although Friedman is more explicit about this). But any survey of the proposals about the growth of science, from Kuhn and Lakatos, in the 1960s must be struck by the differences in the historical episodes to which different authors appeal. Rival suggestions about the terms in which scientific change should be reconstructed seem motivated by attention to details that emerge in different exemplars. Concentrations on the Darwinian revolution or the fate of ether theories or the transition from Newtonian mechanics to the special and general theories of relativity bring rival features into prominence. Perhaps we should see general philosophy of science as a venture in constructing tools, useful for thinking about alternative facets of scientific work. The logical empiricist concept of theories as axiomatic systems is valuable for some purposes, for example recognizing which constituents of theories are independent of others; thinking of theories as families of models is a more adequate way to understand some areas of science, such as the origins and growth of evolutionary theory. Similarly, the loose assemblage of notions that descend from Kuhn's *paradigm* (appeals to cores, axiologies, the relative a priori, significant questions, and the like) might be viewed as parts of a toolkit from which philosophers can draw when attempting to understand scientific practice and scientific change. Not every problem requires a hammer—or a paradigm.

One important feature of Kuhn's specification of the disciplinary matrix was his explicit introduction of values. Certain types of value judgments and value commitments were taken to be essential to normal scientific practice and to be subject to change across scientific revolutions. In "Objectivity, Value Judgment, and Theory Choice" (Kuhn 1977: 320–339), Kuhn offered more detail about the values he had in mind, listing such desiderata as scope, accuracy, precision, and simplicity. Not only are the values weighed differently against one another in alternative normal scientific practices, but there is also variation in how they are articulated with respect to concrete instances of scientific research.

Kuhn's appeal to values was relatively conservative—he found no place within science for the types of social and political values scholars such as Shapin and Schaffer took to play important roles in some episodes of scientific change. Moreover, given his emphasis on fundamental homogeneity within the normal scientific community, his work gave precedence to the question of understanding how consensus is formed. How does a field of science make the transition from one shared set of concepts, factual claims, and value commitments to another? Ironically, Feyerabend, Kuhn's bogeyman twin, had emphasized the opposite issue, viewing dissent as crucial to the flourishing of scientific inquiry and seeking ways to ensure the flowering of alternatives. Laudan (1984) clearly recognized the existence of connected questions—"When is consensus an epistemic goal, and how is it attained?", "When is disagreement an epistemic desideratum and how is it fostered?"—but, challenged by the apparent difficulty of bringing scientists separated by different value commitments into accord with one another, he expended the bulk of his efforts on the issue of understanding consensus.

Despite Kuhn's double conservatism, his focus on epistemic values and his connecting value commitments to apparent difficulties in reaching consensus, this feature of his work, already implicit in *Structure* and explicit in subsequent writings, has left its mark on some of the most important movements in contemporary philosophy of science. Feminist philosophy of science, particularly in the writings of Helen Longino and Evelyn Fox Keller, can be seen as taking rich advantage of the breach Kuhn introduced in the traditional exclusion of values from the scientific mind. Keller's detailed studies of the interactions between value judgments and developments in various scientific fields, recognizing the connections between considerations analysts might subsequently separate as epistemic and as social, revealed both how research might be impoverished by ignoring people with alternative value commitments and how working through value conflicts might advance a scientific field (Keller 1985, 1992, 2000, 2002). Longino has argued for the relevance of social values to judgments about what counts as scientific evidence and has offered a general theory of scientific practice that explicitly recognizes its place within societies composed of people with rival values (Longino 1990, 2002).

Keller and Longino also challenge the Kuhnian assumption that the equilibrium state of any scientific field must be one in which those working in that field are in fundamental agreement. In this they are joined by other pluralists (Dupre 1993; Cartwright 1999; and contributors to Kellert, Longino, and Waters 2006). Pluralists suppose that rival models for some domain of phenomena might reasonably be adopted by a scientific community, not merely as a short-term expedient until some consensus decision about which one should survive but as a permanent feature of practice in the field.

These developments are aligned with another theme, prefigured in *Structure*, namely that of science as a contingently shaped bundle of efforts to introduce pockets of order into a "dappled" or "disorderly" world (Cartwright 1999; Dupre 1993). Instead of seeing science as aimed at producing some unified ("final") theory, John Dupre and Nancy Cartwright take healthy scientific research to issue in ways of coping with, or domesticating, aspects of the natural world that bear on important human concerns. Science should be socially embedded, its research aimed at promoting the collective good (Kitcher 2001, 2011), through supplying tools for intervention and prediction that overcome human problems (Cartwright 2012).

Approaches of this general type fly in the face of many traditional assumptions about science and its goals: that science is primarily directed toward unified theories providing understanding of regularities in nature, that scientific research should be autonomously directed by the scientific community, and that science should be a value-free zone. The last claim, already scrutinized by Keller and Longino, has been decisively challenged by Heather Douglas (2009). Plainly, the route leading from Kuhn to these more recent developments ends at a far remove from his conservatism: the door he left ajar has become a gaping portal. Nevertheless, there are hints in *Structure* of the perspectives I have briefly reviewed. Kuhn's struggles with the concept of scientific progress in the final chapter of the original monograph are tacitly at odds with any notion of a final unified theory toward which the sciences might be advancing. Moreover, in a laconic early sentence, typically unnoticed, he hints at the path dependence and contingency of what becomes accepted as correct science: "An apparently arbitrary element compounded of personal and historical accident, is always a formative ingredient of the beliefs espoused by a scientific community at a given time" (Kuhn 1962: 4).

6 KUHN AFTER KUHN: A RETURN TO PRAGMATISM

None of the issues we have discussed so far corresponds to Kuhn's own conception of the aspects of *Structure* that were most significant and most worthy of further elaboration. Although many philosophers have focused on his notion of methodological incommensurability (and the correlative suggestions about the place of values in science), Kuhn thought of conceptual incommensurability as the core of his contribution. Responding to commentators who offered ways of connecting the languages of rival paradigms, he insisted on the importance of the fact that those languages are not intertranslatable. To explain how Priestley and Lavoisier can specify one another's referents, proceeding token by token, fails to plumb the depths of the situation. Because there is no uniform way to substitute for all tokens of some types (e.g., "phlogiston" and its cognates), simply noting that token-token replacements are available misses the crucial point that what is seen from one perspective as a kind is viewed from the other as a category that is either heterogeneous or incomplete. That point serves as the basis of *Structure*'s claims that scientific revolutions change the world scientists inhabit. Kuhn's presidential address to the Philosophy of Science Association elaborates the connection and defends against charges that he is advocating a form of idealism (Kuhn 2000: ch. 4; originally presented in 1990).

Rival paradigms are conceptually incommensurable in the sense that the words of each language belong to a structured set—the lexicon—in which some terms are marked with particular features (e.g., "acid" is marked as a natural kind term). Intertranslatability breaks down because corresponding expressions in the rival languages are marked differently: the lexicons are structured in cross-cutting ways. In consequence, multilinguals, able to move across the revolutionary divide, "experience aspects of the world differently as they move from one to the next" (Kuhn 2000: 101). Despite this "the world is not invented or constructed" (101). The world in which we live and move is structured by the efforts of previous generations, and, through our own efforts, we restructure it. This "world that changes from one community to another" nevertheless deserves to be called "the real world"—for it "provides the environment, the stage, for all individual and social life" (Kuhn 2000: 102).

The approach to ontology figuring in Kuhn's thinking plainly accords with Quine's famous proposals: to decide what there is, one goes to the best science of the day and offers a logical reconstruction (or regimentation) of it. Kuhn is particularly concerned with differences in the extensions of natural kind terms, seeing these as fundamental to world changes. In principle, there might be linguistic shifts without any significant metaphysical differences. The theoretical notion of the lexicon, however, as central to Kuhn's later work as the concept of paradigm was to *Structure,* serves to mark out those episodes in which ontological views change. Differences in the lexicon will require those who regiment the new science to recognize a modified ontology.

Kuhn's apparently mysterious picture relies on a distinction between two senses of "world," a distinction drawn already by two of the classical pragmatists, William James and John Dewey. In one sense, "the world" denotes whatever is prior to and independent of the cognitive subject. In another, "the world" picks out a structured whole, the world in which people live, containing objects, assorted into kinds, as well as processes with beginnings and ends. For James and Dewey much of this structure depends on us: we determine the boundaries of objects, the clustering into kinds, the interval within which a process unfolds. The structuring and restructuring depends on our cognitive capacities and on the efforts we make to attain our evolving ends.

James presents these ideas in a book to which Kuhn's discussion of changes in worldview—and of worlds—alludes: Kuhn (1962: 113) draws on James (1890/ 1983: 462). Before that passage, James introduces his views about worlds, with an analogy:

> The mind, in short, works on the data it receives very much as a sculptor works on his block of stone. In a sense the statue stood there from eternity. But there were a thousand different ones beside it, and the sculptor alone is to thank for having extricated this one from the rest. (277)

He goes on to suggest that the world "we feel and live in" has been formed in part by the common sensory and psychological capacities (that distinguish our world from that of "ant, cuttlefish or crab") and in part by the "slow cumulative strokes of choice," made by our ancestors (James 1890/1983: 277). Dewey develops the latter theme in his discussion of life experience, which "is already overlaid and saturated with the reflection of past generations and bygone ages" (Dewey 1925/1981: 40).

The pragmatist approach to worlds (and world changes) is not restricted to the style of ordering Kuhn most emphasizes, the taxonomic division into kinds. For example, Hacking (1993) offers a lucid presentation of Kuhn's ideas that situates Kuhn in the tradition of reflecting on natural kinds. Pragmatists also view worlds as constituted by the spatiotemporal boundaries set for objects and for processes. Yet there are passages in *Structure* that point toward this broader conception: the discussion of Galileo's novel perspective on the swinging of heavy bodies (Kuhn 1962: 118–119), for instance, recognizes that world reconstruction can extend to processes.

Kuhn's understanding of progress, already indicated in the final chapter of *Structure* and further developed in his Presidential Address, also expresses an affinity with pragmatism. To see "scientific development" as "a process driven from behind, not pulled from ahead—as evolution from, rather than evolution toward" (Kuhn 2000: 96) is to follow Dewey in generalizing a Darwinian conception to human activities, most centrally to inquiry. Progress is not always measured by the decreasing distance to a goal. It is sometimes a matter of overcoming present difficulties. Medicine makes progress not because it advances toward some ideal of perfect human health (what would that be?) but because physicians learn to cope with some diseases, thereby modifying disease ecology and changing or intensifying problems for their successors. Dewey and Kuhn view scientific progress generally in that way.

The connection between Kuhn and pragmatism (especially in Dewey's version) is not simply a matter of historical coincidence, nor even a recapitulation of Kuhn's own sense of convergence with some strands in neopragmatism (Kuhn 2000: 312–313 notes kinship with Hilary Putnam's ideas of the early 1980s). At the center of Kuhn's thought are the ideas of taxonomic readjustments as world changes and of progress as "progress from"; once we have appreciated this, it becomes natural to combine these themes with the pragmatist tendencies of some of the contemporary movements, briefly reviewed in the previous section. Longino, Dupre, and Cartwright view the sciences as advancing by finding ways to reconstruct and order natural phenomena, so that the problems besetting people and human communities are overcome (see also Keller 1985; Kitcher 2001, 2011). Progressive world changes are those leading to the piecemeal betterment of human lives. Kuhn did not explicitly take that last step, but, from the perspective on his work presented here, it would extend the trajectory he was following. At the end of the line is a very different vision of science, and a very different set of tasks for the philosophy of science, than the ones against which Kuhn was reacting. So the famous opening sentence of *Structure* would be resoundingly vindicated.

7 CODA

Any truly significant, world-changing book allows for many readings. As I have read and taught *Structure* over decades, it has never seemed the same twice, and views of its impact are bound to be equally variable. So my narrative can claim to be only one among a diverse set, a particular slant on the complex intellectual developments of over half a century. Others should write their own narratives, and, though I hope they might overlap with mine, I should be surprised and sorry if there were not important differences. So I end by stealing and twisting Hacking's pithy injunction: "Great books are rare. *Structure* is one. Reread it and you will see."

ACKNOWLEDGMENTS

Many thanks to Paul Humphreys for his constructive suggestions about an earlier draft of this article.

REFERENCES

Barnes, Barry. (1982). *T. S. Kuhn and Social Science* (New York: Columbia University Press).

Bloor, David. (1976). *Knowledge and Social Imagery* (London: Routledge).

Cartwright, Nancy. (1999). *The Dappled World* (Cambridge, UK: Cambridge University Press).

Cartwright, Nancy. (2012). *Evidence-Based Policy* (Oxford: Oxford University Press).

Collins, Harry M. (1985). *Changing Order* London: Sage.

Dewey, John. (1981). *Experience and Nature* (Carbondale: University of Southern Illinois Press). (Original work published 1925)

Doppelt, Gerald. (1978). "Kuhn's Epistemological Relativism: An Interpretation and Defense." *Inquiry* 21: 33–86.

Douglas, Heather. (2009). *Science, Policy, and the Value-Free Ideal* (Pittsburgh: University of Pittsburgh Press).

Dupre, John. (1993). *The Disorder of Things* (Cambridge, MA: Harvard University Press).

Earman, John. (1993). "Carnap, Kuhn, and the Philosophy of Scientific Methodology." In Paul Horwich (ed.), *World Changes* (Cambridge, MA: MIT Press), 9–36.

Feyerabend, P. K. (1962). "Explanation, Reduction, and Empiricism." In *Minnesota Studies in the Philosophy of Science*, Vol. III (Minneapolis: University of Minnesota Press), 28–97.

Feyerabend, P. K. (1965). "Problems of Empiricism." In R. G. Colodny (ed.), *Beyond the Edge of Certainty: Essays in Contemporary Science and Philosophy* (Englewood Cliffs, NJ: Prentice-Hall), 145–260.

Feyerabend, P. K. (1970). "Against Method." In *Minnesota Studies in the Philosophy of Science*, Vol. IV (Minneapolis: University of Minnesota Press), 17–130.

Feyerabend, P. K. (1975). *Against Method* (London: Verso).

Friedman, Michael. (2001). *Dynamics of Reason* (Stanford, CA: CSLI).

Gross, Paul, and Levitt, Norman. (1994). *Higher Superstition* (Baltimore: Johns Hopkins University Press).

Hacking, Ian. (1993). "Working in a New World." In Paul Horwich (ed.), *World Changes* (Cambridge, MA: MIT Press), 275–310.

Hanson, N. R. (1958). *Patterns of Discovery* (Cambridge, UK: Cambridge University Press).

Holmes, F. L. (1985). *Lavoisier and the Chemistry of Life* (Madison: University of Wisconsin Press).

James, William. (1983). *The Principles of Psychology* (Cambridge, MA: Harvard University Press). (Original work published 1890)

Keller, Evelyn. (1985). *Reflections on Gender and Science* (New Haven, CT: Yale University Press).

Keller, Evelyn. (1992). *Secrets of Life/Secrets of Death* (London: Routledge).

Keller, Evelyn. (2000). *The Century of the Gene* (Cambridge, MA: Harvard University Press).

Keller, Evelyn. (2002). *Making Sense of Life* (Cambridge, MA: Harvard University Press).

Kellert, Stephen, Longino, Helen, and Waters, C. Kenneth. (2006). *Scientific Pluralism* (Minneapolis: University of Minnesota Press).

Kitcher, Philip. (1993). *The Advancement of Science* (New York: Oxford University Press).

Kitcher, Philip. (1998). "A Plea for Science Studies." In Noretta Koertge (ed.), *A House Built on Sand* (New York: Oxford University Press), 32–56.

Kitcher, Philip. (2001). *Science, Truth, and Democracy* (New York: Oxford University Press).

Kitcher, Philip. (2011). *Science in a Democratic Society* (Amherst, NY: Prometheus).

Kuhn, T. S. (1962). *The Structure of Scientific Revolutions* (1st ed.) (Chicago: University of Chicago Press).

Kuhn, T. S. (1970). *The Structure of Scientific Revolutions* (2nd ed.) (Chicago: University of Chicago Press).

Kuhn, T. S. (1977). *The Essential Tension* (Chicago: University of Chicago Press).

Kuhn, T. S. (2000). *The Road since Structure* (Chicago: University of Chicago Press).

Kuhn, T. S. (2012). *The Structure of Scientific Revolutions* (50th anniversary ed.) (Chicago: University of Chicago Press).

Lakatos, Imre. (1970). "The Methodology of Scientific Research Programmes." In I. Lakatos and A. Musgrave (eds.), *Criticism and the Growth of Knowledge* (Cambridge, UK: Cambridge University Press), 91–195.

Lakatos, Imre, and Musgrave, Alan. (1970). *Criticism and the Growth of Knowledge* (Cambridge, UK: Cambridge University Press).

Latour, Bruno. (1987). *Science in Action* (Cambridge, MA: Harvard University Press).

Latour, Bruno. (1988). *The Pasteurization of France* (Cambridge, MA: Harvard University Press).

Latour, Bruno. (1992). "One More Turn after the Social Turn." In Ernan McMullin (ed.), *The Social Dimension of Science* (Notre Dame, IN: University of Notre Dame Press), 272–294.

Latour, Bruno. (1993). *We Have Never Been Modern* (Cambridge, MA: Harvard University Press).

Latour, Bruno. (2005). *Reassembling the Social* (New York: Oxford University Press).

Laudan, Larry. (1977). *Progress and its Problems* (Berkeley: University of California Press).

Laudan, Larry. (1984). *Science and Values* (Berkeley: University of California Press).

Laudan, Larry, and Leplin, Jarrett. (1991). "Empirical Equivalence and Underdetermination." *Journal of Philosophy* 88: 449–472.

Longino, Helen. (1990). *Science as Social Knowledge* (Princeton, NJ: Princeton University Press).

Longino, Helen. (2002). *The Fate of Knowledge* (Princeton, NJ: Princeton University Press).

Masterman, Margaret. (1970). "The Nature of a Paradigm." In I. Lakatos and A. Musgrave (eds.), *Criticism and the Growth of Knowledge* (Cambridge, UK: Cambridge University Press), 59–90.

Reisch, George. (1991). "Did Kuhn Kill Logical Empiricism?" *Philosophy of Science* 58: 264–277.

Reisch, George. (2005). *How the Cold War Transformed Philosophy of Science* (Cambridge, UK: Cambridge University Press).

Rudwick, Martin. (1985). *The Great Devonian Controversy* (Chicago: University of Chicago Press).

Scheffler, Israel. (1967). *Science and Subjectivity* (Indianapolis: Bobbs-Merrill).

Sellars, Wilfrid. (1953). "Empiricism and the Philosophy of Mind." In *Minnesota Studies in the Philosophy of Science*, Vol. I (Minneapolis: University of Minnesota Press), 253–329.

Shapere, Dudley. (1984). *Reason and the Search for Knowledge* (Dordrecht, The Netherlands: Reidel).

Shapin, Steven, and Schaffer, Simon. (1985). *Leviathan and the Air-Pump* (Princeton, NJ: Princeton University Press).

Sokal, Alan. (1996). "Transgressing the Boundaries: Towards a Transformative Hermeneutics of Quantum Gravity." *Social Text* 46/47: 217–252.

CHAPTER 31

....

ASTRONOMY AND
ASTROPHYSICS

....

SIBYLLE ANDERL

1 INTRODUCTION

....

MODERN astrophysics,[1] which operates at the "intermediate" scales (i.e., between the physics of the solar system and the cosmology of the entire universe), has rarely been addressed directly within the philosophy of science context. Although there are philosophical subdisciplines in other special sciences, such as the philosophy of biology and the philosophy of chemistry, today's noncosmological astrophysical research, until now, had failed to arouse philosophical interest in any lasting way.

One reason for this lack of focused interest might have been that, when viewed from the outside, astrophysics superficially appears as a special case within the philosophy of physics or as an extension of the existing philosophy of cosmology. However, both fields seem to deal with slightly different core themes than those that come up in reflecting on contemporary astrophysical research. In contrast with much of what gets discussed in the philosophy of physics, most astrophysical research programs are not at all concerned with metaphysical questions. Rather than the development of autonomous mathematical theories that themselves then require specific interpretations, it is the application of existing physical theories to a vast array of cosmic phenomena and different environments that are found in the universe that constitutes the main focus of astrophysical research. Understanding the nonreducible complexity that permeates the universe requires a broad knowledge of statistical samples rather than a detailed or specific understanding of the ontology of theoretical entities. Practical problems in

---

[1] Often, the transition from classical astronomy to today's astrophysics is denoted as the transition from purely descriptive astronomical observations to the explanatory application of physical methods to cosmic phenomena. However, in this article, the terms "astrophysics" and "astronomy" will be used interchangeably.

astrophysical research are so deeply concerned with methodologies that arise from the observational constraints imposed on the empirical part of research that ontological questions are seldom seen to be relevant. More than that, it might not even be clear what a notion of fundamentality could possibly mean with respect to objects of astrophysical research beyond those cases where astrophysics and microphysics meet and where foundational problems might be inherited from there.

Cosmology, on the other hand, evokes its own metaphysical questions, as, for example, when debates center around the nature of space and time or the status of cosmological entities like dark matter and dark energy. Being the study of the origin and the evolution of the universe, cosmology has to face specific problems that arise from the totality of its research object: in studying the global properties of a single, uniquely defined object, possible cosmological knowledge is systematically underdetermined. This underdetermination is theoretical on the one hand, but also empirical on the other hand, because we cannot observe it in its entirety, even in principle. The reason is that light can only reach us from a limited region of the spacetime, given the limited speed of light and the finite age of the universe. However, to the extent that cosmology does not only lay claim to understanding the evolution of the universe as a whole, but also the evolution of its large-scale structures, the transition to noncosmological astrophysical research that is concerned with the understanding of astrophysical phenomena on smaller scales is continuous. A philosophy of (noncosmological) astrophysics and astronomy will therefore necessarily overlap topics from cosmological research. Some of the epistemological questions it faces might, however, be different from those that are classically discussed for the case of cosmology, resulting from a more bottom-up point of view as compared to the top-down case of a cosmology perspective.

Astrophysics and cosmology share a common problem in that they both need to acquire knowledge of their objects of research without directly interacting, manipulating, or constraining them. In reconstructing plausible explanations and evolutionary scenarios of particular, observable objects and processes, the resulting methodology of astrophysics and cosmology resembles the criminology of Sherlock Holmes: the astrophysicist must look for all possible traces and clues that may help to illuminate what has happened in a given region of the universe. However, in contrast to the cosmological study of the present universe, astrophysics goes beyond the understanding of singular events and objects and aims at making general statements about classes of objects like O-type stars, the interstellar medium of galaxies, black holes, or clusters of galaxies. Here, inability to stage real experiments imposes a challenge on the establishment of causal claims that may derive from but must transcend mere correlations. In practice, this can, at least partially, be compensated for by the fact that the universe is a "cosmic laboratory," one that is filled with phenomena and ongoing processes in all manner of evolutionary stages, each of them constrained by different initial conditions and contemporary environments. Accordingly, the laboratory scientist's skill in controlling her experiment has its correspondence in the astrophysicist's ability to statistically sample, analyze, and model this diversity that she observes in the complex process of reaching generalized conclusions about cosmic phenomena. The understanding of the epistemology of modern astrophysical research certainly can profit from the epistemological studies of other scientific fields (such as paleontology, archaeology, and the social sciences) that also cannot easily perform experiments and artificially reduce their subject's complexity. At the same time, the ubiquitous use of simulations and models and the challenges that the ever-increasing generation of the huge amount of data imposes on astrophysics make corresponding discussions from the philosophy of science relevant for the reflection on astrophysical research.

Finally, astrophysical research is an interesting topic with respect to its practical organization and realization, which opens the discussion to historical and sociological directions. The generation of empirical astrophysical knowledge relies on the distribution of observing time at usually internationally operated observatories. The distribution is based on applications by individual researchers or research collaborations and decided on in a peer-review process by a panel of scientists. Because astronomy rarely takes place at small observatories located at universities anymore, extensive division of work in the generation of data and the essential role of international science policy have a growing influence on the practice of research. This evolution of astrophysics not only makes it an interesting topic for social-scientific and philosophical epistemological reflections on their own, but, at the same time, such reflection may also yield important guidance for the challenges and future policy decisions astrophysics has to face in times of increasingly expensive research programs and exponentially growing amounts of generated data.

In Section 2, Ian Hacking's claim of astrophysical antirealism will be taken as a starting point and a motivation to subsequently shed light on the specific methodology of astrophysics in Section 3. In Sections 4 and 5, aspects of the discussions of scientific modeling and data generation, processing, and interpretation that are relevant for astrophysics will be summarized, before the article is concluded in Section 6.

2 ASTROPHYSICAL ANTIREALISM

Astrophysics deals with phenomena and processes that are found occurring in significantly more extreme conditions than anything that can be artificially generated in a terrestrial laboratory. The range of temperatures, pressures, spatial scales, and time scales pertinent to astrophysical phenomena are larger than anything that is commonly accessible to humans by direct experience. Also, the often dominant influence of gravity sets astrophysical processes apart from terrestrial laboratory settings. While applying physical methods to cosmic phenomena, astronomers have to deal with the fact that it is impossible to interact directly with the objects of astrophysical interest. Curiously, this prohibition might not only have practical consequences for the way in which astrophysical research is done, but it may also impact the ontological status of the objects of research themselves.

In his book *Representing and Intervening* (Hacking 1983), Ian Hacking tried to link scientific realism to the ability of experimenters to manipulate the object of their research. By this definition, he cannot attribute the claim of realism to astrophysical objects because direct interaction with cosmic phenomena is impossible (except to a limited degree within the solar system). Accordingly, Hacking (1989) argues for an antirealism when it comes to cosmic phenomena. His exemplary argument cites the ambiguities involved in interpreting the observations of theoretically predicted gravitational lensing events. If, as he claims, the universe is awash with nondetectable microlenses distorting and aberrating light as it traverses space, then all astronomical observations would be inherently untrustworthy. Distorting influences of microlenses on the observed photon fluxes could never be disentangled from intrinsic features of the source. In this context, he revisits the Barnothys' suggestion in 1965 that quasars are not a separate class of cosmic objects but rather they are Seyfert galaxies subject to gravitational lensing.

Although an interaction with the object of research enables systematic tests of possible sources of error in experimental disciplines, this option is prohibited in astrophysics. Instead, astrophysics depends on the use of models brought in line with the observed phenomenon—a fact on which Hacking builds his second argument against astrophysical realism. The limited nature of the modeling approach, together with the fact that cosmic phenomena can be described by a multitude of adequate but mutually contradictory models shows, according to Hacking, that astrophysics cannot make a claim to realism, if realism is understood to entail a convergence of scientific descriptions toward truth. On the basis of these considerations, in a third argument, Hacking then claims a "methodological otherness" for astronomy and astrophysics, as compared with other, realistic scientific disciplines. The transition to a natural science, according to Hacking, is marked by the application of experimental methods. Astronomy has not made that transition. Although astronomical technology has changed drastically since the historical beginnings of this discipline, Hacking sees the underlying methodology as having remained the same: "Observe the heavenly bodies. Construct models of the (macro)cosmos. Try to bring observations and models into line" (Hacking 1989, 577).

There have been several replies to this rather severe view of the status of modern astrophysics, and each of his three arguments has been attacked (e.g., Sandell 2010; Shapere 1993). First, his example of microlenses does not seem to be at all representative of astrophysical research. Moreover, the study of gravitational lensing was still in its infancy at that time. Today, it is quite possible for astrophysicists to observationally decide on the presence (or not) of microlenses. This demonstrates the fact that often theoretical as well as technological progress can eventually resolve problems of apparent underdetermination, as is the case for Hacking's prime example. This point is related to the fact that observational evidence for or against a given theory does not depend on the theory alone, but it also involves and entails auxiliary hypotheses, available instruments, and background assumptions (e.g., Laudan and Leplin 1991). If these auxiliaries are changed, the observational evidence for a theory might change as well and subsequently give rise to differences in the observational support for previously empirically degenerate theories. Similarly, the theory itself may become further developed at a later time and allow for new predictions that can in turn be tested with new observations. For instance, in the case of gravitational microlenses, progress in our detailed theoretical understanding of lensing events in general has suggested that time-dependent brightening and fading of the emission of a microlensed background object can discriminate microlensing from properties intrinsic to the source itself. Compared with the situation Hacking describes, today, we are actually able to predict observable differences between a theory that includes microlenses and one that does not. Although the argument by Laudan and Leplin (1991) does not guarantee that the problem of so-called *contrastive underdetermination*[2] will always be solved sooner or later, it might explain the faith that scientists usually put in "abductive justification" of their science; that is, if a theory is the best possible explanation of the given empirical evidence, it might very likely be true (and if it is not, then they will find out at some later point). Accordingly, Hacking seems to have underestimated the improved understanding of astronomers and the power of their method of research.

Hacking's second argument, referring to the ubiquitous use of models in astrophysics, has implications that reach well beyond astronomy. Models and simulations are crucial tools in basically all fields of modern sciences. Accordingly, applying this argument would mean questioning the realist status of most of today's scientific research. Apart from that, as will be discussed in Section 4, astrophysical models are being continuously refined and improved. Moreover, new and varied observations are brought to bear in resolving ambiguities between different models. In that regard, models in astrophysics are very often critically discussed and reflected on, and scientists are very careful with a realistic interpretation of features of these models and simulations because their limitations are usually obvious.

Hacking's third argument concerning a fundamental difference between experimental and observational disciplines is more general than the two previous arguments and raises two basic questions in response: (1) does the Hacking criterion for scientific realism make sense in general? And (2), are there really any significant epistemic differences between astrophysics and experimental sciences? The first question has been extensively discussed quite independently of its application to astrophysics (e.g., Reiner and Pierson 1995; Resnik 1994), and we refer the reader to that literature. Regarding the second question, a closer look at astrophysical research practice demonstrates that the distinction between experimental and observational sciences is more subtle than it seems at first sight. The crucial part is to clarify what Hacking's experimental argument explicitly means and comprises. In *Representing and Intervening* (1983), he gives several explanations of what it means for an entity to be real. In addition to the "strong

---

[2] Contrastive underdetermination describes the idea that, for each body of possible evidence, there may well exist several different but equally empirically adequate theories. Stanford (2013) contrasts this use of the term underdetermination with "holist underdetermination," which describes the fact that a general theory can always be held even in the light of countervailing evidence. This is because the evidence depends not only on the tested claim, but also on many other auxiliary hypotheses, which could also be equally wrong.

requirement" that direct manipulation of the entity has to be possible, a weaker version of the criterion can be found in Hacking's additional proposal that an entity can be real if its well-understood causal properties can be used to interfere with other parts of nature.

Shapere (1993) has built on this ambiguity of definition and apparent lack of conceptual clarity when he stresses that the "use" of cosmic phenomena in the investigation of others is indeed possible, even when the phenomena under consideration cannot be directly or interactively manipulated. Gravitational lensing, for instance, can and has been widely used for the detection of dark matter and for distance determinations. For Shapere, the scientific method is not so much based on experimental interaction, but rather on the practice of extrapolating from knowledge already obtained to something novel. In this perspective, astronomy is perfectly in line with the other natural sciences. Sandell (2010) argues in the same vein when she suggests that astronomers do in fact carry out experiments, even within Hacking's meaning of manipulating or, equivalently, utilizing the causal powers of phenomena. Astronomical objects also have causal impact, which can be detected by astronomers to generate stable phenomena (e.g., measurement results of a receiver).

Based on these arguments, it would appear that there is in fact no fundamental difference between astrophysics and other scientific disciplines when it comes down to scientific realism.[3] However, Hacking's article raises an interesting epistemological question. Specifically, if his argument on the possible distorting influence of microlenses is read as a variation of the strong underdetermination argument (e.g., the idea that for each body of possible evidence there might exist several different but equally "empirically adequate" theories), the question remains whether experimental sciences have a larger tool box at their disposal to prevent such situations. Could it be the case that astrophysics is particularly prone to intrinsic underdetermination operating on several levels? For instance, is underdetermination found both at the level of astrophysical observations and at the level of astrophysical modeling because of its intrinsic observational basis? In order to investigate this question, a closer look at the astrophysical method, as currently practiced, may well be in order. We do this in the next section.

3 THE ASTROPHYSICAL METHOD

3.1 The Sherlock Holmes Strategy

A significant part of astrophysical research is dedicated to the understanding of singular instances of objects (e.g., the Class 0 protostar NGC 1333-IRAS4B) or a specific process (e.g., the gaseous outflow from the active galactic nucleus in PG1211 + 143). The

---

[3] An interesting but unrelated argument against a realist interpretation of astrophysical objects is based on the ambiguity and interest-dependence of astrophysical classifications. See Ruphy (2010) and note 4.

basic question in such cases is this: what do the observations tell us about the physics and chemistry at work in the observed region? Or, more generally: which circumstances have led to what we observe? These scientific questions resemble corresponding questions in the "historical sciences." Historical research concerns itself with the explanation of existing, natural phenomena in terms of their past (and sometimes distant) causes. The fact that the causes are past means that the causal chain cannot be subject to investigation itself: the investigator finds herself in a "Sherlock-Holmes situation." Classical historical sciences in this sense of investigation include paleontology, archaeology, and some aspects of global climate change. Historical sciences seem to differ from experimental sciences with respect to their restricted mode of evidential reasoning (e.g., Cleland 2002). The experimental activity usually begins with a hypothesis, from which a test condition C is inferred, together with a general prediction about what should happen if C is realized and the hypothesis is true.

For instance, the hypothesis could be that astrophysicists are not interested in philosophy. The test condition C could be a lunch conversation about Kant's "transcendental idealism," and the prediction would be that the astrophysicist will change the topic of discussion or leave after no more than a minute. The series of experiments then contains several experimental tests, in which C is held fixed while other experimental conditions are varied in order to exclude misleading confirmation and disconfirmation. Within the previous example, the test could be performed with a different philosophical speaker because maybe the first speaker is intrinsically boring (independent of the topic), or the test may be rerun with a different philosophical topic than Kant, given that Kant might not be representative overall for philosophy, and the like. This strategy tries to reduce the holistic underdetermination by checking different auxiliaries or confounding variables for their individual influence on the experimental prediction.

In contrast, historical sciences start from a situation that would be the experimental outcome, or, in Cleland's words, they start from the traces of past causes. In the previous example, the evidence would be simply watching the astrophysicist leave the lunch table, and then the problem would be in having to reconstruct what led to this particular behavior. Or, to take a more realistic example, if the question of the extinction of dinosaurs is being investigated, examples of such evidential traces may be a geological layer of sediment containing high levels of (extraterrestrial) iridium and the Chicxulub crater in the Gulf of Mexico, which pointed to a prehistoric asteroid impact. In astrophysics, these traces are in most cases electromagnetic photons that may be created by a wide variety of different processes. For instance, the interaction between the explosion of a supernova and a molecular cloud would lead to highly excited spectral lines that are significantly broadened by the shock interaction between the supernova remnant and the cloud. The scientific task is then to hypothesize a common, local cause for these traces (high excitation, line broadening) and thereby to unify them under one, self-consistent, causal story. In the astrophysical example, broad lines could alternatively be explained by several molecular clouds moving at different velocities along the line of sight. This possible cause of a broad spectral line would, however, not easily explain the existence of highly excited transitions and requires an independent causal story to explain this second trace, whereas a possible shock interaction unifies both.

At first sight, this "Sherlock Holmes" procedure seems like a very uncertain business, given the complexity of possible interactions in a potentially very long causal chain leading up to the remaining traces. However, at least there is a clear methodological direction to go in such a situation—namely, to search for the so-called "smoking guns" (Cleland 2002): traces that are able to discriminate between competing hypotheses, which distinguish one hypothesis as currently being the best explanation. For instance, if a shock interaction is hypothesized, other spectral lines that are predicted to arise in a shock should be observable as well. If they are not seen, alternative explanations of the already observed traces and their respective observable consequences will have to be further investigated. Doing that, the scientist has to rely on what nature has already provided; there is no way to intervene and actively create such "smoking gun" situations.

This fact might give rise to the intuition that historical sciences have to face a methodological disadvantage and are therefore epistemologically inferior to experimental research. However, Cleland claims that the situation is revised by a time asymmetry of nature (Cleland 2002; Lewis 1979) that creates an asymmetry of overdetermination. The laws of nature are directed in time: if an event has occurred, it is very difficult to make things look like nothing had happened. The situation seems analogous to criminal cases, where it is reasonable to hope that the culprit has left some traces that will make it possible to identify him. In this sense, it is difficult to fake past events (i.e., to create all expectable traces artificially without the occurrence of the actual event). Whereas, in an experimental approach, the underdetermination problem makes it difficult to find the "true" (responsible) causes for an observed outcome, the richness of traces in historical science cases can help to distinguish true causes from faked causes. The large number of effects that an event usually creates leads to a so-called *overdetermination of causes* (Lewis 1979). Even if a significant number of traces is erased, the remaining traces might be enough to determine their unifying cause. This, however, does not imply that humans will necessarily have access to these traces: "Traces may be so small, far flung, or complicated that no human being could ever decode them" (Cleland 2002, 488).

Whether this alleged asymmetry really makes it easier to infer past causes (due to causal overdetermination) than to predict the future behavior of a system (due to underdetermination of future events), as Cleland claims, seems less obvious. The large number of potential causally relevant but unidentified auxiliary factors in experimental situations may be faced with a large number of possible explanations that unify the limited set of observed traces in observational situations. Actually, in a thought experiment where an experimental setting is transformed into an observational one (by cutting the possibility of interactions with the experimental setup), it seems possible to objectively cross-link the set of possible alternative explanations with the set of possible causally relevant auxiliary factors. Accordingly, the epistemic situation in terms of underdetermination seems to be very similar for experimental and historical sciences. So far, the argument was based on the "Sherlock Holmes" strategy as found in the historical sciences, which corresponds to the understanding of singular objects and events in astrophysics. However, the astrophysical method is obviously richer than that.

3.2 The Cosmic Laboratory

Although astrophysics shows some resemblances to historical sciences like paleontology or archeology, there is one fundamental difference: astrophysics does not stop at the understanding of the causal history of singular events. Instead, it ultimately tries to find general causal relations with respect to classes of processes or objects, like protostellar molecular outflows, A-type stars, spiral galaxies, or galaxy clusters.[4] This opens new methodological routes to follow: now causal relations can be investigated based on a whole class of phenomena and therefore statistical methods can be applied. In this context, astrophysicists like to refer to the "cosmic laboratory" (e.g., Pasachoff 1977). The universe harbors such a variety of objects in various evolutionary states and conditions that the experimentalist's activity of creating variations of the initial experimental condition might be already set up by the universe itself in all its naturally manifest diversity.

However, the problem remains how to preclude the influence of auxiliary factors or confounding variables on a hypothetical cause–effect relationship. How can the astrophysicist relate a certain behavior or a certain property to a class of objects or processes rather than to the contingent environmental conditions or the specific context of the observation? That is, how can an observed correlation between observational parameters be transformed into a causal relationship?

The classic way to deal with the exploration of effects based on hypothetical causes is the "counterfactual model" (e.g., Shadish, Cook, and Campbell 2002). If a certain effect is observed in a particular situation and ascribed to the existence of one particular factor, the crucial question for the claim of a causal relationship is whether the effect would have also occurred without the respective factor being in play. This method was already briefly described earlier as a way to deal with potentially causally relevant auxiliary factors. The perfect counterfactual inference would compare the occurring effects in two absolutely identical experimental situations that only differ with respect to the factor in question. This is, however, impossible. If an experiment is repeated, something else that was not mirrored by the experimenter might well have been changed and lead to a change in the observed effect that is then incorrectly ascribed to the factor under

[4] Ruphy (2010) points out that it is an interesting topic on its own to study the taxonomies and classifications that are used in astrophysics in order to group entities whose diversity is scientifically investigated. Using the example of stellar taxonomies, she stresses the strong role of epistemic interest in establishing astrophysical classifications, which, for example, usually depend on the particular observational wavelength regime or the observational resolution; these often don't have sharp boundaries when they are based, in fact, on continuous parameters. These properties of astrophysical classifications might give interesting inputs to the monism–pluralism and realism–antirealism debates, as Ruphy shows that the same structural kind–membership condition leads to several possible and interest-dependent groupings of things and that realism about stellar kinds is therefore problematic.

investigation. In the experimental sciences, randomized controlled experiments yield a possible solution to this problem. Two groups of units, the so-called control and the treatment groups, are statistically similar on average but differ with respect to the factor under study. Therefore, the influence of possible confounding variables should, on average, cancel out for both groups and pose no threat to the inferred causal relationship. The challenge for astrophysics, as well as for social scientists, economists, and other scientists who cannot easily perform experiments, is to find a method to deal with the influence of confounding variables that does not rely on experimental interaction with the objects of research.

Two possible methods are usually described in that context (e.g., Dunning 2012; Shadish, Cook, and Campbell 2002): *quasi-experiments* and *natural experiments*. Natural experiments are the direct equivalent of randomized controlled experiments in an observational situation. The idea is that statistically similar groups of units that only differ with respect to one factor may sometimes be created by naturally occurring processes, which accordingly create an "as-if random" assignment. A classic example is the investigation of cholera transmission in nineteenth-century London (Dunning 2012). The question whether cholera could be explained by the theory of "bad air" or was instead transmitted by infected water could be settled due to the particular structure of the London water supply at that time. Two different water companies served most parts of London. One obtained its water from upstream London on the Thames, whereas the other got its water from the Thames inside London, where the water was contaminated by the city's sewage. The anesthesiologist John Snow could then show that the cholera death rate was dependent on the source of water supply. Because both groups of households were statistically equivalent on average, other reasons for the difference in the cholera death rate could be excluded. The main advantage of natural experiments is that their analysis is very simple, without necessarily calling for complicated statistical modeling. If a difference is observed between the control and the treatment group, it is most likely due to the specific factor in which both groups differ. The demanding part, however, is to decide whether the naturally created assignment is indeed "as-if random." This decision usually requires comprehensive qualitative information on the context of the alleged natural experiment. Also, the existence of a natural experiment is a fortunate event that might not be easy to find and that, obviously, is impossible to force.

Quasi-experiments (Campbell and Stanley 1966), in contrast, are experiments in which a control and a treatment group are compared without random assignment being realized. Accordingly, the treatment group may differ from the control group in many ways other than the factor under study. In order to apply the counterfactual inference, all the alternative explanations that rely on other systematic differences between the groups have to be excluded or falsified. The influence of such confounding factors could in turn be investigated by dedicated experiments, but because this is usually a far too complex option to pursue, the confounding factors are often assessed by conventional quantitative methods, such as multivariate regression. Also, numerical models and simulations of the observed phenomenon can be used to explore the possible "covariant" influence of these factors. Another technique to deal with confounding factors is the application of so-called *matching*: the units in comparison are chosen in a way that known confounding variables can be measured and accounted for. The classical example for this method is a medical study that works with twins and can therefore rely on the far-reaching equivalence of the systems under study.

The question now becomes: which of these methods are applied in today's astrophysical research? Does the cosmic laboratory in its vast variety offer as-if randomized assignments that allow for the targeted study of causal relationships? On first sight, a good candidate seems to be the study of evolutionary processes occurring over cosmic time scales. Due to the finite speed of light, astrophysicists are able to observe the past of the universe and thereby populations of cosmic objects in different evolutionary stages. This means that astrophysicists can literally look at the same universe at various earlier points in time. However, the objects in this earlier universe are not statistically equivalent to their present counterparts, apart from their earlier age, because the environmental conditions in the universe have changed in time as well. Therefore, this situation resembles the matching technique, where (hopefully) known confounding factors need to be evaluated separately, rather than it being a truly natural experiment.

In order to evaluate the possibility of the existence of natural experiments in astrophysics more generally, one first needs to identify processes that may create as-if randomized assignments. The answer is not obvious at all, even though cases of so-called *regression-discontinuity designs* (see Dunning 2012) might be conceivable (i.e., cases where two different groups of objects are distinguished relative to a threshold value of some variable). For instance, one could perform a study that compares molecular cloud cores that are just on the brink of gravitational collapse (expressed by the ratio of the gravitational and stabilizing pressure forces) with those cores in the same molecular cloud that have just become gravitationally unstable. However, the existence of natural experiments in an astrophysical context seems to be hindered by the fundamental difficulty involved in obtaining enough qualitative, contextual information on the different groups of objects or processes under study. This contextual information is necessary to make a valid decision on the existence of "as-if randomization." In this example, the application of a threshold criterion is already rather complicated because a decision about which cores are just on the edge of gravitational instability is already very strongly theory laden. Furthermore, astronomy is subject to observational constraints that introduce additional selection effects, thus leading to observational biases. The further the objects are from Earth, the weaker is the received photon flux and the poorer is the spatial resolution of the object for equivalent observations (i.e., if the same telescope, same instrument, same angular resolution, and same signal-to-noise, etc. are used and obtained). If a class of objects is observed at the same distance, it is not clear how representative those particular objects are for other regions of the universe. Also, if objects are observed in different directions, the effects of the ambient interstellar medium along the line-of-sight between the observer and the object needs to be evaluated and compensated for. Factors like this might explain why the concept "cosmic laboratory" is usually not associated with natural experiments and corresponding as-if randomization. The term is instead used in the following sense: astrophysicists can use a multitude of different phenomena provided by the universe in various environments and evolution-ary states to perform/observe quasi-experiments. However, the evaluation of these quasi-experiments requires the application of sophisticated statistical methods and the use of models and simulations, which in turn are often used as substitutes for traditional experiments.

4 MODELS AND SIMULATIONS IN ASTROPHYSICS

Hacking (1989) considered scientific modeling as a particularly important ingredient of astrophysical research: "I suspect that there is no branch of natural science in which modeling is more endemic than astrophysics, nor one in which at every level modeling is more central" (573). Whether one is willing to acknowledge such a special role for astrophysics or not, its scientific practice does appear to be determined by modeling efforts on many different levels. The developmental time scales of many cosmic phe-nomena are so long that directly observing their evolution is simply not possible in a human lifetime. The accessible cosmic sample of objects in various evolutionary states is then reassembled as a self-consistent time series within a given evolutionary model, augmented and aided by simulations.

Models and simulations, however, always rely on idealizations, simplifications, and assumptions about the modeled object itself. Therefore, the question of the reliability and validity of the results obtained by models and simulations is central. In astrophysics, unlike climate science and economics, it can be studied largely independent of political and/or public interest. Because the results of astrophysical simulations cannot be tested experimentally, the only possible empirical test is a comparison with static observations. These usually result in a relatively weak measure of adequacy because exact quantita-tive agreement between simulation and observation is only to be expected in excep-tional cases. Alternatively, different simulations can be compared among each other, and a simulation can be tested for inner consistency and physical correctness (e.g., Sundberg 2010).

The debate concerning models and simulations is of course an independent and vibrant activity within the philosophy of science community. Only a handful of publica-tions, however, explicitly deal with the more philosophical or foundational aspects of modeling and simulations in astrophysics. Numerical models of astrophysical phenom-ena are often developed and refined over a long period of time and can involve a large number and several generations of collaborators. A representative example of a collec-tive and international modeling process, one comprising a long-standing sequence of model modifications, data generation, and investigation of theoretical implications, was reconstructed by Grasshoff (1998). He investigated how the collective of participat-ing researchers interacts despite their great diversity with respect to their experience,

knowledge, and understanding of the overall model. The split-up of the model of a phenomenon into partial submodels, which are easier to handle, has been described by
Bailer-Jones (2000). These submodels are easier to handle but need to be theoretically
and empirically consistent if they are to be embedded at a higher level of application
and generalization. As an important aspect for the recreation of the unity of various
submodels, she identifies the visualization of the phenomenon that supplies concrete
interpretations of these submodels. However, the often contingent choice of submodels and the rigidity of models that grow over generations of modelers are aspects that
calls for critical reflection. Using the example of cosmological simulations and modeling of the Milky Way, Ruphy (2011) stressed that many simulations are theoretically
not well-constrained. There are many possible paths of modeling opened by the necessary choices between different submodels. This yields a potential or also actual plurality of models that all claim to model the same cosmic phenomenon. Typically, different
models are mutually incompatible but still empirically adequate to a similar degree. This
situation creates a problem if it is possible to adjust the models to new empirical data by
retroactively increasing the complexity of the new model while keeping their previous
modeling contents. This occurs in situations where no testing of the ingoing submodels
is even possible, which is often the case in astrophysics. In such a situation, it becomes
impossible to claim an understanding of the real world from the models just on the basis
of their empirical adequacy. At best, these are plausible fits to the limited input data.

That said, the current modeling practice in astrophysics appears highly heterogeneous, depending on the maturity and developmental stage of the astrophysical subdiscipline, modulated by the richness and availability of observational input. Simulations
of extragalactic objects confronted with large uncertainties of ingoing theories and
severe empirical constraints seem to struggle with different problems from those of the
simulations of cosmic phenomena that can be observed within our Milky Way with high
spatial resolution using different information channels.

It is important to note that models also play an important role within astronomical
data acquisition and reduction. For instance, the recording of data using a single dish
radio telescope requires a model of the mechanical and optical properties of the telescope mirror in different positions in order to determine the exact pointing position.
For the calibration of data with respect to atmospheric influences, a model of the Earth's
atmosphere is needed. Flux calibration presupposes models of the individual stars and
planets used in the calibrating observations. As in many complex scientific disciplines,
astrophysics is subject to a "hybrid" solution, where a clear distinction between empirical data and model-based interpretation is becoming increasingly difficult. This situation might challenge elements of the modeling discussion that have occurred so far (e.g.,
Morrison 2009).

Moreover, planned or newly commissioned instruments, such as the ALMA inter-
ferometer, bring about a tremendous improvement to the empirical observational
basis while, at the same time, the need for easy-to-use, standardized models is growing.
Thereby, questions for the verification, validation, and standardization of simulations

gain additional importance, emphasized by a common differentiation of labor between modelers and model-users.

5 ASTRONOMICAL DATA

Astrophysics is fast becoming a science that is confronting many large, multi-wavelength surveys and dealing with huge amounts of data. For example, the planned Square Kilometre Array will create data at a rate of many petabytes per second. Accordingly, in addition to models and simulations, data handling is another central component of astrophysical research that is worthy of philosophical reflection. In order to do so, it, too, can draw on contemporary discussions from the philosophy of science.

Suppes (1962) first spoke of "data models" by pointing at a hierarchy of models that link raw data to theory. According to Suppes, the relationship between a data model and the underlying data is given by a detailed statistical theory of the "goodness of the fit." The semantic concept of models of theories is thereby extended in its application toward models of experiments and models of data. Harris (2003) has illustrated this concept of data models using planetary observations as an example. He shows that bad observations of planetary positions are first dismissed, and the remaining data points are smoothed, such that the finally plotted path of the planet would constitute a data model: on the one hand, the plot is similar to the original data; on the other hand the plot is an idealization that is qualitatively different from the data. At the same time, the very concept of data models underwrites the inevitability of there being different ways to model data. In that sense, the concept of unprocessed "raw data" is of questionable value, surely, because "the process of data acquisition cannot be separated from the process of data manipulation" (Harris 2003, 1512). Even for the simplest instruments, it is necessary for the scientist to possess certain learned skills to properly read the data.

With their distinction between data and phenomena, Bogen and Woodward (1988) drew attention to experimental practice within science, specifically the practice of data analysis and data selection, which had been neglected within the theory-dominated discussions of the philosophy of science. The strategies for the prevention of errors and the extraction of "patterns" that yield the properties of the underlying phenomena thereby constitute an epistemology of the experiment, which has become another active branch within the philosophy of science. It is interesting to apply these strategies to astrophysical data analyses. Although the strategies applied are very similar to experimental physics, knowledge of the details of data generation that is needed to distinguish real features in the data from mere artifacts might not always be easily accessible given the ever-growing differentiation of labor in observational astrophysics (e.g., Anderl 2014).

There are several ways in which astronomical observations might actually be made (Jaschek 1989; Longair, Stewart, and Williams 1986; Zijlstra, Rodriguez, and Wallander 1995). Not all of these modes are necessarily made available at any given observatory, telescope, or instrument due to practicalities, cost factors, and/or institutional policies. The four most common modes are:

1. *Classical observing*: Here, the astronomer carries out the observations herself at the telescope itself, in real time, and she is commonly assisted by a local team consisting of telescope operators, staff scientists, and instrument specialists.
2. *Remote observing*: Here, the local staff actually operate the telescope while the astronomer is not physically present, but she still has the authority and responsibility to decide in real time on the target selection, integration times, filters, and the like.
3. *Queue observing*: In this case, the instrument is operated under preprogrammed and advanced-scheduled remote control. This is generally the case for unmanned telescopes, but it is also a mode undertaken in order to optimally schedule telescopes at many remote facilities.
4. *Service observing*: In this instance, the astronomer provides the technical staff at the telescope with the necessary details needed to perform a complete suite of her observations. The instructions are provided in advance of scheduled observations, they are executed without her being present or online, and the data are sent to her after the observations have been made.

The history seems to indicate that service observing is becoming quite popular as far as ground-based observations are concerned. However, even if the astronomer travels to the site herself, the technical complexity of the instruments involved usually requires specialized technicians to be present who know the specifics of the instrument and how to achieve optimal performance. It is interesting that, in the case of service mode observing, a change of subject takes place with the concomitant possibility of information loss between the recording of raw data and the data reduction and subsequent data analysis. In order to distinguish valid results from instrumental artifacts or blunders, it is necessary that the astronomical instruments are understood at a relatively high level, and it is to be hoped that no relevant information concerning the process of data generation is undocumented or lost. Accordingly, it is a challenge for the design of archival astronomical databases to provide as much of that information on data generation and applied data manipulation as possible.

However, an epistemological analysis of data generation and data treatment has so far not thoroughly been conducted from a philosophical perspective within modern astrophysics. In contrast, the social sciences are home to a broad discussion on topics such as data access, databases, and increasing data intensity of scientific cooperations. This discussion also refers to examples from astrophysics (e.g., Collins 1998; Sands, Borgman, Wynholds, and Traweek 2012; Wynholds, Wallis, Borgman, Sands, and Traweek 2012). A philosophical investigation of data generation, selection, and reduction within astronomy appears particularly interesting because astrophysicists like to stress their status as passive observers: they only gather information that can be received from the universe.

In fact, modern observational methods require much more manipulation and inter-action than the classic picture of the astronomer looking through his or her telescope might suggest. Even before the so-called raw data is generated, many decisions have to be made that depend on the observer's intention with respect to the usage of the data. Furthermore, the calibration of the telescope, the calibration of the data regarding the atmosphere's influence, and the calibration of the receiver used are very complex pro-cesses that rely on assumptions and models. After the user has received the data, he or she is confronted with various tasks: sorting out bad data, data calibration, searching for systematic errors and, if present, correcting for them, and finally visualizing the data so that a scientific interpretation is possible. Depending on the observational tech-nique and instrumentation used, this process of data reduction can become extremely complex.

One extreme case in that respect is given by the technique of interferometry, in which observations acquired by multiple telescopes are pairwise combined in order to simu-late one large telescope having a resolution corresponding to the distance between the two most widely separated elements of the array of telescopes. This technique relies on the measurement of Fourier components of the source distribution on the sky so that the intensity distribution needs to be determined by means of a Fourier transforma-tion. At the same time, information on the underlying spatial intensity in the source plane is lost due to an incomplete sampling of the virtual surface of the simulated large telescope. The central step within the data reduction therefore cannot unambiguously reconstruct the real source distribution, but rather it can provide a plausible reconstruc-tion that is compatible both with the input data and with the real intensity distribution of the source on the sky. This "data inversion" problem might well serve as an example of true "underdetermination" in astrophysics.

In any case, with respect to the complexity of data generation and processing, there seems to be no obvious difference between astronomy and experimental sciences. It is interesting to note that this fact is even acknowledged by Hacking (1989), although he so firmly stresses the difference between astronomy and the experimental sciences.[5]

The more complex the process of data selection and analysis becomes, the more the processing of data also relies on experience. Accordingly, instructions on astronomical data reduction are not found in textbooks and often only briefly in documentations of the respective software packages. The practice of data reduction is instead transferred directly among colleagues in schools or workshops. This practice contains strong ele-ments of what Polanyi (1958) described as "tacit knowledge," distinguished from explicit knowledge (see also Collins 2010). The existence of tacit knowledge can become a prob-lem if it causes a loss of information with respect to data genesis and the context of data processing when data gets transferred. Within modern astrophysical practice, which

[5] "It is sometimes said that in astronomy we do not experiment; we can only observe. It is true that we cannot interfere very much in the distant reaches of space, but the skills employed by Penzias and Wilson [the discoverers of the cosmic microwave background] were identical to those used by laboratory experimenters" (Hacking 1983, 160).

works increasingly based on the division of labor and is directed toward a multiple use of data, the question arises of how databases and corresponding data formats can be organized to prevent a negative impact on the meaningfulness of the data.

6 CONCLUSION

Modern noncosmological astrophysics is a scientific discipline that has so far not been prominently discussed within the philosophy of science. However, the richness, complexity, and extremity of its objects of research, together with its characteristic methodology as an observational science, make astrophysics an interesting field for philosophical analysis. In its attempt to understand the cosmic history of distinct singular objects and processes, it resembles historical sciences such as archaeology or paleontology, whereas its effort to derive general claims on the behavior of classes of objects and processes is reminiscent of social sciences, neither of which are in a position to perform experiments. All these activities, however, are backed up by far-reaching modeling and simulation efforts. The question for the possible scope and inherent limits of such models is pressing in the case of astrophysics because the possibilities of verification and validation of these models are usually limited. Astrophysics is dealing with ever-growing amounts of data that are usually generated by large international observatories and finally delivered to the scientific user. The complexity of data generation, data reduction and analysis, and the reuse of data from large databases calls for a critical reflection on these practices and their epistemic premises. This article is intended to encourage philosophical interest in astrophysical research because astrophysics may offer a wealth of novel, interesting, and as yet still untreated questions and case studies that may yield new impetus to various philosophical discussions.

REFERENCES

Anderl, S. (2014). *Shocks in the Interstellar Medium*. Ph.D. thesis, Rheinische Friedrich-Wilhelms-Universität Bonn. urn:nbn:de:hbz:5n-34622.

Bailer-Jones, D. M. (2000). "Modelling Extended Extragalactic Radio Sources." *Studies in History and Philosophy of Modern Physics* 31(1): 49–74.

Bogen, J., and Woodward, J. (1988). "Saving the Phenomena." *Philosophical Review* 97(3): 303–352.

Campbell, D. T., and Stanley, J. C. (1966). *Experimental and Quasi-experimental Designs for Research* (Chicago: Rand McNally).

Cleland, C. E. (2002). "Methodological and Epistemic Differences Between Historical Science and Experimental Science." *Philosophy of Science* 69: 474–496.

Collins, H. (1998). "The Meaning of Data: Open and Closed Evidential Cultures in the Search for Gravitational Waves." *American Journal of Sociology* 104(2): 293–337.

Collins, H. (2010). *Tacit and Explicit Knowledge* (Chicago: University of Chicago Press).

Dunning, T. (2012). *Natural Experiments in the Social Sciences* (Cambridge: Cambridge University Press).

Grasshoff, G. (1998). "Modelling the Astrophysical Object SS433—Methodology of Model Construction by a Research Collective." *Philosophia Naturalis* 35: 161–199.

Hacking, I. (1983). *Representing and Intervening. Introductory Topics in the Philosophy of Natural Science* (Cambridge: Cambridge University Press).

Hacking, I. (1989). "Extragalactic Reality: The Case of Gravitational Lensing." *Philosophy of Science* 56: 555–581.

Harris, T. (2003). "Data Models and the Acquisition and Manipulation of Data." *Philosophy of Science* 70(5): 1508–1517.

Jaschek, C. (1989). *Data in Astronomy* (Cambridge: Cambridge University Press).

Laudan, L., and Leplin, J. (1991) "Empirical Equivalence and Underdetermination." *Journal of Philosophy* 88: 449–472.

Lewis, D. (1979). "Counterfactual Dependence and Time's Arrow." *Noûs* 13: 455–476.

Longair, M. S., Stewart, J. M., and Williams, P. M. (1986). "The UK Remote and Service Observing Programmes." *Quarterly Journal of the Royal Astronomical Society* 27: 153–165.

Morrison, M. (2009) "Models, Measurement and Computer Simulation: The Changing Face of Experimentation." *Philosophical Studies* 143(1): 33–57.

Pasachoff, J. M. (1977). *Contemporary Astronomy* (New York: Saunders College Publishing).

Polanyi, M. (1958). *Personal Knowledge: Towards a Post-Critical Philosophy* (Chicago: University of Chicago Press).

Reiner, R., and Pierson, R. (1995). "Hacking's Experimental Realism: An Untenable Middle Ground." *Philosophy of Science* 62(1): 60–69.

Resnik, D. B. (1994) "Hacking's Experimental Realism." *Canadian Journal of Philosophy* 24(3): 395–412.

Ruphy, S. (2010). "Are Stellar Kinds Natural Kinds? A Challenging Newcomer in the Monism/Pluralism and Realism/Antirealism Debates." *Philosophy of Science* 77(5): 1109–1120.

Ruphy, S. (2011). "Limits to Modeling: Balancing Ambition and Outcome in Astrophysics and Cosmology" *Simulation & Gaming* 42(2): 177–194.

Sandell, M. (2010). "Astronomy and Experimentation." *Techné* 14(3): 252–269.

Sands, A., Borgman, C. L., Wynholds, L., and Traweek, S. (2012). "Follow the Data: How Astronomers Use and Reuse Data." *Proceedings of the American Society for Information Science and Technology* 49(1): 1–3.

Shadish, W. R., Cook, T. D., and Campbell, D. T. (2002). *Experimental and Quasi-Experimental Design for Generalized Causal Inference* (Boston/New York: Houghton Mifflin).

Shapere, D. (1993). "Astronomy and Antirealism." *Philosophy of Science* 60: 134–150.

Stanford, K. (2013). "Underdetermination of Scientific Theory." In E. N. Zalta (ed.), *The Stanford Encyclopedia of Philosophy (Winter 2013 Edition)*. http://plato.stanford.edu/archives/win2013/entries/scientific-underdetermination/

Sundberg, M. (2010). "Cultures of Simulations vs. Cultures of Calculations? The Development of Simulation Practices in Meteorology and Astrophysics." *Studies in History and Philosophy of Modern Physics* 41: 273–281.

Suppes, P. (1962). "Models of Data." In E. Nagel, P. Suppes, and A. Tarski (eds.), *Logic, Methodology, and Philosophy of Science: Proceedings of the 1960 International Congress* (Stanford, CA: Stanford University Press), 252–261.

Wynholds, L. A., Wallis, J. C., Borgman, C. L., Sands, A., and Traweek, S. (2012). "Data, Data Use, and Scientific Inquiry: Two Case Studies of Data Practices." *Proceedings of the 12th ACM/IEEE-CS Joint Conference on Digital Libraries*: 19–22.

Zijlstra, A. A., Rodriguez, J., and Wallander, A. (1995). "Remote Observing and Experience at ESO." *The Messenger* 81: 23–27.

CHAPTER 32

CHALLENGES TO EVOLUTIONARY THEORY

DENIS WALSH

THE October 9, 2014, issue of *Nature* carried a discussion piece titled "Does Evolutionary Theory Need a Re-Think?" Two groups of authors lined up on either side of the issue, one promoting a radical and immediate root-and-branch revision ("Yes, Urgently"; Laland et al. 2014) and the other offering a reassuring emollient ("No, All Is Well;" Wray et al. 2014). This is one installment in an increasingly common call for the re-evaluation, extension, or outright rejection of the modern synthesis theory of evolution.[1] That this exchange should appear in a mainstream journal such as *Nature*, however, underscores the point that the modern synthesis theory of evolution is now being subject to a level of critical scrutiny the likes of which it has not experienced in its almost 100-year history.

At the heart of the challenge is a dispute about the place of organisms in evolution. Darwin's theory of evolution takes organismal "struggle for life" as the principal cause of evolution; its twentieth-century successor does not. Shortly after its inception, the modern synthesis theory of evolution took a "genetic turn." As early as 1937, Dobzhansky's *Genetics and the Origin of Species* defined evolution as "changes in the genetic structure of a population." The gene became the canonical unit of biological organization, and the study of evolution became the study of gene dynamics.

As the twentieth century gave way to the twenty-first, murmurings of discontent began to be heard: "Perhaps we have hidden behind the Modern Synthesis, and the idea that all the action is in gene frequencies, for too long" (Weiss and Fullerton 2000: 192). Wallace Arthur noted that "the current mainstream theory of evolutionary mechanism, is in several respects lopsided and incomplete" (2000: 56). As the century progresses, calls for reasserting the place of organisms in evolution grow ever more acute. One particularly compelling challenge is raised by the advent of what is being called "evolutionary-developmental biology" ("evo-devo"; Hall 1999, 2012; Carroll, Grenier,

---

[1] See Hall (1999), Müller (2007), Pigliucci (2009), Maienschein and Laubichler (2014).

and Weatherbee 2000; Robert 2004; Müller 2007; Morange 2011; Irschick et al. 2013)—
and its complementary successor developmental-evolutionary biology ("devo-evo";
Laubichler 2009; Maienschein and Laubichler 2014). Other challenges to the modern synthesis have also arisen in the form of developmental systems theory (Oyama,
Griffiths, and Gray 2001; Griffiths and Gray 2001), niche construction theory (Odling-
Smee, Laland, and Feldman 2003), systems biology (Kitano 2004), and reactive genome
approaches (Shapiro 2011, 2013; Noble 2006). This is a motley of alternative takes on
evolution to be sure, but they share a common ground. Each presses the claim that the
heretofore neglected processes occurring within organisms during their development
are crucial to an understanding of evolution.

Together these organism-centered approaches to evolution suggest that gene-
centered conceptions of evolution leave gaps in our understanding. These lacunae occur
precisely where the activities of organisms make a difference to evolution. It is hardly
surprising, then, that the modern synthesis theory should have recently come under
such intense scrutiny. Yet the debate about whether the synthesis should be revised,
rejected, or reaffirmed has been inconclusive. One reason, amply demonstrated by the
exchange in *Nature*, is that it is not entirely clear what would count as a revision, a rejection or a reaffirmation, of the modern synthesis. For all its power and scope, the boundaries of the modern synthesis are a little vague. Throughout its century of dominance,
and in the absence of a serious contender, the fundamental precepts of the modern
synthesis theory have gone largely unarticulated. Before we ask whether these newly
minted organism-centered conceptions of evolution pose a threat to the core commitments of the modern synthesis, we need to know what those commitments are.

1 MODERN SYNTHESIS

The modern synthesis theory of evolution was forged from its Darwinian predecessor
early in the twentieth century. It can be seen as a powerful and elegant solution to a set
of problems that attended Darwin's own evolutionary thinking. Darwin surmised that
when the individuals in a population vary in their capacity to survive and reproduce,
the population will tend to change in its structure. It will come to comprise individuals
increasingly well suited to their conditions of existence. But four conditions must be
met. (i) Individuals must survive long enough to reproduce, of course, and that entails
developing from inception to reproductive age. (ii) In order for those traits that dispose
individuals to thrive in the struggle for life to increase in frequency, they must be heritable. (iii) For population change to be adaptive, there must be some biasing process.
(iv) For adaptive change to continue generation on generation, there must be a constant
source of new evolutionary novelties. Thus Darwin's process of evolution comprises
four component processes: development, inheritance, adaptive population change, and
the origin of evolutionary novelties.

The modern synthesis theory of evolution provides an elegant treatment of each of these component processes and offers a highly compelling account of the relation between them. It does so by positing the gene, rather than the organism, as the canonical unit in each process. The genetic turn enacted by the modern synthesis is dependent on two conceptual innovations: the genotype/phenotype distinction and the fractionation of evolution. Together these secure the centrality of the gene that forms the foundation of the modern synthesis.

1.1 Genotype and Phenotype

The process of inheritance was particularly problematic for Darwin; it appears to make contradictory demands. On the one hand it appears to be blending, in the sense that offspring generally resemble an amalgam or a midpoint between their parents' traits. On the other hand, it is conserving: traits recur generation on generation. They can even disappear from a lineage for a generation or more, only to reappear in an undiluted form. Blending and conserving seem to pull in opposite directions: How can a trait be preserved in a population unchanged if it becomes increasingly rarefied generation after generation?. The blending of inheritance is not just puzzling; it poses a threat to Darwinism. As Fleeming Jenkin (1867) argued, blending would tend to eliminate an exceptional trait from the population through rarefaction faster than it would be promoted by selection. Darwin puzzles:

> The laws governing inheritance are quite unknown; no one can say why the same peculiarity in different individuals of the same species, and in individuals of different species, is sometimes inherited and sometimes not so, why the child often reverts in certain characters to its grandfather or grandmother.(1859/1968: 13)

Three conceptual shifts helped to solve the problems of inheritance. In so doing, they also sowed the seeds for the gene-centered modern synthesis conception of evolution. These are (i) the rediscovery of Mendel's theory of inheritance, (ii) the genotype/phenotype distinction, and (iii) the Weismann doctrine of the continuity of the germline. The rediscovery of Mendel's experiments around the turn of the century introduced the notion that inheritance is underpinned by factors, one copy of each of which was passed unchanged from parent to offspring. To the extent that the offspring is an amalgam of the parents' traits, it is not as a consequence of a dilution or mixing of inherited factors themselves but as a result of their joint expression. Blending occurs in development. So Mendelism opened up a conceptual gap between the processes of inheritance and development. The introduction of the concepts of the genotype and phenotype by Johansen further entrenched the distinction. Genotypes are inherited; phenotypes are not (or only derivatively). Genotypes are turned into phenotypes in the process of development.

In drawing a distinction between inheritance and development, the genotype/phenotype distinction inaugurated a decisive move away from the Darwinian notion of inheritance as a gross pattern of resemblance and difference.

> Such a distinction made it possible to go beneath inheritance as the mere morphological similarity of parent and offspring and investigate the behaviour, and eventually the nature, of the underlying factors (genes) that were transmitted from one generation to the next. (Roll-Hansen 2009: 457)

Inheritance becomes the transmission of factors (later genes or replicators).

The alienation of inheritance from development was further promoted by August Weismann's discovery of the sequestration of the germline. Weismann noticed that an asymmetry emerges early in the development of metazoan embryos—in the blastula stage. There is a separation between the cells that contribute to the germline and those that go on to build the soma. The entire contents of material passed on from parent to offspring in inheritance reside within the cells of the germline. The "sequestered" germline cells are unaffected by any changes that occur within the somatoplasm.

The Weismann doctrine applies only to metazoans. However, there is a further alleged principle of molecular biology that appears to generalize Weismann's sundering of development from inheritance—the central dogma of molecular biology coined by Frances Crick: "It states that such information cannot be transferred back from protein to either protein or nucleic acid" (1970: 561). Taking DNA structure to be the material of inheritance, and protein synthesis to be the process of development, yields something akin to the Weismann doctrine *without* requiring germline sequestration: the material of inheritance is unaffected by the process of development.

> If the central dogma is true, . . ., this has crucial implications for evolution. It would imply that all evolutionary novelty requires changes in nucleic acids, and that these changes—mutations—are essentially accidental and non-adaptive in nature. Changes elsewhere . . . would have no long-term evolutionary effects. (Maynard Smith 1998: 10)

1.2 The Fractionation of Evolution

The genotype/phenotype distinction did not just facilitate the separation of inheritance from development; it also permitted a comprehensive fractionation of evolution into four distinct, quasi-independent component processes: inheritance, development, adaptive population change, and the origin of evolutionary novelties. The great methodological virtue of modern synthesis theory is that it holds these processes to be discrete and to a significant degree mutually independent. Inheritance, for its part, does not affect adaptive population change. Organisms inherit not what would be adaptive for them, but only what their parents give them, adaptive or not. Nor does inheritance contribute to the origin of novelties. Development is affected by inheritance in the sense that organisms just develop what they inherit. But crucially development leaves no trace on the inherited material and introduces no evolutionary novelties. Nor does development impart an adaptive bias into evolution; it is fundamentally conservative. If inheritance, development, and the origin of novelties are not the source of the adaptive bias in evolution, there must be some further, independent process that promotes the adaptiveness of evolutionary change. That process is natural selection.

Evolution may comprise four quasi-independent processes, yet there is one kind of entity that unites them: the gene (or replicator). Genes are units of inheritance, as inheritance is simply just the transmission of genes. Genes are the privileged explanatory units of development. Genes, it is said, embody a "code script," a program or blueprint for building an organism. Development, then, is just the process of translating these genetic instructions into biological form. Genes play a pivotal role in the adaptiveness of evolution. Natural selection is defined and measured in modern synthesis evolution as change in the gene structure of a population. Natural selection tends to eliminate the store of genetic variants, so there must be a source of new genetic variations. Moreover, these variants must be adaptively unbiased. The random mutation of genes is the ultimate source of evolutionary novelties.

Fractionation is the linchpin of modern synthesis evolutionary thinking. It is fractionation that secures the exclusive theoretical privilege enjoyed by genes in modern synthesis evolutionary theory. Furthermore, it is fractionation that renders organismal development virtually irrelevant to evolution. If fractionation goes, so too does gene-centrism. Implicitly, the challenges to modern synthesis theory seem to be challenges to the modern synthesis fractionated picture. The challenges are of two general kinds. Either they contend that genes do not have a privileged role to play in inheritance, development, adaptive population change, or the generation of evolutionary novelties, *or* they seem implicitly to be arguing that the component processes of evolution are not so discrete and independent as the fractionated picture makes out. So, it is worthwhile holding the fractionated picture of evolution up to close examination.

2 THE NATURE OF INHERITANCE

The concept of inheritance has evolved as evolutionary theory has (Gayon 2000; Müller-Wille and Rheinberger 2012). Inheritance has gone from being a pattern of resemblance and difference between lineages in Darwin's theory to the transmission of replicators in the modern synthesis. Given that the concept of inheritance is revisable, beholden to its place in evolutionary theory, it is worth asking whether the distinctly twentieth-century concept of transmission of replicants is still up to the task of accounting for everything our twenty-first-century evolutionary biology needs from the concept of inheritance.

There are two types of critique of inheritance-as-transmission. One suggests that processes other than transmission of replicators contribute to the pattern of resemblance and difference required for evolution. The other line of argument points out that the modern synthesis reconceptualization of inheritance erroneously presupposes that one can differentiate the role of genes from the role of other processes in causing the pattern of resemblance and difference that is necessary for evolution.

2.1 Inheritance Pluralism

Many of those who plump for a revision of the modern synthesis conception of evolution point out that the pattern of resemblance and difference required for evolution is secured by much more than simply the process of gene transmission (Jablonka and Lamb 2004). Control over the *pattern* of inheritance is not located in genes alone. Rather, it is distributed throughout the gene/organism/environment system. One increasingly popular way of giving voice to this idea of a plurality of inheritance mechanisms is to enumerate discrete "channels" through which inheritance is mediated (Lamm 2014).

2.1.1 *Genetic Inheritance*

Of course the pattern of inheritance is partially mediated by the transmission of genes. Genetic inheritance is ideally suited to play the role of a medium of inheritance. Nucleotide structure is stable and copied with high fidelity. It is reliably passed from generation to generation. As DNA is modular, each portion is buffered from the effects of changes elsewhere in the system, and, as it consists in a digital code, it is unlimited in its size and range of possible varieties (Szathmáry 2000). But it is not the only mechanism that secures this pattern, nor does it do so in isolation from extragenetic causes.

2.1.2 *Epigenetic Inheritance*

We know that environmental stress can induce alterations to the methylation pattern of DNA. Methylation occurs when a methyl group is added to a guanine or cytosine base on a strand of DNA. The methyl group regulates the transcription of the gene to which it is attached. A remarkable study of the persistent effects of the extended famine in Holland in the winter of 1944–1945 demonstrated that early exposure of human fetuses to famine conditions is associated with a decreased methylation of insulin-like growth factor genes (Heijmans et al. 2008). These methylation patterns are associated with diminished growth rates, and the diminished growth rates in stressed females are passed on to their offspring (Lumey 1992).

Other studies demonstrate that methylation patterns induced in single-celled organisms, and in somatic cells of the metazoa and metaphyta, can be passed directly to daughter cells. Such epigenetic inheritance of acquired methylation patterns is especially well documented in metaphytans. These look like prima facie cases of the transgenerational stability of environmentally induced alterations that do not require genetic accommodation.

Methylation patterns appear to constitute an alternative mode for the transmission of characters. They provide "additional sources of inheritance" (Echten and Borovitz 2013). Significantly, the characters transmitted from parent to offspring can be acquired during the life of an individual.

> We suggest that environmental induction of heritable modifications in DNA methylation provides a plausible molecular underpinning for the still contentious paradigm of inheritance of acquired traits originally put forward by Jean-Baptiste Lamarck more than 200 years ago. (Ou et al. 2012: e41143)

Increasingly, the importance of transmitted epigenetic markers is being realized.[2] This too certainly looks like a violation of the independence of inheritance from developmentally induced changes to form enshrined in the Weismann doctrine.

The transmission and imprinting of epigenetic methylation "markers" looks like a distinctly nongenetic mechanism for the passing on of crucial developmental resources. This appears to be a separate inheritance channel that operates independently of the replication of DNA.

> Heritable epigenetic variation is decoupled from genetic variation *by definition*. Hence, there are selectable epigenetic variations that are independent of DNA variations, and evolutionary change on the epigenetic axis is inevitable. The only question is whether these variations are persistent and common enough to lead to interesting evolutionary effects. (Jablonka and Lamb 2002: 93)

Once one is prepared to countenance nongenetic channels of inheritance, other modes become readily apparent. A partial list suffices to demonstrate the multiplicity of modes of inheritance.

2.1.3 *Behavioral inheritance*

Behavioral inheritance systems involve those in which offspring learn behaviors from parents or others. A classic example is found in the spread of the ability of great tits in Britain to learn to extract the cream from milk bottles traditionally left on the doorsteps of houses. Further examples include dietary preferences that are transmitted to mammalian fetuses though the placenta.

2.1.4 *Cultural inheritance*

Languages, technological skills, social behaviors, cultural norms, and traditions are all passed from one generation to the next through learning, the maintenance of written records, and enculturation.

---

[2] For example, Guerrera-Bosagna et al. (2013) document the transgenerational inheritance of markers induced by environmental toxicants in mice.

2.1.5 *Environmental inheritance*

The proper development of organisms and the resemblance between parent and off-spring depends heavily on environmental factors. A parent does not only transmit genes and epigenetic resources; it provides a setting for its offspring. The importance of this setting cannot be overestimated (Gilbert and Epel 2009).

Examples of environmentally moderated inheritance abound. Proper growth and development in mammals, for example, requires the correct gut flora (Gilbert and Epel 2009). Failure to acquire the appropriate gut flora can lead to inability to digest food, immunodeficiency, morphological abnormalities, and cognitive deficiencies. Many young acquire their gut flora from their mothers at birth as they pass through the birth canal. Female birds frequently confer on their offspring immunity to specific pathogens by placing specific antibodies in the yolk of their eggs. Hatchlings assimilate the antibodies and hatch with an immunity that has not been acquired in the normal way. Agrawal, Laforsch, and Tollrian (1999) demonstrated the transgenerational transmission of environmentally induced predator defenses in both animals and plants.

Such examples are legion and widely known.[3] They certainly appear to suggest that the pattern of resemblance and difference that constitutes our pretheoretical—or, better, "premodern synthesis"—conception of inheritance is achieved by much more than the transmission of replicators (Jablonka and Lamb 2004).

Inheritance pluralism certainly seems to erode the presumptive primacy of gene transmission in securing the pattern of inheritance. It suggests, for example, that there can be cases that qualify as instances of genuine Darwinian evolution that do not count as modern synthesis evolution. Nevertheless, inheritance pluralism fails to refute the modern synthesis conception of inheritance. Nor does it particularly invalidate the gene as the unit of inheritance. The reason is that the modern synthesis has effectively engineered a *redefinition* of inheritance. Under the auspices of the modern synthesis theory, inheritance is no longer a pattern of resemblance and difference; it has become a process of transmission. Modern synthesis orthodoxy in effect repudiates the intuition upon which inheritance pluralism is based—namely, that any process that produces the relevant pattern of resemblance and difference is a mechanism of inheritance. There is more to this defense than mere doctrinaire intransigence. A successful theory earns the right define its proprietary concepts to suit its needs. The modern synthesis has done so with the concept of inheritance.

2.2 Inheritance Holism

That is not to say that the modern synthesis conception of inheritance is sacrosanct. The gambit of supplanting the pattern of resemblance with the process of gene transmission can be questioned on other grounds. The rationale for privileging genes in inheritance is

---

[3] Gilbert and Epel (2009) provide an impressive compendium of environmentally inherited traits.

predicated on the idea that the pattern of resemblance and difference required for evolution can be decomposed into discrete genetic and extragenetic components. The genetic causes of inheritance can be deemd to be the most significant only if the respective contributions of genes can be differentiated from the contribution of those of other factors.

This kind of causal decomposition seems to be the legacy of classical genetics. The guiding project of classical genetics was to correlate heritable differences between organisms with corresponding differences in chromosome structure. The locations on chromosomes that are indexes of phenotypic differences between lineages came to be considered the principal causes of those differences. As classical genetics matured into molecular genetics, these markers of phenotypic differences became discernible regions of DNA that encoded instructions for building phenotypes (Waters 1994). In the process, these regions—genes—became not just the principal causes of phenotypic *differences*; they also became the principal causes of *phenotypes*.

The mode of inference that moves from differences in effect to causes of those differences (and from causes of differences to principal causes of effects) is based on the method of difference, first articulated by John Stuart Mill (1843). Mill's method of difference allows us to make an inference from the difference in effect in the absence of some causal factor to the causal contribution that that factor makes when it is there. Mill recognizes that his "canon" of causal inference makes certain metaphysical assumptions, specifically, that causes compose in an orderly way that permits their decomposition: "I shall give the name of the Composition of Causes to the principle which is exemplified in all cases in which the joint effect of several causes *is identical with the sum of their separate effects*" (Mill 1843: 243, emphasis added). The canon assumes that in removing one cause, we leave the others more or less unaffected. That is tantamount to saying that the component causes of a system are relatively insensitive to our interventions and are reasonably independent of one another.

The modern synthesis conception of the gene as an individual unit of inheritance was established through the application of Mill's method. As Waters (2007) notes, the difference principle was consciously deployed in the experiments of T. H. Morgan and his colleagues:

> If now one gene is changed so that it produces some substance different from that which it produced before, the end-result may be affected, and if the change affects one organ predominatingly it may appear the one gene alone has produced this effect. In a strictly causal sense this is true, but the effect is produced only in conjunction with all the other genes. In other words, they are all still contributing, as before, to the end-result which is different in so far as one of them is different.(Morgan 1926: 305–306)

The method of difference is an elegant example of what Nancy Cartwright (1999) calls the "analytic method." It involves intervening on systems to isolate the causal contributions of their component parts, under the assumption that the interventions change only the component intervened upon. According to the analytic method, to understand what happens in the world, we take things apart into their fundamental pieces; to control a situation we reassemble the pieces, we reorder them so they will work together to make things happen as we will. You carry the pieces from place to place, assembling them together in new ways and new contexts. But you always assume that they will try to behave in new arrangements as they have tried to behave in others. They will, in each case, act in accordance with their nature. (Cartwright 1999: 83)

The passage quoted from Morgan states the assumptions explicitly. In inferring the activities of a gene one supposes that the other components of the system "are still contributing, *as before*, to the end-result."

However, there is a wrinkle. While Mill is fairly sanguine about the general applicability of his method, he acknowledges that it is inappropriate where the relevant causes do not compose. He is also explicit that composition fails fairly comprehensively in living systems: "it is certain that no mere summing up of the separate actions of those elements will ever amount to the action of the living body itself" (1843: 243). Where compositionality breaks down, so does the power of the method of difference to isolate and differentiate the discrete contribution of a component cause to a complex effect.

The conditions for applying the method of difference are routinely violated in those systems in which genes do their work (Ciliberti, Martin, and Wagner 2007; Davidson 2006; Wagner 2005, 2007, 2011). Typically genes operate as parts of gene regulatory networks. These are complex adaptive systems capable of regulating their own activities in ways that ensure the robustly reliable production of stable outputs across an enormous range of internal (mutational) and external (environmental) perturbations. The architecture of complex adaptive systems, such as gene regulatory networks, exhibits feedback loops, cyclical causation, and multiple regulatory circuits. The components of a complex system regulate the activities of other components in ways that ensure the robust maintenance of the system's state. The regulatory activities of any component of a system ramify throughout the system as a whole and ultimately redound on that component itself. In complex adaptive systems the activities of a component are among the causes of that component's own activities. The activities of genes in a network are neither independent nor additive, nor are they context insensitive. Perhaps paradoxically, the activity of a gene in a network is part of the context of its own activity.

The dynamics of complex adaptive systems violates the assumptions of the analytic method, and this, in turn, has a number of implications for causal inference. Andreas Wagner (1999) describes the several activities of complex adaptive systems as "non-separable." Causes are non-separable when they do not have a constant effect from context to context. The non-separability of causes certainly does not support the classical assumption that after a change to one component of the systems the other components proceed "as before." Moreover, because the contribution of any one component sensitively depends on the activities of every other, the effect of a single gene on a phenotype cannot be isolated from that of any other gene. Furthermore, there is no difference in kind between a change to the dynamics of a complex system that originates within the system and one that originates externally. Any alteration—whether endogenous or exogenous— not only changes the activities of the internal components of the system but also alters the relation of the system to its environment, its capacity to respond to and effect changes in external conditions.

The upshot is that the phenotypic outputs of a gene regulatory network are conjointly caused by the concerted activities of the components and its external influences in such a way that the causal contribution of the each component to the phenotype cannot be disentangled from that of any other. Even if phenotypic differences can be correlated with genetic differences, it does not follow, as classical genetics traditionally assumed, that these differences (or the relevant phenotypes) are principally *caused by* the genes in question.

The orthodox modern synthesis conviction that genes play a privileged, discernible role in inheritance, as we saw, is predicated on the supposition that the contribution of unit genes to the pattern of inheritance are relatively context insensitive and can be isolated from those of extragenetic influences. It further requires that there is a difference in kind between those novelties whose principal causes are genetic (germline) changes and those whose principal causes are extragenetic. The commitment that only genetic mutations are genuinely inheritable further requires the assumption that novelties, or differences in phenotype, that are initiated by genetic mutations are more robustly recurrent than those initiated by environmental changes. But the increased understanding of the dynamics of gene regulatory networks casts serious doubt on these precepts (Noble 2012).

The new gene dynamics suggests that a reconsideration of the modern synthesis redefinition of inheritance is in order. The pattern of phenotypic resemblance and difference that is so crucial to evolution is the joint effect of the complex adaptive dynamics of gene regulatory networks and their extragenetic contexts. So intimate, context sensitive, and commingled are the various causal contributors to this pattern that it generally cannot be decomposed into those portions that are due to genes and those that are due to extragenetic causes. Inheritance is not "genetic" (or "epigenetic," or "environmental," or "cultural"); it is holistic.

Inheritance holism implies that every trait, including every novel trait, is a result of the complex, adaptive causal interaction between genomes, organisms, and environments. As a consequence, there is no difference in kind between those traits initiated by a change in genes (mutation) and those initiated by extragenetic factors. Every phenotypic novelty is the result of the assimilation, regulation, and accommodation of all the causal influences impinging on the reactive, adaptive genome. Any novelty, no matter how it is initiated, may be intergenerationally stable, that is to say, "inherited." Inheritance holism challenges the very coherence of the modern synthesis conviction that inheritance is the *process* of transmitting replicants. It further advocates a reversion to the premodern synthesis conception of inheritance as a gross pattern of resemblance and difference.

There are two crucial implications of holism for the modern synthesis conception of inheritance. The first is that the process of development is intimately involved in the process of inheritance. There is no distinction between them, as the modern synthesis insists (Danchin and Pocheville 2014: 2308). The fractionation of evolution inaugurated by modern synthesis thinking stands in the way of a comprehensive understanding of inheritance. The second is that the modern synthesis notion of inheritance—transmission of genes—is inadequate to explain the pattern of inheritance that evolution by natural selection requires (Mesoudi et al. 2013).

3 THE PLACE OF DEVELOPMENT

If the "genetic turn" overplayed the hand of gene transmission, it had quite the opposite consequence for organismal development. Viktor Hamburger (1980) famously remarked that throughout much of the twentieth century, organismal development was treated as a black box by the modern synthesis. The process is important to evolution insofar as it delivers genes to the arena of selection, but the prevailing line of thought was that the details of development do not make much difference either way, and so they can be left out.

> After the publication of Darwin's *Origin of Species*, but before the general acceptance of Weismann's views, problems of evolution and development were inexplicably bound up with one another. One consequence of Weismann's separation of the germline and the soma was to make it possible to understand genetics, and hence evolution, without understanding development. (Maynard Smith 1982: 6)

On the traditional modern synthesis view, development is a highly conservative process. It introduces no adaptive bias into evolution, nor does it initiate any novel evolutionary traits. Insofar as development is relevant to explaining the fit and diversity of form, it is as a check or constraint on the power of selection. Wagner and Altenberg nicely encapsulate the prevailing position. "For instance developmental constraints frustrate selection by restricting the phenotypic variation selection has to act upon. Adaptations would be able to evolve only to optima within the constrained space of variability" (1996: 973). This traditional view depicts development and selection as two independent forces acting upon a population, one that promotes the adaptation, the other that (where it is discoverable at all) appears as a tendency to oppose adaptive change.

> This has led to a "dichotomous approach" in which constraint is conceptually divorced from natural selection and pitted against it in a kind of evolutionary battle for dominance over the phenotype . . . much of the constraint literature over the last 25 years has explicitly sought to explain evolutionary outcomes as either the result of selection or constraint (Schwenk and Wagner 2004: 392)

The orthodox two-force picture makes two errors: one empirical the other conceptual. These both point progressively toward a reconsideration of the marginal role in which the modern synthesis has cast organismal development.

The empirical error is vividly exposed by the enormous successes of evolutionary developmental biology. Recent advances in the understanding of development demonstrate that it has a much more intimate, varied, and productive role in evolution than merely imposing a brake on the powers of natural selection. In particular, the features of organismal development put in place the conditions required for adaptive evolution. Furthermore, the processes of development bias, initiate, and accommodate evolutionary novelties. The principal feature of development that underwrites its positive contribution is its adaptive flexibility, or plasticity. Organisms are dynamic, responsive, adaptive systems.

> The organism is not robust because it is built in such a manner that it does not buckle under stress. Its robustness stems from a physiology that is adaptive. It stays the same, not because it cannot change but because it compensates for change around it. The secret of the phenotype is dynamic restoration. (Kirschner and Gerhard 2005: 108–109).

This constitutive feature of organisms is most vividly demonstrated in their development.

The significance of robustness for adaptive evolution first became appreciated, perhaps ironically, in the attempt to evolve efficient computer algorithms. Researchers discovered that not just any computer program can be improved by an iterated cycle of random mutation and selection (Wagner and Altenberg 1996). This raised the question: "What makes a system evolvable?" The question applies to organisms just as much as it does to programs.

Whether a system—an organism or a program—is susceptible to adaptive improvement seems to depend on its architecture. Adaptive systems are modular. A modular system comprises a set of integrated modules, decoupled from one another. The internal integration of a module ensures that each module is robust. It can compensate for perturbations. Decoupling ensures that a module is protected against potentially disruptive changes occurring in other modules. Kitano identifies the importance of modularity: "in principle, a modular system allows the generation of diverse phenotypes by the rearrangement of its internal module connections and relatively independent evolution of each module" (2004: 830). Taken together, these properties of modular architecture confer on an organism the capacity robustly and reliably to produce a viable individual typical of its kind and to generate phenotypic novelties as an adaptive response to genetic, epigenetic, or environmental perturbations (Greenspan 2001; Bergman and Siegal 2002; Von Dassow et al. 2000). The architecture of development has a number of implications for the adaptiveness of evolution.

An organism has in its genome a special "tool-kit" of regulatory genes that control development by influencing the timing, or the products of expression, of *other* genes (Carroll, Grenier and Weatherbee. 2000). Changes in the regulatory role of genes have an important place in evolution. Gerhard and Kirschner (2005, 2007) illustrate the role of regulatory evolution by means of a model they call "facilitated variation."

An organism's genome consists of a core of conserved components. These are highly regular structures shared in common across organisms whose latest common ancestor must have been in the Cambrian or Pre-Cambrian. These core components form the basic control units for a startling variety of different phenotypic structures both within organisms and between lineages. For example, the same homeobox gene *lab* found in *Drosophila* and other insects regulates the correct development of the maxillary and mandibular segments on the head. *Hox 4.2* of mammals regulates the development of the atlas/axis complex of the vertebral column. These are highly divergent, nonhomologous structures, but their respective homeobox genes, *lab* and *Hox 4.2*, are virtually identical (Duboule and Dolle 1989). Highly conserved core structures work in concert with more variable, peripheral developmental resources in the production of novel phenotypes.

> Most evolutionary change in the metazoa since the Cambrian has come not from changes of the core processes themselves, but from regulatory changes affecting the deployment of the core processes ... Because of these regulatory changes, the core processes are used in new combinations and amounts at new times and places. (Kirschner and Gerhard 2005: 221–222)

The architectural structure of development confers on organisms a dynamic robustness, the ability to maintain viability through mounting compensatory changes to perturbations. It also imparts to development the capacity to "search" morphological space, alighting upon new, stable, and adaptive structures.

> The burden of creativity in evolution, down to minute details, does not rest on selection alone. Through its ancient repertoire of core processes, the current phenotype of the animal determines the kind, amount and viability of phenotypic variation the animal can produce ... the range of possible anatomical and physiological relations is enormous ... (Gerhard and Kirschner, 2007: 8588)

This suggests an enhanced positive role for development in promoting adaptive evolution.

The new role is secured by the adaptive plasticity of development. Through its capacity to mount adaptive responses to its conditions, an organism's developmental resources are capable of producing novel, stable, and viable forms, *without* the need for genetic mutations (West-Eberhard 2003; Moczek et al. 2011). In fact, Mary Jane West-Eberhard conjectures that most innovations are not induced by mutations but are the result of the adaptive response of development to environmental influences.

> most phenotypic evolution begins with environmentally initiated phenotypic change ... Gene frequency change follows, as a response to the developmental change. In this framework, most adaptive change is accommodation of developmental-phenotypic change. (2003: 157–158)

Andreas Wagner (2011, 2012) and colleagues demonstrate the ways in which development imposes an adaptive bias in evolution. The structure of gene networks permits organisms to produce their characteristic phenotypes while sustaining an enormous amount of genetic mutation. Gene networks thus store up genetic variation that can be utilized in new ways in new conditions. Moreover, the robustness of gene networks facilitate the search of adaptive space, making the origin of adaptive novelties by random mutation much more likely than they would otherwise be (Wagner 2014). As Wagner puts it, "a robust phenotype can help the evolutionary exploration of new phenotypes ... by accelerating the dynamics of change in an evolving population" (2012: 1256).

The robustness of development further promotes adaptive evolution through the orchestration of phenotypic change. The evolution of complex adaptations requires the coordination of all of an organism's developmental systems. An evolutionary change in, for example, the length of a forelimb requires concomitant changes in bone deposition, muscularization, innervation, and circulation. Each of these systems responds in ways that accommodates the changes in the others. If each system required its own independent genetic mutations, adaptive evolution of complex entities would be practically impossible.

> In contrast to the rapid response produced by plasticity, if the production of newly favored phenotypes requires new mutations, the waiting time for such mutations can be prohibitively long and the chance of subsequent loss through drift can be high. (Pfennig et al. 2010: 459–460)

It appears, then, that development contributes to adaptive evolution in (at least) five ways (Pfennig et al. 2010; Hendrikse, Parsons, and Hallgrímsson 2007).

1. It biases the direction of variation. It renders adaptive evolution more likely than it would otherwise be.
2. It regulates the amount of phenotypic variation. The buffering of development against the effects of mutations serves to secure the viable production of phenotypes against the threat of mutations.
3. It coordinates the evolution of organismal subsystems, preventing change in one subsystem from disrupting the viability of an organism.
4. It serves as an evolutionary capacitor, storing up latent variation, and phenotypic repertoire that influences the rate and direction of evolution in the future.
5. It innovates. Evolutionary novelties largely originate in development.

> Responsive phenotype structure is the primary source of novel phenotypes. And it matters little from a developmental point of view whether the recurrent change we call a phenotypic novelty is induced by a mutation or by a factor in the environment. (West-Eberhard 2003: 503)

Evolutionary developmental biology and its successor devo-evo evince a heightened sensitivity to the central place of development in evolution.

> Evo-devo argues that the variational capacities of genomes are functions of the developmental systems in which they are embedded, for example, through their modular organization, the dynamics of their mechanistic interactions and their non-programmed physical properties. (Müller 2007: 4)

It also intimates that natural selection does not have exclusive claim as the source of the adaptive bias into evolution.

> The explanation of adaptive change as a population-dynamic event was the central goal of the modern synthesis. By contrast, evo-devo seeks to explain phenotypic change through the alterations in developmental mechanisms (the physical inter-actions among genes, cells and tissues), whether they are adaptive or not. (Müller 2007: 945–946)
>
> The burden of creativity in evolution, down to minute details, does not rest on selection alone. Through its ancient repertoire of core processes, the current pheno-type of the animal determines the kind, amount and viability of phenotypic variation the animal can produce . . . the range of possible anatomical and physiological rela-tions is enormous . . . (Gerhard and Kirschner, 2007: 8588)

Evolution is adaptive because development is adaptive.

Ontogeny is no longer the poor cousin among the component process of evolution. In the emerging organism-centered biology, it is seen as the cause and facilitator of adap-tive evolution. That, in turn, suggests a wholesale re-evaluation of the modern synthesis approach to evolution.

4 THE PROCESS OF SELECTION

Orthodox modern synthesis thinking pits development and natural selection against one another as antagonistic forces of evolutionary change. One is fundamentally con-servative and the other exclusively adaptive. I mentioned that this picture commits both an empirical error and a conceptual one. We have explored the empirical error. The con-ceptual error resides in a misconstrual of the process of selection.

4.1 Higher Order Effects

In chapter 3 of *The Origin*, Darwin poses the crucial question: "How have all those exqui-site adaptations of one part of the organisation to another part, and to the *conditions of life*, and of one distinct organic being to another being, been perfected? (1859/1968: 114).

His answer, perhaps surprisingly, is not "natural selection." He says "all these things, . . . *follow inevitably* from the struggle for existence" (114). In giving this response Darwin is voicing the (then) rather radical idea that the causes of biological fit and diversity can be observed in the everyday lives of organisms, their struggle for life. No extra causes are needed, no designer, no *vis essentialis*, no extra-organismal guiding hand. One implication, of course, is that if there is no cause over and above those that are manifest in individuals struggling for life, then natural selection cannot be such a process (Walsh 2000). All the causes of evolutionary change are immanent in the processes occurring within individuals: inheritance, development, and the origination of novelties. In fact, in Darwin's theory, natural selection is what we might call a "higher order effect"; it is the effect on a population of all the myriad causes affecting individuals.

If Darwin is right—if natural selection is a higher order effect—then the adaptive bias in evolution does not originate in some extra, population-level cause. In hypostatizing selection as an autonomous causal process, solely responsible for the adaptiveness of evolution, the modern synthesis fractionated picture of evolution radically misrepresents the metaphysics of evolution.

4.2 Interpreting Evolutionary Models

The erroneous view of the nature of selection raises a further question concerning the interpretation of evolutionary models. Modern synthesis evolutionary models, of the sort introduced by Fisher and Wright (but also models such as the Price equation), represent population change as the result of the variation in trait fitness.[4] Trait fitness is a measure of the instantaneous growth rate of a trait type. The standard approach, in keeping with the traditional modern synthesis picture, is to treat this parameter (variation in fitness) as a measure of the efficacy of a discrete force or causal process—selection—that acts upon a population (Sober 1984; Stephens 2004). But if selection is not a force or a discrete autonomous causal process, this interpretation is apt to mislead.

The fact is that when we model the process of selection in this way, we are representing something quite different from the process that Darwin discovered. Darwin discovered that when individuals vary in their heritable capacity to survive and reproduce, a population will change in its *lineage structure*. Some lineages preponderate over others. Nowadays, evolution is defined as change in *trait structure*: in a population undergoing selection, traits change in their relative frequency. Change in lineage structure is neither necessary nor sufficient for change in trait structure (Walsh et al. 2002). Darwin's theory gives us no sure-fire way to predict or quantify the change in trait structure in a population. The modern synthesis theory has devised an elegant and powerful way to do this. Modern synthesis population models do not deal in populations of concrete individual organisms per se. Evolutionary change in trait structure is explained and

---

[4] In the Price equation and the models of quantitative genetics, the inheritability of traits is also represented.

quantified by ascending to an ontology of abstract trait types. These types are assigned growth rates: trait fitnesses. Variation in trait fitness predicts and explains the change in relative frequency of trait types in a population. The point is that the modern synthesis models of population dynamics take on a whole new explanatory project—change in abstract trait structure—and address it by erecting a whole new ontology—populations as ensembles of abstract trait types.

In constructing these models, Fisher (1930) assumed that a population comprises an enormously large number of abstract entities (trait types), whose effects on population growth are typically minute, context insensitive, and wholly independent of one another. With these assumptions in place, Fisher surmised that the statistical distribution of these growth rates could predict and explain the change in trait structure. Margaret Morrison vividly captures the distinctively abstract and statistical nature of Fisher's thinking:

> Essentially, he treated large numbers of genes in a similar way to the treatment of large numbers of molecules and atoms in statistical mechanics. By making these simplifying and idealizing assumptions, Fisher was able to calculate statistical averages that applied to populations of genes in a way *analogous* to calculating the behavior of molecules that constitute a gas. (2002: 58–59; emphasis in original)

These models raise an acute interpretive challenge. The trouble is that these models appear to tell us little about why the change in abstract population structure should eventuate in adapted individuals. It just is not clear what this process has to do with Darwin's process in which, in changing its lineage structure, a population becomes increasingly adapted to its conditions of existence. It has by no means been a trivial issue for theoretical population biology (Orr 2005, 2007). Orr and Coyne discuss the assumptions about the process of adaptive evolution implicit in these models. They conclude that the "neo-Darwinian view of adaptation . . . is not strongly supported by the evidence" (1992: 725). The evolutionary geneticist Richard Lewontin evinces an uncommon sensitivity to the problem:

> A description and explanation of genetic change in a population is a description and explanation of evolutionary change only insofar as we can link those genetic changes to the manifest diversity of living organisms. . . . To concentrate only on genetic change, without attempting to relate it to the kinds of physiological, morphogenetic, and behavioral evolution that are manifest in the fossil record and the diversity of extant organisms and communities, is to forget entirely what it is we are trying to explain in the first place. (1974: 23)

What "we are trying to explain in the first place" is the fit and diversity of biological form. Once we decline to interpret these models as articulating the direction and magnitude of a discrete adaptation-promoting force—selection—that can be distinguished from all the other component processes of evolution, it becomes entirely unclear what they are telling us about fit and diversity.

One increasingly popular line of thought is that these models tell us very little about the causes of adaptive evolutionary change, nor should we expect them to (Walsh, Lewens, and Ariew 2002; Matthen and Ariew 2002; Walsh 2010a). Instead, they provide a sophisticated statistical description of the change in the trait structure of a population. By ascending to an abstract ontology of trait types and the statistical distribution of their growth rates, these models allow us to identify robust regularities discernible only at this level of abstraction. They generate powerful predictions and explanations of the generalized dynamics of populations. They articulate a set of principles that hold across any biological population, no matter what its causes.[5] In doing so they "sacrifice realism for generality" (Levins 1966/1984). This is a venerable strategy of scientific modeling (Levins 1966/1984; Weisberg 2006).

The upshot is that these models do not explicitly represent the causes of population change. Powerful and indispensable as they are, they do not carve evolution at its causal joints in the way that modern synthesis orthodoxy takes them to do. Nor do they explain what makes adaptive evolution adaptive. If our objective is to understand these things, then abstract models of population dynamics are of scant utility. Instead, we should look to the processes occurring within individual organisms. The marginalization of individual organisms precipitated by the modern synthesis emphasis on population dynamics appears to be born of a naïve and misleading interpretation of its models.

5 CONCLUSION

Evolutionary theory has been under the sway of gene-centered modern synthesis thinking for most of the past century. That theory, to be sure, is an extension or an outgrowth of Darwin's own views. But in many ways it is a corruption of Darwin's way of thinking about evolution (Walsh 2010b). In extending Darwin's theories it takes on a battery of specific theoretical commitments. The two most significant are the centrality of genes and the fractionation of evolution into four discrete, quasi-independent processes: (i) inheritance, (ii) development, (iii) adaptive population change, and (iv) genetic mutation. Gene centrism and fractionation go hand-in-hand. It is because the modern synthesis takes the component processes of evolution to be quasi-independent that it can assign to genes a privileged role in each.

No one can gainsay the enormous advances in the understanding of evolution that have been ushered in by the modern synthesis theory. Yet since the advent of this century the modern synthesis has experienced sustained and increasingly critical scrutiny. Recent challenges to modern synthesis orthodoxy are driven by empirical advances in the understanding of inheritance, development, and the origin of evolutionary novelties. Implicitly, or explicitly, the challenges are leveled at the fractionation of evolution

---

[5] In fact, they are so general that they describe the dynamics of nonbiological ensembles too (Walsh 2014).

and the attendant theoretical privilege accorded to genes. They suggest that the component processes of evolution are intertwined and commingled in a way that renders their separation generally impossible. Significantly, the challenges raise telling doubts about the marginalization of organismal development. They suggest that development is intimately involved in inheritance, the generation of evolutionary novelties, and the adaptive bias in evolution. Furthermore, these challenges underscore the conceptual error involved in reifying natural selection as a discrete, autonomous process, independent of inheritance, development, and the origin of novelties—that is the sole source of the adaptive bias in evolution.

Whether the modern synthesis can survive the wave of critical reappraisal remains an unanswered question. But at least we can now see what would count as a satisfactory defense. Those who wish to quell the rising dissent will need to demonstrate how this emerging body of empirical evidence is consistent with—indeed best accounted for by—the fractionated, gene-centered conception of evolution that forms the foundation of the modern synthesis.

References

Agrawal, A., Laforsch, C., and Tollrian, R. (1999). "Transgenerational Induction of Defences in Animals and Plants." *Nature 401*: 60–63.

Arthur, W. (2000). "The Concept of Developmental Reprogramming and the Quest for an Inclusive Theory of Evolutionary Mechanisms." *Evolution and Development 2*: 49–57.

Bergman, A., and Siegal, M. (2002). "Evolutionary Capacitance as a General Feature of Complex Gene Networks." *Nature 424*: 549–552.

Carroll, S. B., Grenier, J. K., and Weatherbee, S. D. (2000). *From DNA to Diversity: Molecular Genetics and the Evolution of Animal Design* (London: Wiley-Blackwell).

Cartwright, N. (1999). *The Dappled World: A Study in the Boundaries of Science* (Cambridge, UK: Cambridge University Press).

Ciliberti, S., Martin, O. C., and Wagner, A. (2007). "Innovation and Robustness in Complex Regulatory Gene Networks." *Proceedings of the National Academy of Sciences 104*(34): 13591–13596.

Crick, F. J (1970). "The Central Dogma of Molecular Biology." *Nature 227*: 561–563.

Danchin, E., and Pocheville, A. (2014). "Inheritance Is Where Physiology Meets Evolution." *Journal of Physiology 592*: 2307–2317.

Darwin, C. (1968). *The Origin of Species* (London: Penguin). (Original work published 1859)

Davidson, E. H. (2006). *The Regulatory Genome: Gene Regulatory Networks in Development and Evolution* (London: Academic Press).

Dobzhansky, T. (1937). *Genetics and the Origin of Species* (New York: Columba University Press).

Duboule, D., and Dolle, P. (1989). "The Structural and Functional Organization of the Murine *Hox* Family Resembles that of Drosophila Homeotic Genes." *Embo 8*: 1497–1505.

Echten, S., and Borovitz, J. (2013). "Epigenomics: Methylation's Mark on Inheritance." *Nature 495*: 181–182.

Fisher, R. A. (1930). *The Genetical Theory of Natural Selection* (Oxford: Clarendon Press).

Gayon, J. (2000). "From Measurement to Organization: A Philosophical Scheme for the History of the Concept of Inheritance." In P. J. Beurteon, R. Falk, and H.-J. Rheinberge (eds.), *The Concept of the Gene in Development and Evolution: Historical and Epistemological Perspectives* (Cambridge, UK: Cambridge University Press), 60–90.

Gerhard, J., and Kirschner, M. (2005). *The Plausibility of Life: Resolving Darwin's Dilemma* (New York: Norton).

Gerhard, J., and Kirschner, M. (2007). "The Theory of Facilitated Variation." *Proceedings of the National Academy of Sciences 104*: 8582–8589.

Gilbert, S., and Epel, D. (2009). *Ecological Developmental Biology* (Sunderland, MA: Sinauer).

Greenspan, R. J. (2001). "The Flexible Genome." *Nature Reviews Genetics 2*: 383–387.

Griffiths, P. E., and Gray, R. (2001). "Darwinism and Developmental Systems." In S. Oyama, P. E. Griffiths, and R. D. Gray (eds.), *Cycles of Contingency: Developmental Systems and Evolution* (Cambridge, MA: MIT Press), 195–218.

Guerrero-Bosagna, C., Savenkova, M., Haque, M. M., Nilsson, E., and Skinner, M. K. (2013). "Environmentally Induced Epigenetic Transgenerational Inheritance of Altered Sertoli Cell Transcriptome and Epigenome: Molecular Etiology of Male Infertility." *PLoS ONE 8*(3): E59922.

Hall, B. K. (1999). *Evolutionary Developmental Biology* (Amsterdam: Kluwer).

Hall, B. K. (2012). "Evolutionary Developmental Biology (Evo- Devo): Past, Present, and Future." *Evolution: Education and Outreach 5*: 184–193.

Hamburger, V. (1980). "Embryology and the Modern Synthesis in Evolutionary Biology." In E. Mayr and W. Provine (eds.), *The Evolutionary Synthesis* (Cambridge, MA: Harvard University Press), 97–112.

Heijmans, B. T., Tobi, E. W., Stein, A. D., et al. (2008). "Persistent Epigenetic Differences Associated with Prenatal Exposure to Famine in Humans." *Proceedings of the National Academy of Sciences 105*: 17046–17049.

Hendrikse, J. L., Parsons, T. E., and Hallgrímsson, B. (2007). "Evolvability as the Proper Focus of Evolutionary Developmental Biology." *Evolution and Development 9*: 393–401.

Irschick, D. J., Albertson, R. C., Brennnan, P., et al. (2013). "Evo-Devo Beyond Morphology: From Genes to Resource Use." *Trends in Ecology and Evolution 28*: 509–516.

Jablonka, E., and Lamb, M. (2002). "The Changing Concept of Epigenetics." *Annals of the New York Academy of Sciences 981*: 82–96.

Jablonka, E., and Lamb, M. (2004). *Evolution in Four Dimensions: Genetic, Epigenetic, Behavioral, and Symbolic Variation in the History of Life* (Cambridge, MA: Bradford Books).

Jenkin, F. (1867). "The Origin of Species." *North British Review* (June). www.victorianweb.org/science/science_texts/jenkins.html

Kirschner, M., and Gerhard, J. (2005). *The Plausibility of Life: Resolving Darwin's Dilemma* (New Haven, CT: Yale University Press).

Kitano, H. (2004). "Biological Robustness." *Nature Reviews Genetics 5*: 826–837.

Laland, K. Uller, T., Feldman, M., et al. (2014). "Does Evolutionary Theory Need a Rethink? [Part 1]: Yes: Urgently." *Nature 514*: 161–164.

Lamm, E. (2014). "Inheritance Systems." In *Stanford Encyclopedia of Philosophy* (Spring 2012 ed.). Stanford, CA: Stanford University. plato.stanford.edu/archives/spr2012/entries/inheritance-systems

Laubichler M. (2009). "Evo-Devo: Historical and Conceptual Reflections." In M. Laublichler and J. Maienschein (eds.), *Form and Function in Developmental Evolution* (Cambridge, MA: Cambridge University Press), 10–46.

Levins, R. (1984). "The Strategy of Model Building in Population Biology." In E. Sober (ed.), *Conceptual Issues in Revolutionary Biology* (Cambridge, MA: MIT Press), 18–27. (Original work published 1966)

Lewontin, R. C. (1974). *The Genetic Basis of Evolutionary Change* (New York: Columbia University Press).

Lumey, L. H. (1992). "Decreased Birthweights in Infants After Maternal in Utero Exposure to the Dutch Famine of 1944–1945." *Pediatric and Perinatal Epidemiology* 6: 240–253.

Maienschein, J., and Laubichler, M. (2014). "Explaining Development and Evolution on the Tangled Bank." In R. P. Thompson and D. M. Walsh (eds.), *Evolutionary Biology: Conceptual, Ethical, and Religious Issues* (Cambridge, UK: Cambridge University Press), 151–171.

Matthen, M., and Ariew, A. (2002). "Two Ways of Thinking about Fitness and Selection." *Journal of Philosophy* 99: 55–83.

Maynard Smith, J. (1982). *Evolution and the Theory of Games* (Cambridge, UK: Cambridge University Press).

Maynard Smith, J. (1998). *Evolutionary Genetics* (Oxford: Oxford University Press).

Mesoudi, A., Blanchet, S., Charmentier, A., et al. (2013). "Is Non-Genetic Inheritance Just a Proximate Mechanism? A Corroboration of the Extended Evolutionary Synthesis." *Biological Theory* 7: 189–195.

Mill, J. S. (1843). *A System of Logic Ratiocinative and Inductive* (London: Harper and Brothers). http://www.gutenberg.org/files/27942/27942-pdf

Moczek, A. P., Sultan, S., Foster, S., et al. (2011). "The Role of Developmental Plasticity in Evolutionary Innovation." *Proceedings of the Royal Society B*, doi:10.1098/rspb.2011.0971.

Morange, M. (2011). "Evolutionary Developmental Biology: Its Roots and Characteristics." *Developmental Biology* 357: 13–16.

Morgan, T. H. (1926). *The Theory of the Gene* (New Haven, CT: Yale University Press).

Morrison, M. (2002). "Modeling Populations: Pearson and Fisher on Mendelism and Biometry." *British Journal for the Philosophy of Science* 53: 339–368.

Müller, G. B. (2007). "Evo-Devo: Extending the Evolutionary Synthesis." *Nature Reviews Genetics* 8: 943–949.

Müller-Wille, S., and Rheinberger, H.-J. (2012). *A Cultural History of Heredity* (Chicago: Chicago University Press).

Noble, D. (2006). *The Music of Life* (Oxford: Oxford University Press).

Noble, D. (2012). "A Theory of Biological Relativity: No Privileged Level of Causation." *Interface Focus* 2: 55–64.

Odling-Smee, F. J., Laland, K., and Feldman, M. (2003). *Niche Construction: The Neglected Process in Evolution* (Princeton, NJ: Princeton University Press).

Orr, H. A. (2005). "The Genetic Theory of Adaptation: A Brief History." *Nature Reviews Genetics* 6: 119–127.

Orr, H. A. (2007). "Theories of Adaptation: What They Do and Don't Say." *Genetica* 123: 3–13.

Orr, H. A., and Coyne, J. A. (1992). "The Genetics of Adaptation Revisited." *American Naturalist* 140: 725–742.

Ou, X., Zhang, Y., Xu, C., et al. (2012). "Transgenerational Inheritance of Modified DNA Methylation Patterns and Enhanced Tolerance Induced by Heavy Metal Stress in Rice (*Oryza Sativa* L.)." *PLoS ONE* 7(9): E41143.

Oyama, S., Griffiths, P. E., and Gray, R. (Eds.). (2001). *Cycles of Contingency* (Cambridge, MA: MIT Press).

Pfennig, D. W., Wund, M., Schlichting, C., et al. (2010). "Phenotypic Plasticity's Impacts on Diversification and Speciation." *Trends in Ecology and Evolution* 25: 459–467.

Pigliucci, M. (2009). "An Extended Synthesis for Evolutionary Biology: The Year in Evolutionary Biology 2009." *Annals of the New York Academy of Sciences* 1168: 218–228.

Robert, J. S. (2004). *Embryology, Epigenesis, and Evolution* (Cambridge, UK: Cambridge University Press).

Roll-Hansen, N. (2009). "Sources of Wilhelm Johannsen's Genotype Theory." *Journal of the History of Biology* 42: 457–493.

Schwenk, K., and Wagner, G. (2004). "The Relativism of Constraints on Phenotypic Evolution." In M. Pigliucci and K. Preston (eds.), *The Evolution of Complex Phenotypes* (Oxford: Oxford University Press), 390–408.

Shapiro, J. (2011). *Evolution: A View from the 21st Century Perspective* (Upper Saddle River, NJ: FT Press Science).

Shapiro, J. A. (2013). "How Life Changes Itself: The Read-Write (RW) Genome." *Physics of Life Reviews* 10: 287–323.

Stephens, C. (2004). "Selection, Drift, and the 'Forces' of Evolution." *Philosophy of Science* 71: 550–570.

Sober, E. (1984). *The Nature of Selection* (Cambridge, MA: MIT Press).

Szathmáry, E. (2000). "The Evolution of Replicators." *Philosophical Transactions of the Royal Society of London, Series B: Biological Sciences* 355: 1669–1676.

Von Dassow, G., Meir, E., Munro, E. M., and Odell, G. M. (2000). "The Segment Polarity Network Is a Robust Developmental Module." *Nature* 406: 188–192.

Wagner, A. (1999). "Causality in Complex Systems." *Biology and Philosophy* 14: 83–101.

Wagner, A. (2005). *Robustness and Evolvability in Living Systems* (Princeton, NJ: Princeton University Press).

Wagner, A. (2007). "Distributional Robustness Versus Redundancy as Causes of Mutational Robustness." *Bioessays* 27: 176–188.

Wagner, A. (2011). *The Origin of Evolutionary Innovations: A Theory of Transformative Change in Living Systems* (Oxford: Oxford University Press).

Wagner, A. (2012). "The Role of Robustness in Phenotypic Adaptation and Innovation." *Proceedings of the Royal Society B* 279: 1249–1258.

Wagner, A. (2014). *The Arrival of the Fittest: Solving Evolution's Greatest Puzzle* (New York: Current Books).

Wagner, G., and Altenberg, L. (1996). "Complex Adaptations and the Evolution of Evolvability." *Evolution* 50: 967–976.

Walsh, D. M. (2000). "Chasing Shadows." *Studies in History and Philosophy of Biological and Biomedical Sciences* 31: 135–153.

Walsh, D. M. (2010a). "Not a Sure Thing." *Philosophy of Science* 77: 147–171.

Walsh, D. M. (2010b). "Two Neo-Darwinisms." *History and Philosophy of the Life Sciences* 32: 317–339.

Walsh, D. M. (2014). "Variance, Invariance, and Statistical Explanation." *Erkkenntnis*, doi:10.1007/s10670-014-9680-3.

Walsh, D. M., Lewens, T., and Ariew, A. (2002). "The Trials of Life." *Philosophy of Science* 69: 452–473.

Waters, C. K. (1994). "Genes Made Molecular." *Philosophy of Science* 61: 163–185.

Waters, K. (2007). "Causes That Make a Difference." *Journal of Philosophy* 104: 551–579.

Weisberg, M. (2006). "Forty Years of 'The Strategy': Levins on Model Building and Idealization." *Philosophy and Biology 21*: 526–645.

Weiss, K. M., and Fullerton, S. M. (2000). "Phenogenetic Drift and the Evolution of Genotype-Phenotype Relationships." *Theoretical Population Biology 57*: 187–195.

West-Eberhard, M. J. (2003). *Developmental Plasticity and Evolution* (Oxford: Oxford University Press).

Wray, G. A., Hoekster, H. E. Futuyma, D. J., et al. (2014). "Does Evolutionary Theory Need a Rethink? [Part 2]: No, All Is Well." *Nature 514*: 161–164.

CHAPTER 33

···

COMPLEXITY THEORY

···

MICHAEL STREVENS

ALMOST everything is a complex system: Manhattan at rush hour, of course, but also, if you know how to look, a rock sitting in the middle of a field. Excited by the heat of the midday sun, the molecules that make up the rock are vibrating madly. Each is pulling at or shoving its neighbors, its parts shifting around its center of mass in a most haphazard way, their next move hinging on a multitude of minute details concerning the many atoms making up the surrounding stone.

The rock and the city—what do they share that makes each in its own way a paradigm of complexity? They are composed of many parts behaving somewhat independently yet interacting strongly. That I take to be the essential recipe for a complex system: sufficiently many parts, independent yet interacting.

Complexity's most salient consequence is intractability. Even supposing that the behavior of every part of a complex system, and thus of the whole, is entirely determined by the exact state of the parts and the fundamental laws of nature, there is little hope of building scientific models capable of representing the system at this fineness of grain, thus little hope of predicting and understanding the behavior of complex systems by tabulating the gyrations of their parts.

You might therefore wonder whether sciences of complex systems are possible. Complexity theory in its broadest sense is the body of work in science, mathematics, and philosophy that aims to provide an affirmative answer: to show how investigators inquire fruitfully into the workings of complex systems and to understand why they so often succeed.

And they do succeed. Our present-day science does not cover everything, but it covers a lot, and a lot of what's covered is complex: rocks, gases, organisms, ecosystems, economies, societies, languages, and minds. Evidently it is possible sometimes to predict, sometimes to explain, sometimes to predict and explain what is going on in a system without having to track the individual histories of its many parts.

Complexity theory looks both retrospectively at the successes, attempting to understand how they were achieved, and prospectively at possible future successes, attempting to formulate ideas that will engender the next generation of discoveries in developmental genetics, economics and sociology, and neuroscience.

This chapter will not discuss two important topics: attempts to define and quantify complexity and the concept of complexity in computer science and mathematics. The latter I take to be outside my remit; as for the former, I am not persuaded that it is a fruitful way to begin the study of complex systems. I will therefore follow Weaver (1948) in relying on a rough-and-ready characterization—complexity as a consequence of numerous independent interacting parts—and a long list of examples from physics, biology, and the social and behavioral sciences. Definitions are left to complexity theory's seventh day.[1]

1   TWO GRAND QUESTIONS

Complex systems exhibit, on a grand scale, two kinds of behavior that call out for explanation: simplicity and sophistication. Sometimes it's one; sometimes it's the other; often, it is a mix of both at different levels.

I call a behavior simple if it conforms at least approximately to a generalization that can be simply expressed—if, for example, it obeys a "law" containing just a few variables, as most gases roughly obey the ideal gas law $PV = kT$ or most organisms obey Kleiber's law according to which metabolic rate increases in proportion to the 3/4 power of body mass.

The simplicity in a complex system typically does not exist at the level of the many interacting parts that make for its complexity—at what I will call the system's *microlevel*. The reason is this: to track a complex system at the microlevel, you need sets of equations that, for each small part, describe its state and prescribe its interactions with the other parts. There is nothing at all simple about such a model. Simplicity in complex systems emerges rather at higher levels of description—at a macrolevel where variables or other terms in the simple description represent aggregate properties of many or all of the microlevel parts, as a gas's temperature represents its molecules' average kinetic energy or an organism's mass represents the total mass of all its cells and other physiological constituents.

The first grand question about complex systems, then, is of how macrosimplicity emerges (when it does) from microcomplexity, of how simplicity in the behavior of the system as a whole is sustained and explained by the convoluted action of the system's fundamental-level constituents.

Consider, for example, gases. Their low-level convolution is typically, for all practical purposes, intractable. We do not have the mental fortitude or the computational power

---

[1] For a helpful overview of attempts to define complexity, see Mitchell (2009), chapter 7. Mitchell also mentions a survey suggesting that many complexity theorists believe that it is too soon to attempt such a definition (299).

to track the behavior of every one of even a small quantity of gas's heptillions of molecules, each colliding with the others billions of times a second. If that were what it took to build a science of gases, it would be beyond our reach.

Happily, there is another possibility. At the macrolevel, the behavior of the same quantity of gas can be characterized by a simple linear function of three variables—the ideal gas equation. The equation is no good for predicting the trajectories of individual molecules, but we are not much interested in that. Predicting changes in temperature, pressure, and volume is good enough for many purposes—good enough to provide a starting point for a science of gases that rises above the intractability of the behavior of the gases' many parts.

The simple behavior of gases and many other complex systems is not only good for complex-system scientists but essential for the rest of us. Gases' macrolevel simplicity means macrolevel stability: the air in a room tends to remain uniformly distributed throughout the available space, rather than surging to and fro from corner to corner.[2] So we can all keep breathing with a minimum of fuss. Numerous other stable behaviors of our environment—both its resources and its inhabitants—also play a critical role in making our continued existence possible. Macrolevel simplicity in the things around us is not sufficient for life, but it is essential.

Many complex systems behave in striking ways that go far beyond the simplicity that provides a stable background for life to flourish: they exhibit what I will call *sophistication*. That label embraces many things: the orchestration of biological development, in which an elaborately structured, highly functional organism is built from a single cell; the organism's subsequent exploitation of its habitat; the operations of the minds of intelligent organisms directing the exploitation; and much more. Frequently, the sophisticated behavior of such systems is evident in the kind of plasticity that we call goal-directed. But there are other varieties of sophistication, to be considered in due course.

The second grand question about complex systems, then, is how sophisticated behavior emerges from the interaction of relatively simplistic parts. Both grand questions are posed by the appearance of something at the macrolevel that appears to belie what is at the microlevel: simplicity or stability from the convoluted interaction of parts; sophistication from the interaction of simplistic or simply behaving parts.[3] The emergence of simplicity makes complex-system science possible by creating behaviors that scientists

---

[2] Simplicity in the sense of compact describability does not guarantee stability—as chaos theory shows, simple laws can generate behavior that looks to us to be highly erratic—but the two go together often enough (and wherever there is stability, there is more or less by definition simplicity).

[3] The second question concerns a kind of complexity emerging from a kind of simplicity; the first question, a slightly different kind of simplicity emerging from a completely different kind of complexity. I keep the two complexities well apart; hence, my different names for the two kinds of complex behavior: *sophistication* and *convolution*. Simplicity in the sense of a dynamics' compact describability is not quite the same thing as simplicity in the sense of a dynamics' lack of sophistication, but the two are close enough that it seems unnecessarily pedantic to encode the difference in separate terms—hence I use *simplicity* for both.

can reasonably hope to capture in a generalization or a model. The emergence of sophistication makes complex-system science interesting; indeed, essential.

The two varieties of emergence may, and often do, coexist. Inside a nerve cell's axon, a great horde of molecules bounces this way and that, the trajectory of each irregular enough to seem utterly random. But the movements of these molecules taken as a whole are regular enough that the cell exhibits a fairly simple, fairly stable behavior: upon receiving the right stimulus, a wave of electric charge runs up the axon. Examined at the microlevel, this wave consists in the individually unpredictable movements of vast numbers of charge-carrying ions; in the aggregate, however, these movements add up to the reliable, predictable pulses by which information is transmitted long distances through the body and brain. The building blocks of thought, in particular, are in large part these simple behaviors that emerge from the convoluted molecular chaos of the cell.

Yet thought itself is not simple; it is sophisticated. Somehow, the relatively simplistic mechanics of individual neurons are, when harnessed together in a brain, capable of producing behavior that is purposive, intelligent, even rational. From the simple but numerous interactions between molecules, then, comes the convoluted behavior of axonal ions; from this convolution comes a higher level simplicity at the cellular level; and from the interactions between neurons behaving relatively simply, something extraordinarily sophisticated from the assembly that is the brain.

The same complementary arrangement of simplicity from convolution and sophistication from simplicity is found in the balance of life in a mature ecosystem. Individual organisms, especially animals, interacting with each other and with their environments can be as haphazard and unpredictable in their movements as colliding molecules. The intersecting life trajectories of such organisms, then, make for a highly convoluted whole.

Within such a whole, however, simple high-level patterns emerge. The success of ecological modeling using equations tracking only aggregate populations suggests that the convolutions in individual lives add up to stable rates of reproduction, predation, and death. Although it may be effectively impossible to predict whether an individual hare will get eaten in the course of a month, it is relatively easy to predict the rate at which hares in general will get themselves eaten: it seems to depend on only a few high-level variables, such as the number of predators in the ecosystem. From convolution, simplicity emerges.

This simplicity in turn gives rise to sophistication: the regularity in rates of birth and death is what makes for stable evolutionary fitness in a trait—for the fact that possession of a trait can result in a determinate increase in the rate of reproduction, in viability, and so on—and thus makes it possible for natural selection to operate with a certain consistency over time periods long enough to result in evolution and adaptation. The adaptedness of things is of course a kind of sophistication, surely the most important kind that we have yet explained.

This chapter is organized around the two grand questions, focusing on attempts to provide answers to those questions of the widest possible scope; that is, "theories of complexity" that attempt to give very general conditions for simplicity's emergence from convolution and sophistication's emergence from simplicity.

Notions such as explanation and emergence can become themselves a topic of debate in these endeavors, but I have put such arguments to one side because they are covered by separate entries in this handbook.

2 FROM CONVOLUTION TO SIMPLICITY

In a convoluted complex system it is almost impossible to model, and so to predict, the movements (or other changes in state) of one of the system's parts—the trajectory of a gas molecule, a lynx's success in hunting rabbits, a soldier's fate on the field of battle. In some cases, the chaos is replicated at the macrolevel: might the outcome of the Battle of Waterloo have hinged on any of a number of small-scale deployments or split-second decisions? But, often enough, the macrolevel is a sea of mathematical calm: somehow, the aggregate properties of the parts are stable or predictable even though the parts themselves are not. How can that be?

2.1 The Sources of Convolution

Let me begin by saying something about the sources of microlevel convolution or unpredictability. I see three:

1. *Convolution of the parts:* In some cases, the internals of the individual parts of a complex system are convoluted and so any given part's behavior is difficult to predict.
2. *Chaos:* The dynamics of the strong interactions between parts is often sensitive to initial conditions (i.e., in a loose sense "chaotic"). Small differences in the states of individual parts can make for big differences in the outcomes of their encounters. These big differences then stand to make still bigger differences down the road.
3. *Combinatorial escalation:* There are many parts in a single system. To keep track of these parts is difficult enough under any circumstances; if their individual behaviors are hard to predict (convolution of the parts) or their interactions depend on small details of their states (chaos), then the complexity of the task is multiplied beyond practical feasibility.

The emergence of some kinds of macrolevel simplicity under these conditions is easy to understand. The weight of a gas at a given time is a simple function of the weight of its constituent molecules; the strong interactions between these molecules, and the ensuing chaos and combinatorial escalation, is simply irrelevant to the determination of weight. (This example is a useful reminder that any complex system has many macrolevel properties. Some may behave simply; some more complexly. Of the simple behaviors, some may be easily explained; some not.)

Most macrolevel properties are not like weight. A confined gas's pressure is constituted by the drumming of its many molecules on its container's walls; the force exerted by the molecules depends on their position and velocity, which is determined in turn by the other molecules with which they collide, a process that can be characterized fully only by a stupendous array of equations. The changes in the population of some organism depend on individual births and deaths, events likewise dictated by a microlevel dynamics so complex that it will never be written down.

In these cases, unlike the case of weight, the microlevel dynamics drives most or all of the changes in the macrolevel properties, yet the resulting macrolevel dynamics is as simple as the microlevel dynamics is convoluted. The convolution of the microlevel is suppressed or dissolved without in any way undercutting the microlevel's causal role. The problem is to understand this suppression, this dissolution.

2.2 Modularity and Near-Decomposability

Some complex systems have a hierarchical or modular structure, Simon (1996) influentially argues: small groups of parts make up larger assemblies—"modules," if you like—from which the whole is built (or from which higher level modules are built). The parts of a module interact strongly, as they must if the system is to be considered complex at all, but because of the modular structure these interactions are felt only weakly if at all outside the module. In a completely "decomposable" system, they are not felt at all. The effects of a module part do not pass through the module's boundaries; as a consequence—since a module consists of nothing but its parts—each module is utterly self-contained, a little universe doing its own thing independently of the others. Complete causal independence of this sort is rare and is in any case hardly a hallmark of complexity. Far more common, Simon observes, is what he calls "near-decomposability": interactions between modules are far weaker in some sense than interactions within modules.

Such near-decomposability can ensure a relatively simple, relatively stable, relatively tractable behavior of the ensemble of modules—of the system as a whole. Simon makes the case, in particular, for systems in which the interaction between modules happens on a much longer time scale than the interaction within modules. The effect of a particular module part, difficult to predict perhaps because of its internal convolutions or because of its strong interactions with the other parts of the same module, will percolate only very slowly from one module to another; consequently, he argued, the effect of one module on another will be dictated not by individual effects but by their long-run average. If this average is stable or has a simple dynamics, then the same should be true for the dynamics of intermodular interactions—in spite of the large number of strongly interacting parts of which the modules are composed. The relatively simple dynamics of intermodular interaction, Simon believed, provided a basis for stability in the system's overall macrolevel behavior—a stability that could be exploited by selection, natural or otherwise, to create sophistication on top of simplicity.[4]

Near-decomposability and the related notion of modularity, although they have remained important in the study of evolvability (Section 2.5) and in evolutionary developmental biology (Section 3.3), are not enough in themselves to explain the emergence of macrolevel simplicity from microlevel convolution. First, such emergence occurs even in systems that have no hierarchical organization to confine the influence of the parts to a single locale; examples include gases, many ecosystems, and some social structures. Second, the appeal to near-decomposability explains macrolevel simplicity only if the behavior of individual modules is, in the aggregate or in the long run, simple. But the parts of many modules are sufficiently numerous to produce convolution even within the module; how, then, is the necessary modular simplicity to be explained?

2.3 The Statistical Approach

The first great formal theories of complex systems were those of statistical physics: first, kinetic theory (Boltzmann 1964; Maxwell 1860, 1867), and then the more general apparatus of statistical mechanics (Gibbs 1902; Tolman 1938). For many complexity theorists writing in the wake of the development of statistical physics, it was natural to apply the same statistical methods to ecosystems (Lotka 1925), to "disorganized" complex systems in general (Weaver 1948), and even to an imagined social science capable of predicting the future history of the galaxy (Asimov's *Foundation* trilogy).

Consider again a gas in a box. The statistical approach, in the version represented by kinetic theory, stipulates a physical probability distribution over the position and velocity of each molecule in the gas. In the simplest case, the distribution is the Maxwell-Boltzmann distribution, which specifies that a molecule is equally likely to be found in any part of the box while imposing a Gaussian (normal) distribution over the components of the molecule's velocity that depends only on the temperature of the gas and the mass of the molecule.

The distributions for different molecules are stochastically independent, from which certain conclusions follow immediately by way of the law of large numbers (for the same reason that, from the probability of one-half that a tossed coin lands heads, it follows that a large number of coin tosses will, with very high probability, produce about one-half heads). The distribution over position implies that the gas is at any time almost certainly

---

[4] See Strevens (2005) for a more detailed account of the decomposition strategy and Bechtel and Richardson (1993) for "decomposition" as a research strategy for understanding complex systems.

distributed evenly throughout the box. The distribution over velocity implies that, if the gas is warmed, the pressure it exerts on the container walls will increase proportionately (Gay-Lussac's law): an increase in temperature results in a proportional increase in average molecular velocity, which almost certainly results in a proportional increase in the force exerted on average against the container walls—that is, an increase in pressure. The probability distributions over microlevel properties, then—over the positions and velocities of individual molecules—can be used to explain stabilities or simplicities in a gas's macrolevel properties, such as its temperature and pressure.

A direct generalization of the statistical approach in kinetic theory explains the simple behavior of convoluted systems in three steps:

1. Probability distributions are placed over the relevant behaviors of a complex system's parts: over the positions of gas molecules, over the deaths of organisms, over decisions to vote for a certain electoral candidate, and so on.
2. The distributions are combined, in accordance with the law of large numbers, to derive a probability distribution over the behavior of aggregate properties of the parts: the distribution of a gas, the death rate of a certain kind of organism, the results of an election, and so on.
3. From this distribution, a simple macrolevel dynamics or prediction is derived.

This schematic approach is what Strevens (2003, 2005) calls *enion probability analysis* (EPA).

Let me elaborate on the assumptions and machinery of EPA. First, the parts of a system over whose states or behaviors the probability distributions range are called *enions*; this distinguishes them from other parts of the system (say, the distribution of vegetation in or the topography of an ecosystem) that are excluded from the analysis—their states either being held fixed or having their variation determined exogenously.

Second, the outcomes over which the enion probability distributions range are those that contribute to the macrolevel aggregates that are under investigation. If you are investigating death rates, you need probability distributions over death. If you are investigating the spatial distribution of a gas, you need probability distributions over molecules' positions. By the same token, no probability distribution need be imposed over outcomes that are not relevant to the aggregates.

Third, the enion probability distributions—the probabilities assigned to enion states or behaviors, such as position or death—should depend only on macrolevel properties of the system. The probability of a hare's death, for example, should depend only on the total number of lynxes in the local habitat and not on the positions of particular lynxes. The reason is this: any variable on which the probabilities depend will tend to find its way into the probability distributions over the macrolevel properties in Step 2. If the probability of a certain hare's death depends, for example, on the positions of particular lynxes, then the death rate as a whole will, if mathematically derived from this distribution, in most circumstances depend on the positions of particular lynxes. But then the macrolevel dynamics derived in Step 3 will depend on these microlevel variables and so will not be a simple dynamics.[5]

Fourth, the enion probability distributions should be stochastically independent. One hare's death by predation should, for example, be left unchanged by conditionalizing on another hare's death by predation, just as one coin toss's landing heads makes no difference to the probability that the next toss also lands heads. It is this assumption that allows you to pass, by way of the law of large numbers, from a probability to a matching long-run frequency—inferring from (say) a 5% probability that any particular hare is eaten in a month that there is a high probability that, over the course of a month, about 5% of a large population of hares will be eaten.

Fifth, as the immediately preceding example suggests, the use of the law of large numbers will tend to give you—provided that each mathematically distinct enion probability distribution is shared by large numbers of enions—probabilities for macrolevel states or behaviors that are close to one. From these probabilities, then, you can derive something that looks much like a definite prediction or a deterministic macrolevel dynamics, with the proviso that there is some chance of deviation. The chance is negligible in the case of a gas with its vast numbers of molecules but rather more noticeable in the case of hares (although in the ecological case there are many other sources of deviance; the assumption that outcomes are independent, for example, holds only approximately).

When the suppositions just enumerated hold, macrolevel simplicity emerges from microlevel convolution. The population dynamics of a lynx–hare ecosystem, for example, can be represented by a relatively simple equation; in the best case, a Lotka-Volterra equation containing little more than variables representing the populations of the two species and parameters (derived from the enion probability distributions) representing rates of reproduction, predation, and so on.

Where does all the convolution go? The population of hares depends on the births and deaths of individual hares, which depend in turn on minute details in position and configuration: whether or not a hare is eaten might hinge on just a few degrees in the angle that a particular lynx's head makes to its body at a particular time. Why is the dependence not passed up the chain?

The answer is that, in the aggregate, these dependences manifest themselves in fluctuations that cancel each other out. The outcome of a tossed coin depends sensitively on the speed with which it is spun: a little faster, and it would have been tails rather than heads. These dependences push the outcomes of coin tosses this way and that, more or less at random. But precisely because of this randomness, they ultimately make little difference to the frequency of heads. There are as many "nudges" in one direction as in any other; consequently, the nudges more or less balance, leaving the frequency of heads to be determined by fixed, underlying features of the coin's material makeup.

---

[5] It is not inevitable that microlevel variables will end up in the macrolevel dynamics: there might be some further mathematical technique by which they can be removed or perhaps aggregated (so that the death rate depends only on the distribution of lynxes). The same goes for all the strictures in this discussion: any general problem might have a tailor-made solution in some particular case.

The same holds, very broadly, in the ecosystem: an individual hare's life turns on a few small details, but for a system of many hares, these details more or less balance, leaving the rate of hare death to be determined by fixed, underlying features of the ecosystem: hare camouflage, lynx eyesight, ground cover, and, of course, the overall number of lynxes in the environs. The system's behavior can therefore be characterized by equations representing, explicitly or implicitly, this fixed background and the handful of variable properties.

EPA is applicable to a wide variety of complex systems and processes: the systems of statistical physics (and, therefore, of physical chemistry); ecosystems (as Lotka hoped) and therefore evolution by both natural selection and genetic drift; various aspects of human societies and economies, such as traffic flow along the highways and through the Internet's tubes. Much of the simplicity and stability we see around us can be accounted for in this way. But there is more.

2.4 Abstract Difference-Making Structures

The statistical approach to explaining the simplicity of macrolevel behavior—what I call EPA—can be understood as a demonstration that the incredibly intricate and involved to-ings and fro-ings of a system's microlevel parts make no difference to its macrolevel behavior: they are fluctuations that, because of their statistical profile, cancel out, leaving macrolevel states and changes in state to be determined by relatively stable and measurable features such as molecular mass, lynx physiology, and so on.

Other methods explaining the emergence of macrosimplicity may be viewed in the same way: they show that certain elements of a system, elements that contribute significantly to its convoluted microlevel dynamics, make no difference to the behavior of its high-level properties, although these properties may themselves be aggregates of the very elements that are seen to make no difference. These methods, including EPA, thus identify certain rather abstract properties of the system in question—abstract in the sense that they may be shared by systems that in many other respects differ significantly—and they show that the abstract properties alone are difference-makers for high-level behavior. A system with those properties will exhibit that behavior, regardless of how the properties are realized.[6]

Boltzmann's kinetic theory, for example, uses an implementation of EPA to show that the physical differences between the geometry of different gas molecules make no difference to gases' tendency to move toward equilibrium and thus to conform to the second law of thermodynamics: all that matters is that, upon colliding, the molecules in a certain sense scatter. This scattering character of collisions is the difference-making property that, however it is realized, secures equilibration.

---

[6] This is not identical, although it is close, to the notion of difference-making that I have developed in my work on scientific explanation (Strevens 2008). This chapter does not require, I think, a formal characterization.

The simplicity of the equilibration dynamics is, I suggest, closely connected to the abstractness of the difference-making property. The connection is not straightforward or directly proportional, but there is a correlation: where you find simplicity, you tend to find abstract difference-makers.

As a consequence, macrosimplicity also goes along with universality: where one complex system behaves in a simple way, many others, often quite different in their constitution, also tend to behave in that same simple way. Looking beyond EPA, then—although the connection holds there, too—the same techniques that explain macrosimplicity tend also to explain universality, in both cases by identifying a highly abstract difference-making structure sufficient, or nearly so, for the phenomenon in question.

Let me give three examples.

Critical point phenomena: A wide variety of complex systems, when their tempera-ture crosses a certain point, undergo phase transitions in which the macrostate of the system changes qualitatively: liquids freeze, unmagnetized solids become magnetized, disordered rod-like molecules align (Yeomans 1992). For what are called continuous phase transitions, the systems' behavior near the point of transformation—the critical temperature—is strikingly similar: certain physical quantities, such as magnetization, conform to an equation of the form

$$F(T) \propto (T - T_c)^{\alpha},$$

where T is the system's temperature (or other relevant variable), $F(T)$ is the value at tem-perature T of the physical quantity in question (such as magnetization), T_c is the critical temperature, and α is an exponent that takes the same value for large classes of systems that otherwise differ greatly in their constitution.

Two things to note: first, these are complex systems with enormous numbers of degrees of freedom, yet their macrolevel behavior near the critical temperature is extremely simple; second, there is great universality to this behavior, with many dissimi-lar systems following the same equation in the critical zone. The explanation of this sim-plicity and universality consists in a demonstration that almost everything about such systems makes no difference to their behavior in the critical zone; what matters is that they have a certain abstract structure, shared by some simple models used to study criti-cal behavior, such as the Ising model.[7]

The neutral theory of biodiversity: Certain features of certain complex ecosystems (such as tropical forests) appear to be the same regardless of the species that constitute the system. One such feature is the distribution of species abundance, that is, the relative numbers of the most abundant species, the second most abundant species, and so on down to the rarest species.

---

[7] Wilson (1979) gives an accessible yet satisfying account of critical point universality, written by the physicist who won the Nobel Prize for its explanation. For a philosophical treatment, see Batterman (2002).

The neutral theory of biodiversity explains the universality of these features using models of ecological dynamics that make no distinction among species and that do not take into account, in particular, the degree of adaptedness of the species to the environment. It is shown that, in the models, the observed abundance curve obtains, and it is claimed that real ecosystems conform to the curve for the same reason (Hubbell 2001). If that is correct, then the patterns of abundance in real ecosystems owe nothing to the fitness of the species in the system but are instead explained as a matter of chance: some species are more plentiful merely because, to put it very simply, they happened to come along at the right time and so managed to establish a decisive presence in the system.

The topology of the World Wide Web: Patterns of linkage in the World Wide Web have, it seems, a certain topology wherever you look: the probability $P(n)$ that a given website has n incoming links from other sites falls off as n increases, following a power law

$$P(n) \propto n^{-\gamma},$$

where γ is a little greater than 2. As a consequence, the topology of the Web is "scale-free": like many fractals, it looks the same in the large and in the small.

The same structure shows up in many other sizable networks with different values for γ: patterns of scientific citation ($\gamma \approx 3$), electrical power grids ($\gamma \approx 4$), and patterns of collaboration among movie actors ($\gamma \approx 2.3$). Barabási and Albert (1999) cite these phenomena and propose to explain them as a result of a preferential attachment process in which the probability of a node's gaining a new connection is proportional to the number of the node's existing connections. If this is correct, then it seems that the content of a website or a scientific paper makes no difference to its chances of being linked to or cited: all that matters is its current degree of connectivity. That the dynamics of connection depend on so little explains at the same time, then, the simplicity and the (near?) universality of the patterns they produce.

There are other ways to explain probabilistic power laws and the resulting scale freedom, however. Adamic and Huberman (2000), focusing on the topology of the Web in particular, argue that preferential attachment predicts a strong correlation between age and connectedness that does not exist; they propose an alternative explanation, also highly abstract but giving a website's content a role in attracting links.[8]

Similar critiques have been made of the neutral theory of biodiversity as well as of other models that purport to explain interesting swathes of universality by treating factors that seem obviously relevant as non–difference-makers (such as Bak's [1996] proposal to use a model of an idealized pile of sand to explain patterns in earthquakes, stock market crashes, and extinctions): there are many different ways to account for a given

---

[8] In a useful survey of "network science," Mitchell (2009, 253–255) covers these and further sources of skepticism, including assertions that power laws are not nearly so widely observed as is claimed.

simple behavior. What explains a power law probability distribution in one system may be rather different from what explains it in the next.

2.5 Evolvability

Macrosimplicity tends to provide, as I remarked earlier, a relatively stable environment in which sophisticated systems may evolve—an environment in which air is distributed uniformly, the same food plants and animals stay around from year to year, and so on. But stability can be more than just a backdrop against which natural selection and other evolutionary processes (such as learning) operate. It can be the very stuff of which sophisticated systems are made.

Simon (1996) identified modularity and the concomitant stability as a key element of evolvable systems. Thinking principally of gradualist natural selection, in which adaptation is a consequence of a long sequence of small changes in a system's behavior, he reasoned that in a convoluted system such changes will be hard to come by since minor tweaks in implementation will tend—at least sometimes—to effect radical transformations of behavior, in which all accumulated adaptation will be lost. What is needed for evolution, then, is a kind of organization that responds to minor tweaks in a proportional way: the system's behavior changes, but not too much.[9]

This amounts to a kind of measured stability. An evolvable system should have many stable states or equilibria; natural selection can then move among them, finding its way step by step to the fittest. Stuart Kauffman describes such systems—perhaps overly dramatically—as occupying the "edge of chaos": they are, on the one hand, not totally inert, their behavior too stubbornly fixed to change in response to small tweaks, but, on the other hand, not too close to chaos, where they would react so violently to small changes that gradual evolutionary progress would be impossible (Kauffman 1993).

The near-decomposability of hierarchical or modular systems described in Section 2.2 is, Simon thought, the key to evolvability. When such systems take a microlevel step in the wrong direction, Simon argued, then at worst a single module is disabled; the hard-won functionality of the rest of the system is preserved.[10]

Kaufmann's best-known work concerns models of genetic regulation (his "NK models") that are in no way modular but that exhibit the same modest pliability: they

---

[9] Qualitatively the same phenomenon is sought by engineers of control systems who want their systems to react to input, but not to overreact—a central part of the subject matter of Norbert Wiener's proposed science of sophisticated complexity, which he called *cybernetics* (Wiener 1965). The additional chapters in the second edition of Wiener's book touch on evolution by natural selection.

[10] At least, that is the upshot of Simon's fable of the watchmakers Tempus and Hora. In real biology, deleterious mutations are often fatal; the evolvability challenge is more to find scope for small improvements than to rule out catastrophic errors. But perhaps it is also a consequence of near-decomposability that a meaningful proportion of tweaks will tend to result in small improvements rather than drastic and almost always disastrous reconfigurations.

typically respond to small random changes in configuration by changing their behavior, but not too profoundly.

Much recent scientific work on evolvability belongs to the new field of evolutionary developmental biology, in which modularity remains an important theme (see Section 3.3).

2.6 Other Approaches

Mathematical derivation has been the preferred tool in the explanations of macrosimplicity and universality surveyed earlier: certain factors are proved not to make a difference to the behavior of certain macrolevel properties (or at least a proof sketch is provided). But complexity theorists have other methods at their disposal.

The first is an empirical search for universal behaviors. Where diverse systems are found exhibiting the same simple macrolevel behaviors, there is at least some reason (perhaps very far from conclusive, as the case of network topology suggests) to think that they share the same abstract difference-making structure.

The second is simulation. Rather than proving that an abstract difference-making structure induces a certain behavior, systems realizing the structure in various ways can be simulated on a computer; if the same behavior appears in each case, there is at least some reason to think that the structure in question is in each case responsible. (The many uses of simulation are treated more generally in a separate entry in this handbook.)

The third is open-ended computer experimentation. Wolfram (2002) advocates the computer-driven exploration of the behavior of cellular automata in the hope of discovering new abstract difference-making structures and developing new ways to model nature.

3 FROM SIMPLICITY TO SOPHISTICATION

Begin with simplicity—often itself emerging from lower level convolution, as in the case of the simple, regular behavior of DNA molecules, neurons, even animal populations. Begin, that is, with an inventory of parts whose dynamics conforms to simple mathematical equations. What kinds of surprising behavior can you expect from these parts?

Even in isolation, a system whose dynamics is simple in the mathematical sense can do unexpected things. The "catastrophe theory" of the 1970s showed that systems obeying one of a family of simple, smooth macrolevel laws that induce fixed-point equilibrium behavior can, when subject to small external perturbations, leap to a new equilibrium point that is quite far away. This is a "catastrophe"; the mathematics of catastrophes has been used to explain the buckling of a steel girder, the radically different development of adjacent parts of an embryo, and the collapse of complex civilizations (Casti 1994).[11]

The "chaos theory" of the 1980s showed that systems obeying simple macrolevel laws can exhibit a sensitivity to initial conditions that renders their long-term behavior all but unpredictable. What's more, holding the law fixed but altering one of its parameters can change the system's behavior from a simple fixed-point equilibrium, through periodic behavior—repeatedly visiting the same set of states—to chaos by way of a "period-doubling cascade" in which the set of states visited repeatedly doubles in number as the parameter increases or decreases in accordance with a pattern that shows remarkable universality captured by the "Feigenbaum constants."[12]

Although interesting and largely unforeseen, these behaviors are not particular to complex systems and are not sophisticated—not goal-directed, not adaptive, not intelligent. For sophistication, it seems, you need to put together many simple parts in the right sort of way.

Two related questions, then. First, what is the "right sort of way?" What structures give you not merely convolution, but sophistication? Second, must there be an architect? Or do some of these structures emerge spontaneously, in circumstances that are not vanishingly rare? Are there parts that tend to "self-organize" into the kinds of structures that give rise to sophisticated behavior?

3.1 The Nature of Sophistication

What is sophisticated behavior? The notion, although expositorily useful, is a loose one. In a liberal spirit, let me consider a range of possible marks of sophistication.

The first is adaptation or plasticity. An adapted system is one whose behavior in some sense fits its environment (relative to a presumed or prescribed goal). A plastic system is one whose behavior is capable of changing in reaction to the circumstances in order to realize a presumed or prescribed goal—finding its way around obstacles, anticipating difficulties, abandoning strategies that prove infeasible for promising alternatives. It might be a plasmodium, a person, or a self-driving car.

The second characterization of sophistication is considerably less demanding. Some behavior is so stable as to be deadly boring—as is the macrolevel behavior of the rock sitting in the field. Some behavior is so convoluted as to be bewildering in a way that is almost equally boring—as is the microlevel behavior of the rock, with its multitude of molecules all vibrating this way and that, entirely haphazardly. Between the two, things get interesting: there is order, but there is variation; there is complexity, but it is regimented in ways that somehow call out to our eyes and our minds (Svozil 2008). Systems

---

[11] For an assessment of catastrophe theory, see Casti (1994, 77–83), which includes a useful annotated bibliography.

[12] Stewart (1989) provides an accessible summary of this work and other results in chaos theory.

that have attracted complexity theorists for these reasons include the hexagonal cells of Bénard convection, the wild yet composed oscillations of Belousov–Zhabotinsky chemical reactions, the more intriguing configurations of the Game of Life, and the sets of equations that generate the Mandelbrot set and other complex fractals.[13]

A third criterion for sophistication is as strict as the previous criterion is lax: sophisticated behavior should exhibit intelligence. Intelligence is, of course, connected to plasticity, but it is a more open-ended and at the same time more demanding kind of sophistication.

My fourth and last criterion attempts to generalize plasticity, not like the previous criterion in the direction of thought, but in the (closely related) direction of animal life. What does that mean? Perhaps having a certain sort of "spontaneity." Perhaps having the characteristics of a typical animal body: appendages, controlled motion with many degrees of freedom, distal senses that control movement, the ability to manipulate objects (Trestman 2013).

These, at least, are some of the behaviors that go beyond simplicity and that complex systems theorists have sought to explain using some variant or other of "complexity theory," understood as a big, broad theory of sophisticated behavior. Two exemplary genres of this sort of complexity theory will be considered here: energetic and adaptive approaches.

3.2 Energetics

The universe is always and everywhere winding down—so says the second law of thermodynamics. But this does not mean everywhere increasing decay and disorder. Life on Earth evolved, after all, in accordance with the second law, yet it is a story (if you look at certain branches) of growing order. Might the increasing energetic disorder prescribed by the second law tend to be accompanied, under the right circumstances, by an increasing order of another sort—increasing complexity, increasing sophistication? The fabulous principle specifying this conjectured tendency to complexity is sometimes called, whether affectionately or incredulously, the "fourth law of thermodynamics." Were it to come to light, it would provide a foundation for a theory of sophisticated complexity with the widest possible scope.

What are the right conditions for the thermodynamic generation of sophisticated structure? Consider, naturally, the planet Earth. Energy pours in from the sun, making things interesting. That energy is then mostly radiated into space in a sufficiently disordered form to comply with the second law. The second law says that systems tend toward equilibrium, as the system containing the sun, Earth, and space very slowly does. But as long as the sun burns, it maintains the surface of the Earth itself in a state very far from

---

[13] These phenomena are discussed in many books on complexity and related fields. For Bénard and Belousov–Zhabotinsky, see, for example, Prigogine (1980); for the Game of Life, Mitchell (2009); for the Mandelbrot set, Stewart (1989).

thermodynamic equilibrium, a state in which there are large temperature differentials and inhomogeneities in the material structure of things.

Other smaller and humbler systems are also maintained far from equilibrium in the same sense: a retort sitting over a Bunsen burner or a thin layer of oil heated from below and pouring that heat into the air above. Taken as a whole, the system obeys the second law. But as long as the heat flows, the system in the middle, between the heat source and the heat sink, can take on various intricate and sophisticated structures. In the thin layer of oil, the hexagonal cells of Bénard convection develop, each transporting heat from bottom to top by way of a rolling motion. Such an arrangement is often called a *dissipative structure*, operating as it does to get energy from source to sink as the second law commands.

That the dissipation is accompanied by the law-like emergence of an interesting physical configuration is what captures the attention of the theorist of complexity and energetics—suggesting as it does that the entropic flow of energy goes hand in hand with sophisticated flow-enabling arrangements. Two laws, then: a "second law" to ensure dissipation; a "fourth law" to govern the character of the complex structures that arise to implement the second law. Followed by a bold speculation: that ecosystems and economies are dissipative structures that can be understood with reference to the fourth law (Prigogine 1980).

What, then, does the fourth law say? In one of the earliest and most lucid presentations, Lotka (1945) proposed a fourth law according to which the throughput of energy in a system maintained far from equilibrium tends to increase. The law says, then, that in systems maintained far from equilibrium, dissipative structures will tend to evolve so as to maximize the rate at which energy passes from source to sink.

Lotka was writing principally about biological evolution, and his reasoning was Darwinian: organisms will evolve, he believed, to maximize as far as possible their uptake and use of energy from the environment, and so, as a second-law corollary, to maximize the amount of energy that they dump back into the environment in disordered form.[14] It is selective pressure, then, that powers the fourth law by choosing, from among many possible physical configurations, those systems that conform to the law.

The fourth law is not, however, confined to biological systems; according to Lotka it holds in any system of "energy transformers" of the right sort, in virtue of something analogous to natural selection. Lotka goes on to characterize the notion of an energy transformer in greater detail, in terms of functional units termed "receptors," "effectors," and "adjusters." He concludes that "a special branch of physics needs to be developed, the *statistical dynamics of systems of energy transformers*" (179).

A tendency for energy flow to increase is not the same thing as a tendency for sophistication to increase. But followers of Lotka, and to a certain extent Lotka himself, have thought there to be a connection: the most effective way to increase the energy flow through a dissipative structure will typically be to make it more sophisticated. So Depew

---

[14] The key underlying assumption is that there is always some way that at least some species in an ecosystem can put additional free energy to use to increase fitness and that natural processes of variation (such as mutation and sexual reproduction) will sooner or later stumble on that way.

and Weber (1988, 337) write[15]: "It is an essential property . . . of dissipative structures, when proper kinetic pathways are available, to self-organize and . . . to evolve over time toward greater complexity."

There are numerous difficulties in formulating a fourth law, especially one that has something to say about the emergence of sophistication. Most importantly, perhaps, writers after Lotka have tended to ignore the significance of his careful characterization of energy transformers. It is nevertheless worth putting aside these objections to focus on another problem—that a fourth law has meager empirical content—because rather similar complaints can be made about every general theory or principle of sophisticated complexity.

What does the fourth law predict? The law is not intended to prescribe any particular variety of sophistication; its empirical content consists instead in its ruling out a certain kind of event, namely, a system's failure to become more complex. On the face of things, this prediction seems to be a bad one: evolutionary stability, in which nothing much changes, is the statistical norm. More generally, dissipative structures, whether thin layers of oil or tropical rainforests, do not tend to become arbitrarily complex. It is for this reason that Depew and Weber hedge their statement of the fourth law by specifying that it holds only "when proper kinetic pathways are available."

What, then, makes a kinetic pathway "proper" or "available"? Not bare physical possibility: a system consisting of many agitated molecules has an infinitesimal probability of doing all sorts of astonishing things—currents in gently heated oil might, for example, form the image of a monkey at a typewriter—yet they do not because of such events' vanishingly low probability. In that case, what is the fourth law saying? That dissipative structures tend to become more complex provided that they are not unlikely to do so? Without the hedge, the fourth law is false; with the hedge, it seems not much more than a bland truism.

3.3 Adaptation

The literature on general theories of complexity in the 1980s and 1990s contained numerous references to "complex adaptive systems." Is it their complexity that allows such systems to adapt? Often, what is meant is something different: they have become complex through adaptation (perhaps after some other process of emergence has provided "evolvable" parts, as discussed in Section 2.5). In Lotka's theory, organisms evolve to maximize their uptake and use of energy from the environment, and so, as a second-law corollary, they maximize the amount of energy that they dump back into the environment in disordered form.[16] It is selective pressure, then, that secures the truth of the

---

[15] To better secure the connection to complexity (in the sense of sophistication), Depew and Weber amend Lotka's fourth law: rather than maximizing energy flow, structures minimize specific entropy—entropy created per unit of energy flow—thus using energy in a certain sense more efficiently.

[16] In some places, Lotka implies that it is an ecosystem as a whole that maximizes the rate of energy throughput; how exactly selection explains this systemic maximization is left unspecified.

fourth law by choosing, from among many possible physical configurations, those systems that conform to the law.

Putting thermodynamics entirely to one side, it has of course been appreciated ever since Darwin that natural selection is capable, given the character of life on Earth, of building ever more sophisticated systems (although, as remarked in the previous section, complexification is only one evolutionary modus operandi among many). So what of novel interest does putting "adaptive" in the middle of "complex systems" accomplish? To what extent is there a theory of the development of sophisticated complexity by way of natural selection that goes beyond Darwin and modern evolutionary biology?

The answer is that attempts at a general theory of "complex adaptive systems" hope to do one or both of two things:

1. Apply broadly Darwinian thinking outside its usual biological scope, using new methods for discovering and understanding the consequences of selection.
2. Find patterns or tendencies common to a large class of systems that includes both those in which adaptation is driven by natural selection and those in which it is effected by other means, such as learning. Gell-Mann (1994a), for instance, provides a very liberal definition of adaptivity and writes that "an excellent example of a [complex adaptive system] is the human scientific enterprise."

The investigation of "artificial life" is an example of the first sort of project. Under this heading, researchers aim to abstract away from the implementation of reproduction and inheritance on Earth—from DNA and RNA and the cellular mechanisms that coordinate their replication and variation—and model, using computer programs or other fabricated constructs, systems in which these things happen by alternative means. In Tom Ray's influential Tierra model, the stuff of life is code itself: the "organisms" are small computer programs that consume CPU time and use it to replicate themselves, sometimes recombining and sometimes mutating (Ray 1992). What happens in Tierra is not the simulation of evolution by natural selection in some other system, then, but the real thing: the scarce resource is CPU time, and survival and reproduction is not represented but rather instantiated in the persistence and replication of units of code. The behavior of these and similar experiments is quite engaging: researchers see the development both of sophisticated complexity (up to a point) and of parasitism. Yet it is as yet unclear whether artificial life has any general lessons to teach about complexity, above and beyond what is already known from mainstream evolutionary biology (Bedau, McCaskill, Packard, Rasmussen, Adami et al. 2000).

When the Santa Fe Institute was founded in 1984 to promote the interdisciplinary study of complex systems, the ideal of a general theory of sophisticated adaptive systems was an important element of its credo. A series of popular books celebrated the promise of this program of research (Gell-Mann 1994b; Lewin 1992; Waldrop 1992). But, by the late 1990s, many figures associated with the theory of sophisticated complexity had begun to worry, like Simon (1996), that "complexity is too general a subject to have much content" (181). Surveying the ghosts of general theories of complexity that have paraded by in the past one hundred years—cybernetics (Wiener 1965), general system theory (von Bertalanffy 1968), autopoiesis (Varela, Maturana, and Uribe 1974), synergetics (Haken 1983), self-organized criticality (Bak 1996)—the inductively inclined non-specialist might well, at least in the case of sophisticated complexity, concur.

That does not mean that an emphasis on complexity in the abstract is not theoretically fruitful. One of the most exciting fields in science in the past few decades has been evolutionary developmental biology, which uses discoveries about the genetic modulation of development to better understand how complex body plans evolved (Gerhart and Kirshner 1997; Raff 1996). An important part of the explanation seems to be, as Simon envisaged, modularity: some important evolutionary steps were taken not by drastically re-engineering modules' internal workings but by tweaking the time and place of their operation.

The feel of the science is typical of the study of many complex systems. On the one hand, the empirical findings, though hard won through great amounts of tedious and painstaking research on particular systems, cry out to be expressed at the highest level, in abstract vocabulary such as "complexity," "modularity," "switching." On the other hand, attempts to generalize, leaving the particular topic of the development and evolution of life on Earth behind to say something broader about the connection between modularity and complexity, seem to produce merely truisms or falsehoods. Thinking about sophisticated complexity in the abstract remains as enticing, as tantalizing as ever—but the best theories of sophisticated complexity turn out to have a specific subject matter.

4 CONCLUSION

Two intriguing features of complex systems have been discussed in this chapter: simple behavior at the high level emerging from convoluted underpinnings and sophisticated behavior at the high level emerging from simple underpinnings. Complexity theory has sometimes concerned itself with the one sort of emergence, sometimes with the other, and sometimes it seems to aim for both at the same time, seeking to explain behaviors that are both surprisingly stable and surprisingly sophisticated.

The default approach to complex systems, registered in the segregation of the university departments, is to tackle one kind of stuff at a time. The term "complexity theory" implies a more interdisciplinary enterprise, an attempt to identify commonalities among complex systems with radically different substrates: to find connections between ecosystems and gases, between power grids and the World Wide Web, between ant colonies and the human mind, between collapsing cultures and species headed for extinction.

With respect to the emergence of simplicity from convolution, I have been (as a partisan) optimistic about the possibility of substantive general theories, although a complete understanding of simplicity's basis in complexity will surely require many distinct ideas at varying levels of abstraction.

With respect to the emergence of sophistication from simplicity, there is less progress on general theories to report and considerable if reluctant skepticism even in friendly quarters—but plenty of interesting work to do on individual systems, with results that will surely continue to fascinate us all.

REFERENCES

Adamic, L. A., and Huberman, B. A. (2000). "Power-Law Distribution of the World Wide Web." *Science* 287: 2115a.

Bak, P. (1996). *How Nature Works: The Science of Self-Organized Criticality* (New York: Copernicus).

Barabási, A.-L., and Albert, R. (1999). "Emergence of Scaling in Random Networks." *Science* 286: 509–512.

Batterman, R. W. (2002). *The Devil in the Details: Asymptotic Reasoning in Explanation, Reduction, and Emergence* (Oxford: Oxford University Press).

Bechtel, W. and Richardson, R. C. (1993). *Discovering Complexity: Decomposition and Localization as Strategies in Scientific Research* (Princeton, NJ: Princeton University Press).

Bedau, M. A., McCaskill, J. S., Packard, N. H., Rasmussen, S., Adami, C., Ikegami, D. G. G. T., Kaneko, K., and Ray, T. S. (2000). "Open Problems in Artificial Life." *Artificial Life* 6: 363–376.

Boltzmann, L. (1964). *Lectures on Gas Theory*. S. G. Brush (trans.) (Berkeley: University of California Press).

Casti, J. (1994). *Complexification* (New York: HarperCollins).

Depew, D., and Weber, B. (1988). "Consequences of Non-equilibrium Thermodynamics for the Darwinian Tradition." In Weber, B., Depew, D., and Smith, J. (eds.), *Entropy, Information, and Evolution* (Cambridge, MA: MIT Press).

Gell-Mann, M. (1994a). "Complex Adaptive Systems." In G. Cowan, D. Pines, and D. Meltzer (eds.), *Complexity: Metaphors, Models, and Reality* (Reading, MA: Addison-Wesley).

Gell-Mann, M. (1994b). *The Quark and the Jaguar: Adventures in the Simple and the Complex* (New York: Henry Holt).

Gerhart, J., and Kirshner, M. (1997). *Cells, Embryos, and Evolution* (Oxford: Blackwell).

Gibbs, J. W. (1902). *Elementary Principles in Statistical Mechanics* (New York: Scribner's).

Haken, H. (1983). *Synergetics, an Introduction: Nonequilibrium Phase Transitions and Self-Organization in Physics, Chemistry, and Biology*, 3rd ed. (Heidelberg: Springer-Verlag).

Hubbell, S. P. (2001). *The Unified Neutral Theory of Biodiversity and Biogeography* (Princeton, NJ: Princeton University Press).

Kauffman, S. (1993). *The Origins of Order* (Oxford: Oxford University Press).

Lewin, R. (1992). *Complexity: Life at the Edge of Chaos* (New York: Macmillan).

Lotka, A. J. (1925). *Elements of Physical Biology* (Baltimore, MD: Williams and Wilkins).

Lotka, A. J. (1945). "The Law of Evolution as a Maximal Principle." *Human Biology* 17: 167–194.

Maxwell, J. C. (1860). "Illustrations of the Dynamical Theory of Gases." *Philosophical Magazine* 19 and 20: 19–32 and 21–37.

Maxwell, J. C. (1867). "On the Dynamical Theory of Gases." *Philosophical Transactions of the Royal Society of London* 157: 49–88.

Mitchell, M. (2009). *Complexity: A Guided Tour* (Oxford: Oxford University Press).

Prigogine, I. (1980). *From Being to Becoming* (New York: W. H. Freeman).

Raff, R. A. (1996). *The Shape of Life: Genes, Development, and the Evolution of Animal Form* (Chicago: University of Chicago Press).

Ray, T. S. (1992). "An Approach to the Synthesis of Life." In C. G. Langton, C. Taylor, J. D. Farmer, and S. Rasmussen (eds.), *Artificial Life II* (Redwood City, CA: Addison-Wesley), 371–408.

Simon, H. A. (1996). *The Sciences of the Artificial*, 3rd ed. (Cambridge, MA: MIT Press).

Stewart, I. (1989). *Does God Play Dice? The Mathematics of Chaos* (Oxford: Blackwell).

Strevens, M. (2003). *Bigger than Chaos: Understanding Complexity through Probability* (Cambridge, MA: Harvard University Press).

Strevens, M. (2005). "How Are the Sciences of Complex Systems Possible?" *Philosophy of Science* 72: 531–556.

Strevens, M. (2008). *Depth: An Account of Scientific Explanation* (Cambridge, MA: Harvard University Press).

Svozil, K. (2008). Aesthetic complexity. http://arxiv.org/abs/physics/0505088.

Tolman, R. C. (1938). *The Principles of Statistical Mechanics* (Oxford: Oxford University Press).

Trestman, M. (2013). "The Cambrian Explosion and the Origins of Embodied Cognition." *Biological Theory* 8: 80–92.

Varela, F. J., Maturana, H. R., and Uribe, R. (1974). "Autopoiesis: The Organization of Living Systems, Its Characterization and a Model. *Biosystems* 5: 187–196.

von Bertalanffy, L. (1968). *General System Theory: Foundations, Development, Applications* (New York: George Braziller).

Waldrop, M. M. (1992). *Complexity: The Emerging Science at the Edge of Order and Chaos* (New York: Simon & Schuster).

Weaver, W. (1948). "Science and Complexity." *American Scientist* 36: 536–544.

Wiener, N. (1965). *Cybernetics, or Control and Communication in the Animal and the Machine*, 2nd ed. (Cambridge, MA: MIT Press).

Wilson, K. (1979). "Problems in Physics with Many Scales of Length." *Scientific American* 241: 158–179.

Wolfram, S. (2002). *A New Kind of Science* (Champaign, IL: Wolfram Media).

Yeomans, J. M. (1992). *Statistical Mechanics of Phase Transitions* (Oxford: Oxford University Press).

CHAPTER 34

....

COMPUTER SIMULATION

....

JOHANNES LENHARD

1 INTRODUCTION

....

LET us start with a surprise. Few people would deny we are in the midst of a major change that is related to computer and simulation methods and that affects many, if not most, scientific fields. However, this does not seem to be a big issue in the philosophy of science. There, simulation had a halting start more than two decades ago, followed by an increasing uptake over the past few years.[1] The philosophy of computer simulation counts as an interesting area for specialists, while its general philosophical significance is not yet agreed upon.

A first observation relativizes the surprise: simulation had a slow start in the sciences too. In disciplines like physics, computer methods were adopted early on but counted as clearly inferior to theory or as minor additions to theory. Simulations were seen as a straightforward extension to mathematical modeling—relevant only because they produced numerical results from already known models. Let us call this stance the amplifier view. According to this view, simulation merely amplifies the signal from a mathematical model that is (conceptually) already there.

This view is not uncommon in philosophy (see, e.g., Frigg and Reiss 2009; Stöckler 2000) and arguably contributes much to putting into doubt the significance of computer simulation. It sees the computer as a logical-mathematical machine, which perfectly suits the long-established trends toward mathematization and formalization. It is well accepted in the history of science that mathematization played a crucial role in the scientific revolution (cf. Dijksterhuis 1961) of the seventeenth century and also in the so-called second scientific revolution of the nineteenth century (cf. Kuhn 1961). For the

---

[1] Recently, computer simulation has received entries in (other) works of reference for philosophy of science; see Parker (2013) and Winsberg (2014). I take the opportunity and concentrate more on an argument than an overview of existing literature.

amplifier view, however, computer simulation is a technical matter that lacks comparable significance and does not produce philosophically interesting novelty.

The main message of this article is that the amplifier view is seriously wrong, because it is based on a misconception—a misconception, however, that has affected and still is affecting part of the sciences and philosophy of science alike. Yes, simulation modeling is a type of mathematical modeling, but—in contrast to the amplifier view—this type of mathematical modeling is a fundamental transformation, rather than a straightforward extension. The challenge for philosophy of science is to characterize this transformation. I make an attempt to meet this challenge in the following, readily admitting that this attempt is preliminary, incomplete, and (hopefully) controversial.

This new type of mathematical modeling entails a number of components, most of which have been discussed in the philosophy of science literature on simulation, among them experimentation, visualization, and adaptability. Whereas these components have been treated in enlightening ways and from different angles, they have mostly been treated separately. What I intend to add in Section 2 is an account that respects how these components are interdependent—how they fit together and build a new type of mathematical modeling. The novelty of simulation, Section 2 concludes, is one of recipe rather than ingredients.

Furthermore, simulation modeling is also exerting considerable impact on fundamental notions and practices in the sciences. Section 3 picks up a suggestion by Paul Humphreys and argues that simulation modeling is best viewed as a new style of reasoning in the sense of Alistair Crombie and Ian Hacking. More precisely, I argue, simulation modeling can be characterized as a *combinatorial* style of reasoning. What does elevate a type of modeling to a style of reasoning? It has to exhibit considerable transformational force. Two examples exemplify such impact: what counts as "understanding phenomena" and what counts as "solution." Both are seminal pieces of traditional mathematical modeling, and both are transformed, if not inverted, in simulation modeling.

Section 4 briefly discusses new challenges for a philosophy of science that wants to develop a meaningful account of computer simulation.

2 ELEMENTS OF SIMULATION MODELING

2.1 Terminology

The article characterizes computer simulation by how simulation models are built; that is, it concentrates on *simulation modeling*. Let us have a brief look at both parts of this term separately. There are several definitions out for "simulation" but no commonly accepted one.[2] A good starting point is Eric Winsberg who reviews a couple of extant proposals and discerns a narrow from a broad definition:

---

[2] The chapter deals with simulation on digital computers as is usual today. In the pioneering days, analogue computing machines like Vannevar Bush's differential analyzer were famous for simulating the

In its narrowest sense, a computer simulation is a program that is run on a computer and that uses step-by-step methods to explore the approximate behavior of a mathematical model. (. . .) More broadly, we can think of computer simulation as a comprehensive method for studying systems. In this broader sense of the term, it refers to an entire process. (2014)

The broader conception is the primarily relevant one for us. Simulation in the narrow sense, however, is part-and-parcel of the broader picture. The process of developing and using simulation in the narrow sense comes with certain requirements and implications of the utmost philosophical relevance. Exploring the behavior of a mathematical model with the help of a computer profoundly changes how and what researchers learn from such models. Moreover, the models themselves also change, since scientists gear their activity toward types of mathematical models that are particularly suitable for computer-based investigations. Seen from this perspective, simulation modeling overlaps with computational modeling, which seems perfectly acceptable to me.

We arrived at the notion of models and the process of modeling. The debate on models arose from the controversy over what role theories and laws play in the natural sciences when it comes to applications. Models, it has now been widely acknowledged, mediate between theories, laws, phenomena, and data.[3] I agree particularly with Margaret Morrison (1999) when she describes models as "autonomous agents." We discuss salient features of simulation models that are taken neither from target systems nor from guiding theories.[4]

The activity of modeling aims at finding a mediating balance. How then does simulation modeling differ from modeling in general? Although the balance itself may look very different in different cases, there are common features of simulation modeling that try to achieve a balance. We can refine the question and ask how simulation modeling differs from mathematical modeling. I agree with Paul Humphreys' view (1991, 2004) according to which simulation is enlarging the realm of tractability, but I try to address the question on a finer resolution. The point is that practices of keeping things tractable with the help of the computer call in further changes regarding how modeling proceeds. Simulation modeling does not merely thrive on computing capacity but differs in philosophically interesting ways from mathematical modeling.

Most accounts of simulation modeling divide their topic according to techniques like Monte Carlo (MC) simulations, finite difference (FD) methods, or agent-based models. I do not follow this route. My aim is to formulate a general account of simulation

---

solution of certain systems of differential equations. For a historical account of analogue computing, cf. Mindell (2002).

[3] Probably the most prominent presentation of this perspective can be found in Mary Morgan and Margaret Morrison's edited book *Models as Mediators* (1999).

[4] The increasing differentiation of the model debate suggests that philosophical, historical, and sociological aspects should be considered together. Gramelsberger (2011), Humphreys and Imbert (2012), Knuuttila, Merz, and Mattila (2006), Lenhard et al. (2006), and Sismondo and Gissis (1999) offer a broad spectrum of approaches to this.

modeling by focusing on what different techniques have in common. In one way or another, they instantiate strategies that coordinate modeling with the capacity of computers to execute iterative procedures. This coordination unavoidably requires that simulation deviates from established strategies of mathematical modeling. Importantly, these deviations might be alien to human epistemology. A necessary part of simulation modeling hence is recreating connections to the established problems, solutions, and tasks. Simulation modeling thus includes the transformation toward the computer plus the compensation of these transformations. I discern four typical components and argue that they are interconnected and make up a new type of modeling.

2.2 Experimentation

Computer simulations work with a special type of experiment—on that, at least, the literature agrees. Indeed, it is precisely this new type of experiment that frequently attracts philosophical attention to simulations.[5] Most authors agree the most important feature is simulation experiments, often combined with viewing simulation as neither (empirical) experiment nor theory. Although it is undisputed that one is dealing with a special concept of experiment here insofar as it does not involve intervention in the material world, agreement does not extend to what constitutes the core of such an experiment. As a result, no standardized terminology has become established, and simulation experiments, computer experiments (described more precisely in English as computational experiments), numerical experiments, or even theoretical model experiments all emphasize different aspects. Overall, a main line of debate in philosophy of simulation is what the relevant similarities and differences between simulation and (ordinary) experiment are. The recent overview articles by Wendy Parker (2013) and Winsberg (2014) provide a useful entry into this discussion.

However, I want to follow a different line. In the following the focus is not on the status of simulation experiments and whether they are rightly called experiments but on their function in simulation modeling. The main point is that relevant properties of simulation models can be known only by simulation experiments. Or, if we want to avoid talking about experiment in this context, these properties can be known only by actually conducting simulations.[6] There are two immediate and important consequences. First, experimentation is unavoidable in simulation modeling. Second, when researchers construct a model and want to find out how possible elaborations of the current version perform, they will have to conduct repeated experiments. Simulation modeling thus will normally involve a feedback loop that includes iterated experiments.

---

[5] A variety of good motivations are given in, for instance, Axelrod (1997), Barberousse, Franceschelli, and Imbert (2009), Dowling (1999), Galison (1996), Humphreys (1994), Hughes (1999), Keller (2003), Morgan (2003), Morrison (2009), Rohrlich (1991), Winsberg (2003).

[6] Mark Bedau (2011) has highlighted properties that can be known only by actually conducting the computational process of a simulation and has aptly called them "weakly emergent."

Let us consider a couple of examples from different simulation techniques that illustrate the main point.

2.2.1 *Finite Difference Method*

Consider a theoretical model is already at hand, for example on the basis of mathematically formulated general laws. This situation is a standard case in many sciences and considered to be the generic one by some philosophers of simulation (see Winsberg 2014). Such models often take on the form of a system of partial differential equations. Learning what the dynamics of the model actually is then requires solving the equations (i.e., integrating the system). In many cases, however, such integration is impossible by analytical mathematical means, which renders the model intractable. Computer simulation opens up a way for treating these models numerically, that is, for finding a substitution for the (generally unknown) solution.

Speaking about numerically solving an already given model tends to obscure that a particular modeling step is necessary, namely replacing the continuous mathematical model by a discrete model that can be processed on a computer. This step is key in simulation modeling, and it normally requires experimentation.

The point surely deserves some illustration. General circulation models in meteorology provide a prime example as they bring to the fore essential features.[7] Historically, they provided the basis for one of the earliest instances of simulation modeling directly after World War II; moreover, it was commonly accepted that the basic theory, namely fluid dynamics, had long been available. In addition, a theoretical model, expressing the known laws in a system of partial differential equations, had been formulated early in the twentieth century. It was not clear at that time whether this system missed important facets. These equations, however, were mathematically intractable and therefore considered without practical use. The newly developed digital computer opened up a new perspective in which this mathematical model appeared as a feasible candidate for simulation.

A group at the Princeton Institute for Advanced Studies, directed by Jule Charney, developed a simulation model whereby the continuous primitive equations were replaced by a stepwise dynamics on a space and time grid. In terms of mathematical description, the differential equations were replaced by FD equations. The model dynamics, consequently, did not follow from integration of the equations but from summing up over many steps. One complex operation thus was replaced by a host of simple operations, a large iterative procedure—infeasible for a human computer but exactly suitable for the capability of the new automatic computing machine.

It is important to recognize that the construction of a discrete model is a major modeling effort. The selection of a space–time grid, the reformulation of the continuous dynamic, the implementation as an executable program—all these are steps of modeling rather than derivation. What, then, makes a particular series of steps adequate? In the

---

[7] For more details and also references, cf. Lenhard (2007).

end, they have to interlock so that the simulation generates salient features of the atmospheric dynamics. This problem is accessible only by experimentation.

In our case it was Norman Phillips, a member of the Charney group, who conducted the simulation, the so-called first experiment." It was this experiment that convinced the community that the simulation model could indeed reproduce important (known) patterns of the global circulation. Only after the experiment did the theoretical equations gain their status as primitive, which they still hold today.

In short, discretization is a key step in simulation modeling. It turns a theoretical model, which is expressed via continuous mathematical equations, into a FD model, which is suitable for the computer. The dynamics of such a discrete model, however, can be investigated only via (computer) experimentation. This insight applies to other techniques of simulation as well.

2.2.2 *Monte Carlo Simulations*

Monte Carlo simulations present a special case insofar as a large number of (pseudo-) random experiments build the core of the method. Right from the start, MC simulations were seen as experiments. For instance, the mathematician Stanislaw Ulam (1952), who contributed decisively to the invention of MC simulation, saw them as a new method for carrying out statistical and mathematical experiments.

A great number of such single runs are carried out on a computer, each run with a parameter value chosen (pseudo)randomly according to a probability distribution. Then, by the law of large numbers, the mean of all the simulation results approximates the expected value of a distribution, which is formally given by an integral. As in FD methods, one has substituted an integral, maybe an analytically inaccessible one, by summing up many single instances.

Philosophical accounts often allot MC a special place. Peter Galison (1996), for example, contributed a study on the historical origin of the MC simulation within the context of the Manhattan Project. He assigns a fundamental philosophical significance to simulation because of the way it is linked to a radically stochastic worldview. While I think the argument about ontology is of a very limited range, my point rather is about methodology, namely about experimentation as a crucial component of simulation modeling. "Statistical" experiments represent an important and philosophically interesting class of computer experiments—concerning the role of experimentation, MC is typical rather than special.

2.2.3 *Cellular Automata*

A number of philosophers and scientists have recognized cellular automata (CA) as the core of what is novel in simulations.[8] Cellular automata investigate systems of cells situated on a neighborhood structure that take on various states dependent on the states of

---

[8] One can count into that group Rohrlich (1991), Keller (2003), and also Wolfram (2002) with his ambitious claim that CA provide the basis for a "new kind of science."

their neighbors. The so-called agent based models, now becoming increasingly popular among philosophers of science, are siblings of CA. They generate specify local interactions between agents and hope to arrive at interesting macro patterns. The point again is that only experimentation can tell what patterns a certain model specification actually generates. Furthermore, if we want to see how global patterns vary with varying model assumptions, we need repeated loops of exploratory experiments.

2.2.4 *Artificial Neural Networks*

Artificial neural networks (ANNs) have a generic structure where nodes (neurons) are ordered in layers and nodes of neighboring layers are connected, resembling the structure of a physiological neural network. Incoming signals to the first layer are processed along the connections until they reach the last layer. Depending on the strength of the single connections (regulating when a neuron "fires"), such ANNs can exhibit universal behavior; that is, they can generate all Turing-computable patterns. By systematically varying parameters (i.e., connection strengths), one can train a network to generate *particular* patterns, like recognizing the speech of a person (audio input, written output). Talking about training a network, however, is another way of talking about using iterated experimentation in a systematic way for modifying model behavior.

What do we take from this somewhat lengthy discussion of simulation techniques? They utilize different strategies for arriving at iterative procedures that then are performed by the computer. Experimentation therefore is a crucial component of the modeling process. Building and modifying a model even requires repeated loops of computer experiments. There are more elements, however, that characterize simulation modeling.

2.3 Visualization

The literature in philosophy, but also art history and media studies, usually debates the qualities of computer visualizations as representational images.[9] They allow a remarkable amount of control in the production of simulation pictures, including the variation of colors, perspectives, textures, and so on. Visualizations hence may appear as direct representations, although they are created by elaborate modeling procedures. The exact nature of the link between data and simulation often is difficult to establish. The question of what images of scanning tunnel microscopes actually show illustrates the point (see Baird, Nordmann, and Schummer 2004). However, the relevant issue here is the way in which visualizations are embedded into the modeling process.

We have already seen that simulation models exploit the capability of computers for performing a large number of simple (logical) calculations. Human beings are not particularly good at foreseeing the outcome of this sort of procedure, therefore

[9] Galison's (1997) monumental *Image and Logic* discusses a range of imaging technologies in physics.

experimentation is needed. However, human beings are equipped with a strong capability to process visual data. This is where visualization comes in. For example, a simulation of a meteorological circulation pattern might provide a large array of multidimensional (simulated) data. Visualization can condense this large array into an image and make graspable how well the simulation fits already known phenomena, like global circulation cells. Visualizations thus create a link back to human cognition.

In general, this observation applies to many kinds of imaging techniques. Nevertheless, it is particularly relevant for simulation modeling. When building a simulation model, researchers must learn how the current model version behaves and how possible modifications would change this behavior. The dynamics of a simulation model thus needs to be accessible to a sufficient degree. In short, visualizations allow scientists to interact with the model dynamics.

Consider the case where a cellular automaton simulates galaxy formation, discussed by Rohrlich (1991). The crucial test for the CA model is whether it can generate the form of galaxies already known from observations (photographs). Additionally, intermediate visualizations are used during the construction process of the CA model. It entails several parameters that control the dynamics of cells but have no direct physical meaning. How should one assign values to such parameters? Their values are chosen so that the simulation in the end matches known aspects of galaxy dynamics, like the speed of rotation, which is judged by (intermediate) visualizations.

Or consider the design process of a certain part in a wind tunnel. Today, design engineers often use fluid dynamics simulations (mostly FD techniques) as a sort of artificial wind tunnel. When optimizing the form of this component part, the engineers modify the form in the model and then look at the visualized streamlines on their computer screen to quickly see whether the modification would be a beneficial one.

When a particular quantitative value is what matters and when this value can be computed, visualization probably is dispensable. But if modelers want to learn something less precise from the simulation, or even want to detect some surprise, then visualization is the main option for accessing the model dynamics.

2.4 Plasticity and Adaptability

A model has the property of plasticity when its dynamical behavior can vary over a wide spectrum while its theoretical structure remains unchanged. A model is said to be adaptable when changing some model parameters can control the model's behavior. Roughly, adaptable models allow exploiting their plasticity in a purposeful way.

Why should this appear in our list of components? Few models are completely determined by their theoretical structure. Filling gaps and adjusting parameters has always been part and parcel of modeling—there is nothing special here about simulation. Furthermore, parameter adjustments and their likes also seem to be ad hoc and of minor importance. Although this line of reasoning is very common, I do not agree with it but would like to maintain instead that such adjustments are characteristic of simulation modeling as they fulfill an important function.

First, for some classes of simulation techniques, plasticity is pivotal. For CA, and to an extreme degree ANNs, plasticity does not appear as a shortcoming that has to be repaired but rather as a systematic pillar for modeling. The structure of the model is deliberately designed to generate a high degree of plasticity. An ANN, for instance, exhibits extremely flexible input-output behavior. It renders necessary an extended process of adaptation (learning), but this is exactly the point. When, for instance, speech recognition is simulated (i.e., when acoustic phrases are mapped on to words), a process of training the ANN is necessary. Training a network means systematically adapting the neural connection parameters. But this does not happen because of a gap in theory or a mistaken implementation of it. Rather, there is no (or little) theory involved in how acoustic phrases relate to words. It is exactly the advantage of ANNs that they exhibit great plasticity so that a suitable learning algorithm may create a workable mapping by stepwise adaptation. Such learning algorithms, in turn, are a matter of theory—a theory, so to speak, about how to exploit plasticity. To a significant degree, the properties of the model are determined during the process of adjustment. The latter thus is an integral part of the modeling process.

The situation with CA is similar. An example mentioned earlier is the simulation of fluid dynamics. What is known are fluid dynamical phenomena and the theory of them, like the Navier-Stokes equations. A CA model splits up the space into neighboring cells and specifies local interactions. How should this be done? Regularly, modelers will make a guess informed by theory and then explore (via experimentation and visualization) whether the resulting global dynamical patterns indeed reproduce the desired behavior. Modelers will try to learn from these experiments and adapt the interaction parameters accordingly so that the desired global patterns might emerge.

Hence the structure of the model is not aimed at completely representing the (or some idealized) structure of the phenomena to be modeled. Based on the model structure further specifications are made during later steps of model building, and only then is the behavior of the simulation model determined. In such a procedure, the specification (or adaptation) of the model, though normally regarded as pragmatic measures of little significance, acquires a pivotal role in modeling.

Is this kind of argument restricted to simulation models with a structure largely independent of the phenomena to be modeled? I do not think so. This brings me to the second argument on the significance of adaptability in simulation modeling.

This time let us look at examples in which a strong theory exists and the simulation model must build on this theory. The circulation of the atmosphere with its primitive equations is an example. A main point in the section on experimentation was that FD methods transform the theoretical equations into a discrete formulation compatible with the computer. Simulation modeling must address a twofold problem. First, the FD dynamics should closely resemble the dynamics of the target phenomena—not merely in the limit of an infinitely refined grid but on the grid actually used. Second, some processes relevant to atmospheric circulation may evolve on a subgrid scale, like the dynamics of clouds, among others. The effects of such unresolved processes, however, should be included in the simulated dynamics. This twofold problem is not depending on the concrete case of clouds and meteorology but rather is generic to simulation modeling based on continuous theory.

Both parts of the problems are normally tackled by introducing parameterizations. Numerical schemes are designed to model the unresolved processes by a condensed description at the grid points.[10] However, since the cloud parameterization cannot be confirmed on its own but only in view of the overall atmospheric model, the parameter values also have to be assigned—or "tuned"—in view of the overall model dynamics. Inevitably, various and partly unknown shortcomings, errors, effects of processes not represented in the model, and so on will compensate when a set of parameters is fitted.

The design of a parameterization scheme can be considered an art. It has to fulfill two conditions that are counteracting each other. The one is to implement tuning knobs into the scheme and to find particular values for the parameters in this scheme so that the simulation *globally* adapts well. This is not an exercise in curve fitting, however, since the second condition requires the scheme has reasonable *local* interpretations (not all parameters have them); only then is the achieved adaptation significant regarding the modeled phenomena. In brief, a good parameterization scheme must have the right kind of plasticity.

2.5 Fictional Components

A more general debate in philosophy of science regards fictions as part of scientific representations. In a sense, all mathematical models are artificial as they are created by human activity and introduce some idealizations. The debate about the character of mathematical representations is an important one. It is also much older than the terminology of fictionalism (for the latter, see, e.g., Barberousse and Ludwig 2009); it has been a part of the discourse on modeling and mathematics since early on.

Since Plato, in his dialogue "Timaios," marveled about Eudoxos' geometrical description of the planetary movements, in what sense a mathematical model can "save the phenomena" has been a much-debated issue. Famously, Ptolemaios' sophisticated model included components (epicycles, equants) that modified the circular movement—not as more accurate representations but to fit the phenomena better. We all know the controversy around Copernicus' *De Revolutionibus* and the foreword by Osiander that explicitly interprets the heliocentric model as merely hypothetical (i.e., working with artificial [non-veridical] hypotheses that are intended to save the phenomena). The challenging metaphysical question is: Can a mathematical model save the phenomena, because it reveals something about the true (structural) character of these phenomena? or get the phenomena saved as a merely pragmatic effect of modeling? Pierre Duhem (1908/1969), for instance, offers a historically sophisticated positivist stance.

---

[10] See Parker (2006) for a discussion in the context of climate modeling.

Here I want to follow a more specific question: Do fictional components assume a particular function in the methodology of simulation modeling? Some philosophers, like Winsberg (see 2014, which also gives further references), argue that fictions do have a special connection to computer simulations. One of his main examples is taken from multiscale modeling in nanoscience. There a model of how cracks propagate in materials connects different scales in one model. A quantum mechanical model is used for covering a small number of atoms near the emerging crack, and a molecular dynamics model is employed to describe the neighborhood. Both models describe atoms differently—how can the models be connected? The modelers create "Silogen" atoms that provide links to both model ranges.

> Silogen atoms are fictions. They are not offered as even a "good enough" description of the atoms at the boundary—their prima facie representational targets. But they are used so that the overall model can be hoped to get things right. Thus the overall model is not a fiction, but one of its components is. (Winsberg 2014)

I do not want to discuss whether "fictional," "instrumental," or "pragmatic" component is the most appropriate term. Instead, I would like to add an argument why there is a tendency in simulation modeling to include such components. Typically, such local entities and their interactions do not carry much ontological commitment, but are negotiable when adapting the global model dynamics.

That means Silogen atoms or other purportedly fictional components function in a way similar to parameterizations discussed previously. They do not bring in properties that have to be respected by other modeling assumptions but the other way round: their main property is that they result in an acceptable overall simulation model behavior given the other parts of the model. Consequently, these components have to be adaptable according to the overall model dynamics, which in turn requires repeated experimentation, because there are no other independent reasons for how such components should work.

Fictional components complement a simulation model for practical (methodological) reasons. Their exact dynamical function is assigned only after repeated loops in the modeling process. The traditional master question asked how mathematical structures represent the world. Any answer presumably would have to include both the world and the mode of mathematical construction. Computer simulations are adding an additional layer of *computational* structures and entities. From the standpoint of human epistemology, these are fictional components.

2.6 A New Type of Modeling

We discussed four components of simulation modeling. It is not claimed that these are "the" components. There are options for framing and naming them differently; maybe experimentation is the only one that would make it on every list. However, the claim is that these components are interdependent and are reinforcing each other in a systematic way. This is why they do not merely form a set of characteristic traits of simulation but constitute a distinct type of mathematical modeling. Although all components we have discussed are not new to science, the type itself is. What matters is the novelty of the recipe more than that of the single ingredients. The *combination* is the point, and the computer is crucial for making such combination feasible.

Plasticity and exploratory modeling complement one another. The discreteness of the model requires artificial and also nonrepresenting elements in the simulation model whose dynamic effects can be determined only in experiments. Simulation modeling is therefore dependent on experimentation. Because of a model's plasticity, exploration of model behavior addresses more than minor pragmatic questions of adaptation; on the other hand, the performance of models with a high plasticity can be investigated and controlled only in an exploratory mode over the course of feedback loops. Visualization, finally, makes model behavior accessible for modelers and hence helps to make the modeling process feasible in practice. It employs experimentation, visualization, adaptation, and again. It is a systematic way of taking advantage of trial and error.

Mathematical models running on a computer require (re-)framing everything—all entities, conditions, and operations—in a discrete, stepwise manner. This imposes strong restrictions, or rather changes, compared to traditional mathematical modeling. At the same time, iterations can be performed in great numbers and at high speed, which, in turn, supports exploration. Modeling hence can proceed via an iterated feedback loop according to criteria of model performance.

I stress again here the significance of the methodological feedback loop. Winsberg, for instance, has proposed a valuable characterization of simulation as downward, motley, and autonomous. All three properties make good points. The first, however, seems to run against the loop character:

> *Downward* EOCS [epistemology of computer simulation] must reflect the fact that in a large number of cases, accepted scientific theories are the starting point for the construction of computer simulation models and play an important role in the justification of inferences from simulation results to conclusions about real-world target systems. The word "downward" was meant to signal the fact that, unlike most scientific inferences that have traditionally interested philosophers, which move *up* from observation instances to theories, here we have inferences that are drawn (in part) from high theory, *down* to particular features of phenomena. (2014)

In view of the previous discussion, we can add that simulation thrives by *combining* downward and upward. Simulation modeling looks more like an oscillation than a unidirectional move.

3 A STYLE OF REASONING

Up to now I have argued about components of simulation modeling and their systematic interconnection, which justifies us speaking about simulation modeling as a new type of mathematical modeling. If pressed to characterize this new type by two notions, I would say "iterative and exploratory"; if pressed to choose just one word, I would vote for "combinatorial," because simulation modeling combines theory, (computer) experimentation, and empirical data in various and intricate ways. However, we have to inquire further how significant this new type of mathematical modeling is from a philosophical perspective.

Here I pick up Humphreys' proposal (2011), which states the concept of "style of reasoning" is appropriate when categorizing simulation. This term refers to Alistair Crombie's (1988) work on how styles of scientific thinking differ (for more detail see Crombie 1994). Ian Hacking (1992) honed the concept and coined the more rigorous notion of "styles of reasoning." The main point is that each style specifies "what it is to reason rightly" (Hacking 1992: 10). Is there a style that fits simulation modeling? Crombie discerns six styles and simulation that would presumably count as "hypothetical modeling," though the category is much broader. According to Peter Galison (1987), there are three independent traditions of research, parallel but largely independent (each with their own life): the theoretical, experimental, and instrumental tradition. Based on the previous discussion, I think the style of simulation modeling combines these three, hence is in a fundamental sense *combinatorial*.[11]

If indeed the term "style of reasoning" should be appropriate, we need some evidence in what respect and to what extent simulation modeling affects "what it is to reason rightly." Indeed, there seem to be a number of fundamental concepts somehow linked to scientific modeling that undergo changes or assume a new meaning in the practices of simulation modeling. One candidate is validation; Winsberg (2014) and Parker (2013) both include this issue in their overviews on simulation.

The remainder of this section concentrates on the conceptions of solution and of (scientific) understanding. The claim is that in simulation modeling each of them is undergoing a significant change. In a nutshell, the established meaning of mathematical solution is transformed into a much broader pragmatic concept of numerical solution. Understanding via mathematical models is transformed from thriving on transparency into dealing with opacity. Let me indicate the arguments that admittedly would deserve further elaboration.

---

[11] The wording here is inspired by Anne Marcovich's and Terry Shinn's (2014) treatment of "combinatorials" in the context of nanoscale research.

3.1 Solution

Simulation is often described as numerical solution. This manner of speaking is equally widespread in scientific and philosophical literature (cf., e.g., Hartmann 1996; Dowling 1999). It suggests a great continuity between ordinary mathematical modeling and simulation modeling. Basically, it is implied, the task remains invariant: instead of solving equations analytically, simulations solve them numerically. Although this viewpoint sounds very plausible, it tends to exaggerate the continuity in a misleading way. I argue for the following thesis: when simulation is characterized as numerical solution, this invokes a pragmatic sense of solution that differs significantly from the usual mathematical sense.

When we talk of solution in the proper sense, we normally refer to the mathematical notion of solution. This is a rather strict notion relevant in a broad class of natural and engineering sciences, namely where the language of differential and integral calculus is employed, and—even more general—where the language of mathematical equations is employed to formulate theoretical models. If we have an equation with one unknown term, the mathematical term that satisfies our equation is a solution.

What counts as a solution in the strict sense solely depends on the mathematical equations that gets solved (including definitions of the underlying mathematical space). A quadratic equation, for example, can be written as

$$a \cdot x^2 + b \cdot x + c = 0$$

and solutions are exactly those numbers x that satisfy this equation (with given numbers a, b, c). This strict sense of a solution easily generalizes to problems more complicated than one single equation. If we have a system of partial differential equations, for instance, a given term will either satisfy (solve) it or not. Talking about the notion of solution, it does not matter whether one can actually calculate the solution by integration. Systems of partial differential equations are endemic in many sciences, but they are usually not analytically solvable.

They often can be tackled by simulation methods, however. This is the motivation for calling simulations numerical solutions. Although this usage seems to be straightforward, it is misleading. The reason is that simulation normally requires additional modeling steps that affect the character of what the simulation actually solves.

Recall our example of the global circulation of the atmosphere. The primitive equations form an analytically intractable system of partial differential equations. While this system has been made tractable by simulation, at the same time, it was transformed. The simulation model required discretization and the discrete model, a FD model, is not mathematically derived from the original theoretical model. There are several options for discretizing the model; most of the plausible ones approach the theoretical model in the limit of an infinitely fine grid. In practice, however, researchers always work on a finite grid. How can they determine whether their (discrete) simulation model is adequate? Normally they will have to compare salient dynamical properties of the simulation with those of the target system.

In general, it is unclear how the simulation is related to the unknown solution of the theoretical system. This is the heart of the matter. In our example of the atmospheric circulation, the primitive equations were remodeled for transforming them into FD equations. It turned out, however, that the simulation model was not numerically stable in the long run. It was a major achievement of simulation modeling to manipulate the discrete model so that it would be numerically stable when run. Researchers accepted that the new model contained components not motivated by the theoretical model but by internal computational reasons (cf. Lenhard 2007).

In short, if one starts with a continuous model and then replaces this model with a discrete simulation model, then in general the solution for the simulation model will not solve the continuous one.

Besides the *strict* sense of solution, there is also a *pragmatic* sense in which one can speak of solving a problem numerically. Such solution then, in a numerical-quantitative sense, would be good enough for solving the problem at hand. For instance, one could search for a number that satisfies an equation up to a certain inaccuracy; that is, if inserted into the previous equation, the resulting value would be close enough to zero. If one specifies what counts as close enough to zero, one has determined what counts as a solution and what does not. In this case, a solution is defined via a condition of adequacy that can be, but need not be, mathematically precise. In general, such a condition could well depend on what works well enough in a certain context.

Such a question is for the pragmatics of modeling and is not one specific to simulation. The point is that the pragmatic sense of solution is the primarily relevant one for simulation. In the meteorological example, the simulation provides a pragmatic (but not a strict) solution of the dynamics of the atmosphere. The simulation generated fluid dynamical patterns that were similar enough to the observed ones, while it was not defined in a quantitative and precise way what counts as adequate.

Computer simulations arc always numerical. If successful, they provide a pragmatic solution and maybe even a strict solution, but this would be an exceptional case. Let us resume that the strict concept of solution is in fact more special:

1. Only in the cases where a theoretical and mathematically formulated model serves as a basis is a strict solution defined at all.
2. Even if a strict solution is defined, simulation modeling regularly will not get it.

The manner of speaking of simulation as involving numerical solutions has to be handled with care, since it mixes up the strict and the pragmatic sense of solution. Speaking in this way supports a common but misleading picture according to which simulation is a straightforward continuation of traditional mathematical modeling. If we want to draw a more adequate picture, we must take into account those modeling steps that come with simulation modeling and that combine theoretical with instrumental components. Therefore, it seems more apt to say the simulation model *imitates*, rather than solves, the dynamics of the theoretical model.

3.2 Understanding

Understanding is a central but somewhat vague and multifaceted notion in epistemology. A couple of decades ago, understanding sometimes was taken to be a mere appendage to explanation. There is a vast literature in philosophy of science dealing with explanation, whereas understanding is covered considerably less. Books such as the one edited by Henk de Regt, Sabina Leonelli, and Kai Eigner (2009), for example, indicate a change—understanding now is on the agenda in philosophy of science.

In our context, it is of particular interest to have a look at mathematical modeling. Does simulation modeling, against this background, involve a change in the notion of understanding? Here I extremely briefly outline how mathematical modeling traditionally is thought to connect to understanding. Mathematical modeling introduces idealizations so that, ideally, a clear and formal picture emerges of how things work (in the model). The physicist Richard Feynman captured the essence when he stated in his lectures that he has understood an equation when he knows what happens without actually calculating. In a nutshell, mathematical modeling creates epistemic transparency, and the latter enables understanding.

The whole conception of simulation modeling seems to run against this view. Instead of reenacting transparency, simulation offers a way for dealing with opacity. There is a series of reasons why epistemic opacity appears in the context of simulation. First, the sheer number of computational steps renders the entirety of a computation unsurveyable. Humphreys 2004: 148; 2009: Section 3.1) emphasizes primarily this aspect when he speaks of epistemic opacity. He is correct, but there are more reasons. Second, the discrete nature of computer models requires artificial or fictional components, which are positively correlated with epistemic opacity. Third, the plasticity of models fosters opacity, because that part of the model dynamics that has been adapted according to performance measures (and not theoretical reasoning) will lead to opacity. Fourth, a variant of the adaptability problem is the import of submodels in software packages. Whether the code is proprietary or simply developed by someone else—both cases will contribute to opacity.

Epistemic opacity is a typical feature of simulations. However, it does not block simulation modeling. Simulation models can provide means to control a dynamics, although these models are epistemically opaque. Researchers and practitioners can employ a series of iterated simulation experiments and visualizations for testing how varying inputs are related to outputs. In this way they can orient themselves in the model—even if the dynamics of the simulation remain (at least partly) opaque. Admittedly, such a kind of acquaintance with the model falls short of the high epistemic standards that usually are ascribed to mathematical models. This lower standard is still sufficient, however, when the aim is controlled intervention.

A typical example is the possible breakdown of the meridional overturning circulation (MOC; e.g., the Gulf Stream). Researchers investigate how the MOC behaves under varying conditions (in the simulation model), like temperature increase. Their goal is to understand how robust it is. But understanding here means the opposite of Feynman's case. Whereas he wanted to know behavior without calculation, gaining a picture of the MOC is based on large amounts of calculations. Similarly, structural engineering has changed its face with computational modeling. Daring constructions can be admired that could not have been planned without calculating their structural stability via computer models. Engineers understand how such constructions behave, but in a very pragmatic sense that does not presuppose epistemic transparency.

Of course one could question whether the pragmatic notion should be called understanding at all. Hence we face two options: First, does simulation eliminate understanding in the practices of sciences and engineering, or second, do simulation practices replace a strong notion of understanding by a weaker, pragmatic notion? If epistemic opacity is unavoidable in complex simulation models and if, nevertheless, possibilities for interventions and predictions exist, will this coexistence lead to a new conception or redefinition of scientific understanding? Devising an answer to this question still is a task for philosophy of science. A promising route is investigating how "acquaintance with model behavior" works in practical examples; see for instance Miles MacLeod and Nancy Nersessian (2015).

My argumentation comes down to the claim that, first, simulations can facilitate orientation in model behavior even when the model dynamics themselves remain (partially) opaque. Second, simulations change mathematical modeling in an important way: theory-based understanding and epistemic transparency take a back seat while a type of simulation-based understanding comes to the fore that is oriented at intervention and prediction rather than targeted at theoretical explanation. Thus simulations circumvent the complexity barrier more than they remove it. They present a kind of surrogate for frustrated analytical understanding.

4 CHALLENGES FOR PHILOSOPHY OF SCIENCE

The computer as an instrument is continuing to reconfigure parts of science and our culture. It affects scientific methodology on a detailed level, as well as in much more general issues like that of human reasoning—think of the cognitive sciences and how they are based on computer instrumentation. Challenges abound on how to conceptually organize this transformation into a philosophically meaningful account. I think at stake is the picture we draw of ourselves and of our position in the world. In my view, when investigating issues of computer simulation, the philosophy of science directly connects to questions of broadest philosophical significance, which is a good thing.

In a kind of ironic twist, the identification of a combinatorial style of reasoning seems to pose a challenge to philosophy of science. The concept of style implies there is a plurality of styles, and, of course, this is motivating Crombie and Hacking to make use of this term. Discerning styles is, in a sense, the counterprogram to elaborating a universal normative framework for "the" science, a program in which earlier philosophy of science was much engaged. This program had assigned mathematics as a role model, in particular for logical and deductive reasoning. Against this background, the style concept insists that reasoning is a much wider concept and that the sciences have in fact developed various styles of reasoning. While this idea has been used against mathematics as the universal role model for science, our investigation into simulation modeling offers an ironic twist. Namely, mathematical modeling is itself not a monolithic block. On the contrary, it harbors different styles of reasoning. To me, the challenge for philosophy of science is to give a nuanced and historically informed account of the pluralistic nature of mathematics and its uses in the sciences.

I made an attempt to characterize simulation modeling against the background of mathematical modeling. Essential properties of mathematical modeling are reversed in simulation, both in methodological and epistemic respects. There are other possible entry points though. One that strikes me as important is (inter) disciplinarity, or perhaps transdisciplinarity. Simulation is going to newly distribute similarities and differences between scientific disciplines. In this context, it is promising to use the concept of "research technology," a term coined by Terry Shinn, and to look at simulation as research technology (cf. Lenhard, Küppers, and Shinn 2006). For example, there is an increasing gap between model developers and model users, since software packages are traveling between groups and even disciplines. This is not merely a sociologically interesting observation but also poses a philosophical challenge.

Philosophy of science has developed and sharpened its analytical instruments with theoretical physics and is currently investing much work for accommodating biology. Computer simulation, however, is somehow cutting across established lines of thinking. It has created new similarities between fields that use computer models, or even the same software packages. One example is computational physics and computational biology. Another example would be quantum chemistry and chemical engineering. In terms of theory, they look quite different. In terms of simulation modeling, however, they look quite comparable. It is likely that computer simulation and the combinatorial style of reasoning create new similarities and dissimilarities that partly question established ways of categorizing disciplines, their objects, and the ways these objects are investigated. I am convinced that the investigation of these issues will help us to make a guess at, or at least understand better, the bigger question of how the computer and simulation methods are changing the sciences.

References

Axelrod, R. M. (1997). "Advancing the Art of Simulation." In R. Conte, R. Hegselmann, and P. Terno (eds.), *Simulating Social Phenomena* (Berlin: Springer), 21–40.

Baird, D., Nordmann, A., and Schummer, J. (Eds.). (2004). *Discovering the Nanoscale* (Amsterdam: IOS Press).

Barberousse, A., Franceschelli, S., and Imbert, C. (2009). "Computer Simulations as Experiments." *Synthese* 169: 557–574.

Barberousse, A., and Ludwig, P. (2009). "Models as Fictions." In M. Suárez (ed.), *Fictions in Science: Philosophical Essays in Modeling and Idealizations* (London: Routledge), 56–73.

Bedau, M. A. (2011). "Weak Emergence and Computer Simulation." In P. Humphreys and C. Imbert (eds.), *Models, Simulations, and Representations* (New York: Routledge), 91–114.

Crombie, A. C. (1988). "Designed in the Mind: Western Visions of Science, Nature and Humankind." *History of Science* 26: 1–12.

Crombie, A. C. (1994). *Styles of Scientific Thinking in the European Traditions*, Vols. I–III (London: Duckworth).

De Regt, H. W., Leonelli, S., and Eigner, K. (Eds.). (2009). *Scientific Understanding: Philosophical Perspectives* (Pittsburgh: University of Pittsburgh Press).

Dijksterhuis, E. J. (1961). *Mechanization of the World Picture* (Oxford: Oxford University Press).

Dowling, D. (1999). "Experimenting on Theories." *Science in Context* 12(2): 261–273.

Duhem, P. (1969). *To Save the Phenomena: An Essay on the Idea of Physical Theory From Plato to Galileo* (Chicago: University of Chicago Press). (Original work published in 1908)

Frigg, R., and J. Reiss (2009). "The Philosophy of Simulation: Hot New Issues or Same Old Stew?" *Synthese* 169(3): 593–613.

Galison, P. (1987). *How Experiments End* (Chicago: University of Chicago Press).

Galison, P. (1996). "Computer Simulations and the Trading Zone." In P. Galison and D. J. Stump (eds.), *The Disunity of Science: Boundaries, Contexts, and Power* (Stanford, CA: Stanford University Press), 118–157.

Galison, P. (1997). *Image and Logic: A Material Culture of Microphysics* (Chicago: Chicago University Press).

Gramelsberger, G. (Ed.). (2011). *From Science to Computational Sciences* (Zürich: Diaphanes).

Hacking, I. (1992). "'Style' for Historians and Philosophers." *Studies in the History and Philosophy of Science* 23(1): 1–20.

Hartmann, S. (1996). "The World as a Process: Simulations in the Natural and Social Sciences." In R. Hegselmann (ed.), *Modelling and Simulation in the Social Sciences from the Philosophy of Science Point of View* (Dordrecht: Kluwer), 77–100.

Hughes, R. I. G. (1999). "The Ising Model, Computer Simulation, and Universal Physics." In M. Morgan and M. Morrison (eds.), *Models as Mediators* (Cambridge, UK: Cambridge University Press), 97–145.

Humphreys, P. W. (1991). "Computer Simulations." In A. Fine, F. Forbes, and L. Wessels (eds.), *PSA 1990*, Vol. 2 (East Lansing, MI: Philosophy of Science Association), 497–506.

Humphreys, P. W. (1994). "Numerical Experimentation." In P. Humphreys (ed.), *Patrick Suppes: Scientific Philosopher*, Vol. 2 (Dordrecht: Kluwer), 103–121.

Humphreys, P. W. (2004). *Extending Ourselves: Computational science, Empiricism, and Scientific Method* (New York: Oxford University Press).

Humphreys, P. W. (2009). "The Philosophical Novelty of Computer Simulation Methods." *Synthese*, 169(3): 615–626.

Humphreys, P. W. (2011). "Computational Science and Its Effects." In M. Carrier and A. Nordmann (eds.), *Science in the Context of Application*. Boston Studies in the Philosophy of Science 274 (New York: Springer), 131–142.

Humphreys, P. W., and Imbert, C. (Eds.). (2012). *Models, Simulations, and Representations* (New York: Routledge).

Keller, E. F. (2003). "Models, Simulation, and 'Computer Experiments.'" In H. Radder (ed.), *The Philosophy of Scientific Experimentation* (Pittsburgh: University of Pittsburgh Press), 198–215.

Knuuttila, T., Merz, M., and Mattila, E. (Eds.). (2006). "Computer Models and Simulations in Scientific Practice." *Science Studies* 19(1): 3–11.

Kuhn, T. (1961). "The Function of Measurement in the Physical Sciences." *Isis* 52(2): 161–193.

Lenhard, J., Küppers, G., and Shinn, T. (Eds.). (2006). *Simulation: Pragmatic Construction of Reality*. Sociology of the Sciences Yearbook 25 (Dordrecht: Springer).

Lenhard, J. (2007). "Computer Simulation: The Cooperation between Experimenting and Modeling." *Philosophy of Science* 74: 176–194.

Marcovich, A., and Shinn, T. (2014). *Toward a New Dimension* (Oxford: Oxford University Press).

Mindell, D. A. (2002). *Between Human and Machine: Feedback, Control, and Computing Before Cybernetics* (Baltimore, MD: Johns Hopkins University Press).

Morgan, M. S., and Morrison, M. (Eds.). (1999). *Models as Mediators* (Cambridge, UK: Cambridge University Press).

Morgan, M. S. (2003). "Experiments without Material Intervention. Model Experiments, Virtual Experiments, and Virtually Experiments." In H. Radder (ed.), *The Philosophy of Scientific Experimentation* (Pittsburgh: University of Pittsburgh Press), 216–235.

Morrison, M. (2009). "Models, Measurement, and Computer Simulation: The Changing Face of Experimentation." *Philosophical Studies* 143: 33–57.

Morrison, M. (1999). "Models as Autonomous Agents." In M. Morgan and M. Morrison (eds.), *Models as Mediators* (Cambridge, UK: Cambridge University Press), 38–65.

MacLeod, M., and Nersessian, N. J. (2015). "Modeling Systems-Level Dynamics: Understanding without Mechanistic Explanation in Integrative Systems Biology." *Studies in History and Philosophy of Science, Part C: Biological and Biomedical Science* 49: 1–11.

Parker, W. S. (2013). "Computer Simulation." In S. Psillos and M. Curd (eds.), *The Routledge Companion to Philosophy of Science*, 2d ed. (New York: Routledge). 135–145.

Parker, W. S. (2006). "Understanding Pluralism in Climate Modeling." *Foundations of Science* 11(4): 349–368.

Rohrlich, F. (1991). "Computer Simulation in the Physical Sciences." In A. Fine, F. Forbes, and L. Wessels (eds.), *PSA 1990*, Vol. 2 (East Lansing, MI: Philosophy of Science Association), 507–518.

Sismondo, S., and Gissis, S. (Eds.). (1999). Special Issue: *Practices of Modeling and Simulation*. *Science in Context* 12(2).

Stöckler, M. (2000). "On Modeling and Simulations as Instruments for the Study of Complex Systems." In M. Carrier, G. J. Massey, and L. Ruetsche (eds.), *Science at Century's End* (Pittsburgh: University of Pittsburgh Press), 355–373.

Ulam, S. (1952). "Random Processes and Transformations." In *Proceedings of the International Congress of Mathematicians 1950* (Providence, RI: American Mathematical Society), 264–275.

Winsberg, E. (2003). "Simulated Experiments: Methodology for a Virtual World." *Philosophy of Science* 70: 105–125.

Winsberg, E. (2014). "Computer Simulations in Science." In *Stanford Encyclopedia of Philosophy* (Fall 2014 ed.). Stanford, CA: Stanford University. http://plato.stanford.edu/archives/fall2014/entries/simulations-science/

Wolfram, S. (2002). *A New Kind of Science* (Champaign, IL: Wolfram Media).

CHAPTER 35

....

DATA

....

AIDAN LYON

1 INTRODUCTION

....

DATA are clearly very important and play a central role in our daily lives. However, this is a fairly recent development. It was not all that long ago that data were mostly confined to lab books and primarily the concern of scientists. But data are now everywhere, and they are increasingly important to everyone, from teenagers who want faster data plans for their cell phones, to photographers who need larger data storage devices, to CEOs who want to know how to use their business data to improve their sales.

Although data have long had a home in science, the relationship between data and science has been changing in recent times. Terms such as "big science" and "data-driven science" refer to sciences that seem to proceed in an increasingly automated way due to their increased access to gigantic datasets—so-called *big data*. Data are now collected, filtered, stored, and even analyzed automatically by electronic devices. The better humans get at building devices that can gather, store, and process data, the less of a role humans seem to have in science (cf. Humphreys 2007).

Since data are so important to our lives and scientific practice, and since big data are apparently changing the way science is done, it is worth putting the concept of data under the philosophical microscope. In this essay, I will review three philosophical issues in relation to data. Section 2 asks what data are and reviews four analyses of the concept that have been discussed in the literature. Section 3 focuses on the big data revolution, its impact on scientific practice, and what issues in the philosophy of science it raises. Finally, Section 4 discusses the data–phenomena distinction (due to Bogen and Woodward 1988) that has been put to work in the philosophy of science.

2 WHAT ARE DATA?

....

The concept of data is clearly an important one, one that appears to be central to the practice of science and the foundation of much of our modern life. Indeed, it seems that more and more things get "data-fied" each year—for example, music, movies, books, and currencies (such as bitcoin). Because data is so important to us, it is worth asking what we mean by "data." When scientists speak of data, do they talk about the same kind of thing as when tech executives speak of our data being stored in the cloud? It is quite possible that the answer is "no" and that we have multiple concepts of the data that should be distinguished before we attempt to analyze them. However, to begin, I will assume that we have one unified concept of data and see if it can be given an interesting and coherent analysis.

Little work has been done on analyzing the concept of data, and the work that has been done shows that it is a concept that is a little trickier to analyze that one might have supposed. In philosophy, the main source of work on this topic has been done by Luciano Floridi, and, in particular, Floridi (2008) has provided a review of some analyses of data that seem initially plausible. In this section, I will discuss these analyses and build on Floridi's work. Floridi identifies four analyses of the concept of data: (1) the epistemic interpretation, (2) the informational interpretation, (3) the computational interpretation, and (4) the diaphoric interpretation, the last of which is his own preferred interpretation.

The epistemic interpretation of data says that data are facts. This simple analyses has some prima facie plausibility, especially when considered in the light of how data can be used in scientific practice. When William Herschel recorded and analyzed his observations of bodies in the night sky, he was recording and analyzing data, which were facts about the positions of the astronomical bodies, and the analysis allowed him to discover new facts, such as the existence of Uranus (Holmes 2008). Similarly, when credit card companies collect data on our purchases, they seem to be collecting facts about our purchases (e.g., the fact that I just bought a coffee with my credit card at the cafe that I'm currently sitting in is now recorded in some far-away database as a datum).

However, as Floridi points out, the epistemic interpretation has at least two problems. The first problem is that it appears to be too restrictive in that it doesn't explain processes that are common to data, such as data compression and data cryptography. Unfortunately, Floridi doesn't elaborate on this problem, and we way wonder why the epistemic interpretation doesn't explain these processes and whether it is a reasonable constraint that any interpretation should explain them (and what kind of explanation is required). One way to understand the objection is if we understand the concept of fact so that facts are not the sorts of things that can be compressed, in which case the epistemic interpretation would entail that data cannot be compressed. Compression, at least in this context, appears to be a process that applies to strings of symbols (e.g., it is common to compress the string "for example" to the string "e.g."). So, however we understand what facts are, if they are not understood as (or involving) strings of symbols, then it seems they cannot be compressed in the appropriate way. The most common understandings of facts understand them in (apparently) nonsymbolic ways (e.g., as propositions, or as events, etc.). If any such analysis of the concept fact is right, then it will turn out that data cannot be facts.

Floridi's second objection to the epistemic interpretation is that it isn't very informative; it simply trades one concept that is difficult to analyze for another. This objection seems correct, in so far as we want an analysis in terms of simpler concepts, but it can be helpful to know that we only have one concept that is difficult to analyze, rather than two. Moreover, since work has already been done in analyzing what facts are (e.g., Carnap 1946; Popper 1959; Lakatos 1970), the identification of data with facts inherits the informativeness of that work.

Nevertheless, it seems that the epistemic interpretation is wrong. In addition to the problem that it seems to entail that data cannot be compressed, it also seems to entail that data cannot be false. The usual understanding of facts is such that if something is a fact, then it is *true*. For example, if it is a fact that 2014 is the hottest year on record, then it is (in fact!) true that 2014 is the hottest year on record. Facts, then, cannot be false. However, it seems that data can be false. Indeed, sometimes scientists are accused of falsifying their data. To take a famous example, Ronald Fisher complained that Mendel's data on trait inheritance was probably falsified:

> [I]t remains a possibility among others that Mendel was deceived by some assistant who knew too well what was expected. This possibility is supported by independent evidence that the data of most, if not all, of the experiments have been falsified so as to agree closely with Mendel's expectations. (Fisher 1936, 132)

Since facts cannot be falsified and data can, data cannot be facts.

This problem with the epistemic interpretation suggests a very natural way for us to modify the epistemic interpretation so that it avoids the problem, and, incidentally, so that it is more deserving of its name: the *epistemic* interpretation. Instead of saying that data are facts, we might try to say that data are *purported* facts. That is, data are the things that *we think* are facts. A false datum, then, is one that we think is a fact but which actually isn't. This modification of the epistemic interpretation also has the advantage that it fits quite comfortably with the original meaning of "datum" in Latin: that which is taken as given or for granted.

However, there are still problems. Whatever data are, they are things that can be stored, compressed, destroyed, encrypted, and hidden. And it seems that purported facts are not the sorts of things that can have all of these things done to them. For example, suppose that we say that a purported fact is a proposition that someone thinks is true or one that enough people are think is true (and let's not worry about how many would be "enough"). The way we normally think about memory sticks and propositions is such that memory sticks don't store propositions, and yet they do store data. Propositions are also not compressed or destroyed or encrypted or hidden.

Another natural modification suggests itself: we might say that data are *representations* of purported facts (or representations of facts, where the representations need not be veridical). This version of the epistemic interpretation renders data as certain kinds of symbols: symbols that are to represent purported facts. Because symbols can be stored, compressed, destroyed, encrypted, and hidden, it seems we have a substantial improvement to the original epistemic interpretation. Because this improvement also says that data are not any old symbols (i.e., because they are symbols that represent purported facts), it is easy to see how they can be false and how they have an epistemic component to them.

Although it seems we are moving in the right direction, there are two problems with this most recent refinement of the epistemic interpretation. The first problem suggests further refinements that could be made, but the second appears to be more problematic. The first problem is that it seems that there can be data that might represent facts but that are not purported by anyone (or intended to exist). Imagine a malfunctioning microphone that is no longer able to record the human voice but is still sensitive enough to record human coughs. Suppose the microphone has been forgotten and accidentally left on in some office, with the recordings being stored on some far-away server. It seems true to say that the server now contains data on the coughing in the office. Indeed, if someone were to ever realize this, they could analyze the coughing and use it for some purpose (e.g., to measure cold and flu activity in the office). However, until someone does realize this, the data on the server do not appear to be purported facts because no one purports them as facts (and no one intended them to represent coughing facts). This problem suggests that the concept of data may live somewhere between being an intentional and representational concept and being a nonintentional informational concept. We therefore might be able to further improve the epistemic interpretation by using the resources from the literature on information and representation (see, e.g., Dretske 1981, Fodor 1987). For example, instead of saying that data are symbols that represent facts that are purported, we might say that they *would* represent the relevant facts if certain conditions obtained or that the relevant facts *would* be purported if certain conditions obtained. I shall leave these considerations here, because the second problem that I alluded to earlier appears to be more daunting.

The second problem with saying that data are representations of purported facts is that it doesn't seem to do justice to a large amount of the data that we use on a daily basis. For example, these days, it is common for music to be stored in the form of data on computers and various portable devices. This music is stored in a way that is fundamentally the same as the way in which, say, the outcomes of a scientific experiment are stored: as a string of 1's and 0's in a file that is encoded in some physical format, such as point variations in magnetism in the magnetic tape of a spinning hard drive. There are many ways in which data are stored physically—solid state hard drives, compact disks, floppy disks—and these details need not concern us here. So, for simplicity, I shall speak as though 1's and 0's are stored directly in some physical medium (e.g., one can think of tiny 1's and 0's written on the magnetic tape). It doesn't seem that any string of 1's and 0's in a music data file represents any purported facts. For example, the first word of the lyrics of *Hey Jude* is "Hey" and was sung by Paul McCartney. I have that song on my computer, and, when I play it, a causal process unfolds that involves a particular string of 1's and 0's on my hard drive that leads to me hearing the word "Hey." However, it doesn't seem correct to say that that string of 1's and 0's represents the purported fact that, say, McCartney sung the word "Hey" (at some particular time and location). Although it is true that we can *infer* from this string of 1's and 0's, along with some other facts, that McCartney did in fact sing the word "Hey," it is nevertheless false that that string represents that fact. Nor does it seem to represent *any* fact. Rather, the string represents something more like a *command* or *input* that is to be passed to and processed by my music player.

Music stored as data is not the only example of this kind. For example, computer programs are also stored in the same way, and they appear to fail to be factive in the same way that music fails to be factive. For example, the first line of one of the programs on my computer is "import numpy," which is a command to import the numpy package, a Python scientific computing package commonly used for data analysis.[1] This command is stored as a string of 1's and 0's somewhere on my computer, but that string doesn't represent any fact about the world. So it seems that not all data are sets of facts. At this point, we might try to say that we have two concepts of data here, one that is associated with the storage of music (and other media) and one that our refinements of the epistemic interpretation were starting to latch onto. This is entirely possible, and the idea cannot be rejected immediately. However, we should try to resist introducing such a distinction, at least to begin with. This is, in part, simply because of parsimony, but also because such a distinction appears ad hoc, at least at a first glance. Along with the *Hey Jude* file on my computer, I have other files, some of which represent purported facts (e.g., outcomes of some experiments), and all these files are stored in fundamentally the same way: as strings of 1's and 0's.[2] So it seems that our first attempt in analyzing the concept of data should be to try to find an analysis that can handle both examples of data.

This brings us to the second interpretation of data that Floridi discusses: the informational interpretation. The informational interpretation of data says that data are information (or bits of information). One advantage to this interpretation is that makes sense of the locution *data mining*. Data mining typically involves running algorithms and queries on a database to obtain some fact that was not stored in an obvious way in the database. One classic example is the retailer Target, which mined its database of customer purchases to find that certain spending habits are indicative of future ones. In particular, Target found that when a woman changes from purchasing scented hand lotion to unscented hand lotion, that is evidence that suggests that the woman has become pregnant. In a famous incident, Target was able to work out that a teenager was pregnant before her parents were able to do so. Data mining, then, is not like coal mining; the goal of data mining is not to get more data, unlike how the goal of coal mining is to get more coal. Rather, the goal of data mining is to obtain *information*. Data mining might be better named "information mining," which suggests that the informational interpretation of data might be on to something.

However, there are serious problems with the informational interpretation. Floridi identifies two problems. The first is that the informational interpretation of data results in a circularity when combined with one of the most common interpretations of information: that information is a certain kind of true and meaningful data (cf. Floridi 2005). The second problem is that sometimes we have data without information. For example, a music CD can contain megabytes of data without containing information about anything (Floridi 2005). This second objection is probably stated a little too

---

[1] http://www.numpy.org

[2] Of course, there are different file formats and encoding systems, but, at the end of the day, all data are stored in the form of 1's and 0's.

strongly, for it might be argued that the music CD at least contains the information that some music was played at some point. At any rate, this second problem is a problem with understanding information as being *necessary* for data. A third problem arises from understanding information as being *sufficient* for data. The low-charge state of a laptop battery contains information—viz., that the battery has low charge—but the low-charge state does not appear to be data (or a datum). In contrast, the line in the battery history file that records the current charge state of the battery is both data and (contains) information. The informational interpretation of data, therefore, doesn't seem very promising.

The third interpretation of data, the computational interpretation, says that data are collections of binary elements (e.g., the etchings in the aluminum foil in compact discs) that are processed electronically. As Floridi states, an advantage to this interpretation is that it "explains why pictures, music files, or videos are also constituted by data" (Floridi 2008, 235). So, the computational interpretation handles the difficulty concerning media files that our refined version of the epistemic interpretation ran into. However, the computational interpretation is surely too limited. Vast amounts of data existed long before the invention of computers or even electricity—recall Herschel's astronomical data. And even today, plenty of data are stored nondigitally and in a nonbinary way (e.g., some DJs still use vinyl records).

This brings us, finally, to the fourth interpretation, the diaphoric interpretation, which has been developed and defended by Floridi (2008). The diaphoric interpretation says that a datum is a lack of uniformity in some domain. Put more formally, the interpretation says that a "datum $= x$ being distinct from y, where x and y are two uninterpreted variables and the domain is left open to further interpretation" (2008, 235). Floridi claims that the diaphoric interpretation captures what is at the core of the concept of data and that it is the most fundamental and satisfactory of the four interpretations.

As stated, the diaphoric interpretation of data lets too many things count as data. Indeed, *any thing x* that is distinct from some other thing y will count as a datum. This is a problem because data often play an epistemic or inferential role that helps us as truth-seeking agents. As I mentioned earlier, the original Latin meaning of "datum" is something that is given or taken for granted, but there also appears to be something of an objective or externalist component to the concept. One of our main uses of data is to acquire knowledge or useful information (recall the Target example), and, as such, there are certain inferences from data that are warranted and some that are not. I won't attempt to articulate what these inferences are here, but it suffices to note that, at the very least, we can infer from them that d is a datum that represents some fact and that that fact is likely—or at least that we have some evidence for that fact. However, the diaphoric interpretation doesn't respect this inferential constraint. For example, let x be the proposition that I will be a billionaire, which is distinct from $y = \neg x$, in that I wish x to be true and I don't wish y to be true. By the diaphoric interpretation, x is a datum. However, it would be crazy for me to infer that x is likely or that I have good evidence for x from the grounds that x is a datum, since x is only a datum (by the lights of the diaphoric interpretation) because I wish x to be true.

This is just one way in which the diaphoric interpretation of data overgenerates, and there will be many others because x and y are left uninterpreted. The natural remedy would be to give x and y an interpretation or to least put some constraints on their interpretation and/or constrain what differences between them are relevant for datum-hood. I'm not sure what interpretations or constraints would be appropriate, but given that the content of the diaphoric interpretation is that, roughly speaking, for x to be a datum, x must differ in some way from something in a domain, it seems that most of the work will be done by specifying these details.

To close this section, I'd like to return to the issue of whether we have multiple concepts of data. With the demise of the refined epistemic interpretation, I said that we should continue the search for an interpretation that treats data as a unified concept. I then reviewed three more interpretations, and they didn't fare any better—indeed, if anything, they fared worse. So it is worth considering the possibility that we have multiple data concepts. I mentioned as a reason against this idea that music files and files that report things like experiment outcomes—let's call them *fact files*—are stored in fundamentally the same way on our devices. However, that could be a red herring because we could have easily decided to store, say, music files using strings of o's, 1's, and 2's. And a reason for introducing the distinction is that we use music files and fact files in very different ways. The roles that they play in our lives and on our computers are rather different, and they seem to differ in a way that is similar to way that works of fiction differ from works of nonfiction. So it may be fruitful to distinguish two concepts of data before we try to give any analyses. For lack of better names, I'll call the two potential concepts *fact data* and *fictional data*. Examples of fact data include scientific data (including falsified data), charge-state data, customer purchasing data, and communications metadata. Examples of fictional data include some music data, some video data, some image data, some book data, and some code data. We may then make progress by taking some of the work done in the philosophy of fiction and using it for an analysis of fictional data. For example, Walton (1990) has analyzed fictional truth as "being a prescription or mandate in some context to imagine something" (39), which is analogous to the idea that music data are instructions or inputs for a music player. We then might be able to give structurally similar analyses of the two concepts. For example, factual data might be (roughly speaking) symbols that represent purported facts, and fictional data might be symbols that represent intended commands.

3 Big Data

According to IBM, humans create 2.5 quintillion bytes of data every day.[3] The volume of data generated around the world each year is increasing exponentially and so has its

---

[3] http://www-01.ibm.com/software/data/bigdata/what-is-big-data.html

importance to businesses.[4] So-called *big data* is booming, and governments and businesses are scrambling to find and/or train enough data scientists so that they can take advantage of it.

Big data aren't limited to government and business, however. The volume and complexity of data that are now available to various scientific fields is so tremendous that some have argued that this has changed the nature of these scientific fields so much so that they have become known as *big sciences* (cf. Borgman 2010). In particular, biology has seen a huge increase of available data in fields such bioinformatics, computational biology, systems biology, and synthetic biology (Callebaut 2012). Physics has also seen an explosion of data, with the digitization of particle colliders that generate far more data than can be currently stored. Astronomy has seen its own data explosion following the digitizations and improvements of telescopes and cameras—indeed, there is now so much data that some data-processing tasks are crowdsourced to regular citizens (Nielsen 2011, ch. 6). Cognitive science and psychology are "getting big," too, with the large synthesis of functional magnetic resonance imaging (fMRI) data such as Neurosynth[5] and the use of crowdsourcing platforms such as Amazon's Mechanical Turk[6] to conduct experiments (Buhrmester, Kwang, and Gosling 2011; Mason and Suri 2012). These are just a few examples, and the trend is expected to continue: the sciences are becoming extremely "data rich." In many cases, the data are collected and analyzed automatically, with very little—if any—manual input by the researchers (Humphreys 2007).

Explicit discussions of issues that big data may raise for epistemology and the philosophy of science are quite rare (exceptions include Callebaut 2012; Floridi 2012*a*, 2012*b*; Humphreys 2007). At first glance, this lack of attention by philosophers might be because "big data" is just a buzz word that doesn't mean much (cf. Floridi 2012*a*) or because big data is just a fad, and so no novel philosophical issues are raised—except for ethical ones (see, e.g., David and Patterson 2012; Ioannidis 2013). Although fields such as data science and analytics are growing rapidly and the sciences are become data richer, it might seem that there is nothing new here of philosophical interest and that there are just interesting technical challenges that need to be solved. It might seem that all that is going on is that we need to develop statistical and processing methods so that we can take advantage of the large amounts of data available to us, but, at bottom, these methods are just methods of inductive inferences. And inductive inferences are still inductive inferences: we just now have much larger total evidence sets.

However, Floridi (2012*a*, 2012*b*) has argued that big data do in fact pose a real epistemological problem, which he calls *small patterns*:

> The real, epistemological problem with big data is small patterns. Precisely because so many data can now be generated and processed so quickly, so cheaply, and on

---

[4] http://www.tcs.com/SiteCollectionDocuments/Trends_Study/TCS-Big-Data-Global-Trend-Study-2013.pdf

[5] http://neurosynth.org

[6] https://www.mturk.com/mturk/welcome

virtually anything, the pressure both on the data nouveau riche, such as Facebook or Walmart, Amazon or Google, and on the data old money, such as genetics or medicine, experimental physics or neuroscience, is to be able to spot where the new patterns with real added value lie in their immense databases and how they can best be exploited for the creation of wealth and the advancement of knowledge.

[W]hat we need is a better understanding of which data are worth preserving. And this is a matter of grasping which questions are or will be interesting. Which is just another way of saying that, because the problem with big data is small patterns, ultimately, the game will be won by those who know how to ask and answer questions (Plato, Cratylus, 390c) and therefore know which data may be useful and relevant, and hence worth collecting and curating, in order to exploit their valuable patterns. We need more and better techniques and technologies to see the small data patterns, but we need more and better epistemology to sift the valuable ones. (Floridi 2012a, 436–437)

So, according to Floridi, the epistemological challenge posed by big data is not to work out how we can process large datasets and draw reliable inferences from them, but rather to work out how we can *best focus* our inferential resources on our large datasets. In big data contexts, there are so many variables between which there may be useful correlations that we cannot simply run our algorithms over all permutations of them when looking for correlations. Instead, we are forced to choose subsets of those permutations for examination, and so we need some way of making such choices as intelligently as possible. A better epistemology could help us guess that variables such as lotion purchases and pregnancies might be correlated and help us not waste time investigating whether, say, light bulb purchases and pregnancies might be correlated (see Section 1).[7]

However, Floridi's epistemological problem doesn't appear to be a new philosophical problem or a new need to make epistemology better. The problem is essentially the problem of scientific discovery, albeit in a new guise. Traditionally, the problem of scientific discovery has been broken down into two components: (1) that of hypothesis, model, theory construction and (2) that of hypothesis, model, theory testing, with the second component receiving the most philosophical attention (e.g., Popper 1959). However, the first component is nevertheless clearly important, and it is what focuses our energies regarding the second component. As Charles Darwin once said:

About thirty years ago there was much talk that geologists ought only to observe and not theorize; and I well remember someone saying that at this rate a man might as well go into a gravel-pit and count the pebbles and describe the colours. How odd it is that any-one should not see that all observation must be for or against some view if it is to be of any service! (Darwin 1861)

Although the problem may not be a new one, big data could still be of philosophical interest for philosophers interested in the problem of scientific discovery, those who may have a lot to learn from big data case studies.

---

[7] For all I know, these may in fact be correlated!

Is that all there is of philosophical interest to big data? Some of the claims that have been made by various authors seem to suggest otherwise. For example, it has been claimed that with big data we have a new paradigm of science:

> The growth of data in the big sciences such as astronomy, physics, and biology has led not only to new models of science—collectively known as the Fourth Paradigm—but also to the emergence of new fields of study such as astroinformatics and computational biology. (Borgman 2010, 2)

And in an article entitled "The End of Theory: The Data Deluge Makes the Scientific Method Obsolete," Chris Anderson, editor-in-chief of *Wired*, argued that when armed with big data, scientists needn't bother with developing theories and building models anymore, and they can just listen to what the data say:

> This is a world where massive amounts of data and applied mathematics replace every other tool that might be brought to bear. Out with every theory of human behavior, from linguistics to sociology. Forget taxonomy, ontology, and psychology. Who knows why people do what they do? The point is they do it, and we can track and measure it with unprecedented fidelity. With enough data, the numbers speak for themselves. (Anderson 2008)

If there is something to such claims, then there may, after all, be something of philosophical interest to big data.[8] Such a dramatic change in scientific methodology seems to be of philosophical interest in and of itself, and perhaps it even ought to affect the work of philosophers—for example, maybe we shouldn't worry so much about scientific explanation.

The quote by Anderson stands in stark contrast to Floridi's epistemological problem posed by big data. Anderson appears to think that we needn't work out what questions to ask; we can, in effect, just ask all of them. Anderson envisages a new scientific methodology based on Google's methodology for analyzing web content:

> Google's founding philosophy is that we don't know why this page is better than that one: If the statistics of incoming links say it is, that's good enough. No semantic or causal analysis is required. That's why Google can translate languages without actually "knowing" them (given equal corpus data, Google can translate Klingon into Farsi as easily as it can translate French into German). And why it can match ads to content without any knowledge or assumptions about the ads or the content. (Anderson 2008)

This methodology is the polar opposite to what philosophers of science have been studying, especially in recent years with the explosion of literature on mechanistic

---

[8] The debate over such claims has, so far, mostly been happening outside of philosophy (e.g., see Shelton, Zook, and Graham [2012] for a response to Anderson).

models and mechanistic explanations (e.g., Bechtel and Abrahamsen 2005; Glennan 1996; Machamer, Darden, and Craver 2000). Indeed, Anderson states this quite explicitly:

> The new availability of huge amounts of data, along with the statistical tools to crunch these numbers, offers a whole new way of understanding the world. Correlation supersedes causation, and science can advance even without coherent models, unified theories, or really any mechanistic explanation at all.
>
>   There's no reason to cling to our old ways. It's time to ask: What can science learn from Google? (Anderson 2008)

Other share similar views; for example.:

> The methods used in data analysis are suggesting the possibility of forecasting and analyzing without understanding (or at least without a structured and general understanding). More specifically, we will argue in Section 2 that understanding occurs when we know how to relate several descriptions of a phenomenon. Instead, these connections are disregarded in many data analysis methodologies, and this is one of the key features of modern data analysis approaches to scientific problems. The microarray paradigm is at work exactly when a large number of measured variables concerning a phenomenon are algorithmically organized to achieve narrow, specific answers to problems, and the connections among different levels of descriptions are not explored. This is the perspective we propose to call agnostic science and at its heart is the methodological principle that we called microarray paradigm. (Panza, Napoletani, and Struppa 2011, 6)

So perhaps the novel philosophical issue raised by big data is not Floridi's epistemological problem but instead the unusual absence of it. Throughout most of the history of science, theories, models, and hypotheses were developed—to steer ourselves away from the gravel-pit—and data were collected to test these products of theorizing. The process of discovery in the context of big data appears to break from this tradition and thus may allow science to proceed without models, theories, and explanations. This is what is sometimes called *data-driven science.*

Anderson's claims appear to be a little stronger than they should be. Take, for example, his allusion to Google's PageRank algorithm, which analyzes the statistics of incoming links. The success of the algorithm depends on the fact that there is a correlation between people's interest in a given page and the number of links going into that page (roughly speaking). The discovery of the PageRank algorithm wasn't an exercise in big data analysis conducted in a background theory/model vacuum. Indeed, Page, Brin, Motwani, and Winograd (1999) explicitly gave an intuitive justification of the algorithm:

> It is obvious to try to apply standard citation analysis techniques to the web's hypertextual citation structure. One can simply think of every link as being like an academic citation. So, a major page like http://www.yahoo.comwill have tens of thousands of backlinks (or citations) pointing to it. (Page et al. 1999, 2)

The algorithm is certainly ingenious, but we can see how even common sense would guide someone to examine whether user interests correlated with something like the numbers of incoming links. So, although Google need not know what the content of a page is to be able to intelligently rank it in response to a search query, it certainly doesn't do this without the use of any theory.

Other examples, however, seem to better support Anderson's claims. For example, Panza et al. describe the use of big data to train neural networks for hurricane forecasting:

> It is remarkable that (cf. Baik and Paek 2000) the use of neural networks gives better forecasts of the intensity of winds than the best available simulations of atmospheric dynamics. In this context, the input variables X_i are sets of measured meteorological data relative to a developing hurricane, and the output variable Y is the intensity of winds of the hurricane after 48 hours. The crucial point of this method is that the structure of neural networks does not express any understanding of the hurricane dynamics. It does not mirror in any understandable way the structure of the atmosphere: the specific problem is solved, but with no new knowledge of the phenomenon. Note moreover that only the ability to access a large quantity of measurements for the hurricane during its development allows this technique to work, in line with the general tenets of the microarray paradigm. (Panza et al. 2011, 21)

They describe other examples as well, such as electroencephalogram data-driven controls of seizures (22) and signal-boosting techniques (24). Although the examples come from different areas of science, what they have in common is that, in each case, "we can see very clearly that no understanding [of] the phenomenon is gained while solving the problem" (19).

Traditional approaches to the philosophy of science have focused a great deal on understanding scientific explanation. These approaches take the sciences as trying to understand and/or explain their target phenomena. In contrast, the data-driven view of science says that we can shed explanations, understandings, theories, and so on because they are pointless: who cares about understanding or explaining some phenomenon if we can predict it with extreme accuracy? One reply might be that by developing models and theories that explain a given phenomenon, we help protect ourselves against making erroneous predictions when something about the target system changes—for example, when we move from predicting what WEIRD (Western, Educated, Industrialized, Rich, and Democratic) people will do to predicting what non-WEIRD people will do (cf. Henrich, Heine, and Norenzayan 2010).[9] However, the proponent of data-driven science could reply that the big data skeptic is not thinking *big enough*. This is just the beginning of big data, and eventually our datasets will include points on all the variables that might ever be relevant, and our analytic tools will be powerful enough so that we will always be shielded against such errant predictions. When something behind the

---

[9] This is the gist of Shelton, Zook, and Graham's (2012) response to Anderson.

phenomena changes, our predictions will change accordingly because we'll have data on the change and algorithms that can analyze the change. We won't know what's going on, but we will nevertheless enjoy the benefits of our accurate and adaptively predictive algorithms.

Will our datasets ever be that large, and will our analytical tools ever be that powerful? It's difficult to say, but, if I had to bet, I would bet that there will always be some surprises and that we will always benefit from some theory. However, although we may never reach the point of a theoryless science, we will probably get a lot closer to that point than we have ever done so before in that more and more science will use less and less theory. Perhaps this will indeed cause a general change in scientific methodology. Science, to date, has been a human activity conducted from a human perspective, despite our best efforts to be "objective observers" (cf. Callebaut 2012). As more science is done automatically by our electronic devices, science becomes less of a human activity, and, so far, that seems to coincide with less theory and explanation (cf. Humphreys 2007). It may be that the only reason that we have been obsessed with causation, explanation, theories, models, and understanding for so long is simply because for so long we have only ever had access to tiny fragments of data and a limited ability to process that data. The notion of causation could simply be a useful fiction created by an evolutionary process for data-limited beings; so perhaps all there really is in the world, after all, is correlation (cf. Hume 1738).

4 DATA AND PHENOMENA

In an influential paper titled "Saving the Phenomena," Bogen and Woodward (1988) introduced a distinction between what they called *data* and *phenomena*, and they used this distinction to argue against a widely held view of science, namely, that scientific theories predict and explain facts about what we observe. Bogen and Woodward argued that scientific theories are in the business of explaining phenomena, which, according to them, are mostly unobservable, and that they are not in the business of explaining data, which are mostly observable (1988, 305–306).

Bogen and Woodward went on to further refine the view in a series of papers (e.g., Bogen and Woodward 1992, 2003; Woodward 1989, 2000, 2011*a*, 2011*b*), and others built upon the view or used it for other purposes (e.g., Basu 2003; Haig 2013; Kaiser 1991; Massimi 2007; Psillos 2004; Suárez 2005) while others criticized it (e.g., Glymour 2000; McAllister 1997; Schindler 2007; Votsis 2010).

To introduce the distinction between data and phenomena and their respective observabilities, Bogen and Woodward use the example of how we (as scientists) would determine the melting point of lead. They first point out that to determine the melting point of lead, one doesn't take a single thermometer reading of a sample of lead that has just melted. Rather, what one does—or should do—is take *many* readings of the thermometer. This is because the readings will tend to vary due to errors in the measurement process. Bogen and Woodward say that these readings constitute *data* (1988, 308). They then note that if certain assumptions about the source of the variations of the data points can be made, we can use the data to construct an *estimate* of the melting point of lead, which is a *phenomenon* (1988, 308). For example, if the data appear to be normally distributed and we have no reason to think that they have systematic errors, then we would estimate the melting point of lead by taking the mean of the data points. In other words, the temperature measurements are the data, and, if all has gone well, the data can be used as evidence for a hypothesis about the melting point of lead, which is the phenomenon in this example.

Bogen and Woodward note that in this process of determining the melting point of lead, we do not actually *observe* the melting point of lead. And whereas we don't observe this phenomenon, we do observe the data—that is, we do observe the thermometer readings. So we see how, at least in this example, data are observed and phenomena are not: we infer unobserved phenomena from observed data. Bogen and Woodward support this generalization with various examples from the history of science. For example, bubble chamber photographs (data) were used to detect weak neutral currents (phenomenon), and test scores for behavioral tasks along with X-ray photographs of human skulls (data) were used to test whether a damaged frontal lobe results in a dysfunction of sensory processing (phenomenon) (1998, 315–316).

So far, this is relatively straightforward and uncontroversial. However, Bogen and Woodward put the distinction to some heavy work. They argue that scientists tend not to develop theories with the goal of explaining the particular data points that they have acquired. Rather, what scientists tend to do is construct theories that explain the phenomena that they have inferred from their data (1998, 309–310). In the example of the melting point of lead, the phenomenon "is explained by the character of the electron bonds and the presence of so-called "delocalized electrons" in samples of this element (Woodward 2011*b*, 166), and this theory doesn't explain the particular data points that were observed. An example that makes the point even more clearly is the theory of general relativity (GR), the phenomena of gravity, and the data of Eddington:

> GR is a theory of gravitational phenomena, not a theory that purports to explain or provide derivations concerning the behavior of cameras and optical telescopes or Eddingtons decisions about experimental design. (Woodward 2011*b*, 166)

Since the data points are often the result of a myriad of causal factors, many of which are peculiar to the local circumstances of the measurements (e.g., the purity of the lead sample, when Eddington decided to let sunlight fall on his photographic plates), any explanation of them would have to take into account those factors. If the theory that is constructed to explain the melting point of lead doesn't account for those factors, it won't be able to explain the data points. Bogen and Woodward claim that this is typical of scientific practice. One consideration that gives this claim plausibility is that phenomena are understood to be robust and repeatable—"features of the world that in principle could recur under different contexts or conditions" (2011*b*, 166)—whereas data are specific and local; for example, Eddington's photographic plates are not going to be created again.[10]

Before moving on, it is worth considering how Bogen and Woodward's distinction between data and phenomena matches the analyses of data that we considered in Section 2. The most promising analysis was the refinement of the epistemic interpretation that said that data are representations of purported facts. This seems to sit fairly well with Bogen and Woodward's distinction. In the 1988 paper, Bogen and Woodward mostly give examples of data (e.g., thermometer readings) and some of the conceptual roles of data (e.g., being evidence for hypotheses about phenomena). In a later paper, Woodward gives a more explicit characterization of data:

> Data are public records (bubble chamber photographs in the case of neutral currents, photographs of stellar positions in the case of Eddington's expedition) produced by measurement and experiment, that serve as evidence for the existence of phenomena or for their possession of certain features. (Woodward 2000, 163; see also Woodward 2011b, 166)

This also fits quite well with the refined epistemic interpretation because, presumably, the public records produced by measurements and experiments are meant to be representative of the outcomes of those measurements and experiments. It also makes it clear that the data are not the states of the measuring devices; instead, they are our recordings of the measuring devices. Data, then, are observable in that public records are observable—for example, you can open up the lab book and observe the numerical symbols.

If Bogen and Woodward are correct that scientific theories typically explain unobservable phenomena, then this would appear to be big—and bad—news for empiricist accounts of science. For example, according to van Fraassen's constructive empiricism, the ultimate criterion by which we judge the quality of a scientific theory is its *empirical adequacy*, which is how well it fits with what has been observed. If a theory fits the observations well, then we need only believe that it is empirically adequate, and there is no need to believe any unobservable entities that the theory apparently posits. However, if Bogen and Woodward are right, then this is the wrong construal of empirical adequacy. Instead, we should say that for a theory to be empirically adequate, it must fit the phenomena. But since phenomena are typically not observable, it is not clear how van Fraassen can maintain that we need not believe in unobservable entities (Bogen and Woodward 1988, 351).

We might worry, then, about these claims of unobservability. Bogen and Woodward (1988) claimed that phenomena "in most cases are not observable in any interesting sense of that term" (306). However, this strong claim isn't crucial to their view, and Woodward (2011b) replaces it with a weaker one:

> [I]t is enough to show that reasoning from data to phenomena can (and not infrequently does) successfully proceed without reliance on theoretical explanation of

---

[10] Of course, *other* photographic plates might be created, but those will be *other* data.

data. It was an unnecessary diversion to claim that this was always or even usually the case. A similar point holds in connection with the claim in our original paper that phenomena, in contrast to data, are typically not observable (in senses of observable closely linked to human perception). As Paul Teller has argued there are plausible examples of observable phenomena, at least if we do not restrict this notion in question-begging ways: for example, various phenomena associated with the operation of the visual system such as color constancy in changing light and pop-out effects in visual attention. Similarly, for some optical phenomena such as the so-called Poisson spot which appears at the center of the shadow cast by a circular object illuminated by a point source. What we should have said (I now think) is that phenomena need not be observable. (Woodward 2011b, 171)

Woodward doesn't comment on the other strong claim of their original paper: that data "for the most part can be straightforwardly observed" (Bogen and Woodward 1988, 305). However, the opposite of this seems to be case, especially given how much data is collected and stored electronically. For example, the Large Hadron Collider (LHC) generates about 200,000 DVDs worth of data per second, and most of that data—about 99.999%—is automatically discarded using triggers.[11] That's a lot of data that cannot be straightforwardly observed. Even the data that are stored do not appear to be straightforwardly observable because it is stored electronically on a server. Bogen and Woodward might respond that we observe the stored data when it is used to create outputs on our computer screens. But their criticism of the view that phenomena can be observed (contrary to their view) appears to preclude that response. For example, they write:

> But to recognize that X played a part in causing Y or to recognize that Y was produced by causes which include X is not to see X. That is why you cannot see our grandfathers, or the midwives who delivered them, by looking at us. No matter how much you know about where we came from, they are not distal stimuli and they are not what you look at when you look at us. (Bogen and Woodward 1988, 346)

Similarly, although the LHC data play a part in causing various outputs on our computer screens, outputs that we do observe, that doesn't mean we observe, or can observe, the LHC data.

At any rate, whether data can for the most part be straightforwardly observed is not crucial to Bogen and Woodward's view. They can retract their claims that phenomena are often unobservable, that data are often observable, and that theories often do not explain data. Even if this weaker and more plausible view is correct, then constructive empiricism appears to be in trouble. Trying to argue that we do observe phenomena that are causally proximate to data—of which, say, the melting point of lead might be an example—doesn't seem like a promising strategy because actual scientific practice

---

[11] See, e.g., http://www.lhc-closer.es/1/3/13/0

sometimes saves phenomena that are extremely far removed from the data. Massimi (2007) argues that the discovery of the J/ψ particle (a flavor-neutral meson made up of a charm quark and a charm anti-quark) is such an example and uses it to present a similar criticism of van Fraassen's constructive empiricism.

As I mentioned earlier, various criticisms of Bogen and Woodward's view have been given, but one of particular interest here, given the discussion of big data in Section 3, is that of Schindler (2007). Schindler argues that Bogen and Woodward have something of an Anderson-style "big data" view of scientific methodology (although he doesn't put it in those words):

> Bogen and Woodward 1988 . . . have argued . . . for a bottom-up construction of sci-
> entific phenomena from data. For them, the construction of phenomena is "theory-
> free" and the exclusive matter of statistical inferences, controlling confounding
> factors and error sources, and the reduction of data. (Schindler 2007, 161)

If Schindler is correct, then we can think of Anderson's view of science as one that adopts Bogen and Woodward's data-phenomena distinction and rejects the need to construct theories to explain the phenomena. On such a picture, theory is not needed at all: it is not needed to explain phenomena because there is no need to explain phenomena, and it is not needed to infer phenomena from the data—if the data is "big."

However, this doesn't seem to be the right way to interpret Bogen and Woodward. Although they argue that theory is not in the business of explaining data, this doesn't entail that theory is not involved in our inferences from data to phenomena. For example, Bogen and Woodward say that, in order to infer phenomena from data, we need to know such as things as whether confounding factors have been controlled for, whether there were systematic sources of error, and statistical arguments of various kinds, such as whether the conditions of central limit theorem (CLT) apply (Bogen and Woodward 1988, 334). To know such things, we often employ theories of various kinds: for example, a theory will tell us whether a factor is a confounding one or not, and a lot of theory can bear upon whether the CLT applies (cf. Lyon 2014 and Woodward 2011a, 797). As Psillos has nicely put it:

> [T]he establishment of the epistemic authority of what is normally called the observ-
> able phenomenon (e.g., the melting point of lead) is a rather complicated process
> which essentially relies on background theory. If all these background theories some-
> how fail to be adequate (or well-confirmed, I should say), the observed phenomenon
> is called into question. Now, this does not imply that before we establish, say, the
> melting point of lead, we need detailed theories of why lead melts at this point. These
> theories will typically be the product of further theoretical investigation. But it does
> imply that establishing the reliability (and hence the epistemic authority) of the data
> as a means to get to stable phenomena relies indispensably on some prior theories.
> So, observation is not epistemically privileged per se. Its epistemic privilege is, in
> a certain sense, parasitic on the epistemic privilege of some theories. (Psillos 2004,
> 406; emphasis in original)

Moreover, in response to Schindler, Woodward clarifies his view:

> Data to phenomena reasoning, like inductive reasoning generally, is ampliative in the sense that the conclusion reached (a claim about phenomena) goes beyond or has additional content besides the evidence on which it is based (data). I believe it is characteristic of such reasoning that it always requires additional substantive empirical assumptions that go beyond the evidence. (Woodward 2011*b*, 172)

So, Bogen and Woodward's data–phenomena distinction and their understanding of the distinction in scientific practice doesn't entail a "bottom-up" approach to phenomena detection.

This last point shows again what is wrong with the big data picture of science that appears to be gaining in popularity. Even (what may seem to be) purely statistical inferences are guided by empirical assumptions that are made on the basis of prior understanding and/or theory of the phenomena being studied. Woodward gives a nice example of what is involved in making an inference from fMRI data:

> [C]onsider the common use of spatial smoothing procedures in the analysis of fMRI data. Each individual voxel measurement is noisy; it is common to attempt to improve the signal to noise ratio by averaging each voxel with its neighbors, weighted by some function that falls with distance. This procedure depends (among other considerations) on the empirical assumption that the activity of each voxel is more closely correlated with nearby spatial neighbors. (Woodward 2011*b*, 173)

If we didn't have the assumption that the voxel activities of nearby spatial neighbors are correlated, we wouldn't know that we can apply the spatial smoothing techniques to the data.

5 Conclusion

In this chapter, I reviewed some of the philosophical issues that arise in connection with the concept of data. In Section 2, I reviewed four analyses of the concept that have appeared in the literature and suggested how one in particular—the epistemic interpretation—can be improved. Since none of the analyses seemed all that promising when considered against all of the uses and instances of data, I suggested that it may be useful to first distinguish two concepts of data and give them separate analyses, rather than trying to treat data as a unified concept. In Section 3, I discussed the role of big data in science and its impact on scientific methodology. Grand claims have been made about the advent of theoryless science in the wake of big data in science. However, these claims appear to be too strong, and less grand ones should replace them. In particular, it appears that big data may be allowing for more science to be done with less theory, but so far it seems that theory is still involved. Finally, in Section 4, I discussed an important distinction between data and phenomena in the philosophy of science and related the distinction to the epistemic interpretation of Section 2 and the role of big data in science discussed in Section 3.

References

Anderson, C. (2008). "The End of Theory: The Data Deluge Makes the Scientific Method Obsolete." *Wired* 16: 108–109.

Baik, J.-J., and Paek, J. -S. (2000). "A Neural Network Model for Predicting Typhoon Intensity." *Journal of the Meteorological Society of Japan* 78(6): 857–869.

Basu, P. K. (2003, June). "Theory-Ladenness of Evidence: A Case Study from History of Chemistry." *Studies in History and Philosophy of Science Part A* 34(2): 351–368.

Bechtel, W., and Abrahamsen, A. (2005, June). "Explanation: A Mechanist Alternative." *Studies in History and Philosophy of Science Part C: Studies in History and Philosophy of Biological and Biomedical Sciences* 36(2): 421–441.

Bogen, J., and Woodward, J. (1988, July). "Saving the Phenomena." *Philosophical Review* 97(3): 303.

Bogen, J., and Woodward, J. (1992). "Observations, Theories and the Evolution of the Human Spirit." *Philosophy of Science* 59(4): 590–611.

Bogen, J., and Woodward, J. (2003). "Evading the IRS." In. M. Jones and N. Cartwright (eds.), *Correcting the Model: Idealization and Abstraction in Science* (Amsterdam: Rodopi Publishers), 233–268.

Borgman, C. L. (2010). "Research Data: Who Will Share What, with Whom, When, and Why? *Fifth China-North America Library Conference*, Beijing 2010. http://www.ratswd.de/download/RatSWD_WP_2010/RatSWD_WP_161.pdf

Buhrmester, M., Kwang, T., and Gosling, S. D. (2011, January). "Amazon's Mechanical Turk: A New Source of Inexpensive, Yet High-Quality, Data?" *Perspectives on Psychological Science* 6(1): 3–5.

Callebaut, W. (2012, March). "Scientific Perspectivism: A Philosopher of Science's Response to the Challenge of Big Data Biology." *Studies in History and Philosophy of Science Part C: Studies in History and Philosophy of Biological and Biomedical Sciences* 43(1): 69–80.

Carnap, R. (1946, December). "Theory and Prediction in Science." *Science* 104(2710): 520–521.

Darwin, C. (1861). Darwin, C. R. to Fawcett, Henry. https://www.darwinproject.ac.uk/letter/entry-3257

David, K., and Patterson, D. (2012). *Ethics of Big Data* (Cambridge: O'Reilly Media).

Dretske, F. (1981). *Knowledge and the Flow of Information* (Cambridge, MA: MIT Press).

Fisher, R. A. (1936, April). "Has Mendel's Work Been Rediscovered?" *Annals of Science* 1(2): 115–137.

Floridi, L. (2005, March). "Is Semantic Information Meaningful Data?" *Philosophy and Phenomenological Research* 70(2): 351–370.

Floridi, L. (2008). Data. In W. A. Darity (ed.), *International Encyclopedia of the Social Sciences*. New York: Macmillan Reference USA, 234–237.

Floridi, L. (2012a). Big Data and Their Epistemological Challenge." *Philosophy & Technology* 25(4): 435–437.

Floridi, L. (2012*b*). "The Search for Small Patterns in Big Data." *Philosophers' Magazine* 59(4): 17–18.

Fodor, J. A. (1987). *Psychosemantics: The Problem of Meaning in the Philosophy of Mind* (Cambridge, MA: MIT Press).

Glennan, S. S. (1996, January). "Mechanisms and the Nature of Causation." *Erkenntnis* 44(1): 49–71.

Glymour, B. (2000, January). "Data and Phenomena: A Distinction Reconsidered." *Erkenntnis* 52(1): 29–37.

Haig, B. D. (2013, May). "Detecting Psychological Phenomena: Taking Bottom-Up Research Seriously." *American Journal of Psychology* 126(2): 135–153.

Henrich, J., Heine, S. J., and Norenzayan, A. (2010, June). "The Weirdest People in the World?" *Behavioral and Brain Sciences* 33(2–3): 61–83.

Holmes, R. (2008). *The Age of Wonder* (New York: HarperCollins).

Hume, D. (1738). *A Treatise of Human Nature*, 2nd ed. L. A. Selby-Bigge, ed. (Oxford: Clarendon Press).

Humphreys, P. (2007). "Twenty-First Century Epistemology." In *Revista Anthropos 214*, 65–70.

Ioannidis, J. P. A. (2013, April). "Informed Consent, Big Data, and the Oxymoron of Research That Is Not Research." *American Journal of Bioethics* 13(4): 40–42.

Kaiser, M. (1991, October). "From Rocks to Graphs—The Shaping of Phenomena." *Synthese* 89(1): 111–133.

Lakatos, I. (1970). "Falsification and the Methodology of Scientific Research Programmes." In Imre Lakatos and Alan Musgrave (eds.), *Criticism and the Growth of Knowledge.* (Cambridge: Cambridge University Press), 91–195.

Lyon, A. (2014). "Why Are Normal Distributions Normal?: *British Journal for the Philosophy of Science* 65(3): 621–649.

Machamer, P., Darden, L., and Craver, C. F. (2000). "Thinking About Mechanisms." *Philosophy of Science* 67(1): 1–25.

Mason, W., and Suri, S. (2012, March). "Conducting Behavioral Research on Amazon's Mechanical Turk." *Behavior Research Methods* 44(1): 1–23.

Massimi, M. (2007, June). "Saving Unobservable Phenomena." *British Journal for the Philosophy of Science* 58(2): 235–262.

McAllister, J. W. (1997, September). "Phenomena and Patterns in Data Sets." *Erkenntnis* 47(2): 217–228.

Nielsen, M. (2011). *Reinventing Discovery: The New Era of Networked Science* (Princeton, NJ: Princeton University Press).

Page, L., Brin, S. Motwani, R., and Winograd, T. (1999). "The PageRank Citation Ranking: Bringing Order to the Web." Technical Report. Stanford: Stanford InfoLab.

Panza, M., Napoletani, D., and Struppa, D. (2011). "Agnostic Science. Towards a Philosophy of Data Analysis." *Foundations of Science* 16(1): 1–20.

Popper, K. R. (1959). *The Logic of Scientific Discovery* (New York: Basic Books).

Psillos, S. (2004). "Tracking the Real: Through Thick and Thin." *British Journal for the Philosophy of Science* 55(3): 393–409.

Schindler, S. (2007). "Rehabilitating Theory: Refusal of the 'Bottom-Up' Construction of Scientific Phenomena." *Studies in History and Philosophy of Science Part A* 38(1): 160–184.

Shelton, T., Zook, M., and Graham, M. (2012). "The Technology of Religion: Mapping Religious Cyberscapes." *Professional Geographer* 64(4): 602–617.

Suárez, M. (2005). "Concepción semántica, adecuación empírica y aplicación" [The Semantic View, Empirical Adequacy, and Application]. *Crítica: revista Hispanoamericana de filosofía* 37(109): 29–63.

Votsis, I. (2010). "Making Contact with Observations." In Mauricio Suárez, Mauro Dorato, and Miklós Rédei (eds.), *Making Contact with Observations. EPSA Philosophical Issues in the Sciences, Volume 2.* (Dordrecht: Springer), 267–277.

Walton, K. L. (1990). *Mimesis as Make-believe* (Cambridge, MA: Harvard University Press).

Woodward, J. (1989). "Data and Phenomena." *Synthese* 79(3): 393–472.

Woodward, J. (2000). "Data, Phenomena, and Reliability." *Philosophy of Science* 67(September): S163–S179.

Woodward, J. (2011a, January). "Data, Phenomena, Signal, and Noise." *Philosophy of Science* 77(5): 792–803.

Woodward, J. F. (2011b). "Data and Phenomena: A Restatement and Defense." *Synthese* 182(1): 165–179.

CHAPTER 36

EMERGENCE

PAUL HUMPHREYS

Modern interest in emergence is usually considered to begin with John Stuart Mill's discussion of the difference between homopathic and heteropathic laws in his monumental work *A System of Logic* (Mill 1843).[1] The publication of this work was followed by an extended period of interest in the topic involving the British Emergentist School, the French emergentists, and some American philosophers. Prominent members of the British group, with representative publications, were G. H. Lewes (1875), Samuel Alexander (1920), C. Lloyd Morgan (1923), and C. D. Broad (1925); of the French group, Henri Bergson (1907) and Claud Bernard; and of the American group William James, Arthur Lovejoy, Stephen Pepper (1926), and Roy Wood Sellars.[2] After the 1930s, interest in emergence declined precipitously, and it was widely regarded as a failed research program, one that was even intellectually disreputable. In recent years, the status of claims about emergent phenomena has changed. The failure to carry through reductionist programs in various areas of science, the appearance of plausible candidates for emergence within complex systems theory and condensed matter physics, what seems to be an ineliminable holistic aspect to certain quantum systems that exist at or near the fundamental level, and the invention of computational tools to investigate model-based claims of emergence have all led to a revival of interest in the topic.

The literature on emergence being vast, some restrictions on the scope of this chapter are necessary. Contemporary philosophical research on emergence tends to fall into two

---

[1] As usual, when assessing genesis claims, one can find earlier origins. William Uzgalis has noted (2009) that the English materialist Anthony Collins argued in the period 1706–08 that there are real emergent properties in the world.

[2] A standard, although partial, history of the British emergentist movement can be found in McLaughlin (1992), and a complementary treatment is Stephan (1992). A more diachronically oriented history is Blitz (1992). A comprehensive account of the French tradition in emergence, less discussed in the Anglophone world, can be found in Fagot-Largeault (2002) and of the influence of Claude Bernard on G. H. Lewes in Malaterre (2007). Works by later American philosophers are cited in the references for this chapter. Some relations between contemporary accounts of emergence and an ancient Indian tradition are presented in Ganeri (2011). Thanks to Olivier Sartenaer for some of these references.

camps, between which there is a widening gap. The first is motivated by scientific interests and is open to the possibility of emergence occurring within physics, chemistry, biology, or other sciences. The second is primarily driven by problems that arise within metaphysics and the philosophy of mind and, for better or for worse, generally lumps all of the nonintentional sciences under the generic category of the physical. In committing itself to this generous form of physicalism, the second camp turns its back on cases of emergence that occur within the natural sciences, focusing almost exclusively on mental phenomena. This is crippling to an understanding of emergence in general, and so this chapter will mostly be concerned with the first research program, although some attention will need to be paid at various points to the second. The question of whether space and time are emergent features of the universe will not be covered.[3]

1 METHODOLOGICAL CONSIDERATIONS

Formulating a unified theory of emergence or providing a concise definition of the concept of emergence is not a feasible project at the current time. The term "emergent" is used in too wide a variety of ways to be able to impose a common framework on the area, and some of those different senses reflect essentially different concepts of emergence, some ontological, some epistemological, and some conceptual. There is another, methodological, reason to be skeptical of any claim to have a unified account of emergence. Compare the situations of causation and emergence. Although there are different theories of causation, and a now widely held view that there is more than one concept of causation, each of those theories can be tested against a core set of examples that are widely agreed to be cases of causation. There does not at present exist an analogous set of core examples of emergence upon which there is general agreement.

As an example of this disagreement, many philosophers deny that the world contains any examples of emergence (Lewis 1986, x), whereas others assert that it is rare and at most occurs in the mental realm (McLaughlin 1997). Yet other authors reject this parsimony and maintain that entangled quantum states or the effects of long-range correlations that result in such phenomena as ferromagnetism or rigidity constitute legitimate examples of ontological emergence (Batterman 2002; Humphreys 1997). Still others insist that some forms of emergence are common and come in degrees (Bedau 1997; Wimsatt 2007). This division appears to be based in part on the extent to which one subscribes to a guiding principle that we can call the *rarity heuristic*: any account of emergence that makes emergence a common phenomenon has failed to capture what is central to emergence. Those early twentieth-century writers who restricted emergence to phenomena that at the time seemed mysterious and little understood, such as life and consciousness, seemed to have been sympathetic to the rarity heuristic, although earlier philosophers, such as Mill, who considered chemical properties to be emergent, would have rejected it.

---

[3] For this last topic see Butterfield and Isham (1999).

Because of the lack of agreement on examples, it would be a useful project for philosophers of science to assemble candidate examples for emergence that can then be assessed for salient characteristics. To illustrate the variety of examples that have been suggested, some are chemical, as in the class of Belousov-Zhabotinsky reactions, one example of which occurs when citric acid is placed in a sulfuric acid solution in the presence of a cerium catalyst. Circular patterns dynamically emerge from the initially homogeneous color, and the colors subsequently oscillate from yellow to colorless and back to yellow. Such are considered to be examples of reactions far from equilibrium, nonlinear oscillators in which self-organizing behavior leads to emergent patterns. Other suggested examples are physical, involving spontaneous symmetry breaking or quantum entanglement. Some examples are biological, such as the existence of flocking behavior in certain species of birds that demonstrates the appearance of self-organizing higher level structure as a result of simple interactions between agents. Still others are computational, as when the probability that a random 3-SAT problem is satisfiable undergoes a phase transition as the ratio of clauses to variables crosses a critical threshold (Monasson, Zecchina, Kirkpatrick, Selman, and Troyansky 1999;[4] for a longer list, see Humphreys 2016). One can see from this brief list of suggested examples the difficulty of identifying properties that are common to all cases of emergence or, conversely, understanding why all of these cases should count as emergent. Too often, especially in the sciences, an account of emergence is developed based on a small set of examples that are assumed to be emergent without any systematic arguments backing that assumption.

In the absence of canonical examples, one approach to understanding emergence is to draw upon broad theoretical principles and frequently used generic characteristics of emergence as a way to gain understanding of what emergence might be. Using a cluster definition approach, it is common to assert that an emergent entity must be novel, that there is some holistic aspect to the emergent entity, and that the emergent entity is in some sense autonomous from its origins. In addition, providing some account of the relation between the original entities and the emergent entity is desirable.

We can stay on reasonably firm taxonomical ground by distinguishing among ontological emergence, epistemological or inferential emergence, and conceptual emergence. All cases of ontological emergence are objective features of the world in the sense that the emergent phenomenon has its emergent features independently of any cognitive agent's epistemic state.

---

[4] The k-SAT problem is: for a randomly chosen Boolean expression in conjunctive normal form such that each clause in the conjunction contains exactly k literals, does there exist a truth value assignment under which it is satisfied? Determining whether such a Boolean sentence has a truth value assignment that makes it true (the satisfaction problem) has been shown to be an NP-complete problem for the 3-SAT case, and various forms of the k-SAT problem exhibit phase transitions. The 2-SAT problem, which has P class complexity, has a phase transition at M/N equals one, where M is the number of clauses and N is the cardinality of the set of variables from which k are randomly drawn. Below that value, almost all formulas are satisfiable, whereas above it almost all are unsatisfiable. Phase transitions also occur for $k = 3, 4, 5, 6$.

Schematically, *A* ontologically emerges from *B* when the totality of objects, properties, and laws present in *B* are insufficient to determine *A*. In the case of epistemological emergence, limitations of our state of knowledge, representational apparatus, or inferential and computational tools result in a failure of reduction or prediction. Schematically, *A* inferentially emerges from *B* when full knowledge of the domain to which *B* belongs is insufficient to allow a prediction of *A* at the time associated with *B*.[5] Conceptual emergence occurs when a reconceptualization of the objects and properties of some domain is required in order for effective representation, prediction, and explanation to take place. Schematically, *A* conceptually emerges from *B* just in case the conceptual frameworks used in the domain to which *B* belongs are insufficient to effectively represent *A*.

These categories are not mutually exclusive. Epistemological emergence is often accompanied by conceptual emergence, and ontological emergence can give rise to epistemological emergence. Discussions of emergence would be less confusing if these modifiers of "ontological," "epistemological," and "conceptual" were regularly used, and I shall use them from now on in this chapter.

There is an alternative and often used terminology of strong and weak emergence but its use is not uniform. "Strong emergence" is sometimes used as a synonym for "ontological emergence" but usually with the additional requirement that downward causation is present, sometimes for any case of emergence in which ontologically distinct levels are present with *sui generis* laws (Clayton 2004), and sometimes to cover cases for which truths about the emergent phenomenon are not metaphysically or conceptually necessitated by lower level truths (Chalmers 2006; McLaughlin 1997).[6]

Regarding weak emergence, a state description is often taken to be weakly emergent just in case it is not deducible—even in principle—from the fundamental law statements and the initial and boundary conditions for the system; in another widely discussed usage (Bedau 1997), it applies to states of a system that are derivable only by means of a step-by-step simulation. Finally, Chalmers (2006) denotes by "weak emergence" phenomena that are unexpected given the principles of a lower level domain. I shall here use the term "weak emergence" only in Bedau's specific sense and "strong emergence" not at all.

2 SYNCHRONIC AND DIACHRONIC EMERGENCE

Examples of emergence fall into two further types: synchronic emergence and diachronic emergence. Within the former category, the emergent entity and the things from

---

[5] If *A* and *B* are atemporal, then the reference to time is vacuous. In other cases, such as global spatiotemporal supervenience, the time covers the entire temporal development of the universe.

[6] In his article, Chalmers uses "nondeducibility" as an expository convenience, although what is meant is lack of necessitation.

which it emerges are contemporaneous; within the latter category, the emergent entity develops over time from earlier entities. The earlier British and French traditions did discuss a form of diachronic emergence that was called "evolutionary emergence," with an interest in how such things as life could emerge from inanimate matter over time, but most of this material was purely speculative. In the contemporary literature, some types of diachronic emergence (Bedau 1997) are compatible with synchronic ontological reduction, but, overall, the philosophical literature on emergence has a pronounced lean away from diachronic emergence, a bias that results from the contrast between emergence and synchronic reduction noted earlier. One can trace the line of descent from the mode of theoretical reduction favored by Ernest Nagel (Nagel 1961, ch. 11). Nagel required that, in the case of inhomogeneous reduction, the situation in which at least one theoretical term t was such that t appeared in T_1, the theory to be reduced, but not in the reducing theory T_2, the reduction of T_1 to T_2 required the formulation of "bridge laws."[7] These provided necessary and sufficient conditions for the elimination of terms specific to T_1, and the task was then to derive any prediction that could be derived from T_1 using only the theoretical apparatus of T_2 plus the bridge laws. Because at that time prediction and explanation were considered to have identical logical structure, this procedure also allowed reductive explanations of phenomena described in the vocabulary of T_1. Keeping in mind that prediction can be synchronic as well as diachronic, the failure of reduction, in a somewhat crude manner, captures one aspect of the primary form of epistemological emergence, which is based on the essential unpredictability of the emergent phenomenon. If biology is Nagel-irreducible to physics and chemistry, then some biological phenomena cannot be predicted from the representational apparatus of those two sciences alone. The need to introduce specifically biological principles, or new biophysical and biochemical principles, in order to represent and predict certain biological phenomena is a reason to consider those biological phenomena as conceptually emergent from physics or chemistry. One can thus see that emergence had not disappeared in the 1960s; it had simply deflated from want of plausible examples and the optimistic outlook toward reduction.

Two other contemporary accounts of ontological emergence are diachronic. Fusion emergence, advocated by Humphreys (1997), rejects the position that there is a set of permanent basic objects, properties, or states in terms of which all others can be generated and argues that emergent entities can occur when two or more prior entities go out of existence and fuse to become a single unified whole. Quantum entangled states are one example; the production of molecules by covalent bonding is another, assuming that at least one model of covalent bonding is veridical. O'Connor and Wong (2005) argue that nonstructural, higher level properties of composite objects are examples of diachronic emergent properties. The key idea concerning nonstructural properties is that the constituents of the system possess latent dispositions that result in a manifest and indecomposable property when the constituents causally interact and the complexity of

---

[7] The term "bridge laws" is potentially misleading because whether these are necessarily true definitions or contingent associations has been the topic of significant disagreement.

the system reaches a certain level. As with a number of other approaches to emergence, O'Connor and Wong's primary examples are taken to fall within the realm of mental properties. Finally, a more theoretically oriented account of diachronic emergence is given in Rueger (2000) that appeals to changes in the topological features of attractors in dynamical systems theory.

3   EMERGENCE AND REDUCTION

Reduction and emergence are usually considered to be mutually exclusive, and, for some authors, the dichotomy is exhaustive. If we hold both positions, then emergence occurs if and only if reduction fails. Yet a failure of reduction cannot be sufficient for emergence. Fundamental entities are irreducible, but few—and on some conceptions of fundamentality, none—are emergent. Such cases tell us what is wrong with the simple opposition between reduction and emergence. Fundamental entities, whether synchronically or diachronically fundamental, have nothing to which they can be reduced and, commensurately, nothing from which they can emerge. Although common parlance simply notes A as emergent, the grammatical form is misleading. An emergent entity A emerges from some other entity B, and the logical form is relational: A emerges from B. When the context makes it clear what B is, it is appropriate to use the contracted version "A is emergent."

There is an additional problem with treating the failure of reduction as sufficient for emergence; physics is not reducible to biology, and the semantic content of Lord Byron's poems is not reducible to the ingredients for egg drop soup, but no emergence is involved because these pairs are not candidates for a reductive relation. So A must be a prima facie candidate for reduction to B in order for the failure of reduction to put emergence in the picture. "Prima facie" is an elusive idea; it is better to drop the view that a failure of reduction suffices for emergence.

Nor is the failure of reduction or of reductive explanation necessary for emergence. An increasing number of authors (Bedau 1997; Wimsatt 2007) have argued that certain types of emergence are compatible with reduction. Wimsatt formulates four conditions for aggregativity, suggests that emergence occurs when aggregativity fails, provides examples to illustrate that aggregativity is much rarer than is often supposed, and argues that reduction is possible even in the absence of aggregativity. Many authors, mostly in the scientific tradition of emergence, have provided accounts of emergence within which emergent phenomena are explainable (e.g., Batterman 2002; Rueger 2000), and, in light of this, the view that emergent phenomena are mysterious and inexplicable should be abandoned.

The simple opposition between reduction and emergence must be amended for another reason. Just as we have different types of emergence, so there are different types of reduction, theoretical, ontological, and conceptual, and we could have a failure of theoretical reduction in the presence of ontological reduction, thus allowing predictive or conceptual emergence in the presence of ontological reduction. This was the message, at least under one interpretation, of Philip Anderson's widely cited article "More Is Different" (Anderson 1972). As with many scientists, Anderson was less than clear in his discussion about whether he was discussing laws or law statements, but, for him, ontological reduction was a given, whereas novel concepts were needed within condensed matter physics in order to allow effective prediction. Anderson himself did not use the term "emergence" in his article, but the common assumption is that this is what he meant and that ontological reduction and conceptual emergence are indeed compatible.

4 EMERGENCE AS NOMOLOGICAL SUPERVENIENCE

Once the difficulties of achieving a complete Nagel reduction of one theory to another were realized, physicalists switched to less stringent types of relation between the two levels, such as logical or nomological supervenience, and realization. Focusing on the first, the idea is that the higher level properties, states, and objects are inevitably present whenever the appropriate lower level properties, states, and objects are present. That is, the "real" ontology lies at the lower level, usually the physical level, and once that is in place, the ontology of the higher levels—chemistry, biology, neuroscience, and so on—automatically and necessarily exists. Suitably configure the right kinds of fermions and bosons and you will, so it is claimed, necessarily get a red-tailed hawk. What kind of necessitation is involved? The most plausible option is conceptual necessitation; the object we call a hawk is ontologically nothing but the configured elementary particles, just reconceptualized as a hawk. There is no ontological emergence here, although if it is impossible to fully reduce the biological representation of hawks to chemical and physical representations, there is scope for epistemological and conceptual emergence. More controversial is nonconceptual metaphysical necessitation—in every possible world where that configuration exists, so does the hawk—but this type of necessitation allows room for the hawk to exist as something different from the configured particles, thus opening the door to ontological emergence. In yet a third approach, if the necessitation is due to a fundamental law of nature that is not among the fundamental laws of the physical and chemical domains, and is thus what Broad called a "transordinal law," the domain of fundamental physics is nomologically incomplete. We thus have the possibility of the ontological emergence of certain parts of biology from physics and chemistry. This is the essence of McLaughlin's account of emergence (McLaughlin 1997).[8] Thus, P is an emergent property of an individual a just in case P supervenes with nomological but not with logical necessity on properties that parts of a have in isolation or in other combinations, and at least one of the supervenience principles is a fundamental

[8] McLaughlin's position is a modification of an earlier position due to van Cleve (1990).

law. McLaughlin suggests that increasing complexity in the base domain could result in such emergence but remained skeptical about whether any such cases exist outside the mental realm.

These necessitation-based approaches maintain an atemporal orientation, even when the positions are couched in the ontological rather than in the linguistic mode. For example, some versions of supervenience—and the more recent grounding—relations are presented as relations between properties rather than relations between concepts or predicates. Properties, whether construed extensionally, intensionally, or platonically, are atemporal, as are relations between them, and this reinforces the tendency toward an atemporal treatment. Even when property instances appear in nonreductive physicalism, the approach is synchronic. The dominance of such approaches for a long period of time was a primary reason why diachronic accounts of emergence were neglected.

5 UNITY AND EMERGENCE

If one could achieve a wholesale ontological reduction of the subject matters of all domains of correct science to that of one domain, usually taken to be that of fundamental physics, that would be conclusive evidence for the ontological unity of our universe. Methodological, representational, and epistemological variety would remain, but, ontologically, all would be physical. Conversely, it would seem that ontological emergence would be evidence for ontological disunity. Yet this overlooks the possibility of upward unification, an oversight that is probably due to a bias toward downward reduction. As Batterman (2002) has argued, the renormalization methods that are widely employed in condensed matter physics provide a considerable amount of theoretical unity via universality classes, capturing in a precise way the higher level unification that is present in cases of multiple realizability. (For more on this approach, see Section 9 on universality). This kind of emergence, which is formulated in mathematical terms, would also provide evidence that otherwise widely successful methods, such as mereological decomposition and experimental causal analysis, are not universally applicable. The scope of some other ontological and methodological positions, among them various "mechanistic" approaches, would also be called into question.

6 EMERGENCE AND DOWNWARD CAUSATION

Ontological emergence is often dismissed on the grounds that it would result in the presence of downward causation, and downward causation is inconsistent with the causal closure of physics. When properly articulated, this objection can be framed in terms of the exclusion argument. The argument has four premises: (1) every physical event E that is caused has a prior sufficient physical cause C;[9] (2) if an event E has a prior sufficient cause C, then no event C^* distinct from C that is not part of the causal chain or process from C to E is causally relevant to E; (3) the realm of the physical is causally closed and so all events in the causal chain or process from C to E are physical events; and (4) no emergent event C^* is identical with any physical event. Therefore, (5) no emergent event C^* is causally relevant to E. Therefore, emergent events are causally excluded from affecting physical events and hence are causally dispensable.

Much has been written about this argument, a clear version of which can be found in Kim (2006), and I shall not attempt to summarize that vast and heterogeneous literature. For the scope of the argument to be determined, however, we must specify the scope of the physical. Originally, "nonphysical" was taken to mean "mental," but this fails to address the possibility of emergent events within the realm of the physical, broadly construed, and it is useful to consider this argument in terms either of "physical" as referring to the domain of physics proper or to the domain of fundamental physics. In the latter case, the argument, if sound, excludes any causally efficacious ontological emergence.

The term "downward causation" tacitly appeals to a hierarchy of levels—the causation is from objects, properties, laws, and other entities occurring or operating at a higher level to objects, properties, laws, and other entities occurring at a lower level. In the case of objects, a necessary condition for the ordering criterion for objects is usually based on a composition relation. Level j is higher than level i only if any object A at level j has objects occurring at level i as constituents. This allows objects at lower levels to be constituents of A. To find a sufficient condition is much harder. One cannot take the just described scenario as a sufficient condition on pain of expanding the number of levels far beyond what is usually considered reasonable. Even this simple ordering has its problems because molecules are not mereological composites of atoms, and only a partial ordering is possible once the subject matter of the social sciences is reached. Once a criterion for objects has been found, the appropriate level for a property or a law is the first level at which the property or law is instantiated. A more satisfactory solution is to avoid the appeal to levels altogether and, adopting a suggestion of Kim, to use the neutral concept of domains instead.

There are two sources of concern about synchronic downward causation. The first worry involves whole-to-part causation. The objection appears to be that if a given system can causally affect its own components via holistic properties that have downward effects, then the cause will include the effect as one of its parts. Yet this concern needs to be better articulated. The situation is not in conflict with the widely held irreflexivity of the causal relation because the part does not, in turn, cause the whole; it is a constituent of it. There is no analogous worry about downward causation when diachronic processes are involved.

The second source of worry about downward causation occurs when the causation is from some higher level entity but the effect does not involve a component of the system

---

[9] The argument can be generalized to include probabilistic causes, but I shall not do that here.

to which the cause is attached. Such a situation is problematical only if one is committed to the causal closure of the domain to which the effect belongs. This commitment is common when the domain in question is that of the physical. What it is for the realm of the physical to be causally closed is often improperly formulated. A common formulation, as in premise (1) of the exclusion argument, is that every physical event has a sufficient prior physical event, but that is inadequate because it allows causal chains to go from the physical domain into the chemical domain and back into the physical domain, thus allowing downward causation without violating the formulation.[10] The appropriate principle for events can be found in Lowe (2000): "Every physical event contains only other physical events in its transitive causal closure."[11]

7   EMERGENCE AS COMPUTATIONAL OR EXPLANATORY INCOMPRESSIBILITY

Weak emergence is a concept designed to capture many of the examples of diachronic emergence found in complex systems. The characteristic feature of systems exhibiting weak emergence is that the only way to predict their behaviors is to simulate them by imitating their temporal development in a step-by-step fashion. In the original formulation (Bedau 1997), the emphasis was on derivation by simulation, but in more recent publications (e.g., Bedau 2008), Bedau has argued that this step-by-step process reflects a parallel causal complexity in the system. That is, successive interactions, either between components of the system or between the system and its environment, preclude predictive techniques that permit computational compressibility. The degree of computational effort needed to calculate the future positions of a body falling under gravity in a vacuum is largely independent of how far into the future we go, whereas, in the current state of knowledge, no such simple formula is available to calculate the future position of a bumper car subject to multiple collisions because a prediction of the car's position requires calculations of each intermediate step in the trajectory of the car. A useful distinction here is between systems that can be represented by functions that have a closed-form solution and those that do not. In the former, the degree of computational effort required to calculate any future state of the system is largely independent of how far into the future that state is. In chaotic systems, among others, the absence of a closed-form solution precludes such computational compression (Crutchfield, Farmer, Packard, and Shaw 1986, 49). With the switch in emphasis within computational theory from abstract definitions of what is computable in principle to measures of degrees of computational complexity and what is feasibly computable, weak emergence is a contemporary version of the unpredictability approach to emergence that provides an explanation for why compressible predictions are unavailable. I note that

---

[10] In the exclusion argument, this is blocked by premise (3).
[11] For an earlier statement of the point, see Humphreys (1997).

weak emergence is a distinctively diachronic concept of emergence that is compatible with synchronic ontological reduction.

8 UNPREDICTABILITY THROUGH UNDECIDABILITY

The unpredictability approach to emergence originally relied on an unarticulated appeal to unpredictability in principle. In the 1920s, that idea was too vague to be of much use, but, with advances in computation theory, sharper accounts are now possible. An approach to knowledge that dates back to Euclid of Alexandria attempts to capture the entire content of a domain, be it mathematical or natural, by an explicitly formulated set of fundamental principles. Usually, this takes the form of an axiomatic system formulated in a specific language. The content of the axioms is used in conjunction with inference rules and definitions to extract all the implicit content of those fundamental principles. This approach also makes sense when the fundamental principles are not representational items but laws of nature that apply to a domain D, the analog of the inference rules are causal or other agencies, and the development of the system under the action of those laws and agencies, together with the initial and boundary conditions of a given system that falls within D, generates all the evolving temporal states of that system. Famously, there is a gap between the representational version of this axiomatic approach and the ontological form when the representational apparatus is too weak to generate every natural state of every system lying within D. One can thus capture a substantial part of the unpredictability approach to emergence by noting that an emergent state is one belonging to a part of D that falls outside the representational resources brought to bear on D. As a result, knowledge of that state must be arrived at by further observations and experiments. This captures the idea, present in many of the early writings on emergence, that the only way to know an emergent state is to experience an instance of it. The domain D is often taken to be that of physics or some subdomain thereof. A sharp case of such representational inadequacy occurs when the theory T involved is undecidable in the standard logical sense that there is some sentence S formulatable within the language of T for which neither S nor $\sim S$ is derivable from T. Gu et al. (Gu, Weedbrook, Perales, and Nielsen 2009), following earlier work by Barahona (1982) showed that there are states of an Ising lattice representing a recognizably macroscopic physical property that are undecidable; hence, they can be known only through observation or, perhaps, by the introduction of a theory predictively stronger than T.[12]

---

[12] The basic technique is to map temporal states of a one-dimensional Ising lattice onto a two-dimensional cellular automaton and to use Rice's theorem to show undecidable states exist in the latter; see Humphreys (2015) for details.

9 EMERGENCE AS UNIVERSALITY

It is a striking fact about the world that it allows for accurate description and manipulation of certain systems without the need to take into account the behavior of that system's components. Such autonomy of a higher level from lower levels is often thought to be a characteristic feature of the higher level having emerged from the lower levels. In philosophy, this corresponds to the situation called multiple realizability and, in physics, to universality.[13] The term "universality" is misleading because the independence from the substratum usually holds only for a restricted set of properties, but it has become standard. This autonomy is one reason why it has seemed natural to divide the world into levels, and it is frequently cited as an argument for the existence of new laws at a higher level (see Laughlin, Pines, Schmalian, Stojkovic, and Wolynes 2000).

Robert Batterman (2002) has argued that renormalization group methods provide a route for understanding why this kind of autonomy occurs in certain types of physical systems.[14] The most important aspect of universality from the present point of view is that phenomena falling within a given universality class U can be represented by the same general principles, and these principles are satisfied by a wide variety of ontological types at the microlevel.[15] This fact needs to be explained. As well as the thermodynamical phenomena such as phase transitions that are discussed by Batterman, this independence from the details of the underlying dynamics is characteristic of many dynamical systems. The similarity is not surprising because both formalisms appeal to attractors, the first using fixed points in the space of Hamiltonians that fall within a basin of attraction; the second, as in the damped nonlinear oscillator example used by Rueger (2000), appeals to fixed points in phase space.[16] However, Batterman's approach is distinctive in that it relies on an appeal to specifically mathematical, rather than ontological, explanations of the universal phenomena and denies that the usual approach of using reductive explanations that start from fundamental laws is capable of explaining the universal behavior.

We can contrast Batterman's approach to emergence with a different approach appealing to physical principles. Order in complex systems is represented by structure that

[13] The theme recurs in complex systems theory as well: "A common research theme in the study of complex systems is the pursuit of universal properties that transcend specific system details" (Willinger, Alderson, Doyle, and Li 2004, 130) The situations considered are different from those that allow us to predict the motion of a brass sphere in a vacuum without knowing the details of the interactions between the copper and zinc atoms that compose the brass. There, aggregating the masses of the constituents suffices. In other cases, such as not needing to know the exact quark configuration underlying donkey physiology in order to predict that the male offspring of a horse and a donkey will be infertile, it is possible that emergent phenomena are present but that the methods underlying universality in physics are unlikely to explain this, and conceptual emergence is clearly present.

[14] See also Auyang (1999, 181–183).

[15] Two systems are assigned to the same universality class if they exhibit the same behavior at their critical points.

[16] There are other similarities, such as the appearance of qualitative changes at critical points.

results from a loss of symmetry in the system, and a preliminary word of caution is necessary regarding the use of the term "symmetry." "Symmetry" is often used in these contexts to refer to a certain kind of homogeneity that is typically produced by a disordered spatial arrangement. Thus, the symmetry in many systems is just the presence of a random spatial arrangement of the values of a given property. In the ferromagnetic state, iron appears to have a great deal of symmetry since all of the spins are aligned, but this state has a lower degree of symmetry than the disordered state. What is called "spontaneous symmetry breaking" is the appearance of order resulting solely from the local interactions between the elements of the system, although there may be some additional influence from a constant external field. A key aspect of spontaneous symmetry breaking is that the aggregate order allows the energy in the system to be minimized.

Consider a ferromagnet the temperature of which can vary. The temperature is a control parameter, a quantity that links the system with its environment, and the net magnetization, which is the thermodynamical average of the spins, is an order parameter. More generally, an order parameter is a variable that determines when the onset of self-organization occurs and serves as a quantitative measure of the loss of symmetry.[17] In a ferromagnet, the order parameter is the net magnetization M.[18] M is a holistic property of the entire system; it emerges through the formation of local areas of uniform spin orientation called Weiss domains and then spreads throughout the system. There is no general method for finding order parameters, and identifying an order parameter is very system-specific.[19] The critical temperature (and, more generally, a critical point) is a value of the control parameter at or near which a phase transition occurs and large-scale structure appears as the result of local interactions. At the critical point, scale invariance of the structures exists.

A phase is a region of the phase space within which the properties are analytic as functions of external variables; that is, they have convergent Taylor expansions. This means that, with small perturbations, the system stays within the phase, and all thermodynamical properties will be functions of the free energy and its derivatives. A point at which the free energy is nonanalytic is a phase transition. Looked at from the perspective of perturbations, the nonanalyticity means that a small perturbation in the control parameter can lead to a switch from one phase to another and corresponding large qualitative changes. Indeed, it is with this reference to nonanalyticity and the associated singularities that the phase transition from paramagentism to ferromagnetism is said to be explained.

There is a dispute about whether the singular infinite limits that play a central role in Batterman's account are necessary for the existence of the critical points that are associated with emergence. Batterman (2011) argues that they are and that the limits are

---

[17] More technically, an order parameter is a monotonic function of the first derivative of the free energy.

[18] Other examples of order parameters are the director D of a nematic liquid crystal, the amplitude P_G of the density wave in a crystal, and the mean pair field in a superconductor $\langle \psi, r \rangle$.

[19] See, e.g., Binney et al. (Binney, Dowrick, Fisher, and Newman 1992, 11).

essential to providing understanding; the need for such limits is denied by Menon and Callender (2011) and by Butterfield (2011). Because this literature depends essentially on theoretical results, the application of those results to real finite systems requires care, and some parts of the physics literature are ambiguous about this issue. For a detailed, necessarily technical, discussion, I refer the reader to Landsman (2013).

10 COMPLEXITY, NONLINEARITY, AND EMERGENCE

A great deal of traditional science has been based on analysis. The classic method of laboratory experimentation, for example, is based on the assumption that it is possible to isolate individual causal influences by controlling other variables and then to combine the results from different experiments. This decomposition and recomposition method is mirrored in many mathematical models using various versions of additivity principles. In recent years, much attention has been paid to the relations among nonlinear systems, complexity theory, and emergence, and it is often claimed that nonlinear systems form a subset of the set of complex systems and that complex systems give rise to emergent phenomena. This section is a brief guide to the origin of those claims.

Some of the early British emergentist literature argued that emergence occurred when the level of complexity of a system passed a certain point, and this tradition has survived both within complexity theory and in the appeal to phase transitions as an example of emergence. One common definition is that a system is complex if it cannot be fully explained by understanding its component parts. Self-organization is often cited as a feature of complex systems, as is nonlinearity (Auyang 1999, 13; Cowan 1994, 1). Complexity theory, if it can fulfill its ambitions, would provide not just a degree of methodological unity, but also a kind of higher level ontological unity.

The property of linearity is best captured through its use in mathematical models. The term "linear system" can then be applied to systems that are correctly described by a linear model. Suppose that an input to the system at time t is given by $x(t)$ and that the temporal development of the system is described by an operator O, so that the output of the system at time t', where t' is later than t, is given by $y(t') = O[x(t)]$. Then, if the system is presented with two distinct inputs $x_1(t)$ and $x_2(t)$, the system is *linear* just in case $O[ax_1(t) + bx_2(t)] = ay_1(t') + by_2(t')$, for any real numbers a and b. This superposition principle—that if f, g are solutions to an equation E, then so is $af + bg$ for arbitrary constants a,b—is the characteristic feature of linear systems, and it represents a specific kind of additive compositionality. One important use of linearity is the ability to take a spatial problem, divide the region into subregions, solve the equation for each subregion, and stitch the solutions together. Linear equations thus parallel the methods of analysis and decomposition. In contrast, as David Campbell (1987, 219) puts it: "one must consider a non-linear problem *in toto*; one cannot—at least not obviously—break the problem into small subproblems and add the solutions." So here we have a glimpse of the holistic feature that is characteristic of emergent phenomena. Linearity also captures the separability of influences on a system in this way: if all of the inputs X_i to the system are held constant except for X_j, then changes in the output of the system depend only upon changes in the value of X_j. Nonlinear systems, in contrast, violate the superposition principle. For a simple, atemporal example, consider the stationary wave function given by $\sin(\theta)$. This describes a nonlinear system because in general, $\sin(a\theta_1 + b\theta_2) \neq a\sin(\theta_1) + b\sin(\theta_2)$. As a result, the system cannot be modeled by decomposing the angle θ into smaller angles θ_1 and θ_2, finding the solutions to $\sin(\theta_1)$ and $\sin(\theta_2)$ and adding the results. When using this definition of a linear system to represent a natural system, one must be careful to specify what the additive procedure "+" represents since it is the failure of the additivity property of linearity that lies behind the oft-quoted and misleading slogan about emergence "the whole is more than the sum of the parts."

There are three characteristic differences between linear and nonlinear systems. The first is that they describe qualitatively different types of behavior. Linear systems are (usually) described by well-behaved functions, whereas nonlinear systems are described by irregular and sometimes chaotic functions. This difference can be seen in the transition from laminar to turbulent motion in fluids for which the former exhibits linear behavior whereas the latter is highly nonlinear. The second characteristic is that, in linear systems, a small change in parameters or a small external perturbation will lead to small changes in behavior. For nonlinear systems, such small changes can lead to very large qualitative changes in the behavior. This feature of linear systems is related to the characteristic of stability under external perturbations that is sometimes considered to be a feature of systems that do not exhibit emergent behaviors. The third characteristic is that linear systems exhibit dispersion, as in the decay of water waves moving away from a central source, whereas the stability of eddies in turbulent flow exhibits nondispersion.

These distinctions are not always sharp. It is often claimed that a linear system can be decomposed into uncoupled subsystems. Strogatz (1994, 8 9), for example, claims that linearity characterizes systems that can be decomposed into parts, whereas nonlinear systems are those in which there are interactions, cooperative or inhibiting, between the parts. However, there are some nonlinear systems that are integrable and can thus be represented in a way that allows a decomposition into a collection of uncoupled subsystems. Moreover, the equation of motion for a nonlinear pendulum can be given a closed-form solution for arbitrary initial conditions (Campbell 1987, 221).

11 HOLISM

The failure of decomposability can lead to holism, one of the most commonly cited characteristics of emergent phenomena. Suppose that we want to capture reduction in the ontological rather than the theoretical mode. A common theme in the emergence literature, stated with great clarity by C. D. Broad but recurring in subsequent literature, is that emergence results from a failure of a certain type of generative principle. It is this:

Anti-holism: Consider the parts of a system S, each part being isolated from the others. Then when each part p of S combines with the other parts to form S as a whole, the properties of the part p within S follow from the properties of p in isolation or in simpler systems than S. Furthermore, any property of S must follow from properties of the parts of S.[20]

This principle attempts to capture the idea that there is nothing that S as a whole contributes to its properties. In contrast, if, for some property P to be present in S, it is necessary for *all* of the components of S to be in place and to stand in the relations that make S what it is, then there is some ontological or epistemic aspect of S that requires S to be considered as a whole, and this runs counter to reductionist approaches. Much depends on what counts as "follows from." Often, this is taken to be "is predictable from"; in other cases, it is taken to be "is determined by." In Broad's characterization, what you need for a mechanistic account is, first, the possibility of knowing the behavior of the whole system on the basis of knowledge of the pairwise interactions of its parts; and, second, knowledge that those pairwise interactions between the parts remain invariant within other combinations. If these two items of knowledge give you knowledge of the features of the whole system, then it is not an emergent system. One way in which anti-holism can fail is when configurational forces appear in a system, and, as Butterfield (2012, 105) has pointed out, configurational forces are those that are not determined by the pairwise interactions between the elements of the system.[21]

12 SKEPTICISM ABOUT EMERGENCE

Perhaps because of unfavorable comparisons with the clarity of various reductive and generative approaches, emergence has been regarded with suspicion by many, either on grounds of obscurantism, lack of clear examples, or a commitment to analytic and constructive methods. It is frequently held that emergence, if it exists at all, is rare and to be found at most in the realm of conscious phenomena (Kim 1999; McLaughlin 1997). One early expression of this skepticism can be found in Hempel and Oppenheim (1948) where it is argued that any evidence in favor of a supposed example

---

[20] For examples of this principle see Broad (1925, 59) and Stephan (1999, 51). Broad's original formulation mentions "in other combinations," rather than "simpler systems," thereby allowing more, as well as less, complex systems than S, but that more general position eliminates some emergent features that result from threshold effects.

[21] Much of McLaughlin's account of British emergentism (McLaughlin 1992) is constructed around the idea of configurational forces.

of emergence is the result of incomplete scientific knowledge and that the evidence will be undermined as science advances. Pairing Hempel's deductive-nomological approach to scientific explanation with Ernest Nagel's contemporaneous deductive approach to theory reduction, this skepticism was an expression of faith that, in time, all apparent examples of emergence would be reductively explained. At the time, there was what seemed to be inductive evidence in favor of that view, but, as confidence in our ability to fully reduce various domains of science to others has eroded and as numerous credible examples of emergence have been discovered, this skepticism has become less persuasive.

Anti-emergentism is a contingent position, and it is important to separate two aspects of the study of emergence. One project is to provide a systematic account of a particular type of emergence, be it inferential, ontological, or conceptual. Such an account will only be plausible if it is motivated by general principles and either has some connection with historical traditions in emergence or makes a persuasive a priori case that the account is indeed one of emergence. The second project is to identify actual cases that satisfy the criteria given by the theoretical treatment of emergence. Assuming that a given account of emergence is consistent, it follows that there is at least one possible example of emergence of that type. So it would be an important discovery about our world if there were no actual instances of that possibility and perhaps one that required an explanation. Would it be because, as Wimsatt (2007) has suggested, our representational practices are biased toward models that possess nonemergent features; would it perhaps be the result of a very general principle that the novel element required for emergence cannot come from a basis that lacks that feature, a variant of the "never something from nothing" principle; or would it be because the laws and other conditions that are present in our universe preclude, or just happen not to have produced, cases of emergence?

13 A FINAL SUGGESTION

Although I began this chapter by noting that there is no unified account of emergence, I conclude by suggesting a framework within which a number of emergentist positions can be accommodated. This is not a definition, but a schema, and no claim is made that all extant positions that address emergence are included—indeed, not all should be, because some are such that it is hard to tell why their subject matter counts as emergent. One can impose some order on the positions discussed by appealing to this, admittedly rough, principle.

Emergence as nonclosure: A is emergent from a domain B just in case A occurs as a result of B-operations on elements of B but falls outside the closure of B under closure conditions C. To give some sense of how this schema is applied, in a typical application to ontological emergence, B is the domain of physics, the closure conditions on B are the laws of physics, and B-operations are physical interactions.

On accounts of ontological emergence in which to be emergent is to be subject to laws that do not apply to the base domain, an emergent chemical phenomenon, for example, would then be one that results from physical interactions between physical entities but is subject to a nonphysical, irreducibly chemical law. Similarly, for any account in which there are novel physico-chemical transordinal laws or in which the nomological supervenience of A upon B is licensed by a physico-chemical fundamental law.

In the case of inferential emergence, a statement A will be epistemologically emergent from the domain of statements B (usually taken to constitute a theory) just in case the statement A can be constructed within the representational apparatus of B using the syntactic or semantic operations licensed by B, but the inferential operations that constitute the closure conditions for B do not permit the derivation of A. This will cover not only cases in which homogeneous Nagel reduction fails but also cases in which adding definitions of terms in A that are not present in B result in a conservative extension of B, yet A still cannot be derived from B. The reader is encouraged to test his or her favorite account of emergence against the schema.

References

Alexander, S. (1920). *Space, Time, and Deity: The Gifford Lectures at Glasgow 1916–1918* (London: Macmillan).

Anderson, P. W. (1972). "More is Different." *Science* 177: 393–396.

Auyang, S. (1999). *Foundations of Complex-Systems Theories.* (Cambridge: Cambridge University Press).

Barahona, F. (1982). "On the Computational Complexity of Ising Spin Glass Models." *Journal of Physics A: Mathematical and General* 15: 3241–3253.

Batterman, R. (2002). *The Devil in the Details: Asymptotic Reasoning in Explanation, Reduction, and Emergence.* (New York: Oxford University Press).

Batterman, R. W. (2011). "Emergence, Singularities, and Symmetry Breaking." *Foundations of Physics* 41: 1031–1050.

Beckerman, A., Flohr, H. and Kim, J. eds. (1992). *Emergence or Reduction?: Essays on the Prospects of Nonreductive Physicalism.* (Berlin: Walter de Gruyter).

Bedau, M. (1997). "Weak Emergence." In J. Tomberlin (ed.), *Philosophical Perspectives: Mind, Causation, and World, Volume 11* (Malden: Blackwell), 375–399.

Bedau, M. (2008). "Is Weak Emergence Just in the Mind?" *Minds and Machines* 18: 443–459.

Bergson, H. (1907). *L'Évolution créatrice.* (Paris: Les Presses universitaires de France). English translation (1911) as *Creative Evolution* (New York: H. Holt).

Binney, J., Dowrick, N., Fisher, A., and Newman, M. (1992). *The Theory of Critical Phenomena.* (Oxford: Clarendon).

Blitz, D. (1992). *Emergent Evolution.* (Dordrecht: Kluwer Academic Publishers).

Broad, C. D. (1925). *The Mind and Its Place in Nature.* (London: Routledge and Kegan Paul).

Butterfield, J. (2011). "Less Is Different: Emergence and Reduction Reconciled." *Foundations of Physics* 41: 1065–1135.

Butterfield, J. (2012). "Laws, Causation and Dynamics." *Interface Focus* 2: 101–114.

Butterfield, J., and Isham, C. (1999). "On the Emergence of Time in Quantum Gravity." In J. Butterfield (ed.), *The Arguments of Time* (Oxford: Oxford University Press), 111-168.

Campbell, D. K. (1987). "Nonlinear Science: From Paradigms to Practicalities." *Los Alamos Science* 15: 218–262.

Chalmers, D. (2006). "Strong and Weak Emergence." In P. Clayton and P. Davies (eds.), *The Re-Emergence of Emergence* (Oxford: Oxford University Press), 244–256.

Clayton, P. (2004). *Mind and Emergence*. (New York: Oxford University Press).

Cowan, G. (1994). "Conference Opening Remarks." In G. A. Cowan, D. Pines, and D. Meltzer (eds.), *Complexity: Metaphors, Models, and Reality. Santa Fe Institute Studies in the Sciences of Complexity Volume XIX*. (Boulder, CO: Westview Press), 1-4.

Crutchfield, J. P., Doyne Farmer, J., Packard, N. H., and Shaw, R. S. (1986). "Chaos." *Scientific American* 255: 46–57.

Fagot-Largeault, A. (2002). "L'émergence." In D. Andler, A. Fagot-Largeault, and B. Saint-Sernin (eds.), *Philosophie des sciences, Volume II*. (Paris: Gallimard), 939–1048. (In French.)

Ganeri, J. (2011). "Emergentisms, Ancient and Modern." *Mind* 120: 671–703

Gu, M., Weedbrook, C., Perales, A., and Nielsen, M. (2009). "More Really Is Different." *Physica D* 238: 835–839.

Hempel, C. G., and Oppenheim, P. (1948). "Studies in the Logic of Explanation." *Philosophy of Science* 13: 135–175.

Humphreys, P. (1997). "How Properties Emerge." *Philosophy of Science* 64: 1–17.

Humphreys, P. (2015). "More Is Different ... Sometimes: Ising Models, Emergence, and Undecidability." In B. Falkenburg and M. Morrison (eds.), *Why More Is Different: Philosophical Issues in Condensed Matter Physics and Complex Systems* (Berlin: Springer), 137–152.

Humphreys, P. (2016). *Emergence*. (New York: Oxford University Press)

Kim, J. (1999). "Making Sense of Emergence." *Philosophical Studies* 95: 3–36.

Kim, J. (2006). "Emergence: Core Ideas and Issues." *Synthese* 151: 547–559.

Landsman, N. P. (2013). "Spontaneous Symmetry Breaking in Quantum Systems: Emergence or Reduction?" *Studies in History and Philosophy of Modern Physics* 44: 379–394.

Laughlin, R. B., Pines, D., Schmalian, J., Stojkovic, B. P., and Wolynes, P. (2000). "The Middle Way." *Proceedings of the National Academy of Sciences of the United States of America* 97: 32–37.

Lewes, G. H. (1875). *Problems of Life and Mind, Volume II*. (London: Kegan Paul, Trench, Turbner, and Company).

Lewis, D. (1986). *Philosophical Papers, Volume II*. (New York: Oxford University Press).

Lowe, E. J. (2000). "Causal Closure Principles and Emergentism." *Philosophy* 75: 571–585.

Malaterre, C. (2007). "Le "néo-vitalisme" au XIXème siècle: une seconde école française de l'émergence." *Bulletin d'histoire et d'épistémologie des sciences de la vie*, 14: 25–44. (In French.)

McLaughlin, B. (1992). "The Rise and Fall of British Emergentism." In A. Beckerman et al. (eds.), *Emergence or Reduction?: Essays on the Prospects of Nonreductive Physicalism*. (Berlin: Walter de Gruyter), 49–93

McLaughlin, B. (1997). "Emergence and Supervenience." *Intellectica* 25: 25–43.

Menon, T., and Callender, C. (2011). "Turn and Face the Strange ... Ch-ch-changes: Philosophical Questions Raised by Phase Transitions." In R. Batterman (ed.), *The Oxford Handbook of the Philosophy of Physics* (New York: Oxford University Press), 189–223.

Mill, J. S. (1843). *A System of Logic: Ratiocinative and Inductive*. (London: Longmans, Green).

Monasson, R., Zecchina, R., Kirkpatrick, S., Selman, B., and Troyansky, L. (1999). "Determining Computational Complexity from Characteristic 'Phase Transitions.'" *Nature* 400: 133–137.

Morgan, C. L. (1923). *Emergent Evolution*. (London: Williams & Norgate).

Nagel, E. (1961). *The Structure of Science. Problems in the Logic of Explanation*. (New York: Harcourt, Brace & World).

O'Connor, T., and Wong, H. Y. (2005). "The Metaphysics of Emergence." *Noûs* 39: 658–678.

Pepper, S. (1926). "Emergence." *Journal of Philosophy* 23: 241–245.

Rueger, A. (2000). "Physical Emergence, Diachronic and Synchronic." *Synthese* 124: 297–322.

Stephan, A. (1992). "Emergence—A Systematic View on Its Historical Facets." In A. Beckerman, H. Flohr, and J. Kim (eds.), *Emergence or Reduction?: Essays on the Prospects of Nonreductive Physicalism*. (Berlin: Walter de Gruyter), 25–48.

Stephan, A. (1999). "Varieties of Emergentism." *Evolution and Cognition* 5: 49–59.

Strogatz, S. (1994). *Nonlinear Dynamics and Chaos*. (Reading, MA: Addison-Wesley).

Uzgalis, W. (2009). "Anthony Collins on the Emergence of Consciousness and Personal Identity." *Philosophy Compass* 4: 363–379.

Van Cleve, J. (1990) "Emergence vs. Panpsychism: Magic or Mind Dust?" In J. E. Tomberlin (ed.), *Philosophical Perspectives, Volume 4* (Atascadero, CA: Ridgeview), 215–226.

Willinger, W., Alderson, D., Doyle, J. C., and Li, L. (2004). "More 'Normal' than Normal: Scaling Distributions and Complex Systems." In R. G. Ingalls, M. D. Rossetti, J. S. Smith, and B. A. Peters (eds.), *Proceedings of the 2004 Winter Simulation Conference* (Washington, D.C.: IEEE Publications), 130-141.

Wimsatt, W. (2007). "Emergence as Non-aggregativity and the Biases of Reductionisms." In W. Wimsatt (ed.), *Re-Engineering Philosophy for Limited Beings: Piecewise Approximations to Reality*. (Cambridge, MA: Harvard University Press), 274-312.

FURTHER READING

Bedau, M., and Humphreys, P. (eds.). (2008). *Emergence: Contemporary Readings in Philosophy and Science* (Cambridge, MA: MIT Press). A collection of seminal contemporary articles on emergence.

Juarrero, A., and Rubino, C. A. (eds.). (2010). *Emergence, Complexity, and Self-Organization: Precursors and Prototypes*. (Litchfield Park, AZ: Emergent Publications). A collection of historical papers from many of the important early contributors to emergence.

An excellent survey article on emergence that has a stronger orientation toward metaphysical issues than does the present article is T. O'Connor and H. Y. Wong, "Emergent Properties," in *The Stanford Encyclopedia of Philosophy* (Summer 2015 Edition), Edward N. Zalta (ed.), http://plato.stanford.edu/archives/sum2015/entries/properties-emergent/.

CHAPTER 37

···

EMPIRICISM AND AFTER[*]

···

JIM BOGEN

1 INTRODUCTION

SCIENCE is usually characterized as an empirical enterprise, and most present-day philosophies of science derive from the work of thinkers classified as empiricists. Hence the term "empiricism" in my title. "And after" is meant to reflect a growing awareness of how little light empiricism sheds on scientific practice.

A scientific claim is credible just in case it is significantly more reasonable to accept it than not to accept it. An influential empiricist tradition promulgated most effectively by twentieth-century logical empiricists portrays a claim's credibility as depending on whether it stands in a formally definable confirmation relation to perceptual evidence. Philosophers in this tradition pursued its analysis as a main task. Section 7 suggests that a better approach would be to look case by case at what I'll call epistemic pathways connecting the credibility of a claim in different ways to different epistemically significant factors. Perceptual evidence is one such factor, but so is evidence from experimental equipment, along with computer-generated virtual data and more. Sometimes perceptual evidence is crucial. Often it is not. Sometimes it contributes to credibility in something like the way an empiricist might expect. Often it does not.

2 EMPIRICISM IS NOT A NATURAL KIND

Zvi Biener and Eric Schliesser observe that "empiricism" refers not to a single view but rather, to

> an untidy heterogeneity of empiricist philosophical positions. There is no body of
> doctrine in early modernity that was "empiricism" and no set of thinkers who self

[*] Thanks to Zvi Biener, Deborah Bogen, Allan Franklin, Dan Garber, Anil Gupta, Michael Miller, Sandra Diane Mitchell, Slobodan Perovic, Lauren Ross, Ken Schaffner, Jim Woodward, and Amy Enrico.

identified as "empiricists" . . . [Nowadays] "empiricism" refers to a congeries of ideas that privilege experience in different ways. (Biener and Schliesser 2014, 2)

The term comes from an ancient use of "empeiria"—usually translated "experience"—to mean something like what we'd mean in saying that Pete Seeger had a lot of experience with banjos. Physicians who treated patients by trial and error without recourse to systematic medical theories were called "empirical" (Sextus Empiricus 1961, 145–146). Aristotle used "empeiria" in connection with what can be learned from informal observations, as opposed to scientific knowledge of the natures of things (Aristotle 1984, 1552–1553). Neither usage has much to do with ideas about the cognitive importance of perceptual experience that we now associate with empiricism.

Although Francis Bacon is often called a father of empiricism, he accused what he called the empirical school of inductive recklessness: Their "premature and excessive hurry" to reach general principles from "the narrow and obscure foundation of only a few experiments" leads them to embrace even worse ideas than rationalists who develop "monstrous and deformed" ideas about how the world works by relying "chiefly on the powers of the mind" unconstrained by observation and experiment (Bacon 1994, 70). Bacon concludes that, just as the bee must use its powers to transform the pollen it gathers into food, scientists must use their powers of reasoning to interpret and regiment experiential evidence if they are to extract knowledge from it (1994, 105).[1] Ironically, most recent thinkers that we call empiricists would agree.

Lacking space to take up questions that Bacon raises about induction, this chapter limits itself to other questions about experience as a source of knowledge. Rather than looking for a continuity of empiricisms running from Aristotle through British and logical empiricisms to the present, I'll enumerate some main empiricist ideas, criticize them, and suggest alternatives.

3 ANTHROPOCENTRISM AND PERCEPTUAL ULTIMACY

Empiricists tend to agree with many of their opponents in assuming:

1. *Epistemic anthropocentrism.* Human rational and perceptual faculties are the only possible sources of scientific knowledge.[2]

---

[1] The reliability of induction from perceptual experience was the main issue that separated empiricists from rationalists according to Leibniz (1949, 44) and Kant (1998, 138).

[2] Michael Polanyi's promotion of emotions and personal attitudes as sources of knowledge (Polanyi 1958, 153–169, 172–197) qualifies him as an exception, but his objections to empiricism are not mine.

William Herschel argued that no matter how many consequences can be inferred from basic principles that are immune to empirical refutation, it's impossible to infer from them such contingent facts as what happens to a lump of sugar if you immerse it in water or what visual experience one gets by looking at a mixture of yellow and blue (Herschel 1966, 76). Given 1, this suggests:

2. *Perceptual ultimacy.* Everything we know about the external world comes to us from . . . our senses, the sense of sight, hearing, and touch, and to a lesser degree, those of taste and smell (Campbell 1952, 16).[3]

One version of perceptual ultimacy derives from the Lockeian view that our minds begin their cognitive careers as empty cabinets or blank pieces of paper, and all of our concepts of things in the world and the meanings of the words we use to talk about them must derive from sensory experiences (Locke 1988, 55, 104–105).

A second version maintains that the credibility of a scientific claim depends on how well it agrees with the deliverances of the senses. In keeping with this and the logical empiricist program of modeling scientific thinking in terms of inferential relations among sentences or propositions,[4] Carnap's *Unity of Science* (UOS) characterizes science as

a system of statements based on direct experience, and controlled by experimental verification . . . based upon "protocol statements" . . . [which record] a scientist's (say a physicist's or a psychologist's) experience . . . (Carnap 1995, 42–43)

Accordingly, terms like "gene" and "electron," which do not refer to perceptual experiences, must get their meanings from rules that incorporate perceptual experiences into the truth conditions of sentences that contain them. Absent such rules, sentences containing theoretical terms could not be tested against perceptual evidence and would therefore be no more scientific than sentences in fiction that don't refer to anything (Schaffner 1993, 131–132). For scientific purposes, theoretical terms that render sentences that contain them untestable might just as well be meaningless. This brings the Lockeian and the Carnapian UOS versions of perceptual ultimacy together.

The widespread and inescapable need for data from experimental equipment renders both 1 and 2 indefensible.[5] Indeed, scientists have relied on measuring and other experimental equipment for so long that it's hard to see why philosophers of science ever put so much emphasis on the senses. Consider for example Gilbert's sixteenth-century use

---

[3] This idea is not dead. Gupta has attempted to defend the claim that experience is the supreme epistemic authority without relying on what I take to be the most objectionable feature of Perceptual Ultimacy" see Gupta (2006, 3, 216, 220 n.3).

[4] See Bogen and Woodward (2005, 233).

[5] For an early and vivid appreciation of this point, seeFeyerabend (1985b).

of balance beams and magnetic compasses to measure imperceptible magnetic forces (Gilbert 1991, 167–168).[6]

Experimental equipment is used to detect and measure perceptibles as well as imperceptibles, partly because it can often deliver more precise, more accurate, and better resolved evidence than the senses. Thus, although human observers can feel heat and cold, they aren't very good at fine-grained quantitative discriminations or descriptions of experienced, let alone actual, temperatures. As Humphreys says,

> [o]nce the superior accuracy, precision, and resolution of many instruments has been admitted, the reconstruction of science on the basis of sensory experience is clearly a misguided enterprise. (Humphreys 2004, 47)

A second reason to prefer data from equipment is that investigators must be able to understand one another's evidence reports. The difficulty of reaching agreement on the meanings of some descriptions of perceptual experience led Otto Neurath to propose that protocol sentences should contain no terms for subjective experiences accessible only to introspection. Ignoring details, he thought a protocol sentence should mention little more than the observer and the words that occurred to her as a description of what she perceived when she made her observation (Neurath 1983a, 93–97). But there are better ways to promote mutual understanding. One main way is to use operational definitions[7] mentioning specific (or ranges of) instrument readings as conditions for the acceptability of evidence reports. For example, it's much easier to understand reports of morbid obesity by reference to quantitative measuring tape or weighing scale measurements than descriptions of what morbidly obese subjects look like.

In addition to understanding what is meant by the term "morbidly obese," qualified investigators should be able to decide whether it applies to the individuals an investigator has used it to describe. Thus, in addition to intelligibility, scientific practice requires public decidability: it should be possible for qualified investigators to reach agreement over whether evidence reports are accurate enough for use in the evaluation of the claims they are used to evaluate.[8] Readings from experimental equipment can often meet this condition better than descriptions of perceptual experience.

The best way to accommodate empiricism to all of this would be to think of outputs of experimental equipment as analogous to reports of perceptual experiences. I'll call the view that knowledge about the world can be acquired from instrumental as well as sensory evidence *liberal empiricism*,[9] and I'll use the term "*empirical evidence*" for evidence from both sources.

Both liberal empiricism and anthropocentrism fail to take account of the fact that scientists must sometimes rely on computationally generated virtual data for information

---

[6] Thanks to Joel Smith for this example.

[7] I understand operational definitions in accordance with Feest (2005).

[8] For a discussion of this and an interesting but overly strong set of associated pragmatic conditions for observation reports, see Feyerabend (1985a, 18–19).

[9] Bogen (2011), 11.

about things beyond the reach of their senses and their equipment. For example, weather scientists who cannot position their instruments to record temperatures, pressures, or wind flows inside evolving thunderstorms may

> examine the results of high-resolution simulations to see what they suggest about that evolution; in practice, such simulations have played an important role in developing explanations of features of storm behavior. (Parker 2010, 41)[10]

Empiricism ignores the striking fact that computer models can produce informative virtual data without receiving or responding to the kinds of causal inputs that sensory systems and experimental equipment use to generate their data. Virtual data production can be calibrated by running the model to produce more virtual data from things that experimental equipment can access, comparing the results to nonvirtual data, and adjusting the model to reduce discrepancies. Although empirical evidence is essential for such calibration, virtual data are not produced in response to inputs from things in the world. Even so, virtual data needn't be inferior to empirical data. Computers can be programmed to produce virtual measures of brain activity that are epistemically superior to nonvirtual data because virtual data

> can be interpreted without the need to account for many of the potential confounds found in experimental data such as physiological noise, [and] imaging artifacts. (Sporns 2011, 164)

By contrast, Sherri Roush argues that virtual data can be less informative than empirical data because experimental equipment can be sensitive to epistemically significant factors that a computer simulation doesn't take into account (Roush forthcoming). But, even so, computer models sometimes do avoid enough noise and represent the real system of interest well enough to provide better data than experimental equipment or human observers.

4 EPISTEMIC PURITY

Friends of perceptual ultimacy tend to assume that

3. In order to be an acceptable piece of scientific evidence, a report must be pure in the sense that none of its content derives from "judgments and conclusions imposed on it by [the investigator]" (Neurath 1983b, 103) [11,12]

---

[10] Cp. Winsberg (2003).
[11] A lot of what I have to say about this originated in conversation with Sandra D. Mitchell and Slobodan Perovic.
[12] Neurath is paraphrasing Schlick (1959, 209–210).

This condition allows investigators to reason about perceptual evidence as needed to learn from it, as long as their reasoning does not influence their experiences or the content of their observation reports. A liberal empiricist might impose the same requirement on data from experimental equipment. One reason to take data from well-functioning sensory systems and measuring instruments seriously is that they report relatively direct responses to causal inputs from the very things they are used to measure or detect. Assuming that this allows empirical data to convey reliable information about its objects, purity might seem necessary to shield it from the errors that reasoning is prone to (cp. Herschel 1966, 83). But Mill could have told the proponents of purity that this requirement is too strong. One can't report what one perceives without incorporating into it at least as many conclusions as one must draw to classify or identify it (Mill 1967, 421).

Furthermore, impure empirical evidence often tells us more about the world than it could have if it were pure. Consider Santiago Ramòn y Cajal's drawings of thin slices of stained brain tissue viewed through a light microscope (DeFelipe and Jones 1988). The neurons he drew didn't lie flat enough to see in their entirety at any one focal length or, in many cases, on just one slide. What Cajal could see at one focal length included loose blobs of stain and bits of neurons he wasn't interested in. Furthermore, the best available stains worked too erratically to cover all of what he wanted to see. This made impurity a necessity. If Cajal's drawings hadn't incorporated his judgments about what to ignore, what to include, and what to portray as connected, they couldn't have helped with the anatomical questions he was trying to answer (DeFelipe and Jones 1988, 557–621).

Functional magnetic resonance imaging (fMRI) illustrates the need for impurity in equipment data. fMRI data are images of brains decorated with colors to indicate locations and degrees of neuronal activity. They are constructed from radio signals emitted from the brain in response to changes in a magnetic field surrounding the subject's head. The signals vary with local changes in the level of oxygen carried by small blood vessels, indicative of magnitudes and changes in electrical activity in nearby neurons or synapses. Records of captured radio signals are processed to guide assignments of colors to locations on a standard brain atlas. To this end, investigators must correct for errors, estimate levels of oxygenated blood or neuronal activity, and assign colors to the atlas. Computer processing featuring all sorts of calculations from a number of theoretical principles is thus an epistemically indispensable part of the production, not just the interpretation, of fMRI data.[13,]

Some data exhibit impurity because virtual data influence their production. Experimenters who used proton-proton collisions in CERN's large hadron collider (LHC) to investigate the Higgs boson had to calibrate their equipment to deal with such difficulties as that only relatively few collision products could be expected to indicate the presence of Higgs bosons, and the products of multiple collisions can mimic Higgs indicators if they overlap during a single recording. To make matters worse, on average,

---

[13]   Cohen (1996), Lazar et al. (2001).

close to a million collisions could be expected every second, each one producing far too much information for the equipment to store (van Mulders 2010, 22, 29–31). Accordingly, investigators had to make and implement decisions about when and how often to initiate collisions and which collision results to calibrate the equipment to record and store. Before implementing proposed calibrations and experimental procedures, experimenters had to evaluate them. To that end, they ran computer models incorporating them and tested the results against real-world experimental outcomes. Where technical or financial limitations prevented them from producing enough empirical data, they had to use virtual data (Morrison 2015, 292-298). Margaret Morrison argues that virtual data and other indispensible computer simulation results influenced data production heavily enough to "cast . . . doubt on the very distinction between experiment and simulation"(2015, 289). LHC experiment outputs depend not just on what happens when particles collide, but also on how often experimenters produce collisions and how they calibrate the equipment. Reasoning from theoretical assumptions and background knowledge, together with computations involving virtual data, exerts enough influence on all of this to render the data triply impure.[14] The moral of this story is that whatever makes data informative, it can't depend on reasoning having no influence on the inputs from which data are generated, the processes through which they are generated, or the resulting data.

5 SCOPE EMPIRICISM

The empiricist assumptions I've been sketching attract some philosophers to what I'll call *scope empiricism*:

4. Apart from logic and mathematics, the most that scientists should claim to know about are patterns of perceptual experiences. The ultimate goal of science is to describe, systematize, predict, and explain the deliverances of our senses.

Bas Van Fraassen supported scope empiricism by arguing that it is "epistemically imprudent" for scientists to commit themselves to the existence of anything that cannot in principle be directly perceived—even if claims about it can be inferred from empirical evidence. Working scientists don't hesitate to embrace claims about things that neither they nor their equipment can perceive, but scope empiricists can dig in their heels and respond that even so, such commitments are so risky that scientists shouldn't make them. Supporting empiricism in this way is objectionable because it disparages what are generally regarded to be important scientific achievements, especially if scope

---

[14] The fact that different theoretical assumptions would call for the production and storage of different data is not to be confused with Kuhn's idea that people who are committed to opposing theories have different perceptual experiences in response to the same stimuli (Kuhn 1996, 63-64, 112-113).

empiricism prohibits commitments to claims that cannot be supported without appeal to impure data.

Van Fraassen knows that scientists make general claims whose truth depends on how well they fit unobserved as well as observed past, present, and future happenings (Van Fraassen 1980, 69). Thus, Snell's law purports to describe changes in the direction of all light rays that pass from one medium into another—not just the relatively few that have been or will be measured. And similarly for laws and lesser generalizations invoked to explain why Snell's law holds (to the extent that it does). Van Fraassen argues that although scientists can't avoid the epistemic risk of commitment to generalizations over unobserved perceptibles, they should avoid the greater risk of commitment to the existence of imperceptibles. To the contrary, it's epistemically imprudent to commit oneself to a generalization without good reason to think that unexamined instances conform to known instances, and, as I've argued elsewhere, the best reasons to accept a generalization over unobserved perceptibles sometimes include claims about unobservable activities, entities, processes, and the like that conspire to maintain the regularity it describes (Bogen 2011, 16–19). Apart from that, it's epistemically imprudent to infer regularities from data unless it's reasonable to believe in whatever the data represent. Anyone who draws conclusions about the brain from Cajal's drawings had better believe in the existence and the relevant features of the neurons he drew. But we've seen that the evidential value of Cajal's drawings depends on details he had to fill in on the basis of his own reasoned judgments. By the same token, the use of LHC data commits one to imperceptible collision products. Here and elsewhere, disallowing commitment to anything beyond what registers on the senses or on experimental equipment undercuts evidence for regularities scientists try to explain and generalizations they rely on to explain them.

6 "Points of Contact with Reality"

As Neurath paraphrased him, Schlick maintained that

> perceptual experiences are the "absolutely fixed, unshakable points of contact between knowledge and reality." (Neurath 1983b, 103)

Perceptual data might seem to be the best candidates for what Schlick had in mind. But the realities that scientists typically try to describe, predict, and explain are phenomena that Jim Woodward and I have argued are importantly different from data. Phenomena are

> events, regularities, processes, etc. whose instances, are uniform and uncomplicated enough to make them susceptible to systematic prediction and explanation. (Bogen and Woodward 1988, 317)

The melting point of lead and the periods and orbital paths of planets are examples (1988, 319–326). By contrast, data (think data points or raw data) correspond to what I've been calling empirical evidence. They are records of what registers on perceptual systems or experimental equipment in response to worldly inputs. They are cleaned up, corrected for error, analyzed, and interpreted to obtain information about phenomena.

The reason scientists seldom try to develop general explanatory theories about data is that their production usually involves a number of factors in elaborate and shifting combinations that are idiosyncratic to specific laboratory or natural settings. For example, the data Bernard Katz used to study neuronal signaling were tracings generated from neuronal electrical activity and influenced by extraneous factors peculiar to the operation of his galvanometers, sensitive to trial-to-trial variations in positions of the stimulating and recording electrodes he inserted into nerves and physiological effects of their insertion, changes in the condition of nerves as they deteriorated during experiments, and error sources as random as vibrations that shook the equipment in response to the heavy tread of Katz's teacher, A.V. Hill walking up and down the stairs outside of the laboratory. Katz wasn't trying to develop a theory about his tracings. He wanted a general theory about postsynaptic electrical responses to presynaptic spikes. No theory of neuronal signaling has the resources to predict or explain effects produced by as many mutually independent influences as Katz's tracings. Katz's data put him into much more direct epistemic contact with the states of his equipment than to the neuronal phenomena he used it to investigate.

Sensory- and equipment-generated data make contact with reality in the sense that they are produced from causal inputs from things that interact with equipment or the senses. But causal contact is not the same thing as epistemic contact. Recall that virtual storm data help bring investigators into epistemic contact with temperatures in storm regions that do not causally interact with the computers that generate them. More to the point, investigators often use data from things that do interact causally with their equipment or their senses to help answer questions about phenomena that do not. Thus, Millikan used data from falling oil drops he could see to investigate something he could not see—the charge on the electron.

7 Tracking Epistemic Pathways as an Alternative to Received Accounts of Confirmation

The last empiricist notion I want to consider is an idea about confirmation:

5. *Two-term confirmation.* A scientific claim is credible only if it is confirmed by perceptual evidence, where confirmation is a special two-term relation between claim and evidence—the same relation in every case.

One major trouble with this is that, as we've seen, many scientific claims owe their credibility to nonperceptual evidence from experimental equipment or computer simulations. Now I want to take up a second difficulty that can be appreciated by considering how two different sorts of evidence were used to support the credibility of general relativity theory (GRT). The first consisted of photographic plates, some of which were exposed to starlight at night, and others during a solar eclipse. The second consisted of records of telescope sightings of Mercury on its way around the sun. Although both kinds of evidence were used to evaluate one and the same theory, the support they provided for it cannot be informatively represented as instantiating one and the same confirmation relation, let alone a two-term one. The starlight photographs were used to compare GRT to Newtonian predictions about the path of light passing near the sun. Their interpretation required measurements of distances between spots on the photographic plates, calculations of the deflection of starlight passing near the sun from differences between relative positions of spots on eclipse and nighttime plates, and comparisons of deflection estimates to predicted values. The telescope data were used to evaluate GRT and Newtonian predictions about the perihelion of Mercury. In both cases, the evidence supported GRT by favoring its predictions. In addition to different calculations involving different background assumptions and mathematical techniques, different pieces of equipment were needed, and different physical adjustments and manipulations were required to promote the reliability of their data.[15] These data bear on the credibility of GRT by way of different, indirect connections. The natures and the heterogeneity of such connections are ignored by hypothetico-deductive and other received general accounts of confirmation. And similarly for a great many other cases.

The pursuit of a two-term relation of confirmation whose analysis will explain for every case what it is for evidence to make a claim credible tends to distract epistemologists from issues analogous to those that are obscured by the pursuit of a single relation whose analysis can distinguish causal from noncausal co-occurrences of events. Among these are questions about intermediate links between causal influences and their effects, differences in strengths and degrees of importance of causal factors in cases where two or more factors contribute to the same effect, and the ways in which some effects can be produced in the absence of one or more of their normal causes. Such questions arise in connection with a great many scientific investigations. To illustrate, shingles presents itself as a collection of symptoms including, most notably, a painful band of rash. Shingles is caused by the chickenpox virus, varicella zoster. The virus typically produces shingles only after remaining dormant in neural tissue for years after the original bout of chickenpox subsides. Instead of closing the books on shingles once they had identified the virus as its cause, investigators looked for intermediate and supplementary causal factors to explain how it survives while dormant, what inhibits and what promotes its activation, and where in the causal process physicians could intervene to control it.[16] This research raises questions about whether or how factors that are present in the virus

---

[15] Bogen and Woodward (2005, 234-236, 242–246) and Earman and Glymour (1980).
[16] See http://www.cdc.gov/vaccines/pubs/pinkbook/downloads/varicella.pdf

from the outset help explain its survival, activation, and rash production, and whether the same account of causality (e.g., interventionist, mechanistic, pragmatic law, etc.) illuminates the explanation of every stage in the process. Such questions are obscured in standard philosophical literature on causality by the analog of idea 5 presented earlier.

Genetic research is another example. Geneticists who know which DNA codons are especially important to the synthesis of a specific protein look for connections between intermediate causal factors at various steps of protein production, including influences of promoter molecules that initiate or encourage and repressors that block or dampen specific subprocesses.[17]

The best way for philosophers to do justice to such research is to construe the effect of interest as the end point of a causal pathway whose steps involve different factors interacting in different ways to contribute to its production.[18] By analogy, I think philosophers interested in questions that empiricists tried to answer should attend to what I'll call **epistemic pathways**.

> An epistemic pathway leads from any epistemically significant factor you choose, e.g., a parcel of empirical or virtual evidence, to the credibility of a scientific claim.

Depending as it does on what it's reasonable to believe, credibility is relative to people, places, and times. What's credible relative to evidence, methods of interpretation, background knowledge, and the like available now need not be credible relative to what was available fifty years ago. What's not credible now may become credible later on. What it's reasonable for one group of people to accept may not be credible for their contemporaries who work with different evidence, background knowledge, methods of analysis, and so forth. For many philosophical purposes, the simplest way to think about credibility is to treat it as a function of logical, probabilistic, and other connections among epistemically relevant factors available at the same time to typical or ideal human cognizers who are proficient in using all the inferential and computational resources we know of.[19]

An epistemic pathway might be represented by a node and edge diagram a little like the ones Judea Pearl (1988) recommended for use in programming machines to draw reasonable conclusions from uncertain premises. For example, if p is uncertain, but has a probability close enough to 1, you might want a machine to infer q from p and (p & p⊃q), even though the probabilities do not rule out ¬q. Furthermore, p can change the probability, and hence the credibility, of q and to different degrees when the premises are taken together with other propositions of different probabilities. Nodes in Pearl graphs are made to correspond to probabilities of propositions, and edges to relations between

---

[17] See Alon (2007, ch. 2) for a description of how the process might be represented, and Burian (2004) passim for an indication of its complexity.

[18] For discussion and examples of this approach, see Ross (forthcoming).

[19] This is analogous to studying credibility in what logical empiricists called the *context of justification*. It would take a time-indexed dynamical treatment to study the development of a claim or its credibility or to answer historical questions about sources of knowledge.

values of the nodes they connect. If there are exactly two nodes, X and Y, connected by a single directed edge running from the former to the latter, then, assuming agreement over how high a probability to require, Bayesian probability theory can decide whether the assignment of a value to X requires the machine to assign a high enough value to Y to warrant its inference. If X and Y are connected indirectly by nodes and edges in a linear chain, the value of each node can be calculated conditional on its immediate predecessor until one gets to Y. In more complicated (hence more realistic) cases, one or more of the nodes will be connected directly or indirectly to other nodes in a branching network with or without feedback loops. In all of these cases, conditional probabilities decide what it would be reasonable for a machine to infer.

Unfortunately, epistemic pathways are not susceptible to any such uniform treatment. For one thing, not all of the factors along typical epistemic pathways are propositions, and not all of them bear on credibility in the same way. To illustrate this, along with my reasons for thinking that node and edge epistemic pathway diagrams can be philosophically useful, here is a very rough description of part of an epistemic pathway leading from experimental results to the credibility of claims, C, about the existence of the Higgs boson.

Although a crucial pathway segment runs from particle collision data to C, the relevance and reliability of the data depend in part on the use of computer models and mathematical techniques to time collisions, calibrate detectors, and select data for storage. Rather than determining probabilities of propositions along the pathway, these factors help decide what data will be produced, the errors the data are liable to, and the extent of their reliability. The same holds for physical facts about the environment and the equipment. For example, the equipment must be buried deep enough to shield it from confounding solar radiation, the detectors must be physically such as to make them sensitive to the relevant particles, and the equipment must be in good working order. The contributions of factors like these are causal rather than inferential or computational. It is better to think of their epistemic significance on analogy to promoters and inhibitors in causal pathways than in terms of probabilistic relations among propositions in a Pearl network. In this case theories, models, and background knowledge contribute in too many ways to too many kinds of inference and derivation to capture in a node and edge diagram where every edge is interpreted in the same way.

The main reasons to look for Higgs bosons came from the need to solve difficulties in standard particle theory (Butterworth 2014, 105–132). But, according to that very theory, Higgs bosons carry no electrical charge and decay too rapidly to register on collision detectors. Fortunately, there was good theoretical reason to think that their decay should release signature arrays of detectable particles. Having chosen to look for instances of a signature pattern featuring electrons and muons, experimenters set up a magnetic inner tracker to deflect charged particles into telltale curves from which they can be identified. A second layer of detectors was lined with tabs that scintillate in response to hadronic particle jets traveling to them through the inner tracker. Calorimeter responses to heat from the scintillators provided indications of jet momenta. Muons pass through these first two layers to an outer layer of detectors that measure their momenta. This arrangement of detectors was used in ATLAS (an acronym for A Toroidal Large hadon collider ApparatuS) experiments.

Elaborate computations involving both empirical and virtual data, along with a variety of mathematical techniques and physics principles, convert ATLAS detector responses to proton collision biproducts products into data. The data are interpreted to reconstruct (i.e., to model) the streams of collision products that first entered the detection equipment. Given the standard theory, along with background knowledge, the data raised the probability of a Higgs boson signature event to a degree that made C credible. Clearly, the factors along the pathway, their interconnections, and their contributions to the credibility of C are not all of the same kind. Heterogeneity is the rule rather than the exception, not just here, but in many other cases.

An example of how much philosophers can learn a lot by considering individual epistemic pathways is the remarkable fact that ATLAS data support C as parts of epistemic pathways in which perceptual experience does little if any justificatory work. Rather than supplying evidence for the existence of Higgs bosons, perceptual experience figures most prominently in paths that lead from displays and reports of ATLAS data and pattern reconstructions to the credibility of perceptual beliefs about the experiment and its results. Apart from helping investigators access the data, and the reports of interpretations and conclusions drawn from it, the main epistemic significance of perceptual evidence derives from its use in running, monitoring, and adjusting the equipment, setting up future runs, and so on.

There are, of course, other cases in which perceptual evidence comes close to being as important as an empiricist would expect. Cajal's drawings are examples. Here, looking at an epistemic pathway from them to the credibility of his anatomical claims can help us appreciate the significance not only of what he saw, but also of features of his stains, his microscopes, the tissues he prepared, the methods he used to guide his drawing, and so on. Among other things, this can help us raise and answer questions about errors and their influence and about how changes in his methods might have affected the credibility of his claims.

8 CONCLUSION

Empiricism was quite right to emphasize the importance of experimental results in the development and evaluation of scientific claims. But it was wrong to think that most, let alone all, experimental results are records of perceptual experience. Ignoring equipment- and computer-generated data as they did, empiricists were also wrong to think, as Herschel put it, that

> [experience is] . . . the great, and indeed only ultimate source of our knowledge of nature and its laws. (Herschel 1966, 77)

Claims like this are correct only in the uninteresting sense that we use our senses to access data and conclusions drawn from it by looking at measurement displays, reading observation reports and published papers, listening to formal and informal verbal presentations, and so on. Finally, it was a mistake to think that scientific credibility can be modeled very informatively in terms of a two-term relation that connects every credible scientific claim to evidence that makes it worthy of belief.

Recognizing the shortcomings of empiricism opens the way to looking for new stories about when and how data and other factors contribute to credibility. I say "stories" because data are produced and in too many different ways to be captured informatively by any single general account. Philosophers who abandon empiricism can confront two major issues concerning the epistemic value of empirical and virtual data. The first has to do with the usefulness of data as indicators of the presence or the magnitude of the item they are used to measure. The second arises from the fact that data about one item are usually produced for use in answering questions about some further item—often one that does not register directly on sensory systems or experimental equipment.

Recall that fMRI images are often used to measure levels of neuronal activity for use in finding out about anatomical or functional patterns of activity that support specific cognitive functions. Examples of the first issue give rise to questions about how specific features of equipment calibration and use, statistical manipulations that are parts of fMRI data production, environmental factors, and the like bear on the resolution, precision, and accuracy of neuronal activity measures. An instance of the second issue arises from the fact that fMRI equipment, calibration, and use often varies from laboratory to laboratory and even from trial to trial in the same laboratory. As J.A. Sullivan points out, such differences raise vehemently interesting questions about which differences bear on the credibility of specific claims about connections between neuronal and cognitive processes and how strongly they affect their credibility.[20] For example, to control for anatomical or functional idiosyncrasies peculiar to specific subjects, neuroscientists often average fMRI data from a number of resting brains and compare the results to averaged data from the same or other brains engaged in a cognitive task. To allow investigators to compare results from different trials, individual and averaged results are projected onto a common atlas constructed from images of individual brains. Because various methods are used to construct atlases and project fMRI measures on to them, there can be appreciable discrepancies between their representations of anatomical landmarks or activity locations (Woods 1996, 333–339). It should be obvious that the epistemic significance of any given epistemically significant factor will depend in part on what is being investigated. For some purposes (e.g., studying fine-grained differences between right and left brain cognitive processing), discrepancies among averaging techniques, atlases, and projection methods can make a big difference. For others (e.g., studying resting neuronal activity in neuronal populations whose positions and functions don't vary much over healthy brains), the differences might not matter very much.

---

[20] Sullivan (2009).

A second example: fMRI data typically consist of grids divided into pixels marked by numbers or colors to represent levels of activity in voxels of neuronal tissue (e.g., 1 mm^3 in volume). Pixels seldom if ever map cleanly onto anatomically or functionally distinct brain regions. The voxels they represent may be large enough to contain neurons doing all sorts of things that aren't especially relevant to the cognitive function of interest. Alternatively, they may be too small to contain a population that supports it. If a cognitive function is supported by neuronal activities distributed over various parts of the brain, there is no reason why those activities should have to be carried out by neuronal populations that fit tidily within the boundaries of any selection of voxels. This raises questions about how to produce fMRI data that are most susceptible to informative agglomeration and interpretation. For instance, what size voxels would best serve the investigators' interests? How many images should be averaged, and how should the averaging be managed to avoid obscuring anatomically or functionally significant details? And so on (Ramsey et al., 2010).

I believe that philosophers who free themselves from the constraints imposed by epistemic anthropocentrism, perceptual ultimacy, epistemic purity, scope empiricism, and two-term confirmation will be in a far better position than empiricists to study and learn from issues like those described herein. And I believe this can shed new light on philosophically important features of scientific practice.

REFERENCES

Alon, U. (2007). *An Introduction to Systems Biology, Design Principles of Biological Circuits* (Boca Raton, FL: Chapman & Hall).

Aristotle. (1984). *Metaphysics*, In Aristotle, J. Barnes (ed.), *Complete Works of Aristotle*, vol. II (Princeton, NJ: Princeton University Press), 1552–1728.

Bacon, F. (1994). *Novum Organum*. P. Urbach, J. Gibson (eds., trans.). (Chicago: Open Court).

Biener, Z., and Schliesser, E. (2014) "Introduction." In Z. Biener and E. Schliesser (eds.), *Newton and Empiricism* (Oxford: Oxford University Press), 1-12.

Bogen, J. (2011). "'Saving the Phenomena' and Saving the Phenomena." *Synthese* 182: 7–22.

Bogen, J. and Woodward, J. (1988) "Saving the Phenomwna." *Philosophical Review* XCVII(3): 303–352.

Bogen, J and Woodward, J., (2005). "Evading the IRS." In M. B. Jones and N. Cartwright (eds.), *Idealization XII: Correcting the Model Idealization and Abstraction in the Sciences* (New York: Rodopi), 233–267.

Burian, R. (2004). "Molecular Epigenesis, Molecular Pleiotropy, and Molecular Gene Definitions." *History and Philosophy of Life Sciences*, 26: 59-80.

Butterworth, J. (2014). *Smashing Physics* (London: Headline).

Campbell, N. (1952). *What Is Science?* (New York: Dover).

Carnap, R. (1995). *Unity of Science*. M. Black (trans.). (Bristol: Theoemmes).

Cohen, M. S. (1996). "Rapid MRI and Functional Applications." In A. W. Toga and J. C. Mazziotta (eds.), *Brain Mapping: The Methods* (New York: Academic Press), 223–252.

DeFelipe, J., and Jones, E. G. (1988). *Cajal on the Cerebral Cortex* (Oxford: Oxford University Press).

Earman, J., and Glymour, C. (1980). "Relativity and Eclipses." *Historical Studies in the Physical Sciences*, 11(1): 41–89.

Feest, Uljana. (2005). "Operationism in Psychology: What the Debate Is About, What the Debate Should Be About." *Journal of History of Behavioral*, 41(2): 131–149.

Feyerabend, P. K. (1985a). "An Attempt at a Realistic Interpretation of Experience." In P. K Feyerabend (ed.), *Philosophical Papers*, vol. 1 (Cambridge: Cambridge University Press), 17–36.

Feyerabend, P. K. (1985b). "Science Without Experience." In P. K. Feyerabend (ed.), *Philosophical Papers*, vol. 1 (Cambridge: Cambridge University Press), 132–136.

Gilbert, W. (1991). *De Magnete*. P. F. Mottelay (trans.). (Mineola, NY: Dover).

Gupta, A. (2006). *Empiricism and Experience* (Oxford: Oxford University Press).

Herschel, J. F. W. (1966). *Preliminary Discourse on the Study of Natural Philosophy* (New York: Johnson Reprint).

Humphreys, P. (2004). *Extending Ourselves: Computational Science, Empiricism, and Scientific Method* (Oxford: Oxford University Press).

Kant, I. (1998). *Critique of Pure Reason*. P. L. Guyer and A. E. Wood (trans.). (Cambridge: Cambridge University Press).

Kuhn, T. (1996). *Structure of Scientific Revolutions*, 3rd ed. (Chicago: University of Chicago Press).

Lazar, N, A., Eddy, W. F., Genovese, C. R., and Welling, J. (2001). "Statistical Issues for fMRI Brain Imaging." *International Statistical Review* 69(1): 105–127.

Leibniz, G. W. (1949). *New Essays on Human Understanding*. A. G. Langley (trans.). (La Salle, IL: Open Court).

Locke, J. (1988). *An Essay Concerning Human Understanding*. P. H. Niddich (ed.). (Oxford: Oxford University Press).

Mill, J. S. (1967). *A System of Logic Ratiocinative and Inductive* (London: Longmans, Green).

Morrison, M. (2015). *Reconstructing Reality: Models, Mathematics, and Simulations* (Oxford: Oxford University Press).

Neurath, O. (1983a). "Protocol Statements." In O. Neurath (ed.), *Philosophical Papers* (Dordrecht: D. Reidel), 91–99.

Neurath, O. (1983b). "Radical Physicalism and the 'Real World.'" In O. Neurath (ed.), *Philosophical Papers* (Dordrecht: D. Reidel), 100–114.

Parker, W. (2010). "An Instrument for What? Digital Computers, Simulation and Scientific Practice". *Spontaneous Generations: A Journal for the History and Philosophy of Science* 4(1). Online http://spontaneousgenerations.library.utoronto.ca/index.php/SpontaneousGenerations/article/view/13765/11198pontaneousgenerations.library.utoronto.ca/index.php/SpontaneousGenerations/article/view/13765/11198.

Pearl, J. (1988). *Probabilistic Reasoning in Intelligent Systems, Networks of Plausible Inference*, revised 2nd printing (San Francisco: Morgan Kaufmann).

Polanyi, M. (1958). *Personal Knowledge* (Chicago: University of Chicago Press).

Ramsey, J. D., Hanson, S. J., Hanson, C., Halchenko, Y. O. Poldrack, R. A., and Glymour, C. (2010) "Six Problems for Causal Infernce from fMRI". *NeuroImage*, 49(2): 1545–1558.

Ross, L. (forthcoming). "Causal complexity in psychiatric genetics."

Roush, S. (forthcoming). The Epistemic Superiority of Experiment to Simulation. Online https://philosophy.berkeley.edu/file/903/Epistemic_superiority_of_experiment_to_simulation_-_2014-_paper_formatted_blinded.pdf

Sextus Empiricus. (1961). *Outlines of Pyrrhonism*. R. G. Bury (trans.). (Cambridge, MA: Harvard University Press).

Schaffner, K. F. (1993). *Discovery and Explanation in Biology and Medicine*. (Chicago: University of Chicago Press).

Schlick, M. (1959). "The Foundation of Knowledge." In A. J. Ayer (ed.), *Logical Positivism* (Glencoe, IL: Free Press) 209-227.

Sporns, O. (2011). *Networks in the Brain* (Cambridge, MA: MIT Press).

Sullivan, J. A. (2009). "The Multiplicity of Experimental Protocols: A Challenge to Reductionist and Non-reductionist Models of the Unity of Neuroscience." *Synthese* 167(3): 511–539.

Van Fraassen, B. C. (1980). *The Scientific Image* (Oxford: Oxford University Press).

Van Mulders, P. (2010). *Calibration of the Jet Energy Scale Using Top Quark Events at The LHC*. Doctoral dissertation: http://www.iihe.ac.be/publications/thesis_petraVanMulders.pdf

Winsberg, E. (2003). "Simulated Experiments: Methodology for a Virtual World." *Philosophy of Science* 70(1): 105–125.

Woods, R. P. (1996). "Correlation of Brain Structure and Function." In A. W. Toga and J. C. Mazziota (eds.), *Brain Mapping: The Methods* (San Diego: Academic Press) 313–339.

CHAPTER 38

....

MECHANISMS AND MECHANICAL PHILOSOPHY

....

STUART GLENNAN

1 INTRODUCTION

....

"MECHANISM" is one of the most widely used words in the scientist's lexicon. A casual review of journals like *Science* or *Nature* will show that the word appears in many article titles and abstracts. Mechanism talk is not confined to one area of science, but shows up in the whole range of natural, life, and social sciences, as well as in applied fields like engineering (of course) and medicine. But, despite this fact, the nature of mechanisms and mechanistic thinking was not until recently a major topic of discussion among philosophers of science. This has changed since the late 1990s, with a rapidly growing literature which seeks to understand the nature of mechanisms and the role of mechanistic thinking in science.

Philosophers have contemplated mechanisms in connection with two different kinds of projects. The first is broadly metaphysical, investigating the nature of mechanisms as an ontological category and exploring how mechanisms may be related to or explain other ontological categories like laws, causes, objects, processes, properties, and levels of organization. The second is methodological and epistemological, exploring how mechanistic concepts and representations are used to explore, describe, and explain natural phenomena. Opinions are divided as to how these two projects are related. Some philosophers see mechanisms as part of the deep structure of the world and ground the methodological importance of mechanistic thinking in this fact. Others have a more instrumentalist view in which mechanistic thinking is an important part of science but is one that is deeply pragmatically tinged.

Section 2 of this chapter offers some remarks about the recent history of thinking about mechanisms, placing it within the broader history of the philosophy of science. Section 3 offers a general account of the nature of mechanisms. Section 4 explores how mechanisms are discovered and represented and how these representations are used in mechanistic explanation. Finally, Section 5 discusses connections between mechanisms and causation.

2 The Emergence of the New Mechanical Philosophy

Much of the recent philosophical discussion of mechanisms began with three publications. The first, Bechtel and Richardson's *Discovering Complexity: Decomposition and Localization as Strategies in Scientific Research* (1993) offered a theory of mechanistic discovery and explanation. The second, Glennan's "Mechanisms and the Nature of Causation" (1996) offered the first explicit characterization in the contemporary philosophy of science literature of what a mechanism was and argued that mechanisms were the truth-makers for both law statements and causal claims. The third, Machamer, Darden, and Craver's "Thinking about Mechanisms" (2000), raised some objections to the first two works but, more importantly, offered a kind of mechanist's manifesto, making a case for the centrality of mechanisms to any adequate account of the natural world and our scientific investigation of it. While not all would agree with their conclusions, MDC (as both the paper and its authors have come to be known) were remarkably successful in their programmatic ambitions. Their paper rapidly became one of the most widely cited papers in the philosophy of science (and, indeed, philosophy generally). MDC, as well as Bechtel, Glennan, Richardson, and some of their collaborators and students, have come to be called "the new mechanists" (or sometimes, either fondly or pejoratively, "mechanistas"). Although the new mechanical philosophy or new mechanism largely started in the philosophy of biology, its reach has extended across the subfields of philosophy of science, and work by new mechanists is often cited in methodological literature by natural and social scientists of a wide variety of stripes.

MDC and the other new mechanists are responsible for spurring recent discussion of mechanisms, but it would be a mistake to see the ideas behind the new mechanism as all that new. The new mechanical philosophy owes a lot to earlier mechanical philosophies, including, of course, the mechanical/corpuscular philosophy of the seventeenth century. But, more importantly, the new mechanical philosophy is part of a broader set of developments within philosophy of science over the past fifty years: developments all in some way connected with the rejection of the legacy of logical empiricism.

One source of the new mechanism lies in the emergence of philosophy of biology in the 1970s, and particularly in the work of William Wimsatt.[1] Wimsatt and other philosophers interested in biology argued that logical empiricist accounts of scientific theories, laws, explanation, intertheoretic reduction, and the like might make sense in the context of physics but were not adequate to characterizing biological

---

[1] Many of Wimsatt's important essays from the 1970s onward are republished in Wimsatt (2007).

phenomena or the activities of biologists. Wimsatt, drawing on work of Herbert Simon (1996), was particularly interested in part–whole relations and levels of organization within complex systems (like cells, organisms, or ecosystems). His emphasis shifted from a logical empiricist approach focusing on logical relations between statements of theory to focusing on causal and compositional relationships between things in the natural world.

A second important strand is the critique of the logical empiricist conception of law and, with it, covering law models of explanation. Cartwright (1983) famously argued that the laws of physics lie, in the sense that the deep explanatory laws of physics don't describe the behavior of actual systems. Cartwright developed an anti-Humean ontological view that emphasizes singular causal powers or capacities over laws. Cartwright, like Wimsatt, has argued that scientific explanation is not about logical relations between claims but involves models that are heuristic and idealized.

Finally, the new mechanism owes much to Wesley Salmon's ideas on causation and causal explanation. After many years of trying to refine covering law and statistical relevance approaches to scientific explanation, Salmon gave up on the covering law tradition and argued for an ontic conception of scientific explanation in which to give an explanation was to situate some phenomenon within the causal structure of the world (Salmon 1984). Salmon's causal/mechanical model of explanation suggested that to explain something required showing how it arose through the propagation of causal influence via a network of singular causal processes and interactions. Salmon saw this move as a return to "the mechanical philosophy."

The new mechanical philosophy is connected to these and other strands within naturalistic post-positivist philosophy of science. We shall elaborate on some of these connections in the sections below.

3   What Is a Mechanism?

The first task for a mechanical philosophy is to say what is and what is not a mechanism. This is not an easy task; the things that scientists, technologists, and ordinary folk have called mechanisms are a diverse lot, and it is not clear whether the term in its various applications refers to a single kind of thing. Perhaps mechanism is a family resemblance concept with no essential defining characteristics. Also, it is unclear what the task of the analysis is. Is the aim semantic—to characterize what is meant by the term "mechanism"—or ontological—to characterize the properties of a kind of thing that exists in the world?

Within the new mechanist literature there have been a number of proposals for how exactly to characterize mechanisms, the most discussed being Bechtel and Abrahamsen (2005), Glennan (1996, 2002), and Machamer et al. (2000). Although there remain disagreements, there are a lot of common features in these definitions, and subsequent discussions suggest that there is considerable consensus at least on a set of necessary conditions for something to count as a mechanism. These can be captured in a characterization I call *minimal mechanism*:

> A mechanism for a phenomenon consists of entities (or parts) whose activities and interactions are organized so as to be responsible for the phenomenon.[2]

Minimal mechanism is an ontological characterization of mechanisms as things in the world rather than an analysis of the meaning of "mechanism." It is minimal in two related senses. First, it is a highly abstract and permissive definition on which many things count as mechanisms. Second, it represents a consensus on minimal (i.e., necessary) conditions for something to be a mechanism. Most of the disagreements about how to characterize mechanisms concern whether further conditions are required. We shall return shortly to the question of the proper scope of mechanicism and how mechanism definitions might be further restricted, but, before doing this, it will be helpful to unpack the various parts of the definition.

To start, a mechanism consists of entities—the mechanism's parts. In the case of mechanical devices of human construction, what is meant by a mechanism's parts is pretty obvious. For a car engine, it includes things like the pistons, the cylinders, and the spark plugs. Living systems also have parts—like the bones in our skeletons or the organelles in a eukaryotic cell. Naturally occurring but nonliving physical systems can also have parts, like the chambers, vents, and magma within a volcano.

The entities that are parts of mechanisms do things. These are their activities. Spark plugs spark, igniting fuel in the chamber, and the expansion of gases from the ignition pushes the piston. Connected bones rotate at joints. When one part acts upon or with another, it is an *interaction*, and these interactions produce changes to these parts.[3]

Mechanisms are responsible for phenomena. This means that mechanisms are always "for" something, and they are identified by what they are for. An engine is a mechanism for turning a drive shaft. Ribosomes are mechanisms for translating bits of messenger RNA into proteins. Volcanoes are mechanisms for ejecting smoke, ash, and lava from their interior. This last example shows the minimal sense in which mechanisms can be said to be for some phenomena. To say that a mechanism is for something is not to imply that it was designed for some purpose or that it is an adaptation that arose via natural selection. It is simply to identify the phenomena for which the mechanism is causally

---

[2] This characterization is taken from Glennan (forthcoming). Minimal mechanism closely resembles a proposal from Illari and Williamson (2012). That paper provides a comprehensive survey of the debates among the new mechanists about how to best define mechanisms.

[3] Within the literature, there has been some controversy about whether "activity" or "interaction" is the best word to characterize the causal relationships in mechanisms, but I do not think that this terminological dispute is of much significance. As I am construing it, "activity" is the more general term, while "interactions" are a class of relational activities that produce change. So, for instance, when a heart contracts, it is an activity, but it is also an interaction with the blood, as the contraction discharges blood from the heart's chambers. Some activities, like daydreaming, may be "solo" activities that change the entity that is acting but don't (immediately at least) affect other entities in its environment.

responsible. Human-built mechanisms certainly are designed to produce the phenomena they do, and many naturally occurring phenomena are the product of mechanisms that were produced by natural selection; but these represent only a subset of mechanisms, and many mechanisms are for phenomena only in the minimal sense.

Mechanisms are said to be responsible for their phenomena, and there are a variety of ways in which this can be so. Craver and Darden (2013, p. 65) have suggested a three-fold classification of the possibilities, represented schematically in Figure 38.1, according to which mechanisms either produce, underlie, or maintain their phenomena. When a mechanism produces a phenomenon, there is a chain or network of interacting entities leading from some initial or input state to a final or outcome state—for instance, when flipping a switch turns on a light or when turning on that light produces a chain of neural firings originating in the eye. When a mechanism underlies a phenomenon, there is some capacity or behavior of a whole system that arises out of the activities and interactions of a system's parts. For instance, a neuron's action potential—the rapid depolarization and repolarization of the cell membrane that travels down the cell's axon—is a phenomenon produced by the whole cell (situated in its environment) in virtue of the operation of the cell's parts (e.g., membranes, ions, and ion channels). A related example is muscle contraction. A muscle is composed of bundles of muscle fibers, which are individual cells (myocytes). The contraction of a muscle is not strictly produced by the contraction of the fibers; instead, the contraction of the fibers *underlies* the contraction of the tissues because the contraction of the tissue just is the collective and coordinated contraction of the fibers. Mechanisms that maintain phenomena are mechanisms that keep some properties of or relationships within a system stable in the face of changing

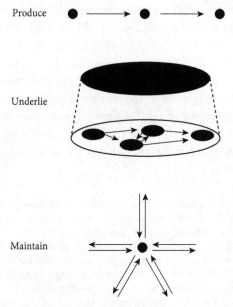

FIGURE 38.1 Relations between Mechanisms and Phenomena.

(Craver and Darden 2013, p. 66)

internal or external conditions. Regulatory mechanisms (as mechanisms of this kind are often called) are a ubiquitous feature of living systems. To name a few examples, cells have mechanisms to maintain concentrations of metabolites, cardiovascular systems have mechanisms to maintain stable blood pressure, or warm-blooded animals have mechanisms to maintain constant body temperature. Many machines and mechanical systems (e.g., heating and cooling systems, steam engines with governors) also have this character.

These three kinds of mechanism–phenomenon relationships highlight an important duality in mechanism talk. On the one hand, mechanisms can be characterized as processes with various stages connecting initial and termination conditions. On the other hand, mechanisms can be characterized as systems—stable collections of parts or components organized in such a way as to have persistent states and dispositions. There is an intimate relationship between these two senses of the term "mechanism" because mechanical systems are organized so as to engage in mechanical processes. A cell is a mechanical system consisting of organized collections of parts that underlie the cell's behavior and maintains its properties in light of environmental fluctuations. At the same time, the way that the cell behaves, including the way that it maintains its stable properties, is through the activities and interactions of its parts. In short, the cell is a mechanical system in which mechanical processes produce, underlie, and maintain the cell's properties and behaviors.

The constitutive and temporal relationships between parts and activities within mechanisms are the most abstract features of mechanical organization—but the actual behavior of mechanisms depends upon much richer organizational features. This is obvious if one considers machines like cars or computers. These machines have many parts, but for them to function as cars and computers, the parts must be connected to each other in a very specific way. The activities of and interactions between these parts must be carefully timed and orchestrated if the mechanism is to exhibit its characteristic phenomena. A pile of car parts is not a car.

The varieties of organization in mechanisms are practically limitless, and there is no easy way to sort out their variety, but one further example—this time from economics—will illustrate some possibilities.[4] Stock markets are mechanisms for trading shares in corporations. In the stock markets, the most important entities are traders, shares, and money, and the primary interaction is trading (i.e., buying and selling). Traders have shares and money and may offer to buy shares with money or sell shares for money. A buyer will buy from the lowest bidder, while a seller will sell to the highest bidder. A more fine-grained description of this mechanism will fill in details about how orders are posted and how buyers are matched with sellers.

Market mechanisms illustrate several important points about mechanistic organization. First of all, the primary mode of organization is causal: the key question is what interacts with what, what properties are changed by those interactions, and when. Other

---

[4] See Glennan (forthcoming) for an attempt at a taxonomy of kinds of organization.

forms of organization (like spatial organization) matter, but only insofar as they affect causal organization. Historically, stock markets needed to have trading floors where traders could see and hear other traders in order to make sales. Now buy and sell orders are managed by computers, so traders can participate from anywhere they have access to the Internet and appropriate software. Second, the stock market shows how much timing matters. Stock prices change in fractions of a second, and how one does in the stock market depends crucially upon the timing of the trade. Third, and related, stock markets involve feedback loops. Investor behavior determines price and volume in trading, and these variables in turn affect investor behavior. Feedback loops are central to explaining market volatility.

A ubiquitous and important aspect of mechanistic organization is its hierarchical character. The parts of mechanisms can themselves be broken into parts, and the activities within mechanisms can be broken down into further activities. Consider again the stock market: a purchase of shares is an interaction between two parts of the market mechanism, a buyer and a seller; but the buyer and seller, whether they be human traders or computers, are themselves decomposable into parts, and it is the organized activities and interactions of the parts of these buyers and sellers that explain their behaviors and dispositions. Similarly, the trade is an interaction that can be decomposed into a series of finer-grained activities—the various verbal, written, or electronic steps in incrementing or decrementing cash and share balances in buyer and seller accounts.

Mechanistic analysis will typically bottom out in some set of entities and activities that are taken to be basic (Machamer et al. 2000). For instance, microeconomists attempt to explain changes in consumption of goods and services within an economy through the theory of consumer choice, whereby consumers are taken to be the bottom-out entities, and their basic activities are choosing bundles of goods and services that maximize their utility relative to their budget constraints. What counts as a bottom-out entity or activity is domain-relative and methodologically motivated, mostly representing a division of labor between scientific disciplines. Whereas the consumer and her utility function are taken as basic in microeconomics, it is understood that consumer behavior is a psychologically complex affair that itself will depend upon psychological mechanisms within the consumer, as well as the social environment in which the consumer is embedded.

In closing, I should make explicit one final point about the ontological status of mechanisms that follows from the preceding characterization: mechanisms are particulars. They consist of entities and activities, which are particular things that are located somewhere in space and time. My heart, for instance is a mechanism for pumping blood that sits in my chest. It came into being during a particular period of time as a result of fetal growth, and, ultimately, it will cease to be. Even though science is generally concerned not with particular mechanisms but with mechanism kinds—for instance, human hearts or hearts generally, as opposed to my heart—what exists in nature are the particular mechanisms. As a consequence of this particularity, it will almost always be the case that the instances of any mechanism kind will be in some respect heterogeneous—for instance, among human hearts there will always be variations.

4 REPRESENTATION, EXPLANATION, AND DISCOVERY

The characterization of minimal mechanism is an ontological one: it supposes that mechanisms, and the entities and activities of which they are made, are things in the world that exist independently of mind and theory. But even if these ontological suppositions are correct (and the supposition can be challenged), discovering, describing, explaining, and controlling mechanisms are scientific activities that necessarily involve representation.

The ontological account of the new mechanists leads naturally to a critique of traditional accounts of scientific theory. What has often been called "the received view of scientific theories" holds that scientific theories should be constructed as sets of general laws. These laws, in conjunction with statements of particular conditions, should deductively entail observations that can be used to test the theory. In the 1970s and 1980s, the received view was largely supplanted by the "semantic view of theories," which saw theories not as consisting of statements, but rather as classes of models that satisfy theory specifications. This is not the place to discuss the challenges to these views or the relationship between them, but suffice it to say that both views see theories as ways to represent knowledge about regularities in the world, and both have a highly formal (either syntactic or semantic) conception of scientific theory.

New mechanists take issue with the received and semantic views on two grounds. First, they are, as we have noted, generally skeptical of the existence—or at least importance—of laws, seeing scientific generalizations as grounded in mechanisms. Second, they are suspicious of formalized accounts of scientific theories in general, holding that to understand the nature of scientific theory philosophers must pay closer attention to the content of scientific theories and the actual means by which scientists express them.[5]

The new mechanists' analysis of scientific representation in fact largely eschews the category of theory. This is in part because of concerns about philosophical accounts of scientific theory; but, more importantly, it stems from the new mechanist's suspicion that the things in science that typically go by the name of theory are not the central vehicles for scientific representation. New mechanists instead focus on models, and particularly models of mechanisms.[6]

Usage is not uniform among scientists or philosophers regarding the terms "model" and "theory," but, typically, theories are seen as more abstract and general, whereas models are seen as more localized—mediators between theories and particular systems or processes in nature. So, for instance, classical mechanics is a general theory of forces that gives rules for composition of forces and relates them to motion of bodies. Using

---

[5] See Craver (2002) for an introductory discussion of the received and semantic views, as well as for a sketch of the mechanist's alternative.

[6] The new mechanist literature on modeling and the views I express here complement and draw upon a rich literature on modeling. See, e.g., Giere (2004), Morgan and Morrison (1999), Teller (2001), Weisberg (2013), and Wimsatt (2007).

mechanics, one can build models of particular mechanical systems, like the solar system or objects on inclined planes.

Ronald Giere characterizes the general relationships among models, modelers, and what is modeled using the following schema:

S uses X to represent W for purposes P.

S represents a scientist or scientific community, X a model, W what is modeled (the target), and P the modeler's purpose (Giere 2004, p. 743). In Giere's account, theories— what he calls "principles"—are resources for model building. Giere's schema brings out several important features of models. First, nothing is intrinsically a model but only becomes a model if it is constructed or used by a modeler to represent something. It might turn out, for instance, that a set of channels and indentations in a rock bore a striking topological resemblance to the London Underground, but that marvelous accident would not make that rock a model (or representation or map) of the London Underground unless someone chiseled the rock for this purpose or at least observed the improbable correspondence and decided to put it to use.

Second, models are constructed to serve the modeler's purpose, and different purposes will lead to different models. Climate and weather models provide an excellent example. Some models represent short-term and geographically localized phenomena (like local temperature or precipitation); others predict the likelihood of severe weather events like tornadoes or hurricanes. Others are focused on longer term patterns like climate change. Some models are principally concerned with prediction of measurable phenomena, while others are designed to explore the causes of those phenomena.

The general relationships among models, modelers, and targets are represented in Figure 38.2. That diagram suggests another important feature of the model target relation, which is that, typically, a single model is used to represent a number of targets. Take, for instance, models of the cell. Biologists construct models of cells that are increasingly specific from models of the most general classes of cells (prokaryotes, eukaryotes) to more specific cells—sex cells, muscle cells, stem cells, blood cells, and the like. But even models of specific kinds of cells will have many targets. For instance, there are many millions of red blood cells within a single human body.

My remarks so far apply to scientific models generally, but not all scientific models are models of mechanisms. What makes a model a mechanical model is that it represents how the entities act and interact in order to be responsible for the mechanism's phenomena. For this reason, we can characterize a mechanical model as having two parts:

> A mechanical model consists of (1) a description of the mechanism's phenomenon (the phenomenal description) and (2) a description of the mechanism that accounts for that phenomenon (the mechanical description).[7]

---

[7] This definition is taken from Glennan (2005), except that I have substituted "phenomenon" and "phenomenal" for "behavior" and "behavioral." I take these terms to be synonymous but have adopted "phenomenon" because it has become more standard within the mechanisms literature.

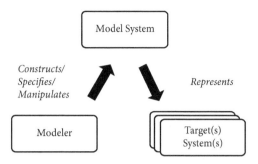

FIGURE 38.2 Modelers, Models, and Targets.

Mechanistic models are typically contrasted with phenomenal models, which characterize a phenomenon without characterizing the mechanism responsible for it.

One of the purposes to which mechanistic models are put is explanation. Scientists use models to explain the mechanisms responsible for certain phenomena. There is, however, some dispute among new mechanists as to whether models (including things like equations, diagrams, and simulations) can themselves be said to be explanations of phenomena. According to the ontic conception of explanation (Craver 2007, 2013; Salmon 1984), it is best to think of explanations not as kinds of representations, but as things in the world. For instance, we might say that sun spots are explained by magnetic fields in the photosphere that reduce convection currents and decrease surface temperature. The sun spots are things in the world, and they are explained by a mechanism that consists of other things in the world. In contrast to this, Bechtel and collaborators (e.g., Bechtel and Abrahamsen 2005) have argued for an epistemic conception in which the models—not the things in the world—are the explanations. For instance, we might say that the Price equation explains the relationship between individual and group selection.

As far as common usage goes, it seems undeniable that we use the word "explanation" in both the ontic and the epistemic sense, but the real philosophical dispute concerns whether the most philosophically important questions are about ontic relationships between mechanisms and the phenomena they are responsible for or about the epistemic characteristics of representations of these mechanisms. Here, I would suggest the middle path advised by Illari (2013), which argues that we should change the debate from the question of whether explanations are ontic or epistemic and instead recognize that good explanations are representations that satisfy both ontic and epistemic norms. Good explanations must refer in appropriate ways to mechanisms in the world, and, at the same time, those representations must have certain features that allow them to elicit understanding.

Much of the discussion of modeling in the mechanisms literature has been associated with accounts of mechanism discovery (Bechtel and Richardson 1993; Craver and Darden 2013). Mechanism discovery is best understood as the process of developing, testing, and accepting, rejecting, or revising models of mechanisms. One of the most influential accounts of this process makes its initial appearance in MDC 2000 and has subsequently been developed in detail by Craver and Darden.

Craver and Darden typically use the term "schema" in place of "model." They define a mechanism schema to be "an abstract description of a type of mechanism" (Machamer et al. 2000, p. 15). They emphasize that abstraction comes in various degrees. To call a description "abstract" is simply to say that it describes only certain features of a thing and leaves out others; all descriptions of mechanisms (or of anything else for that matter) will be abstract to some degree or other. The simplest and most abstract of these schemas they call *mechanism sketches*.

Craver, in his 2007 book, suggests that the process of discovering mechanisms (or mechanistic explanations) proceeds along two axes—from how-possibly models to how actually-models, and from mechanism sketches to ideally complete mechanism schemas. For Craver, these axes are not independent. A mechanism sketch is, in virtue of its sketchiness, only a how-possibly model. Epistemic warrant for a sketch (or model) is accumulated as the sketch is filled in. The more one is able to establish the entities and activities involved in the production of the phenomena, the more confidence one can have that one's model describes how the mechanism actually works.

Craver and Darden's account of mechanism discovery begins with the identification of the phenomenon. All mechanisms are mechanisms for some phenomenon, and without some characterization of the phenomenon, one has not identified a mechanism to search for. (That description might also be called a *phenomenal model*). Once the phenomenon is identified, scientists propose mechanism sketches that seem possible or plausible in light of their existing knowledge of the entities and activities that exist within the domain.

A classic example of such a sketch is Crick's so-called central dogma of molecular biology: "DNA makes RNA which makes protein." The phenomenon in question is the production of proteins in the cell, and the central dogma offers the most basic sketch of the entities and activities involved in the production of proteins. The initial formulation of the central dogma was plausible in light of what Watson and Crick had learned about the structure of DNA and RNA, but further confirmation of the dogma, and, with it, a more complete explanation of protein synthesis, requires the filling in of the specific entities and activities and the details of the timing, location, and other forms of organization that actually determine how DNA in the cell is used to produce specific proteins. As scientists attempt to fill things in, initial descriptions of the phenomena and mechanisms sketches often need to be revised. Developments in molecular genetics illustrate this point. Explorations of the mechanisms of protein synthesis have demonstrated that the central dogma is not just incomplete but in some respects misleading or false. It is now much clearer than it was originally that much DNA does not code for protein but in fact is part of the apparatus of gene regulation, where there are many feedback loops (e.g., in the production of noncoding RNA, which is implicated in regulating the rate of transcription within coding regions).

Craver's formulation of the process of model development and mechanism discovery stems in part from his concern with "boxology" in the functional analysis of cognitive mechanisms. Craver sees the identification of mechanisms as providing normative constraints on the acceptability of functional explanations. Following Cummins (1975), Craver takes functional explanation to involve explaining the capacities of some system in terms of a set of subsidiary capacities (Craver 2001, 2006). A functional analysis can be interpreted as a mechanism sketch in which the exercise of these capacities are the mechanism's activities, but where the entities responsible for them are left unspecified (Piccinini and Craver 2011). Without knowledge of the entities that are responsible for these capacities, the epistemic status of the functional analysis is in doubt. It represents a how-possibly or at best how-plausibly model. Filling in the black boxes will validate the functional analysis and will reveal constraints imposed by the actual mechanism.

Craver and Darden's account of mechanism discovery captures important aspects of the practice of developing mechanistic models, but critics have raised doubts about whether the development of models should always proceed in this way (Gervais and Weber 2013; Levy and Bechtel 2013). Craver and Darden's approach seems to suggest that more is always better. Epistemic progress is made by moving from sketch to schema and ultimately toward an ideally complete model.

But is this always progress? Probably not. There are several issues here: one is whether filling in the details always increases the epistemic warrant for a given mechanism schema. Although no doubt it often does increase warrant, sometimes filling in details of abstract models may actually decrease the plausibility of the model. At any rate, it is possible to gain a high degree of confidence in a model of a mechanism where not all components are localized and some details of realization are left out (Weiskopf 2011). This can be done by experimental manipulation of the target system, as well as by simulations (Bechtel and Abrahamsen 2010; Glennan 2005).

A second issue concerns whether filling in the gaps always improves the explanatory power of a model. One of the sources of explanatory power is in separating the wheat from the chaff—the factors that make a difference from the irrelevant details. Especially in connection with dynamical mechanisms, the argument can be made that many details of timing, state, or initial conditions may be irrelevant to overall systemic outcomes (Kuhlmann 2011). In such cases, more minimal models may in fact be more explanatory. And even if a richer or more concrete model does offer a more complete explanation of a particular system or mechanism, it may do so at the expense of scope. Abstraction, as well as idealization, typically allows a single model to cover a broader range of target phenomena, thus achieving the explanatory aim of unification.

Most generally, the image of the complete model as explanatory ideal fails to acknowledge the inevitable tradeoffs in model building. Models are, as has already been emphasized, used by modelers for different purposes, and, depending upon the purpose, one will choose different models—trading, for instance, generality for realistic detail about mechanisms or for precision of prediction with regard to measurable variables (Levins 1966; Weisberg 2007).

5 Mechanisms and Causation

The concept of mechanism is tightly intertwined with the concept of cause, and our definition of mechanism invokes causal concepts: parts act and interact, mechanisms are responsible for phenomena in a manner that is at least in part causal, and mechanistic explanation is a form of causal explanation. But these observations leave open the question of just how to understand the relationship between mechanisms and causation. Must an account of mechanism be supplemented with an account of causation, and, if so, what accounts of causation is it compatible with? Alternatively, are mechanisms key to understanding the nature of causation?

Advocates of the new mechanism have been divided on this question. I have argued for the possibility of a mechanical theory of causation (Glennan 1996, 2010, 2011), and MDC (along with Jim Bogen) heralded their account of activities as a revisionary theory of the nature of causation (Bogen 2008; Machamer 2004; Machamer et al. 2000). Craver (2007), however, has expressed reservations about these approaches, and Bechtel has avoided the issue, seeing mechanism primarily as an epistemic and explanatory construct.[8]

In this section, I will undertake to sketch a mechanical theory of causation, a theory that takes mechanisms to be the truth-makers for causal claims, and address some of the objections that have been raised to such an approach.[9] To begin, it will be helpful to situate the mechanical theory in the landscape of the literature on causation.

Approaches to causation are sometimes divided into two broad categories: *productivity approaches* suggest that causation is ultimately a singular and intrinsic relation among objects, processes, or events, whereas *difference-making* (or causal relevance) *approaches* suggest that causation is an extrinsic and contrastive relation. The intuition is that for a causal relation to be productive, it must *connect* causes to their effects; whereas, in difference-making relations, causes must *make a difference* to their effects. The distinction can be illustrated by contrasting two widely discussed kinds of causal relationships—causation by omission and causal overdetermination. In causation by omission, an omission (like a failure to put out the campfire) may cause an effect (like a forest fire). The omission makes a difference to the effect, but it can't really be connected to the effect because the omission is not an actual action or event. In causal overdetermination, there are redundant causes, as, for instance, when multiple soldiers in a firing squad shoot a prisoner. The shots are all connected to the prisoner (and the death), but no single shot makes a difference to the death.

Mechanical theories of causation fit within the production tradition: the parts of mechanisms are connected to each other via interactions that exhibit what MDC

---

[8] Bechtel's views are becoming more explicit in some recent unpublished talks about what he is calling "West Coast idealism."

[9] This sketch is a précis of sections of Glennan (forthcoming).

call "productive continuity." But within the mechanical tradition, there are different ways of spelling out the nature of production. Wesley Salmon's (1984) account of causality was an important progenitor of the new mechanist approach, so it would be helpful to begin with a sketch of his account. Salmon argued that a proper understanding of causation required the elucidation of two concepts: production and propagation. If objects interact (like a bat hitting a ball), that interaction will produce changes in the state of the interacting objects, introducing what Salmon called a *mark*. This is production. But for causal influence to be transmitted across space and time, the effects of this marking must be propagated by a continuous causal process. In this case, the path of the ball through the air, carrying its momentum, is that causal process. The image Salmon offers us of the causal structure of the world is of an immense array of causal processes—world lines of objects as they travel through space-time—which, when they cross, produce changes in each other by mark-inducing interactions.

Salmon sees his account as an expression of "the mechanical philosophy." He writes:

> The theory here proposed appeals to causal forks[10] and causal processes; these are, if I am right, the mechanisms of causal production and causal propagation in the universe. (Salmon 1984, p. 239)

Although Salmon's examples of causal processes sometimes veer into the everyday, it is clear that his ultimate target is physical causation (i.e., the sort of causation that occurs in the domain described by physics). All causal processes and interactions depend ultimately on what he calls "fundamental causal mechanisms," of which he thinks there are "a small number" that are subject to "extremely comprehensive laws" (p. 276). Salmon has in mind here physical interactions involving gravitation, electricity and magnetism, nuclear forces, and the like. In subsequent work, in fact, Salmon moved away from his mark transmission criteria and adopted a version of Dowe's (2000) *conserved quantity theory*. According to this theory, causation is a physical relation in which there is an exchange of a conserved quantity like energy or momentum.

What is generally considered the weak point of Salmon's account is that it fails to capture the relation of causal relevance (or difference-making).[11] Suppose a pool player chalks a cue and strikes a cue ball, which in turn strikes the eight ball and sends it into the corner pocket. The striking of the cue ball with the cue, and the cue ball with the eight ball, are interactions that induce marks in the form of changes in momentum. But these strikings may introduce other marks as well. For instance, a bit of chalk may be transferred from the cue to the cue ball and thence to the eight ball. Intuitively, this second kind of marking doesn't make a difference to the end result, but it is productively

---

[10] Salmon's theory discusses two kinds of causal forks, "interactive forks," which explain correlations between causes and effects, and "conjunctive forks," which explain correlated effects of a common cause. Salmon's account of common causes need not concern us here.

[11] This widely discussed example is from Hitchcock (1995).

connected to it. Thus, it is argued, Salmon's account cannot distinguish between relevant and irrelevant causes.

The new mechanists take a rather different approach to understanding interaction—one that has the promise of avoiding the causal relevance trap. Whereas Salmon and Dowe sought a reductive analysis of causal interaction—one that would tell us the one thing causation really is, or at least the small number of fundamental causal mechanisms out of which the causal nexus is constructed—the new mechanists have eschewed a reductive analysis, arguing that there are many different kinds of causal production corresponding to the many different kinds of activities and interactions. Whenever two parts of a mechanism interact, they interact in a rather specific way that produces characteristic changes in the objects: scissors cut paper, wine stains cotton, humans kiss each other. Different kinds of objects have different kinds of characteristic activities and interactions in which they engage and different kinds of objects they can interact with. Scissors can cut paper, and maybe cotton, but not rocks. Scissors can't stain or kiss anything. Wine may stain some things (like cotton or paper) and can't be cut, and so on.

Applying this strategy to Hitchcock's case, we can say that coloring and striking are two distinct activities, one of which produces change in color and the other changes in momentum. Because coloring changes color and not momentum, it is no surprise that coloring is not relevant to the change in the struck ball's direction. Coloring is an activity that chalked balls can engage in, but it is not part of the mechanism that sends the struck ball to the corner pocket.

This answer may strike critics as simply avoiding the problem, for it doesn't give us any criteria for what distinguishes genuinely productive activities from other happenings. But Machamer, at least, thinks we neither have nor need such criteria:

> Activities have criteria that can be given for their identification. We know how to pick out activities such as running, bonding, flowing, etc. . ... One might try to do something more general by giving the conditions for all productive changes, but then one would have to find out what all producings have in common and by what are they differentiated from non-producings. It is not clear that they all have any one thing in common or are similar in any significant way, but neither commonality nor similarity are necessary conditions for an adequate category. (Machamer 2004, p. 29)

Machamer's point is that we come to understand particular activities by empirical (both common-sense and scientific) investigation; philosophers need to accept that there is no one such thing as causing or producing—no secret stuff that constitutes the cement of the universe.[12]

Still, more can be said about what binds these various activities together *as* activities. The two most important points are these: first, activities are the producers of change, and, accordingly, to make changes in the world agents must manipulate and control

---

[12] This position is sometimes called "causal minimalism" (Godfrey-Smith 2009); a view generally thought to originate with Anscombe (1971).

these activities.[13] Second, the fact that a certain activity is productive is seldom or ever a brute fact, for it is almost always the case that the occurrence of an activity or interaction involves the operation of further finer grained mechanisms from which the productive character of the coarse-grained activity derives. When an activity or interaction depends in this way upon finer grained entities and activities, I shall call it a *mechanism-dependent activity or interaction*. Similarly, when the productive capacities of a system or complex object depend upon the organized activities and interactions of its parts, the capacities are mechanism-dependent.

The ubiquity of mechanism dependence is a consequence of the hierarchical organization of mechanical systems and processes, and it explains how productive continuity at lower levels in the mechanistic hierarchy give rise to higher level forms of production. Productive powers of wholes derive from the organization of their parts, and the productive continuity of causal processes derives from productive interactions of parts at various stages in the process. In some sense, this is the same picture drawn by Salmon, but the hierarchical character gives a more informative characterization of production.

Seventeenth-century versions of mechanicism enforced the notion of productive continuity via the stipulation that there was no action at a distance. On standard versions of seventeenth-century mechanical philosophy, action was by contact—pushes and pulls between interlocking corpuscles of various sizes and shapes. In the new mechanicism, the notion of contact is extended. Causal productivity arises from the direct interaction of parts, but what counts as direct contact depends upon the level at which an interaction is described.

To take a simple example, consider the sorts of causal mechanisms by which news spreads. The matriarch of the family dies, an event observed directly by one of her daughters. Her daughter calls her sisters and brothers, who in turn talk to their spouses and children. In such a situation, we distinguish between immediate contact, where one person "speaks directly" to another, and mediated or indirect transmission, where the message is passed to someone via one or more intermediates. But "directness" is not absolute. When the daughter speaks directly to her siblings, she is likely talking to them on the phone, and there is obviously a very elaborate mechanical system with parts whose activities and interactions enable the transmission of her voice to the ears of her geographically distant relations. And even if she tells them "in person," there is obviously an intervening set of mechanisms that allow transmission of the information.[14] There is a transmitter and a receiver, and the transmission requires cognitive and speech

---

[13] Here, we are touching on the relationship between mechanistic theories of causation and manipulability and interventionist theories. Basically, the relationship I propose is this: mechanisms are what allow manipulations to be successful, so mechanisms are the truth-makers for claims about manipulability. So while I would agree with Woodward (2003) that there is a close conceptual connection between manipulability and causation, mechanisms are the causes in the world that account for our capacities to manipulate the world. The position I advocate is not unlike that of Menzies and Price (1993).

[14] In this example, it is natural to characterize the phenomenon in question as information transmission or intentional content, but this distinction between direct and indirect interactions can be found in all mechanisms; moreover, if one is skeptical about information or intentional content, much of this example could be described without reference to these concepts.

production mechanisms at one end, speech recognition and cognitive mechanisms on the other, and physical mechanisms of wave transmission in between.

The mechanistic story is a story about the composition of causes. It shows how the productive capacities of wholes can emerge from the organized activities and interactions of their parts and how effects can be productively related to distal causes through continuous chains of productive activities and interactions. One of its key lessons is that, even though the productive powers of higher level entities and activities depends upon the productive powers of their parts, production at different levels can be genuinely different kinds of production.

Think again about the phone system. This system has countless parts—phones, cables, substations, cell phone towers, and the like, and these parts themselves have parts, which themselves have parts. The causal powers and activities of these parts are extremely various—resistors and capacitors alter current through electrical circuits; fiber optic cable propagates light pulses, batteries convert chemical energy to electrical power. All of these are forms of production, and all of these parts can only exhibit these kinds of productive activities and interactions if they are appropriately situated within a mechanical context. For instance, a battery only provides power if it is connected to a circuit. Moreover, the whole assemblage of the phone system has the capacity to receive sounds from the microphone in one phone and reproduce them on the speaker of the other. Transmitting sound across distances in this way is a wholly different kind of activity than the electrical activities of the components of the phones, cell phone towers, and other parts of the phone system—but it nonetheless mechanistically depends upon these components and their activities. The phone system in turn is a small piece of much larger mechanisms for human communication. A pair of phones, put in the hands of two people who speak the same language and have the relevant technical and cultural knowledge, allows them to transmit their speech, and with it, both information and affect, which will in turn affect individual and collective human behavior.

Hopefully, this sketch has shown how the new mechanist approach to causation differs from the Salmon/Dowe account and how its emphasis on the fact that there are different kinds of production allows it to avoid the causal relevance problem. In the remainder of this chapter, I will briefly discuss two of the most important remaining objections to the mechanistic approach to causation. The first concerns the problem of negative causation, and the second what I shall call the *grounding problem*.

Negative causation is an umbrella term for a variety of apparently causal relationships involving nonhappenings and productive gaps. In causation by omission, an omission "causes" some event, as when my failing to turn off the tap causes the bath to overflow. In prevention, some event stops another event from occurring, as when my turning the tap off prevents the bath from overflowing. Omission and prevention are thought to cause problems for productive accounts of causation because omissions and preventions, as nonoccurrences, cannot be productively linked to anything. The standard (and in my view successful) answer to these questions is to understand omission and prevention as explanatory but not strictly causal relationships (Dowe 2001). When I say that my turning off the tap prevented the overflow, I am simply contrasting an actual causal process with a nonactual one that would have occurred had I not turned off the tap. This counterfactual relationship is like a causal one and is sufficient for most purposes—like the purpose of attributing causal, legal, moral, or financial responsibility for occurrences or nonoccurrences. If the flood was "caused" by my failure to turn off the water rather than a leak in the drain line, it was my fault, not the plumber's.

A more serious threat to the mechanistic account is the problem of causation by disconnection (Schaffer 2000). In causation by disconnection (also called *double prevention*), both the cause and effect are actual events, but the cause does not directly produce the effect. The cause instead removes something that was preventing the effect from happening. A simple example is a burglar alarm. Opening the door breaks a circuit, which was blocking the current to the alarm siren. When the door circuit is disconnected, the block is removed, thus allowing power to flow to the siren. The door opening does not produce the power that turns on the siren; it simply opens a gate that allows power to flow. Causation by disconnection appears to be very widespread in living systems, in engineered mechanisms, and in other parts of nature, and it seems counterintuitive to deny that this is genuine causation.

I think the basic response to this dilemma is not to deny that productivity is required for causation, but to say that it is mechanisms as a whole that are responsible for the production of phenomena, and sometimes part of the wiring of these productive mechanisms involves disconnection. To illustrate, consider a case from pharmacology. Selective serotonin reuptake inhibitors (SSRIs) are a class of psychotropic drugs that are commonly prescribed to combat depression. A very rough sketch of their mechanism of action is as follows: serotonin is a neurotransmitter that facilitates signaling between neurons across synapses. It is synthesized in the presynaptic neuron and is released into the synapse as a result of the firing of that neuron. SSRIs increase (at least in the short term) the amount of serotonin in the synapse, but they do not do so by stimulating increased production of serotonin within the presynaptic neuron. Instead, SSRIs slow the reuptake of serotonin into the presynaptic neuron by inhibiting the action of reuptake pumps in the cell membrane, thus increasing the overall concentration in the synapse. This is clearly a case of causation by disconnection.

I think the correct interpretation of this kind of case is to say that taking an SSRI will cause (in the sense of produce) an increase of serotonin in the synapse, even though the mechanism of action is one of inhibition. To say that one event produces another, one need not insist that it does so on its own. Causal production typically involves the orchestrated activities and interactions of numerous entities. The taking of the SSRI is an essential step in triggering the mechanism that produces the increase in serotonin levels, and hence it contributes to the production of the changed serotonin levels.[15]

The other major objection to a mechanistic account of causation is the grounding problem. If the mechanistic account of causation is an account of how more basic causes are compounded into more complex causes, it would seem that the theory needs in

---

[15] For a more thorough defense of this account of causation by disconnection, see Glennan (forthcoming).

addition an account of the base case. MDC argued that mechanisms begin with bottom-out entities and activities, but this was explicitly a methodological rather than an absolute and metaphysical bottom. The question remains: does the mechanistic theory demand the existence of a metaphysically fundamental level? Moreover, if it does, is this where causation is really located? A serious survey of the scientific and metaphysical issues involved here is beyond the scope of this chapter, so I will here only be able to summarize a few responses that I have argued for elsewhere (Kuhlmann and Glennan 2014).

Suppose there is a level of fundamental entities—the basic building blocks of the universe. We used to call these atoms, even though we now believe that there are subatomic levels; but suppose that these subatomic levels come to an end in the true fundamental entities. If they did, it might be that these fundamental entities would have some basic set of causal powers—powers that would manifest themselves in interactions among these entities. These are the sorts of things that Salmon might have had in mind when he spoke of the "few basic mechanisms" that ground all causal problems. These metaphysically bottom-out entities and activities would solve the grounding problem.

If there is such a fundamental level, some metaphysicians might suggest that this is where we will really find causation: the mechanisms story is only about explanation and methodology. To this, I can only reply that none of the causes that we (scientists, philosophers, or the folk) ever look for and use belong to this fundamental level and that any scientifically informed metaphysical view cannot brush off as unimportant the project of understanding the mechanistic composition of causes.

There are reasons, however, to doubt that there is a fundamental level of the kind described in the previous paragraph. For one thing, it is neither metaphysically nor scientifically obvious that there must be a fundamental level (Schaffer 2003). If there is not, it is possible that the correct answer to the grounding problem is "mechanisms all the way down." To deny the existence of compositional causes because there is no bottom-level cause is no more necessary than to deny the existence of causal chains because there is no first-cause.

For another, even if there does turn out to be a fundamental level, it may not be a fundamental level of "things." There are at least some plausible metaphysical interpretations of quantum mechanics that suggest that "thing-like" entities, a category central to the new mechanistic account, do not exist at levels described by quantum mechanics (Ladyman and Ross 2007). If this is so, what is needed is some account of the emergence of thing-like entities whose localized activities and interactions can be parts of mechanisms. Fortunately, there may be an account of the emergence of such a "fundamental classical level" that does not ignore the challenges posed by nonlocality and quantum entanglement. This may be where the mechanisms stop, and it should provide grounding enough.[16]

---

[16] See Kuhlmann and Glennan (2014). There, we also argue that there are phenomenon that are nonclassical but nonetheless in important respects mechanistic.

REFERENCES

Anscombe, G. E. M. (1971). *Causality and Determination: An Inaugural Lecture* (Cambridge: Cambridge University Press).

Bechtel, W., and Abrahamsen, A. (2005). "Explanation: A Mechanist Alternative." *Studies in the History and Philosophy of Biology and the Biomedical Sciences* 36(2): 421–441.

Bechtel, W., and Abrahamsen, A. (2010). "Dynamic Mechanistic Explanation: Computational Modeling of Circadian Rhythms as an Exemplar for Cognitive Science." *Studies in History and Philosophy of Science* 41(3): 321–333.

Bechtel, W., and Richardson, R. C. (1993). *Discovering Complexity: Decomposition and Localization as Strategies in Scientific Research* (Princeton, NJ: Princeton University Press), 286.

Bogen, J. (2008). "Causally Productive Activities." *Studies in History and Philosophy of Science Part A* 39(1): 112–123.

Cartwright, N. (1983). *How the Laws of Physics Lie* (Oxford: Clarendon Press).

Craver, C. F. (2001). "Role Functions, Mechanisms, and Hierarchy." *Philosophy of Science* 68(1): 53–74.

Craver, C. F. (2002). "Structures of Scientific Theories." In P. Machamer and M. Silberstein (eds.), *The Blackwell Guide to the Philosophy of Science* (Cambridge, MA: Blackwell), 55–79.

Craver, C. F. (2006). "When Mechanistic Models Explain." *Synthese* 153(3): 355–376. doi: 10.1007/sll229-006-9097-x

Craver, C. F. (2007). *Explaining the Brain* (Oxford: Clarendon Oxford Press).

Craver, C. F. (2013). "The Ontic Account of Scientific Explanation." M. I. Kaiser, O. R. Scholz, D. Plenge, and A. Hüttemann (eds.), *Explanation in the Special Sciences: The Case of Biology and History*, Synthese Library. Springer Netherlands, 27–52. doi: 10.1007/978-94-007-7563-3_2

Craver, C. F., and Darden, L. (2013). *In Search of Mechanisms: Discovery across the Life Sciences*. Chicago: University of Chicago Press.

Cummins, R. (1975). "Functional Analysis." *Journal of Philosophy* 72: 741–765. doi: 10.2307/2024640

Dowe, P. (2000). *Physical Causation* (Cambridge: Cambridge University Press).

Dowe, P. (2001). "A Counterfactual Theory of Prevention and 'Causation' by Omission." *Australasian Journal of Philosophy* 79(2): 216–226.

Gervais, R., and Weber, E. (2013). "Plausibility Versus Richness in Mechanistic Models." *Philosophical Psychology* 26(1): 139–152.

Giere, R. N. (2004). "How Models Are Used to Represent Reality." *Philosophy of Science* 71(5): 742–752.

Glennan, S. S. (1996). "Mechanisms and the Nature of Causation." *Erkenntnis* 44(1): 49–71.

Glennan, S. S. (2002). "Rethinking Mechanistic Explanation." *Philosophy of Science* 69(S3): S342–S353. doi: 10.1086/341857

Glennan, S. S. (2005). "Modeling Mechanisms." *Studies in the History of the Biological and Biomedical Sciences* 36(2): 443–464. doi: 10.1016/j.shpsc.2005.03.011

Glennan, S. S. (2010). "Mechanisms, Causes, and the Layered Model of the World." *Philosophy and Phenomenological Research* 81(2): 362–381. doi: 10.1111/j.1933-1592.2010.00375.x

Glennan, S. S. (2011). "Singular and General Causal Relations: A Mechanist Perspective." In P. M. Illari, F. Russo, and J. Williamson (eds.), *Causality in the Sciences* (Oxford: Oxford University Press), 789–817.

Glennan, S. S. (forthcoming). *The New Mechanical Philosophy*. Oxford: Oxford University Press.

Godfrey-Smith, P. (2009). "Causal Pluralism." In H. Beebee, C. Hitchcock, and P. Menzies (eds.), Oxford Handbook of Causation (New York: Oxford University Press), 326–337.

Hitchcock, C. R. (1995). "Discussion: Salmon on Explanatory Relevance." *Philosophy of Science* 62(2): 304–320.

Illari, P. M. (2013). "Mechanistic Explanation: Integrating the Ontic and Epistemic." *Erkenntnis* 78: 237–255. doi: 10.1007/s10670-013-9511-y

Illari, P. M., and Williamson, J. (2012). "What Is a Mechanism? Thinking about Mechanisms across the Sciences." *European Journal for Philosophy of Science* 2(1): 119. doi: 10.1007/s13194-011-0038-2

Kuhlmann, M. (2011). "Mechanisms in Dynamically Complex Systems." P. Illari, F. Russo, and J. Williamson (eds.), *Causality in the Sciences*. Oxford University Press: Oxford, 880–906.

Kuhlmann, M., and Glennan, S. S. (2014). "On the Relation between Quantum Mechanical and Neo-Mechanistic Ontologies and Explanatory Strategies." *European Journal for Philosophy of Science* 4(3): 337–359. doi: 10.1007/s13194-014-0088-3

Ladyman, J., and Ross, D. (2007). *Every Thing Must Go: Metaphysics Naturalized* (Oxford: Oxford University Press).

Levins, R. (1966). "The Strategy of Model Building in Population Biology." *American Scientist* 54(4): 421-431.

Levy, A., and Bechtel, W. (2013). "Abstraction and the Organization of Mechanisms." *Philosophy of Science* 80(2): 241–261.

Machamer, P. (2004). "Activities and Causation: The Metaphysics and Epistemology of Mechanisms." *International Studies in the Philosophy of Science* 18(1): 27–39.

Machamer, P., Darden, L., and Craver, C. F. (2000). "Thinking about Mechanisms." *Philosophy of Science* 67(1): 1–25.

Menzies, P., and Price, H. (1993). "Causation as a Secondary Quality." *The British Journal for the Philosophy of Science* 44(2): 187–203.

Morgan, M. S., and Morrison, M. (1999). *Models as Mediators: Perspectives on Natural and Social Science*, vol. 52 (Cambridge: Cambridge University Press), 401.

Piccinini, G., and Craver, C. F. (2011). "Integrating Psychology and Neuroscience: Functional Analyses as Mechanism Sketches." *Synthese* 183(3): 1–58.

Salmon, W. C. (1984). *Scientific Explanation and the Causal Structure of the World* (Princeton, NJ: Princeton University Press).

Schaffer, J. (2000). "Causation by Disconnection." *Philosophy of Science* 67(2): 285–300.

Schaffer, J. (2003). "Is There a Fundamental Level?" *Noûs* 37(3): 498–517.

Simon, H. A. (1996). *The Sciences of the Artificial*, 3rd ed. (Cambridge, MA: MIT Press).

Teller, P. (2001). "Twilight of the Perfect Model Model." *Erkenntnis* 55(3): 393–415.

Weisberg, M. (2007). "Three Kinds of Idealization." *The Journal of Philosophy* 104(12): 639–659.

Weisberg, M. (2013). *Simulation and Similarity: Using Models to Understand the World* (Oxford: Oxford University Press).

Weiskopf, D. A. (2011). "Models and Mechanisms in Psychological Explanation." *Synthese* 183: 313–338. doi: 10.1007/s11229-011-9958-9

Wimsatt, W. C. (2007). *Re-Engineering Philosophy for Limited Beings* (Cambridge, MA: Harvard University Press).

Woodward, J. (2003). *Making Things Happen* (Oxford: Oxford University Press), 432.

CHAPTER 39

....

PHILOSOPHY AND COSMOLOGY

....

CLAUS BEISBART

1 INTRODUCTION

....

COSMOLOGY is the study of the cosmos or of the universe. The universe, in turn, is, or comprises, all there is. Cosmology is thus a far-reaching endeavor. Ideally, it would take us to the limits of what there is. But we may get stuck within narrower limits of what we can know and of what science can do. The aim of this article is to explore the limits of being, knowledge, and science, insofar as they manifest themselves in cosmology.

When we speak of cosmology today, we mostly refer to a discipline of empirical science, which is sometimes also called "physical cosmology." But cosmology was not always considered a subdiscipline of physics, and questions that were, or may be, called "cosmological" have occupied philosophers quite a bit (cf. Kragh, 2007). The pre-Socratic philosophers such as Thales claimed that water or something else is the *arché* (i.e., the origin or principle) of everything. Plato's dialogue "Timaios" draws a picture of the structure of the visible world. Much later, philosopher Christian Wolff distinguished between an empirical, physical cosmology and a so-called *cosmologia generalis*, which was considered part of metaphysics and supposed to derive truths about the world of material objects from ontology (1731: Sections 1–4).

That philosophers have displayed interest in cosmological questions should not come as a surprise. Clearly, if cosmology is about all there is, we had better be part of the game, particularly because some philosophical disciplines, particularly metaphysics, try to think about everything too, and philosophical questions about what there is and how it is structured naturally lead to an interest in the structure of the material world as investigated by cosmology. Also, for a long time, any inquiry about the structure of the world suffered from the scarcity of observational data and a lack of well-supported physical theories. Consequently, philosophical arguments and assumptions were needed to address the structure of the world.

In the meantime, things have changed. In his "Critique of Pure Reason," Immanuel Kant famously smashed metaphysical cosmology by arguing that questions about the universe lead human reason into a deep problem, which he called the "antinomy of pure reason." Whatever the merits of Kant's arguments, there are few philosophers since then who have tried to do cosmology. Physical cosmology, by contrast, has made great progress after Albert Einstein (1917) first applied the field equations of his general theory of relativity (GTR) to the universe. Today, physical cosmology is a thriving field that is guided by observation and that has made a lot of progress.

It may thus be concluded that cosmology is, and should be, an empirical science and that philosophers cannot make any substantial contribution to answering questions about the universe. Philosophical curiosity and speculation may have been legitimate and perhaps even useful as a source of inspiration in the early days of cosmology when no one could have dreamed of the techniques that present-day observational cosmology employs. But today, there is just no point in philosophers' addressing questions about cosmology, or so the view is. If this is right, then philosophers, if interested in cosmology at all, should content themselves with clarifying the basic concepts of cosmology and scrutinizing the logical structure of our cosmological knowledge but otherwise remain silent.

But this is not what all philosophers did. There is a notable philosophical tradition of questioning our abilities to learn what the universe is like. It is Kant again who set a famous example; his argument in the "Critique of Pure Reason" is not just directed against metaphysical cosmology but is also meant to put doubts on an empirical science of the universe. The same tradition is manifest in more recent claims that science will never be able to determine whether or not the age of the universe is finite (Scriven 1954).

Work of this sort contributes to a more general philosophical project prominently initiated by Socrates (viz., the critical reflection of what we know). In the spirit of this tradition, John Locke, David Hume, and Immanuel Kant tried to systematically determine the limits of human knowledge. Cosmology seems a particularly apt target when we think about these limits because it is our boldest attempt to extend our knowledge, if "extend" is meant in a literal, spatial sense. Attempts to know what the universe is like immediately raise epistemological worries: How can we study the universe, even though it is a unique object and we lack similar objects to which we can compare? And how can we ever ascertain that our studies really address the whole universe?

The aim of this article is to inquire whether physical cosmology can answer the most important questions that we have about the universe. Can we push the limits of scientific knowledge to the limits of being? And what is the role of philosophy? Kant's doubts about the possibility of cosmological knowledge provide a good starting point for the discussion (Section 2). It is then shown that modern physical cosmology escapes Kantian and related worries (Section 3). Our currently best cosmological model, its support, and its problems are reviewed in Section 4, and it is shown that, despite its success, it cannot be used to answer some big cosmological questions, nor can we use possible explanations of why the universe is as it is to settle the issues, or so Section 5 argues.

2 Kant's Challenge

Why did Kant think that both metaphysical and empirical cosmologies are impossible? In his "Critique of Pure Reason," Kant examines a couple of substantive questions about the world (A405–A591, B432–B619). For our purposes, we can focus on the first two:

1. Does the universe have a beginning in time?
2. Is the spatial extension of the universe finite?

According to Kant, any *metaphysical* attempts to answer the questions run into trouble. The reason is that, for each question, both the positive and the negative answers lead to a contradiction with well-established metaphysical assumptions, or so Kant argues.

Kant's arguments can also be understood as identifying problems for *empirical* cosmology. To show this, I translate Kant's arguments from their metaphysical parlance into an epistemic jargon. The arguments are then in fact more convincing. For Kant has often been criticized on the grounds that he ultimately draws conclusions from what we can know to what is. This would be a mistake, unless we succumb to a type of antirealism that denies truths that we cannot verify. To most people, such a position is not plausible, and in the remainder of this article we assume a realist outlook according to which beings and knowables do not necessarily coincide. But there is more to Kant's arguments if we consistently cast them in purely epistemic terms. The arguments may then have a tenable quintessence, namely that both possible answers to our questions raise epistemological worries.

Suppose we answer the first question by saying that the universe has a beginning. According to Kant, we then are left with an explanatory gap. The reason is as follows. Kant assumes that time is independent of the universe and infinite in the past. If the universe had a beginning, we could sensibly ask why the universe originated at the particular instance of time it did. In a similar way, we often ask why a particular object within the universe originated at a particular time. We can explain the latter fact because there are other objects that interact with each other and eventually give birth to the object of interest. But an explanation of this type is impossible when it comes to the universe as a whole because the universe comprises all there is. We can refer to time alone if we are to explain why the universe originated at a particular instance of time. But time is homogeneous, and no instance of time is privileged over the other. We thus lack the resources to explain why the universe originated when it did.

In an analogous manner, Kant argues that a spatially finite universe would lead to an explanatory gap too. Kant takes space to be infinite, and if the universe were finite, there would be no way of explaining why the universe is located in infinite space where it is because the universe is all there is and space is homogeneous.

But for Kant, the assumption of a spatially infinite universe with an infinite past also comes with problems. On his view, we can never know that the universe has an infinite past or that it takes infinite space because we would need infinite evidence to show that. For every event in the past, say a collision between two galaxies, we would have to find an earlier one. Likewise, for every object in our world, say a quasar, we would have to find another one that was even further away. But our evidence does not suffice to do this since it is not infinite.

If all this is correct, then both the positive and the negative answers to our questions lead to severe epistemic problems, and a theoretical dilemma challenges any scientific cosmology.

3 Skeptical Challenges Met

But are we really caught in a dilemma? We certainly are not in the way Kant thought. In arguing this point, we learn a lot more about the foundations of present-day cosmology (cf. Mittelstaedt and Strohmeyer 1990).

Consider first the horns of the dilemma that involve infinities. Kant's arguments presume that we can only know the universe to be infinite in space or in the past if we can prove this on the basis of data. But assuming that the universe is in fact infinite in a spatial or temporal way, isn't there any other way to know this? Couldn't we establish this knowledge in a more indirect way? In particular, if a finite universe does in fact leave us with explanatory gaps, as Kant claims, isn't this good evidence for an infinite universe?

For some reasons, Kant came to think that his worries about a universe that is finite in one or the other way do not indicate that the universe is infinite in this way. But his insistence that an infinity of the universe be proven through data is indeed too strict. What we call "scientific knowledge" is often based on inferences that move beyond what has been observed. Although we have not tested that every electron has the same negative charge of -1.6022×10^{-19} Coulomb, we claim to know this. One promising way to move beyond our observations is to model a system (see, "Models and Theories", Morrison this volume). For instance, in fluid mechanics, water in a basin is modeled as a continuous fluid: Each point in the basin is assigned several numbers that reflect the density, temperature, and pressure of the water at this point. It is true that we have to set infinitely many numbers to specify the model because space is assumed to contain infinitely many points. But this is not a problem because a mathematical function like the sine contains information about infinitely many points.

For the reasons mentioned by Kant, there is no way to prove that such a model gets it right everywhere. But proof seems too strict a requirement on knowledge about the empirical world. In physics, it is common practice to confirm continuum models of fluids using finite data (e.g., about temperature at several heights in a basin filled with water) and to extrapolate the validity of the model.

An analogous extrapolation may be used to establish that the universe is infinite in space or time. True, we have to be cautious because many models do not reflect their target in every respect. Sometimes infinities in models are just assumed as idealizations.

Nevertheless, if we have a very successful model of the universe, if there is no rival that is equally successful, and if the assumption of an infinite extension in space or time is decisive for the success of the model, then it seems fair to conclude that the universe is indeed infinite in this way.

Now to devise a model either in fluid mechanics or in cosmology, we cannot just reason from our data and, for example, connect the dots in a plot with the data. We need theory, and theory too is something that Kant does not properly take into account. The dynamics of a fluid like water is governed by laws expressed in the Navier-Stokes equations, which connect physical characteristics such as density, pressure, and temperature. Good models of water in a basin or of similar systems are thus solutions to these equations.

Physicists have tried to make progress along similar lines in cosmology. But what laws can be used in cosmology? It seems that we are now in trouble. Most laws we know are instantiated many times, and to identify a law, we need several instances. But there is only one universe. Maybe Kant's point was simply that we lack laws and theories about the universe to devise a cosmological model.

But cosmologists have been quite successful in avoiding this problem too. Their strategy is as follows: many physical laws like those that govern the water in the basin are described using partial differential equations. Equations of this type are assumed to hold at every instance of time and at every point that is filled with a certain medium. The Navier-Stokes equations are of this type, and the assumption that they always hold for every point in water has been well confirmed. To get started in cosmology, we choose laws that are supposed to hold everywhere, at every point in the universe. We can test them on our planet and by using observations about systems that are much smaller than the universe (e.g., about star clusters and galaxies). The laws then are extrapolated to every point in the universe. If this approach works out fine (I return to the question of whether this is so later), we do not need any special laws about the universe as such to do cosmology.

We cannot, of course, use a law that applies to water only. The composition of matter in the universe varies to some extent from place to place. The usual approach pioneered by Einstein (1917), among others, is to neglect the variation of matter on small scales and to focus on the very largest scales of the order of 10^{24} meters. These scales exceed those of the largest galaxies (e.g., the Milky Way, which is a system of about 10^{11} stars, among them the sun). Figuratively speaking, the idea is that we look at the universe with glasses through which small-scale variations cannot be discerned anymore; they have been averaged away due to a coarse-graining. What we can then still see looks like a fluid, so cosmologists say that they model the matter in the universe as a cosmic fluid. This fluid is usually assumed to have an isotropic pressure, that is, a pressure that is the same in every direction ("perfect fluid"; Hawking and Ellis 1973: 69–70). Very often, the pressure is simply set to zero ("dust model," 103).

At the scales we are now talking about, only one of the forces well known from terrestrial physics is relevant, namely gravity. Gravity is only sensitive to the mass of a body, so it does not matter that our averaging of the matter of the universe mixes several materials. Cosmologists adopt the currently best theory of gravity, which is very well tested (i.e., Einstein's GTR), and apply it everywhere in space and time.

The usual strategy to average out small-scale fluctuations in the matter distribution has, of course, a downside. The models that we obtain will not reflect entities and processes at smaller scales. As a consequence, cosmology is not really about everything but only about the physics at the largest scales (see later discussion for a qualification). Worries to the effect that our world is just too multifaceted to be characterized by a single model or by a single science are thus pointless. They miss the rather limited scope of models in cosmology.

A more general lesson is looming large here: we cannot define cosmology by saying that it is about everything because everything is a non-topic. (This, by the way, is also the lesson that Kant drew from the alleged antinomy of pure reason.) Every physical cosmology builds upon physical assumptions (often implicit in theories), and the history of science has seen a large number of related proposals that invite different perspectives and lend themselves to different conceptions of cosmology. For instance, for many scholars in ancient and medieval times, the universe was basically the solar system in which the various objects each had their natural places. A cosmology based upon these assumptions will differ greatly from present-day cosmology.

The theory upon which present-day cosmology builds (i.e., GTR) has a number of peculiar characteristics. First, it does not allow for an observer-independent distinction between space and time. Space and time are given up in favor of space-time. This does not matter much for the rest of this article, however, since the standard cosmological models reintroduce a division between space and time (see later discussion). Second, according to the GTR, matter is but one form of energy that gives rise to gravitation. But for reasons of convenience, I simply talk about matter only. Finally, the GTR describes gravitational interactions via curvature of the space-time. Consequently, the geometry of space-time is not fixed in advance as it is in Newtonian physics but rather depends on the matter distribution in a way specified by Einstein's field equations.

Models of relativistic cosmology are solutions to these equations. They not only specify the distribution of matter (which is characterized in terms of the energy-momentum tensor $T_{\mu\nu}$) but also the geometry of space-time (which is described using a four-dimensional differentiable manifold M and a metric tensor field $g_{\mu\nu}$).

We can now show that the other horn of Kant's dilemma is also only apparent. Kant presupposes a specific space-time model well known from Newtonian physics, namely Euclidean space and an independent, infinite time. A finite world with a beginning in time does not properly mirror this space-time structure and thus leads to explanatory gaps. Einstein's field equations preclude the scenarios considered by Kant, and, to this extent, he was right. But, pace Kant, they allow for solutions that do not suffer from the problems feared by Kant. In particular, there is a class of so-called Friedmann-Lemaître models (e.g., Hawking and Ellis 1973: Section 5.4; Wald 1984: Section 5.2; Weinberg 2008: Section 1.5). In all of these models, matter distribution and geometry of space-time are in tune: the matter distribution is homogeneous, as is space-time. Of course the space-time structure in most of the models is quite different from the one Kant assumed a priori. For instance, according to some of the models, space is curved. It then forms a three-dimensional analogue of the surface of a sphere: This very surface constitutes a two-dimensional space that is unbounded but finite in surface area. In an analogous way, the three-dimensional space in which we live may be unbounded but have finite volume. Regarding time, many Friedmann-Lemaître models assume a finite past of the universe. All in all, a universe that is finite in some way does not leave us with the explanatory gaps that Kant feared, simply because space or time may be finite too.

The Friedmann-Lemaître models constitute the standard models of cosmology. They are distinguished by their simplicity and high degree of symmetry: They assume the universe to be spatially homogeneous and isotropic about every location. This is the content of the so-called cosmological principle (e.g., Beisbart and Jung 2006), which implies the so-called Friedmann-Lemaître-Robertson-Walker type of metric (Hawking and Ellis 1973: Section 3.5; Wald 1984: Section 5.2; Weinberg 2008: Section 1.1), which in turn entails a global cosmic time because, under this metric, observers who are comoving with the cosmic fluid all measure the same time. Since our assumptions thus far do not uniquely fix the solutions, we are still talking about a family of models. They can be parameterized using the cosmological parameters (e.g., Weinberg 2008: Section 1.5). The latter reflect the mean mass density of the universe, the mean curvature, and the value of the so-called cosmological constant, which is a free parameter in Einstein's field equations.

Of course the models of this family only fit our universe if it is in fact homogeneous and isotropic on large scales. In the very beginnings of relativistic cosmology, the cosmological principle was simply taken for granted, partly because it would have been difficult to make progress without building on it. Cosmologist George Ellis thus commented that cosmology was heavily reliant on aprioristic, philosophical assumptions (see, e.g., Ellis 1984: 220). But there is empirical support for the principle (e.g., Lahav 2002); in particular, one can make a good case for its validity by drawing on the Ehlers–Geren–Sachs theorem and the so-called Copernican principle (Ellis 2007: 1222–1223). As an alternative, one may in principle construct a cosmological model bottom-up without assuming the Cosmological Principle (Ellis et al., 1985).

All in all then, Kant was wrong on both horns of the alleged dilemma. If we build cosmological models that solve Einstein's field equations and that are supported by data we can escape Kantian and related worries.

4 The Concordance Model

Where has model building of this sort taken us and what do we know about the universe, in particular about its age and its extension?

During the twentieth century, there was much controversy about the question of which model fits the universe best, the most important rivals being various Friedmann-Lemaître models and the so-called steady state cosmology (cf. Kragh 1996). Since the turn of the millennium, however, a consensus in favor of a specific Friedmann-Lemaître model has emerged. According to this model, which is called the concordance model, space is flat but expanding. The expansion of the universe is well confirmed by the cosmological redshift in the spectra of galaxies (see Hubble [1929] for the classic observational study). The precise manner in which the spectra of very distant objects are redshifted significantly constrains the model of the universe (see Riess et al. [1998] and Perlmutter et al. [1999] for recent landmarks). Since the universe is expanding, it was smaller and thus denser and hotter in its past. Indeed, if we move backward in time, the density tends to infinity for a finite temporal distance from now. As an infinite density is supposed to be physically impossible, the space-time has to stop there. In mathematical terms, we are running into an initial space-time singularity, which is called the Big Bang. As a consequence, the universe has a finite age, which happens to be about 13.8 billion years. It became transparent for visible light and for electromagnetic radiation of different wavelengths about 380,000 years after the Big Bang. Photons that became free at that time are still moving through the universe, thus constituting the so-called cosmic microwave background (CMB). The characteristics of the CMB radiation have been measured with high precision and significantly constrain the model of the universe as well (see, e.g., Ellis [2014: 10–11] for more observational support of the concordance model).

In the concordance model, the expansion of the universe is slowed down due to the gravitational attraction between constituents of matter. But this effect is counteracted by a repulsive force that is now often attributed to a dark energy. The repulsion becomes more and more significant as compared to gravity, as the universe expands further. Consequently, according to the model, the universe will expand forever, approaching exponential growth.

Although the concordance model fits many observations very well, there is one mismatch with our current data: the model assumes, by far, more matter than we have detected thus far. The cosmological parameter reflecting the mass density has a value of about .3, while the masses that are directly observed would yield a value of less than .05 (Peebles 1993: 476). This by itself is not too much of a problem. While it is true that all matter gives rise to gravitational interactions—and it is for this very reason that we need a particular mass density in the concordance model—it is not true that any mass is easily detected. Some matter may not emit any radiation and thus escape our attempts to detect it. This matter is called dark matter. The problem, however, is that, by drawing on the concordance model, physicists can tell a plausible story of how light elements arise in the early universe through Big Bang nucleosynthesis: As the universe cools down, protons and neutrons start to combine to form stable nuclei of, for example, helium (Peacock 1999: Section 9.4; Weinberg 2008: Section 3.2). A detailed model of these processes can be brought in good accordance with the abundances of light elements observed in the universe but predicts that most of the dark matter is of a yet unknown type ("non-baryonic"). Even though there are various candidates for such dark matter (e.g., elementary particles predicted by supersymmetric theories), attempts to experimentally detect these particles have been largely unsuccessful (Aprile and Profumo

2009). The problem is not exclusive to cosmological model building though, since some dark matter is also needed to account for the dynamics of galaxies (Peacock 1999: ch. 12) and for the effects of gravitational lenses (Weinberg 2008: ch. 9). If physicists continue to fail in detecting dark matter, then the GTR will be in trouble, because all models that imply yet unseen masses are based on the GTR.

Another problem with the concordance model is the interpretation of the dark energy counteracting gravity. Within the framework of the GTR, the repulsion can be described using a term in Einstein's field equations. The relative weight of this term depends on the value of the cosmological constant, which is not fixed by theory and can thus only be determined using observations. A cosmological constant of the size predicted by the concordance model is compatible with all other successful applications of the GTR, so the term with the constant may describe an effect that is ubiquitous but only shows its head at the largest scales. It seems nevertheless attractive to trace the effect back to an underlying mechanism and to predict the size of the effect on this basis. Attempts to interpret dark energy using the quantum vacuum have failed spectacularly (Rugh and Zinkernagel 2002). An alternative is to postulate a new type of matter or energy that gives rise to a negative pressure, which leads to the repulsion.

Apart from these unsolved puzzles, there are further problems that are now most often avoided by assuming inflation. One of them is the horizon problem: the CMB radiation that we detect is almost the same in every direction, and we find only extremely tiny fluctuations as we change the direction in which we take observations. Given the Friedmann-Lemaître models, this means that, about 380,000 years after the Big Bang, the distribution of matter in the universe was almost perfectly homogeneous. Now if we observe an almost perfectly homogeneous distribution of matter on Earth (e.g., in a gas), we explain the distribution by a process of equilibration. But according to the concordance model, the available time after the Big Bang did not suffice for equilibration. Thus, assuming the model, we cannot causally explain the almost homogeneous distribution of matter manifest in the CMB radiation. This problem can be avoided, however, if we suppose that the very early universe underwent a short phase of rapid, exponential expansion and thus was very quickly "inflated" by a factor of about 10^{26}. As one effect, extremely small regions that had had enough time to equilibrate beforehand became inflated to enormously large scales. The CMB radiation that we detect then is assumed to stem from such an inflated region. Inflation can also account for the extremely tiny fluctuations in the CMB radiation: Under inflationary models, these are inflated quantum fluctuations from the very early universe (cf. Peacock 1999: ch. 11; Weinberg 2008: ch. 4).

The assumption of an epoch of inflation has its costs, however. The rapid expansion itself needs explanation. The most common approach is to postulate a new type of matter or energy, the so-called inflaton field. The field is supposed to give rise to a negative pressure, which, in turn, makes space expand. The inflaton field has never been observed, and there is a lot of freedom to model it. The consequence is that there are rival inflationary scenarios. The generic predictions that these models make (e.g., about the tiny fluctuations manifest in the CMB) are well confirmed. Nevertheless, inflation is not completely uncontroversial. Some philosophers deny that features of the very early universe need explanation in terms of physical processes (Callender 2004). Earman and Mosterin (1999) note and assess other problems about inflation.

Overall, then, our currently best model of the universe does incorporate effects that have otherwise gone unnoticed, such as dark energy or inflation, and these effects call for a more detailed account. But this does not cast into doubt the general strategy of relativistic cosmology, namely to extrapolate the validity of GTR. The effects just mentioned can be accounted for in a general relativistic framework, at least if there is enough dark matter.

Given the success of the concordance model, we can justifiably use it if we try to answer the two questions, which lead to trouble according to Kant: Does the universe have a finite spatial extension, and does it have a beginning?

We turn first to the spatial extension of the universe. According to the concordance model, the universe is flat or almost flat. But flatness does not imply infinite Euclidean space. A three-dimensional flat space can be combined with various topologies that differ from that of infinite space (e.g., with a torus topology that implies finite space; Lachièze-Rey and Luminet 1995). To obtain a flat and three-dimensional torus space, we can cut a cube out of Euclidean space and identify the faces opposite to each other. Since neither Einstein's field equations nor the concordance model fix the topology of our space-time, the only possible option to identify the topology is to use observations.

As it happens, our observations do not suffice either. Since the universe has an age of about 13.8 billion years, all signals had, at most, 13.8 billion years to travel through space. Since signals cannot move faster than the speed of light, there is a maximum distance that signals can have taken from the Big Bang up to now. This distance is called the particle horizon; its size can be calculated from the concordance model. All signals that we detect (e.g., light from distant galaxies) can thus only have been emitted from objects that are closer to us than the particle horizon. The limited region to which our observations are restricted is called the observable universe.

If the universe has a finite spatial size smaller than the particle horizon, some distant objects can be seen more than once, and we expect certain correlations in the CMB radiation. But, so far, there is no evidence for such observational signatures of such a small universe. The question of whether or not the universe is finite is thus wide open. It is either finite, but larger than our particle horizon, or infinite.

That all possible observations only cover some part of the universe has more consequences than that. How can we know what the universe is like in regions outside the observable universe? Although the observable universe looks homogeneous at large scales, it may well be that it is elsewhere markedly different (e.g., that its matter density is higher than we observe). Accordingly, the cosmological principle may extend to the whole universe, or not.

This problem is not peculiar to Friedmann-Lemaître models. Drawing on earlier work (Glymour 1977; Malament 1977), John Manchak (2009) has shown that, for almost every space-time compatible with the GTR and for every observer within it, we can find a markedly different space-time and an observer within it such that the observations of both coincide 100 percent. For each of the observers, the choice between the space-times is thus undetermined by the observations they could have made (cf. Beisbart 2009, Butterfield 2014). This is not too much of a problem if there are nevertheless good reasons to prefer one model to the other. For instance, it may be suggested that the homogeneity of the observed universe provides strong inductive support for extrapolating its characteristics to the whole universe. But this suggestion, as well as other proposals, do not carry much conviction (see Beisbart 2009).

Kant's claim that we cannot know whether or not the universe has a finite spatial extension is right, then. But the problem is not that we are faced with an explanatory gap of the type suggested by Kant, and it does not seem impossible in principle to justify the hypothesis that the universe occupies infinite space. The problem is rather one of underdetermination: Several models that differ on the spatial extension of the universe are compatible with our best theories and the observations we can have.

We turn now to the question of whether the universe had a beginning in time. Within our currently best model of the universe, the answer depends on what exactly it means to say that the universe had a beginning. On the one hand, the past of the universe is finite in the following sense: let us start from our current location in space-time. In the concordance model, this location is occupied by a tiny part of the cosmic fluid. If we follow the motion of this fluid element back in time, we obtain a trajectory with a finite temporal length, which is about 13.8 billion years. A possible observer who moves along with the fluid element would measure this interval of time.

On the other hand, under the model, there was no first moment of time (cf., e.g., Earman 2005). As noted earlier, when we go back in time, the universe becomes denser and denser, eventually approaching an infinite density after a finite interval of time. But there is no part of space-time at which the density is indeed infinite. We can thus most appropriately use positive real numbers excluding zero to characterize the temporal structure of the past: as there is no smallest positive number greater than zero, there is no earliest instance of time.

Both of these results are not specific to the concordance model but rather more generic. Thus even if the whole universe is not well described by the concordance model, our observations and the GTR strongly suggest that at least the observable universe goes back to a singularity and thus has a finite age.

There are nevertheless problems with our answer. Models do not reflect their targets in every respect, and many cosmologists do not think that the concordance model properly represents the first few moments of the universe. The model is a solution to the GTR. Thus only gravitational interactions are taken account of, while other forces are neglected. But since matter is extremely dense in the early universe, physics at small scales then can no longer be neglected and other forces must be taken into account. In addition, we may stretch the GTR beyond its range of legitimate applications, when we think about the beginning of time. The GTR is not a quantum theory, while the forces apart from gravity are characterized in terms of quantum theories. This suggests that gravity too should be described by a quantum theory, maybe even in a grand unified theory, which accounts for all fundamental forces in a unified manner. It is thus often assumed that the GTR breaks down before the so-called Planck time, which is about 10^{-43} seconds after what we take to be the Big Bang. If this is correct, then the implications of the concordance model for the beginning of time may have to be revised. The problem, though, is that we yet lack a convincing quantum theory of gravity (cf. Kiefer 2012). The consequence is that we are at least currently facing a physics horizon, as Ellis (2007: 1232) calls it. There are certainly proposals on the table. A research program called quantum cosmology studies the universe with the methods of quantum mechanics and thus assumes a wave function for the whole universe (Kiefer 2012: ch. 8; Bojowald 2015). But research of this kind is (still?) very speculative. As a consequence, the question of how exactly the universe began is open. But since our notion of time is likely to crumble in a quantum theory of gravity (cf. also Rugh and Zinkernagel 2009), it is still correct to say that the observable universe has a finite past.

All in all, our attempts to answer questions about the spatial extension of the universe and its past have led us to the limits of our current scientific knowledge. The constraint that the particle horizon puts on our observations is in fact a restriction on what we *can* know. But is it really impossible to infer at least something about those parts of the universe that are further away than our horizon? These days, it is not so much philosophers as physicists who try to do so. An interesting idea is that explanations of characteristic features of the observable universe tell us something about other parts of it. And explanation is a task for science anyway. Kant himself insisted that scientists fulfill this task. In his view, we ultimately run into the antinomy of pure reason when we try to provide grounds for, or to explain, phenomena.

5 EXPLANATION IN COSMOLOGY

Let us thus turn to explanation in cosmology and raise questions such as: Why is the space-time structure as it is? And why is matter distributed as it is? If the ideas behind inflation are correct, then any aggregation of matter has arisen from quantum fluctuations. This is extremely difficult to imagine, and we may also ask why the fluctuations were as they were.

Our most urgent quests for explanation do not in fact concern the distribution of matter and the structure of space-time in general but rather our own existence. Why is there life; why are there conscious beings like ourselves who can to some extent know about the world? It may be objected that these questions are not particularly cosmological because they concern scales that are much smaller than those investigated in cosmology. But cosmology is important for the origin of life, not because it gives a full account of life but rather because the possibility of life depends on cosmological conditions. After all, the question of why there is life can be rephrased as the question of why there is life in the universe.

Compare this to the question of why there is life on Earth. This question is most naturally thought to urge for a contrastive explanation: Why is there life on Earth rather than on other planets? The most complete answer to this question would first name sufficient conditions for life that hold on Earth and second mention necessary conditions for life (of the sort we know) that are violated elsewhere. Often, we cannot go as far as this, though, and instead content ourselves with arguments according to which necessary conditions for life are fulfilled here but not there.

But there is also a different question: Why is there any life on Earth rather than not? What is now important is the contrast between there being life and there not being life on Earth. Suppose now that there are laws that entail the emergence of life, given suitable conditions at an earlier time. We can then trace back the origin of life on Earth to conditions at some earlier time. But we may then want to push the explanation one step further and ask why the conditions were as they were. Given suitable laws, we may again be able to answer the quest for explanation by providing yet earlier conditions from which the emergence of the early conditions for life follow and so forth. Since we will plausibly not allow that the time intervals between the conditions considered become arbitrarily small, this sort of explaining conditions in terms of earlier conditions will only come to a halt if we reach something like ultimate initial conditions beyond which we cannot move. These initial conditions must be from the very early universe. If we ask further why the ultimate conditions were as they were, we are led to consider some aspect of the origin of the universe and thus to engage in cosmogony. True, Big Bang models do not have a first moment and a strict initial time. But even then we can ask: Why is the series of states of the universe such that life arises?

Our explanatory questions do not only lead us to prior conditions of life in the matter distribution. When we explain the emergence of life on the basis of prior conditions, we too rely on laws of nature. The validity of some of these laws may be explained by deriving them from other laws, but we will not be able to explain the validity of all laws in this way. The question then is: Why is the universe governed by specific fundamental laws, which cannot be explained in terms of other laws? Why do the laws allow for the existence of life? These questions are cosmological in that they concern laws, which are supposed to hold everywhere in the universe. Maybe the validity of the laws goes back to the creation or genesis of the world because the way the universe has arisen explains why these laws hold.

The question of why the laws of nature allow for life has much been discussed in science and philosophy. A particular emphasis is on constants of nature that appear in laws (e.g., on the speed of light and the gravitational constant). It can be shown that worlds with slightly different laws in which the value of the gravitational constant, say, is slightly different from that in our world do not fulfill necessary conditions for life (see Barrow and Tipler 1986: chs. 4–5). More generally, in the space of possible values for the most important constants of nature, only a very tiny part seems to allow for life. We seem to have been extremely lucky that the universe is characterized by conditions friendly to life. The possibility of life seems extremely improbable, but the universe does render life possible.

Questions of why some very early matter distribution was as it was and why the laws of nature are as they are cannot be meant to be contrastive. There is simply nothing to which we can contrast our universe because there is only our universe. But how then can we answer the questions?

One answer is of course that life was intended. The most common variety of this answer has it that God created the universe and the laws of nature in such a way as to make life possible. If we accept this explanation as the best one and did not believe in the existence of God before, we can infer the existence of God via an inference to the best explanation (cf. Plantinga 1998).

Another explanation that is more popular in contemporary cosmology is as follows: there is not just our universe but rather a multiverse (e.g., a huge plurality of universes, each with its own laws, space-time structure and matter distribution; see Carr 2007). Given this plurality of worlds, it is no coincidence that our universe (i.e., our part of the multiverse) makes life possible, for we could not exist in most of the other universes. Here we are talking about an observation selection effect (see Bostrom 2002). Observation selection effects arise because we only observe what we *can* observe. For instance, at very large distances, we can only observe the very brightest galaxies simply because fainter ones are too faint to be seen by us. This explains why we observe only very bright galaxies at large distances. Likewise, due to observational constraints, we can only observe one universe, and we can obviously only observe a universe in which we can exist. This explains why we observe a universe with constants of nature and initial conditions that make life possible. Reasoning of this kind is often called anthropic, and it appeals to the weak anthropic principle (famously introduced by Carter 1974, see also Barrow and Tipler 1986: ch. 1). According to this principle, the universe that we observe must be so as to make life possible. The "must" here is meant in an epistemic sense: only hypotheses that leave open the possibility of life are compatible with our prior knowledge. According to the strong anthropic principle, by contrast, the universe must enable life in a stronger sense: universes that do not allow for life are metaphysically impossible (Barrow and Tipler 1986: Section 1.2).

A multiverse explanation can be spelled out by using more physics. The basic idea is to use some (known or new) physics to build a model in which various universes arise that differ in other parts of the physics. One idea is that inflation is not the same everywhere; rather, various regions of space-time become inflated in different ways, thus effectively forming a lot of universes, among them our own. This process may continue forever, as is assumed in scenarios of eternal inflation (Weinberg 2008: Section 4.3). Since the scenarios imply that the multiverse looks to most observers as does a Friedmann-Lemaître model, there is no incompatibility with our observations.

There are other ways to obtain a plurality of universes. For instance, one can write down solutions to Einstein's field equations that comprise several disconnected space-time manifolds. We may additionally postulate that they differ in their laws apart from the field equations. Or why not postulate a plurality of causally unconnected worlds that also differ on their laws of gravitation? Maybe every possible world exists, as philosopher David Lewis (1986) famously thought (cf. Tegmark 2004). In any case, if we accept a multiverse explanation as the best, we may again run an inference to the best explanation and conclude that there is a multiverse. This conclusion would extend our cosmological assumptions beyond our horizon.

But do multiverse scenarios really explain well why the universe is friendly to life? Are multiverse scenarios still scientific? There are serious worries that suggest negative answers to these questions (cf. Ellis 2014: 13–17). One problem is that a multiverse explanation comes with huge costs. Very large numbers of other universes are postulated in order to understand some features of ours. (Whatever set of universes is postulated, a multiverse explanation has to assume many of them to ensure that a universe with life is among them.) It is thus not clear that a multiverse explanation is less expensive than one in terms of God. If it is not, it can only qualify as the best explanation if there are other reasons that count strongly in favor of it. Another worry is that multiverse explanations miss the essential point. Our question was why the universe in the sense of everything there is allows for life. As noted previously, this fact cannot be explained in terms of a contrastive explanation of why life has arisen here rather than there. But, at bottom, the multiverse explanation does offer a contrastive explanation of why we observe values of parameters that are friendly to life by contrasting with other universes the existence of which is simply postulated. It is of course correct that we have no reason to suppose that what we observe coincides with everything there is or is typical of everything. But even if there is much more than what we can observe, we may still raise the question of why there is life in everything there is (whatever this is in detail). And the multiverse view does not answer this question. Put differently, the multiverse explanation shifts the explanation only one step further. Our question was why the universe is as it is. The answer given was that what we take to be the universe is not everything there is; rather there is a mechanism that accounts for a plurality of worlds. But we may then ask why this mechanism was as it was, why it produced the worlds it did, and why it allows for the possibility of life somewhere.

If multiverse explanations are not convincing, what are our options? Should we refer to God's action to explain why the universe is as it is? Should we try to argue that our quest for explanation cannot be answered in the way we thought? Is it, maybe, a mistake to think that the observed values of the constants of nature are unlikely because there are no meaningful probabilities for these values (Ellis 2014: 12)? Should we, perhaps, simply stop asking why the universe is as it is?

Whatever the best option, we now have certainly encountered some limits of our empirical science. Explanation is a central task of science, but we have identified a quest for explanation that is not yet sufficiently answered by science. True, many scientists are friendly to the multiverse explanation, but this explanation does not have the empirical support that other scientific explanations entertain. Since most multiverse scenarios do not have any testable consequences, the empirical support is very unlikely to improve.

The issue of an explanation of why the universe is as it is intimately linked with the question of how much being extends. The reason is that our observations are confined to that part of the universe that is observable to us, and the most promising way to draw inferences about other parts of the universe is to run an inference to the best explanation. True, scientific realists claim that we have pushed the limits of our knowledge beyond what is observable to us. They think that atoms and other unobservable entities are at the center of highly successful explanations and that we should take their existence for granted. But multiverse explanations are not as powerful, because they do not predict in much detail what we observe of the universe, nor can they or their rivals detail how the matter distribution is in other parts of the universe. The consequence is that our particle horizon marks some clear limitation to the knowledge we may ever attain on what there is.

Since some of our quests for knowledge and understanding are not answered and, to a large extent, cannot be answered by science, the floor is open for philosophy again, at least if the latter is understood as speculation that is not much supported by evidence. But of course no serious philosopher will engage in speculative model building about the universe. There are nevertheless tasks for philosophers; for example, they can clarify the explanatory issues that are at stake in cosmology (cf. Smeenk 2013). Further, to the extent that we do try to answer the questions we are talking about (not to obtain knowledge, of course, but maybe something like belief that is reasonable given the poor empirical evidence), our answers will depend on our more general philosophical outlook (cf. Ellis 2014), and philosophers can critically examine the answers and their relation to philosophical views.

I lack space and time to delve into these tasks, so let me conclude by returning to Kant. It finally seems that he was right about some limits of science. As indicated, he insisted that science be explanatory, and his central question was: Where do our explanatory questions lead us? If we explain phenomena scientifically, we draw on other things, which are thus prior and which in turn call for explanation. We may either arrive at something that is explanatory of other things but does not need explanation itself because it is self-explaining. Or the chain of explanations may go on forever. For Kant, both options are not satisfactory. Kant rejects the proposal that a phenomenon subject to empirical science explains itself. By contrast, if the chain of explanations is never-ending, then all phenomena seem groundless to Kant. This seems exaggerated, but we can agree with him that a never-ending chain of explanations does not satisfy us either.

ACKNOWLEDGEMENTS

Many thanks to Martina Jakob, Tilman Sauer, Sebastian Schnyder, Michael Spichtig, Francisco Soler-Gil, Andreas Verdun, and particularly the editor for very useful comments.

References

Aprile, E., and Profumo, S. (Eds.). (2009). "Focus on Dark Matter and Particle Physics." *New Journal of Physics* 11(10): 105002–105029.

Barrow, J. D., and Tipler, F. J. (1986). *The Anthropic Cosmological Principle* (Oxford: Oxford University Press).

Beisbart, C. (2009). "Can We Justifiably Assume the Cosmological Principle in Order to Break Model Underdetermination in Cosmology?" *Journal for General Philosophy of Science* 40: 175–205.

Beisbart, C., and Jung T. (2006). "Privileged, Typical, or Not Even That?—Our Place in the World according to the Copernican and the Cosmological Principles." *Journal for General Philosophy of Science* 37: 225–256.

Bojowald M. (2015). "Quantum Cosmology: A Review." *Reports on Progress in Physics* 78: 023901.

Bostrom, N. (2002). *Anthropic Bias: Observation Selection Effects in Science and Philosophy* (New York: Routledge).

Butterfield, J. (2014). "On Underdetermination in Cosmology." *Studies in History and Philosophy of Modern Physics* 46: 57–69.

Callender, C. (2004). "Measures, Explanations and the Past: Should 'Special' Initial Conditions Be Explained?" *British Journal for the Philosophy of Science* 55(2): 195–217.

Carr, B. (2007). *Universe or Multiverse?* (Cambridge, UK: Cambridge University Press).

Carter, B. (1974). "Large Number Coincidences and the Anthropic Principle in Cosmology." In M. S. Longair (ed.), *IAU Symposium 63: Confrontation of Cosmological Theories with Observational Data* (Dordrecht: Reidel), 291–298.

Earman J. (2005). "In the Beginning, At the End, and All in Between: Cosmological Aspects of Time." In F. Stadler and M. Stöltzner (eds.), *Time and History: Proceedings of the 28th International Ludwig Wittgenstein Symposium* (Heusenstamm: Ontos-Verlag), 155–180.

Earman, J., and Mosterin, J. (1999). "A Critical Look at Inflation." *Philosophy of Science* 66: 1–49.

Einstein, A. (1917). "Kosmologische Betrachtungen zur allgemeinen Relativiätstheorie." *Preußische Akademie der Wissenschaften (Berlin) Sitzungsberichte*: 142–152. English translation: "Cosmological Considerations on the General Theory of Relativity" (1952). In H. A. Lorentz, H. Weyl, and H. Minkowski (eds.), *The Principle of Relativity: A Collection of Original Memoirs on the Special and General Theory of Relativity* (Minola, NY: Dover), 175–188.

Ellis, G. F. R. (1984). "Relativistic Cosmology—Its Nature, Aims and Problems." In B. Bertotti, F. de Felice and A. Pascolini (eds.), *General Relativity and Gravitation* (Dordrecht: Reidel), 215–288.

Ellis, G. F. R. (2007). "Issues in the Philosophy of Cosmology." In J. Butterfield and J. Earman (eds.), *Philosophy of Physics* (Amsterdam: Elsevier), 1183–1286.

Ellis, G. F. R. (2014). "On the Philosophy of Cosmology." *Studies in History and Philosophy of Modern Physics* 46: 5–23.

Ellis, G. F. R., Nel, S. D., Stoeger, W., Maartens, R., and Whitman, A. P. (1985). "Ideal Observational Cosmology." *Physics Reports* 124: 315–417.

Glymour, C. (1977). "Indistinguishable Space-Times and the Fundamental Group." In edited by J. Earman, C. Glymour, and J. Stachel (eds.), *Foundations of Space-Time Theories*. Minnesota Studies in the Philosophy of Science 8 (Minneapolis: University of Minnesota Press), 50–60.

Hawking, S. W., and Ellis, G. F. R. (1973). *The Large-Scale Structure of Space-Time* (Cambridge, UK: Cambridge University Press).

Hubble, E. T. (1929). "A Relation between Distance and Radial Velocity among Extra-Galactic Nebulae." *Proceedings of the National Academy of Sciences of the United States of America* 15(3): 168–173.

Kant, I. (1781). *Kritik der reinen Vernunft* (Riga: Hartknoch). English translation: *Critique of Pure Reason*. Translated and edited by P. Guyer (Cambridge, UK: Cambridge University Press 1998).

Kiefer C. (2012). *Quantum Gravity*. 3d ed. (Oxford: Oxford University Press).

Kragh, H. (1996). *Cosmology and Controversy: The Historical Development of Two Theories of the Universe* (Princeton, NJ: Princeton University Press).

Kragh, H. (2007). *Conceptions of Cosmos: From Myth to the Accelerating Universe: A History of Cosmology* (New York: Oxford University Press).

Lachièze-Rey, M., and Luminet, J. P. (1995). "Cosmic Topology." *Physics Reports* 254: 135–214.

Lahav, O. (2002). "Observational Tests of FRW World Models." *Classical and Quantum Gravity* 19(13): 3517–3526.

Lewis, D. (1986). *On The Plurality of Worlds* (Oxford: Blackwell).

Malament, D. (1977). "Observationally Indistinguishable Space-Times." In J. Earman, C. Glymour, and J. Stachel (eds.), *Foundations of Space-Time Theories*. Minnesota Studies in the Philosophy of Science 8 (Minneapolis: University of Minnesota Press), 61–80.

Manchak, J. B. (2009). "Can We Know the Global Structure of Spacetime?" *Studies in History and Philosophy of Modern Physics* 40: 53–56.

Mittelstaedt, P., and Strohmeyer, I. (1990). "Die kosmologischen Antinomien in der Kritik der reinen Vernunft und die moderne physikalische Kosmologie." *Kantstudien* 81: 145–169.

Plantinga, A. (1998). "God, Arguments for the Existence of." In E. Craig (ed.), *Routledge Encyclopedia of Philosophy*, Vol. IV (London: Routledge), 85–93.

Peacock, J. (1999). *Cosmological Physics* (Cambridge, UK: Cambridge University Press).

Peebles, P. J. E. (1993). *Principles of Physical Cosmology* (Princeton, NJ: Princeton University Press).

Perlmutter, S., Aldering, G., Goldhaber, G., et al. (1999). "Measurements of Omega and Lambda from 42 High-Redshift Supernovae." *The Astrophysical Journal* 517(2): 565–586.

Riess, A. G., Filippenko1, A. V., Challis, P., et al. (1998). "Observational Evidence from Supernovae for an Accelerating Universe and a Cosmological Constant." *The Astronomical Journal* 116(3): 1009–1038.

Rugh, S. E., and Zinkernagel, H. (2002). "The Quantum Vacuum and the Cosmological Constant Problem." *Studies in History and Philosophy of Science Part B* 33(4): 663–705.

Rugh, S. E., and Zinkernagel, H. (2009). "On the Physical Basis of Cosmic Time." *Studies in History and Philosophy of Modern Physics* 33: 663–705.

Scriven, M. (1954). "The Age of the Universe." *British Journal for the Philosophy of Science* 5: 181–190.

Smeenk, C. (2013). "Philosophy of Cosmology." In R. Batterman (ed.), *The Oxford Handbook for the Philosophy of Physics* (New York: Oxford University Press), 607–652.

Tegmark, M. (2004). "Parallel Universes." *Scientific American* 288(5): 40–51.

Wald, R. M. (1984). *General Relativity* (Chicago: Chicago University Press).

Weinberg, S. (2008). *Cosmology* (Oxford: Oxford University Press).

Wolff, C. (1731). *Cosmologia generalis* (Leipzig and Frankfurt: n.p.).

FURTHER READING

Leslie J. (Ed.). (1990). *Physical Cosmology and Philosophy* (New York: Macmillan).

Leslie J. (Ed.). (1998). *Modern Cosmology and Philosophy*. Rev. ed. (Amherst, NY: Prometheus Books).

Weinberg, S. (1972). *Gravitation and Cosmology: Principles and Applications of the General Theory of Relativity* (New York: Wiley).

CHAPTER 40

PHILOSOPHY OF NEUROSCIENCE

ADINA L. ROSKIES AND CARL F. CRAVER

1 INTRODUCTION

NEUROSCIENCE is a field of fields that hang together because they share the abstract goals of describing, understanding, and manipulating the nervous system.[1] Different scientists direct their attention to different aspects of the nervous system, yet all of them seem to be concerned in one way or another with relating functions of the nervous system to structures among component parts and their activities. A neuroscientist might focus on large-scale structures, such as cortical systems, or on exceptionally tiny structures, such as the protein channels that allow neurons to conduct electrical and chemical signals. She might consider long time frames, such as the evolution of brain function, or exceptionally short time frames, such as the time required for the release of synaptic vesicles. And, at these different spatial and temporal scales, researchers might investigate, for example, anatomical structures, metabolism, information processing, physiological function, overt behavior, growth, disease, or recovery. These different topics lead scientists to approach the brain with a diverse array of tools: different theoretical background knowledge, different experimental techniques, different methodologies, different model organisms, and different accepted standards of practice.

Likewise, philosophers interested in neuroscience approach the topic considering many entirely distinct questions and problems. Some are puzzled by whether—and if so, how—diverse aspects of our mental lives, from our conscious experiences, to our capacity for understanding, to our ability to make decisions on the basis of reasons, derive from or are otherwise related to the biochemical and electrophysiological workings of

---

[1] For full references, please see online version at http://www.oxfordhandbooks.com/view/10.1093/oxfordhb/9780199368815.013.40

our brains. Others are interested in neuroscience as a distinctive kind of science. Some see it as representative of multidisciplinary sciences that span multiple levels of brain organization. Some see it as distinctive in attempting to forge connections between mental phenomena and biological phenomena. Still others are interested in fundamental concepts in the neurosciences: they seek to clarify the sense in which a neuron carries information, a brain region may properly be said to represent the world, or what it means to claim that a function localizes to a particular brain region. And some philosophers move fluidly among these perspectives.

The diversity of neuroscience makes for a wealth of potential topics that might involve a philosopher, far too much to cover in a single chapter. Here, we map some of the more populated terrain. We begin looking at philosophical problems internal to the practice of neuroscience: neuroscientific explanation and methods. Then we turn to questions in philosophy of science more generally that neuroscience can help inform, such as unification and reduction. Finally, we consider how neuroscience is (and is not) contributing to traditional philosophical discussions about the mind and its place in the natural world.

2 Looking Inward

2.1 Explanation in Neuroscience

The central aims of neuroscience are to predict, understand, and control how the brain works. These goals are clearly interconnected, but they are not the same. Here we focus on what is required of an explanation in neuroscience. We discuss how explanation differs from prediction and why this difference matters for following through on the aim of control. Contemporary debates about explanation in the philosophy of neuroscience have become focused on the nature and limits of causal explanation in neuroscience. Here, we present the background to that debate.

2.1.1 *Explanations as Arguments: Predictivism*

C. G. Hempel's (1965) covering law (CL) model is a common backdrop for thinking about the norms of explanation in the neurosciences. The CL model is the clearest and most concise representative of a predictivist view of explanation. Its failings point the way to a more adequate model.

According to the CL view, scientific explanations are arguments. The conclusions of such arguments (the *explanandum* statements) are descriptions of events or generalizations to be explained, such as the generation of an action potential by a pyramidal cell in the cortex. The *explanans*, the explanation proper, consists of the premises of the argument. The premises in the explanans state the relevant laws of nature (perhaps the Nernst equation and Coulomb's law) and the relevant antecedent and background conditions from which the explanandum follows. The CL model is an epistemic conception of scientific explanation: the central norms of explanation are norms for evaluating inferences. The power of explanation inheres in its capacity to make the diverse phenomena of the world expectable.

2.1.2 *Enduring Challenges to the CL Model*

The central problem with the CL model (and predictivism more generally) is that the norms for evaluating explanations are different from the norms for evaluating arguments. The normative demands of the predictivist CL model are neither sufficient nor necessary for an adequate explanation.

The account is not sufficient because, as Hempel recognized, the model counts every strong correlation as explanatory. This violates scientific practice: scientists searching for explanations design experiments to distinguish causes from correlations, and they recognize a distinction between predictive models and explanatory models. The difference is of extreme practical import: a model that reveals causes reveals the parameters that can be used to control how the system works.

The norms of the CL model are not necessary because one can explain an event without being able to predict it. In the biological sciences, many causal processes operate stochastically, and some of these stochastic causal processes yield their effects only infrequently. For example, only approximately 20% of action potentials cause neurons to release neurotransmitters. Action potentials explain neurotransmitter release, but they do not make release likely or expectable.

2.1.3 *Causal Explanations*

Wesley Salmon's (1984) diagnosis and treatment for the failings of the CL model are equally pithy and compelling: to explain a phenomenon is to discover the mechanisms that underlie, produce, or maintain it. This is the causal mechanical (CM) model of explanation.

The difference between knowing correlations and knowing causes is that knowing causes reveals the buttons and levers by which a system might be controlled. Causes are detected most crisply in well-controlled experiments in which a change induced in the putative cause variable results in a change in the effect variable even when all the other causes of the change in the effect variable have been controlled for (see Woodward 2003). Causal knowledge deserves the honorific title "explanation" because it is distinctively useful: it can be used to control what happens or change how something works. The norms of causal-mechanical explanation, in other words, are tailored not to the ideal of expectation but to the ideal of control.

Salmon identified two forms of CM explanation; both are common in neuroscience. An *etiological explanation* (or causal explanation) explains a phenomenon in terms of its antecedent causes. For example, death of dopaminergic neurons explains Parkinsonian symptoms; a *constitutive explanation* explains a phenomenon in terms of the underlying components in the mechanism, their activities, and their organization. The organization of cells in the hippocampus, for example, is part of the explanation for how it forms spatial maps. Constitutive explanations are inherently interlevel: they explain the behavior of a whole in terms of the organized activities of its parts.

For one who embraces the CM model, not every model of a phenomenon explains the phenomenon. Models can be used to summarize data, infer quantities, and generate predictions without explaining anything. Some models fail as explanations because they are purely descriptive, phenomenal models. Other models fail because they are inaccurate. Still others suffer as explanations because they contain crucial gaps. Consider these kinds of failure in turn.

2.1.4 *Phenomenal Versus Mechanistic Models*

Dayan and Abbott (2001) distinguish descriptive and mechanistic models of neural function. Descriptive models are phenomenal models. They summarize the explanandum. Mechanistic models, in contrast, describe component parts, their activities, and their organization.

Consider an example. Hodgkin and Huxley (1952) used the voltage clamp to experimentally investigate how membranes change conductance to the flow of ions at different voltages. They intervened to raise and lower the voltage across the membrane, measured the resulting ionic current, and used that to infer the membrane's conductance at that voltage. They used these data to fit a curve describing membrane conductance for different ions as a function of voltage. Hodgkin and Huxley insist, however, that these equations fail to explain the changes in membrane conductance. They do not describe the mechanisms and so merely summarize the voltage clamp data. They are descriptive, phenomenal models; they provide expectation without explanation.

2.1.5 *How-Possibly Models*

Some models are representational and true (to an approximation): they are intended to describe how a mechanism works and (to an approximation) they accurately do so. Some models are intended to represent a mechanism but fail because they are false. Other models intentionally distort a system to get at something deep and true about it. Different parts of the same model might be best understood in any of these different ways.

Return again to Hodgkin and Huxley. The conductance equations just described as phenomenal models were components in a powerful, more comprehensive model that describes the total current moving across the cell membrane during an action potential. The conductance equations were prized for their accuracy and were tightly related to the experimental data collected with the voltage clamp. Other parts of the model are not inferred directly from observations but rather apply highly general laws about the flow of electricity through circuits. In order to apply these laws, Hodgkin and Huxley made a number of idealizing assumptions about the neuron. For example, they assume that the axon is a perfect cylinder and that ions are distributed evenly throughout the cytoplasm. Different parts of the model are intended to have different degrees of fidelity to the actual mechanism: some differences are thought to matter; others are not. And this difference of emphasis says something about how Hodgkin and Huxley intended to use the model. Specifically, the model was part of an argument that a mechanism involving diffusion of ions across a semi-permeable membrane is sufficient to account for the diverse features of the action potential. The model is not intended as an accurate description of the mechanism. For their purposes, more or less accurate values for ion concentrations and conductance changes were essential. The precise shape of the neuron was not.

A purely how-possibly model might make accurate predictions for the wrong reasons. Hodgkin and Huxley, for example, temporarily posited the existence of "gating particles" in the membrane that move around and explain how the membrane changes its conductance. But they insisted that this model should not be taken seriously and should be used only for heuristic purposes. The explanatory content of the model is that content that fixes our attention on real and relevant parts, activities, causal interactions, and organizational features of the mechanism. One and the same model of a mechanism might vary internally in how much explanatory information it conveys about different aspects of the mechanism.

2.1.6 Completeness

The goal of explanation in neuroscience is to describe mechanisms. This does not entail that a model must describe everything about a mechanism to count as explanatory. A complete model of a particular phenomenon would not generalize to different situations. And if incomplete models do not explain, then science has never explained anything. The Hodgkin and Huxley model, for example, rises above the gory details (e.g., about the locations of the ions or shapes of the cells) to capture causal patterns relevant to action potentials generally.

Mechanisms can be described from many perspectives and at many different grains; there is no requirement that all of this information be housed in a single model. Yet it is against this ideal of a complete description of the relevant components of a mechanism (an ideal never matched in science) that one can assess progress in knowing how a mechanism or type of mechanism works. For some explanatory projects, very sketchy models suffice. For others, more detail is required.

Not everyone agrees that completeness and correctness are ideals of explanation. Functionalist philosophers argue that explanations need not reveal mechanisms at all to explain how something works. A model might be predictively adequate and fictional yet nonetheless explanatory (see Levy and Bechtel 2013; Weiskopf 2011). This view faces the challenge of articulating a view about explanation that doesn't collapse into the failed predictivism of the CL model (see Kaplan and Craver 2011; Piccinini and Craver 2011; Povich forthcoming). Other philosophers argue that explanations are improved by leaving out mechanistic details (Batterman and Rice 2014): good explanations use idealization and abstraction to isolate the core explanatory features of a system. An advocate of a causal mechanical model will simply acknowledge that such models are useful for getting at core causal features while leaving out other causally and explanatorily relevant details that might well make a difference in a particular case (see Povich forthcoming). Finally, others emphasize the role of optimality explanations in biology (as when one explains a coding scheme by appeal to its optimal efficiency of information transfer

[Chirimuuta 2014; Rice 2013]). Mechanists tend to see such optimality explanations as shorthand for more detailed evolutionary and developmental explanations. The challenge in each case is to articulate a general view of explanation, free of the problems that plague the CL model, according to which the models count as explanatory.

2.2 Neuroscience Methods

One role for philosophy of neuroscience is to reconstruct the logic of experiments and the inferences by which conclusions are drawn from different kinds of data. Neuroscientists operate with a number of assumptions about how to test hypotheses about the relationship between the mind and the brain; philosophy can help make those methodological assumptions explicit and encourage discussion of the merits of those methodological assumptions. Here, we discuss some neuroscientific methods that have attracted philosophical discussion.

2.2.1 *General Approaches*

2.2.1.1 *Animals and Model Systems*

Philosophers tend to focus on human brains and behavior. However, neuroscientists are limited to noninvasive and postmortem methods to study humans directly. For this reason, most of our knowledge about how brains work comes from research on nonhuman animals. The reliance on animal models raises a number of theoretical and practical questions about the similarities and differences between human and nonhuman brains and behaviors and so about the inferences that can be made by comparing them (Shanks, Greek, and Greek 2009). Such inferences are grounded in similarities of structure and function as well as in phylogenetic relationships among species. Such inferences are more warranted in dealing with highly basic, highly conserved functions and in cases where known similarities exist (Bechtel 2009). However, some of the phenomena we most wish to understand are dependent on functions or capacities that nonhuman animals do not share with us (Roskies 2014). Humans and other animals might perform the same task with very different mechanisms. This challenge is especially pressing in the use of animal models to study human neuropsychiatric disorders, where many of the symptoms, such as delusions, emotions, hallucinations, and the like cannot convincingly be established in nonhumans (Nestler and Hyman 2010).

2.2.1.2 *Tasks, Functional Decomposition, and Ontology*

Any technique aimed at revealing the relation between cognitive performance and neural mechanisms relies on methods to measure cognitive performance. Many techniques require tasks specifically designed to engage capacities taken to be real, unitary components in cognitive and/or neural mechanisms. The use of tasks to operationalize psychological constructs relies on a theory of the task according to which the unitary component in question contributes to its performance. Often, these components are derived from elements of an intuitive, background cognitive psychology. However, an abiding worry is that our intuitive functional ontologies do not match the operations by which the brain solves these tasks (Anderson 2015). This issue has been discussed in philosophy both by philosophers of mind who defended folk psychology as an accurate picture of mind (Fodor 1980) and by those who advocated its abandonment (Churchland 1981). Sullivan has emphasized that neuroscientists often use many different task protocols to study the same construct and that it is no trivial matter to show that the different protocols are in fact engaging the same neural mechanisms (Sullivan 2009). Such inferences are especially perilous in dealing with distinct populations (e.g., infants, nonhumans, those with psychiatric disorders) that might exhibit the same behaviors using altogether different mechanisms.

2.2.1.3 *New Methods*

Recent technological developments greatly expand the range of epistemological questions we need to ask. Genomics will undoubtedly provide a much better understanding of the ways in which humans differ from other species at a molecular level and may provide tools for assessing the likelihood of translational success. Behavioral genetics and neurogenetics promise to illuminate the extent and nature of the dependence of neurobiological and behavioral traits on genes. Research has already established that these dependencies are often complex and multigenic and that few traits or diseases are determined by single alleles. In fact, when strong associations have been found, they often are highly modulated by environmental factors (Buckholtz and Meyer-Lindenberg 2008; Caspi et al. 2002). These facts should help dispel the erroneous idea that our genes determine our futures independently from the environment. These techniques also raise a number of questions about the use of "big data" and the assumptions by which one searches for significant effects in a vast sea of comparisons (e.g., Ioannidis 2005).

Advances in such understanding bring with them an inevitable increase in our ability to control such systems. Techniques such as CRISPR provide unprecedented control in gene editing and engineering, opening the possibility for more direct control over brain machinery. Optogenetics, which uses genetic techniques to make specific neural populations functionally responsive to illumination by light of specific frequencies, already allows relatively noninvasive and highly controlled manipulation of specific cell types and is poised to revolutionize our understanding and manipulation of brain function and brain disorders. Some philosophers of neuroscience see the emergence of optogenetics as a case study for considering how science produces a maker's knowledge of how to build and manipulate (rather than represent) brain system (Craver forthcoming). Such technologies also raise pressing ethical questions about how neuroscientific knowledge can and ought to be utilized.

2.2.2 *Lesions*

Behavioral data from brain-lesioned organisms have long provided key evidence about the functional organization of the brain. Experimentally induced lesions in animals have been used to investigate relationships of dependence and independence among cognitive systems, the localization of function in the brain, and the causal structure of the neural systems.

Interpretation of lesion studies is complicated by many factors. Lesions are often imprecise and incompletely damage many structures rather than completely damaging any one structure. Lesions are often also highly variable across individuals. Even assuming that one can find or produce perfectly isolated and replicated lesions, the inference from lesion to the causal structure of the brain is perilous. The brain recovers after the lesion, and redundant systems might take over for the function of the lesioned area; both of these would mask the lesioned area's involvement. Furthermore, the effects of a lesion are not always confined to the area of the lesion. Damage to brain vasculature and far-reaching changes to patterns of network connectivity elsewhere might be responsible for the lesions' effects. Furthermore, a region might contain both intrinsic connectivity and fibers of passage that connect functionally unrelated areas of the brain.

The central form of inference for relating behaviorally measured deficits to brain structure is the dissociation inference. In a single dissociation, it is shown that an individual or group with damage to a particular brain region succeeds on one or more tasks measuring cognitive capacity A while failing on tasks measuring cognitive capacity B. This shows that cognitive capacity B is not necessary for cognitive capacity A. Double dissociations demonstrate the mutual independence of A and B. Double dissociations are commonly understood as powerful arguments for the existence of distinct modules in the brain (Davies 2010).

These inferences, however, face challenges that temper any overly sanguine application of this methodology. First, behavioral patterns reflecting double dissociations have been shown to result from damage to single modules within connectionist models (Plaut 1995). Second, the appearance of double dissociations can be produced if one does not take into account the relative resource demands of the tasks measuring A and the tasks measuring B. If A requires more processing resources (more working memory or faster information processing) than task B, then one might retain the ability to perform the less demanding task while losing the ability to perform the more demanding task even if one and the same system is responsible for both (Glymour 1994). Both defenders and detractors of this methodology should acknowledge that these inferences are not deductive but abductive: we infer to the best explanation for the observed changes in behavior (Davies 2010).

2.2.3 *Neuroimaging*

Dissociation studies provide evidence about whether a cognitive capacity is necessary for a certain kind of task performance. They do not allow one easily to infer what the lesioned structure does in an intact brain. The advent of noninvasive neuroimaging technologies, such as positron emission tomography (PET) and functional magnetic resonance imaging (fMRI), allowed neuroscientists to address this question more directly by studying how blood flow, which correlates with neural activity, changes during the performance of a task. Data are collected from discrete volumes of brain tissue (represented as voxels), allowing levels of activity to be compared across task conditions.

These methods have revolutionized medicine and cognitive neuroscience. They have also generated philosophical controversy (Hanson and Bunzl 2010).

The main philosophical questions are epistemological, concerning what we can learn about brain function from imaging data, but ethical questions about how neuroimages are consumed have also been raised. Brain images are misleadingly simple and compelling relative to the complex procedures and manipulations required to produce them (Weisberg, Keil, Goodstein, Rawson, and Gray 2008). Although the "seductive allure of neuroscience" was widely predicted to bias nonexperts viewing neuroimages (Roskies 2008), several studies in forensic contexts suggests that neuroimages themselves are not biasing (Schweitzer et al. 2011). Because neuroimaging has come to dominate human cognitive neuroscience, we discuss some of these complex concerns in depth.

2.2.3.1 *What Does fMRI Measure?*

The question of what produces the fMRI signal and how it is related to brain function has been an important focus of neuroscientific research. fMRI methods measure the blood oxygen level dependent (BOLD) signal, which is the ratio of oxygenated to deoxygenated hemoglobin in a voxel. This measure has a complicated and indirect relationship to neural activity (Buxton 2009). Studies have corroborated the hypothesis that the fMRI BOLD signal correlates with neural activity and have suggested that the fMRI signal primarily reflects synaptic processing (local field potentials) rather than action potentials (Logothetis and Wandell 2004). However, the BOLD signal has many limitations: it does not distinguish between excitatory and inhibitory neurons, and it has relatively low spatial and temporal resolution. Moreover, because we still do not understand how neurons encode information, it is difficult to know which neural signals are computationally relevant and which ones are irrelevant. Finally, neuroimaging at its best provides only a weak kind of causal information about brain function: namely, that performing a given kind of task causes an increase in activity in given loci. This does not establish that the brain region is actually involved in the performance of the task, or that its involvement is in any way specific to the task. In order to establish the causal relevance of brain areas to particular functions, neuroimaging data must ultimately be supplemented by other information, ideally from more direct interventions on brain function (lesions, stimulation, transcranial magnetic stimulation [TMS], etc.).

2.2.3.2 *What Can Brain Imaging Tell Us About Cognition?*

No brain imager denies the importance of cognitive psychology to the design of task-based imaging experiments. But many psychologists deny the importance of imaging to understanding cognitive function. Philosophical arguments about multiple realizability and the autonomy of the special sciences imply that neuroimaging cannot provide much insight into the nature of mind (Fodor 1974). Arguments from psychology have been similarly dismissive of neuroimaging. For example, some claim that neuroimaging can provide information only about where mental processes occur, not about how they occur, and that mapping (localizing) function is fundamentally uninteresting. Others argue that brain imaging cannot guide theory choice in psychology (Coltheart

2006). Such arguments fail to recognize that empirically testable commitments about structure–function relationships can serve as bridge principles between functional and anatomical data and thus allow neuroimaging to bear on psychological theory choice (Roskies 2009). However, increasingly clever and powerful experimental and analytical techniques have been developed to allow neuroimaging data to more precisely mirror temporal and functional variables, and there are ample illustrations of neuroimaging's evidential relevance to psychology. Even the most vociferous deniers seem to accept now that neuroimaging can shed light on psychological questions (Coltheart 2013). What remains unknown is how much neuroimaging can tell us: what are its scope and limits?

One factor fueling early skepticism about the evidential relevance of neuroimaging came from meta-analyses of neuroimaging studies that showed divergent results even among studies of like phenomena. Over time, however, evidence has amassed for reliable overlap of results among many tasks with similar content or structure, and there has been a growing awareness of the existence of both individual differences and the context sensitivity of brain networks recruited for various tasks. In addition, novel methods of registration provide increasingly precise ways of compensating for functional-anatomical variability across subjects (Haxby, Connolly, and Guntupalli 2014). On the whole, the voices of those who proclaim that neuroimaging results are just noise or are uninterpretable have faded into the past.

Neuroimaging nicely illustrates how our theoretical frameworks and analytical methods shape and are shaped by our interpretations of the neuroscience data. Early neuroimaging was influenced by the pervasive assumption of modularity of brain functions advocated by philosophical and psychological theories alike (Kanwisher, McDermott, and Chun 1997). Studies sought to identify individual brain regions responsible for performing identifiable task-related functions using subtraction methods, in which differences in regional activation across tasks are ascribed to task differences. In early studies, attention focused on the region or regions that showed highest signal change in relation to task manipulations. Some regions of cortex were consistently activated with particular types of tasks or tasks involving particular kinds of stimuli, consistent with the view that the brain/mind is a modular system composed of more or less isolable and independent processing units with proprietary functions (note that this view is distinct from the modularity of mind proposed by Fodor [1983], which was designed to describe only peripheral systems and includes features that are inappropriate for internal processing modules, such as innateness, automaticity, and information encapsulation in any strong sense).

Criticisms were leveled against neuroimaging in virtue of these subtraction methods. Some rightly highlight limitations inherent in the interpretation of single contrasts, underscoring the need for factorial designs (Price, Moore, and Friston 1997) and for what has been called "functional triangulation" (Roskies 2010b). Other criticisms misunderstand subtraction as wedded to assumptions about the sequential nature of neural processing, the absence of feedback, or the localization of functions to single regions of activation (van Orden and Paap 1997). However, subtraction involves no such assumptions (Roskies 2010*b*). Nonetheless, the focus on attributing function to areas with maximum levels of signal change perpetuates a modular picture of brain function that is likely false.

Over the past decade, powerful new methods of analysis have allowed neuroscientists to study distributed processing across large cortical networks. Multivariate techniques allow tasks to be correlated with widespread patterns of brain activity rather than individual regions, thus demonstrating that the brain encodes information in a distributed fashion across large portions of the cortex (Norman, Polyn, Detre, and Haxby 2006). Using these methods, scientists have demonstrated that information in the fMRI signal in different areas of cortex is sufficient to distinguish stimulus identity at various levels of abstraction, task demands, or planned actions. They do not demonstrate whether or how this information is being used (though some of this can be inferred by looking at what information is represented in different parts of the processing hierarchy). Another area of active development is research on (nonfunctional, non–task-related) resting state connectivity in the human brain. In this application, researchers mine data about correlations of slow-wave oscillations in spatially disparate brain regions to form hypotheses about the system-level organization of brain networks (see, e.g., Power et al. 2011). These techniques change the way researchers conceive of cortical representation and functional specificity and vastly improve the predictive power of neuroimaging. The deeper understanding they engender takes seriously the notion that entire brain systems or networks rather than individual regions contribute to task performance.

Novel analytical techniques continue to improve our understanding of how neural representations are encoded throughout cortex. For example, correlations between patterns of neural activity and stimulus features or semantic models allow the prediction of novel activation patterns and evaluation of the models' success (Haxby et al. 2014). Techniques such as representational similarity analysis (RSA) can aid in determining the representational specificity of various brain regions (Kriegeskorte, Mur, and Bandettini 2008). These new methods improve our understanding of how representations are elaborated across cortical areas.

2.2.3.3 *The Logic of Brain Imaging*

People routinely use fMRI to infer the function of regions of activation based on psychological models. Such "forward inferences" are relatively unproblematic (or as unproblematic as the psychological models on which they depend; see Section 2.2.3.4. Cognitive Ontology). Reverse inference, however, has been widely criticized as erroneous. In a reverse inference, the involvement of a psychological capacity in a complex task is inferred from the observation of activity in a brain area previously associated with that capacity. As a deductive inference, the logic is clearly flawed: it assumes that brain regions have the same function in all tasks, and we know that brain regions often subserve multiple functions (Anderson 2010; Poldrack 2006). Many conclude that reverse inference is always illegitimate. However, this view is too strong. Bayesian reasoning can be used to provide probabilistic epistemic warrant for reverse inferences (Poldrack

2011). Taking into account task- or network-related contextual factors can also increase the likelihood that a hypothesis about function based on reverse inference is correct (Klein 2012). In fact, both reverse and forward inferences can be useful, but neither is demonstrative. As in neuropsychology, the interpretation in neuroimaging results involves an inductively risky inference to the best explanation. There are no crucial experiments, not even in neuroimaging.

2.2.3.4 *Cognitive Ontology*

Philosophers have raised the disquieting possibility that our folk psychological understanding of the human mind is mistaken. Perhaps more troubling still is the possibility that neuroimaging is structured so that it can only support and never dislodge these mistaken cognitive ontologies. As Poldrack aptly notes, no matter what functional decomposition or theory of the task one posits, contrasts between tasks will cause something to "light up" (Poldrack 2010). Can neuroimaging (augmented by information from other fields) help to bootstrap our way out of a mistaken ontology? Or are we doomed to tinker with theories that can never converge on the truth? To address this issue, we might develop data-driven, theory-neutral methods for making sense of imaging data. Such work is in its infancy, but early attempts suggest that our intuitive ontologies are not the best explanations of brain data (Anderson 2015).

3 LOOKING OUTWARD: WHAT CAN PHILOSOPHY OF NEUROSCIENCE TELL US ABOUT SCIENCE?

There is no greater villain in the history of philosophy, viewed from the point of view of contemporary neuroscience, than Rene Descartes. This is because Descartes was a dualist, holding that the mind is an altogether distinct substance from the body. Never mind that he held that everything in the non-mental world could be understood ultimately in terms of the lawful, mechanical interactions of minute particles. His villainy results from questioning whether this mechanistic perspective is sufficient to understand the working of the human mind and from his decision that it was not: in addition to physical mechanisms, minds require souls. Although Descartes is often treated as a villain, one might just as easily see him as expressing a thesis about the explanatory limits of a properly "mechanical" conception of the world.

In this section, we consider two sets of questions related to this thesis. First, we ask about the unity of neuroscience, the integration of different scientific disciplines in the study of the mind and brain. Second, we ask about the ontology of the multilevel structures common in neuroscientific models and, in particular, whether higher level things are nothing but lower level things. In our final section, we turn to potential limits of mechanistic, causal explanation when it comes to the domain of the mind.

3.1 Integration and the Unity of Neuroscience

There is, at the moment, no Newton or Darwin to unify our understanding of the nervous system. Is this problematic? Or is the ideal of unification perhaps inappropriate for the neurosciences?

What would a unified theory of the mind-brain even look like? According to a classic picture (Oppenheim and Putnam 1958), the world can be ordered roughly into levels of organization, from high-level phenomena (such as those described by economics) to low-level phenomena, such as those described by quantum mechanics. Things at higher levels are composed of things at lower levels, and theories about higher level phenomena can be explained in terms of theories at lower, and ultimately fundamental, levels. As in the CL model, such explanations would involve derivations of higher-level theories from lower-level theories (e.g., Schaffner 1993). Because theories about different levels are typically expressed using different vocabularies, such deductive unification requires bridge laws linking the two vocabularies. This classic view of the unity of science combines commitments to (1) a tidy correspondence between levels of ontology, levels of theories, and levels of fields, and (2) the idea that all of the levels can be reduced, step by step, to the laws of physics. This caricature of the classic view provides a clear contrast to more contemporary accounts.

Bickle (2003), for example, argues that we can dispense with the idea that mechanisms in neuroscience span multiple levels of organization, as well as with the idea that scientists ought to integrate across such levels. According to his "ruthlessly reductive" view, higher level descriptions of the world are, in fact, merely imprecise and vague descriptions of behavior that guide the search for cellular and molecular mechanisms. The unity of neuroscience is achieved not by a stepwise reduction to the lowest level, but by single explanatory leaps that connect decisions to dopamine or navigation to NMDA receptors. His view fits well with work in many areas of cellular and molecular neuroscience, in which scientists seem to bridge between behaviors and molecules in a single experimental bound. And, once levels are gone, there is no pressing question of identifying things at higher levels with things at lower levels. The pressing question is merely how the higher level sciences fix our attention on the right molecules.

Another alternative, the mechanistic view, is more a conservative elaboration and emendation of the classic view. Designed to fit examples drawn from more integrative areas of neuroscience research, the mechanistic view stresses the need to elaborate the causal structure of a mechanism from the perspective of multiple different levels studied with multiple different techniques and described with multiple theoretical vocabularies (Craver 2007; Machamer, Darden, and Craver 2000). The appropriate metaphor is a mosaic in which individual sciences contribute tiles that together reveal the mechanistic structure of a brain system. Examples of multifield integration are easy to find in neuroscience, but one example concerning the multilevel organization of spatial learning and spatial memory has received particular attention.

For mechanists, the idea that the nervous system has multiple levels is simply a commitment to the idea that the various activities of the nervous system (e.g., forming a spatial map) can be understood by revealing their underlying mechanisms, that the components and activities of these mechanisms can themselves be so decomposed, and so on. Levels thus understood are local, not monolithic. The spatial learning system is at a higher level than spatial map formation in the hippocampus, and spatial map formation is at a higher level than place cell firing. But this does not imply that all things that psychologists study are at the same level, nor that all capacities decompose into the same sets of levels, nor that there is a single theory that covers all phenomena such as learning and memory. Unity in neuroscience, on this view, is achieved capacity by capacity and not by a grand reduction of a "theory of psychology" to a "theory of neuroscience"; local theories about specific capacities are related to the mechanistic parts for that capacity. Viewed from this vantage point, there is little prospect for, and little apparent need for, a unifying theory that encompasses all psychological phenomenon under a concise set of basic laws (as Newton crafted for motion).

Mechanists have also tended to distance themselves from the classical idea that the unity of neuroscience, such as it is, is defined by stepwise reduction among levels. Integration can move up and down across levels, but integration also takes place between fields working on phenomena that are intuitively at the same level. For example, one might combine Golgi staining and electrophysiological recording to understand how electrical signals propagate through the wiring diagram of the hippocampus. In sum, the mechanistic view differs from the classical view in presenting a more local view of levels and a more piecemeal and multidirectional view of how work in different fields, using different techniques and principles, can be integrated without reducing them all to the lowest physico-chemical level (Craver 2007).

To some, however, even this more limited vision of unity is overly optimistic. The ability to combine results from distant corners of neuroscience requires some way of ensuring that everyone is working with the same concepts and constructs. Sullivan (2009) argues that neuroscience is plagued by a multiplicity of experimental protocols and lab norms that makes it difficult, if not impossible, to compare what is discovered in one lab with what is discovered in another. If the same construct is operationalized differently in different labs, one cannot assume that results described with the same terms in fact are really about the same phenomenon. If so, unity might be a good ideal, but there is reason to doubt whether science as currently pursued will genuinely achieve it.

As we write, a novel and potentially unifying theory of brain function is under consideration. Predictive coding models of brain function postulate that error signals are encoded in feedforward neural activity and that feedback activity contains representational information from higher cortical levels (Friston and Kiebel 2009). Incorporating biological data and insights from Bayesianism and other computational approaches, predictive coding promises to unify a wide range of psychological and neurobiological phenomena at a number of levels within a single theoretical framework (Hohwy 2014). Evidence is mounting in accord with predictive coding predictions, yet some psychological constructs still seem difficult to weave into predictive coding accounts (Clark

2013). If the promise of this model is borne out, we may have to revisit the question of unification.

3.2 Ontology

Grant that the capacities of the nervous system can be decomposed into capacities of parts of the nervous system and so on. A second question concerns the ontology of such multilevel structures. Isn't every higher level capacity in some sense just the organized behavior of the most fundamental items? Most philosophers say yes, but they disagree about what, exactly, they are assenting to.

There appears to be broad consensus that wholes can often do things that individual parts cannot do. Cells generate action potentials. A single ion channel cannot generate an action potential. For some, these considerations alone warrant the claim that higher-level phenomena are "more than" their parts: they can do things the parts cannot do, they have properties that are not simple aggregations of the properties of the parts, and they are organized (Wimsatt 1997).

There is still considerable room for disagreement, however, when we ask: is there anything more to the behavior of the mechanism as a whole than the organized activities of its component parts (in context)? For some, the answer is no: the behavior of a type of mechanisms is identical to types of organized entities and activities that underlie them (e.g., Polger 2004). This type-identity thesis explains why cognitive capacities are correlated with activations of physical mechanisms. It also fits with the reductionist component of the classical model of the unity of science: these type-identities are the bridge laws that link the vocabularies describing different levels. Finally, the thesis avoids the problem of top-down causation discussed later.

The type-identity thesis faces the challenge of multiple realizability: behaviors or capacities can typically be realized by an innumerably large number of mechanisms that differ from one another in innumerable ways (see, e.g., Marder and Bucher 2007). The mapping of higher-level to lower-level capacities is one-to-many.

Here, the metaphysical options proliferate. Some hold out for type-identity, claiming that genuine multiple realization is rare (Shapiro 2008) or reflects a failure to match grains of analysis (Bechtel and Mundale 1999; see also Aizawa 2009). Others embrace a form of token identity, holding that each individual manifestation of a higher-level capacity is identical to a particular set of organized activities among lower-level components. Yet it is unclear that the idea of token identity can be expressed coherently without collapsing into the thesis of type identity (see Kim 1996). The discussion of identity and its formulation is well beyond the focus of this essay (see Smart 2007).

Still others have opted for a third way: nonreductive physicalism. According to this view, higher-level properties or capacities are irreducible (not type-identical) to lower-level properties and capacities, and they have causal powers of their own, but they nonetheless are determined by (supervene upon) the most fundamental structure of the world.

Nonreductive physicalism has some virtues. First, it acknowledges the problem of multiple realization and, in fact, uses that as a premise in an argument for the autonomy of higher-level causes. Second, the view preserves the idea that higher-level capacities are real, not dim images of a fundamentally atomic reality (as a Bickle might have it). And third, it preserves a general commitment to physicalism: what goes on at higher levels is ultimately grounded in what happens at the physical level.

Despite the prima facie plausibility of this view, it faces a persistent challenge to explain how higher level phenomena can have causal powers and so a legitimate claim to being real at all (Kim 2000). The challenge might be put as a dilemma: either higher level capacities have effects over and above the effects of their lower level realizers, or they do not. If they do not, then their claim to existence is tenuous at best. If they do, then their effects seem to be entirely redundant with the effects of the physical causes at play. In that case, the nonreductive physicalist is committed to the idea that most events are multiply overdetermined by a set of redundant causes. For some, such as Kim, this is a most unattractive world picture. For others (e.g., Woodward 2003), the sense that multiple overdetermination is unattractive rests on a mistaken view of causation.

What of Bickle's idea that higher level capacity descriptions merely guide our attention to lower level mechanisms? Many of the details about the molecules and their activities don't make any difference at all to how things work at higher levels. The cell is a blooming buzzing confusion; mechanistic order is imposed selectively by focusing on just those lower level interactions and features that make a difference to the higher level capacities (such as spatial learning). If so, higher-level capacities are not pointers to those lowest level mechanisms; rather, they are central to the identity conditions of the mechanism in the first place.

The issues in this section are quite general and have nothing whatsoever to do with the particular topic area of neuroscience. Hierarchical organization and nearly decomposable structure are features of systems generally. For this reason, we leave this discussion behind to consider more direct implications of neuroscience for issues in ethics and the philosophy of mind.

4 LOOKING ACROSS: NEUROSCIENTIFIC APPROACHES TO PHILOSOPHICAL QUESTIONS

Neuroscience has made remarkable progress in uncovering diverse mechanisms of the central nervous system at and across multiple levels of organization. Despite this, there are far fewer solved than unsolved problems. And among these unsolved problems are some that seem especially recalcitrant to a solution. These problems are recalcitrant not so much because we lack the relevant data but because we have no idea what sort of evidence might be relevant. They are recalcitrant not so much because their complexity defies our understanding but because we can't seem to imagine any way that even an arbitrarily complex mechanism could possibly do what it must to explain the phenomenon. They are recalcitrant, finally, not so much because the relevant science has yet to be done but because something about the phenomenon appears to evade formulation within the language and the explanatory building blocks that science offers. It is often with these problems that we find the closest nexus with philosophical puzzles.

4.1 Free Will/Agency

The philosophical problem of free will is generally thought to be tied to the metaphysical question of determinism, although, since Hume, it has been recognized that indeterminism poses an equally troubling background for our commonsense conception of freedom. Nonetheless, libertarian solutions to the problem of free will ground freedom in the existence of indeterministic events in the brain (e.g., Kane 1999). Although some have supposed that neuroscience will answer the question of determinism, it will not (Roskies 2006). At most, it will illustrate what we have long suspected: that neural mechanisms subserve behavior. The hard problems in this area involve assessing the implications of this ever-encroaching mechanistic explanation of human behavior for our ordinary concepts of choice, decision-making, reasons, responsibility, and a whole host of related concepts bound up with the folk conception of human freedom.

Although not specifically aimed at the philosophical problem of freedom, the neuroscience of volition has flourished (Haggard 2008; Roskies 2010*a*). Voluntary (i.e., intentionally guided) action is thought to involve frontoparietal networks, and endogenously generated actions are thought to involve the pre-supplementary motor area (preSMA). Progress is being made on other aspects of agency, such as decision-making, task-switching, conflict-monitoring, and inhibitory processes. Ultimately, however, there is challenging philosophical work to be done to specify the relationship between these subpersonal mechanisms and personal-level decisions and responsibilities.

One thread of research that has attracted philosophical discussion concerns the causal relevance of conscious decision-making to action. The research focuses on an electroencephalographic (EEG) signal, the readiness potential (RP), in the human brain. Libet argued that the RP, which reliably predicts that one is about to spontaneously initiate movement, appears well before subjects report being aware of having decided to move (Libet, Gleason, Wright, and Pearl 1983). He reasoned that the brain "decides" to move before we make the conscious decision to do so, and he concluded that our conscious decisions are utterly inefficacious; so we lack free will. While Libet's basic findings characterizing the RP have been replicated numerous times, many studies question his interpretations. For example, Schlegel et al. have shown that the RP occurs in nonmotor tasks and in tasks that involve preparation but not action (Schlegel et al. 2013). Schurger has shown that averaging signals time-locked to decisions in a drift-diffusion model results in a signal that looks like the RP (Schurger, Sitt, and Dehaene 2012). Thus, the interpretation of the RP as evidence of a subpersonal decision to move cannot be maintained. Libet's method of measuring time of awareness has been criticized, as has the nature of the state that is indexed by subject's report (Roskies 2011). Recordings of single-unit activity during a Libet task show that both movement and time of willing can be decoded from neural data on a single-trial basis and prior to a subject's judgment of having decided to move (Fried, Mukamel, and Kreiman 2011). This study demonstrates that information encoding impending action (and decision) is present in the brain prior to movement but is subject to the same criticisms as the original paradigm with respect to measurement of the time of willing and the nature of the state measured (Roskies 2011).

Neuroimaging studies showing the ability to predict subjects' actions seconds prior to decision are offered in the same spirit as Libet but also fail to successfully demonstrate that we lack flexibility in our decisions or that "our brains decide, we don't." Neuroimaging studies showing that information is present well before decisions show only slightly more predictive power than chance and can be interpreted as merely indicating that decisions we make are sensitive to information encoded in the brain with long temporal dynamics (Soon, Brass, Heinze, and Haynes 2008). Indeed, the finding that brain activity precedes action or even decisions is not surprising. Any materialist view of mind would expect brain activity corresponding to any kind of mental process, and one that evolved in time would likely be measurable before its effects. Thus far, no neuroscientific results demonstrate that we lack (or have) free will.

4.2 Mental Causation

A long-standing philosophical question involves whether mental states can play a causal role in action or whether the mind is epiphenomenal. The current debate is tightly tied to questions of reductionism (see the earlier discussion of Ontology). Neuroscientists have for the most part not grappled with the problem of mental causation in this kind of formulation but have instead begun to work on problems about how executive brain areas affect the functioning of controlled (and so "lower-level" areas, using "level" in an entirely distinct sense than gives rise to the philosophical problem of mental causation). Tasks involving endogenous manipulation of attention and set-switching are frequently used. Attention has been shown to affect lower-level processing, sharpening or potentiating the tuning curves of neurons involved in lower-level task functions. Other work has implicated synchronous neural activity in the modulation of lower-level function. Recent work suggests that the dynamical properties of a system may themselves govern cued set-switching (Mante, Sussillo, Shenoy, and Newsome 2013). This interesting work raises intriguing questions about how such results fit into traditional views of mind and executive function.

4.3 Meaning

The nature of intentionality has been a preoccupation of philosophers since Brentano coined the term. How do mental representations mean anything? Although early thinkers despaired of a naturalistic solution to this problem, later thinkers pursued various naturalistic accounts (Dretske 1981; Fodor 1980; Millikan 1987). The problem of intentionality or meaning has not been solved, but, as neuroscience and psychology continue to articulate the representational vehicles that the brain constructs on the basis of experience and the way in which those are shaped by expectations, the broad outlines of a possible solution to the problem might be taking shape (see, e.g., Clark 2013).

A particular difficulty should be flagged here. When we speak of a neural representation, we often describe a pattern of neural activity as "meaning" or "standing" for the stimulus with which it best correlates. This manner of speaking should not be taken to suggest that neurons or patterns of activity "understand" anything at all; rather, it is often nothing more than the expression of one or more causal or correlational relationships: that the pattern of activity is caused by its preferred stimulus, or with some behavioral output, or that it is correlated with some significant environmental factor in virtue of some less direct causal route (see, e.g., Haugeland 1998). Some have suggested that Brentano's challenge can be met with a purely causal theory of content (e.g., Fodor 1990). Others have appealed to natural selection to explain the normative content of representations (e.g., Garson 2012). Still other philosophers deny that the neural, causal, or correlational sense of representation (understood in causal or correlational terms) has the resources to understand the normative significance of mental content: thus there remains a gap between theories of neural representation and theories of how thinkers have thoughts about things (McDowell 1994).

4.4 Moral Cognition and Responsibility

Neuroimaging began to make direct contact with ethical questions with Greene's publication of brain activity during judgments about moral dilemmas (Greene, Sommerville, Nystrom, Darley, and Cohen 2001). Although a simplistic interpretation of brain activity as mapping to normative ethical positions has been roundly criticized, the results from imaging and other studies have led to a rapid growth in understanding of how the brain represents and processes information relevant to complex decision-making generally (not only moral decisions). Moral judgments appear not to involve special-purpose moral machinery but to recruit areas involved in decision-making and cognition more generally. Insight into moral cognition can be gained by seeing how permanent or temporary disruptions to neural machinery influence judgments. For example, damage to ventromedial cortex leads to moral judgments that deviate from the norm in being more "utilitarian" (Young, Bechara, et al. 2010), whereas disruption of the temporoparietal junction (TPJ), thought to be involved in social cognition, leads to moral judgments that are less sensitive to mental state of the offender (Young, Camprodon, Hauser, Pascual-Leone, and Saxe 2010). Neural data might also explain many aspects of our intuitions about punishment (Cushman 2008). Studies of individuals with episodic amnesia have shown that many cognitive capacities plausibly linked to human choice and moral judgment (such as an understanding of time, valuation of future rewards, executive control, self-knowledge, and the Greene effect of personal scenarios on moral judmgents) remain intact despite global and even life-long deficits in the ability to remember the experiences of one's life and to project one's self vividly into the future (Craver et al. 2014).

Regional patterns of activation are not the only data relevant to understanding moral behavior. Insights from neurochemistry and cellular neurobiology also inform our understanding of morality, indicating how prosocial behavior depends on neural function and perhaps supplying the framework allowing for naturalistic theorizing about the human development of morality (Churchland 2012). Despite some confusion on this issue, no descriptive data from neuroscience have provided a rationale for preferring one normative philosophical theory to another, and none can, without the adoption of some kind of normative assumptions (Berker 2009; Kahane 2012). However, one can see neural foundations that are consistent with prominent elements of all the major philosophical contenders for substantive normative ethical theories: utilitarianism, deontology, and virtue ethics. That said, in probing the neural basis of prudential and moral reasoning, neuroscientists are seeking to reveal the mechanistic basis by which moral decisions are made, but they are also seeking to reveal the mechanistic basis of being a moral decision-maker. As such, they require a number of substantive assumptions about the kind of thing one must be capable of doing in order to count as a maker of decisions, let alone a moral agent, at all.

4.5 Consciousness and Phenomenality

Consciousness, once the province solely of philosophers, has become a respectable topic for research in neuroscience, starting with Crick and Koch's publications on the topic in the 1990s. Their exhortation to find the neural correlates of consciousness has been heeded in several different ways. First, in medicine, novel neuroimaging techniques have been used to provide evidence indicating the presence of consciousness in brain-damaged and unresponsive patients. For instance, a few patients classified as being in a persistent vegetative state showed evidence of command-following in an imagery task (Owen et al. 2006). In further studies, the same group showed that behaviorally unresponsive patients were correctly able to answer questions using the tasks as proxies. Most recently, they contend that normal executive function can be assessed by brain activity during movie viewing and that the presence of this executive function is indicative of consciousness (Naci, Cusack, Anello, and Owen 2014). Their study showed that coherent narratives evoked more frontal activity than did scrambled movies and that this is related to executive function. The authors of these studies have claimed that the results show residual consciousness, yet this claim rests on a strong assumption that differential response to these commands and task situations is indicative of consciousness. The approach raises a host of epistemological questions about how residual mindedness in the absence of behavior might be detected with brain imaging devices and ethical questions about what is owed to persons in this state of being.

A more direct approach contrasts brain activity during liminal and subliminal presentation, with the hopes of identifying the differences in brain activity that accompany awareness (Rees, Kreiman, and Koch 2002). These attempts, although interesting neuroscientifically, are beset by a number of interpretational problems (Kouider and Dehaene 2007). Several other approaches have kinship with Baars's global workspace theory (Baars 1993). One long-standing observation links synchronous and oscillatory activity with binding and consciousness. A number of studies have corroborated the idea that corticothalamic feedback is necessary for conscious awareness of stimuli. Others try to quantify consciousness as a measure of information integration (Oizumi, Albantakis, and Tononi 2014). The available data do not (yet) suffice to distinguish between various theoretical accounts of consciousness, all of which posit central roles for feedback, coherence, oscillations, and/or information-sharing. Moreover, these views may explain functional aspects of consciousness (what Block has called access consciousness) but do not explain the more puzzling aspect of phenomenality.

The "hard" problem of phenomenal consciousness concerns the experiential aspect of human mental life (Chalmers 1995). The term "qualia" is often used for the "raw feelings" that one has in such experiences: the redness of the red, the cumin's sweet spiciness, the piercing sound of a drill through an electric guitar. There is, as Thomas Nagel described it, something it is like to have these experiences; they feel a certain way "from the inside." Why is this problem thought to be so difficult? Neuroscience has made tremendous progress in understanding the diverse mechanisms by which conscious experiences occur. There appears to be nothing in the mechanisms that explains, or perhaps even could explain, the experiences that are correlated with their operation. Chalmers made this point vivid by asking us to consider the possibility of philosophical zombies who have brains and behaviors just like ours, yet lack any conscious experiences. If zombies are conceivable, then all the mechanisms and behaviors might be the same and consciousness experience might be absent. If so, then conscious experience is not identical to any physical mechanism.

Many attempts have been made to respond to this hard problem of consciousness, and it would be impossible to canvass them all here. One could be an eliminativist and reject the hard problem because it asks us to explain something fictional. One could be an optimist and insist that the hard problem seems so hard only because we haven't yet figured out how to crack it. For those committed to a form of physicalism or even to the multilevel mechanistic picture, these problems are exciting challenges for the science of the mind-brain, terra incognita that will fuel interest in the brain for generations to come. How one thinks about the possibility for a science of consciousness is influenced by a host of deep metaphysical commitments about the structure of the world that must be untangled if any scientific progress is to be made.

5 CONCLUSION

Philosophical engagement with the neurosciences is of relatively recent origin, perhaps beginning in earnest with the work of Paul M. Churchland (1989) and Patricia S. Churchland (1989) who predicted that philosophical issues would look different as philosophers began to acquaint themselves with the deliverances of the neurosciences. In the nearly three decades since, this prediction has been borne out. The neurosciences now have a home in the philosophy of science and the philosophy of mind, as well as in moral psychology, ethics, and aesthetics. It is a significant challenge to these philosophers to stay up to date with the latest work in a large and rapidly changing field while, at the same time, maintaining the critical and distanced perspective characteristic of philosophers. Yet this distance must be maintained if philosophy is to play a useful, constructive role in the way that neuroscience proceeds and the way that its findings are understood in light of our need to find a place for ourselves in the natural world. Precisely because philosophers are not primarily caught up in the practice of designing experiments or in the social structure of the sciences, and precisely because they bring the resources of conceptual analysis and a long tradition for thinking about the mind that are not well-known to experimentalists, philosophers have much to contribute to this discussion.

ACKNOWLEDGMENT

ALR would like to thank Jonathan Kubert for research and technical assistance.

REFERENCES

Aizawa, K. (2009). "Neuroscience and Multiple Realization: A Reply to Bechtel and Mundale." *Synthese* 167: 493–510.

Anderson, M. L. (2010). "Neural Reuse: A Fundamental Organizational Principle of the Brain." *Behavioral and Brain Sciences* 33: 245–266.

Anderson, M. L. (2015). "Mining the Brain for a New Taxonomy of the Mind." *Philosophy Compass* 10: 68–77. http://doi.org/10.1111/phc3.12155

Baars, B. J. (1993). *A Cognitive Theory of Consciousness* (New York: Cambridge University Press).

Batterman, R. W., and Rice, C. C. (2014). "Minimal Model Explanations." *Philosophy of Science* 81: 349–376.

Bechtel, W. (2009). "Generalization and Discovery by Assuming Conserved Mechanisms: Cross-Species Research on Circadian Oscillators." *Philosophy of Science* 76: 762–773.

Bechtel, W. and Mundale, J. (1999) "Multiple Realizability Revisited: Linking Cognitive and Neural States" *Philosophy of Science* 66: 175–207.

Berker, S. (2009). "The Normative Insignificance of Neuroscience." *Philosophy and Public Affairs* 37: 293–329.

Bickle, J. (2003). *Philosophy and Neuroscience: A Ruthlessly Reductive Account* (Dordrecht: Kluwer).

Buckholtz, J. W., and Meyer-Lindenberg, A. (2008). "MAOA and the Neurogenetic Architecture of Human Aggression." *Trends in Neurosciences* 31: 120–129. http://doi.org/10.1016/j.tins.2007.12.006

Buxton, R. B. (2009). *Introduction to Functional Magnetic Resonance Imaging: Principles and Techniques*, 2nd ed. (New York: Cambridge University Press).

Caspi, A., McClay, J., Moffitt, T. E., Mill, J., Martin, J., Craig, I. W., and Poulton, R. (2002). "Role of Genotype in the Cycle of Violence in Maltreated Children." *Science* 297: 851–854. http://doi.org/10.1126/science.1072290

Chalmers, D. J. (1995). "Facing up to the Problem of Consciousness." *Journal of Consciousness Studies* 2: 200–219.

Chirimuuta, M. (2014). "Minimal Models and Canonical Neural Computations: The Distinctness of Computational Explanation in Neuroscience." *Synthese* 191: 127–153.

Churchland, P. M. (1981). "Eliminative Materialism and the Propositional Attitudes." *Journal of Philosophy* 78: 67–90. http://doi.org/10.2307/2025900

Churchland, P. M. (1989). *A Neurocomputational Perspective: The Nature of Mind and the Structure of Science* (Cambridge, MA: MIT Press).

Churchland, P. S. (1989). *Neurophilosophy: Toward a Unified Science of the Mind-Brain* (Cambridge, MA: MIT Press).

Churchland, P. S. (2012). *Braintrust: What Neuroscience Tells Us About Morality* (Princeton, NJ: Princeton University Press).

Clark, A. (2013). "Whatever Next? Predictive Brains, Situated Agents, and the Future of Cognitive Science." *Behavioral and Brain Sciences* 36: 181–204. http://doi.org/10.1017/S0140525X12000477

Coltheart, M. (2006). "What Has Functional Neuroimaging Told Us About the Mind (So Far)?" *Cortex* 42: 323–331.

Coltheart, M. (2013). "How Can Functional Neuroimaging Inform Cognitive Theories?" *Perspectives on Psychological Science* 8: 98–103. http://doi.org/10.1177/1745691612469208

Craver, C. F. (2007). *Explaining the Brain: Mechanisms and the Mosaic Unity of Neuroscience* (New York: Oxford University Press).

Craver, C. F. (forthcoming). Thinking About Interventions: Optogenetics and Makers' Knowledge in J. Woodward and K. Waters eds. *Causation in Biology and Philosophy* (Minnesota Studies in the Philosophy of Science).

Craver, C. F. Kwan, D., Steindam, C., and Rosenbaum, R. S. (2014) "Individuals with Episodic Amnesia are not Stuck in Time." *Neuropsychologia* 57: 191–195.

Cushman, F. (2008). "Crime and Punishment: Distinguishing the Roles of Causal and Intentional Analyses in Moral Judgment." *Cognition* 108: 353–380. http://doi.org/10.1016/j.cognition.2008.03.006

Davies, M. (2010). "Double Dissociation: Understanding Its Role in Cognitive Neuropsychology." *Mind and Language* 25: 500–540. http://doi.org/10.1111/j.1468-0017.2010.01399.x

Dretske, F. I. (1981). *Knowledge and the Flow of Information* (Cambridge, MA: MIT Press).

Dayan, P., and Abbott, L. F. (2001). *Theoretical Neuroscience. Computational Modeling of Neural Systems* (Cambridge, MA: MIT Press).

Fodor, J. A. (1974). "Special Sciences (Or: The Disunity of Science as a Working Hypothesis)." *Synthese* 28: 97–115.

Fodor, J. A. (1980). *The Language of Thought* (Cambridge, MA: Harvard University Press).

Fodor (1983). *The Modularity of Mind.* Cambridge, MA: MIT Press.

Fodor, J. A. (1990). *A Theory of Content and Other Essays* (Cambridge, MA: MIT Press).

Fried, I., Mukamel, R., and Kreiman, G. (2011). "Internally Generated Preactivation of Single Neurons in Human Medial Frontal Cortex Predicts Volition." *Neuron* 69: 548–562. http://doi.org/10.1016/j.neuron.2010.11.045

Friston, K., and Kiebel, S. (2009). "Predictive Coding Under the Free-Energy Principle." *Philosophical Transactions of the Royal Society of London B: Biological Sciences* 364: 1211–1221. http://doi.org/10.1098/rstb.2008.0300

Garson, J. (2012). "Function, Selection, and Construction in the Brain." *Synthese* 189: 451–481. doi: 10.1007/s11229-012-0122-y

Glymour, C. (1994). "On the Methods of Cognitive Neuropsychology." *British Journal for the Philosophy of Science* 45: 815–835.

Greene, J. D., Sommerville, R. B., Nystrom, L. E., Darley, J. M., and Cohen, J. D. (2001). "An fMRI Investigation of Emotional Engagement in Moral Judgment." *Science* 293: 2105–2108. http://doi.org/10.1126/science.1062872

Haggard, P. (2008). "Human Volition: Towards a Neuroscience of Will." *Nature Reviews Neuroscience* 9: 934–946. http://doi.org/10.1038/nrn2497

Hanson, S. J., and Bunzl, M. (eds.). (2010). *Foundational Issues in Human Brain Mapping* (Cambridge, MA: MIT Press).

Haugeland, J. (1998). *Having Thought: Essays in the Metaphysics of Mind* (Cambridge: Cambridge University Press).

Haxby, J. V., Connolly, A. C., and Guntupalli, J. S. (2014). "Decoding Neural Representational Spaces Using Multivariate Pattern Analysis." *Annual Review of Neuroscience* 37: 435–456. http://doi.org/10.1146/annurev-neuro-062012-170325

Hempel, C. G. (1965). *Aspects of Scientific Explanation* (New York: Free Press, Macmillan).

Hodgkin, A. L., and Huxley, A. F. (1952). "A Quantitative Description of Membrane Current and Its Application to Conduction and Excitation in Nerve." *Journal of Physiology* 117: 500–544.

Hohwy, J. (2014). *The Predictive Mind* (New York: Oxford University Press).

Ioannidis, J. (2005). "Why Most Published Research Findings Are False." *PLoS Medicine* 2: 124.

Kahane, G. (2012). "On the Wrong Track: Process and Content in Moral Psychology." *Mind and Language* 27: 519–545. http://doi.org/10.1111/mila.12001

Kane, R. (1999). "Responsibility, Luck, and Chance: Reflections on Free Will and Indeterminism." *Journal of Philosophy* 96: 217–240. http://doi.org/10.2307/2564666

Kanwisher, N., McDermott, J., and Chun, M. M. (1997). "The Fusiform Face Area: A Module in Human Extrastriate Cortex Specialized for Face Perception." *Journal of Neuroscience* 17: 4302–4311.

Kaplan, D. M., and Craver, C. F. (2011). "The Explanatory Force of Dynamical and Mathematical Models in Neuroscience: A Mechanistic Perspective." *Philosophy of Science* 78: 601–627.

Kim, J. (1996) *Philosophy of Mind.* Westwood Press.

Kim, J. (2000). *Mind in a Physical World: An Essay on the Mind-Body Problem and Mental Causation* (Cambridge, MA: MIT Press).

Klein, C. (2012). "Cognitive Ontology and Region- Versus Network-Oriented Analyses." *Philosophy of Science* 79: 952–960.

Kouider, S., and Dehaene, S. (2007). "Levels of Processing During Non-conscious Perception: A Critical Review of Visual Masking." *Philosophical Transactions of the Royal Society of London B: Biological Sciences* 362: 857–875. http://doi.org/10.1098/rstb.2007.2093

Kriegeskorte, N., Mur, M., and Bandettini, P. (2008). "Representational Similarity Analysis— Connecting the Branches of Systems Neuroscience." *Frontiers in Systems Neuroscience* 2: 2–24. http://doi.org/10.3389/neuro.06.004.2008

Levy, A., and Bechtel, W. (2013). "Abstraction and the Organization of Mechanisms." *Philosophy of Science* 80: 241–261.

Libet, B., Gleason, C. A., Wright, E. W., and Pearl, D. K. (1983). "Time of Conscious Intention to Act in Relation to Onset of Cerebral Activity (Readiness-Potential)." *Brain* 106: 623–642. http://doi.org/10.1093/brain/106.3.623

Logothetis, N. K., and Wandell, B. A. (2004). "Interpreting the BOLD Signal." *Annual Review of Physiology* 66: 735–769. http://doi.org/10.1146/annurev.physiol.66.082602.092845

Machamer, P., Darden, L., and Craver, C. F. (2000). "Thinking About Mechanisms." *Philosophy of Science* 67: 1–25.

Mante, V., Sussillo, D., Shenoy, K. V., and Newsome, W. T. (2013). "Context-Dependent Computation by Recurrent Dynamics in Prefrontal Cortex." *Nature* 503: 78–84. http://doi.org/10.1038/nature12742

Marder, E., and Bucher, D. (2007). "Understanding Circuit Dynamics Using the Stomatogastric Nervous System of Lobsters and Crabs." *Annual Review of Physiology* 69: 291–316.

McDowell, J. (1994). "The Content of Perceptual Experience." *Philosophical Quarterly* 44: 190–205.

Millikan, R. G. (1987). *Language, Thought, and Other Biological Categories: New Foundations for Realism* (Cambridge, MA: MIT Press).

Naci, L., Cusack, R., Anello, M., and Owen, A. M. (2014). "A Common Neural Code for Similar Conscious Experiences in Different Individuals." *Proceedings of the National Academy of Sciences of the United States of America* 111: 14277–14282. http://doi.org/10.1073/pnas.1407007111

Nestler, E. J., and Hyman, S. E. (2010). "Animal Models of Neuropsychiatric Disorders." *Nature Neuroscience* 13: 1161–1169. http://doi.org/10.1038/nn.2647

Norman, K. A., Polyn, S. M., Detre, G. J., and Haxby, J. V. (2006). "Beyond Mind-Reading: Multi-Voxel Pattern Analysis of fMRI Data." *Trends in Cognitive Sciences* 10: 424–430. http://doi.org/10.1016/j.tics.2006.07.005

Oizumi, M., Albantakis, L., and Tononi, G. (2014). "From the Phenomenology to the Mechanisms of Consciousness: Integrated Information Theory 3.0." *PLoS Computational Biology* 10: e1003588.

Oppenheim, P., and Putnam, H. (1958). "The Unity of Science as a Working Hypothesis." In H. Feigl et al. (eds.), *Minnesota Studies in the Philosophy of Science*, vol. 2. (Minneapolis: Minnesota University Press), 3–36.

Owen, A. M., Coleman, M. R., Boly, M., Davis, M. H., Laureys, S., and Pickard, J. D. (2006). "Detecting Awareness in the Vegetative State." *Science* 313: 1402–1402. http://doi.org/10.1126/science.1130197

Piccinini, G., and Craver, C. (2011). "Integrating Psychology and Neuroscience: Functional Analyses as Mechanism Sketches." *Synthese* 183(3): 283–311.

Plaut, D. C. (1995). "Double Dissociation Without Modularity: Evidence from Connectionist Neuropsychology." *Journal of Clinical and Experimental Neuropsychology* 17: 291–321.

Poldrack, R. A. (2006). "Can Cognitive Processes Be Inferred from Neuroimaging Data?" *Trends in Cognitive Sciences* 10: 59–63. http://doi.org/10.1016/j.tics.2005.12.004

Poldrack, R. A. (2010). "Mapping Mental Function to Brain Structure: How Can Cognitive Neuroimaging Succeed?" *Perspectives on Psychological Science* 5: 753–761. http://doi.org/10.1177/1745691610388777

Poldrack, R. A. (2011). "Inferring Mental States from Neuroimaging Data: From Reverse Inference to Large-Scale Decoding." *Neuron* 72: 692–697. http://doi.org/10.1016/j.neuron.2011.11.001

Polger, T. (2004). *Natural Minds* (Cambridge, MA: MIT Press).

Povich, M. (forthcoming) "Mechansms and Model-Based MRI" Philosophy of Science. *Proceedings of the Philosophy of Science Association.*

Power, J. D., Cohen, A. L., Nelson, S. M., Wig, G. S., Barnes, K. A., Church, J. A., et al. (2011). "Functional Network Organization of the Human Brain." *Neuron* 72: 665–678.

Price, C. J., Moore, C. J., and Friston, K. J. (1997). "Subtractions, Conjunctions, and Interactions in Experimental Design of Activation Studies." *Human Brain Mapping* 5: 264–272. http://doi.org/10.1002/(SICI)1097-0193(1997)5:4<264::AID-HBM11>3.0.CO;2-E

Rees, G., Kreiman, G., and Koch, C. (2002). "Neural Correlates of Consciousness in Humans." *Nature Reviews Neuroscience* 3: 261–270. http://doi.org/10.1038/nrn783

Rice, C. (2013). "Moving Beyond Causes: Optimality Models and Scientific Explanation." Noûs. doi: 10.1111/nous.12042

Roskies, A. L. (2006). "Neuroscientific Challenges to Free Will and Responsibility." *Trends in Cognitive Sciences* 10: 419–423. http://doi.org/10.1016/j.tics.2006.07.011

Roskies, A. L. (2008). "Neuroimaging and Inferential Distance." *Neuroethics* 1: 19–30.

Roskies, A. L. (2009). "Brain-Mind and Structure-Function Relationships: A Methodological Response to Coltheart." *Philosophy of Science* 76: 927–939.

Roskies, A. L. (2010a). "How Does Neuroscience Affect Our Conception of Volition?" *Annual Review of Neuroscience* 33: 109–130.

Roskies, A. L. (2010b). "Saving Subtraction: A Reply to Van Orden and Paap." *British Journal for the Philosophy of Science* 61: 635–665. http://doi.org/10.1093/bjps/axp055

Roskies, A. L. (2011). "Why Libet's Studies Don't Pose a Threat to Free Will." In W. Sinnott-Armstrong (ed.), *Conscious Will and Responsibility* (New York: Oxford University Press), 11–22.

Roskies, A. L. (2014). "Monkey Decision Making as a Model System for Human Decision Making." In A. Mele (ed.), *Surrounding Free Will* (New York: Oxford University Press), 231–254.

Salmon, W. (1984). *Scientific Explanation and the Causal Structure of the World* (Princeton, NJ: Princeton University Press).

Schaffner, K. F. (1993). *Discovery and Explanation in Biology and Medicine* (Chicago: University of Chicago Press).

Schlegel, A., Alexander, P., Sinnott-Armstrong, W., Roskies, A., Tse, P. U., and Wheatley, T. (2013). "Barking Up the Wrong Free: Readiness Potentials Reflect Processes Independent of Conscious Will." *Experimental Brain Research* 229: 329–335. http://doi.org/10.1007/s00221-013-3479-3

Schurger, A., Sitt, J. D., and Dehaene, S. (2012). "An Accumulator Model for Spontaneous Neural Activity Prior to Self-Initiated Movement." *Proceedings of the National Academy of Sciences* 109: E2904–E2913. http://doi.org/10.1073/pnas.1210467109

Schweitzer, N. J., Saks, M. J., Murphy, E. R., Roskies, A. L., Sinnott-Armstrong, W., and Gaudet, L. M. (2011). "Neuroimages as Evidence in a Mens Rea Defense: No Impact." *Psychology, Public Policy, and Law* 17: 357–393. http://doi.org/10.1037/a0023581

Shanks, N., Greek, R., and Greek, J. (2009). "Are Animal Models Predictive for Humans?" *Philosophy, Ethics, and Humanities in Medicine* 4: 1–20.

Shapiro, L. A. (2008). "How to Test for Multiple Realization." *Philosophy of Science* 75: 514–525.

Smart, J. J. C. (2007). "The Mind/Brain Identity Theory", *The Stanford Encyclopedia of Philosophy* (Winter 2014 Edition), Edward N. Zalta (ed.). <http://plato.stanford.edu/archives/win2014/entries/mind-identity/>

Soon, C. S., Brass, M., Heinze, H. -J., and Haynes, J. -D. (2008). "Unconscious Determinants of Free Decisions in the Human Brain." *Nature Neuroscience* 11: 543–545. http://doi.org/10.1038/nn.2112

Sullivan, J. A. (2009). "The Multiplicity of Experimental Protocols: A Challenge to Reductionist and Non-Reductionist Models of the Unity of Neuroscience." *Synthese* 167: 511–539.

van Orden, G. C., and Paap, K. R. (1997). "Functional Neuroimages Fail to Discover Pieces of Mind in the Parts of the Brain." *Philosophy of Science* 64: S85–S94.

Weisberg, D. S., Keil, F. C., Goodstein, J., Rawson, E., and Gray, J. R. (2008). "The Seductive Allure of Neuroscience Explanations." *Journal of Cognitive Neuroscience* 20: 470–477. http://doi.org/10.1162/jocn.2008.20040

Weiskopf, D. A. (2011). "Models and Mechanisms in Psychological Explanation." *Synthese* 183: 313–338.

Wimsatt, W. C. (1997). "Aggregativity: Reductive Heuristics for Finding Emergence." *Philosophy of Science* 64: S372–S384.

Woodward, J. (2003). *Making Things Happen: A Theory of Causal Explanation* (New York: Oxford University Press).

Young, L., Bechara, A., Tranel, D., Damasio, H., Hauser, M., and Damasio, A. (2010). "Damage to Ventromedial Prefrontal Cortex Impairs Judgment of Harmful Intent." *Neuron* 65: 845–851. http://doi.org/10.1016/j.neuron.2010.03.003

Young, L., Camprodon, J. A., Hauser, M., Pascual-Leone, A., and Saxe, R. (2010). "Disruption of the Right Temporoparietal Junction with Transcranial Magnetic Stimulation Reduces the Role of Beliefs in Moral Judgments." *Proceedings of the National Academy of Sciences of the United States of America* 107: 6753–6758. http://doi.org/10.1073/pnas.0914826107

CHAPTER 41

SOCIAL ORGANIZATION OF SCIENCE

MARTIN CARRIER

1 ACCUMULATION OR INTERACTION: SOCIAL ORGANIZATION AND CONFIRMATION

THE role of the scientific community and the importance of its social organization have only slowly ascended the research agenda of philosophy of science and epistemology. In the early stages of the scientific enterprise, the rules individual researchers were supposed to follow were considered sufficient for delineating good epistemic practice. As a consequence, the scientific community was initially conceived as a forum for the accumulation of discoveries and theoretical insights. Later, the interaction among various scientists was recognized as an essential element in providing reliable knowledge.

Francis Bacon (1561–1626) envisaged science as a collective enterprise. He took the cooperation among scientists as a precondition of success. The systematic nature of inquiry and the social collaboration among various researchers provide the chief distinction of the incipient sciences as compared to medieval scholarship. Bacon placed heavy emphasis on the division of epistemic labor. In contradistinction to the "all-in-one" thinkers of the Middle Ages, the new kind of researcher that Bacon heralded and advocated would pursue various projects in parallel and put his results together with those of others eventually. Dividing a research problem into a variety of subtasks and having the latter attacked simultaneously by a number of investigators is the royal road to speeding up the growth of knowledge (Bacon 1620/1863). In this Baconian framework, each scientist is considered a self-sufficient entity. The critical requirement is following Bacon's methodological canon (which includes abandoning prejudices, avoiding overgeneralizations by paying attention to counterinstances, giving heed to particular conditions, or taking distortions introduced by experimental intervention into due account). In other words, the advantage of pursuing scientific inquiry in a community lies in the accumulation of results gained by various scientists and thus in the multiplication of individual forces.

Another example of such a purely additive effect of cognitive achievements is the so-called jury theorem published by Nicolas de Condorcet (1743–94) in 1785. Condorcet proved that if a group of investigators is considered, each of whose members is more likely to hit the truth than not, then the probability that the majority of the group arrives at a truthful judgment increases with the number of members in the group. In other words, if the odds are better than half for each individual to arrive at a correct assessment, then groups have a better chance of passing correct judgments than individuals. This theorem, just as the Baconian view does, addresses the advantages of parallel reasoning, as it were, or of the accumulation of results gained individually, and it does not attend to the possible benefit of communication and debate.

By contrast, the philosophical discussion of the organization of the scientific community in the twentieth century places the interaction among scientists at center stage. The social organization of science is important because new features emerge from the exchange of arguments, results, or commitments among scientists. It is the interrelationship among scientists that distinguishes the scientific community from a collection of individuals. I sketch three aspects of scientific inquiry that are claimed to benefit from this social interaction: namely, test and confirmation, articulation of ideas, and choice of research topics.

First, as to *test and confirmation*, Robert Merton (1919–2003) was the first to analyze science consistently from the angle of the scientific community. He claimed that the interaction among scientists plays a key role in the process of testing and confirming individual knowledge claims. Having such claims checked by others enhances significantly the reliability of scientific knowledge. Science is not primarily a matter of individual geniuses and their lonely discoveries and conceptual innovations. Rather, the community is essential in examining the validity and appropriateness of alleged findings. Experiments are repeated, more sophisticated endeavors are built on their results, and premises are attacked and criticized. Examining such claims is essentially a social process that involves taking up and evaluating contributions of others. The benefit of such social interactions among scientists is reciprocal "policing." This feature serves to encourage scientists to play by the rules and resist the temptation to embellish their findings. Reciprocal policing helps to avoid premature and careless allegations and to anticipate fraudulence. In sum, subjecting scientific claims to the critical scrutiny of other scientists enhances significantly the reliability of scientific knowledge (Merton 1942/1973).

For this reason, the scientific community is the bearer or owner of knowledge. Knowledge is not simply produced by the individuals who conceive and advocate certain ideas but rather by a social body whose members reciprocally check and confirm their ideas. It is the community that transforms such ideas into dependable knowledge.

In this vein, a number of the cultural values that Merton thought to be constitutive of scientific judgment were directed at the social processes in the scientific community. In particular, the Mertonian value of "disinterestedness" is to be understood as a social value. Disinterestedness means that the scientific community does not prefer certain research results over others. This is an institutional imperative, not a psychological factor; it addresses the community, not individual researchers who may well be keen to produce outcome of a certain kind (Merton 1942/1973). The disinterestedness of the scientific community as a whole is epitomized by the struggle between antagonistic approaches. The community is divided and, for this reason, does not express a joint preference as to what the quest at hand should accomplish.

Karl R. Popper (1902–94) occasionally took up Merton's conception of social policing. The *Logic of Scientific Discovery* (1935/2002) leaves it open about who is supposed to criticize a scientist's favorite ideas and sometimes suggests that a self-conscious Popperian needs to be his or her first critic. However, in the context of his later contributions to the dispute about a value-free social science (Popper 1969/1988), Popper emphasized that the proponent of an idea and its critic might be different persons. Popper's core epistemic process of suggesting conjectures and attempting to refute them may be distributed among various scientists.

Merton emphasized that in virtue of the significance of social interaction in the test and confirmation of knowledge claims, the community is the bearer of knowledge. The community owns the knowledge, a feature Merton called "communism" but was later relabeled "communalism." The reason is that social interaction or open communication in the scientific community is an indispensable prerequisite for producing trustworthy knowledge. In Merton's view, taking up a medieval metaphor famously employed by Newton, later scientists look further than earlier scientists because they are dwarfs on the shoulders of giants (Merton 1965). Likewise, the mentioned social understanding of "disinterestedness" addresses only the scientific community as a whole, whereas individual researchers may well proceed in a one-sided fashion. Finally, science as "organized skepticism" does not demand that scientists consistently doubt their own premises. Rather, scientists may take advantage of their colleagues for this purpose, who will be eager to undermine their theories.

The social nature of test and confirmation procedures is stated with particular succinctness and clarity by the psychologist Donald Campbell. As he argues, "the objectivity of physical science does *not* come from the fact that single experiments are done by reputable scientists according to scientific standards. It comes instead from a social process which can be called competitive cross-validation . . . and from the fact that there are many independent decision makers capable of rerunning an experiment, at least in a theoretically essential form. The resulting dependability of reports . . . comes from a social process rather than from dependence upon the honesty and competence of any single experimenter" (Campbell 1971/1988, 302–303).

Second, as to the *articulation of ideas*, Ludwik Fleck (1896–1961) underscored the role of the community in articulating and developing ideas. In particular, it is the interaction between an inner circle of specialists and a wider group around it that transforms ideas that makes theory development an essentially social process. In Fleck's view, the scientific community is composed of an esoteric and an exoteric circle: the group of specialists for a particular field and a wider cluster of more or less competent scientists, stretching eventually to the wider public. The specialists report their discoveries and insights to the larger group, and, in order to be understood, they adjust their mode of expression to the assumed conceptual framework of their audience. Such a process of attuning one's thoughts to the expected intellectual horizon of the addressees is bound to produce changes in these thoughts. As a result, the wandering of ideas through the scientific community, including the associated exoteric circles, cannot help but alter these ideas. The upshot is that the final character an idea assumes is often due to a series of alterations and adaptations and may have little in common with the thoughts of its original author (Fleck 1935/1979, 1936/1986).

Fleck's claim is, accordingly, that the movement of ideas through the scientific community and adjacent social groups modifies these ideas. The articulation of an idea involves a sequence of productive misunderstandings and thus is an achievement of the community. As a result, such adjustments are realized by social processes, which suggests that scientific creativity is, at least in part, of a social nature.

Third, as to the *choice of research topics*, Michael Polanyi (1891–1976) advocated the view that a scientist's selection of problems for scientific inquiry is essentially shaped by the interaction with other scientists. Conceiving projects essentially responds to efforts of others and to results achieved by others. These attempts and findings create a spectrum of opportunities or market niches, as it were, for new endeavors. This mutual adaptation distinguishes coordinated efforts, as characteristic for the division of labor in the scientific community, from completing predesigned subtasks in parallel (as sketched earlier). Polanyi suggested that this mechanism of local adjustment in devising research projects thwarts any central planning of the research process (as realized, for instance, in commissioned research). Any such attempt only serves to mutilate the advancement of science but fails to successfully direct science in a certain direction. It is the interaction between small-scale research projects and, accordingly, the decentralized, local structure of the process of problem choice in science that serves to identify suitable topical niches. In other words, selecting and shaping promising research endeavors is an essentially social process (Polanyi 1962).

2 THE STRUCTURE OF THE SCIENTIFIC COMMUNITY

Important philosophers of the first half of the twentieth century portrayed the scientific community as being open-minded and democratic and as being characterized by transparency, rationality, and approximate equality in terms of epistemic judgment. The scientific community was widely represented as an open society. However, this position was countered by various claims to the limited spread such features actually have in the scientific community.

Merton saw the scientific community as a democratic society (Merton 1938, 1942/1973) because both essentially agree in their chief cultural values and commitments. A cultural value of the scientific community is "universalism," according to which a claim in science is judged by invoking matters of fact (observation and previous knowledge) but no personal features of the author of the claim. Merton argued that this universalism of the scientific community corresponds to the meritocratic nature of democratic societies. The credibility of a scientist and the authority of a citizen are indiscriminately based on his or her achievement, but not on social class or descent. Merton offered only sketchy remarks regarding the claimed correspondence between his remaining cultural values and key features of democratic societies. Fleshing out these remarks may lead to claims such that the scientific community, in virtue of being shaped by communalism, is committed to openness and transparency of communication. The freedom of expression and the freedom of the press can be seen as analogs of this feature in democratic societies. Organized skepticism as a key element of the scientific community can be seen as corresponding to investigative journalism in society at large.

Merton's contentious correspondence claim between the structures of scientific communities and democratic societies was not meant to imply that science can only thrive in a democratic social environment. Rather, Merton's idea is that the latter environment is most conducive to the epistemically relevant social processes in the scientific community. The agreement between the value systems facilitates doing research. In nondemocratic societies, scientists need to switch between their professional and their societal value systems, which is bound to interfere with professional behavior. Merton's claim is, accordingly, that science flourishes best in democratic societies because the scientific community is a democratic society itself with respect to its meritocratic orientation, its openness of discussion, and its appreciation of critical inquiry (Merton 1942/1973).

Independently of Merton and approximately at the same time, Fleck placed the scientific community at the center of his analysis of science. The social structure of the scientific community represents the pivot around which the development of science turns. This community is a "thought collective" that is characterized by a specific "thought style." Such collective moods center on basic ideas that usually entail relations of similarity (such as life and fire) and contrast (as between rest and motion). In addition to taking the scientific community as the bearer of knowledge and to attributing a creative role to it (see Section 1), Fleck also assumed that the thought collective has a democratic structure. An important reason for his claim is that the tendency to establish a fixed and dogmatic corpus of statements, which is inherent to small groups or esoteric circles, is broken up by the transformation of ideas and, correspondingly, the influx of new ideas that arise from interaction with larger exoteric groups. The borders between the community of specialists, on the one hand, and the ambient collection of scientists and scholars as well as society at large, on the other, are permeable to ideas. This is why exoteric circles are able to influence the course of science, and this is tantamount to a democratic impact on science. In addition, test and confirmation practices in science are publicly accessible. The assessment of hypotheses is based on reasons whose viability can be checked by anyone. There are no clandestine circles and no concealed commitments in science; everything lies open before the public eye. These two reasons—the interaction between science and the public and the lack of secret knowledge or allegiance—suggest that the thought collective has a democratic structure (Fleck 1935/1979, 1936/1986).

Popper is the third influential thinker in the 1930s and 1940s who champions the democratic nature of the scientific community. Popper perceives the scientific community as an open-minded, rational, and critical society that is characterized by proceeding in small tentative steps whose appropriateness is tested frequently. This test demands the candid identification and criticism of failures and oversights. Such criticism must not be restrained or blocked by social powers. An open society takes criticism seriously and thereby assumes the ability to adapt to new challenges and to correct mistakes. The tentative implementation of a policy and the critical scrutiny of a hypothesis are similar processes (Popper 1948/1963/2002; see also Popper 1945/1966).

The picture of the scientific community as it emerges from these classic sources highlights four features. First, *open-mindedness* in the sense of a nondogmatic attitude with an emphasis on revisability. Second, *transparency* in the sense of a free and public discussion of all proposals. Third, *rationality* in the sense of relying on experience and argument. Fourth, *equality* in the sense of the prevalence of such rational factors over social characteristics such as institutional standing or reputation.

However, various thinkers have suggested limitations to several of these features. For instance, Polanyi did not consider the scientific community as a democratic association among equals. To be sure, he regarded the scientific community as a free and cooperative association of independent individuals and took this "Republic of Science" as a model of a free society. Yet, this Republic is hierarchically structured. Polanyi advocated an authoritarian structure for the community in which distinguished figures of high reputation lead the way and the rest follows suit. This exemption of the scientific community from democratic equality is justified by the claim that singling out promising pathways of research and assessing the merits of hypotheses requires tacit knowledge and skills that can be transmitted from master to disciple but cannot be fully articulated. It is the scientific elite who need to take the lead if science is to make progress (Polanyi 1958, 1962).

Thomas Kuhn (1922–96) raised objections regarding the open-mindedness of the scientific community. In the periods of so-called normal science, scientists deliberately ignore anomalies (i.e., unexpected empirical counterinstances) and adhere steadfastly to their received conceptual framework or "paradigm." In Kuhn's view, such paradigms enjoy an unquestioned monopoly in normal science. It is the lack of creativity and skillfulness on the part of scientists that is blamed for unresolved difficulties rather than mistaken principles of the paradigm (Kuhn 1962). In fact, as Pierre Duhem (1861–1916) had pointed out in the years around 1900 and as Imre Lakatos (1922–74) forcefully argued in the subsequent Popper–Kuhn debate of the 1960s and 1970s, it is not immediately clear which assumption is called into question by an anomaly. There is ample room left to scientists for digesting seeming refutations by adjusting minor or marginal claims. The refuting power of an anomaly can be shifted at will, at least to a certain degree, and their epistemic threat can be neutralized. As a result, scientists may adhere to their favorite principles in the face of recalcitrant evidence, and Kuhn's claim, strongly backed by Lakatos (1970/1978), is that they often do so. In other words, the scientific community is characterized by a certain amount of dogmatism. This attitude is restricted to certain circumstances (to periods of normal science in Kuhn or to the pursuit of a research program in Lakatos), but it stands in marked contrast to the alleged open-mindedness claimed for the scientific community. In fact, Fleck had already suggested the notion of *Denkzwang* to denote the constraints of thought collectives in recognizing certain features and privileging certain modes of interpretation at the expense of neglecting contrary information or experience.

In spite of Popper's violent protest (Popper 1970), Kuhn is of the opinion that the tenacious adherence to the pillars of the ruling paradigm is among the conditions of the epistemic success of science. Trying hard to reconcile anomalous observations with the received principles rather than quickly abandoning these principles upon the emergence of difficulties is a prerequisite of building powerful theories. Being able to convert apparent defeat into victory is a chief distinction of excellent theories. Accordingly, in Kuhn's view, a certain amount of dogmatism in the scientific community serves to promote the growth of knowledge.

Historical case studies and empirical surveys suggest that scientific communities exhibit a hierarchical structure to a certain degree and that not all voices are listened to with similar attention. Institutional position is important, and opponents tend to be ignored and pushed to the sidelines. The judgment of the leading scientists dominates community assessment and response. However, such sources also reveal that parts of this commitment to an open society are widely recognized. The salient point is that even leading researchers are not relieved of the burden of argument. Even scientists at the top of the pecking order need to give reasons. The benefit of being esteemed by one's colleagues is increased attention. The higher a scientist has climbed up the ladder of reputation, the sooner his argument is noted and his advice followed. Yet everybody needs to argue, even the mandarins (Taubert 2008).

In addition, historical evidence suggests that even the unambiguous verdict of the scientific authorities may fail to force a discipline in a particular direction. Outsiders can overturn a whole field. Daniel Shectman was awarded the Nobel Prize for chemistry in 2011 for his discovery of a crystal structure that had been considered impossible before. His asserted quasiperiodic crystals contradicted the recognized fundamentals of crystallography. Shectman was an outcast for decades. He had been banished from his research group, accused of having brought disgrace upon his laboratory. But the scientific research process is such that ridiculed outsiders can be acknowledged to be right after all. Episodes like this militate against assuming a steep and insurmountable hierarchy in the scientific community.

Likewise, the dogmatism of the scientific community has its limits. Anomalies cannot be ignored consistently, as Kuhn already pointed out. If empirical problems and counterinstances pile up, the pertinent paradigm runs into a crisis in whose course the unquestioned allegiance of the community dissolves.

These features can be taken to delineate the correct core of the classical picture of the scientific community. The nonhierarchical or egalitarian nature of the scientific community is retained to some degree, as is its open-mindedness. The scientific community still exhibits qualities of an open society—but only within limits.

A quite recent development regarding the structure of the scientific community is the rise of a community engaged in "do-it-yourself biology" and biohacking. This development marks the emergence of "garage synthetic biology," which is practiced by alternative research communities. Such informal groups are driven by the spirit of doing research on their own and by being committed to the vision of a community. Such biohacking clusters operate according to their image of a civic science that aspires to liberate science from political and economic powers and help citizens regain control over research. It is their declared aim to render science to small-scale grassroots communities and to thereby restore the democratic appeal science enjoyed prior to the advent of "big science" (Delgado 2013).

3 COMMON GROUND AND DIVERSITY AMONG SCIENTIFIC COMMUNITIES

Important issues regarding the social organization of the scientific community are what keep the community together and define the relations between the scientific community and the ambient society. Kuhn provided an elaborate scheme of what a scientific community shares and what makes it a coherent entity. Scientific communities might be taken to be glued together by their commitment to common scientific principles and methods. Kuhn agrees but argues that additional elements need to come in. Correspondingly, a Kuhnian paradigm is not only composed of substantive tenets and methodological practices. Such commitments alone would not be powerful enough to produce sufficient common ground among the approaches pursued and the judgments passed. Rather, a paradigm also includes a class of shared exemplary solutions to a set of problems. It is one of Kuhn's major contentions that the shared commitment to a collection of exemplars or "paradigmatic" (in the original sense of the term) problem solutions is able to create coherence among procedures and conclusions.

In his "Postscript" of 1969 (Kuhn 1970a), Kuhn saw the scientific community as primarily characterized by social or sociological indicators. In particular, the members of a scientific community have received the same sort of education, they are familiar with the same corpus of literature, they share formal and informal communication networks, and they largely agree in their professional opinions. Kuhn's claim is that such communities, as identified by their sociological features, also share intellectual characteristics, such as a commitment to the same theoretical and metaphysical principles, to the same methodological standards, and to a set of shared core exemplars. Kuhn placed this community structure, which involves an ensemble of institutional and epistemic characteristics, under the new heading of "disciplinary matrix."

A scientific community, in Kuhn's view (Kuhn 1962, 1970b), is an autonomous entity and operates by its own (or "internal") rules. It is a group of specialists whose work is detached from social needs and interests. Scientists select their research topics in accordance with their own aspirations and bring their own value system to bear on the evaluation of hypotheses. A wider social interference would be detrimental to the advancement of science. Kuhn distinguished between scientists in the proper sense and engineers. The former determine their research agenda according to internal considerations, chief among them the expectation that a certain problem can be solved in the first place. The latter are pressed by extra-scientific forces to address topics according to their social urgency and without taking their assumed tractability into account. This difference in solvability explains why scientists produce new solutions more rapidly than engineers (although engineers may come up quickly with technologies that are based on the existing system of knowledge; Kuhn 1962, 164). In virtue of being composed of a distinguished group of specialists, scientific communities have clear borders. The dividing line between scientists, on the one hand, and freaks and amateurs, on the other, is drawn unambiguously and unmistakably. The ideas of the latter are ignored (except in scientific crises).

Kuhnian scientific communities are largely insulated from society at large, and they make their scientific decisions by resorting exclusively to their own yardsticks. Kuhn is sometimes credited with having enriched a narrowly intellectualist philosophy of science with sociological considerations. However, this only applies to his characterization of the scientific community by social features, whereas his picture of scientific research and scientific change is strongly internalist. Kuhn's scientific community proceeds in splendid isolation from social forces.

Bringing these considerations together suggests that Kuhnian scientific communities are identified by *sociological features* (mostly common institutional goals and communication links), they are *autonomous* in the sense of being separated from social demands, and they often enjoy a *monopoly* in their relevant field. This monopoly is a characteristic of Kuhn's normal science, in which a generally adopted paradigm is claimed to rule a field undisputedly. In a scientific crisis, this unanimity dissolves and the strict distinction between specialists and *dilettanti* is broken up. Yet, after the scientific revolution that usually follows this interim period of wavering and doubt, the former relations are re-established. Accordingly, the fourth feature of Kuhn's scientific communities is the *strict demarcation* between experts and laypersons. Scientific communities do not have fuzzy boundaries.

Lakatos was a major player in the Popper–Kuhn controversy that emerged after 1965. Lakatos developed his account (Lakatos 1970/1978, 1971/1978) to uphold what he took to be Popper's normative epistemology while doing justice, at the same time, to what he believed were Kuhn's empirical insights (Carrier 2002, 2012). Lakatos viewed science as a competition among research programs. The identity of such a program is fixed by a "hard core" of theoretical principles that are retained by the program's advocates come what may (together with a shared program heuristics). This tenacity is made possible by the mentioned leeway in taking up and meeting empirical challenges. The shared commitment to the core principles of a program prompt consensual judgments regarding proposed program changes. Yet, this agreement is confined to the proponents of a given program; advocates of rival programs will disagree. In contradistinction to Kuhnian paradigms, research programs, as a rule, fail to gain a monopoly. Research programs remain contested over their life span.

Lakatos defined scientific communities in a narrow, theory-only fashion. In contradistinction to Kuhn, such communities are identified through their commitment to a set of mostly nomological assumptions. This theory-dominated approach is accompanied by a downgrading of the social. The structure and development of a thriving or "progressive" program can be accounted for by appeal to epistemic or internal standards alone. Any intrusion of social considerations in the elaboration of a program indicates that the program has taken a degenerate course. This includes, as in Kuhn, the selection of the research agenda, which is determined by theory-internal (namely, heuristic) considerations in progressive programs. Accordingly, the scientific communities pertaining to thriving programs are autonomous in the sense of being committed to epistemic considerations alone. As in Kuhn, this group of specialists is strictly separate from outsiders (and in particular social forces). Yet, Kuhn's monopolistic community is replaced by a plurality of competing, large-scale camps. As a result, Lakatos agrees with Kuhn regarding the autonomy and strict demarcation of the scientific community but, in contrast to Kuhn, assumes a plurality of such communities and rejects Kuhn's sociological mode of identifying them.

To sum up these classical positions, Merton and Popper claimed that the scientific community operated in harmony with democratic social structures. The scientific community is a meritocratic and open society, just as a democratic society is. Science and democracy are congenial. By contrast, Fleck sees a constant flux of ideas across the intellectual border between science and society. It is the wandering of ideas through esoteric and exoteric circles that contributes to the shaping of scientific knowledge. However, for Kuhn and Lakatos, the scientific community is strictly distinct from society at large. Science may draw some inspirations from other social areas such as technological challenges, but, at bottom, the nature and evolution of the system of knowledge is guided by epistemic standards and aspirations alone.

Positions emerging from the mid-1970s on have contributed to perceiving the boundary between the scientific community and its social environment as being less impervious and more permeable, mostly with respect to ideas but also regarding persons (see the discussion on "lay experts" later in this section). This means that science and its social organization are thought to be rightly susceptible to social forces.

Social constructivism is an early example for the accentuation of the flow of ideas from society to science. David Bloor (b. 1942) opposed Lakatos's primacy of epistemic criteria over social factors and suggested inverting this relation. The interests of scientists as social beings and as parts of society play a major role in their assessment of hypotheses.

In particular, Bloor criticized the view, especially striking in Lakatos, that the prevalence of epistemically privileged (or truthful) theories is explained by appeal to rational judgment (agreement with observation, methodological distinction), whereas the dominance of false beliefs is attributed to social interference. According to this view, Darwinian evolutionary theory flourished because it truthfully captured important mechanisms of the biological world, whereas Lysenkoism fared well temporarily because its claim that hereditary changes could be induced by environmental conditions fitted smoothly into the Stalinist creed of the period (with its aspiration to create a new type of human being by realizing novel social conditions). In contrast, Bloor advanced his "symmetry principle," according to which the same kinds of explanation should account for the emergence of true and false theories, rational and irrational hypotheses, successful and failed suppositions. Social accounts of knowledge generation should be impartial regarding such normative distinctions (Bloor 1976).

The content and viability of this principle has been subject to controversial judgment (see Bloor 1981; Laudan 1981/1996). Its chief relevance for the issue of the social organization of science is the assumption involved in it that social forces legitimately and comprehensively influence the system of knowledge. More modest accounts of such influences assume, for instance, that social factors may well affect or shape the incipient stages of a theory but that successful theories have managed to sift out these factors in the course of their maturation. By contrast, theories that fail to get rid of such factors over time will eventually break down and vanish (McMullin 1982). Yet Bloor insisted that the fingerprint of the social remains visible in theories that are accepted as part of the system of knowledge. Boyle won over Hobbes because the image of an experimental community, as conjured up by Boyle, matched the political idea of a free and still self-disciplined community (Shapin and Schaffer 1985). Pasteur's denial of spontaneous generation gained acceptance in spite of its weak evidential underpinning for theological and political reasons: spontaneous generation tended to reduce life to matter and to interfere with God's creative power (Bloor 1981).

Another instance   if of a quite different nature—of the influence of social forces on the structure and development of science is the important role that application-oriented research has gained in the past decades. Today, large amounts of research are financed by economic companies, are conducted out of commercial interest, or are commissioned by politics. Underlying the industrial funding of research is the idea that science is a primary source of technological development that is in turn viewed as a driving force of economic growth. Public funding of research is mostly due to the same motives. Financial support of academic and industrial research alike is widely understood as a kind of investment whose expected return is economic competitiveness. Research in the natural sciences has become a major economic factor, and scientific goals are increasingly intertwined with economic interests. In some areas, a significant portion of innovative work has moved into industrial laboratories. In a similar vein, university research is directed toward practical goals; the scientific work done at a university institute and a company laboratory tend to approach each other. This convergence is emphasized by strong institutional links. Universities found companies in order to market products based on their research. Companies buy themselves into universities or conclude large-scale contracts concerning joint projects. The commercialization or "commodification" of academic research is one side of a process of reciprocal adaptation, the reverse side of which is the rise of industrial high-tech research. As a result, the scientific community has grown beyond the university and includes parts of the economy.

Several changes in the social organization of science are associated with this institutional transformation. First, the research agenda is set by nonscientific actors according to social demand. Research issues are imposed on science depending on social urgency or economic prospect. This involves a deviation from the mode characteristic of fundamental research in which scientists themselves select research topics according to epistemic interest and solvability. Examples are the identification of suitable measures for combatting climate change; the development of more powerful electrical storage systems; or the discovery of therapies for Alzheimer's disease. As a result, the pathway of research is strongly shaped by social forces.

Second, among the often criticized features of commercial research is its tendency to keep research outcome secret. In order to protect economic interests, such research is often done behind closed doors. Not infrequently, scientists are prohibited by their working contracts from disclosing discoveries they have made. Industrial research projects are intended to produce knowledge that is used exclusively; companies are not eager to pay to produce knowledge that later benefits a competitor (Dasgupta and David 1994; Rosenberg 1990). This secrecy is perceived as doing harm to science in that it restricts intersubjective access to research results and thereby impairs general scrutiny. Traditional commitments to scientific method include that knowledge claims in science should be subject to everyone's judgment. Secrecy may block putting knowledge claims to severe testing. Such claims are neither examined as critically as they would be if the claims or the pertinent evidence were more widely known, nor can the new information be employed in related research projects. Secrecy thus tends to distort the process of test and confirmation in science and the progress of science (Carrier 2008).

Third, intellectual property rights or patents restrict the use of research findings but not general access to them. If the so-called experimental-use privilege—the right to do research with patented material—is interpreted narrowly, as is the case in some legal systems, broad covering of a research field by patents may deter researchers from pursuing certain questions. As a result, such economic features may hurt the development of the system of knowledge. In addition, the whole institution of patenting is in conflict with Mertonian communalism (see Section 1), according to which scientific knowledge is in public possession. Knowledge cannot be owned by a scientist, nor by his or her employer; scientific findings are public property.

It is widely complained that the commercialization of science has interfered with the traditional openness of scientific discourse in which ideas (and materials) were freely shared. Critics conclude that market incentives do not automatically generate an optimal social organization of science. However, some of the detrimental features often highlighted in this connection, such as the unwillingness to share information and material resources, can be attributed to some degree to increased competition among scientists (which equally obtains in academic research). Another widespread complaint regarding commercialized research is its loss of Mertonian disinterestedness (see Section 1). In some fields, such as medical drug research, the community as a whole is claimed to be biased in the sense that only certain assumptions (e.g., assumptions related to marketable products) are considered acceptable in the first place.

Another major shift in the social organization of science concerns the expansion of the scientific community through inclusion of "lay experts" or "experience-based experts." Science has entered the social arena in various fields and is involved in giving policy advice. In such instances, scientific experts are called on to assess specific cases and recommend concrete practical measures. This orientation toward the particular requires the extensive use of more restricted regularities or local knowledge. Local knowledge sometimes acquires a crucial role in conceiving tailor-made solutions to narrow practical problems. Experience-based experts or noncertified experts who have advanced knowledge in virtue of their familiarity with the relevant domain but whose competence is not recognized by university degrees or other certificates may have important factual contributions to make to solving such problems (Collins and Evans 2002).

Local, experience-based knowledge is best included by social participation. Social participation has epistemic and political aspects that should be kept apart. The political requirement calls on scientific experts to listen to and to take up local fears and aspirations because taking stakeholders' views into account improves the odds that the expert view is politically accepted and implemented eventually. It is a quite different matter to include the local perspective as a piece of experience-based expertise, that is, as a means for enhancing the epistemic quality of the scientific analysis or recommendation. Due to their familiarity with the precise circumstances on site, lay people may be able to contribute specific pieces of information. Taking up this local knowledge can improve expert advice. This epistemic contribution of the public is a far cry from advocating stakeholders' interests. Giving heed to local knowledge is rather prone to advance the process of deliberation (Carrier 2010).

4 SOCIAL EPISTEMOLOGY

Social epistemology is based on the assumption that various epistemic features of science are related to social characteristics of the researchers. Topic selection, preferred research strategies, and the practice of examining claims may well depend on the social status and the social role of the pertinent scientist. In other words, the supposition is that social properties (i.e., characteristics of the relevant scientist in society at large) may affect the questions asked, the heuristics followed, and the confirmation procedures applied.

Feminist epistemology is the best elaborated field of social epistemology. On this view, gender is a social factor that has an impact on research. Gender differences are supposed to produce differences in outlook that will bear on topic selection, research strategies, and the examination of claims. As to the questions asked, one of the most prominent examples of feminist epistemology is the change introduced in primate research through the advent of female scientists in the 1970s. Male researchers had focused on social conceptions such as male domination of females or male competition and fighting as chief factors of reproductive success. Female primatologists paid attention to behavior traits such as female cooperation or female–infant care and emphasized their importance for reproductive success. The inclusion of female scientists offered a new perspective on primate interactions that is accepted today also by male researchers (Anderson 1995). A second case in point is feminist archeology. Traditional archeology suffered from an androcentric perspective that neglected the role of women in the prehistoric world. The hypothesis of "man-the-hunter" assumed that the invention of tools grew out of hunting and warfare. Female archeologists in the 1970s opposed this then-prevalent model with their "woman-the-gatherer" account, according to which tools were invented in the context of digging, collecting, and preparing plants for food (Longino 1990, 106–111).

Both examples testify that expanding the social diversity of the scientific community can introduce new conceptual perspectives. The elaboration of such alternative approaches has improved science in its epistemic respect by providing a deeper and more complete understanding of the evidence. Yet, this epistemic benefit was not gained by dropping one-sided approaches and replacing them with a more neutral one. Rather, the alternative feminist approaches involve a social model or political values as well. In consequence, it is claimed that we can make epistemic progress while continuing to bring social-value commitments (or feminist values) to bear (Anderson 1995). Social epistemology maintains that the diversity of social roles is reflected in a diversity of cognitive orientation. Widening the social spectrum of researchers means enhancing the range of theoretical options. The epistemic benefit of such increased diversity becomes clear once it is realized that all scientists take some assumptions for granted. These beliefs appear self-evident and as a matter of course so that they are frequently not recognized as significant principles in the first place. The trouble with such unnoticed or implicit assumptions is that they go unexamined. They are never subjected to critical scrutiny. This means that if one of these seemingly innocuous claims should be mistaken, its falsity is hardly recognized. Predicaments of this sort can be overcome by drawing on the critical force of scientific opponents. These opponents will struggle to expose unfounded principles and try their best to undermine cherished accounts.

Underlying this pluralist approach is the maxim that mistaken assumptions are best discovered from outside; that is, by taking a perspective that does not contain the critical assumption (a precept strongly championed by Paul Feyerabend [Carrier 2012]). Thus, scientific controversies are an appropriate instrument for revealing blind spots, one-sided approaches, and unfounded principles. Such deficiencies are best uncovered by developing an alternative position. For this reason, pluralism is in the epistemic interest of science; it contributes to enhancing the reliability of scientific results.

A notion of objectivity that captures such a pluralist setting is centered on reciprocal criticism and mutual control. Baconian objectivity means abandoning one's prejudices and taking a detached and neutral position. The social notion of scientific objectivity instead focuses on the social interaction between scientists who reciprocally censure their conflicting approaches. Different biases are supposed to keep each other in check. Social objectivity thrives on correcting flaws by taking an opposing stance and thus demands the exchange of views and arguments among scientists (Anderson 1995; Brown 2001; Longino 1990, 1993).

Within this pluralist framework, the objectivity of science needs to be separated from the objectivity of scientists. Individual scientists need not be neutral and detached. They may be anxious to confirm or refute some assumption in order to promote their reputation, support some worldview, or bring about technological novelties. Such divergent values and goals need in no way undermine the objectivity of science. On the contrary, pursuing contrasting avenues and taking one another to task may contribute to the objectivity of science. In the pluralist understanding of objectivity, what matters is not to free scientists from all contentious suppositions but rather to control judgments and interests by bringing to bear opposing judgments and interests (Carrier 2008).

It is among the principles of social epistemology that pluralism in the scientific community conduces to an improved check of assumptions. From this perspective, one of the predicaments of commercial research is that companies share crucial assumptions. All pharmaceutical companies explore similar ways for curing the same set of diseases. However, in some cases, competing economic interests have produced epistemically relevant diversity. For instance, in the early 2000s, scientists articulated worries to the effect that the anti-clotting efficacy of aspirin would decrease over the years. The claim was that a habituation effect occurred. Some years later, it was revealed by a journalist that the whole debate had been launched by a company that was producing alternative anti-clotting agents. Conversely, some of the leading scientists who had opposed this alleged drop in effectiveness had been funded by a competing aspirin manufacturer (Wise 2011). Although the whole debate was manufactured, as it were, and driven by commercial forces, it bears witness to the possible corrective influence of competing economic interests. Competition is able to create a pluralist setting that serves to promote objectivity. In fields in which strong commercial interests meet, this kind of objectivity is often the only one available. Under such conditions, it has proved impossible at times to identify scientists who have detailed expertise but no vested interest. When strong corporate interests are at stake, the neutral, well-informed expert may turn out to be an idle dream.

Diversity in the scientific community is in part produced by the varied cognitive makeups of scientists. Philip Kitcher suggested that, in addition to cognitive variation, a social mechanism contributes to generating diversity. This model is based on the Mertonian supposition that scientists strive for recognition and reputation. A comparative assessment of a scientist's odds of being successful in some scientific endeavor could run like this: if one joins the majority camp, success may be probable, but, if achieved, the recognition would need to be shared with a large number of colleagues. Alternatively, in embarking on a minority approach, the odds of success may be smaller, but, in case of success, one's own share of the recognition could be expected to be higher.

The epistemic optimum for the scientific community as a whole is diversity. The odds of success increase when the community divides itself up into various competing strands. In Kitcher's model, this epistemic optimum is realized by setting up a system of social incentives. Striving for individual reputation generates an epistemically appropriate structure of the scientific community (Kitcher 1993, ch. 9; Kitcher 2011, ch. 8).

In addition to advocating social pluralism in the scientific community, a second related strand of social epistemology concerns requirements for significant test procedures. Social epistemology is characterized by the view that no contrast exists between the rational and the social but that rational procedures of severe test and confirmation are, in fact, of a social nature. They involve the interaction of scientists (Longino 2002; see Section 1). It is characteristic of social epistemology that the relevant requirements address procedural rules as to how to deal with knowledge claims in the scientific community. The Mertonian values of universalism and communalism (in the sense of unrestricted communication; see Section 1) are meant as benchmarks for discussing the merits of hypotheses. In a similar vein, Longino put forward procedural standards that are intended to govern the process of critical examination in the scientific community. One of her demands concerns the need to take up criticism and to respond to objections appropriately. This requirement broadens the Popperian obligation to address anomalies and counterinstances (Popper 1957/1963/2002). The epistemic spirit of science is distinguished by taking challenges seriously and by trying to cope with them. This commitment involves accepting the revisability of scientific statements and thus relinquishing dogmatism. Another one of Longino's procedural standards is "tempered equality of intellectual authority." This community rule is supposed to preclude personal or institutional power playing; arguments should be appreciated independently of institutional hierarchies (Longino 1993, 2002).

Such procedural rules are intertwined with social values of the scientific community. These rules address how to deal with opposing approaches and display a clear epistemic bearing. They are epistemic values of a particular sort and codify the epistemic attitude. Such rules or values are essentially social: they address how to deal with dissenting factions and nonconformist positions. Such procedural rules for addressing diversity are suitable for constraining antagonistic beliefs and driving them toward a common point of view (Carrier 2013).

It is the basic tenet of social epistemology that knowledge production is tied to social presuppositions. The growth of knowledge relies on certain communication networks and on communities of a certain structure. The procedural standards elaborated by social epistemology are meant to be normative. They address what structure of a scientific community is most conducive to gaining scientific knowledge of the kind we think worthwhile.

This normative approach is characteristic of the entire field of the social organization of science as a topic of philosophy of science. Social organization as a philosophical issue is mostly concerned with exploring which kinds of social organization are most beneficial to the epistemic aspirations of science.

REFERENCES

Anderson, E. (1995). "Feminist Epistemology: An Interpretation and a Defense." *Hypathia* 10: 50–84.

Bacon, F. (1620/1863). *The New Organon*, trans. J. Spedding, R. L. Ellis and D. D. Heath, *The Works VIII* (Boston: Taggard and Thompson).

Bloor, D. (1976). "The Strong Programme in the Sociology of Knowledge." In D. Bloor, ed., *Knowledge and the Social Imagery* (London: Routledge and Kegan Paul), 1–19.

Bloor, D. (1981). "The Strengths of the Strong Programme." *Philosophy of the Social Sciences* 11: 199–213.

Brown, J. R. (2001). *Who Rules in Science?* (Cambridge MA: Harvard University Press).

Campbell, D. T. (1971/1988). "The Experimenting Society." In E. S. Overman, ed., *Methodology and Epistemology for Social Science* (Chicago: University of Chicago Press), 290–314.

Carrier, M. (2002). "Explaining Scientific Progress. Lakatos's Methodological Account of Kuhnian Patterns of Theory Change." In G. Kampis, L. Kvasz, and M. Stöltzner, eds., *Appraising Lakatos: Mathematics, Methodology, and the Man (Vienna Circle Library)* (Dordrecht: Kluwer), 53–71.

Carrier, M. (2008). "Science in the Grip of the Economy: On the Epistemic Impact of the Commercialization of Research." In M. Carrier, D. Howard, and J. Kourany, eds., *The Challenge of the Social and the Pressure of Practice: Science and Values Revisited* (Pittsburgh, PA: University of Pittsburgh Press), 217–234.

Carrier, M. (2010). "Scientific Knowledge and Scientific Expertise: Epistemic and Social Conditions of their Trustworthiness." *Analyse und Kritik* 32: 195–212.

Carrier, M. (2012). "Historical Approaches: Kuhn, Lakatos and Feyerabend." In J. R. Brown, ed., *Philosophy of Science: The Key Thinkers* (London: Continuum), 132–151.

Carrier, M. (2013). "Values and Objectivity in Science: Value-Ladenness, Pluralism and the Epistemic Attitude." *Science & Education* 22: 2547–2568.

Collins, H. M., and Evans, R. (2002). "The Third Wave of Science Studies: Studies in Expertise and Experience." *Social Studies of Science* 32: 235–296.

Dasgupta, P., and David, P. A. (1994). "Toward a New Economics of Science." *Research Policy* 23: 487–521.

Delgado, A. (2013). "DIYbio: Making Things and Making Futures," *Futures* 48: 65–73.

Fleck, L. (1935/1979). *Genesis and Development of a Scientific Fact*, transl. F. Bradley and T. J. Trenn (Chicago: Chicago University Press).

Fleck, L. (1936/1986). "The Problem of Epistemology." In R. S. Cohen and T. Schnelle, eds., *Cognition and Fact. Materials on Ludwik Fleck* (Dordrecht: Reidel), 79–112.

Kitcher, P. (1993). *The Advancement of Science. Science without Legend, Objectivity Without Illusions* (New York: Oxford University Press).

Kitcher, P. (2011). *Science in a Democratic Society* (Amherst, MA: Prometheus).

Kuhn, T. S. (1962). *The Structure of Scientific Revolutions* (Chicago: University of Chicago Press).

Kuhn, T. S. (1970a). "Postscript 1969." In T. S. Kuhn, ed., *The Structure of Scientific Revolutions*, 2nd ed., enlarged (Chicago: University of Chicago Press), 174–210.

Kuhn, T. S. (1970b). "Logic of Discovery of Psychology of Research." In I. Lakatos and A. Musgrave, eds., *Criticism and the Growth of Knowledge* (Cambridge: Cambridge University Press), 1–23.

Lakatos, I. (1970/1978). "Falsification and the Methodology of Scientific Research Programmes." In J. Worrall and G. Currie, eds., *The Methodology of Scientific Research Programmes (Philosophical Papers I)* (Cambridge: Cambridge University Press), 8–101.

Lakatos, I. (1971/1978). "History of Science and its Rational Reconstructions." In J. Worrall and G. Currie, eds., *The Methodology of Scientific Research Programmes (Philosophical Papers I)* (Cambridge: Cambridge University Press), 102–138.

Laudan, L. (1981/1996). *Beyond Positivism and Relativism. Theory, Method, and Evidence.* (Boulder CO: Westview), ch. 10.

Longino, H. (1990). *Science as Social Knowledge: Values and Objectivity in Scientific Inquiry* (Princeton, NJ: Princeton University Press).

Longino, H. E. (1993). "Essential Tensions—Phase Two: Feminist, Philosophical, and Social Studies of Science." In L. M. Antony and C. Witt, eds., *A Mind of One's Own. Feminist Essays on Reason and Objectivity* (Boulder, CO: Westview Press), 257–272.

Longino, H. E. (2002). *The Fate of Knowledge* (Princeton, NJ: Princeton University Press).

McMullin, E. (1982). "Values in Science." In P. Asquith and T. Nickles, eds., *PSA 1982 II. Proceedings of the 1982 Biennial Meeting of the Philosophy of Science Association: Symposia* (East Lansing, MI: Philosophy of Science Association), 3–28.

Merton, R. K. (1938). "Science and the Social Order." *Philosophy of Science* 5: 321–337.

Merton, R. K. (1942/1973). "The Normative Structure of Science." In R. K. Merton, ed., *The Sociology of Science. Theoretical and Empirical Investigations* (Chicago: University of Chicago Press), 267–278.

Merton, R. K. (1965). *On the Shoulders of Giants. A Shandean Postscript* (Basingstoke: MacMillan).

Polanyi, M. (1958). *Personal Knowledge. Towards a Post-Critical Philosophy* (Chicago: University of Chicago Press).

Polanyi, M. (1962). "The Republic of Science: Its Political and Economic Theory," *Minerva* 38 (2000): 1–32.

Popper, K. R. (1935/2002). *The Logic of Scientific Discovery* (London: Routledge).

Popper, K. R. (1945/1966). *The Open Society and Its Enemies* (London: Routledge).

Popper, K. R. (1948/1963/2002). "Prediction and Prophecy in the Social Sciences." In K. R. Popper, ed., *Conjectures and Refutations. The Growth of Scientific Knowledge* (London: Routledge), 452–466.

Popper, K. R. (1957/1963/2002). "Science: Conjectures and Refutations." In K. R. Popper, ed., *Conjectures and Refutations. The Growth of Scientific Knowledge.* (London: Routledge), 43–78.

Popper, K. R. (1969/1988). "Die Logik der Sozialwissenschaften." In T. W. Adorno et al., eds., *Der Positivismusstreit in der deutschen Soziologie* (Darmstadt: Luchterhand), 103–123.

Popper, K. R. (1970). "Normal Science and Its Dangers." In I. Lakatos and A. Musgrave, eds., *Criticism and the Growth of Knowledge* (Cambridge: Cambridge University Press), 51–58.

Rosenberg, N. (1990). "Why Do Firms Do Basic Research (with Their Own Money)?" *Research Policy* 19: 165–174.

Shapin, S., and Schaffer, S. (1985). *Leviathan and the Air-Pump. Hobbes, Boyle, and the Experimental Life* (Princeton: Princeton University Press), ch. 8.

Taubert, N. (2008). "Balancing Requirements of Decision and Action: Decision-Making and Implementation in Free/Open Source Software Projects." *Science, Technology and Innovation Studies* 4: 69–88.

Wise, N. M. (2011). "Thoughts on Politicization of Science Through Commercialization." In M. Carrier and A. Nordmann, eds., *Science in the Context of Application. Methodological Change, Conceptual Transformation, Cultural Reorientation* (Dordrecht: Springer), 283–299.

CHAPTER 42

....

SPACES

....

DEAN RICKLES

1 INTRODUCTION

....

WE seem to be rather directly immersed in space, yet, as Frank Arntzenius wryly observes, we can't see it, smell it, hear it, taste it, or touch it (2014, 151). Despite its apparent sensory opacity, the study of space is one of the oldest in philosophy and continues to occupy a central place. It was a key theoretical component of the cosmological investigations of the pre-Socratics. For example, space (or "void") formed one half of a complete ontology in the atomists' worldview, along with the complement of indivisible atoms. Of course, Parmenides had earlier denied the reality of space,[1] but, as the atomists' position indicated, without void, there is nothing to separate one thing from another (no plurality) and no possibility for change between positions (no motion)—an apparent absurdity that Parmenides accepted (and that Zeno attempted to bolster with his paradoxes). Space has, then, been a controversial subject from the very beginnings of our discipline.

Questions that emerged from these early cosmological investigations (most of which continue to perplex) include: is space infinite or finite? Is space continuous or discrete? Why does space have three dimensions? Is space Euclidean? Is space uniform? What is the relationship between space and matter? Why and how is space distinct from (or similar to) time? Is a genuinely empty space (i.e., true vacuum) possible? Was space created? Is space dynamically active or inert? Also emerging from this early period of research on space was the idea that *motion* is deeply entangled with some of these questions so that

---

[1] In this sense, our spatial experience was an "illusion" rather than a representation of reality—we see less drastic examples of mismatch between appearance and reality in Section 3.2, in the context of phenomenal receptor-based spaces and physical space. By contrast, according to, for example, Leibniz's brand of spatial relationalism, space *is* a representation (not an illusion), although of the order of (coexistent) things in the world. As Michael Friedman (1981, 4) notes, in the relationalist's worldview "[m]aterial objects are not, as the *absolutist* imagines, literally embedded in physical space"; instead, physical space is viewed as having a "merely representative . . . function" much like the (abstract) representational spaces we will meet below.

lessons about motion are of direct relevance to questions about space and vice versa.[2] This in turn links up to a close relationship between the geometry (and topology) of a space and the possible motions (and properties) of objects in that space—a feature that will propagate into the more general spaces we discuss later.

But, interesting though such issues are, "space" is a far more general concept than is suggested by these ancient debates and their progeny—hence the plural "spaces" in our title. There are other spaces besides "space out there" (the three-dimensional [3D] stuff of everyday life). There are:

- The curled up, compactified 6D spaces (Calabi-Yau manifolds) of string theory around (and through) which strings and branes wrap (and thread) to generate observable, low-energy physics (Yau and Nadis 2010)
- The parallel but causally disjoint spaces of multiverse theories, which can be viewed (*qua* ensemble) as a space of sorts in its own right (Susskind 2009)
- The internal spaces of particle physics, in which non-spatiotemporal (in the "space out there" sense) symmetries (such as isospin symmetry) operate, with degrees of freedom determining an abstract space (e.g., isospin space) of functions (i.e., fields Ψ mapping the spacetime manifold to some set of values) (Nagashima and Nambu 2010)

Or, and this is more the focus of the present chapter, there are the various *representational spaces* in physics and elsewhere in almost any discipline that demands (or accommodates) a systematic consideration of possibilities for its subject matter. The problem of spaces in general (what they are and *why* they are) demands a wide-angle lens, taking in physics, philosophy, psychology, linguistics, neuroscience, cognitive science, and more. Adopting this liberal approach better reveals the common core of what it is to be a space and highlights the various scientific functions they perform.

We do consider "physical" space *within* this more general approach, ultimately as an additional example of a representational space: there is a strong sense in which both our everyday space and space according to physics are as much representational spaces as the more obviously representational kinds, which we don't ordinarily view as corresponding one-to-one with actually existing things in the world outside our heads, if at all.[3]

---

[2] For example, the principle of Galilean relativity (that the laws of physics can be referred to any uniformly moving reference frame) appears to conflict with the notion of a privileged central location in the Universe, pointing instead to a homogeneous structure.

[3] This is not to say that there *is* no objective, physical space being represented (that the representation is *about*), of course. But that is an inference in need of argument (including scientific evidence)—something I'll return to later. Obviously, however, everyday space is based on a mapping that involves turning information received on 2D surfaces (retinas, skin, etc.) into a 3D representation. We can go beyond the senses in various ways to strengthen our sense of physical space (using instruments, theories, etc.), but these involve representation, too. We are more used to this kind of situation in the case of time (especially the notions of "flow" and "nowness," which most are happy to admit is a mental contribution not shared by the external world).

The remainder of this chapter proceeds as follows. In Section 2 we consider the key features of a space that are shared (to a greater or lesser extent) by the examples we then discuss in Section 3 along with the conceptual issues they raise.[4] Finally, in Section 4, we draw some general lessons about the overall significance of spaces in science.

2 WHEN IS A SPACE A SPACE?

Whether physical or representational (or otherwise), intuitively, a space is a way of ordering and systematizing information, qualities, data, or entities of some kind. A random assortment of items won't do. My sock drawer, the slides of Albert Einstein's brain matter, and the number two does not a space make. The goal is to have the points of the space represent *distinguishable* states or possibilities (which the previous items do satisfy), but with some structure (relations) involved that unifies the elements or members of the collection (which the previous items do not satisfy).[5]

A space allows one to decide when things (such as possibilities) are the same or different. But, again, we need some commonality between the members. If we have a (representation of a) jumble of socks of *every possible* (physically manufacturable) size, color, shape, and fabric, then we are nearing the notion of a space in the sense I want to cover in this chapter. We are *exhausting* what it is to be a plain-colored (unpatterned) sock: any such sock *must* lie somewhere in this jumble. Conversely, any sock represented in this jumble is a *possible* sock, and any sock not thus represented is impossible. If we then use our properties of size, shape, color, and fabric as kinds of "degrees of freedom" for socks and have some ordering[6] for these, then we have ourselves a 4D sock space. An example

[4] The range of spaces considered in Section 3 must be selective, of course, and comprises only a small sample of what exists, although they have been chosen for their coverage of a range of important issues in philosophy of science.

[5] Perhaps we could come up with some such unifying relation, but it would be highly gerrymandered and presumably of no utility. The elements of a space should, then, have some kind of "natural connection" with one another (be *about* or of the same kind of thing).

[6] Sizing has an obvious pre-existing ordering, of course (with several scales). For colors, we have an ordering too. Fabric we could imagine being ordered in terms of "fineness of weave" (denier) or some such. Shape is more difficult, requiring a continuous number of determinations, but one can imagine contriving some such ordering based on what it is possible to manufacture (which could discretize the number of possibilities). One could also consider increasing the dimensionality of this 4D space by adding design patterns to the socks, thus increasing quite radically the number of possible socks (and so the number of distinct points in the space). Or, alternatively, one could decrease the dimensionality (and thereby the number of possible socks) by removing or freezing a value for one of the "spatial axes," say, by only considering color (thus isolating a subspace of the larger sock space). The space of socks would then have the same size as the space of physically printable/dyeable colors. It is also important to note that, depending on what aspects of socks we include, there will be different *equivalence classes* of socks (that are deemed "the same kind of sock"): sameness and difference (in the context of a space) are determined by the "axes" of the space. If we have just one axis (size, say), then any socks of the same size are *the same* (represented by the same point) regardless of any other differences (e.g., in color) that they might have. Hence, adding more dimensions increases the possibilities for *distinguishing* things (and, conversely, reduces the number of things that might be considered the same).

of a point s in sock space would be a quadruple like s = ⟨Euro 42, turquoise, ankle, 85 denier⟩, each point is specified by four values (one coordinate for each magnitude). In this case, by construction, the properties are assumed to be discrete, and so we have a discrete space. This brings with it a natural (counting) measure to order and compare different socks: distance is the number of distinguishable elements separating (e.g., two sock sizes). One could add a temporal dimensional (producing a 5D space) to chart a "trajectory" through sock space (a "sock history" or the socks one has owned).[7] But the space also encodes alternatives: socks one *could* have owned.

This is a silly, although not wholly inaccurate account of what a space is (in the wider sense: in this case, a *quality* or *feature space*). Let's firm things up by isolating the important features of spaces before moving on to a more precise discussion and introducing some of the properties that we will find recurring across the various types of space. We can isolate four features of particular importance:

- *Condensation*: A single point of a space can encode arbitrarily many aspects of a system—where the number of independent aspects or parameters n needed to specify a point corresponds to the dimensionality of the space.[8]
- *Individuation*: Each point represents a unique distinct state or possibility for a system.[9]
- *Exhaustion*: The points of the space together include *all* possibilities (of a certain kind) for a system: any possibility for a system is represented in the space, and any point of the space represents a possibility of the system. This principle of exhaustion allows one to *count* the number of possibilities and ascertain what is *impossible*.
- *Geometrization*: There is some notion of metrical relationships and/or ordering among the elements of the space.[10]

Thus characterized, it is clear that the joint unificatory and abstract nature of such spaces, combined with an ability to speak quantitatively about them (and the totalities of outcomes or features they include), renders them scientifically very useful. Let us next consider some basic formal features of these spaces, beginning with Riemann's early

[7] Unless one has an infinitely busy tailor updating socks during incremental foot growth, this will involve discrete jumps, as mentioned earlier. We can further assume here that one only buys one pair at a time, so that time is also represented as a series of discrete steps.

[8] These aspects may or may not determine genuinely orthogonal (independent) axes. Peter Gärdenfors mentions "color" and "ripeness" qualities of fruit, which will tend to covary (2007, 171).

[9] There is sometimes (e.g., when one has symmetries involved) redundancy in a space such that multiple points represent one and the same possibility. In such cases, we assume that the extra points either serve some representational function (e.g., making a problem more transparent) or else are introduced in error and should be identified as a single point (an equivalence class) in a new smaller space.

[10] In cases of *sensory* spaces, as Gärdenfors points out, "judgments of similarity and difference generate an ordering of the perceptions" (Gärdenfors 2007, 169; cf. Clark 1993, 114). In other spaces (such as phase space), a more complicated interaction of mathematical and physical principles is involved in establishing the geometrical structure.

development of the manifold concept, which laid the foundations for spaces of more general kinds than 3D physical space (which was henceforth viewed as "just one kind of space").[11]

The most minimal idea underlying the notion of a space is that it comprises a set of elements on which can be defined some structure (e.g., an affine connection, giving us parallelism and some metric relations, giving us angles and lengths: two features that lead us to what we ordinarily think of as space in the physical sense—that is, a metric-manifold). In his *Habilitationsvortrag* of 1854, Bernhard Riemann developed the general concept of *manifold* from this basic idea (of a set of elements independent of a geometrical structure), along with the earlier presented notion of condensation (or "*n*-fold extension" in Riemann's terminology: *viz.* the number *n* of coordinates it takes to specify a point of a space).[12] Starting from physical 3-space ("Raum"), he had in mind the generalization to multiply extended $n > 3$ cases ("mannig-faltigkeit") that characterize many abstract representational spaces. As he puts it: "[physical] [s]pace constitutes only a special case of a triply extended quantity" (1854, 313).[13] Given this, however, the differences between physical space and any other entities modeled by triply extended quantities must be found beyond the formal structure of the space itself—in experience for example—and one and the same formal structure can find multiple applications.

Given such a space (with extension alone), the question arises as to the further structure, which might be intrinsic (arising from the internal features of the space's elements,

---

[11]   Note that the modern notion of a manifold (a differentiable manifold, that is) is rather more specialized, amounting to a one-to-one (continuous) mapping from an open set to a real space so that the manifold is locally identical to $\mathbb{R}^n$ (for whatever value of *n* one is interested in: for example, a coordinate patch of a 2-manifold *M* would be mapped to a unique image point in $\mathbb{R}^2$ so that functions on the manifold can be rewritten as functions in the real plane). This structure is chosen to allow continuous and differentiable functions to be defined, without which modern physics would be unimaginable.

[12]   Riemann labels "the essential characteristic" (of an *n*-fold extended manifold) the fact that fixing a location in the manifold is reducible to *n* measurements or "numerical determinations" (e.g., of the three position coordinates, x_1, x_2, x_3 of some point in the manifold, for $n = 3$). The concept of space, and with it geometry, is thus separated off from our direct intuitions of space because such higher dimensional spaces simply cannot be thus intuited. See Scholz (1999) for a thorough historical discussion of the concept of manifold.

[13]   The example of space (at least *position* in space) is somewhat special in another way, however. It is, along with color, as Riemann notes, one of the few "everyday life" examples of a continuous manifold. In a set of unpublished notes, Riemann also mentions the 1D manifold of tones ("the line of sound"): see Scholz (1982), in which Riemann's debt to the philosopher Johann Friedrich Herbart is uncovered, especially the idea of considering spaces of arbitrarily high, even infinite, dimensionality associated with the qualities of everyday objects—though as we have just seen, Riemann disagreed about the ubiquity of such continuous spaces in everyday life. Gärdenfors (2004) has developed a method of representing information based on geometric structures he calls "conceptual spaces." These spaces are constructed from geometrical representations of qualities (based on quality dimensions). There are many similarities between this framework and the work of Herbart—namely, the idea that anything can be viewed as a bundle of qualities each of which determines a qualitative continuum. However, Gärdenfors exploits the idea in a variety of distinct ways, including the notion that learning amounts to an expansion of conceptual space by the accretion of new quality dimensions.

in the case of a discrete space) or extrinsic (imposed externally upon the points, for a continuous space)—see Grünbaum (1973, 563–566) for a discussion of the intrinsic/extrinsic distinction in this context. The most important structural feature for our purposes is the metric. As we saw in our sock space example, qualities with discrete spectra self-determine their own natural "counting measure": one can simply move in steps from one point to another to generate a path and measure the distance. Continuous spaces are more tricky and involve satisfaction of certain conditions of metricity as encoded in a distance function.[14]

The metric is key to the usual understanding of physical space. In the case of a Euclidean space $\mathbb{E}^2$ (i.e., the Euclidean plane) one has a distance function $d(q, p) : \mathbb{E}^2 \to \mathbb{R}$ determining the distance $d = n \in \mathbb{R}$ between any two points of space $p = (x, y)$ and $q = (x', y')$, where $q, p \in \mathbb{E}^2$ (with coordinates (x, y) and (x', y')). This function has to satisfy the following four conditions of metricity:

1. $d(q, q) = 0$
2. $d(q, p) = 0 \supset q = p$
3. $d(q, p) = d(p, q)$
4. $d(q, p) + d(p, r) \geq d(q, r)$

One finds that these conditions are satisfied by the Euclidean (flat) metric $d(q, p) = \sqrt{\Delta x^2 + \Delta y^2}$ (where $\Delta x = (x - x')$ and $\Delta y = (y - y')$), relating distance between points to the difference in the points' coordinates. The set of points M together with this function d is then a metric space (M, d) and can be used to represent a variety of phenomena, including physical space (as in Newtonian and quantum mechanics). There are, of course, alternatives to (positive-definite) Euclidean space, such as those with constant curvature (hyperbolic or elliptical).

In his Erlangen program, Felix Klein proposed to define the geometry of a space in terms of its invariance properties (under some group of transformations): for each group of transformations, there is a corresponding space (e.g., Euclidean space is defined by the six-parameter group of rigid translations and rotations). As with physical space, so with the more general spaces: intrinsic features of the space lead to certain invariances in the phenomena or laws the spaces encode. These invariances will show up as symmetries in the spaces, which can be eliminated (reducing the size of the space) or left intact. In addition, certain redundancies of more general type (multiple representations

---

[14] Riemann writes: "in a discrete manifold the principle of metric relations is already contained in the concept of the manifold, but in a continuous one it must come from something else" (1854, 322). Presciently, in view of the construction of general relativity more than sixty years later, Riemann considered this in relation to physical space, suggesting that the "something else" might involve "binding forces acting upon it" (1854, 322). This suggested that the future study of space should be part of physics, rather than pure mathematics.

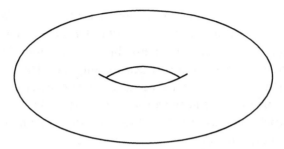

FIGURE 42.1 A torus (or "doughnut") characterized by a multiply connected topology.

of the same state or possibility) can be eliminated via identifications, which will result in a *multiply connected* space.[15] Take the torus, for example (see Figure 42.1).

One can construct this space by rolling up a plane (the fundamental domain) into a cylinder (first identification) and then gluing the ends of the cylinder together (second identification). One could then wind loops around the handle (i.e., threading through the hole), the central hole, or both (these additional kinds of motion are known as "winding numbers"). We can classify the topological invariants of a space by invoking these winding numbers, which count the number of times a loop wraps around the rolled up dimension: in the case of the torus, (m, n) refers to m windings around the hole and n windings around the handle. No winding at all around either would simply be represented by $(0, 0)$ (no motion away from the fundamental domain)—this can be visualized in the "universal covering space" (Figure 42.2).

Non-simply connected spaces can then imply identities between apparently distinct possibilities: the identities will be relative to some relevant relations (such as the way the situations *appear* to an observer or physical laws). Both physical space and the more abstract representational spaces can be multiply connected: the tonal (pitch) spaces of musicology have this form on account of the cyclic nature of the octave,[16] becoming more convoluted depending on other equivalences that one imposes (e.g., chord equivalences through transposition, inversion, and so on). Color space is also non-simply connected with respect to the dimension of Hue, taking on a circular topology, as we see in Section 3.2.

Now that we have a grasp on what a space is and how it might be put together, let's see how such spaces arise in various scientific contexts.

---

[15] There is a standard method of removing this structure (formally at least), if we decide that keeping it is not to our taste, by "quotienting" out the symmetry. What this amounts to, in simple terms, is identifying all of the symmetric possibilities by forming their equivalence classes (relative to the symmetry operation) and treating these classes as points of a new (smaller) space. In the case of non–simply connected spaces, the quotient space is known as an "orbifold" (orbit manifold, where "orbit" refers to the action of the symmetry group).

[16] That is, in terms of harmonic structure, one tone is identical to any tone twelve semitones away in either direction—see Section 3.2.3. In this case, the winding $(0, 0) \rightarrow (1, 0)$ might correspond to moving from middle C to the C above middle C; the motion $(0, 0) \rightarrow (2, 0)$ would involve a movement from middle C to the C *two* octaves higher, and so on.

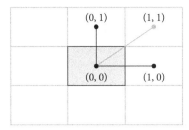

FIGURE 42.2 Unwrapping loops using the universal covering space for the torus. Neighboring horizontal and vertical cells correspond to one winding around the hole (1, 0) and handle (0, 1), respectively. Diagonal cells (1, 1) correspond to single windings around *both* handle and hole. The starting point and the end point correspond to the *same point* in space.

3 KINDS OF SPACE

We can split the class of spaces up into two broad types: those that represent physical targets (such as visual space and spaces associated with other receptors) and those that represent abstract targets (such as phase spaces, in which the points represent *possibilities* for physical targets). As noted earlier, we are not so much concerned with physical space here—inasmuch as we *are* interested, it is with the relation between it and some representational space. The latter often works by assuming a default, ideal model for physical space (e.g., Euclidean geometry) against which the representational space is judged (e.g., for divergence from physical space thus modeled).

3.1 From Physical Space to Possibility Space

When we deal with a complex system of N particles (with large N), it would be a complicated task to deal with all of their paths through 3D space (relative to a time parameter t). Using the principle of condensation, all of this information is bundled into a space of $3N$-dimensions, in which a single point represents the positions (configuration) of all N particles (taking into account each particle's three spatial coordinates in ordinary space) taken at an instant of time.[17] We can then consider paths in this space, which would now correspond to possible histories of the entire system of particles. By adding a temporal component in this way (giving us a $3N + 1$ dimensional space), we can encode the entire dynamical content of a system (the dynamically possible trajectories) according to a

[17] Interestingly, in the context of general relativity, where the dynamical object is spatial geometry itself, the configuration space has as its points Riemannian metrics on a 3-manifold. The dynamics then unfolds in this so-called *superspace* rather than in "real" space. Hence, in a quantum theory of gravity, simplistically construed, the domain space would be this so-called superspace, and the configuration variables (to be quantized) would be 3-metrics (i.e., what we would usually think of as ordinary space).

theory—constraints that the system is under can easily be imposed on the space, which will have the effect of reducing the number of dimensions.

The principle of exhaustion is satisfied in that the space contains *all possible configurations* of the system of N particles in physical space.[18] This space, as a totality, is called "configuration space" Q. Just as the position of a single particle in 3D space is given by a coordinate, so in this new space we have "generalized coordinates" q^i allowing us to label and order the points. However, if we are viewing this space as representing particles in classical mechanics, we face a problem in implementing the individuation aspect of a space since one and the same configuration might represent an infinite range of possibilities, each with different velocities (or momenta) attached to the particles. For this reason, phase space was devised.[19]

In the case of (generalized) velocities q^i, one forms a new manifold called the "tangent bundle" to Q. This can be understood in terms of the tangent vectors d/dt to curves (histories) passing through the points of Q. The set of all such tangent vectors through some point x is the tangent space at x, T_xQ. The tangent *bundle* TQ comprises all of the tangent spaces for all of the points of configuration space Q. Hence, this approach further condenses information into a single space, now including more data (information about velocities). To include these extra data, the new space expands to become $6N$-dimensional (where the extra three dimensions encode information about velocities in the three spatial directions). We can now specify both position and velocity in this space, thus restoring individuation.[20]

If we consider the set of 1-forms[21] at a point of the configuration space, we get a different space: the *cotangent* space at x, T_x^*Q. As previously, bundling together all such spaces, for all points, delivers a new space: the cotangent bundle (or phase space, to physicists), T^*Q. Again, this has $6N$-dimensions, but it is coordinatized by (generalized)

---

[18] In the context of quantum field theory (in the so-called Wilsonian approach), one frequently finds talk of the "space of Lagrangians." Since each Lagrangian defines a theory, this is also thought of as a space of theories (Butterfield and Bouatta, forthcoming). It is, in fact, a kind of configuration space for theories. In this sense, it is another step removed from reality: the target of theory space is a possibility space for theories whose elements are theories that map to the world. A similar kind of theory space occurs in string theory, although there it is often mixed together with discussions of the multiverse, since the theories represent possible universes of string theory. Also in string theory, there have recently emerged some rather surprising symmetries known as *dualities* linking what appear to be radically distinct spaces: one can find large and small spaces identified or even spaces of distinct dimensions (see Rickles 2011, 2013). These often amount to identifying a pair of *theories* in the sense of finding that what looked like a distinct pair of points in theory space map to the same physical possibility (and so lie in an equivalence class).

[19] If we don't have to worry about derivatives, then the individuation aspect is also upheld.

[20] I have not mentioned the existence of a projection map π from the tangent bundle to the pure configuration space Q. This map associates a $3N$-dimensional "slice" (or "fiber") to each point $x \in$ Q and, importantly, allows us to split the six coordinates of TQ into three position and three velocity components.

[21] A 1-form is just a function that takes vectors as arguments and spits out real numbers: $\omega(\mathbf{v}) = scalar\ (n \in \mathbb{R})$.

positions q^i and momenta p_i. We needn't spend too much time over the mathematical details of this space, but we will isolate three important features:

- First, there is a *geometry* on the space, closely related to the metric of ordinary space (though in this case anti-symmetric), which allows one to realize Newton's laws in terms of objects moving in phase space rather than ordinary space.
- Second, one can have *degeneracy* pointing to an unphysical (i.e., purely formal) structure.[22]
- Third, *laws of physics* can be implemented in the geometrical structure of the space. As we see later, the phase space of a theory amounts to the provision of the *possible worlds* of a theory in virtue of a correspondence between trajectories (through points: initial conditions) and world histories.

According to Ismael and van Fraassen (2003), a physical theory is split into two ingredients: theoretical ontology and laws. The former is important for our purposes since Ismael and van Fraassen construe it in terms of metaphysical possibility space, with the laws then acting on this space to generate a smaller space of *physical* (nomological) possibilities. Hence, the theoretical ontology has as its "points" possible worlds of the theory, embodying the kinds of entities, properties, and relations allowed. These possible worlds (being "entire histories") correspond to state space trajectories. Neither of these is yet physical, in the sense of satisfying the dynamical equations of some theory (the laws, that is). Hence, only some of the possible worlds (a subset of the *space* of possible worlds) are associated with the physically possible trajectories. Ismael and van Fraassen speak of the laws of a theory as selecting "a subset of points of the metaphysical possibility space" where "that selection is (or represents) the set of *physical possibilities*" (2003, 374). The worlds isolated in this fashion are, in other words, those that *satisfy* the theoretical laws.

  Thus, one can think of these state space representations as not standing in relation to target systems in the world (at least not directly), but rather to another space: of possible worlds. Hence, the notion of a space is deeply entangled with both the definition of a theory and also interpretation: the provision of the ontology (the worlds) is tantamount to giving an interpretation. Both configuration and phase spaces have been embroiled in recent controversy. Two examples we consider next both concern the "indispensability" of the spaces, although in very different ways: the question of "physicality" of spaces is also involved.

  We have been assuming that configuration space is not isomorphic to physical reality but is instead a mathematical structure that makes a physicist's life easier by condensing information. This is seemingly challenged in the context of quantum mechanics. We represent the quantum state of the world using a wave function $\psi(x)$. The problem is

---

[22] Gauge theories are modeled in this framework (with the inclusion of constraints) since they allow for a transparent geometrical representation of how the unphysical structure appears and how physical structure can be extracted (see Earman 2003).

that, except for the case of a world with a single particle, the domain of this function is not "physical space" (3D space), but $3N$-dimensional configuration space (N duplications of ordinary 3-space: one per particle). That is: the fundamental entity of quantum mechanics (that which obeys the dynamical laws of the theory) is modeled as living in a monstrously high-dimensional space (of three times the number of particles in the universe). Since this is a successful, established theory, we ought to take it seriously ontologically speaking. If we believe in the wave function, then presumably we have to believe in those features that it depends on (that are indispensable).

The point of raising this example is that we have the *choice* of giving configuration space a physical interpretation and viewing the standard 3D "space out there" view as illusory: the line between physical and representational is thus blurred. David Albert (1996) has notoriously defended configuration space realism (see also Lewis 2004).[23] Moreover, he feels, the strange features of quantum mechanics point to the fundamental reality of a higher dimensional space. It's worth spending some time in unpacking the interpretive to-and-fro in the positions of Lewis and Albert because it reveals relevant aspects about the notion of what a "spatial dimension" is.

First, note that both Lewis and Albert point out a further option in how we should think about space: both 3-space *and* $3N$-space exist. This involves some contortions in understanding the dimension concept and trades on an ambiguity in how it can be read, both physical in a sense. The first physical space is ordinary space out there (the 3-space of possible distances between concrete objects). The second is the space in which a *complete state* of the world is defined (in a quantum world, the value of the wave function at every point of the $3N$-dimensional space). So, as Lewis puts it, "the world really is three-dimensional under one reading of 'dimension,' and it really is $3N$-dimensional under the other reading" (2013, 118). Both conceptions of space can, then, be taken as veridical given the disambiguation—although we are hardly left with a fleshed-out picture of how the spaces hang together.

More recently, Lewis has defended spatial realism about ordinary 3-space only and anti-realism about $3N$-space (finding it "illusory"). His approach involves a consideration of a distinction between dimensions and (independent) parameters. As he points out, parameters are not always genuinely spatial in character, so the fact that the wave function depends on $3N$ independent parameters should not be seen to imply $3N$-dimensionality of a space. In support of his claim that parameters and dimensions are not coextensive he gives the following example:

> in evolutionary game theory, the state of a population of organisms can be represented as a function of n parameters, one for each organism, in which each parameter represents the continuum of possible strategies the organism can adopt in

[23] Albert's argument is not purely one of indispensability and, in fact, depends more on the feature alluded to earlier: that phase space has a natural splitting of the coordinates into two groups of three. As Peter Lewis (2013) rightly points out, there are problems in putting too much weight on the grouping into threes: there is, for one, much arbitrariness in the choice of coordinates (even in terms of the grouping into coordinates and momenta), and this points to the choice being an artifact of the representation rather than a fundamental feature of the world (116).

interacting with the others, and the function represents the probability distribution
of over strategy space. (Lewis 2013, 119)

Lewis claims that this tells us nothing about the literal space in which the organisms
live. Although this might not strictly be true (given that the possible interactions and
strategies will be functions of space to some extent), let's suppose it is the case. As he
points out, the configuration space parameters are in fact spatial in his sense (position
coordinates for particles), but still, to reify them as "really spatial" would, he argues, be
tantamount to reifying an "n-organism" in the strategy space example as a result of the
interdependencies between the individual organisms. But this confuses what are repre-
sentationally and functionally different scenarios.

   In the strategy case, each point represents a possible population state for all the organisms. We have no physical dependence relation between such states and some well-
established entity as we do in the case of the quantum mechanical wave function. Hence,
the inference to configuration space realism is grounded in the fundamentality of the
wave function, not simply the fact that the configuration space involves quasi-spatial
parameters. If we had some reason for being indispensably committed to the strategy
space, then we might well think about reifying n-organism states. I'm not suggesting
that this would be a *good* interpretive option, but it would be an option nonetheless.

   Ultimately, Lewis remains committed to configuration space in a sense: the parameters of the space feature just as strongly in quantum mechanics. However, they are
viewed as encoding information about coordinates of all the particles ("they encode
how the wave packet for the particle we are following is correlated with wave packets
for the other particles"; Lewis 2013, 123). But we are left none the wiser about what these
parameters *are* physically (and what the correlations amount to) if they are not spatial: information? If so, is this understood in physical or abstract terms?[24]

   I'm not espousing configuration space realism, but I am pointing out that it remains
an interpretive option to treat what looked abstract (part of the representation alone) as
part of the physical ontology.[25] Indeed, even if we agree with Lewis, there is still a sense
in which the configuration space is representing physical facts (correlations) rather than

---

[24] Jill North (2013) offers a slight twist to configuration space realism by considering "wave function
space realism." Here, the distinction is between a view of the world as containing particles (from whose
properties and relations configuration space is built) and one containing only wave functions. Hence,
the wave function space does not represent a configuration of particles, even though, in order to remain
empirically adequate, the two spaces are isomorphic. This is, then, an interpretive reassignment of one
and the same mathematical space motivated by the view that the fundamental dynamical laws require it.

[25] Just as we can reassign what looks like part of the physical ontology to the representational
machinery, one can do this with the wave function itself, viewing it as an auxiliary mathematical
structure not to be viewed as mapping onto the world—a route taken by Bradley Monton (2013) to reject
configuration space realism, though he is not necessarily against the 3N-dimensional view per se; rather,
he believes that quantum mechanics is false (in the sense of incompatible with facts about gravitation).
We might similarly consider the geometrical structure of the Calabi-Yau spaces to be representational,
viewing them, á la Lewis, as encoding information about "internal degrees of freedom" of spacetime
fields in $\mathbb{R}^4$.

abstract possibilities as Ismael and van Fraassen construe state space representations. This highlights the distinct role played by configuration space in quantum mechanics as opposed to classical mechanics. Of course, the state space of quantum mechanics (Hilbert space) still satisfies Ismael and van Fraassen's claims about the (state space) → (possible world) correspondence.

Interestingly, David Malament (1982) uses this fact—that state spaces represent possible worlds (rather than physical systems)—to launch an argument against nominalism (specifically, Hartry Field's defence in *Science Without Numbers*). The problem, as Lyon and Colyvan put it, is simply that "not only are the points in phase space abstract entities, but they also *represent abstract entities*, viz. possible dynamical states" (2008, 6). Malament explains:

> Suppose Field wants to give some physical theory a nominalist reformulation. Further suppose the theory determines a class of mathematical models, each of which consists of a set of "points" together with certain mathematical structures defined on them [spaces in our sense—DR]. Field's nominalization strategy cannot be successful unless the objects represented by the points are appropriately physical (or non-abstract). In the case of classical field theories the represented objects are space-time points or regions. So Field can argue, there is no problem. But in lots of cases the represented objects are abstract. In particular this is true in all "phase space" theories. (Malament 1982, 533)

The problem here concerns the treatment of abstract versus concrete objects. We can think of a concrete property or structure as having some realization in a particular system, whereas an abstract object is independent of realization: with Platonist hats on, we might say that the former instantiate the latter. For example, we might have some configuration of particles that falls into the shape of a square, which will be represented by some point in a phase space description. What is the relationship between the phase point and the system? Is it a realization of some abstract universal property? The nominalist will want nothing of such abstracta, preferring to say that the abstract property is simply the set of all particular configurations of that sort.

The problems for the nominalist arise because of the principle of exhaustion: there are configurations that are possible for a system of particles, but that will never be realized. Note also that even the world (corresponding to the phase point representing our world) picked out for special treatment as "actual" is also representational because it itself is supposed to *map* to our world—that is, there is still an indirectness in how the phase point represents the real-world target. However, even those representations (such as, e.g., a spacetime manifold) that are taken to map directly to the world do nothing of the sort and map to an intermediary abstract version of reality, which is itself a representation.

In sum: upholding a clear distinction between a representational and a physical space thus comes under pressure from the amount of interpretive freedom available on both sides.

3.2 Receptor-Based Phenomenal Spaces

Phenomenal spaces (visual, color, auditory, etc.) are part of a "reconstruction of the world" (just as theories are), as Patricia Churchland puts it, "from the effects on our nerves" and are "in some measure a product of our brains" (1986, 21)—part, that is, of the inferential process involved in the journey to such representations from the trans-duction of inputs. Of course, it is also a partial product of our physiology (the specific receptors), which imposes constraints on the brain's reconstructions, albeit with pos-sible flexibility in how they are satisfied—Churchland refers to receptors as the "inter-face between world and brain" (1986, 46).[26] The receptors individuate the senses and determine the dimensions of a specific sensory space; so, sensory possibility is relative to them.[27] The ordering of the points of receptor spaces are usually psychologically determined according to the "perceived closeness" of experiences and the hope is that this metric is isomorphic to the source of the perceptions—as we see, it rarely (if ever) is. Here, we have to be careful with our handling of the *individuation* principle because what are distinct *physical* states do not necessarily correspond to *distinguishable* states (i.e., by an observer), in which case the equivalence class machinery comes into play.

Churchland (1986, 427) views the sensory (receptor) space representations of the world as "located" in a (relevant) phase space on which computations are performed (corresponding to mappings between phase spaces). There is, moreover, such a space for each of the sensory modalities. Neuronal arrays are the "medium" for the representa-tion, but, ultimately, these arrays are understood in phase space terms.[28] Churchland presents as an example of a very-high-dimensional space, the "space of faces" involved in face recognition tasks. As with our sock space described earlier, the face space has as its axes some facial features relevant to the individuation of faces. Each point in the space will then represent a unique face by condensing these features. However, the representa-tion supposedly works in virtue of a relatively straightforward *resemblance* between the represented objects and the neuronal configurations (cf. Watson 1995, 107), but, as we see, this is rarely the case.[29]

---

[26] This somewhat masks the fact that the idea of a "receptor" is itself a theoretical entity and also, to a certain extent, part of the brain's representational machinery.

[27] Austen Clark speaks in terms of "*Psychological primaries* for a given domain" that are "members of the smallest set of qualities (a) whose mixtures can match every sensed quality in the domain; but (b) each of which cannot be matched by combinations of the others." These are phenomenologically "pure" qualities: the atomic constituents of other qualities (1993, 126). I take this principle of combination to be included in my four "principles of space-hood."

[28] It is difficult to see what these phase spaces themselves might be, given the extreme physicalism of Churchland. She appears to speak of them as if the neurons (the brain) *embody* such a phase space, but this leaves untouched the modal aspects of such (possibility) spaces. Furthermore, most receptor phase spaces would be so unimaginably complex that it is difficult to see what use they could be in guiding action and cognition.

[29] As Michael Arbib puts it, "[t]he task of perception is *not to reconstruct* a Euclidean model of the surroundings" rather "perception is *action-oriented*, combining current stimuli and stored knowledge to determine a course of action appropriate to the task at hand" (1991, 380).

FIGURE 42.3 Penrose's impossible triangle exposing the fragmented nature of visual space.

In this section, we consider visual, color, and tonal pitch spaces, which each have their own quirks, but all highlight (similar) problems for a naïve realism about the mapping between the spaces and the reality they represent.[30]

3.2.1 *Visual Space*

In Kant's day, it was taken for granted that space had a Euclidean structure. This led to the assumption that our visual space shared this same structure. It was also assumed that this perceptual (phenomenological) space had a global and unified structure, so that the visual scene corresponded to single geometry (rather than local patches or varying geometry)—this latter assumption persisted well into the non-Euclidean days. However, experiments starting in the 1960s chipped away at this idea, showing our perceptual, visual space to be a fragmented affair. This fragmentation is shared at the neurobiological. too, with a distribution of the processes going into visual spatial experience being conducted by distinct areas of the brain (Figure 42.3).[31]

If the visual system has an in-built default geometry, then we ought to be able to detect this in human responses, from which that geometrical structure can be reconstructed: flat (Euclidean), hyperbolic, or elliptical.[32] In a series of famous "alley" experiments from the early twentieth century (devised by Walter Blumenfeld), it was concluded that space perception (visual geometry) is non-Euclidean. The alley

[30] Clark (1993) has argued that these sensory spaces are really "qualia spaces": objective representations of phenomenology. Your tasting some Pinot Noir or smelling some Earl Grey tea means that your cognitive system is occupying some appropriate point in gustatory and olfactory space. Clark's theory includes all such sensory spaces, which are to be given a physical interpretation. This largely matches Churchland's approach mentioned earlier—I have my doubts about Clark's proposal for similar reasons. However, Clark's position along with Churchland's stand as examples of reifying what would ordinarily be viewed as "purely representational spaces" with no literal physical interpretation.

[31] The well-known "impossible figures" can facilitate the exposure of some of this fragmentation (see Figure 42.3). As MacLeod and Willen argue, the Penrose triangle illusion amounts to the attempt to construct a representation from local, fragmented information, which the brain finds impossible to do and ends up switching between various representations. They thus think such illusions indicate "that the phenomenal integrity of visual space is itself an illusion" (1995, 59).

[32] See Wagner (2006) for a detailed overview of the geometries of visual space—the literature on this subject is enormous.

experiments work by distinguishing two kinds of alley: distance and parallel. In the former case, the subjects are shown a pair of (fixed) lights on the horizon positioned either side of them (equidistant from the median plane). A series of test lights are then switched on, positioned along the horizon so that the subject has to adjust her positions to make them equidistant, like the fixed lights. In the parallel case, there are two fixed lights symmetrically placed to either side of the subject, now far away, and the subject was instructed to adjust increasingly closer lights (lying in the same horizontal plane as the fixed lights) so that they appeared to form a parallel track moving toward the distant fixed lights.

The idea is that if visual space were Euclidean, then (modulo some measurement error factor) the settings of lights would match in the two cases. But this was not the result: in the parallel alley scenario, the lights were adjusted closer to the median plane of the observer than in the distance alley scenario. Rudolf Luneberg analyzed these results, concluding that visual geometry is hyperbolic. In the representation of a space of objects (lights) satisfying a Euclidean geometry (i.e., in physical space), the mapping contracts more distant lengths and expands closer lengths.[33]

There are all sorts of methodological problems with this test, spoiling the conclusion's validity, having to do with, for example, the idealized test conditions that have low external validity in terms of application to more complex visual displays (e.g., including more objects, motion, and change; see Howard 2014, 194). It has also been found that the degree of curvature found is highly dependent on the kinds of specific task that a subject is carrying out (e.g., using a tool). Patrick Suppes (1977) suggests that the geometry of visual space is as context dependent as is the geometry of physical space in general relativity: depending on what physical objects are present, the geometry will be different. However, the situation is inverted in how the geometry is affected in the visual and physical cases: more perceptual objects render the visual field more Euclidean, whereas the opposite is true in the case of physical objects in the gravitational field (which is identified with geometry), with more objects (masses) causing greater deviation from flatness. This task dependence supports Suppes's suggestion that the visual geometry is dependent on what is going on in the visual field, although in this case it is more a matter of what is being *attended to* in the visual field.

The geometrical structure of visual space is still not settled (there is no such thing as *the* geometry of visual space), but it does seem to be the case that the direct resemblance between physical/visual cannot be maintained: although visual space looks prima facie like a simple case of direct mapping, this does not fit the facts.

3.2.2 *Color Space*

Colors are extremely useful additions to the world: they provide a new layer of object discrimination (beyond, e.g., shape, size, texture, and so on). Evolutionarily, we use

---

[33] A similar recent experiment by Koenderink and van Doorn (2000), involving large distances and allowing for other visual cues to enter, found that distant geometry was elliptical and near geometry was hyperbolic.

color as an indicator of "safe food," "sweet food," and so on.[34] Colors (as perceived) can be exhaustively described by a combination of hue (H; referring to color class or "pure color"), saturation (S or chroma: the "whiteness" of a color), and value (V; lightness or brightness: how dark a color is). These provide the three axes of an HSV color space. Any variations in perceived color can be represented by altering the values of these parameters. The space of colors perceived by the human eye has the structure of a 3-manifold, call it C, with the dimensions corresponding to three (independent) color receptors (red L-cones, blue S-cones, and green M-cones).

Distances between points of this manifold are taken to map onto subjectively observed differences of colors. The closer the points are together, the closer the colors "seem" to a viewer. However, in rating "closeness" of colors, people often find it difficult to place purples with blues or reds. The solution is to join the colors together by the purples into a wheel, with reds on one side and blues on the other. Psychologically opposed colors then lie opposite one another on the wheel (see Figure 42.4: the so-called "color spindle," in which saturation and value are also represented). This ordering gives the space a non-simply connected topology. We also see here a "preferred basis" of colors (red, yellow, green, and blue) in terms of which all other colors can be described. These are the unique hues.

Isaac Newton developed a color space in terms of the wavelengths of light. A problem with this (in terms of what we *mean* by color) was discovered by Hurvich and Jameson in the 1950s (Hurvich and Jameson 1955): perceived colors are not isomorphic to specific wavelengths (the color of light). Different wavelengths can generate the same color as seen (depending on such factors as illumination). For our purposes, what is important is the distinction between "physical" and "phenomenal" color. Newton discovered that different spectral distributions mapped to one and the same color experience, so that color as perceived is an equivalence class of physical colors—the latter space S is infinite dimensional, whereas the former C is only 3D. The equivalence classes will map to distinct perceived colors in a one-to-one fashion (so that $S \rightarrow C$ is many-to-one), with elements of one and the same equivalence class being "metamers" of some perceived color.

An influential approach (Byrne and Hilbert 2003) to color ontology, familiar from our discussion of symmetries and quotient spaces earlier, involves a commitment to *equivalence classes* of surface-spectral reflectance properties generating the same color experience. This can be viewed as analogous to the move in the context of the "hole argument" (of general relativity) in which one is ontologically committed to the equivalence classes of metrics with respect to the symmetry group of the theory (the *geometry*). Just as the laws of general relativity are blind to changes of metrics lying within the same

---

[34] Joseph Weinberg presents a model of color space in which "Riemann's 'forces' are essentially those of natural selection [such that] [a]ny survival value of color vision must rest upon its help in recognising an object by its reflectance" (1976, 135). A connection is forged between the metric tensor on color space and the various properties of light in nature (plus selection). This model generates an objective (in the sense of independent) metric despite the fact that a subjective (phenomenal) color space is involved.

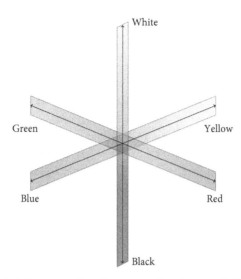

FIGURE 42.4 Simplified (phenomenal) color space showing psychologically opposing hues (red-green) and (yellow-blue), as well as experienced lightness along the vertical dimension. Saturation is represented by distance from the center of the spindle.

equivalence class (i.e., containing diffeomorphic copies), so the eye is blind to metameric transformations.

As will be familiar now, the mapping does not just break down in the (physical space) → (phenomenal space) direction (as a result of the many-to-one relationship). The converse also breaks down as a result of contextual effects that are just as nontrivial in color perception, as with visual perception, so that many phenomenal colors map to one and the same "physical color." One can see this in cases of contrasts with different surrounding colors ("surrounds") that influence how physically identical colors appear (see MacLeod [2003] for examples). But there are everyday examples involving the brain's interference to break the veridical nature of the space. As Ashtekar et al. point out:

> Snow, in particular, looks white even when we are walking in a pine forest, although now it in fact reflects predominantly green light. Similarly, for a large class of illuminations, a patch color can seem to be green in certain surroundings but pale grey when looked at in isolation. A sheet of paper appears white in both bright day light and in a windowless office even though the amount of light reflected from a dark grey sheet in daylight is much more than that reflected by the "white" sheet in a windowless office. In each case, the brain has absorbed not just the three inputs [from the receptors] but a whole range of other "relevant" information to produce a final color perception. (Ashtekar, Corichi, and Pierri 1999, 547–548)

Of course, there are good evolutionary reasons for this persistence of color despite physically different reflectances: color constancy ensures that one's ability to identify objects (including dangers) is stable. However, the non-isomorphism of space and target clearly throws a spanner in the works of most accounts of scientific representation.

3.2.3 *Tonal Space*

The musicological space is a sensory space of sorts because it is so heavily dependent on pitch perception, although many musicologists would baulk at this suggestion. The spaces encode implicit (or explicit for trained musicians) knowledge about musical structure. The structure of the space/s has been analyzed in two ways: "objectively," via psychology and neuroscience, and "subjectively," via music theoretical techniques (often based on a listener's introspective experience of a piece of music). Central to both, however, is the notion of tonality. Musicologist Brian Hyer has claimed that there is a "recurrent tension" in music theory over "whether the term [tonality] refers to the objective properties of the music—its fixed internal structure—or the cognitive experience of listeners" that is "whether tonality is inherent in the music" (2002, 727). There is clear overlap here with the example of color: one can model in terms of the physical or the phenomenal. We see in the case of tonality, too, that the mapping from physical to phenomenal is many-to-one (and, *mutatis mutandis*, one-to-many for certain contextual effects), which suggests the kind of abstraction (and simplification) that characterizes scientific representation in general.

There is a justly famous set of experiments, known as "probe tone tests," designed to examine a listener's representation of tonality (Krumhansl 1990). The experiment involves a priming of the subject with some musical example (from tonal music), providing a context, and then supplying the subject with a probe (a single tone taken from the thirteen chromatic tones of the octave, with the octave tone included). The subject then has to make a judgment about the goodness of fit (using a seven-point scale) with the earlier musical passage. The results suggest a fairly robust tonal geometry that matches musicological practice (as embodied in, e.g., the circle of fifths). As Carol Krumhansl notes, "tones acquire meaning through their relationships to other tones" (1990, 370). In other words, musical context affects the perception and representation of pitch so that multiple mental representations map to one and the same physical source. Tonic tones are perceived as "closest," then diatonic, then nontonic (this generates a tonal hierarchy). It is precisely the interplay of tonal stability and instability that generates musical tension (produced by motions away from the tonal center) and release (produced by motion back to tonal center). The space as constructed through multidimensional scaling techniques on the data from these tests is shown in Figure 42.5.

Krumhansl's (1990, 112–119) space represents the perceptual relations among the various keys. She explicitly appeals to the fact that a spatial-geometric representation can display the various dependencies between pitches, as revealed in the probe tone tests. She also notes that there is something inherently "spatial" about musical attributes: "a tone is described as low or high, an interval as ascending or descending, and so on" (112). A central phenomenological feature that affects the topology of the space is octave equivalence ("the sense of identity between tones separated by octaves"; 113)—this involves frequencies f and $2f$. In identifying tones that are thus linked, they are represented in the space by a single point (so that for any pitch p, any pitches that are integer multiples of twelve away [where each semitone amounts to a unit distance]

C Major Key Context

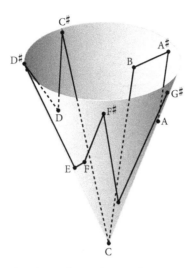

FIGURE 42.5 Multidimensional scaling solution from judgments given in the probe tone test for the case of a C major context—tonic is at the vertex, with distance from the vertex representing goodness of fit: the further away, the worse the perceived fit with the context provided.

(*Source*: Carol Krumhansl (1990) *Cognitive Foundations of Musical Pitch*, Oxford University Press, 128).

are considered "the same").[35] Mathematically, this corresponds to the space $\mathbb{R}/12\mathbb{Z}$ (the reals with the set of multiples of twelve modded out) and is what music theorists call "pitch class space" (cf. Tymoczko 2006, 72).

A space due to Roger Shepard (1982) includes the octave equivalence but keeps the symmetry visible in the space (see Figure 42.6). Shepard implements octave equivalence via a projection down onto the lower disc (what he calls the "chroma" or tone quality), which identifies all octave equivalent tones (such tones share the same chroma, that is).

Dmitri Tymoczko (2011) has recently built these basic ideas up into a detailed geometrical framework for doing musicology. He aims to chart the "shape" of musical spaces, producing a kind of translation manual between musicological ideas and (highly nontrivial) abstract spaces. As described earlier, rather than thinking in terms of pitch space (simply a space whose points are different pitches, ordered in the traditional linear way, such that a musical work traces a path through it), Tymoczko identifies the same pitches (e.g., middle C, C above, C below, and all other Cs), producing "pitch-class" spaces rooted in octave-equivalence. Tymoczko notes that now the corresponding space is an example of an "orbifold" (an orbit manifold, where the manifold has been "quotiented" by octave equivalence, thus identifying certain points). However, Tymoczko generalizes this to *all* intervals. For example, one could go from middle C to E flat by going

[35] Sameness here is couched in terms of "musical structure" (though in such a way as to be related to the listener's experience, such as preserving the character of a chord in a transposition for example).

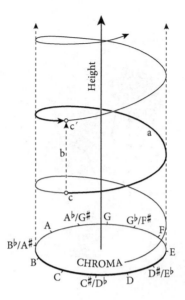

FIGURE 42.6 Helical model of the cyclic nature of tones due to octave equivalence. Pitch height runs vertically in one dimension .

[Source: Roger Shepard (1965) "Approximation to Uniform Gradients of Generalization by Monotone Transformations of Scale," In D. I. Mostofsky (ed.), *Stimulus Generalization*, Stanford University Press, 105].

up or down (and then jump any number of octaves up or down) to get the same interval. Hence, these motions are identified, producing a more complex orbifold. Again, one can apply this to *chords* of any type, with each voice of the chord generating another dimension of the space: the same chord will simply be playable in many different ways, and these are to be identified, too, again producing an orbifold when the redundant elements are eliminated. There are five such musical transformations for quotienting out redundant structure that generate the various equivalence classes of "musical entity" (e.g., chords, chord types, chord progressions, pitch classes, etc.). These are the so-called OPTIC symmetries: *O*ctave shifts, *P*ermutation (reordering), *T*ransposition (the relation between pitches sharing the same succession of intervals, regardless of their starting note), *I*nversion (turning a sequence upside down), and *C*ardinality equivalence (ignoring repetitions: note duplication).

Musical compositions can have their tonal structures modeled in such quotient spaces, and we can judge the success or failure of compositions relative to them. That is, we have a space of musical possibilities and therefore a means of testing that will "work." Musicologists have argued that this geometrical approach kills the phenomenological complexity of musical experience (e.g., Hasegawa 2012). However, as with any scientific modeling, sacrifices of detail have to be made to get at the invariant structure amenable to mathematical analysis. Moreover, as we have seen, the geometrical structure is derived from phenomenological measures.

More recent work has identified a neural basis of such measures, revealed by localized activation patterns in the cortex mapping to relationships among tonal keys. Janata et al.

(2002) claim to have found evidence in functional magnetic resonance imaging (fMRI) experiments, of "an area in rostromedial prefrontal cortex that tracks activation in tonal space" mapping onto the "formal geometric structure that determines distance relationships within a harmonic or tonal space" in the context of Western tonal music. They found that "[d]ifferent voxels [3D pixels] in this area exhibited selectivity for different keys" (2002, 2167; cf. Raffman 2011, 595–596). What this work suggests is that our neural representations of music are relational: on a specific occasion, some family of neurons will fire, say for the key of A minor, whereas on another occasion, that family might fire for the key of C minor. The crucial fact is, however, that the relational structure between the keys is preserved on each occasion despite the fact that keys are not absolutely localized to unique assemblies of neurons:

> what changed between sessions was not the tonality-tracking behaviour of these brain areas but rather the region of tonal space (keys) to which they were sensitive. This type of relative representation provides a mechanism by which pieces of music can be transposed from key to key, yet maintain their internal pitch relationships and tonal coherence. (Janata et al. 2002, 2169)

This strengthens the idea of using a geometric representation. The neural representation also goes some way toward explaining the "musical symmetries" involved (e.g., in transposition), given the encoding of relative structure.

Ultimately, however, as Krumhansl points out, the geometric representation of musical pitch relationships is somewhat limited in what it can do because perceived relations are highly context-sensitive, yet such models make all sorts of assumptions about the fixed structure (e.g., that the relations satisfy the metric axioms). Another contextual factor that matters musically is that the judgments given about similarity and goodness of intervals are not symmetric under permutation of the tones, whereas the symmetry of distance is an axiom of geometric models. Moreover, as Stefan Koelsch points out, one can have situations in which physically quite different frequencies are perceived by listeners as "somewhat identical," whereas frequencies close in physical terms (such as two pitches separated by a semitone) are perceived as "quite different" (2012, 23). Hence, there is no simple mapping between frequency space and tonal pitch space, much as we saw for color and visual space.

4 Why Spaces?

The use of spaces in the sciences is not hard to fathom: after all, they manage to encompass in a single catalog of coordinates totalities of possibilities. So long as one is happy with the structure of the space, one can be confident that nothing will escape the net. It thus provides a complete framework for thinking about whatever it is the space is representing. Obviously, the more general and wide-ranging one makes some framework, the more freedom and flexibility there is likely to emerge. Spaces are like theories and models in this respect: one selects what one thinks are the basic degrees of freedom (ignoring irrelevant features), imposes some structure on these appropriate to the task (in this case as a geometry), and finds encoded in the resulting space what is possible and impossible.

Scientific practice involves essential idealization, abstraction, and simplification of phenomena (think "spherical cows"). All of the spaces we have considered fit this methodology by focusing on a set of core properties that ought to combine to generate a complete possibility space for a phenomenon of interest. Without the elimination of detail, action and analysis based on them would be impossible. Although the sensory spaces are clearly informed by millennia of selection pressures in their elimination of detail, the state spaces are informed by physical assumptions chosen for a variety of reasons (to fit the phenomena, to fit background knowledge, etc.).[36] When a mathematical model (or a representation in general) of a physical phenomenon (including experience) is constructed, then, what is represented will also be highly idealized: we will possess a representation *of a representation* (in which many aspects of the target system have been ignored). Hence, a representation involving a space has (at least) two steps to go to reach the "real world" (the actual target). The initial idealization must preserve some relevant structure of the original, and then the representation must preserve the structure of the idealized version (which we expect will flow down into the target, too, with the ignored details viewed as perturbations).

We saw how the representation can often pick up features that do *not* flow down to target: they are usually taken as having no representational significance themselves but are seemingly indispensable for the representation as a whole to work—for example, the 3*N*-dimensional configuration space of the wave function in quantum mechanics (according to configuration space anti-realists) or, more simply, the use of real or complex numbers to represent values. As Michael Friedman points out,

> Scientists ... distinguish between aspects of theoretical structure that are intended to be taken literally and aspects that serve a purely representational function. No one believes, for example, that the so-called "state spaces" of mechanics—phase space in classical mechanics and Hilbert space in quantum mechanics—are part of the furniture of the physical world. Quantum theorists do not believe that the observable world is literally embedded in some huge Hilbert space.... Rather, the function of Hilbert space is to represent the properties of physical quantities like position and momentum by means of a correlation with Hermitian operators. Similar remarks apply to other auxiliary "spaces" such as "color space," "sound space," etc. (Friedman 1981, 4)

---

[36] Penny Maddy (1997) has argued that the use of continua in our mathematical representations involves either inconsistency or false assumptions so that any such usage in scientific practice is not to be viewed as mapping onto reality. James Robert Brown counters by pointing out that she restricts her focus to physical space, neglecting the various representational spaces: "perhaps physical space and time are not genuine continua, but this need not imply that all our various spaces of representation also fail to be continua" (2013, 143). Of course, many *just are* continua, by construction. Whether they veridically map to the world is another matter and, as we have seen, is not necessary for their useful application. Moreover, we found that the representational spaces *do not* map onto reality in any simple way, although not for reasons having to do with the nature of continua.

As he goes on to say (and as mentioned in Note 1), according to relationalism, physical space serves a purely representational auxiliary function, with physical objects not to be understood as literally embedded in physical space.[37] He also notes that the natural position of a positivist is to treat *all* aspects of theoretical structure in the same way as the relationalist treats space, as "one big auxiliary 'space' for *representing* the observable phenomena" (1981, 5). I would suggest that this is not as outlandish as it sounds, given the evidence from our study of spaces alone, which possess enormous interpretive flexibility: that is, whether a space is considered "auxiliary" or not is an interpretive choice. Moreover, often it is known that some such auxiliary space is *not* faithfully representing observable (physical) structure because other contextual (pragmatic, evolutionary, mathematical, or other) factors veto the mapping that is steering the phenomena from the actual physical structure *although for good reasons*.[38] This leads to ambiguity over what the space is representing and what its function is.

There is more to spaces, it seems, than mapping as closely as possible onto reality, and the notion of the "reality" being mapped to is in any case sometimes up for debate—for example, in the case of Shepard's tonal space, Krumhansl (1990) rejects the choice of geometry on account of a divergent account of how musical/auditory reality is to be understood (namely, as involving *consonance* in a fundamental way that should be represented in the geometry).

Our discussion suggests, then, that Friedman's distinction between "aspects of theoretical structure . . . intended to be taken literally" and "aspects that serve a purely representational function" is hard to uphold given the features of the spaces we discussed.[39] The physicality or "theoreticality" of a space comes into question in each case as a result of contextual effects and/or interpretive choice.

Spaces thus seem to occupy an interesting position in science in that these mapping misalignments (between "space as representation" and "space as reality") are often known and tolerated. Hence, their primary use, in systematizing a truly vast range of features of a theory or phenomenon, also leads to ambiguity: greater generality in representation leads inevitably to difficulties in pinning the spaces onto specific sectors

[37] He also remarks (loc. cit.) that those espousing the many-worlds interpretation might well believe that we do indeed inhabit a huge Hilbert space, again pointing to the flexibility in how we treat spaces in science (analogous to the situation with configuration space realism).

[38] The example of color models is apposite here. We know that our perceptions can't distinguish fine detail in the colors, so we bundle the physical colors into equivalence classes according to how we distinguish them perceptually. But then this space still faces problems in that elements beyond color perception interfere with our color experiences (recall the examples of Ashtekar et al.): that is, although we consider our color model to be a good one, it faces problems modeling both physical color and perceived color!

[39] Friedman himself suggests that unifying power makes the difference between literal and representational function and involves *interaction with other pieces of theoretical structure*. This might explain the willingness to provide literal interpretations of configuration space in the context of quantum mechanics (given its relationship to the wave function) but not in the case of the strategy space example, which does not figure in such a theoretical dependency relationship. There is something to this, as I suggested when considering that example, but I would argue that forging such links simply presents a challenge to interpreters. We have seen how Churchland links phase spaces (a prime example of a purely representational space for Friedman) with neuronal arrays in order to breathe life into them.

of reality. Riemann's great breakthrough in freeing geometry from physical space as standardly conceived, so that the same representational structure is multiply realizable, inevitably leads to such problems. This feature certainly demands the attention of philosophers of science, with a view to seeing where spaces lie relative to other similar, although less extreme, instances of representational machinery (such as thought experiments and simulations).

References

Albert, D. (1996). "Elementary Quantum Mechanics." In J. Cushing, A. Fine, and S. Goldstein (eds.), *Bohmian Mechanics and Quantum Theory: An Appraisal* (Dordrecht: Kluwer), 277–284.

Arbib, M. (1991). "Interaction of Multiple Representations of Space in the Brain." In J. Paillard (ed.), *Brain and Space* (Oxford: Oxford University Press), 379–403.

Arntzenius, F. (2014). *Space, Time, and Stuff* (New York: Oxford University Press).

Ashtekar, A., Corichi, A., and Pierri, M. (1999). "Geometry in Color Perception." In B. R. Lyer and B. Bhawal (eds.), *Black Holes, Gravitational Radiation and the Universe* (Kluwer Academic), 535–549.

Brown, J. R. (2013). *Platonism, Naturalism, and Mathematical Knowledge* (London: Routledge).

Butterfield, J., and Bouatta, N. (forthcoming) "Renormalization for Philosophers." In T. Bigaj and C. Wüthrich (eds.), *Metaphysics in Contemporary Physics* (Poznan Studies in Philosophy of Science, Amsterdam: Brill/Rodopi).

Byrne, D., and Hilbert, D. (2003). "Color Realism and Color Science." *Behavioral and Brain Sciences* 26: 3–21.

Churchland, P. (1986). *Neurophilosophy: Toward a Unified Science of the Mind/Brain* (Cambridge: Bradford Books).

Clark, A. (1993). *Sensory Spaces.* (Oxford: Clarendon Press).

Earman, J. (2003). "Tracking Down Gauge: An Ode to the Constrained Hamiltonian Formalism." In K. Brading and E. Castellani (eds.), *Symmetries in Physics: Philosophical Reflections* (Cambridge: Cambridge University Press), 140–162.

Friedman, M. (1981). "Theoretical Explanation." In R. Healey, *Reduction, Time, and Reality* (Cambridge: Cambridge University Press), 1–16.

Gärdenfors, P. (2004). *Conceptual Spaces* (Cambridge, MA: MIT Press).

Gärdenfors, P. (2007). "Representing Actions and Functional Properties in Conceptual Spaces." In T. Ziemke, J. Zlatev, and R. M. Frank (eds.), *Body, Language, and Mind, Volume 1* (Berlin and New York: Mouton de Gruyter), 167–195

Grünbaum, A. (1973). *Philosophical Problems of Space and Time* (Boston and Dordrecht: D. Reidel).

Hasegawa, R. (2012). "New Approaches to Tonal Theory." *Music and Letters* 93(4): 574–593.

Howard, I. (2014). *Perceiving in Depth, Volume 1: Basic Mechanisms* (Oxford: Oxford University Press).

Hurvich, L. M., and Jameson, D. (1955). "Some Quantitative Aspects of an Opponent-Colors Theory. II. Brightness, Saturation, and Hue in Normal and Dichromatic Vision." *Journal of the Optical Society of America* 45: 602–616.

Hyer, B. (2002). "Tonality." In T. Christensen (ed.), *The Cambridge History of Western Music Theory* (Cambridge: Cambridge University Press), 726–752.

Ismael, J., and van Fraassen, B. (2003). "Symmetry as a Guide to Superfluous Theoretical Structure." In E. Castellani and K. Brading (eds.), *Symmetry in Physics: Philosophical Reflections* (Cambridge: Cambridge University Press), 371–392.

Janata, P., Birk, J. L., van Horn, J. D., Leman, M., and B. Tillmann (2002). "The Cortical Topography of Tonal Structures Underlying Western Music." *Science* 298: 2167–2170.

Koelsch, S. (2012). *Brain and Music* (New York: Wiley).

Koenderink J. J., and van Doorn, A. J. (2000). "Direct Measurement of the Curvature of Visual Space." *Perception* 29: 69–79.

Krumhansl, C. L. (1990). *Cognitive Foundations of Musical Pitch* (Oxford: Oxford University Press).

Lewis, P. (2004). "Life in Configuration Space." *British Journal for the Philosophy of Science* 55: 713–729.

Lewis, P. (2013). "Dimension and Illusion." In A. Ney and D. Z. Albert (eds.), *The Wave Function: Essays on the Metaphysics of Quantum Mechanics* (Oxford: Oxford University Press), 110–125.

Lyon, A., and Colyvan, M. (2008). "The Explanatory Power of Phase Spaces." *Philosophia Mathematica* 16(2): 227–243.

MacLeod, D. I. A. (2003). "New Dimensions in Color Perception." *Trends in Cognitive Sciences* 7(3): 97–99.

MacLeod, D. I. A., and Willen, J. D. (1995). "Is There a Visual Space?" In R. Duncan Luce, et al. (eds.), *Geometric Representations of Perceptual Phenomena* (Mahwah, NJ: Lawrence Erlbaum), 47–60.

Maddy, P. (1997). *Naturalism in Mathematics* (Oxford: Oxford University Press).

Malament, D. (1982). "Review of Field's *Science Without Numbers*." *Journal of Philosophy* 79: 523–534.

Monton, B. (2013). "Against 3N-Dimensional Space." In A. Ney and D. Z. Albert (eds.), *The Wave Function: Essays on the Metaphysics of Quantum Mechanics* (Oxford: Oxford University Press), 154–167.

Nagashima, Y., and Nambu, Y. (2010). *Elementary Particle Physics: Quantum Field Theory and Particles, Volume 1* (Wiley-VCH).

North, J. (2013). "The Structure of the Quantum World." In A. Ney and D. Z. Albert (eds.), *The Wave Function: Essays on the Metaphysics of Quantum Mechanics* (Oxford: Oxford University Press), 184–202.

Raffman, D. (2011). "Music, Philosophy, and Cognitive Science." In T. Gracyk and A. Kania (eds.), *The Routledge Companion to Philosophy of Music* (London: Routledge), 592–602.

Rickles, D. (2011). "A Philosopher Looks at String Dualities." *Studies In History and Philosophy of Modern Physics* 42(1), 54–67.

Rickles, D. (2013). "AdS/CFT Duality and the Emergence of Spacetime." *Studies in the History and Philosophy of Modern Physics* 44(3): 312–320.

Riemann, B. (1854/2012). "On the Hypotheses Which Lie at the Foundations of Geometry." Reprinted in J. McCleary, *Geometry from a Differentiable Viewpoint* (Cambridge: Cambridge University Press), 313–323.

Scholz, E. (1982). "Herbart's Influence on Bernard Riemann." *Historia Mathematica* 9: 413–440.

Scholz, E. (1999). "The Concept of Manifold, 1850–1950." In I. M. James (ed.), *History of Topology* (Amsterdam: Elsevier), 25–65.

Shepard, R. N. (1982). "Geometrical Approximations to the Structure of Musical Pitch." *Psychological Review* 89(4): 305–333.

Suppes, P. (1977). "Is Visual Space Euclidean?" *Synthese* 35: 397–421.

Susskind, L. (2009). "The Anthropic Landscape of String Theory." In B. Carr (ed.), *Universe or Multiverse?* (Cambridge: Cambridge University Press), 247–266.

Tymoczko, D. (2006). "The Geometry of Musical Chords." *Science* 313: 72–74.

Tymoczko, D. (2011). *The Geometry of Music* (Oxford: Oxford University Press).

Wagner, M. (2006). *The Geometries of Visual Space* (Mahwah, N.J.: Lawrence Erlbaum).

Watson, R. A. (1995). *Representational Ideas: From Plato to Patricia Churchland* (Dordrecht: Kluwer Academic).

Weinberg, J. (1976). "The Geometry of Colors." *General Relativity and Gravitation* 7(1): 135–169.

Yau, S.-T., and Nadis, S. (2010). *The Shape of Inner Space: String Theory and the Geometry of the Universe's Hidden Dimensions* (New York: Basic Books).

Index*

* Names are included only for historical figures

relativity, general 119, 178, 360–4, 568, 751,
    788, 889, 897–8
  special 128, 361, 363, 467, 507–9
religion 499–500
renormalization 766, 770
replicators 68–9, 674–5, 678
representation 32, 34, 48, 55–7, 59–61, 80–1,
    96–100, 102–3, 108, 130, 133, 149, 171–3,
    182, 296, 300–1, 323–4, 379–81, 388–94,
    440–459, 463, 475–9, 513–4, 595–6, 726–7,
    740–1, 762–3, 769, 775–6, 803, 839, 846,
    854, 883–9, 892, 895–6, 900, 903–6
  deflationary account 446–7, 454–7
  DDI account 453–5
  substantialism 449–53
representation theorem 193–5, 390, 429–30
research program, progressive 140, 499, 872
resemblance 399–400, 448–9, 674–9
revolution, scientific 511–2, 547–8, 634–47
Riemann, Bernhard 886–7, 906
Russell, Bertrand 123, 368, 370–1, 402–4

Salmon, Wesley 169–70, 531–2, 798,
    809–12, 838
scales, spatial 104–7, 654
science
  aims of 545–6, 559–60, 564
  applied 491
  commercialization of 874
  concept of 485–9
  data-driven 284, 847
  and democracy 872
  and society 625–6
  wars 9, 641–3
scientific method 150, 187, 431–2, 747–50, 874
secrecy 874
selection, natural 67–72, 78–9, 83–5, 313–4,
    675, 682–8, 704, 707, 713, 799–800
  group 805
  unit of 69, 79
self-organization 6, 771–2
Sellars, Wilfrid 101, 636–7
sense data 590
senses, human 243–4, 781–3, 786–7, 895
  individuating 107–9
separability 366, 680, 773
Sextus Empiricus 151–3

Sherlock Holmes strategy 653, 657–660
significance level 204
Silogen atom 727
similarity 75–6, 166–7, 384, 390–1, 404,
    449–51, 496–8, 596–7, 903
Simon, Herbert 700–1, 707, 713–4
simplicity 349–50
simulation. See computer simulation
skepticism 236–40, 275–80, 774–5
Snow, John 661
social
  constructivism 872
  epistemology 875–9
  kind 49–50, 52–3, 56, 59–60, 411, 413–4
  norm 304–6
  ontology 47, 57
  organization of science 863–881
sociobiology 77
sociology
  of knowledge 641
  of science 637, 641–643
spaces 882–908
  color 897–9
  configuration 130, 890–4, 904–5
  physical 357, 370, 886–94, 897–9, 905–6
  possibility 425–6, 889–894
  representational 883, 886, 889
  tonal 888, 900–3
space-time 25–6, 130, 346, 357–369, 371, 822,
    824, 826–31
special consequence condition 190
species 68, 73–8, 358–9, 401–2, 407, 409, 516,
    529, 705–6, 841–2
speech act 53, 454, 525
Spencer, Herbert 66, 70
Standard Model 7, 357, 359–60, 364
statistical mechanics 25, 143, 150, 154, 177–8,
    247–8, 250–3, 447–9, 486, 489–91, 506,
    554, 621, 710, 723, 737–8
statistical relevance 196, 531–2, 798
strategy, pure 296, 310–1, 314
string theory 883
structure
  lineage 687–8
  mathematical 96–7, 118–9, 131, 133, 182,
    592–4, 596, 598, 602, 891–4
  trait 687–9

CPSIA information can be obtained
at www.ICGtesting.com
Printed in the USA
BVHW011050260620
582052BV00010B/6

9 780190 939397